Verilog Digital Computer Design

Algorithms into Hardware

ISBN 0-13-639253-9
90000
9 780136 392538

Verilog Digital Computer Design

Algorithms into Hardware

Mark Gordon Arnold
University of Wyoming

Prentice Hall PTR
Upper Saddle River, NJ 07458
http://www.phptr.com

Editorial/Production Supervision: *Craig Little*
Acquisitions Editor: *Bernard M. Goodwin*
Manufacturing Manager: *Alan Fischer*
Marketing Manager: *Miles Williams*
Cover Design Director: *Jerry Votta*
Cover Design: *Talar Agasyan*

Prentice Hall books are widely used by corporations and government agencies for training, marketing, and resale.

The publisher offers discounts on this book when ordered in bulk quantities.
For more information, contact the Corporate Sales Department at 800-382-3419, fax: 201-236-7141,
email: corpsales@prenhall.com or write

Corporate Sales Department
Prentice Hall PTR
One Lake Street
Upper Saddle River, NJ 07458

Printed in the United States of America
10 9 8 7 6 5 4 3 2 1

ISBN 0-13-639253-9

Prentice-Hall International (UK) Limited, *London*
Prentice-Hall of Australia Pty. Limited, *Sydney*
Prentice-Hall Canada Inc., *Toronto*
Prentice-Hall Hispanoamericana, S.A., *Mexico*
Prentice-Hall of India Private Limited, *New Delhi*
Prentice-Hall of Japan, Inc., *Tokyo*
Simon & Schuster Asia Pte. Ltd., *Singapore*
Editora Prentice-Hall do Brasil, Ltda., *Rio de Janeiro*

Table of Contents

Page

Page

Page

Page

Page

Page

Page

Appendices

Page

Page

List of Figures

Page

Page

Page

Page

Page

APPENDIX C COMBINATIONAL LOGIC BUILDING BLOCKS

Page

Page

Preface

When I started teaching Verilog to electrical engineering and computer science seniors at the University of Wyoming, there were only two books and a handful of papers on the subject, in contrast to the overwhelming body of academic literature written about VHDL. Previously, VHDL had been unsuccessful in this course. For all its linguistic merits, VHDL is too complex for the first-time user. Verilog, on the other hand, is much more straightforward and allows the first-time user to focus on the design rather than on language details. Yet Verilog is powerful enough to describe very exotic designs, as illustrated in chapters 8-11.

As its subtitle indicates, this book emphasizes the algorithmic nature of digital computer design. This book uses the manual notation of Algorithmic State Machine (ASM) charts (chapter 2) as the master plan for designs. This book uses a top-down approach, which is based on the designer's faith that details can be ignored at the beginning of the design process, so that the designer's total effort can be to develop a correct algorithm.

Chapters 2-11 use the same elementary algorithm, referred to as the *childish division algorithm*, for many hardware and software examples. Because this algorithm is so simple, it allows the reader to focus on the Verilog and computer design topics being covered by each chapter. This book is unique in showing the correspondence of ASM charts to *implicit style* Verilog (chapters 3, 5 and 7). All chapters emphasize a feature of Verilog, known as *non-blocking assignment* or *Register Transfer Notation (RTN)*, which is the main distinction between software and synchronous hardware. Except for chapter 6, this book ignores (abstracts away) propagation delay. Instead, the emphasis here is toward designs that are accurate on a clock cycle by clock cycle basis with non-blocking assignment. (Many existing Verilog books either provide too much propagation delay information or are so abstract as to be inaccurate on a clock cycle basis. Appendices C and D motivate the abstraction level used here.)

Chapter 4 gives a novel three-stage design process (behavioral, mixed, structural), which exercises the reader's understanding of many elementary features of Verilog. Chapter 7 explains an automated one hot preprocessor, known as VITO, that eliminates the need to go though this manual three-stage process.

This book defers the introduction of Mealy machines until chapter 5 because my experience has been that the complex interactions of decisions and non-blocking assignments in a Mealy machine are confusing to the first-time designer. Understanding chapter 5 is only necessary to understand chapters 9 and 10, appendix J and sections 7.4 and 11.6.

The goal is to emphasize a few enduring concepts of computer design, such as pipelined (chapters 6 and 9) and superscalar (chapter 10) approaches, and show that these concepts are a natural outgrowth of the non-blocking assignment. Chapter 6 uses ASM charts and implicit Verilog to describe pipelining of a special-purpose machine with only the material of chapter 4. Chapters 8, 9 and 11 use the classic PDP-8 as an illustration of the basic principles of a stored program computer and cache memory. Chapter 8 depends only on the ASM material of chapter 2. Chapter 9 requires an understanding of all preceding chapters, except chapter 7. The capstone of this book, chapter 10 (which depends on chapter 9), uses the elegant ARM instruction set to explore the RISC approach, again with the unique combination of ASMs, implicit Verilog and non-blocking assignment.

Chapters 3-6, 9 and 10 emphasize Verilog *simulation* as a tool for uncovering bugs in a design prior to fabrication. Test code (sometimes called a testbench) that simulates the operating environment for the machine is given with most designs. Chapter 10 introduces the concept of Verilog *code coverage*. Chapters 7 and 11, which are partially accessible to a reader who understands chapter 3, uses specific *synthesis* tools for programmable logic to illustrate general techniques that apply to most vendors' tools. Even in synthesis, simulation is an important part of the design flow. Chapter 11 will be much more meaningful after the reader has grasped chapters 1-9. The designs in chapter 11 have been tested and downloaded (`www.phptr.com`) into Vantis CPLDs using a tool available to readers of this book (appendix F), but these designs should also be usable with minor modifications for other chips, such as FPGAs.

Appendices A, B and G give background on the machine language examples used in chapters 8-11. Appendices C and D give the block diagram notation used in all chapters for combinational logic and sequential logic, respectively. Chapters 1-11 do not use tri-state bidirectional buses, but appendix E explains the Verilog coding of such buses.

This book touches upon several different areas, such as "computer design," "state machine design," "assembly language programming," "computer organization," "computer arithmetic," "computer architecture," "register transfer logic design," "hardware/software trade-offs," "VLSI design," "parallel processing" and "programmable logic design." I would ask the reader not to try to place this book into the pigeon hole of some narrow academic category. Rather, I would hope the reader will appreciate in all these digital and computer design topics the common thread which the ASM and Verilog notations highlight. This book just scratches the surface of computer design and of Verilog. Space limitations prevented inclusion of material on interfacing (other than section 11.6) and on multiprocessing. The examples of childish division, PDP-8 and ARM algorithms were chosen for their simplicity. Sections labeled "Further reading" at the end of most chapters indicate where an interested reader can find more advanced concepts and algorithms, as well as more sophisticated features of Verilog. Appendix F indicates postal and Web addresses for obtaining additional tools and resources. It is hoped that the simple examples of Verilog and ASMs in this book will enable the reader to proceed to these more advanced computer design concepts.

In places, this book states my opinions rather boldly. I respect readers who have differing interpretations and methodologies, but I would ask such readers to look past these distinctions to the unique and valuable approaches in this book that are not found elsewhere. I have sprinkled (somewhat biased) historical tidbits, primarily from the first quarter century of electronic computer design, to illustrate how enduring algorithms are, and how transient technology is. Languages are more algorithmic than they are technological. Just look at the endurance of the COBOL language for business software. Hardware description languages will no doubt change as the twenty-first century unfolds, but I suspect whatever they become, they will include something very much like contemporary implicit style Verilog.

Acknowledgments

The author would like to thank the following people who have contributed to the content of the computer design laboratory at the University of Wyoming: Tighe Fagan, Rick Joslin, Bob Lynn, Susan Taylor McClendon, Philip Schlump, Elaine Sinclair, Tony Wallace and Cao Xu.

The author would also like to thank his colleagues at the University of Wyoming, especially Tom Bailey whose editorial advice is appreciated very much, John Cowles who cheerfully endured a semester of teaching the computer design laboratory, Jerry Cupal who convinced the author to try Verilog and Richard Sandige who has engaged the author in stimulating discussions about digital design.

The author wishes to acknowledge the tremendous contribution of James Shuler of SUNY (Brockport, NY) in the development of the VITO preprocessor described in chapter 7 and appendix F. The author gratefully acknowledges Elliot Mednick of Wellspring Solutions (Salem, NH) for making a limited version of its VeriWell™ Verilog simulator available to everyone. The author is extremely grateful to Kevin Bush of MINC, Inc. (Colorado Springs, CO) for making a limited version of its VerilogEASY synthesis tool available to readers of this book.

The author sincerely thanks Freddy Engineer of MINC, Inc. and Neal Sample of Stanford for proofreading the manuscript. The author also thanks Karolyn Durer, Peggy Hopkins and Phyllis Ranz for their assistance in the manuscript preparation. The author wishes to acknowledge the contribution of his editors at Prentice Hall: Russ Hall, Camille Trentacoste, Diane Spina, Bart Blanken, Craig Little and Bernard Goodwin.

Finally, the author wishes to express deep appreciation to Frank Prosser and David Winkel of Indiana University for the inspiration they provided with *The Art of Digital Design*. It is hoped that the material presented here will contribute a fraction of what *The Art of Digital Design* has.

Acknowledgments

This book is dedicated to the memory of my father,
Gordon William Arnold,
whose encouragement and sense of humor
during the last year of his life stimulated
the writing of this book.
He was a wonderful dad,
and I cherished him.

1. WHY VERILOG COMPUTER DESIGN?

1.1 What is computer design?

A computer is a *machine* that *processes information*. A machine, of course, is some tangible device (i.e., hardware) built by hooking together physical components, such as transistors, in an appropriate arrangement. Processing occurs when the machine follows the steps of a mathematical *algorithm*. Information is represented in the machine by *bits*, each of which is either 0 or 1. This book only considers digital information (i.e., bits) and does not consider analog information. Analog information can be approximated by digital information by using a sufficient number of bits.

Computer design is the thought process that arrives at how to construct the tangible hardware so that it implements the desired algorithm. The goal is to turn an algorithm into hardware. Computer designers have two ways to look at the machines they build: the way they act (known as the *behavioral* viewpoint, which is closely related to algorithms), and the way they are built (known as the *structural* viewpoint, which is like a "blueprint" for building the machine).

1.1.1 General-purpose computers

When you say the word computer today, it brings to mind what we refer to as a *general-purpose computer*, which you can program with *software* to implement any algorithm. With a general-purpose computer it is not necessary to build a new machine to implement each new algorithm. Programming such a general-purpose machine is often done with a conventional high-level language, such as C, C++, Java or Pascal.

1.1.2 Special-purpose computers

If you accept the definition of a computer given in section 1.1, there are many kinds of machines that fit this description in addition to general-purpose computers. We will refer to these other kinds of machines as *special-purpose computers*, which are non-programmable machines that implement one specific algorithm. A general-purpose computer is actually like a special-purpose computer that implements one algorithm, known as *fetch/execute*, that interprets a software program. The fetch/execute algorithm is fairly complex, so it is easier to study computer design by first looking at how simpler algorithms (than fetch/execute) can be transformed into hardware. For example, a traffic light is controlled by a machine that indicates when different colored lights are to be turned on. This machine is not programmable. Once it is designed, it always does the same boring thing: green, yellow, red, green, yellow, red, ... Nevertheless, by the above definition, it is a computer. It follows a particular (although boring) algorithm.

Special-purpose computers are ubiquitous because, in large volumes, it is more economical to manufacture a special-purpose computer that implements one boring algorithm than to purchase a general-purpose computer and waste most of its capabilities on that boring algorithm. However, for small volumes, or for problems where the specifications change frequently (such as tax accounting), the software approach is more economical.

There are only a handful of general-purpose computers on the market, and so there are not many jobs for designers for these popular machines. On the other hand, many non-computer industries use special-purpose computers as parts for the products they manufacture, and so job opportunities exist for designers of special-purpose machines. Also, special-purpose computers play a role in the peripheral devices, such as modems, that attach to general-purpose computers.

1.2 A brief history of computer/digital technology

The history of computer design highlights two things: changing technologies and lasting concepts. It is important to make a distinction between a concept and a technology. Information and algorithms are mathematical concepts that exist regardless of the physical details of their implementation with a particular technology. Many of the algorithms used in computers today were discovered by the great minds of mathematics decades or centuries ago.

Almost four centuries ago, Blaise Pascal (for whom the language is named) built one of the first mechanical calculators (which required a great deal of human intervention to operate). Pascal is remembered today however because he discovered several interesting algorithms, such as "Pascal's triangle," which are still in common use. A century and a half ago, Charles Babbage succeeded in using the technology of his day (precision cams and gears) to build the first fully automatic special-purpose computer for tabulating mathematical functions. Babbage also envisioned a general-purpose machine (with its fetch/execute algorithm) but was unable to complete it due to financial difficulties.

The invention of the vacuum tube was the technological advance that made building computers affordable. For a fraction of the cost of a machine built with cams and gears, a vacuum tube computer could automatically carry out hundreds of algorithm steps in a second. During the 1930s, C. Wynn-Williams in Great Britain built the first binary counter with vacuum tubes and the team of John Atanasoff and Charles Berry at Iowa State University built the first vacuum tube special-purpose computer for solving simultaneous equations. During World War II, several computers were built, including Colossus (in Great Britain), which was used to break coded German messages. In 1945, the mathematician John von Neumann popularized the idea of a general-purpose computer, and his name is often synonymous with a machine that implements the fetch/execute algorithm. The first operational general-purpose computer was the Manchester Mark I, which was a vacuum tube machine built in England that ran its first program in 1948.

In the 1950s, general-purpose vacuum tube computers cost millions of dollars, and only large corporations and governments owned them. The next major technological advance came with the invention of the transistor, which can do the same thing that a vacuum tube can do faster and more economically. Transistors also have the advantages that they run cooler and have a longer life than vacuum tubes.

This, of course, lowered the cost of general-purpose computers so that smaller corporations could own them, but it also made the application of *digital design* practical. Digital designs are special-purpose computers built using electronic circuits that process bits. Devices like digital watches, digital microwave oven timers, digital thermostats, hand-held calculators, etc. are all controlled by special-purpose computers that became economical with the invention of the transistor and related digital electronics.

In the 1960s, it became possible to manufacture hundreds or thousands of transistors on a chip of semiconductor material, known as an integrated circuit, at very low cost. Integrated circuits made it possible to mass-produce general-purpose computers, as well as digital electronic chips. Special- and general-purpose computers are now so powerful and affordable that they are part of almost every complex device built, from children's toys to the space shuttle.

Since the 1960s, there have been continual improvements in semiconductor technologies. It is now possible to get millions of transistors on a single chip. Of course, today's chips cost a fraction of the price of, and run faster and cooler than, their predecessors. But the algorithms that these chips implement are similar to the algorithms implemented with earlier technologies.

1.3 Translating algorithms into hardware

In the beginning, hardware designers were programmers and vice versa. The world of hardware design and software design fragmented into separate camps during the 1950s and 1960s as advancing technology made software programming easier. The industry needs many more programmers than hardware designers and programmers require far less knowledge of the physical machine than hardware designers. Despite this, the role of software designers and hardware designers is essentially the same: solve a problem.

Although many hardware designers realized in the 1960s and 1970s that their primary job was to develop an algorithm that solves a problem and translate that algorithm into hardware, some hardware designers lost sight of this essential truth.

An early notation for describing digital hardware that provides tremendous clarity in this regard is the Algorithmic State Machine (ASM), which was invented in the early 1960s by T.E. Osborne. As the name suggests, the ASM notation emphasizes the algorithmic nature of the machines being designed. Chapter 2 explains the ASM notation, and how it can be used manually to translate an algorithm into hardware. This notation is used throughout the rest of this book.

1.4 Hardware description languages

Unfortunately, hardware designers were inundated with the overwhelming technological changes that occurred with semiconductor electronics. Many hardware designers lost track of the advances in design methodology that occurred in software. Around 1980, as semiconductor technology advanced, it got more and more difficult to design hardware. Up to that time, most hardware design was done manually. Designers realized that the ever-increasing power of general-purpose computers could be harnessed to aid them in designing the next generation of chips. The goal of using the current generation of general-purpose computers to help design the next generation of special- and general-purpose computers required bringing the worlds of hardware and software back together again.

Out of this union was born the concept of the Hardware Description Language (HDL). Being a computer language, an HDL allows use of many of the timesaving software methodologies that hardware designers had been lacking. But as a hardware language, the HDL allows the expression of concepts that previously could only be expressed by manual notations, such as the ASM notation and circuit diagrams.

As technology advances, the details about HDLs will undoubtedly change in the future, but studying an HDL instills fundamental concepts that will endure. These ideas, originally thought of as hardware concepts, are becoming more important in software due to the increased importance of software parallel processing and object-oriented programming. There is a deep theoretic similarity between the concepts in software fields (such as operating systems and data structures) and the concepts in computer design. The growing popularity of HDLs attest to this fact: hardware is becoming more like software, and vice versa.

Chapter 3 discusses a popular HDL, known as Verilog, which is easy to learn because it has a syntax similar to C and Pascal. Verilog was developed in the early 1980s by Philip Moorby as a proprietary HDL for a company that was later accquired by Cadence Design Systems, which put the Verilog standard into the public domain. It is now known as IEEE1364. Verilog is used together with ASM charts in the rest of this book. This book is not just about Verilog or ASM charts for their own sake; this book also describes how these notations illuminate the thought processes of a computer designer. Ultimately, computation takes place on hardware. As children, all of us were inquisitive about everything: "How does this work?" Even if you do not plan on becoming a computer designer, it seems reasonable that you should be able to answer that question about the machines that are at the heart of your chosen career. The power of the Verilog and ASM notations give us insight for answering this question.

1.5 Typography

Fonts are used in this book to distinguish between different kinds of text, as explained below:

Times	is used in the bulk of the text for discussion.
Bold Times	is used to emphasize important or surprising concepts.
Italic	is used for the definition of an important term or phrase.
`Courier`	is used for Verilog text, exactly as it is typed into the file, and for similar notations taken from ASM charts, hardware diagrams, and simulation results. This font is also used for parts of other high level languages, such as C.
`Bold Courier`	is used for parts of Verilog text that are important in the discussion that precedes or follows them.
`Italic Courier`	is used to describe parts of Verilog syntax, such as a *`statement`*, which can be replaced with some particular symbol, such as `while`. Also, it is used to highlight complex simulation results.

1.6 Assumed background

It is assumed that the reader has a reasonable amount of experience programming in a conventional high-level language, such as C, C++, Java or Pascal. Programming experience in assembly language (appendices A, B and G) is very helpful. It is assumed that the reader can understand binary, octal and hexadecimal notations, can convert these to and from decimal and can perform arithmetic in these bases. It is also assumed that the reader is familiar with the common combinational logic gates (AND, OR, NOT, etc.), and that the reader knows about the common digital building blocks used in digital design (appendices C, D and E).

1.7 Conclusion

The few computers built in the nineteenth century were based on classical mechanics (cams and gears visible to the naked eye). Almost all the computers built in the twentieth century have been based on electronics. It is hard to say what technologies will be prevalent for computers in the twenty-first century.

Conventional semiconductor technology will someday reach its limit (based on the minimum size of a transistor and the speed of light). Technologies based on recombinate DNA, photonics, quantum mechanics, superconductivity and nanomechanics (cams and gears built of individual atoms) are all contenders to be the computer technology of the twenty-first century. The point is that it does not matter: technology changes every day, but concepts endure. The intellectual journey you travel by turning an algorithm into hardware illustrates these enduring concepts. I hope you enjoy the journey!

2. DESIGNING ASMs

This chapter explains the graphical notation used throughout the rest of this book. This graphical notation helps a hardware designer working only with pencil and paper. Chapters 4 and above require that the reader understand the notation explained in this chapter. Chapter 3 describes an alternative textual notation, known as Verilog, more suitable for automation (computer-aided design), where software tools help the designer produce a correct machine. The reader of this book will acquire a thorough understanding of both the notation in chapter 3 and the notation in this chapter, but we begin at the beginning with the most important question a hardware designer can ask: "How do I write down the particular algorithm that the hardware is supposed to follow?"

2.1. What is an ASM chart?

An *Algorithmic State Machine* (ASM) chart is a flowchart-like notation that describes the step-by-step operations of an algorithm in **precise** units of time. ASM charts are useful when you want to design hardware that implements some particular algorithm. The ASM chart can describe the behavior of the hardware without having to specify particular hardware devices to implement that algorithm. This allows you to make sure that the algorithm is correct before choosing an interconnection of particular hardware ("a structure") that implements the behavior described by the ASM. The most serious errors in hardware design do not result from connecting wires to the wrong place ("bad structure") but instead are the fault of designers not thinking through their algorithms completely ("bad behavior"). Designing a hardware structure is much more expensive than describing its behavior, and so it is sensible to spend extra time on the behavioral ASM chart before considering how to implement it with a hardware structure.

Although an ASM chart looks similar to a conventional software flowchart, the interpretation of an ASM chart differs from a conventional software flowchart with regard to how the passage of time relates to the operation of the algorithm. In software, the exact amount of time from one algorithm step to the next is not explicitly described by a flowchart. In the ASM chart, each step of the algorithm takes an exact amount of time, known as the *clock period* or *clock cycle*. There are also other time-related distinctions in the ASM chart notation which are described later.

An ASM chart is composed of rectangles, diamonds (or equivalently diamonds can be drawn as hexagons for notational convenience), ovals and arrows that interconnect the rectangles, diamonds and ovals. ASM charts composed only of rectangles and diamonds are said to describe *Moore* machines. ASM charts that also include ovals are said to describe *Mealy* machines. Mealy machines are described in chapter 5, and some of the more advanced concepts in chapter 7 and chapters 9 and above require the reader to understand Mealy notation. At this time, we will ignore the use of ovals and concentrate only on ASM charts for Moore machines.

Each rectangle is said to describe a *state*. A label, such as a number or preferably a meaningful name, can be written on the outside of the rectangle. The term *present state* refers to which rectangle of the ASM chart is active during a particular clock period. The term *next state* indicates which rectangle of the ASM chart will be active during the next clock period. The ASM chart indicates how to determine the next state (given the present state) by an arrow that points from the rectangle of the present state to the rectangle of the next state. Each arrow eventually arrives at one of the rectangles in the ASM chart. Since it has a finite number of rectangles, there is at least one loop in an ASM chart. An ASM chart is said to describe a particular *finite state machine*. Unlike software, there is no way to stop or halt a finite state machine (unless you pull the plug).

There is a relationship between the ASM chart and its behavior. For example, consider the following ASM chart with three states:

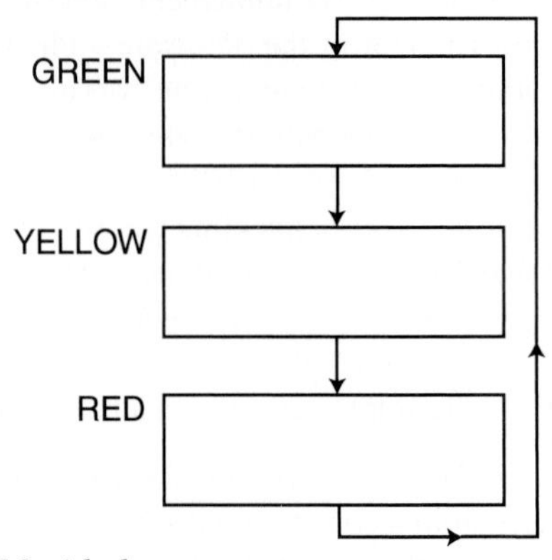

Figure 2-1. ASM with three states.

Assuming that we start in state GREEN, and that the clock has a period of 0.5 seconds, the ASM chart will make the following state transitions forever:

time	present state	next state
0.0	GREEN	YELLOW
0.5	YELLOW	RED
1.0	RED	GREEN
1.5	GREEN	YELLOW
2.0	YELLOW	RED
2.5	RED	GREEN
...	...	...

For example, between 0.5 and 1.0, the ASM is in state YELLOW. It is again in state YELLOW between 2.0 and 2.5.

The following sections explain the commands that can occur in rectangles of an ASM chart, the decisions that can occur in diamonds of an ASM chart, the input and output connections to a machine described by an ASM chart and issues of ASM chart style. Although the examples in the following sections vaguely resemble a traffic light controller, they are not intended to solve such a practical problem. They are instead intended solely to illustrate ASM chart notation and style.

2.1.1 ASM chart commands

Normally, the rectangle for a state is not empty. There are three *command* notations that a designer can choose to put inside the rectangle, which are described in the following sections.

2.1.1.1 Asserting a one-bit signal

A *signal* is a bit (or as explained in section 2.1.1.2, a group of bits) that conveys information. A signal is transmitted via a wire (or similar physical medium). The designer gives a signal (and its corresponding wire) a name to document its purpose. When the name of a signal occurs inside a rectangle of an ASM chart, that signal is asserted when the machine is in the state corresponding to the rectangle in question. In other state rectangles, where that signal is not mentioned, that signal takes on its default value. As an example, assume the default value for the signal STOP is 0. In the following:

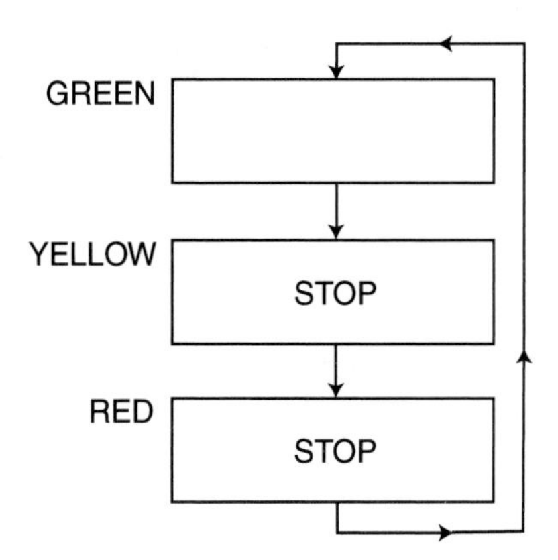

Figure 2-2. ASM with command outputs.

STOP will be 1 when the ASM is in state RED or state YELLOW. STOP will be 0 when the ASM is in state GREEN. The following illustrates this situation:

time	present state	next state	
0.0	GREEN	YELLOW	STOP=0
0.5	YELLOW	RED	STOP=1
1.0	RED	GREEN	STOP=1
1.5	GREEN	YELLOW	STOP=0
2.0	YELLOW	RED	STOP=1
2.5	RED	GREEN	STOP=1
...	...	...	...

2.1.1.2 Outputting a multi-bit command

When the name of a signal is on the left of an equal sign (=) inside a rectangle, that signal takes on the value specified on the right of the equal sign during the state corresponding to the rectangle in question. In other state rectangles, where that signal is not mentioned, that signal takes on its default value.

The following two diagrams show ASM charts that use =. The first of these ASMs is equivalent to the ASM given in section 2.1.1. The second example introduces a two-bit bus SPEED whose default value is 00.

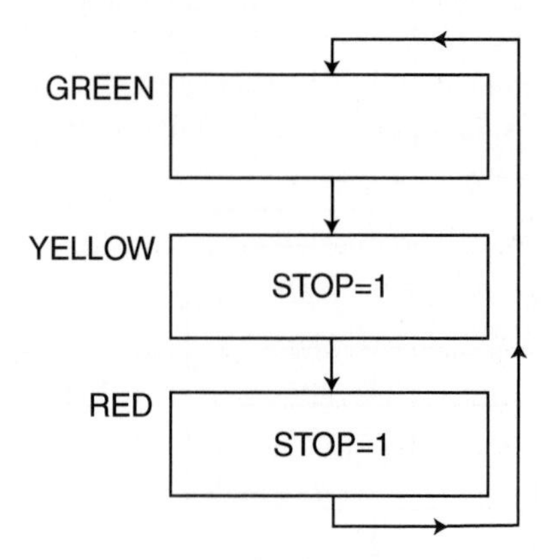

Figure 2-3. Equivalent to figure 2-2.

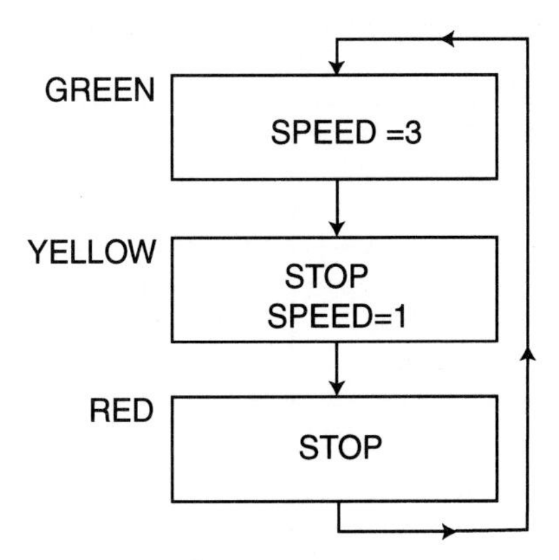

Figure 2-4. ASM with multi-bit output.

In the above, SPEED is 0 in state RED, 1 in state YELLOW and 3 in state GREEN.

time	present state	next state	output
0.0	GREEN	YELLOW	STOP=0 SPEED=11
0.5	YELLOW	RED	STOP=1 SPEED=01
1.0	RED	GREEN	STOP=1 SPEED=00
1.5	GREEN	YELLOW	STOP=0 SPEED=11
2.0	YELLOW	RED	STOP=1 SPEED=01
2.5	RED	GREEN	STOP=1 SPEED=00
...	...	...	... ...

2.1.1.3 *Register transfer*

The last two notations are simply a way of indicating how state names translate into physical signals, such as STOP and SPEED. Although we will eventually find many uses for these two notations, they are by themselves not the most convenient way to describe an algorithm.

Most algorithms manipulate variables that change their values during the course of the computation. It is necessary to have a place to store such values. Eventually, the designer will choose some kind of synchronous hardware register (appendix D) to hold such temporary values. In ASM chart notation, it is not necessary to make this design decision in order to describe an algorithm. Register Transfer Notation (RTN) (denoted by an arrow inside a rectangle) tells what happens to the register on the left of the arrow at the beginning of the next clock cycle. If a particular register is not mentioned on the left of an arrow in a state, the value of that register will remain the same in the next clock cycle. For example,

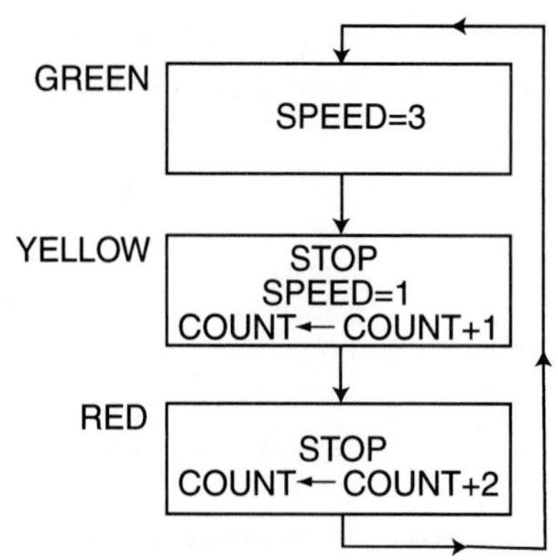

Figure 2-5. ASM with register output.

in the above assume that a three-bit COUNT register is 000 at time 0:

time	present state	next state			
0.0	GREEN	YELLOW	STOP=0	SPEED=11	COUNT=000
0.5	YELLOW	RED	STOP=1	SPEED=01	COUNT=000
1.0	RED	GREEN	STOP=1	SPEED=00	COUNT=001
1.5	GREEN	YELLOW	STOP=0	SPEED=11	COUNT=011
2.0	YELLOW	RED	STOP=1	SPEED=01	COUNT=011
2.5	RED	GREEN	STOP=1	SPEED=00	COUNT=100
...	...	...	...	...	...

Unlike STOP and SPEED, COUNT is not a function of the present state. For example, whenever the ASM is in state GREEN, STOP is 0. The first time in state GREEN, COUNT is 000, but the second time in state GREEN, COUNT is 011.

Notice that the ← causes a delayed assignment, which is different than assignment with conventional software programming languages, such as C. This is an important distinction to keep in mind when designing ASM charts. **One of the central topics that reoccurs throughout this book is the consequence of designing algorithms that use this kind of delayed assignment**. Although at first a novice designer may find ← unnatural and may make mistakes because of a misunderstanding of ←, once the reader masters the concept of this delayed assignment, all of the advanced concepts in later chapters will become much more understandable.

2.1.2 Decisions in ASM charts

One or more diamonds (or hexagons) following a rectangle indicate a decision in an ASM chart. The decision inside the diamond occurs at the **same time** as the operations described in the rectangle. There are two kinds of conditions that a designer can put inside a diamond in an ASM chart. These decisions are described in the following sections.

2.1.2.1 *Relations*

Relational operators (==, <, >, <=, >=, !=) as well as logical operators (&&, ||,!) can occur inside a diamond. It is also permissible to use the shorter bitwise operators (&,|,^,~) inside a diamond when all of the operands are only one-bit wide. When the relation in the diamond involves registers also used in the rectangle pointing to that diamond, the action taken is often different than would occur in software. Because the decision in the diamond occurs at the same time as the operations described in the rectangle, you ignore whatever register transfer occurs inside the rectangle to decide what the next state will be. The register transfer is an independent issue, which will only take effect at the beginning of the next clock cycle. As a illustration of such a decision, consider:

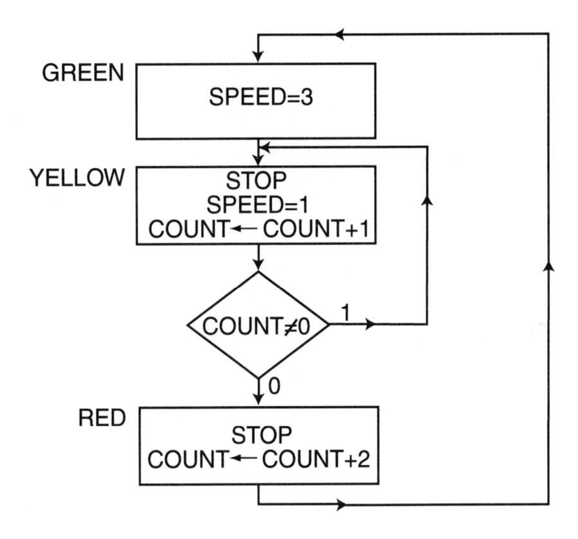

Figure 2-6. ASM with decision.

At first glance, it might appear that the ASM will get stuck in the loop the first time the machine enters state YELLOW because COUNT+1 is 001. However the decision COUNT != 0 is based on the current value of COUNT, which remains 000 until the beginning of the next clock cycle. Therefore the ASM exits from state YELLOW and proceeds to RED. On the second time the machine enters state YELLOW, COUNT is 011, and so it stays in state YELLOW for six clock periods. The only reason the ASM ever leaves state YELLOW is because the three-bit COUNT wraps around from 7 to 0.

```
        present  next
time    state    state
 0.0    GREEN    YELLOW   STOP=0 SPEED=11 COUNT=000
 0.5    YELLOW   RED      STOP=1 SPEED=01 COUNT=000
 1.0    RED      GREEN    STOP=1 SPEED=00 COUNT=001
 1.5    GREEN    YELLOW   STOP=0 SPEED=11 COUNT=011
 2.0    YELLOW   YELLOW   STOP=1 SPEED=01 COUNT=011
 2.5    YELLOW   YELLOW   STOP=1 SPEED=01 COUNT=100
 3.0    YELLOW   YELLOW   STOP=1 SPEED=01 COUNT=101
 3.5    YELLOW   YELLOW   STOP=1 SPEED=01 COUNT=110
 4.0    YELLOW   YELLOW   STOP=1 SPEED=01 COUNT=111
 4.5    YELLOW   RED      STOP=1 SPEED=01 COUNT=000
 5.0    RED      GREEN    STOP=1 SPEED=00 COUNT=001
 ...    ...      ...      ...    ...      ...
```

The highlighted line shows the last time the ASM is in state YELLOW. The next state is RED because COUNT is 000.

2.1.2.2 External status

Many hardware systems are composed of independent actors working cooperatively but in parallel to each other. We use *actor* as an ambiguous term that incorporates other digital hardware (i.e., special- and general-purpose computers) as well as non-digital hardware and people who communicate with the machine described by an ASM chart. From a designer's standpoint, the details of the other actors are normally unimportant.

These actors need to send information to the machine described by the ASM chart. When such external information can be represented in only one bit, it is known as *external status*. (Multi-bit signals can be broken down into several single-bit status signals if desired.) External status signals have names that are simply labels for physical wires connecting the machine that implements the ASM chart to the outside world. By convention, the name of a status signal can occur by itself inside a diamond. The meaning of such a diamond is the same as testing if the status signal is equal to one. For example,

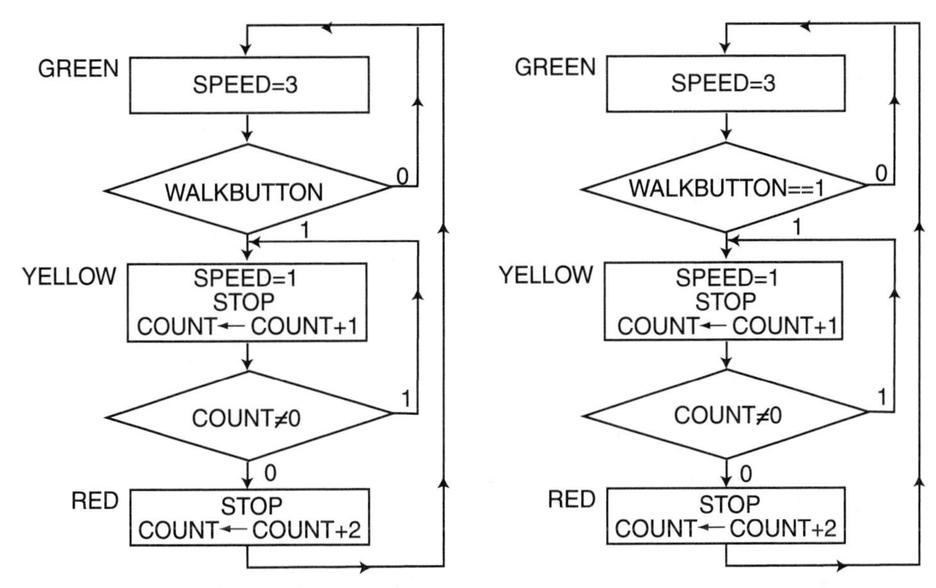

Figure 2-7. ASMs with external status.

The above ASMs are equivalent to each other. They will stay in state GREEN until WALKBUTTON is 1 during the last clock period that the ASM is in state GREEN. When WALKBUTTON is 1 in state GREEN, the next state will be state YELLOW. The machine ignores WALKBUTTON when it is in state YELLOW or in state RED.

2.1.3 Inputs and outputs of an ASM

An ASM chart describes the behavior of a piece of digital hardware without specifying its internal structure. The machine described by an ASM chart is often part of a larger structure. The machine receives inputs from other actors in that larger structure and provides outputs to other actors in that larger structure. In order to be part of such a larger structure, the machine described by an ASM chart must have a limited set of *input ports* (to receive information) and *output ports* (to send information). These ports are either single wires or buses having names that correspond to names used in the ASM chart.

In addition to its ASM chart, we can draw a *block diagram* of a machine. Such a diagram is sometimes called *a black box* because it hides the internal structure that implements the machine. The only thing we know is that the behavior of the machine is described by the corresponding ASM, and that the machine has inputs as specified by arrows pointing into the black box, and outputs as specified by arrows pointing out of the black box. As is standard notation in all hardware structure diagrams, when the input or output ports are more than one bit wide, the width is specified by a slash.

The following is a block diagram of the machine described by the ASM chart in 2.1.2.2:

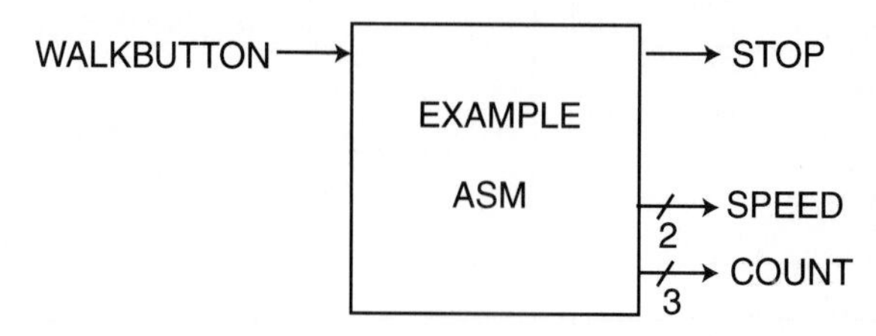

Figure 2-8. Block diagram.

2.1.3.1 ASM inputs

There are two kinds of inputs to a machine described by an ASM chart and black box diagram. It is somewhat arbitrary which of these two approaches a designer uses. The designer is free to choose the way that seems most appropriate for the problem at hand. It is permissible for a designer to mix these two approaches in a particular ASM. Since it plays a role later in the design process, the distinction between these two kinds of inputs is important to note.

2.1.3.1.1 External status inputs

A designer may consider a one-bit input port as an external status input when it is mentioned only by itself in diamond(s). Such status inputs are usually interpreted as providing an answer from the outside world to some yes/no question, such as "has the button been pressed?" As an example, in the block diagram above in section 2.1.3, WALKBUTTON is the only external status input.

2.1.3.1.2 External data inputs

When an input of any width is used only on the right-hand side of register transfers in rectangles and/or only in relational decisions, it is considered an external data input. Such data inputs usually play the same role that input variables play in conventional programming languages.

It is arbitrary whether the designer wishes to consider a one-bit input as a status or data input, as was illustrated in section 2.1.2.2. When a multi-bit input is used only with relational operations in diamonds (and not on the right of register transfers in rectangles), a designer may consider such a multi-bit input as being composed of several status input bits. For example, consider a machine with an external three-bit data input A. At some point in the behavior of the machine, the ASM needs to test if the value being input to the machine from the outside world on the bus A is equal to two. Figure 2-9 shows the block diagram and two equivalent ASM charts. The ASM chart on the left treats A as a single data input, but the ASM on the right treats A as being composed of three status inputs (A[2], A[1] and A[0]).

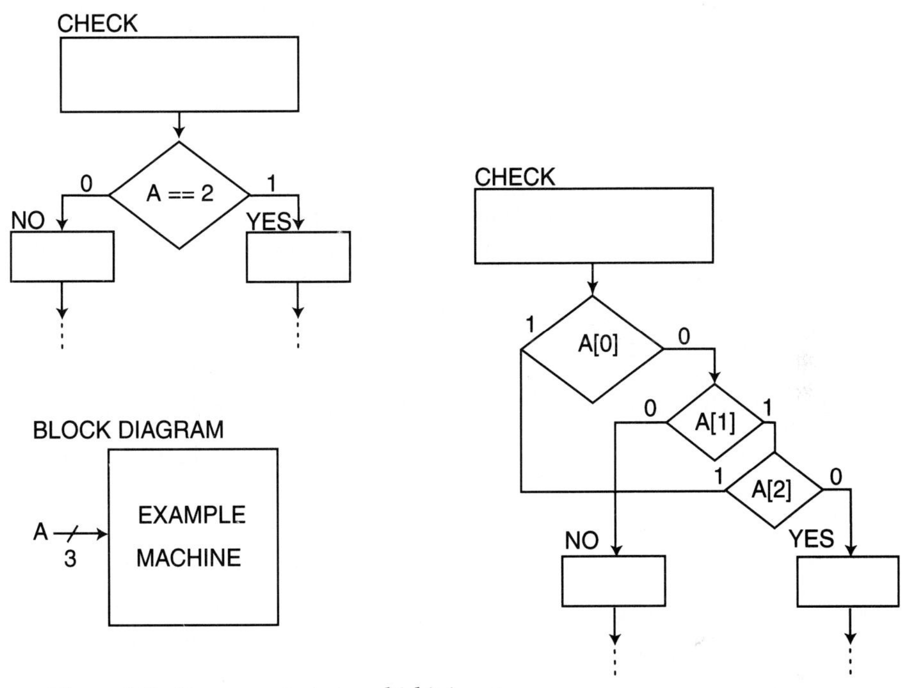

Figure 2-9. Two ways to test multi-bit input.

It is natural for designers to treat yes/no information as status. In most other cases, it is easier for the designer to consider something as a data input than to consider it as a status input, as the above ASMs illustrate. Inputs used on the right of register transfers must be treated as data inputs.

In the block diagram above in figure 2-8, there are no external data inputs.

2.1.3.2 ASM outputs

There are two kinds of outputs from a machine described by an ASM chart and black box diagram. Unlike the two kinds of inputs, with outputs the designer makes a choice based on how the output is generated in the ASM. From that point on, the designer is committed to that kind of output. Since it plays a role later in the design process, the distinction between these two kinds of outputs is important to note.

2.1.3.2.1 External command outputs

External command outputs are generated as described in sections 2.1.1.1 and 2.1.1.2. They are a function only of the present state of the ASM (assuming, as we have been so far, that ovals are not present in the ASM so that it represents a Moore machine). Command outputs do not retain their value when changing from one state to the next. If a particular command output is not mentioned in the next state, it reverts to its default value. In the block diagram given in section 2.1.3, the external command outputs are STOP and SPEED.

2.1.3.2.2 External data outputs

External data outputs are register names mentioned on the left of a register transfer in at least one rectangle. Unlike command outputs, data outputs can retain their value when changing from one state to the next. If a particular data output is not mentioned in the present state, it continues to hold its current value during the next state. Not all registers used in an ASM are necessarily output by the ASM. If a register is not specified as an output from the block diagram, it is an internal register, which the outside world is not allowed to examine (because no bus has been provided to connect that register to the outside world.) The only external data output of the block diagram in section 2.1.3 is COUNT.

2.1.4 Goto-less style

ASM charts allow a designer to specify an arbitrarily complex set of decisions to determine the next state. In theory, the possible next states for a particular state could be any of the rectangles in the ASM chart. Of course, in every particular case (based on register values and status inputs during that clock period), the ASM deterministically describes a particular next state to which the ASM goes in the next clock period. The problem for a careless designer is that the number of possible next states to consider could be quite large if the ASM is large. The flaw here is not in the technical capability of ASM charts (and corresponding hardware) to correctly implement such complex decisions, but rather in the capability of designers to comprehend their designs.

Similar problems were encountered decades ago in software design. At that time, the high-level language `goto` statement was quite popular. Psychological studies have shown that people are only able to keep a few details in their short-term memory at any time. Using `goto` statements correctly requires that a designer remember too many details when the program gets to be of any size. Dijkstra popularized the idea in 1968 that software programmers should avoid the use of `goto` statements in order to make software more readable, and more likely to be correct.

Arbitrary next states in a flowchart are just like `goto`s in software. Therefore, we will mostly use "`goto`-less" style ASM charts. Such ASMs are limited to decisions that are analogous to the high-level language style that is nowadays standard practice in software. In other words, we will try to make our decisions act like high-level language `if` statements, and our loops act like high-level language `while` statements. Although it is technically possible to make an ASM chart look like a plate of spaghetti, the goal of the `goto`-less style is to avoid such a mess. On rare occasions, there may be a compelling need to use an ASM chart which violates the `goto`-less style.

2.1.5 Top-down design

There are two basic approaches to solving problems: bottom up and top down. With the bottom-up approach, the designer begins with some tiny detail of the problem and solves that detail. The designer then goes on to some unrelated tiny detail, and solves it. The designer will eventually have to "glue" the details together to form the complete solution. On large problems, bottom-up designed pieces usually do not "fit" together perfectly. This problem happens because the designer did not view the separate details as fitting into some master plan. An unreasonably large percentage of the designer's time at the end of the project is wasted on integrating the details into a complete system.

The opposite of the bottom-up approach is top-down design. In the top down approach, the designer starts with a master plan. The details come later, and because the designer has a master plan, the details will fit perfectly into the final solution. Top- down design is not natural for novice designers. If you have never built hardware before, it is natural to worry about how the details will work (voltages, wires, gates, etc.). Learning something new is a mostly bottom-up process, but experienced designers use what they have already learned in a top-down fashion. Top-down design is based on **faith**: you have solved details similar to those in your current problem before, and so you can ignore the details when you begin the solution to the current problem. An ASM chart is a useful notation for describing the overall actions of a hardware system without getting into the hardware details. Therefore, the ASM chart makes a good starting point for the top-down design process.

We will take the top-down design process through three stages, briefly described in the following subsections (2.1.5.1 through 2.1.5.3).

2.1.5.1 Pure behavioral

This is the most important stage. It is the stage that most of the examples in later chapters concentrate on. In this stage of the top-down design process, the machine is described with a single ASM chart using primarily RTN and relational decisions. The only differences between the pure behavioral solution and a software solution is that the ASM chart describes the passage of time relating to the hardware system clock (as explained in sections 2.1.1 and 2.1.2) and the machine described by the ASM chart connects to the external world via hardware ports (section 2.1.3). A practical example of taking a simple problem and exploring various solutions using pure behavioral ASM charts is given in section 2.2. The only kind of structure that exists in the pure behavioral stage consists of the input and output ports, as illustrated by the following:

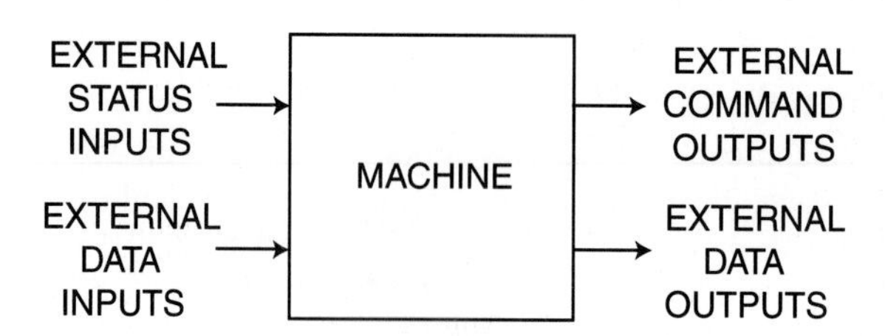

Figure 2-10. Pure behavioral block diagram.

2.1.5.2 Mixed

A pure behavioral ASM chart is merely the statement of an algorithm with precise timing information and includes an indication of which operations occur in parallel. It does not describe precisely what hardware components implement the computation. The goal of computer design is to arrive at a "blueprint" of a physical machine. The pure behavioral ASM chart is merely a description of what the designer wants the machine to do. It does not tell how to connect the physical components together. Software people wonder why the problem is not done upon completing the behavior ASM. After all, we **do** have a solution (an algorithm). Hardware people wonder why we spend so much time with ASM charts. After all, we **do not** yet have a solution (physical hardware). The answer to both groups is: have patience. The pure behavioral stage is important because it enhances the likelihood the designer will produce a **correct** solution. The next stage, which is known as the *mixed* stage, accomplishes part of the transformation from the algorithm into a physical **hardware** structure.

The mixed stage of the top-down design process partitions the problem into two separate but interdependent actors: the *controller* and the *architecture*. The architecture (sometimes called the *datapath*) is the place where physical hardware registers will implement the register transfers originally conceived in the pure behavioral stage. The architecture also contains combinational logic circuits that perform computations required by the algorithm. What the architecture cannot do by itself is sequence events according to the master plan given in the behavioral ASM. This is why the controller exists as an independent actor. The controller tells the architecture what to do during each clock cycle so that the master plan is carried out. Although it may seem the concepts of controller and architecture make things more complicated, in fact working in this fashion simplifies the thought process. In theory, it is possible to design a machine in an extreme way that either has no architecture or has no controller. Such extreme designs are as unnatural to think about as software without variable declarations.

The controller issues commands (as explained in sections 2.1.1.1 and 2.1.1.2) instead of RTN. The architecture receives and acts upon those commands and responds by outputting status. The controller makes decisions based on such status signals received from the architecture (as explained in section 2.1.2.2) instead of relational decisions. It is still possible to draw an ASM chart at this stage of the design, but the ASM chart only describes the independent action of the controller (in terms of commands and status), rather than the complete behavior of the system. This is what top-down design is all about: moving from one master plan (the behavioral ASM) to greater detail on how to carry out the master plan (the mixed ASM). The hardware structure in the mixed stage now has more detail. From the standpoint of the outside world, the mixed stage is identical to the pure behavioral stage, but internally we now see the interconnection of the controller and the architecture.

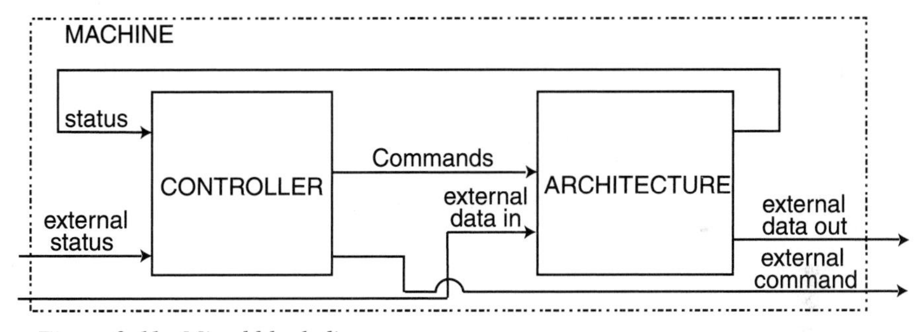

Figure 2-11. Mixed block diagram.

Although, in theory, the architecture could be described by ASM chart(s), it is usually more effective to use a hardware structure diagram. This is because a single ASM chart for the architecture could easily have billions of states (corresponding to all the combinations of values that all the registers in the system could have). Therefore, at the mixed level of abstraction, we use an ASM chart to describe the controller (which still has the same number of states) but use a hardware block diagram to describe the architecture. This stage of the design is known as mixed because it is a mixture of behavior and structure. Examples of translating some of the pure behavioral solutions of section 2.2 into mixed behavioral controller/structural architecture solutions are given in section 2.3.

2.1.5.3 Pure structure

The final stage of the design process is to implement the ASM chart for the controller as a hardware structure. This translation from the mixed stage to the pure structure stage is quite mechanical, and in fact software tools exist that create controller hardware automatically. One can simply describe the controller as a table that says, given the present state and status inputs, what the next state and command outputs will be. Various techniques exist to turn such a table into a hardware structure. For example, such a table can be burned into a Read Only Memory (ROM). The only other hardware required for the controller besides the ROM is a register to hold the present state. Examples of translating some of the mixed ASM charts of section 2.3 into pure structural solutions are given in section 2.4.

2.1.6 Design automation

Synthesis tools exist (chapter 11) that automate much of the final stages of the design process explained above. When using such tools, the designer's job is essentially complete at the end of the pure behavioral stage. Many designers skip over the mixed stage and go straight to the pure structural stage, from which the synthesis tool can automatically create the netlist of gates needed to fabricate an integrated circuit.

Nevertheless, it is important for a designer to understand how all these stages can be carried out manually in order for the designer to know how to create an efficient and correct design. The remainder of this chapter gives manual examples of the three stages described above (pure behavioral, mixed, pure structural). The most important of these stages is the pure behavioral stage because, unless that stage is correct, the rest of the design process (either manual or automated) is pointless.

2.2 Pure behavioral example

To illustrate the design process for a pure behavioral ASM, we need a simple algorithm to implement in hardware. One such algorithm comes from the definition of unsigned integer division. This definition is probably the first thing you ever learned about division when you were a child, and so we will refer to the algorithm that derives from this definition as the *childish division algorithm.*

By way of illustration, suppose you give a child the following problem: "You have seven friends and twenty-one cookies. How can you divide your cookies equally among your friends?" One solution is to give each friend one cookie, and note that each friend received a new cookie. Check to see if there are enough cookies left to give another one to each person. Since there are, repeat this process. When you are done, you will have noted that each person has received three cookies.

The name "childish" may seem pejorative, but this algorithm has a very honored place in the history of computer design (section 8.1). Like a child, this algorithm is simple and unpretentious, yet it raises important issues that also apply to much more complicated algorithms. Variations on the childish division algorithm are used throughout the rest of this book. Even Snoopy can tell whether or not this algorithm has been implemented correctly:

PEANUTS © United Feature Syndicate. Reprinted by Permission.

Of course, high-level software languages are sophisticated enough to have integer division built in. If the variable `x` is the number of cookies, and `y` is the number of friends, `x/y` is the solution to this problem. In hardware, division is seldom implemented as a combinational logic building block (although for small bus sizes this is certainly feasible). This means we need to use an ASM chart to describe a division algorithm. There are much more efficient algorithms than this childish algorithm that are normally implemented in hardware, but the childish division algorithm will allow us to emphasize the properties of ASM charts without having to get into obscure mathematical detail to justify the algorithm. Why this childish algorithm works is obvious.

Before considering the hardware implementation of this childish algorithm, let's consider how to code it in software, such as in the C programming language:

```
r1 = x;
r2 = 0;
while (r1 >= y)
  {
    r1 = r1 - y;
    r2 = r2 + 1;
  }
```

Upon exiting from the loop, `r2` will be `x/y`. This is a slow algorithm when the answer `r2` is of any appreciable size because the loop executes `r2` times.

This software algorithm would still work when the statements inside the loop are interchanged. These two statements are independent of each other (the new value of `r1` does not depend on the old value of `r2` and vice versa):

```
r1 = x;
r2 = 0;
while (r1 >= y)
  {
    r2 = r2 + 1;
    r1 = r1 - y;
  }
```

2.2.1 A push button interface

Before we can put the childish division algorithm into an ASM chart, we need to consider how the eventual hardware will communicate with other actors. In this case, we will assume the only other actor is a person, referred to as the user, who will give our ASM data inputs for `x` and `y`, and who will wait for the machine to answer back with `x/y`. Software designers have sophisticated user interfaces to choose from (using keyboard and mouse software drivers) that make it easy for people to interact with software. Such software solves two fundamental problems that always arise when two actors try to communicate: what data is being communicated and when should the data transfer occur? These problems arise because the actors operate independently in parallel to each other. In hardware, these same problems also arise, but it is the responsibility of the designer of the ASM chart to solve them.

A simple user interface scheme that will work here is to assume that we have a *friendly user* working with a push button that produces a status signal, `pb`, that our ASM can use. When the user pushes the button, `pb` will be 1 for exactly a single clock cycle (the design of such a push button is actually another exercise in working with ASM charts which we will ignore for the moment).

Also, we will assume that the division machine produces a READY signal. When READY is 1, the machine is waiting for the user to push the button. When READY is 0, the machine is busy computing the quotient. When READY becomes 1 again, the machine will output the valid quotient. We will assume that the user is patient enough to wait at least two clock cycles before pushing the button again. The user sets the value of the buses `x` and `y` in binary on switches before pushing the button. The user leaves `x` and `y` alone during the computation of the quotient. We will not consider how the machine would malfunction if an unfriendly user who does not obey these assumptions tries to use the machine.

The overall appearance of the ASM chart and hardware structure block diagram will be:

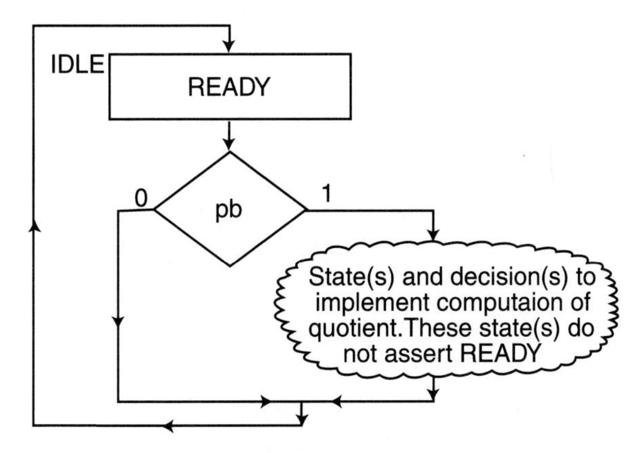

Figure 2-12. ASM for friendly user interface.

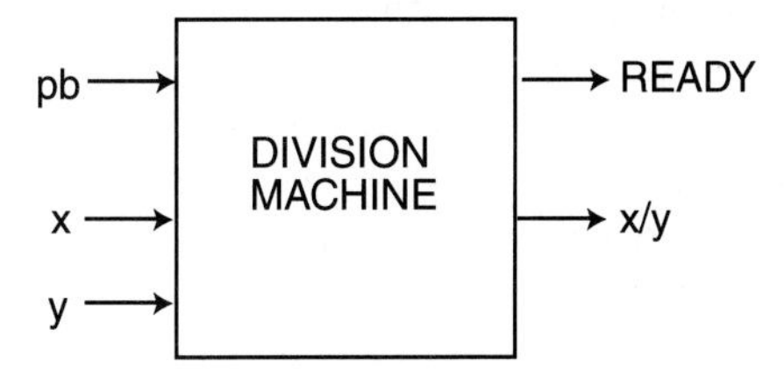

Figure 2-13. Block diagram.

The cloud will be replaced by one or more rectangles and one or more diamonds required to implement the childish division algorithm. None of the rectangles that replace the cloud will assert the READY signal. This means that on the next clock cycle after the user pushes `pb`, READY will become 0. READY only becomes 1 when the states in the cloud have finished and the ASM loops back to state IDLE.

In the following sections, we will examine several different ways to implement the childish division algorithm hidden in the cloud. An experienced hardware designer would not need to go through so many alternatives to arrive at our final solution. The reason we will look at so many different ways to do the same thing is to illustrate important properties of ASM charts that are somewhat different than conventional software. In this discussion, we will see that certain ASM charts that look reasonable to one familiar with software are actually incorrect and that some ASM charts that look somewhat strange are actually correct. Later, in section 2.3, we will use some of the pure behavioral ASMs we develop in this section as the starting point for the mixed stage of the top-down design process. These examples will also be used in later chapters.

2.2.2 An ASM incorporating software dependencies

The software paradigm used by conventional programming languages, such as C, can be described as each statement completes whatever action it is meant to accomplish before the software proceeds to execute the next statement in the program. Ultimately, all such software programs execute on some kind of hardware, therefore, it must be possible to describe this software paradigm using the ASM chart notation. Although it is often **inefficient**, a software algorithm can always be translated correctly into an ASM with the following rules:

1. Each assignment statement is written by itself in RTN in a unique rectangle that is not followed by a diamond.[1]
2. Each `if` or `while` is translated into an empty rectangle with a diamond to implement the decision.

With this approach, either of the following ASMs correctly implements the software algorithm for division given earlier:

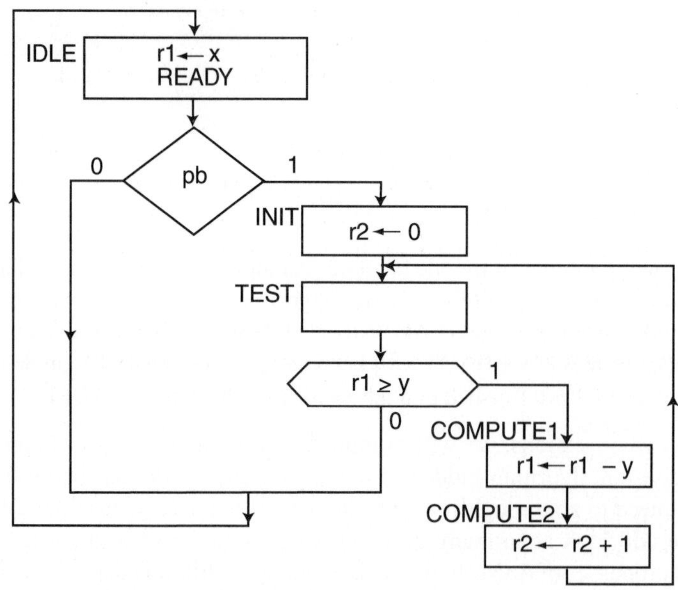

Figure 2-14. ASM for software paradigm (COMPUTE1 at top).

[1] Although in the following example there is a diamond in state IDLE involving an external status signal, the original software algorithm does not mention this status signal (`pb`), and so the software paradigm is preserved in this example.

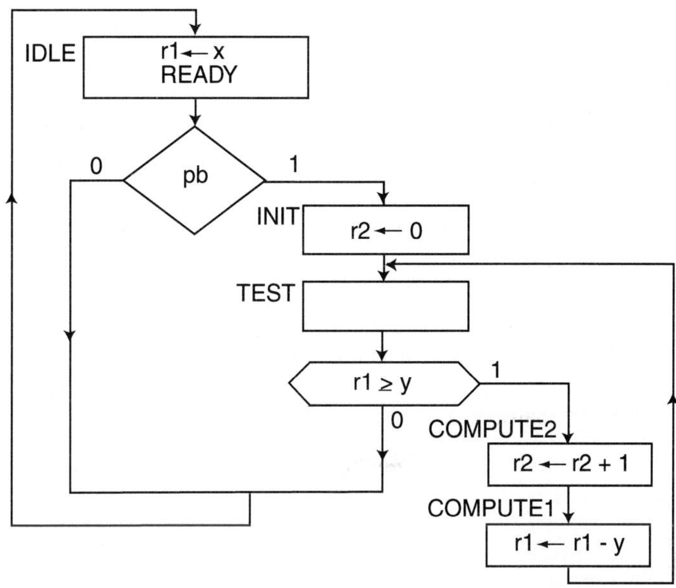

Figure 2-15. ASM for software paradigm (COMPUTE1 at bottom).

The only difference between these two ASMs is whether state COMPUTE1 is at the top of the loop or at the bottom of the loop. Since these ASMs exactly model the way software executes one statement at a time (one software statement per ASM rectangle), whether `r1` or `r2` gets a value assigned first is irrelevant, because this was also irrelevant in software.

The value of `x` is assigned to the register `r1` in state IDLE. Although this could have been done in an additional state, since we have assumed (see section 2.2.1) that the user waits at least two clock cycles when READY is 1 before pushing `pb`, the initialization of `r1` can occur here. The value of `x` will not be loaded into `r1` until the second of these two clock cycles. If `pb` is true, the ASM proceeds to state INIT, which will eventually cause `r2` to change. If `pb` is false, as would be the case most of the time, state IDLE simply loops to itself. Since state IDLE leaves `r2` alone and `r2` typically contains the last quotient, this user interface allows the user as much time as required to view the quotient. The user interface, not the division algorithm, requires that `r2` be assigned after the `pb` test.

State INIT makes sure that `r2` is 0 at the time the ASM enters state TEST. State TEST checks if `r1>=y`, just as the `while` statement does in software. States COMPUTE1 and COMPUTE2 implement each software assignment statement as RTN commands in separate clock cycles.

Both of these ASMs work when `x < y`. For example, the following shows how the ASMs proceed when `x` is 5 and `y` is 7 (all values are shown in decimal for ease of understanding):

```
IDLE        r1=   ? r2=   ? pb=0 ready=1
IDLE        r1=   5 r2=   ? pb=1 ready=1
INIT        r1=   5 r2=   ? pb=0 ready=0
TEST        r1=   5 r2=   0 pb=0 ready=0
IDLE        r1=   5 r2=   0 pb=0 ready=1
```

The way each of the above ASMs operates is slightly different when `x >= y`. The following shows how the ASM with COMPUTE1 at the top of the loop proceeds when `x` is 14 and `y` is 7:

```
IDLE        r1=   ? r2=   ? pb=0 ready=1
IDLE        r1=  14 r2=   ? pb=1 ready=1
INIT        r1=  14 r2=   ? pb=0 ready=0
TEST        r1=  14 r2=   0 pb=0 ready=0
COMPUTE1    r1=  14 r2=   0 pb=0 ready=0
COMPUTE2    r1=   7 r2=   0 pb=0 ready=0
TEST        r1=   7 r2=   1 pb=0 ready=0
COMPUTE1    r1=   7 r2=   1 pb=0 ready=0
COMPUTE2    r1=   0 r2=   1 pb=0 ready=0
TEST        r1=   0 r2=   2 pb=0 ready=0
IDLE        r1=   0 r2=   2 pb=0 ready=1
IDLE        r1=   ? r2=   2 pb=0 ready=1
```

The time to compute the quotient with this ASM includes at least two clock periods in state IDLE, a clock period in state INIT, and the time for the loop. The number of times through the loop is the same as the final quotient (`r2`). Since there are three states in the loop, the total time to compute the quotient is at least `3+3*quotient`.

Here is what happens with the ASM that has COMPUTE2 at the top of the loop:

```
IDLE        r1=   ? r2=   ? pb=0 ready=1
IDLE        r1=  14 r2=   ? pb=1 ready=1
INIT        r1=  14 r2=   ? pb=0 ready=0
TEST        r1=  14 r2=   0 pb=0 ready=0
COMPUTE2    r1=  14 r2=   0 pb=0 ready=0
COMPUTE1    r1=  14 r2=   1 pb=0 ready=0
TEST        r1=   7 r2=   1 pb=0 ready=0
COMPUTE2    r1=   7 r2=   1 pb=0 ready=0
COMPUTE1    r1=   7 r2=   2 pb=0 ready=0
TEST        r1=   0 r2=   2 pb=0 ready=0
IDLE        r1=   0 r2=   2 pb=0 ready=1
IDLE        r1=   ? r2=   2 pb=0 ready=1
```

The latter ASM illustrates the need for an empty rectangle in state TEST. State COMPUTE1 schedules a change in register `r1` (for example, from 7 to 0), but the change does not take effect until the beginning of the next clock cycle. Therefore the decision cannot be part of state COMPUTE1 but instead needs the empty rectangle (state TEST).

2.2.3 Eliminating state TEST

The empty rectangle for state TEST was introduced only to allow a mechanical translation from software to an ASM. In many instances, a decision like this can be merged in with other states. Remember with an ASM that a non-empty rectangle having a diamond following it means the computation in the rectangle and the decision in the diamond take place in parallel. It is inappropriate to merge a decision onto states doing computation when the outcome of the decision (in the software paradigm) could depend on the computation. Consider the following modified version of the ASM (figure 2-15) that has COMPUTE2 at the top of the loop:

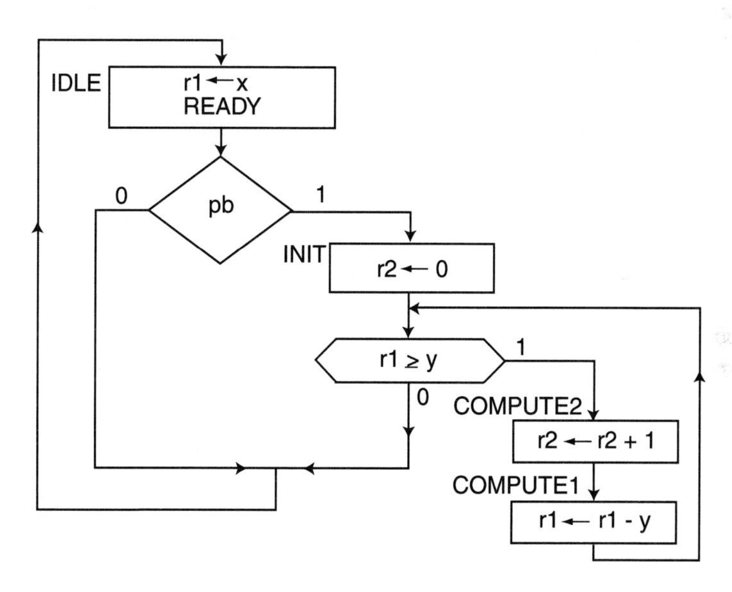

Figure 2-16. Incorrect four-state division machine.

The only difference here is that state TEST has been eliminated. Although this works for `x<y`, it fails to compute the correct quotient for `x>=y`. As an illustration of this error, assume that `r1` is twelve bits and consider when `x` is 14 and `y` is 7:

```
IDLE        r1=   ? r2=   ? pb=0 ready=1
IDLE        r1=  14 r2=   ? pb=1 ready=1
INIT        r1=  14 r2=   ? pb=0 ready=0
COMPUTE2    r1=  14 r2=   0 pb=0 ready=0
COMPUTE1    r1=  14 r2=   1 pb=0 ready=0
COMPUTE2    r1=   7 r2=   1 pb=0 ready=0
COMPUTE1    r1=   7 r2=   2 pb=0 ready=0
COMPUTE2    r1=   0 r2=   2 pb=0 ready=0
COMPUTE1    r1=   0 r2=   3 pb=0 ready=0
IDLE        r1=4089 r2=   3 pb=0 ready=1
IDLE        r1=   ? r2=   3 pb=0 ready=1
```

The decision `r1>=y` actually occurs separately in two states: INIT and COMPUTE1. In state INIT, the only computation involves `r2`, and so the decision (14 is >= 7) proceeds correctly. The problem exists in state COMPUTE1 because the computation changes `r1`, and the decision is based on `r1`. The second time in state COMPUTE1, `r1` is still 7, although it is scheduled to become 0 at the beginning of the next clock cycle. The decision is based on the current value (7), and so the loop executes one more time than it should and the incorrect value of `r2` (3) results. The mysterious decimal 4089 is the side effect of 12-bit underflow ($4089+7=2^{12}$).

Although it is incorrect to remove state TEST in the last example, what about removing state TEST in the other ASM (figure 2-14, with COMPUTE1 at the **top** of the loop)?

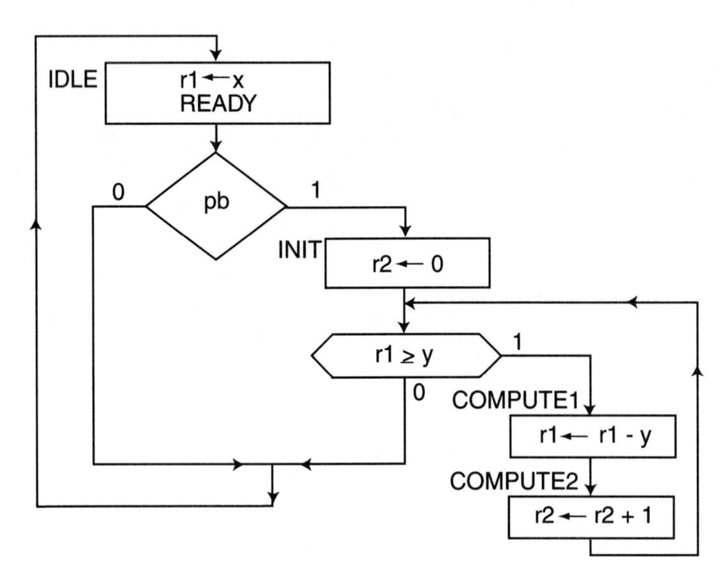

Figure 2-17. Correct four-state divison machine.

This ASM has the decision r1>=y happening in two different states: INIT and COMPUTE2. The difference here is that the decision is not dependent on the result of the computation in state COMPUTE2. Therefore, this ASM is correct. As an illustration, consider when x is 14 and y is 7:

```
IDLE        r1=   ? r2=   ? pb=0 ready=1
IDLE        r1=  14 r2=   ? pb=1 ready=1
INIT        r1=  14 r2=   ? pb=0 ready=0
COMPUTE1    r1=  14 r2=   0 pb=0 ready=0
COMPUTE2    r1=   7 r2=   0 pb=0 ready=0
COMPUTE1    r1=   7 r2=   1 pb=0 ready=0
COMPUTE2    r1=   0 r2=   1 pb=0 ready=0
IDLE        r1=   0 r2=   2 pb=0 ready=1
IDLE        r1=   ? r2=   2 pb=0 ready=1
```

The second time in state COMPUTE1 schedules the assignment that changes r1 from 7 to 0. This takes effect at the beginning of the clock cycle when the ASM enters state COMPUTE2 for the second time. The decision, which is now part of COMPUTE2, is based on the correct value (0). This means the loop goes through the correct number of times and the quotient in r2 is correct. As was the case with the earlier ASMs, r2 will remain unchanged until pb is pushed again.

Although the ASMs in section 2.2.2 are also correct, this ASM has the advantage that it executes faster as it requires only 3+2*quotient clock cycles.

2.2.4 Eliminating state INIT

In addition to being able to describe a decision and a computation that occur in parallel, the ASM chart notation can describe multiple computations that occur in parallel. Consider eliminating state INIT by merging the assignment of zero to r2 into the rectangle for state IDLE:

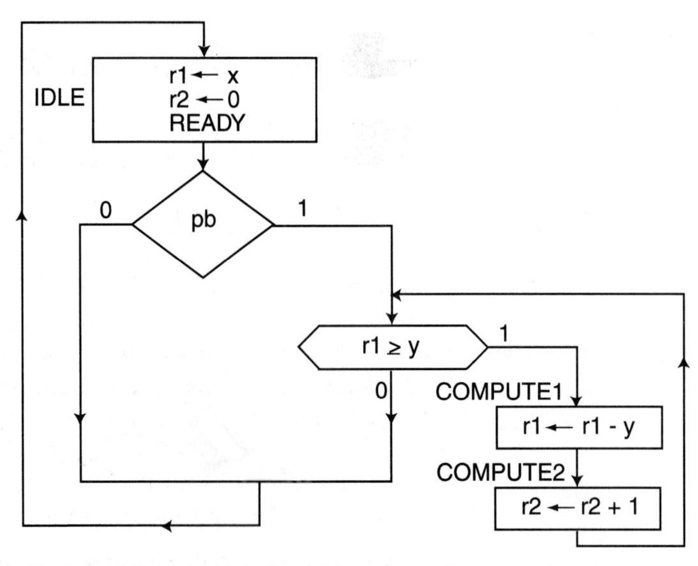

Figure 2-18. Incorrect user interface (throws quotient away).

You may have as many RTN assignments occurring in parallel within a state as you want as long as each left-hand side within that state is unique. In this instance, `r1` and `r2` are scheduled to have new values assigned at the beginning of the next clock cycle. Since we have assumed that the user will ensure that the ASM stays in state IDLE while `x` remains constant for at least **two** clock cycles, `r1` and `r2` will be properly initialized before entering the loop. This ASM will correctly compute the quotient and leave the loop after the proper number of times for the same reason. To illustrate what this ASM does, consider the same example as the other ASMs (when `x` is 14 and `y` is 7):

```
IDLE        r1=   ? r2=   0 pb=0 ready=1
IDLE        r1=  14 r2=   0 pb=1 ready=1
COMPUTE1    r1=  14 r2=   0 pb=0 ready=0
COMPUTE2    r1=   7 r2=   0 pb=0 ready=0
COMPUTE1    r1=   7 r2=   1 pb=0 ready=0
COMPUTE2    r1=   0 r2=   1 pb=0 ready=0
IDLE        r1=   0 r2=   2 pb=0 ready=1
IDLE        r1=   ? r2=   0 pb=0 ready=1
```

There is a new problem with this ASM that we have not seen before: the quotient (2) exists in `r2` for only one clock cycle. This ASM throws it away because the assignment of 0 to `r2` is in state IDLE. From a mathematical standpoint, this ASM is correct, but from a user interface standpoint, it is unacceptable.

2.2.5 Saving the quotient

One way to overcome the user interface problem in section 2.2.4 is to introduce an extra register, `r3`, that saves the quotient in a new state COMPUTE3:

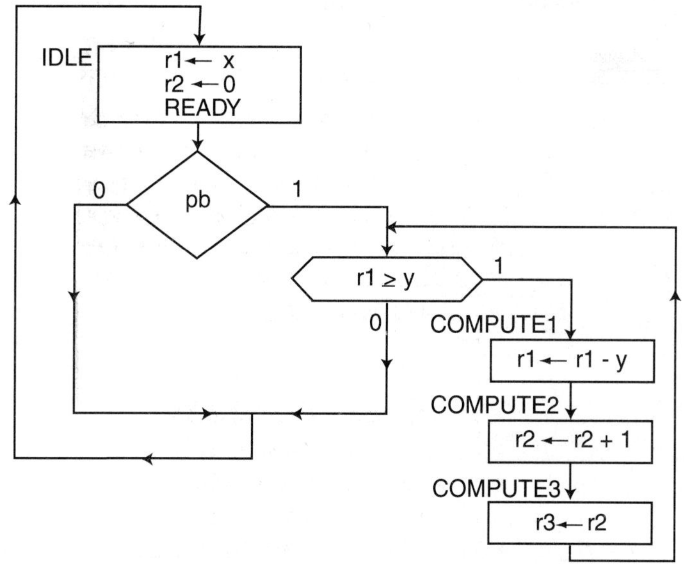

Figure 2-19. Saving quotient in `r3`.

This ASM works for `x>=y` (the quotient is now in `r3`, not `r2`). For example, when `x` is 14 and `y` is 7:

```
IDLE        r1=   ? r2=   0 r3=   ? pb=0 ready=1
IDLE        r1=  14 r2=   0 r3=   ? pb=1 ready=1
COMPUTE1    r1=  14 r2=   0 r3=   ? pb=0 ready=0
COMPUTE2    r1=   7 r2=   0 r3=   ? pb=0 ready=0
COMPUTE3    r1=   7 r2=   1 r3=   ? pb=0 ready=0
COMPUTE1    r1=   7 r2=   1 r3=   1 pb=0 ready=0
COMPUTE2    r1=   0 r2=   1 r3=   1 pb=0 ready=0
COMPUTE3    r1=   0 r2=   2 r3=   1 pb=0 ready=0
IDLE        r1=   0 r2=   2 r3=   2 pb=0 ready=1
IDLE        r1=   ? r2=   0 r3=   2 pb=0 ready=1
```

Unfortunately, there is a subtle error in the above ASM: when the answer is supposed to be zero, `r3` is left unchanged instead of being cleared. This occurs because the only assignment to `r3` is inside the loop, but the loop never executes when the quotient is zero. One way to overcome this problem is to include an extra decision in the ASM to test for the special case that `x<y` (which can be done by testing if `r1>=y` is false):

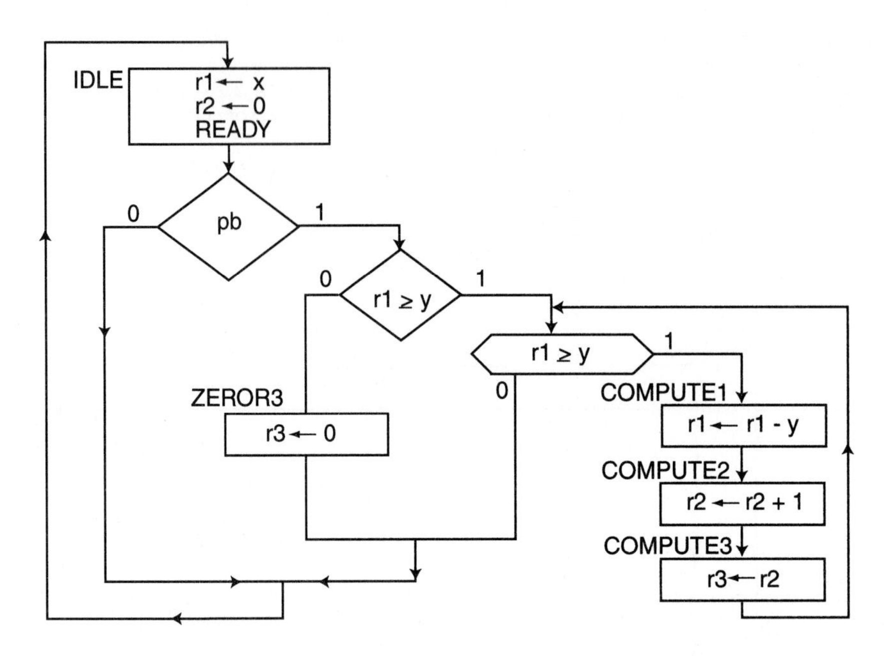

Figure 2-20. Handling quotient of zero.

Of course, this has the disadvantage of taking longer (`2+3*quotient` clock cycles), but sometimes a designer must consider a slower solution to eventually discover a faster solution.

2.2.6 Variations within the loop

Let's take the final ASM of section 2.2.5 and consider some variations of it inside the loop that will make it incorrect. Our eventual goal is to find a faster solution that is **correct**, but for the moment, let's just play around and see how we can break this ASM.

One incorrect thing to do would be to assign to `r3` before incrementing `r2`:

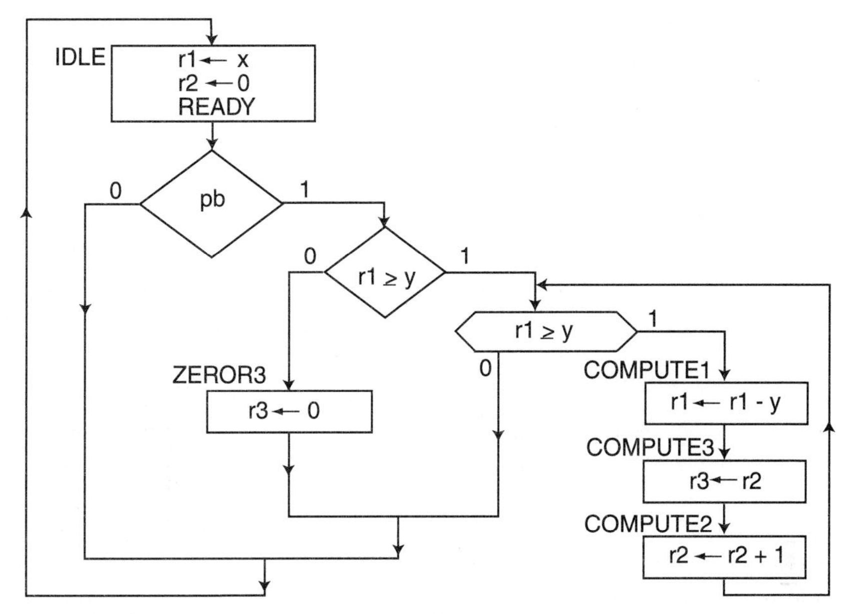

Figure 2-21. Incorrect rearrangement of states.

Here is an example when x is 14 and y is 7 of what kind of error occurs:

```
IDLE        r1=  ? r2=  0 r3=  ? pb=0 ready=1
IDLE        r1= 14 r2=  0 r3=  ? pb=1 ready=1
COMPUTE1    r1= 14 r2=  0 r3=  ? pb=0 ready=0
COMPUTE3    r1=  7 r2=  0 r3=  ? pb=0 ready=0
COMPUTE2    r1=  7 r2=  0 r3=  0 pb=0 ready=0
COMPUTE1    r1=  7 r2=  1 r3=  0 pb=0 ready=0
COMPUTE3    r1=  0 r2=  1 r3=  0 pb=0 ready=0
COMPUTE2    r1=  0 r2=  1 r3=  1 pb=0 ready=0
IDLE        r1=  0 r2=  2 r3=  1 pb=0 ready=1
IDLE        r1=  ? r2=  0 r3=  1 pb=0 ready=1
```

The value in r3 is one less than it should be since it was assigned too early.

Another thing to try (which unfortunately will also fail for similar reasons) is to merge states COMPUTE2 and COMPUTE3 into a single state COMPUTE23:

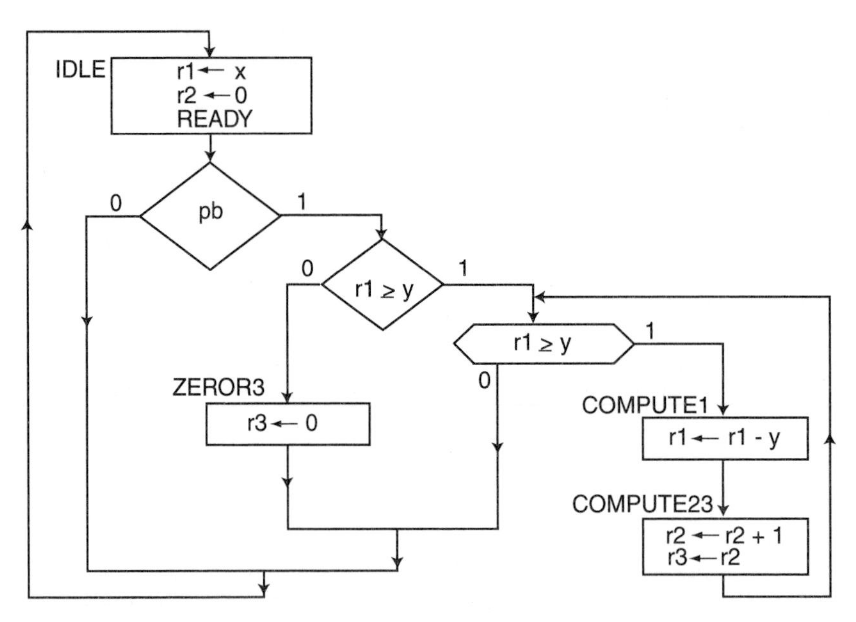

Figure 2-22. Incorrect parallelization attempt.

Here is an example when x is 14 and y is 7 of what kind of error occurs:

```
IDLE        r1=  ? r2=  0 r3=  0 pb=0 ready=1
IDLE        r1= 14 r2=  0 r3=  0 pb=1 ready=1
COMPUTE1    r1= 14 r2=  0 r3=  0 pb=0 ready=0
COMPUTE23   r1=  7 r2=  0 r3=  0 pb=0 ready=0
COMPUTE1    r1=  7 r2=  1 r3=  0 pb=0 ready=0
COMPUTE23   r1=  0 r2=  1 r3=  0 pb=0 ready=0
IDLE        r1=  0 r2=  2 r3=  1 pb=0 ready=1
IDLE        r1=  ? r2=  0 r3=  1 pb=0 ready=1
```

Even though inside the rectangle the assignment to r2 is written above the assignment to r3, they happen in parallel. **The meaning of a state in an ASM is not affected by the order in which a designer writes the commands inside the rectangle**. Since there is a dependency between the commands in state COMPUTE23, this ASM is not equivalent to the correct solution of section 2.2.5 but is instead equivalent to the incorrect solution given a moment ago. After the second time in state COMPUTE23, r2 is incremented (from 1 to 2), but r3 changes to the old value of r2 (1), which is not what we want.

Although all of the above variations may seem hopeless, there is in fact a correct and faster solution if we press on with this kind of variation. Let's merge all three commands into a single state COMPUTE:

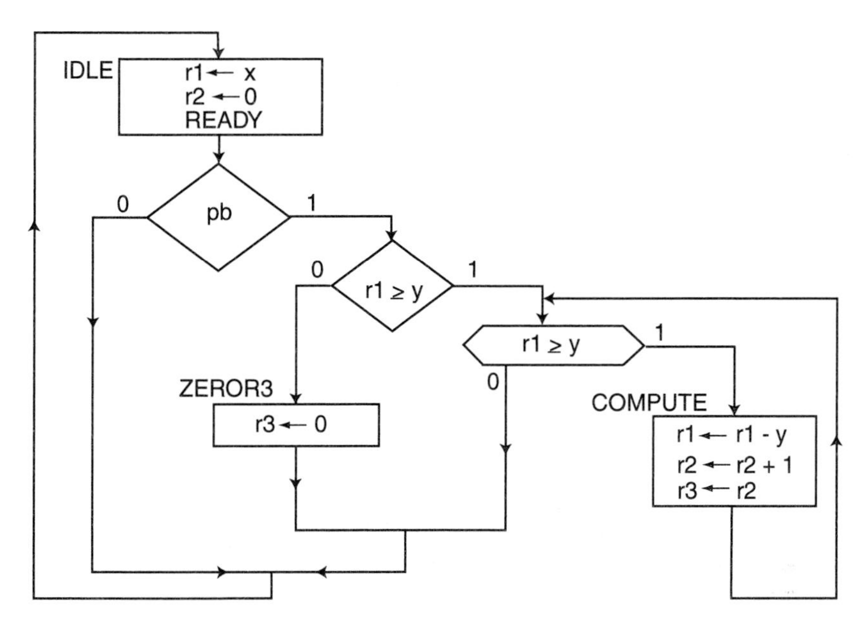

Figure 2-23. Correct parallelization.

This ASM is correct, as illustrated by the example used before (when `x` is 14 and `y` is 7):

```
IDLE       r1=    ? r2=   0 r3=    ? pb=0 ready=1
IDLE       r1=   14 r2=   0 r3=    ? pb=1 ready=1
COMPUTE    r1=   14 r2=   0 r3=    ? pb=0 ready=0
COMPUTE    r1=    7 r2=   1 r3=    0 pb=0 ready=0
COMPUTE    r1=    0 r2=   2 r3=    1 pb=0 ready=0
IDLE       r1= 4089 r2=   3 r3=    2 pb=0 ready=1
IDLE       r1=    ? r2=   0 r3=    2 pb=0 ready=1
```

The decision involving `r1>=y` is part of state COMPUTE (as well as being part of state IDLE), and `r1` is affected in state COMPUTE. Also there is the interdependence of `r2` and `r3` observed earlier. The reason why state COMPUTE works here is that all of these things occur at the same time in parallel. We have now totally left the sequential software paradigm of section 2.2.1 (one statement at a time; no dependency within a state). We are now using the dependency in the algorithm with parallelism to get the correct result much faster.

Although in this ASM `r3` still serves as the place where the user can observe the quotient when the ASM returns to state IDLE, `r3` accomplishes something even more important. It compensates for the fact that the loop in state COMPUTE executes one more time than the software loop would. Even though r2 becomes one more than the correct quotient, r3 is loaded with the old value of r2 each time through the loop. On the last time through the loop, r3 is scheduled to be loaded with the correct quotient.

The loop in state COMPUTE is interesting because it has a property that software loops seldom have: it either does not execute or it executes at least two times. This is because the decision r1>=y is part of both states IDLE and COMPUTE. To illustrate this, consider when x is 7 and y is 7:

```
IDLE       r1=    ? r2=   0 r3=    ? pb=0 ready=1
IDLE       r1=    7 r2=   0 r3=    ? pb=1 ready=1
COMPUTE    r1=    7 r2=   0 r3=    ? pb=0 ready=0
COMPUTE    r1=    0 r2=   1 r3=    0 pb=0 ready=0
IDLE       r1=4089 r2=   2 r3=    1 pb=0 ready=1
IDLE       r1=    ? r2=   0 r3=    1 pb=0 ready=1
```

You can see that r1 is 7 in state IDLE, and so the ASM proceeds to state COMPUTE. In state COMPUTE, r1 is scheduled to change, but it remains 7 the first time in state COMPUTE; thus the next state is state COMPUTE (it loops back to itself). Only on the second time through state COMPUTE has the scheduled change to r1 taken place; thus the next state finally becomes IDLE.

As with earlier ASMs, this ASM works for x<y only because of state ZEROR3. For example, consider when x is 5 and y is 7:

```
IDLE       r1=    ? r2=   0 r3=    ? pb=0 ready=1
IDLE       r1=    5 r2=   0 r3=    ? pb=1 ready=1
ZEROR3     r1=    5 r2=   0 r3=    ? pb=0 ready=0
IDLE       r1=    5 r2=   0 r3=    0 pb=0 ready=1
IDLE       r1=    ? r2=   0 r3=    0 pb=0 ready=1
```

The time required for this ASM is 3+quotient clock cycles.

2.2.7 Eliminate state ZEROR3

If the loop in state COMPUTE could execute one or more (rather than two or more) times, it would be possible to eliminate state ZEROR3. This would work because r2 is already 0, and the assignment of r2 to r3 would achieve the desired effect of clearing r3.

One way to describe this in ASM chart notation is to note that pb is true when making the transition from state IDLE to state COMPUTE (the first time into the loop), but pb remains false until the quotient is computed (by our original assumption about a friendly user). Let's change the decision so that it ORs the status signal pb together with the result of the r1>=y:

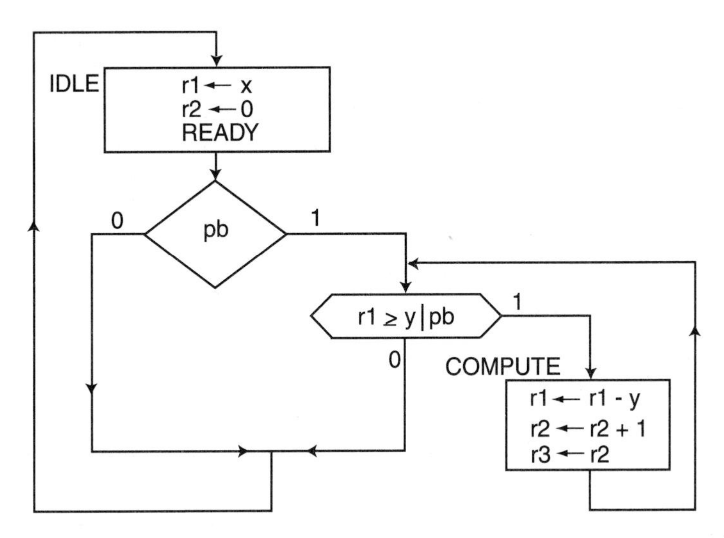

Figure 2-24. Goto-less two-state childish division ASM.

Since pb is true for only one clock cycle, this only causes the loop to go in the first time. Whether to repeat the loop subsequent times depends **only** on r1. Since for all values x>=y the loop occurs at least twice anyway, ORing pb in the decision does not change the operation of the ASM for x>=y. ORing pb only affects what happens when x<y. Rather than executing state COMPUTE zero times, ORing pb forces it to execute once. In this case, state COMPUTE will not execute more than once because x (r1) is not >= y.

For example, consider when x is 5 and y is 7:

```
IDLE        r1=    ? r2=    0 r3=    ? pb=0 ready=1
IDLE        r1=    5 r2=    0 r3=    ? pb=1 ready=1
COMPUTE     r1=    5 r2=    0 r3=    ? pb=0 ready=0
IDLE        r1=4094 r2=    1 r3=    0 pb=0 ready=1
IDLE        r1=    ? r2=    0 r3=    0 pb=0 ready=1
```

The fact that r1 and r2 are different after executing state COMPUTE than they were in the earlier ASM after executing state ZEROR3 is irrelevant since the user only looks at data output r3.

The time required for this ASM is also 3+quotient clock cycles.

The above ASM was arranged to follow the goto-less style mentioned in section 2.1.4. In essence, there is a while loop (testing r1>=y|pb) nested inside an if statement (testing pb). In order to describe this ASM in the goto-less style, pb is tested twice coming out of state IDLE. Such ASMs with redundant tests (in the same clock cycle) can be simplified into shorter equivalent ASM notation. Although this equivalent ASM is truly identical and would be implemented with the same hardware, it does not follow a style that can be thought of in terms of `ifs` and `whiles`:

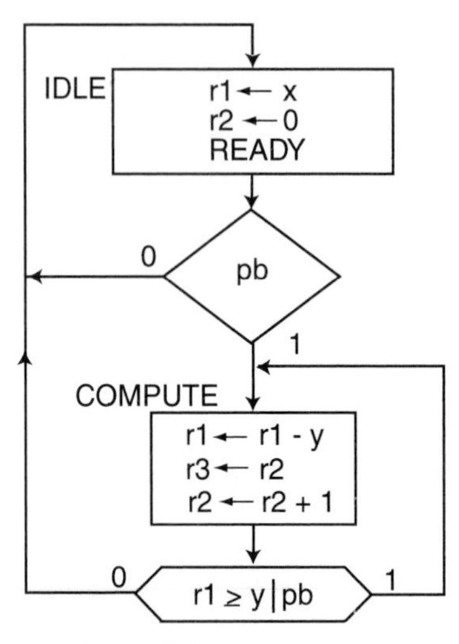

Figure 2-25. Equivalent to figure 2-24.

Also, in the above, the order of the statements within state COMPUTE were rearranged for ease of understanding. As mentioned earlier, changing the order with a rectangle does not change the meaning. Which way you draw the ASM is both a matter of personal taste and also a matter of how you intend to use it. We will see examples where both forms of this ASM prove useful. At this stage it is important for you to be comfortable that these two ASMs mean **exactly** the same thing because under **all** possible circumstances they cause the same state transitions and computations to occur.

2.3 Mixed examples

The three stages of the top-down design process were discussed in section 2.1.5. Section 2.2 gives several alternative ways to describe the childish division algorithm in the first stage of the top-down design process (as a pure behavioral ASM). This section continues this same example into the second stage of the top-down design process. In the second stage, we partition the division machine into a controller and an architecture.

2.3.1 First example

We could use any of the ASM charts in section 2.2 as an example of translating a pure behavioral ASM into a structural architecture and a behavioral controller. For example, consider the ASM chart described in section 2.2.7, which is the simplest correct ASM chart for the division machine. We can, for the moment, ignore pb and READY because they are external ports (and will remain the same in the mixed stage). In the mixed stage, we need to eliminate RTN commands and relational decisions.

Consider the RTN commands in this ASM chart. In state IDLE, r1 is scheduled to be loaded with x, and in parallel r2 is scheduled to be cleared. In state COMPUTE, all three registers are scheduled to change. Of course, these scheduled changes take effect at the beginning of the next clock cycle.

There are many possible hardware structures that could implement these RTN commands. The designer makes an arbitrary decision (based on speed, cost, availability, ease, personal prejudice, etc.) about what hardware components to use in the architecture. The only requirement is that interconnection of the chosen components can correctly implement all RTN transfers with the precise timing indicated by the original behavioral ASM chart. The easiest (but not necessarily best) way to accomplish this is to choose register components (like those in appendix D) that internally take care of as many of the required RTN commands as possible. For example, if the designer chooses a counter register for r2, the counter can internally take care of clearing r2 (as is required in state IDLE) and also take care of incrementing r2 (as is required in state COMPUTE). We will be able to eliminate the RTN commands (such as r2 ← r2+1) from the ASM chart and replace them with internal command signals (such as incr2).

If the designer were to choose a non-counter register for r2, the designer would have to provide for these actions with additional combinational devices (like those in appendix C) in the architecture. It is not wrong to choose a non-counter register for r2; it would just make the designer work harder. To keep this example as simple as possible, we will choose a counter register for r2.

On the other hand, registers r1 and r3 are loaded with values that must come from outside the register (unlike a simple counter). Therefore, it is sensible to use the simplest register component possible (the enabled register) for r1 and r3. Additional hardware will be required to make available the new values to be loaded into r1 and r3.

Having decided on the kind of registers to use in the architecture, we need to consider how those registers are interconnected. For a moment, let's concentrate only on the RTN in state COMPUTE. In this state, r1 will be loaded with a difference, r2 increments, and r3 gets loaded with the old value of r2. All three of these actions occur in parallel. This means, for example, that the difference must be computed by a dedicated combinational device (such as a subtractor). Such a combinational device is always computing the difference between r1 and y (even though that difference is loaded into r1 only when the controller is in state COMPUTE).

Loading r3 with the old value of r2 is easy. The output port of the r2 counter register is simply connected via a bus to the input port of the r3 enabled register. If the only state to mention r1, r2 or r3 were state COMPUTE, we would have the following architecture:

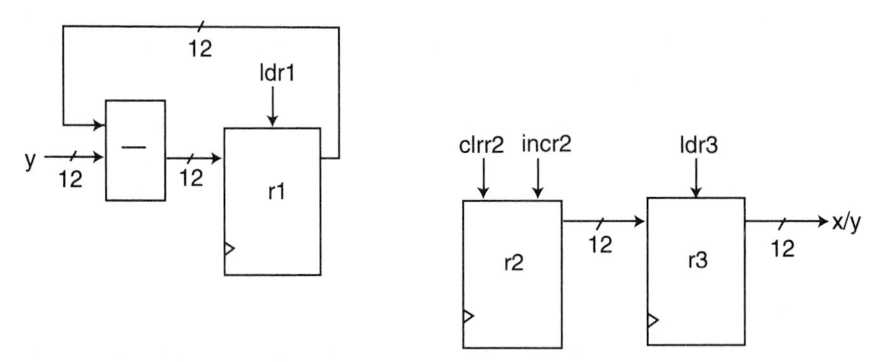

Figure 2-26. Architecture using subtractor.

but the above architecture fails to implement the RTN of state IDLE. The above architecture provides no way for r1 to be loaded with x.

One approach that often allows an architecture to deal with different kinds of RTN in different states is to use an Arithmetic Logic Unit (ALU), which is capable of many different operations, instead of a dedicated combinational device (such as a subtractor). Also, there is often a need for one or more muxes so that the proper information can be routed to the ALU. In this particular ASM, there are only two different results that might be loaded into r1: either the difference of r1 and y or passing through x unchanged. This means the ALU must be capable of at least two different operations: computing the difference of the ALU's two data inputs and passing through the ALU's second input unchanged. The ALU is commanded to do these operations by particular bit patterns on the six-bit aluctrl bus. Symbolically, we will refer to these bit patterns as 'DIFFERENCE and 'PASS. The grave accent ('), which is also known as backquote or tick, indicates a symbol that is replaced by a particular bit pattern.

On one hand, the ALU should be able to subtract y; on the other hand, the ALU should be able to pass x. To accomplish this requires a mux which can select either x or y. The output of this mux is connected as the second input of the ALU. Input 0 of this mux is connected to the external bus x. Input 1 of this mux is connected to the external bus y.

When `muxctrl` is 0, the output of the mux is the same as `x`. When `muxctrl` is 1, the output of the mux is the same as `y`. Using the mux and ALU, the architecture now appears as follows:

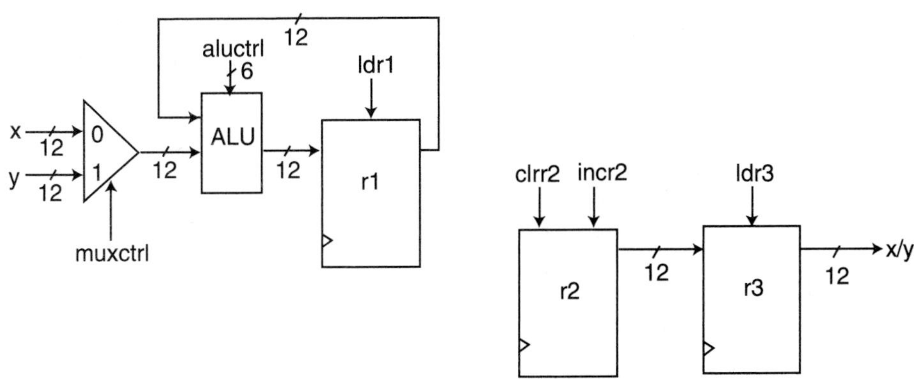

Figure 2-27. Architecture using ALU.

Although the above architecture implements all the RTN of the ASM chart in section 2.2.7, it does not consider the relational decision `r1>=y`. The simplest way to translate relational decisions into the mixed stage is to dedicate a combinational device (usually a comparator) for calculating an *internal status* signal that indicates the result of the relational comparison. In the mixed ASM, this internal status signal will be tested instead of referring to the relational decision. This is why ultimately the ASM uses only status signals. In this particular instance, we will use a comparator whose inputs are `r1` and `y`.

There are three outputs of a comparator: the strictly <, the exactly == and the strictly >. There is no >= output, but we can obtain that output, since it is the complement of the strictly < output. We will use the strictly < output as the input of an inverter, whose output is the internal status signal `r1gey`.

At last, we have an architecture which can correctly implement all the RTN and relational decisions of the ASM chart in section 2.2.7. Now it will be a mechanical matter to translate the pure behavioral ASM chart of section 2.2.7 into an equivalent mixed ASM chart. The purpose of this translation will be to use command signals (such as `incr2`) instead of RTN, and to use status signals (such as `r1gey`) instead of relational decisions. The ← in RTN translates to a command signal (such as `ldr1, clrr2, incr2` or `ldr3`) corresponding to the register on the left of the arrow. The computation on the right of the arrow may or may not require additional commands directed to the combinational logic units, such as the ALU and mux.

This translation from pure behavioral ASM to mixed ASM always relates to a particular architecture that the designer has in mind. Although many architectures might have been chosen for one pure behavioral ASM, each architecture will have a distinct mixed ASM. The following shows the particular architecture we have just developed and the corresponding translation of the pure behavioral ASM of section 2.2.7 into the particular mixed ASM required by this architecture. Finally, we give a system block diagram showing the interconnection of this particular controller (as described by the mixed ASM) and this particular architecture:

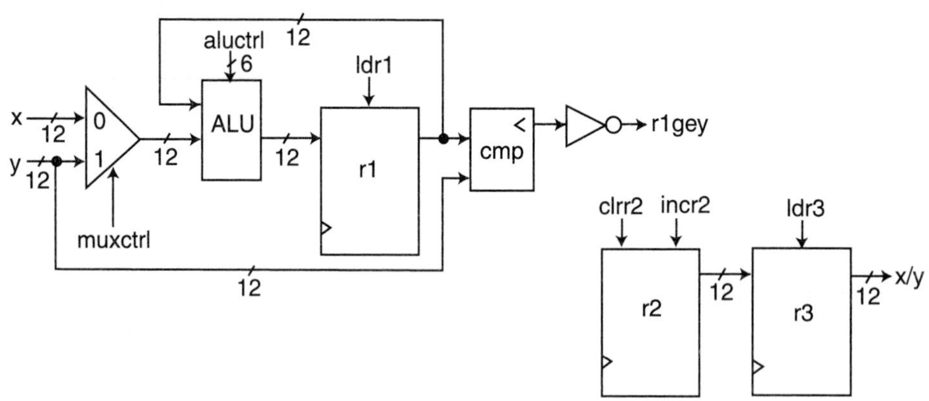

Figure 2-28. Methodical architecture.

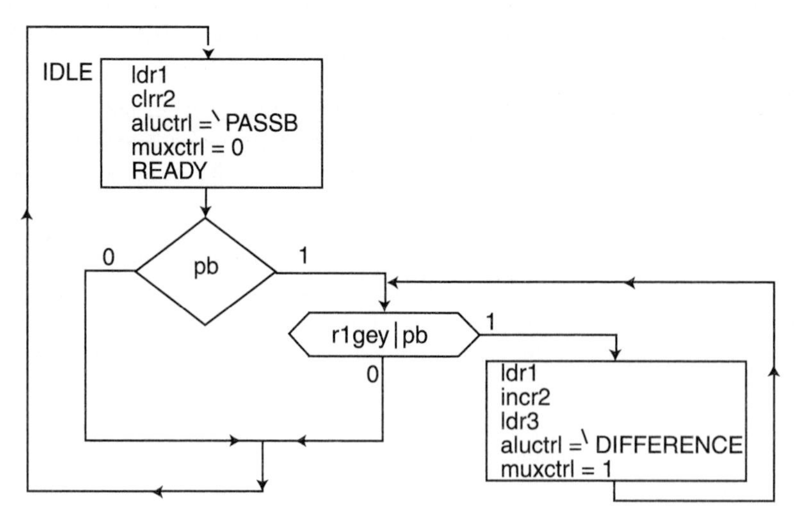

Figure 2-29. Mixed ASM corresponding to figures 2-24 and 2-28.

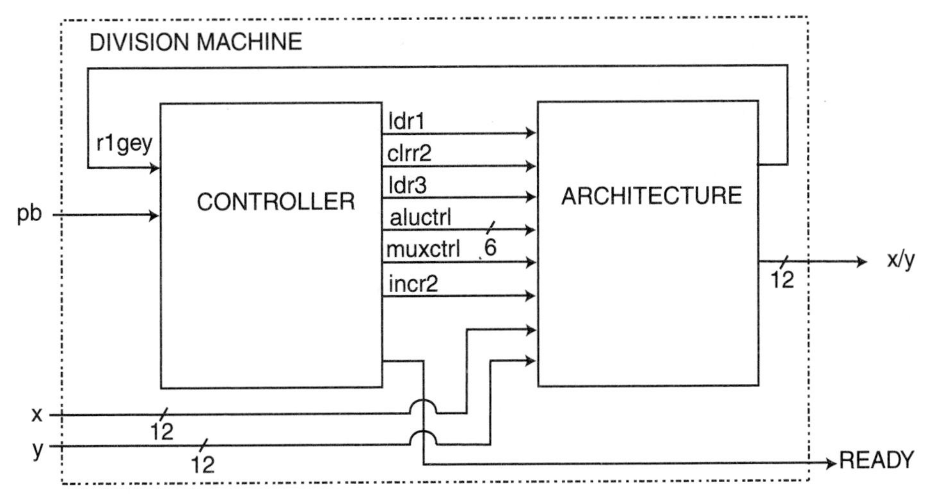

Figure 2-30. System diagram.

The external command READY and the external status `pb` are not affected by the translation to the mixed stage.

2.3.2 Second example

It happens in this example that the architecture just developed could have been used to implement any of the ASM charts in section 2.2. (It is rare for the same methodical architecture to work with different ASMs.) In general, this would not be the case, but it just happens that the ASM chart of section 2.2.7 requires maximal parallelism of the same register transfers as the other ASMs. All the other ASM charts of section 2.2 implement the same RTN commands with less parallelism, and so an architecture designed for maximal parallelism can implement an ASM that demands less parallelism. For example, the first ASM of section 2.2.2 could be implemented using the same architecture and system block diagram. The register `r3` is not used by this ASM chart, thus the data output of the machine should be `r2` instead of `r3`. Figure 2-31 is the mixed ASM chart that corresponds to this architecture and the ASM of section 2.2.2:

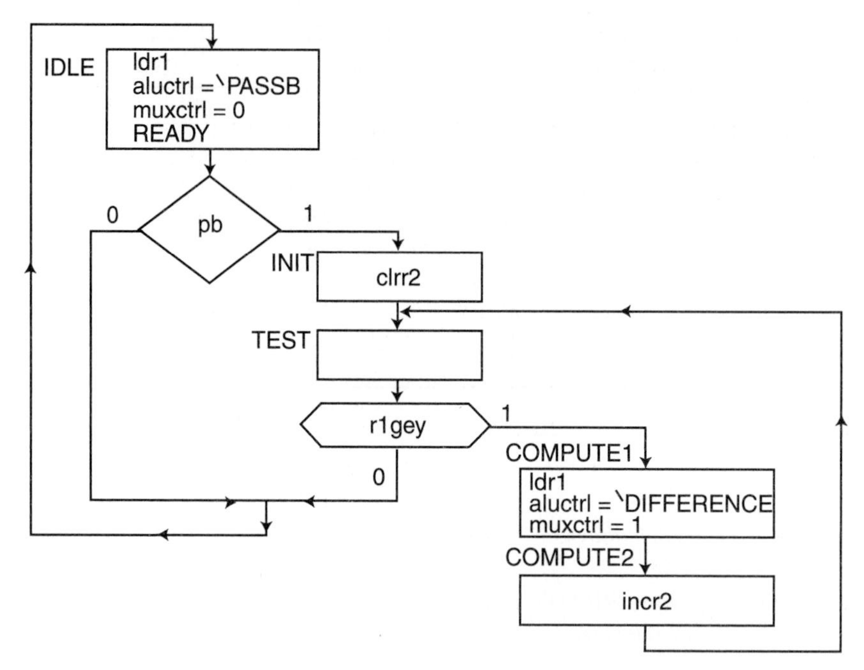

Figure 2-31. Mixed ASM corresponding to figures 2-14 and 2-28.

2.3.3 Third example

If we are content with the slower ASM chart of section 2.2.2, perhaps we can find a cheaper architecture (with less potential for parallelism) that can correctly implement the RTN given in section 2.2.2. Of course, such an architecture could not implement ASM charts, such as that in section 2.2.7, which require more parallelism. One way to reduce cost (at the expense of speed) is to use the ALU as a central unit that can do many different operations. In particular, it can output the value zero (when `aluctrl` is 'ZERO) and it can increment (when `aluctrl` is 'INCREMENT). Since the ALU is used for everything, the mux must have enough inputs to provide anything required by the ALU. In this instance, there is a three-input mux (with a two-bit `muxctrl`). This allows the mux to select `x`, `r1` or `r2` to be output to the bus which is the a input of the ALU.

It is no longer necessary for `r2` to be a counter register since the ALU can increment its input, and the mux can provide the value of `r2` to the ALU. The output of the ALU must be available on a central bus, from which both `r1` and `r2` can be loaded. (This ASM does not use `r3`.)

The carry out (`cout`) status signal output from the ALU is available at no cost. It can be used to determine the result of the `r1>=y` test. This is permissible because the ASM of section 2.2.2 has an empty rectangle for state TEST. In the mixed ASM, this rectangle will not be empty, although no registers will be loaded. The ALU is simply commanded to compute a difference without issuing a register load signal. As a side effect of computing the difference, `~cout` indicates `r1>=y`. The following shows the architecture and mixed ASM chart for the third example:

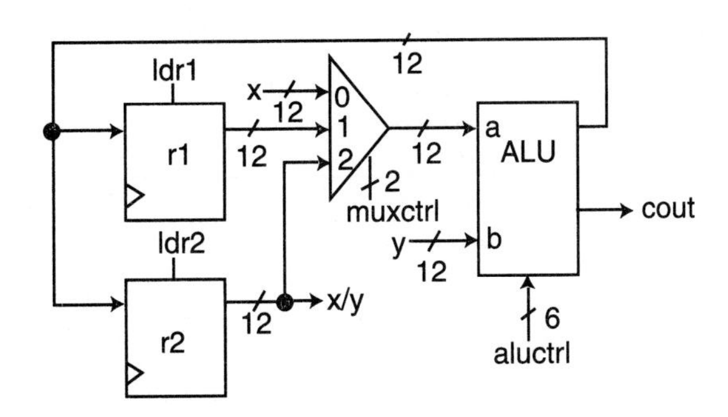

Figure 2-32. Central ALU architecture.

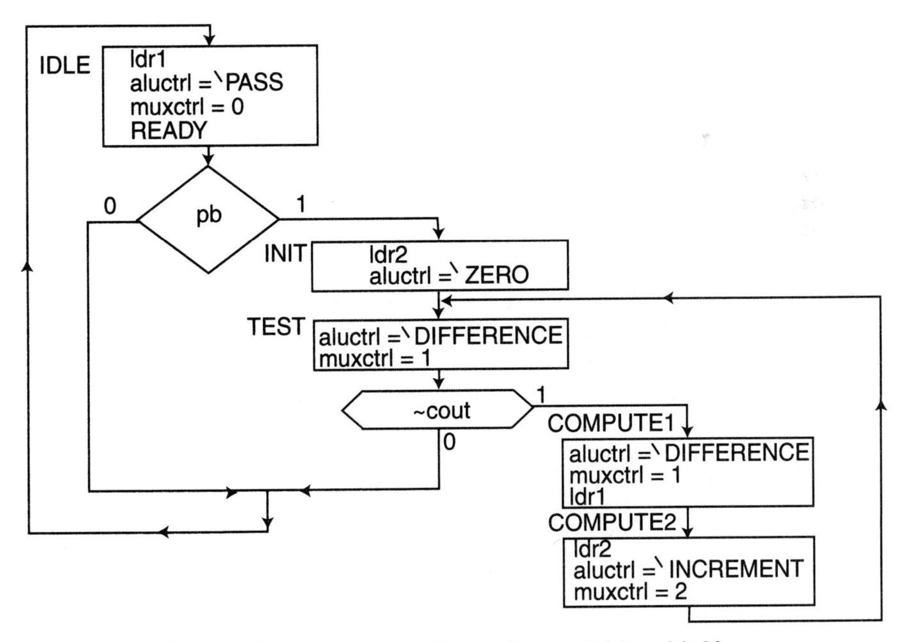

Figure 2-33. Mixed ASM corresponding to figures 2-14 and 2-32.

2.3.4 Methodical versus central ALU architectures

There is a spectrum of possible ways to choose an architecture at the mixed stage of the design process. At one extreme is the *central ALU approach*, illustrated in section 2.3.3, where one ALU does all the computation required by the entire algorithm. At the other extreme are *methodical approaches*, illustrated by sections 2.3.1 and 2.3.2, where each computation in the algorithm is done by a different hardware unit. The central ALU approach typically uses less hardware but only works with certain kinds of ASMs. For example, the ASM that works with the methodical approach in section 2.3.1 cannot work with the central ALU approach because that ASM performs more than one computation per clock cycle. The following table highlights the differences between these two approaches:

	Central ALU	Methodical
What does computation?	one ALU	registers themselves or registers tied to dedicated muxes and ALUs
What ALU output connects to?	every register	only one register
What kind of register?	enabled	all kinds[2]
Number of ← per clock cycle	one	many
Speed	slower	faster
Cost	lower	higher
Example	2.3.3	2.3.1 & 2.3.2
Figures	2-32, 2-33	2-28, 2-29, 2-31

In the methodical architecture of section 2.3.2, the output of the ALU only connects to `r1`, but in the central ALU architecture of section 2.3.3, the output of the ALU connects to both `r1` and `r2`. In the ASM implemented with the methodical architecture of

[2] Including customized registers (other than those described in appendix D) built using muxes and combinational logic that are tailored to the specific algorithm. See section 7.2.2.1 for an example.

section 2.3.1, there are multiple $\leftarrow$ per clock cycle. The central ALU approach is slower because it takes an ASM that uses more clock cycles to accomplish similar register transfers.

2.4 Pure structural example

In the mixed stage (described in section 2.3), the division machine was partitioned into a controller (described by a mixed ASM chart) and a structural architecture. This section continues with the same example (variations of the childish division algorithm) into the third stage (of the three stages for the top-down design process explained in section 2.1.5).

The third stage involves converting the mixed ASM chart into a hardware structure that implements the behavior described by the mixed ASM chart. At the end of the third stage, the top level of the system is completely described in terms of structure. The top level of the system is no longer described in terms of what it does (behavior) but is instead described in terms of how to build it (structure). The designer has enough information to wire together the hardware components into an operational physical machine. The algorithm has become a working piece of hardware.

2.4.1 First example

To translate the mixed ASM chart into hardware we must assign each symbolic state name a specific bit pattern. The bit patterns used are completely arbitrary. The only requirement is that the designer be consistent. One approach that is easy for the designer (when the number of states in the ASM chart is small) is to use a binary code. For an ASM with two states, a one-bit code suffices. For an ASM with three or four states, a two bit code will do. For an ASM with five to eight states, a three-bit code does the job. For an ASM with nine to sixteen states, the designer needs a four-bit code. In general, an ASM with `n` states requires a `ceil(log2(n))`-bit code.

The mixed ASM of section 2.3.1 is quite simple. It requires only two states. Let us say that being in state IDLE is represented by 0, and being in state COMPUTE is represented by 1.

We also need to know the bit patterns that control the ALU. As explained in appendix C, for an ALU inspired by the 74xx181, 'PASSB is 101010 and 'DIFFERENCE is 011001.

The structural controller is composed of two parts: the present state register, and the next state combinational logic. The status and the present state are the inputs of the combinational logic. The next state and the commands are the output of the combinational logic. The next state is the input to the present state register:

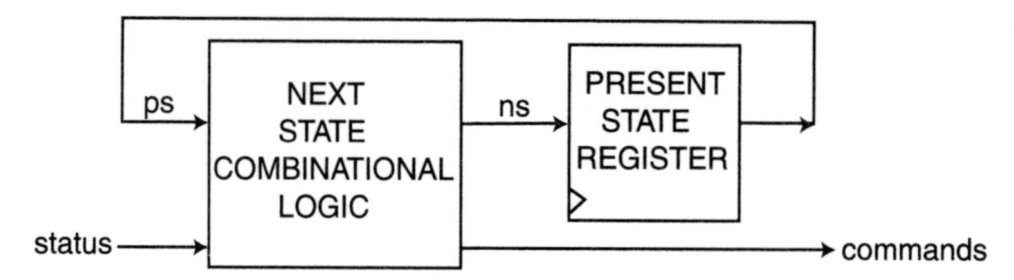

Figure 2-34. Controller.

We can describe the next state logic with a table. For the ASM chart of section 2.3.1, the corresponding table is:

inputs			outputs							
ps	pb	r1gey	ns	ldr1	clrr2	incr2	ldr3	muxctrl	aluctrl	ready
0	0	0	0	1	1	0	0	0	101010	1
0	0	1	0	1	1	0	0	0	101010	1
0	1	0	1	1	1	0	0	0	101010	1
0	1	1	1	1	1	0	0	0	101010	1
1	0	0	0	1	0	1	1	1	011001	0
1	0	1	1	1	0	1	1	1	011001	0
1	1	0	1	1	0	1	1	1	011001	0
1	1	1	1	1	0	1	1	1	011001	0

In the above, `ps` stands for the representation of the present state, and `ns` stands for the representation of the next state.

One possible hardware implementation of this table is a ROM. The above table can be used as is to "burn" the ROM. Since there are three bits of address input to the ROM (one bit for the present state, and two bits for the status), there are eight words (each 13-bits wide) stored in the ROM for this controller.

Another approach would be to use the above table to derive minimized logic equations for each bit of output and then use the logic equations to arrive at a structure composed of AND/OR gates. For example, the following logic equations are equivalent to the above table:

```
ns = ~ps&pb|ps&(r1gey|pb)
ldr1 = 1
clrr2 = ~ps
incr2 = ps
ldr3 = ps
muxctrl = ps
aluctrl[5] = ~ps
aluctrl[4] = ps
aluctrl[3] = 1
aluctrl[2] = 0
aluctrl[1] = ~ps
aluctrl[0] = ps
ready = ~ps
```

For a more complicated controller, using Karnaugh maps (or other logic minimization tools) could be helpful. Other approaches exist. Turning a table, such as the above, into actual hardware is a task that can be automated. Many software tools exist to aid in this job.

Several of the entries in the above table are identical to each other. For example, 000 and 001 are identical because the transition from IDLE to COMPUTE depends only on `pb` and not on `r1gey`. (The second ASM chart in section 2.2.7 makes this clear.) An abbreviated form of the table:

inputs			outputs							
ps	pb	r1gey	ns	ldr1	clrr2	incr2	ldr3	muxctrl	aluctrl	ready
0	0	-	0	1	1	0	0	0	101010	1
0	1	-	1	1	1	0	0	0	101010	1
1	0	0	0	1	0	1	1	1	011001	0
1	1	0	1	1	0	1	1	1	011001	0
1	-	1	1	1	0	1	1	1	011001	0

shows a "don't care" as a hyphen for those status inputs that do not affect a particular state transition. This table means exactly the same thing as the longer form of the table given earlier.

2.4.2 Second example

Assuming that the five states in the ASM chart of section 2.3.2 are represented as follows:

```
IDLE      000
INIT      001
TEST      010
COMPUTE1  011
COMPUTE2  100
```

the following table describes the controller:

inputs			outputs						
ps	pb	r1gey	ns	ldr1	clrr2	incr2	muxctrl	aluctrl	ready
000	0	-	000	1	0	0	0	101010	1
000	1	-	001	1	0	0	0	101010	1
001	-	-	010	0	1	0	0	101010	0
010	-	0	000	0	0	0	0	101010	0
010	-	1	011	0	0	0	0	101010	0
011	-	-	100	1	0	0	1	011001	0
100	-	-	010	0	0	1	0	101010	0
101	-	-	000	0	0	0	0	101010	0
11-	-	-	000	0	0	0	0	101010	0

Using the "don't care" form of the table is useful because otherwise the table would be 32-lines long. The values of `muxctrl` and `aluctrl` in states INIT, TEST and COMPUTE2 are arbitrary. There are three extra state encodings (101, 110 and 111) that should never occur in the proper operation of the machine. On power up, the physical hardware might find itself in one of these states. To avoid problems of this kind, these states go to state IDLE but otherwise do nothing.

2.5 Hierarchical design

Even though upon completion of the third stage, the top level of the system is described with pure structure, some devices (actors) that make up part of the architecture or controller may still be described in terms of behavior. For example, if an architecture needs an adder, the block diagram for the architecture would simply show a device with two inputs (let's label them `a` and `b`) and one output (`sum`). There are many possible hardware structures that could implement such an adder. At the top level of the system we will view the black box of an adder as being part of a structure, even though we have not specified what internal structure (of AND/OR gates for instance) implements the adder. Whatever internal structure is used, we can describe the behavior of the adder with a one-state ASM chart. ASM charts such as this that have one state correspond to combinational logic. The following shows the black box for a two- bit adder, the ASM chart for the adder, and one of many possible internal structures (circuits) for the adder:

The above circuit diagram is *flattened*, which means it does not show any hierarchy. A flattened circuit diagram, composed entirely of logic gates, each producing a single bit of output, is equivalent to a *netlist*. A netlist is a list of gates with the corresponding

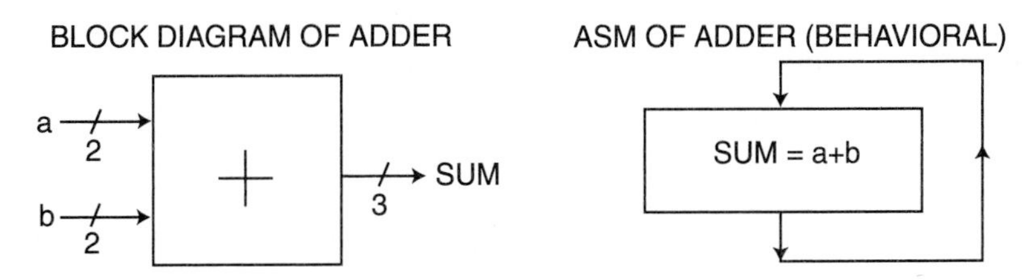

Figure 2-35. Block diagram and behavioral ASM for adder.

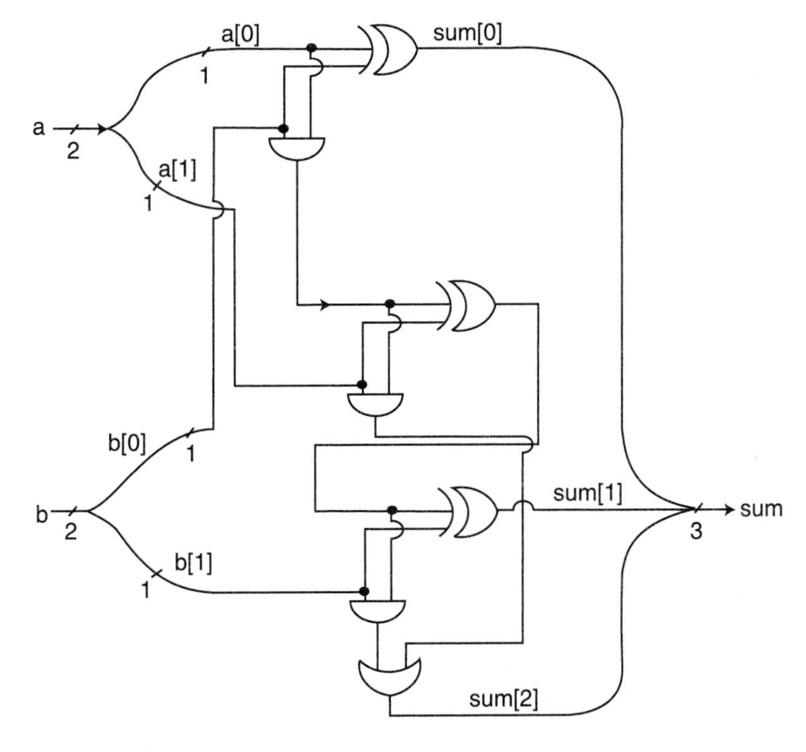

Figure 2-36. Flattened circuit diagram for adder

one-bit wires (nets) that each gate is connected to. A netlist can be submitted to a *silicon foundry* to fabricate an actual integrated circuit, or it can be used manually as directions (to a low-skilled worker) on how to hook together copper wire to form the desired machine. Although a flattened circuit diagram (or its equivalent netlist) is ultimately what is used to build a machine, it does not express much of the thought process the designer uses to arrive at the final circuit diagram.

The term *hierarchical* design refers to keeping a hierarchy of design components in the final design. Hierarchical design applies only to hardware structures, and not to behavioral ASMs. In hierarchical designs, the designer documents how to arrive at the final circuit diagram. Instead of just saying how gates are interconnected, the designer defines *modules*. For example, the two-bit adder block diagram is a module. The definition of a module occurs by *instantiating* other modules. In this example, the adder is composed of a full-adder and a half-adder, and so the designer instantiates a full-adder and also instantiates a half-adder in the definition of the adder.

The designer has to define a separate module for the full-adder. The full-adder is composed of two half-adders and an OR gate. The two half-adders that the designer instantiates in the definition of the full-adder are like tract houses in a boring suburb. Such houses are build from identical blueprints and therefore look the same. But different people live in each house. So it is for the two half-adders that compose the full-adder. They are instantiated from an identical "blueprint" (the definition of the half-adder will be discussed in the next paragraph), but different data is processed by each half-adder.

In the final circuit, the half-adder is instantiated three times. Each of the three half-adders is in turn composed of an exclusive-OR gate and an AND gate. The following diagrams show the adder module definition, full-adder module definition, half-adder module definition and a circuit diagram with dotted lines showing the instantiation of these definitions inside each other.

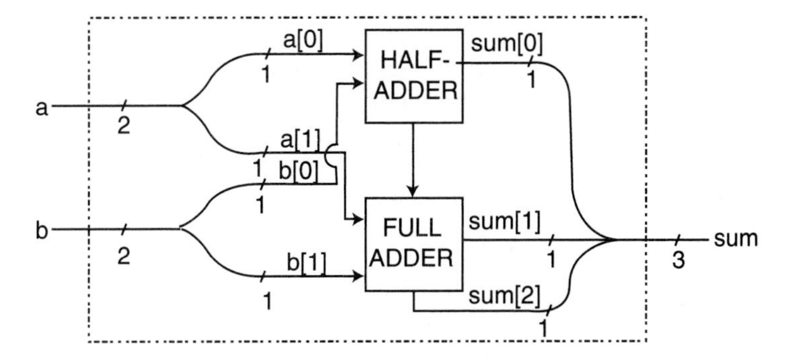

Figure 2-37. Definition of the adder module.

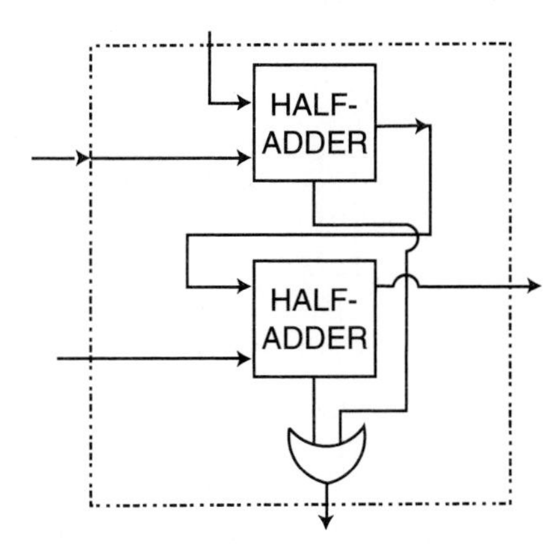

Figure 2-38. Definition of the full-adder module.

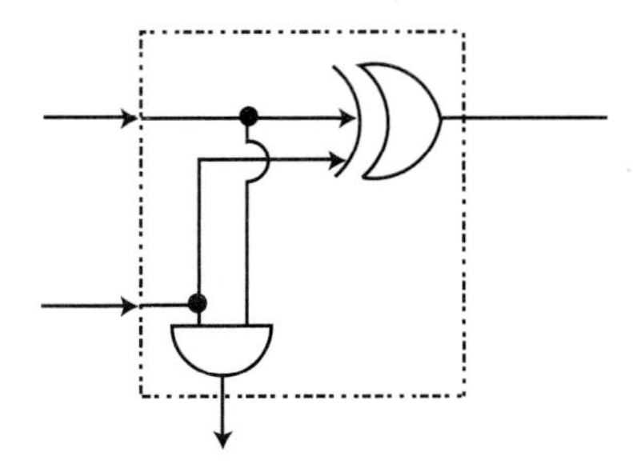

Figure 2-39. Definition of the half-adder module.

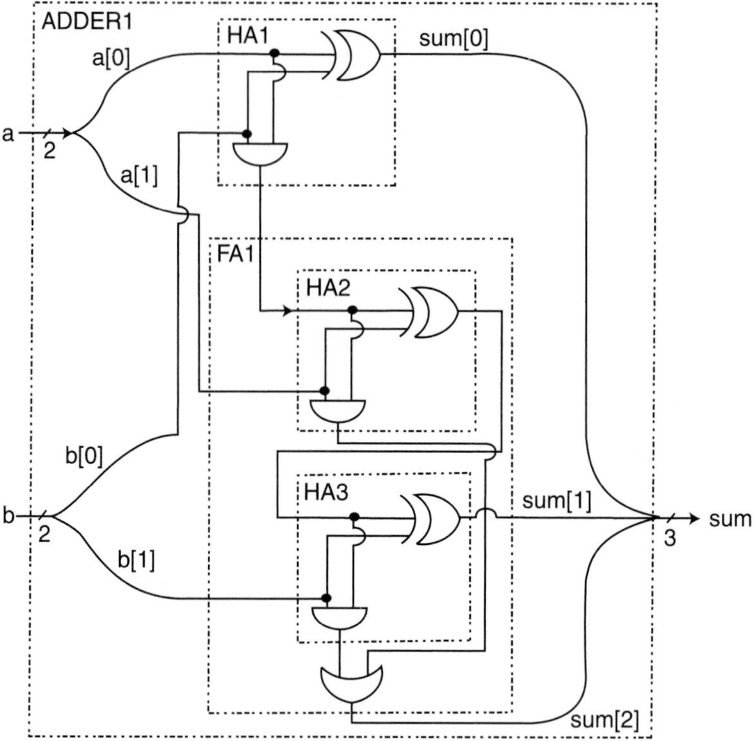

Figure 2-40. Hierarchical instantiation of modules.

Although hierarchical design can be used with either bottom-up or top-down design, it is most important with top-down design. In top-down design, upon completion of the third stage, the designer may apply the same three stages over again on any components (actors) that are not standard building blocks. Building blocks such as adders are well understood, and the designer is not normally concerned about their internal structure. Other problem-specific building blocks (such as the push button in the division machine) would be dealt with at the end of the third stage for the top-level system.

2.5.1 How pure is "pure"?

The terms "pure behavioral" and "pure structural" can be a little confusing. No matter at what level of abstraction you view a hardware system, there is some irreducible structure and some irreducible behavior. In the "pure" behavioral stage, the input and output ports are part of some larger (unspecified) structure. In the "pure" structural stage, the nature of the black boxes instantiated in the hardware diagram is known only by their behavior. Even if the designer takes the hierarchy all the way down to the gate level, the nature of each gate is known only by its behavior.

2.5.2 Hierarchy in the first example

Sections 2.2.7, 2.3.1 and 2.4.1 discuss the details for the three stages for the example two-state version of the division machine. The following three diagrams illustrate the increasing amount of structure and decreasing amount of behavior as the designer progresses through the three stages. In each of these diagrams, black boxes (and architectural devices, which are in fact black boxes themselves) represent aspects of the system whose internal nature is known only by behavior.

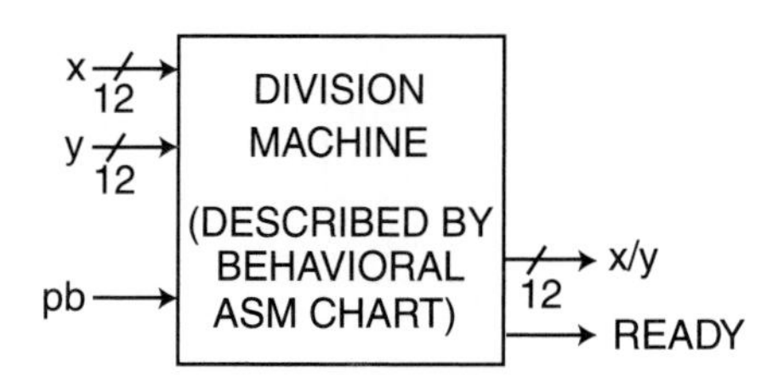

Figure 2-41. "Pure" behavioral block diagram.

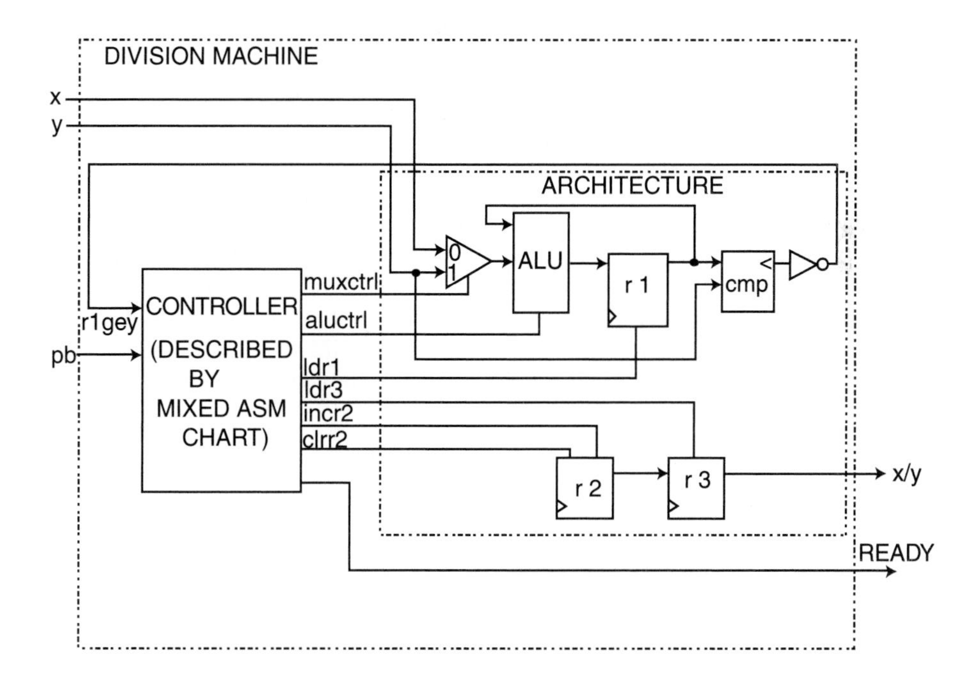

Figure 2-42. Mixed block diagram.

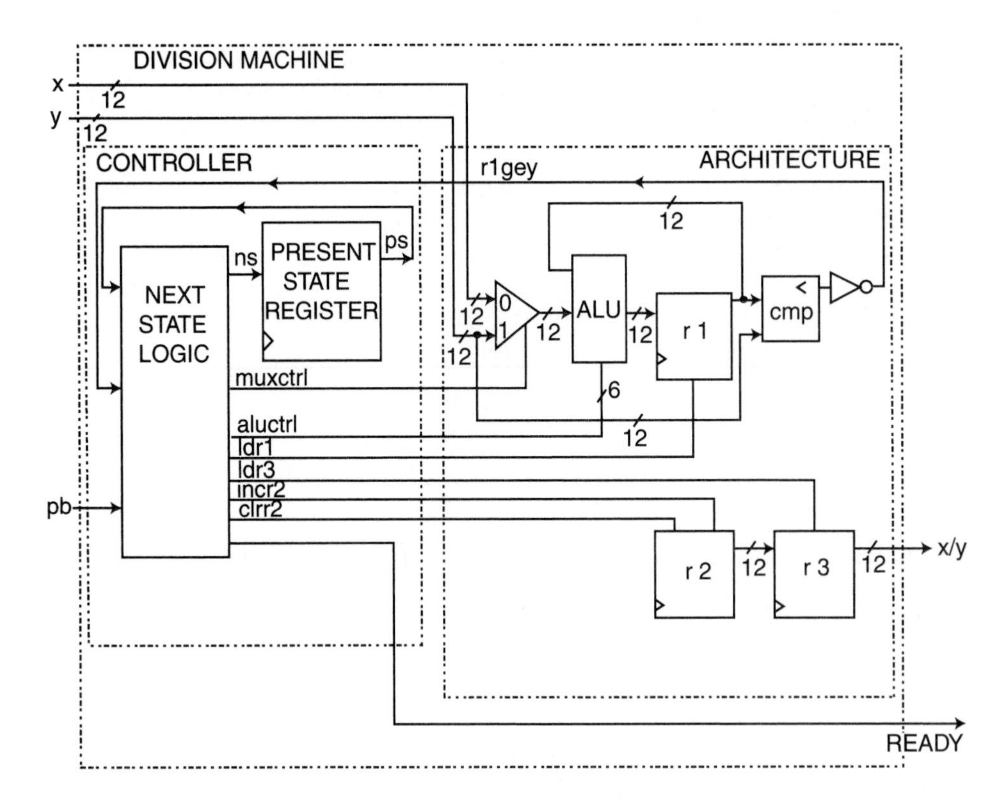

Figure 2-43. "Pure" strucural block diagram.

The first diagram (figure 2-41), which illustrates the "pure" behavioral stage, has a single black box. This box represents the complete division machine. The machine has a unified behavior which can be described by a single ASM chart (see section 2.2.7). The only structure in the first stage is the port structure that allows the machine to communicate with the outside world.

The second diagram (figure 2-42) shows the "mixed" stage. Instantiated inside the division machine are the controller and the architecture. In the mixed stage, the structure of the controller remains a mystery (i.e., a black box) which is described only in behavioral terms by the ASM chart of section 2.3.1. This ASM chart refers only to status and command signals. The architecture, on the other hand, is visible; however architectural devices (such as the ALU) remain as black boxes, known only by their behavior.

The third diagram (figure 2-43) shows the "pure" structural stage. Here the controller includes the next state logic and the present state (`ps`) register. The architecture remains the same as in the "mixed" stage. Even though this is the "pure" structural stage, the internal structure of the black boxes (next state logic, mux, ALU, comparator, inverter, `ps`, `r1`, `r2` and `r3` registers) remains hidden.

Here is where the power of hierarchical design comes to aid us in our top-down approach. We do not need to worry about the gate-level details of, say, the ALU. As the earlier full-adder example illustrates, such behavioral black boxes can be decomposed down to a gate-level netlist. For our purposes, we will be content with the third diagram. The reason we do not have to work our way down to reach the gate level is that automated tools now exist to do this dirty work.

2.6 Conclusion

This chapter illustrates two manual *graphical* notations: the ASM chart (to describe behavior) and the block diagram (to describe structure). There are three stages in the top-down design process to turn an algorithm (behavior) into hardware (structure): pure behavioral, mixed and pure structural. The mixed and pure structural stages partition the machine into a controller and an architecture.

In addition, this chapter describes instantiating structural modules (hierarchical design). This allows the pure structural stage to be described in an understandable way, without having to descend to the extreme gate-level detail of a netlist.

The next chapter introduces an automated *textual* notation that allows us to express behavioral, structural and hierarchical design in a unified notation.

2.7 Further Reading

Clair, C. R., *Designing Logic Systems Using State Machines,* McGraw-Hill, New York, 1973. This short but influential book was the first to explain the ASM chart notation, which T. E. Osborne had invented in the early 1960s at Hewlett Packard.

Gajski, Daniel D., *Principles of Digital Design*, Prentice Hall, Upper Saddle River, NJ, 1997. Chapter 8 describes ASM charts. This book uses the term datapath to mean what is called an architecture here and uses = rather than ← for RTN.

PROSSER, FRANKLIN P. and DAVID E. WINKEL, *The Art of Digital Design: An Introduction to Top Down Design*, 2nd ed., PTR Prentice Hall, Englewood Cliffs, NJ, 2nd ed., 1987. Chapters 5-8 give several examples of ASM charts using RTN. This book uses the term architecture the way it is used here.

2.8 Exercises

2-1. Give a pure behavioral ASM for a factorial machine. The factorial of `n` is the product of the numbers from 1 to `n`. This machine has the data input, `n`, a push button input, `pb`, a data output, `prod`, and an external status output, READY. READY and `pb` are similar to those described in section 2.2.1. Until the user pushes the button, READY is asserted and `prod` continues to be whatever it was. When the user pushes the button, READY is no longer asserted and the machine computes the factorial by doing no more than one multiplication per clock cycle. For example, when `n=5` after an appropriate number of clock cycles `prod` becomes `120 == 1*1*2*3*4*5 == 1*5*4*3*2*1` and READY is asserted again.

Use a linear time algorithm in the input `n`, which means the exact number of clock cycles that this machine takes to compute `n!` for a particular value of `n` can be expressed as some constant times `n` plus another constant. (All of the childish division examples in this chapter are linear time algorithms in the quotient.) For example, a machine that takes `57*n+17` clock cycles to compute `n!` would be acceptable, but you can probably do better than that.

2-2. Design an architecture block diagram and corresponding mixed ASM that is able to implement the algorithm of problem 2-1 assuming the architecture is composed of the following building blocks: up/down counter registers, multiplier, comparator and muxes. Give a system diagram that shows how the architecture and controller fit together, labeled with appropriate signal names.

2-3. Give a table that describes the structural controller for problem 2-2.

2-4. Give a pure behavioral ASM similar to problem 2-1, but use repeated addition to perform multiplication. For example, `13*14 == 0 +13 +13 +13 +13 +13 +13 +13 +13 +13 +13 +13 +13 +13 +13`. Direct multiplication in a single cycle is not allowed. The algorithm should be suitable for implementation with the central ALU approach. This will be a quadratic time algorithm in `n` because of nested loops.

2-5. Design an architecture block diagram and corresponding mixed ASM for problem 2-4 assuming the following building blocks: enabled registers, muxes, comparator, and the ALU described in section C.6. Give a system diagram that shows how the architecture and controller fit together, labeled with appropriate signal names. Label the `a` and `b` inputs of the ALU.

2-6. Give a table that describes the structural controller for problem 2-5.

2-7. Give a pure behavioral ASM similar to problem 2-4, but use a shift and add algorithm to perform multiplication. Direct multiplication in a single cycle is not allowed. Here is an example of multiplying 14 by 13 using the shift and add algorithm with 4-bit input representations, and an 8-bit product.

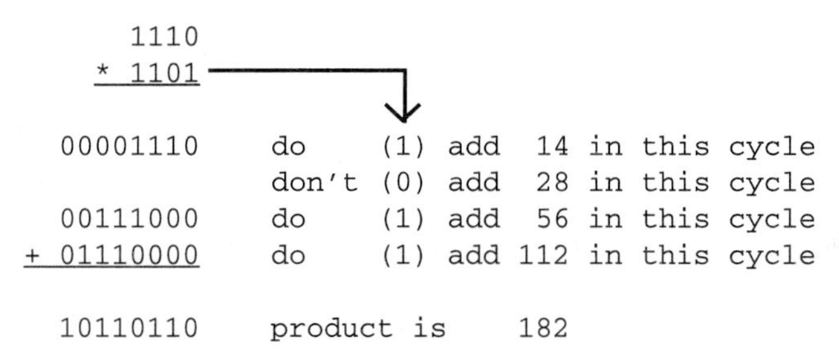

1110
* 1101
00001110 do (1) add 14 in this cycle
don't (0) add 28 in this cycle
00111000 do (1) add 56 in this cycle
+ 01110000 do (1) add 112 in this cycle
10110110 product is 182

The number of cycles to perform a single multiplication by `n` is proportional to the number of bits used to represent `n`, which is roughly the logarithm of `n`. But you have to perform `n` such multiplications, and so this factorial algorithm is what is called an `n log n` time algorithm, which takes more clock cycles than a linear time algorithm but fewer clock cycles than a quadratic time algorithm when `n` is large. (Note that unlike the linear time algorithm of problem 2-1, this approach does not require an expensive multiplier.) You should use a methodical approach that exploits maximal parallelism.

2-8. Design an architecture block diagram and corresponding mixed ASM that is able to implement problem 2-7 assuming the following building blocks: enabled registers, counter registers, shift registers, muxes, adder, comparator. Give a system diagram that shows how the architecture and controller fit together, labeled with appropriate signal names.

2-9. Give a table that describes the structural controller for problem 2-8. (See section D.9 for details about controlling a shift register.)

2-10. For each of the following ASMs, draw a timing diagram. `x`, `y` and `z` are 8-bit registers, whose values should be shown in decimal.

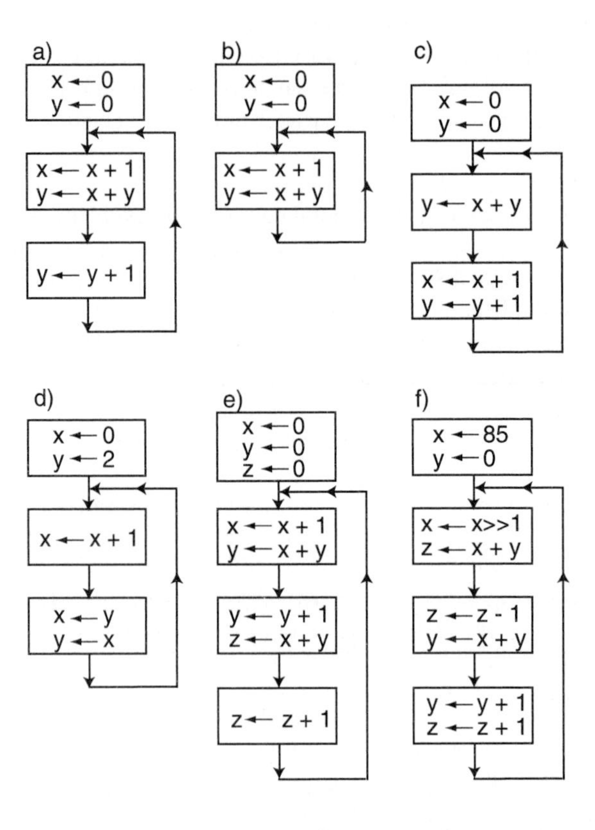

a)
x ← 0
y ← 0
x ← x + 1
y ← x + y
y ← y + 1
b)
x ← 0
y ← 0
x ← x + 1
y ← x + y
c)
x ← 0
y ← 0
y ← x + y
x ← x + 1
y ← y + 1
d)
x ← 0
y ← 2
x ← x + 1
x ← y
y ← x
e)
x ← 0
y ← 0
z ← 0
x ← x + 1
y ← x + y
y ← y + 1
z ← x + y
z ← z + 1
f)
x ← 85
y ← 0
x ← x>>1
z ← x + y
z ← z - 1
y ← x + y
y ← y + 1
z ← z + 1

2-11. Design an architecture block diagram and corresponding mixed ASM that is able to implement problem 2-10, part a assuming the following building blocks: two 8-bit counter registers, one 8-bit adder, and any number of any kind of mux. Give a system diagram that shows how the architecture and controller fit together, labeled with appropriate signal names.

2-12. Like 2-11, except use: one 8-bit counter register, one 8-bit enabled register, one 8-bit ALU and any number of any kind of mux. Label the `a` and `b` inputs of the ALU.

2-13. Design an architecture block diagram and corresponding mixed ASM that is able to implement problem 2-10, part b assuming the following building blocks: two 8-bit enabled registers, one 8-bit adder, one 8-bit incrementor and any number of any kind of mux. Give a system diagram.

2-14. Design an architecture block diagram and corresponding mixed ASM that is able to implement problem 2-10, part c assuming the following building blocks: one 8-bit counter register, one 8-bit enabled register, one 8-bit ALU (section C.6) and any number of any kind of mux. Give a system diagram. Label the a and b inputs of the ALU.

2-15. Design an architecture block diagram and corresponding mixed ASM that is able to implement problem 2-10, part d assuming the following building blocks: two 8-bit enabled registers, one 8-bit ALU (see section C.6) and one 8-bit two-input mux. Give a system diagram. Label the `a` and `b` inputs of the ALU.

2-16. Design an architecture block diagram and corresponding mixed ASM that is able to implement problem 2-10, part e assuming the following building blocks: two 8-bit counter registers, one 8-bit enabled register, one 8-bit ALU (section C.6) and any number of any kind of mux. Give a system diagram. Label the `a` and `b` inputs of the ALU.

2-17. Like 2-16, except use: one 8-bit incrementor, three 8-bit enabled registers, one 8-bit adder, and any number of any kind of mux.

2-18. Design an architecture block diagram and corresponding mixed ASM that is able to implement problem 2-10, part f assuming the following building blocks: one 8-bit shift register (section D.9), one 8-bit counter register, one 8-bit up/down counter register (section D.8) and one 8-bit adder. You may not use any muxes. Give a system diagram.

2-19. Like 2-18, except use: one 8-bit incrementor, one 8-bit decrementor, three 8-bit enabled registers, one 8-bit adder, one 8-bit combinational shifter, and any number of any kind of mux.

3. VERILOG HARDWARE DESCRIPTION LANGUAGE

The previous chapter describes how a designer may manually use ASM charts (to describe behavior) and block diagrams (to describe structure) in top-down hardware design. The previous chapter also describes how a designer may think hierarchically, where one module's internal structure is defined in terms of the instantiation of other modules. This chapter explains how a designer can express all of these ideas in a special hardware description language known as Verilog. It also explains how Verilog can test whether the design meets certain specifications.

3.1 Simulation versus synthesis

Although the techniques given in chapter 2 work wonderfully to design small machines by hand, for larger designs it is desirable to automate much of this process. To automate hardware design requires a Hardware Description Language (HDL), a different notation than what we used in chapter 2 which is suitable for processing on a general-purpose computer. There are two major kinds of HDL processing that can occur: simulation and synthesis.

Simulation is the interpretation of the HDL statements for the purpose of producing human readable output, such as a timing diagram, that predicts approximately how the hardware will behave before it is actually fabricated. As such, HDL simulation is quite similar to running a program in a conventional high-level language, such as Java Script, LISP or BASIC, that is interpreted. Simulation is useful to a designer because it allows detection of functional errors in a design without having to fabricate the actual hardware. When a designer catches an error with simulation, the error can be corrected with a few keystrokes. If the error is not caught until the hardware is fabricated, correcting the problem is much more costly and complicated.

Synthesis is the compilation of high-level behavioral and structural HDL statements into a flattened gate-level netlist, which then can be used directly either to lay out a printed circuit board, to fabricate a custom integrated circuit or to program a programmable logic device (such as a ROM, PLA, PLD, FPGA, CPLD, etc.). As such, synthesis is quite similar to compiling a program in a conventional high-level language, such as C. The difference is that, instead of producing object code that runs on the same computer, synthesis produces a physical piece of hardware that implements the computation described by the HDL code. For the designer, producing the netlist is a simple step (typically done with only a few keystrokes), but turning the netlist into physical hardware is often costly, especially when the goal is to obtain a custom integrated circuit from a commercial silicon foundry. Typically after synthesis, but before the physical fabrication, the designer simulates the synthesized netlist to see if its behavior matches the original HDL description. Such post-synthesis simulation can prevent costly errors.

3.2 Verilog versus VHDL

HDLs are textual, rather than graphic, ways to describe the various stages in the top-down design process. In the same language, HDLs allow the designer to express both the behavioral and structural aspects of each stage in the design. The behavioral features of HDLs are quite similar to conventional high-level languages. The features that make an HDL unique are those structural constructs that allow description of the instantiation and interconnection of modules.

There are many proprietary HDLs in use today, but there are only two standardized and widely used HDLs: Verilog and VHDL. Verilog began as a proprietary HDL promoted by a company called Cadence Data Systems, Inc., but Cadence transferred control of Verilog to a consortium of companies and universities known as Open Verilog International (OVI). Many companies now produce tools that work with standard Verilog. Verilog is easy to learn. It has a syntax reminiscent of C (with some Pascal syntax thrown in for flavor). About half of commercial HDL work in the U.S. is done in Verilog. If you want to work as a digital hardware designer, it is important to know Verilog.

VHDL is a Department of Defense (DOD) mandated language that is used primarily by defense contractors. Although most of the concepts in VHDL are not different from those in Verilog, VHDL is much harder to learn. It has a rigid and unforgiving syntax strongly influenced by Ada (which is an unpopular conventional programming language that the DOD mandated defense software contractors to use for many years before VHDL was developed). Although more academic papers are published about VHDL than Verilog, less than one-half of commercial HDL work in the U.S. is done in VHDL. VHDL is more popular in Europe than it is in the U.S.

3.3 Role of test code

The original purpose of Verilog (and VHDL) was to provide designers a unified language for simulating gate-level netlists. Therefore, Verilog combines a structural notation for describing netlists with a behavioral notation for saying how to test such netlists during simulation. The behavioral notation in Verilog looks very much like normal executable statements in a procedural programming language, such as Pascal or C. The original reason for using such statements in Verilog code was to provide *stimulus* to the netlist, and to test the subsequent *response* of the netlist. The pairs of stimulus and response are known as *test vectors.* The Verilgo that creates the stimulus and observes the response is known as the *test code* or *testbench.* Snoopy's "woof" in the comic strip of section 2.2 is analougus to the role of the test codes warning us that the expected response was not observed. For example, one way to use simulation to test whether a small machine works is to do an *exhaustive test*, where the test code provides each possible combination of inputs to the netlist and then checks the response of the netlist to see if it is appropriate.

For example, consider the division machine of the last chapter. Assume we have developed a flattened netlist that implements the complete machine. It would not be at all obvious whether this netlist is correct. Since the bus width specified in this problem is small (twelve bits), we can write Verilog test code using procedural Verilog (similar to statements in C) that does an exhaustive test. A reasonable approach would be to use two nested loops, one that varies x through all its 4096 possible values, and one that varies y through all its 4095 possible values. At appropriate times inside the inner loop, the test code would check (using an `if` statement) whether the output of the netlist matches `x/y`. Verilog provides most of the integer and logical operations found in C, including those, such as division, that are difficult to implement in hardware. The original intent was not to synthesize such code into hardware but to document how the netlist should automatically be tested during simulation.

Verilog has all of the features you need to write conventional high-level language programs. Except for file Input/Output (I/O), any program that you could write in a conventional high- level language can also be written in Verilog. The original reason Verilog provides all this software power in a “hardware” language is because it is impossible to do an exhaustive test of a complex netlist. The 12-bit division machine can be tested exhaustively because there are only 16,773,120 combinations with the 24 bits of input to the netlist. A well-optimized version of Verilog might be able to conduct such a simulation in a few days or weeks. If the bus width were increased, say to 32-bits, the time to simulate all 2^{64} combinations would be millions of years. Rather than give up on testing, designers write more clever test code. The test code will appear longer, but will execute in much less time. Of course, if a machine has a flaw that expresses itself for only a few of the 2^{64} test patterns, the probability that our fast test code will find the flaw is usually low.

3.4 Behavioral features of Verilog

Verilog is composed of modules (which play an important role in the structural aspects of the language, as will be described in section 3.10). All the definitions and declarations in Verilog occur inside a module.

3.4.1 Variable declaration

At the start of a module, one may declare variables to be `integer` or to be `real`. Such variables act just like the software declarations `int` and `float` in C. Here is an example of the syntax:

```
integer x,y;
real Rain_fall;
```

Underbars are permitted in Verilog identifiers. Verilog is case sensitive, and so `Rain_fall` and `rain_fall` are distinct variables. The declarations `integer` and `real` are intended only for use in test code. Verilog provides other data types, such as `reg` and `wire`, used in the actual description of hardware. The difference between these two hardware-oriented declarations primarily has to do with whether the variable is given its value by behavioral (`reg`) or structural (`wire`) Verilog code. Both of these declarations are treated like `unsigned` in C. By default, `reg`s and `wire`s are only one bit wide. To specify a wider `reg` or `wire`, the left and right bit positions are defined in square brackets, separated by a colon. For example:

```
reg [3:0] nibble,four_bits;
```

declares two variables, each of which can contain numbers between 0 and 15. The most significant bit of `nibble` is declared to be `nibble[3]`, and the least significant bit is declared to be `nibble[0]`. This approach is known as *little endian notation.* Verilog also supports the opposite approach, known as *big endian notation*:

```
reg [0:3] big_end_nibble;
```

where now `big_end_nibble[3]` is the least significant bit.

If you store a signed value[1] in a `reg`, the bits are treated as though they are unsigned. For example, the following:

```
four_bits = -5;
```

is the same as:

```
four_bits = 11;
```

[1] In order to simplify dealing with twos complement values, many implementations allow integers with an arbitrary width. Such declarations are like `regs`, except they are signed.

Verilog supports concatenation of bits to form a wider `wire` or `reg`, for example, `{nibble[2], nibble[1]}` is a two bit `reg` composed of the middle two bits of `nibble`. Verilog also provides a shorthand for obtaining a contiguous set of bits taken from a single `reg` or `wire`. For example, the middle two bits of `nibble` can also be specified as `nibble[2:1]`. It is legal to assign values using either of these notations.

Verilog also allows arrays to be defined. For example, an array of reals could be defined as:

```
real monthly_precip[11:0];
```

Each of the twelve elements of the array (from `monthly_precip[0]` to `monthly_precip[11]`) is a unique real number. Verilog also allows arrays of `wire`s and `reg`s to be defined. For example,

```
reg [3:0] reg_arr[999:0];
wire[3:0] wir_arr[999:0];
```

Here, `reg_arr[0]` is a four-bit variable that can be assigned any number between 0 and 15 by behavioral code, but `wir_arr[0]` is a four-bit value that cannot be assigned its value from behavioral code. There are one thousand elements, each four bits wide, in each of these two arrays. Although the `[]` means bit select for scalar values, such as `nibble[3]`, the `[]` means element select with arrays. It is **illegal** to combine these two uses of `[]` into one, as in `if(reg_arr[0][3])`. To accomplish this operation requires two statements:

```
nibble = reg_arr[0];
if (nibble[3]) ...
```

3.4.2 Statements legal in behavioral Verilog

The behavioral statements of Verilog include[2] the following:

```
var = expression;

if (condition)
  statement
```

[2] There are other, more advanced statements that are legal. Some of these are described in chapters 6 and 7.

Continued

```
if (condition)
  statement
else
  statement

while (condition)
  statement

for (var=expression;condition;var=var+expression)
  statement

forever
  statement

case (expression)
  constant: statement
  ...
  default: statement
endcase
```

where the italic *`statement`*, *`var`*, *`expression`*, *`condition`* and *`constant`* are replaced with appropriate Verilog syntax for those parts of the language. A *`statement`* is one of the above statements or a series of the above statements **terminated by semicolons inside** `begin` and `end`. A *`var`* is a variable declared as `integer`, `real`, `reg` or a concatenation of `reg`s. A *`var`* cannot be declared as `wire`.

3.4.3 Expressions

An *`expression`* involves constants and variables (including `wire`s) with arithmetic (`+`, `-`, `*`, `/`, `%`), logical (`&`, `&&`, `|`, `||`, `^`, `~`, `<<`, `>>`), relational (`<`, `==`, `===`, `<=`, `>=`, `!=`, `!==`, `>`) and conditional (`?:`) operators. A *`condition`* is an expression. A *`condition`* might be an expression involving a single bit, (as would be produced by `||`, `&&`, `!`, `<`, `==`, `===`, `<=`, `>=`, `!=`, `!==` or `>`) or an expression involving several bits that is checked by Verilog to see if it is equal to 1. Except for `===` and `!==`, these symbols have the same meaning as in C. Assuming the result is stored in a 16-bit `reg`,[3] the following table illustrates the result of these operators, for example where the left operand (if present) is ten and the right operand is three:

[3] Some results are different if the destination is declared differently.

<table>
<tr><th>symbol</th><th>name</th><th>example</th><th>16-bit unsigned result</th></tr>
<tr><td>+</td><td>addition</td><td>10+3</td><td>13</td></tr>
<tr><td>-</td><td>subtraction</td><td>10-3</td><td>7</td></tr>
<tr><td>-</td><td>negation</td><td>-10</td><td>65526</td></tr>
<tr><td>*</td><td>multiplication</td><td>10*3</td><td>30</td></tr>
<tr><td>/</td><td>division</td><td>10/3</td><td>3</td></tr>
<tr><td>%</td><td>remainder</td><td>10%3</td><td>1</td></tr>
<tr><td><<</td><td>shift left</td><td>10<<3</td><td>80</td></tr>
<tr><td>>></td><td>shift right</td><td>10>>3</td><td>1</td></tr>
<tr><td>&</td><td>bitwise AND</td><td>10&3</td><td>2</td></tr>
<tr><td>|</td><td>bitwise OR</td><td>10|3</td><td>11</td></tr>
<tr><td>^</td><td>bitwise exclusive OR</td><td>10^3</td><td>9</td></tr>
<tr><td>~</td><td>bitwise NOT</td><td>~10</td><td>65525</td></tr>
<tr><td>?:</td><td>conditional operator</td><td>0?10:3</td><td>3</td></tr>
<tr><td></td><td></td><td>1?10:3</td><td>10</td></tr>
<tr><td>!</td><td>logical NOT</td><td>!10</td><td>0</td></tr>
<tr><td>&&</td><td>logical AND</td><td>10&&3</td><td>1</td></tr>
<tr><td>||</td><td>logical OR</td><td>10||3</td><td>1</td></tr>
<tr><td><</td><td>less than</td><td>10<3</td><td>0</td></tr>
<tr><td>==</td><td>equal to</td><td>10==30</td><td></td></tr>
<tr><td><=</td><td>less than or equal to</td><td>10<=3</td><td>0</td></tr>
<tr><td>>=</td><td>greater than or equal</td><td>10>=3</td><td>1</td></tr>
<tr><td>!=</td><td>not equal</td><td>10!=3</td><td>1</td></tr>
<tr><td>></td><td>greater than</td><td>10>3</td><td>1</td></tr>
</table>

3.4.4 Blocks

All procedural statements occur in what are called *blocks* that are defined inside modules, after the type declarations. There are two kinds of procedural blocks: the `initial` block and the `always` block. For the moment, let us consider only the `initial` block. An `initial` block is like conventional software. It starts execution and eventually (assuming there is not an infinite loop inside the `initial` block) it stops execution. The simplest form for a single Verilog `initial` block is:

```
module top;

 declarations;

 initial
   begin
     statement;
     ...
     statement;
   end

endmodule
```

The name of the module (`top` in this case) is arbitrary. The syntax of the *`declarations`* is as described above. All variables should be declared. Each *`statement`* is terminated with a semicolon. Verilog uses the Pascal-like `begin` and `end`, rather than { and }. There is no semicolon after `begin` or `end`. The `begin` and `end` may be omitted in the rare case that only one procedural statement occurs in the `initial` block.

Here is an example that prints out all 16,773,120 combinations of values described in section 3.3:

```
module top;
 integer x,y;
 initial
   begin
     x = 0;
     while (x<=4095)
       begin
         for (y=1; y<=4095; y = y+1)
           begin
             $display("x=%d y=%d",x,y);
           end
         x = x + 1;
       end
   end
   $write("all ");
   $display("done");
endmodule
```

The loop involving x could have been written as a `for` loop also but was shown above as a `while` for illustration. Note that Verilog does not have the ++ found in C, and so it is necessary to say something like `y = y + 1`. This assignment statement is just like its counterpart in C: it is instantaneous. The variable changes value before the next statement executes (unlike the RTN discussed in the previous chapter). The `$display` is a *system task* (which begin with $) that does something similar to what `printf("%d %d \n",x,y)` does in C: it formats the textual output according to the string in the quotes. The system task `$write` does the same thing as `$display`, except that it does not produce a new line:

```
x=    0 y=    1
x=    0 y=    2
      ...      ...
x= 4095 y= 4094
x= 4095 y= 4095
all done
```

The above code would fail if the declaration had been:

```
reg [11:0] x,y;
```

because, although twelve bits are adequate for the hardware, the test code requires that `x` and `y` become 4096 in order for the loop to stop.

Since infinite loops are useful in hardware, Verilog provides the syntax `forever`, which means the same thing as `while(1)`. In addition, the `always` block mentioned above can be described as an `initial` block containing only a `forever` loop. For simulation purposes, the following mean the same:

```
initial               initial
  begin                 begin
    while(1)              forever       always
      begin                 begin         begin
        ...                   ...           ...
      end                   end           end
  end                   end
```

For synthesis, one should use the `always` block form only. The statement `forever` is not a block and cannot stand by itself. Like other procedural statements, `forever` must be inside an `initial` or `always` block.

3.4.5 Constants

By default, constants in Verilog are assumed to be decimal integers. They may be specified explicitly in binary, octal, decimal, or hexadecimal by prefacing them with the syntax `'b`, `'o`, `'d`, or `'h`, respectively. For example, `'b1101`, `'o15`, `'d13`, `'hd`, and 13 all mean the same thing. If you wish to specify the number of bits in the representation, this proceeds the quote: `4'b1101`, `4'o15`, `4'd13`, `4'hd`.

3.4.6 Macros, `include` files and comments

As an aid to readability of the code, Verilog provides a way to define macros. For example, the `aluctrl` codes described in 2.3.1 can be defined with:

```
`define DIFFERENCE  6'b011001
`define PASSB       6'b101010
```

Later in the code, a reference to these macros (preceded by a backquote) is the same as substituting the associated value. The following `if`s mean the same:

```
if (aluctrl == `DIFFERENCE)          if (aluctrl == 6'b011001)
 $display("subtracting");             $display("subtracting");
```

Note the syntax difference between variables (such as `aluctrl`), macros (such as `` `DIFFERENCE ``), and constants (such as `6'b011001`). Variables are not preceded by anything. Macros are preceded by backquote. Constants may include one forward single quote.

You can determine whether a macro is defined using `` `ifdef `` and `` `endif ``. This preprocessing feature should not be confused with `if`. For example, the following:

```
`ifdef DIFFERENCE
   $display("defined");
`endif
```

prints the message regardless of the value of `` `DIFFERENCE ``, as long as that macro is defined. The message is not printed only when there is not a `` `define `` for `` `DIFFERENCE ``.

Verilog allows you to separate your source code into more than one file (just like `#include` in C and `{$I}` in Pascal). To use code contained in another file, you say:

```
`include "filename.v"
```

There are two forms of comments in Verilog, which are the same as the two forms found in C++. A comment that extends only for the rest of the current line can occur after `//`. A comment that extends for several lines begins with `/*` and ends with `*/`. For example:

```
        /*  a multi line comment
            that includes a declaration:
reg a;
            which is ignored by Verilog
        */
reg b;  //  this declaration is not ignored
```

3.5 Structural features of Verilog

Verilog provides a rich set of built-in logic gates, including `and`, `or`, `xor`, `nand`, `nor`, `not` and `buf`, that are used to describe a netlist. The syntax for these structural features of Verilog is quite different than for any of the behavioral features of Verilog mentioned earlier. The outputs of such gates are declared to be `wire`, which by itself describes a one-bit data type. (Regardless of width, an output generated by structural Verilog code must be declared as a `wire`.) The inputs to such gates may be either declared as `wire` or `reg` (depending on whether the inputs are themselves computed by structural or behavioral code). To instantiate such a gate, you say what kind of gate you want (`xor` for example) and the name of this particular instance (since there may be several instances of `xor` gates, let's name this example `x1`). Following the instance name, inside parentheses are the output and input ports of the gate (for example, say the output is a `wire` named `c`, and the inputs are `a` and `b`). The output(s) of gates are always on the left inside the parentheses:

```
module easy_xor;
   reg a,b;
    wire c;
    xor x1(c,a,b);
    ...
endmodule
```

People familiar with procedural programming languages, like C, mistakenly assume this is "passing `c,` `a` and `b` and then calling on `xor`." **It is doing no such thing**. It simply says that an `xor` gate named `x1` has its output connected to `c` and its inputs connected to `a` and `b`. If you are familiar with graph theory, this notation is simply a way to describe the edges (`a,b,c`) and vertex (`x1`) of a graph that represents the structure of a circuit.

3.5.1 Instantiating multiple gates

Of course, there is an equivalent structure of `and/or` gates that does the same thing as an `xor` gate (recall the identity `a^b == a&(~b)|(~a)&b`):

```
module hard_xor;
  reg a,b;
  wire c;
  wire t1,t2,not_a,not_b;

  not i1(not_a,a);
  not i2(not_b,b);
  and a1(t1,not_a,b);
  and a2(t2,a,not_b);
  or  o1(c,t1,t2);
  ...
endmodule
```

The order in which gates are instantiated in structural Verilog code does not matter, and so the following:

```
module scrambled_xor;
  reg a,b;
  wire c;
  wire t1,t2,not_a,not_b;

  or  o1(c,t1,t2);
  and a1(t1,not_a,b);
  and a2(t2,a,not_b);
  not i1(not_a,a);
  not i2(not_b,b);
  ...
endmodule
```

means the same thing, because they both represent the interconnection in the following circuit diagram:

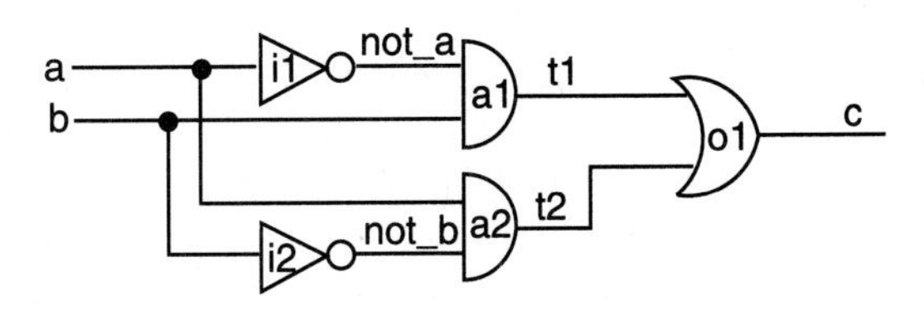

Figure 3-1. Exclusive or built with ANDs, OR and inverters.

3.5.2 Comparison with behavioral code

Structural Verilog code does not describe the order in which computations implemented by such a structure are carried out by the Verilog simulator. This is in sharp contrast to behavioral Verilog code, such as the following:

```
module behavioral_xor;

   reg a,b;
   reg c;
   reg t1,t2,not_a,not_b;

   always ...
      begin
        not_a = ~a;
        not_b = ~b;
        t1 = not_a&b;
        t2 = a&not_b;
        c = t1|t2;
      end
endmodule
```

which is a correct behavioral rendition of the same idea. (The ellipses must be replaced by a Verilog feature described later.) Also, `c`, `t1`, `t2`, `not_a` and `not_b` must be declared as `regs` because this behavioral (rather than structural) code assigns values to them.

To rearrange the order of behavioral assignment statements is incorrect:

```
module bad_xor;

   reg a,b;
   reg c;
   reg t1,t2,not_a,not_b;

   always ...
     begin
       c = t1|t2;
       t1 = not_a&b;
       t2 = a&not_b;
       not_a = ~a;
       not_b = ~b;
     end
endmodule
```

because `not_a` must be computed before `t1` by the Verilog simulator.

3.5.3 Interconnection errors: four-valued logic

In software, a bit is either a 0 or a 1. In properly functioning hardware, this is usually the case also, but it is possible for gates to be wired together incorrectly in ways that produce electronic signals that are neither 0 nor 1. To more accurately model such physical possibilities,[4] each bit in Verilog can be one of four things: `1'b0`, `1'b1`, `1'bz` or `1'bx`.

Obviously, `1'b0` and `1'b1` correspond to the logical 0 and logical 1 that we would normally expect to find in a computer. For most technologies, these two possibilities are represented by a voltage on a wire. For example, active high TTL logic would represent `1'b0` as zero volts and `1'b1` as five volts. Active low TTL logic would represent `1'b0` as five volts and `1'b1` as zero volts. Other kinds of logic families, such as CMOS, use different voltages. ECL logic uses current, rather than voltage, to represent information, but the concept is the same.

3.5.3.1 High impedance

In any technology, it is possible for gates to be miswired. One kind of problem is when a designer forgets to connect a wire or forgets to instantiate a necessary gate. This means that there is a wire in the system which is not connected to anything. We refer to this as *high impedance*, which in Verilog notation is `1'bz`. The TTL logic family will normally view high impedance as being the same as five volts. If the input of a gate to which this wire is connected is active high, `1'bz` will be treated as `1'b1`, but if it is active low, it will be treated as `1'b0`. Other logic families treat `1'bz` differently. Furthermore, electrical noise may cause `1'bz` to be treated spuriously in any logic family. For these reasons, it is important for a Verilog simulator to treat `1'bz` as distinct from `1'b0` and `1'b1`. For example, if the designer forgets the final `or` gate in the example from section 3.5.1:

```
module forget_or_that_outputs_c;
  reg a,b;
  wire c;
  wire t1,t2,not_a,not_b;

  not i1(not_a,a);
  not i2(not_b,b);
  and a1(t1,not_a,b);
  and a2(t2,a,not_b);
  ...
endmodule
```

[4] Verilog also allows each bit to have a strength, which is an electronic concept (below gate level) beyond the scope of this book.

there is no gate that outputs the wire c, and therefore it remains 1'bz, regardless of what a and b are.

3.5.3.2 Unknown value

Another way in which gates can be miswired is when the output of two gates are wired together. This raises the possibility of *fighting outputs*, where one of the gates wants to output a 1'b0, but the other wants to output a 1'b1. For example, if we tried to eliminate the or gate by tying the output of both and gates together:

```
module tie_ands_together;
  reg a,b;
  wire c;
  wire t1,t2,not_a,not_b;

  not i1(not_a,a);
  not i2(not_b,b);
  and a1(c,not_a,b);
  and a2(c,a,not_b);
  ...
endmodule
```

the result is correct (1'b0) when a and b are the same because the two and gates both produce 1'b0 and there is no fight. The result is incorrect (1'bx) when a is 1'b0 and b is 1'b1 or vice versa, because the two and gates fight each other. Fighting gates can cause physical damage to certain families of logic (i.e., smoke comes out of the chip). Obviously, we want to be able to have the simulator catch such problems before we fabricate a chip that is doomed to blow up (literally)!

3.5.3.3 Use in behavioral code

Behavioral code may manipulate bits with the four-valued logic. Uninitialized regs in behavioral code start with a value of 'bx. (As mentioned above for structural code, disconnected wires start with a value of 'bz.) All the Boolean operators, such as &, | and ~ are defined with the four-valued logic so that the usual rules of commutativity, associativity, etc. apply.

The four-valued logic may be used with multi-bit wires and regs. When all the bits are either 1'b1 or 1'b0, such as 3'b110, the usual binary interpretation (powers of two) applies. When any of the bits is either 1'bz or 1'bx, such as 3'b1z0, the numeric value is unknown.

Arithmetic and relational operators (including == and !=) produce their usual results only when both operands are composed of `1'b0`s and `1'b1`s. In any other case, the result is `'bx`. This relates to the fact the corresponding combinational logic required to implement such operations in hardware would not produce a reliable result under such circumstances. For example:

```
if ( a == 1'bx)
   $display("a is unknown");
```

will never display the message, even when `a` is `1'bx`, because the result of the == operation is always `1'bx`. `1'bx` is not the same as `1'b1`, and so the `$display` never executes.

There are two special comparison operators (=== and !==) that overcome this limitation. === and !== cannot be implemented in hardware, but they are useful in writing intelligent simulations. For example:

```
if ( a === 1'bx)
  $display("a is unknown");
```

will display the message if and only if `a` is `1'bx`.

To help understand the last examples, you should realize that the following two `if` statements are equivalent:

```
if(expression)              if((expression)===1'b1)
  statement;                  statement;
```

The following table summarizes how the four-valued logic works with common operators:

a	b	a==b	a===b	a!=b	a!==b	a&b	a&&b	a|b	a||b	a^b
0	0	1	1	0	0	0	0	0	0	0
0	1	0	0	1	1	0	0	1	1	1
0	x	x	0	x	1	0	0	x	x	x
0	z	x	0	x	1	0	0	x	x	x
1	0	0	0	1	1	0	0	1	1	1
1	1	1	1	0	0	1	1	1	1	0
1	x	x	0	x	1	x	x	1	1	x
1	z	x	0	x	1	x	x	1	1	x
x	0	x	0	x	1	0	0	x	x	x
x	1	x	0	x	1	x	x	1	1	x
x	x	x	1	x	0	x	x	x	x	x
x	z	x	0	x	1	x	x	x	x	x
z	0	x	0	x	1	0	0	x	x	x
z	1	x	0	x	1	x	x	1	1	x
z	x	x	0	x	1	x	x	x	x	x
z	z	x	1	x	0	x	x	x	x	x

This table was generated by the following Verilog code:

```
module xz01;
  reg a,b,val[3:0];
  integer ia,ib;

  initial
    begin
      val[0] = 1'b0;
      val[1] = 1'b1;
      val[2] = 1'bx;
      val[3] = 1'bz;
      $display
        ("a b  a==b a===b  a!=b a!==b   a&b a&&b  a|b a||b  a^b");

      for (ia = 0; ia<=3; ia=ia+1)
        for (ib = 0; ib<=3; ib=ib+1)
          begin
            a = val[ia];
            b = val[ib];

            $display
     ("%b %b  %b  %b    %b    %b     %b     %b     %b     %b    %b ",
           a,b,a==b,a===b,a!=b,a!==b,a&b,a&&b,a|b,a||b,a^b);
          end
    end
endmodule
```

3.6 `$time`

A Verilog simulator executes as a software program on a conventional general-purpose computer. How long it takes such a computer to run a Verilog simulation, known as *real time*, depends on several factors, such as how fast the general-purpose computer is, and how efficient the simulator is. The speed with which the designer obtains the simulation results has little to do with how fast the eventual hardware will be when it is fabricated. Therefore, the real time required for simulation is not important in the following discussion.

Instead, Verilog provides a built-in variable, `$time`, which represents simulated time, that is, a simulation of the actual time required for a machine to operate when it is fabricated. Although the value of `$time` in simulation has a direct relationship to the physical time in the fabricated hardware, `$time` is not measured in seconds. Rather, `$time` is a unitless integer. Often designers map one of these units into one nanosecond, but this is arbitrary.

3.6.1 Multiple blocks

Verilog allows more than one behavioral block in a module. For example:

```
module two_blocks;
  integer x,y;

  initial
    begin
      a=1;
      $display("a is one");
    end

  initial
    begin
      b=2;
      $display("b is two");
    end
endmodule
```

The above *simulates* a system in which `a` and `b` are **simultaneously** assigned their respective values. This means, from a simulation standpoint, `$time` is the same when `a` is assigned one as when b is assigned two. (Since both assignments occur in `initial` blocks, `$time` is 0.) Note that this does not imply the sequence in which these assignments (or the corresponding `$display` statements) occur.

3.6.2 Sequence versus `$time`

In software, we often confuse the two separate concepts of time and sequence. In Verilog, it is possible for many statements to execute without `$time` advancing. The sequence in which statements within one block execute is determined by the usual rules found in other high-level languages. The sequence in which statements within different blocks execute is something the designer cannot predict, but that Verilog will do consistently. The advancing of `$time` is a different issue, discussed in section 3.7.

If you change the `wires` to be `regs`, a structural Verilog netlist is equivalent to several `always` blocks, where each `always` block computes the result output by one gate. If the design is correct, the sequence in which such `always` blocks execute at a particular `$time` is irrelevant, which helps explain why the order in which you instantiate gates in structural Verilog is also irrelevant. With Verilog, you can simulate the parallel actions of each gate or module that you instantiate, as well as the parallel actions of each behavioral block you code.

3.6.3 Scheduling processes and deadlock

Like a multiprocessing operating system, a Verilog simulator schedules several processes, one for each structural component or behavioral block. The `$time` variable does not advance until the simulator has given each process that so desires an opportunity to execute at that `$time`.

If you are familiar with operating systems concepts, such as semaphores, you will recognize that this raises a question about how Verilog operates: what are the atomic units of computation, or in other words, when does a process get interrupted by the Verilog simulator?

The behavioral statements described earlier are uninterruptible. Although it is nearly correct to model an exclusive OR with the following behavioral code:

```
module deadlock_the_simulator;
  reg a,b,c;
  always
    c = a^b;
  ... other blocks ...
endmodule
```

the Verilog simulator would never allow the other blocks to execute because the block computing `c` is not interruptible. Overcoming this problem requires an additional feature of Verilog, discussed in the next section.

3.7 Time control

Behavioral Verilog may include *time control* statements, whose purpose is to release control back to the Verilog scheduler so that other processes may execute and also tell the Verilog simulator at what `$time` the current process would like to be restarted. There are three forms of time control that have different ways of telling the simulator when to restart the current process: #, @ and `wait`.

3.7.1 # time control

When a statement is preceded by # followed by a number, the scheduler will not execute the statement until the specified number of `$time` units have passed. Any other process that desires to execute earlier than the `$time` specified by the # will execute before the current process resumes. If we modify the first example from section 3.6:

```
module two_blocks_time_control;
  integer x,y;


  initial
    begin
      #4
      a=1;
      $display("a is one at $time=%d",$time);
    end

  initial
    begin
      #3
      b=2;
      $display("b is two at $time=%d",$time);
    end

endmodule
```

the above will assign first to `b` (at `$time`=3) and then to `a` one unit of `$time` later. The order in which these statements execute is unambiguous because the # places them at a certain point in `$time`.

There can be more than one # in a block. The following nonsense module illustrates how the # works:

```
module confusing;
  integer a;


  initial
    begin
         a = 10;
      #2 a = 20;
      #5 a = 30;
    end

  initial
    begin
      #1 a = 40;
      #3 a = 50;
      #4 a = 60;
    end
endmodule
```

In the above code, a becomes 10 at `$time` 0, 40 at `$time` 1, 20 at `$time` 2, 50 at `$time` 4, 30 at `$time` 7 and 60 at `$time` 8. The interaction of parallel blocks creates a behavior much more complex than that of each individual block.

3.7.1.1 Using # in test code

One of the most important uses of # is to generate sequences of patterns at specific `$times` in test code to act as inputs to a machine. The # releases control from the test code and gives the code that simulates the machine an opportunity to execute. Test code without some kind of time control would be pointless because the machine being tested would never execute.

For example, suppose we would like to test the built-in `xor` gate by stimulating it with all four combinations on its inputs, and printing the observed truth table:

```
module top;
  integer ia,ib;
  reg a,b;
  wire c;

  xor x1(c,a,b);

  initial
    begin
      for (ia=0; ia<=1; ia = ia+1)
        begin
          a = ia;
          for (ib=0; ib<=1; ib = ib + 1)
            begin
              b = ib;
              #10 $display("a=%d b=%d c=%d",a,b,c);
            end
        end
    end
endmodule
```

The first time through, `a` and `b` are initialized to be 0 at `$time` 0. When #10 executes at `$time` 0, the `initial` block relinquishes control, and `x1` is given the opportunity to compute a new value (0^0=0) on the wire `c`. Having completed everything scheduled at `$time` 0, the simulator advances `$time`. The next thing scheduled to execute is the `$display` statement at `$time` 10. (The simulator does not waste real time computing anything for `$time` 2 through 9 since nothing changes during this `$time`.) The simulator prints out that "`a=0 b=0 c=0`" at `$time` 10 and then goes through the inner loop once again. While `$time` is still 10, `b` becomes 1. The #10 relinquishes control, `x1` computes that `c` is now 1 and `$time` advances. The `$display` prints out that "`a=0 b=1 c=1`" at `$time` 20. The last two lines of the truth table are printed out in a similar fashion at `$times` 30 and 40.

3.7.1.2 Modeling combinational logic with #

Physical combinational logic devices, such as the exclusive OR gate, have propagation delay. This means that a change in the input does not instantaneously get reflected in the output as shown above, but instead it takes some amount of physical time for the change to propagate through the gate. Propagation delay is a low-level detail of hardware design that ultimately determines the speed of a system. Normally, we will want to ignore propagation delay, but for a moment, let's consider how it can be modeled in behavioral Verilog with the #.

The behavioral exclusive OR example in section 3.6.3 deadlocks the simulator because it does not have any time control. If we put some time control in this `always` block (say a propagation delay of #1), the simulator will have an opportunity to schedule the test code instead of deadlocking inside the `always` block:

```
module top;
  integer ia,ib;
  reg a,b;
  reg c;

  always #1
    c = a^b;

  initial
    begin
      for (ia=0; ia<=1; ia = ia+1)
        begin
          a = ia;
          for (ib=0; ib<=1; ib = ib + 1)
            begin
              b = ib;
              #10 $display("a=%d b=%d c=%d",a,b,c);
            end
        end
      $finish;
    end
endmodule
```

As in the last example, `a` and `b` are initialized to be 0 at `$time` 0. When #10 executes at `$time` 0, the `initial` block relinquishes control, which gives the `always` loop an opportunity to execute. The first thing that the `always` block does is to execute #1, which relinquishes control until `$time` 1. Since no other block wants to execute at `$time` 1, execution of the `always` block resumes at `$time` 1, and it computes a new value (0^0=0) for the reg `c`. Because this is an `always` block, it loops back to the #1. Since no other block wants to execute at `$time` 2, execution of the `always` block resumes at `$time` 2, and it recomputes the same value for the reg `c` that it just computed at `$time` 1. The `always` block continues to waste real time by unnecessarily recomputing the same value all the way up to `$time` 9.

Finally, the `$display` statement executes at `$time` 10. The test code prints out "`a=0 b=0 c=0`" and goes through its inner loop once again. While `$time` is still 10, `b` becomes 1. The #10 relinquishes control, and the `always` block will have another ten chances to compute that `c` is now 1. The remaining lines of the truth table are printed out in a similar fashion.

There is an equivalent structural netlist notation for an `always` block with # time control. The following behavioral and structural code do similar things in `$time`:

```
reg c;                        wire c;
always #2                     xor #2 x2(c,a,b);
  c = a^b;
```

Both model an exclusive OR gate with a propagation delay of two units of `$time`. On many (but not all) implementations of Verilog simulators, the structural version is more efficient from a real-time standpoint. This is discussed in greater detail in chapter 6.

3.7.1.3 Generating the system clock with # for simulation

Registers and controllers are driven by some kind of a clock signal. One way to generate such a signal is to have an `initial` block give the clock signal an initial value, and an `always` block that toggles the clock back and forth:

```
reg sysclk;

initial
  sysclk = 0;

always #50
  sysclk = ~sysclk;
```

The above generates a system clock signal, `sysclk`, with a period of 100 units of `$time`.

3.7.1.4 Ordering processes without advancing `$time`

It is permissible to use a delay of #0. This causes the current process to relinquish control to other processes that need to execute at the current `$time`. After the other processes have relinquished control, but before `$time` advances, the current process will resume. This kind of time control can be used to enforce an order on processes whose execution would otherwise be unpredictable. For example, the following is algorithmically the same as the first example in 3.7.1 (`b` is assigned first, then `a`), but both assignments occur at `$time` 0:

```
module two_blocks_time_control;
  integer x,y;
  initial
    begin
      #0
      a=1;
      $display("a is one at $time=%d",$time);
    end
  initial
    begin
      b=2;
      $display("b is two at $time=%d",$time);
    end
endmodule
```

3.7.2 @ time control

When an @ precedes a statement, the scheduler will not execute the statement that follows until the event described by the @ occurs. There are several different kinds of events that can be specified after the @, as shown below:

```
@(expression)
@(expression or expression or ...)
@(posedge onebit)
@(negedge onebit)
@ event
```

When there is a single expression in parenthesis, the @ waits until one or more bit(s) in the result of the `expression` change. As long as the result of the `expression` stays the same, the block in which the @ occurs will remain suspended. When multiple expressions are separated by `or`, the @ waits until one or more bit(s) in the result of any of the `expressions` change. The word `or` is not the same as the operator |.

In the above, `onebit` is single-bit `wire` or `reg` (declared without the square bracket). When `posedge` occurs in the parenthesis, the @ waits until `onebit` changes from a 0 to a 1. When `negedge` occurs in the parenthesis, the @ waits until `onebit` changes from a 1 to a 0. The following mean the same thing:

```
reg a,b,c;          reg a,b,c;
@(c) a=b;           @(posedge c or negedge c) a=b;
```

An `event` is a special kind of Verilog variable, which will be discussed later.

3.7.2.1 Efficient behavioral modeling of combinational logic with @

Although you can model combinational logic behaviorally using just the #, this is not an efficient thing to do from a simulation real-time standpoint. (Using # for combinational logic is also inappropriate for synthesis.) As illustrated in section 3.7.1.2, the `always` block has to reexecute many times without computing anything new. Although physical hardware gates are continuously recomputing the same result in this fashion, it is wasteful to have a general-purpose computer spend real time simulating this. It would be better to compute the correct result once and wait until the next time the result changes.

How do we know when the output changes? Recall that perfect combinational logic (i.e., with no propagation delay) by definition changes its output whenever **any of its input(s) change**. So, we need the Verilog notation that allows us to suspend execution until any of the inputs of the logic change:

```
module top;
  integer ia,ib;
  reg a,b;
  reg c;

  always @(a or b)
    c = a^b;

  initial
    begin
      for (ia=0; ia<=1; ia = ia+1)
        begin
          a = ia;
          for (ib=0; ib<=1; ib = ib + 1)
            begin
              b = ib;
              #10 $display("a=%d b=%d c=%d",a,b,c);
            end
        end
      $finish;
    end
endmodule
```

At the beginning, both the initial and the always block start execution. Since neither a nor b have changed yet, the always block suspends. The first time through the loops in the initial block, a and b are initialized to be 0 at $time 0. When #10 executes at $time 0, the initial block relinquishes control, and the always block is given an opportunity to do something. Since a and b both changed at $time 0, the @ does not suspend, but instead allows the always block to compute a new value (0^0=0) for the reg c. The always block loops back to the @. Since there is no way that a or b can change anymore at $time 0, the simulator advances $time. The next thing scheduled to execute is the $display statement at $time 10. (Like the example in section 3.7.1.1, but unlike the example in section 3.7.1.2, the simulator does not waste real time computing anything for $time 1 through 9 since nothing changes during that $time.) The simulator prints out that "a=0 b=0 c=0" at $time 10, and then goes through the inner loop once again. While $time is still 10, b becomes 1. The #10 relinquishes control, and the always block has an opportunity to do something. Since b just changed (though a did not change), the @ does not suspend, and c is now 1. After $time advances, the $display prints out that "a=0 b=1 c=1" at $time 20. The last two lines of the truth table are printed out in a similar fashion at $times 30 and 40.

Since this is a model of combinational logic, it is very important that **every input to the logic be listed after the @**. We refer to this list of inputs to the physical gate as the *sensitivity list.*

3.7.2.2 Modeling synchronous registers

Most synchronous registers that we deal with use rising edge clocks. Using @ with posedge is the easiest way to model such devices. For example, consider an enabled register whose input (of any bus width) is din and whose output (of similar width as din) is dout. At the rising edge of the clock, when ld is 1, the value presented on din will be loaded. Otherwise dout remains the same. Assuming din, dout, ld and sysclk are taken care of properly elsewhere in the module, the behavioral code to model such an enabled register is:

```
always @(posedge sysclk)
  if (ld)
    dout = din;
```

Similar Verilog code can be written for a counter register that has clr, ld, and cnt signals:

```
always @(posedge sysclk)
  begin
    if (clr)
      dout = 0;
    else
      if (ld)
        dout = din;
      else
        begin
          if (cnt)
            dout = dout + 1;
        end
  end
```

Note that the nesting of `if` statements indicates the priority of the commands. If a controller sends this counter a command to `clr` and `cnt` at the same time, the counter will ignore the `cnt` command. At any `$time` when this `always` block executes, only one action (clearing, loading, counting or holding) occurs. Of course, improper nesting of `if` statements could yield code whose behavior would be impossible with physical hardware.

3.7.2.3 Modeling synchronous logic controllers

Most controllers are triggered by the rising edge of the system clock. It is convenient to use `posedge` to model such devices. For example, assuming that `stop`, `speed` and `sysclk` have been dealt with properly elsewhere in the module, the second ASM chart in section 2.1.1.2 could be modeled as:

```
always
  begin
    @(posedge sysclk)  //this models state GREEN
      stop = 0;
      speed = 3;
    @(posedge sysclk) //this models state YELLOW
      stop = 1;
      speed = 1;
    @(posedge sysclk)  //this models state RED
      stop = 1;
      speed = 0;
  end
```

There are several things to note about the above code. First, the indentation is used only to promote readability. Assuming the code for generating `sysclk` given in section

3.7.1.3, the `stop = 0` and `speed = 3` statements execute at `$time` 50, 350, 650, ... because there is no time control among them. The indentation simply highlights the fact that these two statements execute atomically, as a unit, without being interrupted by the simulator.

The second thing to note is that the = in Verilog is just a **software assignment statement**. (The variable is modified at the `$time` the statement executes. The variable will retain the new value until modified again.) This is different than how we use = in ASM chart notation. (The command signal is a function of the present state. The command signal does not retain the new value after the rising edge of the system clock but instead returns to its default value.) Another way of saying this is that there are no default values in standard Verilog variables as there are for ASM chart commands. Despite the distinction between Verilog and ASM chart notation, we can model an ASM chart in Verilog by fully specifying every command output in every state. For those states where a command is not mentioned in an ASM chart, one simply codes a Verilog assignment statement that stores the default value into the Verilog variable corresponding to the missing ASM chart command. The `stop=0` and `speed=0` statements above were not shown in the original ASM chart but are required for the Verilog code to model what the hardware would actually do.

The third thing is the names of the states are not yet included in the Verilog code. (The comments are of course ignored by Verilog.) Eventually, we will find a way of including meaningful state names in the actual code.

The fourth thing is that this ASM chart does not have any RTN (i.e., it is at the mixed stage). We will need an additional Verilog notation to model ASM charts that use RTN. This notation is discussed in section 3.8.

3.7.2.4 @ for debugging display

@ can also be used for causing the Verilog simulator to print debugging output that shows what happens as actions unfold in the simulation. For example,

```
always @(a or b or c)
  $display("a=%b b=%b c=%b at $time=%d",a,b,c,$time);
```

The above block would eliminate the need for the designer to worry about putting `$display` statements in the test code or in the code for the machine being tested.

With clocked systems, it is often convenient to display information shortly after each rising edge of the clock:

```
always @(posedge sysclk)
 #20 $display("stop=%b speed=%b at $time=%d",
 stop,speed,$time);
```

3.7.3 `wait`

The `wait` statement is a form of time control that is quite different than # or @. The `wait` statement stands by itself. It does not modify the statement which follows in the way that @ and # do (i.e., there must be a semicolon after the `wait` statement). The `wait` statement is used primarily in test code. It is not normally used to model hardware devices in the way @ and # are used. The syntax for the `wait` statement is:

```
wait(condition);
```

The `wait` statement suspends the current process. The current process will resume when the condition becomes true. If the condition is already true, the current process will resume without `$time` advancing.

For example, suppose we want to exhaustively test one of the slow division machines described in chapter 2. The amount of time the machine takes depends on how big the result is. Furthermore, different ASM charts described in chapter 2 take different amounts of `$time`. Therefore, the best approach is to use the `ready` signal produced by the machine:

```
module top;
  reg pb;
  integer x,y;
  wire [11:0] quotient;
  wire sysclk;
  ...
  initial
    begin
      pb= 0;
      x = 0;
      y = 0;
      #250;
      @(posedge sysclk);
      while (x<=4095)
        begin
          for (y=1; y<=4095; y = y+1)
            begin
              @(posedge sysclk);
              pb = 1;
```

Continued

```
              @(posedge sysclk);
              pb = 0;
              @(posedge sysclk);
              wait(ready);
              @(posedge sysclk);
              if (x/y === quotient)
                $display("ok");
              else
               $display("error x=%d y=%d x/y=%d  quotient=%d",
                               x,y,x/y,quotient);
            end
          x = x + 1;
        end
      $stop;
    end
endmodule
```

This test code (based on the nested loops given in section 3.4) embodies the assumptions we made in section 2.2.1. The first two @s in the loop produce the `pb` pulse that lasts exactly one clock cycle. The third @ makes sure that the machine has enough time to respond (and make `ready` 0). The `wait(ready)` keeps the test code synchronized to the division machine, so that the test code is not feeding numbers to the division machine too rapidly. The fourth @ makes sure the machine will spend the required time in state IDLE, before testing the next number.

The ellipsis shows where the code for the actual division machine was omitted in the above. The `quotient` is produced by this machine which is not shown here. The design of this code will be discussed in the next chapter.

3.8 Assignment with time control

The # and @ time control, discussed in sections 3.7.1 and 3.7.2, precede a statement. These forms of time control delay execution of the following statement until the specified `$time`. There are two special kinds of assignment statements[5] that have time control **inside the assignment statement**. These two forms are known as *blocking* and *non-blocking procedural assignment.*

[5] Assignment with time control is not accepted by some commercial synthesis tools but is accepted by all Verilog simulators. Since there are problems with intra-assignment delay (section 3.8.2.1), some authors recommend against its use, but when used as recommended later in this chapter (section 3.8.2.2), it becomes a powerful tool. Chapter 7 explains a preprocessor that allows all synthesis tools to accept the use proposed in this book.

3.8.1 Blocking procedural assignment

The syntax for blocking procedural assignment has the # or @ notation (whose syntax is described in sections 3.7.1 and 3.7.2) after the = but before the expression. For example, three common forms of this are:

```
var = # delay expression;
var = @(posedge onebit) expression;
var = @(negedge onebit) expression;
```

Other variations are also legal. What distinguishes this from a normal instantaneous assignment is that the expression is evaluated at the `$time` the statement first executes, but the variable does not change until after the specified delay. For example, assuming `temp` is a `reg` that is not used elsewhere in the code and that `temp` is declared to be the same width as `a` and `b`, the following two fragments of code mean the same thing:

```
                                         initial
initial                                    begin
  begin                                      ...
    ...                                     temp = b;
    a = @(posedge sysclk) b;                @(posedge sysclk) a = temp;
    ...                                      ...
  end                                      end
```

Blocking procedural assignment is almost what we need to model an ASM chart with RTN. The one problem with it, as its name implies, is that it blocks the current process from continuing to execute additional statements at the same `$time`. We will not use blocking procedural assignment for this reason.

3.8.2 Non-blocking procedural assignment

The syntax for a non-blocking procedural assignment is identical to a blocking procedural assignment, except the assignment statement is indicated with <= instead of =. This should be easy to remember, because it reminds us of the ← notation in ASM charts. For example, the most common form of the non-blocking assignment used in later chapters is:

```
var <= @(posedge onebit) expression;
```

Typically, `onebit` is the `sysclk` signal mentioned in section 3.7.1.3. Although other forms are legal, the above `@(posedge onebit)` form of the non-blocking assignment is the one we use in almost every case for ← in ASM charts.[6]

The expression is evaluated at the `$time` the statement first executes and further statements execute at that same `$time`, but the variable does not change until after the specified delay. For example, assuming `temp` is a `reg` that is not used elsewhere in the left-hand code and that `temp` is declared to be the same width as `a` and `b`, the following two fragments of code mean nearly the same thing:

```
                                              always @(posedge sysclk)
                                                 #0 a = temp;

initial                                       initial
  begin                                         begin
     ...                                           ...
     a <= @(posedge sysclk) b;                       temp = b;
     ...                                           ...
   end                                          end
```

Note that, all by itself, the effect of the non-blocking assignment is like having a parallel `always` block to store into `a`. An advantage of the <= notation is that you do not have to code a separate `always` block for each register.

A subtle detail is that the right-hand `always` block is the last thing to execute (#0) at a given `$time`. Similarly, the <= causes the `reg` to change only after every other block (including the one with the <=) has finished execution. This subtle detail causes a problem, which is discussed in the next section, and which is solved in section 3.8.2.2.

3.8.2.1 Problem with <= for RTN for simulation

An obvious approach to translating RTN from an ASM chart into behavioral Verilog is just to put <= for each ← in the ASM chart. For example, assuming `stop`, `speed`, `count` and `sysclk` are taken care of properly elsewhere, one might think that the ASM chart from section 2.1.1.3 could be translated into Verilog as:

[6] The exceptions are when the left-hand side of the ← is a memory being changed every clock cycle, in which case `@(negedge onebit)` is appropriate, as explained in section 6.5.2, and for post-synthesis behavorial modeling of logic equations, in which case # is appropriate, as explained in section 11.3.3.

```
always
  begin
    @(posedge sysclk)          //this models state GREEN
      stop = 0;
      speed = 3;

    @(posedge sysclk)          //this models state YELLOW
      stop = 1;
      speed = 1;
      count <= @(posedge sysclk) count + 1;

    @(posedge sysclk)          //this models state RED
      stop = 1;
      speed = 0;
      count <= @(posedge sysclk) count + 2;
  end
```

However, when one runs this code on a Verilog simulator, the following incorrect result is produced (assuming the debugging `always` block shown in section 3.7.2.4):

```
stop=0    speed=11    count=000 at    $time=     70
stop=1    speed=01    count=000 at    $time=    170
stop=1    speed=00    count=001 at    $time=    270
stop=0    speed=11    count=010 at    $time=    370
stop=1    speed=01    count=010 at    $time=    470
stop=1    speed=00    count=011 at    $time=    570
```

Recall from section 2.1.1.3 that at `$time` 370, `count` should be three instead of two. The underlying cause of this error is the subtle detail mentioned above: The <= causes the `reg` to change only after every other block (including the one with the <=) has finished execution.

The above Verilog starts to execute the statements for state YELLOW at `$time` 150. The last of these statements evaluates `count+1` at `$time` 150 and schedules the storage of the result. Since count is still 3'b000 at `$time` 150, the result scheduled to be stored at the end of `$time` 250 is 3'b001. The `@(posedge sysclk)` that starts state RED causes the always block to suspend until `$time` 250. The problem shown above occurs at `$time` 250 because the assignment initiated by the <= at `$time` 150 will be the last thing that occurs at `$time` 250. Prior to the assignment, the process will resume and execute the three statements, including `count <= @(posedge sysclk) count + 2`. Since `count` is still 3'b000, this <= schedules 3'b010 to be assigned at `$time` 350, which is not what happens in an ASM chart. As soon as the assignment of 3'b010 has been scheduled at `$time` 250, 3'b001 will be stored into `count` (as a result of the first <=).

3.8.2.2 *Proper use of <= for RTN in simulation*

To overcome the problem described in the last section, you need to use a non-zero delay after each `@(posedge sysclk)` that denotes a rectangle of the ASM chart. For example, here is the complete Verilog code to model (in a primitive way) the ASM chart from section 2.1.1.3:

```
module top;
  reg stop;
  reg [1:0] speed;
  reg sysclk;
  reg [2:0] count;

  initial
    sysclk = 0;
  always #50
    sysclk = ~sysclk;

  always
    begin
      @(posedge sysclk) #1        //this models state GREEN
        stop = 0;
        speed = 3;
      @(posedge sysclk) #1 //this models state YELLOW
        stop = 1;
        speed = 1;
        count <= @(posedge sysclk) count + 1;

      @(posedge sysclk) #1        //this models state RED
        stop = 1;
        speed = 0;
        count <= @(posedge sysclk) count + 2;
    end

  always @(posedge sysclk)
    #20 $display("stop=%b speed=%b count=%b at $time=%d",
          stop,speed,count,$time);

  initial
    begin
      count = 0;
      #600 $finish;
    end
endmodule
```

Let's analyze the reason why each block is required in this module. The first `initial` block is required to give `sysclk` a value other than 1'bx at `$time` 0. The next block toggles `sysclk` so that the clock period is 100. If `sysclk` were not initialized at `$time` 0, it would stay `1'bx` forever (`~1'bx` is `1'bx`).

The only new thing in the `always` block that models the ASM chart is the addition of #1 after each `@(posedge sysclk)`. The `always` block that follows it displays `stop`, `speed` and `count` during each state.

The test code in the final `initial` block simply initializes count to be 3'b000. (In a real machine, this would occur in a state of the ASM, but instead here it is part of the test code for the purposes of illustration only.) The test code schedules a `$finish` system task to be called at `$time` 600. This is required because the `always` blocks would otherwise tell the simulator to go on forever.

With the #1 after each @, the Verilog simulator produces the following correct output:

```
stop=0   speed=11   count=000   at   $time=    70
stop=1   speed=01   count=000   at   $time=   170
stop=1   speed=00   count=001   at   $time=   270
stop=0   speed=11   count=011   at   $time=   370
stop=1   speed=01   count=011   at   $time=   470
stop=1   speed=00   count=100   at   $time=   570
```

3.8.2.3 Translating `goto`-less ASMs to behavioral Verilog

This book concentrates on several design techniques that all begin by expressing an ASM with behavioral Verilog. Since Verilog is a `goto`-less language, only certain kinds of ASMs can be translated in this fashion. Chapters 5 and 7 explain how arbitrary ASMs can be translated into Verilog, but this section will concentrate only on ASMs that adhere to this highly desirable `goto`-less style.

3.8.2.3.1 Implicit versus explicit style

The approach of expressing a state machine with high-level statements (like `if` and `while`) is known as *implicit style* because the next state of the machine is described implicitly through the use of `@(posedge sysclk)` within the statements of an `always` block. Implicit style is the opposite of the *explicit style* table (illustrated in section 2.4.1) that requires the designer to say what state the machine goes to under all possible circumstances.

Experienced hardware designers who are new to Verilog may find the implicit style approach confusing because it requires thinking about a state machine in a different way. The implicit style is much more like software concepts, such as the distinction between `if` and `while`. On the other hand, experienced software designers may also find this approach difficult at first because the timing relationship between <= and decisions in Verilog is different than in conventional software languages. The following sections go through a series of examples that illustrate some typical kinds of ASM constructs and how they translate into implicit style Verilog.

3.8.2.3.2 Identifying the infinite loop

Unlike software, all ASMs have at least one infinite loop. Implicit style behavioral Verilog is defined by an `always` block. Many times this `always` block can also serve to implement the infinite loop of the ASM. In the following ASM, the transitions from states FIRST, SECOND, THIRD and FOURTH are implicit. The designer does not have to say anything about their next states. The transition from FIFTH to FIRST occurs because of the `always`:

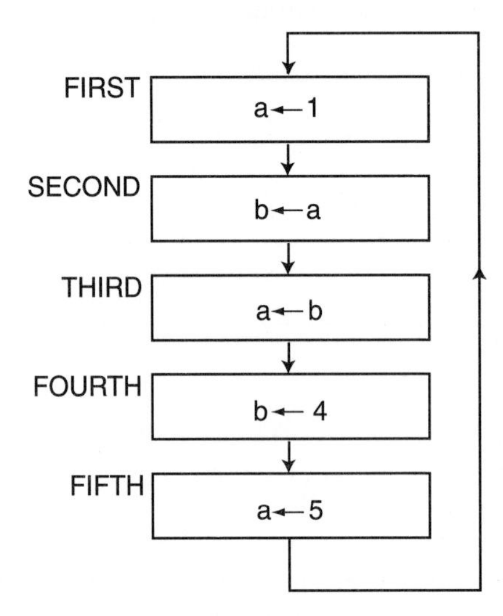

Figure 3.2 Every ASM has an infinite loop.

Inside the `always`, there is a one to one mapping of rectangles into `@(posedge sysclk)` statements. In this example, the ASM has five states, so the `always` uses five `@(posedge sysclk)`:

```
module top;
  //Following are actual hardware registers of ASM
 reg [11:0] a,b;


 //Following is NOT a hardware register
 reg sysclk;

 //The following always block models actual hardware
```

Continued

```
always
  begin
    @(posedge sysclk) #1;              // state FIRST
     a <= @(posedge sysclk) 1;
    @(posedge sysclk) #1;              // state SECOND
     b <= @(posedge sysclk) a;
    @(posedge sysclk) #1;              // state THIRD
     a <= @(posedge sysclk) b;
    @(posedge sysclk) #1;              // state FOURTH
     b <= @(posedge sysclk) 4;
    @(posedge sysclk) #1;              // state FIFTH
     a <= @(posedge sysclk) 5;
  end

//Following initial and always blocks do not correspond to
// hardware. Instead they are test code that shows what
// happens when the above ASM executes

always #50 sysclk = ~sysclk;
always @(posedge sysclk) #20
  $display("%d a=%d b=%d ", $time, a, b);

initial
  begin
    sysclk = 0;
    #1400 $stop;
  end
endmodule
```

The above is slightly more primitive than what will be used in later chapters, but the emphasis of this example is to show how an ASM translates into Verilog. In the above, there are three `always` blocks, but only the first one corresponds to hardware. The other two `always` blocks and the `initial` block are necessary for simulation (in later chapters these other blocks will be moved to other modules).

3.8.2.3.3 Recognizing `if else`

Most ASMs have decisions. Decisions in implicit Verilog are described either with the `if` statement (possibly followed by `else`) or with the `while` statement. For hardware designers without extensive software experience, determining whether the `if` or the `while` is appropriate for a particular decision can seem confusing at first.

The following ASM is an example where the `if else` construct is appropriate:

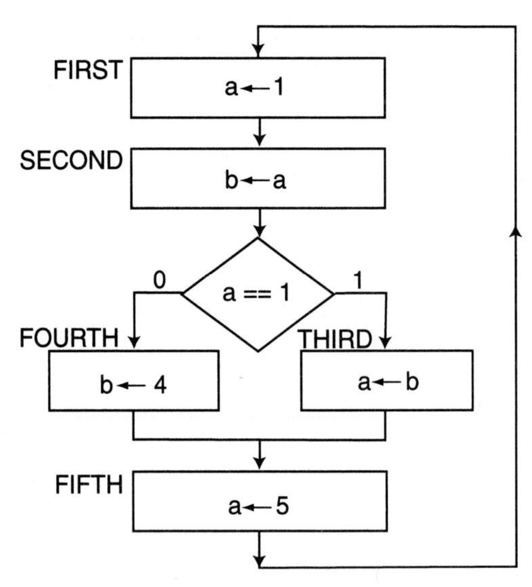

Figure 3-3. ASM corresponding to `if else`.

For brevity, only the `always` block that corresponds to the actual hardware is shown:

```
always
  begin
    @(posedge sysclk) #1;                // state FIRST
     a <= @(posedge sysclk) 1;
    @(posedge sysclk) #1;                // state SECOND
     b <= @(posedge sysclk) a;
     if (a == 1)
       begin
         @(posedge sysclk) #1;           // state THIRD
           a <= @(posedge sysclk) b;
       end
     else
       begin
         @(posedge sysclk) #1;           // state FOURTH
           b <= @(posedge sysclk) 4;
       end
    @(posedge sysclk) #1;                // state FIFTH
     a <= @(posedge sysclk) 5;
  end
```

The `if else` is appropriate here because only one of the states (THIRD or FOURTH) will execute. Because `a` is one in state SECOND, state THIRD will execute. In the following very similar Verilog, state FOURTH rather than state THIRD will execute:

```
always
  begin
    @(posedge sysclk) #1;              // state FIRST
     a <= @(posedge sysclk) 1;
    @(posedge sysclk) #1;              // state SECOND
     b <= @(posedge sysclk) a;
     if (a != 1)
       begin
         @(posedge sysclk) #1;         // state THIRD
           a <= @(posedge sysclk) b;
       end
     else
       begin
         @(posedge sysclk) #1;         // state FOURTH
           b <= @(posedge sysclk) 4;
       end
    @(posedge sysclk) #1;              // state FIFTH
     a <= @(posedge sysclk) 5;
  end
```

3.8.2.3.4 Recognizing a single alternative

Often, it is appropriate to omit the `else`, as in the following ASM:

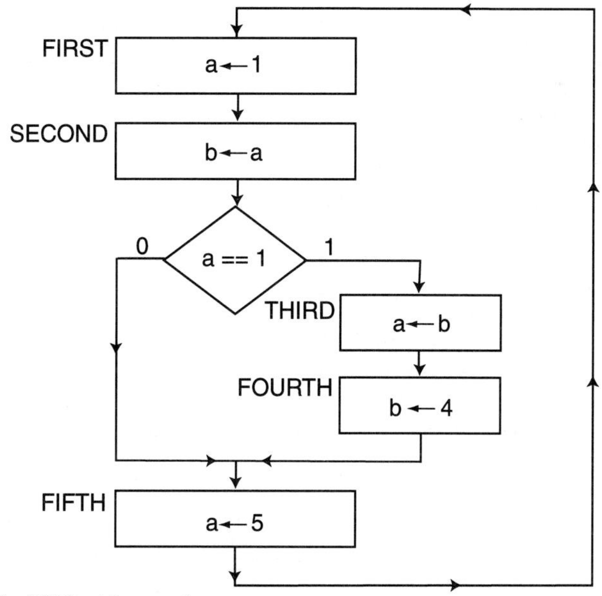

Figure 3-4. ASM without `else`.

which translates to the following Verilog:

```
always
  begin
    @(posedge sysclk) #1;               // state FIRST
     a <= @(posedge sysclk) 1;
    @(posedge sysclk) #1;               // state SECOND
     b <= @(posedge sysclk) a;
     if (a == 1)
       begin
         @(posedge sysclk) #1;          // state THIRD
           a <= @(posedge sysclk) b;
         @(posedge sysclk) #1;          // state FOURTH
           b <= @(posedge sysclk) 4;
       end
    @(posedge sysclk) #1;               // state FIFTH
     a <= @(posedge sysclk) 5;
  end
```

In the above, both state THIRD and state FOURTH will execute because a is one in state SECOND. The following very similar Verilog skips directly from state SECOND to state FIFTH:

```
always
  begin
    @(posedge sysclk) #1;               // state FIRST
     a <= @(posedge sysclk) 1;
    @(posedge sysclk) #1;               // state SECOND
     b <= @(posedge sysclk) a;
     if (a != 1)
       begin
         @(posedge sysclk) #1;          // state THIRD
           a <= @(posedge sysclk) b;
         @(posedge sysclk) #1;          // state FOURTH
           b <= @(posedge sysclk) 4;
       end
    @(posedge sysclk) #1;               // state FIFTH
     a <= @(posedge sysclk) 5;
  end
```

3.8.2.3.5 Recognizing `while` *loops*

The following two ASMs describe the same hardware. The first of the following two ASMs is very similar to the one in section 3.8.2.3.4, except that state FOURTH does not necessarily go to state FIFTH . Instead, state FOURTH goes to a decision which determines whether to go to state THIRD or state FIFTH. The second of the following two ASMs is a much less desirable way to describe the identical hardware. It is undesirable because the `a==1` test is duplicated; however, its meaning is exactly the same as the first of the following two ASMs:

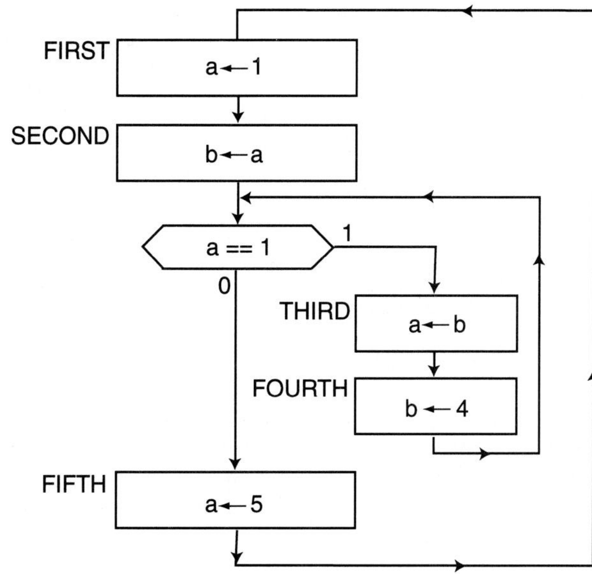

Figure 3-5. ASM with `while`.

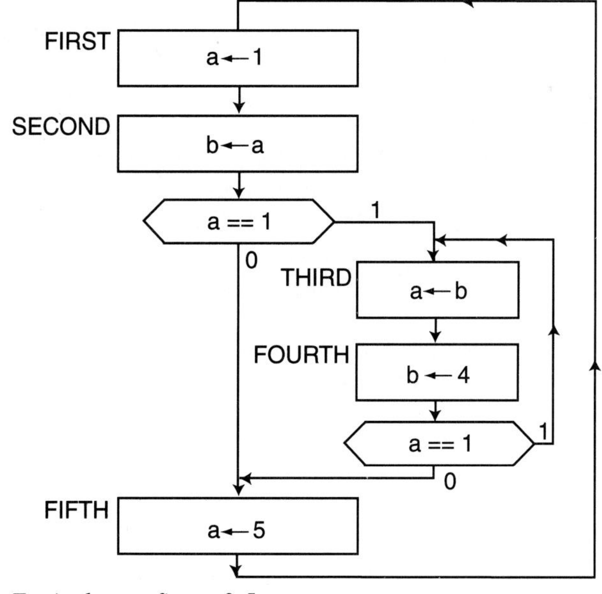

Figure 3-6. Equivalent to figure 3-5.

The reason the first of the ASMs is preferred is because it is more obvious that it translates into a `while` loop in Verilog:

```
always
  begin
    @(posedge sysclk) #1;              // state FIRST
     a <= @(posedge sysclk) 1;
    @(posedge sysclk) #1;              // state SECOND
     b <= @(posedge sysclk) a;
     while (a == 1)
       begin
         @(posedge sysclk) #1;         // state THIRD
           a <= @(posedge sysclk) b;
         @(posedge sysclk) #1;         // state FOURTH
           b <= @(posedge sysclk) 4;
       end
    @(posedge sysclk) #1;              // state FIFTH
     a <= @(posedge sysclk) 5;
  end
```

In fact, the only syntactic difference between the above Verilog and the Verilog in section 3.8.2.3.4 is that the word `if` has been changed to `while`. The advantage of looking at this particular ASM as a `while` loop is that the decision `a==1` is shared by both state SECOND and state FOURTH. With the `while` loop, the designer does not have to worry that the decision is actually part of two states. Many practical algorithms that produce useful results (as illustrated in chapter 2) demand a loop of this style. The `while` in Verilog makes this easy.

3.8.2.3.6 Recognizing `forever`

Sometimes machines need initialization states that execute only once. Since synthesis tools only accept behavioral Verilog defined with `always` blocks, such ASMs still begin with the keyword `always`. However, the looping action of the `always` is not pertinent. (If the designer only wanted to simulate the machine, `initial` would work just as well as `always`, but ultimately the synthesis tool will demand `always`.)

In order to describe the infinite loop that exists beyond the initialization states, the designer must use `forever`. For example, consider the following ASM:

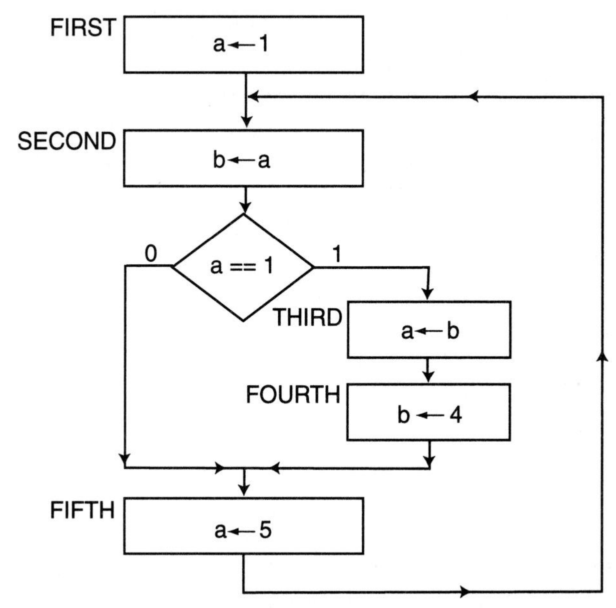

Figure 3-7. ASM needing `forever`*.*

It is almost identical to the one in section 3.8.2.3.4, except that state FIFTH forms an infinite loop to state SECOND instead of going to state FIRST. The corresponding Verilog implements this using `forever`:

```
always
  begin
    @(posedge sysclk) #1;              // state FIRST
     a <= @(posedge sysclk) 1;
     forever
       begin
         @(posedge sysclk) #1;         // state SECOND
           b <= @(posedge sysclk) a;
           if (a == 1)
             begin
               @(posedge sysclk) #1;     // state THIRD
                 a <= @(posedge sysclk) b;
               @(posedge sysclk) #1;     // state FOURTH
                 b <= @(posedge sysclk) 4;
             end
         @(posedge sysclk) #1;           // state FIFTH
           a <= @(posedge sysclk) 5;
       end
  end
```

3.8.2.3.7 Translating into an `if` at the bottom of `forever`

The following two ASMs are equivalent. Many designers would think the one on the left is more natural because it describes a loop involving only state THIRD. As long as `a==1`, the machine stays in state THIRD. The noteworthy thing about this machine is that state THIRD also forms the beginning of a separate infinite loop. (Such an infinite loop might be described with an `always` or in this case a `forever`.) Because of this, it is preferred to think of this ASM as an `if` at the bottom of a `forever`, as illustrated by the ASM on the right:

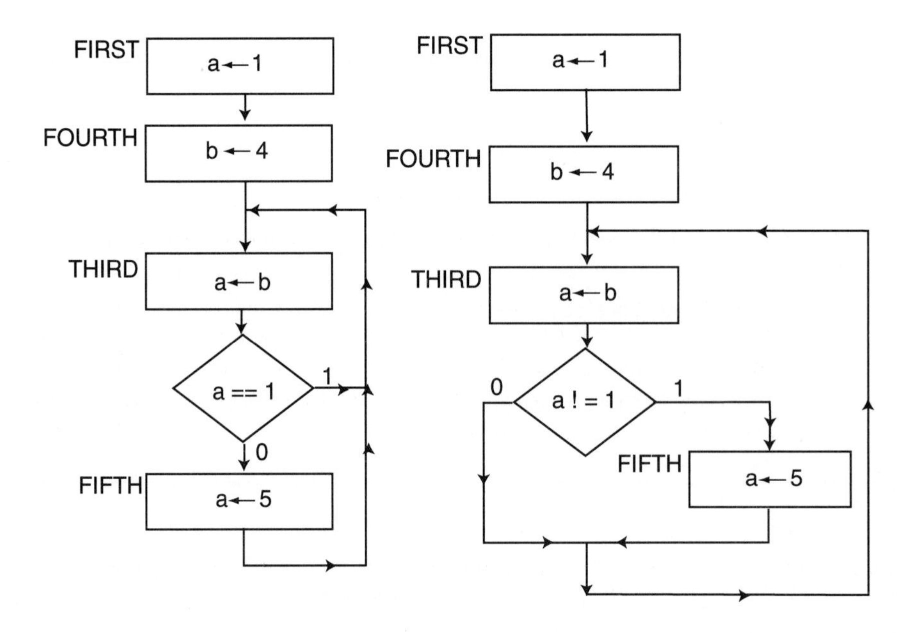

Figure 3-8. Two ways to draw `if` at the bottom of `forever`.

The ASM on the right tests if `a != 1` to see whether to leave the loop involving only state THIRD and proceed to state FIFTH. The reason the ASM on the right is preferred is that its translation into Verilog is obvious:

```
always
  begin
    @(posedge sysclk) #1;              // state FIRST
      a <= @(posedge sysclk) 1;
    @(posedge sysclk) #1;              // state FOURTH
      b <= @(posedge sysclk) 4;
      forever
        begin
          @(posedge sysclk) #1;        // state THIRD
            a <= @(posedge sysclk) b;
            if (a != 1)
              begin
                @(posedge sysclk) #1;  // state FIFTH
                  a <= @(posedge sysclk) 5;
              end
        end
  end
```

In software, an `if` never implements a loop. This is also true in Verilog of an isolated `if`, but the combination of an `if` at the bottom of `forever` or `always` has the effect of nesting a non-infinite loop inside an infinite loop. It is the `forever` or `always` that forms the looping action, not the `if`. This example illustrates a kind of implicit behavioral Verilog that sometimes causes novice Verilog designers confusion. It is suggested that the reader should fully appreciate this example before proceeding to later chapters. Designers need to be careful not to confuse `if` with `while`.

3.9 Tasks and functions

In conventional software programming languages, it is common for a programmer to use functions and procedures (known as void functions in C) to break an algorithm apart into manageable pieces. There are two main motivations for using functions and procedures: they make the top-down design of a complex algorithm easier, and they sometimes allow reuse of the same code. Verilog provides tasks (which are like procedures) and functions, which can be called from behavioral code.

3.9.1 Tasks

The syntax for a task definition is:

```
task name;
     input arguments;
     output arguments;
     inout arguments;
     ...
     declarations;
     begin
       statement;
       ...
     end
   endtask
```

This task definition must occur inside a module. The task is usually intended to be called only by `initial` blocks, `always` blocks and other tasks within that module. Tasks may have any behavioral statements, including time control.

Verilog lets the designer choose the order in which the `input`, `output` and `inout` definitions are given. (The order shown above is just one possibility.) The order in which `input`, `output` and `inout` definitions occur is based on the calling sequence desired by the designer. The sequence in which the formal arguments are listed in some combination of `input`, `output` and/or `inout` definitions determines how the actual arguments are bound to the formal definitions when the task is called.

The purpose of an `input` argument is to send information from the calling code into the task by value. An `input` argument may include a width (which is equivalent to a `wire` of that width) or it may be given a type of `integer` or `real` in a separate declaration. An `input` argument may not be declared as a `reg`.

The purpose of an `output` argument is to send a result from the task to the calling code by reference. An `output` argument must be declared as a `reg`, `integer` or `real` in a separate declaration.

An `inout` definition combines the roles of `input` and `output`. An `inout` argument must be declared as a `reg`, `integer` or `real` in a separate declaration.

3.9.1.1 Example task

Consider the following nonsense code:

```
integer count,sum,prod;
initial
  begin
    sum = 0;
    count = 1;

    sum = sum + count;
    prod = sum * count;
    count = count + 2;
    $display(sum,prod);

    sum = sum + count;
    prod = sum * count;
    count = count + 3;
    $display(sum,prod);

    sum = sum + count;
    prod = sum * count;
    count = count + 5;
    $display(sum,prod);

    sum = sum + count;
    prod = sum * count;
    count = count + 7;
    $display(sum,prod);

    $display(sum,prod,count);
  end
```

After initializing `sum` and `count`, there is a great similarity in the following four groups (each composed of four statements). Using a task allows this `initial` block to be shortened:

```
integer count,sum,prod;
initial
  begin
    sum = 0;
    count = 1;
    example(sum,prod,count,2);
    example(sum,prod,count,3);
    example(sum,prod,count,5);
    example(sum,prod,count,7);
    $display(sum,prod,count);
  end
```

The definition of the task `example` is:

```
task example;
  inout sum_arg;    //1st positional argument
  output prod_arg; //2nd positional argument
  inout count_arg; //3rd positional argument
  input numb_arg;   //4th positional argument

  integer count_arg,numb_arg,sum_arg,prod_arg;

  begin
   sum_arg = sum_arg + count_arg;
   prod_arg = sum_arg * count_arg;
   count_arg = count_arg + numb_arg;
   $display(sum_arg,prod_arg);
 end
endtask
```

Because the formal `inout sum_arg` is defined first, it corresponds to the actual `sum` in the `initial` block. Similarly, the formal `output prod_arg` corresponds to `prod`, and the formal `inout count_arg` corresponds to `count`. In order to pass different numbers each time to `example`, the formal `numb_arg` is defined to be `input`. The order in which the arguments are declared (in this case with the `integer` type) is irrelevant. The `$display` statements produce the following:

```
          1           1
          4          12
         10          60
         21         231
         21         231          18
```

3.9.1.2 `enter_new_state` *task*

The translation of the ASM chart from section 2.1.1.3 into Verilog given in section 3.8.2.2 is correct but could be improved in two ways. First, this translation did not include state names as part of the Verilog code (they were only in the comments). Second, this translation did not automatically provide default values for states where command signals were not mentioned, as occurs in ASM chart notation.

To overcome both of these limitations, we will define a task, which is arbitrarily given the name `enter_new_state`. The purpose of this task is to do things that occur whenever the machine enters any state. This includes storing into `present_state` a representation of a state (which is passed as an input argument, `this_state`), doing the #1 (which is legal in a task) to allow the <= to work properly and giving default values to the command outputs. In order to use this task, the designer needs to define several arbitrary bit patterns for the state names, define the `present_state` as a `reg` and indicate the number of bits in the `present_state`:

```
`define NUM_STATE_BITS     2
`define GREEN              2'b00
`define YELLOW             2'b01
`define RED                2'b10

  ...

  reg [`NUM_STATE_BITS-1:0] present_state;

  ...
```

The `always` block that implements the ASM chart is similar to the one given in section 3.8.2.2:

```
always
  begin
    @(posedge sysclk) enter_new_state(`GREEN);
      speed = 3;

    @(posedge sysclk) enter_new_state(`YELLOW);
      stop = 1;
      speed = 1;
      count <= @(posedge sysclk) count + 1;

    @(posedge sysclk) enter_new_state(`RED);
      stop = 1;
      count <= @(posedge sysclk) count + 2;
  end
```

The only differences are that the state names are passed as arguments to `enter_new_state`, and default values do not have to be mentioned. For example, state GREEN uses the default value 0 for `stop`, and state RED uses the default value 0 for `speed`.

The task that accomplishes these things for this particular ASM is:

```
task enter_new_state;
   input ['NUM_STATE_BITS-1:0] this_state;
   begin
     present_state = this_state;
     #1 stop = 0;
        speed = 0;
   end
 endtask
```

Even though default values are assigned for every state, since no time control occurs in this task after the assignment of default values, those states where non-default values are assigned work correctly. For example, assume the machine enters state GREEN at `$time` 50. At that `$time`, `present_state` will be assigned 2'b00. At `$time` 51, `stop` and `speed` will assigned their defaults of 0, but since there is no more time control, the `always` block which called on the task is not interruptable. At the same `$time` 51 `speed` changes to 3. Any other module concerned about `speed` at `$time` 51 would only observe a change to a value of 3. To understand this, we need to distinguish between sequence and `$time`. Because the task was called, two changes occurred to `speed` in sequence, but since they happened at the same `$time`, the outside world can only observe the last change. This creates exactly the effect we want. We are now ready to model ASM charts that do practical things with behavioral Verilog. Examples of translating ASM charts into Verilog using tasks like this are given in chapter 4.

3.9.2 Functions

The syntax for a function is similar to a task:

```
function type name;
     input arguments;
     ...
     declarations;
     begin
       statement;
       ...
       name = expression;
     end
   endfunction
```

except only input arguments are allowed. In the function definition, *`type`* is either `integer`, `real` or a bit width defined in square brackets. The statement(s) in a **function never include any time control**. The name of the function must be assigned the result to be returned (like the syntax of Pascal). These restrictions on functions exist so that every use of a function could, in theory, be synthesized as combinational logic.

3.9.2.1 *Real function example*

Verilog does not provide built-in trigonometric functions, but it is possible to define a function that approximates such a function using a polynomial:

```
function real sine;
    input x;
    real x;
    real y,y2,y3,y5,y7;
    begin
      y = x*2/3.14159;
      y2 = y*y;
      y3 = y*y2;
      y5 = y3*y2;
      y7 = y5*y2;
      sine = 1.570794*y - 0.261799*y3 +
               0.0130899*y5 - 0.000311665*y7;
    end
  endfunction
```

Such a function might be useful if a designer needs to test the Verilog model of a machine, such as a math coprocessor, that implements an ASM to approximate transcendental functions.

3.9.2.2 *Using a function to model combinational logic*

A more common use of a function in Verilog is as a behavioral way to describe combinational logic. For example, rather than being described by the logic gates given in section 2.5, a half-adder can also be described by a truth table:

inputs		output	
a	b	c	s
0	0	0	0
0	1	0	1
1	0	0	1
1	1	1	0

Such a table can be written in Verilog as a function defined with a `case` statement. Since the result of the function is composed of more than one bit, the function is better documented by using local variables (`c` and `s` in this example), which are concatenated to form the result:

```
function [1:0] half_add;
    input a,b;
    reg c,s;  //local for documentation

    begin
      case ({a,b})
        2'b00:  begin
                  c = 0;
                  s = 0;
                end
        2'b01:  begin
                  c = 0;
                  s = 1;
                end
        2'b10:  begin
                  c = 0;
                  s = 1;
                end
        2'b11:  begin
                  c = 1;
                  s = 0;
                end
        default:begin
                  c = 1'bx;
                  s = 1'bx;
                end
      endcase
      half_add = {c,s};
    end
  endfunction
```

So `half_add(0,0)` returns 2'b00 and `half_add(1,1)` returns 2'b10. Both `half_add(1,0)` and `half_add(0,1)` return 2'b01. All other possibilities, such as `half_add(1'bx,0)` return 2'bx. In order to use this function to model the combinational logic of a half-adder, the designer would define an `always` block with @ time control as explained in section 3.7.2.1:

```
reg C,S;
...
always @(A or B)
  {C,S} = half_add(A,B);
```

The actual argument A in the `always` block is bound to the formal `a` in `half_add`, and the actual argument B is bound to the formal `b`. The locals `c` and `s` are concatenated to form a two-bit result (hence the [1:0] declaration for the function.) This two bit result is stored in the two-bit concatenation `{C,S}`.

3.10 Structural Verilog, modules and ports

The preceding sections have covered many behavioral and a few structural (built-in gate), features of Verilog. This section discusses the most central aspect of Verilog: how the designer can define and instantiate Verilog modules to achieve hierarchical design.

Verilog code is composed of one or more modules. Each module is either a *top-level module* or an *instantiated module.* A top-level module is one (like all the earlier examples in this chapter) which is not instantiated elsewhere in the source code. There is only one copy of a top-level module. The definition of a top-level module is the same as the code that executes. The `regs` and `wires` in a top-level module are unique.

An instantiated module, on the other hand, is a unique executable copy of the definition. There may be many such copies. The definition is a "blueprint" for each of these instances. For example, section 2.5 illustrates an adder that needs three instances of a half-adder. It is only necessary to define the half-adder once. It can be instantiated as many times as required. Each instance of an instantiated module has its own copy of the `regs` and `wires` specified by the designer. For example, the value stored in a particular `reg` in one instance of a module need not be the same as the value stored in the `reg` of the same name in another instance of that module.

Instantiated modules should have ports that allow outside connections with each instance. It is this interconnection (i.e., structure) with the system external to the instance that gives each instance its unique role in the total system. Normally, each instance is internally identical to other instances derived from the same module definition, and how an instance is connected within the system gives that instance its characteristics.

The syntax for a module definition with ports is:

```
module name (port1,port2, ... );
  input ...  ;
  output ... ;
  inout ... ;
  declarations;
  structural instance;
  ...
  behavioral instance;
  ...
  tasks
  ...
  functions
  ...
endmodule
```

An example of a `structural instance` is given using built-in gates in section 3.5.1. Examples of designer supplied (rather than built-in) `structural instances` will be given later in section 3.10.6. A `behavioral instance` is either an `always` or `initial` block, as explained in section 3.4. (Tasks and functions are local to a module, and may be called by a `behavioral instance`, but are not by themselves `behavioral instances`.) The `declarations` include specifying either `wire` or `reg` of an appropriate width for each port listed in parentheses, as well as any local variables used internally within the module.

The order in which ports appear in the parentheses on the first line of the module definition is the order which matters elsewhere when this module is instantiated. Every one of the ports listed in the parentheses must be defined as one of the following: `input`, `output` or `inout`. Unlike tasks, the order in which the ports of a module appears in the `input`, `output` or `inout` definitions themselves is irrelevant. Although there is some vague similarity, the meaning of the words `input`, `output` and `inout` for a module is quite different than for a task. The designer makes the choice among these three alternatives based on the direction of information flow relative to the module in question. When making this decision, the designer looks at the system from the viewpoint of this one module.

3.10.1 `input` ports

An `input` port is one through which information comes into the module in question from the outside world. An `input` port must be declared within the module to have a size, or else Verilog will treat the `input` port as a one-bit `wire`, which is often incorrect. There are two ways to declare the size: either as a `wire` of some size (regardless of whether the module uses a *`behavioral instance`* or a *`structural instance`*) or with the `input` definition.[7]

Failure to declare an `input` port as a `wire` will cause it to be treated as a single-bit `wire`.

3.10.2 `output` ports

An `output` port is one through which information goes out of the module in question to the outside world. When the module in question uses a *`behavioral instance`* to produce the `output` port, the `output` port **must** be declared as a `reg` of some size. When the module in question uses a *`structural instance`*, the `output` port should be declared as a `wire` of some size. In other words, whether to declare an `output` port to be a `wire` or `reg` depends on whether it is generated by structural or behavioral code within the module in question.

3.10.3 `inout` ports

An `inout` port is one that is used to send information both directions. The advantage of an `inout` port is that the same port can do two separate things (at different times). The Verilog code for using an `inout` port is more complex than for simple `input` and `output` ports. An `inout` port corresponds to a hardware device known as a *tri-state buffer*. The details of `inout` ports and tristate buffers are discussed in appendix E.

3.10.4 Historical analogy: pins versus ports

Consider the analogy that "ports are like the doors of a building." For buildings like a store in a shopping center, some doors are labeled "IN," meaning that customers who wish to enter the store in question should go through that door. Those who are finished shopping leave through a different door labeled "OUT." It would be possible to look at the world from the viewpoint of the parking lot, but it is more convenient to look at things relative to the store in question (since there may be many stores in the shopping center to choose from).

There is another analogy for ports: ports are like the pins on an integrated circuit. Some pins are inputs and some pins are outputs. This is a very good analogy, but it is a little dangerous because when a large design is fabricated by a modern silicon foundry, most of the ports in the design do not correspond to a physical pin on the final integrated circuit.

---

[7] Some synthesis tools require that the `input` definition have the size.

To understand this pin analogy, let's digress for a moment and look at the history of hierarchical design and integrated circuit technology. Before the mid-1960s, all digital computers were built using *discrete electronic devices* (such as relays, vacuum tubes or transistors). It takes several such devices, wired together by hand in a certain structure, to make a gate, and of course, as we have seen in section 2.5, it takes many such gates to make anything remotely useful. In the early 1960's, photographic technologies became practical to mass-produce entire circuits composed of several devices on a wafer of semiconductor material (typically silicon). The wafer is sliced into "chips," which are mounted in epoxy (or similar material) with metal pins connecting the circuitry on the chip to the outside. There are several standard sizes for the number and placement of pins. For example, one of the oldest and smallest configurations is the 16-Pin Dual Inline Package (DIP). It is a rectangle with seven data pins on each side, and no pins on the top or bottom. (Two pins are reserved for power and ground.) A notch or dot at the top of the chip indicates where pin one is.

Designers in the 1960s and 1970s were limited by the number of devices that fit onto the chip and also by the number of pins allowed in these standard sizes. Realizing the power of hierarchical design, these designers built chips that contain standard building blocks that fit within the number of pins available. An example is a four-bit counter in one chip, TTL part number 74xx163, which is still widely used. Whenever designers needed a four-bit counter, they could simply specify a 74xx163, without worrying about its internal details. This, of course, is hierarchical design and provides the same mental simplification as instantiating a module. Physically, the pins of the 74xx163 chip would be soldered into the final circuit.

The relationship between these early integrated circuits and hierarchical design is not perfect, hence the danger of saying ports are like pins. If a design needs one 13-bit counter, a designer in the 1970s would have to specify that four 74xx163s be soldered into the final circuit to act as a single counter. There is an interconnection between these four chips so that they collectively count properly. From a hierarchical standpoint, we want to see only one black box, with a 13-bit bus, but this counter is fabricated as four 74xx163s wired together. Some of the physical pins (connected to another one of the 74xx163s) have nothing to do with the ports of a 13-bit counter.

With modern silicon fabrication technologies, the limitations on the number of devices on a chip have been eased, but the limitations on physical pins have become even more severe. Although chips can contain millions of gates, the number of pins allowed is seldom more than a few hundred. Hierarchical design should be driven by the problem being solved (which is the fundamental principle of all top-down design) and not by the limitations (such as pins) of the technology used. Every physical pin on a chip is (part of) a Verilog port, but not every Verilog port necessarily gets fabricated as a physical pin(s). Even so, the **analogy** is a good one: ports are **like** pins.

3.10.5 Example of a module defined with a behavioral instance

Section 2.5 defines an adder several ways. The simplest way to explain what an adder does is to describe it behaviorally. Since an adder is combinational logic, we can use the @ time control technique discussed in section 3.7.2.1 to model its behavior. However, since an adder is used in a larger structure, we should make the `always` block that models the adder's behavior part of a module definition. Those ports (`a` and `b`) that are physical inputs to the fabricated adder will be `input` ports to this module, and are exactly the variables listed in the sensitivity list. The port that is a physical output (`sum`) is, of course, defined to be an `output` port. Since this module computes `sum` with behavioral code, `sum` is declared to be a `reg`. (There are no "registers" in combinational logic, but a Verilog `reg` is used in a behavioral model of combinational logic. A `reg` is not a "register" as long as the sensitivity list has all the inputs listed.) As in the example of section 2.5, the widths of `a` and `b` are two bits each, and the width of `sum` is three bits:

```
module adder(sum,a,b);
  input [1:0] a,b;
  output [2:0] sum;
  wire [1:0] a,b;
  reg [2:0] sum;

  always @(a or b)
    sum = a + b;
endmodule
```

The widths shown on input and output definitions are optional for simulation purposes.[8]

To exhaustively test this small adder, test code similar to section 3.7.2.1 enumerates all possible combinations of `a` and `b`:

[8] The width will not be shown on later examples in this chapter, although describing the width on `input` and `output` definitions would be legal in simulation. The width might be required to overcome the limitations of some commercial simulation tools.

```
module top;
  integer ia,ib;
  reg [1:0] a,b;
  wire [2:0] sum;

  adder adder1(sum,a,b);

  initial
    begin
      for (ia=0; ia<=3; ia = ia+1)
        begin
          a = ia;
          for (ib=0; ib<=3; ib = ib + 1)
            begin
              b = ib;
              #1 $display("a=%d b=%d sum=%d",a,b,sum);
            end
        end
    end
endmodule
```

The important thing in this top-level test module is that `adder` (the name of the module definition) is instantiated in `top` with the name `adder1`. In the top-level module, `a` and `b` are `reg`s because, within this module (`top`), `a` and `b` are supplied by behavioral code. On the other hand, `sum` is supplied by `adder1`, and so `top` declares `sum` to be a `wire`. The syntax for instantiating a user defined module is similar to instantiating a built-in gate. In this example, the local `sum` of `top` corresponds to the output port (coincidentally named `sum`) of an instance of module `adder`. If the names (such as `sum`) in module `adder` were changed to other names (such as `total`), the module would work the same:

```
module adder(total,alpha,beta);
  input alpha,beta;
  output total;
  wire [1:0] alpha,beta;
  reg [2:0] total;

  always @(alpha or beta)
    total = alpha + beta;
endmodule
```

It is the position within the parentheses, and not the names, that matter[9] when the module is instantiated in the test code.

3.10.6 Example of a module defined with a structural instance

Of course, in hierarchical design, we need a structural definition of the module. As described in section 2.5, the module `adder` can be defined in terms of instantiation of an instance of a `half_adder` (which we will call `ha1`) and an instance of a `full_adder` (which we will call `fa1`):

```
module adder(sum,a,b);
  input a,b;
  output sum;
  wire [1:0] a,b;
  wire [2:0] sum;

  wire c;

  half_adder ha1(c,sum[0],a[0],b[0]);
  full_adder fa1(sum[2],sum[1],a[1],b[1],c);
endmodule
```

Since the adder is defined with two *`structural instances`* (named `ha1` and `fa1`), all of the ports, including the `output` port, `sum`, are `wires`. The local wire `c` sends the carry from the half-adder to the full-adder. Of course, we need identical test code as in the last example, and we also need module definitions for `full_adder` and `half_adder`.

3.10.7 More examples of behavioral and structural instances

Even though `half_adder` and `full_adder` are instantiated structurally in section 3.10.6, they can be defined either behaviorally or structurally. For example, a behavioral definition of these modules is:

[9] Verilog provides an alternative syntax, described in chapter 11, that allows the name, rather than the position, to determine how the module is instantiated.

```
module half_adder(c,s,a,b);
  input a,b;
  wire a,b;
  output c,s;
  reg c,s;

  always @(a or b)
    {c,s} = a+b;
endmodule

module full_adder(cout,s,a,b,cin);
  input a,b,cin;
  wire a,b,cin;
  output cout,s;
  reg cout,s;

  always @(a or b or cin)
    {cout,s} = a+b+cin;
endmodule
```

Once again, notice that the outputs are `regs`. Concatenation is used on the left of the = to make the definition of the module simple. `{cout,s}` is a two-bit `reg` capable of dealing with the largest possible number (`2'b11`) produced by `a+b+cin`.

An alternative would be to define the `half_adder` and `full_adder` modules with *`structural instances`*, which means all outputs are `wires`:

```
module half_adder(c,s,a,b);
  input a,b;
  wire a,b;
  output c,s;
  wire c,s;

  xor x1(s,a,b);
  and a1(c,a,b);
endmodule

module full_adder(cout,s,a,b,cin);
  input a,b,cin;
  wire a,b,cin;
  output cout,s;
  wire cout,s;
  wire cout1,cout2,stemp;

  half_adder ha2(cout1,stemp,a,b);
  half_adder ha3(cout2,s,cin,stemp);
  or         o1(cout,cout1,cout2);
endmodule
```

There are two instances of `half_adder` (`ha2` and `ha3`). The only difference between these two instances is how they are connected within `full_adder`. There are three local wires (`cout1`, `cout2` and `stemp`) that allow internal interconnection within the module.

At this point, we have reduced the problem down to Verilog primitive gates (`and, or, xor`) whose behavior is built into Verilog.

3.10.8 Hierarchical names

Although ports are intended to be the way in which modules communicate with each other in a properly functioning system, Verilog provides a way for one module to access the internal parts of another module. Conventional high-level languages, like C and Pascal, have *scope rules* that absolutely prohibit certain kinds of access to local information. Verilog is completely different in this regard. The philosophy of Verilog for accessing variables is very similar the philosophy of the NT or UNIX operating systems for accessing files: if you know the path to a file (within subdirectories), you can access the file. Analogously in Verilog: if you know the path to a variable (within modules), you can access the variable.

For example, using the definition of `adder` given in section 3.10.6, and the instance `adder1` shown in the test code of section 3.10.5, `adder1` has a local wire `c` that is not accessible to the outside world. The following statement **in the test code** would allow the designer to observe this wire, even though there is no port that outputs `c`:

```
$display(adder1.c);
```

A name, such as `adder1.c` is known as a *hierarchical name*, or *path*.

The following statement allows the designer to observe `cout2` from the test code:

```
$display(adder1.fa1.cout2);
```

which happens to be the same as:

```
$display(adder1.fa1.ha3.c);
```

The parts of a hierarchical name are separated by periods. Every part of a hierarchical name, except the last, is the name of an instance of a module. The names of the corresponding module definitions (`adder`, `full_adder` and `half_adder` in the above example) **never** appear in a hierarchical name.

3.10.9 Data structures

The term "structure" has three distinct meanings in computer technology. Elsewhere in this book, "structure" takes on its hardware meaning: the interconnection of modules using wires. But you have probably heard of the other two uses of this word: "structured programming," and "data structures." The concept of "structured programming" is a purely behavioral software concept which is closely related to what we call goto-less programming (see section 2.1.4). "Data structures" are software objects that allow programmers to solve complex problems in a more natural way.

The period notation used in Verilog for hierarchical names is reminiscent of the notation used in conventional high-level languages for accessing components of a "data structure" (`record` in Pascal, `struct` in C, and `class` in C++). In fact, you can create such software "data structures" in Verilog by defining a portless module that has only data, but that is intended to be instantiated. Such a portless but instantiated module is worthless for hardware description, but is identical to a conventional software "data structure." Such a module has no behavioral instances or structural instances. For example, a data structure could be defined to contain payroll information about an employee:

```
module payroll;
  reg [7:0] id;
  reg [5:0] hours;
  reg [3:0] rate;
endmodule
```

Suppose we have two employees, `joe` and `jane`. Each employee has a unique instance of this module:

```
payroll joe();
payroll jane();

  initial
    begin
      joe.id=254;
      joe.hours=40;
      joe.rate=14;
      jane.id=255;
      jane.hours=63;
      jane.rate=15;
    end
```

The empty parentheses are a syntactic requirement of Verilog. In this example, the fields of `jane` contain the largest possible values.

Data structures usually have a limited set of operations that manipulate the fields of the data. For example, the `hours` and `rate` fields can be combined to display the corresponding total pay. This operation is defined as a local task of the module. However, since there are no behavioral instances in this module, this task sits idle until it is called from the outside (using a hierarchical name):

```
module payroll;
  reg [7:0] id;
  reg [5:0] hours;
  reg [3:0] rate;

  task display_pay;
    integer pay;  //local
    begin
      if (hours>40)
        pay = 40*rate + (hours-40)*rate*3/2;
      else
        pay = hours*rate;
      $display("employee %d earns %d",id,pay);
    end
  endtask
endmodule

module top;
  payroll joe();
  payroll jane();
  initial
    begin
      joe.id=254;
      joe.hours=40;
      joe.rate=14;
      joe.display_pay;
      jane.id=255;
      jane.hours=63;
      jane.rate=15;
      jane.display_pay;
    end
endmodule
```

This is very close to the software concept of *object-oriented programming* in languages like C++, except the current version of Verilog lacks the inheritance feature found in C++.

Data structures are a powerful use of hierarchical names, but they are somewhat afield from the central focus of this book: hardware structures. Application of hierarchical names are useful in test code, and so it is important to understand them. Also, the above example helps illustrate what instantiation really means in Verilog.

3.10.10 Parameters

Verilog modules allow the definition of what are known as parameters. These are constants that can be different for each instance. For example, suppose you would like to define a module behaviorally that models an enabled register of arbitrary width:

```
module enabled_register(dout, din, ld, sysclk);
  parameter WIDTH = 1;
  input din,ld,sysclk;
  output dout;
  wire [WIDTH-1:0] din;
  reg [WIDTH-1:0] dout;
  wire ld,sysclk;

  always @(posedge sysclk)
    if (ld)
      dout = din;
endmodule
```

By convention, we use capital letters for parameters, but this is not a requirement. Note that parameters do not have a backquote preceding them.

If you instantiate this module without specifying a constant, the default given in the `parameter` statement (in this example, 1) will be used as the `WIDTH`, and so the instance `R1` will be one bit wide:

```
wire ldR1,sysclk;
wire R1dout,R1din;
enabled_register      R1(R1dout,R1din,ldR1,sysclk);
```

To specify a non-default constant, the syntax is a # followed by a list of constants in parentheses. Since there is only one parameter in this example, there can be only one constant in the parentheses. For example, to instantiate a 12-bit register for `R12`:

```
wire ldR2,sysclk;
wire [11:0] R12dout,R12din;
enabled_register #(12) R12(R12dout,R12din,ldR12,sysclk);
```

Verilog requires that the width of a wire that attaches to an output port match the reg declaration within the module. In this example, R12dout is a wire twelve bits wide, the parameter WIDTH in the instance R12 is twelve, and the corresponding output port, dout, is declared as reg[WIDTH-1:0], which is the same as reg [11:0].

Since there is only one constant in the parentheses above, it is legal to omit the parentheses:

```
enabled_register #12   R12(R12dout,R12din,ldR12,sysclk);
```

Sometimes, you need more than one constant in the definition of a module. For example, a combinational multiplier has two input buses, whose widths need not be the same:

```
module multiplier(prod,a,b);
  parameter WIDTHA=1,WIDTHB=1;
  output prod;
  input a,b;
  reg [WIDTHA+WIDTHB-1:0] prod;
  wire [WIDTHA-1:0] a;
  wire [WIDTHB-1:0] b;

  always @(a or b)
    prod = a*b;
endmodule
```

Here is an example of instantiating this:

```
wire [5:0] hours;
wire [3:0] rate;
wire [9:0] pay;

multiplier #(6,4) m1(pay,hours,rate);
```

3.11 Conclusion

Modules are the basic feature of the Verilog hardware description language. Modules are either top-level or instantiated. Top-level modules are typically used for test code. Instantiated modules have ports, which can be defined to be either input, output or inout. Constants in modules may be defined with the parameter statement. A module is either defined with a *behavioral instance* (always or initial block(s) or with a `structural instance` (built-in gates or instantiation of other designer-provided modules). Behavioral and structural instances may be mixed in the same module.

Variables produced by behavioral code, including outputs from the module, are declared to be `regs`. Behavioral modules have the usual high-level statements, such as `if` and `while`, as well as time control (#, @ and `wait`) that indicate when the process can be suspended and resumed. The `$time` variable simulates the passage of time in the fabricated hardware. Verilog makes a distinction between algorithmic sequence and the passage of `$time`. The most important forms of time control are # followed by a constant, which is used for generating the clock and test vectors; `@(posedge sysclk)`, which is used to model controllers and registers; and @ followed by a sensitivity list, which is used for combinational logic. Verilog provides the non-blocking assignment statement, which is ideal for translating ASM charts that use RTN into behavioral Verilog. Verilog also provides tasks and functions, which like similar features in conventional high-level languages, simplify coding.

Structural modules have a simple syntax. They may instantiate other designer-provided modules to achieve hierarchical design. They may also instantiate built-in gates. The syntax for both kinds of instantiation is identical. All variables in a structural module, including outputs, are `wires`.

Hierarchical names allow access to tasks and variables from other modules. Use of hierarchical names is usually limited to test code.

The next chapter uses the features of Verilog described in this chapter to express the three stages (pure behavioral, mixed and pure structural) of the design process for the childish division machine designed manually in chapter 2. The advantage of using Verilog at each of these stages is that the designer can simulate each stage to be sure it is correct before going on to the next stage. Also, the final Verilog code can be synthesized into a working piece of hardware, without the designer having to toil manually to produce a flattened circuit diagram and netlist.

3.12 Further reading

LEE, JAMES M., *Verilog Quickstart*, Kluwer, Norwell, MA, 1997. Gives several examples of implicit style.

PALNITKAR, S., *Verilog HDL: A Guide to Digital Design and Synthesis,* Prentice Hall PTR, Upper Saddle River, NJ, 1996. An excellent reference for all aspects of Verilog.

SMITH, DOUGLAS J., *HDL Chip Design: A Practical Guide for Designing, Synthesizing, and Simulating ASICs and FPGAs Using VHDL or Verilog*, Doone Publications, Madison, AL, 1997. A Rosetta stone between Verilog and VHDL.

STERNHEIM, ELIEZER, RAJVIR SINGH and YATIN TRIVEDI, *Digital Design with Verilog HDL*, Automata Publishing, San Jose, CA, 1990. Has several case studies of using Verilog.

THOMAS, DONALD E. and PHILIP R. MOORBY, *The Verilog Hardware Description Language*, Third edition, Kluwer, Norwell, MA., 1996. Explains how a simulator works internally.

3.13 Exercises

3-1. Design behavioral Verilog for a two-input 3-bit wide mux using the technique described in section 3.7.2.1. The port list for this module should be:

```
module mux2(i0, i1, sel, out);
```

3-2. Design a structural Verilog module (`mux2`) equivalent to problem 3-1 using only instances of `and`, `or`, `not` and `buf`.

3-3. Modify the solution to problem 3-1 to use a parameter named `SIZE` that allows instantiation of an arbitrary width for `i0`, `i1` and `out` as explained in section 3.10.10. For example, the following instance of this device would be useful in the architecture drawn in section 2.3.1:

```
wire muxctrl;
wire [11:0] x,y,muxbus;
mux2 #12 mx(x,y,muxctrl,muxbus);
```

3-4. Given the instance (`mx`) of the module (`mux2`) shown in problem 3-3, what hierarchical names are equivalent to `x`, `y`, `muxctrl` and `muxbus`?

3-5. Design behavioral Verilog for combinational incrementor and decrementor modules using the technique described in section 3.7.2.1. Use a parameter named `SIZE` that allows instantiation of an arbitrary width for the ports as explained in section 3.10.10.

3-6. Design behavioral Verilog for an up/down counter (section D.8) using the technique described in section 3.7.2.2. The port list for this module should be:

```
module updown_register(din,dout,ld,up,count,clk);
```

3-7. Modify the solutions to problem 3-6 to use a parameter named `SIZE` that allows instantiation of an arbitrary width for the ports as explained in section 3.10.10.

3-8. Design behavioral Verilog for a simple D-type register (section D.5) using the technique described in section 3.7.2.2. Use a parameter named `SIZE` that allows instantiation of an arbitrary width for the ports as explained in section 3.10.10. The port list for this module should be:

```
module simpled_register(din,dout,clk);
```

3-9. Design a structural Verilog module (`updown_register`) equivalent to problem 3-7 using only instances of the modules defined in problems 3-3, 3-5 and 3-8.

3-10. For each of the ASM charts given in problem 2-10, translate to implicit style Verilog using non-blocking assignment for ← and `@(posedge sysclk)#1` for each rectangle, as explained in section 3.8.2.3.1. As in that example, there should be one `always` that models the hardware, one `always` for the `$display` and an `always` and `initial` for `sysclk`. Compare the result of simulation with the manually produced timing diagram of problem 2-10.

3-11. Without using a Verilog simulator, give a timing diagram for the machine described by the ASM chart of section 3.8.2.3.3. Show the values of `a` and `b` in the first twelve clock cycles, and label each clock cycle to indicate which state the machine is in. Next, run the **original** implicit style Verilog code equivalent to the ASM and make a printout of the .log file. On this printout, write the name of the state that the machine is in during each clock cycle. The manually created timing diagram should agree with the Verilog .log file. Finally, modify the following:

```
@(posedge sysclk) #1;                // state FIRST
  a <= @(posedge sysclk) 1;
```

to become:

```
@(posedge sysclk) #1;                   // state FIRST
  a = 1;
```

Run the modified Verilog code and make a printout of its .log file. On this printout, circle the differences, if any, that exist between the correct timing diagram and the .log file for the modified Verilog. In no more than three sentences, explain why there are or are not any differences between = and <=.

3-12. Without using a Verilog simulator, give a timing diagram for the machine described by the ASM of section 3.8.2.3.4. Show the values of `a` and `b` in the first twelve clock cycles, and label each clock cycle to indicate which state the machine is in. Next, run the **original** implicit style Verilog code equivalent to the ASM and make a printout of the .log file. On this printout write the name of the state that the machine is in during each clock cycle. The manually created timing diagram should agree with the Verilog .log file. Finally, modify the code to change the `if` to a `while`. Run the modified Verilog code and make a printout of its .log file. On this printout, circle the differences, if any, that exist between the correct timing diagram and the .log file for the modified Verilog. In no more than three sentences, explain why there are or are not any differences between `if` and `while`.

3-13. Without using a Verilog simulator, give a timing diagram for the machine described by the ASM of section 3.8.2.3.5. Show the values of `a` and `b` in the first twelve clock cycles, and label each clock cycle to indicate which state the machine is in. Next, run the **original** implicit style Verilog code equivalent to the ASM and make a printout of the .log file. On this printout write the name of the state that the machine is in during each clock cycle. The manually created timing diagram should agree with the Verilog .log file. Finally, modify the code to eliminate all #1s. Run the modified Verilog code and make a printout of its .log file. On this printout, circle the differences, if any, that exist between the correct timing diagram and the .log file for the modified Verilog. In no more than three sentences, explain why there are or are not any differences between using and omitting #1s.

4. THREE STAGES FOR VERILOG DESIGN

The design of the childish division machine described in chapter 2 can be expressed in Verilog using the language features discussed in chapter 3. This chapter uses variations on the childish division machine as a simple but practical example of how to design hardware with Verilog.

As described in sections 2.1.5.1 through 2.1.5.3, the three-stages of the top-down design process are: pure behavioral, mixed behavior/structure and pure structural. Because Verilog allows both behavioral and structural constructs, one can transform a design through these three stages by minor editing of the Verilog source code. Section 4.1 gives several examples of how the pure behavioral ASMs of chapter 2 can be written in pure behavioral Verilog using the implicit style of section 3.8.2.3. Section 4.2 uses one of these behavioral examples to illustrate translation into the mixed stage. Section 4.3 translates the mixed example from section 4.2 into the pure structural stage (the "explicit style" often used by Verilog designers). Section 4.4 shows that, having completed the three phases of our design process, Verilog allows additional structure (in this case, dealing with the controller) to be instantiated in place of behavior using the hierarchical design process described in section 2.5.

Chapter 7 and appendix F describe an alternate way (which can be automated) for translating pure behavioral (implicit style) Verilog directly to a form that can be synthesized into physical hardware. The manual technique in this chapter is more intricate and involved than the one in chapter 7, but an understanding of the three-stage technique in this chapter will give the reader a better appreciation for the behavioral and structural aspects of Verilog simulation.

4.1 Pure behavioral examples

This section gives examples of modeling various ASM charts for the childish division algorithm with pure behavioral Verilog.

4.1.1 Four-state division machine

Let's consider translating the second ASM chart of section 2.2.3 (the one that has four-states, with state COMPUTE1 at the top of the loop) into Verilog, using the `enter_new_state` approach described in section 3.9.1.2.

4.1.1.1 Overview of the source code

First, we have to define the arbitrary bit patterns that represent the states and indicate the number of bits required:

```
`define NUM_STATE_BITS 2
`define IDLE         2'b00
`define INIT         2'b01
`define COMPUTE1     2'b10
`define COMPUTE2     2'b11
```

These definitions occur outside any modules. Next, we need to include the definition of a module that generates the clock in a fashion similar to 3.7.1.3, except the clock is output as a port of the module:

```
module cl(clk);
     parameter TIME_LIMIT = 110000;
     output clk;
     reg clk;

     initial
       clk = 0;

     always
       #50 clk = ~clk;

     always @(posedge clk)
       if ($time > TIME_LIMIT) #70 $stop;
endmodule
```

The purpose of the parameter TIME_LIMIT is to stop the simulation after a certain amount of `$time` has elapsed. Without scheduling such a `$stop` (or `$finish`) statement, the simulation could continue forever.

Third, the module that implements the behavioral simulation of the actual hardware needs to be defined with an appropriate portlist:

```
module slow_div_system(pb,ready,x,y,r2,sysclk);
  input pb,x,y,sysclk;
  output ready,r2;
  wire ...
  reg...
  ...

endmodule
```

This module corresponds to the first block diagram in section 2.5.2. At this stage, we are leaving the details of this module ambiguous. The discussion of the details inside this module will continue in section 4.1.1.2.

Finally, the last module that needs to be defined is the top-level test code. Using a technique involving #, @ and `wait` that is similar to the one described in section 3.7.3, the test code checks if the simulated hardware can divide values of `x` that vary from 0 to 14 properly when `y` is held fixed at 7:

```
module top;
  reg pb;
  reg [11:0] x,y;
  wire [11:0] quotient;
  wire ready;
  integer s;
  wire sysclk;

  cl #20000 clock(sysclk);
  slow_div_system  slow_div_machine(pb,ready,x,y,
                                   quotient,sysclk);
  initial
    begin
      pb= 0;
      x = 0;
      y = 7;
      #250;
      @(posedge sysclk);
      for (x=0; x<=14; x = x+1)
        begin
          @(posedge sysclk);
          pb = 1;
          @(posedge sysclk);
          pb = 0;
          @(posedge sysclk);
          wait(ready);
          @(posedge sysclk);
          if (x/y === quotient)
           $display("ok");
          else
           $display("error x=%d y=%d x/y=%d quotient=%d",
                    x,y,x/y,quotient);
        end
      $stop;
    end
endmodule
```

Even for a small machine like this, an exhaustive test (such as was given in section 3.7.3) would be too time consuming to use as an illustration.

In the above code, note that slow_division_system is instantiated with the instance name slow_division_machine. The first port in the module, the input pb, corresponds to a reg in the test code of the same name. The second port in the module, the output ready, corresponds to a wire of the same name in the test code. The third and fourth ports of the module, the inputs x and y, correspond to 12-bit regs of the same names in the test code. The fifth port of the module, an output named r2, corresponds to a 12-bit wire whose name is quotient in the test code. The final port of the module, an input named sysclk, corresponds to a wire of the same name in the test code.

The reason sysclk is a wire is because it happens to be an output port of the instance of the cl module named clock. This means clock, rather than the test code, supplies sysclk. The test code therefore has a little bit of structure connecting clock to slow_div_machine. The variables in the test code that are regs are so declared because the test code must supply them to slow_division_machine using the behavioral =. The remaining wires are so declared because the behavioral test code does not supply them.

4.1.1.2 *Details on* **slow_division_system**

Let's return to the definition of the slow_division_system module. We need to declare the types of the inputs and outputs. Also, we need to declare local variables, such as those that model the physical registers of the hardware (r1 and r2), and also the present_state:

```
module slow_div_system(pb,ready,x,y,r2,sysclk);
  input pb,x,y,sysclk;
  output ready,r2;
  wire pb,sysclk;
  wire [11:0] x,y;
  reg ready;
  reg [11:0] r1,r2;
  reg ['NUM_STATE_BITS-1:0] present_state;
  ...
endmodule
```

These declarations were constrained by the portlist, how the module was instantiated in the test code and by the description of the problem given in chapter 2.

As described in section 3.9.1.2, we should define a task that simplifies the code for the sequence of statements that must occur when the machine enters each state:

```
task enter_new_state;
  input ['NUM_STATE_BITS-1:0] this_state;
  begin
    present_state = this_state;
    #1 ready=0;
  end
endtask
```

The definition of this task will be nearly identical for every pure behavioral ASM. The only distinction from one problem to another is the list of external command outputs (see section 2.1.3.2.1) specific to the particular machine. In this case, the only external command output is `ready`. It has a default value of 0; thus this task must initialize it at the beginning of every clock cycle.

Having defined the above task within the `slow_division_system` module, it is possible to translate the ASM from section 2.2.3 into Verilog:

```
always
  begin
   @(posedge sysclk) enter_new_state('IDLE);
    r1 <= @(posedge sysclk) x;
    ready = 1;
    if (pb)
      begin
       @(posedge sysclk) enter_new_state('INIT);
        r2 <= @(posedge sysclk) 0;
        while (r1 >= y)
          begin
           @(posedge sysclk) enter_new_state('COMPUTE1);
            r1 <= @(posedge sysclk) r1 - y;
           @(posedge sysclk) enter_new_state('COMPUTE2);
            r2 <= @(posedge sysclk) r2 + 1;
          end
      end
  end
```

The only other thing that would be desirable to put in this module is a debugging display, as described in section 3.7.2.4:

```
always @(posedge sysclk) #20
    $display("%d r1=%d r2=%d pb=%b ready=%b",
                 $time, r1,r2, pb, ready);
```

The net effect of the other modules defined in section 4.1.1.1 and all the details inside the module given above is to produce the following simulation output from Verilog:

```
                  70 r1=    x r2=    x pb=0 ready=1
                 170 r1=    0 r2=    x pb=0 ready=1
                 270 r1=    0 r2=    x pb=0 ready=1
                 370 r1=    0 r2=    x pb=1 ready=1
                 470 r1=    0 r2=    x pb=0 ready=0
                 570 r1=    0 r2=    0 pb=0 ready=1
ok
                 670 r1=    0 r2=    0 pb=0 ready=1
                 770 r1=    1 r2=    0 pb=1 ready=1
                 870 r1=    1 r2=    0 pb=0 ready=0
                 970 r1=    1 r2=    0 pb=0 ready=1
ok
...
                6670 r1=   12 r2=    1 pb=0 ready=1
                6770 r1=   13 r2=    1 pb=1 ready=1
                6870 r1=   13 r2=    1 pb=0 ready=0
                6970 r1=   13 r2=    0 pb=0 ready=0
                7070 r1=    6 r2=    0 pb=0 ready=0
                7170 r1=    6 r2=    1 pb=0 ready=1
ok
                7270 r1=   13 r2=    1 pb=0 ready=1
                7370 r1=   14 r2=    1 pb=1 ready=1
                7470 r1=   14 r2=    1 pb=0 ready=0
                7570 r1=   14 r2=    0 pb=0 ready=0
                7670 r1=    7 r2=    0 pb=0 ready=0
                7770 r1=    7 r2=    1 pb=0 ready=0
                7870 r1=    0 r2=    1 pb=0 ready=0
                7970 r1=    0 r2=    2 pb=0 ready=1
ok
```

The `regs` `r1` and `r2` are not initialized at `$time` 0, and so the value `12'bx` is printed by the `$display` simply as `x` (not to be confused with the variable `x`). Each time the machine returns to state IDLE (`ready=1`), the outputs of the machine from `r2` (which is the same as `quotient`) are highlighted above. These are the values that the test code uses to determine that everything is "`ok`" each time.

4.1.2 Verilog catches the error

The first ASM chart in section 2.2.3 (with state COMPUTE2 at the top of the loop) has an error. An advantage of using Verilog to simulate such a machine while it is still in the behavioral stage is that Verilog can usually catch such errors before they become costly. In this case, the Verilog code would be identical to section 4.1.1.2, except the ASM chart would be translated as:

```
always
  begin
     ...
     while (r1 >= y)
       begin
        @(posedge sysclk) enter_new_state(`COMPUTE2);
         r2 <= @(posedge sysclk) r2 + 1;
        @(posedge sysclk) enter_new_state(`COMPUTE1);
         r1 <= @(posedge sysclk) r1 - y;
       end
    end
  end
```

The output from the Verilog simulator makes the problem obvious:

```
    ...
          2670 r1=    5 r2=    0 pb=0 ready=1
          2770 r1=    6 r2=    0 pb=1 ready=1
          2870 r1=    6 r2=    0 pb=0 ready=0
          2970 r1=    6 r2=    0 pb=0 ready=1
    ok
          3070 r1=    6 r2=    0 pb=0 ready=1
          3170 r1=    7 r2=    0 pb=1 ready=1
          3270 r1=    7 r2=    0 pb=0 ready=0
          3370 r1=    7 r2=    0 pb=0 ready=0
          3470 r1=    7 r2=    1 pb=0 ready=0
          3570 r1=    0 r2=    1 pb=0 ready=0
          3670 r1=    0 r2=    2 pb=0 ready=0
          3770 r1=4089 r2=    2 pb=0 ready=1
    error x=    7 y=    7 x/y=    1 quotient=    2
    ...
          8670 r1=   13 r2=    2 pb=0 ready=1
          8770 r1=   14 r2=    2 pb=1 ready=1
          8870 r1=   14 r2=    2 pb=0 ready=0
          8970 r1=   14 r2=    0 pb=0 ready=0
          9070 r1=   14 r2=    1 pb=0 ready=0
          9170 r1=    7 r2=    1 pb=0 ready=0
```

Continued.

```
                9270 r1=    7 r2=    2 pb=0 ready=0
                9370 r1=    0 r2=    2 pb=0 ready=0
                9470 r1=    0 r2=    3 pb=0 ready=0
                9570 r1=4089 r2=    3 pb=0 ready=1
           error x=   14 y=     7 x/y=    2 quotient=    3
```

Because of the error (which causes the loop to execute an extra time), the time to complete the test is longer. The `wait` statement in the test code compensates for this; thus the test code is checking `r2` via `quotient` at the proper time, but when `y>=7`, `quotient` is just plain wrong.

Rather than spending thousands of dollars actually fabricating a faulty computer, the designer can observe the problem simply from the behavioral Verilog code.

4.1.3 Importance of test code

Do not be **deluded** that because Verilog **can sometimes** catch errors such as described in section 4.1.2 that it **will always** catch such errors. Just because a designer uses an automated tool like a Verilog is not the same as saying that the designer does not have to think. In fact, using such tools requires a higher level of thought process. The designer is responsible not just for trying to design a machine that works, but also for designing test code that aggressively checks to see if the design fails.

When we began with this example in section 2.2.1, we stated in informal English some assumptions about the environment in which this machine operates. We assumed two things: there is a friendly user (who provides inputs in a specific sequence), and that this user needs the output of the machine to remain constant when the machine is in state IDLE (because the user is a person who cannot perceive events that happen within a single clock cycle, which are typically less than a millionth of a second).

The test code given in section 4.1.1.1 satisfies the assumption that the inputs are provided in the order demanded in section 2.2.1 (in particular the test code lets the machine stay in state IDLE two clock cycles before the button is pressed). There is, however, a problem with the above test code with regard to the second assumption, as is illustrated below.

Consider the ASM chart of section 2.2.4. The Verilog code to simulate this machine is identical to the previous examples, except for the following:

```
always
  begin
   @(posedge sysclk) enter_new_state(`IDLE);
    r1 <= @(posedge sysclk) x;
    r2 <= @(posedge sysclk) 0;
    ready = 1;
    if (pb)
      begin
       while (r1 >= y)
         begin
          @(posedge sysclk) enter_new_state(`COMPUTE1);
           r1 <= @(posedge sysclk) r1 - y;
          @(posedge sysclk) enter_new_state(`COMPUTE2);
           r2 <= @(posedge sysclk) r2 + 1;
         end
      end
  end
```

When Verilog simulates this with the test code given earlier, it detects no errors:

```
                  70 r1=    x r2=    x pb=0 ready=1
                 170 r1=    0 r2=    0 pb=0 ready=1
                 270 r1=    0 r2=    0 pb=0 ready=1
                 370 r1=    0 r2=    0 pb=1 ready=1
                 470 r1=    0 r2=    0 pb=0 ready=1
                 570 r1=    0 r2=    0 pb=0 ready=1
ok
...
                6070 r1=   12 r2=    0 pb=0 ready=1
                6170 r1=   13 r2=    0 pb=1 ready=1
                6270 r1=   13 r2=    0 pb=0 ready=0
                6370 r1=    6 r2=    0 pb=0 ready=0
                6470 r1=    6 r2=    1 pb=0 ready=1
ok
                6570 r1=   13 r2=    0 pb=0 ready=1
                6670 r1=   14 r2=    0 pb=1 ready=1
                6770 r1=   14 r2=    0 pb=0 ready=0
                6870 r1=    7 r2=    0 pb=0 ready=0
                6970 r1=    7 r2=    1 pb=0 ready=0
                7070 r1=    0 r2=    1 pb=0 ready=0
                7170 r1=    0 r2=    2 pb=0 ready=1
ok
```

But as was discussed in section 2.2.4, it is unacceptable from a user interface standpoint because it throws away the correct answer after only one clock cycle. For example, at `$time` 6470, the correct answer 1 is present in `r2`, and the test code notes this. But at `$time` 6570, the machine has thrown away the correct answer even though `ready` still indicates that the correct answer should be displayed.

The proper approach is to take all details in the informal English specification and make these details part of the test code. The test code should be a formal specification of what the machine is supposed to do under all circumstances. This raises the interesting, and somewhat unsolvable dilemma: how does the designer test the test code?[1] Nevertheless, it is important to put reasonable effort into creating robust test code.

4.1.4 Additional pure behavioral examples

The first ASM chart of 2.2.5 can be translated into Verilog as:

```
always
  begin
   @(posedge sysclk) enter_new_state(`IDLE);
    r1 <= @(posedge sysclk) x;
    r2 <= @(posedge sysclk) 0;
    ready = 1;
    if (pb)
      begin
       while (r1 >= y)
         begin
          @(posedge sysclk) enter_new_state(`COMPUTE1);
           r1 <= @(posedge sysclk) r1 - y;
          @(posedge sysclk) enter_new_state(`COMPUTE2);
           r2 <= @(posedge sysclk) r2 + 1;
          @(posedge sysclk) enter_new_state(`COMPUTE3);
           r3 <= @(posedge sysclk) r2;
         end
      end
  end
```

[1] See J. Cooley, *Integrated System Design*, July 1995, pp. 56-60 for a description of a Verilog contest where the test code provided to the contestants was erroneous. The "winning design" would not actually work correctly because the test code could not detect a flaw in the design.

where everything else is the same as earlier examples, except the states are represented as:

```
`define NUM_STATE_BITS 2
`define IDLE        2'b00
`define COMPUTE1    2'b01
`define COMPUTE2    2'b10
`define COMPUTE3    2'b11
```

and the portlist now has `r3` rather than `r2` as the output:

```
module slow_div_system(pb,ready,x,y,r3,sysclk);
  input pb,x,y,sysclk;
  output ready,r3;
  wire pb;
  wire [11:0] x,y;
  reg ready;
  reg [11:0] r1,r2,r3;
  reg [`NUM_STATE_BITS-1:0] present_state;
```

Also, `r3` must be mentioned in the debugging `$display` statement. As was described in chapter 2, this machine fails for `x<y` (the quotient is unknown):

```
         70 r1=    x r2=    x r3=x pb=0 ready=1
        170 r1=    0 r2=    0 r3=x pb=0 ready=1
        270 r1=    0 r2=    0 r3=x pb=0 ready=1
        370 r1=    0 r2=    0 r3=x pb=1 ready=1
        470 r1=    0 r2=    0 r3=x pb=0 ready=1
        570 r1=    0 r2=    0 r3=x pb=0 ready=1
error x=    0 y=    7 x/y=    0 quotient=x
...
```

although it does work for larger values:

```
...
      3070 r1=   6 r2=   0 r3=   x pb=0 ready=1
      3170 r1=   7 r2=   0 r3=   x pb=1 ready=1
      3270 r1=   7 r2=   0 r3=   x pb=0 ready=0
      3370 r1=   0 r2=   0 r3=   x pb=0 ready=0
      3470 r1=   0 r2=   1 r3=   x pb=0 ready=0
      3570 r1=   0 r2=   1 r3=   1 pb=0 ready=1
ok
...
```

The second ASM chart of section 2.2.5 can be translated as:

```
always
  begin
   @(posedge sysclk) enter_new_state(`IDLE);
    r1 <= @(posedge sysclk) x;
    r2 <= @(posedge sysclk) 0;
    ready = 1;
    if (pb)
      begin
       if (r1 >= y)
         while (r1 >= y)
           begin
            @(posedge sysclk) enter_new_state(`COMPUTE1);
             r1 <= @(posedge sysclk) r1 - y;
            @(posedge sysclk) enter_new_state(`COMPUTE2);
             r2 <= @(posedge sysclk) r2 + 1;
            @(posedge sysclk) enter_new_state(`COMPUTE3);
             r3 <= @(posedge sysclk) r2;
           end
       else
         begin
          @(posedge sysclk) enter_new_state(`ZEROR3);
           r3 <= @(posedge sysclk) 0;
         end
      end
  end
```

where the states are represented as:

```
`define NUM_STATE_BITS 3
`define IDLE        3'b000
`define COMPUTE1    3'b001
`define COMPUTE2    3'b010
`define COMPUTE3    3'b011
`define ZEROR3      3'b100
```

which does work correctly for all values, including x<y:

```
        70 r1=   x r2=    x r3=    x pb=0 ready=1
       170 r1=   0 r2=    0 r3=    x pb=0 ready=1
       270 r1=   0 r2=    0 r3=    x pb=0 ready=1
       370 r1=   0 r2=    0 r3=    x pb=1 ready=1
       470 r1=   0 r2=    0 r3=    x pb=0 ready=0
       570 r1=   0 r2=    0 r3=    0 pb=0 ready=1
ok
...
```

The second ASM of section 2.2.6 can be translated as:

```
always
  begin
   @(posedge sysclk) enter_new_state(`IDLE);
    r1 <= @(posedge sysclk) x;
    r2 <= @(posedge sysclk) 0;
    ready = 1;
    if (pb)
      begin
       if (r1 >= y)
         while (r1 >= y)
           begin
            @(posedge sysclk) enter_new_state(`COMPUTE);
             r1 <= @(posedge sysclk) r1 - y;
            @(posedge sysclk) enter_new_state(`COMPUTE23);
             r2 <= @(posedge sysclk) r2 + 1;
             r3 <= @(posedge sysclk) r2;
           end
       else
         begin
          @(posedge sysclk) enter_new_state(`ZEROR3);
```

Continued

```
          r3 <= @(posedge sysclk) 0;
        end
     end
  end
```

The above correctly models the design error due to inappropriate use of parallelism in state COMPUTE23 that causes `r3` to be assigned a value too early:

```
...
     3070 r1=   6 r2=   0 r3=   0 pb=0 ready=1
     3170 r1=   7 r2=   0 r3=   0 pb=1 ready=1
     3270 r1=   7 r2=   0 r3=   0 pb=0 ready=0
     3370 r1=   0 r2=   0 r3=   0 pb=0 ready=0
     3470 r1=   0 r2=   1 r3=   0 pb=0 ready=1
error x=    7 y=    7 x/y=    1 quotient=    0
...
```

The corrected ASM chart of 2.2.6 that has all three computations happening in parallel in state COMPUTE123 can be translated into Verilog as:

```
always
  begin
   @(posedge sysclk) enter_new_state(`IDLE);
    r1 <= @(posedge sysclk) x;
    r2 <= @(posedge sysclk) 0;
    ready = 1;
    if (pb)
      begin
       if (r1 >= y)
         while (r1 >= y)
           begin
            @(posedge sysclk) enter_new_state(`COMPUTE123);
             r1 <= @(posedge sysclk) r1 - y;
             r2 <= @(posedge sysclk) r2 + 1;
             r3 <= @(posedge sysclk) r2;
           end
       else
         begin
          @(posedge sysclk) enter_new_state(`ZEROR3);
           r3 <= @(posedge sysclk) 0;
         end
      end
  end
```

where the states are defined as:

```
`define NUM_STATE_BITS 2
`define IDLE       2'b00
`define COMPUTE123 2'b01
`define ZEROR3     2'b11
```

The simulator shows that this version works correctly:

```
...
      3070 r1=   6 r2=   0 r3=    0 pb=0 ready=1
      3170 r1=   7 r2=   0 r3=    0 pb=1 ready=1
      3270 r1=   7 r2=   0 r3=    0 pb=0 ready=0
      3370 r1=   0 r2=   1 r3=    0 pb=0 ready=0
      3470 r1=4089 r2=   2 r3=    1 pb=0 ready=1
ok
...
```

4.1.5 Pure behavioral stage of the two-state division machine

The best correct design proposed in chapter 2 for the division machine is described by the ASM chart in section 2.2.7. It has the advantage that it takes only one clock cycle each time it goes through the loop, and it only needs an ASM with two states. Here is how this ASM chart can be translated into a pure behavioral module, similar to the earlier examples:

```
`define NUM_STATE_BITS 1
`define IDLE           1'b0
`define COMPUTE        1'b1

`include "clock.v"

module slow_div_system(pb,ready,x,y,r3,sysclk);
  input pb,x,y,sysclk;
  output ready,r3;
  wire pb;
  wire [11:0] x,y;
  reg ready;
  reg [11:0] r1,r2,r3;
  reg [`NUM_STATE_BITS-1:0] present_state;

  always
    begin
```

Continued

```
      @(posedge sysclk) enter_new_state('IDLE);
       r1 <= @(posedge sysclk) x;
       r2 <= @(posedge sysclk) 0;
       ready = 1;
       if (pb)
         begin
          while (r1 >= y | pb)
            begin
             @(posedge sysclk) enter_new_state('COMPUTE);
              r1 <= @(posedge sysclk) r1 - y;
              r2 <= @(posedge sysclk) r2 + 1;
              r3 <= @(posedge sysclk) r2;
            end
         end
     end

   task enter_new_state;
     input ['NUM_STATE_BITS-1:0] this_state;
     begin
       present_state = this_state;
       #1 ready=0;
     end
   endtask

   always @(posedge sysclk) #20
     $display("%d r1=%d r2=%d r3=%d pb=%b ready=%b",
               $time, r1,r2,r3, pb, ready);
endmodule
```

For brevity, the `c1` module has been placed in the `"clock.v"` file. Here is the Verilog simulation that shows it working:

```
...
      2670 r1=    5 r2=    0 r3=    0 pb=0 ready=1
      2770 r1=    6 r2=    0 r3=    0 pb=1 ready=1
      2870 r1=    6 r2=    0 r3=    0 pb=0 ready=0
      2970 r1=4095 r2=    1 r3=    0 pb=0 ready=1
ok
      3070 r1=    6 r2=    0 r3=    0 pb=0 ready=1
      3170 r1=    7 r2=    0 r3=    0 pb=1 ready=1
      3270 r1=    7 r2=    0 r3=    0 pb=0 ready=0
      3370 r1=    0 r2=    1 r3=    0 pb=0 ready=0
      3470 r1=4089 r2=    2 r3=    1 pb=0 ready=1
```

Continued

```
ok
...
        6570 r1=   13 r2=   0 r3=   1 pb=0 ready=1
        6670 r1=   14 r2=   0 r3=   1 pb=1 ready=1
        6770 r1=   14 r2=   0 r3=   1 pb=0 ready=0
        6870 r1=    7 r2=   1 r3=   0 pb=0 ready=0
        6970 r1=    0 r2=   2 r3=   1 pb=0 ready=0
        7070 r1=4089 r2=   3 r3=   2 pb=0 ready=1
ok
```

This two-state machine will be the basis for the examples that show how to translate from pure behavioral Verilog into mixed Verilog (section 4.2) and into pure structural Verilog (section 4.3). This example is also used to illustrate the hierarchical refinement of the controller to become a netlist (section 4.4).

4.2 Mixed stage of the two-state division machine

As was explained in section 2.3.1, to translate an algorithm into hardware eventually requires that the designer decide upon a particular architecture. The mixed stage is the point in the design process when the designer decides how the registers and combinational logic devices of the architecture are to be interconnected. The only constraint is that this interconnection be able to implement all of the RTN commands used in the pure behavioral stage at the times required.

4.2.1 Building block devices

Verilog only provides built-in primitives for elementary gates, such as `and`. In order to model the mixed stage of the design, we need to define modules that simulate the bus width devices that are instantiated in the architecture. The devices outlined here are a few of the common ones from appendixes C and D.

The details of just exactly how these modules work internally need not concern us now. We just need to know the (arbitrary) order and definition of ports in the portlist for devices we want to use in the architecture. We will assume these devices are fully defined in a file, "`archdev.v`," with behavioral code.

All of these modules have a parameter, SIZE, that indicates the bus width of the data inputs and outputs of the module. (The widths of command and status ports, if any, on the device are determined by the nature of the device.)

4.2.1.1 `enabled_register` *portlist*

This module has a data input bus, `di`, of a chosen SIZE (the default is 1), and a similar sized data output bus, `do`. (See section D.6 for a description of the hardware being modeled by this module.) It also has the `enable` input, which, when it is 1, causes the current value of `di` to be loaded into the register (thereby changing `do`) at the next rising edge of `sysclk`:

```
module enabled_register(di,do,enable,clk);
     parameter SIZE = 1;
     input di,enable,clk;
     output do;
     reg [SIZE-1:0] do;
     wire [SIZE-1:0] di;
     wire enable;
     wire clk;
     ...
endmodule
```

The ellipsis indicates where the behavioral definition of the register (see sections 3.7.2.2 and 3.10.10) goes. This module was inspired by the 74xx377 (eight-bit enabled register) and 74xx378 (six-bit enabled register) TTL chips, which have an active low enable signal. This means, in a clock cycle where the physical 74xx377 is scheduled to change value at the next rising edge, the 74x377 will have zero volts representing the `1'b1` used in the simulation for `enable`. Other than this minor detail, any architecture instantiating the above module can be constructed from these chips.

4.2.1.2 `counter_register` *portlist*

This module has a data input bus, `di`, and a similar sized data output bus, `do`. (See section D.7 for a description of the hardware being modeled by this module.) It has an output, `tc`, that is 1 when the current output is at its maximal value. It also has `load`, `count` and `clr` inputs that determine what the value of the counter will be after the next rising edge of `sysclk`:

```
module counter_register(di,do,tc,load,count,clr,clk);
     parameter SIZE = 1;
     input di,load,count,clr,clk;
     output do,tc;
     reg [SIZE-1:0] do;
     wire [SIZE-1:0] di;
     reg tc;
```

Continued

```
      wire load,count,clr;
      wire clk;
      ...
endmodule
```

This module was inspired by the 74xx163 (4-bit up counter), which has active low `clr` and `load` signals (see the discussion in section 4.2.1.1.) Also, this chip has two inputs that must both simultaneously be one to cause counting. The reason for having two inputs, rather than just the one `count` shown above, is to simplify the connections required to cascade the four-bit chip to form larger counters. Since at this stage of the design we are not at all concerned with such physical details, the above module was simplified to have a single `count` signal.

4.2.1.3 `alu181` *portlist*

This module models a combinational ALU inspired by the 74xx181. (See section C.6 for a description of the hardware being modeled by this module.) It has two data inputs, `a` and `b`, and a similar sized data output bus, `f`. It also has status outputs, `cout` (1 when addition and similar operations produce a carry) and `zero` (1 when `f` is zero). It is controlled by the commands: `s`, `m` and `cin`:

```
module alu181(a,b,s,m,cin,cout,f,zero);
      parameter SIZE = 1;
      input a,b,s,m,cin;
      output cout,f,zero;
      wire [SIZE-1:0] a,b;
      wire  m,cin;
      wire [3:0] s;
      reg [SIZE-1:0] f;
      reg cout,zero;
      ...
endmodule
```

In chapter 2, the ALU was considered to have a six-bit command input, `aluctrl`. When this module is instantiated, this input should be subdivided in the following fashion:

```
  alu181 #size instancename(a,b,aluctrl[5:2],
                     aluctrl[1],aluctrl[0],cout,f,zero);
```

4.2.1.4 comparator portlist

This module models a comparator inspired by the 74xx85. (See section C.7 for a description of the hardware being modeled by this module.) It has two data inputs, `a` and `b`, and three status outputs, `a_lt_b, a_eq_b, a_gt_b`. At any time, only one of the outputs will be 1, depending on `a` and `b`:

```
module comparator(a_lt_b, a_eq_b, a_gt_b, a, b);
     parameter SIZE = 1;
     output a_lt_b, a_eq_b, a_gt_b;
     input a, b;
     wire [SIZE-1:0] a,b;
     reg a_lt_b, a_eq_b, a_gt_b;
     ...
endmodule
```

4.2.1.5 mux2 portlist

This module models a multiplexor inspired by the 74xx157. (See section C.4 for a description of the hardware being modeled by this module.) It has two data inputs, `i0` and `i1`, and a similarly sized data output, `out`. When the command input, `sel`, is 0, the output is `i0`. When `sel` is 1, the output is `i1`:

```
module mux2(i0, i1, sel, out);
     parameter SIZE = 1;
     input i0, i1, sel;
     output out;
     wire [SIZE-1:0] i0, i1;
     wire sel;
     reg [SIZE-1:0] out;
     ...
endmodule
```

4.2.2 Mixed stage

As discussed in chapter 2, the system is no longer described simply in terms of its behavior. Instead, in the mixed stage, there is a specific structure that interconnects the controller and the architecture:

```
module slow_div_system(pb,ready,x,y,r3,sysclk);
  input pb,x,y,sysclk;
  output ready,r3;
  wire pb;
  wire [11:0] x,y;
  wire ready;
  wire [11:0] r3;
  wire sysclk;

  wire [5:0] aluctrl;
  wire muxctrl,ldr1,clrr2,incr2,ldr3,r1gey;

  slow_div_arch a(aluctrl,muxctrl,ldr1,clrr2,
             incr2,ldr3,r1gey,x,y,r3,sysclk);
  slow_div_ctrl c(pb,ready,aluctrl,muxctrl,ldr1,
             clrr2,incr2,ldr3,r1gey,sysclk);
endmodule
```

This version of `slow_div_system` replaces the behavioral version of this module discussed in section 4.1. The test code that instantiates `slow_div_system` should not notice any difference between this mixed stage module and the earlier pure behavioral stage. Note that all ports and locals in this module are now declared to be `wire` since this module is composed simply of two structural instances, and there are no behavioral assignment statements.

4.2.3 Architecture for the division machine

At the end of section 2.3.1, a particular architecture for the division machine was chosen that handles all the RTN and decisions required by the division machine. To aid in understanding the Verilog that is equivalent to this architecture, figure 4-1 shows the same architecture from section 2.3.1 redrawn with the names to be used in the Verilog code:

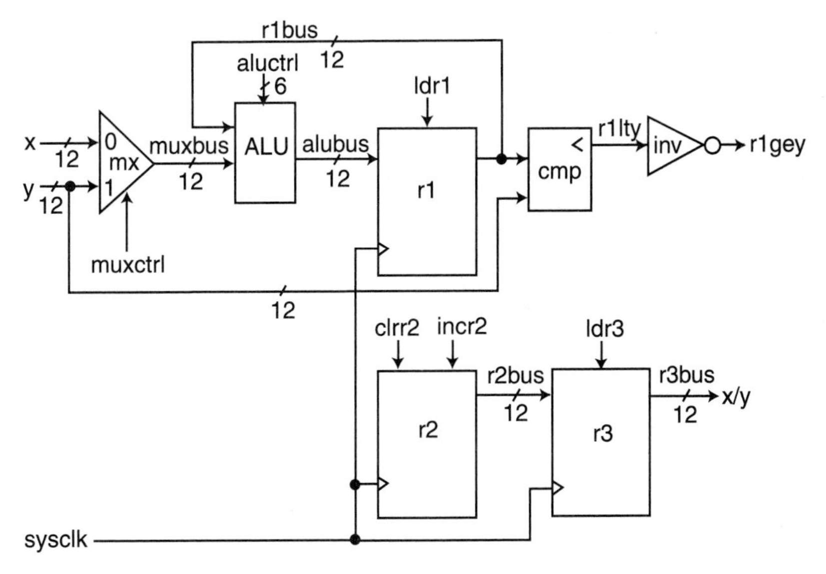

Figure 4-1. Architecture with names used for Verilog coding.

Here is how this architecture can be described as a structural Verilog module:

```
module slow_div_arch(aluctrl,muxctrl,ldr1,
              clrr2,incr2,ldr3,r1gey,x,y,r3bus,sysclk);
 input aluctrl,muxctrl,ldr1,clrr2,incr2,ldr3,x,y,sysclk;
 output r1gey,r3bus;
 wire [5:0] aluctrl;
 wire muxctrl,ldr1,clrr2,incr2,ldr3,r1gey,sysclk;
 wire [11:0] x,y,r3bus;
 wire [11:0] muxbus,alubus,r1bus,r2bus;

 enabled_register #12 r1(alubus,r1bus,ldr1,sysclk);
 mux2             #12 mx(x,y,muxctrl,muxbus);
 alu181           #12 alu(r1bus,muxbus,aluctrl[5:2],
                      aluctrl[1],aluctrl[0],,alubus,);

 comparator       #12 cmp(r1lty,,,r1bus,y);
 not                  inv(r1gey,r1lty);

 counter_register #12 r2(,r2bus,,1'b0,incr2,clrr2,sysclk);
 enabled_register #12 r3(r2bus, r3bus,ldr3,sysclk);
```

Continued

```
  always @(posedge sysclk) #20
   begin
     $display("%d r1=%d r2=%d r3=%d pb=%b ready=%b", $time,
              r1bus,r2bus,r3bus,
              slow_div_machine.pb,slow_div_machine.ready);
     $write("            %b       %b        %b",
              ldr1,{clrr2,incr2},ldr3);
     $display("     muxbus=%d alubus=%d",muxbus,alubus);
     $write("                              ");
     $display("     muxctrl=%b   aluctrl=%b",
              muxctrl,aluctrl);
     $write("                              ");
     $display("     x=%d r1gey=%b",x,r1gey);
   end
endmodule
```

The portlist for this module includes the commands that are input to this architecture (that were output from the controller). These commands include the six-bit `aluctrl` as well as `muxctrl`, `ldr1`, `clrr2`, `incr2` and `ldr3`. Also, the portlist has the status output `r1gey`. The portlist has the twelve bit data inputs `x` and `y` and the 12-bit data output `r3bus`. Of course, since there are clocked registers in the architecture, they must be supplied with `sysclk`. The order in this portlist matches the order where this is instantiated in section 4.2.2.

The first three structural instances (`r1`, `mx` and `alu`) define the portion of the block diagram that relates to register `r1`. This name is no longer a `reg` as it was in the pure behavioral stage, but is instead the instance name for an `enabled_register`, whose portlist is defined in section 4.2.1.1. This instance is for a twelve bit wide register (because the parameter is instantiated with 12). The input to this `enabled_register` comes from `alubus`, which is described below. The output from this `enabled_register` is known as `r1bus`. Of course both `alubus` and `r1bus` are `wires` since this module is defined only with structure. The load signal for `r1` is `ldr1`, and as is necessary in synchronous design, `r1` is connected to `sysclk`.

There is an instance (named `mx`) of `mux2` (see section 4.2.1.5) instantiated to be 12 bits wide. It selects the data input `x` when `muxctrl` is 0, and `y` when `muxctrl` is 1. Its output is `muxbus`. All of these buses are of course 12-bits wide. The instance of `alu181` (see section 4.2.1.3) named `alu` takes its inputs from `r1bus` and `muxbus`. The `aluctrl` is provided to the appropriate ports. The `cout` and `zero` ports of `alu181` are left disconnected. The `f` output connects to the `alubus` (mentioned in the last paragraph) that provides the input to `r1`.

The next two structural instances (cmp and inv) produce r1gey. The instance of comparator (see section 4.2.1.4) produces the output r1lty when r1bus is less than y. The a_eq_b and a_gt_b outputs of comparator are left disconnected. The built-in not produces r1gey from r1lty.

There is a 12-bit instance of counter_register (see section 4.2.1.2) named r2. Its data input is left disconnected. Normally this is a bad idea to leave an input (as opposed to an output) disconnected. However, in this case, the counter is only cleared and incremented, so the data input is not needed. The output of this counter is r2bus. The tc output of this register is not used. The load port of this instance is not utilized, but it **must** be specified as being tied to 1'b0. (In the physical TTL active low logic of the 74xx163, this would be equivalent to tying this to the five-volt supply. In active high logic, this would be equivalent to tying it to ground.) The load port **cannot be left disconnected**. The commands incr2 and clrr2 are provided to r2, as is sysclk.

Finally, there is another instance (r3) of enabled_register whose input comes from r2bus and whose output is r3bus, which is the data output of this entire module. The command ldr3 is provided to r3, as is sysclk.

At the bottom of this module are debugging $display statements. Some of the names have changed from those used in the pure behavioral version of this machine. We now refer to r1bus, r2bus and r3bus. Also, this module does not have pb or ready in its portlist, and so to produce the same style output as earlier simulations, the $display statement must use hierarchical names. (Since the $display statement does not correspond to synthesized hardware, and is only there for the convenience of the designer, it would be inappropriate to use ports to access such information. Hierarchical names are exactly what we need for a situation like this.) Two extra $display statements show signals that were not present in the pure behavioral stage. The command signals (ldr1, {clrr2, incr2} and ldr3) are displayed directly below the values of the registers they affect. Also, the values of internal buses are shown. Directly below them are the command signals that affect them. Also, the status signal output to the test code (r1gey) and the data input from the test code (x) are displayed. The other data input (y) is not displayed because in this test code it remains unchanged at 7.

4.2.4 Controller for the division machine

Although instantiating the components for the architecture as described in section 4.2.3 is a creative task that could have been done in a variety of ways, translating the pure behavioral stage into the mixed stage controller is a straightforward process that does not involve any creativity. For a particular pure behavioral machine, and for a chosen architecture, there is only one correct controller. To arrive at this controller,

1. Define a module (slow_div_ctrl in this example) for the controller with appropriate declarations that match how it was instantiated in the mixed system module (in this example, it is instantiated inside slow_div_system as shown in section 4.2.2).
2. Initialize all the internal command signals (aluctrl, muxctrl, ldr1, clrr2, incr2 and ldr3 in this example) used by the architecture to their default values in the enter_new_state task.
3. Move the always block from the pure behavioral system (in this example, slow_div_system of section 4.1.5) to inside the controller module, and comment out all non-blocking assignment statements (using //).
4. Following each non-blocking assignment that is commented out, put assignment statement(s) that assert the equivalent command signals. For example, //r2 <= @(posedge sysclk) 0; is followed by clrr2 = 1.
5. Replace relational conditions, such as r1 >= y, with status signals, such as r1gey.

Here is the module that is instantiated in section 4.2.2 and corresponds to pure behavioral Verilog of section 4.1.5, and that is equivalent to the mixed ASM chart of section 2.3.1:

```
module slow_div_ctrl(pb,ready,aluctrl,muxctrl,ldr1,
                     clrr2,incr2,ldr3,r1gey,sysclk);
  input pb,r1gey,sysclk;
  output ready,aluctrl,muxctrl,ldr1,clrr2,incr2,ldr3;

  reg ['NUM_STATE_BITS-1:0] present_state;
  wire pb;
  reg ready;
  reg [5:0] aluctrl;
  reg muxctrl,ldr1,clrr2,incr2,ldr3;
  wire r1gey,sysclk;

  always
    begin
     @(posedge sysclk) enter_new_state('IDLE);
      //r1 <= @(posedge sysclk) x;
      //r2 <= @(posedge sysclk) 0;
      ready = 1;
      aluctrl = 'PASSB;
      muxctrl = 0;
      ldr1 = 1;
      clrr2 = 1;
      if (pb)
```

Continued

```
        begin
         while (r1gey | pb)
           begin
            @(posedge sysclk) enter_new_state('COMPUTE);
             ready = 0;
             //r1 <= @(posedge sysclk) r1 - y;
             //r2 <= @(posedge sysclk) r2 + 1;
             //r3 <= @(posedge sysclk) r2;
             aluctrl = 'DIFFERENCE;
             muxctrl = 1;
             ldr1 = 1;
             incr2 = 1;
             ldr3 = 1;
           end
        end
    end

   task enter_new_state;
    input ['NUM_STATE_BITS-1:0] this_state;
    begin
     present_state = this_state;
     #1{ready,aluctrl,muxctrl,ldr1,clrr2,incr2,ldr3}=0;
    end
   endtask
endmodule
```

The boldface above shows the editing done to transform the pure behavioral Verilog into this mixed stage. Of some interest is the fact that the pure behavioral `while`, which has the condition `((r1>=y)|pb)`, is translated above into `(r1gey | pb)`. Use of single bit `&` and `|` (or perhaps more clearly `&&` and `||`) is permitted inside a mixed controller. This notation is not a data computation that must occur in the architecture (although the designer could have chosen to put a single `or` gate in the architecture to accomplish this). It is important to distinguish this decision-making use of `|` from a data manipulation use of `|`, such as `r1|y`, which should be performed by combinational logic (such as the ALU) in the architecture. In the case of `((r1>=y)|pb)`, there are two reasonable ways to translate this into the mixed stage: the way that was shown above, and the way that requires introducing an extra signal in the architecture to represent the `or` of `r1gey` and `pb`. (`pb` would then be classified as an external data input to the architecture, in addition to being an external status input to the controller.) Since we would like to minimize the number of wires that interconnect the controller to the architecture, we chose the former approach where `pb` is simply an external status signal.

Of course, the bit patterns for controlling the ALU must be defined outside this module:

```
`define DIFFERENCE  6'b011001
`define PASSB       6'b101010
```

The test code is the same as the pure behavioral system. Here is the output from the completed mixed stage:

```
  70 r1=    x r2=    x r3=    x pb=0 ready=1
        1       10        0     muxbus=    0 alubus=    0
                                muxctrl=0    aluctrl=101010
                                x=    0 r1gey=x
 170 r1=    0 r2=    0 r3=    x pb=0 ready=1
        1       10        0     muxbus=    0 alubus=    0
                                muxctrl=0    aluctrl=101010
                                x=    0 r1gey=0
 270 r1=    0 r2=    0 r3=    x pb=0 ready=1
        1       10        0     muxbus=    0 alubus=    0
                                muxctrl=0    aluctrl=101010
                                x=    0 r1gey=0
 370 r1=    0 r2=    0 r3=    x pb=1 ready=1
        1       10        0     muxbus=    0 alubus=    0
                                muxctrl=0    aluctrl=101010
                                x=    0 r1gey=0
 470 r1=    0 r2=    0 r3=    x pb=0 ready=0
        1       01        1     muxbus=    7 alubus=4089
                                muxctrl=1    aluctrl=011001
                                x=    0 r1gey=0
 570 r1=4089 r2=    1 r3=    0 pb=0 ready=1
        1       10        0     muxbus=    0 alubus=    0
                                muxctrl=0    aluctrl=101010
                                x=    0 r1gey=1
ok
...
6570 r1=   13 r2=    0 r3=    1 pb=0 ready=1
        1       10        0     muxbus=   14 alubus=   14
                                muxctrl=0    aluctrl=101010
                                x=   14 r1gey=1
6670 r1=   14 r2=    0 r3=    1 pb=1 ready=1
        1       10        0     muxbus=   14 alubus=   14
                                muxctrl=0    aluctrl=101010
                                x=   14 r1gey=1
```

Continued

```
6770 r1=  14 r2=   0 r3=   1 pb=0 ready=0
        1      01       1    muxbus=   7 alubus=   7
                             muxctrl=1   aluctrl=011001
                             x=  14 r1gey=1
6870 r1=   7 r2=   1 r3=   0 pb=0 ready=0
        1      01       1    muxbus=   7 alubus=   0
                             muxctrl=1   aluctrl=011001
                             x=  14 r1gey=1
6970 r1=   0 r2=   2 r3=   1 pb=0 ready=0
        1      01       1    muxbus=   7 alubus=4089
                             muxctrl=1   aluctrl=011001
                             x=  14 r1gey=0
7070 r1=4089 r2=   3 r3=   2 pb=0 ready=1
        1      10       0    muxbus=  14 alubus=  14
                             muxctrl=0   aluctrl=101010
                             x=  14 r1gey=1
ok
```

4.3 Pure structural stage of the two state division machine

Translating from the mixed stage to the "pure" structural stage is an easy and mechanical (although somewhat tedious) process. All modules except the controller remain the same. As explained in section 2.4.1, the controller module becomes a structure composed of a present state register (which is an instance of an actual register module, and not a `reg`) and the next state logic.

In the "pure" structural stage, the definition of the next state logic may remain as behavioral code (a function) that is a transformation of the code inside the `always` block of the mixed stage. In section 4.4, we will see how the next state logic could also be defined in terms of built-in gates, using hierarchical design. Fortunately, it is not normally necessary to worry about the details given later in section 4.4, because synthesis tools exist that can automatically transform the behavioral next state function described in this section into a netlist. For this reason, we consider this section to be the final step that the designer has to be involved with. Section 4.4 is presented later only to motivate the kind of transformations that synthesis tools are capable of.

4.3.1 The pure structural controller

The structure of the controller is quite simple. The instance name of the next_state_logic is nsl. As shown in the diagram in section 2.4.1, r1gey and pb are the inputs to nsl. Also, next_state, ldr1, incr2, clrr2, ldr3, muxctrl, aluctrl, ready and next_state are outputs of nsl. The input to ps_reg is next_state and its output is present_state. The portlist of the controller module is identical to the mixed stage, except the outputs are declared to be wires:

```
module slow_div_ctrl(pb,ready,aluctrl,muxctrl,ldr1,
                     clrr2,incr2,ldr3,r1gey,sysclk);
  input pb,r1gey,sysclk;
  output ready,aluctrl,muxctrl,ldr1,clrr2,incr2,ldr3;

  wire ['NUM_STATE_BITS-1:0] present_state;
  wire pb;
  wire ready;
  wire [5:0] aluctrl;
  wire muxctrl,ldr1,clrr2,incr2,ldr3;
  wire r1gey,sysclk;

  next_state_logic  nsl(next_state,
                         ldr1,incr2,clrr2,ldr3,
                         muxctrl,aluctrl,ready,
                         present_state, r1gey, pb);
  enabled_register #('NUM_STATE_BITS) ps_reg(next_state,
                         present_state,1'b1,sysclk);
endmodule
```

For simplicity, we are using an enabled_register for ps_reg with its enable tied to 1'b1.

4.3.2 next_state_logic module

The combinational logic that computes the next state needs to be defined. Following the technique outlined in section 3.7.2.1, there is an always block with an @ sensitivity list which has all the inputs to this module. Since the calculation of the next state and corresponding outputs is quite lengthy, this calculation is isolated in a function, state_gen, that is defined in the file "divbookf.v":

```
module next_state_logic(next_state,
                        ldr1,incr2,clrr2,ldr3,
                        muxctrl,aluctrl, ready,
                        present_state, r1gey, pb);
  output next_state,ldr1,incr2,clrr2,ldr3,muxctrl,
         aluctrl,ready;
  input present_state, r1gey, pb;
  reg ['NUM_STATE_BITS-1:0] next_state;
  reg ldr1,incr2,clrr2,ldr3,muxctrl,ready;
  reg [5:0] aluctrl;
  wire ['NUM_STATE_BITS-1:0] present_state;
  wire r1gey,pb;

  'include "divbookf.v"

  always @(present_state or r1gey or pb)
   {next_state,ldr1,clrr2,incr2,ldr3,muxctrl,aluctrl,
     ready} = state_gen(present_state, pb, r1gey);
endmodule
```

4.3.3 `state_gen` function

To create the file that contains the `state_gen` function is a simple matter of editing a portion of the mixed code:

1. Create a function header that returns the proper number of bits and has the appropriate input arguments. The number of bits to be returned is `'NUM_STATE_BITS-1` plus the number of command output bits. The input arguments of the function are the same as the input ports of `next_state_logic`. (In this example, the inputs are `ps`, `pb` and `r1gey`.)
2. The output `regs` of the mixed controller become local `regs` in the function. (In this example, `ldr1`, `incr2`, `clrr2`, `ldr3`, `muxctrl`, `aluctrl` and `ready` are local `regs` in the function.) A local `ns reg` is also defined to hold the next state within the function. Remember that local `regs` of a function would never be synthesized as physical registers. They are `regs` simply because the function uses behavioral assignment.
3. The assignment of default values that occurs in the `enter_new_state` task of the mixed stage becomes the first executable statement of the function. Also, `ns` is also given a default value (the starting point of the algorithm).
4. The next executable statement of the function is a `case` statement based on the present state, `ps`. This `case` statement is equivalent to a truth table, as explained

in section 3.9.2.2. The advantage of the case statement for this purpose is that it is more compact than a truth table, and it documents some of the thought process of the earlier stages through meaningful identifiers (such as aluctrl).

5. The statements that follow each @(posedge sysclk) inside the always block of the mixed controller are moved to a place within the case statement that corresponds to that state. The @ and the call to enter_new_state are eliminated. The commented-out non-blocking assignment statements are retained for documentation.
6. In each block of code that corresponds to a state, ns is computed. This computation, in effect, acts like a goto. It says which state in the case statement will execute when this function is called during the next clock cycle (after ns becomes ps). Unlike the mixed stage, the order in which the designer types the states into the case statement has no effect on the order in which the states execute. The ns computation determines the order in which they execute. By putting the ASM into a function, we have lost the perfect correspondence to the goto-less style that we had in the mixed stage.
7. The next state and the outputs are concatenated to be returned from this function in the order needed (see section 4.3.2).

Here is the state_gen function for the two-state division machine:

```
function ['NUM_STATE_BITS-1+12:0] state_gen;
  input ['NUM_STATE_BITS-1:0] ps;
  input pb,r1gey;
  reg ready;
  reg [5:0] aluctrl;
  reg muxctrl,ldr1,clrr2,incr2,ldr3;
  reg ['NUM_STATE_BITS-1:0] ns;

  begin
    {ns,ready,aluctrl,muxctrl,ldr1,clrr2,incr2,ldr3}=0;
    case (ps)
      'IDLE:      begin
                    //r1 <= @(posedge sysclk) x;
                    //r2 <= @(posedge sysclk) 0;
                    ready = 1;
                    aluctrl = 'PASSB;
                    muxctrl = 0;
                    ldr1 = 1;
                    clrr2 = 1;
                    if (pb)
                      if (r1gey|pb)
```

Continued

```
                              ns = 'COMPUTE;
                            else
                              ns = 'IDLE;
                          else
                            ns = 'IDLE;
                        end
        'COMPUTE: begin
                          ready = 0;
                          //r1 <= @(posedge sysclk) r1 - y;
                          //r2 <= @(posedge sysclk) r2 + 1;
                          //r3 <= @(posedge sysclk) r2;
                          aluctrl = 'DIFFERENCE;
                          muxctrl = 1;
                          ldr1 = 1;
                          incr2 = 1;
                          ldr3 = 1;
                          if (r1gey|pb)
                            ns = 'COMPUTE;
                          else
                            ns = 'IDLE;
                        end
      endcase
      state_gen = {ns,ldr1,clrr2,incr2,ldr3,
                    muxctrl,aluctrl,ready};
    end
  endfunction
```

Here boldface shows some changes that were made to make this work as a function.

4.3.4 Testing `state_gen`

Since `state_gen` is isolated in a file by itself, we can test `state_gen` by writing some trivial Verilog, unrelated to any of the earlier code:

```
 'define DIFFERENCE  6'b011001
 'define PASSB       6'b101010

 'define NUM_STATE_BITS 1
 'define IDLE           1'b0
 'define COMPUTE        1'b1

module test;
```

Continued.

```
'include "divbookf.v"


integer i;
reg ps,pb,r1gey;


initial
  begin
    for (i=0;i<=7;i=i+1)
      begin
        {ps,pb,r1gey} = i;
        $display("%b %b %b %b",ps,pb,r1gey,
                    state_gen(ps,pb,r1gey));
      end
  end
endmodule
```

This produces an output which agrees with the manual derivation given in section 2.4.1. The bit patterns used in the function must be defined, as shown above, because they are not defined in `"divbookf.v"`.

4.3.5 It *seems* to work

Having tested the function to see that it behaves as expected, we can now put the controller code from sections 4.3.1 through 4.3.3 together with the architecture code from section 4.2.3 to obtain the pure structural version of the two-state division machine. The simulation of this produces the same output as that of the mixed stage, shown in section 4.2.4.

There is one additional detail, described in the next section, that we will want to consider in all future designs. Since the code (including this additional detail) that the designer develops in the pure structural stage can be run through a synthesis tool, it is not necessary for the designer to manually transform Verilog code after reaching the pure structural stage. In the next section, we will see how the synthesis tool continues to transform the Verilog code for the designer automatically.

4.4 Hierarchical refinement of the controller

Except for one small physical problem to be explained in this section, the Verilog code in section 4.3 can be submitted to a synthesis tool, which produces the netlist that can be used to fabricate the chip. Assuming we recognize and fix this little problem, we do not need to write any more Verilog. The synthesis tool can do the rest of the job of creating the netlist.

On the other hand, since the two-state division machine, with its childish algorithm, is so simple, it makes a good example to illustrate what a synthesis tool does. Also, by looking at the netlist, we will discover this small physical problem alluded to earlier. This problem, which occurs with most controllers, has a simple solution; however, it was previously hidden from view because behavioral Verilog usually does not model all the physical details of a circuit. The power of top-down design is to hide those details until the last moment. We have reached this moment of truth when we get down to gates!

This section illustrates the need to do post-synthesis simulation prior to fabrication. All three-stages of design (including the pure structural stage) have some behavioral aspects that are well above the gate level. To predict more accurately what the fabricated circuit will do, we need to simulate the synthesized netlist. With this simple machine, the netlist for the controller is simple enough that we can generate it manually, but in most machines, an automatically produced netlist would be incomprehensible to the designer.

4.4.1 A logic equation approach

The `state_gen` function defined in section 4.3.3 can be replaced by a series of logic equations representing the low-level behavior of `ns1`, as explained in section 2.5:

```
module next_state_logic(next_state,
                        ldr1,incr2,clrr2,ldr3,
                        muxctrl,aluctrl, ready,
                        present_state, r1gey, pb);
  output next_state,ldr1,incr2,clrr2,ldr3,muxctrl,
         aluctrl,ready;
  input present_state, r1gey, pb;
  reg ['NUM_STATE_BITS-1:0] next_state;
  reg ldr1,incr2,clrr2,ldr3,muxctrl,ready;
  reg [5:0] aluctrl;
  wire ['NUM_STATE_BITS-1:0] present_state;
  wire r1gey,pb;

  always @(present_state or r1gey or pb)
```

Continued

```
      begin
        next_state =
         ~present_state&pb|present_state&(r1gey|pb);
        ldr1 = 1;
        clrr2 = ~present_state;
        incr2 = present_state;
        ldr3 = present_state;
        muxctrl = present_state;
        aluctrl[5] = ~present_state;
        aluctrl[4] = present_state;
        aluctrl[3] = 1;
        aluctrl[2] = 0;
        aluctrl[1] = ~present_state;
        aluctrl[0] = present_state;
        ready = ~present_state;
      end
  endmodule
```

Even though this is a rather tedious way to describe the controller, it is still a behavioral description as explained in section 3.7.2.1. This module has an `always` block with @ followed by a sensitivity list naming all the inputs to the combinational logic, and the outputs are defined as `regs`. Although using logic equations in this fashion is perfectly legal, this is not the preferred style of Verilog coding for three reasons: it is not a netlist, a designer can easily make a mistake when deriving the logic equations manually, and even if the logic equations are correct, they are usually meaningless to the designer. One reason why we practice top-down and hierarchical design is to minimize our exposure to details, especially details as tedious as a page full of logic equations.

Algorithms to manipulate Boolean equations are some of the most studied aspects of computer design. Even prior to the introduction of Verilog, software existed to automatically produce logic equations from truth tables. It would be a giant leap backward to use Verilog with manually produced logic equations. We will not consider writing modules of this style again.

4.4.2 At last: a netlist

Even though logic equations should not normally be your first choice when designing behavioral code, it is important for you to be aware of the properties of Boolean algebra. Ultimately, at the lowest levels, all computation carried out on digital computers are the result of iterative application of Boolean equations.

It was popular to use logic equations manually before the introduction of HDLs because there is a one-to-one mapping between logic equations and a netlist. For the same reason, a synthesis tool internally manipulates logic equations as it explores the vast space of possible hardware structures that correctly implement a combinational function. When the synthesis tool decides what the optimum logic equation is, it is trivial for it to produce the netlist. Using the logic equations from section 4.4.1, we have at last gotten all the way down to the netlist:

```
module next_state_logic(next_state,
                        ldr1,incr2,clrr2,ldr3,
                        muxctrl,aluctrl, ready,
                        present_state, r1gey, pb);
  output next_state,ldr1,incr2,clrr2,ldr3,muxctrl,
         aluctrl,ready;
  input present_state, r1gey, pb;
  wire ['NUM_STATE_BITS-1:0] next_state;
  wire ldr1,incr2,clrr2,ldr3,muxctrl,ready;
  wire [5:0] aluctrl;
  wire ['NUM_STATE_BITS-1:0] present_state;
  wire r1gey,pb;

  buf b0(ldr1,aluctrl[3],1'b1);
  buf b1(aluctrl[2],1'b0);
  buf b2(incr2,ldr3,muxctrl,aluctrl[4],aluctrl[0],
         present_state[0]);
  not i1(not_ps,clrr2,aluctrl[5],aluctrl[1],ready,
         present_state[0]);
  and a1(not_ps_and_pb,not_ps,pb);
  and a2(ps_and_or,present_state[0],r1gey_or_pb);
  or  o1(r1gey_or_pb,r1gey,pb);
  or  o2(next_state[0],not_ps_and_pb,ps_and_or);
endmodule
```

Figure 4-2 shows the corresponding circuit diagram.

Of course, many other possible solutions exist that produce the same truth table. The built-in Verilog gate `buf` (non-inverting buffer) passes through its last port unchanged to all the other ports, which are outputs. The only difference between `buf` and `not` is that the latter inverts its outputs.

We will ignore discussing the netlist for the architectural devices, such as `mux2`, since this is a trivial but tedious task. The synthesis tool would do this identically to the way the controller was synthesized.

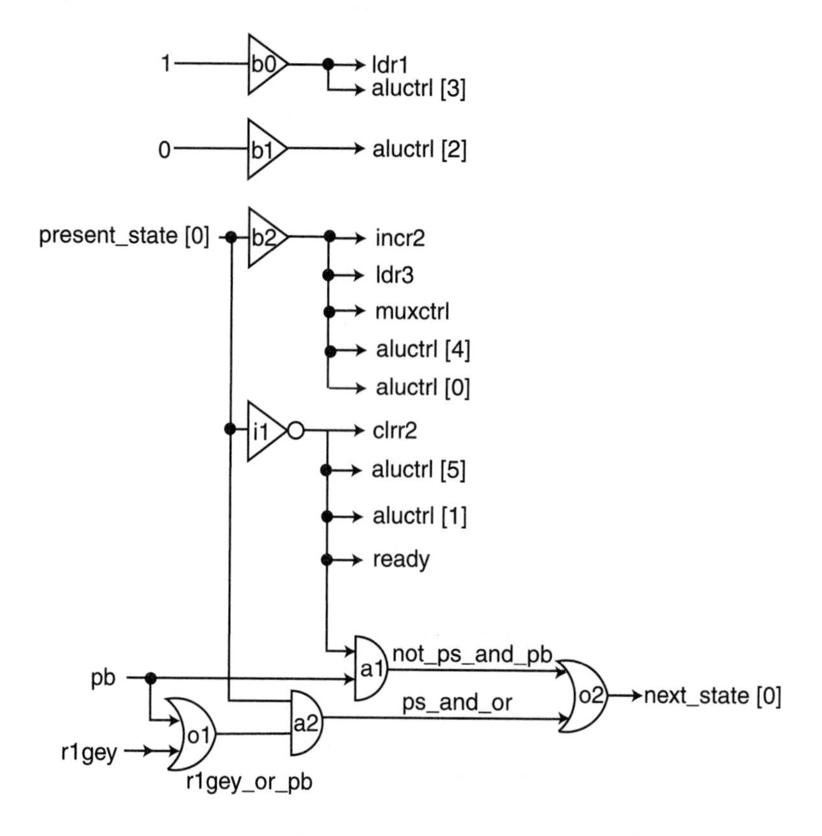

Figure 4-2. Netlist for the childish division controller.

4.4.3 Post-synthesis simulation

Even though up to this point all the simulation results indicate that our synthesized division machine should work, it is wise to conduct post-synthesis simulation prior to fabricating the hardware. This is why Verilog was developed in the first place: to simulate gate level netlists. Synthesis from behavioral code is now the predominate use of Verilog, but synthesis tools appeared later in the history of Verilog than simulators for netlists.

So let's use Verilog in the way it was originally intended to be used and simulate the netlist given in section 4.4.2. When this simulation runs, we get a very interesting but discouraging output:

```
 70 r1=x r2=x r3=x pb=0 ready=x
      1      xz        z      muxbus=x alubus=x
                              muxctrl=z   aluctrl=xz10xz
                              x=   0 r1gey=x
170 r1=x r2=x r3=x pb=0 ready=x
      1      xz        z      muxbus=x alubus=x
                              muxctrl=z   aluctrl=xz10xz
                              x=   0 r1gey=x
270 r1=x r2=x r3=x pb=0 ready=x
      1      xz        z      muxbus=x alubus=x
                              muxctrl=z   aluctrl=xz10xz
                              x=   0 r1gey=x
...
```

that continues like this for as long as you are willing to let the simulator run. This is the physical problem alluded to earlier: as the controller is currently interconnected, the gate level netlist does not seem to work. The simulator's output is splattered with `1'bx`s and `1'bz`s.

Although you might think something is wrong with the logic equations (given in section 4.4.1) or the equivalent netlist (given in section 4.4.2), there is not. The logic equations and equivalent netlist are correct. What's the problem?

To understand the problem, you need to remember the intent behind having the four-valued logic system. When `1'bx`s or `1'bz`s appear where you were expecting a 1 or a 0, this is an indication of some flaw in the design. Although major interconnection errors can cause this (see section 3.5.3), more subtle problems can cause this as well. Since everything is interconnected properly in this netlist, we need to understand what the `1'bx`s and `1'bz`s are trying to tell us here.

At `$time` 0, all `reg`s start as `'bx`s and all wires start as `'bz`s. If the simulation does not change these values, that is how they will stay. The `ps_reg` of the controller has an internal `reg` that holds the present state. At `$time` 0, it is `1'bx`. The next state that the machine computes from `ps_reg` is also unknown. A Boolean function of `'bx` is usually `'bx`. Therefore, the `ps_reg` is reloaded with `1'bx`, rather than the proper sequence of states. The four-valued logic of the simulation has detected a potential flaw in the design: we do not know what state the controller starts out in, so we cannot predict what happens next.

Why didn't the pure structural version (see section 4.3.3) detect this problem? The reason is found in the definition of the `state_gen` function. The first statement of this function initializes `ns` (which is what becomes `next_state`) to be `1'b0`:

```
{ns,ready,aluctrl,muxctrl,ldr1,clrr2,incr2,ldr3}=0;
case (ps)
  ...
endcase
```

Unfortunately, this also means that even though `ps` is `'bx` at `$time` 0, `ns` will become 0 after the first rising edge. This is a disguised way of saying:

```
{ready,aluctrl,muxctrl,ldr1,clrr2,incr2,ldr3}=0;
if (ps === 1'bx | ps === 1'bz)
  ns = 1'b0;
else
case (ps)
  ...
endcase
```

The reason this is cheating is because hardware cannot implement the === operation. Each physical wire can only carry one bit of information. Each simulated wire tested with === carries two bits of information (to represent 0, 1, `1'bx` and `1'bz`). At `$time` 0, all `regs` are intialized to `'bx`. As soon as the above function detects `'bx`, it turns it into 0. In this indirect way, the function is informed when `$time` is 0. The fabricated hardware has no way to know when `$time` is 0, because a wire is simply either 0 or 1.

We can make the pure structural stage act more like the netlist by omitting `ns` from the default initialization:

```
{ready,aluctrl,muxctrl,ldr1,clrr2,incr2,ldr3}=0;
case (ps)
  ...
```

By omitting `ns` here, the simulation using `state_gen` fails similarly to the netlist.

4.4.4 Resetting the present state

What post-synthesis simulation discovered is a problem that all state machines exhibit, and that happily has a simple solution. When the power is first turned on, we do not know what state the machine will be in. The pure behavioral and mixed stages use position within behavioral code (rather than the ps_reg) to indicate the current state, so they start at the top of the code at $time 0. As explained above, the way we did the pure structural stage (with the state_gen function) tricked Verilog into disguising the problem. At the netlist level, the problem cannot be hidden any longer.

If you are at all familiar with state machines, you probably spotted this problem a long time ago. We need an asynchronous reset for the ps_reg that is activated soon after $time 0, and left inactive thereafter. Here is a behavioral model of such a simple D-type resettable register:

```
module resetable_register(di,do,reset,sysclk);
  parameter SIZE=1;
  input di,reset,sysclk;
  output do;
  wire [SIZE-1:0] di;
  reg [SIZE-1:0] do;
  wire reset,sysclk;

  always @(posedge sysclk or posedge reset)
    begin
      if (reset)
        do = 0;
      else
        do = di;
    end
endmodule
```

The above is patterned after the 74xx175 (six-bit resettable D-type register), except as is typical with TTL logic, the reset signal on the 74xx175 is active low.

This is the first and only time that we will admit an asynchronous signal into our design. *Asynchronous* means that a change happens in a register at a $time other than the rising edge of sysclk. Notice the difference between the clr signal used in the synchronous counter_register (described in sections 3.7.2.2 and 4.2.1.2) and the reset signal described here. Although both signals cause the register to become zero at some point in $time, the clr signal simply schedules the change to happen at the next rising edge, but the reset signal causes the clearing to happen instantly. The register is continually rezeroed for as long as reset is asserted because of the if , even should a rising edge of the clock occur. Without the posedge reset, the register would be a synchronous, clearable D-type register.

The ps_reg needs to have an asynchronous reset so that it is zero prior to the arrival of the first rising edge of sysclk. The reset signal must be provided by our friendly user, which means for the Verilog simulation that reset becomes a port of several modules and must be provided by the test code. It is a input port of the controller:

```
module slow_div_ctrl(pb,ready,aluctrl,muxctrl,ldr1,
              clrr2,incr2,ldr3,r1gey,reset,sysclk);
  input pb,r1gey,sysclk,reset;
  output ready,aluctrl,muxctrl,ldr1,clrr2,incr2,ldr3;

  wire ['NUM_STATE_BITS-1:0] present_state;
  wire pb;
  wire ready;
  wire [5:0] aluctrl;
  wire muxctrl,ldr1,clrr2,incr2,ldr3;
  wire r1gey,sysclk,reset;

  next_state_logic  nsl(next_state,
                        ldr1,incr2,clrr2,ldr3,
                        muxctrl,aluctrl,ready,
                        present_state, r1gey, pb);
  resetable_register #('NUM_STATE_BITS) ps_reg(next_state,
                        present_state,reset,sysclk);
endmodule
```

and of the system that instantiates the controller:

```
module slow_div_system(pb,ready,x,y,r3,reset,sysclk);
  input pb,x,y,sysclk,reset;
  output ready,r3;
  wire pb;
  wire [11:0] x,y;
  wire ready;
  wire [11:0] r3;
  wire sysclk,reset;

  wire [5:0] aluctrl;
  wire muxctrl,ldr1,clrr2,incr2,ldr3,r1gey;

  slow_div_arch a(aluctrl,muxctrl,ldr1,clrr2,
             incr2,ldr3,r1gey,x,y,r3,sysclk);
```

Continued

```
   slow_div_ctrl c(pb,ready,aluctrl,muxctrl,ldr1,
                 clrr2,incr2,ldr3,r1gey,reset,sysclk);
endmodule
```

It must also appear in the test code:

```
module top;
   reg pb;
   reg [11:0] x,y;

   wire [11:0] quotient;
   wire ready;

   integer s;
   wire sysclk;
   reg reset;

   cl #20000 clock(sysclk);
   slow_div_system
slow_div_machine(pb,ready,x,y,quotient,reset,sysclk);

   initial
     begin
       pb= 0;
       x = 0;
       y = 7;
       reset = 0;
       #30 reset = 1;
       #10 reset = 0;
       #210;
       ...
endmodule
```

The test code issues a reset pulse that lasts for 30 units of `$time`, which causes the present state to become zero. When the netlist for the controller (section 4.4.2) is re-simulated with the above, it produces the same correct answers we obtained for the mixed stage.

4.5 Conclusion

The three stages of top-down design can be expressed in Verilog. The pure behavioral stage requires writing a single system module to model the machine. The only structure at this stage is the portlist of the module. During this stage, the designer also develops test code that instantiates the pure behavioral module. The test code is important because it is a specification of what the machine is supposed to do. By translating the ASM that describes the machine with the `enter_new_state` approach, the pure behavioral Verilog is organized so that it will be easy to translate to the mixed stage.

In the mixed stage, the designer develops structural Verilog code for the architecture. The modules instantiated inside the architecture may themselves be defined behaviorally. The architecture and a controller are instantiated as the system module. The controller is behavioral code derived from the pure behavioral stage. The <= statements are commented out and replaced with appropriate command signals for the chosen architecture, and relational conditions are replaced with appropriate status signals. The default values for the command signals are indicated in the `enter_new_state` task.

In the pure structural stage, the controller is edited to become a structural module that instantiates a next state generator and a resettable register. The test code and portlists must be edited to include a `reset` signal. The next state generator has an `always` block modeling combinational logic which calls a function defined with a `case` statement. The cases in this statement are copied from the mixed controller code. Additional statements must be provided in each state to describe the calculation of the next state. Default values are indicated prior to the `case` statement.

The next state logic and all of the architectural building blocks of the pure structural stage could be refined down to the gate level by using hierarchical design. However, synthesis tools can take a pure structural description (with all of its instantiated modules still defined behaviorally) and produce a gate-level netlist. In most cases, the pure structural stage is equivalent to working hardware. It is important to do post-synthesis simulation before fabrication to insure the netlist solves the problem correctly because it is cheaper to find flaws before fabrication.

4.6 Exercises

4-1. Use a simulator to take problems 2-1, 2-2 and 2-3 through the three-stages in Verilog.

4-2. Use a simulator to take problems 2-4, 2-5 and 2-6 through the three stages in Verilog.

4-3. Use a simulator to take problems 2-7, 2-8 and 2-9 through the three-stages in Verilog.

5. ADVANCED ASM TECHNIQUES

Although the ASM techniques illustrated in chapter 2 and the corresponding Verilog notation given in chapter 4 are adequate to solve any problem, they may yield a hardware solution that is not optimal in terms of speed (number of clock cycles) and cost (number of gates.) Despite all the marketing hype one hears, neither speed nor cost should be the primary concern of the designer. The primary responsibility of the designer is producing a **correct** design. (Intel Corporation illustrated the wisdom that before one produces a fast chip one ought to design a correct chip when they sold a version of the Pentium in 1994 whose division algorithm was incorrect.)

Despite the fact speed should not be our first concern, in many problems a correct solution demands that an algorithm find its answer by a certain time. Consider, for example, the onboard computers of the space shuttle. They need to compute the correct result in a timely enough fashion that the shuttle may correct its course. In such a context, a machine that computes a correct answer too late is not a solution at all.

Our search for a faster solution should always begin in the abstract world of algorithms, and not in the gruesome world of gates. The best way to speed up a machine is not to do some trickery with gates; instead the best way is to describe a better algorithm that solves the same problem. Chapter 2 illustrates this point with several different variations on the division machine with the final solution being three times faster than the slowest solution. One difficulty is that certain faster algorithms cannot be expressed with the notations discussed in chapters 2 and 4. This chapter discusses an additional ASM feature that helps us describe more efficient algorithms. Also, this chapter explains how any ASM can be written in Verilog, including those that use such notations, as well as those that have complex branches that do not follow the `goto`-less style we adhered to in earlier examples.

5.1 Moore versus Mealy

Recall that an ASM chart is composed of diamonds, rectangles and ovals connected by arrows. Chapter 2 ignored the use of ovals. Such ASM charts that do not have ovals are referred to as Moore machines. All the ASM charts discussed previously were for Moore machines. ASM charts that also include ovals are known as Mealy machines. Mealy ASM charts provide a way to express algorithms that are faster (and in some instances less costly) than Moore ASM charts.

Commands (both RTN and signal names) that occur inside a rectangle are known as *unconditional* commands, because the commands are issued regardless of anything when the machine is in the state corresponding to the rectangle. Ovals are used to describe *conditional* commands, that are sometimes (but not always) issued when a machine is in a particular state. Ovals are not by themselves a state, but rather they are the children of some parent state that corresponds to a rectangle in the ASM chart. Because ovals represent a conditional command, they must occur after one or more diamond(s). If you follow the arrows of a Mealy ASM, you would first come to a rectangle, then you would come to one or more diamond(s) and finally you would come to the oval. After the oval, the arrow might go to another diamond or a rectangle.

The actions in one rectangle and all ovals and diamonds that are connected (without any intervening rectangles) to that one rectangle occur in the same clock cycle. In essence, a combination of diamonds and ovals allows the designer to implement an arbitrarily nested `if else` construct that executes in a single clock cycle. Large numbers of decisions can be carried out in parallel by such ASM charts, allowing some algorithms to be sped up considerably.

5.1.1 Silly example of behavioral Mealy machine

Suppose we take the silly ASM of section 2.1.2.1 and include two conditional command signals: STAY and LEAVE. STAY is supposed to be 1 while the machine stays in state YELLOW, and LEAVE is 1 during the last cycle that the machine is in state YELLOW. STAY and LEAVE are never asserted at the same time. Here is the ASM chart:

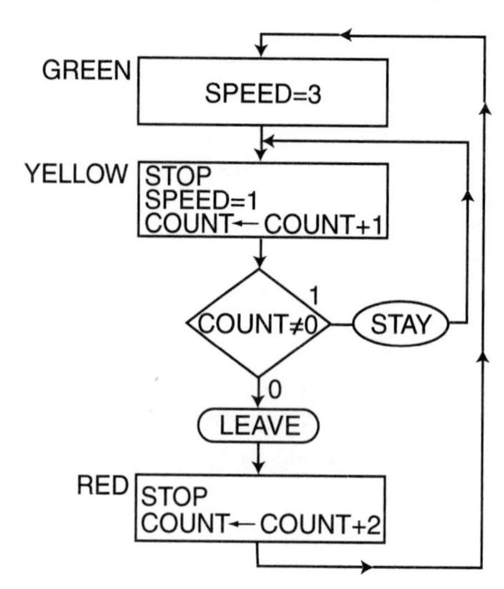

Figure 5-1. Behavioral Mealy ASM.

Assuming, as was the case for the example in section 2.1.2.1, that the period of the clock is 0.5 seconds, the following illustrates what would be observed physically when the hardware corresponding to this ASM operates:

present time	state					
0.0	GREEN	stop=0	speed=11	count=000	stay=0	leave=0
0.5	**YELLOW**	**stop=1**	**speed=01**	**count=000**	**stay=0**	**leave=1**
1.0	RED	stop=1	speed=00	count=001	stay=0	leave=0
1.5	GREEN	stop=0	speed=11	count=011	stay=0	leave=0
2.0	**YELLOW**	**stop=1**	**speed=01**	**count=011**	**stay=1**	**leave=0**
2.5	YELLOW	stop=1	speed=01	count=100	stay=1	leave=0
3.0	YELLOW	stop=1	speed=01	count=101	stay=1	leave=0
3.5	YELLOW	stop=1	speed=01	count=110	stay=1	leave=0
4.0	YELLOW	stop=1	speed=01	count=111	stay=1	leave=0
4.5	**YELLOW**	**stop=1**	**speed=01**	**count=000**	**stay=0**	**leave=1**
5.0	RED	stop=1	speed=00	count=001	stay=0	leave=0
5.5	GREEN	stop=0	speed=11	count=011	stay=0	leave=0
6.0	YELLOW	stop=1	speed=01	count=011	stay=1	leave=0
...	...	...	...	...	...	...

Between 0.5 and 1.0, the machine is in state YELLOW, but because COUNT is zero, the decision does not go on the path that loops back to state YELLOW but instead goes on the path where the next state is state RED. On this path is an oval that asserts the LEAVE signal. This conditional signal is asserted during the entire clock cycle, just as the unconditional signal STOP is asserted during the same time. The signal STAY, which is on a different path, is not asserted during this clock cycle.

Between 2.0 and 2.5, the machine again is in state YELLOW, but because COUNT is non-zero, the decision goes on the path that loops back to the same state. On this path is the oval that asserts the STAY signal. The signal LEAVE is not asserted in this clock cycle.

Because COUNT is three bits, COUNT is zero again between 4.5 and 5.0, and so this is the last clock cycle that the machine loops in state YELLOW. This means that STAY is not asserted but that LEAVE is asserted.

5.1.2 Silly example of mixed Mealy machine

An architecture that implements the silly example of section 5.1.1 is a counter register attached to an adder. One input of the adder is tied to the constant two. There is a comparator to see when COUNT is equal to zero. Here are the architecture and corresponding mixed ASM:

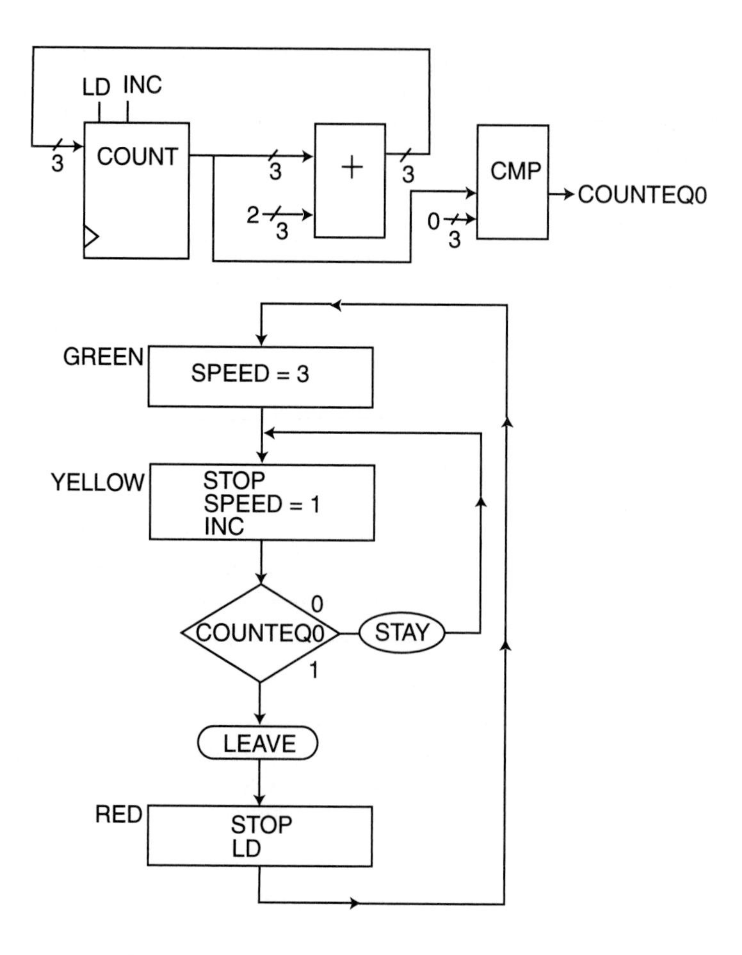

Fig 5-2. Mixed Mealy ASM.

5.1.3 Silly example of structural Mealy machine

The generic diagram of the pure structural controller given in section 2.4.1 applies to any machine, whether it is a Mealy or Moore machine. The next state combinational logic will be a little different when the machine is a Mealy machine than when it is a Moore machine. With a Moore machine, only the next state bits (and not the command bits) are a function of both the present state and the status inputs. With a Moore machine, the commands are a function of the present state only. In other words, for a Moore machine, every line of the truth table where `ps` is the same has the same command outputs.

A Mealy machine is completely general. The commands as well as the next state are a function of both the present state and the status. To illustrate this, consider the truth table of the next state combinational logic for the machine described by the ASM chart and architecture of section 5.1.2:

ps	COUNTEQ0	ns	STOP	SPEED	**LEAVE**	**STAY**	INC	LD
00	-	01	0	11	0	0	0	0
01	0	10	1	01	**1**	**0**	1	0
01	1	01	1	01	**0**	**1**	1	0
10	-	00	1	00	0	0	0	1

SPEED, STOP, INC and LD are unconditional commands that are a function of `ps` only. LEAVE and STAY are a function of both `ps` and COUNTEQ0. The conditional signals LEAVE and STAY are the only things here that make this a Mealy machine.

5.2 Mealy version of the division machine

Section 2.2 gives many variations of Moore ASMs that implement the childish division algorithm. This section describes how this algorithm can be improved by including ovals in the ASM.

5.2.1 Eliminating state INIT again

Section 2.2.3 describes a correct four-state version of the division machine that uses only two registers (`r1` and `r2`) in the architecture. Section 2.2.4 describes an unsuccessful attempt to remove state INIT from this ASM. Register `r3` was introduced in section 2.2.5 to compensate for the user interface bug that exists in the ASM of section 2.2.4.

The problem in section 2.2.4 is that the assignment to `r2` was written as an unconditional command in state IDLE. By using a Mealy ASM, it is possible to eliminate state INIT without destroying the contents of register `r2` when the machine is waiting in state IDLE for `pb` to be pressed. Only when the machine is leaving state IDLE to begin computing the quotient does `r2` get cleared. In other words, when the machine makes a transition from state IDLE to state COMPUTE1 is the time when `r2` becomes zero. Here is the ASM:

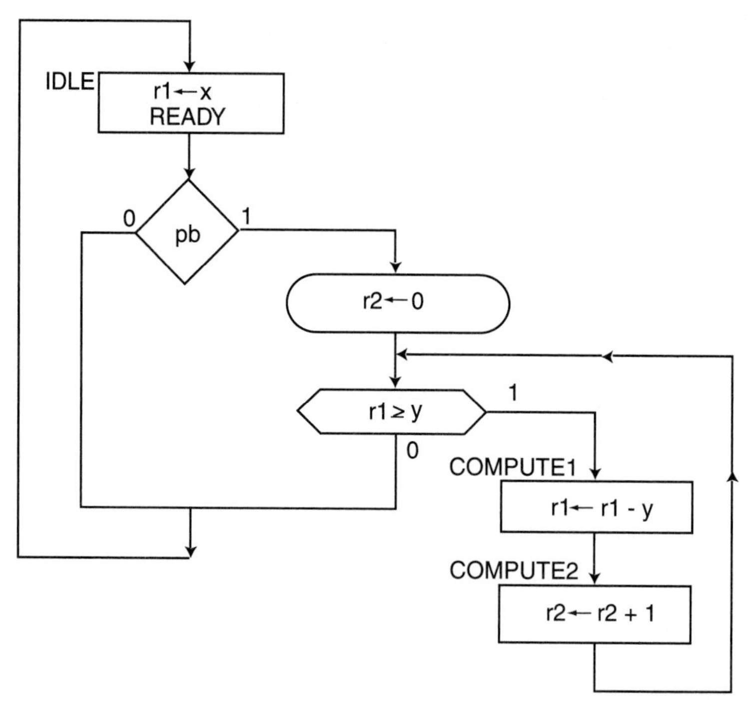

Figure 5-3. Mealy division machine with two states in loop.

Here is an example that shows the machine works when x=14 and y=7:

```
IDLE        r1=   ?   r2=     ?   pb=0    ready=1
IDLE        r1=   ?   r2=     ?   pb=0    ready=1
IDLE        r1=  14   r2=     ?   pb=1    ready=1
COMPUTE1    r1=  14   r2=     0   pb=0    ready=0
COMPUTE2    r1=   7   r2=     0   pb=0    ready=0
COMPUTE1    r1=   7   r2=     1   pb=0    ready=0
COMPUTE2    r1=   0   r2=     1   pb=0    ready=0
IDLE        r1=   0   r2=     2   pb=0    ready=1
IDLE        r1=   ?   r2=     2   pb=0    ready=1
```

The highlighted line shows where the conditional command to clear `r2` occurs. This takes effect at the next rising edge of the clock, which is when the machine enters state COMPUTE1 (`r2=0` on the next line is also highlighted to illustrate this).

Based on the assumptions used throughout all of the chapter 2 examples, the above ASM executes in `2+2*quotient` clock cycles, which is one clock cycle faster than the correct ASM of section 2.2.3.

5.2.2 Merging states COMPUTE1 and COMPUTE2

The above ASM requires about twice as long as the best solution discussed in chapter 2. To achieve the same kind of speed up with the Mealy ASM, we need to do the same thing we did in chapter 2: the operations in the loop need to occur in parallel. Consider the following **incorrect** ASM:

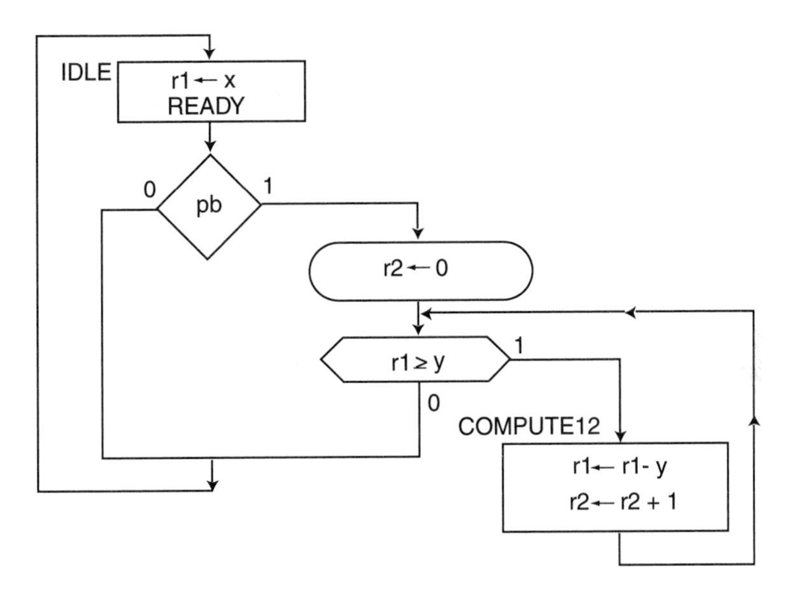

Figure 5-4. Incorrect Mealy division ASM.

To illustrate how this ASM fails, consider when x=14, and y=7:

```
IDLE        r1=     ?    r2=   ?   pb=0    ready=1
IDLE        r1=     ?    r2=   ?   pb=0    ready=1
IDLE        r1=    14    r2=   ?   pb=1    ready=1
COMPUTE12   r1=    14    r2=   0   pb=0    ready=0
COMPUTE12   r1=     7    r2=   1   pb=0    ready=0
COMPUTE12   r1=     0    r2=   2   pb=0    ready=0
IDLE        r1=  4089    r2=   3   pb=0    ready=1
IDLE        r1=     ?    r2=   3   pb=0    ready=1
```

By the point when the machine returns to state IDLE, `r2` has been incremented one time too many. In section 2.2.5, this problem was solved by using the `r3` register to save the correct quotient. However, since we are striving for a faster and cheaper solution here, it would be better to avoid introducing the `r3` register in this design.

5.2.3 Conditionally loading **r2**

To solve the bug illustrated in section 5.2.2, we need to load `r2` only when the machine stays in the loop, and to keep the old value of `r2` when the machine leaves the loop to return to state IDLE. This of course requires another oval in the ASM:

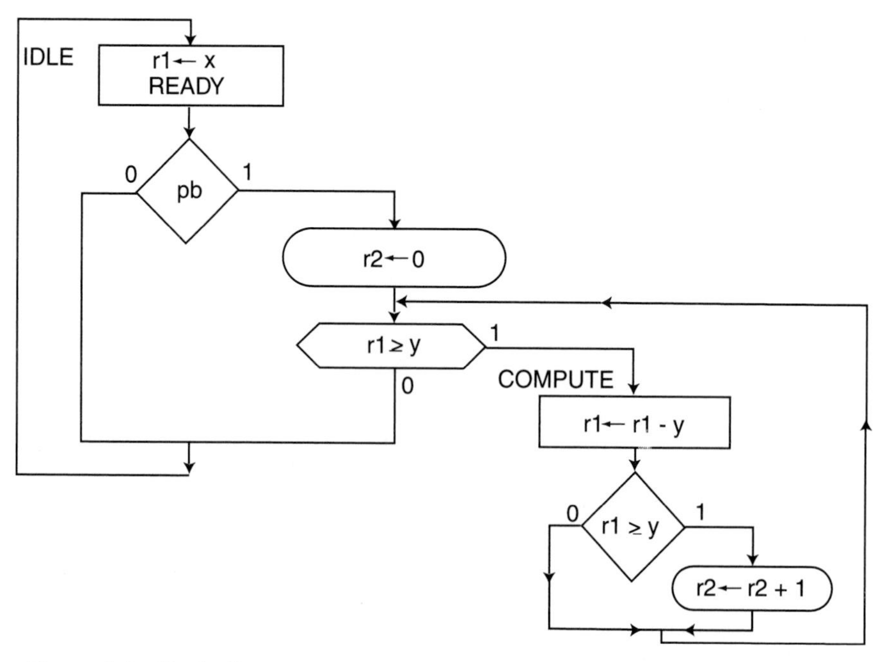

Figure 5-5. Mealy division ASM with conditional load.

To illustrate that this ASM works correctly, consider the case we looked at in the last section:

```
IDLE      r1=     ?   r2=   ?   pb=0   ready=1
IDLE      r1=    13   r2=   ?   pb=0   ready=1
IDLE      r1=    14   r2=   ?   pb=1   ready=1
COMPUTE   r1=    14   r2=   0   pb=0   ready=0
COMPUTE   r1=     7   r2=   1   pb=0   ready=0
COMPUTE   r1=     0   r2=   2   pb=0   ready=0
IDLE      r1=  4089   r2=   2   pb=0   ready=1
IDLE      r1=     ?   r2=   2   pb=0   ready=1
```

This machine can achieve the correct result in `3+quotient` clock cycles using only two (instead of three) registers. Therefore, it is as fast as the fastest Moore machine in chapter 2 using fewer registers.

5.2.4 Asserting READY early

The reason that the Mealy machine in section 5.2.3 is no faster than the the Moore machine in 2.2.7 is because of the assumption that the user waits at least two clock cycles while the machine asserts READY. In the Moore machines of chapter 2, asserting READY was the same as being in state IDLE, but with a Mealy machine, it would be possible to assert READY one clock cycle earlier. There are two reasons why asserting READY one clock cycle early works. First, during the last clock cycle of state COMPUTE in the ASM of section 5.2.3, `r2` already contains the correct quotient, and so `r2` is not scheduled to be incremented again. Second, the user is unaware of what state the machine is in and instead relies on READY to indicate the proper time when `pb` can be pressed again.

The following machine asserts READY in state IDLE and in the last clock cycle of state COMPUTE. When the machine is in state COMPUTE, the `r1 >= y` test at the bottom of the loop is evaluated at the same time as the `r1 >= y` test at the top of the loop. When the machine stays in the loop another time, `r2` is incremented. When the machine will leave the loop, READY is asserted instead.

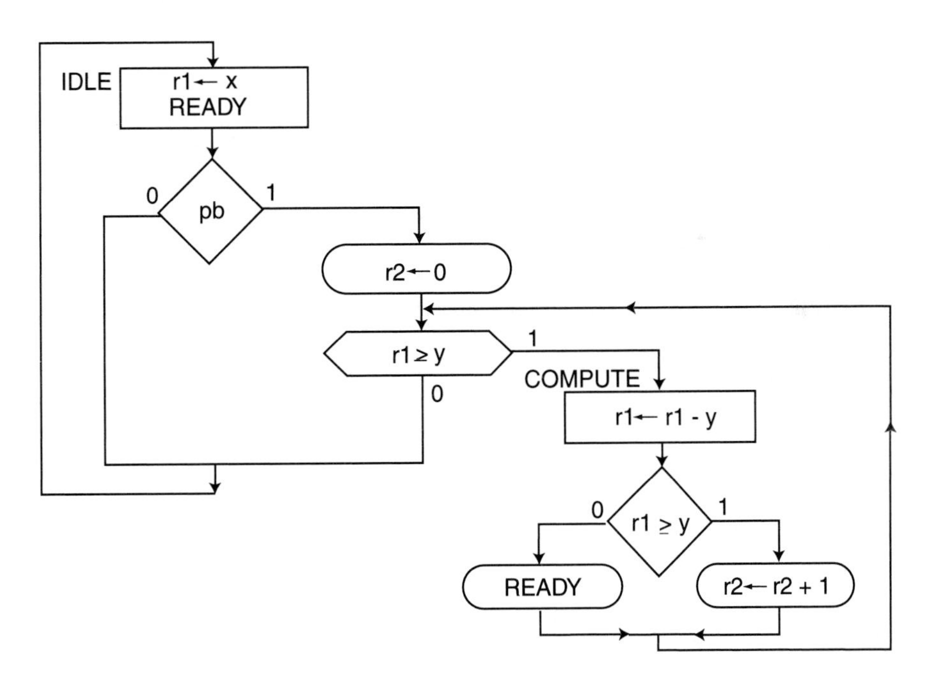

Figure 5-6. Mealy division ASM with conditional READY.

It would not be legal to attempt to assert READY after leaving the loop because READY is already asserted in state IDLE. The conditional assertion of READY as shown above is legal because it can only happen in state COMPUTE. The following illustrates how READY is asserted for the example when x=14 and y=7:

```
IDLE       r1=     ?   r2=    ?   pb=0   ready=1
IDLE       r1=    14   r2=    ?   pb=1   ready=1
COMPUTE    r1=    14   r2=    0   pb=0   ready=0
COMPUTE    r1=     7   r2=    1   pb=0   ready=0
COMPUTE    r1=     0   r2=    2   pb=0   ready=1
IDLE       r1=  4089   r2=    2   pb=0   ready=1
```

The last time in state COMPUTE, `r2` is already the correct quotient, and the conditional signal READY is asserted. The user can begin counting clock cycles from this moment, rather than having to wait an extra clock cycle.

This machine can achieve the correct result in `2+quotient` clock cycles using only two (instead of three) registers. Therefore, this Mealy machine is cheaper and faster than any of the Moore machines given in chapter 2.

5.3 Translating Mealy ASMs into behavioral Verilog

Pure behavioral Mealy ASMs that use only RTN can usually be translated into Verilog quite easily. Often it is as simple as using an `if` statement, where the first statement inside the `if` is not time control. (A Moore machine that has an `if` decision must have time control as the first statement inside the body of the `if`. Of course nested `if`s are allowed in both Moore and Mealy machines.) The statements inside an `if` without intervening time control correspond to ovals in a Mealy ASM. For example, consider how the ASM of section 5.2.1 translates into Verilog:

```
always
  begin
     @(posedge sysclk) enter_new_state(`IDLE);
       r1 <= @(posedge sysclk) x;
       ready = 1;
       if (pb)
         begin
           r2 <= @(posedge sysclk) 0;
           while (r1 >= y)
             begin
           @(posedge sysclk) enter_new_state(`COMPUTE1);
        r1 <= @(posedge sysclk) r1 - y;
```

Continued

```
@(posedge sysclk) enter_new_state(`COMPUTE2);
   r2 <= @(posedge sysclk) r2 + 1;
    end
      end
  end
```

Although there is time control later inside the `if`, the statement `r2 <= @(posedge sysclk) 0` occurs directly after the `if` with no intervening time control, and so is conditional during state IDLE.

As another example, consider translating the Mealy ASM of section 5.2.3 into Verilog:

```
  always
    begin
         @(posedge sysclk) enter_new_state(`IDLE);
            r1 <= @(posedge sysclk) x;
            ready = 1;
            if (pb)
              begin
                 r2 <= @(posedge sysclk) 0;
                 while (r1 >= y)
                   begin
                     @(posedge sysclk) enter_new_state(`COMPUTE);
                        r1 <= @(posedge sysclk) r1 - y;
                        if (r1 >= y)
                        r2 <= @(posedge sysclk) r2 + 1;
                   end
            end
  end
```

The condition (`r1 >= y`) always produces identical results in the `if` and in the `while` because no `$time` passes from when it is evaluated by the `if` and when it is later reevaluated by the `while`.

As a final example, consider translating the Mealy ASM of section 5.2.4:

```
  always
     begin
         @(posedge sysclk) enter_new_state(`IDLE);
            r1 <= @(posedge sysclk) x;
            ready = 1;
            if (pb)
              begin
```

Continued.

```
                r2 <= @(posedge sysclk) 0;
                while (r1 >= y)
                  begin
                     @(posedge sysclk) enter_new_state(`COMPUTE);
                      r1 <= @(posedge sysclk) r1 - y;
                      if (r1 >= y)
                        r2 <= @(posedge sysclk) r2 + 1;
                      else
                         ready = 1;
                  end
         end
end
```

The conditional command simply translates into an `else`.

5.4 Translating complex (`goto`) ASMs into behavioral Verilog

Section 2.1.4 discusses the `goto`-less style for ASM charts, where every decision is described in terms of high-level `while`, `if` and `case` constructs. Since Verilog has statements that correspond to these constructs, it is usually straightforward to translate such an ASM chart into behavioral Verilog, regardless of whether it is a Moore or Mealy ASM.

On the other hand, because Verilog does not provide a `goto` statement, there are three situations when translating an ASM chart into Verilog is more difficult. First, translation is difficult when an ASM chart uses a bottom testing loop construct, similar to the `repeat ... until` of Pascal or `do ... while()` of C. Second, translation is difficult when an ASM chart has intervening time control before the loop exit decision (as in the ASM of section 2.2.2). Third, translation is difficult when the decision can only be described with `goto`s.

The general solution to these difficulties involves using the `present_state` variable inside `if`s and `while`s. In the behavioral Verilog model of an ASM, the `present_state` variable indicates which algorithmic state the ASM is currently performing. By testing the `present_state` inside `if`s and `while`s with the `!==` operator, it is possible to implement arbitrary (`goto`-like) decisions without needing a `goto` statement. Such tests are not part of what the hardware does. Mentioning `present_state` in an ASM chart is unnecessary since an ASM chart allows arbitrary `goto`s to any state. Such decisions are required only to overcome a limitation of Verilog, and so using `!==` (rather than `!=`) is appropriate. The need for using `!==`

comes from the fact that present_state may be 'bx. The following three sections illustrate the three kinds of difficulties, and how testing present_state with !== can solve the problem.

5.4.1 Bottom testing loop

A bottom testing loop is, technically speaking, "goto-less," but such a loop is still difficult to translate because Verilog does not provide a bottom testing loop construct in the language. In essence, since such a construct does not exist, the decision at the bottom of the loop has to be thought of as a conditional goto that branches to the top of the loop. (In the pure structural stage, this is how the loop would be implemented by the computation of the next state in the state_gen function.) In the pure behavioral stage, since Verilog lacks a goto statement, the only choice is to describe such a loop using a while.

As an illustration, consider the nonsense ASM chart from section 2.1.2.1. Suppose the states are assigned the following representations:

```
`define         NUM_STATE_BITS  2
`define         GREEN    2'b00
`define         YELLOW   2'b01
`define         RED      2'b10
```

It is **incorrect** to translate the loop involving state YELLOW using just a while:

```
always
  begin
    @(posedge sysclk) enter_new_state(`GREEN);
       stop = 0;
       speed = 3;
       while(count != 0)
         begin
           @(posedge sysclk) enter_new_state(`YELLOW);
              stop = 1;
              speed = 1;
              count <= @(posedge sysclk) count + 1;
         end
    @(posedge sysclk) enter_new_state(`RED);
      stop = 1;
      speed = 0;
      count <= @(posedge sysclk) count + 2;
  end
```

because the `count != 0` test in the while occurs as part of both states GREEN as well YELLOW, but in the original ASM the test is part of state YELLOW only. To illustrate this problem, consider the simulation of the above while loop:

```
$time=      70   ps=00   stop=0   speed=11   count=000
$time=     170   ps=10   stop=1   speed=00   count=000
$time=     270   ps=00   stop=0   speed=11   count=010
$time=     370   ps=00   stop=1   speed=01   count=010
$time=     470   ps=01   stop=1   speed=01   count=011
$time=     570   ps=01   stop=1   speed=01   count=100
$time=     670   ps=01   stop=1   speed=01   count=101
$time=     770   ps=01   stop=1   speed=01   count=110
$time=     870   ps=01   stop=1   speed=01   count=111
$time=     970   ps=01   stop=1   speed=01   count=000
$time=    1070   ps=10   stop=1   speed=00   count=001
...
```

The condition in the `while` is evaluated in state GREEN at `$time` 51, which is the problem. At `$time` 170, the machine has gone directly to state RED, rather than to state YELLOW where it is supposed to be. This error occurs because the `while` loop has inserted an extra (incorrect) test whether `count != 0` in state RED. Because `count` is zero, the machine avoids state YELLOW altogether, which is not the desired behavior.

It is an unavoidable feature of Verilog that the behavioral construct that we have to implement this loop is the `while` statement. A `while` loop always tests at the top of the loop, but in hardware, as is the case here, we often want to test at the bottom of the loop. To implement the ASM of section 2.1.2.1 correctly for Verilog simulation requires nullifying the fact that this `while` loop executes both in state GREEN and in state YELLOW. To overcome this problem, the correct code ORs the original ASM condition with a `!==` test that mentions the bottom state of the loop. In this example, because there is only the state YELLOW in the loop, the `while` condition ORs `count != 0` with `present_state !== 'YELLOW`:

```
always
  begin
    @(posedge sysclk) enter_new_state('GREEN);
    speed = 3;
    while(count != 0 | present_state !== 'YELLOW)
      begin
        @(posedge sysclk) enter_new_state('YELLOW);
          stop = 1;
          speed = 1;
          count <= @(posedge sysclk) count + 1;
      end
```

Continued.

```
    @(posedge sysclk) enter_new_state('RED);
      stop = 1;
      count <= @(posedge sysclk) count + 2;
end
```

When the machine is in the bottom state of the loop (YELLOW in this example), the !== will be false, and the original ASM condition will be the only thing that decides whether the `while` loop continues. When the machine is in any other state, such as state GREEN, the `!==` will be true (even if present_state is `'bx`), and so the ORed condition is true. The loop will begin execution, regardless of the original ASM condition. It does not matter whether `count != 0`; the loop will execute at least once. The simulation of this code shows that it correctly models the behavior of the original ASM:

```
$time=    70   ps=00   stop=0   speed=11   count=000
$time=   170   ps=01   stop=1   speed=01   count=000
$time=   270   ps=10   stop=1   speed=00   count=001
$time=   370   ps=00   stop=0   speed=11   count=011
$time=   470   ps=01   stop=1   speed=01   count=011
$time=   570   ps=01   stop=1   speed=01   count=100
$time=   670   ps=01   stop=1   speed=01   count=101
$time=   770   ps=01   stop=1   speed=01   count=110
$time=   870   ps=01   stop=1   speed=01   count=111
$time=   970   ps=01   stop=1   speed=01   count=000
$time=  1070   ps=10   stop=1   speed=00   count=001
...
```

At `$time` 170, the machine has entered state YELLOW as it should. Since during that clock cycle `count` is zero, the machine proceeds to state RED.

5.4.2 Time control within a decision

An ASM chart that is translated directly from a software paradigm `while` loop (see section 2.2.2 for an example) would appear to be `goto`-less, but in fact it is not. The problem with such an ASM chart is that it must have @ time control between the algorithmic top of the loop and the place where the decision occurs. Despite the fact that Verilog has a `while` loop, the testing of the condition required by a software "`while`" loop occurs in the middle of the loop (after the @), rather than at the algorithmic top of the loop (as required by Verilog's `while` construct). The reason for this @ time control is to give the algorithm `$time` to make the decision before any computations occur that could effect the outcome of the decision.

Again, it is possible to overcome this kind of problem using a present_state !== test. For example, consider translating the Moore ASM chart of section 2.2.2 into behavioral Verilog:

```
module slow_div_system(pb,ready,x,y,r2,sysclk);
  input pb,x,y,sysclk;
  output ready,r2;
  wire pb;
  wire [11:0] x,y;
  reg ready;
  reg [11:0] r1,r2;
  reg ['NUM_STATE_BITS-1:0] present_state;

  always
    begin
      @(posedge sysclk) enter_new_state('IDLE);
        r1 <= @(posedge sysclk) x;
        ready = 1;
        if (pb)
          begin
           @(posedge sysclk) enter_new_state('INIT);
            r2 <= @(posedge sysclk) 0;
            while ((r1 >= y)|present_state !=='TEST)
             begin
               @(posedge sysclk)enter_new_state('TEST);
                 if (r1 >= y)
                  begin
                    @(posedge sysclk enter_new_state('COMPUTE1);
                      r1 <= @(posedge sysclk) r1 - y;
                    @(posedge sysclk) enter_new_state('COMPUTE2);
                      r2 <= @(posedge sysclk) r2 + 1;
                  end
             end
          end
    end
task enter_new_state;
 input ['NUM_STATE_BITS-1:0] this_state;
 begin
   present_state = this_state;
   #1 ready=0;
 end
endtask
```

Continued

```
 always @(posedge sysclk) #20
   $display("%d ps=%b r1=%d r2=%d pb=%b ready=%b",
      $time, present_state,r1,r2, pb, ready);
endmodule
```

where the states are represented as:

```
`define    NUM_STATE_BITS 3
`define    IDLE         3'b000
`define    INIT         3'b001
`define    TEST         3'b010
`define    COMPUTE1     3'b011
`define    COMPUTE2     3'b100
```

The troublesome state here is state TEST. There is a Verilog `while` loop whose body includes state TEST and an `if` statement that includes the other states of the ASM loop. Three situations can occur with the Verilog `while` loop: It is possible that the `while` loop is being entered for the first time from state INIT, it is possible that the `while` loop is to be reexecuted from state COMPUTE2, and it is possible that the `while` loop is to exit from state TEST. In each of these three situations, the condition inside the Verilog `while` loop is evaluated. The only one of these three situations in which the Verilog loop body does not proceed to execute is when the ASM loop exits from state TEST. The other two situations (from state INIT and state COMPUTE2) are guaranteed to stay inside the Verilog `while` loop. Therefore, the `present_state !== 'TEST` condition makes sure that the next thing to execute in both of those two situations will be the algorithmic top of the Verilog `while` loop (state TEST).

In order to allow the Verilog `while` loop to exit at the identical `$time` that the ASM loop exits, there is a nested `if` inside the Verilog `while` loop, after state TEST. This `if` uses the same ASM condition (`r1 >= y`) that was also mentioned in the Verilog `while` loop. In the situation when this condition is false, no `$time` has elapsed before the Verilog `while` condition `((r1 >= y)|present_state !== 'TEST)` is re-evaluated. Since the present state is state TEST and ASM condition (`r1 >= y`) remains false since no `$time` has elasped, the Verilog `while` loop exits properly. Here is a simulation for `x`=14 and `y`=7:

```
    9370    ps=000  r1=   13   r2=   1  pb=0   ready=1
    9470    ps=000  r1=   14   r2=   1  pb=1   ready=1
    9570    ps=001  r1=   14   r2=   1  pb=0   ready=0
    9670    ps=010  r1=   14   r2=   0  pb=0   ready=0
    9770    ps=011  r1=   14   r2=   0  pb=0   ready=0
    9870    ps=100  r1=    7   r2=   0  pb=0   ready=0
    9970    ps=010  r1=    7   r2=   1  pb=0   ready=0
   10070    ps=011  r1=    7   r2=   1  pb=0   ready=0
   10170    ps=100  r1=    0   r2=   1  pb=0   ready=0
   10270    ps=010  r1=    0   r2=   2  pb=0   ready=0
   10370    ps=000  r1=    0   r2=   2  pb=0   ready=1
```

5.4.3 Arbitrary gotos

It is poor style to use arbitrary `gotos`. Therefore such an example is not presented here. Nevertheless, regardless of how messy the ASM, some combination of `ifs` and `whiles` that use `===` and `!==` tests with the `present_state` can implement the ASM in Verilog.

5.5 Translating conditional command signals into Verilog

To translate a Mealy ASM, such as the one in section 5.1.1, that has conditional command signals (rather than conditional RTN), `enter_new_state` must include all the Mealy command signal outputs. In the example ASM of section 5.1.1, the task has four commands to initialize, each of which has a default of zero:

```
task enter_new_state;
   input ['NUM_STATE_BITS-1:0] this_state;
   begin
     present_state = this_state;
     #1 stop =0;
        speed = 0;
        stay = 0;
        leave = 0;
   end
 endtask
```

Initializing such conditional command signals is important because in many situations the Mealy command is explicitly mentioned only on certain paths through the ASM. By describing the default values for all outputs (whether they are Mealy or Moore) in `enter_new_state`, the behavioral Verilog will be a one-to-one mapping of the corresponding ASM chart.

The Mealy ASM chart of section 5.1.1 then can be translated into Verilog as shown below (using the !== technique described in section 5.4.1):

```
always
  begin
    @(posedge sysclk) enter_new_state('GREEN);
      speed = 3;
      while(count!=0 | present_state !== 'YELLOW)
        begin
          @(posedge sysclk) enter_new_state('YELLOW);
            stop = 1;
            speed = 1;
            count <= @(posedge sysclk) count + 1;
            if (count!=0)
              stay = 1;
        end
      leave = 1;
    @(posedge sysclk) enter_new_state('RED);
      stop = 1;
      count <= @(posedge sysclk) count + 2;
  end
```

The diamond and oval inside the loop simply translate into an `if` statement followed by the `stay = 1` statement, with no intervening time control. Therefore, there is no time control between the return from `enter_new_state` and the execution of the `if` (and the possible consequent execution of `stay=1`.) Suppose that `count` is non-zero, which means `stay` becomes one at the same `$time` that `speed` and `count` become one. Since `leave` is not mentioned inside the loop, it retains its default value of zero.

On the other hand, suppose `count` is zero inside the loop. This means `stay=1` does not execute, and so `stay` retains its default value (of zero) given to it by `enter_new_state`. No `$time` passes at the point where the `while` retests whether `count !=0`. Since `count` is zero, the `while` is guaranteed to exit, but still no `$time` has elasped. This means that the `leave=1` statement executes at the same `$time` as the final call to `enter_new_state('YELLOW)` returns back to the loop body. Therefore the last cycle in which the machine is in state YELLOW will output `leave` as one, but `stay` as zero.

Since this is a correct translation of a bottom testing loop, the only way the machine exits from the `while` loop is from state YELLOW. (It is not possible to get directly from state GREEN to the exit of the `while` because of the `present_state !== 'YELLOW`.) Therefore, this Verilog is a one to one mapping of the ASM.

In the above Verilog, the states are represented as:

```
`define    NUM_STATE_BITS  2
`define    GREEN    2'b00
`define    YELLOW   2'b01
`define    RED      2'b10
```

and the following declarations occur at the beginning of the module:

```
reg stop;
reg [1:0] speed;
reg [2:0] count;
reg [`NUM_STATE_BITS-1:0] present_state;
reg stay,leave;
```

5.6 Single-state Mealy ASMs

As discussed in section 2.4.1, an ASM with four rectangles requires two bits to represent those states. An ASM with two rectangles requires only one bit to represent those states. What about an ASM with only one state? As explained in section 2.5, an ASM with only one rectangle represents pure combinational logic and therefore needs zero bits to represent the state. (The machine is always in that one state, and so it is not necessary to record which state the machine is in.)

The oval notation allows such an ASM to have an arbitrarily complex decision happening in that one state. For example, a decoder, whose input is a two-bit bus, `inbus`, and whose outputs are `o0`, `o1`, `o2` and `o3` can be described as:

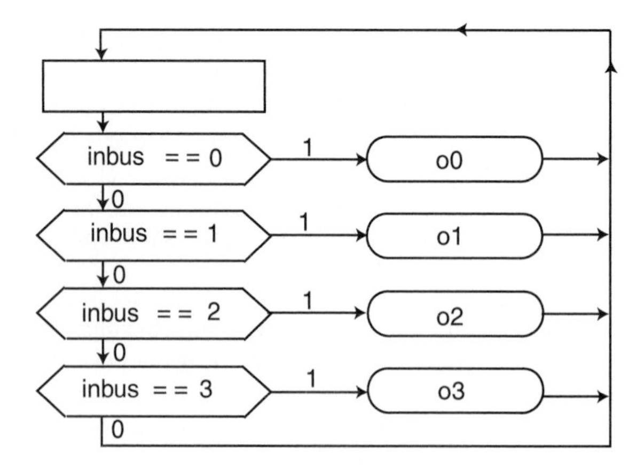

Figure 5-7. ASM for combinational logic (decoder).

Such an ASM bears a close resemblance to the way in which a combinational device is modeled in Verilog:

```
      wire [1:0] inbus;
      reg o0,o1,o2,o3;
      always @(inbus)
        begin
          o0 = 0; o1 = 0; o2 = 0; o3 = 0;   //defaults
          case (inbus)
            0: o0 = 1;
            1: o1 = 1;
            2: o2 = 1;
            3: o3 = 1;
          endcase
        end
```

The only difference between the ASM and the Verilog is that the Verilog needs the proper time control for combinational logic, which is @ followed by a sensitivity list, rather than any mention of the system clock.

5.7 Conclusion

Moore machines have commands that occur when the machine is in a particular state. Mealy machines allow commands in a particular state to occur based on status. This chapter shows how Mealy ASMs allow a designer to express faster and better algorithms. Like Moore ASMs, Mealy ASMs have unconditional commands in rectangles. Unlike Moore ASMs, Mealy ASMs have conditional commands in ovals that follow diamonds. The conditional commands in the ovals happen at the same time as the unconditional commands in the rectangle and the decisions in the diamonds.

Translating a Mealy ASM into behavioral Verilog is usually simple, typically involving an `if` statement with no intervening time control. When the ASM involves command signals (rather than RTN), as would be the case at the mixed stage, the `enter_new_state` task must initialize the conditional commands. Some ASMs (both Moore and Mealy) cannot be expressed in the `goto`-less style with simple `whiles` and `ifs`. Such ASMs need to be translated into Verilog using !== tests of `present_state`. A common example of an ASM that must be translated into Verilog with a `present_state !==` test is a bottom testing loop. These techniques work only for simulation. See chapter 11 for synthesis techniques.

Single-state Mealy ASMs are a general notation to describe combinational logic in a behavioral fashion. As such, they are closely related to the behavioral Verilog description of combinational logic.

6. DESIGNING FOR SPEED AND COST

Chapter 2 uses ASM charts as a description of how to carry out each step of an algorithm in an arbitrarily chosen unit of physical time, known as the clock cycle. Chapter 3 introduces Verilog's `$time` variable, whose incrementation by the Verilog scheduler simulates the passage of physical time in the fabricated hardware. The example Verilog clock (section 3.7.1.3) used in chapters 3 through 5 has an arbitrarily chosen clock period of 100 units of `$time`. The exact amount of physical time that this 100 units of Verilog `$time` relates to is not specified nor is it of any concern in chapters 3 through 5. Up to this point, the emphasis has been on designing correct algorithms and implementing them properly in hardware. The only attempt to increase speed in chapters 2 through 5 is by doing more steps in parallel, thereby requiring fewer clock cycles. The number of clock cycles required by an ASM chart is a mathematical property of the algorithm and is separate from its physical implementation.

The total physical time required by a machine to compute an answer is the number of clock cycles required by the algorithm multiplied by the physical time of the clock period used in the hardware implementation. Physical clocks are often measured in frequency, rather than time. There is a reciprocal relationship between clock period and clock frequency. For example, what is the total time required to divide 14 by 7 using the machine described in section 2.2.7 when that machine is clocked at 200 MHz? According to the analysis there, the number of clock cycles required by the algorithm (including the time for two clock cycles in state IDLE) is `3+quotient=5` clock cycles. The clock period in this example is 5 ns, and so the total time is 25 ns.

In the first stages of design, algorithmic correctness and speed only in terms of clock cycles are the primary focus of the designer. The harsh physical reality of time should enter into the designer's thinking only after the design has been synthesized. This chapter shows how the `$time` features of a Verilog simulator allow a designer to experimentally determine the speed of a synthesized design without having to fabricate it. This chapter also illustrates three alternative design techniques that allow a trade-off between speed and cost should a synthesized design fail to meet its speed requirements.

6.1 Propagation delay

Every physical combinational logic device takes an amount of physical time, known as propagation delay, to compute its answer. Except in section 3.7.1.2, we have modeled combinational logic devices in Verilog as having no propagation delay. There are three reasons for modeling perfect (delayless) combinational logic. First, it is the only acceptable way to model combinational logic for synthesis. (The synthesis tool, rather than the designer, chooses the propagation delay of the synthesized hardware. Only after synthesis can the designer simulate the netlist to see how fast it is.) Second, the delayless style (section 3.7.2.1) is more efficient to simulate in the early stages of design, when the designer is more concerned with algorithmic correctness than with speed. Third, the delayless style is easier.

The purpose of top-down design is to defer (but not forget) as many details for as long as possible. Though it is simpler to ignore propagation delay, if we want to fabricate a practical machine, we must ultimately confront these details. Real-world machines must meet certain criteria of speed and cost. Even if a machine is algorithmically correct, if it is too slow or too expensive, it will not be practical to build. Running a machine at a higher clock frequency increases the speed of the machine, but there is a cost associated with operating at higher clock frequencies.

6.2 Factors that determine clock frequency

Synchronous devices (registers and the controller) are clocked by a clock signal of a particular frequency. If that frequency is low enough (the clock period is long enough), the machine will behave as predicted by the ASM chart and the Verilog simulations. For as long as the machine does not malfunction, increasing the clock frequency will proportionately increase the speed of the machine without having to redesign it. Unfortunately, there is a limit on how fast the clock can be. Therefore, to get the most out of the hardware, the designer wants to know what the maximum permissible clock frequency is.

The propagation delay of every combinational device and every gate in a machine potentially plays a role in determining the actual speed of the machine. There must be enough time in each clock cycle for every combinational logic path to stabilize on its correct value. A combinational device whose interconnection of gates is shallow can have its result ready to be clocked into a register faster than an equivalent device whose interconnection of gates is deep. Unfortunately, combinational devices whose internal interconnections are shallow tend to be more costly. Simulation and synthesis tools help the designer decide on a trade-off between speed and cost.

The factors that determine the maximum clock frequency include the delay of signals on each wire, the propagation delay of each gate, the way such gates and wires are interconnected to form building blocks (such as adders) and the way such building blocks are interconnected to form the problem-specific architecture and the controller. The designer does not have much influence on the first three factors. The implementation technology (not the designer or the synthesis tool) determines the delays of wires, gates and devices. (Of course, the designer can choose to use a more expensive technology to achieve higher speed. For example, by fabricating custom silicon with smaller chip dimensions, the propagation delays of each gate are correspondingly reduced.) At best, the designer can only give hints to the synthesis tool to favor either lower-cost building blocks or higher-speed building blocks. If speed is essential, the designer can manually create a netlist for a shallow (high-speed) building block, but this circumvents the advantages of the synthesis tool.

When using a synthesis tool, the only major factor to increase speed over which a designer has much control is the way building blocks are interconnected to form the architecture and the controller. The end of this chapter discusses various methods that the designer has at the architectural level to increase speed at minimum cost. But first, the next sections look at netlist level propagation delay, which is the underlying cause of this difficulty.

6.3 Example of netlist propagation delay

Putting propagation delays at the netlist level is easy. You simply instantiate the built-in gate with a parameter, which is the delay in units of `$time`. (This works only for built-in gates, since parameters in user-defined modules take on whatever meaning the user desires.)

Section 2.5 explains how a two-bit adder black box can be decomposed down to the gate level using hierarchical design. Sections 3.10.6 and 3.10.7 give the equivalent structural (hierarchical) Verilog modules for this adder assuming no propagation delay. As an example of modeling propagation delay, assume `and` and `or` gates have a delay of one unit of `$time`, and the more complicated `xor` gates have a delay of two units of `$time`. A slight change to the modules from sections 3.10.6 and 3.10.7 will provide a much more realistic model of what the actual hardware does:

```
module half_adder(c,s,a,b);
   input a,b;
   wire a,b;
   output c,s;
   wire c,s;
   xor #2 x1(s,a,b);
   and #1 a1(c,a,b);
 endmodule

 module full_adder(cout,s,a,b,cin);
   input a,b,cin;
   wire a,b,cin;
   output cout,s;
   wire cout,s;
   wire cout1,cout2,stemp;
   half_adder ha2(cout1,stemp,a,b);
   half_adder ha3(cout2,s,cin,stemp);
   or  #1     o1(cout,cout1,cout2);
 endmodule

 module adder(sum,a,b);
   input a,b;
   output sum;
   wire [1:0] a,b;
   wire [2:0] sum;
   wire c;
   half_adder ha1(c,sum[0],a[0],b[0]);
   full_adder fa1(sum[2],sum[1],a[1],b[1],c);
 endmodule
```

Assuming this module is instantiated as before:

```
            adder adder1(sum,a,b);
```

the following diagram illustrates the interconnection of gates described by the above Verilog:

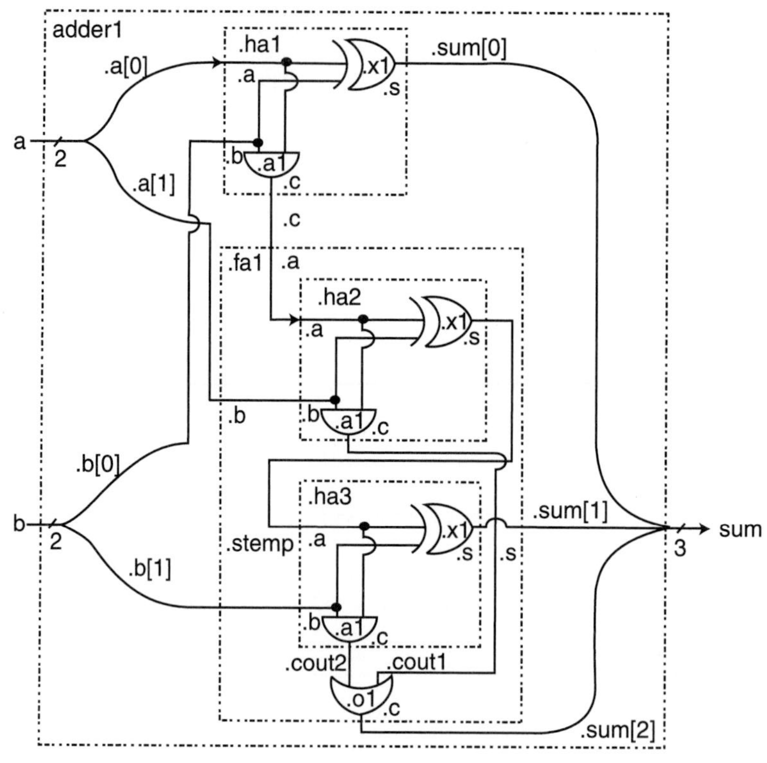

Figure 6-1. Adder with names used in structural Verilog.

6.3.1 A priori worst case timing analysis

Without using simulation, we could determine how long it takes in the worst case for the above modules to stabilize. There are many possible paths through the gates that have to be considered to determine the longest path. For example, there is a dependency of `sum[2]` on `a[0]`. A change in `a[0]` could cause an incorrect result for `sum[2]` until the effects of the change in `a[0]` can propagate through the path to `sum[2]`. The following shows where this change has to propagate, and how much `$time` is required:

```
Verilog name                total delay thus far

a[0]                             0
adder1.a[0]                      0
adder1.ha1.a                     0
adder1.ha1.a1                    1
adder1.ha1.c                     1
adder1.c                         1
adder1.fa1.a                     1
adder1.fa1.ha2.a                 1
adder1.fa1.ha2.a1                2
adder1.fa1.ha2.c                 2
adder1.fa1.cout1                 2
adder1.fa1.o1                    3
adder1.fa1.c                     3
adder1.sum[2]                    3
sum[2]                           3
```

There is also a dependency of `sum[1]` on `a[0]`. The following shows where the change in `a[0]` has to propagate, and how much `$time` is required:

```
Verilog name                total delay thus far

a[0]                             0
adder1.a[0]                      0
adder1.ha1.a                     0
adder1.ha1.a1                    1
adder1.ha1.c                     1
adder1.c                         1
adder1.fa1.a                     1
adder1.fa1.ha2.a                 1
adder1.fa1.ha2.x1                3
adder1.fa1.ha2.s                 3
adder1.fa1.stemp                 3
adder1.fa1.ha3.a                 3
adder1.fa1.ha3.x1                5
adder1.fa1.ha3.s                 5
adder1.sum[1]                    5
sum[1]                           5
```

There are other similar delay paths, but none of them are longer than five units of `$time`. Therefore, whatever code instantiates `adder` must wait more than five units of `$time` after changing `a` and `b` before using `sum`.

The test module in section 3.10.5 did not do this, so it would print out incorrect values for sum. If you simulate adder1 with the test code from section 3.10.5, most of the results printed out will be wrong. This is because the #1 in the test code is an inadequate amount of $time for adder1 to stabilize on the correct answer.

6.3.2 Simulation timing analysis

In circuits more complex than this trivial example, it is difficult to do a priori timing analysis, especially on a synthesized netlist. Also, since a priori analysis finds the worst case propagation delay, such analysis may be overly pessimistic. Therefore, designers often use post-synthesis simulation to obtain timing information about the design.

In this example, according to the a priori timing analysis, the test code for the adder module described in section 6.3.1 needs to wait longer than #5. If we use simulation to do the timing analysis, we would take a guess and check if the guess is an adequate amount of delay. Here we try #7:

```
module top;
```

```
integer ia,ib;
reg [1:0] a,b;
wire [2:0] sum;
reg [2:0] oldsum;

adder adder1(sum,a,b);

always #1
 begin
  #0 if (a+b==sum)
      $display("a=%d b=%d sum=%d CORRECT      $time=%d",
                a,b,sum,$time);
     else
      if (sum==oldsum)
       $display("a=%d b=%d sum=%d WRONG LAG    $time=%d",
                 a,b,sum,$time);
      else
       $display("a=%d b=%d sum=%d WRONG GLITCH $time=%d",
                 a,b,sum,$time);
     oldsum = sum;
 end
initial
```

Continued

```
  begin
   for (ia=0; ia<=3; ia = ia+1)
    begin
     a = ia;
     for (ib=0; ib<=3; ib = ib + 1)
      begin
          b = ib;
       #7 if (a + b === sum)
           $display("                        tested CORRECT");
          else
           $display("                        tested WRONG");
      end
    end
    $finish;
  end
endmodule
```

6.3.3 Hazards

In addition to the `initial` block in the test code of section 6.3.2, it is helpful to have an `always` block that monitors the change in `sum` at every unit of `$time`. At the end (#0) of each unit of `$time`, the `always` block checks if the current output of the adder (`sum`) is equal to `a+b`. If it is, it prints out the `"CORRECT"` message. If it is not, it prints out a message explaining the reason why. There are two possible explanations for a `"WRONG"` value of `sum`. The first is that the current value of `sum` is the same as what `sum` used to be at the previous unit of `$time`. In other words, `sum` is lagging behind the change in `a` or `b`. The other possible error is that `sum` has changed (due to the change in `a` or `b`) to an incorrect value. Such a momentary incorrect value from combinational logic with propagation delay is known as a *hazard* (also known as a *glitch*). Hazards occur when combinational logic internally has different path delays.

Here is a partial output of this simulation:

```
a=0 b=0 sum=X WRONG GLITCH $time=          1
a=0 b=0 sum=X WRONG GLITCH $time=          2
a=0 b=0 sum=X WRONG GLITCH $time=          3
a=0 b=0 sum=0 CORRECT      $time=          4
a=0 b=0 sum=0 CORRECT      $time=          5
a=0 b=0 sum=0 CORRECT      $time=          6
                               tested CORRECT
a=0 b=1 sum=0 WRONG LAG    $time=          7
a=0 b=1 sum=0 WRONG LAG    $time=          8
a=0 b=1 sum=1 CORRECT      $time=          9
a=0 b=1 sum=1 CORRECT      $time=         10
a=0 b=1 sum=1 CORRECT      $time=         11
a=0 b=1 sum=1 CORRECT      $time=         12
a=0 b=1 sum=1 CORRECT      $time=         13
                               tested CORRECT
...
a=2 b=2 sum=3 WRONG LAG    $time=         70
a=2 b=2 sum=3 WRONG LAG    $time=         71
a=2 b=2 sum=6 WRONG GLITCH $time=         72
a=2 b=2 sum=6 WRONG LAG    $time=         73
a=2 b=2 sum=4 CORRECT      $time=         74
a=2 b=2 sum=4 CORRECT      $time=         75
a=2 b=2 sum=4 CORRECT      $time=         76
                               tested CORRECT
...
```

In the above, for cases such as `a=0 b=1`, the output of `sum` simply retains its old value until `sum` makes a single change to the correct value. In essence, in these cases, it is like describing the adder with the following behavioral block:

```
module adder(sum,a,b);
parameter DELAY=1;
output sum;
input a,b;
reg [2:0] sum;
wire [1:0] a,b;
  always (a or b)
    # DELAY sum=a+b;
endmodule
```

where DELAY is an integer propagation delay. Although the above is an attractive way of viewing propagation delay, it does not describe the more complex behavior that occurs in other cases. For example, in the simulation of the adder given in section 6.3, for cases such as `a=2 b=3`, at first (like the other cases) the output makes no change (since the input change has not yet propagated to the output). Later, the output changes to an incorrect result that is different from the earlier value of `sum`. Finally, the output stabilizes on the correct result.

Although the a priori analysis using the circuit diagram indicates that more than #5 would always be safe, we could use simulation to see if #4 would be enough. Here is a partial output of this simulation:

```
a=0 b=0 sum=X WRONG GLITCH $time=            1
a=0 b=0 sum=X WRONG GLITCH $time=            2
a=0 b=0 sum=X WRONG GLITCH $time=            3
                                    tested WRONG
a=0 b=1 sum=0 WRONG GLITCH $time=            4
a=0 b=1 sum=0 WRONG LAG    $time=            5
a=0 b=1 sum=1 CORRECT      $time=            6
a=0 b=1 sum=1 CORRECT      $time=            7
                                    tested CORRECT

...
a=2 b=2 sum=3 WRONG LAG    $time=           40
a=2 b=2 sum=3 WRONG LAG    $time=           41
a=2 b=2 sum=6 WRONG GLITCH $time=           42
a=2 b=2 sum=6 WRONG LAG    $time=           43
                                    tested WRONG

...
```

Although four units of `$time` was enough for the adder to stabilize in many cases, it failed to stabilize for all of them. Therefore, a longer period is required for completely correct behavior for any possible input. On the other hand, if we knew before we build our machine that the inputs would *always* be among those cases where the combinational logic stabilizes early, we could run the machine faster. For an adder, such a situation is unlikely, but for other kinds of combinational logic, we might be able to use Verilog simulation to determine that our machine can run faster than a priori worst case analysis would predict.

6.3.4 Advanced gate-level modeling

Verilog provides three additional features for modeling gate-level delays: rising/falling delays, minimum/typical/maximum delays and `specify` blocks. The first of these allows us to model the fact that, for many electronic gate technologies, the time for a gate to change its output to one is not the same as the time for the gate to change its output to zero. For example, suppose the `xor` in the half-adder of section 6.3 takes two units of `$time` when its output changes to a one, but three units of `$time` when its output changes to a zero:

```
xor #(2,3) x1(s,a,b);
```

Note that the xor in section 6.3 could have been described as:

```
xor #(2,2) x1(s,a,b);
```

The second of these advanced gate delay features allows us to model that there are certain variations in the fabrication process that affect the speed of supposedly identical gates at the time of physical manufacturing. Even though the gates are supposed to be identical, these minor variations mean some of the gates will be slower than others. The maximum speed possible would be obtained if there were no variation during fabrication. Using statistical quality control methods, manufacturers determine the typical speed expected given random variations and determine a minimum acceptable speed by discarding parts that do not obtain this speed. Many Verilog simulators allows resimulation at each of these three speeds without recompilation by specifying these three delays separated by colons. For example, the following:

```
xor #(1:2:3) x1(s,a,b);
```

indicates a minimum delay of 1, a typical delay of 2 and a maximum delay of 3. The way in which this is used depends on a particular simulator, and not all simulators implement this feature.

The third of these advanced gate delay features, known as the specify block, allows us to model delays within modules without having to indicate delays on individual gates. For example, the following module is equivalent to the one given in section 6.3:

```
module half_adder(c,s,a,b);
  input a,b;
  wire a,b;
  output c,s;
  wire c,s;

  specify
    (a >= c) = 1;
    (a >= s) = 2;
    (b >= c) = 1;
    (b >= s) = 2;
  endspecify
```

Continued

```
  xor x1(s,a,b);
  and a1(c,a,b);
endmodule
```

Not all simulators support `specify` blocks. For more information, check the documentation for your simulator, or see the book by Palnitkar mentioned at the end of this chapter.

6.4 Abstracting propagation delay

As the previous sections illustrate, once a design has been synthesized down to the gate level, Verilog can provide a fairly accurate model of propagation delay. A problem arises if one wishes to estimate propagation delay before synthesis. For a given technology, manufacturers usually publish a priori estimates of worst case propagation delays for bus-width building blocks (such as adders). We would like to be able to use such worst case estimates to simulate the propagation delay of an architecture when it is still at the mixed stage (block diagram). The problem is that the propagation delay of a physical bus-width device exhibits itself only as specific hazards (like those illustrated in section 6.3.3) that require a synthesized netlist to be simulated.

This section illustrates how Verilog can be used to model abstractly the propagation delay of a bus-width device. The correct Verilog code for doing this uses some relatively advanced features of Verilog. To motivate the need for these features, we will first consider some incorrect attempts at modeling propagation delay.

6.4.1 Inadequate models for propagation delay

The simplest Verilog code for a bus-width device that includes some notation of propagation delay is similar to the code given in 6.3.3, except that the port sizes are defined by the first parameter, and the propagation delay is defined by the second parameter:

```
module adder(s,a,b);
parameter SIZE = 1, DELAY = 0;
output s;
input a,b;
reg [SIZE-1:0] s;
wire [SIZE-1:0] a,b;
  always @(a or b)
    # DELAY s=a+b;
endmodule
```

As explained in section 6.3.3, this code is deficient because it does not model cases where there is a hazard but instead always models the error as a lag.

How should a hazard be represented abstractly? The specific value that presents itself when a hazard occurs can only be predicted from the synthesized netlist. Instead, at the abstract level, we will use `'bx` to represent the hazard.

```
module adder(s,a,b);
parameter SIZE = 1, DELAY = 0;
output s;
input a,b;
reg [SIZE-1:0] s;
wire [SIZE-1:0] a,b;
  always @(a or b)
    begin
      s = 'bx;
      # DELAY s=a+b;
    end
endmodule
```

Although this is an improvement, the above still has a flaw. To see why it is deficient, consider the following design which instantiates the above `adder` twice:

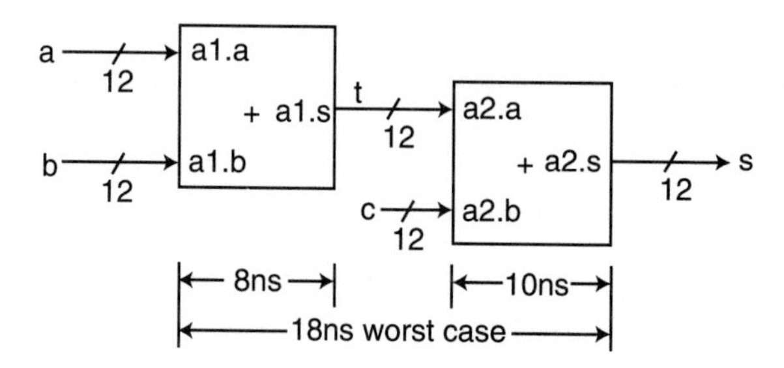

Figure 6-2. Two instantiations of `adder`.

The following test code gives an example how the last definition of `adder` gives a misleading result from simulation:

```
module test;
  reg [11:0] a,b,c;
  wire [11:0] t,s;
  adder #(12,8)  a1(t,a,b);
  adder #(12,10) a2(s,t,c);
  initial
    begin
          a = 0;
          b = 0;
          c = 0;
      #30 a = 100;
          b = 20;
          c = 3;
      #40 $finish;
    end
  always @(s)
    $display("$time=%d s=%d",$time,s);
  always @(t)
    $display("$time=%d t=%d",$time,t);
endmodule
```

This is illustrated by the simulation produced from the above test code:

```
$time=          0 t=x
$time=          0 s=x
$time=          8 t=   0
$time=         10 s=   0
$time=         30 t=x
$time=         30 s=x
$time=         38 t= 120
$time=         40 s= 123
```

The test code instantiates two adders. The first instance, `a1`, has a propagation delay of 8, and the second instance, `a2`, has a delay of 10. At `$time` 30, the test code causes `a`, `b` and `c` to change. A worst case analysis indicates that it should take 18 additional `$time` units (`$time`=48) to produce the sum of 100+20+3; however the simulation shows the correct sum in only 10 `$time` units (`$time`=40).

This flaw exists because the `always` block for `a2` is still delaying (#10) when the change in `t` (also known as `a2.a`) occurs at `$time`=38. Rather than delaying an additional 10 units of `$time` from `$time`=38, Verilog simply returns to the @ time control at `$time` 40. Since `a2.a` is stable at `$time` 40, the algorithmically correct answer 123 is available too soon. This Verilog model allows us to conclude that the machine could run faster than is physically possible.

6.4.2 Event variables in Verilog

Verilog provides special variables, known as `event` variables, that are helpful in overcoming the problems shown in the last section. Variables declared as events are used only in two places: First, in a triggering statement:

```
        event e;
        ...
        ->e;
```

and, second, in time control:

```
        always @ e
          ...
```

Note: There are no parentheses around the variable in the @ time control. The `->` triggers the corresponding @ to be scheduled. For example, the following prints "10" and "30":

```
event e;
 initial
   begin
     #10;
     -> e;
     #20
     -> e;
   end
 always @ e
   $display($time);
```

Here is an example of how an event could be used to model the adder:

```
module adder(s,a,b);
  parameter SIZE = 1, DELAY = 0;
  output s;
  input a,b;
  reg [SIZE-1:0] s;
  wire [SIZE-1:0] a,b;
  event change;

  always @(a or b)
    start_change;

  always @change
    # DELAY s=a+b;

  task start_change;
    begin
      s = 'bx;
      -> change;
    end
  endtask
endmodule
```

Unfortunately, the above produces the same incorrect model of the adder (the correct result 123 is available too soon) for the same reasons discussed in the previous section.

6.4.3 The `disable` statement

This statement causes Verilog to cease execution of a labeled block. For `always` blocks, it causes them to restart at the top. (It has another use, explained in section 7.5.) An optional label is given after a colon on a `begin` statement. Here is how the `disable` statement overcomes the problem shown in the previous section:

```
module adder(s,a,b);
  parameter SIZE = 1, DELAY = 0;
  output s;
  input a,b;
  reg [SIZE-1:0] s;
  wire [SIZE-1:0] a,b;
  event change;

  always @(a or b)
       start_change;
```

Continued

```
   always @change
         begin : change_block
           # DELAY s=a+b;
         end

   task start_change;
         begin
           s = 'bx;
           disable change_block;
           #0;
           -> change;
         end
      endtask
endmodule
```

The task `start_change` can be called many times from the first `always` block without `$time` advancing. This way, every change in the inputs will be noticed by the Verilog scheduler. The only # control in `start_change` is #0. This is required so the `disable` statement can take effect. After `change_block` has been disabled, the `change` event is retriggered. This, in turn, causes the full # DELAY before the output changes from `'bx`.

The #DELAY is in the block (`change_block`) which can be disabled. There is no way that changes that occur in the middle of a #DELAY will be missed. Therefore, instantiating a series of these `adders` will produce a correct model of the propagation delay. For example, here is the simulation using the same test code as section 6.4.1:

```
$time=          0 t=x
$time=          0 s=x
$time=          8 t=   0
$time=         18 s=   0
$time=         30 t=x
$time=         30 s=x
$time=         38 t= 120
$time=         48 s= 123
```

Note that `s` is `'bx` from `$time` 30 until `$time` 48, as is predicted by worst case timing analysis.

6.4.4 A clock with a PERIOD parameter

The reason we wish to simulate propagation delays is ultimately to design faster machines that are clocked at higher frequencies. Therefore, it is desirable to have a clock with PERIOD as a parameter:

```
module cl(clk);
     parameter TIME_LIMIT = 110000,
               PERIOD = 100;
     output clk;
     reg clk;
     initial
       clk = 0;
     always
       begin
         #(PERIOD/2) clk = ~clk;
         #(PERIOD-PERIOD/2) clk = ~clk;
       end
     always @(posedge clk)
       if ($time > TIME_LIMIT) #(PERIOD-1) $stop;
endmodule
```

Note that if PERIOD is omitted in the instantiation, it will default to 100, as has been the situation in all earlier simulations.

6.4.5 Propagation delay in the division machine

Suppose the propagation delays are 70 for the ALU, 25 for the mux, 20 for the comparator and 10 for the inverter. To backannotate this in the original code of section 4.4.5, simply include the propagation delay parameter with the instantiation:

```
module slow_div_arch(aluctrl,muxctrl,ldr1,clrr2,
                     incr2,ldr3,r1gey,x,y,r3bus,sysclk);
  input aluctrl,muxctrl,ldr1,clrr2,incr2,ldr3,x,y,sysclk;
  output r1gey,r3bus;
  wire [5:0] aluctrl;
  wire muxctrl,ldr1,clrr2,incr2,ldr3,r1gey,sysclk;
  wire [11:0] x,y,r3bus;
  wire [11:0] muxbus,alubus,r1bus,r2bus;

  enabled_register #12  r1(alubus,r1bus,ldr1,sysclk);
  mux2             #(12,25)  mx(x,y,muxctrl,muxbus);
  alu181           #(12,70) alu(r1bus,muxbus,aluctrl[5:2],
                             aluctrl[1],aluctrl[0],,alubus,);
  comparator       #(12,20)  cmp(r1lty,,,r1bus,y);
  not              #10       inv(r1gey,r1lty);
```

```
   counter_register #12  r2(,r2bus,,1'b0,incr2,clrr2,sysclk);
   enabled_register #12  r3(r2bus, r3bus,ldr3,sysclk);
   ...
endmodule
```

When this is simulated with a clock period of 100, it works:

```
   cl #(20000,100) clock(sysclk);
   slow_div_system  slow_div_machine(pb,ready,x,y,
           quotient,reset,sysclk);
```

as is illustrated by the following timing diagram:

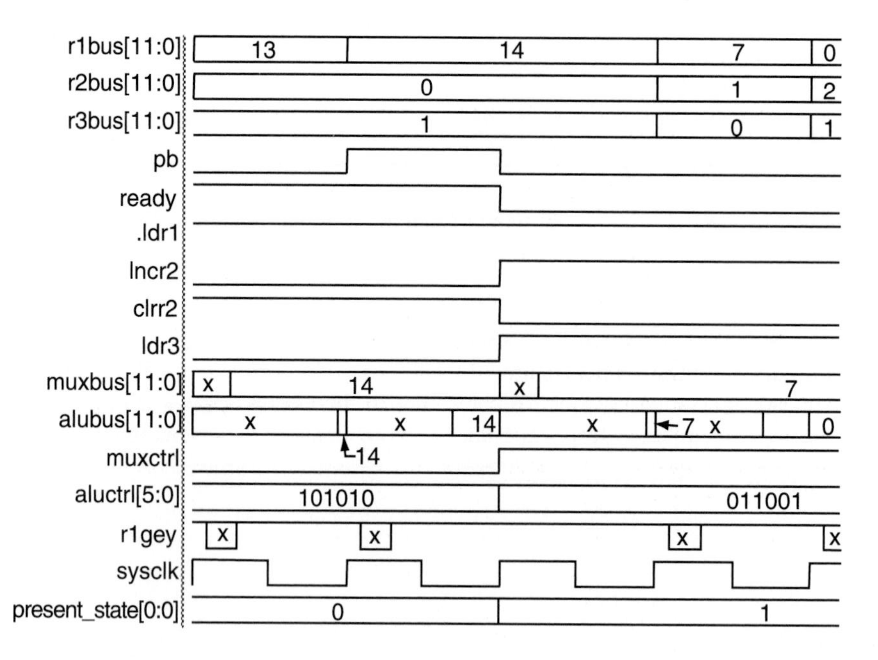

Figure 6-3. Timing diagram for division machine with abstract propagation delay.

A designer might want to experiment to see if the clock can be speeded up to have a period of, say, 90:

```
cl #(20000,90) clock(sysclk);
slow_div_system  slow_div_machine(pb,ready,
   x,y,quotient,reset,sysclk);
```

The test code will detect an error:

```
      6365 r1=x r2=    2 r3=     1 pb=0 ready=1
              1    10        0      muxbus=x alubus=x  x= 14 r1gey=x
                                    muxctrl=0   aluctrl=101010
error x=   14 y=     7 x/y=    2 quotient=   1
```

because the ALU will not have had a chance to stabilize by the time of the rising edge of the clock.

6.5 Single cycle, multi-cycle and pipeline

The solution to many problems involves performing the same kind of computation on large amounts of independent data values. The term *independent* means that the result of doing computation on one data value does not affect nor is affected by doing the computation on any of the other data values. For example, three-dimensional computer graphics (such as occur in virtual reality systems) usually require evaluating the same formulae at millions of points. The order in which such points are processed does not matter. You will get the same answer if you start processing on the lower-left-hand side of the screen as you will get if you start processing on the upper-right-hand of the screen, or if you process in any other order that you might choose. In problems like this that have complete data independence, many possible hardware solutions exist. All such hardware solutions are correct in that they all eventually arrive at the desired result. These hardware solutions differ in terms of their speed and cost.

If cost were not a constraint, problems with totally independent data values could be solved by building one combinational logic machine for each data value to be processed. Each such machine could compute its answer in parallel to all the other machines. Although this kind of massively parallel approach is sometimes used, it is not practical in many situations due to cost constraints.

Because practical problems with perfectly independent data are commonplace where cost is as or more important than speed, three standard techniques have been developed that allow the designer to choose the trade-off between speed and cost. These three techniques are known as the single-cycle, pipelined, and multi-cycle approaches. What these three techniques share in common is that no more than one complete result is produced per clock cycle.

At one extreme is the single-cycle approach. With the *single-cycle* approach, the result for one of the independent values is completely computed (start to finish) in a single clock cycle. The single-cycle approach is perhaps the most natural approach to think about, but it is usually not the most efficient.

At the opposite extreme is the multi-cycle approach. With the *multi-cycle* approach, the result for one of the independent values requires several clock cycles to be computed. Thinking about the multi-cycle approach is quite analogous to thinking about the software paradigm (section 2.2.2). The multi-cycle approach is usually the slowest, but it often requires minimum hardware because it can be implemented with a central ALU (section 2.3.4).

In between the single-cycle approach and the multi-cycle approach is the pipelined approach. The *pipelined* approach usually requires more hardware than the other approaches but often is the fastest and most efficient. In order to understand the pipelined approach, it is necessary to investigate the two other approaches first.

As discussed earlier in this chapter, the total time required by a machine is the number of clock cycles multiplied by the clock period. The three approaches discussed in this section differ both in terms of the number of clock cycles required and the clock period. We can understand the algorithmic distinctions among these three approaches at the behavioral stage and even predict the number of clock cycles required at the behavioral stage; however, we cannot predict which approach will be fastest at the behavioral stage. This is because the clock period is determined by the propagation delay in the architecture, which we cannot predict until the mixed stage, or when the hardware has been synthesized.

6.5.1 Quadratic polynomial evaluator example

The quadratic polynomial `a*x*x + b*x + c` is a simple example of a formula that a machine might evaluate many times with different values of `x`, but the same values of `a`, `b` and `c` (which remain unchanged for a suitable period before, during and after the quadratic evaluations). For each unique `x` value, the computation of the quadratic formula is independent of the computation for other values of `x`. Although the formulae used for practical problems, such as computer graphics, are more complex than this familiar old quadratic, the nature of the formulae in such practical problems is very similar to this quadratic.

Although a practical machine would probably store the `x` values in a synchronous memory, for the sake of simplicity in this example, assume the values of `x` are contained in a ROM. The goal of the machine is to evaluate the quadratic polynomial for each of these `x` values and store the corresponding `y` values into a synchronous memory (see section 8.2.2.3.1 for how a synchronous memory can be implemented). The address where each y value is stored should be the same as the address in the ROM from which the corresponding x value was fetched.

Suppose a is 1, b is 2 and c is 3. If the contents of x are as shown on the left in decimal, when the machine is done, the contents of the y memory will be as shown on the right:

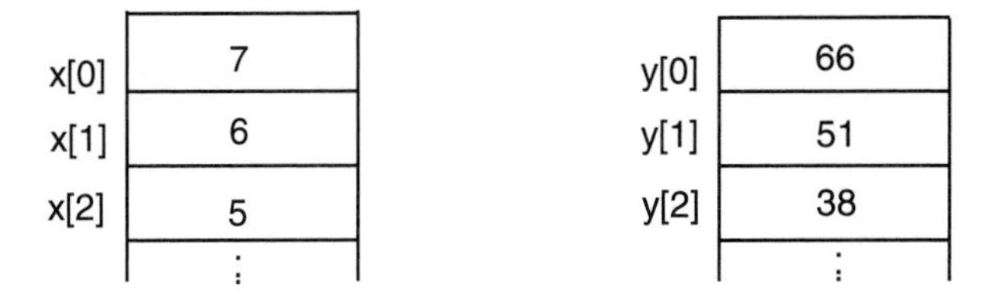

Figure 6-4. Example contents of x and y for quadratic machine.

For example, `y[2] = a*x[2]*x[2] + b*x[2] + c = 25+10+3 = 38.`

The machine will use a push button interface, similar to the one described in section 2.2.1. The machine will wait in state IDLE until pb is pressed. While the machine is in state IDLE, it leaves the contents of y alone. Some time after the machine leaves state IDLE it will begin to fill y with the correct results.

The following sections will look at behavioral ASMs to illustrate how this machine can be implemented with the single-cycle, multi-cycle and pipelined approaches. Several incorrect versions of pipelining will be presented before the final correct pipelined solution is shown in section 6.5.7.

6.5.2 Behavioral single cycle

The ASM chart for the single-cycle approach is quite simple and obvious. The machine only needs to have a memory address (ma) register. The ma register in the single- cycle approach provides the address used by both the x and y during each clock cycle. In each clock cycle, the content of the ROM is fetched from the address indicated by ma, the quadratic is evaluated using the value of x fetched from the ROM and the result is stored into y at the same address indicated by ma.

Suppose the maximum memory address is the constant MAXMA. The first of the following two ASM charts (figure 6-5) describes the single-cycle approach in the simplest possible fashion. The computation of the quadratic actually involves several multiplications and additions. The second (equivalent) ASM chart (figure 6-6) makes this clearer:

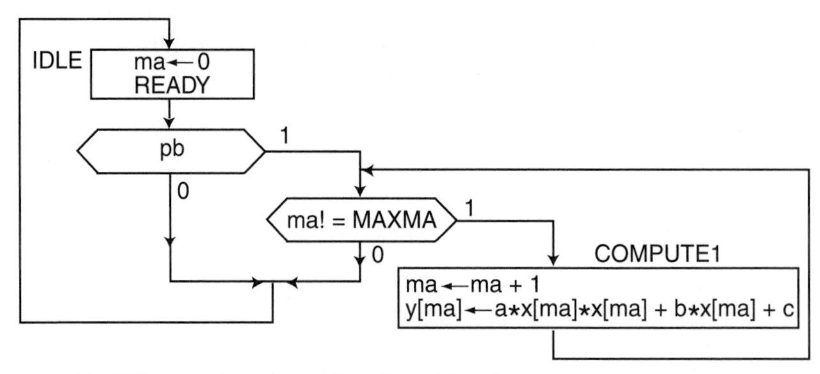

Figure 6-5. Behavioral single cycle ASM with only ←.

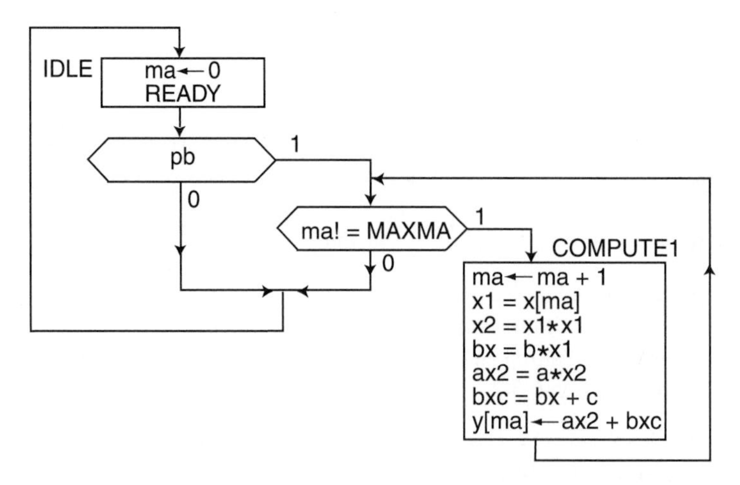

Figure 6-6. Equivalent single cycle ASM with = for combinational logic.

Note the use of = rather than ← for the intermediate results (`x1,x2,bx,ax2` and `bxc`). As discussed in chapter 2, the = means that combinational logic computes all of these values in one clock cycle. Note that `x2` and `bx` are dependent on `x1`; `ax2` is dependent on `x2`; and `bxc` is dependent on `bx`. This means that the minimum clock period for the single-cycle approach must allow enough time for the computations of all of these intermediate results to stabilize. The amount of time it takes for the combinational logic to finish computing these intermediate values is not something we can predict at the behavioral stage. Up until this chapter, we have neglected such propagation delays, but later in this chapter, we will estimate what these delays will be.

Since we do not know how fast the machine can be clocked, let us assume that the clock period is 100 units of Verilog `$time` for the purpose of the following and later simulations. Also, for reasons to be explained later, we will assume that each word in `y` is initialized to `'bz` prior to execution of this ASM. In the following partial simulation output, the `$time` and registers are printed on one line with the contents of `y` on the following line:

```
 349 ps=000 ma= 0 x1=x     x2=x  bx=x  ax2=x  bxc=x
        z   z   z   z   z   z   z   z
 449 ps=001 ma= 0 x1=    7 x2=49 bx=14 ax2=49 bxc=17
        z   z   z   z   z   z   z   z
 549 ps=001 ma= 1 x1=    6 x2=36 bx=12 ax2=36 bxc=15
       66   z   z   z   z   z   z   z
   ...
1149 ps=001 ma= 7 x1=    0 x2= 0 bx= 0 ax2= 0 bxc= 3
       66  51  38  27  18  11   6   z
1249 ps=000 ma= 8 x1=x     x2=x  bx=x  ax2=x  bxc=x
       66  51  38  27  18  11   6   3
```

In state COMPUTE1 at `$time` 449, the machine obtains the value (7) of `x1` from the ROM. After getting this from the ROM, but during the same clock cycle, the machine computes the square (49), and the product (49) of `a` and the square. Also in this clock cycle, the machine computes the product (`bx=14`) of `b` and `x1`. After computing `bx`, but before the end of the clock cycle, the machine computes the sum (`bxc=17`) and the sum (`ax2+bxc`). This final result (66) is scheduled to be stored in the memory. This value appears at the correct place (`y[0]`) by `$time` 549.

The number of clock periods required for this single cycle ASM to complete is MAXMA+1, because one result is produced each clock cycle.

Here is the behavioral Verilog code used to produce the above simulation:

```
`define NUM_STATE_BITS 3
`define IDLE           3'b000
`define COMPUTE1       3'b001
module poly_system(pb,a,b,c,ready,sysclk);
 input pb,a,b,c,sysclk;
 output ready;
 wire pb,sysclk;
 wire [11:0] a,b,c;
 reg [11:0] x[`MAXMA:0],y[`MAXMA:0];
 reg ready;
 reg [11:0] ma;
```

Continued

```
 reg [11:0] x1,x2,bx,ax2,bxc;
 reg ['NUM_STATE_BITS-1:0] present_state;
 integer i;
 initial
  begin
   for (i=0;i<='MAXMA;i=i+1)
    begin
     x[i]='MAXMA-i;
     y[i]='bz;
    end
  end

 always
  begin
   @(posedge sysclk) enter_new_state('IDLE);
    ma <= @(posedge sysclk) 0;
    ready = 1;
    if (pb)
     begin
      while (ma != 'MAXMA)
       begin
        @(posedge sysclk) enter_new_state('COMPUTE1);
         ma <= @(posedge sysclk) ma + 1;
         x1 = x[ma];
         x2 = x1*x1;
         bx = b*x1;
         ax2 = a*x2;
         bxc = bx + c;
         y[ma] <= @(negedge sysclk) ax2 + bxc;
       end
     end
  end
endmodule
```

Note that the order of the intermediate computations (=) matters in Verilog.

The non-blocking assignment to the memory location, `y[ma]`, uses `@(negedge sysclk)` rather than the `@(posedge sysclk)` typical for non-blocking assignment to ordinary registers (see section 3.8.2). The problem here arises because new values are stored into *distinct* elements of `y` during *every clock cycle.* Some simulators will do the proper thing in a situation like this even if you were to use:

```
ma <= @(posedge sysclk) ma + 1;
...
 y[ma] <= @(posedge sysclk) ax2 + bxc; // not portable
```

However, `@(negedge sysclk)` is necessary to produce the correct result on other Verilog simulators. To understand why, remember that there are two separate concepts in Verilog: sequence and `$time`. A non-blocking assignment by itself will cause the new value to be stored as the last event at the specified `$time`. All Verilog simulators save the right-hand expressions (`ax2 + bxc` and `ma + 1` in this example) until the specified clock edge. For left-hand values that are not arrays, this is sufficient to model the behavior of synchronous registers. But when you are dealing with a synchronous memory (`y[ma]`), a Verilog simulator must also save the address (`ma`) for the left-hand memory until the next clock edge. The sequence in which a simulator will update `ma` and `y[ma]` is not defined. The Verilog standard is ambiguous on this issue, and some vendors, for reasons of efficiency, have chosen not to save the address.

To overcome this problem at the pure behavioral stage, we need to remember that the statements in a particular state will execute one unit of `$time` after the rising edge. The falling edge of the system clock will occur prior to the next rising edge, but after all non-blocking assignments of this state have been scheduled. Therefore, `ma` will still have the correct value, but changing `y[ma]` at the falling edge will not disturb some dependent computations[1] within this state.

This is another illustration of choosing an appropriate level of abstraction (see section C.1). In the pure behavioral stage, we are only interested in the values of registers and memories at the moment of the rising edge. The `negedge` memory approach simulates the values at the rising edge properly on *all* Verilog simulators. How a particular simulator arranges to simulate between the rising edges is irrelevant as the designer explores different algorithmic possibilities. Of course, in the actual architecture, a synchronous memory will change its values only at the rising edge, *synchronously* to the ordinary registers in the design. At the mixed or later stages of design, such details are significant to determine the proper clock frequency, but since the mixed stage removes all <=, there is no problem. Before the mixed stage, the designer should not worry about clock frequency; thus the `negedge` memory approach is perfectly acceptable.

[1] There are no dependent computations in this example.

6.5.3 Behavioral multi-cycle

The single-cycle approach discussed in the previous section does all of the intermediate computations in one clock cycle. The multi-cycle approach, on the other hand, does each intermediate computation in a separate clock cycle. This of course means it takes several clock cycles to produce one result. In the multi-cycle approach, each intermediate result is stored in a register, thus the ASM uses the ← notation. The multi-cycle approach is like the software paradigm (section 2.2.2), where each intermediate computation occurs in a separate rectangle by itself. Here is an ASM chart for the multi-cycle version of the quadratic evaluator:

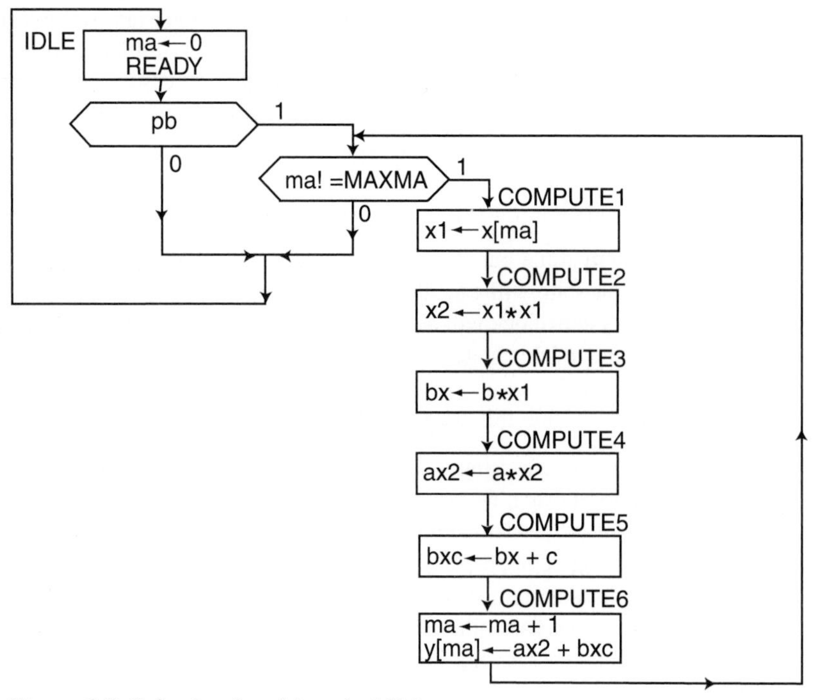

Figure 6-7. Behavioral multi-cycle ASM.

This machine has six registers (`ma, x1, x2, bx, ax2` and `bxc`), and has six states inside the loop. Here is a partial simulation, again assuming a clock period of 100 (which may be much longer than is actually required):

```
 349 ps=000 ma= 0 x1=x  x2=x  bx=x  ax2=x  bxc=x
           z   z   z   z   z   z   z   z
 449 ps=001 ma= 0 x1=x  x2=x  bx=x  ax2=x  bxc=x
           z   z   z   z   z   z   z   z
 549 ps=010 ma= 0 x1= 7 x2=x  bx=x  ax2=x  bxc=x
           z   z   z   z   z   z   z   z
 649 ps=011 ma= 0 x1= 7 x2=49 bx=x  ax2=x  bxc=x
           z   z   z   z   z   z   z   z
 749 ps=100 ma= 0 x1= 7 x2=49 bx=14 ax2=x  bxc=x
           z   z   z   z   z   z   z   z
 849 ps=101 ma= 0 x1= 7 x2=49 bx=14 ax2=49 bxc=x
           z   z   z   z   z   z   z   z
 949 ps=110 ma= 0 x1= 7 x2=49 bx=14 ax2=49 bxc=17
           z   z   z   z   z   z   z   z
1049 ps=001 ma= 1 x1= 7 x2=49 bx=14 ax2=49 bxc=17
          66   z   z   z   z   z   z   z
 ...
5249 ps=000 ma= 8 x1= 0 x2= 0 bx= 0 ax2= 0 bxc= 3
          66  51  38  27  18  11   6   3
```

In state COMPUTE1 around `$time` 449, `x1` is scheduled to be loaded with `x[ma]=x[0]=7`. This change in `x1` shows up by `$time` 549. In state COMPUTE2 around `$time` 549, the square (49) of this value is scheduled to be loaded into `x2`, which shows up by `$time` 649. In state COMPUTE3 around `$time` 649, the product (14) of `x1` and `b` is scheduled to be loaded into `bx`, which shows up by `$time` 749. In state COMPUTE4 around `$time` 749, the product (49) of `a` and `x2` is scheduled to be loaded into `ax2`, which shows up by `$time` 849. In state COMPUTE5 around `$time` 849, the sum of `bx` and `c` (17) is scheduled to be loaded into `bxc`, which also shows up by `$time` 949. Finally, in state COMPUTE6 around `$time` 949, the sum of `ax2` and `bxc` is scheduled to be loaded into `y`, and this shows up by `$time` 1049.

The number of clock periods required for this multi-cycle ASM to complete is 6*(MAXMA+1). Although at first glance, this appears much slower than the single-cycle approach of section 6.5.2, it need not be that much slower. Later we will be able to predict the propagation delay of the architecture for the multi-cycle approach (which determines the maximum clock frequency). Since there is less computation being done in each clock cycle, it should be possible to clock the multi-cycle machine faster than the single-cycle machine. The relative performance of these two machines is something we can only predict given a structural architecture.

Here is the behavioral Verilog code used to produce the above simulation:

```
`define NUM_STATE_BITS 3
`define IDLE           3'b000
`define COMPUTE1       3'b001
`define COMPUTE2       3'b010
`define COMPUTE3       3'b011
`define COMPUTE4       3'b100
`define COMPUTE5       3'b101
`define COMPUTE6       3'b110
...
 always
  begin
   @(posedge sysclk) enter_new_state(`IDLE);
    ma <= @(posedge sysclk) 0;
    ready = 1;
    if (pb)
     begin
      while (ma != `MAXMA)
       begin
        @(posedge sysclk) enter_new_state(`COMPUTE1);
         x1 <= @(posedge sysclk) x[ma];
        @(posedge sysclk) enter_new_state(`COMPUTE2);
         x2  <= @(posedge sysclk) x1*x1;
        @(posedge sysclk) enter_new_state(`COMPUTE3);
         bx  <= @(posedge sysclk) b*x1;
        @(posedge sysclk) enter_new_state(`COMPUTE4);
         ax2 <= @(posedge sysclk) a*x2;
        @(posedge sysclk) enter_new_state(`COMPUTE5);
         bxc <= @(posedge sysclk) bx + c;
        @(posedge sysclk) enter_new_state(`COMPUTE6);
         ma <= @(posedge sysclk) ma + 1;
         y[ma] <= @(negedge sysclk) ax2 + bxc;
       end
     end
  end
```

6.5.4 First attempt at pipelining

The single-cycle approach puts all the computation steps into one clock cycle but uses = (corresponding only to combinational logic) for the intermediate results. The multi-cycle approach spreads the computation steps across separate clock cycles, but uses $\leftarrow$ (corresponding to registers) for the intermediate results. The pipelined approach is half-way between these two approaches.

The pipelined approach puts all the computation steps into one clock cycle and uses ← (corresponding to registers) for the intermediate results. This means, unlike the other two approaches, each intermediate computation in the pipelined approach occurs in parallel to the other intermediate computations. The only reason that the intermediate computations can occur in parallel is that a machine like this is processing a large amount of independent data in an identical fashion.

A pipelined machine is very much like a factory assembly line. Factories are efficient because they mass produce many copies of an identical item. At each point in time, each worker in the factory does one thing to a partially assembled item on the production line. For example:

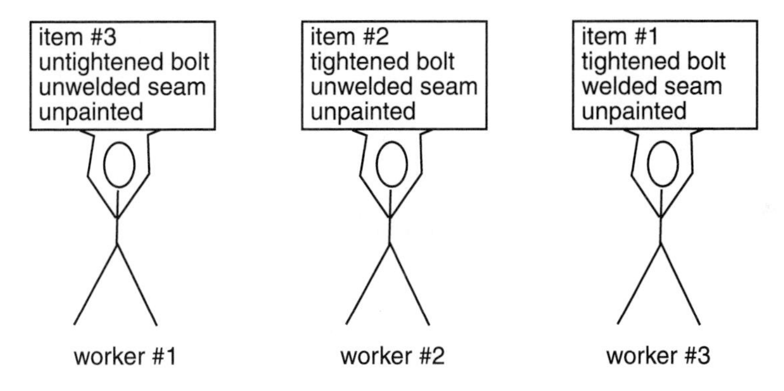

Figure 6-8. Analogy to factory.

Worker #1 might tighten a bolt, worker #2 might weld a seam and worker #3 might paint the item. Each worker acts in parallel to the other workers. In the above picture, worker #1 is tightening the bolt on item #3 at the same time that worker #2 is welding item #2 (which already has its bolt tightened) and that worker #3 is painting item #1 (which has it bolt tightened and which has been welded). Each item has experienced the correct sequence in order (tightening, welding and painting), but the tightening, welding and painting that happens at any given instance occurs to independent items.

With this analogy in mind, we can understand that each of the intermediate computations (←) in the pipelined quadratic evaluator produces an intermediate result derived from a different `x` value. Here is a first (somewhat flawed) attempt to describe the factory-like operation of this pipelined system:

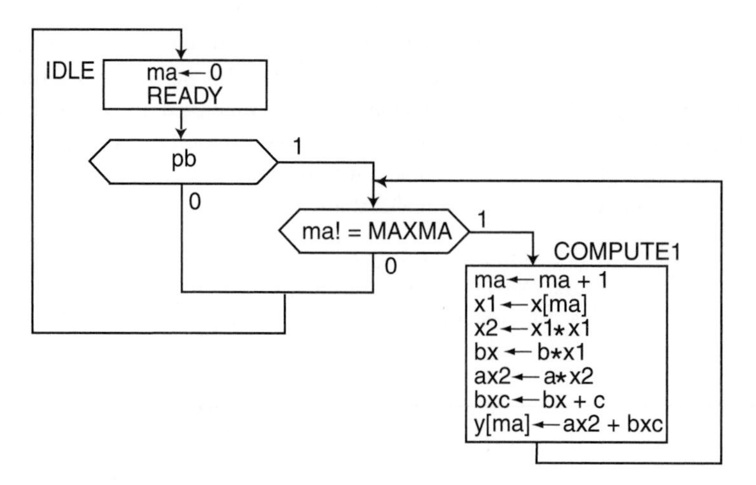

Figure 6-9. Incorrect pipelined ASM.

Here is a simulation showing how this ASM malfunctions:

```
 349 ps=000 ma= 0 x1=x  x2=x  bx=x    ax2=x    bxc=x
       z   z   z   z   z   z   z   z
 449 ps=001 ma= 0 x1=x  x2=x  bx=x    ax2=x    bxc=x
       z   z   z   z   z   z   z   z
 549 ps=001 ma= 1 x1= 7 x2=X  bx=4094 ax2=4095 bxc=x
       x   z   z   z   z   z   z   z
 649 ps=001 ma= 2 x1= 6 x2=49 bx=  14 ax2=   1 bxc= 1
       x   x   z   z   z   z   z   z
 749 ps=001 ma= 3 x1= 5 x2=36 bx=  12 ax2=  49 bxc=17
       x   x   2   z   z   z   z   z
 849 ps=001 ma= 4 x1= 4 x2=25 bx=  10 ax2=  36 bxc=15
       x   x   2  66   z   z   z   z
 949 ps=001 ma= 5 x1= 3 x2=16 bx=   8 ax2=  25 bxc=13
       x   x   2  66  51   z   z   z
1049 ps=001 ma= 6 x1= 2 x2= 9 bx=   6 ax2=  16 bxc=11
       x   x   2  66  51  38   z   z
1149 ps=001 ma= 7 x1= 1 x2= 4 bx=   4 ax2=   9 bxc= 9
       x   x   2  66  51  38  27   z
1249 ps=000 ma= 8 x1= 0 x2= 1 bx=   2 ax2=   4 bxc= 7
       x   x   2  66  51  38  27  18
```

There are several problems with this, but before discussing what is wrong, let us consider what is almost right. In state COMPUTE1 around `$time` 449, `x1` is scheduled to be loaded with `x[ma]=x[0]=7`. This change in `x1` shows up by `$time` 549. Around `$time` 549, the square (49) of this value is scheduled to be loaded into `x2`, which shows up by `$time` 649. In parallel, the product (14) of `x1` and `b` is scheduled to be loaded into `bx`, which also shows up by `$time` 649. Around `$time` 649, the product (49) of `a` and `x2` is scheduled to be loaded into `ax2`, which shows up by `$time` 749. In parallel, the sum of `bx` and `c` (17) is scheduled to be loaded into `bxc`, which also shows up by `$time` 749. Finally, around `$time` 749, the sum of `ax2` and `bxc` is scheduled to be loaded into `y`, and this shows up by `$time` 849.

The problem at `$time` 849 is that although 66 is the correct value, it shows up at the wrong address for `y`. This is because `ma` has necessarily been incremented in each of the clock cycles. This machine has forgotten that the 66 is supposed to stored at y[0]. Instead it stores it at y[3]. Aside from the fact that the addresses to `y` are offset by three, the machine continues to compute a correct result each clock cycle. By `$time` 949, 51 is stored into `y`. By `$time` 1049, 38 is stored into `y`, and so forth.

In addition to storing the correct results at the wrong addresses, this machine also has another flaw. It does not finish the complete job (storing 11, 6 and 3). The intermediate results needed to produce 11, 6 and 3 are left frozen in the pipeline when the machine returns to state IDLE.

Another, less obvious, flaw is that garbage values (`'bx`, `'bx` and `2`) are stored into the memory during the first clock cycles (449, 549 and 649). Initializing `y` to `'bz` rather than `'bx` to highlights this flaw.

6.5.5 Pipelining the ma

The major problem with the ASM chart of section 6.5.4 is that the memory address used to store into `y` does not correspond to the value being stored. To overcome this problem, we can introduce three additional registers, `ma1`, `ma2` and `ma3`, that will save the memory addresses from the previous three clock cycles. In a given clock cycle, `ma1` is the value of `ma` one clock cycle ago, `ma2` is the value of `ma` two clock cycles ago, and `ma3` is the value of `ma` three clock cycles ago.

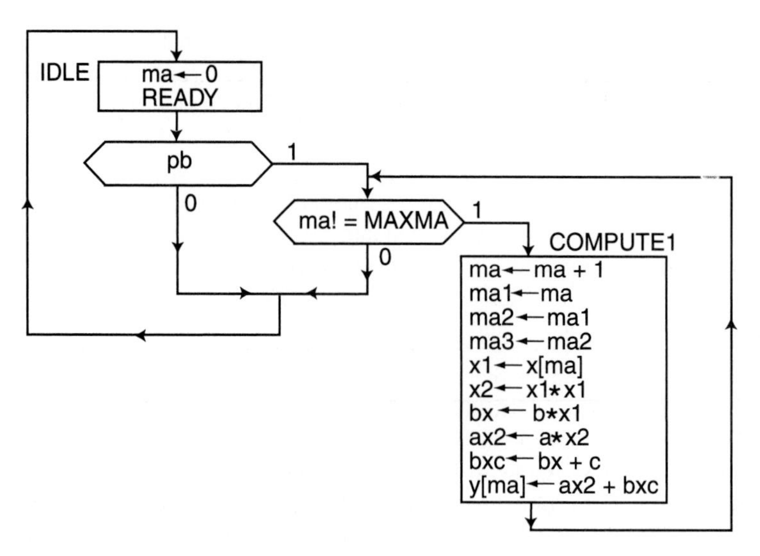

Figure 6-10. Pipelined ASM with multiple addresses but without flush.

Here is a simulation that shows how addresses flow through the `ma` pipeline:

```
349 ps=000 ma=   0 x1=x x2=x bx=x
           ma1=   0 ax2=x bxc=x ma2=    0 ma3=    0
     z   z   z   z   z   z   z   z
449 ps=001 ma=   0 x1=x x2=x bx=x
           ma1=   0 ax2=x bxc=x ma2=    0 ma3=    0
     z   z   z   z   z   z   z   z
549 ps=001 ma=   1 x1=    7 x2=X bx=4094
           ma1=   0 ax2=4095 bxc=x ma2=    0 ma3=    0
     x   z   z   z   z   z   z   z
649 ps=001 ma=   2 x1=    6 x2=   49 bx=   14
           ma1=   1 ax2=    1 bxc=    1 ma2=    0 ma3=    0
     x   z   z   z   z   z   z   z
749 ps=001 ma=   3 x1=    5 x2=   36 bx=   12
           ma1=   2 ax2=   49 bxc=   17 ma2=    1 ma3=    0
     2   z   z   z   z   z   z   z
849 ps=001 ma=   4 x1=    4 x2=   25 bx=   10
           ma1=   3 ax2=   36 bxc=   15 ma2=    2 ma3=    1
    66   z   z   z   z   z   z   z
949 ps=001 ma=   5 x1=    3 x2=   16 bx=    8
           ma1=   4 ax2=   25 bxc=   13 ma2=    3 ma3=    2
    66  51   z   z   z   z   z   z
```

Continued.

```
1049 ps=001 ma=    6 x1=    2 x2=     9 bx=     6
              ma1=    5 ax2=   16 bxc=    11 ma2=     4 ma3=     3
      66   51   38    z    z    z    z    z
1149 ps=001 ma=    7 x1=    1 x2=     4 bx=     4
              ma1=    6 ax2=    9 bxc=     9 ma2=     5 ma3=     4
      66   51   38   27    z    z    z    z
1249 ps=000 ma=    8 x1=    0 x2=     1 bx=     2
              ma1=    7 ax2=    4 bxc=     7 ma2=     6 ma3=     5
      66   51   38   27   18    z    z    z
```

Although the addition of `ma1`, `ma2` and `ma3` solves the addressing problem, the revised ASM still does not finish the complete job (storing 11, 6, and 3). As was the case in the ASM of section 6.5.4, the intermediate results needed to produce 11, 6 and 3 are left frozen in the pipeline when the machine returns to state IDLE. Also, garbage values (`'bx, 'bx` and `2`) are still stored into the memory during the first clock cycles, although now they are stored in `y[0]` each time.

6.5.6 Flushing the pipeline

In order to prevent the final values from being frozen in the pipeline, there need to be some additional clock cycles spent "flushing" those values out of the pipeline. Returning to the factory analogy, when the factory is about to cease production of a particular model item, worker #1 can stop work earliest, but the other workers must finish their tasks on the last item worker #1 tightened. Similarly, worker #2 can stop before worker #3. So it is with flushing the pipeline.

The ASM needs three states, FLUSH1, FLUSH2 and FLUSH3, that perform the required computations on the valid data in the pipeline. For those registers that have valid data, the computations are identical to those in state COMPUTE1. At each successive flushing state, there are fewer registers in the pipeline that contain valid data; thus each successive state has fewer computations to perform.

6.5.7 Filling the pipeline

The reason that garbage values have been stored by all the previous pipeline attempts is because of the assignment to `y[ma3]` in state COMPUTE1. During the first clock cycles when state COMPUTE1 executes, `ma3`, `ax2` and `bxc` do not have legitimate values. Therefore, to store `ax2+bxc` at address `ma3` is illegitimate. The situation is the opposite problem from flushing the pipeline. In this case, the pipeline must be filled prior to storing the first result in y. The following is a completely correct pipelined ASM that accomplishes this operation by introducing states FILL1, FILL2 and FILL3:

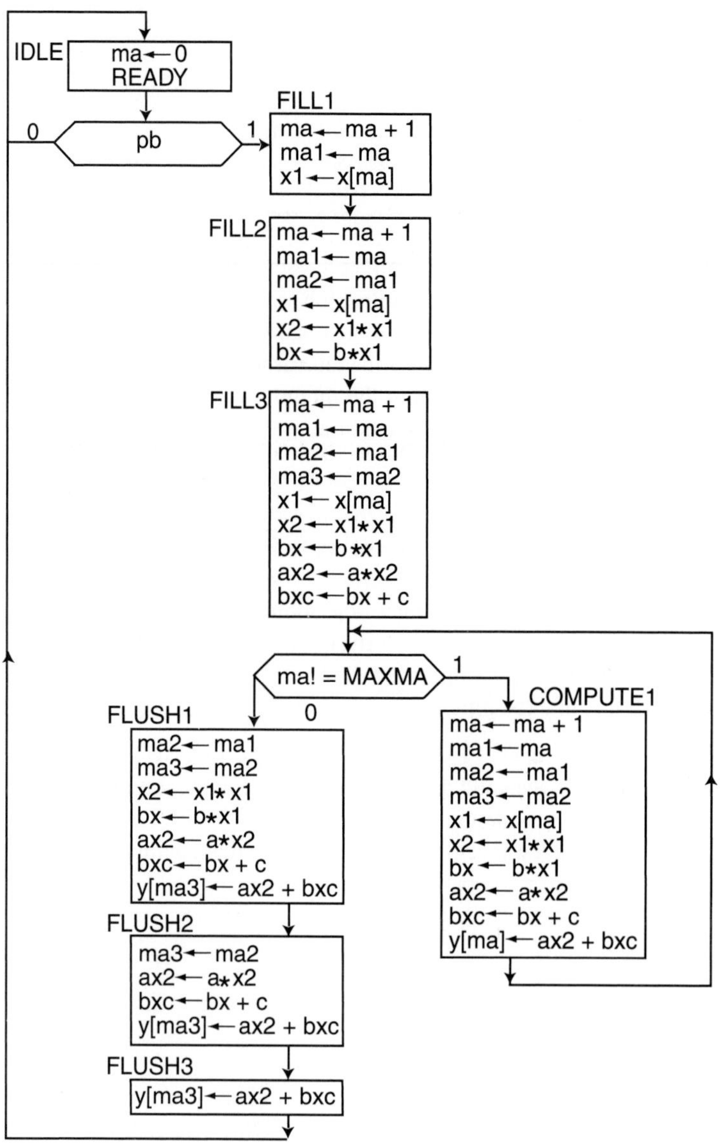

Figure 6-11. Correct pipelined ASM that fills and flushes.

Here is a simulation that shows the proper values filling the pipeline:

```
349 ps=000 ma=   0 x1=x x2=x bx=x
           ma1=   0 ax2=x bxc=x ma2=    0 ma3=    0
     z   z   z   z   z   z   z   z
449 ps=010 ma=   0 x1=x x2=x bx=x
           ma1=   0 ax2=x bxc=x ma2=    0 ma3=    0
     z   z   z   z   z   z   z   z
549 ps=011 ma=   1 x1=    7 x2=x bx=x
           ma1=   0 ax2=x bxc=x ma2= 0 ma3= 0
     z   z   z   z   z   z   z   z
649 ps=100 ma=   2 x1=    6 x2=49 bx=14
           ma1=   1 ax2=x bxc=x ma2= 0 ma3= 0
     z   z   z   z   z   z   z   z
749 ps=001 ma=   3 x1=    5 x2=   36 bx=   12
           ma1=   2 ax2=49 bxc=17 ma2= 1 ma3= 0
     z   z   z   z   z   z   z   z
849 ps=001 ma=   4 x1=    4 x2=   25 bx=   10
           ma1=   3 ax2=36 bxc=15 ma2= 2 ma3= 1
    66   z   z   z   z   z   z   z
    ...
```

The rest of the simulation is similar to the previous example. Here is the behavioral Verilog code used to produce the simulation of the correct pipelined machine:

```
`define NUM_STATE_BITS 3
`define IDLE           3'b000
`define COMPUTE1       3'b001
`define FLUSH1         3'b101
`define FLUSH2         3'b110
`define FLUSH3         3'b111
`define FILL1          3'b010
`define FILL2          3'b011
`define FILL3          3'b100
...
always
 begin
  @(posedge sysclk) enter_new_state(`IDLE);
   ma <= @(posedge sysclk) 0;
   ready = 1;
   if (pb)
    begin
     @(posedge sysclk) enter_new_state(`FILL1);
      ma <= @(posedge sysclk) ma + 1;
      ma1 <= @(posedge sysclk) ma;
      x1  <= @(posedge sysclk) x[ma];
     @(posedge sysclk) enter_new_state(`FILL2);
      ma <= @(posedge sysclk) ma + 1;
```

Continued.

```
      ma1 <= @(posedge sysclk) ma;
      ma2 <= @(posedge sysclk) ma1;
      x1  <= @(posedge sysclk) x[ma];
      x2  <= @(posedge sysclk) x1*x1;
      bx  <= @(posedge sysclk) b*x1;
     @(posedge sysclk) enter_new_state('FILL3);
      ma <= @(posedge sysclk) ma + 1;
      ma1 <= @(posedge sysclk) ma;
      ma2 <= @(posedge sysclk) ma1;
      ma3 <= @(posedge sysclk) ma2;
      x1  <= @(posedge sysclk) x[ma];
      x2  <= @(posedge sysclk) x1*x1;
      bx  <= @(posedge sysclk) b*x1;
      ax2 <= @(posedge sysclk) a*x2;
      bxc <= @(posedge sysclk) bx + c;
      while (ma != 'MAXMA)
       begin
        @(posedge sysclk) enter_new_state('COMPUTE1);
         ma <= @(posedge sysclk) ma + 1;
         ma1 <= @(posedge sysclk) ma;
         ma2 <= @(posedge sysclk) ma1;
         ma3 <= @(posedge sysclk) ma2;
         x1  <= @(posedge sysclk) x[ma];
         x2  <= @(posedge sysclk) x1*x1;
         bx  <= @(posedge sysclk) b*x1;
         ax2 <= @(posedge sysclk) a*x2;
         bxc <= @(posedge sysclk) bx + c;
         y[ma3] <= @(negedge sysclk) ax2 + bxc;
       end
     @(posedge sysclk) enter_new_state('FLUSH1);
      ma2 <= @(posedge sysclk) ma1;
      ma3 <= @(posedge sysclk) ma2;
      x2  <= @(posedge sysclk) x1*x1;
      bx  <= @(posedge sysclk) b*x1;
      ax2 <= @(posedge sysclk) a*x2;
      bxc <= @(posedge sysclk) bx + c;
      y[ma3] <= @(negedge sysclk) ax2 + bxc;
     @(posedge sysclk) enter_new_state('FLUSH2);
      ma3 <= @(posedge sysclk) ma2;
      ax2 <= @(posedge sysclk) a*x2;
      bxc <= @(posedge sysclk) bx + c;
      y[ma3] <= @(negedge sysclk) ax2 + bxc;
     @(posedge sysclk) enter_new_state('FLUSH3);
      y[ma3] <= @(negedge sysclk) ax2 + bxc;
   end
end
```

As described in section 6.5.2, the use of `negedge` memory modeling is necessary with many Verilog simulators.

6.5.8 Architectures for the quadratic evaluator

Only by proceeding to the mixed stage of the top-down design process can the maximum speed of the quadratic evaluator be estimated for the single-cycle, multi-cycle and pipelined versions. The mixed stage for each version requires choosing an architecture appropriate for each algorithm. Since the computations (to be performed by combinational logic) in each version are identical, the combinational logic devices (adders and multipliers) in these three versions will be identical. The three versions differ only with respect to whether and when intermediate computations are saved in registers. The following sections describe the architectures for these three versions.

6.5.8.1 Single-cycle architecture

The behavioral single-cycle ASM charts in section 6.5.2 describe all of the computations that must occur in one clock cycle before a result can be loaded into `y`. The following architecture implements these computations:

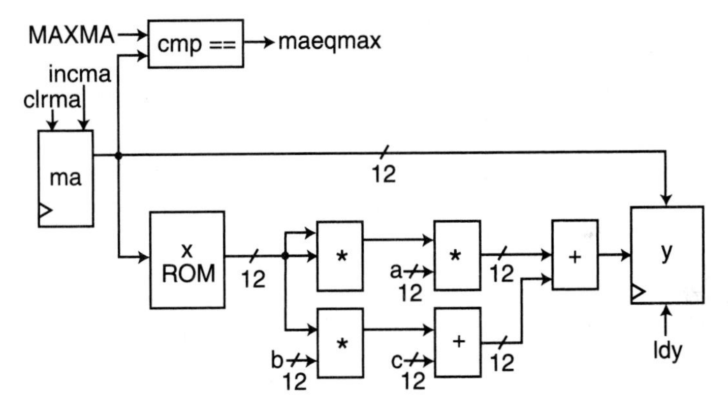

Figure 6-12. Single-cycle architecture.

The `ma` register provides the same address to `x` and `y`. The `mabus` is also compared against MAXMA to produce the status signal `maeqmax`.

There are three multipliers in the architecture. The `ma` register selects a particular word from the ROM. This value is fed to the first two multipliers. One of these multipliers produces the square, and the other multiplies this value by `b`. There is a third multiplier that multiplies the square by `a`.

There are also two adders in the architecture. The first adder produces the sum of `bx` and `c`. The other adder produces the final result. The final result is loaded into `y[ma]` when the `ldy` signal is asserted.

The following Verilog code shows the definition of this single-cycle architecture, along with the corresponding mixed controller:

```
module poly_ctrl(pb,maeqmax,ldy,incma,clrma,ready,sysclk);
 input pb,maeqmax,sysclk;
 output ldy,incma,clrma,ready;
 wire pb,maeqmax,sysclk;
 reg ldy,incma,clrma,ready;
 reg ['NUM_STATE_BITS-1:0] present_state;
 always
  begin
   @(posedge sysclk) enter_new_state('IDLE);
    //ma <= @(posedge sysclk) 0;
    ready = 1;
    clrma = 1;
    if (pb)
     begin
      while (~maeqmax)
       begin
        @(posedge sysclk) enter_new_state('COMPUTE1);
         //ma <= @(posedge sysclk) ma + 1;
         //x2 = x[ma]*x[ma];
         //bx = b*x[ma];
         //ax2 = a*x2;
         //bxc = bx + c;
         //y[ma] <= @(negedge sysclk) ax2 + bxc;
         incma = 1;
         ldy = 1;
       end
     end
  end
...
endmodule


module poly_arch(maeqmax,ldy,incma,clrma,a,b,c,sysclk);
 output maeqmax;
 input ldy,incma,clrma,a,b,c,sysclk;
 wire maeqmax,ldy,incma,clrma,sysclk;
 wire [11:0] a,b,c;
 wire [11:0] x2,bx,ax2,bxc,xbus,ydibus,mabus;
```

Continued

```
counter_register #12   ma(,mabus,,1'b0,incma,clrma,sysclk);
comparator       #12   cmp(,maeqmax,,mabus,`MAXMA);
rom              #(12,23)  x(mabus,xbus);

multiplier       #(12,24)  m1(x2,xbus,xbus);
multiplier       #(12,24)  m2(bx,b,xbus);
multiplier       #(12,24)  m3(ax2,a,x2);
adder            #(12,25)  a1(bxc,bx,c);
adder            #(12,25)  a2(ydibus,ax2,bxc);

ram              #12   y(ldy,mabus,ydibus,,sysclk);
endmodule
```

In the above, it is assumed that the propagation delays of the ROM, multipliers and adders are 23, 24 and 25 units of `$time`, respectively. `CLOCK_PERIOD is 100, which is just barely long enough for all the combinational logic to stabilize before a result is clocked into `y`, as illustrated by the following timing diagram produced by a Verilog simulator:

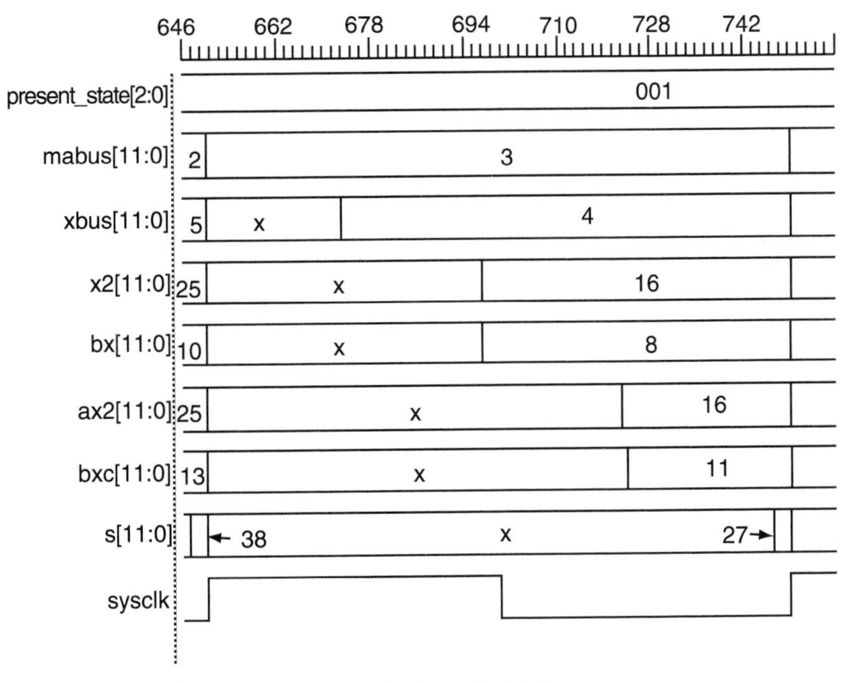

Figure 6-13. Timing diagram for single-cycle ASM.

6.5.8.2 *Multi-cycle architecture*

The behavioral multi-cycle ASM chart in section 6.5.3 can be implemented by many different possible architectures. Some of these possible architectures could be considerably cheaper than the architecture presented in this section; however, the architecture in this section was chosen for its consistency with the architecture in the previous section:

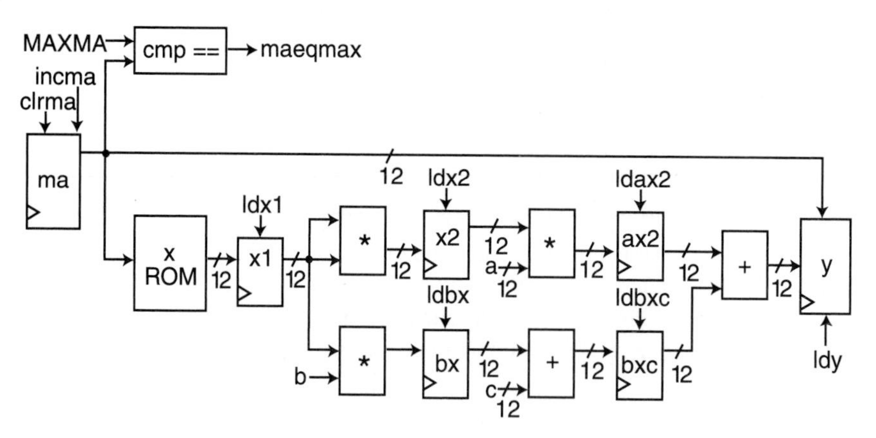

Figure 6-14. Multi-cycle architecture.

The only difference between this architecture and the previous architecture is the insertion of registers for `x1`, `x2`, `bx`, `ax2` and `bxc`. As indicated by the ASM chart, it takes six clock cycles for each computation to travel through this architecture. In the first cycle, only `ldx1` is asserted. In the second cycle, only `ldx2` is asserted. In the third cycle, only `ldbx` is asserted. In the fourth cycle, only `ldax2` is asserted. In the fifth cycle, only `ldbxc` is asserted. In the sixth cycle, finally `ldy` and `incma` are asserted.

As mentioned above, this is not a particularly efficient architecture for the multi-cycle approach because in any given clock cycle, five-sixths of the architecture is not performing any useful computation. Nevertheless, the insertion of the registers allows this architecture to be clocked considerably faster than the architecture in section 6.5.8.1.

The following Verilog code shows the definition of this multi-cycle architecture, along with the corresponding mixed controller:

```
always
 begin
  @(posedge sysclk) enter_new_state(`IDLE);
   //ma <= @(posedge sysclk) 0;
   clrma = 1;
   ready = 1;
   if (pb)
    begin
     while (~maeqmax)
      begin
       @(posedge sysclk) enter_new_state(`COMPUTE1);
        //x1 <= @(posedge sysclk) x[ma];
        ldx1 = 1;
       @(posedge sysclk) enter_new_state(`COMPUTE2);
        //x2  <= @(posedge sysclk) x1*x1;
        ldx2 = 1;
       @(posedge sysclk) enter_new_state(`COMPUTE3);
        //bx  <= @(posedge sysclk) b*x1;
        ldbx = 1;
       @(posedge sysclk) enter_new_state(`COMPUTE4);
        //ax2 <= @(posedge sysclk) a*x2;
        ldax2 = 1;
       @(posedge sysclk) enter_new_state(`COMPUTE5);
        //bxc <= @(posedge sysclk) bx + c;
        ldbxc = 1;
       @(posedge sysclk) enter_new_state(`COMPUTE6);
        //ma <= @(posedge sysclk) ma + 1;
        //y[ma] <= @(negedge sysclk) ax2 + bxc;
        incma = 1;
        ldy = 1;
      end
    end
 end
...
 counter_register #12   ma(,mabus,,1'b0,incma,clrma,sysclk);
 comparator       #12   cmp(,maeqmax,,mabus,`MAXMA);
 rom              #(12,23)  x(mabus,xbus);
 enabled_register #12       x1(xbus,x1bus,ldx1,sysclk);
 multiplier       #(12,24)  m1(x2dibus,x1bus,x1bus);
 enabled_register #12       x2(x2dibus,x2dobus,ldx2,sysclk);
 multiplier       #(12,24)  m2(bxdibus,b,x1bus);
 enabled_register #12       bx(bxdibus,bxdobus,ldbx,sysclk);
 multiplier       #(12,24)  m3(ax2dibus,a,x2dobus);
 enabled_register #12       ax2(ax2dibus,ax2dobus
                                       ldax2,sysclk);
 adder            #(12,25)  a1(bxcdibus,bxdobus,c);
```

Continued

```
enabled_register #12         bxc(bxcdibus,bxcdobus
                             ldbxc,sysclk);
adder             #(12,25)   a2(ydibus,ax2dobus,bxcdobus);
ram               #12   y(ldy,mabus,ydibus,,sysclk);
```

In the above, it is assumed that the propagation delays are the same as in the single-cycle approach of section 6.5.8.1. With the multi-cycle approach, 'CLOCK_PERIOD can now be 26 in this example, which is nearly four times faster than the single-cycle approach. The faster clock is possible because there is less logic that has to stabilize before each intermediate result is clocked into one of the registers. The following timing diagram illustrates this:

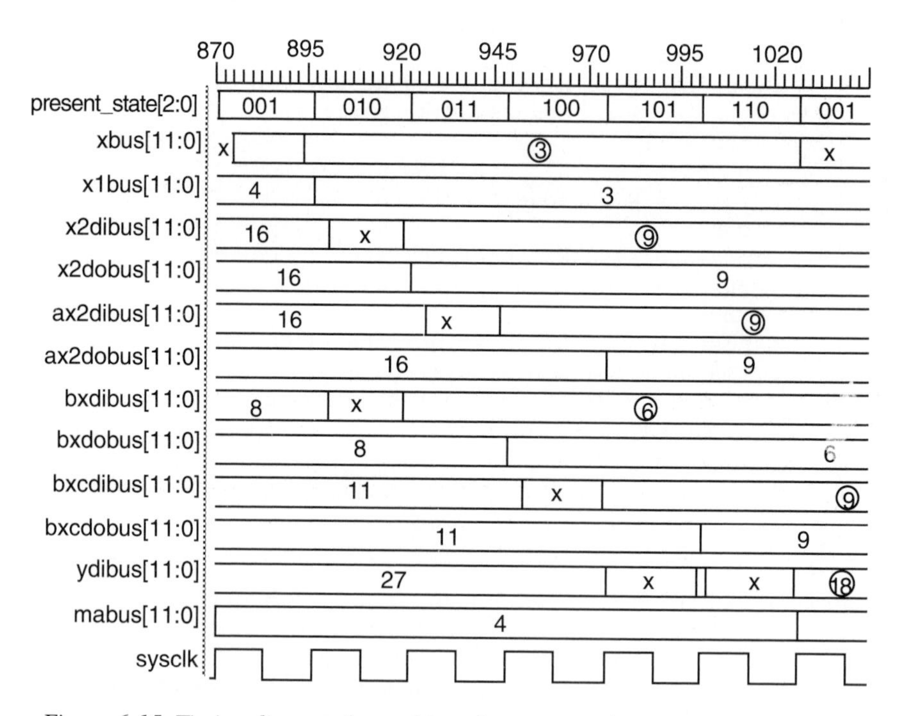

Figure 6-15. Timing diagram for multi-cycle.

6.5.8.3 *Pipelined architecture*

The correct behavioral ASM for the pipelined method given in section 6.5.7 requires three additional registers: `ma1`, `ma2` and `ma3`. In a pipelined design, all of the registers inserted in the single cycle architecture to make it a pipelined architecture are referred to as *pipeline registers*. In this architecture, every register, except `ma`, is a pipeline register.

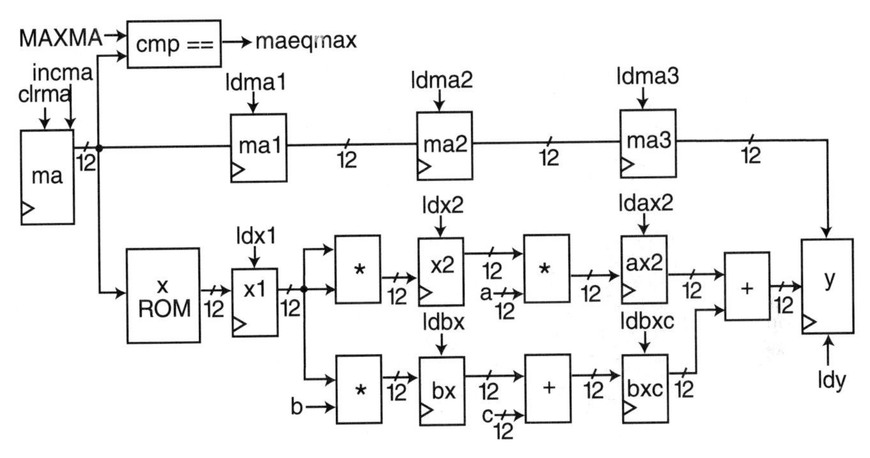

Figure 6-16. Pipelined architecture.

Notice that the pipeline registers are drawn in columns. The `ma1` and `x1` registers are drawn in the leftmost pipeline register column. The `ma2`, `x2` and `bx` registers are in the next pipeline register column. The `ma3`, `ax2` and `bxc` registers are in the third pipeline register column. In between each pipeline register column are buses and combinational logic only. Such a column of combinational logic to the left of a pipeline register is known as a *pipeline stage*. For example, the first pipeline stage is the ROM. The second pipeline stage involves two multipliers and a bus (that passes `ma1`). The third stage involves an adder and a multiplier. The fourth stage involves just an adder. (This architecture assumes that a value can be clocked into the appropriate word of `y` as though it just a clocked register. See section 8.2.2.3.1 for details.)

The following Verilog code shows the definition of this pipelined architecture, along with the corresponding mixed controller:

```
always
 begin
  @(posedge sysclk) enter_new_state(`IDLE);
   //ma <= @(posedge sysclk) 0;
   ready = 1;
   clrma = 1;
   if (pb)
    begin
     @(posedge sysclk) enter_new_state(`FILL1);
      //ma <= @(posedge sysclk) ma + 1;
      //ma1 <= @(posedge sysclk) ma;
      //x1  <= @(posedge sysclk) x[ma];
      ldx1 = 1;
      incma = 1;
      ldma1 = 1;
     @(posedge sysclk) enter_new_state(`FILL2);
      //ma <= @(posedge sysclk) ma + 1;
      //ma1 <= @(posedge sysclk) ma;
      //ma2 <= @(posedge sysclk) ma1;
      //x1  <= @(posedge sysclk) x[ma];
      //x2  <= @(posedge sysclk) x1*x1;
      //bx  <= @(posedge sysclk) b*x1;
      ldx1 = 1;
      ldx2 = 1;
      ldbx = 1;
      incma = 1;
      ldma1 = 1;
      ldma2 = 1;
     @(posedge sysclk) enter_new_state(`FILL3);
      //ma <= @(posedge sysclk) ma + 1;
      //ma1 <= @(posedge sysclk) ma;
      //ma2 <= @(posedge sysclk) ma1;
      //ma3 <= @(posedge sysclk) ma2;
      //x1  <= @(posedge sysclk) x[ma];
      //x2  <= @(posedge sysclk) x1*x1;
      //bx  <= @(posedge sysclk) b*x1;
      //ax2 <= @(posedge sysclk) a*x2;
      //bxc <= @(posedge sysclk) bx + c;
      ldx1 = 1;
      ldx2 = 1;
      ldbx = 1;
      ldax2 = 1;
      ldbxc = 1;
      incma = 1;
      ldma1 = 1;
      ldma2 = 1;
      ldma3 = 1;
```

Continued

```
      while (~maeqmax)
       begin
        @(posedge sysclk) enter_new_state('COMPUTE1);
         //ma <= @(posedge sysclk) ma + 1;
         //ma1 <= @(posedge sysclk) ma;
         //ma2 <= @(posedge sysclk) ma1;
         //ma3 <= @(posedge sysclk) ma2;
         //x1  <= @(posedge sysclk) x[ma];
         //x2  <= @(posedge sysclk) x1*x1;
         //bx  <= @(posedge sysclk) b*x1;
         //ax2 <= @(posedge sysclk) a*x2;
         //bxc <= @(posedge sysclk) bx + c;
         //y[ma3] <= @(negedge sysclk) ax2 + bxc;
         ldx1 = 1;
         ldx2 = 1;
         ldbx = 1;
         ldax2 = 1;
         ldbxc = 1;
         incma = 1;
         ldma1 = 1;
         ldma2 = 1;
         ldma3 = 1;
         ldy = 1;
       end
      @(posedge sysclk) enter_new_state('FLUSH1);
       //ma2 <= @(posedge sysclk) ma1;
       //ma3 <= @(posedge sysclk) ma2;
       //x2  <= @(posedge sysclk) x1*x1;
       //bx  <= @(posedge sysclk) b*x1;
       //ax2 <= @(posedge sysclk) a*x2;
       //bxc <= @(posedge sysclk) bx + c;
       //y[ma3] <= @(negedge sysclk) ax2 + bxc;
       ldx2 = 1;
       ldbx = 1;
       ldax2 = 1;
       ldbxc = 1;
       ldma2 = 1;
       ldma3 = 1;
       ldy = 1;
      @(posedge sysclk) enter_new_state('FLUSH2);
       //ma3 <= @(posedge sysclk) ma2;
       //ax2 <= @(posedge sysclk) a*x2;
       //bxc <= @(posedge sysclk) bx + c;
```

Continued

```
        //y[ma3] <= @(negedge sysclk) ax2 + bxc;
        ldax2 = 1;
        ldbxc = 1;
        ldma3 = 1;
        ldy = 1;
       @(posedge sysclk) enter_new_state(`FLUSH3);
        //y[ma3] <= @(negedge sysclk) ax2 + bxc;
        ldy = 1;
      end
  end
...
 counter_register #12  ma(,mabus,,1'b0,incma,clrma,sysclk);
 enabled_register #12  ma1(mabus,ma1bus,ldma1,sysclk);
 enabled_register #12  ma2(ma1bus,ma2bus,ldma2,sysclk);
 enabled_register #12  ma3(ma2bus,ma3bus,ldma3,sysclk);
 comparator       #12  cmp(,maeqmax,,mabus,`MAXMA);
 rom              #(12,23)   x(mabus,xbus);
 enabled_register #12       x1(xbus,x1bus,ldx1,sysclk);
 multiplier       #(12,24)  m1(x2dibus,x1bus,x1bus);
 enabled_register #12       x2(x2dibus,x2dobus,ldx2,sysclk);
 multiplier       #(12,24)  m2(bxdibus,b,x1bus);
 enabled_register #12       bx(bxdibus,bxdobus,ldbx,sysclk);
 multiplier       #(12,24)  m3(ax2dibus,a,x2dobus);
 enabled_register #12       ax2(ax2dibus,ax2dobus
      ldax2,sysclk);
 adder            #(12,25)  a1(bxcdibus,bxdobus,c);
 enabled_register #12       bxc(bxcdibus,bxcdobus
      ldbxc,sysclk);
 adder            #(12,25)  a2(ydibus,ax2dobus,bxcdobus);
 ram              #12  y(ldy,ma3bus,ydibus,,sysclk);
```

In the above, it is assumed that the propagation delays are the same as in the single- and multi-cycle approaches. With the pipelined approach, ‘CLOCK_PERIOD can usually be about as fast as in the multi-cycle approach (26 in this example), but the number of such clock cycles is nearly the same as in the single-cycle approach. Unlike the multi-cycle approach, where five sixths of the combinational logic is unproductive during each clock cycle, when the pipeline is full, all of the combinational logic is doing productive work. The following timing diagram illustrates how the pipelined approach gets the complete job done faster than the single-cycle or multi-cycle approach:

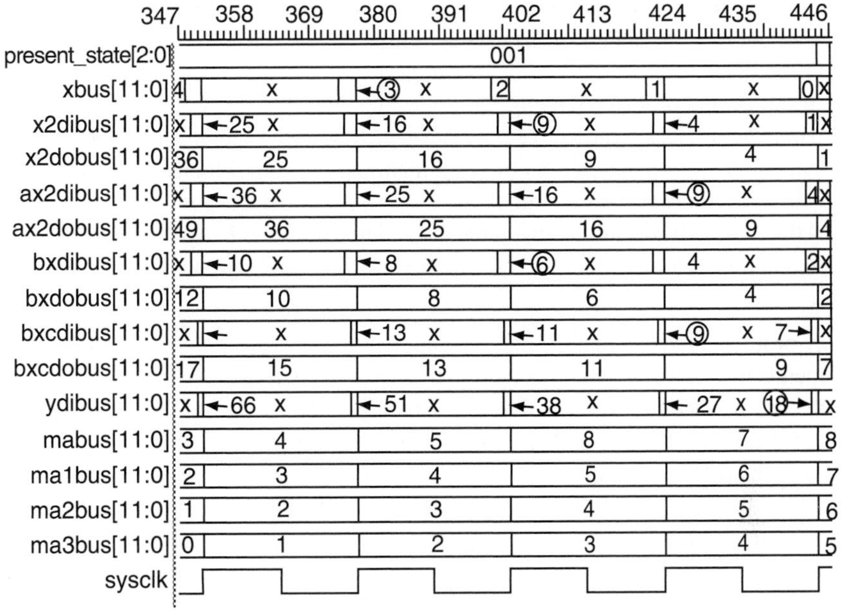

Figure 6-17. Timing diagram for pipelined ASM.

6.6 Conclusion

The first duty of a designer is to produce a correct design. The top-down design process explained in earlier chapters helps organize a designer's thinking to achieve this goal. Often, in addition to being algorithmically correct, a practical design must meet the criteria of speed and cost. This chapter explains how Verilog can help a designer determine if a design meets its speed goals. This chapter also explains different design alternatives that allow a designer to trade off speed and cost.

The speed of an algorithm implemented in hardware depends on two factors: the clock period and the number of clock cycles. The algorithm itself determines how many clock cycles are required, but the limiting factor on how fast the clock period can be is gate-level propagation delay. Synthesizable Verilog cannot have propagation delay, but once a design is synthesized, it is easy to annotate the built-in gates of the netlist with propagation delays. (Some synthesis tools automatically backannotate the netlist for post-synthesis simulation.) Gates with delays create the possibility of spurious wrong outputs, known as hazards.

There are many ways that a building block, such as an adder, can be synthesized into a gate level netlist. Each such unique netlist may give rise to unique patterns of hazards that can only be simulated in detail after synthesis. Despite the fact that we cannot know the details of a hazard prior to synthesis, it is possible to abstractly model a hazardous period of a signal using Verilog events, the `disable` statement and `'bx`. Such models are not synthesizable but instead provide more accurate timing simulation prior to synthesis. Instantiations of combinational building blocks defined this way provide an accurate worst case model of the propagation delay for bus-width devices. From the `$time` the inputs to any of the instantiated devices change until such changes propagate through all of the instantiated devices, the final output of the collection is `'bx`.

Many problems require that the same computation be performed on large amounts of independent data. There are three common algorithmic alternatives to solve such problems, known as single-cycle, multi-cycle, and pipelined. In the single-cycle approach, the computation on each independent piece of data is begun and finished in one clock cycle. In the multi-cycle approach, the computation on each independent piece of data requires several clock cycles to complete before another piece of data can be processed. The pipelined approach, like an assembly line in a factory, does different aspects of the computation with different pieces of independent data at the same time. Although it still takes several clock cycles to complete the computation for a particular piece of data, once the pipeline is filled, it produces one result per clock cycle.

What the multi-cycle and pipelined approaches have in common is that they both can be clocked by "faster" clocks, determined by the worst case delay of a single building block. The single-cycle approach demands a "slower" clock, determined by the delay path through several devices. The single-cycle approach produces exactly one result per clock cycle, and the pipelined approach usually produces almost one result per clock cycle (because the time to fill the pipeline is usually negligible compared to how much independent data is to be processed). The multi-cycle approach needs several clock cycles to produce each result. Therefore, the pipelined approach is usually fastest and most efficient because it can be clocked fast and it produces nearly one result per clock cycle.

Although in recent years pipelining has become important in the design (and marketing) of personal computers, pipelining is not a new concept. It has been used since the 1960s to design general-purpose computers (chapter 9), but its use with special-purpose computers has a much longer history. Pipelining is one of those algorithmic concepts that endure. It was first applied to computer design by Babbage in the 1820s. On Babbage's machine, the clock cycle was generated when one turned a crank. To avoid muscle strain, Babbage designed for speed and cost and chose a pipelined design. Despite almost unimaginable technological change in two centuries, many designers since then have followed in Babbage's algorithmic footsteps and have chosen pipelined designs. One does not have to be a genius like Babbage to understand pipelining today, because modern tools like Verilog simulators make these intricate `$time` related concepts much easier to understand.

6.7 Further reading

GAJSKI, DANIEL D., *Principles of Digital Design*, Prentice Hall, Upper Saddle River, NJ, 1997. Chapter 8 discusses how pipelining can be applied to both the controller and the architecture (datapath).

PALNITKAR, S., *Verilog HDL: A Guide to Digital Design and Synthesis,* Prentice Hall, PTR,Upper Saddle River, NJ, 1996. Chapters 5 and 10 explain about sophisticated gate-level delay modeling in Verilog.

PATTERSON, DAVID A. and JOHN L. HENNESSY, *Computer Organization and Design: The Hardware/Software Interface,* Morgan Kaufmann, San Mateo, CA, 1994. Chapters 5 and 6 discuss the trade-offs of the single-cycle, multi-cycle and pipelined approaches.

6.8 Exercises

6-1. A complex number, *X*, can be represented inside a machine as two integers: the real part, `xr`, and the imaginary part, `xi`. Mathematicians say that *X* = `xr+i*xi`, where `i` is the square root of minus one. (Some electrical engineers use the symbol `j` instead of `i`.) To add two complex numbers, *X* and *Y*, simply requires adding the real and imaginary parts separately. To multiply two complex numbers, *X* and *Y*, requires computing `xr*yr-xi*yi` and `xr*yi+xi*yr`. Suppose that a machine has four ROMs: `xr[ma]`, `xi[ma]`, `yr[ma]` and `yi[ma]`. Design a multi-cycle behavioral ASM suitable for a central ALU architecture that computes the sum of the products of the complex values in *X* and *Y* ROMs. This computation has many practical applications in the field of digital signal processing, such as filtering out unwanted noise in a telephone conversation. Note that there is no need for a memory in this problem because the desired answer is a single complex sum composed of `sumr` and `sumi`. You may assume the ALU can do either an integer addition, subtraction or multiplication in a single cycle.

6-2. Implement a pure behavioral Verilog simulation and test code that verifies your design in problem 6-1.

6-3. Draw a block diagram for the architecture of problem 6-1. Assume the propagation delays (in nanoseconds) of building blocks are the same as in section 6.5.8.1. How many seconds will it take for your machine to compute the sum assuming there are ten million words in each ROM?

6-4. Implement a mixed Verilog simulation that verifies your architecture for problem 6-3. Assume the propagation delays of building blocks are the same as in section 6.5.8.1.

6-5. Design a single-cycle behavioral ASM for problem 6-1. The architecture that will eventually implement the register transfers of this single-cycle machine may have as many integer adders, subtractors and multipliers as necessary.

6-6. Implement a pure behavioral Verilog simulation and test code that verifies your design in problem 6-5.

6-7. Draw a block diagram for the architecture of problem 6-5. Assume the propagation delays (in nanoseconds) of building blocks are the same as in section 6.5.8.1. How many seconds will it take for your machine to compute the sum assuming there are ten million words in each ROM?

6-8. Implement a mixed Verilog simulation that verifies your architecture for problem 6-7. Assume the propagation delays of building blocks are the same as in section 6.5.8.1.

6-9. Design a pipelined behavioral ASM for problem 6-1. The architecture that will eventually implement the register transfers of this pipelined machine may have as many pipeline stages and as many integer adders, subtractors and multipliers as necessary.

6-10. Implement a pure behavioral Verilog simulation and test code that verifies your design in problem 6-9.

6-11. Draw a block diagram for the architecture of problem 6-9. Assume the propagation delays (in nanoseconds) of building blocks are the same as in section 6.5.8.1. How many seconds will it take for your machine to compute the sum assuming there are ten million words in each ROM?

6-12. Implement a mixed Verilog simulation that verifies your architecture for problem 6-11. Assume the propagation delays of building blocks are the same as in section 6.5.8.1.

7. ONE HOT DESIGNS

The manual process of translating an ASM (or the equivalent Verilog) into hardware is quite involved. The final step of creating a structural controller is tedious because we have to determine what the next state is in every situation. There is an alternative way to create a structural controller directly from the behavioral Verilog or from the mixed ASM without the need to consider the next state logic and the present state register. The final hardware structure that is created by this alternative method is slightly more expensive than what is created by the process described in chapter 4, but it is much easier to understand.

This approach is known as a *one hot* method. The one hot controller uses as many flip flops as there are states in the ASM. As described in appendix D, a D-type flip flop is a one-bit register, whose output Q is simply its input D, delayed by one clock cycle.

The reason the one hot method is preferred is that the translation from behavior to structure is much more straightforward than the process described in chapter 4. Given a behavioral ASM or the equivalent behavioral Verilog, there is a one-to-one mapping to the circuit diagram for the one hot controller. (There is a tool described in appendix F that automates this process.)

7.1 Moore ASM to one hot

As explained in chapter 2, a Moore ASM is composed of three symbols: rectangles describing states, diamonds describing decisions and arrows describing control flow. With the one hot technique, each of these behavioral symbols (or the corresponding Verilog) translates directly into a physical piece of hardware.

7.1.1 Rectangle/flip flop

A rectangle in an ASM [or the equivalent `@(posedge sysclk)` in Verilog] translates into a flip flop. This technique is known as one hot because it is assumed that only one of the flip flops is hot (i.e., contains a one) in any clock cycle. The rest of the flip flops will be cold (contain 0). The rules of the one hot technique ensure that if this assumption is true shortly after `$time` 0, the one hot interconnections will guarantee this one hot property will remain in effect from then on.

7.1.2 Arrow/wire

An arrow in an ASM (or the implicit flow of control in Verilog) corresponds to a physical wire in a one hot controller. When that wire is hot it means that the corresponding statement in Verilog is active during the current clock cycle. Several statements (that execute in parallel) might be hot in a particular clock cycle, but only one flip flop (corresponding to a state) is hot in that clock cycle.

7.1.3 Joining together/OR gate

It is common for two or more arrows to join together in an ASM. This joining together occurs because in different clock cycles there are different paths to arrive at the same next state. Because they could fight (and produce a `1'bx` in Verilog simulation), it is illegal to tie together two wires that are each connected to an output. Therefore, when arrows in an ASM join together, the corresponding physical wires in the one hot design must be ORed together.

7.1.4 Decision/demux

A decision (diamond in an ASM or an equivalent `if` or `while` in Verilog) translates into a one bit wide demux. Recall from appendix C that the combinational logic for a demux is very different from that of a mux. The following truth table describes the outputs (`out0` and `out1`) of the demux, given its two inputs (`in` and `cond`):

in	cond	out1	out0
0	0	0	0
0	1	0	0
1	0	0	1
1	1	1	0

Notice when `in` is cold (i.e., 0), both outputs are cold. When `in` is hot (i.e., 1), only one of its two outputs is hot; hence it preserves the one hot property.

7.1.5 Summary of the one hot method

The following diagram illustrates the above four rules for translating a Moore ASM into a one hot design:

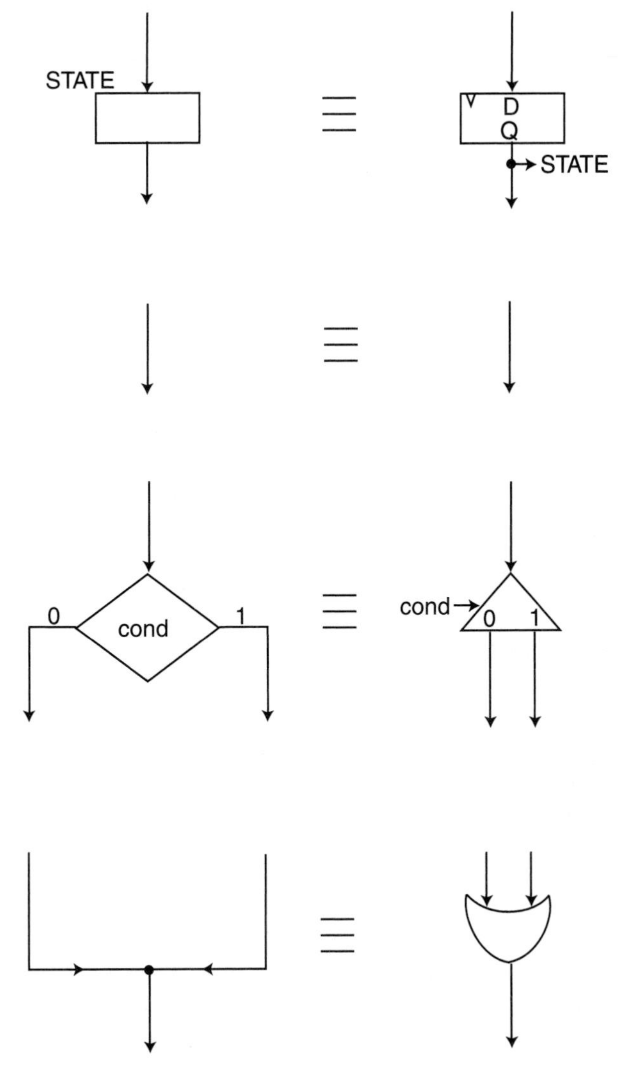

Figure 7-1. Moore ASMs and corresponding components of one hot controllers.

The one hot method uses more flip flops than the method shown in chapter 4. The number of flip flops of the present state register in chapter 4 is approximately the base two logarithm of the number of states. The following table shows how many flip flops are required in each method:

number of states	Ch 4	one hot	one hot + power on
2	1	2	3
3	2	3	4
4	2	4	5
5	3	5	6
6	3	6	7
7	3	7	8
8	3	8	9

The need for the extra flip flop for "power on" will be explained in the next section.

7.1.6 First example

Consider the first ASM given in section 2.2.2 (and the mixed ASM in section 2.3.3), which has five states. The diagram below shows how to translate the ASM from section 2.2.2 into one hot hardware. There are five states in this ASM; therefore there are five flip flops in the corresponding one hot controller. For clarity, the wires connected to the Q outputs of these five flip flops are labeled with states that correspond to them. Later we will use a numeric labeling scheme, but for now using the state names as wire names will emphasize how the one hot method works.

Coming out of state IDLE, there is a decision based on `pb`. If `pb` is false, there is a path that eventually leads back to state IDLE. If `pb` is true, there is a path that leads to state INIT. In the one hot controller, this decision corresponds to a demux whose input is the Q output of the flip flop for IDLE. The `out0` output of the demux goes to OR gates that in turn provide the D input for the flip flop for IDLE. The `out1` output of the demux connects to the D input for the flip flop for INIT. In other words, the ASM chart and the circuit diagram have an identical structure.

In the ASM, there are two paths that lead to state TEST. One comes from state INIT and the other comes from state COMPUTE2. In the one hot controller, this corresponds to another OR gate. In state TEST there is a decision based on `r1gey` (which tells if `r1 >= y`). If this condition is true, the ASM proceeds to state COMPUTE1; otherwise the ASM proceeds to state IDLE. This decision corresponds to a demux whose input is the Q output of the flip flop for state TEST. The `cond` input for this demux is `r1gey`. The `out1` output of this demux connects to the D input of the flip flop for state COMPUTE1. The `out0` output of this demux connects to the OR gate that leads back to the flip flop for state IDLE.

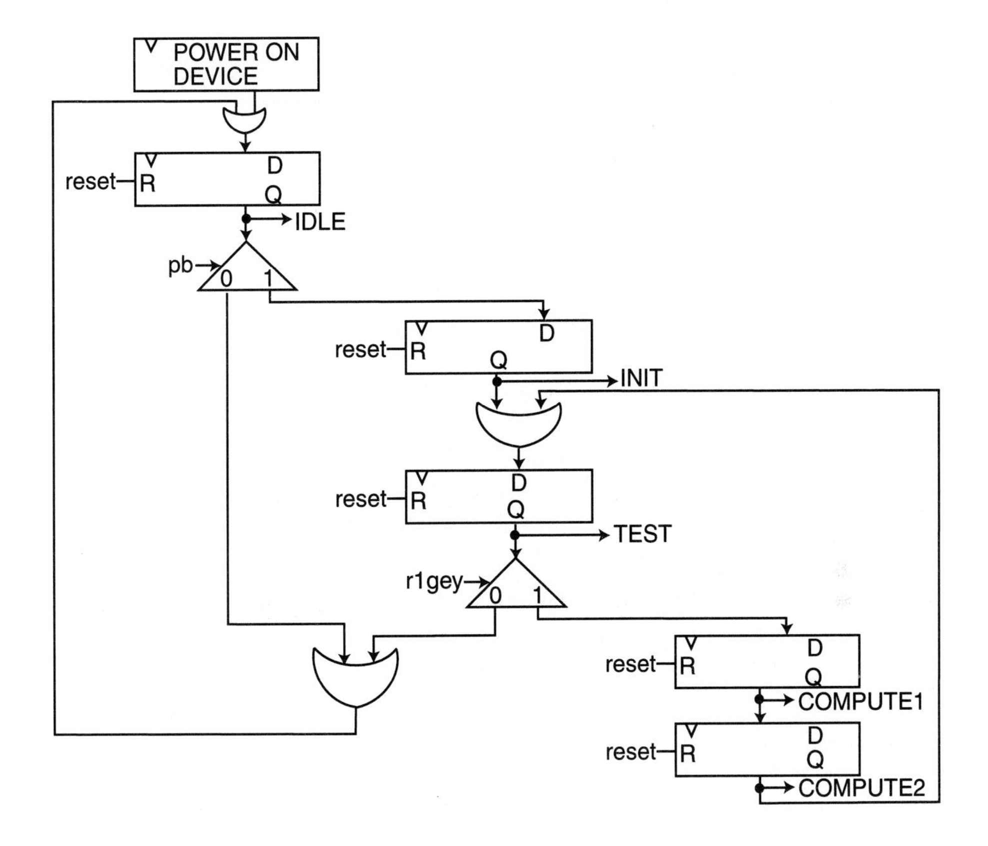

Figure 7-2. One hot controller for ASMs of sections 2.2.2 and 2.3.3.

In order to guarantee that the one hot property holds at `$time` 0, all of the flip flops are connected to an asynchronous `reset` signal. In physical hardware, shortly after `$time` 0 is when this `reset` signal ceases to be active. It is never used again.

By itself, just reseting the flip flops that correspond to the states in a particular ASM would have the effect of making all those flip flops cold at `$time` 0. Exactly one of these flip flops (the one for state IDLE in this example) needs to become hot at the first rising edge of the clock. To accomplish this, we need to OR an additional wire on the path to the flip flop for state IDLE.

This extra wire will be the output of a power-on device that will be hot only between `$time` 0 until the first rising edge of the clock. After the first rising edge of the clock, this power on device will will be cold thereafter.

This power-on device is constructed as a D flip flop with its D input tied to a one. The Q output of this flip flop is complemented to form the signal that "ignites" the first flip flop of the actual controller.

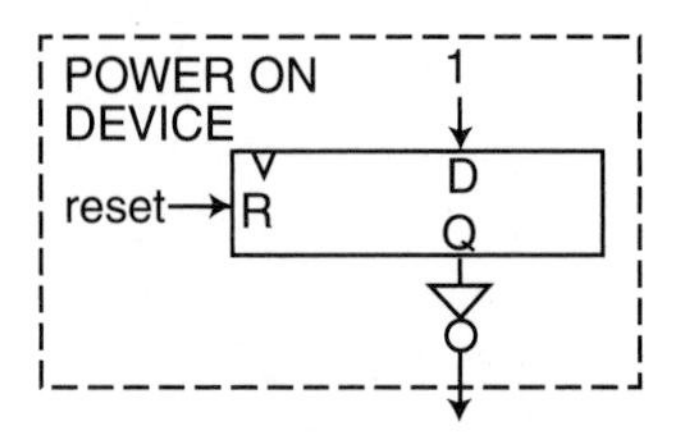

Figure 7-3. Power-on device for one hot controllers.

The next diagram shows the translation of the ASM from sections 2.2.7 and 2.3.1 (or the corresponding Verilog from section 4.1.5) into a one hot controller. Like any other controller, this also needs to generate the command outputs required for the architecture. This is very similar to the netlist given in section 4.4.2, except instead of `present_state[0]` and `~present_state[0]`, we have IDLE and COMPUTE as wires.

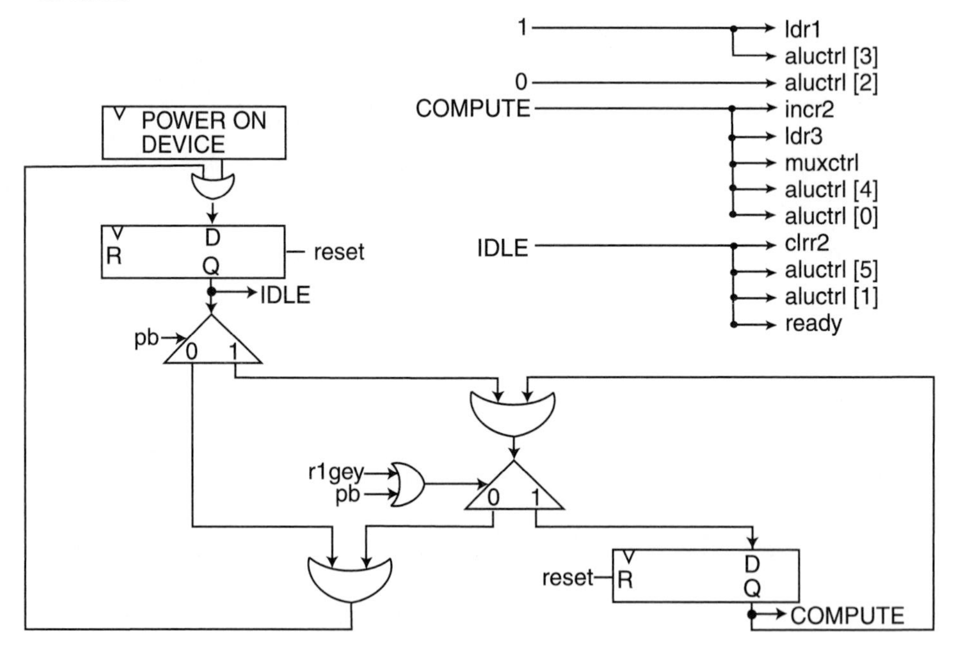

Figure 7-4. One hot controller for ASMs of sections 2.2.7 and 2.3.1.

7.2 Verilog to one hot

It usually takes a little more hardware to implement a machine with the one hot method, so why is this method worthwhile? The one hot method has a tremendous advantage: not only is there a one-to-one mapping of ASMs into one hot designs—there is also a similar correspondence between an implicit style behavioral Verilog block and the one hot circuit. Because of this direct translation process, it is relatively simple to write software that translates such Verilog into a one hot design without the need for the designer to go through the lengthy process described in chapter 4.

VITO is an automated preprocessor tool that performs the translation from implicit style behavioral Verilog into a one hot design. How to use VITO is described in appendix F. In order to understand the approach used in VITO, it is necessary to appreciate the one-to-one mapping between Verilog and the one hot circuit. Such an appreciation is best developed by working through a few examples. The following examples describe manual translation of implicit style Verilog directly into a one hot circuit. Rather than instantiating built-in gates, the following examples will use what is called *continuous assignment.*

7.2.1 Continuous assignment

Continuous assignment is a shorthand way of describing combinational logic. It is equivalent to defining a module without having to declare the ports and so forth that would otherwise be required. Let's consider an example of a continuous assignment:

```
module test;
  reg ff_1,ff_2;
  ...  /code that deals with ff_1 and ff_2

  wire s_3;
  assign s_3 = ff_1 | ff_2;
endmodule
```

There also is an additional shorthand for continuous assignment that allows the wire declaration to occur on the same line. For example, the above is equivalent to:

```
wire s_3 = ff_1 | ff_2;
```

Note that continuous assign is not a behavioral statement. The left-hand side is a `wire`, not a `reg`. Clearly, using continuous assign shortens the code considerably compared to the "`hidden_module`" shown below:

```
module test;
  reg ff_1,ff_2;
  ...  /code that deals with ff_1 and ff_2

  wire s_3;
  hidden_module h1(s_3, ff_1, ff_2);
endmodule
```

This "hidden_module" defines combinational logic in the usual way with a reg for the output port and a sensitivity list for the input wires:

```
module hidden_module(s_3, ff_1, ff_2);
  output s_3;
  input ff_1, ff_2;
  reg s_3;
  wire ff_1, ff_2;
  always @(ff_1 or ff_2)
    s_3 = ff_1 | ff_2;
endmodule
```

The computation, ff_1 | ff_2, is the same as given in the continuous assign. The power of the continuous assign is that it allows arbitrarily complicated expressions (of arbitrary width) to be evaluated. For example, the following:

```
module test;
  reg [11:0] a,b;
  ...  /code that deals with a and b

  wire [11:0] sum;

  assign sum = a + b;
endmodule
```

is equivalent to instantiating a hidden_adder:

```
module test;
  reg [11:0] a,b;
  ...  /code that deals with a and b
```

Continued

```
    wire [11:0] sum;

    hidden_adder h2(sum, a, b);
endmodule
module hidden_adder(sum, a, b);
  input a,b;
  output sum;
  wire [11:0] a,b;
  reg [11:0] sum;
  always @(a or b)
    sum = a + b;
endmodule
```

An advantage of the continuous assignment is that you do not have to specify the widths of `sum`, `a` and `b` multiple times—their previous declarations are sufficient. In contrast, with the hidden adder approach you have to duplicate the declaration of their widths inside the declaration of `hidden_adder`.

Also, continuous assignment allows the use of the conditional operator (`? :`). For example, the following:

```
module test;
  reg [11:0] a,b;
  reg sel;
  ...  /code that deals with a,b,sel

  wire [11:0] muxout;

  assign muxout =  sel ? b : a;
endmodule
```

is equivalent to instantiating a hidden instance of a `mux2`, whose portlist is given in section 4.2.1.5.

```
        mux2 #(12) h3(a, b, sel, muxout);
```

7.2.2 One hot using continuous assignment

The wires that implement the combinational logic of a one hot circuit can be described with continuous assignment. This is done as a notational convenience because continuous assignments are equivalent to structural instances but are much more concise. For example, the adder and mux in the last section could have been of any width, but the syntax of the actual continuous assignment would have been the same. Synthesis tools available from many different vendors are able to translate continuous assignments into the structural instances (netlist) required to fabricate hardware. Each flip flop required by the one hot circuit will be described by a separate one-bit `reg`. Such `regs` are also synthesizable. The names of these `wires` and `regs` will relate to the statement numbers of the lines in the Verilog `always` block from which they derive.

7.2.2.1 One hot with `if else`

The following example Verilog (taken from section 3.8.2.3.3) illustrates implicit style behavioral Verilog with an `if else` statement. In this example `@(posedge sysclk) #1` occurs on lines 3, 5, 9, 14 and 17, so the names of the five flip flops for the one hot controller will be `ff_3`, `ff_5`, `ff_9`, `ff_14` and `ff_17`:

```
 1:always
 2: begin
 3:  @(posedge sysclk) #1;    //FIRST  is ff_3
 4:   a <= @(posedge sysclk) 1;
 5:  @(posedge sysclk) #1;    //SECOND is ff_5
 6:   b <= @(posedge sysclk) a;
 7:   if (a == 1)
 8:    begin
 9:     @(posedge sysclk) #1;//THIRD  is ff_9
10:      a <= @(posedge sysclk) b;
11:    end
12:   else
13:    begin
14:     @(posedge sysclk) #1;//FOURTH is ff_14
15:      b <= @(posedge sysclk) 4;
16:    end
17:  @(posedge sysclk) #1;    //FIFTH  is ff_17
18:   a <= @(posedge sysclk) 5;
19: end
```

It is easier to give each flip flop a name that relates to what statement number the `@(posedge sysclk) #1` occurs on than to use the name from the original ASM. The reason that we do not use the names FIRST, SECOND, THIRD, FOURTH and

FIFTH for the flip flops is that those names were inside comments, which are ignored by Verilog. [The reason we do not use the `enter_new_state` task (sections 3.9.1.2 and chapter 4] to give each state a name is that the VITO preprocessor does not support tasks.) The example in section 7.1 of translating from an ASM to a one hot circuit used the state names given in the ASMs as the names of the flip flops only for the purpose of illustrating the nature of the one hot method. Since this translation will now be automated, the names do not matter. In the automated process, the designer will seldom notice what name is given to each `wire`.

Every statement also has a `wire` associated with it that is active when the corresponding statement executes. (In the earlier example, these names were also the original ASM state names. In general this need not be the case, and so separate names are appropriate for an automated tool.) In this example, there are nineteen `wires` (`s_1` through `s_19`) that correspond to statements in the original implicit style Verilog code. Of these `wires`, five act as command signals to the architecture:

```
wire      action in architecture when wire is active
s_4                 a <= @(posedge sysclk) 1;
s_6                 b <= @(posedge sysclk) a;
s_10                a <= @(posedge sysclk) b;
s_15                b <= @(posedge sysclk) 4;
s_18                a <= @(posedge sysclk) 5;
```

The other `wires` (`s_1`, `s_2`, `s_3`, `s_5`, `s_7`, `s_8`, `s_9`, `s_11`, `s_12`, `s_13`, `s_14`, `s_16`, `s_17` and `s_19`) are used to define the rest of the one hot controller. Some of those `wires` are synonymous with each other. For example `s_11` (an `end` statement) is synonymous with the `s_10` wire that precedes it.

Although there are nineteen `wire` names in the one-hot controller, only the above five are sent to the architecture. Using the methodical approach (such as in sections 2.3.1 and 8.4.1) for designing the architecture, we sort the above list according to the left-hand side of the `<=`, and separate them according to these destinations:

```
s_4                 a <= @(posedge sysclk) 1;
s_10                a <= @(posedge sysclk) b;
s_18                a <= @(posedge sysclk) 5;

s_6                 b <= @(posedge sysclk) a;
s_15                b <= @(posedge sysclk) 4;
```

Of course, there are many architectures that could implement the above register transfers. In earlier chapters we have focused on using standard building blocks, such as enabled registers. In the approach of this chapter, we instead choose an architecture that is easier to describe using continuous assignment. For this reason, we will use simple (non-enabled) D-type registers that have no command inputs whatsoever. In other words, such registers can be described simply as:

```
reg [11:0] a,b;
wire [11:0] new_a,new_b;
always @(posedge sysclk)
  a = new_a;
always @(posedge sysclk)
  b = new_b;
```

All of the actions normally encapsulated inside a register building block (of the kind described in appendix D) now have to be given with the combinational logic that computes `new_a` and `new_b`. From the sorted list above, one approach would be to use three muxes for computing `new_a` and two muxes for computing `new_b`:

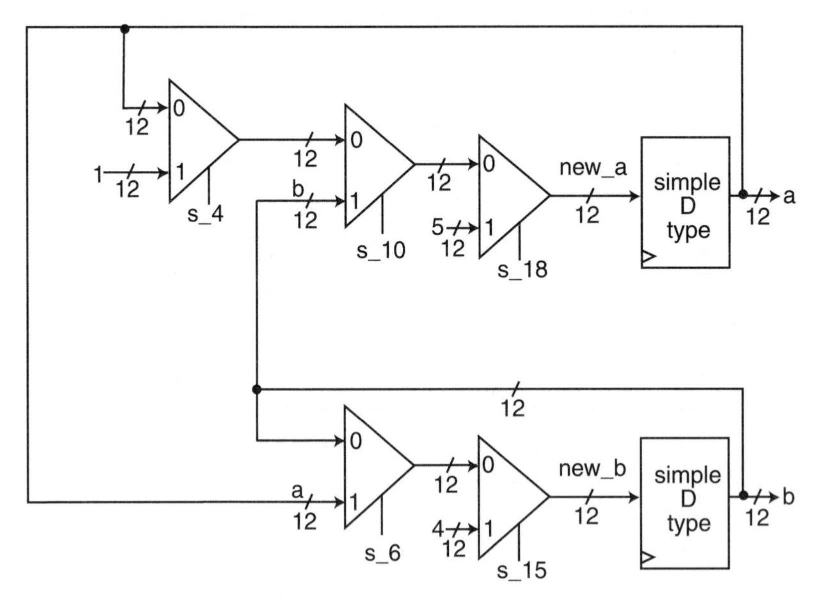

Figure 7-5. Architecture generated from implicit Verilog of sections 7.2.2.1 and 3.8.2.3.3.

The combinational logic in the above diagram can be expressed as two continuous assignments:

```
assign new_a =
  (s_18) ? 5 :
  (s_10) ? b :
  (s_4)  ? 1 :
           a;

assign new_b =
  (s_15) ? 4 :
  (s_6)  ? a :
           b;
```

Because of the nature of correct one hot designs, it is guaranteed that `s_18+s_10+s_4 <= 1` and `s_6+s_15 <= 1`. In other words, within each group (dealing with the same destination register), no more than one of the command signals will be active in any clock cycle. This means there are several permutations of the muxes that would also suffice, such as:

```
assign new_a =
  (s_4)  ? 1 :
  (s_10) ? b :
  (s_18) ? 5 :
           a;

assign new_b =
  (s_6)  ? a :
  (s_15) ? 4 :
           b;
```

Notice that the architecture was created by a *textual transformation* of the original Verilog. The block diagram given above was shown only as an aid to understand how the Verilog continuous assignment works. The preprocessor produces similar Verilog by just rearranging the original text of the Verilog.

Having defined the architecture, all that remains is to define the controller. The following circuit diagram shows how each implicit style behavioral statement translates into hardware:

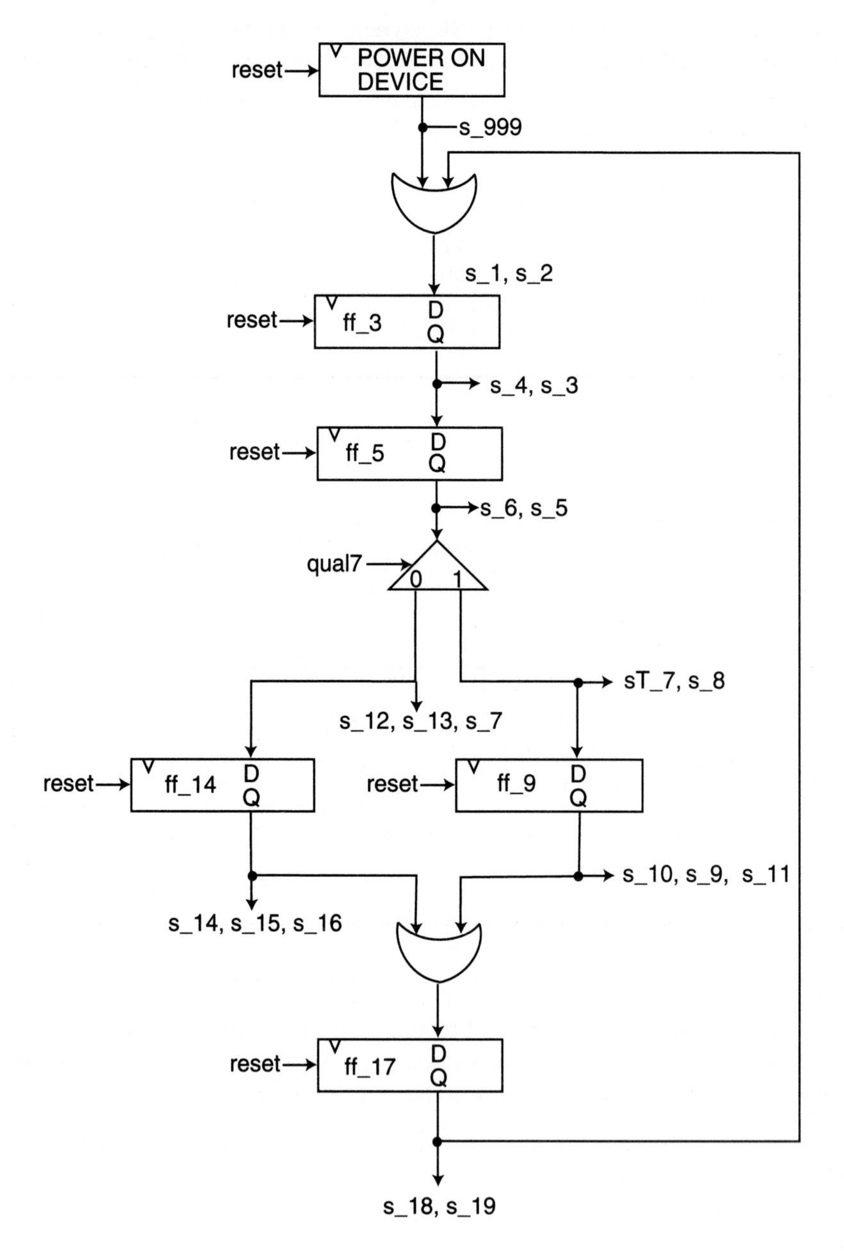

Figure 7-6. Controller generated from implicit Verilog of sections 7.2.2.1 and 3.8.2.3.3.

Again, VITO does not draw such a circuit diagram. The above diagram is provided only to help explain the textual transformations that the preprocessor performs. Starting from the original Verilog, the preprocessor generates the many continuous assignments required to describe the above circuit.

7.2.2.2 One hot with `if`

The following example is taken from the implicit style example in section 3.8.2.3.4. It illustrates a slightly different one hot controller than the last example because the original Verilog uses an `if` without an `else`:

```
 1:always
 2: begin
 3:  @(posedge sysclk) #1;     //FIRST   is ff_3
 4:   a <= @(posedge sysclk) 1;
 5:  @(posedge sysclk) #1;     //SECOND  is ff_5
 6:   b <= @(posedge sysclk) a;
 7:   if (a == 1)
 8:    begin
 9:     @(posedge sysclk) #1;//THIRD   is ff_9
10:      a <= @(posedge sysclk) b;
11:     @(posedge sysclk) #1;//FOURTH  is ff_11
12:      b <= @(posedge sysclk) 4;
13:    end
14:  @(posedge sysclk) #1;     //FIFTH   is ff_14
15:   a <= @(posedge sysclk) 5;
16: end
```

The `if` statement translates to a demux whose input, `qual7`, comes from the comparator for statement 7 that implements the condition `a == 1`. The 1 output, `sT_7`, corresponds to when this condition is true at the `$time` the `if` executes. The 0 output, `s_7` corresponds to when this condition is false at the `$time` the `if` executes.

The preprocessor generates the following one hot controller. In the following, only some of the `wire` names are shown:

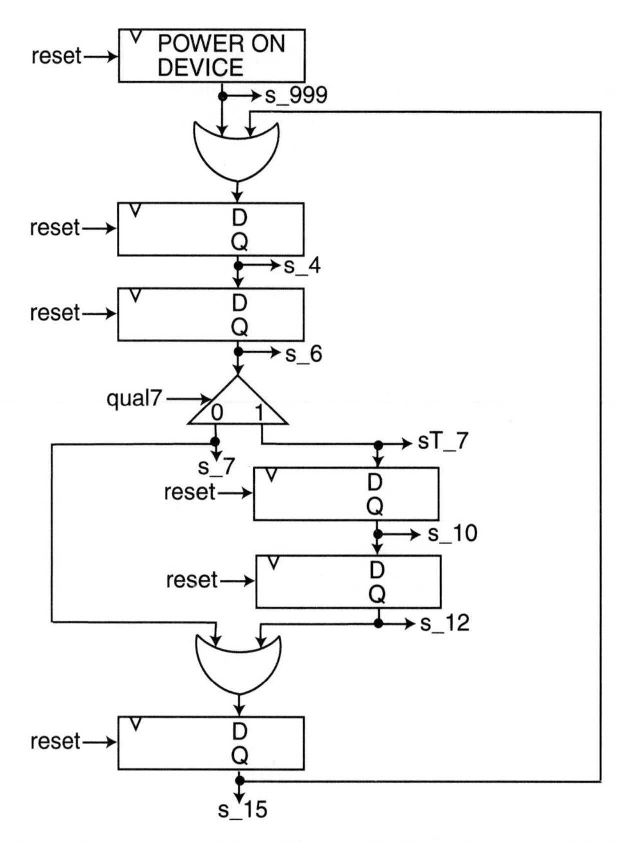

Figure 7-7. Controller generated from implicit Verilog of sections 7.2.2.2 and 3.8.2.3.4.

7.2.2.3 One hot with `while`

The following example is taken from the implicit style example in section 3.8.2.3.5. It is similar to the last example, except it uses a `while` loop (and a corresponding OR gate in the one hot controller to indicate the two paths that lead to the top of the `while`):

```
1:always
2: begin
3:  @(posedge sysclk) #1;    //FIRST  is ff_3
4:   a <= @(posedge sysclk) 1;
5:  @(posedge sysclk) #1;    //SECOND is ff_5
6:   b <= @(posedge sysclk) a;
7:   while (a == 1)
8:    begin
9:     @(posedge sysclk) #1;//THIRD  is ff_9
```

Continued

```
10:        a <= @(posedge sysclk) b;
11:       @(posedge sysclk) #1;//FOURTH is ff_11
12:        b <= @(posedge sysclk) 4;
13:      end
14:   @(posedge sysclk) #1;    //FIFTH  is ff_14
15:     a <= @(posedge sysclk) 5;
16: end
```

The preprocessor generates the following one hot controller:

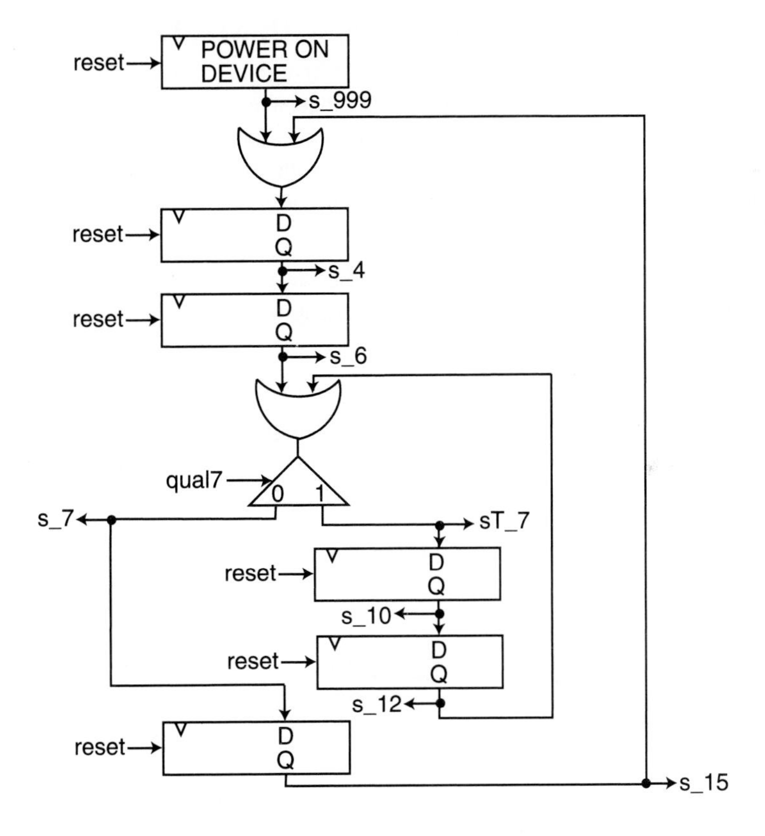

Figure 7-8. Controller generated from implicit Verilog of sections 7.2.2.3 and 3.8.2.3.5.

7.3 Mealy commands in a one hot machine

Chapter 5 describes Mealy machines, where a command can be conditional. Conditional commands, shown as ovals in ASM charts, are not guaranteed to execute simply because the machine is in a particular state. In order for a Mealy command to execute, some specified condition must also be true. In a one hot controller, this condition corresponds to an output `wire` from the proper demux. In a Mealy one hot controller, some of the wires associated with the statements that compose a state may not necessarily be active when the machine is in that state. This is why the preprocessor creates a `wire` for every statement: any statement in a Mealy machine might be conditional.

7.4 Moore command signals with Mealy <=

The VITO preprocessor only permits <= . At first glance, this might appear to prevent implementation of mixed ASMs or of ASMs that have external command outputs, such as `ready` in the ASM of section 2.2.7. In fact, as long as such command signals are unconditional (Moore), they can be described using Mealy <=. By doing so, the cost of the architecture will increase by some extra flip flop(s); however this is usually a small fraction of the total cost.

In addition, using the technique described below ensures that the command signal is hazard-free,[1] which is necessary in certain situations, such as interfacing to asynchronous memories (section 8.2.2.3.2).

7.4.1 Example to illustrate the technique

As an example to explain this technique, consider the following machine that asserts an external command signal, `comm`, when the machine is in state BOT:

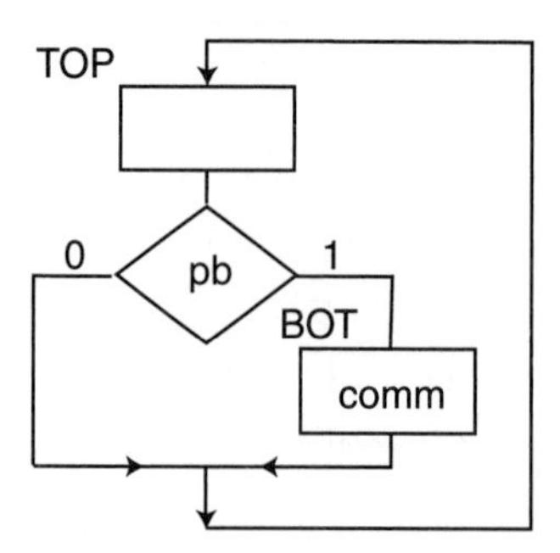

Figure 7-9. Example with Moore command signal.

[1] The physical cause for hazards is explained in section 6.3.3.

This can be described with the following implicit style behavioral Verilog:

```
always
  begin
    @(posedge sysclk) #1;       //TOP
      comm = 0;
      if (pb)
        begin
          @(posedge sysclk) #1;//BOT
            comm = 1;
        end
  end
```

As explained in section 3.9.1.2, the `comm = 0` statement can been hidden inside the `enter_new_state` task so that 0 is the default value for `comm`:

```
always
  begin
    @(posedge sysclk) enter_new_state(`TOP);
      if (pb)
        begin
          @(posedge sysclk) enter_new_state(`BOT);
            comm = 1;
        end
  end

task enter_new_state;
  input [`NUM_STATE_BITS-1:0] this_state;
  begin
    present_state = this_state;
    #1 comm =0;
  end
endtask
```

Since the VITO preprocessor only allows <=, we need to describe a machine without using = whose behavior will be identical to the above after the first clock cycle.

One of the essential ideas used throughout this entire book is the meaning of the non-blocking assignment. It computes a value now but assigns that value to a register at **the next rising edge of the clock**. Since the above Verilog is a Moore machine, the command is synonymous with the machine being in a particular state. As described in sections 2.4 and 4.4.5, such Moore commands can be generated by combinational logic that is part of the next state logic:

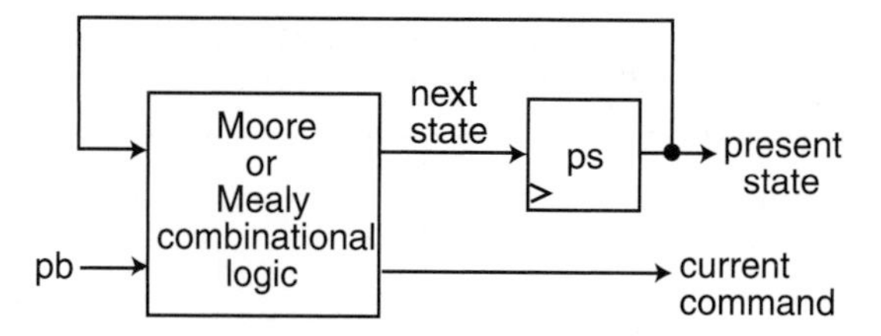

Figure 7-10. Current command approach suitable for Moore or Mealy controller.

Although Mealy machines must be defined using the above, we can look at a Moore machine such as this example in a different way. We know what the next state is going to be, and we know that there is a command synonymous with being in that next state. Instead of using combinational logic for the current command as we have done in previous chapters, we can instead use a register that will contain the *next command:*

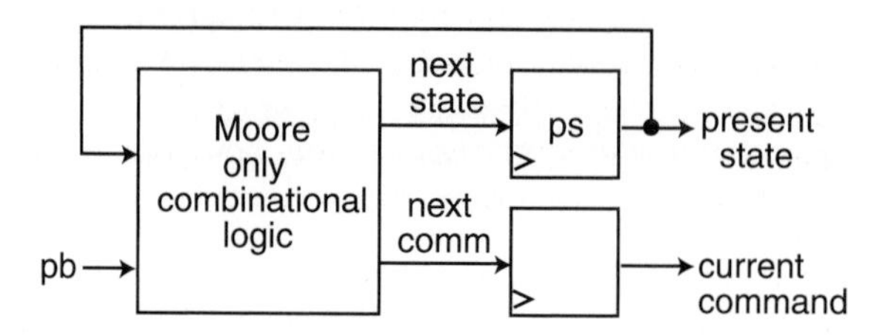

Figure 7-11. Next command approach suitable only for Moore controller.

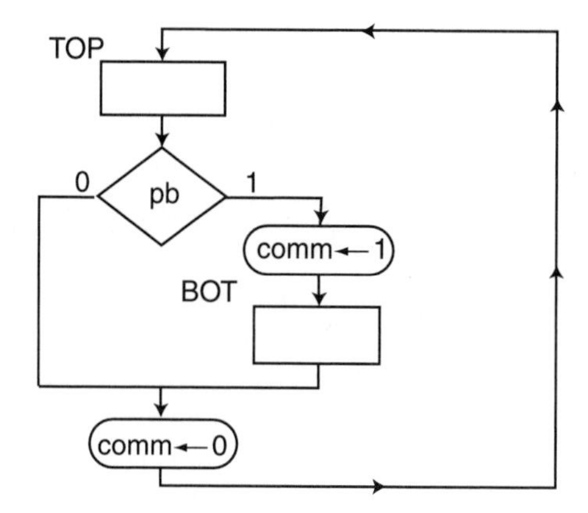

Figure 7-12. Behavioral ASM with ← for next command approach.

In the implicit style behavioral Verilog corresponding to figure 7-12, we use `<=` to assign values to the next command (`comm`) register:

```
always
  begin
    @(posedge sysclk) #1;       //TOP
      if (pb)
        begin
            comm <= @(posedge sysclk) 1;
          @(posedge sysclk) #1;//BOT
        end
      comm <= @(posedge sysclk) 0;
  end
```

Except for the time prior to the second full clock cycle (during which time `comm` is `1'bx`), the above machine has the same behavior as the original Verilog code that used =. Note that the non-blocking assignments are pure behavioral Mealy commands, and so the assignment is conditional. The machine only schedules the assignment of 1 to `comm` when the machine is already on the path where the next state will be state BOT. In other words, we only know that the next command will be 1 when we already know that the next state will be state BOT. The only way to get to state BOT is conditionally in state TOP when `pb == 1`.

Likewise, the machine only schedules the assignment of 0 to `comm` when the machine is already on the path where the next state will be state TOP. The next command will be 0 only when we know that the next state will be state TOP. There are of course two ways that we could know that the next state is state TOP: conditionally in state TOP when `pb == 0`, and unconditionally in state BOT. The non-blocking assignment, `comm <= 0`, only has to be described once because it was given after the `if` statement.

By rearranging the `comm <= @(posedge sysclk) 0` to the top of the `always` loop, the following is identical to the original Moore ASM, including the first *full* clock cycle:

```
always
  begin
      comm <= @(posedge sysclk) 0;
    @(posedge sysclk) #1;       //TOP
      if (pb)
        begin
            comm <= @(posedge sysclk) 1;
          @(posedge sysclk) #1;//BOT
        end
  end
```

This works on the assumption[2] that the power on circuit can reliably provide s_999 as the command signal that clears comm prior to that first full clock cycle.[3] When this assumption holds, the above example illustrates a simple rule for converting a Moore machine with external commands into a Mealy machine using <=: put the <= for the next commands just prior to the @(posedge sysclk) that marks the beginning of the corresponding next state. All commands must be described, including those to take on default values.

7.4.2 Pure behavioral two-state division example

Section 4.1.5 shows how to translate the two-state ASM chart for the childish division algorithm (section 2.2.7) into implicit style behavioral Verilog in a way that is suitable for simulation only (using the enter_new_state task). With the technique illustrated in the last section, it is very easy to modify this source code so that you can synthesize physical hardware directly from the implicit style (pure behavioral) Verilog without having to go through the tedious "mixed" and "pure structural" stages (sections 4.2 and 4.3). The boldface in the following shows the changes to the code from section 4.1.5:

```
module slow_div_system(pb,ready,x,y,r3,sysclk);
 input pb,x,y,sysclk;
 output ready,r3;
 wire pb;
 wire [11:0] x,y;
 reg ready;
 reg [11:0] r1,r2,r3;
 //reg ['NUM_STATE_BITS-1:0] present_state;
 wire sysclk;
 always
  begin
    ready <= @(posedge sysclk) 1;
   @(posedge sysclk) #1;              //IDLE
    r1 <= @(posedge sysclk) x;
    r2 <= @(posedge sysclk) 0;
    //ready = 1;
```

[2] Some synthesis tools might not produce a reliable circuit under these circumstances, and so the former method (assigning 0 to comm at the bottom of the always) might be preferred. The latter Verilog code is logically correct, but its physical implementation may be unreliable, depending on the clock frequency and the physical properties of the technology.

[3] Actually, the signal that clears the register is s_3, which is the OR of s_999 and the wire from the bottom of the always loop.

Continued

```
    if (pb)
     begin
      while (r1 >= y | pb)
       begin
         ready <= @(posedge sysclk) 0;
        @(posedge sysclk) #1;        //COMPUTE
         r1 <= @(posedge sysclk) r1 - y;
         r2 <= @(posedge sysclk) r2 + 1;
         r3 <= @(posedge sysclk) r2;
       end
     end
  end
  // task enter_new_state; ...
endmodule
```

7.4.3 Mixed two-state division example

Although the code in section 7.4.2 alone is enough to create physical hardware using appropriate synthesis tools (perhaps with the help of the VITO preprocessor described in appendix F), sometimes (for reasons of availability, speed or cost), the designer may want to create the architecture manually, as described in sections 2.3.1 and 4.2.3. The reader of chapter 4 may be left with the impression that in such a case, the designer must also create the controller manually. In fact, as long as the controller is a Moore machine, and the designer is willing to expend a few extra flip flops for the controller, it is possible to go straight from the mixed stage to physical hardware, without going through the tedious details of the "pure" structural stage.

Section 4.2.4 shows how to translate the mixed Moore ASM of the two-state division machine, (which generates command signals for a specific architecture that the designer has selected) into mixed Verilog. Using the techniques described in the preceding sections, here is equivalent implicit style Verilog acceptable to the synthesis preprocessor (bold indicates differences from section 4.2.4):

```
module slow_div_ctrl(pb,ready,aluctrl,muxctrl,ldr1,
               clrr2,incr2,ldr3,r1gey,sysclk);
 input pb,r1gey,sysclk;
 output ready,aluctrl,muxctrl,ldr1,clrr2,incr2,ldr3;

 //reg ['NUM_STATE_BITS-1:0] present_state;
 wire pb;
 reg ready;
 reg [5:0] aluctrl;
 reg muxctrl,ldr1,clrr2,incr2,ldr3;
 wire r1gey,sysclk;
```

Continued

```
always
 begin
   ready <= @(posedge sysclk) 1;
   aluctrl <= @(posedge sysclk) `PASSB;
   muxctrl <= @(posedge sysclk) 0;
   ldr1 <= @(posedge sysclk) 1;
   clrr2 <= @(posedge sysclk) 1;
   incr2 <= @(posedge sysclk) 0;
   ldr3 <= @(posedge sysclk) 0;
  @(posedge sysclk) #1;          //IDLE
   //r1 <= @(posedge sysclk) x;
   //r2 <= @(posedge sysclk) 0;
   //ready = 1;
   //aluctrl = `PASSB;
   //muxctrl = 0;
   //ldr1 = 1;
   //clrr2 = 1;
   if (pb)
    begin
     while (r1gey | pb)
      begin
        ready <= @(posedge sysclk) 0;
        aluctrl <=@(posedge sysclk)`DIFFERENCE;
        muxctrl <= @(posedge sysclk) 1;
        ldr1 <= @(posedge sysclk) 1;
        clrr2 <= @(posedge sysclk) 0;
        incr2 <= @(posedge sysclk) 1;
        ldr3 <= @(posedge sysclk) 1;
       @(posedge sysclk) #1;    //COMPUTE
        //ready = 0;
        //r1 <= @(posedge sysclk) r1 - y;
        //r2 <= @(posedge sysclk) r2 + 1;
        //r3 <= @(posedge sysclk) r2;
        //aluctrl = `DIFFERENCE;
        //muxctrl = 1;
        //ldr1 = 1;
        //incr2 = 1;
        //ldr3 = 1;
      end
    end
 end
endmodule
```

7.5 Bottom testing loops with `disable` inside `forever`

Although `while` loops are sufficient to implement any algorithm, and are preferable for many mathematical problems, there are situations when a bottom testing loop is more convenient than a `while` loop. Chapter 5 describes how the `enter_new_state` approach allows bottom testing loops to be simulated. The problem is that the technique used in chapter 5 is not acceptable to synthesis tools including the VITO preprocessor. There is another approach for bottom testing loops, involving the `disable` statement inside a `forever` loop that the preprocessor accepts.

The `disable` statement in Verilog has two main uses: stopping execution of a parallel block in a simulator (explained in section 6.4.3) and implementing bottom testing loops for synthesis (explained below).

As an example of a bottom testing loop, consider the following ASM. It is supposed to go through the bottom testing loop (consisting of states TOP and BOT) five times before returning to state OUTSIDE:

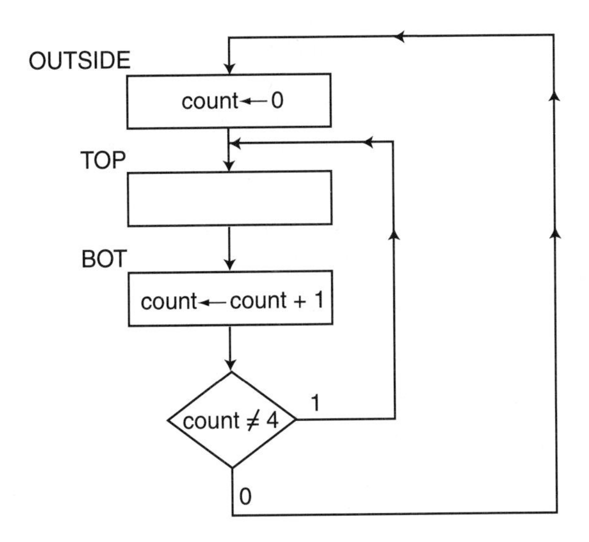

Figure 7-13. Example bottom testing loop.

Using the simulation technique of section 5.4.1, this could be translated to Verilog as follows:

```
always
 begin
  @(posedge sysclk) enter_new_state(`OUTSIDE);
   count <= @(posedge sysclk) 0;
   while (count!=4 & present_state !== `BOT);
    begin
     @(posedge sysclk) enter_new_state(`TOP);
     @(posedge sysclk) enter_new_state(`BOT);
      count <= @(posedge sysclk) count + 1;
    end
 end
```

The above simulates correctly. However, the above cannot be synthesized into a one hot machine using the VITO preprocessor.

An alternative way to implement a bottom testing loop is to use a `forever` statement with a `disable` statement inside. Using a `disable` statement requires an extra block that has a label to surround the `forever`. The `forever` by itself would never exit, and so the `disable` statement causes a `goto` the `end` that matches the labeled `begin`. For example, the above ASM could be translated into:

```
always
 begin
  @(posedge sysclk) #1;           //OUTSIDE
   count <= @(posedge sysclk) 0;
   begin : looplab
    forever
     begin
      @(posedge sysclk) #1;       //TOP
      @(posedge sysclk) #1;       //BOT
       count <= @(posedge sysclk) count + 1;
       if(count==4)
        begin
         disable looplab;
        end
     end
   end
 end
```

The above works correctly with the VITO preprocessor. However, this will not simulate properly on most Verilog simulators because the `disable` statement will also disable the non-blocking assignment. Putting `#1` in front of the `disable` may help on some simulators, but on many simulators there seems to be no way to use `disable` in this way properly. Therefore, the Verilog you choose for a bottom testing loop will be quite different if you want to simulate than if you want to synthesize. This book has avoided using bottom testing loops in most examples in order that simulation may agree with synthesis, but there are situations where hardware designers prefer bottom testing loops.

7.6 Conclusion

One hot encoding provides a more natural way of translating complex algorithms into hardware than the binary encoded approach described in earlier chapters. Because of this, a preprocessor tool is available that directly translates an algorithm written in implicit style behavioral Verilog into a one hot circuit. There is a one to one mapping between the Verilog (or the equivalent ASM) and the one hot controller.

There are *graphical* software tools that can automatically translate an ASM chart into Verilog, but the use of such tools is often ill advised. The use of such tools locks the designer into a proprietary file format. Although manually drawn ASM charts are useful to a designer in the early stages of design, they lack the expressive power of Verilog to hide the details of a design with good notation. Instead, this book uses graphical **ASM charts** only as the **master plan** for the design. The **details** of the design occur in *textual* form as **implicit style** behavioral **Verilog**. With one of several commercial synthesis tools and perhaps the VITO synthesis preprocessor described in appendix F (that uses the one-hot techniques given in this chapter), implicit style Verilog alone is often enough to create operational hardware.

The central concept of this book is that algorithms can be described using pure behavioral ASM charts (with RTN) or the equivalent pure behavioral Verilog (with implicit style `whiles` and `ifs` together with the non-blocking assignment). This approach is different than traditional software because of the potential for parallel processing and because of the idea of the system clock. This approach is different than traditional hardware because of the emphasis on algorithms and behavior. Such implicit style behavioral Verilog algorithms (or their equivalent ASM charts) describe in an abstract fashion the operations carried out by some specific synchronous architecture. Chapters 4 and 5 show how you can manually design such architectures using Verilog, but the Verilog to one hot preprocessor (explained in this chapter) eliminates the need for such manual translation.

7.7 Further reading

ARNOLD, MARK G. and JAMES D. SHULER, "A Preprocessor that Converts Implicit Style Verilog into One-hot Designs," *6th International Verilog HDL Conference,* Santa Clara, CA, March 31-April 3, 1997, pp. 38-45. Gives more information about the VITO preprocessor.

PROSSER, FRANKLIN P. and DAVID E. WINKEL, *The Art of Digital Design: An Introduction to Top Down Design*, Prentice Hall PTR, Englewood Cliffs, NJ, 2nd ed., 1987. Chapter 5 gives examples of the one hot technique.

7.8 Exercises

7-1. Draw a circuit diagram for a one hot controller corresponding to the Moore ASM given in section 2.2.4. Label the output of each flip flop with the name of the state. Assume the command and status signals of the architecture are the same as in sections 2.3.1 and 4.2.3.

7-2. Draw a circuit diagram for a one hot controller corresponding to the Mealy ASM given in section 5.2.4. Label the output of each flip flop with the name of the state. Assume the command and status signals of the architecture are the same as in sections 2.3.1 and 4.2.3.

7-3. Draw a block diagram using muxes and combinational logic which is equivalent to the following continuous assignment (assume that the 12-bit a and b are defined elsewhere):

```
wire [11:0] new_a;
assign new_a =
  (s_10) ? a+b :
  (s_20) ? 2*a-b :  a;
```

7-4. Rewrite the pure behavioral Verilog of section 4.1.3 into the implicit style form suitable for the VITO preprocessor. (Eliminate the `enter_new_state` task and convert `ready` to <= as described in section 7.4.2.) Use the preprocessor to produce the continuous assignments that are equivalent to the one hot design. Draw a circuit diagram for the one hot controller labeled with the names used in the output of the preprocessor. Also draw a block diagram for the architecture constructed only with combinational logic, muxes and simple (non-enabled) D-type registers corresponding to the `? :` in the output of the preprocessor.

7-5. Rewrite the pure behavioral Verilog of section 5.4.2 into the implicit style form suitable for the VITO preprocessor. (Eliminate the `enter_new_state` task and convert `ready` to <= as described in section 7.4.2. Also, use the `disable` statement in a different way than was described in this chapter.) Use the preprocessor to produce the continuous assignments that are equivalent to the one hot design. Draw a circuit diagram for the one hot controller labeled with the names used in the output of the preprocessor. Also draw a block diagram for the architecture constructed only with combinational logic, muxes and simple (non-enabled) D-type registers corresponding to the `? :` in the output of the preprocessor.

8. GENERAL-PURPOSE COMPUTERS

8.1 Introduction and history

The machines described earlier in this book have each implemented a single algorithm that solves a specific problem, such as the childish division algorithm given in chapter 2. We use the term *special-purpose computer* to describe such machines, which are designed to solve only one problem. The designer of a special-purpose computer transforms the algorithm which solves just that one problem into the hardware structure that directly performs the computations required by that specific algorithm. The history of automation is filled with examples of such machines. Prior to the electronic age, Blaise Pascal's 1642 calculator, Jacquard's automated loom, Charles Babbage's 1823 difference engine, Herman Hollerith's[1] 1887 electromechanical punch card counter (which tabulated the 1890 U.S. census), Leonardo Torres y Quevedo's 1911 electromechanical chess playing machine as well as all of the calculators and business equipment of the early twentieth century are illustrations of special-purpose computers.

It is not surprising that the first "computers"[2] implemented with electronic (vacuum tube) technologies were also special-purpose machines. C. E. Wynn-Williams' 1932 binary counter (for nuclear particles), John V. Atanasoff and Clifford Berry's 1938 computer (later dubbed the ABC machine) at Iowa State University for solving simultaneous equations and the British Colossus machines of World War II (that cracked German coded messages) are all illustrations of successful vacuum tube implementations of special-purpose machines that followed specific algorithms.

In contrast to such specialized machines, a *general-purpose computer* is designed to solve any problem, limited only by the size of the machine. The idea of a general-purpose computer is independent of any technology. Babbage and Augusta Ada (Lady Lovelace) envisioned a machine that could create its own programs. Alan Turing published a theoretical paper in 1936 which proved there are mathematical functions that cannot be computed mechanically. To do this, he envisioned a (technologically inefficient but plausible) machine programmed via a "tape" that could be both read and written. The theoretical machines envisioned by Babbage and Turing were "universal" because they would have the capability of self-modification.

[1] Hollerith started a company that later became IBM.

[2] The term "computer" did not develop its current meaning as a machine that processes information until the 1950s. Previously, a "computer" was a person hired by a scientist to carry out an algorithm manually.

Governments on both sides during World War II focused more resources on the design of computers than had ever occurred before. Although at first many such machines were justified because they solved some important special-purpose problem, such as ballistics, the huge expense required to build and maintain such machines motivated several independent groups to design machines that could be reused to solve different problems. These wartime machines were not fully general-purpose in the modern sense, but were programmable via punched tape. The tape moved in one direction past a reader, and the holes told the machine what to do. Although on most such machines looping was not possible (because the tape moved in only one direction) and self-modification was not possible (because once a tape was punched, it could not be repunched), such machines made it easy to change the program by changing the tape.

Konrad Zuse filed a patent in Germany in 1936 on such a tape-controlled machine and built several versions of this machine, the first of which became operational in 1941. Colossus, in fact, was tape-controlled due to the flexibility required by British mathematicians (including Turing) who sought to break ever-changing German codes. George R. Stibitz and others at Bell Labs built several tape-controlled relay computers, some of which remained in use for over a decade after the war. In 1943 Howard Aiken and others at Harvard, with the help of engineers from IBM, built the tape-controlled Harvard Mark I, which was used by U.S. Navy personnel (including the later to become famous Admiral Grace Hopper). Near the end of the war, IBM started to build the SSEC, which was unique among the tape-controlled computers of the war because it had some limited ability for the type of self-modification alluded to by Babbage and Turing (and was therefore almost a true general-purpose computer).

John P. Eckert, John W. Mauchly and others in the Moore School at the University of Pennsylvania built ENIAC from 1943 to 1945 for ballistic computations required by the U. S. Army. It was the largest computer built during the war, constructed with an order of magnitude more vacuum tubes (nearly 20,000) than any of the other machines. Unlike other machines of the era, it was not programmed via a tape, but instead it had to be rewired (via a plugboard) to solve a different problem. (Designing a "program" for the ENIAC was similar to designing the controller and architecture as illustrated in chapter 2). This made ENIAC far more specialized and inconvenient than the tape-controlled machines. Recognizing this inconvenience, people at the Moore School (notably John von Neumann) proposed building EDVAC, which would represent programs in the same memory as data, rather than on tape or with a plugboard.

Although EDVAC was not the first general-purpose computer to become operational, von Neumann's 1945 proposal was profoundly influential. To this day, his name is synonymous with general-purpose computers that store their programs in the same memory as their data and that use what we now call the fetch/execute algorithm. The hardware implementation of the actual EDVAC machine did not become operational until 1951, in part because von Neumann left the Moore School to join Princeton's Institute for Advanced Studies where he designed the IAS machine.

The first general-purpose computer to become operational was a prototype (known as "Baby Mark I") built by Frederic C. Williams and Tom Kilburn at the University of Manchester in England. Although small (less than one thousand vacuum tubes) and limited (only 32 words of memory and no hardware support for division), it ran its first software program (division using the childish algorithm described in chapter 2) on June 21, 1948. A later version of this machine, the Manchester Mark I, became operational in 1949 with 128 words of memory. A commercial version of the Mark I was produced in Britain a few years later by Ferranti. Also in Britain during 1949, Maurice Wilkes and others at Cambridge completed the EDSAC, which had 512 words of memory.

In the U.S., the first operational general-purpose computer was BINAC with 512 words of memory, built in 1949 by Eckert and Mauchly after they left the Moore School to start their own company. Their company later produced the UNIVAC, which was the first general-purpose computer that was commercially successful in the U.S. (more than 20 were sold).

Since the early 1950s, thousands of slightly different implementations of general-purpose computers have proliferated worldwide. Although they differ quite significantly in the details, all of them implement essentially the same algorithm in hardware: fetch/execute, which is the subject of this chapter.

8.2 Structure of the machine

Since the Manchester Mark I, almost all general-purpose machines have had the same overall top-level structure, illustrated by the following diagram:

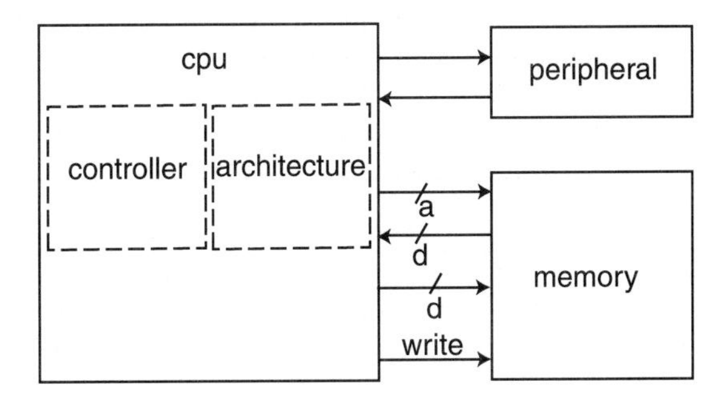

Figure 8-1. Block diagram of typical general-purpose computer.

8.2.1 CPU, peripheral and memory

There are typically three components of a general-purpose machine:

1. The Central Processing Unit (CPU) is composed of a controller and an architecture. As discussed in chapter 2, for special-purpose computers, the architecture is where the ALU(s), register(s) and other computational elements reside.
2. The peripheral(s) are distinct special-purpose computer(s) that interface the CPU to outside actors, such as a keyboard. Each interface typically has its own controller and architecture, including synchronizers as required.
3. Memory is a special kind of device that stores the bits that can represent both programs and data.

Sections 8.3 and 8.4 of this chapter will illustrate how the same techniques used for special-purpose computers in chapter 2 can implement the CPU of a general-purpose computer. The next section describes memory.

8.2.2 Memory: data and the stored program

The one component that has not been discussed in detail previously is memory.[3] From a behavioral standpoint, a *memory* is simply an array of words. The subscript to this array is known as an *address*. We will refer to the number of bits required for the address as `a`. The designer can choose how many bits (`d`) are in each word, and also how many words (2^{a}) are in the memory. For example, Williams and Kilburn's first machine had a word size of 32 bits, and there were 32 such words in the memory (five address bits). Therefore, their memory had $32*2^5 = 1024$ bits total. Later in this chapter we will design a machine with 4096 words (12-bit address), and a wordsize of 12-bits. This machine will require a memory with $12*2^{12} = 49{,}152$ bits. Most machines typically have memories that hold billions of bits, but that is a detail that is irrelevant to learning the essential ideas of this chapter.

In this chapter and in appendices A and B, we will indicate both address and contents of memory in octal. As an abbreviation, the address will be shown on the left, and the contents will be shown on the right, with a slash separating the address from the contents. For example,

```
0123/4567
```

[3] The memory we are talking about can be used both to store and retreive bits. It is refered to as "RAM" by some people.

means the address is 0123_8 == 000001010011_2 == 83 and the contents is 4567_8 == 100101110111_2 == 2423, or more succinctly in array notation, `memory[83] == 2423`. We will sometimes abbreviate even further to say `m[83] == 2423`.

There are five independent issues that can be used to categorize memory: unidirectional versus bidirectional (section 8.2.2.1), deterministic versus non-deterministic access time (section 8.2.2.2), synchronous versus asynchronous (section 8.2.2.3), static versus dynamic (section 8.2.2.4) and volatile versus non-volatile (section 8.2.2.5).[4]

8.2.2.1 Unidirectional buses versus a bidirectional bus

There are two common variations on how a memory device connects to the rest of the system. One approach uses simple unidirectional buses of the type seen in chapter 2, and the other combines two data buses together into what is known as a bidirectional bus.

The simplest form of memory has two input buses and one output bus. This simple type of memory with only unidirectional buses is what we will use in this chapter. In this type of memory, the `d`-bit-wide `din` bus is an input to the memory device, and the `d`-bit-wide `dout` bus is the output of the memory device. Also, the `a`-bit wide `addr` bus is another input to the memory device. There must be additional input(s) to the memory device, which are described in later sections.

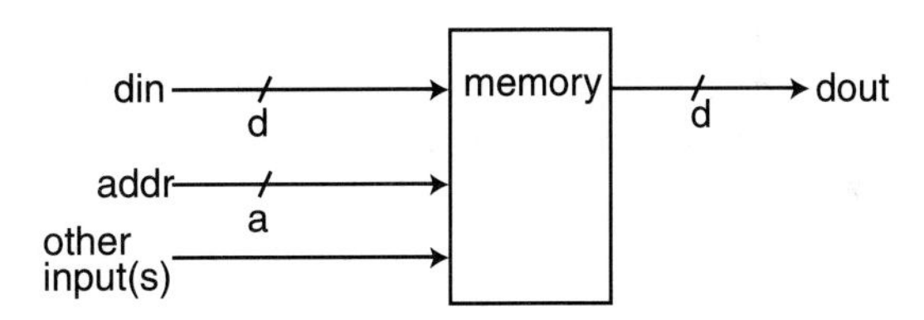

Figure 8-2. Symbol for memory with unidirectional data buses.

A *bidirectional* bus is one that is used to send information two ways. In the following diagram, a bidirectional bus is indicated by an arrow that points both ways. In this case, bidirectionality allows combining the `din` and `dout` buses, into a single `data` bus as illustrated in the following:

[4] A different issue related to memory (not discussed in this chapter) is how many ports the memory has. Multi-ported memory is discussed in section 9.8, but in this chapter all memory is assumed to have only one read port and one write port, as illustrated in the following sections.

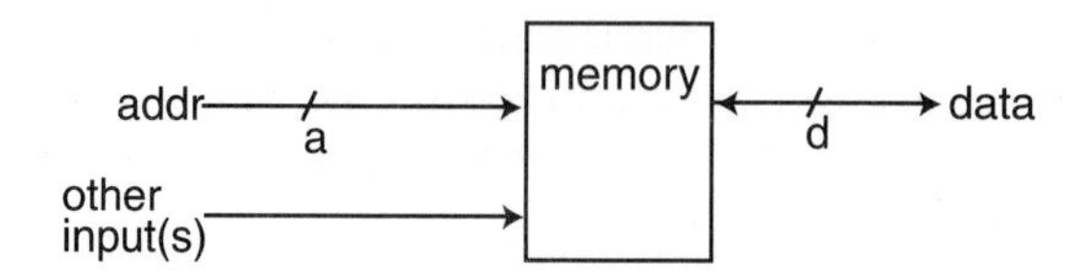

Figure 8-3. Symbol for memory with bidirectional data bus.

The advantage of bidirectionality is that there are fewer wires connecting the memory device to the rest of the system, however, interfacing to such a device is more complicated. This requires the use of tri-state buffers. Except for such tri-state buffers, the internal structure of this memory is identical to that of a memory with separate `din` and `dout` buses. We will not consider memory devices with bidirectional buses in this chapter.

8.2.2.2 Deterministic versus non-deterministic access time

Like all other physical devices, the actual memory hardware is not instantaneous. Although at the early behavioral stages of the design we prefer to ignore the time it takes to use the memory (known as the *access time*), ultimately, the speed of the memory will have a major influence on the speed of a general-purpose computer (since both the program and the data have to be obtained from the same memory). In the final stages of design, the designer must consider memory timing.

When the access time is *deterministic*, the time to obtain an arbitrary bit from memory does not vary significantly as a function of the address.

Almost all primary memories used today have deterministic access times. Almost all secondary memories (e.g., disk drives) have non-deterministic access times.

8.2.2.3 Synchronous versus asynchronous memory

There are two different ways in which the memory timing can occur: synchronous and asynchronous. The difference is whether or not the memory uses the system clock.

8.2.2.3.1 Synchronous memory

This kind of memory is the fastest, most expensive and simplest for the designer to incorporate. This kind of memory is commonly used where speed is important, such as in pipelined systems (section 6.5.8.3) or RISC computers (chapter 10). Because of its cost, it has not generally been used for the primary memory of a stored program computer, although recently, as clock speeds have increased, synchronous primary memories have become more commonplace.

In addition to the address and data buses, there must be a command input, `ldmem`, that tells the memory what to do. In order for the memory to be synchronous with the rest of the machine, it also needs a clock input. The top-level structure of a synchronous memory with separate `din` and `dout` is shown below:

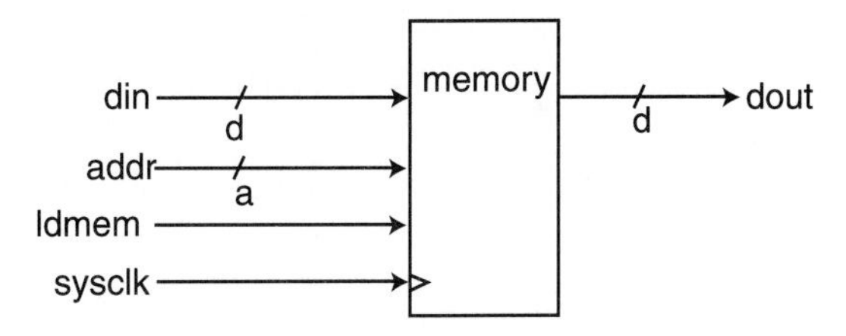

Figure 8-4. Symbol for synchronous memory.

There are two things that the device does. Given enough time, the output of the device reflects the contents of the memory at the word indicated by the address bus. In other words, neglecting the propagation delay (i.e., neglecting the access time),

```
dout = memory[addr]
```

The second thing that the memory can do is based on the `ldmem` command signal. On the next rising edge after `ldmem` becomes true, the word in memory indicated by the address bus changes to become the value of the data input bus,

```
memory[addr] ← din
```

Note that almost instantly after this change takes effect, `dout` will also change. At most one word in memory can be changed in one clock cycle when a memory is single-ported.[5] If `ldmem` is not true, memory remains unchanged.

A synchronous memory device can be thought of as a bank of registers. Although it is not usually the most efficient way to build a memory, the following diagram shows a structure that achieves this goal:

[5] Chapters 9 and 10 describe multi-ported memories that allow more than one memory operation per clock cycle.

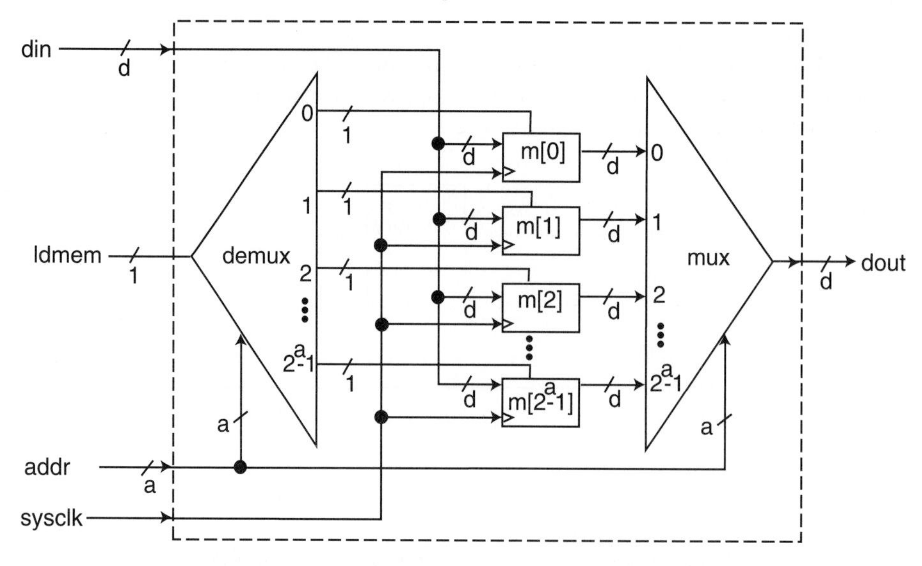

Figure 8-5. Implementation of synchronous memory.

This diagram uses 2^a enabled registers, each containing `d` bits. It also has a $1*2^a$ demux and a `d`$*2^a$ mux.

Let's ignore the left-hand side of this diagram (the demux that connects to the `ldmem` signal) for the moment. For example, assume the proper data has already been placed in each register, and the user wishes to obtain the contents of one of the memory locations. The user provides the address of the desired memory cell on the `addr` bus. The mux selects the output of the corresponding register, and outputs that on the `dout` bus.

To understand how a user is able to write new data into memory, you need to recall what the combinational logic of a demux does. When `ldmem` is 0, the demux will produce 0s on all of its 2^a outputs, so none of the registers would change. When `ldmem` is 1, the demux will output a 1 on **exactly** one of its outputs and 0s on the others. The output that is 1 will be determined by the value of the `addr` bus. Therefore if `ldmem` is 1, only the contents of the register corresponding to the current `addr` bus will change at the next rising edge of the clock.

Notice that the above implementation has a deterministic access time (essentially the propagation delay of the mux). It is possible to build a synchronous clocked memory based on shift registers, where the access time varies depending on how many clock cycles are required to shift the desired bit out. However, there is seldom any advantage to such a memory.

8.2.2.3.2 Asynchronous memory

A significant portion of the cost of the memory shown in 8.2.2.3.1 is due to the clock being provided to each register. All of the cheaper memory technologies invented have been asynchronous, which means they do not use the system clock. If the designer can cope with a memory that is asynchronous, the cost of memory can be reduced.

The block diagram for such an asynchronous memory (with separate `din` and `dout` ports) is:

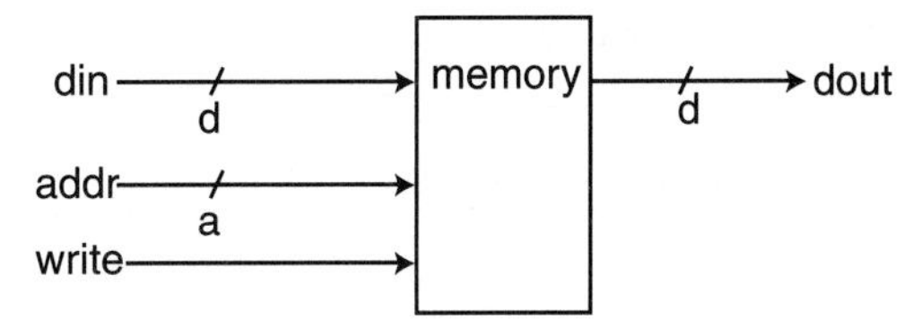

Figure 8-6. Symbol for asynchronous memory.

Here the `write` signal combines the roles of the `ldmem` signal and the `sysclk` signal. Asynchronous memory may also have a bidirectional data bus instead of two unidirectional buses.

In general, asynchronous design is highly unsafe, and should only be attempted by expert designers. Proper asynchronous design involves consideration of much lower (electronic) details than is the case for synchronous design. With the introduction of HDLs, the vast majority of design (such as CPUs) in industry is synchronous because synchronous designs are much more likely to be synthesized correctly. Asynchronous design is beyond the scope of this text, and so we will not consider the internal structure that implements this memory (although it is similar in concept to the diagram in 8.2.2.3.1).

Fortunately, since memory is such an important commodity, electronic experts have hidden most of the asynchronous ugliness inside commonly available memory chips. To cope safely with such devices, there are only three extra things that the designer has to do:

1. Choose a conservative clock speed for the rest of the system relative to the access time of the memory. In other words, the access time of the memory should be a small fraction of the clock period. Some memories have a different time for read and write, and so you should choose the larger of these.

2. Hold `addr` and `din` constant for at least one clock cycle before and during the cycle `write` is active. This means both `addr` and `din` should come from registers in the architecture that are not changed during this time.

3. Make sure there are no hazards in `write` (see section 6.3.3). As long as the controller is a Moore machine, this is easy to guarantee (section 7.4).

Sometime after the memory receives the `write` pulse but before the access time has elapsed, the asynchronous memory will latch the new value into memory. In this way, the asynchronous `write` pulse combines the role of the system clock and the `ldmem` signal.

8.2.2.4 *Static versus dynamic memory*

Static memory retains its contents regardless of whether it is used or not. The only limitation on using static memory is that the clock speed can be no faster than the access time. (More precisely, the maximum clock speed needs to consider all propagation delays, including access time. Examples of calculating the maximum possible clock speed are given in chapter 6.)

Dynamic memory places an additional requirement on the designer. Not only is there a maximum possible clock speed, but there is also a minimum clock speed. This is because every word in dynamic memory needs to be *refreshed*. Dynamic memory technology is usually based on very cheap electrical devices known as capacitors, which store charge. Over time, the charge leaks away. Unless the capacitors are refreshed, the information will disappear.

Dynamic memory is the cheapest kind of fast memory that is currently available. For problems where the machine will continually use all memory addresses over and over, there is no extra inconvenience to use dynamic memory. For most other problems, where this cannot be guaranteed, it is best to use a *dynamic memory controller* between the dynamic memory and the rest of the system.

8.2.2.5 *Volatile versus non-volatile*

A memory device is *volatile* if it loses its contents when the power is turned off. A memory device is *non-volatile* if it can retain its contents without consuming any power. A memory technology that is inherently non-volatile would be desirable because it allows a program to remain in memory when the power is turned off; however with most technologies, it is more cost-effective to provide a backup battery to preserve memory contents when the power to the rest of the system is removed.

8.2.3 History of memory technology

One of the recurring themes of this book is that *technology changes, but algorithms endure.* This means the time you spend honing your problem solving skills will benefit you throughout your career because such skills do not become outdated. Learning about the fetch/execute algorithm in particular is important because it is the algorithm which makes your career possible.[6]

From the historical account in section 8.1, you might wonder why it took so long for the general-purpose computer to be realized. The reason is a corollary to the italicized phrase above: some algorithms have to wait to be implemented until technology changes enough so the hardware is affordable and practical. To see how this postponed implementation of general-purpose computers until after World War II, consider the three components of a general-purpose machine: CPU, peripherals and memory.

As the ABC, Colossus and ENIAC illustrate, vacuum tube technology was available by the start of World War II to implement CPUs and peripherals. The technological problem from the late 1930s until the early 1950s was memory. Although the pioneers were aware of techniques like section 8.2.2.1.1 and used small memories of this sort for data access, the cost of storing programs in such memory was prohibitive. Currently, memories of this kind are commonly used, but not as the primary memory for stored program computers. It takes about six switching devices (relays, vacuum tubes, transistors or whatever technology is in vogue) to construct an enabled flip flop, so it would take $6*d*2^{a}$ switching devices to build the registers. It takes approximately $a*2^{a}$ switching devices to construct the demux, and $d*a*2^{a}$ switching devices for the mux. This makes the total about $((a+6)*d + a)*2^{a}$ switching devices to construct a working memory unit along the lines shown in section 8.2.2.3.1.

For Williams and Kilburn's 32 word memory (which, even in 1948, was considered too small for practical programming), this would require $((5+6)*32 + 5)*32 = 11{,}424$ vacuum tubes, which is more than an order of magnitude more tubes than was required for their entire CPU. (The ENIAC used about 20,000 vacuum tubes because it stored all its **data** in vacuum tubes. Also, storing a **program** in vacuum tubes would have been unrealistic.)

In order to build their machines, the pioneers had to invent technologies for memory that were more efficient and reliable than simple vacuum tubes. Zuse invented a binary mechanical memory. Atanasoff invented a rotating drum using capacitors (which is conceptually similar to the dynamic memory chips in widespread use today). Although creative, neither of these technologies would have been reasonable for a general-purpose computer in the 1940s.

The breakthrough came when Williams and Kilburn invented a TV-like tube for storing bits in the Mark I. Using the terminology defined above, the Williams tube was bidirectional, asynchronous, dynamic and volatile. Most importantly, the Williams tube was the first affordable technology that had the same kind of deterministic access time

---

[6] For example, modern Verilog simulators and synthesis tools are possible only because of large and fast general-purpose computers.

provided by the mux in section 8.2.2.3.1. The Williams tube also had the desirable side effect that a programmer could actually see every bit in memory (since they were actually stored as glowing dots of electric charge on the phosphor screen).

The only other cost-effective memory technology available before 1953 was the delay line, which had been used during the war for (analog) radar signals. These were also bidirectional, asynchronous, dynamic and volatile. The problem with delay lines is that they do not have deterministic access times. A delay line memory recirculates the same data over and over again. One has to wait until the desired data comes out of the delay line before it can be used. Therefore, there is a range of possible access times, the longest of which is quite slow. Wilkes as well as Eckert and Mauchly used such delay lines for the memory on their first general-purpose machines. Most computers of the early 1950s used delay lines, including IBM's first general-purpose machine, the model 701.

In 1953, Jay W. Forrester at MIT invented a new memory technology that stores each bit on a small doughnut-shaped magnetic *core*. Wires were woven through the cores to interface them to the system. By selecting which wire electric current flows through, the system could selectively magnetize or demagnetize each core (corresponding to storing a 1 or a 0). Core memory is fast compared to earlier technologies. It is also nearly ideal from a designer's viewpoint. It is unidirectional, deterministic, asynchronous, static and non-volatile. It is far less expensive to construct a core memory of a given size than to construct a comparable memory (of the kind outlined in sections 8.2.2.3.1 and 8.2.2.3.2) using vacuum tubes or transistors. Core memory dominated the computer industry until the late 1970s so much so that the term "core" became synonymous with the primary memory of a general-purpose computer. The first practical application of core memory was in Forrester's general-purpose Whirlwind computers, used by the U. S. Air Force for strategic defense for decades. Although it is not as cheap as the technology that replaced it, the military continues to use core memory because it is non-volatile, and it can retain its contents better than any other technologies when in close proximity to a nuclear explosion.

The final technological change for memory occurred at the end of the 1960s when Robert Noyce, Gordon Moore and others at Intel developed semiconductor integrated circuit memories that had all of the hardware for a memory similar to the one described in 8.2.2.3.1 on a single chip of silicon.

Integrated circuit memories come in many varieties. Today there are many competing manufacturers worldwide of interchangeable memory chips. A designer is able to choose from many different chips in a trade-off between speed, cost and ease of design.

Some chips are unidirectional, but others are bidirectional. Since the number of pins on an integrated circuit tends to be more of a constraint than the number of devices that can fit on the chip, larger integrated circuit memories tend to be bidirectional.

Almost all integrated circuit memories have deterministic access times. In the 1970s, research occurred in magnetic "bubble" memories with non-deterministic access times, however these memories did not succeed in the marketplace.

Some integrated circuit memories are synchronous, but many others are asynchronous. Larger memories have tended to be asynchronous because that enables more bits to be packed onto a chip of comparable area. The difference is between about six switching devices per bit versus only two switching devices per bit.

Many integrated circuit memories are static, which allows designers to observe them operate slowly enough for the details to be intelligible. Such memories are ideal for a computer design lab. Also, static memories tend to be faster than comparable dynamic memories. Despite these desirable properties, larger memories are dynamic. The difference is between two switching devices per bit versus one switching device and a capacitor per bit. Synchronous dynamic memories offer high speed at low cost.

Almost all existing integrated circuit memories are volatile. Successful research into ferrous semiconductors that would in essence put core memory on a chip occurred in the early 1990s. Whether such memories will be successfully commercialized remains to be seen at the time of this writing.

One principal limitation of integrated circuit memories is the number of pins that connect the memory to the CPU. In the mid 1990s, attempts were made by Cray Computer and others to overcome this restriction by fabricating multiple CPUs on the same chip as memory. Whether such "smart memories" will be successfully commercialized remains to be seen at the time of this writing.

For the first quarter of a century of the computer age, the physical appearance of memory devices changed radically as technology improved. For the second quarter of a century, memory looked basically the same: a silicon chip. As bit densities increased, packaging changed to hold more bits, but the semiconductor electronics that store each bit have remained essentially the same. By the mid 1990s, Single In-line Memory Modules (SIMMs) and Dual In-line Memory Modules (DIMMs) that can fit in the palm of your hand and that can contain billions of bits became a common way to package several dynamic memory chips.

What the preferred memory technology will look like by the end of the 21st century is, of course, unclear. Although visibility of bits was simply a side effect of the properties of the phosphor in the Williams tube, the idea of using light to store information has not gone away. In the late 1990s, prototype photochemical and holographic memories accessed using lasers were demonstrated that have the potential of storing orders of magnitude more bits than semiconductor memories. Daydreaming farther into the future, perhaps some nanomechanical computer designer in the 21st century might even pursue Zuse's memory designs at the atomic scale!

What is certain is that the cost, speed and capacity of integrated circuit memory has improved radically in the last quarter century. It is likely these improvements will continue well into the 21st century. That these technological factors have improved exponentially is in large part responsible for the success of the general-purpose computer, which needs to store both its programs and its data in memory.

8.3 Behavioral fetch/execute

The three components of a general-purpose computer described in section 8.2 (CPU, peripherals and memory) act together as a unified system that implements the fetch/execute algorithm. This section describes how to model the behavior of this unified system with an ASM chart, without regard to the structural interconnection of the hardware. This section explains what is referred to in chapter 2 as the "pure" behavioral stage of the top-down design process. Later, in section 8.4, the "mixed" stage of the top-down design process shows some of the structural interconnections for the CPU and memory.

This section focuses on the algorithm that makes the general-purpose possible: fetch/execute. Although the details of the fetch/execute algorithm vary widely among the thousands of general-purpose machines designed and built since 1948, the fundamental operations of the fetch/execute algorithm have remained essentially the same:

1. Fetch the current instruction from memory
2. If needed, fetch data from memory
3. Prepare for fetching the next instruction
4. Decode and execute the current instruction
    a) Interpret what the current instruction means
    b) Carry out the operation asked for by the current instruction, possibly modifying memory

Steps 2 and 4 have details that are machine specific. It may be possible to rearrange the order in which steps 2, 3 and 4 occur, depending on these machine-specific details.

A general-purpose computer can modify its instructions without programmer intervention because it uses the same memory to store instructions as it uses to store data. In other words, it can treat instructions as though they were data. This characteristic of universal machines, known as self-modification, is difficult for programmers to use effectively. However, this capability is the key to the success of the general-purpose computer. The ability for self-modification allows software (known as compilers and assemblers) to translate programs automatically from an easy to understand high-level language (C, Java, Pascal, Verilog, etc.) to the much more tedious machine language that is specific to the hardware.

For readers without intimate experience with low-level programming, appendix A gives a short introduction to machine and assembly language (and how they relate to high-level language) using an example of adding three numbers. This example will also be used in later sections of this chapter.

8.3.1 Limited instruction set

Although the fetch/execute algorithm is similar on all general-purpose computers, the machine-specific details depend on the instruction set being implemented. The *instruction set* is the set of machine language bit patterns that the hardware can interpret. All software on a particular machine is eventually translated to such instructions. The hardware is only capable of executing instructions that are in its unique instruction set. Although conceptually similar, a different model computer probably has an entirely different instruction set.

8.3.1.1 The PDP-8

We need a simple yet concrete example of an instruction set so that we can go through the stages of the top-down process, starting at the abstract algorithm for fetch/execute (which has remained essentially unchanged for half a century) and concluding with a unique hardware structure that implements those instructions. The instruction set that we will use as an example in this chapter is a subset of the PDP-8's instruction set. The PDP-8 is a classic example of what is called *a single-accumulator, one-address instruction set.* (All early stored program machines, including the Manchester Mark I, had this simple kind of instruction set.)

8.3.1.2 History of the PDP-8

The PDP-8, which was designed by C. Gordon Bell and Ed DeCastro at DEC in 1965, is pivotal in the history of general-purpose computers. It was the first computer that cost only a few thousand rather than hundreds of thousands of dollars. Bell was able to achieve this with core memory and transistor technology by keeping the instruction set simple and the memory small. The PDP-8 continued to be manufactured (with improved technologies) into the 1990s due to the simplicity and elegance of its instruction set. These attributes also make it an ideal example of the fetch/execute algorithm.

8.3.1.3 Instruction subset

The complete PDP-8 instruction set, which is described in appendix B, has about thirty instructions. Even though the PDP-8 has one of the simplest and smallest instruction sets ever designed, attempting to concentrate on all thirty of these instructions at once would distract from our primary goal: understanding the enduring fetch/execute algorithm. Therefore, for the example in this section, we will implement a machine that executes only the following four PDP-8 instructions (that are also used in the example of summing three numbers in appendix A):

mnemonic	octal machine language	what the mnemonic stands for
TAD	1xxx	add memory to accumulator (Twos complement ADd)
DCA	3xxx	Deposit accumulator in memory and Clear Accumulator
CLA	7200	CLear Accumulator
HLT	7402	HaLT

This subset does not contain enough of the PDP-8's instruction set for practical programming, but it provides a good introduction to fetch/execute. The capitalized letters explain what the mnemonic stands for. The first two instructions (TAD and DCA) are memory reference instructions, which require the machine to calculate an *effective address* (the address of the data in memory that the instruction is going to reference). Although, like most other instruction sets, the PDP-8 has several variations (known as *addressing modes*) on how to calculate the effective address, we will only consider the simplest one of these, known as *direct* page zero addressing. (Two bits in the instruction register determine which addressing mode the machine uses.) With direct page zero addressing mode, the effective address is simply the low-order seven bits of the instruction, denoted as xxx in the octal machine language above. The reason the PDP-8 is known as a one-address machine is because each instruction uses at most a single effective address.

The next instruction (CLA) manipulates the accumulator register without referencing memory; therefore it does not need an effective address. The final instruction (HLT) causes the machine to stop executing a program and instead proceed to a special state where the machine waits until an external signal tells it to run another program.

8.3.1.4 Registers needed for fetch/execute

The pure behavioral ASM for a general-purpose computer uses register transfer notation, similar to that of a special-purpose computer, as explained in chapter 2. Therefore, we need to determine what registers will be manipulated in the behavioral fetch/execute ASM.

Some of the registers are specified by the specific instruction set. The details of these registers are machine dependent. In the case of the PDP-8, the 12-bit accumulator, `ac`, is the primary register that the machine language programmer uses. (There are a few other registers, such as the `link`, that are specific to the PDP-8. As was done in appendix A, we will ignore these for the moment in order to keep this example simple.) Other machines, such as the Pentium, have different registers that the programmer can manipulate. We refer to the registers that are visible to the programmer as the *programmer's model*. Some people refer to these as the *computer architecture*; however we do not use this term since the registers in the programmer's model are not everything contained in the internal architecture of the CPU.

In addition to the registers required to implement a specific machine language, the fetch/execute algorithm requires the hardware to have two registers: the program counter, `pc`, and the instruction register, `ir`. Typically, the `pc` contains the address in memory of the next instruction to execute, and the `ir` contains the current instruction which is about to execute. If the machine did not have an HLT instruction, the machine would simply loop forever doing the four phases of the algorithm:

1. fetch the instruction from `m[pc]` into `ir`
2. calculate the effective address
3. increment the `pc` (prepare for fetch of *next* instruction)
4. decode and execute the instruction in the `ir`

where `m` refers to memory array. Most machines, including the PDP-8, have some form of HLT instruction. In order to keep track of whether the machine has halted or not, there needs to be an additional one-bit register, `halt`. When the machine has not executed an HLT instruction, `halt` is 0. When the machine has just executed an HLT instruction, `halt` becomes 1. The fetch/execute algorithm proceeds only when `halt` is 0.

The machine needs a register to hold the effective address of data in memory to be manipulated by an instruction. This register, which may be used for other purposes at different times, is known as the memory address register, `ma`. It will be convenient to have an additional register, known as the memory buffer register, `mb`, to contain the data that was in memory at the effective address prior to the execution of the instruction.

In later stages of the top-down design process, it will be convenient to have `ma` as the sole source providing the `addr` input to the memory device. At the "pure" behavioral stage, we can ensure this is possible by restricting the use of the memory array. All references to memory must be `m[ma]`, rather than the somewhat more natural references, `m[pc]`. Also it will be convenient to have `mb` as the sole source providing the `din` input to the memory device. In the restricted behavioral ASM, the only way to store something into memory is by saying `m[ma]` $\leftarrow$ `mb`. This will require that the behavioral ASM have states that initialize `mb` properly.

8.3.1.5 ASM for fetch/execute

The following is an ASM that implements the fetch/execute algorithm for the tiny instruction set described in section 8.3.1.3:

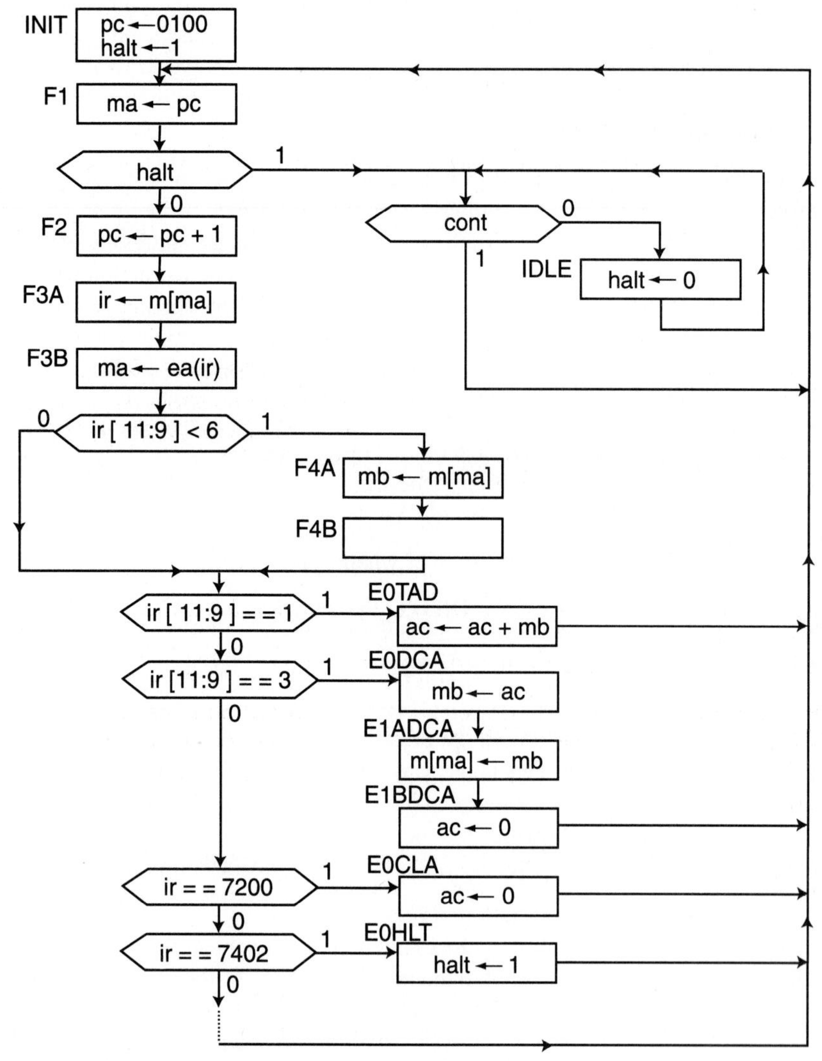

Figure 8-7. ASM implementing four instructions of PDP-8.

For ease of notation, the four-digit constants in the above ASM (0100, 7200 and 7402) are in octal. Smaller constants are shown in decimal. Although the names are arbitrary, the first letter in the names chosen for most of the states indicates the role of the state in the fetch/execute algorithm. States whose names begin with the letter "F" primarily have to do with the fetch aspect of the algorithm. States whose names begin with the letter "E" have to do with execution of machine language instructions. States whose names begin with other letters (e.g., "I") have to do with aspects of the machine besides the fetch/execute algorithm, such as interfacing to the programmer.

When the machine is first powered on, it goes through the INIT state once, where `halt` is initially set to 1, and `pc` is set to an arbitrary address (0100 octal in this example) where we assume the program resides. (Later we will make the location of the program more flexible, but for this limited example, assume the program always starts at 0100.)

When `halt` is 1, the machine proceeds to the IDLE state. When the machine is in the IDLE state, it waits for an external signal, known on the PDP-8 as `cont` ("continue"), that tells it to run another program. When `cont` is true in state IDLE, the machine proceeds to the beginning of the actual fetch/execute algorithm, which starts at state F1. Since state IDLE clears the `halt` register, the machine will not return to state IDLE until the fetch/execute algorithm (F1, F2, F3A, F3B, ...) has executed an HLT instruction.

State F1 is necessary because of the restriction on the use of memory described at the end of section 8.3.1.4. Since `pc` was set up to contain the address of the next instruction that should be executed (either by state INIT for the first time through the fetch/execute algorithm, or by some state inside the algorithm for later instructions), it would be natural to say something like `ir` ← `m[pc]`; however this violates the aforementioned restriction. At the time the machine enters state F1, the `ma` contains no useful information, and so it is possible to copy the program counter into `ma`. This will be used by a later state (F3A) to actually fetch the current instruction into the instruction register.

State F1 also needs to check the halt register to see if the previous instruction that just finished executing was an HLT instruction or if the machine has just been powered up (the machine just came from state INIT). If `halt` is true, the machine proceeds to state IDLE as explained above.

If `halt` is not true, the machine proceeds to state F2. This state prepares the machine for executing the next instruction after the current one by incrementing the program counter. The placement of state F2 is somewhat arbitrary since a copy of the original program counter has been preserved for the moment in the memory address register. (For example, state F2 could be placed after state F3B, which would more closely match the description given in appendix A.)

The next state after state F2 is state F3A. State F3A fetches the instruction stored in memory pointed to by the original program counter, which is now in the memory address register. In the behavioral ASM, we use the same register transfer notation for dealing with memory, `ir←m[ma]`, as is used for dealing with other registers. In later stages of the design, the timing of memory may be somewhat more difficult than that of the internal CPU registers, but at this early stage, we will ignore these details.

The next state after state F3A is state F3B. State F3B calculates the effective address using the information in the instruction register. This calculation is denoted by a function referred to as `ea(ir)`. For example, if the `ir` is 1107_8, `ea(ir)` is 0107_8. If the `ir` is 3111_8, `ea(ir)` is 0111_8. In the later stages of the top-down design process, the `ea` function will be realized using combinational logic. Appendix A assumes there is an additional register for the effective address, but this is not necessary here since state F3B uses the existing `ma` register to hold the effective address. (To implement the complete instruction set of the PDP-8 described in appendix B, more complicated combinational logic is required for `ea`. This is because the `ea` function for some of the addressing modes not implemented in this chapter needs an additional argument.)

State F3B has a decision to determine what state occurs next. If the instruction is a Memory Reference Instruction (MRI), the next state after F3B will be F4A. If the instruction is not an MRI, the next state after F3A will be one that implements the operation requested by the instruction ("E"xecute it). Even though in this section the only MRIs in our PDP-8 subset start with 1 and 3, we will describe how to test for any MRI. In the complete PDP-8 instruction set (given in appendix B), an instruction is MRI if the high order octal digit (three bits) of the instruction is between 0 through 5 inclusive. There are several ways one could write this test. It could be written as `ir < 6000`, however, this does not emphasize that only the high-order three bits of the instruction register determine the outcome. The test could be written as `ir/1000`$_8$ `< 6` or `ir>>9 < 6` to emphasize that the outcome is based on the high-order three bits, but neither one of these tests is the most succinct way to express this. We need a notation that clearly says "just look at these bits." Although the material in this section does not depend on any knowledge of Verilog, Verilog does indeed have such a bit selection notation: `ir[11:9]` says form a three-bit value using bits 11 through 9 of `ir`, which is roughly equivalent to `(ir >> 9) & 7`, which in this case is equivalent to `ir >> 9` since `ir` is 12-bits. For example, if `ir` is 1107_8, `ir[11:9]` is 1. If `ir` is 3111_8, `ir[11:9]` is 3. We will use this aspect of Verilog notation in our ASM because it clearly documents what the hardware will do, which of course is the goal of a behavioral ASM.

If the instruction in the `ir` is MRI, the machine proceeds to state F4A. State F4B loads `mb` with the data that the machine may need to use to execute the memory reference instruction. For example, in the program of appendix A, when the memory reference instruction 1107 is fetched by state F3A, the machine schedules 0107 to be loaded into `ma` at the next rising edge of the clock. Since `ir[11:9]` is 1 (which is less than 6), at that next clock edge the machine proceeds to state F4A. In state F4A, `ma` has just become 0107, and so the contents of memory at that effective address, `m[ma]`, can be loaded into mb. In this example, one clock cycle later (when the machine is in state F4B), `mb` becomes 0152.

State F4B is not necessary. It is included here as a placeholder for operations needed in the ASM to implement the complete instruction set of the PDP-8, including features not described yet.

The bottom of the ASM has a series of decisions that determines which instruction is currently in the instruction register.

This series of decisions is known as the *decoding* portion of the fetch/execute algorithm. For MRI, the decoding decisions happen in state F4B. For non-MRI, the decoding decisions happen in state F3B. Since we are implementing only four instructions in this instruction subset, there are only four decisions required to decode these instructions. The more complex the instruction set, the more difficult it is to do decoding. Most machines, including the complete PDP-8 have a long string of decisions to implement decoding. Notice, from a high-level view, decoding occurs as a series of `if ... else if ... else if ...` style decisions, since each instruction has a unique machine language code.

The remaining states of the machine perform certain actions required during the execution of each specific instruction.

If `ir[11:9]` is 1 in state F4B, the instruction is what the programmer calls a "Twos complement ADd," and so the machine proceeds to state E0TAD. In this state, the machine adds the data from memory at the effective address to the accumulator. In a complete implementation of the PDP-8, other operations are involved with a TAD, but we will ignore those details for the moment. After performing the addition in state E0TAD, the machine has completely executed the TAD instruction and is ready to fetch another instruction. Therefore, the next state after state E0TAD is state F1.

If `ir[11:9]` is 3 in state F4B, the instruction is what the programmer calls a "Deposit and Clear Accumulator" (DCA) and so the machine proceeds to state E0DCA. Although TAD and DCA are both MRIs, the operations involved for the DCA are more complex because the DCA instruction stores the accumulator in memory and then clears the accumulator. It takes three clock cycles to accomplish all the operations required by the DCA instruction. State E0DCA occurs during the first of these three clock cycles. The restrictions on the use of memory described at the end of section 8.3.1.4 require anything that is to be stored in memory be placed in the memory buffer register. State E0DCA schedules that the memory buffer register be assigned a copy of the value in the accumulator at the next rising edge of the clock. This is done in preparation for the next state, which is state E1ADCA.

State E1ADCA stores `mb` (which is now the same as the value of the accumulator) into memory at the effective address. After state E1ADCA, the next state is state E1BDCA.

State E1BDCA schedules that zero be assigned to the accumulator at the next rising edge of the clock. After scheduling the accumulator to be cleared in state E1BDCA, the machine has completely executed the DCA instruction and is ready to fetch another instruction. Therefore, the next state after state E0DCA is state F1.

If `ir` is 7200 in state F3B, the instruction is what the programmer calls a "CLear Accumulator"(CLA) and so the machine proceeds to state E0CLA. Note that since 7200 is not MRI, the decoding occurs earlier (in state F3B) than for the MRI examples above. Also note that for non-MRI-like 7200, all twelve bits of the instruction register must be tested, since there is no effective address. In state E0CLA, the machine schedules that zero be assigned to the accumulator at the next rising edge of the clock. After this, the machine is ready to fetch another instruction, and so the next state after state E0CLA is state F1.

If `ir` is 7402 in state F3B, the instruction is what the programmer calls a "HaLT," and so the machine proceeds to state E0HLT. Again all twelve bits of the instruction register must be tested in state F3B to decode this instruction. In state E0HLT, the machine schedules that one be assigned to the `halt` register at the next rising edge of the clock. The next state after state E0HLT is state F1, not because the machine is going to fetch another instruction, but instead because state F1 is where the test of the `halt` register occurs. (As mentioned above, when `halt` is one in state F1, the machine proceeds to state IDLE.) The final sequence of states through which the machine goes near the end of a program will be F1, F2, F3A, F3B, E0HLT, F1, IDLE, IDLE, IDLE ....

8.3.1.6 Example machine language program

Assume that the following machine language program:

```
0100/7200
0101/1106
0102/1107
0103/1110
0104/3111
0105/7402
0106/0112
0107/0152
0110/0224
0111/0510
```

is present in memory starting at address 0100 when power is turned on. (For an explanation of this program, see appendix A.) The following shows how the ASM proceeds when the external input `cont` is not asserted:

```
INIT    ma=????  mb=????  pc=????  ir=????  halt=?  ac=????
F1      ma=????  mb=????  pc=0100  ir=????  halt=1  ac=????
IDLE    ma=0100  mb=????  pc=0100  ir=????  halt=0  ac=????
```

The question marks indicate an unknown value in registers when power is first turned on.[7] State INIT initializes the `halt` flag so that the machine goes straight from F1 to IDLE. State INIT is also initialized to the starting address of our sample program. The machine stays in IDLE until the external input `cont` is asserted. When it is asserted, the following happens:

```
IDLE    ma=0100  mb=????  pc=0100  ir=????  halt=0  ac=????
F1      ma=0100  mb=????  pc=0100  ir=????  halt=0  ac=????
F2      ma=0100  mb=????  pc=0100  ir=????  halt=0  ac=????
F3A     ma=0100  mb=????  pc=0101  ir=????  halt=0  ac=????
F3B     ma=0100  mb=????  pc=0101  ir=7200  halt=0  ac=????
E0CLA   ma=0000  mb=????  pc=0101  ir=7200  halt=0  ac=????
F1      ma=0000  mb=????  pc=0101  ir=7200  halt=0  ac=0000
```

In state F1, the program counter (0100) is saved in the memory address register. In state F2, the program counter is scheduled to be incremented to become 0101 (as can be seen in state F3A) in preparation for fetching the next instruction four clock cycles later. In state F3A, the instruction register is scheduled to be loaded from memory address 0100. In state F3B, this instruction (7200) becomes available in the instruction register, and since `ir[11:9] >= 6`, the instruction decoding takes place. State E0CLA schedules zero to be loaded into the accumulator. Having completed the fetching and execution of the CLA instruction, the machine performs similar operations to fetch the second instruction. This time, the program counter is 0101 in state F1. The following shows how the fetching and execution of the second instruction proceeds:

[7] In other chapters, a similar idea is denoted with the "x" value in Verilog.

```
F2        ma=0101  mb=????  pc=0101  ir=7200  halt=0  ac=0000
F3A       ma=0101  mb=????  pc=0102  ir=7200  halt=0  ac=0000
F3B       ma=0101  mb=????  pc=0102  ir=1106  halt=0  ac=0000
F4A       ma=0106  mb=????  pc=0102  ir=1106  halt=0  ac=0000
F4B       ma=0106  mb=0112  pc=0102  ir=1106  halt=0  ac=0000
E0TAD     ma=0106  mb=0112  pc=0102  ir=1106  halt=0  ac=0000
F1        ma=0106  mb=0112  pc=0102  ir=1106  halt=0  ac=0112
```

In state F2 the saved program counter (0101) is visible in the memory address register at the same time the program counter is scheduled to be incremented to become 0102 (as can be seen in state F3A). In state F3A, the instruction register is scheduled to be loaded from memory address 0101. In state F3B, this instruction (1106) becomes available in the instruction register, but unlike the above non-MRI, instruction decoding does not take place in state F3B. Instead, state F3B schedules the memory address register to be loaded with the effective address (0106), derived from the instruction register. Since `ir[11:9] < 6` in state F3B, the machine proceeds to state F4A, where the memory buffer register is scheduled to be loaded with the contents of memory (0112) at that effective address, as can be seen in state F4B. In state F4B, instruction decoding takes place. Since, `ir[11:9] == 1`, the machine proceeds to state E0TAD, where the 0112 in the memory buffer register is added to the zero in the accumulator. The remaining two TAD instructions execute in a similar fashion:

```
F2        ma=0102  mb=0112  pc=0102  ir=1106  halt=0  ac=0112
F3A       ma=0102  mb=0112  pc=0103  ir=1106  halt=0  ac=0112
F3B       ma=0102  mb=0112  pc=0103  ir=1107  halt=0  ac=0112
F4A       ma=0107  mb=0112  pc=0103  ir=1107  halt=0  ac=0112
F4B       ma=0107  mb=0152  pc=0103  ir=1107  halt=0  ac=0112
E0TAD     ma=0107  mb=0152  pc=0103  ir=1107  halt=0  ac=0112
F1        ma=0107  mb=0152  pc=0103  ir=1107  halt=0  ac=0264
F2        ma=0103  mb=0152  pc=0103  ir=1107  halt=0  ac=0264
F3A       ma=0103  mb=0152  pc=0104  ir=1107  halt=0  ac=0264
F3B       ma=0103  mb=0152  pc=0104  ir=1110  halt=0  ac=0264
F4A       ma=0110  mb=0152  pc=0104  ir=1110  halt=0  ac=0264
F4B       ma=0110  mb=0224  pc=0104  ir=1110  halt=0  ac=0264
E0TAD     ma=0110  mb=0224  pc=0104  ir=1110  halt=0  ac=0264
F1        ma=0110  mb=0224  pc=0104  ir=1110  halt=0  ac=0510
```

The accumulator now contains the sum of the three numbers (0510). The following shows the execution of the DCA instruction:

```
F2        ma=0104  mb=0224  pc=0104  ir=1110  halt=0  ac=0510
F3A       ma=0104  mb=0224  pc=0105  ir=1110  halt=0  ac=0510
F3B       ma=0104  mb=0224  pc=0105  ir=3111  halt=0  ac=0510
F4A       ma=0111  mb=0224  pc=0105  ir=3111  halt=0  ac=0510
F4B       ma=0111  mb=0000  pc=0105  ir=3111  halt=0  ac=0510
E0DCA     ma=0111  mb=0000  pc=0105  ir=3111  halt=0  ac=0510
E1ADCA    ma=0111  mb=0510  pc=0105  ir=3111  halt=0  ac=0510
E1BDCA    ma=0111  mb=0510  pc=0105  ir=3111  halt=0  ac=0510
F1        ma=0111  mb=0510  pc=0105  ir=3111  halt=0  ac=0000
```

In state F2 the saved program counter (0104) is visible in the memory address register at the same time the program counter is scheduled to be incremented to become 0105 (as can be seen in state F3A). In state F3A, the instruction register is scheduled to be loaded from memory address 0105. In state F3B, this instruction (3111) becomes available in the instruction register. State F3B schedules the memory address register to be loaded with the effective address (0111), derived from the instruction register. Since `ir[11:9] < 6` in state F3B, the machine proceeds to state F4A, where the memory buffer register is scheduled to be loaded with the contents of memory (0000) at that effective address, as can be seen in state F4B. In state F4B, instruction decoding takes place. Since, `ir[11:9] == 3`, the machine proceeds to state E0DCA, where the memory buffer register is scheduled to be assigned the value from the accumulator (0510).[8] In state E1ADCA, this value is stored in memory. State E1BDCA schedules the accumulator to be cleared. Finally, the HLT instruction executes:

```
F2        ma=0105  mb=0510  pc=0105  ir=3111  halt=0  ac=0000
F3A       ma=0105  mb=0510  pc=0106  ir=3111  halt=0  ac=0000
F3B       ma=0105  mb=0510  pc=0106  ir=7402  halt=0  ac=0000
E0HLT     ma=0002  mb=0510  pc=0106  ir=7402  halt=0  ac=0000
F1        ma=0002  mb=0510  pc=0106  ir=7402  halt=1  ac=0000
IDLE      ma=0106  mb=0510  pc=0106  ir=7402  halt=1  ac=0000
IDLE      ma=0106  mb=0510  pc=0106  ir=7402  halt=0  ac=0000
```

The value in the memory address register calculated by state F3B (this value, 0002, becomes visible in state E0HLT) is irrelevant. In hardware, unnecessarily doing a harmless computation sometimes is more efficient than having a decision avoid the computation when it is unwanted.[9] It does not slow the machine, and it is simpler always to load these bits from the instruction register into the memory address register, even, as

[8] Having the ASM proceed through F4A and F4B was unnecessary in this case since state E0DCA does not use the value loaded into memory buffer register by state F4A in the same way E0TAD does. This is harmless but slower than required and was done to simplify the explanation of state F3B.

[9] As the last footnote indicates, whether it is efficient depends on whether extra states, like F4A, are involved or not. Here there are no extra states involved.

in this case, when they are not needed. State E0HLT schedules the `halt` flag to become zero, which causes the machine to go to F1 and then back to IDLE, where the machine will stay (unless `cont` is pressed again).

8.3.2 Including more in the instruction set

The machine described by the ASM in section 8.3.1.5 is rather useless. It was presented only to introduce the essential aspects of the fetch/execute algorithm. Rather than implement a useless subset of instructions in hardware, let's include more of the PDP-8's instructions in our instruction set. For the extended example in this section, we will implement a machine that executes the following PDP-8 instructions:

mnemonic	octal machine language	what the mnemonic stands for
AND	0xxx	AND memory with accumulator
TAD	1xxx	add memory to accumulator (Two's Complement Add)
DCA	3xxx	Deposit accumulator in memory and Clear Accumulator
JMP	5xxx	goto a new instruction (JuMP)
CLA	7200	CLear Accumulator
CLL	7100	CLear Link
CMA	7040	CoMplement Accumulator
CML	7020	CoMplement Link
IAC	7001	Increment ACcumulator
RAL	7004	Rotate Accumulator and link Left
RAR	7010	Rotate Accumulator and link Right
CLACLL	7300	CLear Accumulator and CLear Link
SZA	7440	Skip next instruction if Zero is in Accumulator
SNA	7450	Skip next instruction if Non-zero value is in Accumulator
SMA	7500	Skip next instruction if Minus (negative) value is in Accumulator
SPA	7510	Skip next instruction if Positive (non-negative) value is in Accumulator
SZL	7430	Skip next instruction if Zero is in Link
SNL	7420	Skip next instruction if one (Non-zero) is in Link
HLT	7402	HaLT
OSR	7404	Or Switch "Register" with accumulator

These instructions are explained more fully in appendix B. The first four instructions (AND, TAD, DCA and JMP) are memory reference instructions. As in section 8.3.1, we will only consider direct page zero addressing.

The next eight mnemonics (CLA, CLL, CMA, CML, IAC, RAL, RAR, CLACLL) describe instructions that manipulate the accumulator and the link registers without referencing memory. (The link register was not considered in section 8.3.1.5 because it requires some details that are discussed in section 8.3.2.2.) Although the full PDP-8 instruction set allows for 256 combinations of these operations, we will only consider the eight listed here.

The skip instructions allow conditional execution of the following instruction. If the condition is met, the following instruction does not execute. If the condition is met, the skip acts like a NOP.

The HLT instruction causes the machine to stop executing a program and instead proceed to states that allow the machine to interface with its programmer. Unlike the example in section 8.3.1.5, the ASM in this section will include interface states after the HLT instruction that allow an arbitrary program to be loaded at an arbitrary address any time the programmer wishes. The programmer communicates with the halted machine using an external 12-bit input, `sr`. In the original PDP-8 documentation, the `sr` is known as the switch "register"; however `sr` is not a register. `sr` is an external input bus, very much like the buses `x` and `y` in the division machine of chapter 2. In the physical realization of the PDP-8, the `sr` is simply a set of twelve switches (one for each bit).

The OSR instruction is an unusual kind of input instruction unique to the PDP-8, which is ideal for our purposes in this section. The OSR instruction ORs input coming from the external `sr` bus with the contents of the accumulator. This allows a discussion here of software input without the need for machine language instructions 6xxx.

Even though the PDP-8 is one of the simplest instruction sets ever designed, and we have still chosen to implement only about half of it, you may have a feeling of panic about whether you will ever be able to design such a machine. Have faith—top-down design will come to the rescue.

8.3.2.1 ASM implementing additional instructions

Here is the ASM for the improved machine that implements the instructions listed above:

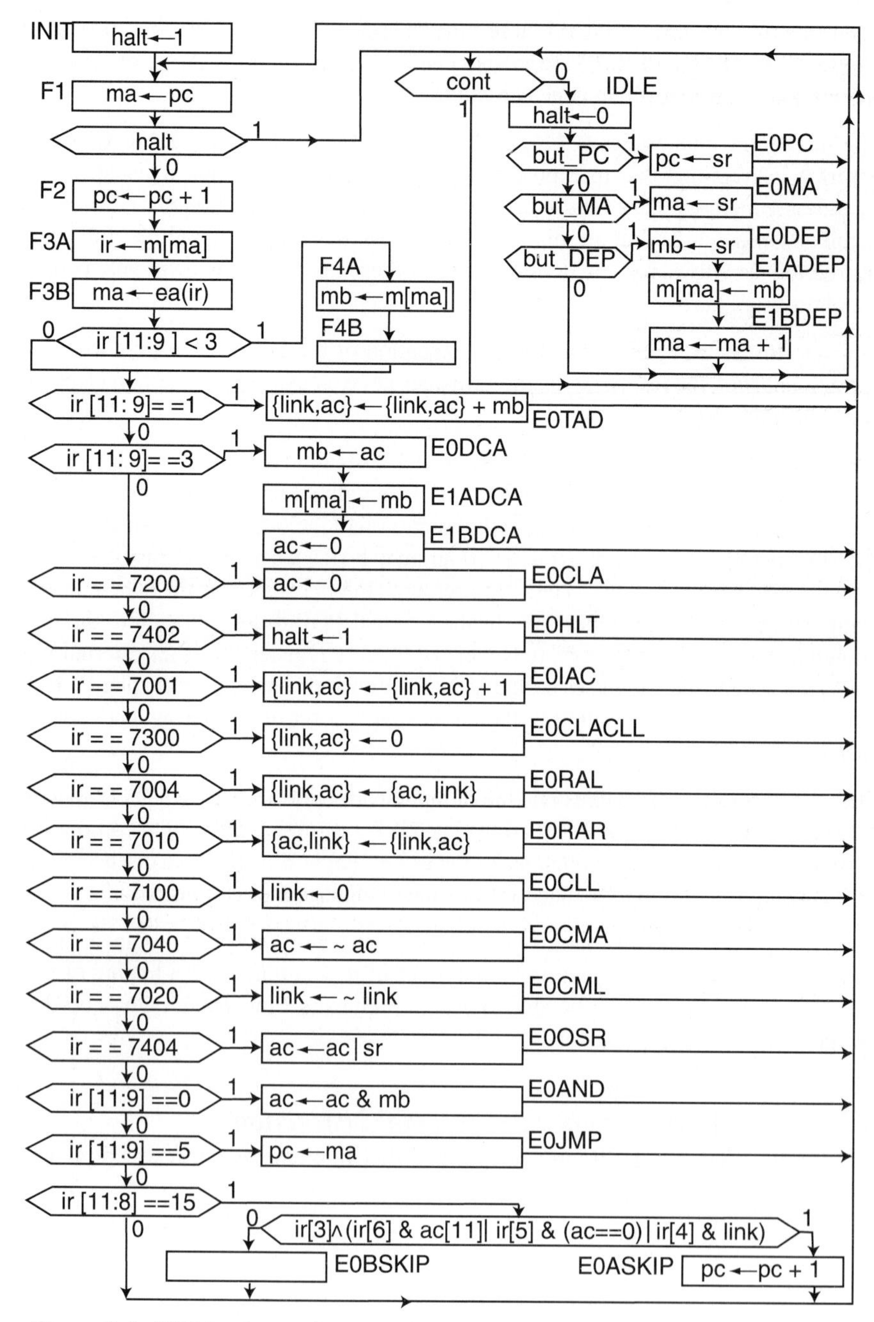

Figure 8-8. ASM implementing more instructions of the PDP-8.

Section 8.3.2.2 describes the states required to execute these additional instructions. Section 8.3.2.3 describes the additional states that allow the programmer a manual interface when the machine is halted.

8.3.2.2 States for additional instructions

Many of the execute states (E0CLA, E0HLT, E0DCA, E1ADCA and E1BDCA) of the improved ASM in section 8.3.2.1 are identical to the states with the same names of the useless ASM in section 8.3.1.5. Therefore, these states will not be discussed here.

Also, the fetch states (F1, F2, F3A, F3B, F4A and F4B) of the improved ASM in section 8.3.2.1 are identical to the states with the same names in the useless ASM of section 8.3.1.5, except the decision whether to go to state F4A is different. As the DCA instruction (3111) used in the example machine language program of section 8.3.1.6 illustrates, it is not necessary for some memory reference instructions to have the memory buffer register initialized. Although it is harmless to do so, it slows the machine down. Therefore, in this section, we will only perform state F4A if it is required. For direct addressing, only AND(0xxx), TAD (1xxx), ISZ (2xxx) and JMS (4xxx) require state F4A. Since JMS and ISZ are not part of the instruction subset implemented in this section (JMS and ISZ are left as exercises), the condition can be restated as `ir[11:9] < 3`. If the full PDP-8 instruction set with all addressing modes were implemented, this condition would be more complicated.

8.3.2.2.1 Instruction described with concatenation

One place where there is a noticeable difference between the useless ASM and the improved ASM is in state E0TAD. This difference is due to the fact that the TAD instruction treats the `link` and the `ac` together as a 13-bit entity. One way to describe this is to say that value inside the CPU is `4096*link + ac`, but the CPU never performs such a wasteful computation. Instead, inside the CPU the 12-bit bus coming out of the `ac` and the one-bit bus coming out of the `link` are joined together to form a 13-bit bus. We need a notation to describe this joining together, technically known as concatenation. Although the material in this section does not depend on any knowledge of Verilog, Verilog does indeed have such a concatenation notation: `{link,ac}` is a 13-bit value. The most significant bit (bit 12) of `{link,ac}` is `link`, the next to most significant bit (bit 11) is `ac[11]`, ... and the least significant bit (bit 0) is `ac[0]`. As a different illustration of concatenation, note that `ir[11:9]` is the same as `{ir[11],ir[10],ir[9]}`.

State E0TAD of the improved ASM properly shows that the TAD instruction affects and is affected by the link. It is perfectly legal for a concatenation to be on the left-hand side of a register transfer. The operation `{link,ac}` ← `{link,ac} + mb` means that the 12-bit `mb` is extended to have 13 bits (by implicitly concatenating a 0 on the left). This is added to the 13-bit {link,ac}. The respective portions of the 13-bit sum are stored back into link and ac. The following table shows four examples of before and after state E0TAD:

before link	ac	mb	after link	ac	mb
0	0040	4001	0	4041	4001
0	4040	4002	1	0042	4002
1	0040	4003	1	4043	4003
1	4040	4004	0	0044	4004

For instance, the last line shows $\{1,4040_8\}+4004_8$ == 1100000100000_2 + 0100000000100_2 == 10000000100100_2 == 20044_8, which is too big to fit in 13 bits, and so the result is 0 in the link and 0044_8 in the accumulator.

There are four other instructions (IAC, CLACLL, RAL and RAR) that are most easily described using concatenation. If the instruction register contains 7001 in state F3B, the instruction is what the programmer calls "Increment ACcumulator," and so the machine proceeds to state E0IAC, which is similar to E0TAD, except one rather than mb is added to {link,ac}. If the instruction register is 7300 in state F3B, the instruction is what the programmer calls a "CLear Accumulator, CLear Link," and so the machine proceeds to state E0CLACLL, where a 13-bit zero is assigned to {link,ac}.

If the instruction register is 7010 in state F3B, the instruction is what the programmer calls a "Rotate Accumulator and link Right," and so the machine proceeds to state E0RAR, where the 13-bit {link,ac} is rotated right. Similarly, if the instruction register is 7004 in state F3B, the instruction is what the programmer calls a "Rotate Accumulator and link Left," and so the machine proceeds to state E0RAL, where the 13-bit {link,ac} is rotated left. Concatenation is the simplest way to describe rotation. Recall that:

```
{link,ac}==
{link,ac[11],ac[10],ac[9],ac[8],ac[7],ac[6],ac[5],ac[4],ac[3],ac[2],ac[1],
ac[0]}
```

and

```
{ac,link} ==
{ac[11],ac[10],ac[9],ac[8],ac[7],ac[6],ac[5],ac[4],ac[3],ac[2],ac[1],ac[0],link}
```

The single 13-bit wide register transfer:

```
{link,ac} ← {ac,link}
```

is a more succinct way to describe 13 separate register transfers, each one bit wide:

```
link   ← ac[11]
ac[11] ← ac[10]
ac[10] ← ac[9]
ac[9]  ← ac[8]
       ...
ac[2]  ← ac[1]
ac[1]  ← ac[0]
ac[0]  ← link
```

Observe that, except for the link, the bits are shifted over one place to the left. The old value of the link "rotates" around to the least significant bit of the accumulator. The following table illustrates examples what is in the link and the accumulator before and after state E0RAL:

before link	before ac	after link	after ac
0	1001	0	2002
0	2002	0	4004
0	4004	1	0010
1	1001	0	2003
1	2002	0	4005
1	4004	1	0011

Although the software uses of the previous instructions were fairly obvious, the RAL instruction may seem a bit strange. In fact, RAL has two uses: arithmetic and logical. The first three lines above illustrate the arithmetic use: if the programmer has previously cleared the link, RAL is like multiplication by two (with overflow in the link). The remaining examples above illustrate the logical use: to rearrange bits without losing any information.

RAR is the inverse of RAL, and the concatenation notation makes this clear:

```
{ac,link} <- {link,ac}
```

In the following, the second, third and last three lines use data from the previous (RAL) table to illustrate that RAR is the inverse of RAL (the rotates do not lose information, they simply rearrange it):

before		after	
link	ac	link	ac
0	1001	1	0400
0	2002	0	1001
0	4004	0	2002
1	1001	1	4400
1	2002	0	5001
1	4004	0	6002
0	2003	1	1001
0	4005	1	2002
1	0011	1	4004

The first three examples above illustrate the arithmetic use of RAR: if the programmer clears the link, RAR is like unsigned division by two (with the remainder in the link).

8.3.2.2.2 Additional non-memory reference instructions

If the instruction register is 7100 in state F3B, the instruction is what the programmer calls a "CLear Link," and so the machine proceeds to state E0CLL, where zero is assigned only to the link (the accumulator is left alone). If the instruction register is 7040 in state F3B, the instruction is what the programmer calls a "CoMplement Accumulator," and so the machine proceeds to state E0CMA, where `~ac` is assigned only to the accumulator (the link is left alone). If the instruction register is 7020 in state F3B, the instruction is what the programmer calls a "CoMplement Link," and so the machine proceeds to state E0CML, where `~link` is assigned only to the link (the accumulator is left alone).

If the instruction register is 7404 in state F3B, the instruction is what the programmer calls "Or Switch Register," and so the machine proceeds to state E0OSR, where the external `sr` input is ORed with the current value of the accumulator. Here is a typical use of this instruction:

```
0002/7402
0003/7200
0004/7404
0005/3100
```

Assume the machine executed instructions, prior to 0002, which are irrelevant to this discussion. When the program wants input from the user, it halts (by executing the HLT instruction, 7402). The machine will proceed to state IDLE, but the program counter remains at 0003. While the machine is halted, the user is free to enter in whatever value is desired on the switches. When the user pushes the `cont` button, the fetch/execute algorithm proceeds with the instruction at 0003, which is a CLA instruction (7200). This is done to get rid of any extraneous value in the accumulator in preparation for the next instruction. The next instruction in fact is the OSR (7404), which ORs zero in the accumulator with the desired value from the external `sr` input. Because zero is the identity for OR (i.e., `0 | sr == sr`), the input value from the switches is loaded into the accumulator. Finally, a DCA instruction stores the input value into memory. The OSR instruction, in conjunction with IDLE state and the `cont` input, provides a simple user interface that will work nicely for the software in this chapter.\

8.3.2.2.3 Additional memory reference instructions

There are six memory reference instructions in the instruction set of the PDP-8. Two of these (TAD and DCA) were described earlier. Two of these (JMS and ISZ) are left as exercises. The other two (AND and JMP) are described in this section.

If `ir[11:9]` is 0 in state F4B, the instruction is what the programmer refers to as "AND," and so the machine proceeds to state E0AND. This state is similar to E0TAD, except the AND instruction only changes the accumulator. (AND leaves the link register alone.) Recall that & is the bitwise AND operator, and so the register transfer:

```
ac ← ac & mb
```

is equivalent to:

```
ac[11] ← ac[11] & mb[11]
ac[10] ← ac[10] & mb[10]
ac[9]  ← ac[9]  & mb[9]
         ...
ac[2]  ← ac[2]  & mb[2]
ac[1]  ← ac[1]  & mb[1]
ac[0]  ← ac[0]  & mb[0]
```

If `ir[11:9]` is 5 in state F4B, the instruction is what the programmer calls a "JuMP," and so the machine proceeds to state E0JMP. All general-purpose computers have some kind of jump (sometimes known as branch) instruction. The purpose of a jump instruction is to modify the program counter. The jump instruction allows high-level language features (such as loops and decisions) to be translated into machine language.

Although the jump instruction of the PDP-8 is categorized as a memory reference instruction, it does not actually reference memory. It simply takes the effective address (from the memory address register) and uses this as the new value of the program counter.

8.3.2.2.4 Skip instructions

High-level language programs are composed of statements like `if` and `while`. The JMP instruction by itself is not enough to translate such programs. For this reason, the PDP-8 instruction set includes several "skip" instructions. These instructions test to see whether the value in the accumulator (or link) meets certain conditions. If it does, the next instruction will be skipped. If the condition is not met, the next instruction will execute normally. The following table illustrates several skip instructions, and how they are encoded in machine language.

mnemonic	octal ir	ir[6]	ir[5]	ir[4]	ir[3]	
SMA	7500	1	0	0	0	Skip if Minus (negative) Accumulator
SZA	7440	0	1	0	0	Skip if Zero Accumulator
SNL	7420	0	0	1	0	Skip if Non-zero Link
SPA	7510	1	0	0	1	Skip if Positive Accumulator
SNA	7450	0	1	0	1	Skip if Non-zero Accumulator
SZL	7430	0	0	1	1	Skip if Zero Link

If `ir[11:8]` is 15 in state F3B, the instruction is one of the above skip instructions. If the condition is met, the machine proceeds to state E0ASKIP, where the program counter is scheduled to be incremented an extra time. If the condition is not met, the machine proceeds to state E0BSKIP, where the machine leaves the program counter the way it was.

The condition is determined by `ir[6:3]`. `ir[3]` is a bit that reverses the meaning of the instruction; hence `ir[3]` is 0 for SMA, SZA and SNL, but `ir[3]` is 1 for SPA, SNA and SZL. (If you think about it, you will realize SMA, SZA and SNL, respectively, are the opposites of SPA, SNA and SZL.) `ir[6]` is 1 for SMA and SPA. `ir[5]` is 1 for SZA and SNA. `ir[4]` is 1 for SNL and SZL. Therefore, the condition that decides whether to proceed to state E0ASKIP is:

```
ir[3]^(ir[6]&ac[11]|ir[5]&(ac==0)|ir[4]&link)
```

where ^ is the exclusive OR, which reverses the meaning of the parenthesized expression when `ir[3]` is one. Note: `ac[11]` is the "sign" bit of the accumulator (the bit that indicates 12-bit negative twos complement values).

As an illustration of how a programmer uses the skip instructions in conjunction with the other instructions, consider the unsigned greater than or equal decision. Suppose `r1` (stored at 0032) and `y` (stored at 0101) are software variables that contain 12-bit unsigned numbers. Should the high-level language programmer wish to test (in either an `if` or a `while`) to see whether `r1` is greater than or equal to `y`, there are several equivalent ways to write this (given that the following is performed in 13-bit twos complement arithmetic):

```
r1                >= y
r1                >= {0,y}
-{0,y}      + r1 >= 0
{~0,~y}+1  + r1 >= 0
```

The last of these can be performed with the instructions described earlier. The final signed 13-bit result in `{link,ac}` can be tested with the SZL instruction, as shown below:

```
0014/7200    CLA
0015/7100    CLL       /{link,ac} = {0,0}
0016/1101    TAD y     /{link,ac} = {0,y}
0017/7040    CMA       /{link,ac} = {0,~y}
0020/7020    CML       /{link,ac} = {~0,~y}
0021/7001    IAC       /{link,ac} = {~0,~y}+1
0022/1032    TAD r1    /{link,ac} = {~0,~y}+1 + r1
0023/7430    SZL       /test whether {~0,~y}+1 + r1 >= 0
0024/5xxx    JMP xxx   /if {~0,~y}+1 + r1 < 0, goto xxx
             ...       /if {~0,~y}+1 + r1 >=0, execute here
```

8.3.2.3 Extra states for interface

The ASM shown in section 8.3.2.1 has three additional states (E0MA, E0PC and E0DEP) that allow the programmer to interface with the machine using the `sr` input when the machine is not performing the fetch/execute algorithm. These states only occur when the programmer pushes buttons (`but_MA`, `but_PC` and `but_DEP`, respectively) during the time that the machine is in state IDLE.

State E0PC allows the programmer to load the the program counter with a value previously placed on the switches. This allows a program to reside anywhere in memory, unlike the nearly useless ASM given in section 8.3.1. For this reason, state INIT no longer initializes the program counter, and instead the programmer is responsible for pushing `but_PC` appropriately prior to pushing `cont`.

The programmer uses `but_MA` and `but_DEP` together to load a program into memory at an arbitrary address. First, the programmer enters the address on the switches to indicate where in memory the programmer desires to place the program or data. Then the programmer pushes `but_MA`, which causes state E0MA to occur, where the `sr` input is assigned to the memory address register. Next, the programmer enters the first word to go into memory onto the switches, and pushes `but_DEP`, which causes state E0DEP to occur. State E0DEP assigns the `sr` input to the memory buffer register, and then the machine proceeds to state E1ADEP. In state E1ADEP the machine deposits the memory buffer (containing the programmer's desired word) into memory at the memory address. Finally, state E1BDEP increments the memory address (in case the programmer has more words to deposit). The programmer may enter as many successive words as desired with this technique. Finally, the programmer uses `but_PC` and `cont` as described above.

8.3.2.4 Memory as a separate actor

At this point, we have described the behavior of the complete general-purpose computer system. Now, we need to consider what the external inputs and outputs of this system are:

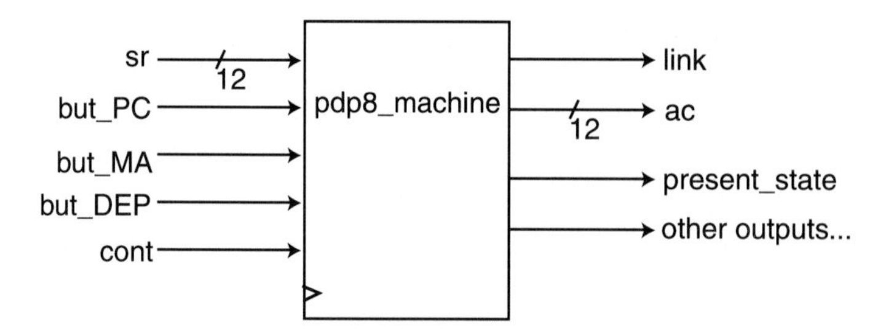

Figure 8-9. Block diagram for the PDP-8 system.

The present state and the other outputs (such as the memory buffer register, memory address register, program counter, instruction register and the halt register) are sent out from the machine primarily to allow the programmer to observe the internal operation of the machine.

8.3.2.4.1 Top-level structure of the machine

In fact, the machine hidden inside the last diagram is composed of three components: the CPU, the peripherals and the memory. Since in this chapter we are ignoring the peripherals, that leaves two separate components that must be interconnected to form the system: the CPU and the memory. Although one could consider the memory as just another component of the CPU's architecture, this is normally not done. As described in section 8.2, there are many different technologies for memory, and often the technology to implement memory is different than the technology used to implement the CPU. Therefore, we would like to physically separate the memory from the CPU and design the CPU independently from the memory. This means that memory is an independent actor, as illustrated in the following diagram:

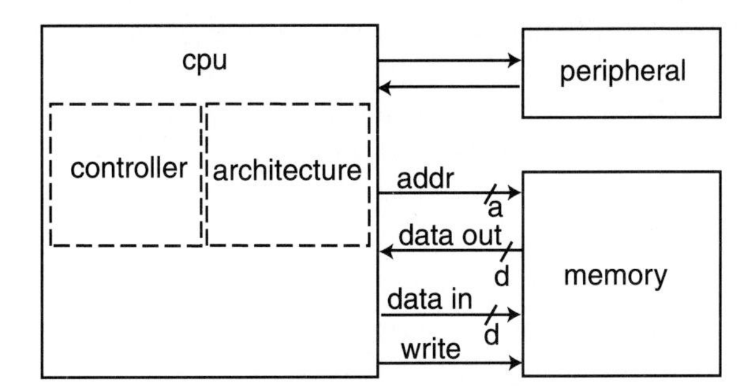

Figure 8-10. System composed of processor (controller and architecture) with memory as a separate actor.

Let's assume that we will implement this machine using an asynchronous, volatile, static, deterministic access time memory with separate data input and data output. The choice of this kind of memory device simplifies the design in several ways. First, since this memory is static, there is no need to refresh it. Second, since the access time is known, proper functioning is easily guaranteed by choosing a sufficiently slow clock. Third, since this memory has separate buses for data input and data output, there is no need to introduce the complexity of tri-state buffers.

The one design complexity that must be dealt with is the fact that this memory is asynchronous. The reason for choosing an asynchronous memory device is cost and availability. The problem with doing so is that extra care must be taken in providing the inputs to the memory device. In particular, the register transfer `m[ma]` ← `mb` must be implemented by asserting a hazard-free external (to the CPU) command output, `write`, as described in section 8.2.2.1. The memory address and memory buffer outputs of the CPU provide the `addr` and `din` inputs to the memory device. The `dout` of the memory device provides the `membus` input to the CPU.

8.3.2.4.2 Pure behavioral ASM with memory as a separate actor

The ASM of section 8.3.2.1 can be rewritten to reflect that memory is a separate actor. Every place (states F3A and F4A) where `m[ma]` is used on the right-hand side of a register transfer in section 8.3.2.1, the revised ASM will use `membus` instead. All other places (states E1ADEP and E1ADCA) that mention `m[ma]` are of the form `m[ma]` ← `mb`. These can be replaced with an assertion of the `write` signal, as illustrated in figure 8-11.

8.3.2.5 Our old friend: division

This book uses the childish division algorithm (first described in section 2.2) in most chapters to illustrate various ways that hardware can be designed. This algorithm is ideal as a learning example because it is simple. Although the operations used to implement this algorithm are typical of the most sophisticated algorithms, it is so elementary that any child can perform it. Unlike faster division algorithms, why it works is obvious.

There is another reason why division is the centerpiece of this book. Division has played an interesting role in the history of general-purpose computers. As mentioned previously, the very first program ever run on a general-purpose computer was the childish division algorithm. Much faster algorithms than the childish algorithm exist for division, but they are very complex and hard to understand. Many general-purpose computers throughout history (from the BINAC in 1949 up to the Pentium half a century later) have provided "divide" instructions that implement much more sophisticated division algorithms in hardware than our old friend, the childish algorithm. Despite this, many computers, including many PDP-8s,[10] have shunned division in hardware in favor of division in software. The irony of this is that a flaw in the hardware divide instruction of the Pentium general-purpose computer caused Intel great embarrassment in the mid 1990s.

[10] Some models of the PDP-8 had an optional hardware feature, known as EAE, that assisted in performing division.

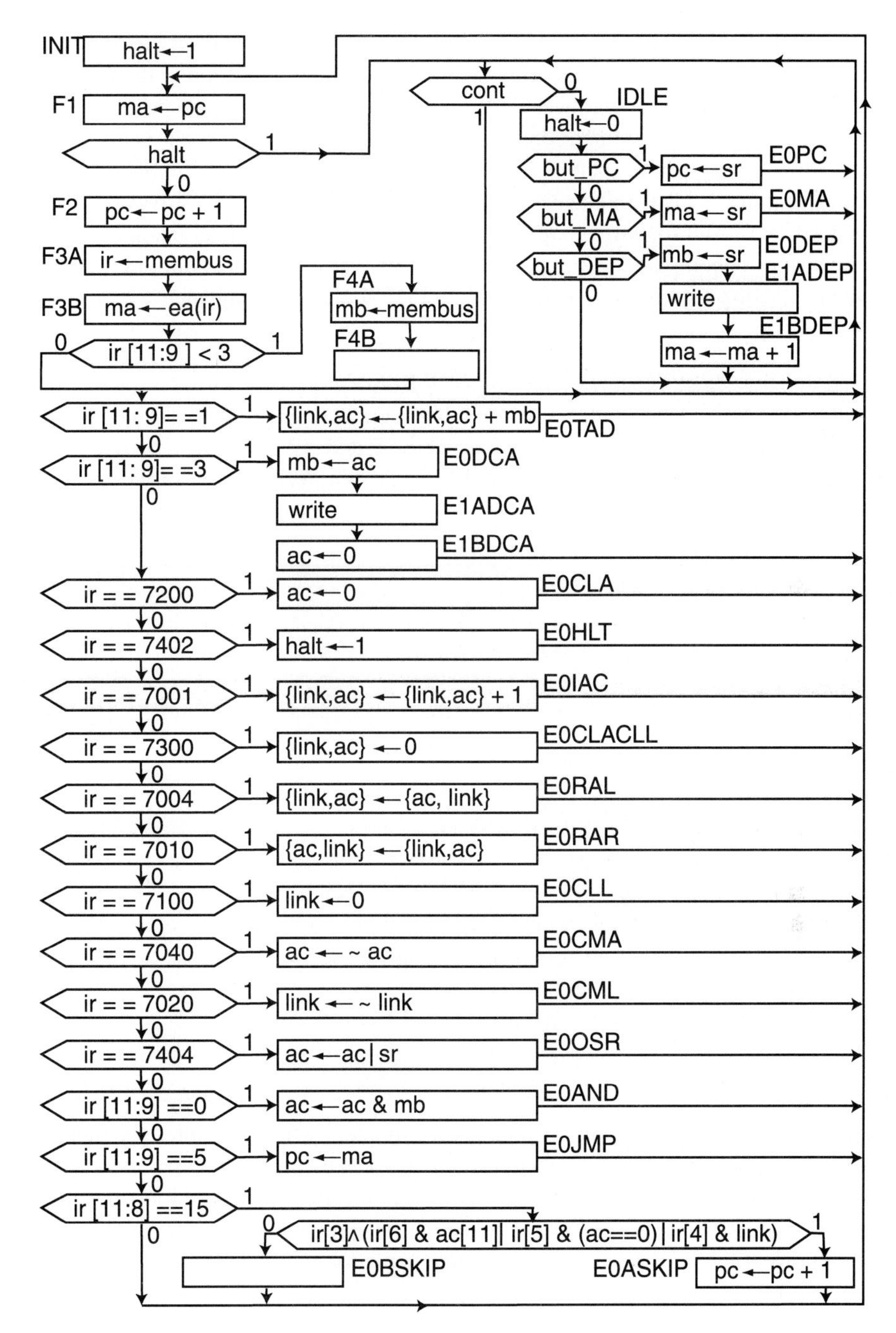

Figure 8-11. ASM with memory as a separate actor.

A major theme of this book is that speed is not the primary concern of the designer—**correctness** is. An algorithm implemented in software might be slower than the same algorithm implemented in hardware, but you should not worry about speed. The most important thing is to implement the algorithm properly, whether in hardware or software. (Sometimes speed is part of the specification of a correct design, but even then, as the Pentium incident indicates, the rest of the design must be correct before the speed matters.) And so, with an appreciation of the important role division has played in the history of computer design, let us consider how to implement our old friend, the childish division algorithm, in software with PDP-8 machine language.

8.3.2.5.1 Complete childish division program in C

A complete software program written in a high-level language typically has some input/output formatting statements, such as `scanf` and `printf` in C or READ and WRITE in Pascal (or `$readmemb` and `$display` in Verilog). For instance, a complete C program to implement the the childish division algorithm from section 2.2 might appear as follows:

```
main ()
  {
    unsigned x,y,r1,r2;

    while (1)
      {
        scanf("%04o",&x);
        scanf("%04o",&y);
        r1 = x;
        r2 = 0;
        while (r1 >= y)
          {
            r1 = r1 - y;
            r2 = r2 + 1;
          }
         printf("%04o\n",r2);
      }
  }
```

Since octal is a convenient notation that is used frequently in this chapter, the input and output are shown formatted as four-digit octal numbers by `scanf` and `printf`. Translating statements like `scanf` and `printf` into the PDP-8 instruction set requires using 6xxx machine language instructions. We have avoided implementing the 6xxx instructions of the PDP-8 because their implementation requires concepts not covered in this chapter.

8.3.2.5.2 User Interface for the software

The purely hardware implementations of the childish division algorithm in chapter 2 avoided these input/output formatting problems with:

a) two separate external data input buses (`x` and `y`), presumably connected to switches, so that the "friendly" user can toggle in the binary values desired to be input into the hardware.

b) an external status input (`pb`), presumably connected to a push button, so the "friendly" user can indicate the proper time for the hardware to look at the external input buses.

c) an external data output bus (`r2`), presumably connected to lights, so the "friendly" user can observe the computation of the quotient in binary as it progresses. (If the clock is fast enough, the user will not notice anything except the result.)

d) an external command output (READY), presumably connected to a light, so the "friendly" user can know when `r2` has become the correct quotient.

This approach requires that the person who uses the hardware described in chapter 2 be "friendly," someone who is willing and able to adhere to several rules that describe how to use the machine properly. Some of these rules relating to the timing of the push button were described in chapter 2. Although chapter 2 explained the operation of this machine in decimal notation, the use of switches as a physical input medium and lights as a physical output medium additionally demands a user who is comfortable with the binary number system.

To translate the childish division algorithm into PDP-8 machine language, we are going to assume a similar "amicable" user. This "amicable" user is willing to toggle binary values into the `sr` to provide inputs to the algorithm (as described in section 8.3.2.2.2) and observe the binary result in the accumulator. The main difference between the "amicable" user of the algorithm implemented in PDP-8 machine language and the "friendly" user of the algorithm implemented in the hardware given in chapter 2 is how the user must operate the machine. The PDP-8 software version (which has two HLT instructions) re-uses the single `sr` input bus so that the "amicable" user must press the `cont` button twice. The hardware in chapter 2 uses two separate input buses (`x` and `y`) so that the "friendly" user only presses the `pb` button once.

The following summarizes the hardware of the PDP-8 that is utilized for crude input and output by the software implementation of the childish division algorithm:

a) one external data input bus (`sr`), connected to twelve switches, so the "amicable" user can toggle in the desired binary input values.

b) an external status input (`cont`), connected to a push button, so the "amicable" user can push `cont` after toggling each separate value in on the `sr`.

c) an external data output bus (ac), connected to twelve lights, so the "amicable" user can observe the computation of the quotient (and other things) in binary as it progresses. (If the clock is fast enough, the user will not notice anything except the result.)

d) an external command output (present_state), connected to lights, so the "amicable" user can know when ac has become the correct quotient (i.e., when present_state becomes IDLE).

8.3.2.5.3 *Childish division program in machine language*

On the left side of the following is the PDP-8 machine language code for the childish division algorithm. On the right side is the corresponding assembly language mnemonics and symbolic operands, in the style explained in appendix A, with comments following the slash. Only the machine language resides in memory. The commented assembly language is shown only to clarify how the program operates:

```
          / Childish division algorithm in PDP-8 machine language
          /
          / The following 2 instructions allow the user to observe
          / the previous result (r2) in the ac

0000/7300 L0, CLACLL      /{link,ac} = {0,0}
0001/1033     TAD r2      /{link,ac} = {0,r2}
          /
          / The following 4 instructions wait for the user to toggle in
          / the first value on sr (while still displaying r2 on ac)
          / When the user presses cont, this first value toggled
          / in on sr is stored into x

0002/7402     HLT         /wait for user to toggle first value
0003/7200     CLA         /           {link,ac} = {0,0}
0004/7404     OSR         /           {link,ac} = {0,0|sr}
0005/3100     DCA x       /x = sr;    {link,ac} = {0,0}
          /
          / The following 3 instructions wait for the user to toggle in
          / the second value on sr (ac is now cleared)
          / When the user presses cont, this second value toggled
          / in on sr is stored into y

0006/7402     HLT         / wait for user to toggle second value
0007/7404     OSR         /           {link,ac} = {0,0|sr}
0010/3101     DCA y       /y = sr;    {link,ac} = {0,0}
          /
          / The following 3 instructions initialize r1 and r2
          / prior to the while loop
```

Continued

```
0011/1100     TAD x       /          {link,ac} = {0,0+x}
0012/3032     DCA r1      /r1 = x;   {link,ac} = 0
0013/3033     DCA r2      /r2 = 0;
         /
         / The following 9 instructions implement
         /  while (r1>=y)

0014/7200     CLA
0015/7100     CLL         /          {link,ac} = {0,0}
0016/1101     TAD y       /          {link,ac} = {0,y}
0017/7040     CMA         /          {link,ac} = {0,~y}
0020/7020     CML         /          {link,ac} = {~0,~y}
0021/7001     IAC         /          {link,ac} = {~0,~y}+1
0022/1032     TAD r1      /          {link,ac} = {~0,~y}+1 + r1
0023/7430     SZL         / test whether {~0,~y}+1 + r1 >= 0
0024/5000     JMP L0      / if {~0,~y}+1 + r1 < 0, exit while (goto L0)
                          / if {~0,~y}+1 + r1 >=0, stay in while loop

         /The following 5 instructions implement body of the while loop

0025/3032     DCA r1          /r1 = r1 - y;  i.e. r1=~y+1+r1; ac=0
0026/1033     TAD r2          /              {link,ac} = {0,0+r2}
0027/7001     IAC             /              {link,ac} = {0,0+r2}+1
0030/3033     DCA r2          /r2 = r2+1
0031/5014     JMP L1      /continue while loop
         /
         / The following 2 words store data manipulated by
         / the childish division algorithm
         /
0032/0000 r1, 0000
0033/0000 r2, 0000
         /
         / The following 2 words store data input from the sr
         /
0100/0000 x,  0000
0101/0000 y,  0000
```

8.3.2.5.4 Analysis of childish division software

The following table summarizes how many clock cycles it takes for each part of the childish division program given in section 8.3.2.5.3 to execute:

High-level Operations	Before while	During while	After while
r1 = x;	14		
r2 = 0;	7		
while (r1 >= y)		44	44
{			
r1 = r1 - y;		7	
r2 = r2 + 1		24	
}			5
display r2 in accumulator and halt			18
Total Clock Cycles	21	75	67

The times are listed in three columns. The left column indicates operations that execute just once, before the `while` loop begins. The right column indicates operations that execute just once, upon exiting from the `while` loop. The middle column indicates operations that occur **each** time through the loop. The `while` statement itself involves formation of the 13-bit twos complement (32 cycles) and testing (12 cycles). The entry 44 (32+12) occurs both in the middle and right columns because this machine code occurs each time through the loop as well as the final time when the condition `r1>=y` becomes false. Just as in chapter 2, the number of times the loop executes is proportionate to the quotient. Neglecting how long it takes for the user to toggle in the inputs, from the time the program actually starts computing the quotient (when the program counter was 0011) until the machine returns to state IDLE is `21 + 67 + 75*quotient` clock cycles.

8.3.2.5.5 Comparison with special-purpose implementation

This table compares different implementations of the childish division algorithm:

	section	clock cycles	12-bit registers	12-bit memory words	ctrl states
	2.2.7	3+quotient	3	0	2
special	2.2.3	2+2*quotient	2	0	4
purpose	2.2.2	3+3*quotient	2	0	5
hardware	2.2.5	2+3*quotient	3	0	5
PDP-8 software	8.3.2	88+75*quotient	5	30	31

The "section" column indicates where the ASM (and in the case of the PDP-8 software, also the machine language) is defined. The "clock cycle" column indicates how long it takes to compute the quotient, neglecting the time for the user to toggle in the binary inputs. The "12-bit registers" indicates how many **hardware** registers of this size are required (in the case of the PDP-8, this is the accumulator, instruction register, memory address register, memory buffer register and program counter). We neglect one bit registers such as halt and link as being insignificant in the total cost. We also neglect the cost of the combinational logic that interconnects the registers within the architecture (since this is the subject of section 8.4). The "12-bit memory words" indicates how many words the machine language version requires for both program and data. Special-purpose implementations of this algorithm do not need memory because the hardware registers continually hold the data, and the controller implements the algorithm. The "ctrl states" indicates how many states are required by the **hardware** ASM.

Any way you look at the above table, software appears to be a real loser. Compared to the fastest special-purpose implementation listed above (section 2.2.7), the software approaches being about seventy-five times slower for a large `quotient`:

```
  Lim                    (88+75*quotient)/(3+quotient) = 75
quotient  →  ∞
```

For the particular case traced above and in chapter two (quotient=14/7), the ratio is 238/5, or about 47 times slower. One reason why the hardware implementation in section 2.2.7 makes the software look so bad is because the hardware does the equivalent of three high-level operations (test `r1>=y`, `r1=r1-y` and `r2=r2+1`) in parallel during **each** clock cycle. The childish division algorithm has the potential for this parallelism, and so we ought to exploit this.

On the other hand, if we wanted to handicap the hardware to make the contest seem more sporting, the ASM of section 2.2.2 is the closest to the software implementation because it only does one high-level operation at a time. For very large `quotient`, the software approaches being about 25 times slower than section 2.2.2:

```
  Lim                    (88+75*quotient)/(3+3*quotient) = 25
quotient  →  ∞
```

For the particular case traced above and in chapter two (`quotient`=14/7), the ratio is 238/9, or about 26 times slower.

Even when the hardware only does one thing at a time (as in section 2.2.2), the software appears much slower. There are two reasons for this. First, it takes several PDP-8 instructions to do the equivalent of one high-level language statement (which is most noticeable in implementing the `while`). Second, the way we have implemented the ASM for the PDP-8, it takes several clock cycles (either five or seven) for each instruction to execute.

Software requires general-purpose hardware in order to run. The PDP-8 is about as simple as a general-purpose computer can be, but even so, it requires five registers. Software also requires memory for programs and data. Because of technological differences explained in section 8.1, the cost to store a bit in memory is usually several times lower than to store a bit in a register. For the sake of argument, say that the cost for a 12-bit word in memory is five times cheaper than for a 12-bit register. The storage costs are then 2*5 for section 2.2.2 hardware, 3*5 for section 2.2.7 hardware, and 5*5+30 for the PDP-8 implementation (assuming we only pay for the memory actually used to implement the childish division program). Therefore, section 2.2.2 storage cost is about one fifth that of the PDP-8 implementation, and section 2.2.7 storage cost is about one-quarter that of the PDP-8 implementation.

8.3.2.5.6 Trade-off between hardware and software

One cannot draw sweeping conclusions having examined only a single algorithm in hardware and software, and having examined the software on a single implementation of a single instruction set. The difference between hardware and software may be less pronounced when the algorithm is more complicated or when the instruction set is more capable. In particular, algorithms that require memory for storage of data structures, such as arrays, may show software performance closer to that of special-purpose hardware. However, for the childish division algorithm, we can conclude the software solution gives lower performance and costs more.

Would you pay more to buy something slower? Paradoxically, in most instances, you probably would because hardware speed and cost are often not the primary concern. Certainly in this case, speed is unimportant when you consider the problem that the childish division algorithm solves. It interactively obtains two 12-bit inputs, divides them in a very inefficient way,[11] and displays the answer. It is going to take the user several seconds to toggle in the inputs, and several more for the user to comprehend the output. Since the largest 12-bit quotient is 4095, the maximum total time for the PDP-8 implementation is 88+75*4095 = 307213 clock cycles. Although this seems awful in comparison to the 4098 clock cycles required by the hardware implementation of section 2.2.7, it is less than the blink of an eye when the clock period is 100 nanoseconds

[11] Regardless of the underling implementation (hardware or software), there are much better algorithms than the childish division algorithm if you really want to divide fast.

(a very moderate clock speed using current integrated circuit technology). The user will only see a brief flash before the correct answer appears. Occasionally, the specification of a problem has a real-time aspect to it. For example, if instead of our friendly user, the input came from another machine that needed to divide two thousand numbers per second, only the hardware of section 2.2.7 would be able to keep up.

In most instances however, the factor that matters more than hardware speed and cost is design speed and cost. In other words, how long does it take and how much does it cost for the designer to produce a correct design? Designers are willing to use hardware which is, in a technological[12] sense, more costly and slower than is theoretically necessary because in doing so they obtain the benefit of rapid debugging. When a designer finds an error, it is easier to change a few bits in the memory of a general-purpose computer than it is to fabricate a corrected version of a special-purpose computer. Also, many design changes occur not because of a designer's mistake but instead are required due to changing specifications. Productivity tools for general-purpose computers, such as compilers, assemblers, linkers, editors, debuggers, etc., make the software designer's task of coping with bugs and changing specifications much easier than the above machine language examples.

The situation that has existed for the last half century is designers have had the choice between using a general-purpose computer or building a special-purpose computer. If the market price (in dollars, rather than in gates) of the general-purpose computer is within budget and its speed is adequate (not the fastest, just adequate) and otherwise meets physical constraints (size, weight, power consumption, ruggedness) for the intended application, the designer typically chooses the general-purpose computer because of the advantages of rapid debugging. Although most algorithms work adequately on general-purpose computers, some demand special-purpose hardware. This has created two different economic phenomena.

First is the emergence of the general-purpose computer industry, composed of only a handful of companies worldwide that actually design CPUs. All together, these companies employ only a few hundred computer designers at best, and so few of the readers of this book will ever be employed as general-purpose computer designers. These designers face a daunting challenge: they design machines that will be used for tasks that no one has yet conceived. Programmers in the future will think of new things to do with the general-purpose machines that designers are working on today. Why does knowing how the machine will be used assist the designer? Speed is not the primary concern of a designer solving a specific problem because the designer can easily tell if the machine is fast enough. A special-purpose computer does not have to be the fastest computer in existence—it just has to be fast enough, and, of course, do its job correctly. General-purpose computer designers do not have the luxury of knowing what is fast

[12] Here "technological" means measuring cost in in terms of registers, gates, chip area, etc., rather than in dollars.

enough. Because of the market pressures created by this uncertainty, they have developed more efficient (but intricate) variations on the fetch/execute algorithm that allow software to approach the speed of a special-purpose computer. This has come at the cost of increased hardware by using sophisticated techniques, such as pipelining (chapters 6 and 9), and is why there is such variety among the instruction sets (chapter 10).

Second, a more recent phenomenon is the emergence of hardware description languages (chapter 3) running on general-purpose computers that allow the debugging (simulation) and synthesizing of efficient special-purpose computers to be almost as easy as the programming of software. It is the theme of this book that the worlds of hardware and software are converging. You will need to be aware of both of these to prosper in the next half century of the computer age.

8.4 Mixed fetch/execute

In order to illustrate that there is nothing mysterious about the design of a general-purpose computer once the details of the fetch/execute algorithm are specified, let's translate the pure behavioral ASM (section 8.3.2.4.2) for the PDP-8 instruction subset into the mixed stage of the top-down design process. Recall from section 2.1.5.2, the mixed stage consists of two hardware structures: a controller and an architecture.

There are many possible architectures that can implement a given pure behavioral ASM. The more complicated the ASM, the more room there is for creativity in the design of the architecture. Once the designer decides upon an architecture, the design of the controller is a relatively mechanical process.

8.4.1 A methodical approach for designing the architecture

When an ASM uses more than a handful of registers and/or states, it becomes difficult to keep track of all of the details in your head. In such an instance, it is wise to take a methodical approach to designing the architecture. To begin with, note all of the register transfers that occur in each state. Write down this information grouped together by destination. Since in section 8.3.2.4.2 there are six possible destinations (left-hand sides of ←, excluding the memory, which in section 8.3.2.4.2 is a separate actor), there are six groups to note:

a) register transfers to the accumulator and/or lind. (These are together in one group since `{link,ac}` often acts as a 13-bit register, and so modifications to the accumulator by itself or to the link by itself should be considered as modifications to `{link,ac}`);

b) register transfers to the halt flag;
c) register transfers to the instruction register;
d) register transfers to the memory address register;
e) register transfers to the memory buffer register; and
f) register transfers to the program counter.

It is wise to write down the state(s) in which each transfer occurs so that you can refer back to the ASM as necessary. (When both the right-hand sides and left-hand sides of register transfers in two or more states are identical, note the names of all such states.) The following table illustrates this for the ASM in section 8.3.2.4.2:

RTN	State(s)
ac ← 0	E0CLA, E1BDCA
ac ← ac & mb	E0AND
ac ← ac | sr	E0OSR
ac ← ~ac	E0CMA
link ← 0	E0CLL
link ← ~link	E0CML
{ac,link} ← {link,ac}	E0RAR
{link,ac} ← {ac,link}	E0RAL
{link,ac} ← 0	E0CLACLL
{link,ac} ← {link,ac} + 1	E0IAC
{link,ac} ← {link,ac} + mb	E0TAD
halt ← 0	IDLE
halt ← 1	E0HLT,INIT
ir ← membus	F3A
ma ← ea(ir)	F3B
ma ← ma + 1	E0BDEP
ma ← pc	F1
ma ← sr	E0MA
mb ← ac	E0DCA
mb ← membus	F4A
mb ← sr	E0DEP

Continued

RTN	State(s)
pc ← ma	E0JMP
pc ← pc + 1	F2, E0ASKIP
pc ← sr	E0PC

Note that implicitly, the `{link,ac}` group should be thought of as implementing the following register transfers:

{link,ac} ← {link,0}	E0CLA, E1BDCA
{link,ac} ← {link,ac & mb}	E0AND
{link,ac} ← {link,ac | sr}	E0OSR
{link,ac} ← {link,~ac}	E0CMA
{link,ac} ← {0,ac}	E0CLL
{link,ac} ← {~link,ac}	E0CML
{ac,link} ← {link,ac}	E0RAR
{link,ac} ← {ac,link}	E0RAL
{link,ac} ← 0	E0CLACLL
{link,ac} ← {link,ac} + 1	E0IAC
{link,ac} ← {link,ac} + mb	E0TAD

These two ways of describing link and accumulator register transfers are equivalent. The former is easier for the designer to comprehend. The latter is important in the next step the designer takes.

8.4.2 Choosing register types

Here is where the creative part occurs. Whatever hardware structure the designer chooses must be capable of implementing each of the above register transfers during the state(s) indicated. The controller will take care of making sure the states happen at the proper times, so we do not have to worry about that. Our concern now is that the architecture can manipulate the data as listed above.

The first decision the designer must make is what kind of structural device will implement each register. One possibility would be to use enabled registers for every variable in the ASM (other than memory); however, this will typically cause the architecture to be more complex than if other types of registers are selected. A better approach is to look at each group (corresponding to transfers to a particular register) individually and note those register transfers where the right-hand side consists only of constants and/or the variable (or concatenated variables) on the left-hand side.

For the link and accumulator group, there are several such register transfers:

```
{link,ac} ← {link,0}            E0CLA, E1BDCA
{link,ac} ← {link,~ac}          E0CMA
{link,ac} ← {0,ac}              E0CLL
{link,ac} ← {~link,ac}          E0CML
{link,ac} ← 0                   E0CLACLL
{ac,link} ← {link,ac}           E0RAR
{link,ac} ← {ac,link}           E0RAL
{link,ac} ← {link,ac} + 1       E0IAC
```

For the halt flag, both of the possible register transfers are of this kind:

```
halt ← 0                        IDLE
halt ← 1                        E0HLT,INIT
```

For the memory address register, only one of the register transfers meet this criteria:

```
ma ← ma + 1                     E0BDEP
```

Similarly, for the program counter, there is only one register transfer that uses pc on both sides:

```
pc ← pc + 1                     F2, E0ASKIP
```

For the instruction register and memory buffer register, there are no such register transfers.

The reason for identifying such register transfers is that, in theory, such transfers can be implemented internally within a register device without the need for any external data interconnection. Although such devices may be slightly more expensive, the intellectual simplification they provide to the architecture is usually worth the added cost.

For registers where no such transfers occur, it is clear that the designer should use enabled registers. Therefore, to implement an architecture for the ASM of section 8.3.2.4.2, the instruction register and the memory buffer register should be enabled registers. For these registers, whatever new data is loaded always comes from outside the enabled register.

In the case of the memory address register and the program counter, it is obvious from the above that an up counter register is the most appropriate choice. For the halt flag, a clearable enabled register (or its equivalent) is a reasonable choice because this allows the `halt` ← `0` transfer to occur internally (leaving only the `halt` ← `1` to be provided externally).

The choice of the register type for `{link,ac}` is less clear. In theory, one could imagine a device that is capable internally of doing all the operations listed for the link and accumulator. The problem is that such a contrived device is not one of the standard register building blocks discussed in appendix D. The intellectual simplification of register building blocks occurs not only because they hide details internally (hierarchical design) but also that their behavior is widely understood in industry and they can be concisely explained in a single cohesive sentence. (An up counter can hold, load, clear and increment its data. These operations are no more and no less than what is required to "count up.") It would not be wrong to build a device that does everything for the link and accumulator. (An automated synthesis tool, such as the one described in chapter 7, might take such an approach.) As a matter of good style for a manually synthesized design and out of consideration to others who attempt to understand the architecture, we will instead choose standard register building blocks of the kind described in appendix D.

Of the building blocks described in appendix D, there are two possible choices for the link and accumulator: the up counter and the shift register. If the designer chooses an up counter, the following register transfers can be implemented internally by the device:

`{link,ac}` ← `{link,0}`	E0CLA, E1BDCA
`{link,ac}` ← `{0,ac}`	E0CLL
`{link,ac}` ← `0`	E0CLACLL
`{link,ac}` ← `{link,ac} + 1`	E0IAC

If the designer chooses a shift register, a different set of register transfers can be implemented internally by the device:

```
{ac,link} ← {link,ac}          E0RAR
{link,ac} ← {ac,link}          E0RAL
```

Of the complete group of link and accumulator register transfers, the ones that **cannot** be implemented by **either** of these building blocks include:

```
{link,ac} ← {link,~ac}         E0CMA
{link,ac} ← {~link,ac}         E0CML
{link,ac} ← {link,ac & mb}     E0AND
{link,ac} ← {link,ac | sr}     E0OSR
{link,ac} ← {link,ac} + mb     E0TAD
```

The best way to implement such computations that must occur outside the register building block is with a combinational ALU, such as the one described in section 2.3.4, that is capable of doing addition and logical operations. Since the ALU can add an arbitrary number to the accumulator (as required in state E0TAD), it can also increment the accumulator (as required in state E0IAC). The ALU can perform sixteen different logical (bitwise) operations, including AND, OR and NOT (as required in states E0AND, E0OSR, E0CMA and E0CML). The ALU can output zero, and so the clearing operations (as required in states E0CLA, E1BDCA, E0CLL and E0CLACLL) can be accomplished at no added cost. Since the ALU is suitable for either design alternative (`{link,ac}` as an up counter or `{link,ac}` as a shift register) but the ALU can do the incrementing that would otherwise require a counter, an appropriate design decision is to use a shift register for `{link,ac}`.

Here are the register types chosen above:

```
{link,ac}       13-bit shift register
mb              12-bit enabled register
ma              12-bit up counter register
ir              12-bit enabled register
pc              12-bit up counter register
halt             1-bit clearable enabled register
```

8.4.3 Remaining register transfers

Having decided on each register type, we can eliminate those register transfers that occur internally within the register device, which leaves the following:

```
ac ← 0                                  E0CLA, E1BDCA
ac ← ac & mb                            E0AND
ac ← ac | sr                            E0OSR
ac ← ~ac                                E0CMA
link ← 0                                E0CLL
link ← ~link                            E0CML
{link,ac} ← 0                           E0CLACLL
{link,ac} ← {link,ac} + 1               E0IAC
{link,ac} ← {link,ac} + mb              E0TAD

halt ← 1                                E0HLT,E0INIT

ir ← membus                             F3A

ma ← ea(ir)                             F3B
ma ← pc                                 F1
ma ← sr                                 E0MA

mb ← ac                                 E0DCA
mb ← membus                             F4A
mb ← sr                                 E0DEP

pc ← ma                                 E0JMP
pc ← sr                                 E0PC
```

The remaining `{link,ac}` transfers are listed above in their original form to provide documentation that more closely matches the original ASM. For example, `ac ← 0` is more concise than `{link,ac} ← {link,0}`.

8.4.4 Putting the architecture together

In choosing the shift register, we also determined that every one of the remaining `{link,ac}` transfers (listed in section 8.4.3) can be implemented by the ALU. One of the inputs to the ALU will be the 13-bit `{link,ac}`. The other will be a 13-bit mux that selects between {0,sr}, {0,mb} and the constant one. It is important to note that although the {link,ac} is a unified thirteen-bit shift register, the link and accumulator portions are controlled separately. Therefore, it is possible to load just the accumulator, just the link or both of them. The controls for the {link,ac} are as follows:

link_ctrl bits	link_ctrl symbol	ac_ctrl bits	ac_ctrl symbol	action
00	`HOLD	00	`HOLD	do nothing
00	`HOLD	11	`LOAD	ac ←alubus[11:0]
11	`LOAD	00	`HOLD	link ← alubus[12]
11	`LOAD	11	`LOAD	{link,ac} ← alubus
10	`LEFT	10	`LEFT	{link,ac} ← {ac,link}
01	`RIGHT	01	`RIGHT	{ac,link} ← {link,ac}

The default (when link_ctrl and ac_ctrl are not mentioned in a state) is to hold the accumulator and link as they are.

The ea(ir) function can be implemented by trivial combinational logic. We leave this as a separate device since there are other addressing modes not implemented here that are described in appendix B and that are left as exercises.

There is only one register transfer left for the halt flag, and so its input is a constant one. Similarly, there is only one register transfer for the instruction register, and so its input is the memory bus (which provides m[ma] to the architecture from the external memory device).

The remaining register transfers can be provided for by placing muxes on the inputs of the appropriate registers. The input to the memory buffer register is a 12-bit mux that selects among sr, the accumulator and memory bus. The input to the memory address register is a 12-bit mux that selects among sr, ea(ir) and the program counter. The input to the program counter is a 12-bit mux that selects between the sr and the memory address register.

Here is the block diagram of the architecture that was just derived for the subset PDP-8:

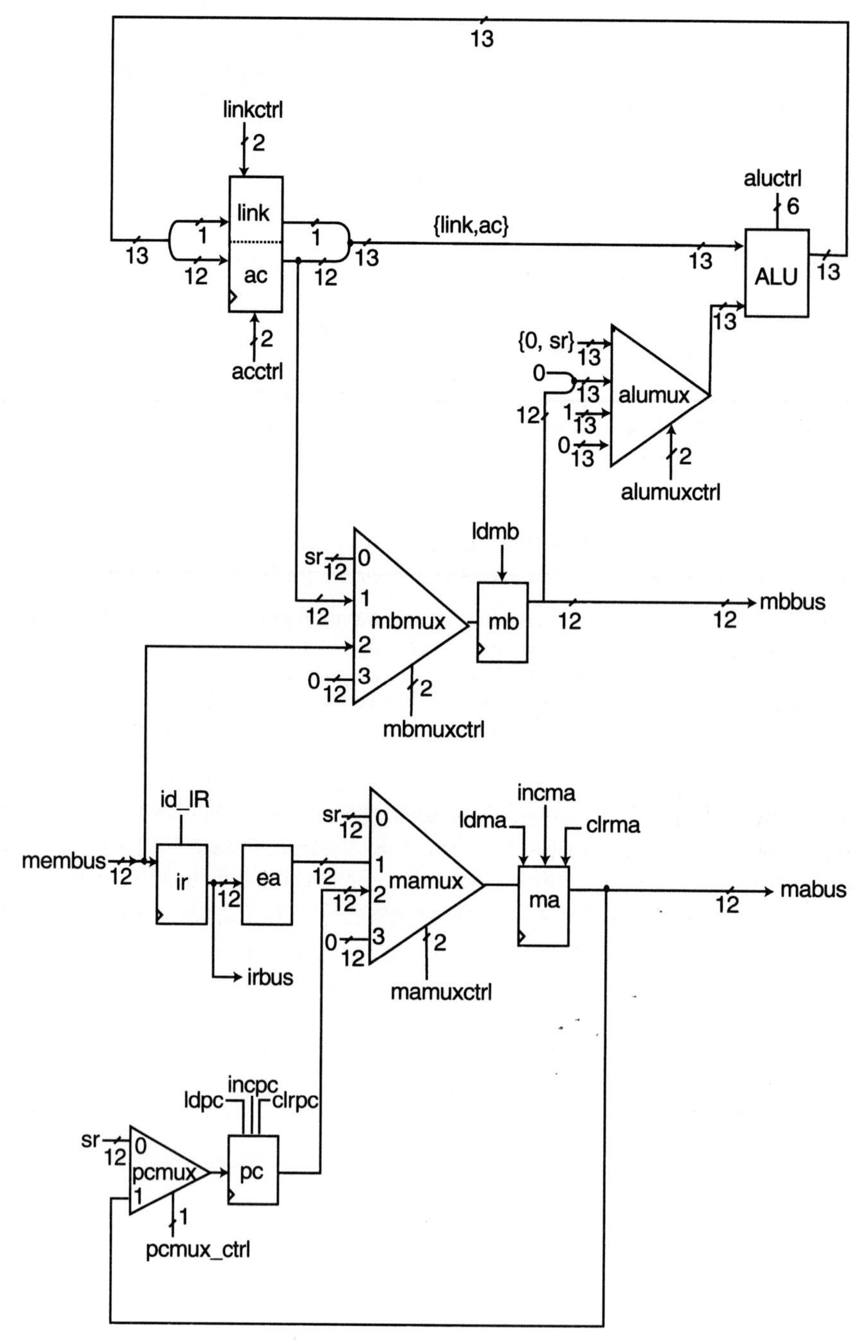

Figure 8-12. Architecture for PDP-8 subset.

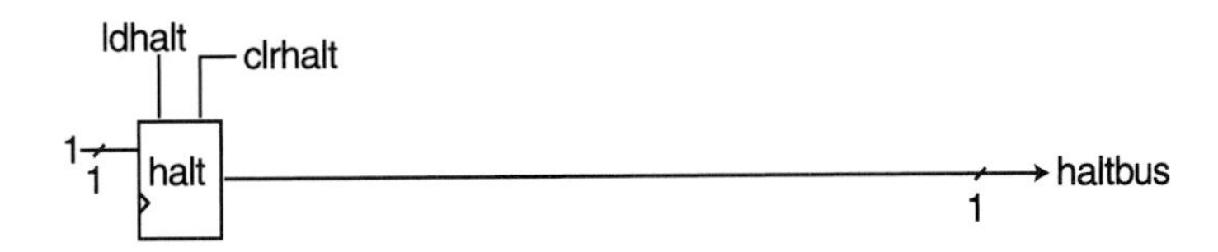

Figure 8-12. Architecture for PDP-8 subset (continued).

The fourth inputs to the memory address and memory buffer muxes are not required and therefore tied to zero. It is left as an exercise to show that these fourth inputs will help to implement more of the instructions given in appendix B.

8.4.5 Implementing the decisions

The ASM of section 8.3.2.4.2 has several decisions. Some of these test external status inputs (`cont`, `but_PC`, `but_MA` and `but_DEP`) that have nothing to do with the architecture. The remaining decisions test data contained in the registers of the architecture (`link`, `ac`, `ir` and `halt`). Although it would be possible to implement each of these decisions using a comparator, another easier way to implement these decisions exists.

Recall from section 2.1.3.1.2 that a multi-bit external status signal which is only tested against constants can be rewritten as a nested series of decisions that test the individual bits of the status. Using this approach, the internal status inputs to the controller are simply `link`, `ac`, `ir` and `halt` with no need for comparators in the architecture. In particular, since the instruction register is used in so many decisions, it is prudent to make it an input to the controller.

8.4.6 Mixed ASM

Here is the mixed ASM for the architecture of section 8.4.4 that implements the register transfers of section 8.3.2.4.2:

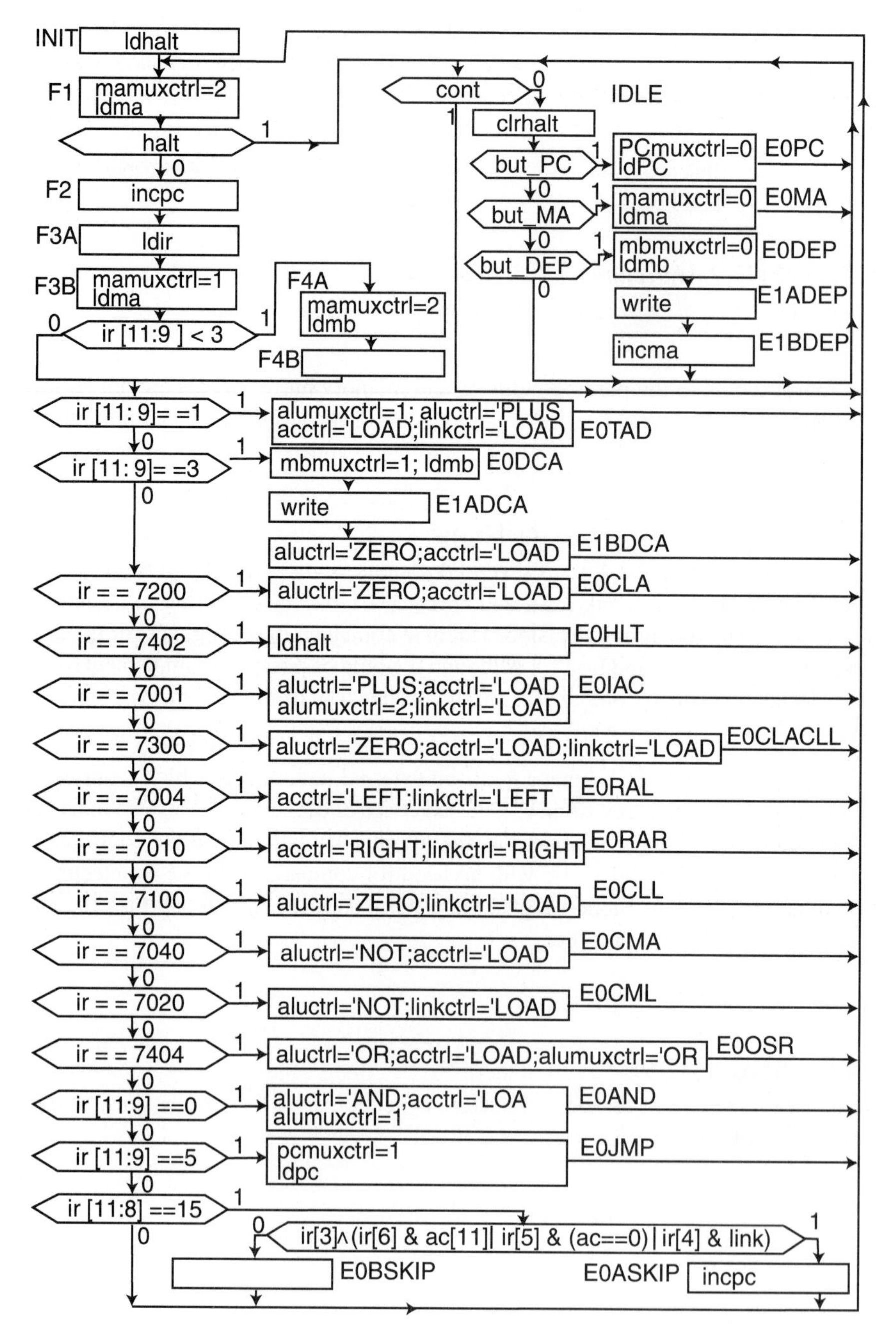

Figure 8-13. Mixed ASM for PDP-8 subset.

As discussed in the previous section, the decisions, which are implemented entirely in the controller, can (and should) be documented in the most understandable way. This means using `ir == 7200` rather than the equivalent twelve individual bit decisions that the controller actually uses. As is shown elsewhere in this book, modern synthesis tools can translate decisions like `ir == 7200` into the details required in the controller.

8.4.7 Block diagram

The following block diagram shows how to put the entire structure together:

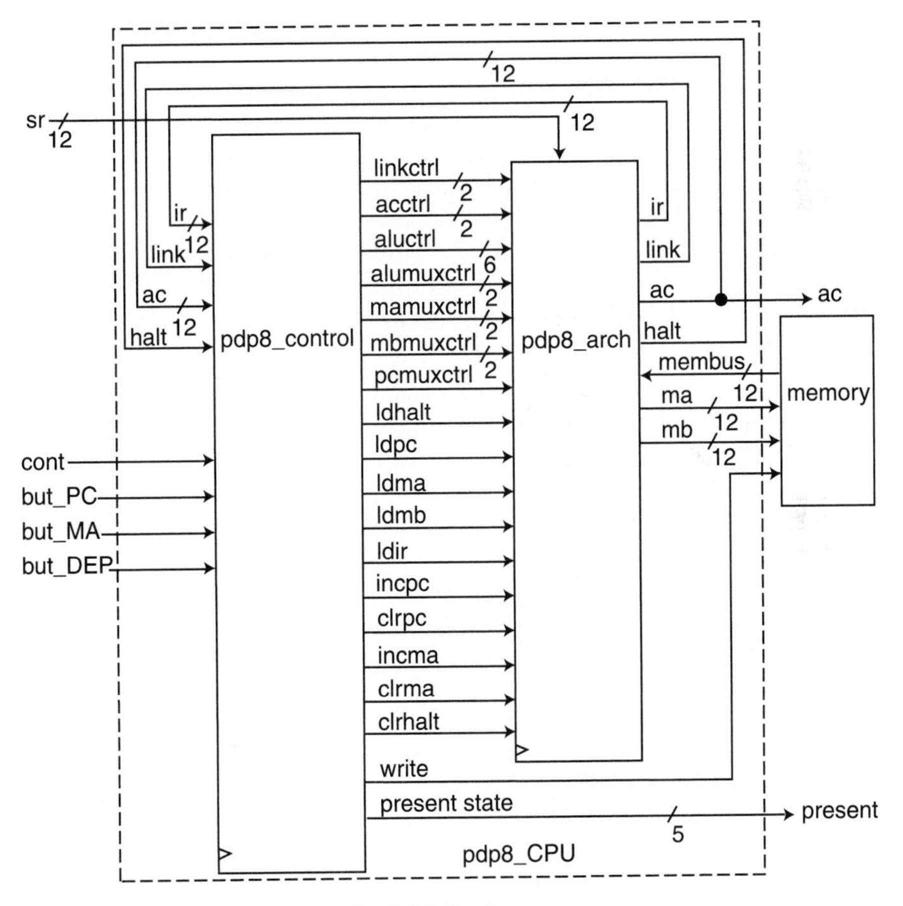

Figure 8-14. Block diagram for PDP-8 subset.

It is a simple, but tedious, matter to use hierarchical design to fill in the details of the controller from the ASM of section 8.4.6. Happily, synthesis tools can also aid the designer with this process.

8.5. Memory hierarchy

As described in section 8.2.3, the design of large and fast memories has been a challenge since the time of the earliest electronic computers. By the end of the twentieth century, these issues became of increasing concern because improving silicon technologies[13] allowed general-purpose processors to run at ever-higher clock frequencies. Large low-cost memory, such as dynamic memory chips, was unable to keep up with increasing processor speeds.

If general-purpose computers accessed memory in a completely random and haphazard fashion such that we could not make any kind of accurate prediction for which word in memory the processor would access next, this mismatch of processor and memory speed would be unsolvable. Happily, because of the nature of the fetch/execute algorithm and the nature of most machine language programs interpreted by the fetch/execute algorithm, we can predict, with reasonably good odds, what word the processor might fetch next. This solution to the mismatch between processor and memory speed has been recognized since 1962, when Kilburn and others at the University of Manchester designed the Atlas computer to take advantage of the fact that not all words in memory are accessed with the same frequency.

Kilburn's solution, which has endured with minor variations for more than a third of a century, is to design a hierarchy of memories of different speeds, sizes and costs. The hierarchy might have several different levels, each containing a different memory technology. The lowest level has a memory technology that costs the least per bit. This memory will have the largest number of words since we can afford to buy quite a lot of such cheap memory. Such inexpensive memory necessarily has a slow access time. Each higher level in the hierarchy has a kind of memory which is faster than lower levels in the hierarchy. Because the faster memories are more expensive per bit than the memories in the lower levels in the hierarchy, we can only afford smaller memory sizes in the upper levels of the heriarchy.

[13] Primarily smaller chip dimensions which mean lower propagation delays.

The memory hierarchy is usually effective because, statistically speaking, most memory accesses occur to words that have already been accessed before. If the system keeps the few words that are more likely to be accessed in the fast but small-sized higher levels of the hierarchy, and all the other words that are less likely to be accessed in the lower levels, we observe two benefits. First, the cost of the system is not significantly higher than if it were built entirely of slow cheap memory. Second, the speed of the system is not significantly slower than if it were built entirely of fast expensive memory. In essence, we almost get the best of both alternatives. However, this good cost and performance occurs only in a statistical sense: the "average" program will on "average" execute almost as fast as if the system used a fast memory. The program you are interested in may actually execute considerably slower, depending on the pattern in which that program accesses memory for the particular data you give it and depending on the details of the memory hierarchy you use.

There are two common kinds of memory hierarchy. The first of these is known as *cache memory*, which is discussed in the next section. The second of these, which is what Kilburn used, is known as *virtual memory*. The idea of virtual memory is to keep less frequently used parts of memory on disk. The access time for the data on disk is many orders of magnitude slower than for data in semiconductor memory. It is also very non-deterministic because of the unpredictable distance the disk has to rotate to be positioned on the proper data.[14] Although conceptually, virtual memory is very similar to cache memory, its implementation requires complicated hardware and software. Hardware implementation of virtual memory requires a disk controller, and the management of virtual memory is usually intertwined with the software details of an operating system. Since hardware disk controllers and software operating systems are beyond the scope of this book, we will not consider virtual memory.

8.5.1 Cache memory

Cache memory is the fastest part of the memory hierarchy. It is built out of several components. The cache needs its own controller, which we will ignore for the moment. Of course, the cache needs high-speed memory for the data to be stored in the cache, but the cache also needs a *tag memory* which indicates the address associated with each portion of the cache. The data and tag memories of the cache are usually composed of expensive high-speed static memory that can be accessed in significantly less than one clock cycle. When the propagation delay of the rest of the system is considered, this still allows data to be fetched from the cache in one clock cycle.

[14] Also, there is the chance the disk head has to move, which can take a significant fraction of a second.

The tag memory is needed because a particular part of the cache may be associated with more than one address at different times during the operation of the cache. In contrast, a particular part of an ordinary (main) memory will always be associated with one particular constant address. As explained in section 8.2.2.3.1, such a main memory can be thought of as a mux which selects one of several register values. Each cell in a main memory is always associated with its particular address because that address specifies the port of the mux to which the corresponding register is wired.

There are two common approaches to designing a cache. In the *direct mapped* approach, there is only one tag memory and one data memory. In the multi-way set associative approach, there are several parallel tag and corresponding data memories. The direct mapped approach is simpler and therefore allows a faster access time. On the other hand, the direct approach is often not as successful in keeping the appropriate words in the cache as the multi-way approach, and so even though the access time of the multi-way approach is slower, it may be faster overall for some programs than the direct approach. This section, however, will concentrate on the direct mapped approach, which is easier to comprehend.

The typical cache memory uses the low-order bits of the address bus to select information out of both the data and tag portions of the cache. In order for a memory access to be fast, the information fetched from the tag memory must match the address bus.[15] If it does not, the cache must be updated from some lower level of the memory hierarchy. Commercial computer systems often have more than one level of cache. In such systems, the first level is often on the same chip as the processor to maintain the highest (single clock) speed. The second level (referred to as L2) is contained on separate chips that allow access in a small number of cycles. The main memory is composed of dynamic memory, with an access time of many clock cycles. In this section, however, there will only be two levels in the memory hierarchy: the direct mapped cache and the main memory.

In this chapter, we will assume each element of the cache content memory is a single word. Often, in commercial systems, each element of the cache content memory is a group of several contiguous words, known as a *line*. Using a line composed of several words may improve the performance of the cache, but including such details here would obscure the idea being discussed in this section: how a cache is a cost-effective way to improve the performance of a general-purpose computer.

For example, assume a cache size of four words[16] with the following simple program that goes through a loop eight times producing nine values[17] (7760, 7762, ... 7776 and 0000) in the accumulator:

[15] In an actual implementation, only the high-order bits need to be stored in the tag memory and checked against the high-order bits of the address bus, but we will ignore this detail for now.

[16] This is too small for practical use but will illustrate how a cache works.

[17] These are the nine decimal values -16, -14, ... -2 and 0.

```
0000/7300          CLACLL      // ac = -16
0001/1006          TAD A
0002/1011   L,     TAD B       // ac = ac + 2
0003/7510          SPA         // if ac>=0, halt
0004/5002          JMP L       // if ac<0, stay in loop
0005/7402          HLT
0006/7760   A,     7760        // equivalent to decimal -16
                   *0011
0011/0002   B,     0002        // +2
```

Assuming the instructions of this program are loaded into memory in the same order as listed above, at the time the fetch/execute cycle begins, the cache will contain:

```
        cache                  main memory
   tag        data              0000/7300
 0/0004    0/5002               0001/1006
 1/0011    1/0002               0002/1011
 2/0006    2/7760               0003/7510
 3/0003    3/7510               0004/5002
                                0005/7402
                                0006/7760
                                0011/0002
```

The words shown in bold for the main memory are the ones currently in the cache. When the processor fetches the first instruction, the memory access will be slow because address 0000 is not currently in the cache. This is known as a *cache miss*. The cache has to bring in this word (7300) from the main memory, and so the cache now looks like:

```
        cache                  main memory
   tag        data              0000/7300
 0/0000    0/7300               0001/1006
 1/0011    1/0002               0002/1011
 2/0006    2/7760               0003/7510
 3/0003    3/7510               0004/5002
                                0005/7402
                                0006/7760
                                0011/0002
```

Fetching the next instruction (1006) causes another cache miss:

cache tag	cache data	main memory
0/0000	0/7300	**0000/7300**
1/0001	1/1006	**0001/1006**
2/0006	2/7760	0002/1011
3/0003	3/7510	**0003/7510**
		0004/5002
		0005/7402
		0006/7760
		0011/0002

However, when this TAD instruction is executed, the cache already has the data 7760 required by the processor. This is known as a *cache hit*. The second memory access during this instruction is fast because it is a cache hit.

Fetching and executing the next instruction (1011) causes two cache misses:

cache tag	cache data	main memory
0/0000	0/7300	**0000/7300**
1/0011	1/0002	0001/1006
2/0002	2/1011	**0002/1011**
3/0003	3/7510	**0003/7510**
		0004/5002
		0005/7402
		0006/7760
		0011/0002

Fetching and executing the SPA instruction (7510) causes a cache hit, and so this memory access is fast. Since the accumulator is negative, the skip does not occur, and the processor needs to fetch the next (5002) instruction. This causes another cache miss:

cache tag	cache data	main memory
0/0004	0/5002	0000/7300
1/0011	1/0002	0001/1006
2/0002	2/1011	**0002/1011**
3/0003	3/7510	**0003/7510**
		0004/5002
		0005/7402
		0006/7760
		0011/0002

From this point on, as long as the program stays inside this three-instruction loop (TAD; SPA; JMP), all of the instruction and data accesses are cache hits. The final cache miss occurs when the program halts:

```
      cache                    main memory
  tag       data                0000/7300
0/0004   0/5002                 0001/1006
1/0005   1/7402                 0002/1011
2/0002   2/1011                 0003/7510
3/0003   3/7510                 0004/5002
                                0005/7402
                                0006/7760
                                0011/0002
```

In total, there are six cache misses[18] and 29-cache hits in this example. With the given value of A, this is a 17% miss rate and an 83% hit rate, although the hit rate would increase for values of A that are more negative.[19]

The good performance that the above program exhibits using this little cache depends heavily on how the instructions and data are arranged. For example, if B were located at address 0007, there would be 20 cache misses and only 15 cache hits, which is a 57% miss rate and 43% hit rate. A larger cache size will often improve performance. If the program with B at address 0007 runs on a machine with a cache size of eight, the hit rate becomes 100% because this entire tiny program can reside in the cache. If the program with B at address 0011 runs on a machine with a cache size of eight, the hit rate is 94% (two misses) because the program cannot all fit in the cache at once (0001 and 0011 cannot reside in a direct mapped cache of size eight at the same time).

8.5.2 Memory handshaking

Regardless of whether a machine uses cache memory, virtual memory or both in its memory hierarchy, one thing is clear: the access time is non-deterministic. Although we expect the majority of memory accesses to occur in a single cycle, some accesses will take additional cycles. The ASM chart for fetch/execute given in section 8.3.2.4.2 assumes that every memory access can occur in one cycle, which is not the case for a memory hierarchy. A more sophisticated ASM is required that waits for the memory

[18] The number of misses is the same in this program regardless of the value of A and therefore of how many times the loop executes.

[19] The number of hits depends on how many times the loop executes.

hierarchy to provide the requested instruction or data. To coordinate the operation of the memory hierarchy with the CPU requires using what is called a *partial handshaking* protocol.

In the partial handshake protocol, there is an extra command signal, `memreq`, that the CPU sends to the memory when the CPU requests a particular word of the memory hierarchy. Unlike the simple memory described in section 8.2.2.3.1, the memory hierarchy might ignore the address bus when `memreq` is not asserted. Only when `memreq` is asserted does the memory take notice of the address bus and respond accordingly.

In the partial handshake protocol, there is also an extra status signal, the memory read acknowledge (`memrack`), that the memory sends back to the CPU to acknowledge that the memory hierarchy has obtained the word desired by the CPU. The CPU must continue to assert its memory request until the memory responds with its acknowledge signal. If the desired word is already in the cache, the memory hierarchy will instantly[20] assert `memrack`. Having the memory hierarchy assert `memrack` within the same cycle that the CPU first asserts `memreq` means that only one cycle is spent on the memory access. If the desired word is not already in the cache, the memory hierarchy will wait however long is necessary before asserting `memrack`. The ASM for the CPU must stay in a wait state prior to when the memory hierarchy asserts the `memrack` status signal. For example, consider the portion of the ASM from section 8.3.2.4.2 (consisting of states F3B, F4A and F4B) shown on the left:

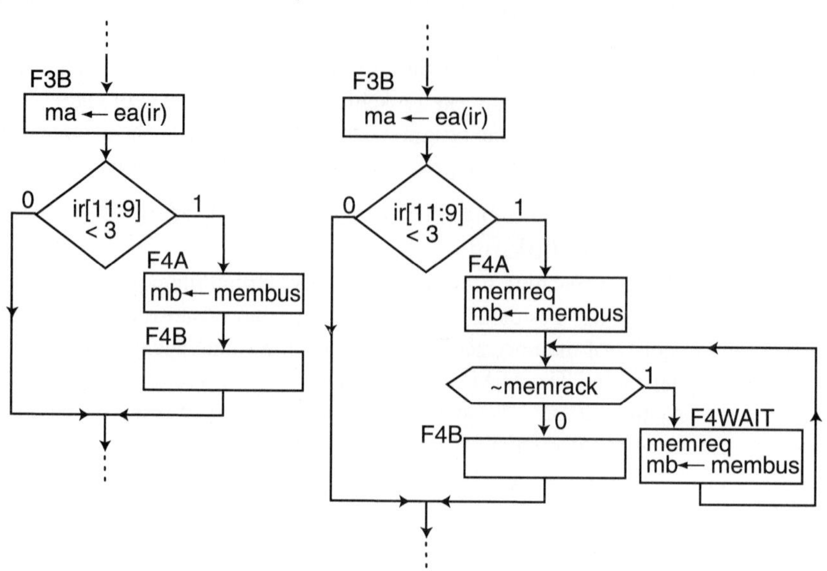

Figure 8-15. ASMs without and with memory read handshaking.

[20] Ignoring a trivial amount of propagation delay, as was done in earlier portions of this chapter.

As explained in section 8.3.1.5, these states fetch the operand for PDP-8 instructions, such as TAD, on the assumption that every memory access can occur in one clock cycle. In state F3B, the memory address register is scheduled to change to the effective address of the instruction. As with all ←, the change does not take effect until the next rising edge of the clock. Assuming the instruction is a TAD, the CPU will be in state F4A when that next rising edge occurs. In state F4A, the memory buffer register is scheduled to be loaded with the corresponding contents of memory that comes via the memory bus. In this ASM, there is never more than one clock cycle for the memory to give the correct data to the CPU.

In contrast, the more complicated ASM on the right uses handshaking to adapt to the speed of each particular memory access. This requires introducing an extra state, F4WAIT, and asserting `memreq` in both states F4A and F4WAIT. Also, there is a decision involving `memrack` that occurs in both states F4A and F4WAIT. If the memory access is fast, the CPU never goes to state F4WAIT, and so the state transitions of the left ASM are identical to the state transitions of the right ASM. On the other hand, if the memory access is slow, the machine goes to state F4WAIT where it will loop until the memory hierarchy asserts `memrack`. Note the data that the memory hierarchy provides on `membus` must be valid **before** the hierarchy can assert `memrack`.

A similar handshaking approach is required for memory writes, except the memory hierarchy responds back with a memory write acknowledge (`memwack`):

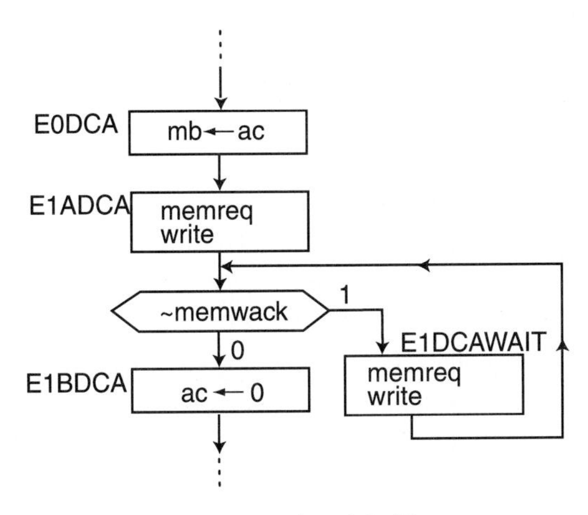

Figure 8-16. ASM with memory write handshaking.

8.5.3 Architecture for a simple memory hierarchy

The memory hierarchy is a separate actor from the CPU, and so the memory hierarchy needs its own controller and architecture. Assuming we keep the names the same as in the earlier sections of this chapter, the memory hierarchy and the CPU communicate address and data using `mabus`, `mbbus` and `membus`. More importantly, the CPU tells the memory hierarchy what it needs done to memory with `write` and `memreq`, and the memory responds back when the requested operation is complete using either `memrack` or `memwack`. Both the CPU and the memory hierarchy are fed the system clock.[21] Here is a diagram that illustrates this interconnection:

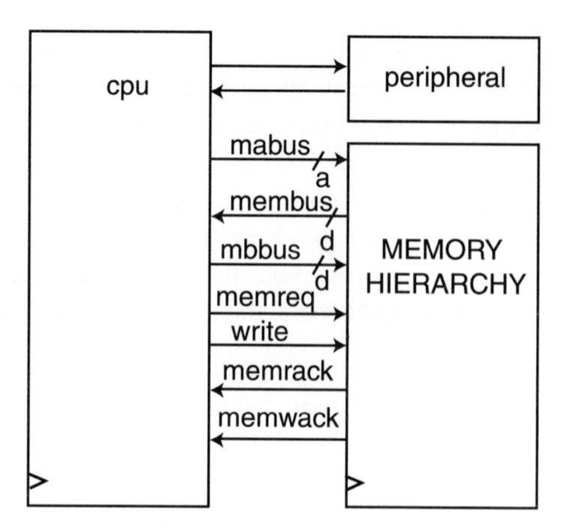

Figure 8-17. Connection of processor to memory hierarchy.

We are going to design a memory hierarchy consisting of a main memory and a cache. There are many choices available to the designer of a cache. Although the cache could be either set associative or direct mapped, we will use the direct mapped technique since it is easier to understand. Also, there is a choice about how the cache treats writes: either a *write-through cache* or a *write-back cache* .

A write-back cache waits until it is necessary to write data back into the main memory. This has the advantage that operations on values such as loop counters do not have to wait for the slower access time of the main memory. The problem with the write-back approach is that the main memory and the cache can become inconsistent with each other. Because *cache consistency* is not guaranteed at all times with the write-back approach, a request from the CPU for a memory read may also cause a write to the main memory (that restores cache consistency). This makes a write-back cache considerably harder to design than a write-through cache. It may even make the write-back

[21] This is a requirement of the partial handshake protocol.

cache slower than a comparable write-through cache.[22] The complexity of write-back caches are even more pronounced when multiple CPUs share the same main memory but have distinct caches.

In contrast, write-through caches are simple enough[23] that the design of an elementary one can be described using only the mixed Moore ASM notation (section 2.3) or using the equivalent Verilog (section 4.2). The essential idea of a write-through cache is that when the CPU requests the memory hierarchy to do a write operation, the memory hierarchy will store the data into both the main memory and the cache. A write-through cache is guaranteed to be consistent with the main memory.

We are going to skip over the pure behavioral stage of the design, and proceed straight to the mixed stage so that we may focus on the architecture of the memory hierarchy.[24] This architecture consists of the main memory, the cache content memory and the cache tag memory. The main memory is asynchronous with a deterministic access time bounded by a known number of clock cycles. Like the one and only memory shown previously in section 8.4.7, the main memory of the hierarchy has its data input connected to `mbbus` and its address input connected to `mabus`. The distinctions between the main memory of the hierarchy and the memory of section 8.4.7 are the main memory of the hierarchy receives its `mainwrite` signal from the internal cache controller (rather than from the CPU) and the data output of the main memory of the hierarchy does not connect directly to the CPU. Instead, the data output of the main memory connects to the data input of the cache content memory.

The cache content and cache tag memories are synchronous memories that can be accessed within one clock cycle. The address inputs to both the cache content and cache tag memories come from the low order `j` bits of the CPU's memory address bus. We will use 'CACHE_SIZE to indicate the number of words that can reside in the cache, which is the same as 2^j. The data output of the cache content register (`cache_content[mabus % 'CACHE_SIZE]`) is connected to the memory bus that goes back to the CPU. The data output of the cache tag register (`cache_tag[mabus % 'CACHE_SIZE]`) is connected to a comparator. The other input of the comparator is the memory address bus from the CPU.[25] The output of the comparator is the memory read acknowledge signal (`memrack`). This signal is sent to

---

[22] Whether write-through or write-back is faster depends on several factors, including the pattern in which the particular program accesses memory.

[23] A write-back cache requires the Mealy notation of chapter 5.

[24] We are also avoiding the pure behavioral stage now because, even on this simple write-through cache, the pure behavioral ASM requires the Mealy notation of chapter 5. The mixed ASM does not need the Mealy notation because the `memrack` signal is generated by the architecture, and not the controller.

[25] Many implementations would only use the high order `a-j` bits, but for simplicity, we use all `a` bits.

both the CPU (where it controls the duration of the wait loops described in section 8.5.2) and to the cache controller. The cache controller generates the `ldcont` and `ldtag` commands for the cache content and cache tag memories. When asserted, these commands indicate that the cache content and cache tag memories will be loaded with new information at the next rising edge of the clock.

The cache controller also generates the memory write acknowledge (`memwack`) signal after the main memory, the cache tag memory and the cache content memory have been updated as a result of a `write` signal from the CPU. The following shows the architecture for the memory hierarchy (using the write-through direct mapped cache described above) and the corresponding mixed Moore ASM for the cache controller:

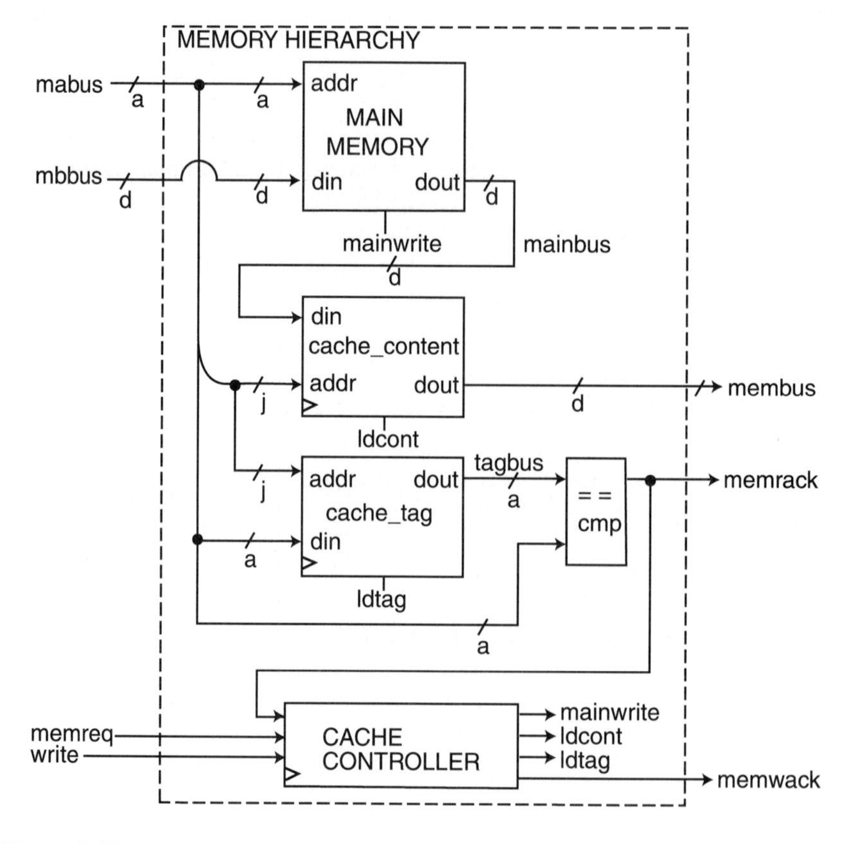

Figure 8-18. Memory hierarchy architecture with direct mapped write-through cache.

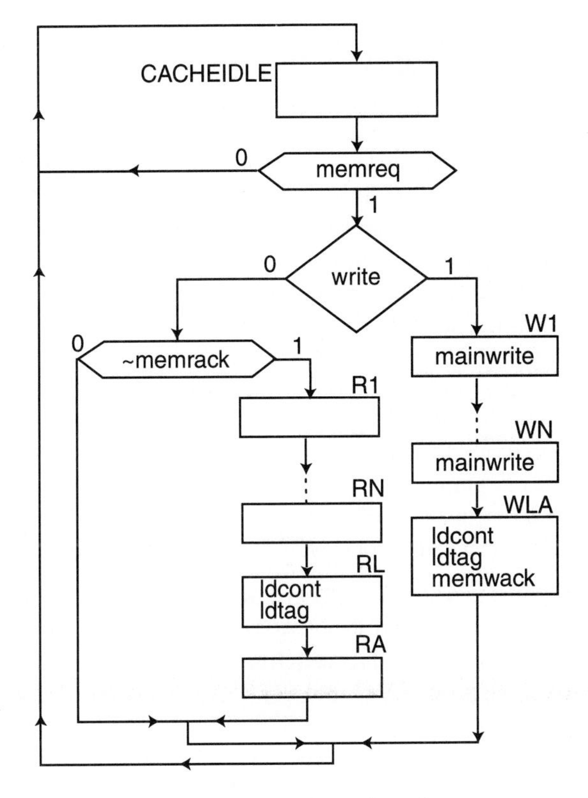

Figure 8-19. ASM for direct mapped write-through cache memory controller.

The ASM stays in state CACHEIDLE unless a memory request occurs. There are three possibilities for a memory request. Two of these possibilities are when the memory request is for a read operation (i.e., `write` is zero): either the requested data is in the cache or the requested data is not in the cache. The third possibility is a memory write request from the CPU (regardless of whether it is in the cache).

The first possibility is when the data being read by the CPU is already in the cache. In this case, `memrack` will be true during the first clock cycle that `memreq` is asserted. Because the `memrack` signal comes straight from the combinational logic comparator, both the ASM for the CPU and the ASM for the cache controller proceed without delay states. The ASM for the CPU makes a transition such as from state F4A to F4B, and the ASM for the cache controller makes a transition from CACHEIDLE back to that same state.

The second possibility is when the data being read by the CPU is not in the cache during the first clock cycle that `memreq` is asserted. During this clock cycle `memrack` will be false. It will stay false for as long as the output of the cache tag memory does not equal the memory address bus from the CPU. In a case like this, when `memrack` is false, the ASM for the CPU makes a transition such as from state F4A to F4WAIT, and the ASM for the cache controller makes a transition from CACHEIDLE to state R1. The ASM has an appropriate number of empty delay states (not shown) to allow for the read access time of the asynchronous main memory. Then, in state RL, the cache controller issues the `ldcont` command. This causes the cache content memory to be loaded at the next rising edge of the clock with the data obtained from the slow main memory. Also in state RL, the cache controller issues the `ldtag` command. This causes the cache tag memory to be loaded at the next rising edge of the clock with the address being provided by the CPU. Because of this change to the tag memory, when the cache controller proceeds to the empty state, RA, the architecture will for the first time assert `memrack`. The one empty state, RA, is all that is necessary to allow the CPU to make a transition such as from state F4WAIT to F4B. Of course, the cache controller makes a transition during that same clock cycle from state RA back to CACHEIDLE.

The third possibility is when the CPU makes a memory write request. The ASM for the cache controller proceeds from state CACHEIDLE to state W1 during the same clock cycle that the ASM for the CPU proceeds from a state such as E1ADCA to E1DCAWAIT. The ASM has an appropriate number of delay states (not shown) that each assert `mainwrite`. This allows for the write access time of the asynchronous main memory.[26] Finally, in state WLA, the cache controller asserts `ldcont`, `ldtag` and `memwack`. The assertion of `ldcont` and `ldtag` is not necessary for this write operation but is required for any future read operations to be fast. Therefore, a separate empty state for write acknowledgement is not necessary here as was the case for read acknowledgement. Because `memwack` is asserted in state WLA, at the same time that the ASM for the cache controller makes a transition from state WLA to state CACHEIDLE, the ASM for the CPU makes a transition such as from state E1DCAWAIT to E1BDCA.

The following example is a program that adds two numbers together and stores the sum in memory. Both state machines (CPU and memory controllers) cooperate to fetch instructions and data and to store results back in memory. This example illustrates each of the three possibilities explained above. The first two instructions, as well as the first word of data fetched, are already in the cache. In such an instance (shown in bold) the cache state remains in CACHEIDLE and the CPU does not need a wait state. This situation is signaled by `memreq` and `memrack` both being one during the same clock cycle.

[26] The read and write access times need not be the same.

CPU state	ma	mb	pc	ir	h	ac	CACHE state	memreq	memrack	memwack
F1	0002	1006	0000	xxxx	0	xxxx	CACHEIDLE	0	0	0
F2	0000	1006	0000	xxxx	0	xxxx	CACHEIDLE	0	1	0
F3A	**0000**	1006	0001	xxxx	0	xxxx	CACHEIDLE	**1**	**1**	0
F3B	0000	1006	0001	**7300**	0	xxxx	**CACHEIDLE**	0	1	0
E0CLACLL	0100	1006	0001	7300	0	xxxx	CACHEIDLE	0	0	0
F1	0100	1006	0001	7300	0	0000	CACHEIDLE	0	0	0
F2	0001	1006	0001	7300	0	0000	CACHEIDLE	0	1	0
F3A	**0001**	1006	0002	7300	0	0000	CACHEIDLE	**1**	**1**	0
F3B	0001	1006	0002	**1006**	0	0000	**CACHEIDLE**	0	1	0
F4A	**0006**	1006	0002	1006	0	0000	CACHEIDLE	**1**	**1**	0
F4B	0006	**1000**	0002	1006	0	0000	**CACHEIDLE**	0	1	0
E0TAD	0006	1000	0002	1006	0	0000	CACHEIDLE	0	1	0
F1	0006	1000	0002	1006	0	1000	CACHEIDLE	0	1	0

As execution proceeds, the next instruction causes a cache miss, which causes the memory controller to leave CACHEIDLE and causes the CPU to enter a wait state (shown in italic):

F2	0002	1000	0002	1006	0	1000	CACHEIDLE	0	0	0
F3A	**0002**	1000	0003	1006	0	1000	CACHEIDLE	**1**	**0**	0
F3WAIT	*0002*	*1000*	*0003*	*1000*	*0*	*1000*	*R1*	*1*	*0*	*0*
F3WAIT	*0002*	*1000*	*0003*	*1000*	*0*	*1000*	*R2*	*1*	*0*	*0*
F3WAIT	*0002*	*1000*	*0003*	*1000*	*0*	*1000*	*R3*	*1*	*0*	*0*
F3WAIT	*0002*	*1000*	*0003*	*1000*	*0*	*1000*	*RL*	*1*	*0*	*0*
F3WAIT	*0002*	*1000*	*0003*	*1000*	*0*	*1000*	*RA*	**1**	*1*	*0*
F3B	0002	1000	0003	**1007**	0	1000	**CACHEIDLE**	0	1	0

Fetching the operand of this instruction causes a cache hit, but fetching the following instruction causes a cache miss:

F4A	**0007**	1000	0003	1007	0	1000	CACHEIDLE	**1**	**1**	0
F4B	0007	**2000**	0003	1007	0	1000	**CACHEIDLE**	0	1	0
E0TAD	0007	2000	0003	1007	0	1000	CACHEIDLE	0	1	0
F1	0007	2000	0003	1007	0	3000	CACHEIDLE	0	1	0
F2	0003	2000	0003	1007	0	3000	CACHEIDLE	0	0	0
F3A	**0003**	2000	0004	1007	0	3000	CACHEIDLE	**1**	**0**	0
F3WAIT	*0003*	*2000*	*0004*	*2000*	*0*	*3000*	*R1*	*1*	*0*	*0*
F3WAIT	*0003*	*2000*	*0004*	*2000*	*0*	*3000*	*R2*	*1*	*0*	*0*
F3WAIT	*0003*	*2000*	*0004*	*2000*	*0*	*3000*	*R3*	*1*	*0*	*0*
F3WAIT	*0003*	*2000*	*0004*	*2000*	*0*	*3000*	*RL*	*1*	*0*	*0*
F3WAIT	*0003*	*2000*	*0004*	*2000*	*0*	*3000*	*RA*	***1***	***1***	*0*
F3B	0003	2000	0004	**3005**	0	3000	**CACHEIDLE**	0	1	0

Because this example uses a write-through cache, executing the DCA instruction causes the memory controller to leave state CACHEIDLE and causes the CPU to enter a wait state:

E0DCA	0005	2000	0004	3005	0	3000	CACHEIDLE	0	0	0
E1ADCA	0005	3000	0004	3005	0	3000	CACHEIDLE	**1**	0	**0**
E1DCAWAIT	*0005*	*3000*	*0004*	*3005*	*0*	*3000*	*W1*	*1*	*0*	*0*
E1DCAWAIT	*0005*	*3000*	*0004*	*3005*	*0*	*3000*	*W2*	*1*	*0*	*0*
E1DCAWAIT	*0005*	*3000*	*0004*	*3005*	*0*	*3000*	*W3*	*1*	*0*	*0*
E1DCAWAIT	*0005*	*3000*	*0004*	*3005*	*0*	*3000*	*W4*	*1*	*0*	*0*
E1DCAWAIT	*0005*	*3000*	*0004*	*3005*	*0*	*3000*	*WLA*	***1***	*0*	***1***
E1BDCA	0005	3000	0004	3005	0	3000	**CACHEIDLE**	0	1	0
F1	0005	3000	0004	3005	0	0000	CACHEIDLE	0	1	0

Fetching the final instruction causes another cache miss:

F2	0004	3000	0004	3005	0	0000	CACHEIDLE	0	0	0
F3A	**0004**	3000	0005	3005	0	0000	CACHEIDLE	**1**	**0**	0
F3WAIT	*0004*	*3000*	*0005*	*7300*	*0*	*0000*	*R1*	*1*	*0*	*0*
F3WAIT	*0004*	*3000*	*0005*	*7300*	*0*	*0000*	*R2*	*1*	*0*	*0*
F3WAIT	*0004*	*3000*	*0005*	*7300*	*0*	*0000*	*R3*	*1*	*0*	*0*
F3WAIT	*0004*	*3000*	*0005*	*7300*	*0*	*0000*	*RL*	*1*	*0*	*0*
F3WAIT	*0004*	*3000*	*0005*	*7300*	*0*	*0000*	*RA*	***1***	***1***	*0*
F3B	0004	3000	0005	7402	0	0000	**CACHEIDLE**	0	1	0
E0HLT	0002	3000	0005	7402	0	0000	CACHEIDLE	0	1	0

8.5.4 Effect of cache size on the childish division program

There are many alternatives that a designer must choose from when implementing a cache. It is often hard to predict manually what effect these choices will have on the speed of the system when it is running a particular program. This is a case where simulation is essential to allow the designer to estimate the effects different design decisions will have on the overall performance of the system. For example, one could simulate to observe the effect of cache size. One of the reasons HDLs such as Verilog have become popular is because designers need to conduct such simulations before building their machines.

The ASMs for the PDP-8 (section 8.3.2.4.2) and the cache controller (section 8.5.3) were translated into Verilog code (not shown), and the childish division program (section 8.3.2.5.3) was run for `x=14` and `y=7` with various cache sizes. In each case, there are 53 read accesses and seven write accesses. In this simulation, all write accesses and any cache misses cause the CPU to wait for five clock cycles. Here are hit and miss ratios for reads in this simulation:

cache size	clocks	#miss	miss ratio	hit ratio
8	541	51	96%	4%
16	396	22	58%	41%
32	296	2	4%	96%

8.6 Conclusion

General-purpose computers implement the fetch/execute algorithm, which in turn allows the hardware to interpret other algorithms coded in machine language that is stored in memory. Memory is the critical component for a general-purpose computer to be useful, and various technologies have been used to implement memory during the last half century. Static memories are fast, but dynamic memories are cheaper. Memory hierarchies that include a cache offer the best compromise between speed and cost. From an abstract behavioral viewpoint, all memory technologies can be thought of as arrays of binary words, but in reality, memory devices are independent actors that operate in parallel to the CPU. When the access time is non-deterministic, there must be handshaking between the memory and the CPU so that the CPU can adjust its speed to that of the memory.

The design process for a general-purpose CPU is similar to that of special-purpose hardware. The example used in this chapter of the PDP-8 was implemented at the behavioral and mixed stages of the design process. A methodical architecture was presented, and a variation using a direct mapped cache was considered. These designs were benchmarked using the childish division program to show that software running on the CPU designed in this chapter is slower and less efficient than when the childish division algorithm is implemented in special-purpose hardware. The next chapter will look at how this performance discrepancy can be diminished.

8.7 Further reading

BELL, C. GORDON and A. NEWELL, *Computer Structures: Readings and Examples*, McGraw-Hill, New York, NY, 1971. Chapter 5 is the definitive description of the PDP-8 from the man who also invented the first HDL (a language known as ISP).

BELL, C. GORDON, J. C. MUDGE and JOHN E. MCNAMARA, *Computer Engineering: A DEC View of Hardware Systems Design*, Digital Press, Bedford, MA, 1978. Chapter 8.

LAVINGTON, S., *Early British Computers: The Story of Vintage Computers and the People Who Built Them*, Digital Press/Manchester University Press, Bedford, MA, 1980. Describes the work of Kilburn, Williams, Turing, Wilkes and other British pioneers.

The Origins of Digital Computers: Selected Papers, 2nd ed., Edited by B. Randell, Springer-Verlan, Berlin, 1982. Reprints of original papers by computer pioneers.

PATTERSON, DAVID A. and JOHN L. HENNESSY, *Computer Organization and Design: The Hardware/Software Interface,* Morgan Kaufmann, San Mateo, CA, 1994. Chapter 7 explains virtual memory and multi-way set associative caches.

PROSSER, FRANKLIN P. and DAVID E. WINKEL, *The Art of Digital Design: An Introduction to Top down Design*, 2nd ed., Prentice Hall PTR, Englewood Cliffs, NJ, 1987. Chapter 7 describes an elegant central ALU architecture for the complete PDP-8 instruction set.

SLATER, ROBERT, *Portraits in Silicon*, MIT Press, Cambridge, MA, 1987. Gives biographies of several important pioneers including Babbage, Zuse, Atanasoff, Turing, Aiken, Eckert, Mauchly, von Neumann, Forrester, Bell and Noyce.

WAYNER, P., "Smart Memory," *BYTE*, June 1995, p. 190.

WOLF, WAYNE, *Modern VLSI Design: A Systems Approach*, 2nd ed., Prentice Hall PTR, Englewood Cliffs, NJ, 2nd ed., 1994, p. 356-370. Shows how to layout a VLSI chip that implements a PDP-8 architecture.

8.8 Exercises

8-1. Revise the ASM of section 8.3.2.1 to include the ISZ instruction described in appendix B.

8-2. Revise the architecture of section 8.4.7 to correspond to problem 8-1.

8-3. Revise the mixed ASM of 8.4.6 to correspond to problem 8-2.

8-4. Revise the ASM of section 8.3.2.1 to include the JMS instruction described in appendix B.

8-5. Revise the ASM of problem 8-4 to include all the addressing modes described in appendix B.

8-6. Revise the ASM of problem 8-5 to include the interrupt instructions ION and IOF and associated hardware described in appendix B.

8-7. Revise the architecture of section 8.4.7 to correspond to problem 8-6.

8-8. Revise the mixed ASM of 8.4.6 to correspond to problem 8-7.

8-9. Suppose a direct mapped write-through cache of size four contains the contents of addresses 0004, 0001, 0002 and 0003 when starting to run the following PDP-8 program:

```
0000/7200
0001/1004
0002/3006
0003/7402
0004/1000
```

a) How many cache read hits occur?
b) How many cache read misses occur?
c) What will be in the cache_tag and cache_contents when the program halts?

8-10. Translate the ASM of figure 8-8 into behavioral Verilog. Use test code that loads and runs the childish division program of section 8.3.2.5.3 using the `sr`.

8-11. Translate the architecture of figure 8-12 together with the mixed ASM of figure 8-13 into Verilog. Use the same test code as problem 8-10.

8-12. Modify problem 8-11 to include the direct mapped cache designed in section 8.5. Assume it takes five clock cycles to read or write to the main memory. Use similar Verilog test code.

9. PIPELINED GENERAL-PURPOSE PROCESSOR

The fetch/execute algorithm described in section 8.3.1.5 typically requires five clock cycles to execute each instruction. In the terminology of chapter 6, that ASM uses a multi-cycle approach. The clock is fast because its frequency is determined by the maximum propagation delay of a single combinational unit, most likely the ALU. On the other hand, the effective speed is approximately one-fifth of what could be achieved if pipelining were used instead.

In order to pipeline an algorithm that makes decisions (as fetch/execute must do in order to decode instructions), we need to use a Mealy ASM with ovals. (See chapter 5 for details about Mealy ASMs.) A Mealy approach is required because the pipeline will process different stages of independent instructions at the same time. Later stages depend upon the completion within one clock cycle of the earlier stages. In a Mealy ASM, a conditional computation begins instantly and is ready one clock cycle after the decision. In a Moore ASM, a conditional computation cannot begin until one clock cycle after the decision, and the result is not ready until two clock cycles after the decision. This would be too late for a pipelined fetch/execute, and so a Mealy ASM is required to describe the overall behavior of a pipelined general-purpose computer.

The existence of the NOP instruction (7000) is important to the design of the pipelined fetch/execute. By putting a NOP in the pipeline when none existed in the original program, it will be possible to cope with several special situations. The essential goal of the pipelined machine is to end up with the same answer in memory and the accumulator as would be obtained from a non-pipelined version. Since a NOP leaves both the accumulator and memory alone, NOP provides for a safe way to stall later stages of the pipeline while earlier stages of the pipeline are being filled. This is quite advantageous, since it can eliminate the need for "FILL" and "FLUSH" states of the kind described in chapter 6.

9.1 First attempt to pipeline

The following is a somewhat flawed attempt to design a pipelined ASM that is equivalent to the multi-cycle ASM of 8.3.1.5. This ASM is for a three-stage pipeline consisting of instruction fetch, operand fetch and instruction execution. Ideally, in each clock cycle independent instructions are being fetched, having an operand fetched and being executed. It is important to understand that what is being pipelined is the fetch/execute algorithm itself and not the software algorithm implemented by the machine language program (which may not even be possible to pipeline). The efficiency of a software algorithm running on a pipelined general-purpose computer will be better than the same software running on a multi-cycle general-purpose computer, but not as good as the efficiency of a special-purpose computer that implements the same algorithm in hardware rather than in software.

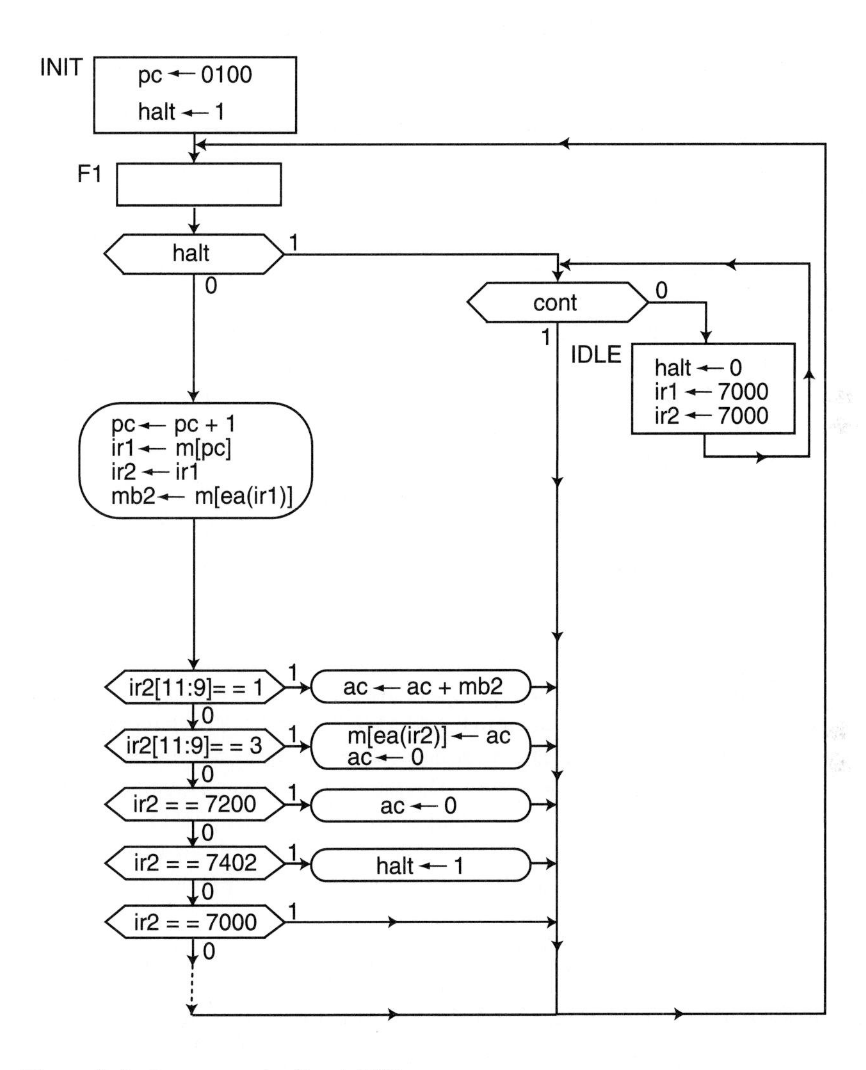

Figure 9-1. Incorrect pipelined ASM.

Here is a portion of the implicit style Verilog corresponding to this ASM:

```
  ...
  forever
    begin
     @(posedge sysclk) enter_new_state('F1);
       if (halt)
         ...
       else
         begin
           pc <= @(posedge sysclk) pc + 1;
           ir1 <= @(posedge sysclk) m[pc];
           ir2 <= @(posedge sysclk) ir1;
           mb2 <= @(posedge sysclk) m[ea(ir1)];
           if      (ir2[11:9] == 1)
             ac <= @(posedge sysclk) ac + mb2;
           else if (ir2[11:9] == 3)
             begin
               m[ea(ir2)] <= @(negedge sysclk) ac;
               ac <= @(posedge sysclk) 0;
             end
           else if (ir2 == 12'o7200)
             ac <= @(posedge sysclk) 0;
           else if (ir2 == 12'o7402)
             halt <= @(posedge sysclk) 1;
           else if (ir2 == 12'o7000)
        ...
        end
    end
```

In this ASM, the operations of states F2, F3A, F3B, F4A, F4B, E0CLA, E0TAD, E0DCA, E1ADCA, E1BDCA and E0HLT from the ASM of 8.3.1.5 have been merged into state F1 using the Mealy notation. The most noticeable change is that the set of registers used is somewhat different than before. The memory address register has been eliminated altogether so that three separate things can be done in parallel to memory during the same clock cycle: an instruction can be fetched, an operand can be fetched and the accumulator can be stored. A single memory address register would not allow all of these to happen in parallel.

The instruction register of section 8.3.1.5 is now replaced by two registers, `ir1` and `ir2`, so that the machine can distinguish the instructions as they travel through the pipeline. The memory buffer register is no longer used for writing to the memory, and instead is used only to hold an operand fetched from memory in the previous clock cycle. The memory buffer register has been renamed as `mb2`, reminding us that it is the operand for the instruction (`ir2`) in the final stage of the pipeline.

State INIT is identical to the one in the multi-cycle ASM. Since state INIT makes `halt` equal to one, when state F1 executes for the first time, the machine goes to state IDLE, where it waits for `cont` to become true. Note that state F1 does not do anything unconditionally, thus the program counter remains the way state INIT initialized it.

In state IDLE, `halt` becomes zero. There is an additional detail in state IDLE which was not present in section 8.3.1.5. The instruction pipeline (`ir1` and `ir2`) gets initialized to NOPs. After `cont` becomes true, it will take two clock cycles for the pipeline to fill with actual instructions from the program before the first instruction can execute. By putting NOPs in `ir1` and `ir2`, the machine can execute the imaginary NOPs harmlessly while the pipeline is filling with actual instructions. This eliminates the need for "FILL" states of the kind discussed in chapter 6.

When the machine is not halted in state F1, there are three separate things that occur in parallel:

a) The youngest instruction is fetched into `ir1` (`pc` ← `pc + 1` and `ir` ← `m[pc]` occur in parallel),
b) The operand for the middle aged instruction is fetched as that instruction moves down the pipeline (`ir2` ← `ir1` and `mb2` ← `m[ea(ir1)]` occur in parallel)
c) The oldest instruction is decoded and executed (decisions similar to section 8.3.1.5 but involving `ir2`)

The execution of each instruction must be described in a Mealy oval. When `ir2` contains a TAD instruction, the accumulator is scheduled to be updated by adding it to the operand fetched in the previous stage. When `ir2` contains a DCA instruction, the accumulator is scheduled to be stored (`m[ea(ir2)]` ← `ac`) in parallel to scheduling that the accumulator be cleared.

9.2 Example of independent instructions

The ASM of section 9.1 is only able to execute certain PDP-8 programs correctly. By "correctly," we mean that the pipelined version produces (in fewer clock cycles) the same result that the multi-cycle version (section 8.3.1.5) produces in more clock cycles. Since the multi-cycle and the pipelined versions proceed differently, we have to wait until both machines are halted to check if the results are the same. The limitation on the kind of machine language program that figure 9-1 will execute properly is that each instruction is independent of the others. In other words, there are no data dependencies. (This is the only kind of pipelining discussed in chapter 6.) An example of such a program is the one given in appendix A, which is used with the multi-cycle ASM in section 8.3.1.6:

```
0100/7200
0101/1106
0102/1107
0103/1110
0104/3111
0105/7402
0106/0112
0107/0152
0110/0224
0111/0000
```

Here is what happens when the first pipelined ASM executes this program:

```
INIT   pc=xxxx ir1=xxxx mb2=xxxx ir2=xxxx ac=xxxx h=x     59
F1     pc=0100 ir1=xxxx mb2=xxxx ir2=xxxx ac=xxxx h=1    159
IDLE   pc=0100 ir1=xxxx mb2=xxxx ir2=xxxx ac=xxxx h=1    259
F1     pc=0100 ir1=7000 mb2=xxxx ir2=7000 ac=xxxx h=0    359
F1     pc=0101 ir1=7200 mb2=xxxx ir2=7000 ac=xxxx h=0    459
F1     pc=0102 ir1=1106 mb2=xxxx ir2=7200 ac=xxxx h=0    559
F1     pc=0103 ir1=1107 mb2=0112 ir2=1106 ac=0000 h=0    659
F1     pc=0104 ir1=1110 mb2=0152 ir2=1107 ac=0112 h=0    759
F1     pc=0105 ir1=3111 mb2=0224 ir2=1110 ac=0264 h=0    859
F1     pc=0106 ir1=7402 mb2=0000 ir2=3111 ac=0510 h=0    959
F1     pc=0107 ir1=0112 mb2=xxxx ir2=7402 ac=0000 h=0   1059
F1     pc=0110 ir1=0152 mb2=xxxx ir2=0112 ac=0000 h=1   1159
IDLE   pc=0110 ir1=0152 mb2=xxxx ir2=0112 ac=0000 h=1   1259
IDLE   pc=0110 ir1=7000 mb2=xxxx ir2=7000 ac=0000 h=0   1359
```

In the above, bold shows how the first and third TAD instructions travel through the pipeline, and italics show how the CLA, the second TAD, and the DCA instructions travel through the pipeline. In the first clock cycle after leaving IDLE (`$time` 359), `ir1` and `ir2` contain NOPs, so nothing happens to the accumulator. In the next clock cycle, `ir1` contains the first instruction (7200), but `ir2` still contains a NOP. Only in the third clock cycle after leaving IDLE (`$time` 559) does an actual instruction from the program execute—in this case the accumulator is scheduled to be cleared. This action becomes visible at `$time` 659. At that same time the first TAD instruction is ready to execute. In the previous clock cycle, the operand (0112) needed for this TAD instruction was scheduled to be loaded into `mb2`. Therefore, at `$time` 659 the `ac` ← `ac + mb2` can be scheduled. The sum (0000+0112) becomes visible at `$time` 759. The remaining TAD instructions have filled the pipeline, so they can execute one per clock cycle. This is possible because the operands (0152 available at `$time` 759 and 0224 available at `$time` 859) have also been fetched. At `$time` 959, the correct sum (0510) is stored into memory at address 0111.

9.3 Data dependencies

What happens if the instructions are not independent of each other? For software to do practical things, often one instruction needs to depend on results computed by previous instructions. This is known as a *data dependency*. For example, a slight variation of the program from appendix A:

```
0100/7200
0101/1106
0102/1107
0103/3111   ← this is different from appendix A
0104/1111   ← this is also different
0105/7402
0106/0112
0107/0152
0110/0224
0111/0000
```

illustrates the problem that the above ASM has with instructions that are not independent. In this program, instead of doing a third TAD at 0103, the DCA (3111) occurs. This is followed by a TAD (1111) from this same location. The TAD instruction at 0104 is dependent on the DCA instruction at 0103. Here is the wrong result that figure 9-1 produces:

```
INIT   pc=xxxx ir1=xxxx mb2=xxxx ir2=xxxx ac=xxxx h=x      59
F1     pc=0100 ir1=xxxx mb2=xxxx ir2=xxxx ac=xxxx h=1     159
IDLE   pc=0100 ir1=xxxx mb2=xxxx ir2=xxxx ac=xxxx h=1     259
F1     pc=0100 ir1=7000 mb2=xxxx ir2=7000 ac=xxxx h=0     359
F1     pc=0101 ir1=7200 mb2=xxxx ir2=7000 ac=xxxx h=0     459
F1     pc=0102 ir1=1106 mb2=xxxx ir2=7200 ac=xxxx h=0     559
F1     pc=0103 ir1=1107 mb2=0112 ir2=1106 ac=0000 h=0     659
F1     pc=0104 ir1=3111 mb2=0152 ir2=1107 ac=0112 h=0     759
F1     pc=0105 ir1=1111 mb2=0000 ir2=3111 ac=0264 h=0     859
F1     pc=0106 ir1=7402 mb2=0000 ir2=1111 ac=0000 h=0     959
F1     pc=0107 ir1=0112 mb2=xxxx ir2=7402 ac=0000 h=0    1059
F1     pc=0110 ir1=0152 mb2=xxxx ir2=0112 ac=0000 h=1    1159
IDLE   pc=0110 ir1=0152 mb2=xxxx ir2=0112 ac=0000 h=1    1259
IDLE   pc=0110 ir1=7000 mb2=xxxx ir2=7000 ac=0000 h=0    1359
```

In the above, italics show how the instruction at 0104 travels through the pipeline. Everything looks fine until `$time` 959. The `mb2` register (shown in bold italics) should contain the operand needed in the next clock cycle for the TAD (1111) instruction. Unfortunately, at `$time` 859 when `mb2` was scheduled to be loaded, memory at address 0111 still contains the zero put there originally. The DCA (3111) instruction that is going to put the correct value (0264) into memory has not yet finished executing. By `$time` 1059, this error in the accumulator (also shown in bold italics) becomes obvious. The accumulator is supposed to contain 0264, but instead it contains 0000.

9.4 Data forwarding

The problem illustrated in the last section is known as data dependency. *Data dependency* means that the machine needs an operand that has not yet been stored in memory because a previously fetched instruction has not yet finished executing. To overcome this data dependency, we can introduce the idea of *data forwarding* into the following improved version of the pipelined ASM. This ASM is nearly identical to the earlier one, except how `mb2` is computed depends on what is in the pipeline. Under most situations, `mb2` comes from memory as it did earlier (`mb2` ← `m[ea(ir1)]`). In one special situation, the current value of the accumulator is "forwarded" to the `mb2` register. This situation occurs when the oldest instruction (`ir2`, the one currently executing) is a DCA and the effective address of that instruction is the same as the effective address of the instruction (`ir1`) that will execute in the next clock cycle. The ASM in figure 9-2 uses data forwarding.

The following shows in bold how the Verilog must be changed to implement the data forwarding given in the ASM:

```
if (halt)
  ...
else
  begin
    pc <= @(posedge sysclk) pc + 1;
    ir1 <= @(posedge sysclk) m[pc];
    ir2 <= @(posedge sysclk) ir1;
    if ((ir2[11:9] == 3)&&(ea(ir1)==ea(ir2)))
      mb2 <= @(posedge sysclk) ac;
    else
      mb2 <= @(posedge sysclk) m[ea(ir1)];
    ...
```

The following shows how data forwarding solves this problem:

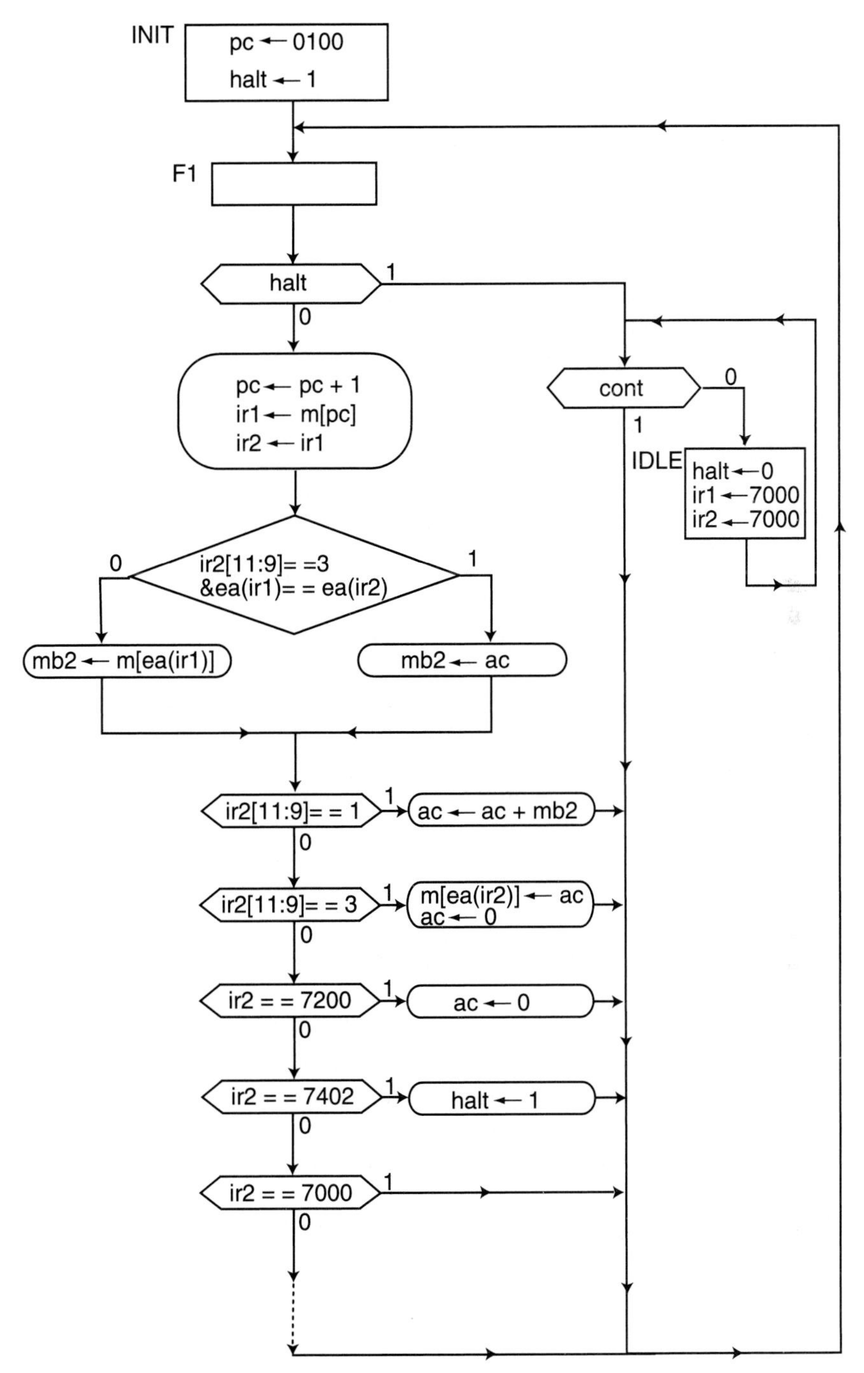

Figure 9-2. Pipelined fetch/execute with data forwarding.

```
INIT   pc=xxxx ir1=xxxx mb2=xxxx ir2=xxxx ac=xxxx h=x    59
F1     pc=0100 ir1=xxxx mb2=xxxx ir2=xxxx ac=xxxx h=1   159
IDLE   pc=0100 ir1=xxxx mb2=xxxx ir2=xxxx ac=xxxx h=1   259
F1     pc=0100 ir1=7000 mb2=xxxx ir2=7000 ac=xxxx h=0   359
F1     pc=0101 ir1=7200 mb2=xxxx ir2=7000 ac=xxxx h=0   459
F1     pc=0102 ir1=1106 mb2=xxxx ir2=7200 ac=xxxx h=0   559
F1     pc=0103 ir1=1107 mb2=0112 ir2=1106 ac=0000 h=0   659
F1     pc=0104 ir1=3111 mb2=0152 ir2=1107 ac=0112 h=0   759
F1     pc=0105 ir1=1111 mb2=0000 ir2=3111 ac=0264 h=0   859
F1     pc=0106 ir1=7402 mb2=0264 ir2=1111 ac=0000 h=0   959
F1     pc=0107 ir1=0112 mb2=xxxx ir2=7402 ac=0264 h=0  1059
F1     pc=0110 ir1=0152 mb2=xxxx ir2=0112 ac=0264 h=1  1159
IDLE   pc=0110 ir1=0152 mb2=xxxx ir2=0112 ac=0264 h=1  1259
IDLE   pc=0110 ir1=7000 mb2=xxxx ir2=7000 ac=0264 h=0  1359
```

As in the last example, italics show how the instruction at 0104 travels through the pipeline. In this case, data forwarding only occurs at `$time` 859, because `ea(1111)==ea(3111)&ir2[11:9]==3`. The underlining emphasizes the parts of `ir1` and `ir2` that must be identical for data forwarding to occur. During that clock cycle, the accumulator (shown in non-italic bold) contains 0264. The effect of the data forwarding becomes visible at `$time` 959, when `mb2` (shown in italic bold) becomes 0264, which is correct. At `$time` 1059, we see that the accumulator (shown in italic bold) has the correct value because of this data forwarding.

9.5 Control dependencies: implementing JMP

The multi-cycle ASM of section 8.3.2.1 implemented several additional instructions. Of these, the JMP (5xxx) instruction presents a problem to implement with a pipelined ASM. If we do not do something special, two of the instructions that follow the JMP will execute prior to executing the instruction being jumped to. To avoid this error, the following ASM does not fetch these instructions after the JMP, but instead it puts two NOPs in the instruction pipeline:

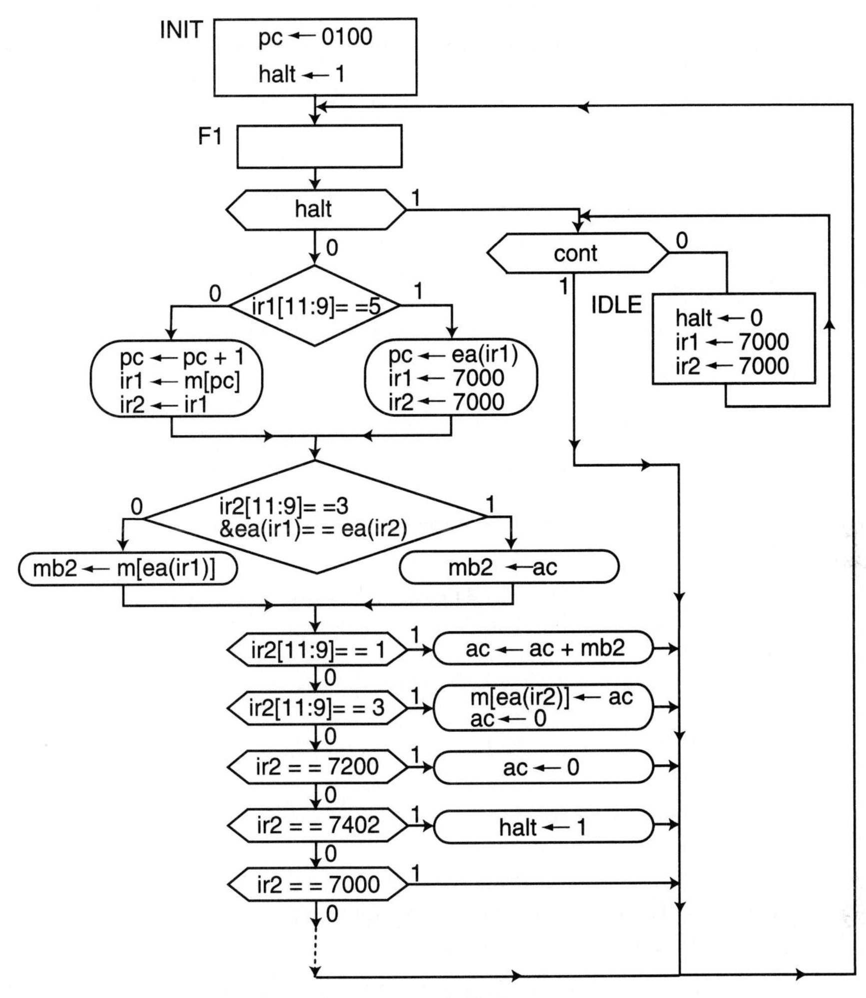

Figure 9-3. Pipelined fetch/execute with JMP.

The following shows in bold how the Verilog must be changed to implement JMP properly for the pipelined ASM:

```
    if (halt)
      ...
    else
      begin
        if (ir1[11:9] == 5)
          begin
            pc <= @(posedge sysclk) ea(ir1);
```

Continued

```
        ir1 <= @(posedge sysclk) 12'o7000;
        ir2 <= @(posedge sysclk) 12'o7000;
      end
    else
      begin
        pc <= @(posedge sysclk) pc + 1;
        ir1 <= @(posedge sysclk) m[pc];
        ir2 <= @(posedge sysclk) ir1;
      end
    if ((ir2[11:9] == 3)&&(ea(ir1)==ea(ir2)))
      ...
```

This occurs when the instruction in `ir1` is a JMP (rather than waiting until `ir2` contains the JMP). To illustrate how this works, consider the following variation of the program in appendix A:

```
0100/7200
0101/1106
0102/5105  ←  This is different from appendix A
0103/1110
0104/3111
0105/7402
0106/0112
0107/0152
0110/0224
0111/0000
```

Instead of a TAD instruction at address 0102, there is a JMP (5105) instruction that avoids executing the TAD (1110) instruction at address 0103 and the DCA instruction at 0104. The following shows how figure 9-3 executes this program correctly:

```
INIT  pc=xxxx ir1=xxxx mb2=xxxx ir2=xxxx ac=xxxx h=x     59
F1    pc=0100 ir1=xxxx mb2=xxxx ir2=xxxx ac=xxxx h=1    159
IDLE  pc=0100 ir1=xxxx mb2=xxxx ir2=xxxx ac=xxxx h=1    259
F1    pc=0100 ir1=7000 mb2=xxxx ir2=7000 ac=xxxx h=0    359
F1    pc=0101 ir1=7200 mb2=xxxx ir2=7000 ac=xxxx h=0    459
F1    pc=0102 ir1=1106 mb2=xxxx ir2=7200 ac=xxxx h=0    559
F1    pc=0103 ir1=5105 mb2=0112 ir2=1106 ac=0000 h=0    659
F1    pc=0105 ir1=7000 mb2=7402 ir2=7000 ac=0112 h=0    759
F1    pc=0106 ir1=7402 mb2=xxxx ir2=7000 ac=0112 h=0    859
F1    pc=0107 ir1=0112 mb2=xxxx ir2=7402 ac=0112 h=0    959
F1    pc=0110 ir1=0152 mb2=xxxx ir2=0112 ac=0112 h=1   1059
IDLE  pc=0110 ir1=0152 mb2=xxxx ir2=0112 ac=0112 h=1   1159
IDLE  pc=0110 ir1=7000 mb2=xxxx ir2=7000 ac=0112 h=0   1259
```

At `$time` 559, the pipeline is filled with instructions as before. At `$time` 659, the first value (0000) becomes visible in the accumulator, and the operand (0112) is in `mb2` for the first TAD instruction (1106). What is different at `$time` 659 is that `ir1` contains a JMP (5105) instruction. Instead of incrementing the program counter as was done in the earlier ASMs, the program counter is loaded with the effective address (0105) of the JMP instruction. The pipeline cannot contain any fetched instruction at $time 759. Therefore, the same decision at `$time` 659 that changes the program counter must also schedule the instruction pipeline to be loaded with NOPs. At `$time` 759, the accumulator (0000+0112) contains the correct sum from the previous TAD instruction, but it will take two clock cycles before another instruction from the program can execute. At `$time` 959, the HLT instruction is ready to execute.

9.6 Skip instructions in a pipeline

The conditional skip instructions of the PDP-8, such as SPA (Skip on Positive Accumulator, 7510) and SMA (Skip on Minus Accumulator, 7500), were described as incrementing the program counter of the multi-cycle implementation given in chapter 8. To implement these instructions with a pipelined machine requires a different approach because the program counter changes during **every clock cycle** of pipelined execution.

One of the important ideas of this book is the meaning of the non-blocking assignment. Regardless of what kind of machine you are building, in any clock cycle there can be only **one** non-blocking assignment to a particular register. It is impossible for two values to be stored in the same register during the same clock cycle. Therefore, we cannot describe a skip instruction in a pipelined implementation as incrementing the program counter yet again. The program counter at that stage is preparing to fetch the instruction **after the one to be skipped**, which will execute regardless of the skip. It is already too late to increment the program counter by two at the time we realize that the next instruction is to be skipped because the instruction to be skipped has already been fetched into `ir1`. At the time it would have been appropriate to increment the program counter by two, we would not yet know whether the accumulator will be positive or negative. We need a different way to think about the skip instruction.

The overall effect of incrementing the program counter yet again in the multi-cycle implementation is to *nullify* the instruction that follows the skip. In a pipelined implementation, we can accomplish the same thing by replacing the instruction to be nullified with a NOP (7000):

```
always
 begin
  @(posedge sysclk) enter_new_state(`INIT);
   pc <= @(posedge sysclk) 12'o0100;
   halt <= @(posedge sysclk) 1;
   forever
    begin
     @(posedge sysclk) enter_new_state(`F1);
      if (halt)
       begin
        $stop;
        while (~cont)
         begin
          @(posedge sysclk) enter_new_state(`IDLE);
           halt <= @(posedge sysclk) 0;
           ir1 <= @(posedge sysclk) 12'o7000;
           ir2 <= @(posedge sysclk) 12'o7000;
         end
       end
      else
       begin
        if ((ir2 == 12'o7510) && (~ac[11]) ||
            (ir2 == 12'o7500) && (ac[11]))
         begin
          pc <= @(posedge sysclk) pc + 1;
          ir1 <= @(posedge sysclk) m[pc];
          ir2 <= @(posedge sysclk) 12'o7000;
         end
        else
         if (ir1[11:9] == 5)
          begin
           pc <= @(posedge sysclk) ea(ir1);
           ir1 <= @(posedge sysclk) 12'o7000;
           ir2 <= @(posedge sysclk) 12'o7000;
          end
         else
          begin
           pc <= @(posedge sysclk) pc + 1;
           ir1 <= @(posedge sysclk) m[pc];
           ir2 <= @(posedge sysclk) ir1;
          end
         if ((ir2[11:9] == 3)&&(ea(ir1)==ea(ir2)))
          mb2 <= @(posedge sysclk) ac;
         else
          mb2 <= @(posedge sysclk) m[ea(ir1)];
```

Continued

```
         if      (ir2[11:9] == 1)
          ac <= @(posedge sysclk) ac + mb2;
         else if (ir2[11:9] == 3)
          begin
           m[ea(ir2)] <= @(negedge sysclk) ac;
           ac <= @(posedge sysclk) 0;
          end
         else if (ir2 == 12'o7200)
          ac <= @(posedge sysclk) 0;
         else if (ir2 == 12'o7402)
          halt <= @(posedge sysclk) 1;
         else if (ir2 == 12'o7041)
          ac <= @(posedge sysclk) -ac;
         else if (ir2 == 12'o7001)
          ac <= @(posedge sysclk) ac + 1;
         else if (ir2 == 12'o7000)
          ;
         else if (ir2 == 12'o7510)
          ;
         else if (ir2 == 12'o7500)
          ;
         else
          $display("other instructions...");
       end
    end
end
```

The decision whether to nullify the instruction that follows the skip must occur at the top of the algorithm. This is because each register, such as `ir2`, can only have one value stored into it during each clock cycle. The normal behavior of the pipeline (transferring `ir1` into `ir2`) cannot occur when the next instruction is to be nullified. Similarly, if that next instruction (in `ir1`) is a JMP (as is likely), the skip needs to take precedence over the JMP. Therefore the precedence of the decisions at the top of the algorithm is:

a) a skip instruction in `ir2` that is to be taken
b) a JMP instruction in `ir1`
c) normal pipelined behavior

Any other precedence would be incorrect. At the time the algorithm makes this decision, `ir2` already contains the skip instruction. Therefore, the bottom of the algorithm (which executes in parallel) needs to treat the 7500 or 7510 as a NOP, regardless of whether or not the following instruction will be nullified.

The above also includes the IAC (Increment ACcumulator, 7001) and CIA (Complement and Increment Accumulator, 7041) instructions. These non-memory reference instructions are similar to the CLA (7200) instruction in that the pipeline follows its normal behavior. To achieve simple pipelined behavior here with the CIA instruction, we assume that the ALU can form the twos complement negation of the accumulator in a single clock cycle.

9.7 Our old friend: division

The recurring example in this book is the childish division algorithm, introduced in section 2.2. It is used in chapter 2 to illustrate Moore ASMs, used in chapter 3 to illustrate Verilog test code, used in chapter 4 to illustrate behavioral, mixed and structural Verilog, used in chapter 5 to illustrate Mealy ASMs, used in chapter 6 to illustrate propagation delay and used in chapter 8 to benchmark the multi-cycle general-purpose PDP-8 against the special-purpose hardware of earlier chapters. The conclusion in chapter 8 is that special-purpose hardware implementations of the childish division algorithm were considerably faster and cheaper than the same algorithm running as software on the multi-cycle implementation of the general-purpose PDP-8. Yet most algorithms are implemented in software rather than hardware because software is easier to design and maintain. Pipelining allows a designer to create a more expensive general-purpose computer where the speed of its software comes closer to that of special-purpose hardware.

To illustrate what we have achieved by pipelining the PDP-8 as described in the previous sections, recall the description of the childish division algorithm in C:

```
r1 = x;
r2 = 0;
while (r1 >= y)
  {
     r1 = r1 - y;
     r2 = r2 + 1;
  }
```

For simplicity, we will assume x and y already have their values stored in memory, and that these values are less than 2048.[1]

As has been illustrated many times in earlier chapters, the `while` loop serves two roles: it avoids entering the loop and thus keeps r2 zero when x<y, or otherwise it stops the loop when it has repeated the proper number of times. In chapter 8, this was implemented as a skip and JMP at the top of the software loop and an unconditional JMP at the bottom. Such an approach is the easiest way to translate to machine language, but it has the cost of requiring additional instructions to execute each time through the loop. We need to find as good a machine language translation of this algorithm as we can. Such a machine language program will make the best use of the pipelined machine. The following uses an SPA instruction at the top to cause the loop to be entered the first time, and an SMA instruction at the bottom to cause the loop to exit:

```
               *0100
0100/7200       CLA
0101/1126       TAD X          // ac = 0+x
0102/3124       DCA R1         // r1 = x
0103/3125       DCA R2         // r2 = 0
0104/7200       CLA
0105/1127       TAD Y          // ac = 0+y
0106/7041       CIA            // ac = -y
0107/1124       TAD R1         // ac = r1-y
0110/7510       SPA            // if (r1-y >= 0) goto L1
0111/5123       JMP L2         //            else goto L2
0112/3124 L1,   DCA R1         //    r1 = r1-y
                               //     depends on ac containing r1-y
                               //     on both paths to this inst.
0113/1125       TAD R2         //    ac = 0+r2
0114/7001       IAC            //    ac = r2+1
0115/3125       DCA R2         //    r2 = r2+1
0116/1127       TAD Y          //    ac = 0+y
0117/7041       CIA            //    ac = -y
0120/1124       TAD R1         //    ac = r1-y
0121/7500       SMA            //    if (r1-y < 0) goto L2
0122/5112       JMP L1         //              else goto L1
0123/7402 L2,   HLT            // done
                               //
0124/0000 R1,   0000
0125/0000 R2,   0000
0126/0016 X,    0016           //  These must be < 2048 (3777 octal)
0127/0007 Y,    0007
```

[1] Since we have not implemented the link register of the PDP-8 in this pipelined version, larger values of x and y could cause the program to malfunction.

The execution of the above software on the pipelined PDP-8 illustrates how the skip instructions work:

```
INIT   pc=xxxx ir1=xxxx mb2=xxxx ir2=xxxx ac=xxxx h=x      59
F1     pc=0100 ir1=xxxx mb2=xxxx ir2=xxxx ac=xxxx h=1     159
IDLE   pc=0100 ir1=xxxx mb2=xxxx ir2=xxxx ac=xxxx h=1     259
F1     pc=0100 ir1=7000 mb2=xxxx ir2=7000 ac=xxxx h=0     359
F1     pc=0101 ir1=7200 mb2=xxxx ir2=7000 ac=xxxx h=0     459
F1     pc=0102 ir1=1126 mb2=xxxx ir2=7200 ac=xxxx h=0     559
F1     pc=0103 ir1=3124 mb2=0016 ir2=1126 ac=0000 h=0     659
F1     pc=0104 ir1=3125 mb2=0000 ir2=3124 ac=0016 h=0     759
F1     pc=0105 ir1=7200 mb2=0000 ir2=3125 ac=0000 h=0     859
F1     pc=0106 ir1=1127 mb2=xxxx ir2=7200 ac=0000 h=0     959
F1     pc=0107 ir1=7041 mb2=0007 ir2=1127 ac=0000 h=0    1059
F1     pc=0110 ir1=1124 mb2=xxxx ir2=7041 ac=0007 h=0    1159
F1     pc=0111 ir1=7510 mb2=0016 ir2=1124 ac=7771 h=0    1259
F1     pc=0112 ir1=5123 mb2=7510 ir2=7510 ac=0007 h=0    1359
F1     pc=0113 ir1=3124 mb2=7402 ir2=7000 ac=0007 h=0    1459
F1     pc=0114 ir1=1125 mb2=0016 ir2=3124 ac=0007 h=0    1559
F1     pc=0115 ir1=7001 mb2=0000 ir2=1125 ac=0000 h=0    1659
F1     pc=0116 ir1=3125 mb2=xxxx ir2=7001 ac=0000 h=0    1759
F1     pc=0117 ir1=1127 mb2=0000 ir2=3125 ac=0001 h=0    1859
F1     pc=0120 ir1=7041 mb2=0007 ir2=1127 ac=0000 h=0    1959
F1     pc=0121 ir1=1124 mb2=xxxx ir2=7041 ac=0007 h=0    2059
F1     pc=0122 ir1=7500 mb2=0007 ir2=1124 ac=7771 h=0    2159
F1     pc=0123 ir1=5112 mb2=7200 ir2=7500 ac=0000 h=0    2259
F1     pc=0112 ir1=7000 mb2=3124 ir2=7000 ac=0000 h=0    2359
F1     pc=0113 ir1=3124 mb2=xxxx ir2=7000 ac=0000 h=0    2459
F1     pc=0114 ir1=1125 mb2=0007 ir2=3124 ac=0000 h=0    2559
F1     pc=0115 ir1=7001 mb2=0001 ir2=1125 ac=0000 h=0    2659
F1     pc=0116 ir1=3125 mb2=xxxx ir2=7001 ac=0001 h=0    2759
F1     pc=0117 ir1=1127 mb2=0001 ir2=3125 ac=0002 h=0    2859
F1     pc=0120 ir1=7041 mb2=0007 ir2=1127 ac=0000 h=0    2959
F1     pc=0121 ir1=1124 mb2=xxxx ir2=7041 ac=0007 h=0    3059
F1     pc=0122 ir1=7500 mb2=0000 ir2=1124 ac=7771 h=0    3159
F1     pc=0123 ir1=5112 mb2=7200 ir2=7500 ac=7771 h=0    3259
F1     pc=0124 ir1=7402 mb2=3124 ir2=7000 ac=7771 h=0    3359
F1     pc=0125 ir1=0000 mb2=xxxx ir2=7402 ac=7771 h=0    3459
```

At `$time` 1259, when `ir1` gets loaded with the first skip instruction (7510), we do not yet know whether the accumulator will be positive or negative, so the pipeline continues filling normally. At `$time` 1359, there is a decision that must be made because `ir1` contains the JMP instruction (5123) and `ir2` contains the SPA instruction

(7510).[2] As described above, the skip is given precedence over the JMP. Therefore, whether the next instruction (currently in `ir1`) will be nullified is based on `ac[11]`. In this case, `ac[11] == 0`, so the SPA will nullify the following instruction. At `$time` 1459, `ir2` has become NOP (7000), but 3124 was fetched normally into `ir1` so that the algorithm can proceed sequentially.

A different situation occurs at `$time` 2259. Here the SMA (7500) does not nullify the JMP instruction (5112) because the accumulator is not negative, so the behavior described in section 9.5 occurs. Both `ir1` and `ir2` are loaded with NOPs (7000), as is visible at `$time` 2359. The machine does not start executing useful instructions after the JMP until `$time` 2559 because of the time required to fill the pipeline.

Finally, at `$time` 3259, the SMA (7500) does nullify the JMP instruction (5112) because the accumulator is negative, so only `ir2` has a NOP (7000) at `$time` 3359. This allows sequential execution of the HLT (7402) at `$time` 3459.

Between `$time` 359 and `$time` 3459 are 32 clock cycles. In general, if the `quotient >= 1`, the number of clock cycles is `12+10*quotient`. The following table summarizes implementations of the childish division algorithm given in this and earlier chapters:

```
max  pipe kind hardware software
int       of ASM section  section    clock cycles

4095  n   Moore  2.2.7    n/a        3 +    quotient
4095  n   Moore  2.2.3    n/a        2 + 2*quotient
4095  n   Moore  2.2.2    n/a        3 + 3*quotient
4095  n   Moore  2.2.5    n/a        2 + 3*quotient
4095  n   Mealy  5.2.1    n/a        2 + 2*quotient
4095  n   Mealy  5.2.3    n/a        3 +    quotient
4095  n   Mealy  5.2.4    n/a        2 +    quotient
4095  n   Moore  8.3.2.1 8.3.2.5.3  88+75*quotient
2047  n   Moore  8.3.2.1 9.7        55+55*quotient
2047  y   Mealy  9.6      9.7        12+10*quotient
```

The first seven lines above are for special-purpose computers whose ASMs implement the childish division algorithm. The last three lines are for general-purpose computers (whose ASMs implement fetch/execute) that need a machine language program to implement division. The "max int" column shows the maximum allowable integer input, which is 2047 for the software given in this section. The "pipe" column indicates whether the hardware is pipelined. The "kind of ASM" indicates whether the ASM uses condi-

[2] The 7510 in mb2 is sheer coincidence.

tional commands (Mealy) or not (Moore). The "hardware section" column indicates where the ASM is described. The "software section" applies only to general-purpose computer implementations and describes the machine language for a version of the childish division algorithm. The "clock cycle" column indicates how long it takes to compute the `quotient`, neglecting input/output time, if possible. [The "friendly user" assumptions cause the special purpose machines to do useful work (clearing `r2`) during this time that cannot be neglected. On the other hand, the software results ignore these times.]

The next to the bottom line shows how long the software given in this section takes to run on the multi-cycle hardware designed in section 8.3.2.1. This is shown to make a fair comparison of the effects of pipelining given on the bottom line. As `quotient` gets large, the speedup of the software running on the pipelined PDP-8 versus the same software running on the multi-cycle machine of section 8.3.2.1 approaches 55/10=5.5. But still, the speed of the special-purpose hardware in chapter 2 can be up to ten times faster than the speed of the pipelined PDP-8. As is discussed in the next section, the speed of the pipelined PDP-8 comes at the cost of a special kind of memory, known as multi-port memory, which is several times more expensive than the single-ported memory described in section 8.2.2.

9.8 Multi-port memory

In order to realize the pipelined fetch/execute ASM in hardware, it must be possible to do three things simultaneously with memory: fetch an instruction, fetch data and store data. The memory devices discussed in section 8.2.2 would not allow this to happen, because they are restricted to at most one read or write per clock cycle. To allow multiple operations per clock cycle in memory, we need a multi-port memory, which is shown as a letter "E" on its side.

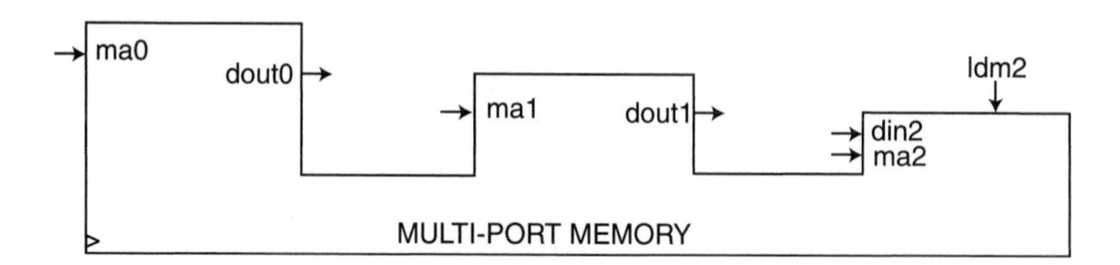

Figure 9-4. Symbol for multi-port memory.

The reason for this unconventional figure is to illustrate the fact that the multi-port memory acts like several independent devices that share a common foundation (the contents of memory). In each clock cycle, three separate operations occur in parallel:

```
always @(m[ma0])
  dout0 = m[ma0];

always @(m[ma1])
  dout1 = m[ma1];

always @(posedge sysclk)
  begin
    if (ldm2)
      m[ma2] = din2;
  end
```

so that the architecture that instantiates the multi-port memory may do three things to memory in parallel.

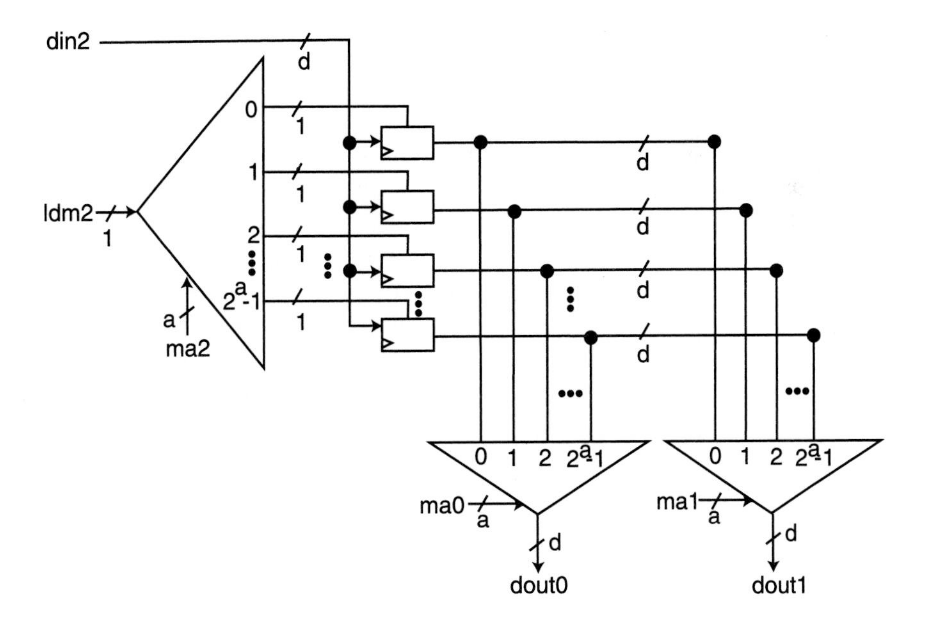

Figure 9-5. Implementation of multi-ported memory.

Figure 9-5 shows a block diagram for a synchronous multi-port memory using a demux, two muxes and enabled registers. This block diagram is a generalization of the single-port memory shown in section 8.2.2.3.1. The distinction with the multi-port memory is that there are two muxes, each of which can access the memory cells independently.

9.9 Pipelined PDP-8 architecture

Figure 9-6 shows an architecture for the pipelined PDP-8 that uses the multi-port memory. The program counter, `pc`, is a counter with `clrpc`, `incpc` and `ldpc` command signals. The other registers (`ir1`, `ir2`, `mb2` and the accumulator) are enabled registers with load signals (`ldir1`, `ldir2`, `ldmb2`, `ldac`). There are two muxes that allow `ir1` and `ir2` to be loaded with NOPs (7000). There is another mux that allows for data forwarding from the accumulator to `mb2`. Also, there is a comparator that detects when `ea(ir1)` equals `ea(ir2)`.

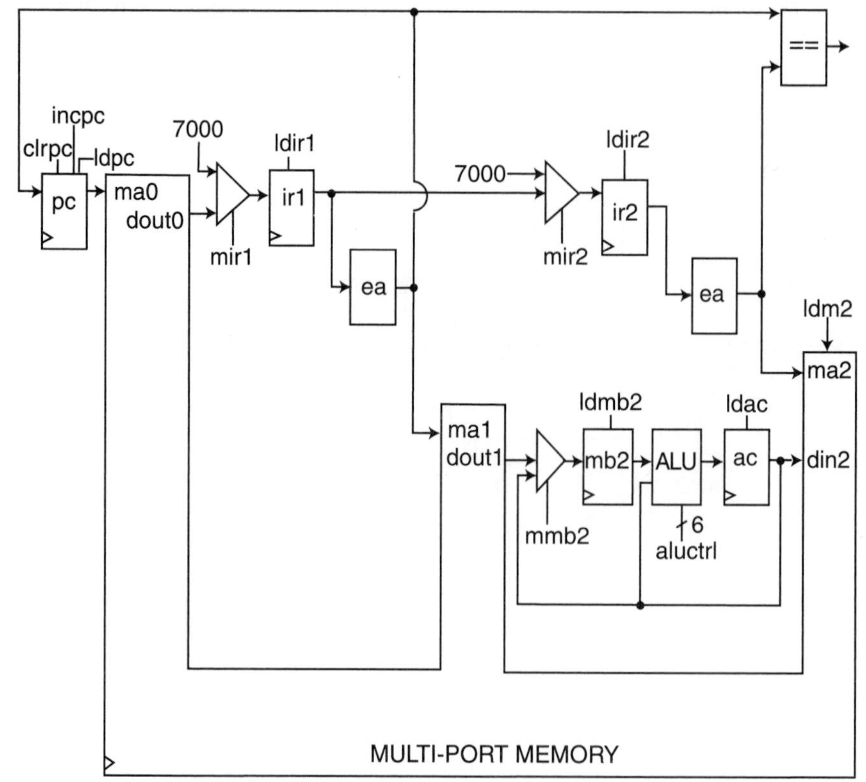

Figure 9-6. Architecture for pipelined PDP-8.

9.10 Conclusion

The pipelined PDP-8 designed in this chapter can run software in some situations about five times faster than the multi-cycle PDP-8 given in the last chapter. Because the propagation delays (which determine the clock frequency) in the pipelined and multi-cycle versions are nearly identical, there are two other factors that determine the speed. First is the number of clock cycles per instruction. (In chapter 8, most instructions take five clock cycles, but in this chapter instructions other than JMP take only one cycle.) Second is the the mix of instructions in the program, such as how frequently JMPs occur. (The example here is the childish division algorithm, which may or may not be representative of how the algorithm you want to implement will perform.)

The major cost of the pipelined approach in this chapter is the multi-port memory, which allows simultaneous access to memory for instructions and data. The problem is that even with pipelining, this approach provides one-tenth the speed of the specialized hardware for the childish division algorithm.

When you consider both cost and speed, special-purpose hardware is much better than software running on a pipelined PDP-8, at least for this example. Although the relative performance of other algorithms might be different, this example points out that other techniques beyond pipelining of the PDP-8 are going to be required if software speed is going to approach that of special purpose hardware. The next chapter illustrates some of these techniques.

9.11 Further reading

PATTERSON, DAVID A. and JOHN L. HENNESSY, *Computer Organization and Design: The Hardware/Software Interface,* Morgan Kaufmann, San Mateo, CA, 1994. Chapter 6 gives more details about implementing data forwarding with an instruction set more complicated than the PDP-8.

STERNHEIM, ELIEZER, RAJVIR SINGH and YATIN TRIVEDI, *Digital Design with Verilog HDL,* Automata Publishing, San Jose, CA, 1990. Chapter 3 gives a different approach to modeling a pipelined general-purpose computer in Verilog.

9.12 Exercises

9-1. Modify the behavioral design in section 9.6 to include the ISZ instruction of the PDP-8 (described in appendix B). Including an ISZ instruction in a program should only increase its execution time by one clock cycle for each time the ISZ is executed. Simulate the modified design with the following programs:

```
0000/7200
0001/2007
0002/2007
0003/1007
0004/2007
0005/5003
0006/7402
0007/7774
```

9-2. Using the ISZ instruction, it is possible to implement a version of the childish division program that is about twice as fast as the one given in section 9.7. Implement such a program and use test code with the behavioral Verilog for problem 9-1 to measure how long it takes to divide `x` by seven, as the test code varies `x` from 0 through 28. Derive a mathematical formula for clock cycles comparable to those listed in section 9.7 which generalizes the data observed by the Verilog test code. Hints: The machine language program needs to precompute `-y`, and `r1` should reside in the accumulator rather than in memory. The skipping action of the ISZ is irrelevant to this program.

9-3. Draw a modified architecture for problem 9-1. The only extra devices needed are two input muxes, a comparator and an incrementor.

9-4. Modify the behavioral design in section 9.6 to include the JMS instruction of the PDP-8 (described in appendix B). Simulate your modified design.

9-5. Draw a modified architecture for problem 9-4.

9-6. Modify the behavioral design in problem 9-4 to include the indirect page zero addressing mode of the PDP-8 (described in appendix B). You may assume there is an additional port to memory. How many stages will be in your pipeline?

9-7. Draw a modified architecture for problem 9-6.

9-8. Modify the behavioral design in problem 9-4 to include interrupts.

9-9. The behavioral design of section 9.4 does not execute self-modifying programs properly. Modify the design to process properly the two kinds of dependencies that are possible in such programs.

9-10. Draw a modified architecture for problem 9-9.

10. RISC PROCESSORS

All general-purpose computers implement fundamentally the same algorithm in hardware: fetch/execute. A multi-cycle implementation of fetch/execute, which is described in chapter 8 using the PDP-8, takes several clock cycles per instruction. Since the PDP-8 has a simple instruction set with a single-accumulator and a single memory address, it often takes several instructions to execute a single high-level language statement, as illustrated in section 8.3.2.5.3. It is possible to pipeline fetch/execute, as was shown in chapter 9, so that most instructions appear to execute in a single clock cycle, but a single-accumulator machine still requires several instructions to perform typical high-level language statements. The speed of a pipelined single-accumulator machine is several times slower than the special-purpose hardware of chapter 2, at least for the division example used throughout this book.

In the beginning, most general-purpose computers, such as the Manchester Mark I, adhered to this single-accumulator, single-address style of instruction set. The PDP-8 is one of the best and purest illustrations of this very simple approach. The reason for introducing it in the preceding chapters is to illustrate how the same design process used for special-purpose computers also works for general-purpose computers. The simplicity of the PDP-8 allows us to focus on using ASM charts and Verilog in the design process without having to worry about excessive complications that exist in other instruction sets.

The problem is that, after pipelining the PDP-8, we have about reached the limits of performance from a single-accumulator, single-address instruction set. To make software closer to the speed of special-purpose hardware will require specifying a different kind of machine language. The central concept of fetch/execute remains the same, but the way the machine uses the bits in the instruction register will have to be different.

10.1 History of CISC versus RISC

One attempt to increase performance of general-purpose processors that became popular in the 1970s is the idea of a Complex Instruction Set Computer (CISC). In essence, the idea is to merge a simple general-purpose machine together with special hardware (and special registers) that solve certain specific computations. The thought was that this would give the user the best of both worlds (special-purpose and general-purpose computers). To activate each special hardware unit requires including a new instruction in the instruction set. Rather than the handful of machine language instructions described in appendix B for the PDP-8, a CISC machine might have thousands of distinct instructions. Fitting all these instructions into a reasonable sized instruction register requires that some instructions occupy multiple words, which is known as a *variable length instruction set*. Such machines are aptly named CISC because the fetch/execute algorithm, although fundamentally the same, has much more complex details with a variable length instruction set. This is especially true if the machine is to be pipelined.

Two factors led to the popularity of CISC processors. First, improved fabrication technologies allowed ever-increasing amounts of hardware to fit on a chip. Second, instruction set designers had a mistaken belief that programmers and compilers would be able to utilize all this special-purpose hardware effectively.

By the early 1980s, several empirical studies had shown that CISC processors did not make effective use of all of their special-purpose hardware. As a result of these studies, several groups designed Reduced Instruction Set Computers (RISC). Like the PDP-8, RISC machines have fixed length instruction sets. This simplifies pipelined implementation. RISC instruction sets are chosen with pipelining of fetch/execute in mind, while CISC instruction sets make pipelining fetch/execute difficult. Unlike the PDP-8, RISC processors have several features that allow higher performance than is possible on a single-accumulator machine. Although CISC processors remained popular through the end of the twentieth century (the Pentium II is a CISC processor), the momentum in computer design shifted to the RISC philosophy.

10.2 The ARM

In the early 1980s, Acorn Computers, Ltd. designed an inexpensive computer for teaching computer literacy in conjunction with a BBC television program in Great Britain. The machine was originally dubbed the "Acorn RISC Microprocessor" (ARM). Several years later, Acorn entered into a consortium with more than a dozen manufacturers, including DEC[1] (the company that manufactured the PDP-8) and Apple (which uses the ARM in its Newton PDA), to promote the ARM worldwide. The ARM acronym was redefined to mean "Advanced RISC Microprocessor." The ARM is probably the most elegant RISC processor ever marketed. Its instruction set is simpler than most of the other RISC processors with which it can be compared. Although, as explained below, it does not have the performance bottlenecks of a single-accumulator machine, its superb simplicity is in some respects reminiscent of the PDP-8. This chapter will use only a small subset of ARM instructions to introduce some key ideas in choosing an instruction set for maximal performance. Appendix G explains how to access the official ARM documentation for the complete instruction set. In particular, the ARM supports several different modes of operation. We will only be concerned with what is called *user mode*.

---

[1] DEC sold its rights to the StrongARM to Intel in late 1997, but the other members of the ARM consortium were not part of that agreement and continue to produce various versions of the ARM.

10.3 Princeton versus Harvard architecture

Like all other general-purpose ("stored program") computers, commercial versions of the ARM use the same memory to store both machine language programs and data. This approach is sometimes referred to as a *Princeton architecture.* The ARM, like all other popular general-purpose computers, requires memory reference instructions, akin to those of the PDP-8, to bring data into and out of the central processing unit. (The mnemonics of these ARM instructions, given in appendix G, are LDR, STR, LDM, STM and SWP.) For the moment let us ignore these memory reference instructions, and instead assume that only programs reside in memory. A machine where programs reside in a memory exclusively for programs is sometimes called a *Harvard architecture.*[2] Although the ARM is not actually a Harvard machine, it will simplify the discussion if we assume it is.

10.4 The register file

If, at least for the moment, data is not going to reside in the same memory as programs, where is it going to be? To achieve software performance that approaches that of special-purpose hardware, there is only one plausible answer: put the data in hardware registers. In contrast to the PDP-8, with its single accumulator, a RISC processor like the ARM needs many registers for storing data.

When you design a special-purpose machine, like those in chapter 2, it is usually fairly clear how to interconnect the registers to implement the transfers required by the algorithm. The designer of a general-purpose computer does not have the luxury of knowing how registers might need to be interconnected because the register transfers will be determined by software. Therefore, the registers of a RISC processor need to be lumped together into what is called a *register file*. The register file is really a small and fast synchronous multi-port memory. The ARM has sixteen registers available in its register file in user mode.[3] We will refer to these sixteen registers using the Verilog array notation `r[0]` through `r[15]`. In assembly language, the programmer refers to these as R0 through R15. Each one of these registers contains a 32-bit value.

The program counter on the ARM is actually synonymous with `r[15]`. We can improve the clarity of our Verilog description of the ARM using:

[2] Harvard architectures usually have a separate memory for data, which we will ignore. Commercial versions of the ARM actually have Princeton architectures that share the same memory for data. It is an oversimplification to think of the ARM with a Harvard architecture.

[3] There are several other registers available for so-called supervisor modes, but we will ignore these for simplicity.

```
`define PC    r[15]
```

10.5 Three operands are faster than one

The most common operations in typical algorithms are things like addition and subtraction. For example, in the childish division algorithm implemented in section 9.7, we need to compute a difference, `d = r1 - y`. On the PDP-8, `r1` and `y` are data that residesin memory. The PDP-8's accumulator contains partial results as the following four instructions execute:

```
0104/7200      CLA
0105/1127      TAD Y
0106/7041      CIA
0107/1124      TAD R1
```

As with the examples in chapters 8 and 9, the above shows the address and corresponding machine language in octal. Upon completion of the second TAD instruction, the PDP-8's accumulator contains `d`.

To perform a similar computation with the ARM requires that all data reside in registers. Let us assume that the ARM's `r[0]` register takes on the role served by the PDP-8's accumulator (to contain the difference, `d`), that the ARM's `r[1]` register serves the same role as the R1 location in the PDP-8's memory and that the ARM's `r[4]` register contains the value of `y`.

Since there are only sixteen registers to choose from, the ARM makes it possible to specify both operands of the subtraction (`r[1]` and `r[4]`) as well as the destination register (`r[0]`) within a single 32-bit instruction. For example, given the above assumptions, the following single ARM instruction is equivalent to the four PDP-8 instructions shown earlier:

```
0000000c/e0510004    SUBS   R0,R1,R4

            ^^   ^
            ||   |
            ||   +- specifies r[4]
            |+--- specifies r[0]
            +--- specifies r[1]
```

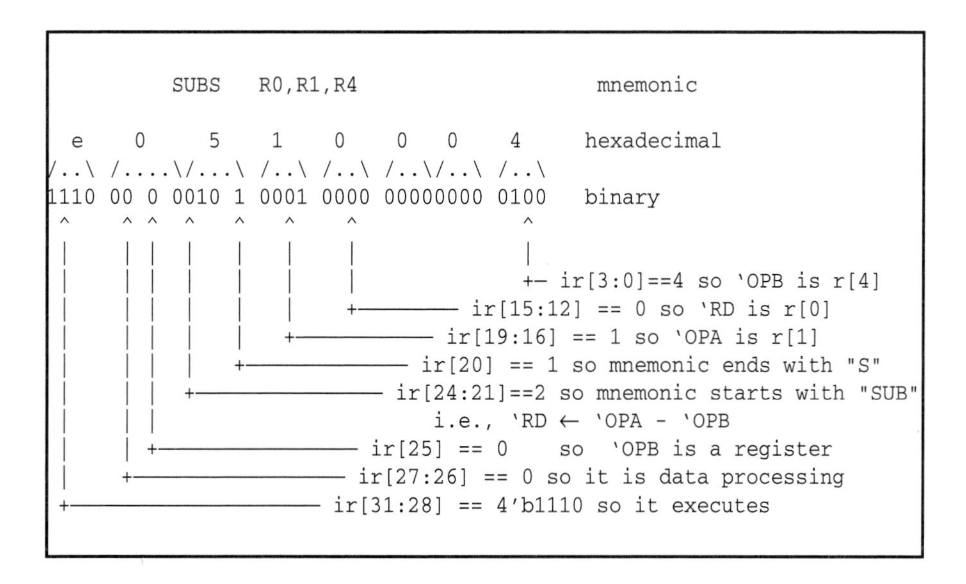
SUBS R0,R1,R4
mnemonic
e 0 5 1 0 0 0 4
hexadecimal
1110 00 0 0010 1 0001 0000 00000000 0100
binary
ir[3:0]==4 so 'OPB is r[4]
ir[15:12] == 0 so 'RD is r[0]
ir[19:16] == 1 so 'OPA is r[1]
ir[20] == 1 so mnemonic ends with "S"
ir[24:21]==2 so mnemonic starts with "SUB"
i.e., 'RD ← 'OPA - 'OPB
ir[25] == 0 so 'OPB is a register
ir[27:26] == 0 so it is data processing
ir[31:28] == 4'b1110 so it executes

10.6.2 Branch instruction

The second major instruction category we will use for our subset of the ARM instruction set is the *branch instruction.* When instruction register bits 27 down to 25 are 3'b101, the ARM categorizes the instruction as a branch. There are two mnemonics for this category: B (branch) and BL (branch and link). Instruction register bit 24 distinguishes between the simple branch instruction and the branch and link (zero means simple branch). We will not implement the branch and link instruction here, although it is straightforward. (It utilizes R14 to save a return address, quite analogously to the way the JMS instruction on the PDP-8 uses a memory location to save a return address.)

The branch instruction on the ARM is very similar to the JMP instruction of the PDP-8, with three differences. First, the branch instruction of the ARM uses a relative addressing mode, rather than the direct addressing mode of the PDP-8's JMP instruction. In a relative addressing mode, the offset field of the branch instruction is added (as a signed twos complement value) to the program counter (rather than being moved to the program counter as occurs on the PDP-8). Second, the branch instruction of the ARM refers to the offset in terms of 32-bit words, but the program counter of the ARM refers to eight-bit bytes. (The value in the ARM's program counter is always divisible by four, and so the offset field of the branch instruction is one-quarter the value needed to add to the program counter.) Third, since the offset field is only 24 bits wide, it must be sign extended before it is added to the program counter.

The following example branch instruction forms an infinite loop by branching back to itself. Because of the relative addressing mode, this same machine language instruction will work identically regardless of the location where it occurs in a program:

```
              L2 B      L2              mnemonic (L2 is a label)

  e    a     f   f   f   f   f   e      hexadecimal
/..\ /...\ /..\/..\/..\/..\/..\/..\
1110 101 0 111111111111111111111110     binary
  ^    ^  ^              ^
  |    |  |              |
  |    |  |              +---------    two's complement -2 offset
  |    |  +-----------------           ir[24] ignored here
  |    +---------------------          ir[27:25] == 5 so it branches
  +------------------------            ir[31:28] == 4'b1110 so it executes
```

It may seem a little strange, but the -2 indicates branching back to the same instruction. In other words, the new value of the program counter is the value of the program counter at the time the instruction is fetched plus `4*offset+8`, where the `offset` is a sign extended version of instruction register bits 23 down to 0. The reason the ARM designers chose to make -2 mean branching back to itself will become clear later in this chapter.

10.6.3 Program status register

Another detail in which the ARM is different than the PDP-8 is the way in which the ARM tests for conditions, such as testing for negative numbers. On the PDP-8, since the accumulator is the only place where a number to be tested can reside, the hardware simply uses the most significant bit of the accumulator to determine whether that number is negative or not. On the ARM, there are sixteen different registers that a programmer might choose to test, and so there are sixteen different sign bits that the hardware might need to use, which would not be economical. Instead, the ARM allows the programmer to specify a one as bit 20 of the instruction register for a data processing instruction ("S" suffix on the mnemonic). When bit 20 is a one, certain critical information about the result of the data processing instruction is saved in the *program status register*. (The "S" suffix means set the PSR.) In this chapter, we will consider bit 31 of the program status register (PSR), which is known as the "N" (negative) flag. The N flag stores the sign bit of the result of the most recent data processing instruction with an "S" suffix mnemonic.

For example, suppose `r[1] == 32'h00000007` and `r[4] == 32'h00000007` prior to the execution of the SUBS R0,R1,R4 instruction (e0510004) given in section 10.6.1. Because the result is not negative (`r[0] == 32'h00000000`) and bit 20 of the instruction register is set, the N flag becomes zero.

As a different example, suppose `r[1] == 32'h00000000` and `r[4] == 32'h00000007` prior to the second execution of the same SUBS R0,R1,R4 instruction (e0510004). Because the result is negative seven (`r[0] == 32'hfffffff9`) and bit 20 of the instruction register is set, the N flag becomes one.

If bit 20 of a data processing instruction is zero, the PSR remains unchanged. For example, suppose `r[1] == 32'h00000007` and `r[4] == 32'h00000007` prior to the execution of a SUB R0,R1,R4 instruction (e0410004) similar to the SUBS except that bit 20 of the instruction register is zero. Even though the result is not negative (`r[0] == 32'h00000000`), the N flag remains what it was (one) because bit 20 of the instruction register is zero.

There are several other bits in the PSR which we will not implement here. For example, bit 30 of the PSR is the "Z" flag, which indicates whether the result (of the most recent data processing instruction with an "S" suffix mnemonic) was equal to zero. Bit 29 of the PSR is the "C" flag, which indicates whether the result (of the most recent data processing instruction with an "S" suffix mnemonic) produced a carry (analogous to the LINK of the PDP-8). Bit 28 of the PSR is the "V" flag, which indicates whether the result (of the most recent data processing instruction with an "S" suffix mnemonic) caused a signed overflow (what should be a negative number appears positive or vice versa).

10.6.4 Conditional execution

One of the most interesting and useful features of the ARM is that every instruction can be conditional, that is, if a certain condition recorded in the PSR is not satisfied, the instruction is treated as a NOP. If that condition is satisfied, the instruction executes normally. The condition is indicated by bits 31 through 28 of the instruction register. Although there are sixteen different conditions that the actual ARM recognizes (as shown in appendix G), we will only consider the following four in the subset implemented here:

decoding	cond name	mnem suffix	instruction executes if	acts like NOP if
ir[31:28]==4'b0100	minus	MI	psr[31]==1	psr[31]==0
ir[31:28]==4'b0101	plus	PL	psr[31]==0	psr[31]==1
ir[31:28]==4'b1110	always	none	1	0
ir[31:28]==4'b1111	never	NV	0	1

As illustrated in earlier sections, most instructions have `4'b1110` for instruction register bits 31 down to 28 so that execution does not depend on the `psr`. Although the ARM documentation discourages it, for our subset, we will treat f0000000 as a NOP. (There are many other ways to form a NOP on this machine.)

Using a condition suffix like PL or MI for an instruction on the ARM is very analogous to preceding an instruction on the PDP-8 with an SMA or SPA, respectively. The only difference on the ARM is that since the condition is **part of each instruction**, only one instruction needs to be fetched, rather than two. For example, the childish division program given in section 9.7 uses an SPA prior to a JMP for the special case when the quotient is zero:

```
0110/7510      SPA
0111/5123      JMP L2
```

These two PDP-8 instructions are analogous to the BMI instruction (it branches when the PSR indicates minus, so it nullifies (treats like a NOP) the instruction when the PSR indicates plus):

```
00000010/4a000003      BMI     L2
```

As another example, the PDP-8 program in section 9.7 also uses an SMA prior to a JMP for deciding whether to go through the loop another time:

```
0121/7500      SMA
0122/5112      JMP L1
```

The analogous ARM instruction is BPL:

```
00000020/5affffb     BPL     L1
```

10.6.5 Immediate operands

The only practical way to put a constant value into the PDP-8's accumulator is to use a memory reference instruction. This means that two memory locations must be accessed: the one that contains the instruction and the one that contains the data.

Although the ARM does actually have memory reference instructions (which we are ignoring for now), the ARM provides a different way of working with constant values, known as immediate operands, that only requires one memory access. This is possible because the constant is part of the instruction. In a data processing instruction, when instruction register bit 25 is a one, OPB is an immediate constant, rather than the value of a register. Assuming that instruction register bits 11 down to 8 are zeros, the value of the immediate constant is given by instruction register bits 7 down to 0. (We will ignore the rotation that the full-fledged ARM does when instruction register bits 11 down to 8 are non-zero.) For example, consider the ARM instruction that adds the constant one to the R2 register without setting the PSR:

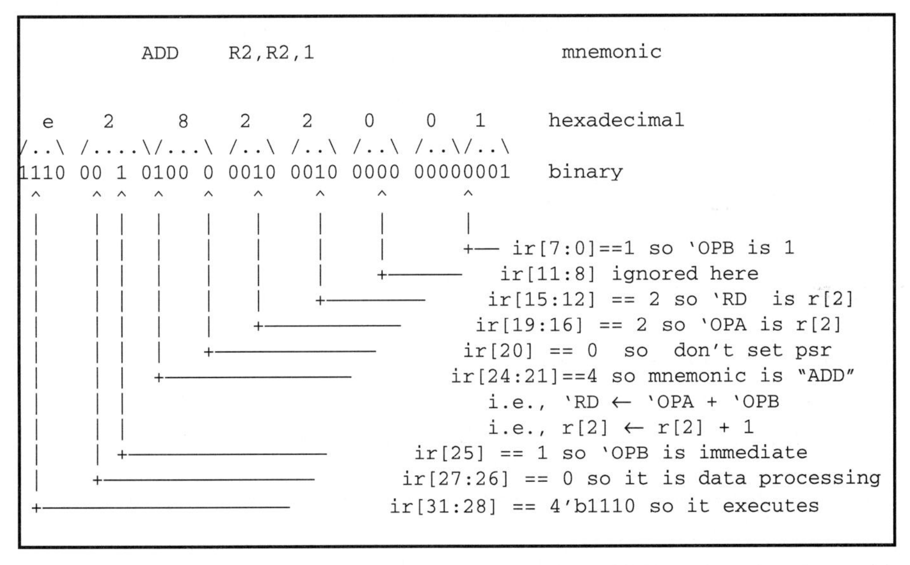

As another example, consider the ARM instruction that initializes the R1 register with the decimal constant fourteen:

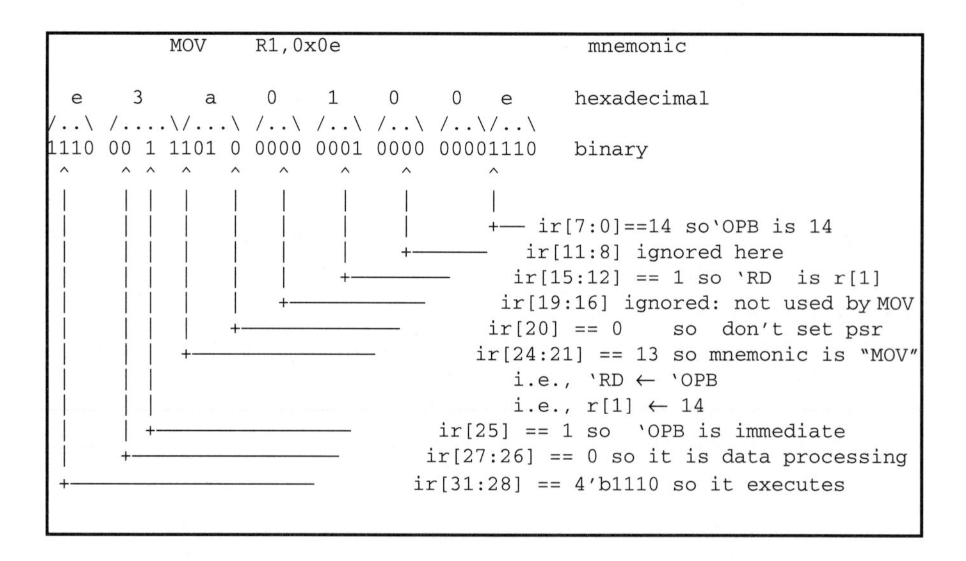
MOV R1,0x0e
mnemonic
e 3 a 0 1 0 0 e
hexadecimal
1110 00 1 1101 0 0000 0001 0000 00001110
binary
ir[7:0]==14 so 'OPB is 14
ir[11:8] ignored here
ir[15:12] == 1 so 'RD is r[1]
ir[19:16] ignored: not used by MOV
ir[20] == 0 so don't set psr
ir[24:21] == 13 so mnemonic is "MOV"
i.e., 'RD ← 'OPB
i.e., r[1] ← 14
ir[25] == 1 so 'OPB is immediate
ir[27:26] == 0 so it is data processing
ir[31:28] == 4'b1110 so it executes

10.7 Multi-cycle implementation of the ARM subset

The multi-cycle ASM in figure 10-1, which follows the basic outline of the PDP-8's ASM given in section 8.3.1.5, implements the fetch/execute algorithm for the ARM instruction set described in section 10.6.

The state names are the same as the ones in the PDP-8's ASM, except for the execute states. In the ASM for the ARM, state E0DP occurs when a data processing instruction (such as ADD or SUB) executes, and state E0B occurs when a branch instruction (B) executes.

10.7.1 Fake SWI as a halt

The actual ARM does not have a halt instruction. Instead, it has a software interrupt (SWI) instruction (ef000000) which changes the mode from user mode to a supervisor mode. Since we are ignoring the issue of modes, and since it is helpful to keep this ASM as similar as possible to the ASM in chapter 8 for the purpose of Verilog test code, we will treat the SWI as a halt. The operation of the SWI on the actual ARM is much more complicated, as explained in appendix G.

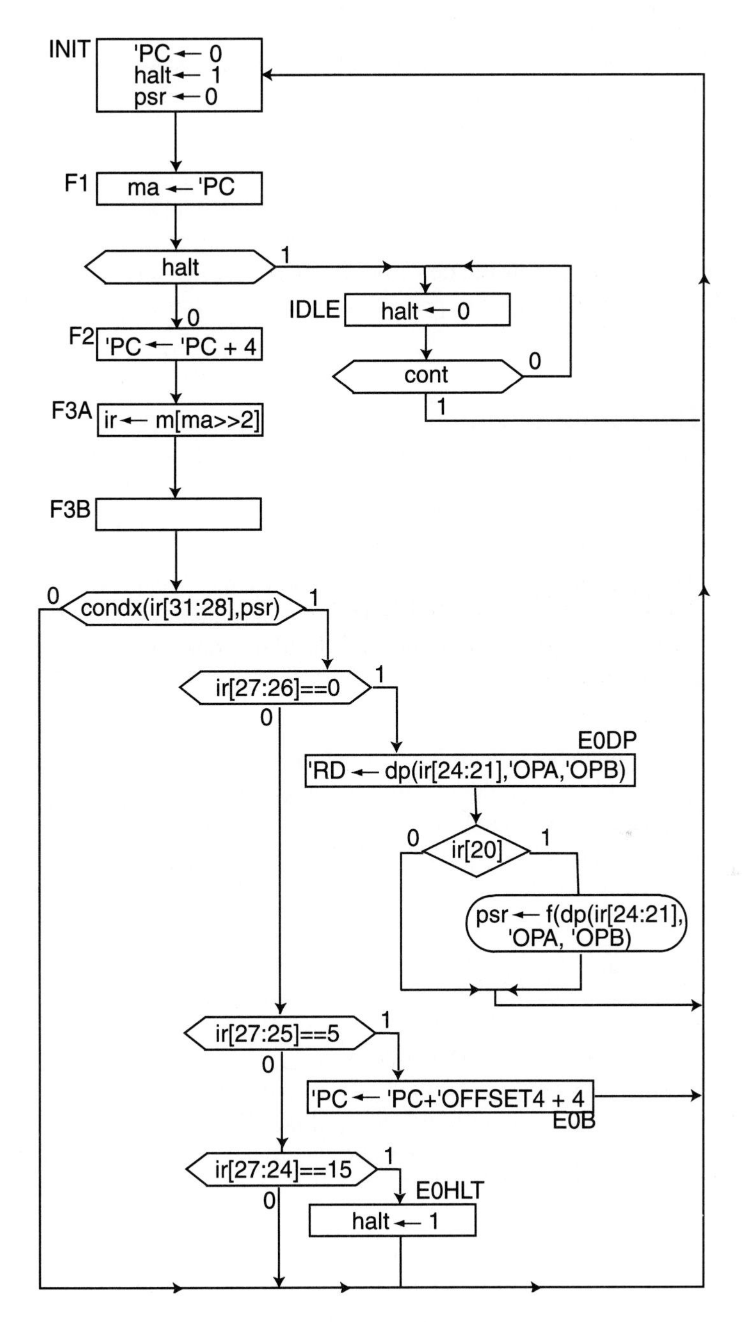

Figure 10-1 Multi-cycle ASM for ARM subset.

10.7.2 Fetch states

State INIT initializes the program counter, halt and program status registers. The machine will then proceed to state F1 and to state IDLE.

When a program executes, the normal sequence is to proceed through states F1, F2, F3A, F3B and one of the execute states. State F2 increments the program counter by four (rather than by one) because the program counter refers to an address in terms of eight-bit bytes but each 32-bit instruction is actually four bytes long. In a related way, when state F3A fetches an instruction from memory, the memory address is shifted over two bits to the right because the program counter is four times the required memory address.[4]

10.7.3 The `condx` function

The decoding (F3B) and executing (E0DP, E0B or E0HLT) states for the ARM are quite different than the analogous states for the PDP-8. First of all, every instruction on the ARM has the potential of being conditional, which is why instruction register bits 31 down to 28 are reserved for this purpose. The first decision that occurs in state F3B is whether the instruction should be nullified or not. On the actual ARM, this decision involves sixteen possibilities. Even though we are only going to implement four of these (4, 5, e and f), it is prudent to isolate this detail in a function which we will refer to as `condx(ir[31:28],psr)`.

In the actual hardware, there will be some combinational logic that implements this function. The important observation is that whether an instruction is executed or is nullified depends only on two things: instruction register bits 31 down to 28 and the current information in the program status register (which, in this implementation, only contains the N flag). Because these details have been isolated inside the `condx` function, the other twelve conditions (0-3, 6-d) not considered here could be implemented fairly easily without having to change this ASM.

After recognizing that the condition for the instruction has been satisfied, state F3B proceeds to decode the instruction. If it is a data processing instruction (instruction register bits 27 and 26 equal zero), the ASM proceeds to state E0DP. If it is a branch instruction (instruction register bits 27 down to 25 equal 5), the ASM proceeds to E0B. If it is a SWI instruction, the machine proceeds to the PDP-8 like state E0HLT for the purpose of communicating with the Verilog top-level module that will test this machine. (As mentioned above, the actual ARM would do something more complicated for SWI.)

---

[4] The reason for this inconsistency only becomes apparent with some of the instructions we are ignoring, such as LDR and STR, that use byte-sized data in memory.

10.7.4 Data processing

State E0DP has two actions that it must perform. First, it needs to perform the requested data processing using the operands (‘OPA and ‘OPB) and to store this result into the destination register (‘RD). Second, it needs to deal with the program status register.

10.7.4.1 The `dp` function

Which kind of data processing occurs in state E0DP depends upon instruction register bits 24 down to 21. In this implementation, we are only considering three data processing operations (ADD, SUB and MOV). Once again, it is advisable to isolate such details in a function, which we will refer to here as dp, so that the implementation of the other 13-data-processing operations will be straightforward. The important observation is that the result of the data processing only depends on ‘OPA, ‘OPB and instruction register bits 24 down to 21.

10.7.4.2 Conditional assignment of `psr`

Using instruction register bit 20, the programmer can choose whether or not a data processing instruction will modify the program status register. If instruction register bit 20 is zero, state E0DP leaves the program status register the way it is. On the other hand, if instruction register bit 20 is a one (meaning the mnemonic has an “S” suffix), state E0DP has a conditional oval which assigns a new value to the program status register. This new program status register value is a function (`f`) of the unconditional data processing that occurs in this state (`dp(ir[24:21],`OPA,`OPB)`). Again, isolating details in a function will make it easier to implement the full capabilities of the ARM, should you choose to do so. For now, `f` simply masks off the sign bit of the result to record the N flag. This assignment to the program status register must be conditional in state `E0DP because the information to compute the new program status register properly might not exist after that clock cycle (i.e., `RD might refer to the same register as `OPA or `OPB).

10.7.4.3 Use of macros

The use of macros (‘PC, ‘RD, ‘OPA and ‘OPB) helps conceal many of the tedious details required to implement the ARM. For example:

```
`define RD   r[ir[15:12]]
```

allows us to describe the destination register for a data processing instruction without having to mention the instruction register bits.

More importantly, macro definitions allow simple descriptions of the operands of the data processing instructions. For example, from the short explanation in section 10.6.1, one might think that 'OPA would simply be defined as `r[ir[19:16]]`, but there are some other details to consider. For example, when 'OPA is supposed to be what the programmer refers to as R15, this is in fact the program counter. According to the specification of the ARM instruction set (appendix G), when R15 is the first operand of a data processing instruction, the value of R15 used in the computation will be eight larger than the value the program counter contained when the instruction was fetched. At the time 'OPA is evaluated, the program counter has already been incremented by four in state F2. Therefore, if the programmer wants to use R15 as 'OPA, the value used in state E0DP should be `r[15]+4`. On the other hand, if the programmer wants to use a different register, its value should not be incremented.

To distinguish between R15 or another register inside a macro requires using the Verilog conditional operator (`? :`). This feature of Verilog, which acts like the similar feature of C, works with three values: the first value occurs before the question mark, the second value occurs between the question mark and the colon, and the third value occurs after the colon. If the first value is equal to one, the result is the second value; otherwise the result is the third value. From a hardware standpoint, the Verilog conditional operator is like a mux, where the select is the first value. In this particular situation, the expression `ir[19:16]!=15` chooses between two different results:

```
`define OPA  (ir[19:16]!=15?r[ir[19:16]]:r[ir[19:16]]+4)
```

As in C, parentheses are a good idea to avoid creating precedence problems when Verilog substitutes such a complicated macro.

The definition of 'OPB is even more involved because instruction register bit 25 allows the programmer to choose between an immediate value or a register value. The same problem with R15 mentioned above also must be considered:

```
`define OPB  (ir[25]?ir[7:0]:(ir[3:0]!=15?  r[ir[3:0]]:r[ir[3:0]]+4))
```

In fact, there are other issues about 'OPB that we are ignoring here. (The actual ARM allows rotation of 'OPB, which would require a more complicated expression for 'OPB.)

10.7.5 Branch

State E0B performs the relative branch by adding four plus four times the signed offset (from the low-order 24 bits of the instruction register) to the program counter.

10.7.5.1 *Multiplying by four*

The reason for multiplying by four is that instructions on the ARM are required to reside at byte addresses that are divisible by four. The low-order two bits are not recorded in the branch instruction in order to maximize the distance that a programmer can choose to branch with the available bits of the instruction register.

10.7.5.2 *Adding four in the multi-cycle implementation*

The reason for adding four in this multi-cycle implementation is to adhere to the specifications referred to in appendix G. The ARM documentation specifies that an offset of -2 in a branch instruction (eafffffe) means branching back to itself. Remember that the -2 will be multiplied by 4 to yield -8. By state E0B, the program counter has been incremented by 4. To leave the program counter the same as it was when the branch instruction (eafffffe) was fetched, an additional 4 must be added to the program counter (-8 + 4 + 4 == 0).

10.7.5.3 *Sign extension macro*

The offset times four needs to be treated as a signed 32-bit value. The offset in the instruction register is only 24-bits wide (bits 23 down to 0). This means the sign bit (`ir[23]`) must be duplicated several times on the left, in a process commonly referred to as *sign extension*. The following macro performs both the sign extension and the multiplication by four using the Verilog concatenation operator:

```
`define OFFSET4  {ir[23],ir[23],ir[23],ir[23],ir[23], ir[23],ir[23:0],2'b00}
```

10.7.6 Verilog for the multi-cycle implementation

Throughout this book, algorithms have been described using ASM charts and Verilog. For simple machines, like the childish division examples of chapter 2, the ASM and the Verilog are equivalent, and either notation gives a completely accurate description of the hardware. It is a theme of this book that both ASM charts and Verilog are important to understand, and that each notation offers the designer useful insights.

The last section illustrates the proper use of an ASM in a complex design: to provide documentation so people can see the "big picture." The last section actually uses quite a bit of Verilog notation to achieve the proper level of abstraction. This section shows that Verilog implements not only what the ASM describes but also fills in the details that should be omitted from the ASM. Verilog can do all this reasonably because it is a **textual** language.

The previous sections use many paragraphs to describe the data processing function, `dp`. Such a description is informal and therefore cannot be synthesized into hardware. Although such details could have been put formally into the ASM, they would have made the ASM considerably more complicated. Ultimately, such a complex ASM would lead us down the wrong path for the performance issues (such as pipelined and/or superscalar design) described later in this chapter. In Verilog, we can be precise but still set aside such details because they can occur in a different part of the source code (the function definition). Although such thinking is commonplace in modern software design, hardware designers are only beginning to realize the power of the notation available in Verilog.

In the case of the `dp` function, we are only implementing three of the sixteen possible operations:

```
function [31:0] dp;
  input [3:0] opcode;
  input [31:0] opa,opb;
  begin
    if      (opcode == 4'b0010)
      dp = opa - opb;
    else if (opcode == 4'b0100)
      dp = opa + opb;
    else if (opcode == 4'b1101)
      dp = opb;
    else
      begin
        dp = 0;
        $display("other DP instructions...");
      end
  end
endfunction
```

This function formally describes the SUB, ADD and MOV instructions. Except for the `$display` statement, this function could be synthesized into the combinational ALU required in the actual hardware. (The `$display` statement warns us if we attempt to execute a data processing instruction that is one of the 13 not implemented here.) This function can be reused as we improve the performance of the design. Because these details have been isolated into a function, it is easy for a designer to know where to modify the Verilog code in order to implement the remaining 13 operations.

In a similar way, we can define the Verilog function that determines whether the condition for a particular instruction to execute is true:

```
function condx;
  input [3:0] condtype;
  input [31:0] psr;
  begin
    if (condtype == 4'b1110)
      condx = 1;
    else if (condtype == 4'b0100)
      condx = psr[31];
    else if (condtype == 4'b0101)
      condx = !psr[31];
    else
      condx = 0;
  end
endfunction
```

Again, isolating this in a function makes it easy to know how to implement the remaining operations. Also, as will be shown later, defining this function will prove extremely helpful as we use more sophisticated techniques to improve performance.

For our subset of the ARM, we only implement the N flag in the program status register. The function which creates this information from the result of the ALU is trivial:

```
function [31:0] f;
  input [31:0] dpres;
  begin
    f = dpres & 32'h80000000;
  end
endfunction
```

Generating all bits of the program status register is considerably more complicated, but isolating it here helps some future designer whose job might be to do so.

A great deal of the abstraction needed for this design comes from the Verilog macros mentioned in the last section. For this multi-cycle implementation, these are:

```
`define PC       r[15]
`define RD       r[ir[15:12]]
`define OPA      (ir[19:16]!=15?r[ir[19:16]]:r[ir[19:16]]+4)
`define OPB      (ir[25]?ir[7:0]:(ir[3:0]!=15?r[ir[3:0]]:r[ir[3:0]]+4))
`define OFFSET4 {ir[23],ir[23],ir[23],ir[23],ir[23],ir[23],ir[23:0],2'b00}
```

For example, the definition of 'OPA says if the register number given in instruction register bits 19 down to 16 is not the program counter (`!=15`), use the value of that register; otherwise (`==15`) use the program counter + 4. This decision is required here because of the definition of the ARM instruction set. The Verilog conditional operator allows for compact, if somewhat cryptic, code for the decision. Recall that 'OPA will be substituted back where required in the code automatically by Verilog, and so the designer using Verilog can usually ignore these details. A designer using only an ASM would have been required to put all this detail in the ASM.

Another advantage of using Verilog here is that as we proceed to a more sophisticated technique for higher performance (such as pipelining), we can change the definition of the macros to match the more sophisticated technique, but keep the macro name the same. This is good, because the concept behind the macro is the same, even though its implementation will be different in later sections of this chapter. Given the above macros and functions, the actual translation of the ASM to implicit style Verilog is simple and is left as an exercise.

10.8 Pipelined implementation

The problem with the multi-cycle implementation is that it requires five cycles per instruction. To improve this performance, we can use a pipelined approach. There are several reasons why a pipelined implementation of our ARM instruction subset will be easier than the pipelined PDP-8 discussed in chapter 9. First, the ARM has a RISC instruction set which was designed to be pipelined. Second, we are neglecting memory reference instructions, and so the issues of operand fetch and data forwarding may be ignored here. Third, we can reuse the functions defined above without modification. Fourth, the Verilog macros given earlier can easily be redefined to match the needs of the pipelined implementation.

10.8.1 ASM for three-stage pipeline

The original versions of the ARM (referred to as the ARM1-ARM6) use a three-stage pipeline, but some of the more recent versions of the ARM use a five-stage pipeline. We will implement a three-stage pipeline because it is easier, and is similar to the three-stage examples in chapter 9. Also, since the original ARM instruction set was designed for a three-stage pipeline, implementing a pipeline of that size is most natural. The three stages are referred to as fetch, decode and execute. For our subset, decoding only considers instruction register bits 27 down to 24. The logic for decoding our subset is trivial and hardly warrants its own separate pipeline stage. The full ARM instruction set includes eleven categories, and so full decoding is rather involved, especially considering the memory reference instructions. Therefore, for our simple ARM subset, we will implement a three-stage pipeline but with the middle stage doing nothing. Later in this chapter, the middle pipeline stage will take on an important role.

As in chapter 9, it will be necessary to have multiple instruction registers (`ir1` and `ir2`). The youngest instruction is in memory, the middle-aged instruction is in `ir1` and the oldest instruction is in `ir2`. Also, as in chapter 9, we need to use NOPs to deal with the filling of the pipeline after a branch instruction. There are many ways to describe a NOP in ARM machine language. The simplest[5] is f0000000. Here is the pipelined ASM for our ARM subset:

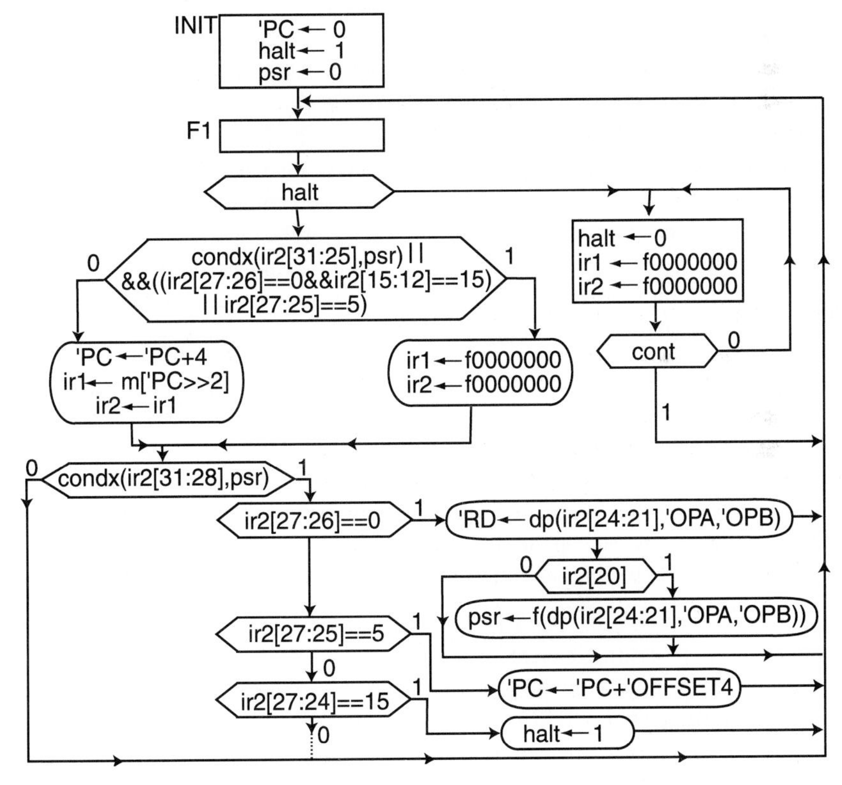

Figure 10-2. Pipelined ASM for ARM subset.

[5] Although the ARM's designers may someday redefine the meaning of this machine code to be something other than NOP, f0000000 is convenient since it is easy to recognize

Rather than translate this ASM as is into Verilog, the following sections discuss some interesting issues that it raises. These issues, such as Mealy ASMs and the non-blocking assignment, have been touched upon in earlier chapters, but with an example of this size these issues become more important.

10.8.2 Mealy ASM

There are several reasons why the pipelined ARM needs to be designed using a Mealy ASM. As in the examples in chapter 9, a decision has to be made and an assignment has to be scheduled in a single cycle because the result of the assignment must be available in the next clock cycle. With the PDP-8, this typically arises with assigning a value to the accumulator having decided already that a TAD instruction is supposed to execute. On the ARM, this same issue arises with assigning a value to the `RD macro (which could be any of the sixteen user registers), having decided a data processing instruction is supposed to execute. The decision (that recognizes the instruction) and the resulting assignment occur in the same clock cycle.

The ARM's conditional execution feature offers an additional reason why a Mealy approach is required. Often, a data processing instruction that assigns a new value to the program status register (such as SUBS) will execute just before a subsequent instruction that depends on this result in the program status register (such as BMI). The decision whether to execute the subsequent instruction cannot occur earlier than the clock cycle when that instruction is in the final stage of the pipeline because, until that stage, the program status register does not reflect the result of the preceding instruction. If the condition is satisfied, the subsequent instruction must schedule whatever assignment(s) are associated with its execution in this same clock cycle; therefore a Mealy approach is mandatory.

10.8.3 Two parallel activities

There are two parallel activities that occur in state F1. The first determines what will be in the instruction registers and the program counter in the next clock cycle. The second, which is similar to the corresponding part of the ASM in section 10.7, deals with decoding and executing the instruction in the final stage of the pipeline.

The first parallel activity in state F1 (determination of what will be in the instruction registers and the program counter) has two cases. Let's refer to these as the "B/R15" case (for branch or modifying the program counter) and the "normal" case (increment program counter). The "B/R15" case is when the instruction in the final stage of the pipeline is an instruction that will execute (as indicated by `condx(ir2[31:28],psr)`) and that is either a branch instruction (bits 27 down to 25 equal to 5) or a data processing instruction that modifies `r[15]` (since `r[15]` is the same as the program counter, such an instruction is effectively like a branch instruction). The "B/R15" case is sim-ilar to the JMP instruction on the pipelined PDP-8 given in section 9.5, except that on the ARM the branch instruction cannot be executed until it reaches the last stage of the pipeline. On the PDP-8, it is possible to begin execution of the JMP when it reaches the middle stage of the pipeline because the JMP on the PDP-8 is unconditional. Since the branch on the ARM is conditional, the bits in the program status register must be valid before the decision to branch can occur. In general, the program status register will not be valid for a particular instruction until that instruction reaches the final stage of the pipeline. This is because the program status register might have been changed by the preceding instruction, which only completed its execution after the preceding clock cycle. If the "B/R15" instruction executes, the instruction registers are filled with NOPs.

The "normal" case decides what will be stored in the instruction registers and the program counter for situations other than the "B/R15" case. In the "normal" case the program counter is incremented by four, and the instructions move down the pipeline.

Besides the part of the ASM that decides what will be stored in the instruction registers and the program counter, there is the part of the ASM that decodes and executes the instruction in the final stage of the pipeline. In this part of the ASM there are five cases:

```
"nullify"     psr prevents this instruction from executing
"dp set"      execute data processing instruction that modifies psr
"dp no set"   execute data processing instruction that leaves psr alone
"ir2 is B"    modify R15
"SWI"         set halt flag
```

Except for the "ir2 is B" case, these are identical to the multi-cycle ASM given in section 10.7. In the "ir2 is B" case, four times the sign extended offset (`OFFSET4) is added to the program counter. Here is where we see that the ARM was designed to work with a three-stage pipeline. The reason that an offset of -2 means branch back to itself is that by the time the branch instruction has reached the final stage of the pipeline, the program counter will already have been incremented twice, i.e., it is eight greater than when the branch was fetched. When the offset is -2, `OFFSET4 is -8 and so adding it to the program counter in this case puts the program counter back to where the same branch instruction will be fetched again.

10.8.4 Proper use of <=

One of the main themes of this book is the proper use of the non-blocking assignment statement. A common mistake with non-blocking assignment is to attempt to assign more than one value to a register during one clock cycle. To avoid making this mistake, a designer needs to check all possible paths through the ASM. Since there are two paths through one-half of the ASM and there are five paths through the other half that executes in parallel, there are, in theory, ten paths for the designer to check, but of these, two are contradictory. It is impossible for `ir2` to contain a branch or data processing instruction that modifies R15 in the same clock cycle that it contains an SWI instruction. Also, when "ir2 is B," the "normal" case cannot occur. When these cases are eliminated, we are left with eight cases to consider. The "B/R15" case might occur in parallel with either the "nullify," "dp set," "dp no set" or "ir2 is B" case. Alternatively, the "normal" case might occur together with either the "nullify," "dp set," "dp no set" or "SWI" case.

The "B/R15" and "normal" cases are the only places where the instruction registers are scheduled to be assigned, and so there is no problem with them. The "dp set" case is the only place where the program status register is scheduled to be assigned a value, and so it is fine. Also, the "SWI" case is the only place where the halt flag is scheduled to be assigned a value; thus we do not need to be concerned with it. The danger arises with the program counter and `RD, since `RD could be `r[15]`, which is the program counter. To avoid this danger, we must leave the program counter alone in the "B/R15" case, because the program counter is modified in parallel by the "dp no set," "dp set" or "ir is B" cases of this Mealy ASM.

10.8.5 Verilog for the pipelined implementation

Here is a partial listing of the Verilog translation for the ASM from section 10.8.1:

```
...
forever
 begin
  @(posedge sysclk) enter_new_state(`F1);
   ...
   else
    begin
     if (condx(ir2[31:28],psr) &&
      ((ir2[27:25]==3'b101)
      || (ir2[27:26]==2'b00&&ir2[15:12]==4'b1111)))
      begin // "B/R15"
       ir1 <= @(posedge sysclk) 32'hf0000000;
       ir2 <= @(posedge sysclk) 32'hf0000000;
      end
```

Continued

```
        else
         begin // "normal"
          'PC <= @(posedge sysclk) 'PC + 4;
          ir1 <= @(posedge sysclk) m['PC>>2];
          ir2 <= @(posedge sysclk) ir1;
         end
        if (condx(ir2[31:28],psr))
         begin
          if       (ir2[27:26] == 2'b00)
           begin // "dp set" or "dp no set"
            'RD <= @(negedge sysclk)
                         dp(ir2[24:21]'OPA'OPB);
            if (ir2[20]) //"dp set"
             psr <= @(posedge sysclk)
                          f(dp(ir2[24:21],'OPA,'OPB));
           end
          else if (ir2[27:25] == 3'b101) //"ir2 is B"
           'PC <= @(posedge sysclk) 'PC + 'OFFSET4;
          else if (ir2[27:24] == 4'b1111)//"SWI"
           halt <= @(posedge sysclk) 1;
          else
           $display("other instructions...");
         end
       end
    end
```

Some of the macros need to be redefined to take into account that `ir2` is the final stage of this pipeline:

```
'define RD       r[ir2[15:12]]
'define OPA      r[ir2[19:16]]
'define OPB      (ir2[25] ? ir2[7:0] : r[ir2[3:0]])
'define OFFSET4 {ir2[23],ir2[23],ir2[23],ir2[23],ir2[23],ir2[23],ir2[23:0],2'b00}
```

Interestingly, because the original ARM was designed with a three-stage pipeline in mind, the definition of 'OPA and 'OPB are simpler than for the multi-cycle implementation. This simplification occurs since `r[15]` does not have to be explicitly mentioned. The value of `r[15]` at the time the instruction is in the final stage of the pipeline is, by definition, the correct value to use.

Execution of a data processing instruction involves non-blocking assignment to `RD, which is a macro that substitutes the subscripted Verilog array, `r[ir2[15:12]]`. This non-blocking assignment therefore uses `negedge` rather than `posedge` to be portable for the reasons explained in section 6.5.2. (Remember that, in this pipelined implementation, `ir2` changes every clock cycle.)

10.9 Superscalar implementation

The pipelined implementation given in the last section has a speed that approaches (but never quite reaches) one clock cycle per instruction. Because ARM data processing instructions have three register operands (`RD, `OPA and `OPB), one basic computation, such as incrementing `r[2]`, can be performed per clock cycle. Although this can be up to three times faster than the pipelined single-accumulator design described in chapter 9, it still is certain to be no better than the slowest special-purpose designs in chapter 2 (such as section 2.2.2). Even for a simple algorithm like childish division, it is often possible for more than one computation to occur in parallel (e.g., incrementing `r[2]` in parallel with subtracting from `r[1]`). A pipelined general-purpose processor only works because of quite a bit of parallel activity in the implementation of fetch/execute. Even so, a pipelined general-purpose computer cannot exploit the parallelism in an algorithm. Such parallelism can be exploited by special-purpose hardware (such as section 2.2.7).

Since the designer of a general-purpose computer can never be certain how fast is "fast enough," it would be desirable if the general-purpose computer could execute more than one instruction in parallel. Such an approach, known as a *superscalar* implementation, is an extension to the pipelined approach. Superscalar implementation is considerably more complex than the pipelined approach because the hardware itself must take seemingly sequential instructions and recognize when it is permissible for them to execute in parallel. In essence, some of the intelligence and skill of the hardware designer (as illustrated by the design alternatives of chapter 2) must be placed inside the hardware itself. Because the hardware of a superscalar general-purpose computer will never have as much information about the software algorithm as the designer of a special-purpose computer has about the ASM, a superscalar general-purpose machine will not be as fast as the best special-purpose hardware. Also, the complexities of superscalar design means its hardware cost may be many times the cost of the equivalent but faster special-purpose machine. However, the economies of scale for general-purpose computers have made superscalar processors viable.

10.9.1 Multiple-port register file

From a structural standpoint, a superscalar general-purpose processor can be distinguished from the slower design alternatives given earlier in this chapter (multi-cycle and pipelined) because the superscalar machine needs multiple ALUs for executing multiple instructions per clock cycle. The simplest case is to imagine that we can afford to have two ALUs, and therefore, under the best circumstances, two instructions can execute per clock cycle.

A consequence of having multiple ALUs is that the register file must be more sophisticated. If there are two ALUs, each of which might have to be fed two independent operands in each clock cycle, we need a register file with four read ports. From a behavioral standpoint, we will refer to the operands of the final stage of the pipeline the way we did in the last section (‘OPA and ‘OPB). However, sometimes another instruction will be executing in parallel. The operands of this parallel instruction will be referred to as ‘POPA and ‘POPB.

The two results of the two ALUs need two write ports into the register file. From a behavioral standpoint, we will refer to these as ‘RD and ‘PRD.

A register file that has four read ports and two write ports is considerably more expensive than the register file used in the pipelined implementation. There will be additional complexities with `r[15]` because it serves the role of the program counter. We will see later that the program counter in a superscalar design does some non-intuitive things.

10.9.2 Interleaved memory

In order to keep a superscalar processor going at full speed, it is necessary to provide it with as many new instructions as it is capable of executing per clock cycle. For example, if our machine is to execute two instructions per clock cycle, it will be necessary to load both `ir1` and `ir2` with instructions from memory addresses ‘PC+4 and ‘PC, respectively.

The single-port memory shown in figure 10-3, which can be used for the multi-cycle and pipelined implementations, does not allow more than one instruction to be fetched per clock cycle:

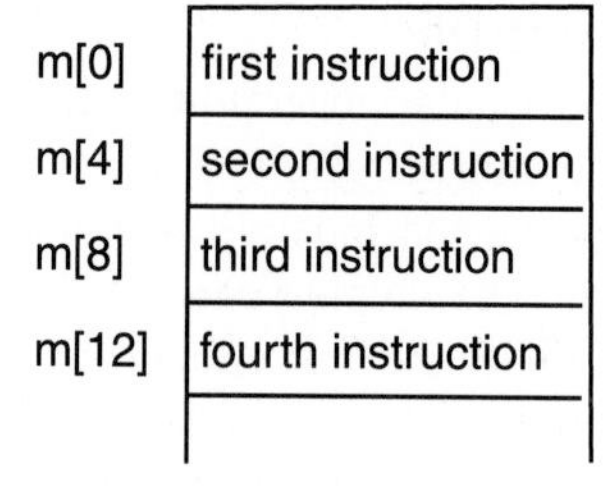

Figure 10-3. Non-interleaved memory.

Although a dual-ported memory for instructions would allow fetching of two instructions per clock cycle, such a memory is expensive. A cheaper alternative is to use an *interleaved* memory. A simple interleaved memory stores half of the instructions in one bank and the adjacent instructions in another:

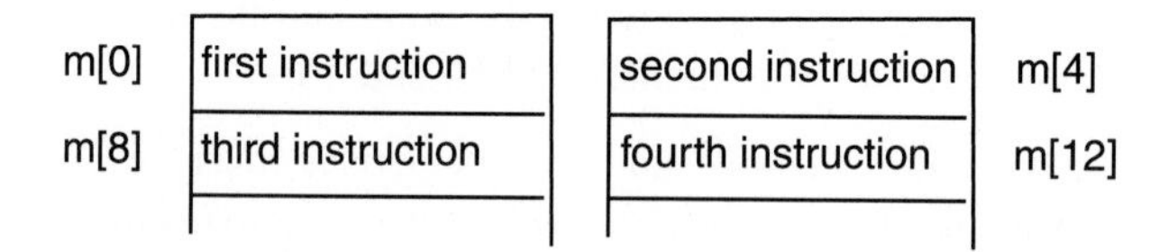

Figure 10-4. Interleaved memory.

In other words, two separate memories act as one. This approach is sufficient only because when the superscalar fetch/execute algorithm wants two instructions, they will always reside in separate banks. One of the instructions comes from an address divisible by eight; the other instruction will be plus or minus four of that address. From a behavioral standpoint, we will simply use the same kind of Verilog array notation for the two instructions that are fetched in parallel: `m[('PC+4)>>2]` and `m['PC>>2]`.

10.9.3 Examples of dependencies

If all instructions in a program were independent of each other, such as:

```
SUB R1,R1,R4
ADD R2,R2,1
```

designing a superscalar machine would be fairly easy. For example, the above two instructions could be fetched from the interleaved memory in parallel, presented to the two separate ALUs (for subtraction and addition, respectively) in parallel and their results could be written back to the register file in parallel.

Unfortunately, in real programs, instructions are often dependent on each other. For example:

```
SUB R2,R1,R4
ADD R2,R2,1
```

It might appear that data forwarding (of R1 minus R4) could be helpful here. Such an approach would be algorithmically correct but would be slow. To make these instructions execute in parallel, the clock period would have to be slow enough allow enough time for both the ADD and the SUB:

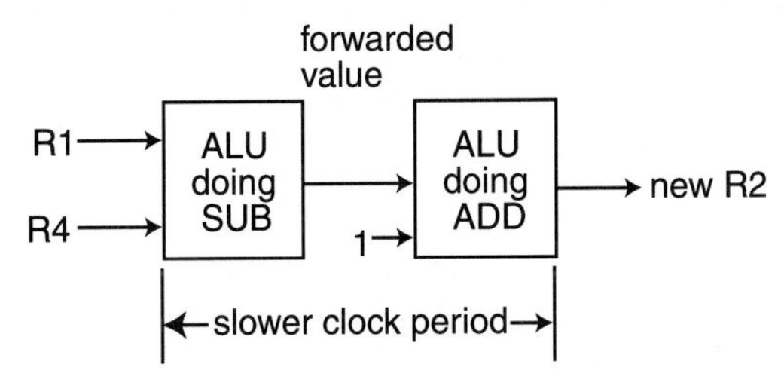

Figure 10-5. Forwarding results of dependent instructions slows clock.

Instead of data forwarding in a situation like this, it is better for the machine to execute only one instruction per clock cycle. At least this way, the clock cycle remains fast. In other words, it behaves like the simple pipeline approach of section 10.8. The hope is that after executing these two instructions sequentially, the machine will fetch some independent instructions (like the ones shown earlier) that it can execute in parallel.

Some programs have combinations of instructions that simply cannot be executed in parallel:

```
SWI
ADD R2,R2,1
```

The machine is supposed to halt (in our subset, at least) before the ADD instruction executes. In such a situation, we have to revert back to a one instruction per clock cycle (simple pipeline) approach, which allows the machine to process the SWI in exactly the order the programmer intends. On a machine that actually implements interrupts (unlike our subset), exact processing of interrupts and similar issues are significant.

A very common problem that occurs with superscalar design is that we cannot be certain, at the time when we have the hardware resources for doing so, whether we are supposed to execute an instruction (such as ADDPL):

```
SUBS R1,R1,R4
ADDPL R2,R2,1
```

The SUBS instruction will modify the program status register, but the ADDPL instruction needs to know that new program status information to decide whether to execute. If these instructions only executed one per clock cycle, there would be no difficulty. Also, if it were not for the "S" and "PL" suffixes on the mnemonics, there would be no problem with executing them in parallel during the same clock cycle. Although we could revert back to a simple pipeline approach (one instruction per clock cycle), the point of superscalar design is to maximize speed.

It might be tempting to try "program status forwarding" in a case like this. Such an approach would be algorithmically correct, but would have the undesirable side effect of doubling the propagation delay (the ADD cannot start until the SUB completes). This would mean the clock cycle of the machine would be twice as long, which would more than negate any advantage of our attempt at superscalar design.

10.9.4 Speculative execution

In contrast to such a flawed approach, the typical superscalar implementation uses a technique, known as *speculative execution*, to solve the problem of not knowing whether an instruction that could execute in parallel is supposed to execute. For most instructions on a RISC machine, the only irreversible consequence of executing that instruction is the storage of its result back in the register file. Speculative execution means we compute the result of an instruction at a time when it is uncertain whether or not that instruction will execute, but at that time we do not store the result back in the register file.

10.9.5 Register renaming

Instead, we put the result of an instruction that is being executed speculatively in a *rename register*. Such a register has the ability, at a later time, to take on the role of any of the user registers in the machine. In a later clock cycle, if the machine discovers that the instruction that executed speculatively in the previous clock cycle was not supposed to execute, the contents of the rename register can simply be discarded, and it will be as though the instruction never executed. If the machine discovers that the instruction was supposed to execute, the rename register takes on the role of the destination register. For our simple implementation, this will be very much like data forwarding of a single value. In a more complicated superscalar design, the renaming process could be much more sophisticated because every register in the file might have been renamed. Our simple implementation guarantees that at most one register will be renamed in any given cycle.

To implement this, there will have to be three components of the renameable register: `reg_val`, which indicates its 32-bit value, `reg_tag`, which indicates its identity if speculative execution succeeds (since it could be any of the registers) and `reg_cond`, which indicates the condition upon which the instruction is supposed to execute. In Verilog, these are declared as:

```
reg [31:0] reg_val;
reg [4:0] ren_tag;
reg [3:0] ren_cond;
```

It is interesting to note that `ren_tag` is five, rather than four, bits wide. This is required because, in addition to the sixteen user registers, we need to indicate when the renamed register is not valid. To do so, the following constant is defined:

```
`define INVALID 16
```

When an instruction cannot be executed speculatively (as in the SWI example from section 10.9.3), the machine assigns 'INVALID to `ren_tag`. In the next clock cycle, this will cause `ren_val` to be ignored.

On the other hand, when an instruction can be executed speculatively, the machine assigns the destination register number to `ren_tag`, the condition upon which that assignment succeeds to `ren_cond` and the potential new value of that register to `ren_val`.

10.9.5.1 First special-purpose renaming example

Even though register renaming and speculative execution may appear difficult to understand at first, they are simple extensions to the idea of the non-blocking assignment which has been discussed in earlier chapters. It is still true that only one assignment can be scheduled for a particular register during a particular clock cycle and that such assignments do not take effect until the next clock cycle. In order to see how register renaming and speculative execution relate to concepts in earlier chapters, let us set aside our goal of implementing the general-purpose ARM for a moment and consider some simple **special-purpose** machines that illustrate these same concepts. In other words, we are going to describe a special-purpose machine that only executes one (nonsensical) algorithm, which we will state in terms of ARM mnemonics:

```
            SUBS R2,R1,R4    ;sets psr
            ADDPL R3,R3,R3 ;speculative
            ADD R3,R3,1      ;R3 same as last inst
            NOP              ;NOP to simplify discussion
```

Designing such a special-purpose machine is easy if all we wanted to do is to carry out the same register transfers as would occur when the above instructions execute on the pipelined implementation of the general-purpose ARM given in section 10.8.1:

```
@(posedge sysclk) #1;
  r[2] <= @(posedge sysclk) r[1] - r[4];
  psr <= f(r[1] - r[4]);
@(posedge sysclk) #1;
  if (psr[31]==0)
    r[3] <= @(posedge sysclk) r[3] + r[3];
@(posedge sysclk) #1;
  r[3] <= @(posedge sysclk) r[3] + 1;
@(posedge sysclk) #1;
```

In the first clock cycle, the sign bit of the difference is scheduled to be stored in the `psr`. In the next clock cycle, after this assignment has taken effect, the `psr` determines whether or not the doubling of `r[3]` will occur (which makes this a Mealy machine). In any event, incrementation of `r[3]` does not occur until the third clock cycle, at which `$time` the appropriate value (either doubled or not) will be available. The fourth clock cycle does nothing because of the NOP in the algorithm.

It is possible to cut the number of states in half by combining two actions per state; however, there is a difficulty. SUBS sets the `psr`, but the ADDPL that we desire to execute in parallel depends on that `psr`. Here is where speculative execution and register renaming come into play. The register transfers of the special-purpose machine described by the Verilog below are similar to those carried out when the equivalent instructions execute on the general-purpose superscalar ARM given in section 10.9.6; however, the following is much simpler because it only considers actions that occur related to the specific instructions:

```
@(posedge sysclk) #1;
 r[2] <= @(posedge sysclk) r[1] - r[4];
 psr <= f(r[1] - r[4]);
 ren_val <= @(posedge sysclk) r[3] + r[3];
 ren_tag <= @(posedge sysclk) 3;
 ren_cond <= @(posedge sysclk) 'PL;
@(posedge sysclk) #1;
 ren_cond <= @(posedge sysclk) 'NV;
 if ((ren_cond == 'PL)&&(psr[31]==0) ||
     (ren_cond == 'MI)&&(psr[31]==1) ||
     (ren_cond == 'AL))
  r[3] <= @(posedge sysclk) ren_val+1; //renamed
 else
  r[3] <= @(posedge sysclk) r[3]+1;     //not renamed
```

In parallel to the subtraction during the first clock cycle, the doubling of `r[3]` occurs before the machine can know whether the difference will be positive. Therefore, the machine saves the doubled value in `ren_val`, and at the same `$time` makes note in `ren_cond` of the condition ('PL) under which this speculative doubled result is to be renamed as `r[ren_tag]`. In the second clock cycle, after the `psr` resulting from the subtraction is valid, the machine makes a decision whether or not renaming occurs. If it does not, incrementation of `r[3]` occurs based on the value already in the register file from two or more clock cycles ago. If renaming does occur, there is a literal substitution of `ren_val` for `r[3]` in this clock cycle. Regardless of whether renaming occurs in the second clock cycle, `ren_cond` is set to 'NV because the NOP will not cause any renaming in the third cycle (not shown).

10.9.5.2 Second special-purpose renaming example

Let's consider a second example, similar to the last one, except the destination of the third instruction (shown in bold) is not the same as the destination of the ADDPL instruction that executes speculatively:

```
      SUBS R2,R1,R4   ;//sets psr
      ADDPL R3,R3,R3  ;//speculative
      ADD R6,R3,1     ;//R3 same but not dest
      NOP             ;//NOP to simplify discussion
```

Again, there is no problem when all we want to do is to carry out the same register transfers as would occur when the above instructions execute on the pipelined implementation of the general-purpose ARM given in section 10.8.1:

```
@(posedge sysclk) #1;
  r[2] <= @(posedge sysclk) r[1] - r[4];
  psr <= f(r[1] - r[4]);
@(posedge sysclk) #1;
  if (psr[31]==0)
    r[3] <= @(posedge sysclk) r[3] + r[3];
@(posedge sysclk) #1;
  r[6] <= @(posedge sysclk) r[3] + 1;
@(posedge sysclk) #1;
```

Of course, things get more interesting when we use speculative execution and register renaming. The register transfers of the special-purpose machine below are similar to those carried out when the equivalent instructions execute on the general-purpose superscalar ARM given in section 10.9.6:

```
@(posedge sysclk) #1;
 r[2] <= @(posedge sysclk) r[1] - r[4];
 psr <= f(r[1] - r[4]);
 ren_val <= @(posedge sysclk) r[3] + r[3];
 ren_tag <= @(posedge sysclk) 3;
 ren_cond <= @(posedge sysclk) `PL;
@(posedge sysclk) #1;
 ren_cond <= @(posedge sysclk) `NV;
 if ((ren_cond == `PL)&&(psr[31]==0) ||
     (ren_cond == `MI)&&(psr[31]==1) ||
     (ren_cond == `AL))
  begin
   r[ren_tag] <= @(posedge sysclk) ren_val; //renamed
   r[6] <= @(posedge sysclk) ren_val+1;
  end
 else
   r[6] <= @(posedge sysclk) r[3]+1; //not renamed
```

The first clock cycle is identical to the speculative example in section 10.9.5.1; thus the speculative doubling of `r[3]` occurs before the machine knows whether the difference of `r[1]` and `r[4]` will be positive. Again, `ren_val` will contain the doubled value and `ren_cond` will indicate the condition (`PL) when `ren_val` is to be renamed as `r[ren_tag]`. In the second clock cycle, after the `psr` resulting from the subtraction is valid, the machine makes a decision whether or not renaming occurs. If it does not, the assignment to `r[6]` occurs based on the value of `r[3]` already in the register file from two or more clock cycles ago. If renaming does occur, the situation is quite different than in the example of section 10.9.5.1. In this example, the destination of the third instruction (`r[6]`) is different than the destination of the speculative instruction (`r[3]`). There is still a literal substitution of `ren_val` for `r[3]`, but there must also be storage

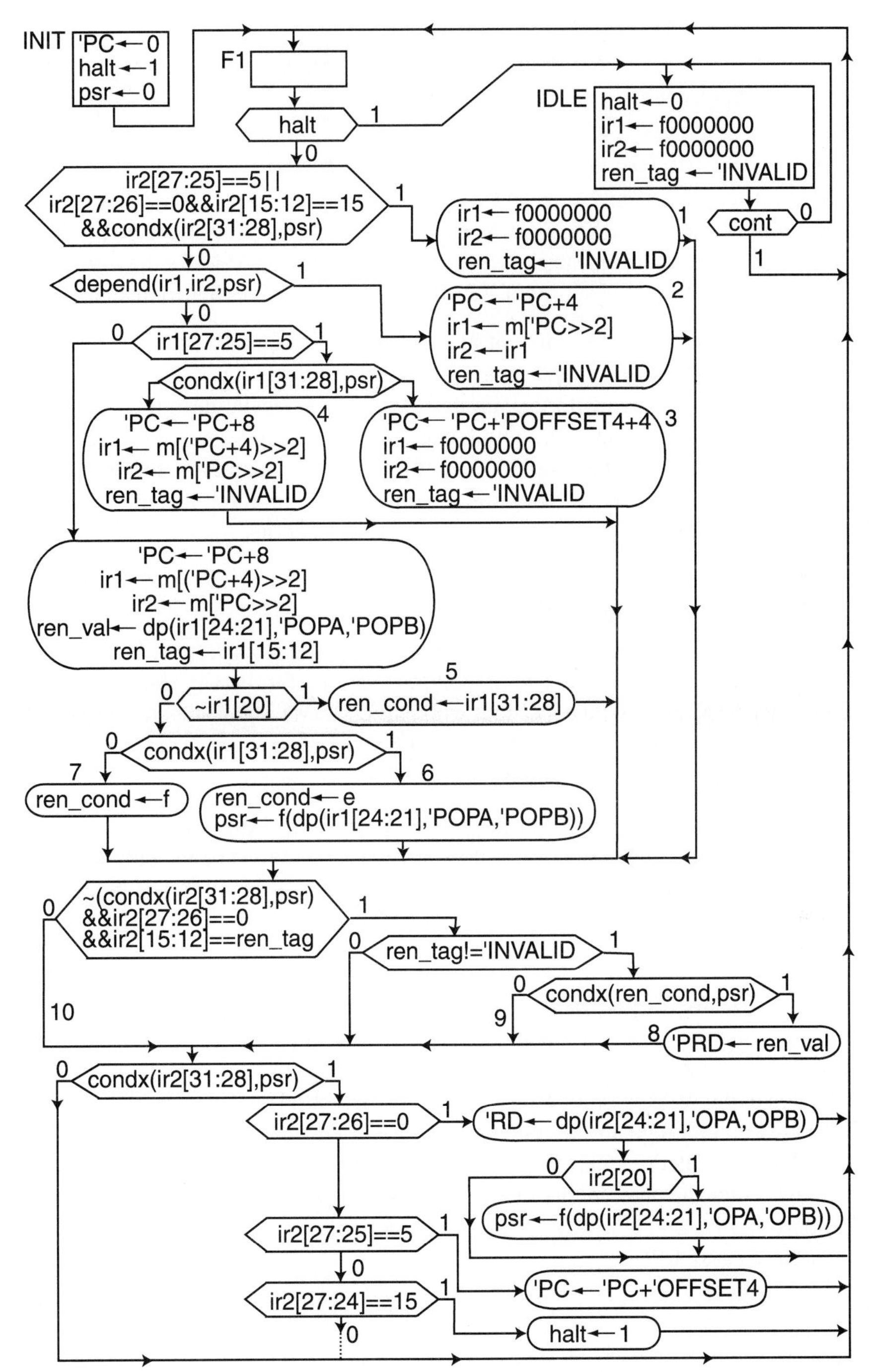

Figure 10-6. Superscalar ASM for ARM subset.

of `ren_val` in `r[ren_tag]` so that `r[3]` will contain the doubled value for future clock cycles (not shown).

10.9.6 ASM for the superscalar implementation

The preceding examples only considered speculative execution in the very limited context of a single algorithm. We want to use this in the general context of the fetch/ execute algorithm for a superscalar processor. Figure 10-6 is the ASM for the superscalar implementation of the same subset of ARM instructions used earlier in this chapter.

Although the reader may not realize it at first, much of the motivation for the ARM subset used throughout this chapter was to keep this intricate ASM as simple as possible. Also, this ASM deals with the simplest case of superscalar design (two instructions per clock cycle). Common commercial superscalar processors are much more complicated than this trivial example.[6] Even so, this ASM is something to behold! The observation is that superscalar design is significantly more complicated than multicycle or pipelined design, even for a tiny subset of instructions.

10.9.7 Three parallel activities

The ASM in section 10.9.6 shows three parallel activities that occur in each clock cycle when the machine is in state F1. The first parallel activity (arbitrarily shown at the top of this Mealy ASM only to be consistent with earlier ASMs in this chapter) deals with the instruction registers in the pipeline and the program counter. This first parallel activity also deals with speculative execution.

The second parallel activity (arbitrarily show in the middle) deals with register renaming. The final parallel activity (arbitrarily shown at the bottom) deals with executing instructions sequentially in the final stage of the pipeline. This portion of the ASM is identical to the pipelined ASM given in section 10.8.1.

10.9.7.1 Pipeline, parallel and speculative execution

The first parallel activity deals with several different interacting components of the superscalar machine. This portion of this ASM is doing analogous work to what the first portion of the ASM of section 10.8.1 does. The essential goal is to decide what will be in the instruction register pipeline (`ir1` and `ir2`) for the next clock cycle. In the pipelined implementation, this was easy: either put NOPs into the pipeline or move the instructions down the pipe. In this superscalar implementation, there are seven distinct cases, only one of which will occur in any clock cycle. For ease of discussion, let's number these cases 1-7:

---

[6] As of 1997, despite its suitability for superscalar implementation, ARM had not yet introduced such a version of its processor, instead focusing on low-cost, low-power versions that use only pipelining.

1. `ir2` has a branch or data processing instruction that changes R15.
2. There is a dependency between `ir1` and `ir2`.
3. `ir1` has a branch that changes R15.
4. `ir1` has a branch that is nullified.
5. `ir1` has a data processing instruction that does not affect `psr`.
6. `ir1` has a data processing instruction that does affect `psr`.
7. `ir1` has a data processing instruction that is nullified.

Each one of these cases corresponds to an oval in the Mealy ASM. In all of these cases, it is guaranteed that an instruction in `ir2` will execute (as described later at the bottom the ASM). The determination of which of the above seven cases applies here is based upon whether `ir1` can be executed in parallel to `ir2`. We must know that to decide how much to increment the program counter and how to load the instruction registers for the next clock cycle. The more instructions we can execute in parallel, the more new instructions have to be fetched.

Cases 1 and 2 describe situations when it is not possible for `ir1` to execute in parallel with `ir2`. Cases 3-7 describe situations when it is possible to do something with `ir1` in parallel to `ir2`

Case 1 is similar to the analogous case of the ASM of section 10.8.1 ("B/R15"—put NOPs in the instruction register pipeline), except case 1 must also make the renamed register invalid. There is no result being computed in parallel by this case.

Case 2 is similar to the only other analogous case of the ASM of section 10.8.1 ("normal"—instructions travel down the pipeline), except case 2 must also make the renamed register invalid. There is no result being computed in parallel by this case. The decision that causes case 2 to occur is extremely intricate; thus we will defer those details until we get to the Verilog function. (The motivation here is similar but hopefully even more persuasive than earlier in this chapter: both ASMs and Verilog have their place in the toolset of the designer.) The name of this Verilog function is `depend`, and it decides whether there is a dependency (i.e., case 2) based only on the information in `ir1`, `ir2` and the program status register.

Of the remaining cases, the only one in which truly speculative execution occurs for the instruction in `ir1` is case 5. In cases 3, 4, 6 and 7, it is known whether the instruction in `ir1` will execute. The reason this is known is because of some details in the definition of the `depend` function for case 2. In other words, if you are not in case 1, 2 or 5, the machine has enough information to say with certainty whether the instruction in `ir1` will execute. (It is the responsibility of the designer to make sure this property holds, but let's ignore the details of that for a moment.)

Case 5 is interesting because it is the reason for using a renamed register. The value scheduled to be assigned to the renamed register is the result from the parallel ALU. The function computed by this ALU is based on `ir1[24:21]`. The other ALU uses `ir2[24:21]`. The condition that says whether the value in the renamed register will actually be used in the next clock cycle comes from `ir1[31:28]` in this clock cycle. The tag for the renamed register is scheduled to become the register specified as the destination in this instruction (`ir1[15:12]`). At the next rising edge of the clock after case 5 occurs, `ren_tag` will be the register number that will be modified if this speculatively executed instruction actually executes; `ren_cond` will indicate whether the register indicated by `ren_tag` should change based on the program status register in this next clock cycle and `ren_val` will be that new value.

There is a hidden detail in the `depend` function that relates to case 5. The `depend` function prevents parallel execution if both the instructions in `ir1` and `ir2` set the program status register. Because of this, distinguishing between case 5 versus cases 6 and 7 is simply a matter of looking at `ir1[20]`. If `ir1[20]` and `ir2[20]` indicate both instructions will modify the program status register, case 2 applies, and the instructions will execute sequentially. The reason for this is that both instructions cannot modify the program status register in the same clock cycle, but it is acceptable for each of them to modify the program status register in sequence.

If by the point of the decision `ir1[20]` indicates that this instruction will modify the program status register, we know that `ir2` will not. This means the program status register in the current clock cycle (rather than in the next clock cycle as was the situation for case 5) accurately reflects the information needed to decide whether `ir1` will execute. Therefore, the decision to choose between cases 6 and 7 is `condx(ir1[31:28],psr)`. Note once again the advantage of being able to reuse this Verilog function. If it is known in this clock cycle (case 6) that the instruction will execute, `ren_cond` will become `4'b1110` (always) in the next clock cycle, rather than whatever condition was present in `ir1[31:28]`. If it is known in this clock cycle (case 7) that the instruction will not execute, `ren_cond` will become `4'b1111` (never) in the next clock cycle. This way we can use the same hardware that implements speculative execution also to handle cases 6 and 7.

The reason we cannot use speculative execution here (i.e., making `ren_cond` be `ir1[31:28]`) is that case 6 changes the program status register. If a conditional instruction that changes the program status register (such as ADDPLS) executes due to the current program status information, it is possible register renaming will fail to happen in the next clock cycle because the condition is no longer true. That would prevent an instruction that is supposed to execute from actually executing. Therefore, cases 6 and 7 evaluate `condx` with the current program status register and communicate this unambiguously into the next clock cycle with the `4'b1110` or `4'b1111`.

Cases 5, 6 and 7 have quite a few things in common. In each case, two data processing instructions execute in parallel. This means the program counter needs to increment by eight rather than four. Also two instructions need to be fetched in parallel from the interleaved memory. In each case, `ren_val` is computed, whether or not it will actually be used later. In theory, for case 7, `ren_val` need not be computed, but it is easier (and harmless) to do so.

Cases 3 and 4 deal with a branch instruction in `ir1`. If we reach case 3 or 4, we know (because of the `depend` function) that the instruction in `ir2` will not affect the program status register. (If it does, case 2 applies instead.) Therefore, the decision whether to take the branch can be based on `condx(ir1[31:28],psr)`. The reason is analogous to the decision for cases 6 and 7. If the branch instruction is nullified (case 4), the program counter is incremented by eight and two instruction are fetched (as in cases 5, 6 and 7). If the branch instruction in `ir1` occurs (case 3), the instruction pipeline fills with NOPs and the program counter changes (by adding 'POFFSET4 + 4, similar to the 'OFFSET+4 in multi-cycle implementation). In either case 3 or case 4, the branch instruction does not modify a user register; thus the tag for the renamed register becomes 'INVALID.

10.9.7.2 Dealing with register renaming

The second parallel activity of this ASM is to decide whether the value in the renamed register set up in the previous clock cycle (perhaps as the result of speculative execution of an instruction by case 5 of that clock cycle) should take effect permanently in the register file.

There are three conditions that cause the value in the renamed register to be discarded without being written back into the register file:

a. `ir2` has a data processing instruction that stores a newer value into that user register than what is in `ren_val` (refered to as case 10).
b. `ren_tag` indicates 'INVALID.
c. evaluation of `ren_cond` in this clock cycle fails (refered to as case 9 of this clock cycle which relates to cases 5 and 7 of the previous clock cycle).

If none of these apply, `ren_val` is written back into the register indicated by `ren_tag` (case 8). The parallel write port through which this occurs can be defined in behavioral Verilog as:

```
`define PRD        r[ren_tag]
```

It will also be necessary to define the macros for the read ports so that they forward `ren_val` during this clock cycle:

```
`define OPA  ((condx(ren_cond,psr)&&ir2[19:16]==ren_tag)?ren_val:r[ir2[19:16]])
`define OPB  (ir2[25]?ir2[7:0]:((condx(ren_cond,psr)&&ir2[3:0]==ren_tag)?ren_val:r[ir2[3:0]]))
`define POPA ((condx(ren_cond,psr)&&ir1[19:16]==ren_tag)?ren_val:r[ir1[19:16]])
`define POPB (ir1[25]?ir1[7:0]:((condx(ren_cond,psr)&&ir1[3:0]==ren_tag)?ren_val:r[ir1[3:0]]))
```

Here the only conditions that must be satisfied are that the operand comes from the register described by `ren_tag` (which is guaranteed not to be 'INVALID if `ren_tag` matches the register number) and `ren_cond` evaluates to true in this clock cycle. Even though the renamed register might be discarded after this clock cycle, if the above conditions are satisfied, the renamed register should be forwarded in this clock cycle.

10.9.8 Verilog for the superscalar ARM

This example illustrates the many advantages of using ASMs together with Verilog. It is simply not possible to put all the details of the superscalar design into a one-page ASM. Some of these details need to be placed into Verilog functions or macros. Most of the functions required, such as `condx`, were defined earlier for the multi-cycle implementation, and the fact that we can reuse them is very helpful.

10.9.8.1 The depend function

There is one function that is unique to this superscalar implementation: `depend`. This function recognizes those situations where it is not possible to execute two instructions in parallel:

a. `ir1` operand (`POPA) same as `ir2` destination ('RD).
b. `ir1` operand (`POPB) same as `ir2` destination ('RD).
c. `ir1` is conditional non-dp and `ir2` sets `psr`.
d. `ir1` and `ir2` both set `psr`.
e. `ir1` is not branch or dp.
f. `ir2` is not branch or dp.
g. `ir1` has R15 as operand.

The first three of these are a form of hazard known as Read After Write (RAW). If these instructions executed sequentially, the older instruction (`ir2`) would write a value into a register ('RD or `psr`) which the next instruction (`ir1`) must read. To attempt to execute these instructions in parallel would mean `ir2` would read the wrong value.

The fourth condition above is a form of hazard known as Write After Write (WAW). This is a situation that has been warned against throughout this entire book: you cannot have two non-blocking assignments to the same register in the same clock cycle. As explained in section 10.9.7.1, the ASM is designed with the understanding that this situation will never occur in case 5; thus the `depend` function must cause the ASM to handle such situations in case 2 (i.e., `ir2` and `ir1` will execute sequentially).

The final three situations deal with instructions for which the ASM was not designed to execute in parallel. Here is the Verilog function that detects these seven conditions that cause the ASM to proceed to case 2:

```
function depend;
 input [31:0] ir1,ir2,psr;
 begin
 depend=(ir2[15:12] == ir1[19:16]
      && ir2[27:26] == 2'b00 && ir1[27:26] == 2'b00
      && condx(ir2[31:28],psr))//POPA bad (RAW)
   || (ir2[15:12] == ir1[3:0] && ir1[25]==0
      && ir2[27:26] == 2'b00 && ir1[27:26] == 2'b00
      && condx(ir2[31:28],psr))//POPB bad (RAW)
   || (ir2[20]
      && ir2[27:26] == 2'b00
      && condx(ir2[31:28],psr)
      && ir1[31:28] != 4'b1110
      && ir1[27:26] != 2'b00)  //psr bad(RAW)non-dp
   || (ir1[20] && ir1[27:26] == 2'b00
      && ir2[20] && ir2[27:26] == 2'b00
      && condx(ir2[31:28],psr))//psr bad(WAW) dp
   || ((ir1[27:26] != 2'b00)
      &&(ir1[27:25] != 3'b101))//ir1 not dp or branch
   || ((ir2[27:26] != 2'b00)
      &&(ir2[27:25] != 3'b101))//ir2 not dp or branch
   || (ir1[27:26] == 2'b00      //ir1 has PC as ALUop
      && ((ir1[3:0] == 4'b1111 && ir1[25]==1'b0)
            ||ir1[15:12] == 4'b1111
            ||ir1[19:16] == 4'b1111));
 end
endfunction
```

Since the goal is to execute as many instructions in parallel as can be executed correctly, it is useful to ignore instructions that are known will be nullified. Since we know with certainty whether `ir2` will be nullified (based on the current program status register), conditions a-d (which mention `ir2`) can be ANDed with condx(ir2[31:28],psr). This means the depend function only slows the machine to one instruction per clock cycle when it is actually necessary. For example, the following two instructions

```
ADDPL  R1,R1,1
ADD    R2,R1,1
```

can be processed in parallel if the ADDPL is nullified but must execute sequentially if the ADDPL is not nullified.

10.9.8.2 Translating the ASM to Verilog

Once all the macros and functions are defined, it is easy to translate the ASM to Verilog. For example, the following is the beginning portion of the Verilog code corresponding to the first parallel activity of state F1 (parallel and speculative execution):

```
...
begin
 if (condx(ir2[31:28],psr) &&
    ((ir2[27:25] == 3'b101)
    || (ir2[27:26] == 2'b00 &&
        ir2[15:12] == 4'b1111)))
  begin
   ir1 <= @(posedge sysclk) 32'hf0000000;
   ir2 <= @(posedge sysclk) 32'hf0000000;
   ren_tag <= @(posedge sysclk) `INVALID;
   `ifdef DEBUG
     $display(
      " 1. ir2 branch or R15 prevents ||",$time);
     cover(1);
   `endif
  end
 else ...
```

10.9.8.3 Code coverage

The power of Verilog is twofold. First, Verilog allows the designer to express the behavior of the hardware abstractly, without having to consider too many details. Second, Verilog allows a designer to test the design using simulation. It would be an underutilization of Verilog to synthesize a design without having ever simulated it. So, the designer has two responsibilities. The first responsibility, of course, is to design the hardware. The second responsibility, which is more important but sometimes neglected by careless designers, is the test code. This code, sometimes referred to as the *testbench*, exercises the Verilog that simulates the hardware. Ideally, we would like to try every possible case. For tiny special-purpose machines, such as the 12-bit childish division test code example in section 3.7.3, this is **barely** possible. For a **general-purpose** machine, it is **impossible to test everything**. The usefulness of simulation, however, depends on how completely the Verilog code that simulates hardware has been tested by the Verilog test code. It does not do any good to test the same correct Verilog statement a million times but ignore another statement that has a bug in it. The advantage of Verilog is that its software-like statements can be used to warn the designer that parts of the Verilog code that simulates hardware has not been tested.

The superscalar implementation given in the last section is far more complex than any of the earlier designs in this book. It is not feasible to test every possible program to see if it works correctly. We will create several programs, but then Verilog will inform us whether all of the cases we are interested in have been tested. The reason for doing this is that the designer will more than likely make a mistake in guessing what cases a moderately complex program will test. The operation of the superscalar machine is so counterintuitive (even on a small program) that it is better for Verilog to keep track of what is being tested.

10.9.8.4 Using **`'ifdef`** *for the* **`cover`** *task*

The Verilog code, of which a portion was shown in section 10.9.8.2, has several statements that compile conditionally, such as:

```
'ifdef DEBUG
  $display(
   " 1. ir2 branch or R15 prevents ||",$time);
  cover(1);
'endif
```

What this means is that if the macro is defined:

```
'define DEBUG
```

the call to the system task, `$display`, and the `cover` task (described below) will be compiled with the rest of the Verilog. If this `'define` were omitted from the top of the file, it would be as though the `$display` and `cover` tasks were not there. The reason for using `'ifdef` is that once simulation is correct, the macro can be undefined, and the Verilog will no longer exhibit these test actions.

Note that `'ifdef` is different than an `if` statement (where the tasks would be compiled, but might not execute). In particular, `'ifdef` can be used to alter which control statements are compiled into the code. For example, cases 9 and 10 of the renaming parallel activity do nothing:

```
'ifdef DEBUG
else
 begin
  $display("10. dp overwrites renamed r%d",
    ren_tag,$time);
  cover(10);
 end
'endif
```

If 'DEBUG is not defined, there is no need for the `else begin ... end` to be compiled. The above shows how the scope of statements that are conditionally compiled can cross `begin end` boundaries. This is possible because the substitution occurs at compile time.

In addition to calling on these tasks, the `cover` task has to be defined. We only want to define it if the 'DEBUG macro is defined. Therefore, the task and everything associated with it will be enclosed in the `'ifdef`:

```
  'ifdef DEBUG
    reg ['MAX_CASE_NO:0] coverage_set;
    task cover;
      input case_no;
      integer case_no;
      begin
        coverage_set = coverage_set |
           ((1 << 'MAX_CASE_NO) >> case_no);
      end
    endtask

    initial
      begin
        coverage_set = 0;
        wait(halt === 1'b0);
        wait(halt === 1'b1);
        $display("coverage=%b",
          coverage_set['MAX_CASE_NO-1:0]);
      end
  'endif
```

Notice how this is completely different from an `if` statement. An `if` can only exist inside a behavioral block or task. This `` `ifdef `` is outside the task and the `initial` block. If 'DEBUG is not defined, the `reg`, the `cover` task and the `initial` block will not be defined.

Each case of interest in the code has a number between 1 and 'MAX_CASE_NO, which is used to identify that case in the call to `cover`. (There is no case 0.) What `cover` does is to OR the corresponding bit of the coverage set with one.

After the program has halted, the initial block prints out the coverage set. The more cases of the Verilog code that were covered by the program, the more ones there will be in the coverage set. We will run several programs in order to obtain complete coverage. (In reality, we have not considered enough test cases here to have total confidence in this design, but this task can be expanded to cover an arbitrary number of cases.)

10.9.9 Test programs

The special-purpose machines in chapters 4 and 5 are easier to test than a general-purpose machine because the test code simply has to supply test data to the machine being tested. A special-purpose machine is supposed to follow some algorithm that manipulates the data, and it is often easy to tell if the result is correct. A general-purpose machine implements a (sometimes intricate) variation of fetch/execute which in turn interprets a program that manipulates the data. It is much harder to tell if a general-purpose machine is correct.

10.9.9.1 A test of R15

One of the details of the ARM instruction set that takes special consideration in all the implementations is R15. R15 is, in fact, just another name for the program counter. The Verilog macros for all the above implementations have to take this into account. Additionally, on the pipelined implementation, R15 as a destination causes the pipeline to fill with NOPs just as though a branch had occurred. On the superscalar implementation, there are many places where special consideration is given to R15.

Therefore, we need an ARM program that exercises at least some of this Verilog code involving `r[15]`. Here is such a program:

```
`ifdef PROGRAM1 //test R15 source and destination
 arm7_machine.m[0]=32'he3b00000; //MOVS R0,0
 arm7_machine.m[1]=32'he08f1000; //ADD  R1,R15,R0
 arm7_machine.m[2]=32'he080200f; //ADD  R2,R0,R15
 arm7_machine.m[3]=32'he08f3000; //ADD  R3,R15,R0
 arm7_machine.m[4]=32'he080400f; //ADD  R4,R0,R15
```

Continued

```
  arm7_machine.m[5]=32'he3b0e0ff; //MOVS R14,0xff
  arm7_machine.m[6]=32'he2400008; //SUB  R0,R0,8
  arm7_machine.m[7]=32'he1a0f001; //MOV  R15,R1
`endif
```

A Verilog macro, 'PROGRAM1, is defined when this is the program we want to use to test the machine with. This test code can be used with any of the implementations. For example, the pipelined implementation produces the following:

```
...
PC=00000024 IR1=eaffffe IR2=e1a0f001 N=0          1251
 r0=fffffff8 r1=0000000c r2=00000010 r3=00000014 r4=00000018
PC=0000000c IR1=f0000000 IR2=f0000000 N=0          1351
 r0=fffffff8 r1=0000000c r2=00000010 r3=00000014 r4=00000018
PC=00000010 IR1=e08f3000 IR2=f0000000 N=0          1451
 r0=fffffff8 r1=0000000c r2=00000010 r3=00000014 r4=00000018
PC=00000014 IR1=e080400f IR2=e08f3000 N=0          1551
 r0=fffffff8 r1=0000000c r2=00000010 r3=00000014 r4=00000018
PC=00000018 IR1=e3b0e0ff IR2=e080400f N=0          1651
 r0=fffffff8 r1=0000000c r2=00000010 r3=0000000c r4=00000018
```

Notice the contents of the registers at `$time` 1251 and the NOPs in the pipeline at `$time` 1351. Also notice the contents of the registers at `$time` 1651. On the other hand, the superscalar implementation produces equivalent results in an entirely different way:

```
...
PC=00000024 IR1=eaffffe IR2=e1a0f001 N=0              1051
r0=fffffff8 r1=0000000c r2=00000010 r3=00000014 r4=00000018
1.ir2 branch or R15 prevents ||                       1051
other DP instructions...
5.use || ALU noS ir1=f0000000 A=fffffff8 B=fffffff8  1151
PC=0000000c IR1=f0000000 IR2=f0000000 N=0             1151
r0=fffffff8 r1=0000000c r2=00000010 r3=00000014 r4=00000018
PC=00000014 IR1=e080400f IR2=e08f3000 N=0
  cond=f,ren_R 0 =00000000                            1251
r0=fffffff8 r1=0000000c r2=00000010 r3=00000014 r4=00000018
2.depend prevents ||                                  1251
9.nullify renamed r 0 ||dp                            1251
6.use || ALU S ir1=e3b0e0ff A=fffffff8 B=000000ff    1351
PC=00000018 IR1=e3b0e0ff IR2=e080400f N=0             1351
r0=fffffff8 r1=0000000c r2=00000010 r3=0000000c r4=00000018
PC=00000020 IR1=e1a0f001 IR2=e2400008 N=0
```

Continued

```
  cond=e,ren_R14 =000000ff                                1451
r0=fffffff8 r1=0000000c r2=00000010 r3=0000000c r4=00000010
2.depend prevents ||                                      1451
8.writeback renamed r14 ||dp                              1451
```

Notice the values of the registers at `$time` 1051 and the NOPs in the pipeline at `$time` 1151. Also notice the value of the registers at `$time` 1451. These correspond to what was visible in the pipelined implementation at `$time` 1251, 1351 and 1651. This program does not execute much faster on the superscalar implementation than on the pipelined implementation because the superscalar implementation properly prevents execution of more than one instruction per clock cycle when R15 is involved. This program is important as a means of testing the Verilog code because it illustrates that case 2 happens due to the following portion of the `depend` function:

```
|| (ir1[27:26] == 2'b00
  && ((ir1[3:0] == 4'b1111 && ir1[25]==1'b0)
    ||ir1[15:12] == 4'b1111
    ||ir1[19:16] == 4'b1111));//ir1 has PC as ALUop
```

The coverage set for this program is 1100110101; in other words, cases 1, 2, 5, 6, 8 and 10 are covered. Notice how helpful the output from the `$displays` is in annotating why these cases occur.

10.9.9.2 Our old friend: division

The last program has no practical value other than that of testing the operation of certain ARM instructions. The remaining programs that we will consider will be based on the childish division algorithm used in earlier chapters. Like the PDP-8, the ARM does not have a divide instruction in the hardware; thus to do division requires some software. Although if speed were essential one would choose a faster algorithm for division than the childish algorithm, this simple algorithm illustrates many of the properties that more sophisticated algorithms also have. This is the reason why it has been implemented many times in this book, in both hardware and software. Here again is the childish division algorithm in C:

```
r1=x;
r2=0;
while (r1>=y)
  {
    r1 = r1 - y;
    r2 = r2 + 1;
  }
```

Since we have not implemented the memory reference instructions of the ARM, the values of x and y must be constants. We will use immediate addressing with x and y. For example, let us assume that x is 14 and y is 7, which is a typical test case used with this algorithm in chapters 2, 4, 5, 8 and 9. It is natural for the r1 and r2 high-level variables to reside in R1 and R2, respectively. R1 and R2 are the registers that the assembly language programmer refers to, but these are r[1] and r[2] in Verilog. Also, it will be convenient for y to reside in R4. The implementations of this algorithm given in chapters 8 and 9 made use of the accumulator of the PDP-8 to contain the difference. We need to have a similar register on the ARM. In the following program, let us use R0 to serve the same role as as the accumulator. This illustrates an important property of all RISC machines (not just the ARM): there is nothing special about R0—we could have chosen any other available register to hold the difference. The following is an ARM program that implements this algorithm in the most straightforward way possible:

```
00000000/e3a0100e      MOV   R1,0x0e
00000004/e3a04007      MOV   R4,0x07
00000008/e3a02000      MOV   R2,0x00
0000000c/e0510004  L1  SUBS  R0,R1,R4
00000010/4a000002      BMI   L2
00000014/e1a01000      MOV   R1,R0
00000018/e2822001      ADD   R2,R2,0x01
0000001c/eaffffffa     B     L1
00000020/ef000000  L2  SWI
```

The above is analogous to the PDP-8 program given in section 8.3.2.5.3. The first three MOV instructions set up R1, R4 and R2 to their initial values of x (14), y (7) and zero, respectively. The SUBS is the only instruction that sets the program status register. The purpose of the SUBS instruction is twofold: to compute the difference and to see if R1>=R4. The BMI makes use of this program status information. As long as R1 >= R4, the BMI is nullified and the loop continues. The difference would then be moved from R0 to R1, and R2 is incremented. The unconditional branch to the label L1 causes the test at the top of the loop to happen again. This loop repeats while the difference (in R0) is greater than zero. The BMI branches to the label L2 when the difference in R0 is negative, at which point the SWI causes the Verilog test code to finish (using the halt flag that would not exist on a an actual ARM).

By defining 'PROGRAM5 in the test code, appropriate Verilog assignment statements (not shown) would place the above machine language instructions into memory. This program can be used with any of the implementations of the ARM. For example, the pipelined implementation produces the following:

```
...
PC=00000014 IR1=4a000002 IR2=e0510004 N=0        2251
 r0=00000000 r1=00000000 r2=00000002 r3=xxxxxxxx r4=00000007
PC=00000018 IR1=e1a01000 IR2=4a000002 N=1        2351
 r0=fffffff9 r1=00000000 r2=00000002 r3=xxxxxxxx r4=00000007
PC=00000020 IR1=f0000000 IR2=f0000000 N=1        2451
 r0=fffffff9 r1=00000000 r2=00000002 r3=xxxxxxxx r4=00000007
PC=00000024 IR1=ef000000 IR2=f0000000 N=1        2551
 r0=fffffff9 r1=00000000 r2=00000002 r3=xxxxxxxx r4=00000007
PC=00000028 IR1=xxxxxxxx IR2=ef000000 N=1        2651
 r0=fffffff9 r1=00000000 r2=00000002 r3=xxxxxxxx r4=00000007
```

The loop executes two times. The third execution of the SUBS (`$time` 2251) produces a negative number (fffffff9), which sets the N flag. This in turn causes the BMI to branch. Not show earlier, the BMI had been nullified. On the other hand, running 'PROGRAM5 on the superscalar implementation produces equivalent results in an entirely different way:

```
...
2.depend prevents ||                                      1351
10.dp overwrites renamed r 0                              1351
PC=00000014 IR1=4a000002 IR2=e0510004 N=0
  cond=f,ren_R 0 =00000000                                1351
r0=00000000 r1=00000000 r2=00000002 r3=xxxxxxxx r4=00000007
PC=00000018 IR1=e1a01000 IR2=4a000002 N=1                 1451
r0=fffffff9 r1=00000000 r2=00000002 r3=xxxxxxxx r4=00000007
1.ir2 branch or R15 prevents ||                           1451
other DP instructions...
5.use || ALU noS ir1=f0000000 A=fffffff9 B=fffffff9       1551
PC=00000020 IR1=f0000000 IR2=f0000000 N=1                 1551
r0=fffffff9 r1=00000000 r2=00000002 r3=xxxxxxxx r4=00000007
PC=00000028 IR1=xxxxxxxx IR2=ef000000 N=1
  cond=f,ren_R 0 =00000000                                1651
r0=fffffff9 r1=00000000 r2=00000002 r3=xxxxxxxx r4=00000007
2.depend prevents ||                                      1651
9.nullify renamed r 0 ||dp                                1651
```

The contents of the registers in the superscalar implementation at `$time` 1651 are the same as the registers in the pipelined implementation at `$time` 2651. The output from the `$display` statements explain the intricate way in which the superscalar machine was able to produce the correct answer in less time. The `depend` function only slows the machine down (case 2) three times. For example, at `$time` 1351, `depend` detects the conditional branch (BMI in this example) following the data processing instruction that sets the program status register (SUBS in this case). At `$time` 1641, `depend` detects SWI. The Verilog coverage set for the superscalar run of 'PROGRAM5 is 1110110111. This means all but cases 4 and 7 were exercised.

10.9.9.3 Faster childish division

In section 9.7, a variation of the childish division algorithm was given that illustrates a different way of implementing a `while` loop in software (testing at both the top and bottom of the loop). The effect of this is to reduce the number of times the branch penalty is incurred and to reduce the number of nullified branch instructions. The following ARM program implements this approach by using both the BPL and BMI instructions:

```
00000000/e3a0100e      MOV    R1,0x0e
00000004/e3a02000      MOV    R2,0x00
00000008/e3a04007      MOV    R4,0x07
0000000c/e0510004      SUBS   R0,R1,R4
00000010/4a000003      BMI    L2
00000014/e1a01000  L1  MOV    R1,R0
00000018/e2822001      ADD    R2,R2,0x01
0000001c/e0510004      SUBS   R0,R1,R4
00000020/5afffffb      BPL    L1
00000024/ef000000  L2  SWI
```

Running on the pipelined implementation, this program (let's refer to it as 'PROGRAM4) produces at `$time` 2051 the same results that 'PROGRAM5 produces (also running on the pipelined implementation) at `$time` 2651. This illustrates that to make good use of a pipelined machine, a good compiler is essential. Manually created assembly language programs are often not as effective as the automatically created machine language from compilers. Running on this superscalar implementation, this program produces at `$time` 1451 the same results that 'PROGRAM5 (also running on the superscalar implementation) produces at `$time` 1651.

The Verilog coverage set for the superscalar run of 'PROGRAM4 is 1100110110. 'PROGRAM4 does not add to the coverage of the Verilog code provided by 'PROGRAM5 (1110110111); thus we need an additional test program to cover cases 4 and 7.

10.9.9.4 Childish division with conditional instructions

One of the big advantages of the ARM instruction set is that any instruction can be made to execute conditionally. In the previous program, only the branch instructions executed conditionally, but with the ARM, the programmer is free to specify that any instruction, such as a data processing instruction, should execute only when a certain condition is true. The importance of this is that conditional execution does not incur any branch penalty.

One technique that compilers use to improve the performance of high-level software running on a pipelined and/or superscalar processor is *loop unrolling*. In C programs using `for` loops, such as:

```
for(i=0;i<3;i++)
  {
    r1 = r1 - y;
     r2 = r2 + 1;
  }
```

the compiler knows a priori how many times the loop will execute; thus the unrolled code looks like:

```
r1 = r1 - y;
r2 = r2 + 1;
r1 = r1 - y;
r2 = r2 + 1;
r1 = r1 - y;
r2 = r2 + 1;
```

which can be implemented without branch penalty:

```
SUB    R1,R1,R4
ADD    R2,R2,0x01
SUB    R1,R1,R4
ADD    R2,R2,0x01
SUB    R1,R1,R4
ADD    R2,R2,0x01
```

If the program has a `for` loop that repeats too many times for it to be practical to unroll the loop completely, it can be partially unrolled. For example:

```
for(i=0;i<3000;i++)
  {
    r1 = r1 - y;
    r2 = r2 + 1;
  }
```

is the same as:

```
for(i=0;i<1000;i++)
  {
    r1 = r1 - y;
    r2 = r2 + 1;
    r1 = r1 - y;
    r2 = r2 + 1;
    r1 = r1 - y;
    r2 = r2 + 1;
  }
```

which would incur the branch penalty one-third as often.

The difficulty with the childish division algorithm (and with many practical programs) is that we do not know before we run the program how many times the loop will execute. (In the case of childish division, the number of times the loop will execute is the answer we are trying to compute.)

Here is where conditional data processing instructions come in handy. Assuming we do not care about the result in `r1`, the childish division algorithm can be partially unrolled as the following C code:

```
r1=x;
r2=0;
do
  {
    r1 = r1 - y;
    if (r1>=0) r2 = r2 + 1;
    if (r1>=0) r1 = r1 - y;
    if (r1>=0) r2 = r2 +1;
  } while (r1>=0)
```

Each time through this loop is equivalent to two times through the loop in the original version of the childish division program. This program (refered to as 'PROGRAM3) incurs the branch penalty half as often provided that the three `if` statements are translated into conditional data processing instructions:

```
00000000/e3a0100e     MOV    R1,0x0e
00000004/e3a04007     MOV    R4,0x07
00000008/e3a02000     MOV    R2,0x00
0000000c/e0511004 L1  SUBS   R1,R1,R4
00000010/52822001     ADDPL  R2,R2,0x01
00000014/50511004     SUBPLS R1,R1,R4
00000018/52822001     ADDPL  R2,R2,0x01
0000001c/5affffa      BPL    L1
00000020/ef000000 L2  SWI
```

The reason that the ARM provides the ability to either set the program status register (bit 20 equal to one) or leave the program status register alone (bit 20 equal to zero) is so that several instructions can be made conditional on the same condition. In this program, the SUBS (and possibly the SUBPLS) determine whether R1 >= 0. The ADDPL and SUBPLS instructions use this program status information to decide whether to execute. Since the pipelined and superscalar implementations allow execution of conditional data processing instructions without branch penalties, such techniques can often speed up a program.

The Verilog coverage set for the superscalar run of 'PROGRAM3 is 0110110110. 'PROGRAM3 does not add to the coverage of the Verilog code provided by 'PROGRAM5 (1110110111); thus we need to do a different test to cover cases 4 and 7. One such test is identical to 'PROGRAM3, except R1 is loaded with 6 rather than 14 as the value of `x`.

```
00000000/e3a01006     MOV    R1,0x06
```

The coverage set for the superscalar run of this modified program is 0101111110, which covers cases 4 and 7. Therefore, all of the ten cases identified in the source code have been tested at least once. This is not to say that the overall design is correct, but at least we know we have checked all the Verilog statements translated from the original ASM. Verilog has helped us make sure that all the code has been covered.

10.10 Comparison of childish division implementations

Determining how long it takes for each of the above division programs (‘PROGRAM3, ‘PROGRAM4 and ‘PROGRAM5) to execute is tedious, especially on the pipelined and superscalar versions of the ARM. A better approach is to let Verilog measure the time for a range of input values:

```
`ifdef DIVTEST
  cont = 0;
  t = 0;
  for (x=0; x<=42; x = x + 1)
    begin
      arm7_machine.m[0] =
        (arm7_machine.m[0] & 32'hffffff00) + x;
      arm7_machine.r[15] = 0;
      #200 cont = 1;
      #100 cont = 0;
      #400 wait(arm7_machine.halt);
      if (arm7_machine.r[2] != x/7)
        $display("error");
      $display("x=%d cl=%d r2=%d %d",x,
        ($time-t)/100,arm7_machine.r[2], $time);
      t = $time;
    end
  $finish;
`endif
```

The above Verilog modifies the MOV immediate at address 0 to initialize different values of x, that range from 0 to 42 and causes the arm7_machine to run the modified program. If the quotient (in r[2]) is not erroneous, the Verilog code simply prints the number of clock cycles (of period 100) elapsed since the machine language program started running for the given value of x. To use the above code, ‘DIVTEST, as well as the macro for the desired machine language program (‘PROGRAM3, ‘PROGRAM4 or ‘PROGRAM5), must be defined. When each of these three programs is run on each of the three implementations (multi-cycle, pipelined and superscalar), we obtain the following data:

run	specific clock cycles, for given quotient 0	1	2	3	4	5 ...	upper bound for quotient > 0
M 5	33	57	81	105	129	153 ...	24*quotient + 33
M 4	33	51	71	91	111	131 ...	20*quotient + 31
M 3	44	46	69	71	94	96 ...	12.5*quotient + 44
P 5	13	20	27	34	41	48 ...	7*quotient + 13
P 4	13	15	21	27	33	39 ...	6*quotient + 9
P 3	14	14	21	21	35	35 ...	3.5*quotient + 14
S 5	9	13	17	21	25	29 ...	4*quotient + 9
S 4	10	11	15	19	23	27 ...	4*quotient + 7
S 3	9	9	13	13	17	17 ...	2*quotient + 9

The column on the left ("run") indicates which program (3, 4 or 5) was run on which machine ("M" for multi-cycle, "P" for pipelined, or "S" for superscalar). For example, "S 3" indicates `PROGRAM3 was run on the superscalar implementation. The column on the right is an equation of an upper bound on this data for `quotient > 0`. Let us look back at the interesting journey we have traveled with our old friend, the childish division algorithm. The following table summarizes implementations of the childish division algorithm given in this and earlier chapters:

register f/e	data	bit	P	S	kind	hard sect	soft sect	upper bound on clock cycles
0	2	12	n	n	Mealy	5.2.4	n/a	2 + quotient
0	2	12	n	n	Mealy	5.2.3	n/a	3 + quotient
0	3	12	n	n	Moore	2.2.7	n/a	3 + quotient
0	2	12	n	n	Moore	2.2.3	n/a	2 + 2*quotient
0	2	12	n	n	Mealy	5.2.1	n/a	2 + 2*quotient
4	15	31	y	y	Mealy	10.9.6	10.9.9.4	9 + 2*quotient
0	3	12	n	n	Moore	2.2.5	n/a	2 + 3*quotient
0	2	12	n	n	Moore	2.2.2	n/a	3 + 3*quotient
3	15	31	y	n	Mealy	10.8.1	10.9.9.4	14+ 3.5*quotient
4	15	31	y	y	Mealy	10.9.6	10.9.9.2	9 + 4*quotient
4	15	31	y	y	Mealy	10.9.6	10.9.9.3	7 + 4*quotient
3	15	31	y	n	Mealy	10.8.1	10.9.9.3	9 + 6*quotient
3	15	31	y	n	Mealy	10.8.1	10.9.9.2	13+ 7*quotient
4	1	11	y	n	Mealy	9.6	9.7	12+ 10*quotient

Continued

3	15	31	n	n	Mealy	10.7	10.9.9.4	44+12.5*quotient
3	15	31	n	n	Mealy	10.7	10.9.9.3	31+ 20*quotient
3	15	31	n	n	Mealy	10.7	10.9.9.2	33+ 24*quotient
4	1	11	n	n	Moore	8.3.2.1	9.7	55+ 55*quotient
4	1	12	n	n	Moore	8.3.2.1	8.3.2.5.3	88+ 75*quotient

The "register" columns indicate how many "data" registers and how many "f/e" (fetch/execute) registers are used. (The number of "f/e" registers is 0 for special-purpose hardware; the number of "data" registers is 1 for a single-accumulator machine like the PDP-8 and much larger (e.g., 15) for a RISC machine like the ARM.) The "bit" column shows maximum number of bits the implementation allows for `x` or `y`. (For software implementations that use the sign bit (the PDP-8's SPA/SMA or the ARM's N flag), this is one less than the register size.) The "S" column indicates whether the hardware is superscalar. The "P" column indicates whether the hardware is pipelined. (Remember a superscalar implementation also uses pipelining.) The "kind" column indicates whether the ASM uses conditional commands (Mealy) or not (Moore). The "hard sect" column indicates where the ASM is described. (The ASM implements the childish division algorithm for special-purpose hardware but the ASM implements fetch/execute for general-purpose hardware.) The "soft sect" applies only to general-purpose computer implementations and describes the machine language for the childish division algorithm. The "upper bound on clock cycles" column indicates how long it takes to compute the quotient. This table is sorted by the linear coefficient of the upper bound; thus for large `quotient`, the order in this table indicates the relative speed of the machines, assuming that the clock frequency of each machine is the same. (The clock frequencies may be different due to different propagation delays in each architecture, as described in chapter 6, but we will assume the clock frequency is the same here.)

There are several interesting things to note in the above table. First, special-purpose hardware is cheaper (number of registers) than general-purpose hardware, especially for the faster kinds of general-purpose hardware (pipelined and superscalar). Second, special-purpose hardware is faster than software running on general-purpose machines except that the superscalar ARM running ‘PROGRAM3 is faster than the special-purpose hardware described in sections 2.2.2 and 2.2.5. Third, the expensive superscalar implementation is competitive with cheap special-purpose hardware only for ‘PROGRAM3 (with its loop unrolling). This illustrates that to capitalize on sophisticated general-purpose hardware requires a good compiler. Fourth, all things being equal, Mealy machines tend to take fewer clock cycles than Moore machines. Fifth, a single-accumulator multi-cycle general-purpose machine (PDP-8) is slower than a RISC multi-cycle general-purpose machine (ARM) because the latter needs fewer instructions to carry out the algorithm. Sixth, pipelining improves the speed of a general-purpose machine. Seventh, pipelining the single-accumulator PDP-8 makes it faster than the multi-cycle ARM but slower than the pipelined ARM. The equivalent of 'PROGRAM4 on the multi-cycle PDP-8 takes `55+55*quotient` clock cycles while the same program on the pipelined PDP-8 takes `12+10*quotient` clock cycles.

10.11 Conclusions

This chapter has compared three different implementations for a RISC instruction set, using a small subset of the ARM as the example hardware and the childish division algorithm as the example software. A multi-cycle implementation requires several cycles to execute an instruction. A pipelined implementation requires one cycle to execute an instruction, except for an instruction such as branch. A superscalar implementation attempts to execute more than one instruction per clock cycle whenever possible.

A RISC machine provides a large set of registers available to the programmer, and an instruction set that allows three register operands to be specified in a single instruction. In comparison to a single-accumulator machine (like the PDP-8), this tends to reduce the number of instructions required to implement an algorithm and to enhance the chance that adjacent instructions will be independent of each other. This latter property makes the design of superscalar general-purpose machines feasible.

Superscalar implementations often use speculative execution, where the result of an instruction is computed before it is known whether that instruction will actually execute. Rather than storing the result in the actual register specified by the instruction, the physical register where this speculative result resides will be renamed to act as the destination if and only if the corresponding instruction actually executes. The superscalar example in this chapter is highly simplified. Commercial superscalar machines are often much more aggressive with speculative execution, using techniques such as *branch prediction*, where the machine executes instructions before it is known whether the branch to those instructions will actually occur, and *out of order execution*, where more instructions are issued (fetched into the pipeline) than can be retired (executed) per clock cycle. The beauty of the ARM's conditional instructions is that they allow us to illustrate the same principles of speculative execution with much simpler hardware.

The design of a superscalar processor is considerably more complex than the design of a pipelined or multi-cycle processor. Because of this, use of a hardware description language such as Verilog is helpful. Through the use of macros and functions, Verilog source code allows the designer to hide unnecessary details early in the design process, yet have those details fully specified in the final source code. Through the use of tasks, the Verilog designer can make sure that the test code covers all cases the designer considers important.

10.12 Further reading

VLSI TECHNOLOGY, INC., *Acorn RISC Machine (ARM) Family Data Manual*, Prentice Hall, Englewood Chiffs, NJ, 1990. Provides documentation on an early version known as the ARM2, which had a three-stage pipeline. How to access documentation about more current versions is given in appendix G.

10.13 Exercises

10-1. The following Verilog code for a special-purpose machine describes the register transfers carried out by four ARM instructions followed by two NOPs run on the superscalar general-purpose ARM. What are these four instructions?

```
@(posedge sysclk) #1;
 r[2] <= @(posedge sysclk) r[1] + r[4];
 psr <= f(r[1] + r[4]);
 ren_val <= @(posedge sysclk) r[3] - r[4];
 ren_tag <= @(posedge sysclk) 3;
 ren_cond <= @(posedge sysclk) `PL;
@(posedge sysclk) #1;
 ren_val <= @(posedge sysclk) r[5] + 5;
 ren_tag <= @(posedge sysclk) 5;
 if ((ren_cond == `PL)&&(psr[31]==0) ||
     (ren_cond == `MI)&&(psr[31]==1) ||
     (ren_cond == `AL))
  begin
   r[6] <= @(posedge sysclk) ren_val;
   r[ren_tag] <= @(posedge sysclk) ren_val;
  end
 else
   r[6] <= @(posedge sysclk) r[3];
@(posedge sysclk) #1;
 if ((ren_cond == `PL)&&(psr[31]==0) ||
     (ren_cond == `MI)&&(psr[31]==1) ||
     (ren_cond == `AL))
   r[5] <= @(posedge sysclk) ren_val;
```

10-2. In problem 10-1, which registers are involved with renaming?

10-3. In problem 10-1, which of the seven cases described in section 10.9.7.1 applies to each state of the special-purpose machine?

10-4. In problem 10-1, which of the instructions is executed speculatively?

10-5. The register file for the multi-cycle and pipelined ARM is a multi-port memory with two read ports and one write port (similar to that described in section 9.8), except the program counter (`r[15]`) must be able to be incremented (by 4) or loaded (with `r[15]` plus an externally supplied 26-bit signed value) independently of the operations that occur on the other ports. Also, there must be a port supplying `r[15]` as the address to memory. Assume that the register file has command inputs `ldPC` and `incPC` to deal with these special operations:

ldPC	incPC	action
0	0	r[15] depends on write port
0	1	r[15]←r[15]+4
1	0	r[15]←r[15]+external
1	1	impossible

The external input is 'OFFSET4 in the pipelined version and 'OFFSET4+4 in the multi-cycle version. Draw a block diagram that implements this synchronous register file.

10-6. Using a register file of the kind given in problem 10-5, design an architecture for the multi-cycle ARM subset given in section 10.7, and give the corresponding mixed ASM.

10-7. Using a register file of the kind given in problem 10-5, design an architecture for the pipelined ARM subset given in section 10.8.1, and give the corresponding mixed ASM.

10-8. The register file for the superscalar ARM is a multi-port memory with four read ports and two write ports. The program counter (`r[15]`) must be able to be incremented (by 4 or 8) or loaded (with `r[15]` plus an externally supplied 26-bit signed value plus either 0 or 4) independently of the operations that occur on the other ports. Assume that the register file has command inputs `ldPC`, `incPC` and `plus4PC` to deal with these special operations:

ldPC	incPC	plus4PC	action
0	0	0	r[15] depends on write ports
0	1	0	r[15]←r[15]+4
0	1	1	r[15]←r[15]+4+4
1	0	0	r[15]←r[15]+'OFFSET4
1	0	1	r[15]←r[15]+'OFFSET4+4
1	1	-	impossible

Draw a block diagram that implements this synchronous register file.

10-9. The interleaved memory described in section 10.9.2 has two conventional memories. For this problem, since the program does not change during execution, we will replace these memories with ROMs (`odd_m` and `even_m`). One of the ROMs is for words whose `addr/4` is odd. The other ROM is for words whose `addr/4` is even. The problem is we cannot predict whether the CPU will need the odd and even instructions fetched into `ir1` and `ir2` or vice versa. Give a block diagram for the interleaved memory that overcomes this problem using three muxes and an incrementor in addition to the ROMs.

10-10. Using a register file of the kind given in problem 10-8 and an interleaved memory of the kind described in problem 10-9, design an architecture for the superscalar ARM subset given in section 10.9.6, and give the corresponding mixed ASM.

10-11. As explained in appendix G, the ARM is actually a Princeton machine, which stores its program and data in the same memory. Like many other RISC machines, the ARM does not allow computation on values in memory. Rather, it only allows load and store instructions. The two most important instructions of this kind are LDR (`ir[27:26]==1&ir[20]==1`) and STR (`ir[27:26]==1&ir[20]==0`). There are several addressing modes available, but for this problem only consider the simple indexed addressing mode (`ir[24:21]` can be ignored in this problem) that accesses `m['OPA+'OPB]`. Assuming a single-port memory of the kind described in section 8.2.2.3.2, give multi-cycle behavioral Verilog to implement such LDR and STR instructions along with the other instructions described in 10.7. Create appropriate test code.

10-12. Assuming a multi-port memory of the kind described in section 9.6, modify the pipelined behavioral Verilog of section 10.8.7 to implement the LDR and STR instructions described in problem 10-11. Create appropriate test code. Unlike chapter 9, operand fetch does not occur until the execution stage of the pipeline, because the ARM has a *load* instruction (LDR), rather than the addition instruction (TAD) of the PDP-8 which required an extra stage to complete. This is important because 'OPA or 'OPB may not be available until that final clock cycle. For the same reason, a STR followed by a LDR from the same address will not require forwarding .

10-13. Rework problem 10-7 to support the instructions of problem 10-12. Note that for the STR instruction there will need to be a mux that provides `ir2[15:12]` to one of the read ports of the register file.

10-14. Using a multi-port memory like that in section 9.6 for problem 10-12 may be too expensive. Design a memory hierarchy, consisting of two direct mapped caches (section 8.5) and a main memory (that takes five cycles per access). One cache is for data manipulated by LDR and STR instructions, and uses the `read`, `write`, `memreq`, `memrack` and `memwack` signals described in section 8.5.3. The other cache is only for instructions being fetched, and uses `ireq` (which combines the roles of `read` and `memreq` for this cache) and `imemack`. You may assume that no machine language instruction will be modified during the execution of the program so that there is no need for write-through with the instruction cache.

10-15. Draw a pure behavioral ASM chart which combines problems 10-12 and 10-14. In the event that an instruction is not in the instruction cache, let NOP(s) enter the pipeline to stall until `imemack` is asserted. In the event that an LDR executes when the operand is not in the data cache, use a wait loop similar to those in section 8.5.2. For STR instructions, use a *write buffer*, which consists of registers that hold the memory address and contents while it is being written. The second of two successive STR instructions will go to a wait state only if the first is still being processed by the memory hierarchy. Because of the write buffer an STR followed by a LDR from the same address creates a dependency that will require forwarding.

10-16. Assuming a powerful multi-port memory of some kind exists, modify the superscalar behavioral Verilog of section 10.9.8.2 to implement the LDR and STR instructions described in problem 10-11. Create appropriate test code.

10-17. Modify the multi-cycle behavioral Verilog of section 10.7.6 to implement the remaining data-processing instructions described by appendix G. Give test code.

10-18. Modify the pipelined behavioral Verilog of section 10.8.7 to implement the remaining data-processing instructions described by appendix G. Give test code.

10-19. Modify the superscalar behavioral Verilog of section 10.9.8.2 to implement the remaining data processing instructions described by appendix G. Give test code.

10-20. The ARM has two multiplication instructions which are identified by `ir[27:22]==0 && ir[7:4]==9`, MUL (`ir[21]==0`) and MLA(`ir[21]==1`):

```
MUL r[ir[19:16]]<-r[ir[3:0]]*r[ir[11:8]]

MLA r[ir[19:16]]<-r[ir[3:0]]*r[ir[11:8]]+r[ir[15:12]
```

Assume that the ALU does not include a combinational multiply operation. Modify the multi-cycle behavioral Verilog of section 10.7.6 to implement the MUL and MLA instructions using a shift and add algorithm such as the one explained in problem 2-7. Test with code that computes a quadratic polynomial, `a*x*x+b*x+c`.

10-21. Modify the pipelined behavioral Verilog of section 10.8.7 to implement the MUL and MLA instructions assuming a combinational multiplier can produce one product per clock cycle. Use the same test code as problem 10-20.

10-22. Modify the superscalar behavioral Verilog of section 10.9.8.2 to implement the MUL and MLA instructions. Use the same test code as problem 10-20.

10-23. Modify `condx` and `f` to allow for all sixteen conditions. Hint: `f` will need a 33-bit input to detect overflow. Give a written justification why your test code is adequate.

10-24. Modify `OPB to include shift and rotate.

11. SYNTHESIS

There are two common uses of Verilog: simulation and synthesis. Chapters 3, 5 and 6 describe various features of Verilog that are useful for simulation, which is the interpretation of Verilog source code on a general-purpose computer. This chapter gives examples of synthesis, which is the automated process of transforming a subset of Verilog statements into a netlist of gates whose interconnnections perform the algorithm specified by the Verilog source code.

11.1 Overview of synthesis

There are two main vehicles for the implementation of synthesized designs: custom integrated circuits (sometimes called Application-Specific Integrated Circuits or ASICs) and programmable logic. Custom integrated circuits are created by transforming the synthesized netlist for a particular design (i.e., the equivalent of gate-level structural Verilog) into a specific geometric arrangement of metal, semiconductor and insulator materials on an integrated circuit. To manufacture such circuits, an automated tool draws the physical layout of the circuit on what is called a *mask*. The mask is then used with photolithography or similar processes to mass-produce the circuit on chips.

Programmable logic is fabricated in a similar way, but the masks used by the manufacturer do not represent some specific design. Instead, a programmable logic chip consists of many building block devices together with a programmable interconnection network. After the programmmable logic chip is manufactured and sold to the designer, bits are transferred into the chip which customize the programmable logic for a specific design. Thus, the same physical hardware might be used by two different designers to implement two completely different designs. Because by itself programmable logic lacks the ability for self-modification, it is not quite general purpose in the same sense as a stored program machine (such as the Manchester Mark I, the PDP-8 or the Pentium II). Historically, the concept of rewiring a fixed set of building block units to solve different problems can be traced back to the ENIAC in the early 1940s. Modern technology now allows interconnections inside a programmable logic chip to be reconfigured by simply changing the bits, rather than having to pull out and plug in wires as was the situation with the ENIAC's plugboards. As a further convienence, Verilog synthesis tools allow the modern designer (who may be ignorant of the wiring bit patterns) to reconfigure the programmmable logic by simply changing source code. Thus, in many instances, using programmable logic with a synthesis tool provides a viable alternative to software.

Regardless of whether the designer is targeting custom integrated circuits or programmable logic, the Verilog used for synthesis is similar, except the cost of synthesizing to programmable logic is considerably cheaper than synthesizing to custom integrated circuits.[1] This chapter concentrates on programmable logic because the necessary tools should be accessible to interested readers.

11.1.1 Design flow

Figure 11-1 gives the *design flow*, which shows how various automated tools interact with one another. The design flow shows how to synthesize and test a design. The design starts as Verilog source code. It gets translated and downloaded into a kind of programmable logic, known as a Complex Programmable Logic Device (CPLD). Figure 11-1 illustrates the files (shown as rectangles) that are produced as output of various tools (shown as circles), and how, in turn, these files are used as input to other tools. The tools shown in figure 11-1 include an optional synthesis preprocessor, the main synthesis tool, a post-synthesis place and route tool targeting programmable logic and a download tool that transfers the design into the programmable logic chip. Also, a standard Verilog simulator plays an important part in the design flow.

The designer only creates two or three files: one file to be synthesized, one file for the test code and one optional file for the physical (pin number) information. The remaining files of figure 11-1 are created automatically. The file containing test code can be used with the Verilog simulator to verify the operation of the file to be synthesized and/or the results of synthesis. The file to be synthesized, which is input to the synthesis tool (or possibly the preprocessor), contains one or more Verilog module(s) using behavioral and/or structural features of Verilog. For pure behavioral synthesis, there is only one module, which, of course, is the highest level module of the file. For structural synthesis, there could be multiple modules defined that are instantiated hierarchically inside the highest level module of the file.

Only the portlist of this highest level module determines how the physical pins of the synthesized chip will be used. The portlists of lower level modules, if any, only deal with the internal connections within the chip and therefore have no influence on the physical pins of the chip.[2]. The `input` and `output` definitions for the highest level module should include the width of each port, since some synthesis tools require this, but all synthesis tools and simulators accept this syntactic variation. Each bit of each port of the highest level module corresponds to a distinct physical pin of the synthe-

[1] Assuming the design will be manufactured only in small quanities.

[2] Some tools, such as PLDesigner-XL, allow a design to be partitioned onto multiple chips, but even in such a case, the partitioning is automatic and not influenced by the Verilog ports of lower level modules.

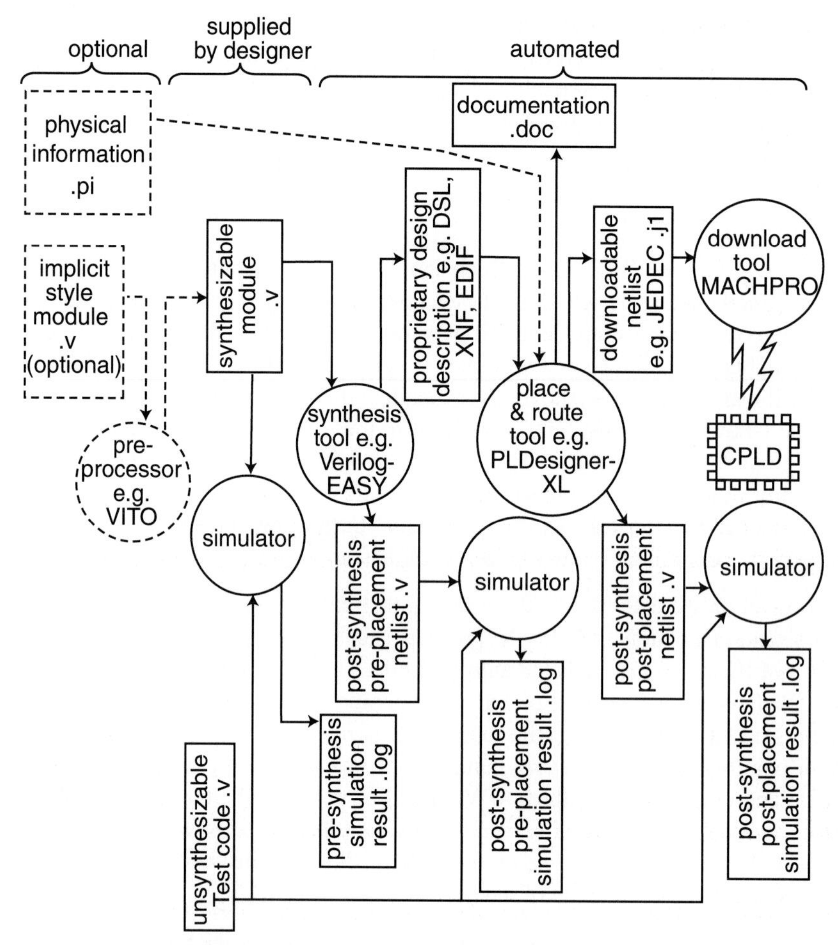

Figure 11-1 Design flow for CPLD synthesis. VITO, VerilogEASY, MACHPRO and PLDesigner-XL are specific tools discussed in section 11.1.3.

sized chip. The easiest and often most efficient approach is to let the synthesis tool choose how to connect these bits to the physical pins. In some cases, such as using a circuit board where the programmable logic chip already has its `sysclk` and similar signals soldered to specific pins, it is necessary for the synthesis tool to use specific pins. There is no standard syntax in Verilog to indicate pin numbers in the file (`.v`) that contains the highest level module, but many synthesis tools allow the designer to force the tool to connect specific bits to specific pins with the physical information file (section 11.3.6) or a similar approach.

11.1.2 Testing approaches

The file to be synthesized does not contain test code, since test code usually contains unsynthesizable statements such as `$display`. On the other hand, test code is essential for using a synthesis tool properly. Since the Verilog file used for synthesis contains a module with a portlist, running that file by itself on a simulator would produce erroneous results since the ports would be disconnected (`'bz`). When a designer does synthesis of a Verilog file, there needs to be another Verilog file with test code that does an `` `include `` of the synthesizable Verilog.

There are several reasonable strategies for using a simulator to test synthesizable Verilog:

a) before synthesis
b) after synthesis but before place and route
c) after place and route
d) all of the above.

What is not reasonable is to neglect testing altogether. Designers who think they are saving time by not writing test code for use in simulation are fooling themselves. It is far easier to detect bugs in simulation than after the synthesized design has been implemented in a physical chip.

The most thorough strategy is to simulate at each step in the design flow. Most vendors support some kind of backannotated output from the synthesis and/or place and route tools. This output is typically some kind of structural Verilog that gives the gate-level netlist produced by the synthesis tool. By using this output together with the test code, the designer may verify that the synthesized result behaves similarly to the original unsynthesized Verilog. The *backannotation* provides timing information to the simulator that allows the designer to predict the speed of the design without having to test the physical chip.

Although somewhat less desirable, the strategy of simulating only after synthesis makes sense when the designer needs to explore many different algorithms and design alternatives to find an acceptable solution to a problem. In such a case, the first issue is to discover which algorithms fit within the available hardware resources. If the designer makes a small mistake with an algorithm that fits within the desired chip, it is likely that the corrected version of the algorithm will also fit. After having discovered an algorithm that fits within the constraints of the hardware, it is important that the designer use simulation to determine whether there are any such bugs in the algorithm. It is much harder to debug the algorithm after it is embodied in physical hardware than when it is given only as Verilog code for simulation, and so simulation should be part of the design process.

When the physical chip will not be operated near its maximum frequency, there often is no need to simulate the back annotated Verilog resulting from synthesis.[3] In such a case, simulating only before synthesis may be reasonable. For example, the clock used in this chapter is slow enough that propagation delay is not a concern with the designs discussed below. We will do post-placement simulation only to illustrate the logical correctness of the process, and not out of concern for speed. In commercial design, speed is often an important issue, but correctness is always the first concern.

11.1.3 Tools used in this chapter

There are five specific software packages that are used as example tools in the design flow of this chapter. The details would differ slightly if other vendors' tools were used instead, but the basic principles of the design flow would be similar. The first software package, which is used only with the later examples in this chapter, is the Verilog Implicit To One hot (VITO) synthesis preprocessor described in chapter 7 and appendix F. The second software package is a synthesis tool, known as PLSynthesizer, which is available from a company called MINC Incorporated.[4] The output of PLSynthesizer is in a proprietary HDL known as DSL. PLSynthesizer also outputs structural Verilog which is logically equivalent to the DSL for post-synthesis simulation. The third software package is a place and route tool, known as PLDesigner, also from MINC, that converts the output of PLSynthesizer into what is called *JEDEC format*, which is a standard file format (`.jl`) used by several different vendors for downloading to programmable devices. PLDesigner may optionally use a physical information (`.pi`) file to indicate pin numbers. PLDesigner outputs the equivalent of the JEDEC file as a structural Verilog netlist for post-placement simulation. In addition, PLDesigner creates a documentation file (`.doc`) which indicates pin numbers and logic equations (in DSL syntax). The fourth software package is a download tool known as MACHPRO, from a CPLD manufacturer known as Vantis,[5] that reads the JEDEC file and sends the desired configuration to the programmable logic chip. The fifth software package is a simulator, such as VeriWell.

11.1.4 The M4-128/64 CPLD

The M4-128/64 is a CPLD chip manufactured by Vantis that comes in a 100-pin package, of which 64 pins are available for the designer to use.[6] Each of these 64 I/O pins may be used as one bit of an `input`, `output` or `inout` port. Internally, the M4-128/

---

[3] Provided that the subset of Verilog used means the same thing in both synthesis and simulation.

[4] MINC has made a restricted version of this technology known as VerilogEASY available to readers of this book. See appendix F for details.

[5] Vantis is a spinoff from Advanced Micro Devices (AMD), and the M4-128/64 used to be known as the AMD Mach445.

[6] The restricted version of VerilogEASY only allows 40 of these pins to be used.

64 contains 128 units, known as *macrocells*. Each macrocell contains a single OR gate, and a single optional flip flop. The output of the OR gate either feeds the flip flop or is directly available to other macrocells or I/O pins as combinational logic. The macrocell's OR gate receives its inputs from a series of AND gates. Put another way, a particular macrocell, `m`, can either implement sequential logic:

```
reg m;
always @(posedge sysclk)
m =  t1_1 & t1_2 & t1_3 & ...
    | t2_1 & t2_2 & t2_3 & ...
        ...
```

or alternatively, that macrocell can implement combinational logic:

```
wire m =  t1_1 & t1_2 & t1_3 & ...
        | t2_1 & t2_2 & t2_3 & ...
            ...
```

where the (optionally complemented) terms `t1_1`, `t1_2`, etc. are either input `wires` from the I/O pins or are the outputs of macrocells. In a particular macrocell, the designer does not have to use all the terms possible, but there are fairly complex internal constraints on how many and which terms may be used in particular macrocells. Because of the internal complexity of the CPLD, it is necessary for the designer to use a tool. This is true even if the designer were to create a netlist manually because the *place and route* tool must transform the original netlist into one that fits within the complex constraints of the CPLD.

Each I/O pin of the M4-128/64 has an optional flip flop, which the synthesis tool may choose to disconnect (for a combinational logic function of the input). Considering the macrocells (64 bonded to I/O pins and 64 hidden) and the I/O pins, the total number of flip flops that the M4-128/64 contains is 192. When all its macrocells are fully in use, the M4-128/64 is the equivalent of about 5000 gates.

Vantis makes a printed circuit board, known as a *demo board,* that has one M4-128/64 mounted on it together with additional hardware, such as a 1.8432MHz oscillator[7] that produces the `sysclk` signal. Although many similar types of devices exist,[8] the reason

[7] When the inputs to every macrocell only come from the internal flip flops, the M4-128/64 may be clocked up to 125 MHz. When macrocells are cascaded together to form complex combinational logic, the maximum frequency is lower. The 1.8432 MHz is slow enough to be safe for most designs.

[8] Of which the Field Programmable Gate Array (FPGA) is perhaps the most common. The FPGA uses a table rather than the AND/OR structure of a CPLD, but such details are seldom important to a designer using a synthesis tool. In the 1990s, companies such as Xilinx and Altera were leading suppliers of FPGAs.

for describing the M4-128/64 demo board here is that it is well suited for small synthesis experiments. The M4-128/64 demo board connects to a personal computer via the parallel (printer) port of that computer. This personal computer runs the synthesis tools (such as PLSynthesizer and PLDesigner) and also the MACHPRO software, which downloads the configuration of hardware determined by the synthesis tools into the M4-128/64. The downloading process changes which terms are connected to which macrocells. If a designer makes a mistake, it is a simple matter to download a corrected version of the design because the internal technology of the CPLD is similar to an EEPROM.

11.2 Verilog synthesis styles

Regardless of whether the designer wants programmable logic or custom integrated circuits, and regardless of which vendors' tools are involved, there are five basic styles of Verilog code used in synthesis: behavioral registers, behavioral combinational logic, behavioral implicit style state machines, behavioral explicit style state machines and structural instantiation. Often a particular design contains a combination of these styles.

11.2.1 Behavioral synthesis of registers

As described in sections 3.7.2.2 and 4.4.4, the synthesizable model for a register is an instantaneous assignment statement inside a block with a single time control syntax, such as `@(posedge sysclk)`, `@(posedge sysclk or posedge reset)` or `@(posedge sysclk or negedge reset)` time control. All synthesis vendors support this Verilog construct. Registers synthesize to a group of flip flops, typically D-type flip flops. Often there is combinational logic associated with a register. An example of synthesizing a register is given in section 11.3.

11.2.2 Behavioral synthesis of combinational logic

There are two ways to describe combinational logic using behavioral Verilog that all synthesis tools accept: the continuous `assign` statement (section 7.2.1) and an `always` block with a sensitivity list composed of all the variables in the block that are **not** on the left of any of the =s inside the block (section 3.7.2.1).[9] All synthesis vendors support both of these constructs. Combinational logic synthesizes to the primitive combinational units of the target hardware which are AND/OR gates for CPLDs, lookup tables for FPGAs and ROMs and arbitrary combinational gates for custom logic. An example of synthesizing combinational logic is given in section 11.4.

[9] With the additional requirement that none of the variables on the left of the =s occur on the right of the =s.

11.2.3 Behavioral synthesis of implicit style state machines

One of the main themes of this book is the advantage of solving problems at the highest level possible. Implicit style state machines provide this high-level approach for designing hardware. Most of the examples in this book only consider the "pure behavioral ASM" (section 2.1.5.1) or its equivalent coding with implicit style Verilog (section 3.8.2.3). Although, as noted in chapter 1, this high level of design has a history that goes back for decades before the introduction of Verilog, the implicit style has only recently begun to capture the attention of many Verilog designers. The reason that designers have become interested in this style is that it allows them to produce correct designs in less time. Unfortunately, not all synthesis vendors support this style, and there are some restrictions on the support provided by those that do. The preprocessor described in appendix F and chapter 7 allows a designer to use a reasonable subset of the implicit style even when the synthesis tool does not support it. Synthesizable implicit style consists of multiple `@(posedge sysclk)` inside an `always`, with a few additional syntax restrictions that were not considered in earlier chapters. Examples are given in sections 11.5, 11.6 and 11.9.

11.2.4 Behavioral synthesis of explicit style state machines

In contrast to the implicit style, the explicit style requires the designer to specify the present state register and next state combinational logic, as explained in section 4.3. All synthesis vendors support this style. An example is given in section 11.7.

11.2.5 Structural synthesis

The most primitive form of synthesis, which all tools accept, involves structural instances. If all modules being synthesized use only structural Verilog, the result of synthesis is simply to flatten the netlists (section 2.5) and fit the design within the constraints of the chip. More typically, some of the modules being instantiated have some of the kind(s) of behavioral code described above. In this case, the behavioral code is synthesized appropriately before the netlist is flattened. An example is given in section 11.8.

11.3 Synthesizing `enabled_register`

As described in section D.6, the enabled register is one of the most important sequential building blocks used in computer design. Suppose we wish to synthesize a two-bit-wide enabled register. The behavioral Verilog for this is identical to section 4.2.1.1, except that we substitute the literal `[1:0]` to indicate the two-bit width since many synthesis tools do not work properly with parameters and do not work properly unless the size is mentioned in the `input` and `output` declarations:

```
module enabled_register(di,do,enable,clk);
 input [1:0] di;
 input enable,clk;
 output [1:0] do;
 reg [1:0] do;
 wire [1:0] di;
 wire enable, clk;
 always @(posedge clk)
  begin
   if (enable)
     do = di;
  end
endmodule
```

The above Verilog should work with any synthesis tool. If we synthesize the above with PLSynthesizer targeting the Vantis M4-128/64 chip mentioned above, we get the following preplacement structural Verilog netlist:

```
module LPM_DFF_2_x(Clk,ClkEn,D,Q);
 input Clk,ClkEn,D; output Q;
 wire net0, net1, net2, net3, net4;
 NAN2 I_2_NAN2(.I0(net0),.I1(net1),.O(net2));
 NAN2 I_3_NAN2(.I0(Q),.I1(net3),.O(net1));
 NAN2 I_4_NAN2(.I0(ClkEn),.I1(D),.O(net0));
 INV I_1_INV(.I0(ClkEn),.O(net3));
 DFF I0(.CLK(Clk),.D(net2),.Q(Q),.QBAR(net4));
endmodule
module LPM_DFF_1_x(Clk,ClkEn,D,Q);
 input Clk,ClkEn; input[1:0]D; output[1:0]Q;
 LPM_DFF_2_x I0(.Clk(Clk),.ClkEn(ClkEn),
               .D(D[1]),.Q(Q[1]));
 LPM_DFF_2_x I1(.Clk(Clk),.ClkEn(ClkEn),
               .D(D[0]),.Q(Q[0]));
endmodule
module enabled_register(di,do,enable,clk);
 input [1:0] di;  output [1:0] do;
 input enable,clk;
 LPM_DFF_1_x dox_x(.Clk(clk),.ClkEn(enable),
                  .D(di), .Q(do));
endmodule
```

11.3.1 Instantiation by name

There are two kinds of structural instantiation syntax that are legal in Verilog. The first kind is instantiation by position, as described in section 3.10. Had instantiation by position been used above, the Verilog shown in bold would have been written as:

```
LPM_DFF_1_x dox_x(clk,enable,di,do);
```

There is no other way to write this with the positional syntax of section 3.10. The other kind of syntax that is legal in Verilog is instantiation by name, which is illustrated above in bold. Like many synthesis tools, PLSynthesizer uses this alternative syntax because the modules generated by the tool may have lengthy portlists. The advantage of instantiation by name is that the ports may be rearranged in any order and the meaning is the same. For example, the following:

```
LPM_DFF_1_x dox_x(.D(di), .Q(do),
                  .Clk(clk),.ClkEn(enable));

LPM_DFF_1_x dox_x(.Q(do), .D(di),
                  .ClkEn(enable),.Clk(clk));
```

are among the twenty-four permutations that mean the same thing.

11.3.2 Modules supplied by PLSynthesizer

The modules in section 11.3.1 (such as NAN2, INV and DFF) could contain detailed gate-level timing information, but this netlist has not yet been placed. After placement inside the M4-128/64 CPLD, the netlist is likely to be considerably different than the one in section 11.3.1. Rather, the netlist in section 11.3.1 is primarily of use to show the logical correctness of the transformation carried out by the synthesis tool. To illustrate this transformation, we will define idealized versions of the modules it instantiates:

```
module NAN2(I0,I1,O);
 input I0,I1;output O;nand g1(O,I0,I1);
endmodule
module INV(I0,O);
  input I0;output O;not g2(O,I0);
endmodule
```

Continued

```
module DFF(CLK,D,Q,QBAR);
  input CLK,D;output Q,QBAR;
  assign QBAR = ~Q;
  always @(posedge CLK)Q = D;
endmodule
```

11.3.3 Technology specific mapping with PLDesigner

In addition to structural Verilog, PLSynthesizer produces the same netlist in a proprietary form, known as DSL. The place and route tool, PLDesigner, uses the DSL to generate a netlist that is fitted within the constraints of the M4-128/64 CPLD. The output of PLDesigner includes the JEDEC netlist and an equivalent post-placement Verilog netlist. Such post-placement structural Verilog more accurately reflects the result of place and route than the netlist produced by PLSynthesizer. For this example, the resulting structural Verilog[10] of the enabled register is:

```
//Model automatically generated by Modgen Version 3.8
`timescale 1ns/100ps
enabledo00(dol0r,dol1r,dil1r,enable,dil0r,clk);
 output dol0r, dol1r;
 input dil1r, enable, dil0r, clk;  supply0 GND;
 wire pin_8,pin_11,pin_12,pin_13,pin_93,pin_94,tmp12,
tmp14,tmp15,tmp16,tmp17,tmp18,tmp19,tmp20,tmp21,tmp22;
portin PI1(pin_8,dil1r); portin PI2(pin_11,enable);
portin PI3(pin_12,dil0r); portin PI4(pin_13,clk);
portout PO1(dol0r,pin_93); portout PO2(dol1r,pin_94);
mbuf B1(tmp12,pin_13); and A1(tmp15,pin_12,pin_11);
not I1(tmp17,pin_11); and A2(tmp16,pin_93,tmp17);
or O1(tmp14,tmp15,tmp16);
dffarap DFF1(pin_93, tmp12, tmp14, GND, GND);
mbuf B2(tmp18,pin_13); and A3(tmp20,pin_8,pin_11);
not I2(tmp22,pin_11); and A4(tmp21,pin_94,tmp22);
or O2(tmp19,tmp20,tmp21);
dffarap DFF2(pin_94, tmp18, tmp19, GND, GND);
endmodule
```

[10] This Verilog was edited slightly for brevity.

Continued

```
module enabledo(do, di, enable, clk);
    output [1:0] do;
    input [1:0] di;
    input enable, clk;
enabledo00 U1(.do10r(do[0]), .do11r(do[1]),
              .di11r(di[1]), .enable(enable),
              .di10r(di[0]), .clk(clk));
endmodule
```

Although much of the syntax used in the above code should be familiar from chapter 3, there are a few features of Verilog used in the above code (shown in bold) that were not mentioned previously. First is `'timescale` which allows the simulator to attach the proper meaning to `$time` that corresponds to the actual physical hardware. Second is the `supply0` declaration, which is a shorthand for a continuous assignment (section 7.2.1) of the one-bit wire `GND` to 0, which models the connection to electrical ground inside the M4-128/64. Third are some user-defined modules (`portin`, `portout`, `mbuf` and `dffarap`) supplied by PLDesigner and explained in section 11.3.4 that model hardware resources of the M4-128/64. The name of the top level module generated by PLDesigner derives from the file name (`enabledo.v`) rather than the behavioral module name (`enabled_register`), and the order of the ports of this module may differ from that of the original behavioral Verilog which is why instantiation by name is done.

In addition to the post-placement structural Verilog, PLDesigner produces a documentation file (`.doc`) which summarizes the logic equations implemented by the netlist:

```
do[1].D=do[1]*/enable+di[1]*enable;
do[1].CLK=clk;
do[0].D=do[0]*/enable+di[0]*enable;
do[0].CLK=clk;
```

This is a much more primitive language than Verilog. The `.D` and `.CLK` notations indicate the macrocells are being used as D-type flip flops. To put the above in the more understandable Verilog form, the notation for Boolean operations ('`*`', '`+`', '`/`') must be rewritten into the corresponding Verilog notation ('`&`', '`|`', '`~`'). The following manual translation is the equivalent behavioral Verilog:

```
always @(posedge clk)
 begin
  do[1]<= #0((di[1]&enable)|(~enable&do[1]));
  do[0]<= #0((di[0]&enable)|(~enable&do[0]));
 end
```

The assignment statements must be non-blocking (with time control of #0) and must be listed inside an always block with a single @(posedge clk) as the time control. This non-blocking assignment is somewhat different than the one used in earlier chapters. It is used above so that the order in which the Verilog statements occur will not effect the result. Since the non-blocking assignments use #0, the effect is almost the same as plain =, except all of the right-hand values will be evaluated before any of the left-hand values are changed. Because of the single @(posedge clk) at the beginning of the always block, do will only change at the rising edge of the clock. The only reason to manually rewrite these logic equations back into Verilog is to describe the meaning of the .doc file. This file explains the transformation that PLSynthesizer has performed on the original behavioral Verilog more succinctly than the netlist.

11.3.4 Modules supplied by PLDesigner

Like most other place and route tools, PLDesigner allows for backannotation of timing information in the netlist after the place and route phase. Such post-placement information is more accurate than post-synthesis preplacement information because the place and route tool knows how signals will be routed through the actual chip. The details of how such information gets inserted into the structural Verilog output from a place and route tool varies among different vendors. In the case of PLDesigner, the modules used in section 11.3.3, such as portin, portout and mbuf, can contain detailed gate-level timing information, such as specify blocks, to model the interconnect delays that occur between macrocells in the M4-128/64. PLDesigner also generates an .sdf file which includes actual min/typ/max timing information for the routed circuit. For our purposes, we are not concerned about such detailed timing information but are rather only interested in the logical correctness of the transformation carried out by the place and route tool. To illustrate these transformations, we will define idealized versions of these modules. The first three of these are simply buffers that pass the input (i) through unchanged as the output (o):

```
module portout(o,i);
 input i; output o; buf b(o,i);
endmodule
module portin(o,i);
 input i; output o; buf b(o,i);
endmodule
module mbuf(o,i);
 input i; output o; buf b(o,i);
endmodule
```

The reason PLDesigner uses all three of these is that, in the actual backannotated Verilog, these might have different delays associated with them because they correspond to different hardware units within the M4-128/64. `portin` corresponds to routing a signal from an I/O pin used as an `input` on a `wire` that connects to an internal macrocell. `portout` corresponds to taking the output of an internal macrocell and routing it to an I/O pin to be used an `output`. `mbuf` corresponds to an internal connection between macrocells. Logically, all three of the above are equivalent. The only difference would be timing, which we are ignoring in this example.

Although Verilog provides built-in gates for combinational logic, such as `and`, `or` and `not`, Verilog does not provide built-in gates for sequential logic.[11] Therefore a place and route tool must supply modules for the sequential logic resources of the target technology, in this case the M4-128/64. Recall that each macrocell contains a D-type flip flop, which can be modeled as:

```
module dffarap(Q,CLK,D,AR,AP);
  output Q;
  input  CLK,D,AR,AP;
  reg Q;
    always @(posedge CLK or
        posedge AP or posedge AR)
      begin
        if (AP)
          Q = 1;
        else if (AR)
          Q = 0;
        else
          Q = D;
      end
endmodule
```

The above models a flip flop with an asynchronous reset (`AR`) and an asynchronous preset (`AP`). Such asynchronous signals are typically only used to initialize a controller when it is first powered up (see sections 4.4.4 and 7.1.6). In this example, these asynchronous signals are not used, and so they are instantiated with a connection to `GND`.

11.3.5 The synthesized design

As explained in figure D-17, an enabled register can be described with a block diagram consisting of a mux and a simple D-type register. Since figure D-17 is going to be synthesized into a physical component, either the designer or the synthesis tool must

[11] For a CPLD, there is no delay attributed to a particular AND or OR gate. Rather the delay is associated with the macrocell. For this reason, PLDesigner-XL uses built-in delayless `and`, `or` and `not`. Place and route tools for FPGAs or custom logic may take a different approach.

choose which signals will go in and out of the pins of the chip. In this instance, there are four bits of information being input and two bits of information being output, as illustrated in figure 11-2.

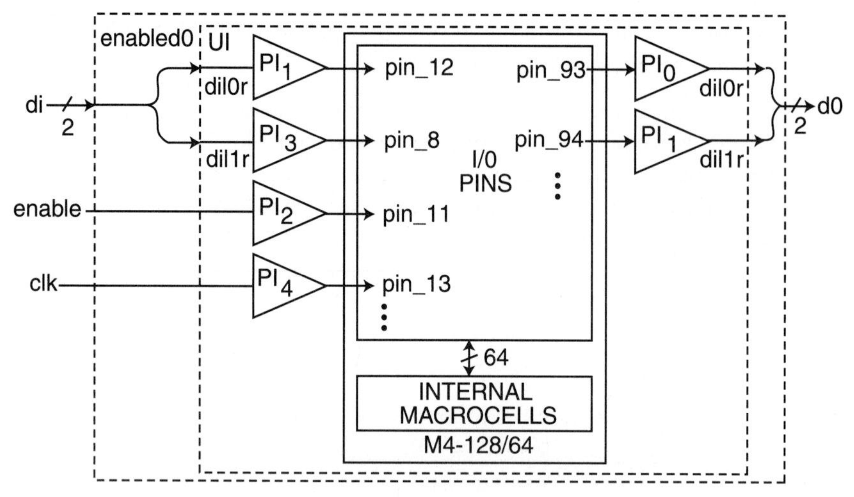

Figure 11-2. Physical pins of M4-128/64 used for two-bit enabled register.

The structural Verilog refers to the internal wires that connect to the I/O pins with the prefix `pin_`. This could be a little confusing since, for example, `pin_13` is not actually the physical pin 13 but rather is the signal from that pin after it has been buffered internally by the M4-128/64. In this design, `pin_13` is logically the same as the `clk` port, which presumably would connect to the global `sysclk` signal. The instance `U1` of `enabledo00` separates the individual one-bit nets from the multi-bit ports. There are buses (which are oversimplified in this diagram) that connect the I/O pins to the macrocells. The actual implementation of the synthesized design occurs in the macrocells.

The synthesis tool has *bit blasted* the design into individual one-bit-wide *bit slices*, each one of which fits into a single macrocell, as shown in figure 11-3.

The circuit in figure 11-3 is a literal transcription of the Verilog produced by the synthesis tool. Notice how each bit slice of the mux has turned into an AND/OR gate arrangement. When `enable` (`pin_11`) is asserted, the outputs of the A2 and A4 AND gates will be zero. Thus, `di[0]` (`pin_12`) and `di[1]` (`pin_8`) will pass through their respective OR gates (O1 and O2) to become the new values of their respective flip flops (DFF1 and DFF2) at the next rising edge of `clk` (`pin_13`). When `enable` is not asserted, the old values (`pin_93` and `pin_94`) will be reloaded into their respective flip flops (DFF1 and DFF2) at the next rising edge of `clk`.

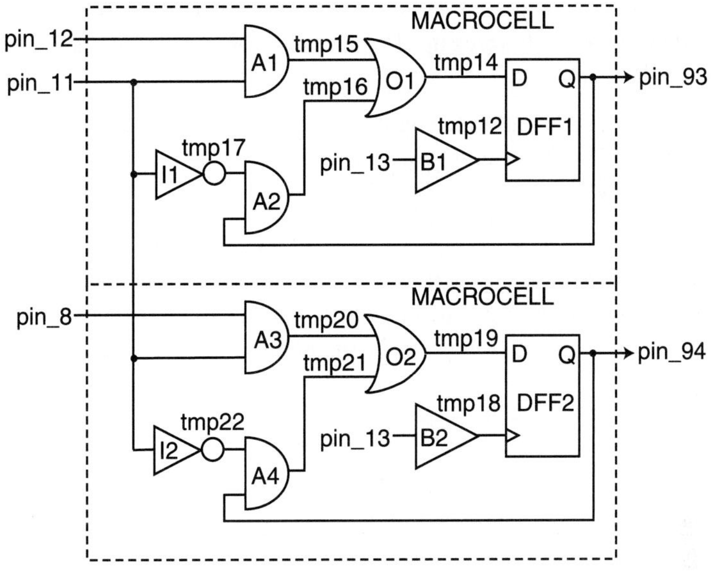

Figure 11-3. Macrocells in M4-128/64 implementing enabled register bit slices.

11.3.6 Mapping to specific pins

The physical pin numbers shown in figure 11-2 were chosen by PLDesigner. Often a designer wishes to override the choices automatically made by the place and route tool. For example, on the M4-128/64 demoboard, certain pins are attached to other hardware soldered on the board:

```
a1:93   a2: 5   a3:19   a4:31   sysclk:13
b1:94   b2: 6   b3:20   b4:32    reset: 4
c1:95   c2: 7   c3:21   c4:33      sw3:18
d1:96   d2: 8   d3:22   d4:34      sw2:54
e1:97   e2: 9   e3:23   e4:35      sw1:63
f1:98   f2:10   f3:24   f4:36      sw0:68
g1:99   g2:11   g3:25   g4:37
```

The `wires` whose names above begin with "a" through "g" are for Light Emitting Diodes (LEDs) in seven-segment displays. For example, the active low `a1 ... g1` signals control the leftmost digit. These seven segments are labeled clockwise, with `a1` at the top and `g1` at the center; thus `b1`, `c1`, `f1` and `g1` should be 0 and `a1`, `d1` and `e1` should be 1 to display the digit "4." The 1.8432MHz clock is available as `sysclk`, and a debounced push button provides the active low reset, which is also activated when the demoboard is powered up. There are four input DIP switches (sw0 Ö sw3) available on the demoboard.

These input pins can be named anything the designer wishes. For example, in the enabled register of section 11.3, it might be reasonable to take the enable from the switch on pin 54, and the di bus from the switches on pins 63 and 68. The two do bits might directly drive the a1 and b1 LED segments[12] on pins 93 and 94. Note that because of the active low nature of the LEDs, the light will not illuminate when the bit is a one, but it will light up when the bit is a zero. The following file, whose name must be similar to the name of the file that contains the module to be synthesized but with the extension .pi, is required to indicate the pin numbers to PLDesigner:

```
{MAX_SYMBOLS 0,MAX_PTERMS 0,POLARITY_CONTROL TRUE,
 MAX_XOR_PTERMS 0,XOR_POLARITY_CONTROL FALSE};
device target 'part_number amd MACH445-12YC';
 OUTPUT do[1]:93;OUTPUT do[0]:94;INPUT clk:13;
 INPUT enable:54;INPUT di[1]:63;INPUT di[0]:68;
end device;
```

11.4 Synthesizing a combinational adder

As described in section C.3, the adder is one of the most important combinational building blocks used in computer design. There are many ways (sections 3.10.5 through 3.10.7) to code an adder in Verilog, both behaviorally and structurally. Of these, the behavioral description is the easiest for the designer:

```
module addpar(s,a,b);
  output [3:0] s;
  input [3:0] a,b;
  reg [3:0] s;
  wire [3:0] a,b;
  always @(a or b)
    s = a + b;
endmodule
```

As in the last example, the input and output definitions need a size (four bits in this case). When the above is synthesized similarly to the last example, PLDesigner produces a .doc file that describes a series of logic equations for each bit of s. The following is a manual translation of this back into Verilog:

[12] Which just happen to be the same as the last example.

```
module addpar(s,a,b);
output [3:0] s; input [3:0] a,b;
wire [3:0] s,a,b; wire [3:3] c;
assign s[0] = ((a[0]&~b[0])|(~a[0]&b[0]));
assign s[1] = ((a[1]&~b[1]&~b[0])|(a[1]&~a[0]&~b[1])
|(a[1]&a[0]&b[1]&b[0])|(~a[1]&a[0]&~b[1]&b[0])
|(~a[1]&~a[0]&b[1])|(~a[1]&b[1]&~b[0]));
assign c[3]=((a[1]&a[0]&b[2]&b[0])|(a[2]&b[2])|(a[0]
&b[2]&b[1]&b[0])|(a[2]&a[1]&b[1])|(a[1]&b[2]&b[1]))
|(a[2]&a[1]&a[0]&b[0])|(a[2]&a[0]&b[1]&b[0]);
assign s[2] = ((a[2]&~b[2]&~b[1]&~b[0])
|(a[2]&~a[1]&~b[2]&~b[1])|(a[2]&~a[1]&~a[0]&~b[2])
|(a[2]&~a[1]&~b[2]&~b[0])|(~a[2]&~a[1]&b[2]&~b[1])
|(a[2]&a[1]&a[0]&b[2]&b[0])|(~a[2]&a[1]&~b[2]&b[1])
|(a[2]&a[0]&b[2]&b[1]&b[0])|(~a[2]&b[2]&~b[1]&~b[0])
|(~a[2]&a[1]&a[0]&~b[2]&b[0])|(a[2]&a[1]&b[2]&b[1])
|(~a[2]&~a[1]&~a[0]&b[2])|(~a[2]&~a[1]&b[2]&~b[0])
|(a[2]&~a[0]&~b[2]&~b[1])|(~a[2]&~a[0]&b[2]&~b[1])
|(~a[2]&a[0]&~b[2]&b[1]&b[0]));
assign s[3] = ((a[3]&~b[3]&~c[3])|(a[3]&b[3]&c[3])
|(~a[3]&~b[3]&c[3])|(~a[3]&b[3]&~c[3]));
endmodule
```

Of course, the designer would probably use the backannotated output from PLDesigner. This lengthy output, which is equivalent to the above `assign` statements, has been omitted for brevity. In this output, the internal name for the one-bit carry `wire` varies depending on how the module is synthesized. The name might be something like `LPM_ADD_SUB_1_x_1_n0002`. It might also be just `c[3]` as shown above.

This result from synthesis is quite a bit more complicated than one might expect when solving the same problem manually using full-adders. The above is complex because the place and route tool utilizes the wide AND/OR gates that exist in each macrocell of the M4-128/64. In the classical ripple carry adder (section 2.5), there needs to be a distinct carry signal input to each full-adder. Here the tool has eliminated the carry for all but the most significant bit by merging the logic equations for several full-adders together in a process known as *node collapsing*. This has the effect of lowering the propagation delay.

11.4.1 Test code

In any event, the designer needs to test the adder. Here is test code (sections 3.10.5 and 6.3.2) that does an exhaustive test:

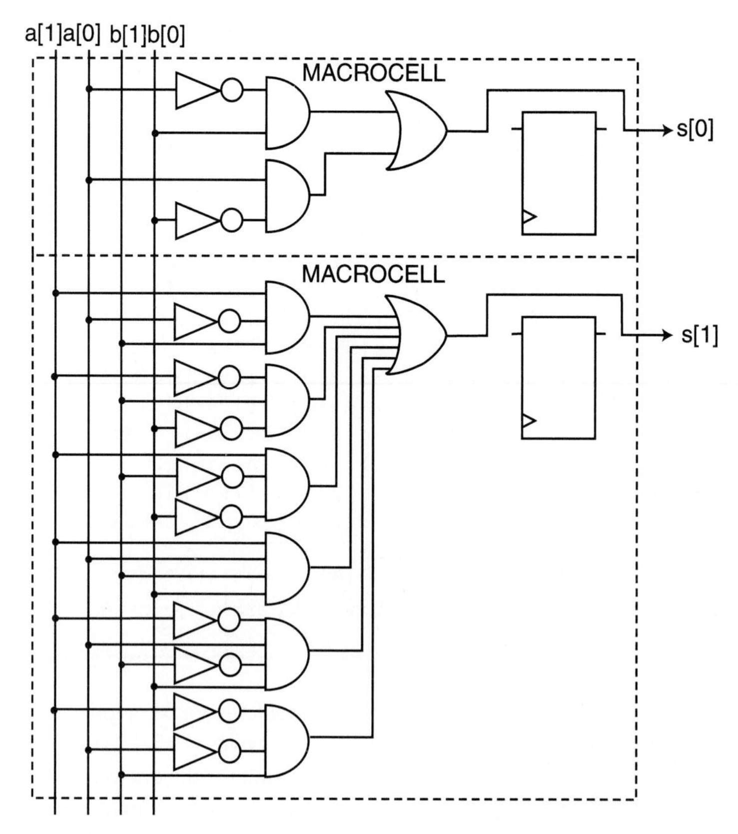

Figure 11-4. Macrocells in the M4-128/64 for low-order two-bit slices of adder

```
module test;
  integer ia,ib,numerr;
  reg [3:0] a,b;  wire [3:0] sum;
  addpar a1(sum,a,b);
  initial
   begin
    numerr = 0;
    for (ia=0; ia<=15; ia=ia+1)
     for (ib=0; ib<=15; ib=ib+1)
      begin
       a=ia; b=ib;
       #1 $display("%b %b %b",a,b,sum);
       if ((ia+ib)%16 !== sum)
        begin
         $display("error");numerr=numerr+1;
```

Continued.

```
          end
        end
      $display("numerr=",numerr);
    end
  endmodule
```

The original behavioral adder, the preplacement netlist and the post-placement netlist all produce the correct results for the 256 combinations of inputs. If the width of the inputs were much larger, such an exhaustive test would be impossible.

As in the last example, we are ignoring the back annotated delay by supplying delayless modules for `portin`, `portout` and `mbuf`. If the backannotation capability were used, the #1 would have to be changed to an appropriate delay longer than the longest propagation delay of the synthesized design.

11.4.2 Alternate coding with `case`

An alternate way to describe this adder is to use behavioral statements that express the mathematics behind the ripple carry approach:

```
module addpar(s,a,b);
  output [3:0] s;
  input [3:0] a,b;

  reg [3:0] s;
  wire [3:0] a,b;
  reg [3:0] c;


 function car;
   input a,b,c;
   begin
     case ({a,b,c})
       3'b000: car = 0;
       3'b001: car = 0;
       3'b010: car = 0;
       3'b011: car = 1;
       3'b100: car = 0;
       3'b101: car = 1;
       3'b110: car = 1;
       3'b111: car = 1;
     endcase
```

Continued

```
    end
  endfunction

  function sum;
    input a,b,c;
    begin
      case ({a,b,c})
        3'b000: sum = 0;
        3'b001: sum = 1;
        3'b010: sum = 1;
        3'b011: sum = 0;
        3'b100: sum = 1;
        3'b101: sum = 0;
        3'b110: sum = 0;
        3'b111: sum = 1;
      endcase
    end
  endfunction

  always @(a or b)
    begin
      c[0] = 0;
      s[0] = sum(a[0],b[0],c[0]);
      c[1] = car(a[0],b[0],c[0]);
      s[1] = sum(a[1],b[1],c[1]);
      c[2] = car(a[1],b[1],c[1]);
      s[2] = sum(a[2],b[2],c[2]);
      c[3] = car(a[2],b[2],c[2]);
      s[3] = sum(a[3],b[3],c[3]);
    end
endmodule
```

Here `car` is a function that models the carry required for the next higher bit position when adding three bits, and `sum` is the corresponding result in the current bit position. These functions may be coded several ways.

The `case` statement approach used above is a direct expression of the truth table for a full-adder. For synthesis, we do not consider `'bx` and `'bz` values in the cases the way that might be necessary for simulation. This is because the synthesis tool implements the `case` statement using == rather than ===, which is all that is physically possible in hardware:

```
if      ({a,b,c}==3'b000) car = 0;
else if ({a,b,c}==3'b001) car = 0;
else if ({a,b,c}==3'b010) car = 0;
else if ({a,b,c}==3'b011) car = 1;
else if ({a,b,c}==3'b100) car = 0;
else if ({a,b,c}==3'b101) car = 1;
else if ({a,b,c}==3'b110) car = 1;
else if ({a,b,c}==3'b111) car = 1;
```

Either the `case` or the `if` statement approach is acceptable because for the three bits of input, all 2^3 possible cases are listed. Such a situation is known as a *full case*. The problem is that without a `default` clause in the `case` or an equivalent `else` at the end of the nested `if`s, the synthesis tool will not synthesize combinational logic properly except for a full case. For example, the following `case`, which only lists the ones of the function, is not full:

```
case ({a,b,c})
  3'b001: sum = 1;
  3'b010: sum = 1;
  3'b100: sum = 1;
  3'b111: sum = 1;
endcase
```

A `case` like this that is not full will synthesize to what is known as a *latch,* which is an asynchronous sequential circuit, rather than the desired combinational logic. To make the `case` full requires using `default:sum=0;` in the above or using a `full case` synthesis directive. A *synthesis directive* is a comment which would be ignored by a simulator, but which causes the synthesis tool to alter its operation. Use of synthesis directives such as `full case` is common, but is dangerous because it may cause synthesis to disagree with simulation. It is better to make the `case` statement be full by supplying the appropriate `default` since that acts the same in both synthesis and simulation. Another common but dangerous directive is `parallel case`, which changes how synthesis interprets the `case` to be like `if`s without `else`s.

An alternative approach to the `case` statement would have been to use logic equations inside the functions, such as `sum=a^b^c`.

In any event, the combinational logic is defined using an `always` block having the same sensitivity list as the example in the last section that invokes the functions. Another way this could have been defined is with eight separate continuous assignment statements rather than the one `always` block:

```
      assign c[0] = 0;
      assign s[0] = sum(a[0],b[0],c[0]);
      assign c[1] = car(a[0],b[0],c[0]);
      assign s[1] = sum(a[1],b[1],c[1]);
      assign c[2] = car(a[1],b[1],c[1]);
      assign s[2] = sum(a[2],b[2],c[2]);
      assign c[3] = car(a[2],b[2],c[2]);
      assign s[3] = sum(a[3],b[3],c[3]);
```

Regardless of which of these variations we choose, the result is isomorphic to the original. This is because the `wire [3:0] c` is internal, and the synthesis tool can optimize it away, just as it did when synthesizing directly from `a+b`. The only distinction is the name chosen for `c[3]`, but otherwise the result is an identical netlist. In this example, all the extra coding of the `sum` and `car` functions did not change the actual structure of the synthesized circuit. The details of the synthesized logic equations were dictated more by the capabilities of the CPLD exploited by the place and route tool than by anything that the designer codes. The main responsibility of the designer is to write correct Verilog. Usually, the designer should choose the modeling style which is easiest to understand (`a+b` in this example) and trust the synthesis tool to choose the logic equations that fit into the target device.

11.5 Synthesizing an implicit style bit serial adder

Rather than worrying about gate-level details, the designer should consider algorithmic alternatives. Although addition of two binary numbers usually is implemented as combinational logic, there are other approaches. The dependent sequence of calls to the `sum` and `car` functions inside the module `addpar` of section 11.4.2 makes it clear that the conventional ripple carry adder is the combinational logic required to implement one of the algorithmic variations explained in chapter 6: the single-cycle approach. Assuming we have a single register to load the sum at the next rising edge of the clock, the ripple carry adder computes in a single clock cycle all the information needed to form the next sum. In earlier chapters, we have assumed such a building block whenever we need to add. This approach for the module `addpar` is sometimes known as a *bit parallel* adder because all of the bits of the sum are available in parallel by the end of a single clock cycle.

Chapter 6 also describes other algorithmic alternatives besides the single-cycle approach. One of these is the multi-cycle approach. In the multi-cycle approach, each step in the dependent sequence is scheduled to occur in a different clock cycle. It is often possible to clock a multi-cycle machine faster than is possible with the single-cycle approach because less computation occurs per clock cycle. The multi-cycle approach takes several of these faster clock cycles to achieve the same result that the single-cycle approach achieves in one slower clock cycle.

11.5.1 First attempt at a bit serial adder

In order for the designer to have the freedom to explore such algorithmic variations easily, the synthesis tool should support the implicit style of Verilog. (The preprocessor described in chapter 7 and appendix F is available for those synthesis tools that do not support the implicit style.) With the implicit style, whether something occurs in parallel during a single cycle or in series during multiple cycles is simply a question of how many `@(posedge sysclk)`s the designer uses. In the case of the bit parallel addition algorithm given in section 11.4.2, it is a trivial matter with implicit style Verilog to change it to what is called a *bit serial* addition algorithm, which produces only one bit of the sum at a time. To do this, the `assign`s become non-blocking assignments, and the designer inserts `@(posedge sysclk)` at appropriate places inside the implicit `always` block:

```
`define CLK @(posedge sysclk)
`define ENS #1

 always
  begin
     ...
   @(posedge sysclk) `ENS;
     c[0] <= `CLK 0;
   @(posedge sysclk) `ENS;
     s[0] <= `CLK sum(a[0],b[0],c[0]);
     c[1] <= `CLK car(a[0],b[0],c[0]);
   @(posedge sysclk) `ENS;
     s[1] <= `CLK sum(a[1],b[1],c[1]);
     c[2] <= `CLK car(a[1],b[1],c[1]);
   @(posedge sysclk) `ENS;
     s[2] <= `CLK sum(a[2],b[2],c[2]);
     c[3] <= `CLK car(a[2],b[2],c[2]);
   @(posedge sysclk) `ENS;
     s[3] <= `CLK sum(a[3],b[3],c[3]);
     ...
  end
```

11.5.2 Macros needed for implicit style synthesis

In order for the implicit style to be practical, the result of simulation of implicit style Verilog before synthesis must agree with the result of simulation after synthesis (and, of course, the behavior of the physical hardware). Some synthesis tools are restricted as to the use of time control, but as discussed in section 3.8.2.1, simulators need # time control to simulate non-blocking assignment properly inside implicit style blocks. Therefore, in order that simulation agree with synthesis, it is recommended that all of the time control required for simulation be coded as macros. Only the time control needed by the synthesis tool (the `@(posedge sysclk)` that denotes a state boundary outside a non-blocking assignment) is written without a macro. The other two forms of time control [the #1 and the `@(posedge sysclk)` inside the non-blocking assignment] are written using macros (`` `ENS `` and `` `CLK ``). This way, they can simulate properly when the macros are defined as shown above, but they can be synthesized properly when the macros are defined as empty.

11.5.3 Using a shift register approach

A disadvantage of the code in section 11.5.1 is that it performs similar computations on different bits of the data. The synthesis tool will either have to duplicate the hardware to implement the `sum` and `car` functions multiple times, or use muxes to allow *resource sharing*, in a way analogous to the central ALU approach. To avoid this problem, we can use a shift register approach:

```
reg c;
reg [3:0] r1,r2;
  ...
@(posedge sysclk) `ENS;
 r2 <= `CLK y; r1 <= `CLK x;
 c <= `CLK 0;
@(posedge sysclk) `ENS;
 r1 <= `CLK {sum(r1[0],r2[0],c),r1[3:1]};
 c  <= `CLK car(r1[0],r2[0],c);
 r2 <= `CLK r2 >> 1;
@(posedge sysclk) `ENS;
 r1 <= `CLK {sum(r1[0],r2[0],c),r1[3:1]};
 c  <= `CLK car(r1[0],r2[0],c);
 r2 <= `CLK r2 >> 1;
@(posedge sysclk) `ENS;
 r1 <= `CLK {sum(r1[0],r2[0],c),r1[3:1]};
 c  <= `CLK car(r1[0],r2[0],c);
 r2 <= `CLK r2 >> 1;
@(posedge sysclk) `ENS;
 r1 <= `CLK {sum(r1[0],r2[0],c),r1[3:1]};
 c  <= `CLK car(r1[0],r2[0],c);
 r2 <= `CLK r2 >> 1;
```

A synthesis tool can produce a more efficient netlist from the above than from the earlier example in section 11.5.1 because the same computations, `sum(r1[0], r2[0], c)` and `car(r1[0], r2[0], c)`, occur in each state. Resource sharing can now occur at no added cost. Also, in the above code there is no need to use a four-bit `wire` for `c` since a single-bit `c` can be reused in each state. A single-bit `c` variable suffices here because we are going to discard the carries in the end, anyway.

With this shift register approach, `r2` starts out with the value of `y`. During each clock cycle, `r2` is shifted over one position to the right. Therefore, during the first clock cycle, `r2[0]` is `y[0]`. During the second clock cycle, `r2[0]` is `y[1]`. During the third clock cycle, `r2[0]` is `y[2]`, etc. In other words, the least significant bit is processed first, and greater significant bits are processed later. This order is essential to the bit serial technique.

The role of `r1` is somewhat more complicated: `r1` starts out with the value of `x`, but `r1` is reused not just for holding the original bits of `x` but also for holding bits of the result. As a low-order bit of `x` shifts out of `r2`, the high-order bit of `r2` is scheduled to become the corresponding `sum(r1[0],r2[0],c)` bit. Although in the beginning, such bits are to the left of where they need to be, by the completion of the process, the result bits will have been shifted over to the proper position.

11.5.4 Using a loop

There is still room to improve the code given in section 11.5.3 because it takes many states (and therefore many flip flops in a one hot controller) to produce the answer. Although the bit serial approach necessarily takes a number of clock cycles proportionate to the number of bits in the word, the size of the controller should not also have to be proportional to the word size.

Here is where the flexibility of the implicit style is useful. With the implicit style, the designer can roll up the related computations that occur in separate states into one state inside a `while` loop. (Rolling up identical computations into a loop is the opposite of the loop unrolling explained in section 10.9.9.4 used by some optimizing compilers for RISC machines.) Although rolling all of these states into a single-loop state does not increase the speed of the machine, it usually will reduce the number of gates required to implement the machine. There is however, an added complication. There needs to be a loop counter that determines how many times the machine should repeat the loop state.

In previous chapters, it would have been natural to use a binary counter. Instead, here we will use a shift register (`r3`) to count in a unary code because with this it will be easier to understand the resulting logic equations after synthesis:

```
 1: always
 2:  begin
 3:    ready <= `CLK 1;
 4:   @(posedge sysclk) `ENS; //ff_4
 5:    r2 <= `CLK y;
 6:    r3 <= `CLK 1;
 7:    c <= `CLK 0;
 8:    if (pb)
 9:     begin
10:       ready <= @(posedge sysclk) 0;
11:      @(posedge sysclk)`ENS; //ff_11
12:       r1 <= `CLK x;
13:        while (~r3[3])
14:         begin
15:          @(posedge sysclk) `ENS; //ff_15
16:           r1 <= `CLK{sum(r1[0],r2[0],c),r1[3:1]};
17:           c  <= `CLK car(r1[0],r2[0],c);
18:           r2 <= `CLK r2 >> 1;
19:           r3 <= `CLK r3 << 1;
20:         end
21:     end
22:  end
```

When the most significant bit of `r3` becomes one, the loop stops. In other words, `r3` contains the unary values 0001, 0010, 0100 and 1000 in successive clock cycles. The effect is similar to what would happen by counting 0, 1, 2 and 3. Since the computation only depends on the number of **times** the loop repeats, and not on the **value** of `r3`, the above unary code is just as reasonable as a binary code. A binary code might produce a somewhat smaller synthesized netlist, but the unary code will produce a synthesized circuit that typically runs faster and is easier to understand.

The above Verilog includes the friendly user interface described in sections 2.2.1 and 7.4.2. The signal `ready` is asserted when the machine is able to accept inputs. The user pulses `pb` for exactly one clock cycle to cause the machine to compute the sum, which will be available in `r1` when the machine exits from the `while` loop.

11.5.5 Test code

The implicit style block of section 11.5.4 together with the function definitions from section 11.4.2 can be placed inside the module to be synthesized:

```
module vsyadd1(pb,ready,x,y,r1,r2,reset,sysclk);
  input [3:0] x,y; input pb,reset,sysclk;
  output ready; output [3:0] r1,r2;
  wire reset,sysclk,pb;    reg ready,c;
  wire [3:0] x,y; reg [3:0] r1,r2,r3;
  ...
endmodule
```

Prior to synthesis, it is prudent to test whether the algorithm is correct. To do such a test, we use the following test code in a different file. The test code has an instance of the module to be synthesized. Unlike the test code for the combinational logic of section 11.4.1, there needs to be a `wait` statement (section 3.7.3) so that the test code can adapt to the speed of the bit serial addition.

```
module top;
 reg [3:0] x,y;  wire [3:0] sum,r2;
 wire ready,sysclk; reg reset,pb;
 integer numerr; time t1,t2;
 cl #52000 clock(sysclk);
 vsyadd1 slow_add_machine(pb,ready,
          x,y,sum,r2,reset,sysclk);
 initial
  begin
   numerr = 0; pb= 0; x = 0; y = 0; reset = 1;
   #30 reset = 0;  #10 reset = 1;
   #210; @(posedge sysclk);
   for (x=0; x<=7; x = x+1)
    for (y=0; y<=7; y = y+1)
     begin
      @(posedge sysclk) pb = 1;
      @(posedge sysclk) pb = 0; t1 = $time;
      @(posedge sysclk) wait(ready); t2 = $time;
      @(posedge sysclk);
      if (x + y === sum)
       $display("ok %d",t2-t1);
      else
       begin
        $display("error x=%d y=%d x+y=%d sum=%b",
         x,y,x+y,sum);  numerr = numerr + 1;
       end
     end
   $display("number of errors=",numerr);
   $finish;
  end
```

Continued

```
  always @(posedge sysclk) #20
   $display("%d r1=%d r2=%d pb=%b ready=%b",
    $time, sum,r2, pb, ready);
endmodule
```

The active low `reset` signal is necessary for the VITO preprocessor described in chapter 7 and appendix F. The test code detects no errors, so it is reasonable to synthesize `vsyadd1`.

11.5.6 Synthesizing

Since PLSynthesizer does not support the implicit style, the first step in synthesizing `vsyadd1` is to use the VITO preprocessor.[13] VITO passes through the module definitions and functions unchanged, which allows use of these names in the code generated by VITO. VITO generates a one hot controller using continuous assignment and one bit `regs` according to the principles described in chapter 7. VITO uses the line number in the names of the `wires` and `regs` generated. In this particular machine, the states correspond to `ff_4`, `ff_11` and `ff_15`. When the code generated by VITO is run through PLSynthesizer, logic equations are formed that describe the inputs to these macrocell flip flops. PLSynthesizer and PLDesigner will eliminate most of the redundant `wire` names created by VITO. The following is the manual translation of the `.doc` file into Verilog for the logic equations of the one hot controller:

```
always @(posedge sysclk or negedge reset)
 begin
  if (~reset)
   {ff_999,ff_4,ff_11,ff_15} = 0;
  else
   begin
    ff_999 <= #0 1;
    ff_4 <= #0(~ff_999|(~pb&ff_4)
      |(r3[3]&ff_15)|(r3[3]&ff_11));
    ff_11 <= #0 (pb&ff_4);
    ff_15 <= #0((~r3[3]&ff_11)|(~r3[3]&ff_15));
   end
 end
```

[13] The preprocessor is not necessary with synthesis tools, such as Synopsys, that support the implicit style.

If ff_999, ff_4, ff_11 and ff_15 are not listed in the portlist of VITO's output, PLSynthesizer will choose cryptic names for them. The above logic equations describe the conditions under which state transitions occur to the particular states. For example, a transition to state ff_11 only occurs when the machine is in state ff_4 and pb is true. There are several ways in which a transition to state ff_4 occurs: when the machine is powered up, when the machine loops back to state ff_4 because pb is false, or when the most significant bit of r3 is one and the machine is in either state ff_15 or state ff_11. The machine makes a transition to state ff_15 when the most significant bit of r3 is zero and the machine is in either state ff_15 or state ff_11. Transitioning to state ff_15 from state ff_11 corresponds to entering the while loop for the first time. Transitioning to state ff_15 from state ff_15 corresponds to remaining in the while loop for an additional cycle.

In addition to the one hot controller, the VITO preprocessor also generates an architecture composed of combinational logic, muxes and simple D-type registers in the style described in section 7.2.2.1. For example, here is what VITO generates corresponding to r2:

```
assign new_r2 = s_5 ? y : s_18 ? r2>>1 : r2;
always @(posedge sysclk) r2 = new_r2;
```

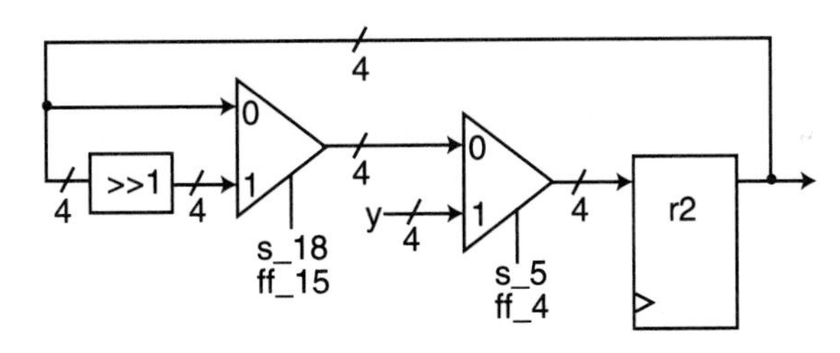

Figure 11-5. Block diagram for r2 portion of architecture.

The above Verilog is equivalent to figure 11-5, which is a kind of specialized shift register that is loadable (in state ff_4 which includes statement s_5) and only shifts right (in state ff_15 which includes statement s_18). Again, logic equations are given in the .doc file that describe the inputs to each macrocell flip flop. The following is the manually translated Verilog for the logic equations that correspond to r2:

```
always @(posedge sysclk)
 begin
  r2[0] <= #0 ((r2[0]&~ff_15&~ff_4)
    |(y[0]&ff_4)|(r2[1]&ff_15&~ff_4));
  r2[1] <= #0 ((r2[1]&~ff_15&~ff_4)
    |(y[1]&ff_4)|(r2[2]&ff_15&~ff_4));
  r2[2] <= #0 ((r2[2]&~ff_15&~ff_4)
    |(y[2]&ff_4)|(r2[3]&ff_15&~ff_4));
  r2[3] <= #0 ((r2[3]&~ff_15&~ff_4)|(y[3]&ff_4));
 end
```

Although it might have appeared from figure 11-5 that there would be two macrocells of delay (for each mux), the synthesis tool merged the logic equations of the two muxes together into a single macrocell per bit slice. Except for `r2[3]`, each bit slice is similar to the others. For example, there are three cases to consider for `r2[0]`. First, when `ff_4` is active, two terms of the logic equation, `r2[0]&~ff_15&`**`~ff_4`** and `r2[1]&ff_15&`**`~ff_4`**, are guaranteed to be zero. This leaves only `y[0]&`**`ff_4`**, which passes through the proper bit of `y` into the input of the `r2[0]` flip flop. Second, when `ff_15` is active, we know (because of the nature of one hot controllers) that `ff_4` could not be active, but the synthesis tool did not know this. Therefore, the tool generates `r2[1]&ff_15&~ff_4`. The `~ff_4` is not necessary considering the total one hot system but is necessary to achieve the mux behavior shown in figure 11-5. Because the other two terms of the logic equation, `r2[0]&`**`~ff_15`**`&~ff_4` and `y[0]&`**`ff_4`**, are guaranteed to be zero in this case (`ff_15` active and `ff_4` inactive), the remaining term, `r2[1]&`**`ff_15`**`&~ff_4`, passes through the right-shifted bit (`r2[1]`) into the input of the `r2[0]` flip flop. Third, the last possibility is that neither `ff_4` nor `ff_15` is active. In this case, `r2[0]&`**`~ff_15&~ff_4`** holds the former value of the `r2[0]` flip flop.

Of course, as mentioned earlier, it is tedious to have to manually translate the non-standard `.doc` file back into Verilog. The designer would probably prefer to use the structural Verilog automatically generated by PLDesigner. The instance and `wire` names shown in figure 11-6 and in the following may vary slightly, depending on tool- specific details:

```
mbuf B5(tmp85,pin_13);and A17(tmp90,tmp91,tmp92,pin_12);
and A16(tmp87,ff_15,tmp88,pin_46);not I19(tmp88,ff_4);
not I20(tmp91,ff_15);and A18(tmp93,ff_4,pin_25);
not I21(tmp92,ff_4);or O5(tmp86,tmp87,tmp90,tmp93);
dffarap DFF5(pin_12,tmp85,tmp86,GND,GND);
mbuf B6(tmp94,pin_13);and A20(tmp98,tmp99,tmp100,pin_46);
and A19(tmp96,ff_15,tmp97,pin_44);not I22(tmp97,ff_4);
not I24(tmp100,ff_4);and A21(tmp101,ff_4,pin_23);
not I23(tmp99,ff_15);or O6(tmp95,tmp96,tmp98,tmp101);
dffarap DFF6(pin_46,tmp94,tmp95,GND,GND);
...
```

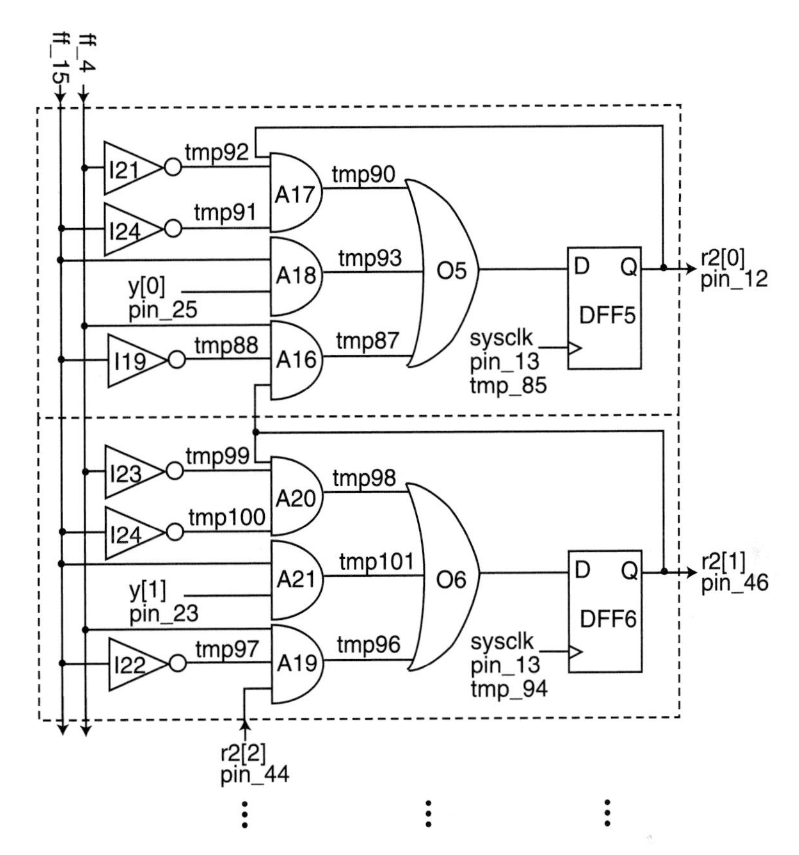

Figure 11-6. Macrocells implementing low-order bit slices of `r2` *.*

11.6 Switch debouncing and single pulsing

To have a useful machine, it is often necessary for the machine to receive information from a person. In many situations, the most economical way to design such a machine is to use mechanical switches and buttons. In previous chapters, we have assumed the existence of ideal switches and buttons. For example, in section 2.2.1, we assumed a push button would assert its output for exactly one clock cycle when it is pushed. The problem is that mechanical switches and buttons are not perfect. They are neither synchronous nor is their output reasonable for use with a machine being clocked millions of times per second. This is because mechanical switches exhibit an undesirable property, known as *bounce*.

Ideally, when a person flips a switch on, we would hope that the output of the switch would become and remain one until the person flips the switch off. Unfortunately, real switches do not behave this way, as is illustrated by the following timing diagram:

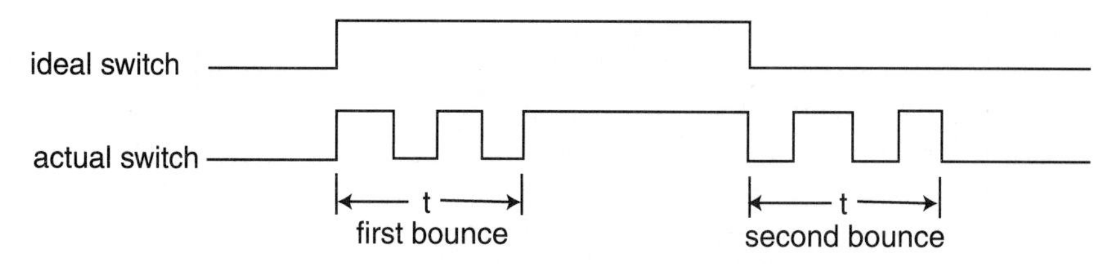

Figure 11-7. Ideal versus actual switch behavior shows need for debouncing.

Happily, real switches bounce for less than a constant time `t` seconds. For example, even the very awkward DIP switches[14] soldered onto the M4-128/64 demoboard bounce for less than a quarter of a second.

One solution to the bounce problem is to design a debounce machine[15] whose input is the actual switch, and whose output is the idealized `pb` signal needed by many of the designs in this book. Most of the time, the actual switch is quiet; thus the debounce machine continually reassigns 0 to `pb`. The debounce machine does something different when the actual switch makes its first transition to a one. During this first `t` second period when bounce occurs, we assume that the output of the actual switch will eventually stabilize to 1. Therefore, the number of clock cycles when the actual switch could be zero during this first bounce period is less than `t` times the clock frequency. After the first bounce period but before the second bounce period, the actual switch continually reads as a one. A second bounce period occurs when the switch is released.

The total number of clock cycles during which the actual switch reads as a zero from the time of the first transition to one until the final transition to zero is less than twice `t` times the clock frequency. The designer precomputes this constant, which will be loaded into a counter when the machine first detects that the actual switch has become a one. For example, with the M4-128/64, two times 0.25 seconds times 1.8432 MHz is approximately one million. Since 0.25 is an overestimation of `t`, the exact number of clock cycles is not too important, as long it is near one million. A convenient number around this size is 2^{20}-1.

[14] People often use pencils to move these tiny switches, which aggravates the bounce problem. The constant `t` tends to be smaller for switches that are easier for people to manipulate, but the underlying cause of bounce is always electrical.

[15] The design here assumes that a single-pole single-throw switch is used and that the debounce machine must be completely digital. Much more economical solutions are possible that either use a few analog components, such as a capacitor and a resistor, or that use a single-pole double-throw switch. In the case of the M4-128/64 demoboard, neither alternative is possible without external components.

In addition to debouncing the switch, we need to make sure that the pb output of the debounce machine lasts for exactly one clock cycle. Otherwise, it would be as though the user is making millions of requests for computation, when in fact the user makes just one request. The following implicit style module solves both the debouncing and single pulsing aspects of this problem:

```
module debounce(sw3,pb,cnt,sysclk,reset);
  input sw3,sysclk,reset;
  output pb;
  output [19:0] cnt;
  wire sw3,sysclk,reset;
  reg pb;
  reg [19:0] cnt;
  always
   begin
    @(posedge sysclk) `ENS;
    pb <= `CLK 0;
    if (sw3 == 1)
     cnt <= `CLK 20'hfffff;
    else
     while (cnt[19:1] != 0)
      begin
       @(posedge sysclk) `ENS;
        if (sw3 == 0)
         cnt <= `CLK cnt - 1;
        if (cnt[19:1] == 0)
         pb <= `CLK 1;
      end
   end
endmodule
```

Assuming cnt is zero and the actual switch, sw3, is zero, the machine leaves cnt alone and therefore does not enter the while loop. The first time the machine detects that sw3 is one, the machine assigns the constant to cnt. Eventually, sw3 becomes zero again during the first bounce period. Since cnt now contains the constant, the machine enters the while loop. Inside the while loop, cnt is decremented only when sw3 is zero. The while loop exits when all bits of cnt other than the least significant are zero (i.e., cnt is 1). During this last clock cycle in the while loop, cnt might or might not be decremented one last time (hence the reason for ignoring the least signifcant bit). In that same clock cycle, pb is scheduled to become one for a single cycle. (pb will be scheduled to return to zero in the next clock cycle when the machine returns to the top state.) Therefore, the above code allows us to use the rather primitive DIP switch, sw3, as an ideal push button, pb.

11.7 Explicit style switch debouncer

As explained in chapter 4, a pure behavioral design can be broken down manually into an architecture and a controller. For example, the controller can be written in the explicit style, where each state transition must be given explictly, using a `case` statement:

```
`define TOP 0
`define BOT 1
module deboun_control(sw3,cnteq0_1,clrpb,
          ldpb,ldcnt,deccnt,sysclk,reset);
  input sw3,cnteq0_1,sysclk,reset;
  output clrpb,ldpb,ldcnt,deccnt;
  wire sw3,cnteq0_1,sysclk,reset;
  reg clrpb,ldpb,ldcnt,deccnt,ps,ns;
  always @(posedge sysclk or negedge reset)
    begin
      if (~reset)
        ps = 0;
      else
        ps = ns;
    end
  always @(ps or sw3 or cnteq0_1)
    begin
      clrpb = 0;ldcnt = 0;deccnt = 0;ldpb = 0;
      case (ps)
      `TOP: begin
              clrpb = 1;
              if (sw3 == 1)
                begin
                  ldcnt = 1;
                  ns = `TOP;
                end
              else
                if (cnteq0_1)
                  ns = `TOP;
                else
                  ns = `BOT;
            end
      `BOT: begin
              if (sw3 == 0)
                 deccnt = 1;
              if (cnteq0_1)
                begin
                  ldpb = 1;
                  ns = `TOP;
```

Continued

```
                    end
                  else
                    ns = `BOT;
                end
        endcase
      end
    endmodule
```

The above code corresponds to what was called the pure structural stage in chapter 4, but for brevity, the above uses only behavioral statements. (The present state register and next state logic are not given in separate modules as was done in chapter 4.) Although similar in operation to the implicit style design given in section 11.6, the explicit style design is much more tedious to understand. Also, the designer must give a Verilog architecture (not shown) consisting of a counter (controlled by `ldcnt` and `deccnt`) and an enabled register (controlled by `ldpb` and `clrpb`). Finally, the designer must instantiate the controller and architecture to make a module that is identical to section 11.6:

```
module debounce(sw3,pb,cnt,sysclk,reset);
  input sw3,sysclk,reset;
  output pb;
  output [19:0] cnt;
  wire sw3,sysclk,reset;
  wire [19:0] cnt;
  wire pb,cnteq0_1,clrpb,ldpb,ldcnt,deccnt;
  deboun_arch architec(pb,cnteq0_1,cnt,
    clrpb,ldpb,ldcnt,deccnt,sysclk);
  deboun_control controller(sw3,cnteq0_1,
    clrpb,ldpb,ldcnt,deccnt,sysclk,reset);
endmodule
```

In this case, the binary encoding makes only a slight savings in macrocells (3%) compared to the one hot encoding used by VITO. As in many other designs, the majority of the macrocells are devoted to the architecture. Those macrocells must be present, regardless of whether the original Verilog was implicit or explicit style. All of the extra manual coding required for the explicit style was not worth the effort.

11.8 Putting it all together: structural synthesis

A typical design often uses a combination of the above techniques. For example, consider a machine activated by the debounced sw3 DIP switch that takes a three bit binary number from the other DIP switches ({sw2,sw1,sw0}) and does bit serial addition of this to a four-bit accumulator, r1, whose output is displayed in hexadecimal on the LEDs a1 .. g1. In order to reuse the code given above, the designer needs structural instances of vsyadd1 and debounce:

```
module mach445(sw3,sw2,sw1,sw0,sysclk,reset,
            a1,b1,c1,d1,e1,f1,g1);
  input sw3,sw2,sw1,sw0,sysclk,reset;
  output a1,b1,c1,d1,e1,f1,g1;
  wire sw3,sw2,sw1,sw0,sysclk,reset;
  reg a1,b1,c1,d1,e1,f1,g1;
  function [7:0] seven_seg;
    input [3:0] i;
    ...
  endfunction
  wire pb,ready;
  wire [3:0] r1,r2;
  reg [3:0] y;
  wire [19:0] cnt;
  vsyadd1 v1(pb,ready,r1,y,r1,r2,reset,sysclk);
  debounce deb1(sw3,pb,cnt,sysclk,reset);
  always @(sw2 or sw1 or sw0)
    y = {sw2,sw1,sw0};
  always @(r1)
    {a1,b1,c1,d1,e1,f1,g1} = ~seven_seg(r1);
endmodule
```

The vsyadd1 and debounce module definitions are given in the same file as the above module. In the above, y is simply another name for {sw2,sw1,sw0}. Note that r1 connects both to the v1.r1 output as well as the v1.x input for the instance of vsyadd1. In other words, r1 plus y will eventually replace the old value of r1.

The function seven_seg (whose case statement definition is not shown) takes a four-bit binary input, i, and outputs the seven bits required to drive one LED digit in hexadecimal. This combinational logic output is complemented to accommodate the active low requirements of the LEDs.

The .pi file must be defined using the pin numbers given in section 11.3.6. When synthesized and downloaded to the M4-128/64 demoboard, the above design will operate properly.

11.9 A bit serial PDP-8

All the designs in chapters 8 through 10 use bit parallel arithmetic to illustrate concepts about general-purpose computers. In contrast, many early general-purpose computers, including the Manchester Mark I, used bit serial arithmetic because it required less hardware. Most modern general-purpose computers are designed with bit parallel arithmetic because it is faster and easier. As a concluding synthesis example, however, let us build a bit serial PDP-8. This allows the CPU to fit within one M4-128/64 chip, and it simplifies the connections to an external memory chip, which must be wired manually to the demoboard.

The PDP-8 subset chosen for this example is the same as section 9.6 (CLA, TAD, DCA, HLT, JMP, SPA, SMA and CIA), with the addition of the SNA, SZA, CMA and IAC instructions described in appendix B. The link as well as additional instructions are not implemented in this example. This subset is sufficient for the childish division program given in section 9.7. Bit serial arithmetic is necessarily a multi-cycle approach, and so the multi-cycle PDP-8 ASM of section 8.3.1.3 is a good starting point for the design, but there are several algorithmic variations required for the CPU to fit into the M4-128/64.

First, bit serial addition loops are used for incrementing `pc` (the user interface and states F3A and E1ASKIP), incrementing `ac` (state E0IAC) and adding to `ac` (state E0TAD). Second, bit parallel comparisons, such as `ir==12'o7200` for CLA, need to be replaced with comparisons of only the appropriate bits, such as `ir[11:9]==7 & ir[8]==0 & ir[7]==1` for CLA. Third, like the original PDP-8 (but unlike chapters 8 and 9), combined instructions (e.g., CMA and IAC to form CIA) are allowed at no extra cost because the bits of the instruction register are tested individually. Fourth, memory accesses occur one bit at a time with a one-bit-wide `mb` register wired to the data in pin of the memory chip and a one-bit-wide `membus` wired to the data out pin of the memory chip. Fifth, like section 8.3.2.4 and figure 8-11, memory must be a separate actor so that it can be physically wired to the M4-128/64. Sixth, the `write` signal is active low for the memory chip used here, which is the opposite of figure 8-11. Seventh, since the number of bits in a memory chip is a power of two but the number of bits in the PDP-8's memory is a multiple of twelve, the simplest approach is to disregard four out of every sixteen bits from the one-bit-wide memory chip. In other words, `bitmem[0]` through `bitmem[11]` form the twelve-bit `m[0]`, and `bitmem[16]` through `bitmem[27]` form `m[1]`. Eighth, in addition to the memory address register, `ma`, the bit serial approach needs a bit address register, `ba`, which provides the low-order four bits of the address going to the memory chip. At any time, the bit from the memory chip currently being processed by the CPU is `bitmem[{ma,ba}]`. Ninth, `ba` also serves as a binary counter for bit serial arithmetic loops, rather than the unary `r3` counter described in section 11.5.4. Tenth, because this subset only implements the direct page zero addressing mode (and not the full set of addressing modes described in appendix B), the memory address register only needs to be seven bits wide (a reduction which saves several macrocells). Eleventh, the user interface of chapter 8 (`but_DEP`, `but_PC`, `but_MA`, `cont` and the twelve-bit switch register) has been replaced with a simpler but workable scheme using four undebounced switches and a push button, `cont`, that must be externally debounced. Twelfth, `swin`, which is the concatenation of the four switches, determines the user interface action taken when `cont` is pressed:

```
swin action
0000 ba ← 0
001- ba ← 0; pc ← {swin[0],pc[11:1]}
010- bitmem[{pc,ba}] ← swin[0]; Advance {pc,ba}
011- Advance {pc,ba}
1000 Execute
```

where advancing `{pc,ba}` means incrementing just `ba`, except in the case when `ba==4'b1011`. In that special case, `pc` is incremented and `ba` becomes zero.

11.9.1 Verilog for the bit serial CPU

In the following implicit style Verilog, the comments indicate names of states similar (but not identical) to those of figure 8-11. Many of the states, especially those for the user interface, have no direct correspondence to figure 8-11:

```
always
 begin
  @(posedge sysclk) `ENS; //INIT
   halt <= `CLK 1; write <= `CLK 1;
   forever
    begin
     @(posedge sysclk) `ENS; //F1
      ma <= `CLK pc; ba <= `CLK 0;
      c <= `CLK 1;
      if (halt)
       begin
        while (~(cont&swin[3]))
         begin
          @(posedge sysclk) `ENS; //IDLE
           halt <= `CLK 0; ma <= `CLK pc;
           mb <= `CLK swin[0]; c <= `CLK 1;
           if      (cont&(swin[3:2] == 2'b00))
            begin
             @(posedge sysclk) `ENS;
```

Continued

```
          ba <= 'CLK 0;
          if (swin[1])
           begin
            @(posedge sysclk) 'ENS;
              pc <= 'CLK {swin[0],pc[11:1]};
           end
       end
      else if (cont&(swin[3:2] == 2'b01))
       begin
        @(posedge sysclk) 'ENS;
          write <= 'CLK swin[1];
        @(posedge sysclk) 'ENS;
          write <= 'CLK 1;
        @(posedge sysclk) 'ENS;
          ba <= 'CLK ba + 1;
          if (ba == 11)
           begin
            @(posedge sysclk) 'ENS;
             ba <= 'CLK 0;
             while (ba != 11)
              begin
               @(posedge sysclk) 'ENS;
                pc <= 'CLK
                 {sum(pc[0],0,c),pc[11:1]};
                c <= 'CLK car(pc[0],0,c);
                ba <= 'CLK ba + 1;
              end
            @(posedge sysclk) 'ENS;
              ba <= 'CLK 0;
           end
       end
    end
  end
 else
  begin
   @(posedge sysclk) 'ENS; //F2
    while (ba != 11)
     begin
      @(posedge sysclk) 'ENS; //F3A
       ir <= 'CLK {membus,ir[11:1]};
       pc <= 'CLK {sum(pc[0],0,c),pc[11:1]};
       c <= 'CLK car(pc[0],0,c);
       ba <= 'CLK ba + 1;
     end
```

Continued

```
        @(posedge sysclk) `ENS; //F3B
         ma <= `CLK ea(ir); ba <= `CLK 0;
         mb <= `CLK ac[0]; c <= `CLK ir[11];
         if      (ir[11:9] == 1)
          begin
           while (ba != 11)
            begin
             @(posedge sysclk) `ENS; //E0TAD
              ac <= `CLK {sum(ac[0],membus,c),ac[11:1]};
              c <= `CLK car(ac[0],membus,c);
              ba <= `CLK ba + 1;
            end
          end
         else if (ir[11:9] == 3)
          begin
           while (ba != 11)
            begin
             @(posedge sysclk) `ENS;//E0DCA
              ac <= `CLK {1'b0,ac[11:1]};
              write <= `CLK 0;
             @(posedge sysclk) `ENS;//E1ADCA
              write <= `CLK 1;
             @(posedge sysclk) `ENS;//E1BDCA
              ba <= `CLK ba + 1; mb <=`CLK ac[0];
            end
          end
         else if (ir[11:9] == 5)
          begin
           @(posedge sysclk) `ENS; //E0JMP
            pc <= `CLK ma;
          end
         else if (ir[11:9] == 7)
          begin
           if (ir[8])
            begin
             if (ir[1])
              begin
               @(posedge sysclk) `ENS; //E0HLT
                halt <= `CLK 1;
              end
             if (ir[3]^(ir[6]&ac[11]|ir[5]&(ac==0)))
              begin //SPA,SZA,SMA,SNA
               while (ba != 11)
                begin
```

Continued

```
                  @(posedge sysclk) `ENS; //E0ASKIP
                   pc <= `CLK {sum(pc[0],0,c),pc[11:1]};
                   c <= `CLK car(pc[0],0,c);
                   ba <= `CLK ba + 1;
                 end
               end
             end
            else
             begin
              if (ir[7])
               begin
                @(posedge sysclk) `ENS; //E0CLA
                 ac <= `CLK 0;
               end
              if (ir[5])
               begin
                @(posedge sysclk) `ENS; //E0CMA
                 ac <= `CLK ~ac;
               end
              if (ir[0])
               begin
                while (ba != 11)
                 begin
                  @(posedge sysclk) `ENS; //E0IAC
                   ac <= `CLK {sum(ac[0],0,c),ac[11:1]};
                   c <= `CLK car(ac[0],0,c);
                   ba <= `CLK ba + 1;
                 end
               end
             end
           end
        end
     end
end
```

11.9.2 Test code

For a design as complicated as this, it is important to simulate before synthesis. Even a tiny bug could prevent the fabricated hardware from operating at all and give no trace as to the cause. In order to simulate the above, we need a non-synthesizable model of the memory chip that will be connected to the fabricated CPU:

```
module mem(mabus,babus,mbbus,membus,write);
 input mabus,babus,mbbus,write;
 output membus;
 wire [11:0] mabus;
 wire [3:0] babus;
 wire mbbus, write;
 reg membus;
 reg [11:0] m[0:127];
 reg [11:0] temp;
 always @(mabus or babus)
  begin
   temp = m[mabus]; membus = temp[babus];
  end
 always @(negedge write)
  begin
   #50 membus = mbbus; temp = m[mabus];
   temp[babus] = membus; m[mabus] = temp;
  end
endmodule
```

The above models memory as twelve-bit words but interfaces to the CPU one bit at a time. An attempt to access one of the four unused bits will result in `1'bx` because of the way Verilog treats bit selects that are out of bounds. The above must be instantiated together with the CPU:

```
module  pdp8_system(swin,cont,halt,sysclk,reset);
 input swin,cont,sysclk,reset;
 output halt;
 wire cont,sysclk,reset,halt,mb, membus, write;
 wire [3:0] swin,ba;
 wire [11:0] ma;
 pdp8_cpu cpu(swin,write,membus,cont,
                 ba,ma,mb,halt,reset,sysclk);
 mem memory(ma,ba,mb,membus,write);
endmodule
```

Assuming `pdp8_system` is instantiated as `pdp8_machine`, the test code can initialize a memory location using a twelve-bit word refered to with hierarchical reference to the array `pdp8_machine.memory.m[ ... ]`. In order to simulate the pushing of `cont`, a task is helpful:

```
task push;
  input [3:0] sw;
  begin
    swin = sw; #200 cont = 1;
    #100 cont = 0; #300;
    case (swin[3:2])
      0:   #200;
      1:   #2000;
      2,3: #100 wait(halt);
    endcase
    #300;
  end
endtask
```

The time control in the task depends upon what `swin` selection was requested. For example, for the test code to set the program counter to `12'o0100` and then execute a program, the task waits 200 units of `$time` for each bit shifted into the program counter and then waits until the CPU halts:

```
push(4'b0010);push(4'b0010);push(4'b0010);//0
push(4'b0010);push(4'b0010);push(4'b0010);//0
push(4'b0011);push(4'b0010);push(4'b0010);//1
push(4'b0010);push(4'b0010);push(4'b0010);//0
push(4'b1000);//Execute until HLT
```

11.9.3 Our old friend: division

In running this simulation with the childish division program of section 9.7, we observe that this bit serial implementation takes 558 cycles when 1 is the quotient, 827 cycles when 2 is the quotient and 1096 cycles when 3 is the quotient. Let us put this in perspective with running the childish division software on the other PDP-8 implementations discussed earlier:

section	kind	arithmetic	clock cycles
11.9.1	multi	serial	289+269*quotient
8.3.2.1	multi	parallel	55+55*quotient
9.6	pipe	parallel	12+10*quotient

Assuming the same clock period, the bit serial approach is about five times slower than the multi-cycle bit parallel approach of chapter 8, which in turn is about five times slower than the pipelined bit parallel approach of chapter 9. To execute one instruction, it takes on average about one cycle for the pipelined bit parallel machine of section 9.6, five cycles for the multi-cycle bit parallel machine of section 8.3.2.1 and twenty-seven cycles for the multi-cycle bit serial machine of section 11.9.1. In the latter case, it takes twelve cycles to fetch the instruction, twelve cycles to fetch the data and three cycles for the other typical states (i.e., F1, F2 and F3B).

11.9.4 Synthesizing and fabricating the PDP-8

This design will occupy the majority of the macrocells in the M4-128/64. After synthesis with VITO and PLSynthesizer, it is necessary to let PLDesigner choose the pins where the signals are routed. If the designer provides complete `.pi` information at first, it is likely that PLDesigner would be unable to fit this design into a single M4-128/64. Instead, the designer should only constrain critical pins. This design does not make use of any of the hardware on the demoboard, other than `sysclk` and `reset`. The only other critical pins are 18, 54, 63 and 68, which should not be used since these are tied to the DIP switches. Instead, `swin` will come from external switches. Once the design does get placed in a single chip, the pins selected by PLDesigner should be put in a `.pi` file so that future minor modifications of the design will not require physical rewiring of the memory chip to the demoboard:

```
membus:10   sysclk:13   contin:19   ma[5]:22
ma[4]:23    ma[6]:24    swin[1]:37  swin[3]:38
mb:43       reset:4     ba[3]:56    ba[2]:58
ba[1]:59    ba[0]:60    ma[0]:62    swin[0]:70
swin[2]:73  halt:74     write:84    ma[3]:93
ma[1]:96    ma[2]:98
```

Each I/O pin of the M4-128/64 is attached to a pin on one of two headers (JP4 or JP5) soldered to the demoboard. A small, low-cost static memory chip that can be used is the 2102, which is arranged as 1 x 1024 bits. Using wirewrap wire and a sixteen-pin dual in-line wirewrap socket, the memory can be attached to the demoboard as follows:

PDP-8 signal	M4 header	2102 pin	PDP-8 signal	M4 header	2102 pin
ba[0]	JP5-27	1	GND	JP4-2	9
ba[1]	JP5-25	2	Vcc	solder	10
write	JP5-12	3	mb	JP5-1	11
ba[2]	JP5-23	4	membus	JP4-27	12
ba[3]	JP5-19	5	GND	JP4-2	13
ma[0]	JP5-31	6	ma[3]	JP4-1	14
ma[1]	JP4-7	7	ma[4]	JP4-26	15
ma[2]	JP4-11	8	ma[5]	JP4-28	16

It is desirable that the `ma` and `ba` signals also be attached to external LEDs to provide feedback to the user. (The onboard LEDs cannot be used because of the place and route limitations of the M4-128/64.) The five-volt power supply (`Vcc`) to the memory chip must be soldered on the demoboard power connection. In addition, the following external switches must be connected: `contin` (externally debounced) to `JP4-34`, `swin[1]` to `JP4-6`, `swin[3]` to `JP4-4`, `swin[0]` to `JP5-32` and `swin[2]` to `JP5-26`.

11.10 Conclusions

Five kinds of synthesizable Verilog were considered in this chapter: behavioral registers, behavioral combinational logic, behavioral implicit style state machines, behavioral explicit style state machines and structural instantiation. Of these, the implicit style is the best choice because it has such a close relationship to the behavioral ASMs discussed in earlier chapters. Often a designer must use some of the other kinds of Verilog, such as combinational logic, to create a complete design, but implicit style should be the first choice for synthesizing hardware.

This chapter has used the M4-128/64 CPLD with VITO, PLSynthesizer and PLDesigner. Although the details of performing synthesis using chips and software from different vendors may vary somewhat from those described here, the design flow for Verilog synthesis is similar. Simulation is a critical part of this design flow. Even though simulation takes some effort by the designer, in most cases, a bug discovered during simulation will be much less expensive than one that remains hidden until after the hardware is fabricated. Synthesis as well as place and route tools output structural Verilog netlists, which can be used with test code to verify the operation of the synthesized design.

11.11 Further reading

PALNITKAR, S., *Verilog HDL: A Guide to Digital Design and Synthesis,* Prentice Hall PTR,Upper Saddle River, NJ, 1996. Chapter 14.

11.12 Exercises

11-1. Give the synthesizable `seven_seg` function used in section 11.8.

11-2. Synthesize a 3-bit childish division machine based on the Verilog given in section 7.4.2 that will work with the hardware resources of the M4-128/64 demoboard. The code should be modified so that `x` is a register (rather than a bus) that is loaded with `y={sw2,sw1,sw0}` when the debounced `sw3` generates the first `pb` pulse. The second `pb` pulse starts the computation of `x/y`, which will be displayed in hexadecimal on the seven-segment display. Use the `debounce` module of section 11.6 and a top-level module similar to section 11.8 with the function from problem 11-1. Use test code that verifies the design after each step in the design flow.

11-3. Synthesize a factorial machine based on problem 2-4 that will work with the hardware resources of the M4-128/64 demoboard. `{sw2,sw1,sw0}` is the 3-bit value of `n` which is used when the debounced `sw3` generates the `pb` pulse. The 13-bit factorial of `n` will be displayed in hexadecimal on the LEDs. Use the `debounce` module of section 11.6 and a top-level module similar to section 11.8 with the function from problem 11-1. Use test code that verifies the design after each step in the design flow.

11-4. Modify the design of section 11.9 to include the link and the CLL, CML, RAR and RAL instructions (appendix B) in a way that allows the design to fit in the M4-128/64. Make appropriate changes to other instructions. Use test code based on the machine language program in section 8.3.2.5.3. Hint: because of the restrictions on <= in VITO, you need to define a 13-bit `lac` register, rather than separate `link` and `ac` registers.

11-5. Modify the design of section 11.9 to include the ISZ instruction (appendix B) in a way that allows the design to fit in a single M4-128/64. Use appropriate test code, such as the machine language code from problem 9-2.

11-6. Give Verilog for the architecture of the `debounce` module in section 11.7.

A. MACHINE AND ASSEMBLY LANGUAGE[1]

Most people use programs written in *high-level languages*. High-level languages are hardware-independent, complex languages that are relatively easy to use. *Hardware independent* means that programs written in high-level languages will run on nearly any general-purpose computer. Examples of high-level languages include Pascal, Verilog and C.

In contrast, low-level languages are simple in form and closer to how computers actually operate. This makes them harder for the programmer to use. Low-level languages are hardware dependent and have one statement per machine operation. Hardware dependent means that low-level languages are designed for a specific computer's hardware. Each statement is called a mnemonic. Mnemonics are easily memorized symbols that represent each fundamental computer operation in a textual form for the programmer's use. An instruction is a binary word that represents these fundamental operations in a form the computer can process.

Low level languages include machine language and assembly language. Assembly language is made up of instructions represented by mnemonics. Machine language consists of the instructions represented in binary. Assembly language has four major parts:

1. labels - symbolic names for places in memory (where variables are stored).
2. mnemonics - indications of what the computer will do.
3. operand - the data operated on by the instruction.
4. comments - a guide to the program that are ignored by the computer.

One statement in a high-level language program often corresponds to many assembly language and machine language instructions. For example, the machine language file of a program written in C and the machine language file of the same program written in assembly language are basically equivalent. But, the assembly language version is much longer than the C program. Consider the following very simple program:

```
/* Total tuition for three classes*/
int tuit,engl=74,cosc=106,math=148;
main(){tuit = engl + cosc + math;}
```

[1] This appendix was written by Susan Taylor McClendon and Mark G. Arnold.

This is equivalent to the following assembly language program written for the PDP-8, a simple general-purpose computer used as an example in chapters 8, 9 and 11:

```
label  mnemonic operand comment
       *0100              /starting addr
        CLA               /put zero in AC
        TAD      ENGL     /add ENGL to 0
        TAD      COSC     /add COSC to ENGL
        TAD      MATH     /add MATH to COSC+ENGL
        DCA      TUIT     /store in TUIT, clear AC
        HLT               /halt
ENGL,   0112              /74 dollars
COSC,   0152              /106 dollars
MATH,   0224              /148 dollars
TUIT,   0000
}
```

The *0100 indicates the starting address of the program in octal. The mnemonics indicate what each instruction does. The operand refers to a label defined later in the program. The following shows this example program translated to PDP-8 machine language code:

```
0100/7200
0101/1106
0102/1107
0103/1110
0104/3111
0105/7402
0106/0112
0107/0152
0110/0224
0111/0000
```

The four digits on the right of the "/" indicate a memory address in octal. The four digits on the left indicate the contents which show the octal values of the bit patterns representing the machine language equivalent of each mnemonic. Starting at address 0106_8 the contents are data values, not instructions.

TAD performs a *T*wo's complement *AD*dition of the operand to the contents held in the AC. DCA, *D*eposit and *C*lear the *A*C, deposits the value held in the AC into memory and then clears the AC. `CLA` and `HLT` are *non-memory reference instructions*. The CLA instruction *CL*ears the *A*C and the `HLT` instruction causes the fetch/execute algorithm to stop. The machine language code for `CLA` is 7200_8 and for `HLT` is 7402_8. More details about these and other instructions of the PDP-8 are given in appendix B.

B. PDP-8 COMMANDS[1]

The commands listed below are the Memory Reference Instructions (MRI) and the non-memory reference instructions of the PDP-8. The bits referred to below are given in little endian notation.

Memory reference instructions

1. TAD ($1xxx_8$) - *T*wo's complement *AD*d contents of memory address xxx_8 to the `link` and `ac`.
2. DCA ($3xxx_8$) - *D*eposit contents of `ac` at memory address xxx_8 and then *C*lear `ac`.
3. AND ($0xxx_8$) - Logical *AND* of `ac` with contents of memory address xxx_8.
4. JMP ($5xxx_8$) - *JuMP* to memory address xxx_8 so that the fetch/execute cycle will process the instruction stored there instead of the next sequential instruction. The PC is simply loaded with xxx_8.
5. ISZ ($2xxx_8$) - *I*ncrement (add 0001_8) to contents of memory address xxx_8 and *S*kip next instruction if contents become *Z*ero.
6. JMS ($4xxx_8$) - *JuM*p to *S*ubroutine located at memory address xxx_8. The JMS instruction saves the return address at memory addess xxx_8 and then the PC becomes xxx_8 + 0001_8. (The return address is the value of the PC indicating which instruction would have otherwise executed next.)

The "xxx_8" in the MRI instructions indicates a memory address used by the instruction. There are four addressing modes of the PDP-8: direct page zero, indirect page zero, direct current page and indirect current page. There is also a variation of the indirect addressing mode known as *autoincrement*.

Why do we need other addressing modes? One reason lies in the number of addresses we can represent using the page addressing bits. The page addressing bits are bits 6-0 of the `ir`. Only 2^7 or 128_{10} addresses (starting at address 0_{10}) can be represented by these seven bits. To represent the other 3968_{10} memory locations possible with the 12-bit address bus, the PDP-8 subdivides the 4096_{10} memory locations into 32_{10} *pages* (starting at page zero) of 128_{10} memory locations (4096_{10} DIV $128_{10} = 32_{10}$). To access a particular page, the PDP-8 uses two types of addressing modes: *direct* and *indirect*. Bit eight indicates either direct or indirect addressing mode and bit seven indicates

[1] This appendix was written by Susan T. McClendon and Mark G. Arnold.

either *page zero* or *current page*. Page zero is normally used for global variables and constants and the current page (01_8-37_8) is normally used for local data and corresponding code. The following lists the combinations of bits seven and eight for each possible addressing mode:

```
ir[8] ir[7] Addressing Mode         Effective Address
  0     0   Direct Page Zero        ir[6:0]
  0     1   Direct Current Page     {pc[11:7],ir[6:0]}
  1     0   Indirect Page Zero      m[ir[6:0]]
  1     1   Indirect Current Page m[{pc[11:7],ir[6:0]}]
```

Direct page zero computes the Effective Address (EA) as simply the low-order seven bits of the instruction register. This is the only addressing mode used in appendix A. Direct current page computes EA as the high-order five bits of the program counter concatenated to the low-order seven bits of the instruction register. This is useful for programs that do not fit in the 128 words of page zero. Indirect page zero computes EA as the contents of memory pointed to by the low-order seven bits of the instruction register. Similarly, indirect current page computes EA as the contents of memory pointed to by the concatenation of the high-order five bits of the program counter and low-order seven bits of the instruction register. These indirect addressing modes are useful when the address of data varies during runtime, and also in conjunction with the JMP instruction to return from a subroutine (called by a JMS instruction) or from an interrupt service routine.

The indirect addressing modes are slower, but more powerful, than the direct addressing modes since the EA comes from memory. First, the machine obtains the address of the EA from the instruction register (and possibly the program counter). Next, it accesses memory to obtain the EA. Finally, it accesses memory to obtain the data.

Autoincrement occurs on the PDP-8 with indirect addressing when the address of the EA (not the EA itself) is between 0010_8 and 0017_8. In these eight cases, the EA in memory is incremented *prior* to execution of the instruction. For example, the instruction 1417_8 increments the word at `m[0017`$_8$`]`, and then adds `m[m[0017`$_8$`]]` to the accumulator.

Non-memory reference instructions

Group 1 microinstructions

1. CLA (7200_8) - *CL*ear the *A*ccumulator, bit 7 on. This instruction sets the `ac` to 0000_8.
2. CLL (7100_8) - *CL*ear the `link`, bit 6 on. This instruction sets the `link` to 0.
3. CMA (7040_8) - *CoM*plement the *A*ccumulator, bit 5 on. This instruction complements (sets all 1's to 0's and 0's to 1's) the `ac`.
4. CML (7020_8) - *CoM*plement the `link`, bit 4 on. This instruction complements the `link`.
5. RAR (7010_8) - *R*otate the *A*ccumulator and `link` *R*ight, bit 3 on. This instruction shifts bit 11 through bit 0 one position to the right. The `link` shifts to bit 11 and bit 0 shifts to the `link`. All other bits shift one position to the right.
6. RTR - (7012_8) - *R*otate the accumulator and `link` *T*wice *R*ight, bit 3 and 1 on. Bit 0 shifts to bit 11, the `link` shifts to bit 10 and bit 1 shifts to the `link`. All other bits shift two positions to the right.
7. RAL (7004_8) - *R*otate the Accumulator and `link` *L*eft, bit 2 on. This instruction shifts bit 10 through 0 one position to the left. The `link` shifts to bit 0 and bit 11 shifts to the `link`. All other bits shift one position to the left.
8. RTL (7006_8) - *R*otate the accumulator and `link` *T*wice *L*eft, bit 2 and 1 on. Bit 11 shifts to bit 0, the `link` shifts to bit 1 and bit 10 shifts to the `link`. All other bits shift two positions to the left.
9. IAC - (7001_8) - *I*ncrement the *AC*cumulator, bit 0 on. Adds 1 to the contents of the `ac`. If the `ac` is 7777_8, the `link` will be complemented (as in the CML instruction). This allows the `link` and `ac` to act together as a 13- bit counter register.
10. NOP (7000_8) - *N*o *OP*eration, bits 0-7 off.

The Group 1 Microinstructions can be combined together. For example CLA CLL is 7300_8.

Group 2 microinstructions

1. SMA (7500_8) - *S*kip on *M*inus *A*ccumulator, bit 6 is 1_2 and bit 3 is 0_2. Normally used with signed data. Skips the next instruction if the value in the `ac` is negative.
2. SPA (7510_8) - *S*kip on *P*ositive *A*ccumulator, bit 6 is 1_2 and bit 3 is 1_2. Normally used with signed data. Skips the next instruction if the value in the `ac` is positive.
3. SZA (7440_8) - *S*kip on *Z*ero *A*ccumulator, bit 5 is 1_2 and bit 3 is 0_2. Skips the next instruction if the value in the `ac` is equal to zero.
4. SNA (7450_8) - *S*kip on *N*on-zero *A*ccumulator, bit 5 is 1_8 and bit 3 is 1_2. Skips the next instruction if the value in the `ac` is not equal to zero.
5. SZL (7430_8) - *S*kip on *Z*ero `link`, bit 4 is 1_2 and bit 3 is 1_2. Skips the next instruction if the `link` is 0_2.
6. SNL (7420_8) - *S*kip on *N*on-zero `link`, bit 4 is 1_2 and bit 3 is 0_2. Skips the next instruction if the `link` is not equal to zero.
7. SKP (7410_8) - *SK*i*P* unconditionally, bit 3 is 1_2. Skips the next instruction.
8. HLT (7402_8) - *H*a*LT*s the computer. Implemented by setting the HALT bit.
9. OSR (7404_8) - Inclusive *O*r of the *S*witch *R*egister with the `ac`. The result is left in the `ac` and the original content of the `ac` is destroyed.

Note that all the memory reference instructions begin with 0_8 to 5_8, and that all the non-memory reference instructions (group 1 and group 2 microinstructions) begin with 7_8. The I/O (Input/Output) instructions are not given here, but they all begin with 6_8.

Interrupts are external signals that cause temporary suspension of the fetch/execute cycle. On the PDP-8, there are two instructions, ION (6001_8) and IOF (6002_8) that control whether interrupts are ignored. ION sets the interrupt enable flag, and IOF clears it. On the PDP-8, an interrupt is ignored unless the last instruction was not 6001_8 and interrupt enable flag is 1. If these conditions are met, the interrupt causes the same action as executing the instruction 4000_8 without fetching such a machine code from memory. The interrupt also causes the interrupt enable flag to become 0. At that point, the fetch/execute cycle resumes. At the end of the interrupt service routine, the programmer must put an ION instruction followed by a JMP indirect instruction.

C. COMBINATIONAL LOGIC BUILDING BLOCKS

Combinational logic (also known as *combinatorial logic*) is the term used to describe the kind of digital hardware whose output depends only on its inputs. Combinational logic is critical to the operation of all computers; however, by itself, combinational logic has no memory. Because of this inability for combinational logic to "remember" previous results, combinational logic, by itself, is insufficient to implement non-trivial algorithms. Combinational logic needs to be combined with sequential logic (see appendix D) to implement such algorithms. The early chapters of this book explain the design process by which an algorithm is transformed into a structure composed of a mixture of combinational and sequential logic. This appendix provides a review of the combinational logic building blocks used throughout this book, and the notations used to describe them.

This appendix does not focus on circuit diagrams or netlists. Instead it describes things at a higher level, known as block diagrams. As described in section 2.5, hierarchical design (the relationship between circuit diagrams and block diagrams) allows us to look at a design at several levels of detail. Designers should work at the highest possible level, which means the notation used should conceal as much of the detail as possible. Lower levels of detail (such as the gate level, circuit diagram or netlist levels) must be dealt with at some point. Modern approaches, such as Verilog synthesis tools, have largely eliminated the need for designers to manipulate these lower level details manually. The bottom-up skills traditionally taught in an introductory digital design course (dealing with optimization of gates) are precisely those manipulations that nowadays are carried out automatically. This appendix assumes the reader has had enough exposure to such details previously to believe that they can be carried out automatically. Instead, sections C.2 through C.11 focus on a top-down approach, based on combinational logic building blocks. First, section C.1 discusses how detailed we might want to be when describing these devices.

C.1 Models of reality

All scientific and engineering disciplines use simplified and idealized models of reality that are easier to describe with mathematics than the reality itself. The role of the computer scientist or engineer is to create a useful product and get it to market rapidly. Such a designer does this by applying a model of reality to a practical problem. By using a model of reality that is too complicated, the designer will be burdened with unnecessary details, and the product will be late to market. The designer needs to choose a model of reality that is appropriate, as illustrated by the following planetary analogy. A planetary model that says that the sun orbits around the earth in a perfect circle every twenty-four hours is an acceptable model of reality for everyday problems. The mathematical simplicity of a circle is compelling, but there are problems where the highly simplified model is insufficient. A more accurate model would say that the earth orbits the sun in an elliptical path as the earth itself rotates. Although not as simple as a circle, an ellipse is still fairly straightforward to describe with simple mathematics. For some problems the elliptical model would also be insufficient, and a very complex model, considering lunar interaction, etc., might be required.[1]

C.1.1 Ideal combinational logic model

Speed and cost are not the first concerns of the designer. Producing a design that implements a correct algorithm is the top priority. For this reason, it will be convenient to think of combinational logic as being instantaneous. Such idealized combinational logic cannot exist in the physical world and is analogous to saying the sun orbits around the earth. Although an idealized model may seem too simple, it is the proper model for automatic Verilog synthesis, which helps ensure that the designer gets the product to market on time. Most of this book (with the primary exception of chapter 6) assumes idealized combinational logic.

C.1.2 Worst case delay model

Just as in the planetary analogy, sometimes the problem will demand something more accurate. As illustrated in chapter 6, there are problems where the computer designer must meet certain speed and cost constraints. Rather than jumping from no detail to every detail, it would be nice to have a simple, but reasonably accurate, model, analogous to the elliptical model of planetary motion. In computer design, the worst case propagation delay model satisfies this need. Worst case propagation indicates the maximum number of gates through which a signal change must pass in the worst case. An assumption commonly used in this model is that each gate has a delay of one unit of `$time` in a Verilog simulation.

C.1.3 Actual gate-level delay model

The delay in a combinational logic device depends on the values being processed by that device. Sometimes, the delay may be shorter than predicted by the worst case model. As described in chapter 6, it is difficult to consider all the possible paths through

[1] In fact, such complex planetary models are only practical because of electronic computers that can simulate these complex interactions.

the gates that compose the device. A Verilog simulator is a tool that allows the designer to try out many combinations of values to see how long it takes for the simulated combinational device to process the information under different circumstances.

C.1.4 Physical delay model

The most accurate but cumbersome model considers all the physical and geometric factors that compose the machine. Such a model considers physical laws that govern analog electronics, including factors such as the speed of light, capacitance, inductance, etc. Although there are times when computer designers must confront these harsh realities, the goal of a good top-down technique is to insulate the designer from physical reality to as great an extent as possible. This book never considers this level of detail.

C.2 Bus

The fundamental building block of all combinational logic is the bus. A bus is a device that transmits information from a source to a destination. The symbol for a bus is a line with a slash drawn through it. Next to the slash is a number, which indicates the number of bits that the bus transmits at any instant.

C.2.1 Unidirectional

A bus is either unidirectional or bidirectional. Most of the buses used in this book are unidirectional. A unidirectional bus is drawn as a line with an arrow pointing in one direction. The arrow indicates the direction in which information flows through the bus. For example, the following is a four-bit unidirectional bus:

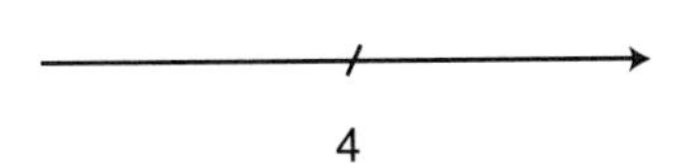

Figure C-1. Symbol for a four-bit unidirectional bus.

Being four bits wide, this bus can only transmit numbers that range from 0 to 15.

All but a few experimental computers have been built with electronics. Although other technologies besides electronics (such as the relatively new field of photonics) can implement these abstractions, modern synthesis tools and the entire computer industry are oriented toward the following electronic approach. In this electronic approach, the abstraction of an n-bit-wide bus is physically implemented as n "wires" running conceptually in parallel to each other. For example, the above four-bit bus would actually be four "wires":

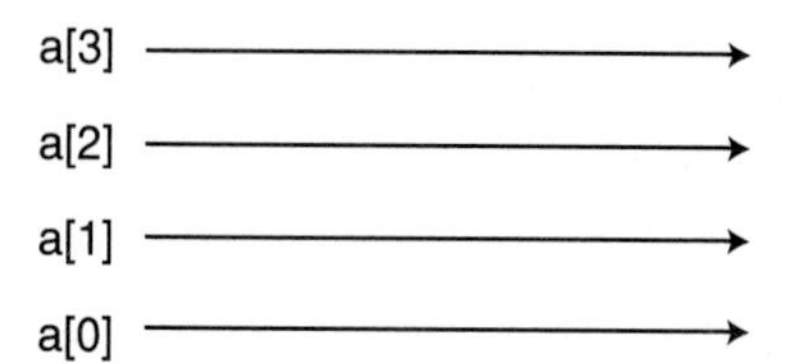

Figure C-2. Implementation of a four-bit bus.

Each "wire" transmits one bit of binary information from the source. For example, to send the number a=15 from the source to the destination, all four transmit a one:

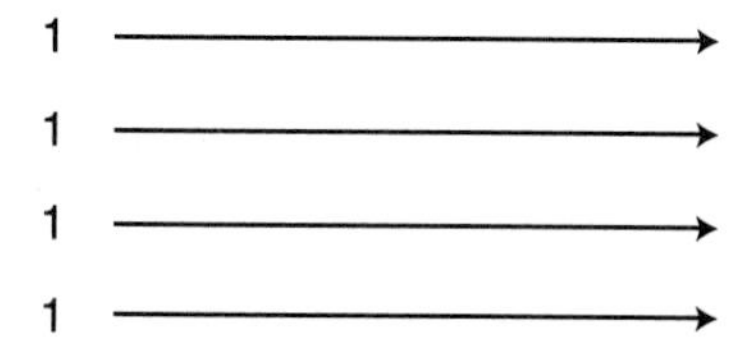

Figure C-3. Transmitting 15 on a four-bit bus.

On the other hand, to send the number a=7, the most significant "wire" instead transmits a zero:

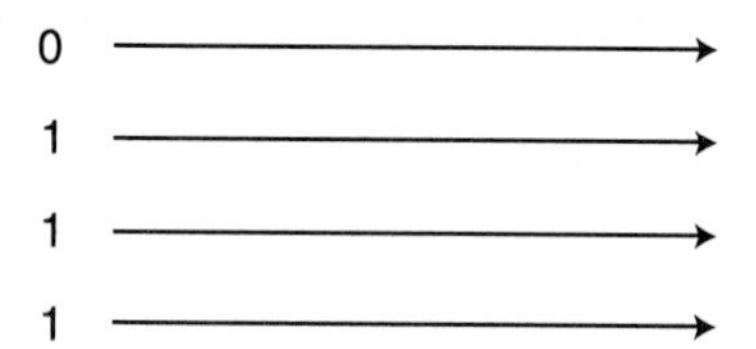

Figure C-4. Transmitting 7 on a four-bit bus.

C.2.2 Place and route

Whether or not the physical implementations of these four "wires" (composed perhaps of a connection between insulated wire dangling in the air, copper plating on a circuit board and traces within an integrated circuit) actually run geometrically parallel to each other is irrelevant. For example the following:

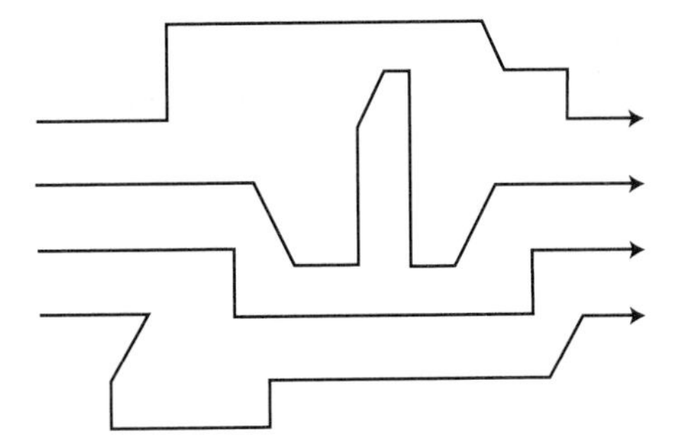

Figure C-5. One possible routing of a four-bit bus.

is equivalent to figure C-4. Such geometric details need not concern us because tools (known as place and route) determine this automatically.

C.2.3 Active high versus active low

At any instant, each "wire" will be at one of two voltages, known as *high* and *low*. The physical values of these voltages seldom concern the designer.[2]

Voltage by itself is not information. The goal of this book is to describe how to design machines that process binary information. Additional abstractions are necessary to relate physical voltages to the binary information being processed by an algorithm. On each wire, one of two possible abstractions is chosen (perhaps by the synthesis tool rather than the human designer) to forever describe how a voltage on that wire translates into binary information. These two abstractions are *active high* and *active low*.

In the active high abstraction, the high voltage means 1 and the low voltage means 0. In the active low abstraction, the opposite holds.[3] The easiest approach is to assume all wires are active high, which is the approach used in this book in order to avoid confusion. Beware that with actual physical chips it is common that some wires will be

---

[2] The numeric values of these voltages vary depending on the technology used. Typically, the lower the voltage, the faster the machine. For the rugged TTL logic families commonly used in educational labs, high is five volts, and low is zero volts. Faster, more modern but less rugged chips based on CMOS use lower voltages, such as 3.3 volts. Slow vacuum tube machines of the 1950s used around +50 volts and -25 volts.

[3] When all the signals are active high, the system is know as positive logic. When all the signals are active low, the system is known as negative logic. When the system is a mixture of both, it is known as mixed logic (not to be confused with the very different concept in Verilog of mixed behavioral structural design, as described in chapter 4).

active low while other are active high. If you use test equipment to observe the operation of an actual physical chip, you must understand the active low abstraction; however, during the design process, you can ignore this confusing issue.

C.2.4 Speed and cost

In this book, we assume an ideal bus, even if other combinational devices in the design are not ideal. Such an ideal bus, which would transmit a signal change from the source to the destination in zero seconds, cannot exist in the physical world.

Unlike the other devices described later in this appendix, the speed of a unidirectional bus does not depend on gate-level propagation delay. Therefore, in the worst case model, the speed of a bus is also instantaneous. As described in chapter 6, Verilog allows a designer to simulate propagation delay on other devices, but buses in Verilog typically have no delay.

The physical speed of a unidirectional bus can be determined by dividing its geometric length by a constant which describes how fast a change in a signal travels along the bus. For most electronic buses, this constant is approximately the speed of light, which is roughly one third of a meter (one foot) per nanosecond.

The reason unidirectional buses have been preferred is that they are extremely cheap and fast compared to the other combinational devices described later that are built out of gates. Bidirectional buses (appendix E) have gate-level propagation delay, like any other combinational device.

Cost is usually related to how much area a device takes on a chip. The area of a bus is at least the width of a wire times its geometric length times the number of bits in the bus. The area of a bus may be larger when problems occur in place and route.

As technology has improved, the relative speed and cost advantage of buses have diminished. In modern "deep submicron" silicon fabrication, interconnection delay, particularly between chips, is a significant factor.

C.2.5 Broadcasting with a bus

Another advantage of unidirectional buses is that they allow broadcasting of information from a single source to a reasonably large number of destinations[4] at little additional cost. It is a common misconception among novice hardware designers that to send the same information to two places in a design requires some special device:

---

[4] How many destinations is determined by the fanout of the logic family.

Figure C-6. Unnecessary device.

This misconception is understandable since to make b and c synonymous with a in software requires two explicit (and possibly expensive) steps:

```
b=a;
c=a;
```

Accomplishing the same thing in hardware is essentially free. You simply run the bus two different places, and refer to the same physical bus by different names at these new locations:

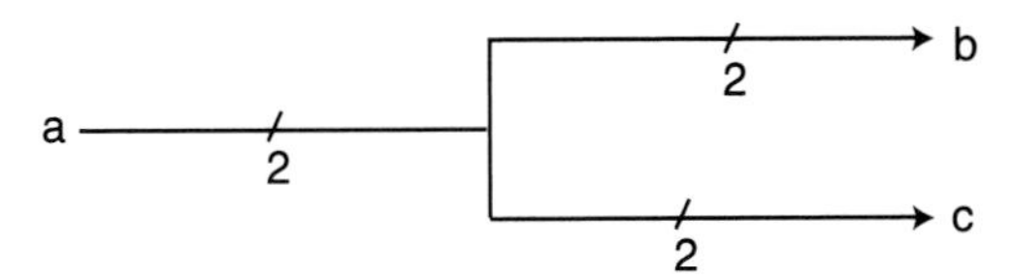

Figure C-7. Transmitting on one bus to multiple destinations for free.

One of many geometric arrangements of "wires" that can accomplish this is:

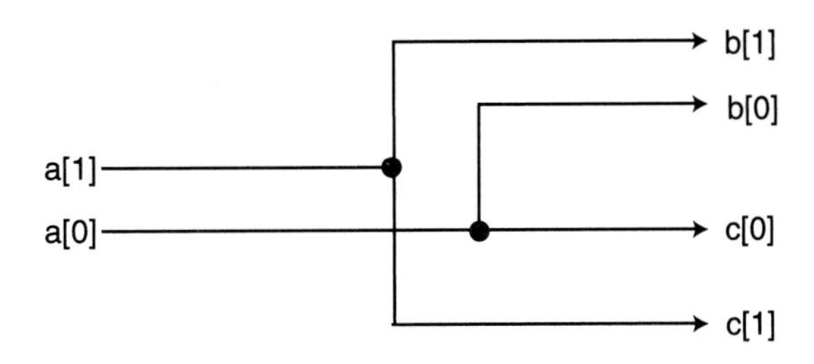

Figure C-8. Implementation of figure C-7.

Note there is no connection between a[1] and a[0], although there are connections between a[0], b[0] and c[0] and between a[1], b[1] and c[1]. Within the time it takes light to travel the physical distance of the bus, the voltages at b[1] and c[1] will be the same as a[1], which means that bit of information has been transmitted to those two locations.

It is common for a designer to use different names for the same bus, but doing so can be confusing. It would be better whenever possible to use the same name at both the source and destinations:

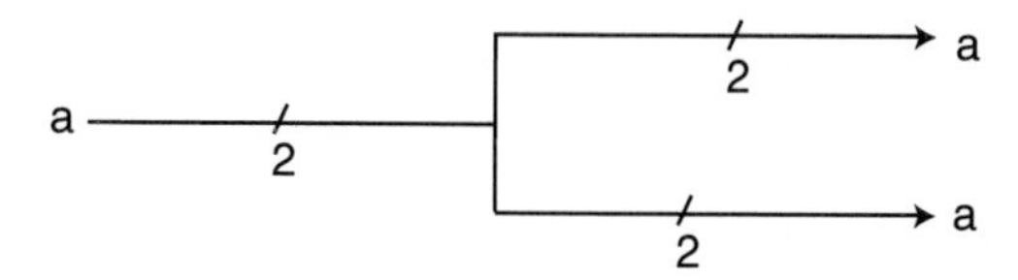

Figure C-9. Using the same name at every node.

since this more accurately reflects physical reality. Nevertheless, there will be times when it is advantageous to re-label the same bus with different names. A rose by any other name is just as sweet, and a bus by any other name is just as cheap.

C.2.6 Subbus

There are certain operations in the binary number system that are trivial to implement. For example, unsigned division by two, b=a/2 (also known as shift right, b=a>>1) appears to require some special device:

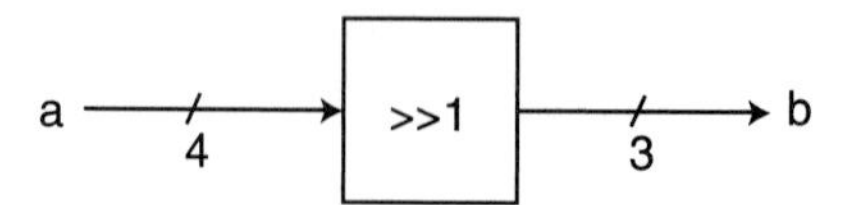

Figure C-10. Combinational device to divide by two (three-bit output).

but in fact can be implemented at no cost simply by rearranging how a subset of the wires of the bus a is connected to b:

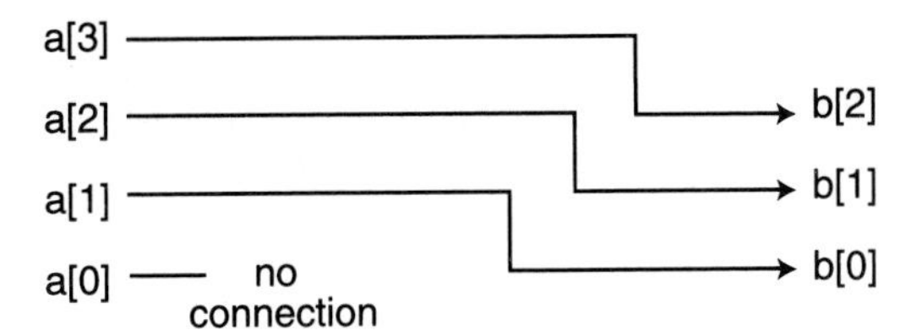

Figure C-11. Implementation of figure C-10.

This subset bus is known as a subbus. A designer can select any bits of the source bus to form a subbus. The notation we use for this is the concatenation syntax of Verilog, which separates the name of the individual wires with commas inside `{}`. For example, the bus `b` can be described as `{a[3],a[2],a[1]}`. Since subbuses that take a continuous group of bits from the source are common, there is another notation, known as bit select, that can be used: `a[3:1]` means the same as `{a[3],a[2],a[1]}`.

If the destination was also supposed to be four bits:

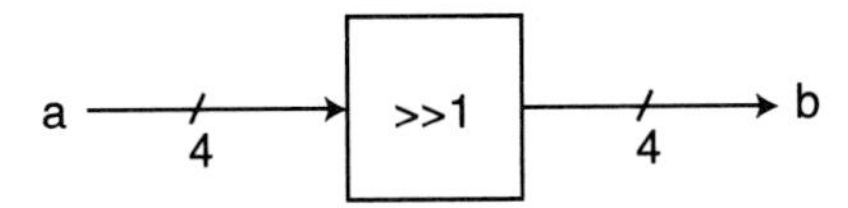

Figure C-12. Combinational device to divide by two (four-bit output).

`b[3]` would have to be tied to a constant 0:

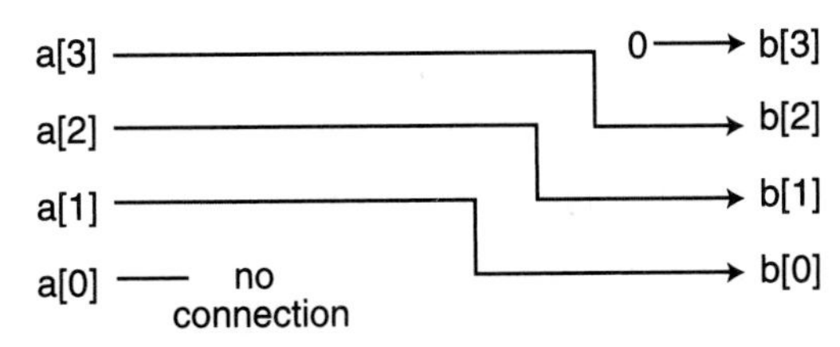

Figure C-13. Implementation of figure C-12.

In concatenation notation, this[5] would be `{0,a[3],a[2],a[1]}` or simply `{0,a[3:1]}`

[5] In correct Verilog notation, the constant 0 would have to be described as `1'b0`.

C.3 Adder

Many algorithms, even those that are not primarily mathematical, often need to do addition of binary numbers. One way to accomplish this is to provide a combinational logic unit that performs this computation:

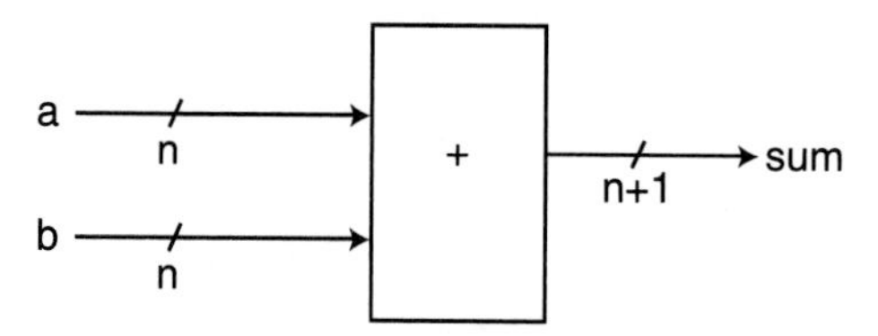

Figure C-14. Combinational device to add two n-bit values (n+1 bit output).

The block diagram symbol for an adder is simply a rectangle with a "+" or the word "adder" inside it. The number of bits in the output bus is one more than the number of bits in the larger of the input buses to allow for the largest possible sum. Note that `a`, `b` and `sum` are typically unsigned. (The low-order `n` bits of `sum` are also valid when they are signed twos complement; however there are complications with signed values beyond the scope of the discussion here.)

C.3.1 Carry out

It is common for the extra bit in the sum to be broken into a separate carry out (`cout`) signal, with `wordsum` being a subbus:

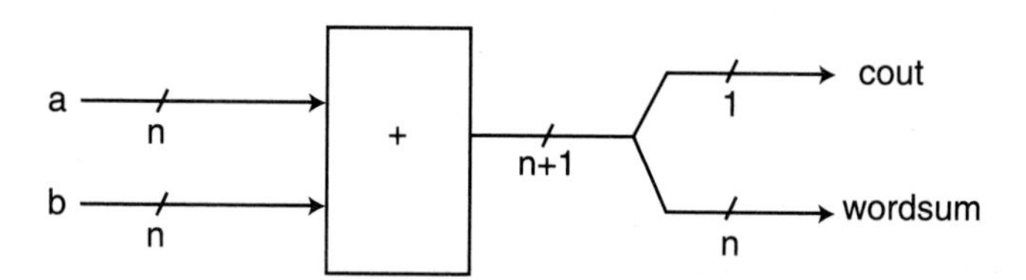

Figure C-15. Treating the high-order bit as carry out.

where `sum={cout,wordsum}`. The above is often drawn as:

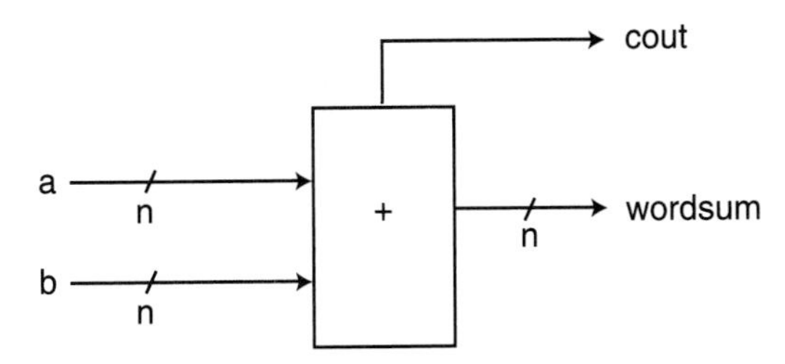

Figure C-16. Alternate symbol for figure C-15.

where the n-bit wordsum is a valid result that fits within the same sized word as a and b only when cout is 0. When cout is 1, an *overflow* error is said to have occurred. The physical implementation of this approach is identical to the earlier view of the adder. The advantage of this view is that all buses are the same width, which often simplifies the design of a larger system. The disadvantage is that cout must be observed while the system is in operation to detect the possibility of an error. Sometimes, however, the designer has a priori knowledge that wordsum is small, and so cout can be ignored.

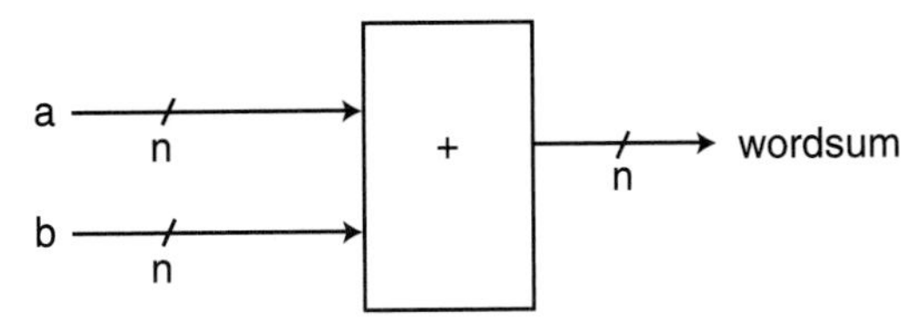

Figure C-17. Adder without carry out.

C.3.2 Speed and cost

There are many ways that adders can be implemented physically. One of the most common techniques is the ripple carry approach, which requires a circuit composed of about 3*n OR gates (each having two inputs) and 6*n AND gates (each having two inputs). Another way to state this is that it requires n full-adder modules (as described in sections 2.5 and 3.10.6). The worst case propagation delay for a ripple carry adder is proportional to n (as illustrated in section 6.3).

Faster techniques exist that require more gates. Some commonly used techniques include carry lookahead and carry skip.

C.4 Multiplexer

The multiplexer (commonly referred to as a mux) is the most important combinational logic building block next to the bus. Its purpose is to select one of its inputs to be its output and to ignore its other inputs. This "out of many, choose one" behavior is symbolized in this book as a triangle, whose tip is the one chosen output:

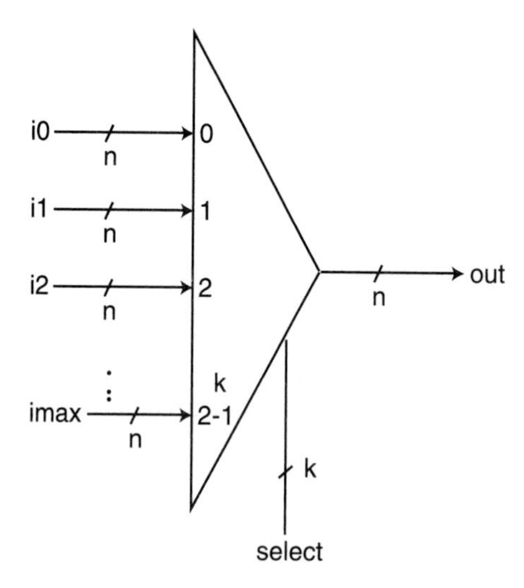

Figure C-18. Symbol for multiplexer.

Some people draw muxes as a rectangle with the word "mux" written inside. The mux has a select input, which is `k` bits wide. The mux also has (at most) 2^k other buses that are data inputs (`i0, i1, i2, ... imax`), each `n` bits wide. The mux has one data output which is `n` bits wide. If any input bus has fewer bits, assume zeros are concatenated on the left.

C.4.1 Speed and cost

There are several ways a mux can be implemented physically. In the most common approach, the mux shown above would be implemented using `n` OR gates (each having 2^k inputs), $n*2^k$ AND gates (each having `k` inputs) and `k` inverters. This approach needs only three stages of propagation delay. Sometimes it is possible to reduce this down to two stages (by eliminating the need for the inverters[6]) and so muxes implemented this way are quite fast.

C.5 Other arithmetic units

Although addition is the arithmetic operation for which we typically use a combinational building block, other operations can be implemented similarly. This section describes several arithmetic operations for which it is reasonable to fabricate specialized combinational logic.

C.5.1 Incrementor

One of the most common operations is adding one to a number:

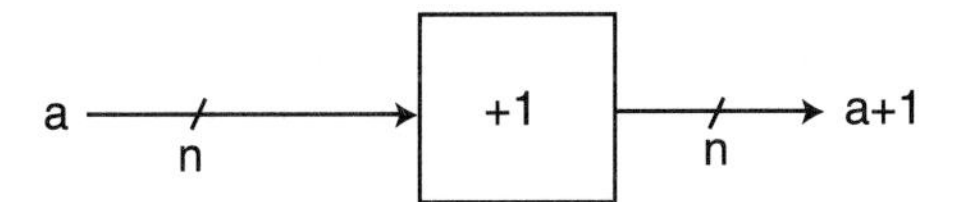

Figure C-19. Symbol for incrementor.

Although conceptually this could be implemented as:

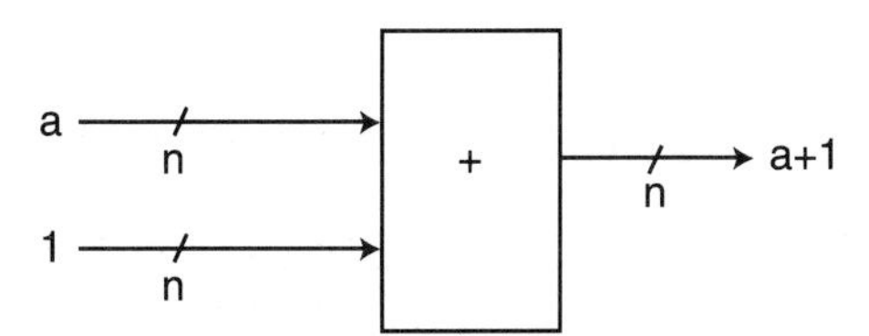

Figure C-20. Inefficient implementation of incrementor.

it is better to specify an incrementor if that is all the problem needs. Using a general adder is inefficient both in terms of speed and cost.

[6] The need for inverters can be eliminated in so-called *dual rail* designs, where every signal is provided in both active high and active low form. The reason the inverters are not needed is because certain devices, such as flip flops, naturally provide both active high and active low versions of the same signal at no extra cost.

C.5.1.1 Speed and cost

Although ripple carry addition of two arbitrary numbers requires `n` stages of worst case propagation delay, incrementation can be done in only two stages of propagation delay using `n-1` OR gates (each with two inputs), `2*n-2` AND gates (each with two inputs), `n-1` AND gates (of various sizes) and some inverters.

C.5.2 Ones complementor

The ones complement, `-a-1` (also known as bitwise not, `~a`):

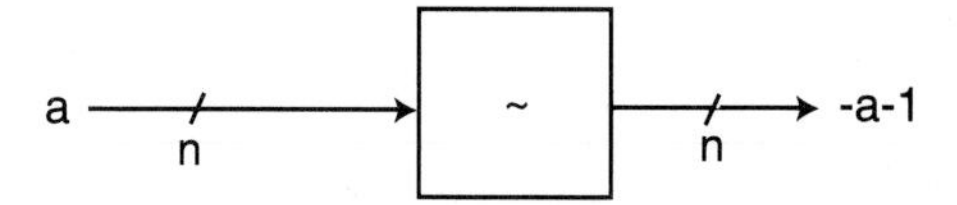

Figure C-21. Symbol for ones complementor.

is often part of a larger computation.

C.5.2.1 Speed and cost

The ones complement only takes `n` inverters, and one unit of propagation delay.

C.5.3 Twos complementor

Forming the negative of a signed number is necessary in many algorithms:

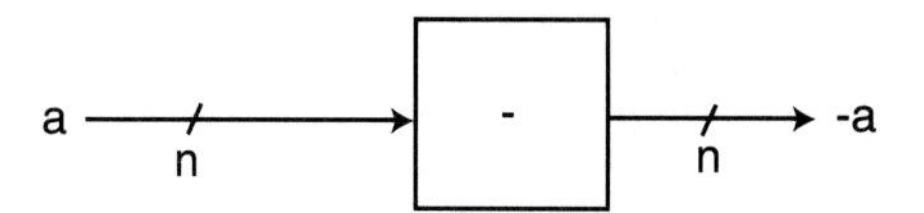

Figure C-22. Symbol for twos complementor.

This can be implemented as

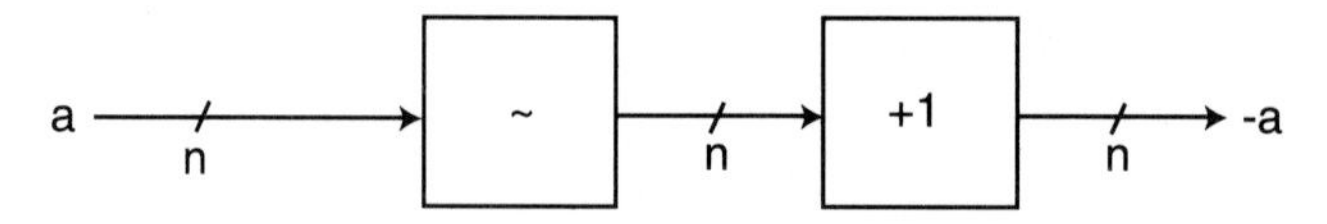

Figure C-23. Possible implementation of twos complementor.

C.5.4 Subtractor

The building block for a combinational logic subtractor is analogous to addition.

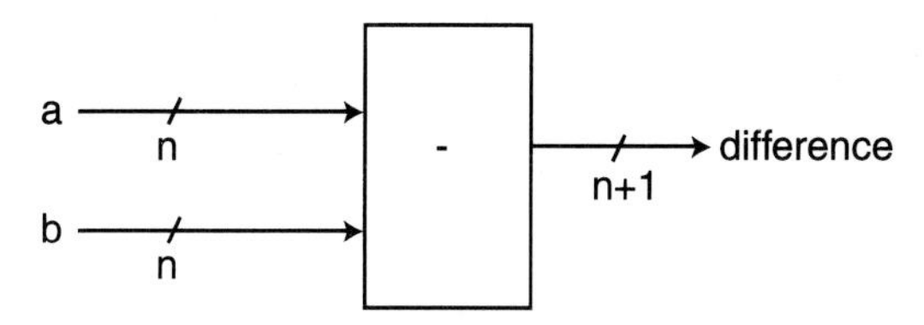

Figure C-24. Symbol for subtractor.

Although `a` and `b` are unsigned, `difference` is a signed twos complement value. The additional bit in the output indicates whether the difference is positive or negative.

One approach to implement a subtractor would be to use an adder and a twos complementor. A more efficient but less common approach would be to derive specialized logic for subtraction ("full subtractors").

C.5.5 Shifters

Multiplication and division by constant powers of two can be accomplished at essentially no cost through subbusing and concatenation. For example, multiplication by 4 (shifting left two places):

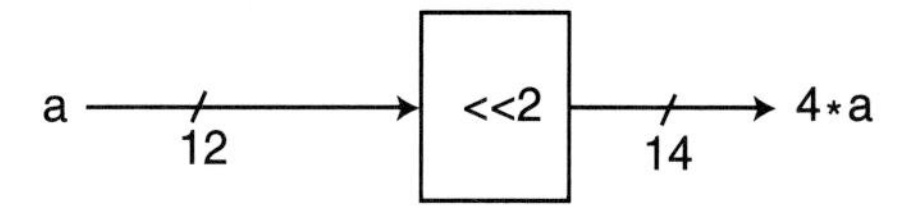

Figure C-25. Symbol for shifter.

simply concatenates `a` to two bits that are zero on the right. The reason this does not cost anything is because the power of two is a constant.

Sometimes the shifter has another input, known as the shift in (`si`), that allows the designer to specify what the least significant bits are:

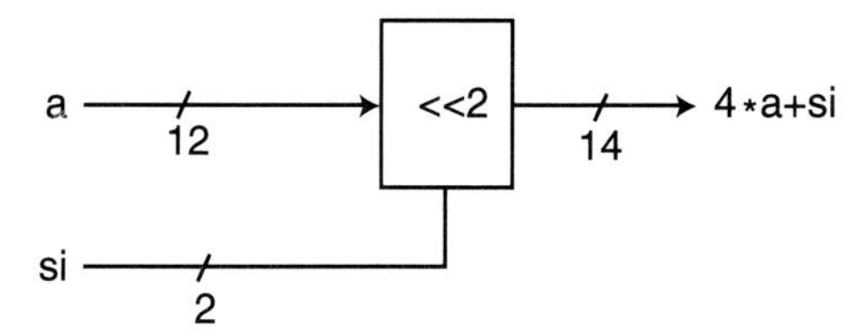

Figure C-26. Symbol for shifter with shift input.

Again, such a device is essentially free because it is implemented as the concatenation of `a` to `si`.

A barrel shifter allows a variable number of places for the shift. The number of places to shift is given by a `k`-bit shift count (`sc`) bus:

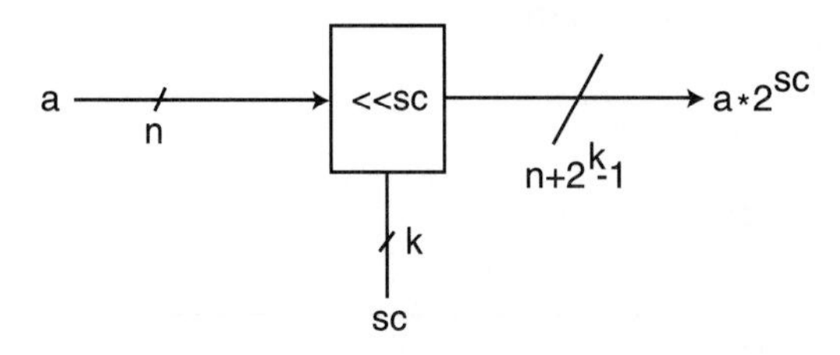

Figure C-27. Symbol for barrel shifter with shift count input.

This can be implemented in two (or three) levels of worst case propagation delay as constant shifters and a mux:

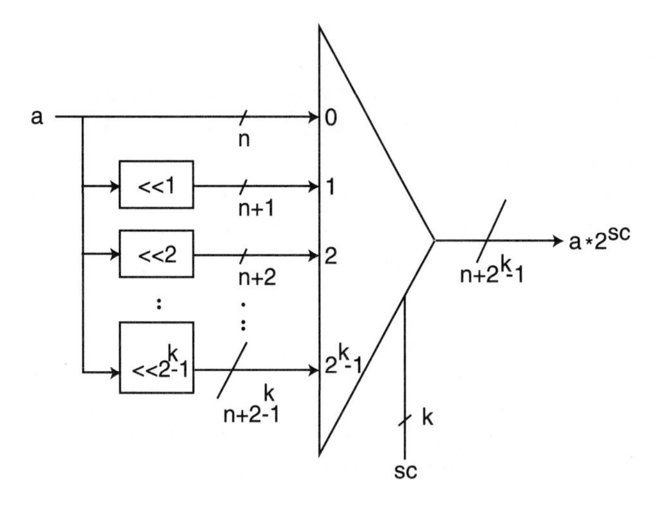

Figure C-28. Possible implementation of barrel shifter.

An alternative implementation which is slower but less costly uses `k` muxes, each with two inputs.

A similar right shifter can be implemented for division by variable powers of two. Barrel shifters can be arranged to allow for both multiplication or division by arbitrary powers of two, and to allow for arbitrary shift input (rather than concatenation with zeros).

C.5.6 Multiplier

The left barrel shifter in the last section only allows multiplication by a power of two. Many algorithms need multiplication by variables that are not powers of two. The multiplier is a fairly costly hardware device that allows two arbitrary numbers to be multiplied:

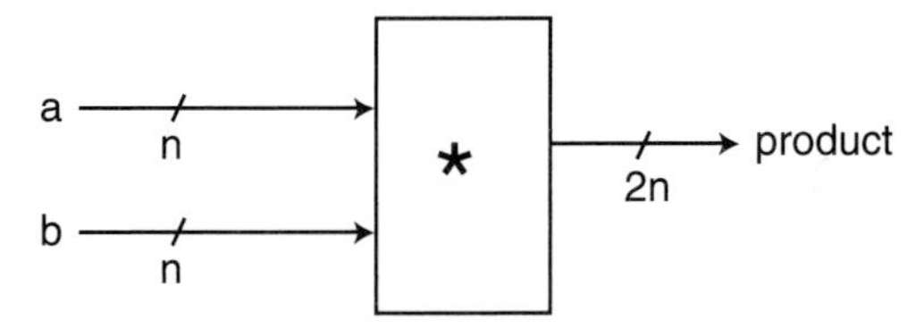

Figure C-29. Symbol for multiplier.

Note that the product has twice as many bits as the input buses. We will normally assume that `a`, `b` and `product` are unsigned. It takes a physically different device to multiply signed numbers.

C.5.6.1 Speed and cost

There are many ways to do multiplication; however the most commonly used techniques require on the order of n^2 gates, and at least twice the worst case propagation delay of addition. Because the cost and speed of a combinational multiplier is so high compared to the devices discussed in this appendix, a slower but cheaper approach involving sequential logic (appendix D) and ASM charts (chapter 2) is often used to generate a product.

C.5.7 Division

Division (by non-powers of two) is even more costly than multiplication when implemented as a combinational logic building block. Division is seldom implemented as combinational logic. Most of this book uses an example of one simple way that division can be implemented using sequential logic and ASM charts.

C.6 Arithmetic logic unit

In many problems, the same building block needs to compute different mathematical functions under different circumstances. A single unit that can handle most of the functions needed for a system is known as an Arithmetic Logic Unit (ALU). A designer can choose to put whatever functionality in an ALU as is appropriate for a particular problem, however it may be convenient[7] to use an ALU that has already been designed, such as the 74xx181.

Regardless of what details are inside the ALU that a designer chooses, the basic principle of how a combinational logic ALU operates is the same. There is a `k`-bit bus, `aluctrl`, that customizes the ALU for the particular function that needs to be computed.

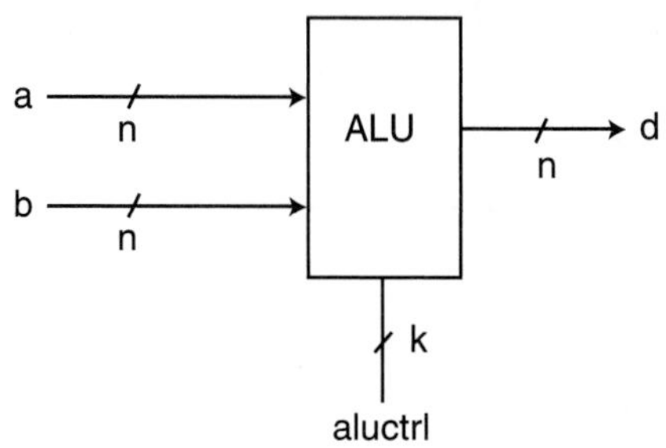

Figure C-30. Symbol for Arithmetic Logic Unit (ALU).

Conceptually, an ALU could be implemented as a mux which selects from the various combinational functions which that particular ALU is capable of performing:

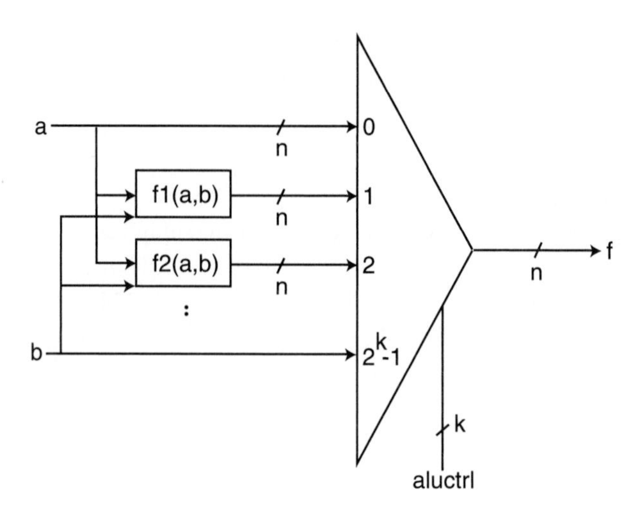

Figure C-31. Possible implementation of ALU.

[7] Especially in an educational lab setting.

Note that ALUs typically allow the passing through of a or b unmodified, and so an ALU can also serve the role of a mux. In physical reality, the implementation of an ALU might be quite different than a mux, but conceptually an ALU is equivalent to the above.

C.6.1 Logical operations

As its name implies, an ALU can perform both arithmetic and logical operations. Logical operations are those in which each bit of f depends only on the corresponding bit in a and also possibly on the corresponding bit in b. For example, consider the operation of a four-bit ALU doing the bitwise 'AND operation, f=a&b:

```
a     0011
b     1001

f     0001
```

In other words, f[3]=a[3]&b[3], f[2]=a[2]&b[2], f[1]=a[1]&b[1] and f[0]=a[0]&b[0]. These bitwise operations are the only dependencies in computing 'AND. For example, in computing f[1], there is no dependence on a[0] or b[0]. Breaking a bitwise operation such as a&b apart into separate single-bit logic equations as shown above is known as *bit blasting*.[8] Bit blasting is one of the many trivial but tedious details of hardware design that designers seldom need be concerned with because Verilog synthesis tools do such things automatically. Of course, the designer needs to understand that an n-bit-wide operation like a&b eventually becomes n separate AND gates operating independently. From this knowledge, it is easy to understand the worst case propagation delay of a&b is only one unit of gate delay, regardless of how many bits are in a and b.

Mathematically, there are only sixteen primitive bitwise logical operations involving no more than two variables. All other formulas involving no more than two variables and involving only combinations of these sixteen primitive operations can be simplified by the laws of Boolean algebra to one of these sixteen operations. The following is a table of the sixteen primitive logical operations:

[8] This quite descriptive term was coined by Synopsys, the pioneering vendor of Verilog synthesis tools in the early 1990s.

mnemonic	aluctrl	alternative aluctrl	operation
'NOT	000010		f = ~a;
'NOR	000110		f = ~(a|b);
'ANDNA	001010		f = (~a)&b;
'ZERO	001110		f = 0;
'NAND	010010		f = ~(a&b);
'NOTB	010110		f = ~b;
'XOR	011010		f = a^b;
'ANDNB	011110		f = a&(~b);
'ORNA	100010		f = (~a)|b;
'EQU	100110		f = ~(a^b);
'PASSB	101010		f = b;
'AND	101110	101101	f = a&b;
'ONES	110010		f = -1;
'ORNB	110110		f = a|(~b);
'OR	111010	000100	f = a|b;
'PASS	111110	000000	f = a;

The mnemonic column gives arbitrary names to these sixteen operations which will be used throughout this book.

When designing an ALU for a particular problem, it may not be necessary to include all of the sixteen mathematically possible operations inside the ALU. Omitting some of these operations may economize the total area required for the ALU and also may reduce the number of bits required for `aluctrl`.

The 74xx181 is an ALU that implements all sixteen of the possible logical operations. Since it also implements many arithmetic operations, it needs a six-bit `aluctrl`. In the above table, the `aluctrl` and alternative `aluctrl` columns indicate the six-bit pattern that must be input to the 74xx181 in order to perform the desired operation. For certain operations, such as 'AND, 'OR and 'PASS, more than one bit pattern can be used to produce the desired result.

C.6.2 Arithmetic operations

In contrast to logical operations, arithmetic operations are those where a change in one bit position of `a` or `b` potentially affects several bit positions of `f`. For example, consider addition ('PLUS) with the same ALU as the last example using the same values for `a` and `b`:

```
a     0011
b     1001

f     1100
```

The fact that a[0] and b[0] are both one in this example ultimately affects f[0], f[1] and f[2]. This ripple effect is why addition has a worst case propagation delay proportional to n.

The following table shows some of the most useful arithmetic operations available in the 74xx181 ALU:

'INCREMENT	000001		f = a+1
'DECREMENT	111100		f = a-1
'PLUS	100100		f = a+b
'DIFFERENCE	011001		f = a-b
'DOUB	110000		f = 2*a
'DOUBINCR	110001		f = 2*a+1

Because the 74xx181 is a low-cost ALU which is readily available for educational laboratory experiments, it does not implement multiplication or division. Section 2.3.1 shows how to use one of these ALUs to implement division using a slow but simple algorithm.

C.6.3 Status

An ALU commonly outputs extra information besides just the n-bit-wide result, f. For example, the 74xx181 has two status outputs that provide information about the computation currently being performed by the ALU. The first of these, cout, comes from the ALU's internal adder. It may be used to detect overflow (a+b>=2^n) when 'PLUS is being performed, and to detect a<b when 'DIFFERENCE is being performed.

The second status output, zero, detects whether f==0. It may be used to detect whether a==b when 'DIFFERENCE is being performed.

C.7 Comparator

There are six mathematical relational operators (==, !=, <, >, <= and >=). The vast majority of useful algorithms use one or more of these to make decisions that determine how the algorithm proceeds. Although the status outputs of an ALU may be able to answer such questions, it is often not efficient to use an ALU to do so. Instead, a specialized combinational building block, known as a comparator, is used instead. A comparator has two n-bit-wide input buses, a and b. At most, a comparator has three bits of output:

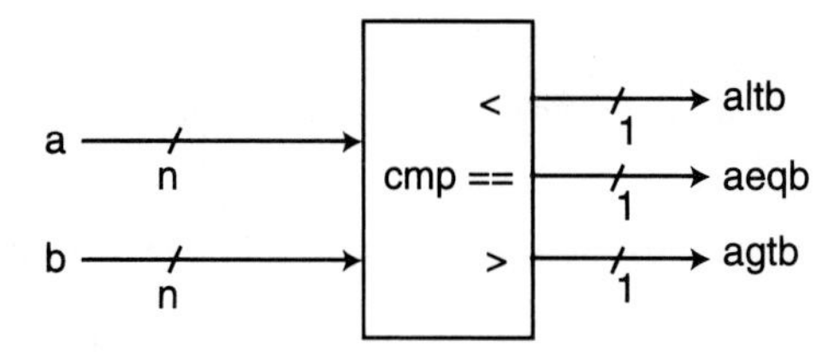

Figure C-32. Symbol for comparator.

From these three outputs, the other three conditions can be derived, for example `ageb=agtb|aeqb`. Many problems only need an equality test:

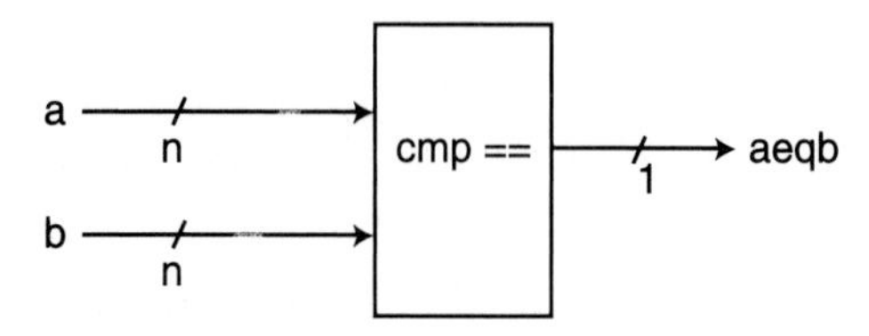

Figure C-33. Symbol for equality comparator.

C.7.1 Speed and cost

A comparator that only provides the equality output is much cheaper than one that also provides inequalities, such as `altb`. Such an equality only comparator needs `2*n` inverters, `2*n` AND gates (each having two inputs), `n` OR gates (each having two inputs) and one AND gate (having `n` inputs). The propagation delay for an equality only comparator is four units of gate delay under these assumptions. The cost is even lower when one input is a constant.

A comparator that also provides inequality outputs will have a worst case propagation delay proportional to `n`. It will also use considerably more area than an equality only comparator.

C.8 Demux

The demultiplexor (demux) is a specialized combinational building block which is not used in the early chapters of this book. It plays an important role in implementing concepts found in later chapters.

As the name implies, the demux is the opposite of a mux. The symbol used in this book for a demux reflects this. It is a triangle where the only input to the demux is connected to the tip of the triangle. The many outputs of the demux are drawn on the opposite side:

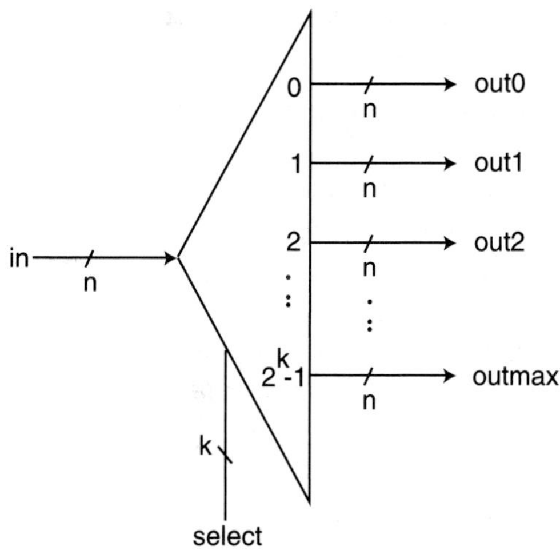

Figure C-34. Symbol for demux.

Like the mux, the demux has a `k`-bit input bus known as `select`. Some people may draw the demux as a box. All but one of the `n`-bit output buses will be zero. The selected output bus will pass through unchanged the value on the input bus.

C.8.1 Speed and cost

Demuxes are simply a large collection of AND gates that operate independently. The demux shown above requires $n*2^k$ AND gates (each having `k+1` inputs) and `k` inverters. Such an implementation would have a worst case propagation delay of only two gates. Sometimes, the inverters can be eliminated, in which case the propagation delay is only one gate.

C.8.2 Misuse of demuxes

Novice hardware designers often use demuxes where they are unnecessary. For example, suppose part of the time a machine needs to increment a number, `a`, and part of the time the machine needs to multiply it by two, but the machine never needs to increment and double the number at the same time. Novice designers often put a demux in such a design:

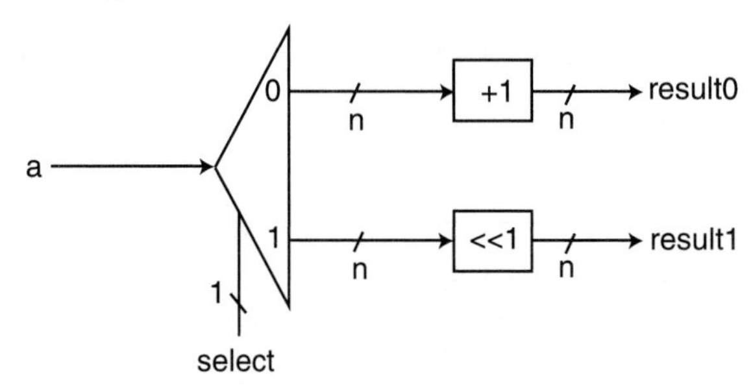

Figure C-35. Misuse of demux.

In the above, `result0` is `a+1` if `select==0` but is `0` otherwise. Also, `result1` is `a*2` if `select==1` but is `0` otherwise. Since it was assumed that `a+1` and `2*a` do not need to be simultaneously available, the above might work, but it would be considered a bad design for three reasons. First, the demux is an unnecessary and expensive (both in terms of speed and cost). Second, if the problem changes so that `result1` is supposed to be `2*a` simultaneously with `result0` being `a+1`, the above design would be completely wrong. Third, even if `2*a` and `a+1` are never needed simultaneously, the designer is burdened with providing the proper `select`.

It is understandable that novice designers make this mistake. In software, the programmer only specifies the operation required (either `a+1` or `2*a`) based on `select`. But in hardware, as was discussed in section C.2.5, it is easier to route a bus to every place where it is needed. The cost of doing this is usually quite low, and certainly less than using a demux:

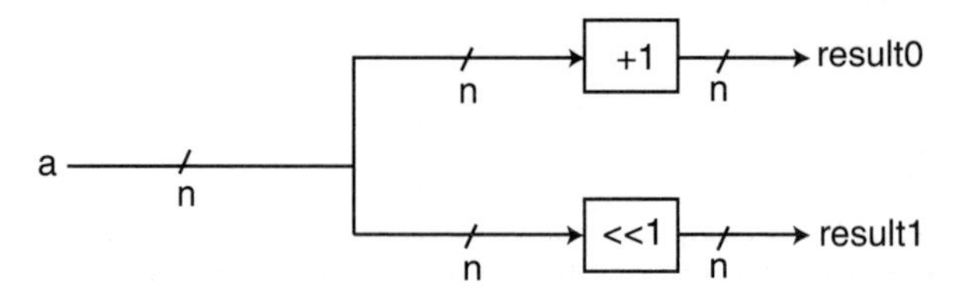

Figure C-36. Proper design omits demux.

There is no harm in `result1` being `2*a` simultaneously with `result0` being `a+1`. **In hardware**, it is often **more economical to compute everything you might need and ignore those results that are not pertinent** under particular circumstances.

Therefore, demuxes are not needed in the early chapters of this book. Demuxes are important in more advanced design topics. Demuxes are important in the design of memory systems (section 8.2.2.3.1) and in the implementation of one-hot controllers (chapter 7).

C.9 Decoders

A decoder is a specialized combinational device that converts from a binary code to some other code. The most common decoder converts from binary to what is called a unary code. The following table lists these codes for the numbers between 0 and 7

value	binary	unary out
0	000	00000001
1	001	00000010
2	010	00000100
3	011	00001000
4	100	00010000
5	101	00100000
6	110	01000000
7	111	10000000

Such a decoder can be thought of in two ways. First, it can be thought of as a building block that simply takes `k` bits of binary input, and produces 2^k bits of unary output:

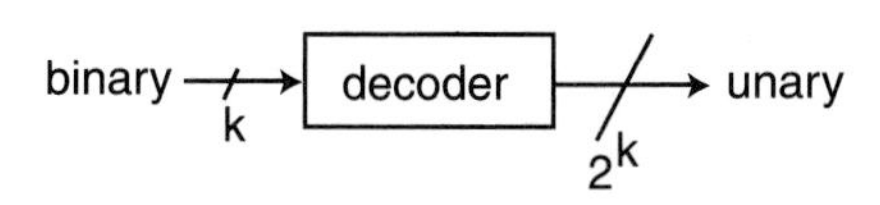

Figure C-37. Symbol for binary to unary decoder.

Second, a binary to unary decoder can be thought of as a building block composed of 2^k comparators. The output of each comparator provides one of the bits of the unary code. The second input of each comparator is connected to a binary constant. Each comparator is comparing against a different `k`-bit binary constant (from 0 to 2^k-1):

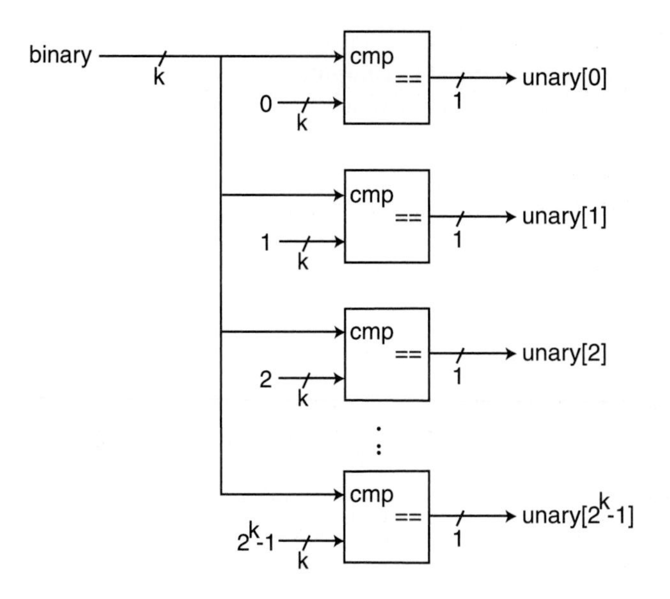

Figure C-38. Possible implementation of decoder.

Finally, an alternative way of looking at a binary to unary decoder is as a demux whose one-bit-wide input bus is tied to 1:

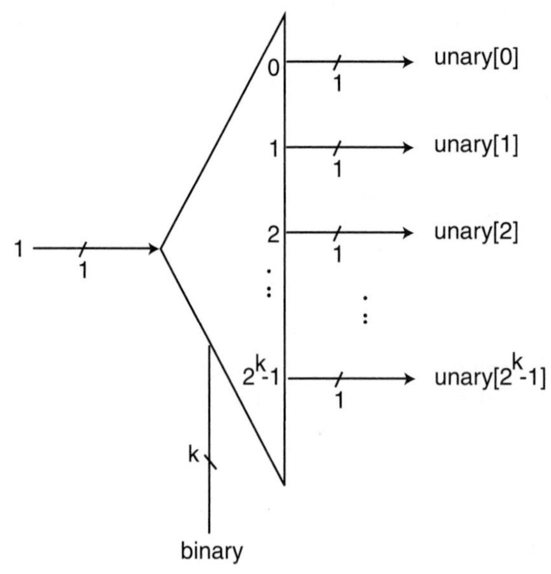

Figure C-39. Alternate implementation of decoder.

C.9.1 Speed and cost

The speed and cost of a binary to unary decoder is similar to a demux.

C.9.2 Other kinds of decoders

Decoders exist that involve other codes besides unary, such as those for seven segment displays. Such decoders are more specialized, and not widely used in computer design.

C.10 Encoders

Sometimes, a designer needs to convert from a unary code to binary. A combinational building block that performs this conversion is known as an encoder. If the designer could be sure the input were always a proper unary code (with exactly one bit that is one), the encoder could be implemented simply with k OR gates. But there are only 2^k valid unary codes out of the very large number (two *raised to the* k) of bit patterns that might appear on the input.

Instead, designers use a priority encoder.

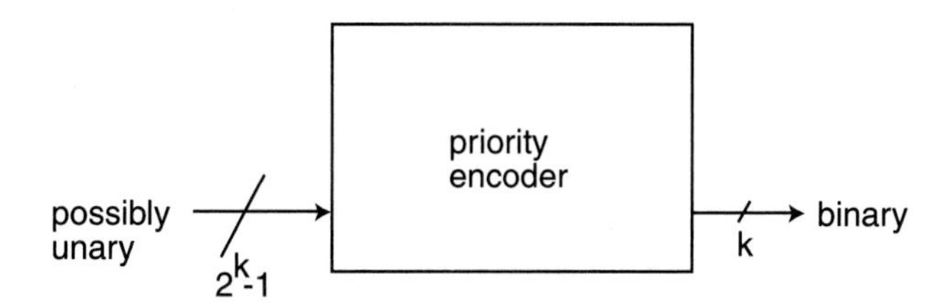

Figure C-40. Symbol for priority encoder.

It outputs the binary code corresponding to the bit position of the least significant leading zero.

```
000   00
001   01
010   10
011   10
100   11
101   11
110   11
111   11
```

The priority encoder is useful for counting how many leading bits of a number are zero. This is a computation that is necessary to implement floating-point arithmetic. Priority encoders are also often used so that a general -purpose computer can select which one of several external interrupts has the highest priority.

C.11 Programmable devices

Almost any imaginable mathematical function can be realized as a combinational building block if it involves a small enough number of bits of input. With Verilog synthesis tools available since the mid 1990s, functions involving around sixteen or fewer bits of input are routinely converted into combinational logic without the designer having to worry about their technological or gate-level implementation. The synthesis tool produces a file that can be downloaded into one of many kinds of programmable devices. The process of transferring the design into a programmable device is known as *programming* it. Such programming is a mechanical process, which does not require human intervention or creativity. The term *burning* is sometimes used to mean the same thing as programming. This use of the term programming should not be confused with its use in software (chapter 8), where the term programming means the same thing as design, which, of course, requires lots of creativity.

There are many kinds of programmable devices available, including Programmable Logic Arrays (PLAs), simple and Complex Programmable Logic Devices (CPLDs), Field Programmable Gate Arrays (FPGAs) and Read Only Memories (ROMs). CPLDs and FPGAs also have provision for sequential logic (see appendix D), but ROMs are pure combinational logic. The combinational logic implemented by all ROMs and by many FPGAs are based on truth tables without the need for expressing logic equations. In contrast, PLAs and CPLDs are based on logic equations (sum of products) rather than truth tables. Synthesis tools automatically produce truth tables or logic equations, depending on the target technology the designer selects.

C.11.1 Read only memory

Automatic synthesis of combinational logic for functions involving more than about sixteen bits depends on the complexity of the function. A simple function like addition can be implemented for an arbitrarily large number of input bits with combinational logic because the function decomposes into smaller combinational logic units, e.g., full adders in the case of addition. The synthesis tool is well aware of the properties of commonly used functions like addition. The decomposition of more complicated functions (whose properties are not built into the synthesis tool) is often less obvious. Synthesis tools explore many possible implementations for the combinational logic requested by the designer, however; as the number of input bits increases, the number of design alternatives grows exponentially. It becomes difficult for the synthesis tool to derive the logic equations needed for technologies such as CPLDs. ROMs tend to be the most practical approach for complex functions as the number of input bits grow. This is because with ROMs, all the designer has to do is tabulate the desired behavior, rather than find logic equations that produce that behavior. This avoids using a synthesis tool that has to explore exponential possibilities.

There are several ways of describing a ROM. The usual viewpoint of the designer is that the ROM is a black box, specialized for computing some particular function:

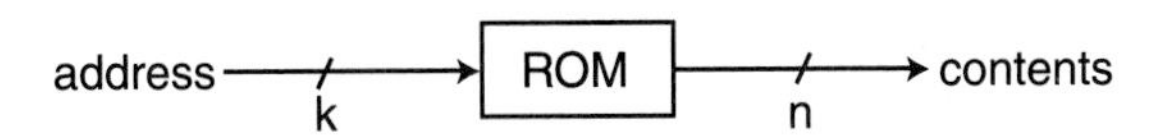

Figure C-41. Symbol for a Read Only Memory (ROM).

The number of input bits, `k`, and output bits, `n`, need not be the same, although often `k==n`. The input bus to the ROM is known as the address. The output bus of the ROM is known as the contents. This address and contents terminology is borrowed from memory systems (section 8.2.2.3.1). However, such a ROM is not truly a "memory" because once a value is burned into a ROM, it cannot be changed.[9]

Normally, the designer will indicate more than just the word "ROM" inside the box, since the ROM could be programmed to implement any function. For instance, the designer might need a "square root ROM," or something like that. The designer is then responsible for providing a table of the contents that need to be burned into that particular ROM.

Another viewpoint of the ROM is to describe it in terms of the combinational logic that it implements. A ROM is simply a mux whose data inputs are connected internally to the constants (`c0, c1, c2, ... cmax`) that the designer has burned into the ROM.

[9] So-called Electrically Erasable Programmable ROMs (EEPROMs) are not truly ROMs when the system in which they are used controls their erasure, such as when they are used as memory in general-purpose computers. In the terminology of chapter 8, such EEPROMs are non-volatile memory with slow access time. EEPROMs are however truly ROMs when their erasure is activated only by a separate development system under the control of the designer and not the hardware being designed. Put another way, if the designer does not use an EEPROM's erasure property in the design of the system itself, then the EEPROM is acting like a ROM, which is to say, the EEPROM implements some combinational logic function.

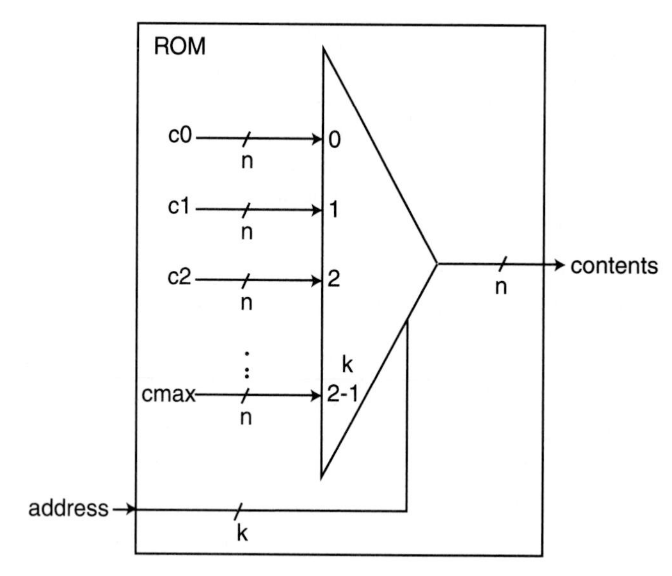

Figure C-42. Possible implementation of a ROM.

C.11.2 Complex PLDs

An alternative to using a ROM is to use a CPLD. The internal structure of a CPLD is too complex to describe here. A designer is seldom concerned with such details. Instead, a synthesis tool takes care of the details for the logic equations required inside the CPLD.

C.12 Conclusions

Combinational logic is an important building block for computer design. The distinguishing characteristic of combinational logic is that it lacks memory: its output is a function of its input. Common combinational building blocks include muxes, demuxes, incrementors, shifters, adders, ALUs, comparators, encoders and decoders. Some devices, such as buses and fixed position shifters, can be implemented at zero cost. Others, such as multipliers, are quite expensive. Ideally, we model combinational logic as a pure mathematical function having no propagation delay, but in reality, different approaches to implementing combinational building blocks have different propagation delays and costs.

Read Only Memories (ROMs) are not actually memory because they do not have the ability to forget. ROMs are simply a different, more convenient, approach for implementing combinational logic. The use of ROMs as well as the use of programmable logic with Verilog synthesis tools has made the design of specialized combinational logic relatively easy.

C.13 Further reading

GAJSKI, DANIEL D., *Principles of Digital Design*, Prentice Hall, Upper Saddle River, NJ, 1997. Chapter 5.

PROSSER, FRANKLIN P. and DAVID E. WINKEL, *The Art of Digital Design: An Introduction to Top Down Design*, 2nd ed., Prentice Hall PTR, Englewood Cliffs, NJ, 1987. Chapter 3.

C.14 Exercises

Using the combinational logic building block devices listed in each of the following problems, give a block diagram that implements the more complex combinational building block described by the data output(s). The buses in these problems should be interpreted as unsigned binary integers.

C-1. Control Inputs: CTRL (3 bits)
Data Inputs: A (32 bits), B (32 bits), C (32 bits), D (32 bits), E (32 bits)
Data Output: F (32 bits)
Devices: one 32-bit adder, one 32-bit 2-input mux, one 32-bit 4 input mux

CTRL	Data outputs
000	F=A
001	F=B
010	F=C
011	F=D
100	F=A+E
101	F=B+E
110	F=C+E
111	F=D+E

C-2. Control Inputs: CTRL (3 bits)
Data Inputs: A (32 bits), B (32 bits), C (32 bits), D (32 bits), E (32 bits)
Data Output: F (32 bits)
Devices: one 32-bit adder, four 32-bit 2-input muxes

CTRL	Data outputs
000	F=A
001	F=B
011	F=D
100	F=A+E
101	F=B+E
110	F=C+E
111	F=D+E

C-3. Control Inputs: CTRL (3 bits)
Data Inputs: A (32 bits), B (32 bits), C (32 bits), D (32 bits), E (32 bits)
Data Output: F (32 bits)
Devices: one 32-bit 8 input mux, four 32-bit incrementors.

CTRL	Data outputs
000	F=A
001	F=B
010	F=C
011	F=D
100	F=A+1
101	F=B+1
110	F=C+1
111	F=D+1

C-4. Control Inputs: none
Data Inputs: A (8 bits), B (8 bits), C (8 bits)
Data Output: A+B+C+2 (10 bits)
Devices: one 8-bit adder, one 9-bit adder

C-5. Control Inputs: none
Data Inputs: A (32 bits), B (32 bits)
Data Output: max(A,B), min(A,B)
Devices: one 32-bit comparator, two 32-bit 2-input muxes

C-6. Control Inputs: none
Data Inputs: array of four unsorted 32-bit integers
Data Output: same integers in sorted order
Devices: five of the devices from problem C-5

Hint: This is hierarchical design. Do not draw any muxes or comparators.

C-7. Control Inputs: CTRL (3 bits)
Data Inputs: A (32 bits), B (32 bits), C (32 bits)
Data Outputs: D (32 bits), E (32 bits)
Devices: three 32-bit adders, one 32-bit 2 input mux, one 32-bit 8-input mux

CTRL	Data outputs	
000	D=A;	E=A
001	D=B;	E=B
010	D=C;	E=0
011	D=C;	E=0
100	D=A+C;	E=A+C
101	D=B+C;	E=B+C
110	D=A+B;	E=0
111	D=A+B;	E=0

C-8. Control Inputs: ALUCTRL (6 bits), CTRL(1 bit),
Data Inputs: H (8 bits), L (8 bits), M(8 bits)
Data Outputs: F (8 bits), G (8 bits)
Devices: two 8-bit integer ALUs (74LS181),
one 8-bit 2 input mux

ALUCTRL	CTRL	Data	Output
100100	0	F=H+L;	G=H+2*L
100100	1	F=H+M;	G=H+L+M
101101	0	F=H&L;	G=H&L
101101	1	F=H&M;	G=H&L&M
000100	0	F=H|L;	G=H|L
000100	1	F=H|M;	G=H|L|M

Hint: It is a theorem of Boolean algebra that L&L==L.

C-9. Control Inputs: CTRL(2 bits)
Data Inputs: X (4 bits), Y (4 bits), Z (8 bits),
Data Output: W (9 bits)
Devices: one 8-bit adder, one 4-bit multiplier, any number of 8-bit 2-input muxes

CTRL	Data output
00	W=X
01	W=X+Z
10	W=X*Y
11	W=X*Y+Z

C-10. Control Inputs: CTRL(2 bits)
Data Inputs: X (4 bits), Y (4 bits), Z (8 bits)
Data Output: W (9 bits) Same as in problem C-9.
Devices: any number of adders of any width (specify), any number of shifters, any number of 2-input muxes of any width (specify)

Hint: With these building blocks, you need to implement the 4-bit multiplier using the shift and add algorithm for multiplication, which is analogous to the pencil and paper algorithm for decimal multiplication:

```
  six times thirteen
      0110
    * 1101 ──────────┐
                     ↓
      0110    do     (1) select 6
              don't  (0) select 18=12+6, instead pass 6
    011000    do     (1) select 30=24+6
+  0110000    do     (1) select 78=48+30

  01001110    product is    78
```

D. SEQUENTIAL LOGIC BUILDING BLOCKS

Although combinational logic (appendix C) is useful for implementing mathematical functions in hardware, combinational logic has no memory of values that were computed previously. Most practical algorithms make use of old values to compute new ones. Therefore, combinational logic by itself is insufficient to implement interesting algorithms. In addition to combinational logic building blocks, interesting machines must include sequential logic building blocks, commonly referred to as registers, that allow the hardware to remember old values. This appendix reviews several important synchronous sequential logic building blocks.

D.1 System clock

The term synchronous means that all of the sequential building blocks are connected to a single signal, known as the system clock or `sysclk` for short. The place where the system clock is connected is shown as a wedge in the lower-left corner of each synchronous building block:

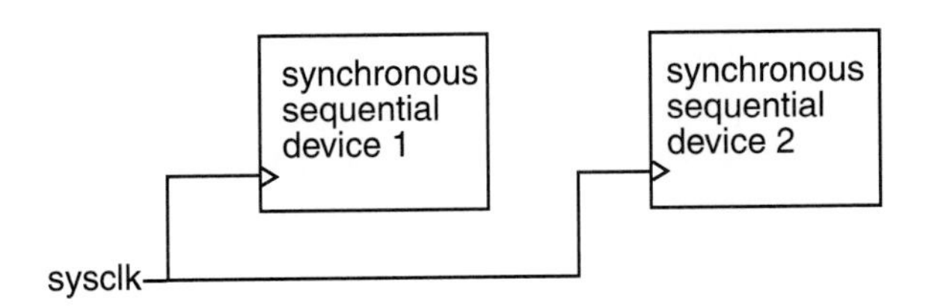

Figure D-1. Universal connection to system clock signal shown.

This connection need not be drawn because it is understood that all synchronous sequential devices connect to this same signal. For example, the following is understood to mean the same as the above:

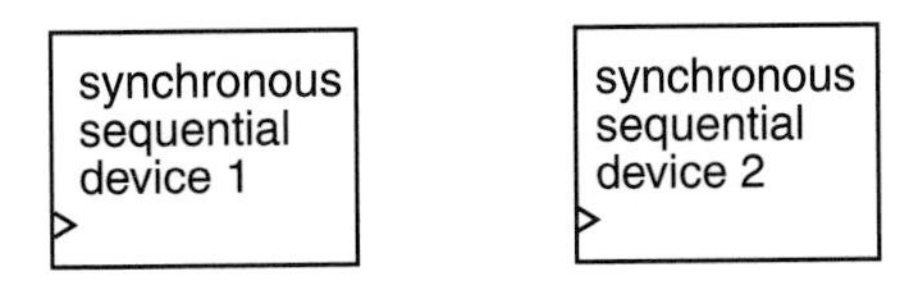

Figure D-2. Universal connection to system clock signal assumed.

D.2 Timing Diagrams

To describe the behavior of sequential logic, it is often helpful to use timing diagrams. A timing diagram plots values against time. For single-bit signals, like `sysclk`, this plot appears similar to some kind of graph you might have drawn in an algebra class.

The concept of a timing diagram originated with the display produced by an oscilloscope. Computers operate at speeds too fast to be observed by the unaided eye. When testing (or repairing) an actual physical computer, a computer designer needs some kind of test equipment to observe signals inside the computer down to a time resolution of about a nanosecond. The oscilloscope is a kind of test equipment that plots voltage versus time on a phosphor screen. The earliest electronic computer designers half a century ago used primitive oscilloscopes, and modern versions of oscilloscopes are still used by computer designers today.[1] For example, if you were to connect an oscilloscope to the `sysclk` signal, you might see:

Figure D-3. An analog waveform for the system clock signal.

which shows how the analog voltage (vertical axis) on the `sysclk` wire varies with time (horizontal axis). Physical properties, like capacitance and inductance, affect the ragged shape of the analog voltage shown on the oscilloscope. Computer designers are not concerned with analog voltages, and so this rather messy physical reality is abstracted to an idealized square wave:

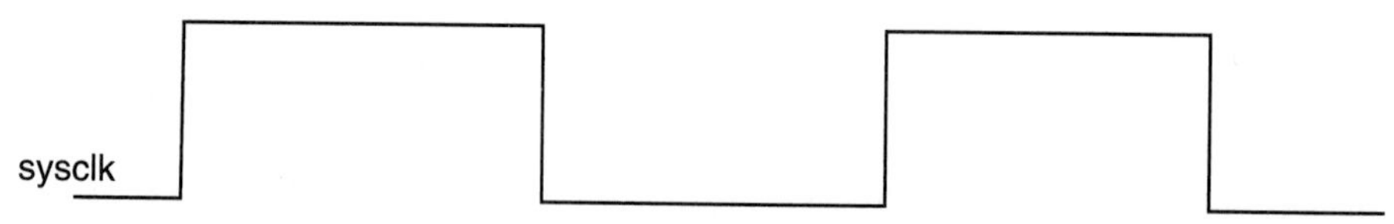

Figure D-4. A digital abstraction of the system clock signal.

Such a square wave is not physically possible; however, as explained in section C.1.1, computer designers often use models of reality that are physically unrealistic because such simplified models emphasize only those things which are algorithmically important.

[1] Computer designers now often use more sophisticated kinds of test equipment.

In the case of the `sysclk` signal, the only thing that is important is that it subdivides time into equal-sized intervals, known as clock periods or clock cycles:

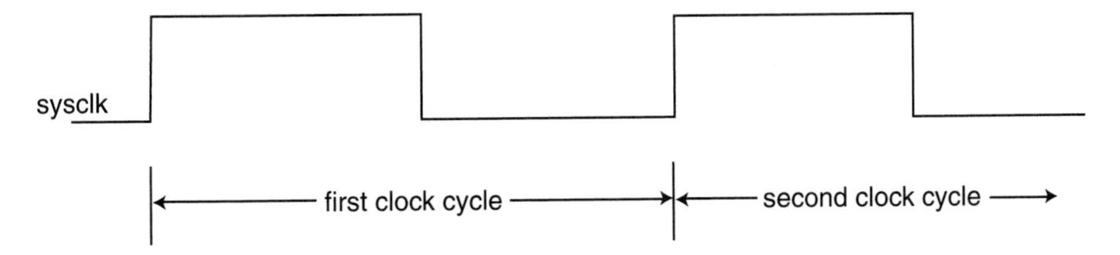

Figure D-5. The system clock divides time into cycles.

Each clock period begins and ends on the rising edge of `sysclk`. (Some kinds of sequential logic use the falling edge; however in this book all synchronous sequential building blocks use the rising edge.)

D.3 Synchronous Logic

Synchronous logic is a restriction on physical reality where changes in the values shown in a timing diagram occur only at the exact instant of the rising edge of `sysclk`. For example, in the following timing diagram, one bit of data is being manipulated by an algorithm:

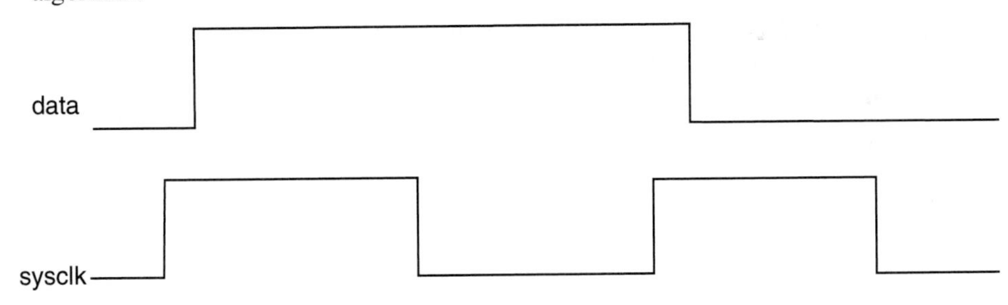

Figure D-6. An ideal synchronous timing diagram.

The above is a valid synchronous timing diagram because the changes in data occur only at the rising edge of `sysclk`. In physical reality, the changes in data occur slightly later than the actual instant of the rising edge due to propagation delay of the circuits used to generate the `data` signal:

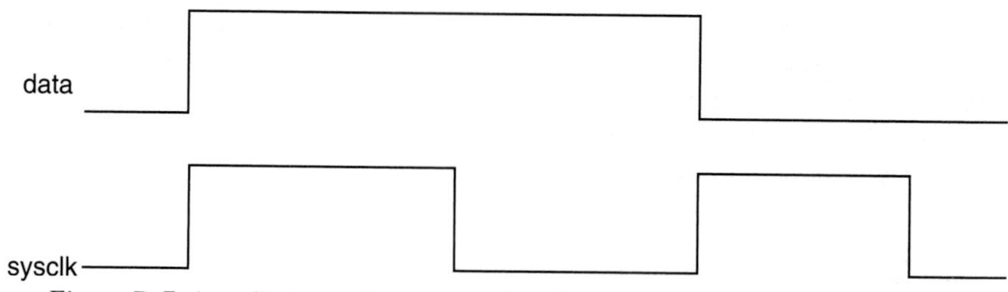

Figure D-7. A realistic synchronous timing diagram with propagation delay.

but as discussed in section C.1.1, we normally ignore propagation delay. At the beginning of the design process, the primary concern of the designer is getting the algorithm right. Worrying about physical reality is a distraction from the designer's most important mission—ensuring that the algorithm is correct.

The following diagram is not synchronous. It is known as *asynchronous* because the data pulse might occur at any time with respect to `sysclk`:

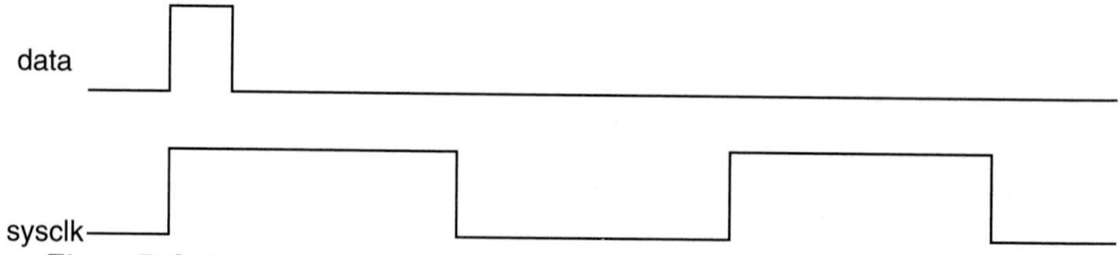

Figure D-8. An asynchronous timing diagram.

With only one exception that happens when a machine is first turned on (described in sections 4.4.5 and 7.1.6), we will not use such asynchronous logic.

Synchronous design is safe and easy. Asynchronous design is hard and dangerous. Commercial synthesis tools concentrate on synchronous design. Therefore, synchronous design is widely used in industry.

D.4 Bus timing diagrams

Digital computers represent values other than zero and one using a group of bits on a bus with the binary number system. The physical reality is that each wire in a bus represents a separate bit of information. But from an algorithmic viewpoint, the designer wants to look at the bus as containing a single binary value. Suppose the value of a variable, `v`, goes through the sequence 0, 1, 2, 3, 0, 1, 2, 3, 0 .... This could be shown on a timing diagram as two separate bits that change synchronously with `sysclk`:

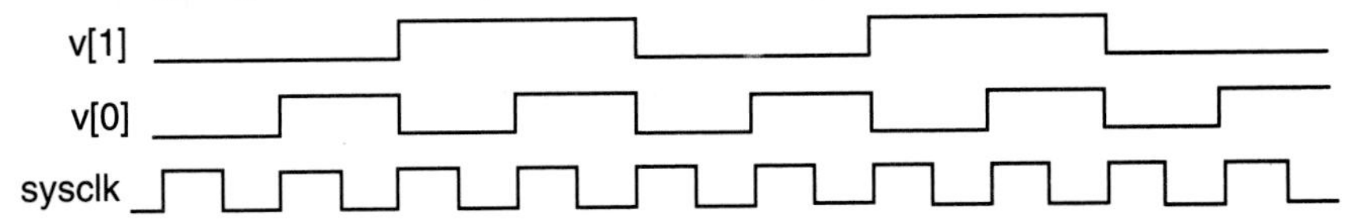

Figure D-9. Timing diagram showing individual bits of a bus.

However dealing with separate bits is quite tedious. Instead timing diagrams usually show the numeric value of the complete bus during each clock cycle:

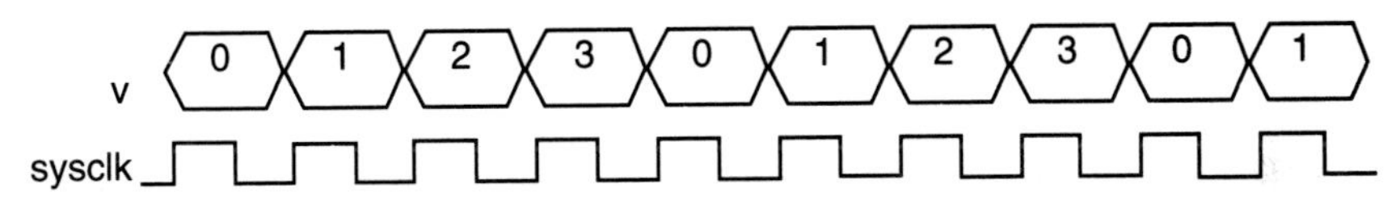

Figure D-10. Timing diagram showing numeric values on a bus in decimal.

In timing diagrams, the notation shown in figure D-11:

Figure D-11. Notation used for bus timing diagrams.

shows the instant in time (a particular rising edge of `sysclk`) when the numeric value of the bus changes. It is only necessary for one bit of the bus to change for the numeric value of the bus to be completely different.

D.5 The D-type register

The simplest sequential building block is the D-type (or delay type) register:

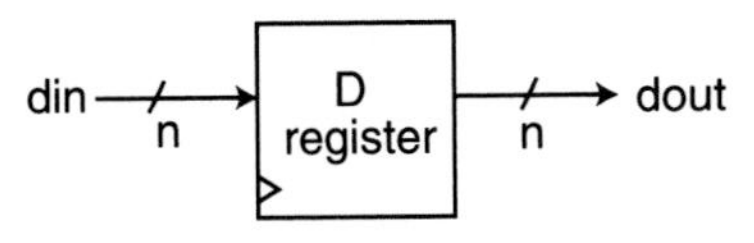

Figure D-12. Symbol for D-type register.

Some people refer to `din` as the D input and `dout` as the Q output. When `n=1`, this device is referred to as a D-flip flop. In fact, an `n`-bit D-type register is usually built from `n` D-type flip flops.

In the D-type register, `dout` is simply a delayed version of `din`. Put another way, `dout` in the present clock cycle is the same as `din` in the previous clock cycle. Suppose that `din` just happens to be going through the binary sequence:

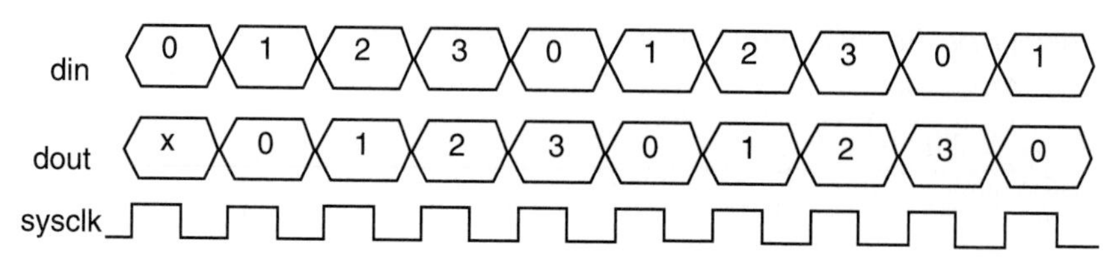

Figure D-13. Example timing diagram for D-type register.

`dout` will also go through the same sequence, but it will lag by one clock cycle. In the above, `x` means unknown (see section 3.5.3 for details on how `'bx` is used in Verilog simulation), because there is not enough information to predict what is in the register at the beginning.

As another example, consider what happens when `din` is somewhat more random:

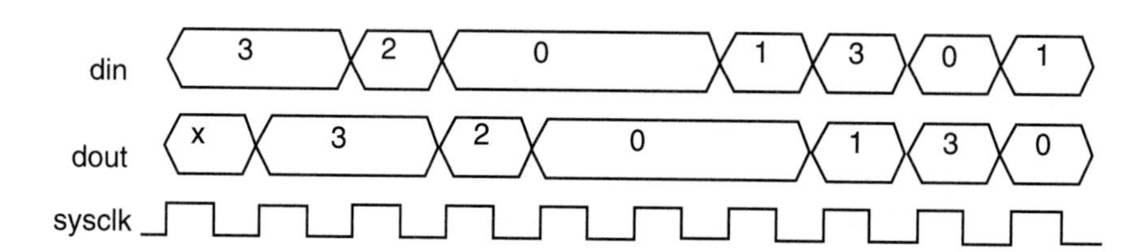

Figure D-14. Another timing diagram for D-type register.

The D-type register is not used by itself in computer design very often. The two most common uses of the D-type register are synchronizers and the present state register for controllers (see sections 2.4.1 and 7.1.1).

All of the more useful registers described below can be constructed from a simple D-type register combined with combinational logic, but it is often not helpful to think of things that way. It is usually better to think in terms of one of the more sophisticated building blocks described below. On the other hand, D-type registers together with specialized logic are often included in designs created by synthesis tools (see section D.11).

The reason the D-type register by itself is often inadequate for many problems is that the D-type register only remembers the old value for one clock cycle. Most algorithms have variables that must remain unchanged for multiple clock cycles. This requires a more sophisticated kind of register, discussed in the next section.

D.6 Enabled D-type register

Algorithms are the starting point for the hardware design approach described throughout this book. Algorithms are composed of steps that manipulate variables at certain moments in time. The rising clock edge determines when those moments occur. The vast majority of algorithms manipulate their variables in complicated ways, so that the variables do not change at **every** rising clock edge. For this reason, we need a kind of register building block that can hold its former contents for multiple clock cycles as well as being able to load itself with new contents. The building block that has such a capability is known as an enabled D-type register, or just simply an enabled register.

In order to allow the designer to choose between these two different actions, the enabled register has a command input. This command input is sometimes known as the load signal or the enable signal. In this book, this command input is typically abbreviated with a name like `ld`.

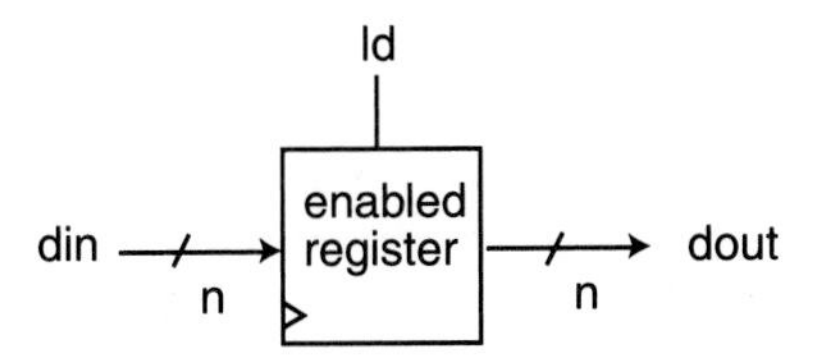

Figure D-15. Symbol for enabled D-type register.

The following action table describes what the enabled register does based on the `ld` input:

ld	action
0	hold
1	load

An action table is not a truth table, because unlike a truth table, an action table includes the concept of time.

For example, suppose the following `din` and `ld` signals are provided to an enabled register:

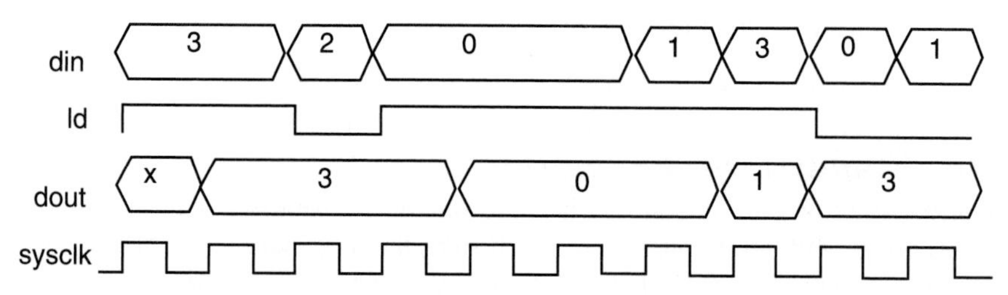

Figure D-16. Example timing diagram for enabled D-type register.

In this example, `ld` happens to be 0 at certain times when `dout` happens to be 3. This means in the clock cycle after `ld` is 0, `dout` will continue to hold the value 3, regardless of what `din` happens to be. On the other hand, when `ld` is 1, the enabled register acts just like a simple D-type register.

The enabled register can be implemented as a mux connected to a simple D-type register:

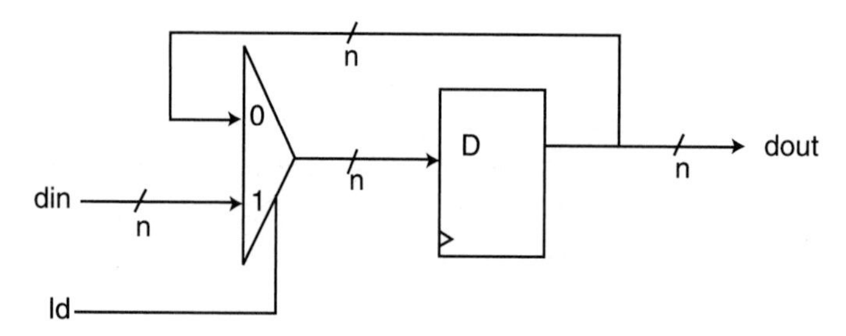

Figure D-17. Implementation of enabled D-type register using simple D-type and mux.

When `ld` is 0, the mux passes through the old value of `dout` to be reloaded into the simple D-type register. When `ld` is 1, the mux passes through the new `din` value to be loaded into the simple D-type register.

Other arrangements of hardware not based on the simple D-type register can also implement an enabled register. Therefore, in the top-down approach, designers typically specify an enabled (or loadable) register without concern for how it is implemented.

In the TTL logic family, the 74xx377 (for n=8) and 74xx378 (for n=6) chips implement the same actions as the above.[2]

D.7 Up counter register

When combined with combinational logic, the enabled D-type register is sufficient to implement any algorithm, however, certain register operations occur so frequently that these operations deserve special implementation as sequential building blocks in their own right. The distinguishing characteristic of these operations is that the register has within itself all the information necessary to perform the operation.

Perhaps the most important of these special operations is counting. Most algorithms include steps that involve counting. In fact, the very first practical machine ever built with digital electronics (by Wynn-Williams in 1932) was a counter used to count alpha particles for a physics experiment conducted by Lord Rutherford. Since that time, billions of counters have been fabricated.

There are many variations on how to build a counter. In this book, we will concentrate only on the two most important kinds of counters: the synchronous loadable binary up counter (described in this section), and the synchronous loadable binary up/down counter (described in section D.8). We will refer to these more simply as the up counter and the up/down counter, respectively. When the word *counter* is used by itself in this book, it means the synchronous loadable binary up counter.

The up counter has three command inputs. The `ld` command signal is the same as it is in an enabled register. The `clr` command signal causes the counter to become zero at the next rising edge of the clock. The `count` command signal (sometimes referred to as the `inc` command signal) causes the counter to increment at the next rising edge of the clock.

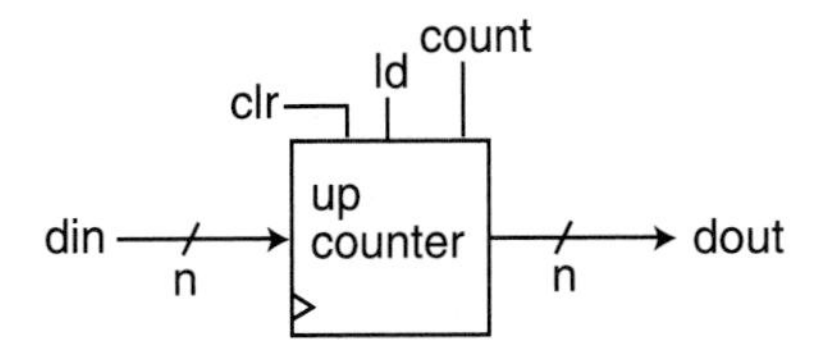

Figure D-18. Symbol for up counter register.

[2] Except `ld` is active low, which has the apparent effect of reversing the 0 and 1 inside the mux of the TTL chip.

The behavior of the up counter is summarized by the following action table:

ld	clr	count	action
0	0	0	hold
0	0	1	increment
0	1	0	clear
0	1	1	clear
1	0	0	load
1	0	1	load
1	1	0	load
1	1	1	load

Note that the `ld` signal has a higher priority than `clr` and `count`. Also `clr` has a higher priority than `count`. An up counter can be constructed from a simple D-type register, three muxes and an incrementor:

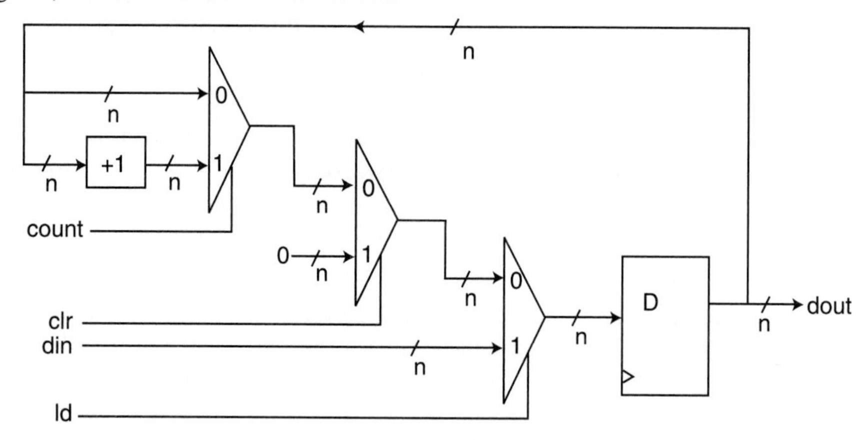

Figure D-19. Implementation of up counter register using simple D-type register and combinational logic.

Recall that the combinational logic incrementor (section C.5.1) is considerably faster than an adder. Even so, there are other more efficient ways of constructing a counter than the technique shown above. For example, in the TTL logic family, the 74xx163 chip provides for `n`=4 the same actions[3] as the above using fewer gates and less propagation delay.

[3] Except that `clr` and `ld` are active low.

D.8 Up/down counter

Some algorithms involve both incrementing and decrementing the same variable. For such algorithms, the use of an up/down counter may be appropriate. The up/down counter has three command inputs. The `ld` command signal is the same as in the earlier registers. The `count` command signal causes the counter to increment or decrement at the next rising edge of the clock, depending on the `up` command signal. If `up` is 1 when `count` is 1, the counter increments. If `up` is 0 when `count` is 1, the counter decrements.

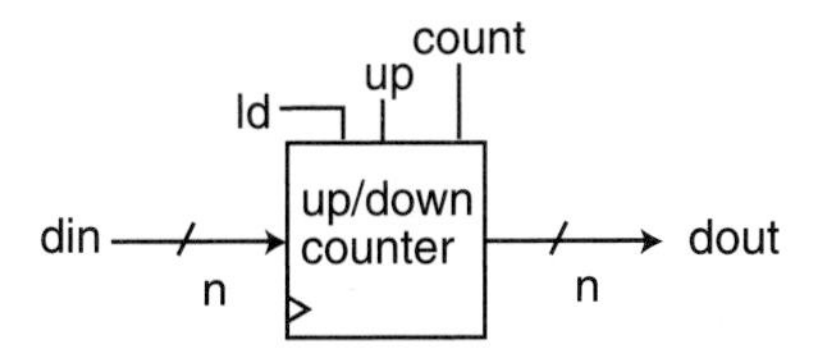

Figure D-20. Symbol for up/down counter register.

The behavior of the up/down counter is summarized by the following action table:

ld	count	up	action
0	0	0	hold
0	0	1	hold
0	1	0	decrement
0	1	1	increment
1	0	0	load
1	0	1	load
1	1	0	load
1	1	1	load

An up/down counter can be constructed from a simple D-type register, two muxes, a combinational logic incrementor and a combinational logic decrementor:

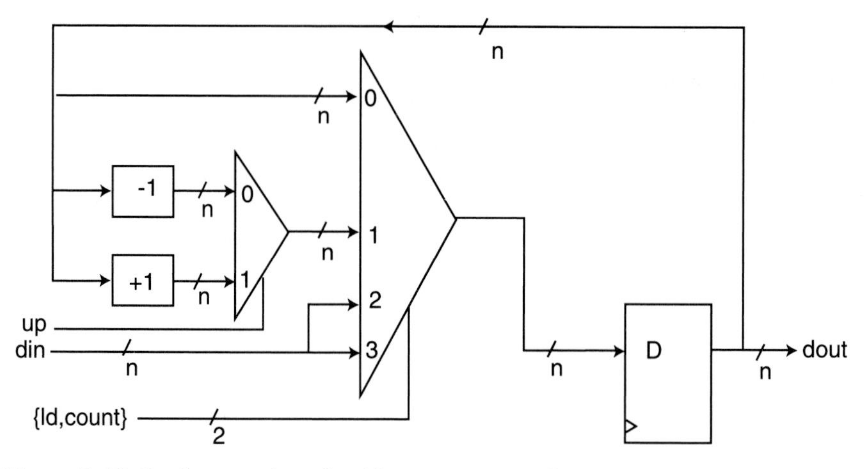

Figure D-21. Implementation of up/down counter register.

There are other ways of constructing this than the technique shown above. For example, in the TTL logic family, the 74xx669 chip provides for `n`=4 the same actions as the above using fewer gates and less propagation delay.

D.9 Shift register

Like counting, multiplication and division by two, as well as the related operations of rotation, can be implemented within a specialized device. Shift registers are sequential building blocks that implement these operations internally. There are many kinds of shift registers. The kind used in this book is a synchronous parallel loadable left/right shift register, with left and right shift (serial) inputs. This device is referred to simply as a shift register in this book.

The shift register has a `clr` signal (similar to the up counter) and a two-bit `shiftctrl` signal. The action table for this shift register is:

clr	shiftctrl	action
0	00	hold
0	01	right
0	10	left
0	11	load
1	00	zero
1	01	zero
1	10	zero
1	11	zero

In addition to the n-bit-wide din bus that all synchronous registers have, the shift register has two inputs, rsi and lsi, each one bit wide, that only play a role when the shift register is shifting:

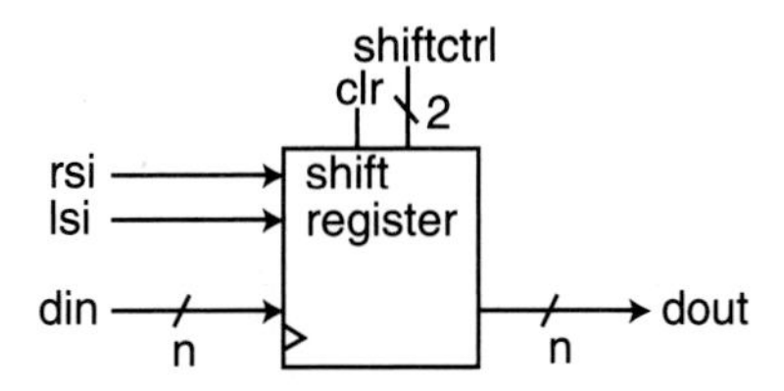

Figure D-22. Symbol for shift register.

The one-bit input rsi is ignored except when the register is shifting right (shiftctrl=01), in which case rsi determines the value of the most significant bit of dout for the next clock cycle. Similarly, lsi is ignored except when the register is shifting left (shiftctrl=10), in which case lsi determines the value of the least significant bit of dout for the next clock cycle.

This can be implemented using two combinational logic shifters, two muxes and a simple D-type register.

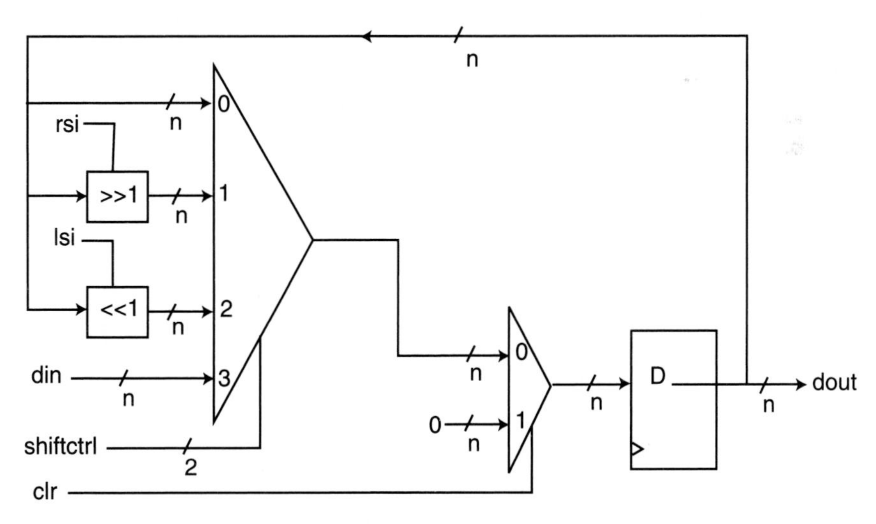

Figure D-23. Implementation of shift register.

Recall that the combinational logic shifters do not cost anything. There are other ways of constructing this than the technique shown above. For example, in the TTL logic family, the 74xx194 chip provides for `n=4` the same actions[4] as the above using fewer gates and less propagation delay.

D.10 Unused inputs

Sometimes a designer needs more capability than an enabled register, but not as much as is offered by one of the other register building blocks described above. For example, a designer may need a register that omits any one of the three command inputs of an up counter:

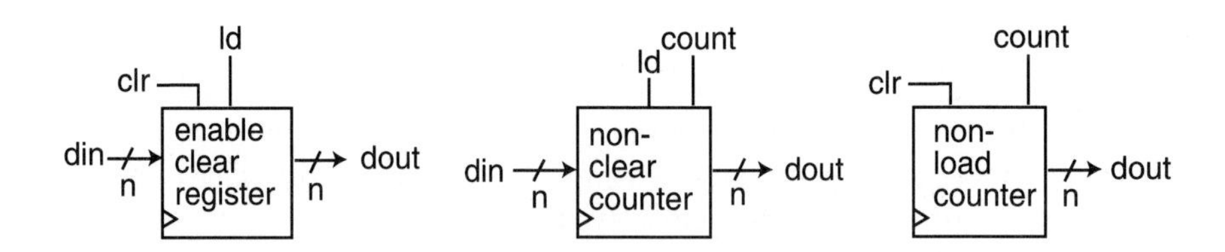

Figure D-24. Symbols for other registers.

The register on the left omits the count signal and is therefore not truly a counter. The register on the left is known as a enabled clearable register. The register in the middle is a counter that does not ever need to be cleared but that instead is loaded with `din`. The register on the right is a counter that never has to be loaded and therefore does not need a `din` bus.

All three of these are specializations of the up counter described in section D.7. They can be implemented by tying one of the three command inputs of an up counter to 0:

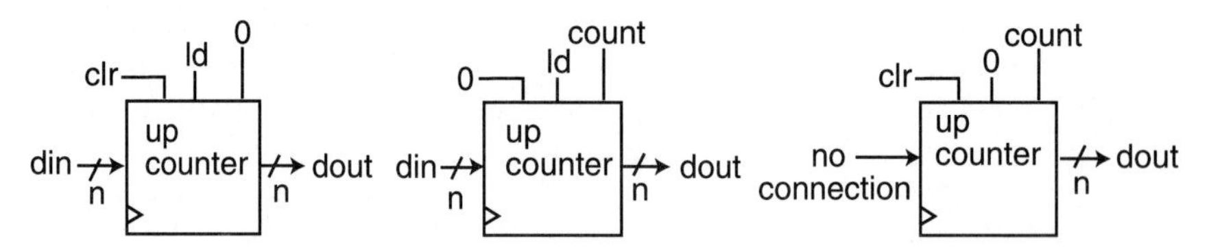

Figure D-25. Implementations for these registers using a loadable clearable up counter.

[4] Except that `clr` is active low.

It is important to understand the distinction between a block diagram and a circuit diagram. A circuit diagram is a detailed description used by people (or more likely automated manufacturing equipment) that put together a machine with no understanding of how the machine was designed. A block diagram is an abstract description used by designers as they think through various design alternatives.

The guiding philosophy for drawing block diagrams is how well the diagram describes the thoughts of the designer. A block diagram should be as simple as possible. Even if the circuit diagram will eventually use a counter with one of its command inputs tied to zero, it is easier for designers to communicate with each other by simply omitting that detail. Designers understand that one way of implementing the following:

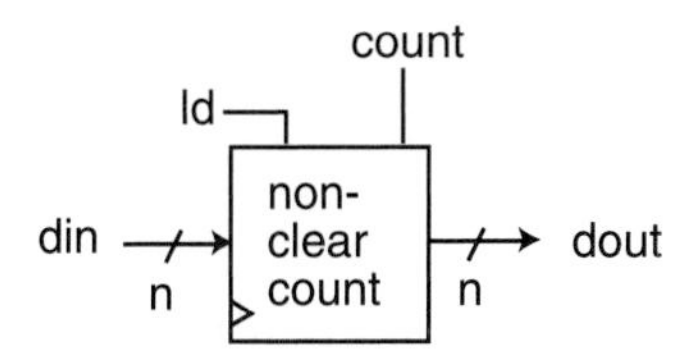

Figure D-26. Symbol for a non-clearable up counter.

is as:

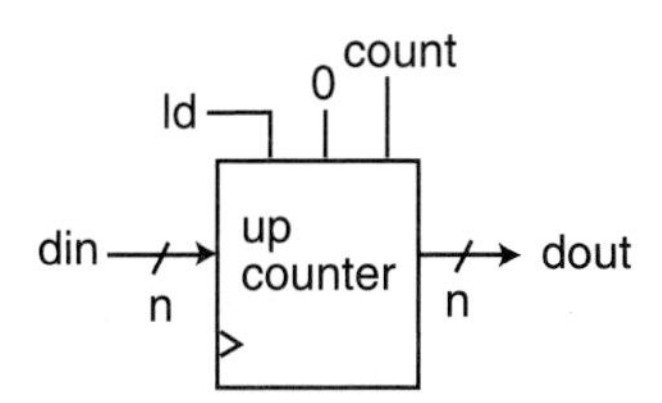

Figure D-27. Possible implementation using a clearable up counter.

although there is probably a more efficient way. Rather than overspecifying a block diagram with details, the designer only shows what is essential to the problem being solved. This is the same philosophical reason why we omit drawing the connection to `sysclk`, ground and Vcc: we know they have to be there,[5] and so why clutter the block diagram with a detail that adds nothing to our understanding?

In a similar way, it is common to use a shift register that never needs to be cleared:

[5] Vcc and ground supply power to a chip. The chip will not operate without these connections. Likewise, synchronous devices will not operate without a connection to `sysclk`.

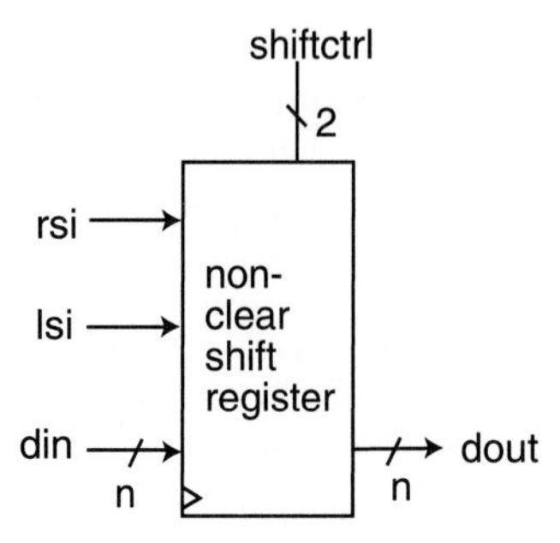

Figure D-28. Symbol for a non-clearable shift register.

This can be implemented as:

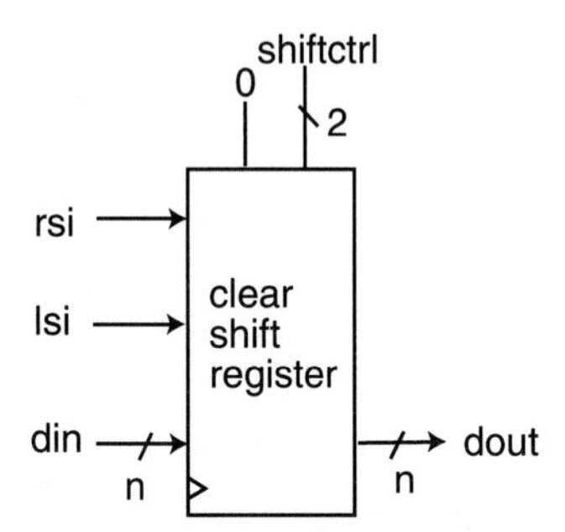

Figure D-29. Possible implementation using a clearable shift register.

D.11 Highly specialized registers

The registers shown above are but a small sample of the ones that are theoretically possible. A designer can create a specialized building block just for a particular problem if the value of `dout` in the next clock cycle can usually be computed as a combinational function of the current `dout`. As with the registers shown above, such specialized building blocks are typically implemented with a simple D-type register combined with muxes and other combinational logic. With the introduction of Verilog synthesis tools in the mid 1990s, designers may start conceptualizing a problem in terms of the building blocks given in earlier sections, only to have the synthesis tool convert those building blocks into some more efficient specialized one which is specific to their particular algorithm.

From a theoretical viewpoint, every computer can be thought of as a single very big register, whose value is meaningless to the human mind. In essence, this theoretical approach treats this one register as the concatenation of every piece of information the computer needs to remember. Mathematicians like to conceptualize things this way, but such an approach is an oversimplification that does not help a practical designer.

The building blocks given earlier are at the right level of abstraction for practical use. They are available as isolated chips (74xx377, 74xx378, 74xx163, 74xx669 and 74xx194) suitable for laboratory experiments which build the confidence of novice designers. They are commonly used by synthesis tools, even though synthesis tools may sometimes do something more sophisticated. In order to understand the more sophisticated things that synthesis tools do, one must already be familiar with the building blocks given in the earlier sections of this appendix.

D.12 Further Reading

GAJSKI, DANIEL D., *Principles of Digital Design*, Prentice Hall, Upper Saddle River, NJ, 1997. Chapter 6.

PROSSER, FRANKLIN P. and DAVID E. WINKEL, *The Art of Digital Design: An Introduction to Top Down Design*, 2nd ed., Prentice Hall PTR, Englewood Cliffs, NJ, 1987. Chapter 4.

D.13 Exercises

D-1. Complete the following timing diagram to show `dout`, given an enabled register with a 4-bit `din`, and a control input `ld`:

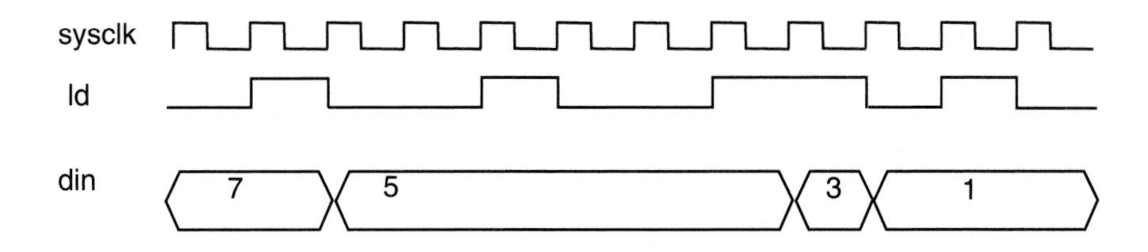

D-2. Complete the following timing diagram to show `dout`, given a shift register with a 4-bit `din`, a 2-bit control input `shf` and inputs `clr`, `rsi` and `lsi`:

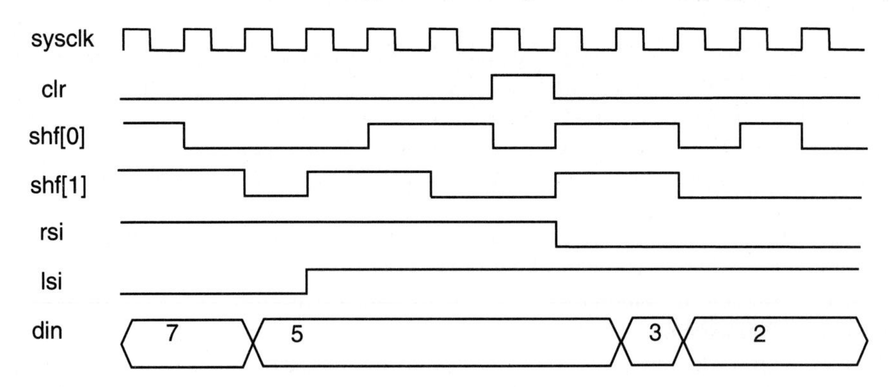

D-3. Complete the following timing diagram to show `dout`, given an up counter register with a 4-bit `din`, and control inputs `clr`, `ld` and `inc`:

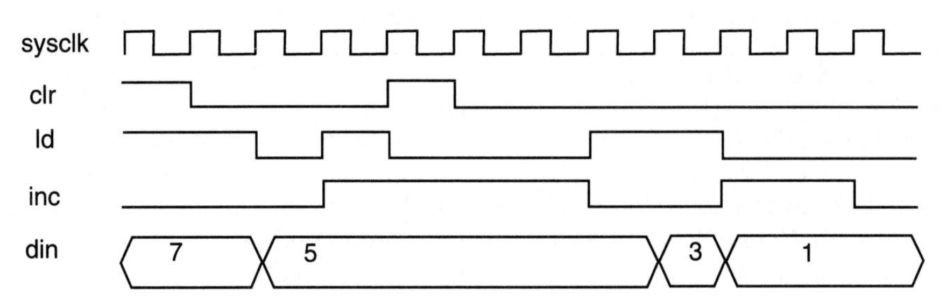

D-4. Complete the following timing diagram to show `dout`, given an up/down counter register with a 4-bit `din`, and control inputs `count`, `ld` and `up`:

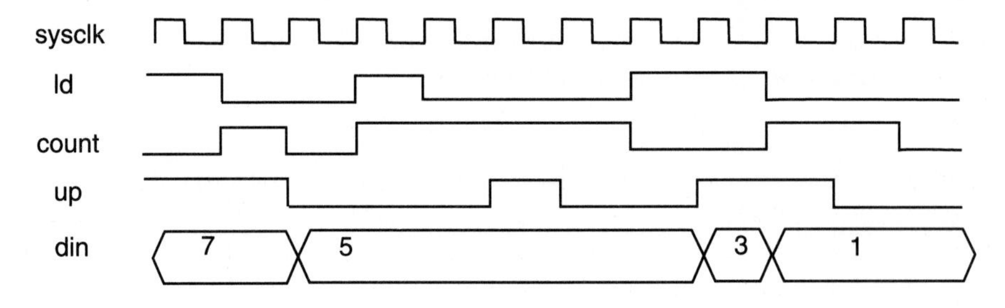

E. TRI-STATE DEVICES

A *tri-state device* is a special kind of combinational building block that has the ability to disconnect its output logically from the bus to which that output is physically connected. For simplicity, the combinational devices defined in appendix C and used throughout most of this book do not have tri-state capabilities, although many actual chips do. This appendix describes what tri-state devices are, and shows two common uses for them.

E.1 Switches

As explained in section C.2.1, a bus is composed of several wires that run in parallel to each other. The bit transmitted on each wire of the bus originates at the output of some gate (such as an AND gate), and is received at the input(s) of other gate(s). Although computer designers normally prefer to abstract away the electronic details of how a gate operates, some understanding of how a non-tri-state device operates is necessary to understand the extra feature provided by a tri-state device.

Each non-tri-state gate is actually composed of several simpler switching devices, such as transistors. Although the details in the operation of these switching devices depend upon the technology family used (CMOS, TTL, etc.), the effect they have on the gate's output is partly analogous to the effect that a wall switch has on the voltage across the filament of a light bulb. When the wall switch is open, the light is turned off because the voltage at point `a` is independent of the voltage at point `b`:

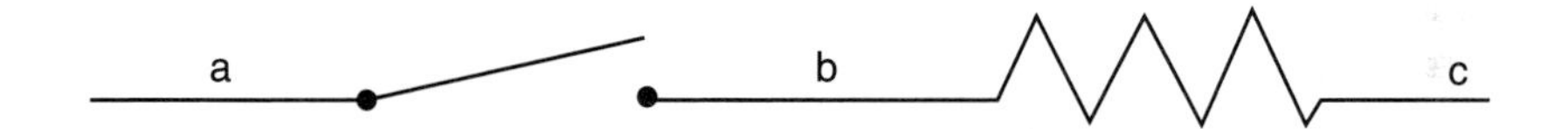

Figure E-1. An open switch causes a light to be off.

Saying that the switch is open is the same as saying `a` is disconnected from `b`. Since the filament of an ordinary light bulb is really just a wire that is a poor conductor (a resistor), the voltage at `b` will be the same as at `c`. For this reason, the filament is cool, and the light does not shine. On the other hand, when the switch is closed, the light is turned on because the voltage at `a` is identical to the voltage at `b`.

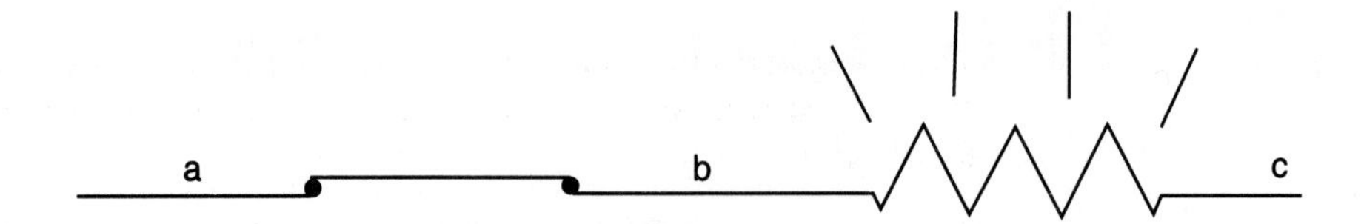

Figure E-2. A closed switch causes the light to be on.

Saying that the switch is closed is the same as saying a is connected to b.

E.1.1 Use of switches in non-tri-state gates

Non-tri-state gates are more complicated than light switches in two ways. First, the gate has to compute the desired output bit (which may require switching devices not described here). Second, the gate has to connect the output wire to the proper voltage.

In most technologies, connecting the output wire to the proper voltage requires two switches: the top switch connects the output wire to the voltage[1] for the bit 1, and the bottom switch connects the output wire to the voltage[2] for the bit 0. For example, suppose the gate needs to output the bit 0. To do this, the "1" switch is open and the "0" switch is closed:

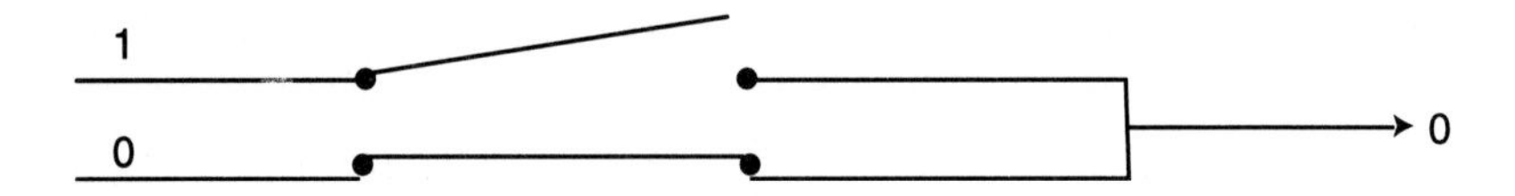

Figure E-3. A gate producing 0 as output.

The only other possibility for a non-tri-state gate is that the gate needs to output the bit 1. To do this, the "1" switch is closed and the "0" switch is open:

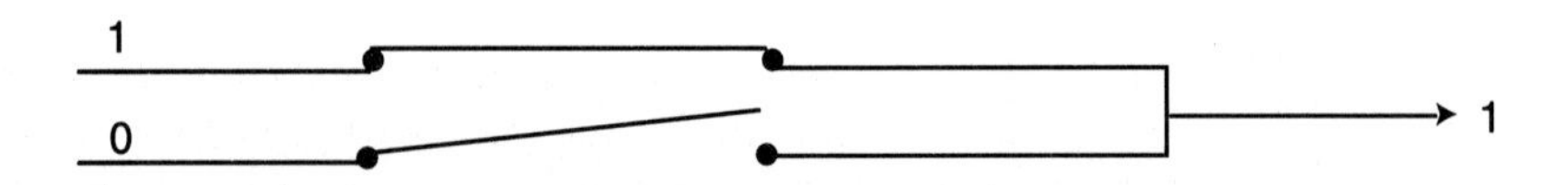

Figure E-4. A gate producing 1 as output.

[1] For active high TTL, 5 volts.
[2] For active high TTL, 0 volts.

A non-tri-state gate is always in one of these two configuration ("1" switch open and "0" switch closed or vice versa).

E.1.2 Use of switches in tri-state gates

The electronic distinction between a non-tri-state gate and a tri-state gate is that a tri-state gate allows a third configuration[3] (both the "1" switch and the "0" switch open):

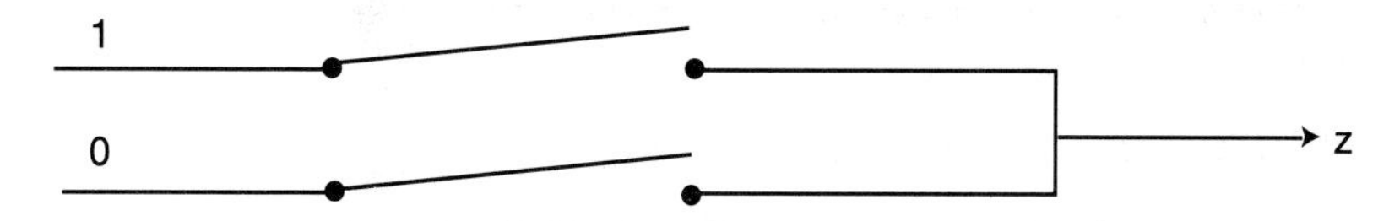

Figure E-5. A gate producing z as output.

The output wire is logically disconnected from the part of the gate the computes an answer. The voltage on the output wire will not be determined by this gate (but could be determined by some other gate). To denote this situation symbolically, we say that the output bit is z (`1'bz` in proper Verilog notation), which stands for high impedance.

E.2 Single bit tri-state gate in structural Verilog

A tri-state gate has two inputs, `enable` and `in`, and one output, `out`:

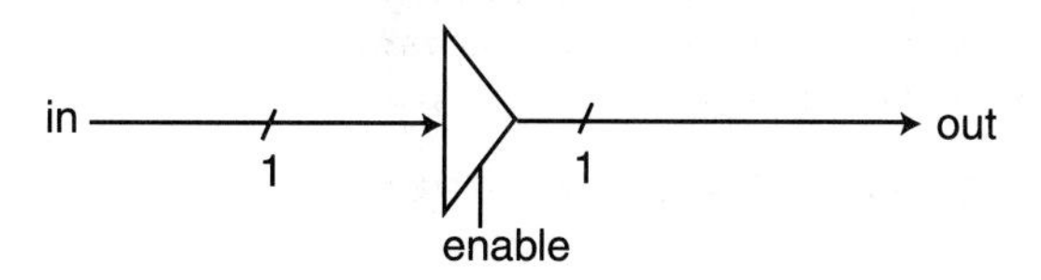

Figure E-6. A tri-state gate.

The behavior of this gate can be described by the following truth table:

enable	in	out
0	0	z
0	1	z
1	0	0
1	1	1

[3] Hence the name tri-state.

In other words, the tri-state driver gate is really nothing more than an electronically controlled switch. When `enable` is 1, the switch is closed:

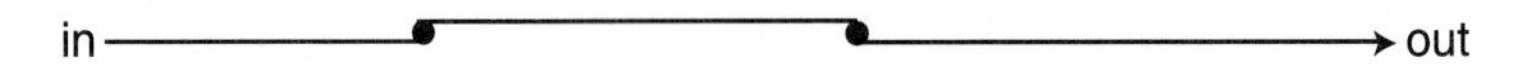

Figure E-7. Effect of tri-state gate when enable is 1.

When `enable` is 0, the switch is open:

Figure E-8. Effect of tri-state gate when `enable` *is 0.*

There is a Verilog built-in gate, known as `bufif1`, that implements this. For example, the following instance:

```
wire out,in,enable;
bufif1 b1(out,in,enable);
```

is equivalent to the single-bit tri-state gate shown above.

As described in section 6.3.4, Verilog allows you to indicate the propagation delay of a built-in gate, such as the `bufif1`:

```
wire out,in,enable;
bufif1 #10 b1(out,in,enable);
```

Also, Verilog allows you to indicate different delays for the (rising) time required to change to a one and the (falling) time required to change to a zero. For built-in gates such as `bufif1`, there is a third separate time that may be of interest in some designs, the *turn off* delay, which is how long it takes when the output changes to `1'bz`. For example:

```
wire out,in,enable;
bufif1 #(10,20,30) b1(out,in,enable);
```

takes 10 units of `$time` if `out` becomes one, 20 units of `$time` if `out` becomes zero and 30 units of `$time` if `out` becomes `1'bz`.

Also Verilog provides other forms of tri-state gates, such as `bufif0`, which has an active low enable signal. For example, the following:

```
wire out,in,enable,enable_low;
not i1(enable_low,enable);
bufif0 #(10,20,30) b1(out,in,enable_low);
```

is functionally identical to the above, as explained by the following:

enable_low	in	out
0	0	0
0	1	1
1	0	z
1	1	z

E.3 Bus drivers

It is the computer designer's job to avoid details such as the ones given above. In order to work with tri-state devices without having to get down to the gate level, we need a more abstract model of what a tri-state device does. Such an abstract model will allow us to use bus-width tri-state devices without having to be concerned with how the tri-state gates switch on and off for each individual bit. This abstract model describes the actions of the tri-state device in terms of a non-tri-state device connected to a special device, known as a tri-state bus driver:

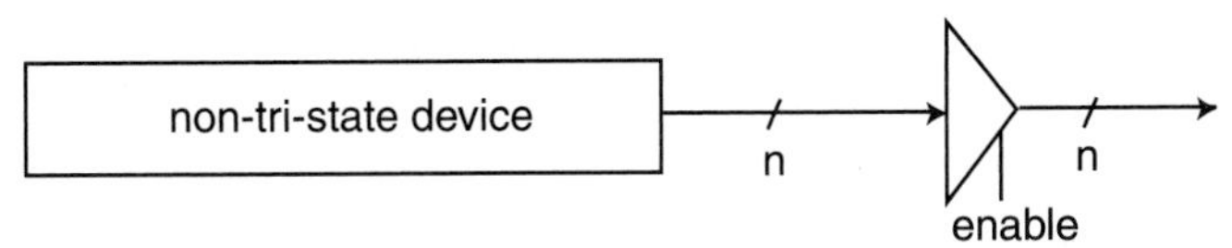

Figure E-9. Tri-state bus driver.

The symbol for a tri-state bus driver looks like a mux, except there is only one input bus (which is n bits wide). Since a mux always has at least two input busses, there should be no reason to confuse these two devices, both of which are symbolically represented as triangles.

Physically, the tri-state bus driver is composed of n independent tri-state driver gates, each one of which is physically a bufif1 instance. Like all other gate-level features of Verilog, working with bufif1 gates is not easy, and so it is better to think of an n-bit-wide tri-state bus driver like any other bus-width building block device, using the combinational logic modeling technique described in section 3.7.2.1:

```
module tristate_buffer(out,in,en);
  parameter SIZE=1;
  output out;
  input in,en;
  reg [SIZE-1:0] out;
  wire [SIZE-1:0] in;
  wire en;
  always @(in or en)
    begin
       if (en === 1)
         out = in;
       else if (en === 0)
         out = 'bz;
       else
         out = 'bx;
     end
endmodule
```

The 'bz provides as many 1'bz values as is required by SIZE.

E.4 Uses of tri-state

There are two main uses of tri-state devices: replacement for muxes and bidirectional buses.

E.4.1 Tri-state buffers as a mux replacement

The first primary use of tri-state bus drivers is to create a structure that is a replacement for a mux. For example, the following:

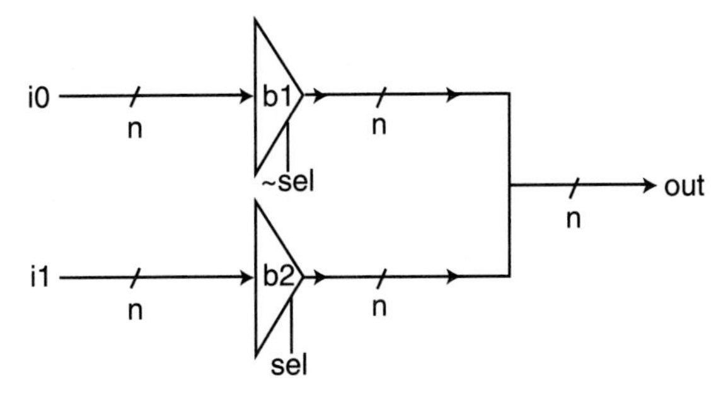

Figure E-10. Using tri-state bus drivers to form a mux.

serves the same role as a two-input mux. The above can be described in Verilog by using two instances of the `tristate_buffer` defined in the last section:

```
module silly_mux(out,i0,i1,sel);
  parameter SIZE=1;
  output out;
  input i0, i1, sel;
  wire [SIZE-1:0] out,i0,i1;
  wire sel;
  wire nsel;

  not n1(nsel, sel);
  tristate_buffer #SIZE b1(out, i0, nsel);
  tristate_buffer #SIZE b2(out, i1, sel);
endmodule
```

E.4.1.1 How Verilog processes four-valued logic

Section 3.5.3 describes the four-valued logic (`0, 1, 1'bz, 1'bx`) used for each bit of Verilog `wires` and `regs`. The need for the binary values 0 and 1 is obvious. The value `1'bx` is often the result of a misconnection of gates. In this appendix, the reason for the fourth value, `1'bz`, should now become clear.

If it were not for the high-impedance value, `1'bz`, it would never make sense for the outputs of two devices to be tied together, such as shown above in the diagram of section E.4.1. Because of `1'bz`, smoke does not come out of the chip when the two tri-state buffers are wired together.

There is an algorithm built into Verilog that models the physical behavior of a wire, based upon the output port(s) of instantiated modules to which that wire is connected. When there is only one output port connected to a wire, the value of the wire in question reflects the value of that single-output port. When that single-output port changes, the wire connected to it is instantaneously and automatically changed. This is the situation that occurs throughout most of the structural examples this book.

The situation is more complicated when there are two or more output ports connected to the same wire. In this example, the output ports of b1 and b2 both drive the same wire. In hierarchical naming (section 3.10.8) the output ports are b1.out and b2.out, and the wire they both drive is simply out. The following table describes what Verilog computes automatically as a particular bit of the wire out, given the corresponding bits of b1.out and b2.out:

	b2.out	0	1	z	x
b1.out					
0		0	x	**0**	x
1		x	1	**1**	x
z		**0**	**1**	**z**	x
x		x	x	x	x

If we guarantee either that every bit of either b1.out is 1'bz or that every bit of b2.out is 1'bz, we can be certain that no bit of out will be 1'bx (see bold above). This is precisely what the two tri-state drivers do for us. When sel is 1, every bit of b1.out is tri-stated, but when sel is 0, every bit of b2.out is tri-stated.

*E.4.1.2 The **tri** declaration*

Verilog provides an alternative to declaring a wire when tri-state drivers are in use, known as tri. The following would also have been legal inside the declaration of silly_mux:

```
wire [SIZE-1:0] i0,i1;
tri [SIZE-1:0] out;
```

The wire and tri declarations do the same thing, and so which one to use is a matter of personal taste.

E.4.2 Bidirectional buses

Although most of this book assumes that a `wire` is essentially free in the fabricated hardware, in fact a wire does cost something. The cost is fairly reasonable when the wire remains hidden inside a physical chip, but the cost is quite high when that wire must be routed outside the chip.

Each bit of a Verilog wire that must be routed outside a chip requires a physical pin. One of the most severe limitations in hardware design is the number of pins available for a chip. Therefore, hardware designers often wish to make maximum utilization of the pins that are available. A *bidirectional bus* is one which sends information both ways:

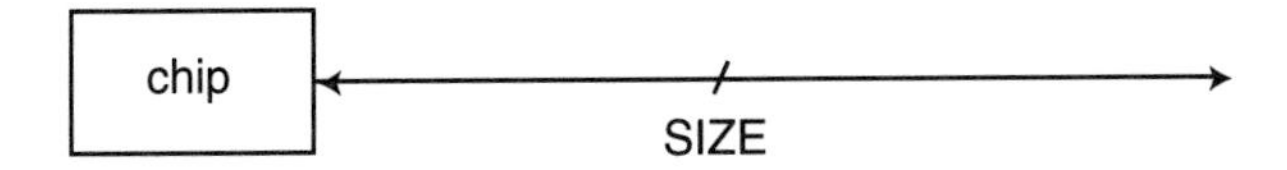

Figure E-11. One bidirectional bus.

Routing a bidirectional bus off chip requires half the number of pins that routing two unidirectional buses requires:

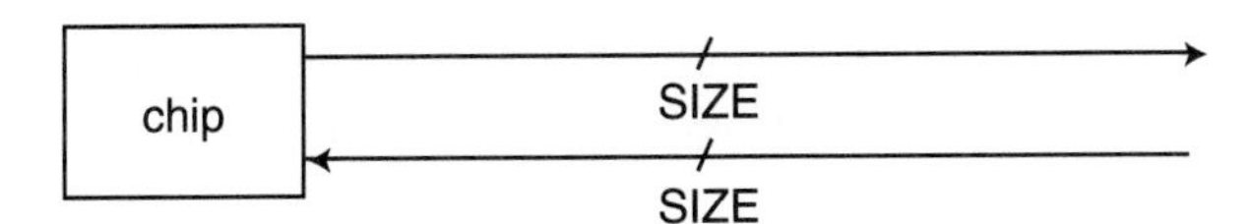

Figure E-12. Two unidirectional buses.

Bidirectional buses are especially important in the design of memory systems (section 8.2.2.1).

E.4.2.1 The `inout` declaration

In order for a Verilog module to use a bidirectional bus, the port for the bidirectional bus must be declared as `inout`. An `inout` port must be declared as a `wire` (or `tri`). Therefore, an `inout` port cannot be directly given its value by the behavioral code of the module in which it is declared. The `inout` port means the wire connecting the instantiated and instantiating modules is physically tied together, without an intervening buffer.[4]

[4] Although it is not necessary, an `input` port could have an intervening buffer into the chip, and an `output` port could have an intervening buffer out of the chip.

The algorithm Verilog uses to determine the value on an `inout` port combines the value outside the module together with the value inside the module, according to the table given in section E.4.1.1 (except the names will be at a different point in the hierarchy than `b1.out` and `b2.out`). The distinction between an `input` port and an `inout` port is not visible within the instantiated module (containing the `inout` declaration). This distinction is only visible within some other instantiating module (which instantiates the module having the `inout` port).

E.4.2.2 A read/write register

To illustrate how a bidirectional bus can reduce the number of pins on a chip, consider a register whose values can be read and written using a single bidirectional bus:

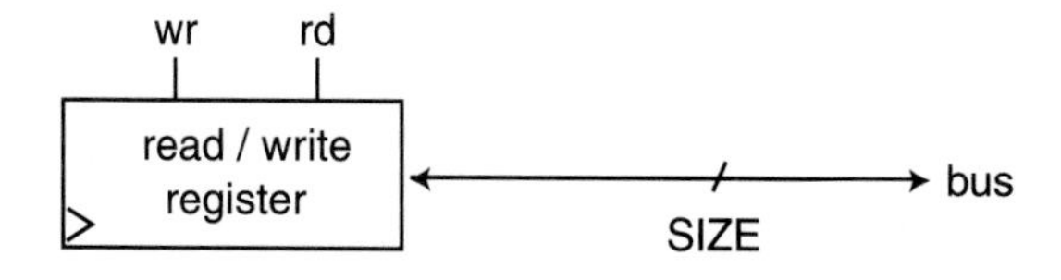

Figure E-13. A read/write register with a bidirectional bus.

If this device were fabricated on a single chip, it would require 5+SIZE pins (including the clock and power). In comparison, the enabled register using unidirectional buses (described in sections D.6 and 4.2.1.1) would require 4+2*SIZE pins, which is almost twice as many.

In order for the bidirectional bus to do double duty, there must be two command inputs: `rd` and `wr`. When `rd` is one, this device drives the bus (provides output) to show the current contents of the register. When `wr` is one, this device leaves the bus alone (`'bz`) and instead the bus provides the input which the register will load at the next rising edge of the clock. Here is the internal structure of this register:

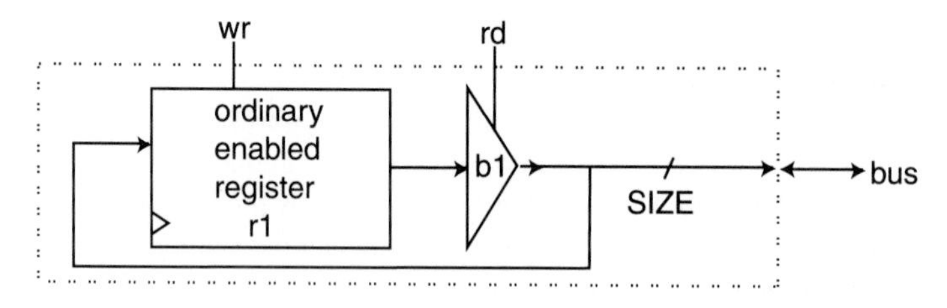

Figure E-14. Implementation of figure E-13.

and here is how this can be described in Verilog:

```
module rw_register(bus,rd,wr,clk);
  parameter SIZE = 1;
  inout bus;
  input rd,wr,clk;
  wire [SIZE-1:0] bus;
  wire [SIZE-1:0] do;
  wire rd,wr;

  enabled_register #SIZE r1(bus,do,wr,clk);
  tristate_buffer #SIZE b1(bus,do,rd);
endmodule
```

Here is an example of using two instances of the read/write register defined above:

```
reg r1rd,r1wr,r2rd,r2wr;
wire [3:0] bus1;

rw_register #4 r1(bus1, r1rd, r1wr, sysclk);
rw_register #4 r2(bus1, r2rd, r2wr, sysclk);
```

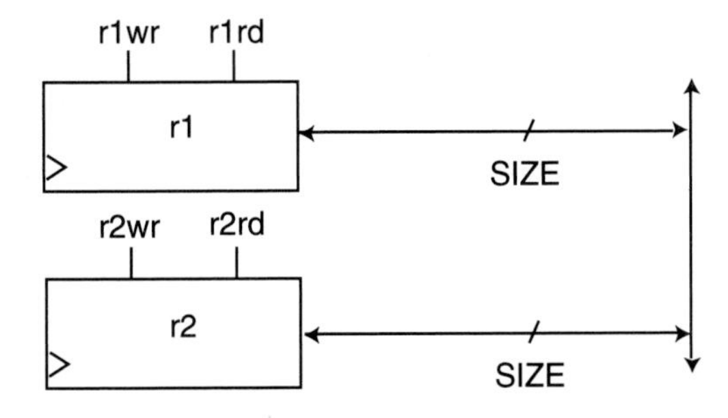

Figure E-15. Instantiation of two read/write registers.

Unlike the `silly_mux` example, there is nothing in the above to guarantee that `bus1` will avoid becoming `'bx`. Instead, it is the responsibility of the designer to ensure that `r1rd` and `r2rd` are never simultaneously one.

For example, to implement the register transfer `r1` $\leftarrow$ `r2` requires generating the commands:

```
                    r1rd = 0;
                    r1wr = 1;
                    r2rd = 1;
                    r2wr = 0;
```

E.5 Further Reading

PALNITKAR, S., *Verilog HDL: A Guide to Digital Design and Synthesis,* Prentice Hall PTR, Upper Saddle River, NJ, 1996. Chapter 5.

E.6 Exercises

E-1. Revise the architecture of the two-state division machine (whose Verilog code is given in section 4.2.3) so as to eliminate the instance of `mux2` and instead use two instances of the `tri-state_buffer` defined in section E.3. Use the test code given in section 4.1.1.1.

E-2. Define a behavioral Verilog module (`bi_mem`) for an asynchronous bidirectional memory (section 8.2.2.1) consisting of 4096 words, each 12-bits wide. The ports are a 12-bit `addr` bus, a 12-bit `data` bus and the commands `write` and `enable`. The following table describes the actions of this memory:

enable	write	action
0	-	data = 12'bz
1	0	data = m[addr]
1	1	m[addr] ← data

E-3. Show how the ASM of section 8.4.6 and the architecture of section 8.4.4 need to be modified to work with the memory defined in problem E-2.

E-4. Define the Verilog corresponding to problem E-3.

E-5. One of the reasons why the tri-state approach is attractive for memory system design is that it allows multiple memory modules to be connected together to form a larger memory without the need for a mux. Using eight instances of the memory defined in problem E-2 together with a decoder having a 3-bit input, give a block diagram that implements a memory of 32,768 12-bit words.

E-6. Give the structural Verilog for problem E-5.

E-7. Pins are so limited in many memory packaging technologies that industry has resorted to several contorted techniques to minimize pin count. One common approach with dynamic memories is to transmit the address in two parts: the *row* address and the *column* address. Internally, this makes for a square geometric arrangement consisting of rows and columns of identical memory cells. Externally, this cuts the number of address pins in half by (approximately) doubling the time to access a word. To distinguish the use of the half-size `addr` bus, there are two input signals: `ras` and `cas`. If `ras` is asserted when `cas` is not, the `addr` bus indicates the row. When both are asserted, the `addr` bus indicates the column. For simplicity, assume `ras` and `cas` signals are synchronous to the `sysclk` input which is provided to this chip. (Often such chips are asynchronous with much stricter timing constraints than are shown below). The following timing diagram illustrates reading a word from such a memory:

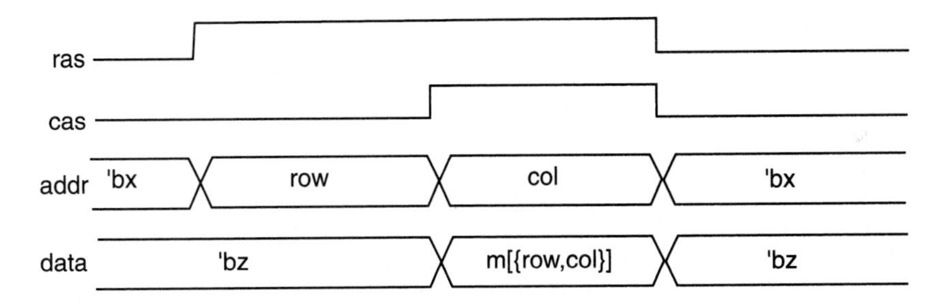

Writing to such a memory is similar, except a `write` signal is asserted and the new content is provided to the chip on the data bus during the entire time. Define a structural Verilog module for a memory containing 16 twelve-bit words using twenty instances of the `rw_register` defined in section E.4.2.2 together with additional combinational logic.

F. TOOLS AND RESOURCES

There are several Verilog example files used in this book. There are also several design automation software packages (tools) used with these files. In addition, there are several information resources that may be helpful on the Internet. This appendix briefly describes how to obtain and use these tools and resources. The details are subject to change, and the respective Web sites should contain the most up-to-date information.

F.1 Prentice Hall

Selected Verilog examples can be downloaded from the Prentice Hall Web site, `www.phptr.com`.

F.2 VeriWell Simulator

Most of the examples in this book have been tested using the Verilog simulator from Wellspring Solutions, Inc, known as VeriWell. At the time of this writing, this excellent software package is available at no charge by downloading it from `www.wellspring.com`. The downloaded version has limits on the size of Verilog source files that it accepts and does not provide graphical (timing diagram) output. The downloaded version is available for MSDOS (command-line), Windows 95/NT (GUI), Macintosh (GUI) and several UNIX (command-line) dialects. For the command-line versions, simply type:

```
veriwell file1.v file2.v ...
```

which will produce the output of `$display` commands both on the screen and in a file known as `veriwell.log`. For the GUI versions, you need to create a "project file" by selecting Project (Alt P) New (Alt N) and choose a name (`.prj`) for the project file. Then select Project (Alt P) Add file (Alt F) to specify the `.v` file name(s). To run the simulator, select Project (Alt P) rUn (Alt U).

Most of the designs in this book are able to simulate on the free version. Wellspring Solutions sells a hardware key that removes the limitations of the free version and also sells a separate package for graphical output:

Wellspring Solutions
7 Tudor Drive, Suite 300
Salem, NH 03079
(603) 898-1100

F.3 M4-128/64 demoboard

Documentation for the M4-128/64 CPLD chip used in chapter 11 can be downloaded from `www.vantis.com`. The demoboard, power supply, download cable and MACHPRO software can be obtained from:

Vantis
Box 3755
Sunnyvale, CA 94088

F.4 Wirewrap supplies

To build the CPU described in chapter 11 requires wirewrap wire, a wirewrap tool (such as an "all in one" tool that strips, wraps and unwraps) and a wirewrap socket. It also requires a memory ("RAM") chip, such as the 2102. Some of these may be available at local electronics stores, but there are several mail-order companies, such as Jamesco (`www.jamesco.com`), that carry a complete selection of such supplies.

F.5 VerilogEASY

The synthesis package used in chapter 11, known as VerilogEASY, is sold by MINC, Inc. (`www.minc.com`). VerilogEASY comes in several versions, each targeting different vendors' programmable logic. A limited version of VerilogEASY that targets the M4-128/64, but that is restricted on the number of inputs and outputs, will be available to readers of this book in the last quarter of 1998. There is no charge for downloading this limited version, but MINC requires that people downloading their software register at their Web site. VerilogEASY accepts the common synthesizable subset of Verilog other than implicit style. VerilogEASY produces two output files: `.src` (in the proprietary DSL language) and `.v` (structural Verilog netlist). To fabricate working hardware requires other tools, described in sections F.3 and F.6. MINC also sells a full version of VerilogEASY and an even more powerful synthesis tool, known as PLSynthesizer:

MINC, Inc.
6755 Earl Drive
Colorado Springs, CO 80918-1039
(719) 590-1155
`info@minc.com`

F.6 PLDesigner

The place and route tool used in this book for CPLDs, such as the Vantis M4-128/64, is known as PLDesigner. It runs on Windows 95/NT. It is not possible to fabricate a design for the M4-128/64 without using this tool. PLDesigner can be purchased from MINC. The following directions apply to PLDesigner: At the PLDesigner menu, choose File (Alt F) Open (Alt O), and enter the name of the `.src` file created by VerilogEASY. Do a File (Alt F) eXit (Alt X). Select Device (Alt D) Parameters (P) and choose the M4-128/64 (MACH445) and say OK. Select Settings (Alt S) Options (Alt O) and be sure Timing Models are set only to generic Verilog. Select Project (alt P) Build all (Alt B) to create the JEDEC file (`.j1`). To create the back annotated Verilog, select Project (Alt P) Generate Timing Model (Alt T), which will put the `.v` file in a `model` subdirectory (since a similarly named `.v` file (the input to VerilogEASY) will already exist).

F.7 VITO

The Verilog Implicit To One hot (VITO) preprocessor is a freely available synthesis preprocessor written by James D. Shuler and Mark G. Arnold. It may be downloaded from the Prenticite Hall Web site. It can also be downloaded from `www.cs.brockport.edu/~jshuler` or `plum.uwyo.edu/~vito`. UNIX and MSDOS versions are available at those Web sites. The theory of how this tool operates is discussed in chapter 7. It is a command-line program, and the following is a typical use:

```
vito -t implicit.v >explicit.v
```

where `implicit.v` is the name of a file consisting of one or more modules that have implicit style state machines. The `-t` option generates comments that explain the transformation. The output of VITO is redirected to another file (`explicit.v`), which would then be used as the input to VerilogEASY (or another synthesis tool). The designer is free to choose other file names.

Each module must include a `sysclk` and `reset` port. The names of these are also given in a special file known as `vito.rc`. For the M4-128/64 (with its active low `reset`), the `vito.rc` file should contain:

```
vito.out
vito.stmt
vito.arch

s_
sT_
tmp_
join_
ff_
qual_
reset
sysclk
@(posedge sysclk)
new_
@(posedge sysclk or negedge reset)
~reset
vito.tail
```

The other names in this file are the prefixes of wire names that VITO will generate, and the temporary files VITO uses. This file is based on position, and so extra blank lines are not allowed.

F.8 Open Verilog International (OVI)

The independent organization that developed the Verilog standard (IEEE 1364) is known as Open Verilog International (`www.ovi.org`). OVI is co-sponsor of the International Verilog Conference (`www.hdlcon.org`) held each spring at the Santa Clara Convention Center. OVI sells a language reference manual, which is the authoritative source for questions of Verilog syntax and semantics:

Open Verilog International
15466 Los Gatos Blvd.
Suite 109-071
Los Gatos, CA 95032
(408) 353-8899
`ovi@netcom.com`

F.9 Other Verilog and programmable logic vendors

Here is a partial list of other Verilog and vendors' Web sites: www.altera.com, www.avanticorp.com, www.cadence.com, www.fintronic.com, www.sunburst-design.com, www.synopsys.com, www.simucad.com, www.synplicity.com, www.veribest.com, www.xilinx.com and www.eg.bucknell.edu/~cs320/1995-fall/verilog-manual.html.

F.10 PDP-8

Additional resources relating to the PDP-8 can be found at strawberry.uwyo.edu, www.in.net/~bstern/PDP8/pdp8.html, www.faqs.org/faqs/dec-fac and www.cs.uiowa.edu/~jones/pdp8. A portion of this information is also available at the Prentice Hall website.

F.11 ARM

Additional resources relating to the ARM can be found at www.arm.com.

G. ARM INSTRUCTIONS[1]

1. Efficient instruction set

The foundation of all processor architectures is the instruction set. When designing it there are two contradictory aims: high code density and easy instruction decoding. The ARM instruction set strikes an optimal balance between these. The instructions are powerful so programs in ARM are short, saving memory and speeding execution because of reduced bus bandwidth requirements. Yet because the instructions decode so easily the ARM processor is small and cheap, consumes very little power and runs at high speed.

In general RISC processors code less densely than CISCs because of their very nature - yet ARM code is generally as dense as code for 32-bit CISC processors, which is significantly better than other 32-bit RISC processors. With the 16-bit Thumb extension, ARM code density is the best in the business.

Features of the ARM fundamental to easy decoding include:

- A small number of highly flexible instruction types
- Consistent instruction data formats

Features implemented for high code density include:

- Barrel shifter to perform arbitrary shifts within the same cycle, at no speed penalty
- Conditional execution on every instruction to eliminate many branches
- Load and store multiple instructions for rapid context switching and memory transfer

2. Instruction set summary

The ARM instruction set is a good target for compilers of many different high-level languages. Where required, though, assembly code programming in ARM is straightforward and enjoyable. The instructions are flexible and orthogonal, the memory model is flat and there are no complicated instruction interdependencies as there are for some RISC processors. Because a whole line of C code can often be performed within one or two instructions, the instructions correspond closely to natural program steps. See the ARM Code Examples which demonstrate the true power and magic of ARM machine code.

The ARM7 instruction set comprises 10 basic instruction types

- Two of these make use of the on-chip arithmetic logic unit, barrel shifter and multiplier to perform high-speed operations on the data in the 16 visible 32-bit registers.
- three classes of instruction control the transfer of data between main memory and the register bank, one optimized for flexibility of addressing, another for rapid context switching and the third for managing semaphores.
- two instructions control the flow and privilege level of execution.
- three types of instruction are dedicated to the control of external coprocessors which allow the functionality of the instruction set to be extended in an open and uniform way.

The ARM7 instruction set is summarised in the table below:

Instruction Type	Instruction bit format, 31...0
Data Processing	cond 0 0 I Opcode S Rn Rd Operand 2
Multiply	cond 0 0 0 0 0 0 A S Rd Rn Rs 1 0 0 1 Rm
Single Data Swap	cond 0 0 0 1 0 B 0 0 Rn Rd 0 0 0 0 1 0 0 1 Rm
Single Data Transfer	cond 0 1 I P U B W L Rn Rd Offset
Undefined	cond 0 1 1 xxxx 1 xxxx
Block Data Transfer	cond 1 0 0 P U S W L Rn Register List
Branch	cond 1 0 1 L offset
Copro Data Transfer	cond 1 1 0 P U N W L Rn CRd CP# Offset
Copro Data Operation	cond 1 1 1 0 CP OpcP CRn CRd CP# CP 0 CRm
Copro Regester Transfer	cond 1 1 1 0 CPOpcP L CRn CRd CP# CP 1 CRm
Software Interrupt	cond 1 1 1 1 Ignored by processor

Register Model

The processor has a total of 37 registers made up of 31 general 32 bit registers and 6 status registers. At any one time 16 general registers (R0 to R15) and one or two status registers are visible to the programmer. The visible registers depend on the processor mode and the other registers (the *banked registers*) are switched in to support rapid interrupt response and context switching.

Below is a list of the visible registers for each of the processor modes. The banked registers are in italics.

User32	FIQ32	Svc32	Abort32	IRQ32	Undef32
R0	R0	R0	R0	R0	R0
R1	R1	R1	R1	R1	R1
R2	R2	R2	R2	R2	R2
R3	R3	R3	R3	R3	R3
R4	R4	R4	R4	R4	R4
R5	R5	R5	R5	R5	R5
R6	R6	R6	R6	R6	R6
R7	R7	R7	R7	R7	R7
R8	*R8_fiq*	R8	R8	R8	R8
R9	*R9_fiq*	R9	R9	R9	R9
R10	*R10_fiq*	R10	R10	R10	R10
R11	*R11_fiq*	R11	R11	R11	R11
R12	*R12_fiq*	R12	R12	R12	R12
R13	*R13_fiq*	*R13_svc*	*R13_abt*	*R13_irq*	*R13_und*
R14	*R14_fiq*	*R14_svc*	*R14_abt*	*R14_irq*	*R14_und*
R15 (PC)	R15 (PC)	R15 (PC)	R15 (PC)	R15 (PC)	R15 (PC)
CPSR	CPSR	CPSR	CPSR	CPSR	CPSR
–	*SPSR_fiq*	*SPSR_svc*	*SPSR_abt*	*SPSR_irq*	*SPSR_und*

More complete documentation can be found at `www.arm.com`.

H. ANOTHER VIEW ON NON-BLOCKING ASSIGNMENT

There are other forms of the non-blocking assignment besides the RTN form (<= `@(posedge sysclk)`) used earlier in this book:

```
var <= expression;
var <= #delay expression;
```

The semantics of <= and <=`#delay` are more subtle than the extra `always` analogy explained in section 3.8.2, which only applies to <= `@(posedge sysclk)`. In general, non-blocking expressions are evaluated immediately and put into a simulator queue to be stored after all blocking assignments at the `$time` given by the specified `#delay`. Clifford Cummings of Sunburst Design, Inc. (`www.sunburst-design.com`) gave a very informative presentation on the use of such non-blocking assignment statements at the 1998 International Verilog Conference. He suggested that the blocking assignment, =, should be primarily limited to modeling combinational logic, as in:

```
always @(a or b)
  sum = a + b;
```

H.1 Sequential logic

According to Cumming's guidelines, sequential logic, such as a simple D type register, should use the non-blocking assignment:

```
always @(posedge sysclk)
  dout <= din;
```

The <= without time control above has the same meaning as <= `#0` (sections 11.3.3 and 11.5.6). It causes the simulator to put the assignment into a special "non-blocking event queue" that stores a new value after all = and =`#0` have finished but before `$time` advances. The <= without time control is useful for situations where an explicit style `module` uses the same `reg`s on opposite sides (example in bold) of different assignments that model distinct sequential devices, as in:

```
always @(posedge sysclk)
  r2 <= r1;
always @(posedge sysclk)
  r1 <= r2;
```

Assuming `r1` and `r2` were initialized (not shown), the above exchanges `r1` and `r2`, regardless of the order in which the simulator schedules the two `always` blocks. Using = instead of <= above would have the incorrect effect of duplicating the value of one of the registers (seemingly chosen at random) into both of them. To use = in this situation requires intervening combinational logic:

```
always @(posedge sysclk)
  r2 = new_r1;
always @(r1)
  new_r1 = r1;   //identity
```

with similar code for `r2`, which is hard for designers to remember when the combinational logic is simply the identity function. This problem does not occur when the interacting `always` blocks are in separate `module`s because the port(s) act like the intervening combinational logic.

Most existing explicit style designs, including examples in this book (sections 3.7.2.2, 7.2.2.1, 11.3 and 11.7), use = properly (with intervening combinational logic or ports) rather than <= . Probably many designers stumble onto correct sequential logic using = without understanding why it is correct. Even more alarming, some incorrect sequential logic using = may appear to be correct because of the arbitrary order in which the

Verilog simulator schedules the assignments. Cummings suggested designers use only <= for sequential logic to guarantee correct operation without making the designer remember the intervening combinational logic or ports.

H.2 `$strobe`

Cummings also suggested using the `$strobe` system task, which works like `$display`, but shows the result of non-blocking assignment at the same `$time` the assignment is made. For example, instead of the `$display` code with delay used in many examples in this book, as typified by section 3.8.2.3.2:

```
always @(posedge sysclk) #20
  $display("%d a=%d b=%d ", $time, a, b);
```

Cummings would recommend:

```
always ? posedge sysclk)
  $strobe("%d a=%d b=%d ", $time, a, b);
```

which has the advantage that the values that will take effect during a particular clock cycle will be displayed at the actual `$time` of the rising edge. With `$display`, there must be at least #1 delay (#20 in this example) beyond `@(posedge sysclk)` to view the values changed by non-blocking assignments.

H.3 Inertial versus transport delay

An interesting contrast between blocking and non-blocking assignment that Cummings illustrated is the difference between *transport delay*, which retains all the values `signal` has, regardless of how briefly they exist:

```
always @(signal)
  dela <= #3 signal;
```

and *inertial* delay, which filters out certain values of `signal` when values change more rapidly than a specified amount of `$time` (3 in this example):

```
always @(signal)
  delb = #3 signal;
always @(signal)
  #3 delc = signal;

$time  0  1  2  3  4  5  6  7  8  9
signal 1  1  1  1  2  3  3  3  3  3 ...
dela   x  x  x  1  1  1  1  2  3  3 ...
delb   x  x  x  1  1  1  1  2  2  2 ...
delc   x  x  x  1  1  1  1  3  3  3 ...
```

Of course, transport delay is more realistic of how physical signals behave.

H.4 Sequence preservation

Because of their queued implementation, non-blocking assignments at the same `$time` occur in the sequence the <=s executed. For example, `a<=1` followed by `a<=2` stores 2 into `a`. As emphasized in section 9.6, hardware registers (described in implicit style code with `<= @(posedge sysclk)`) cannot store multiple values during a single clock cycle. Even though Verilog allows it, it is inappropriate for implicit style code to have more than one `<= @(posedge sysclk)` to a given `reg` during a particular clock cycle. On the other hand, Cummings pointed out that it is useful in explicit style code for a plain <= to give a default values to the output of a state machine, which a later <= can modify at the same `$time`.

H.5 Further reading

CUMMINGS, CLIFFORD E., "Verilog Nonblocking Assignments Demystified", *7th International Verilog HDL Conference,* Santa Clara, CA, Mar 16-19, 1998, pp. 67-69.

I. GLOSSARY

The following include terms used in computer design. Terms marked with * are unique to this book. Synonyms for terms not used in this book are also given. In addition, the following includes Verilog features (courier font), some of which are not described elsewhere in this book. See the references given at the end of chapter 3 for details about Verilog features not described in this book.

Access time: The *propagation delay* of a *memory*.

Active high: A *pin* of a physical chip where 1 is represented as a high voltage.

Active low: A *pin* of a physical chip where 1 is represented as a low voltage.

***Actor**: A machine or person that interacts with the machine being designed.

Address bus: A *bus* used to indicate which word of a *memory* is selected.

Algorithmic State Machine, *see* ASM

Architecture: 1. The hardware of a machine that manipulates data, as opposed to the *controller*. Is present in *mixed(1)* and *pure structural* designs. Also known as a datapath. 2. The *programmer's model* and *instruction set* of a *general-purpose computer*. See also *computer architecture*. 3. A feature of *VHDL* that provides greater abstraction of *instantiation* than *Verilog* does.

ALU (Arithmetic Logic Unit): *Combinational* logic capable of computing several different functions of its input based on a command signal. Typically, the functions include arithmetic operations, such as addition, and bitwise (logical) operations, such as AND.

ASM (Algorithmic State Machine): A graphic notation for finite state machines consisting of *rectangles(1)*, *diamonds* (or equivalently *hexagons*), and possibly (for *Mealy machines*) *ovals*. A *pure behavioral* ASM is equivalent to *implicit style* Verilog with *non-blocking assignment*. Moore *mixed(1)* ASMs can be implemented as implicit style Mealy Verilog.

Asynchronous: Logic which has memory but which does not use the *system clock*.

Backannotation: Recording the *propagation delay* in a *netlist* after *synthesis*.

Behavioral: Code which describes what a machine does, rather than how to build it. *see also pure behavioral*.

Big endian notation: The most significant bit (byte, word, etc.) is labeled as 0.

Blocking procedural assignment: A Verilog statement (=) that evaluates an expression now, causes the process to delay for a specified time and then stores the result.

Bottom testing loop: A loop that is guaranteed to execute at least once. Such loops are difficult to code in *implicit style* Verilog. For *simulation*, use `enter_new_state` task with !== to the bottom state. For synthesis, use `disable` inside `forever`.

Bus: A groups of wires that transmit information.

Bus driver: A *tri-state* device that passes through its input when it is enabled, but outputs `'bz` when it is not.

Cache: A *memory* that allows faster access to words used most frequently.

casex: A variation of `case` that treats `1'bz` or `1'bx` as a don't care. For example, the 13 bits of output for the truth table given in section 2.4.1 could be coded as:

```
always @(ps or pb or r1gey)
 begin
   casex ({ps,pb,r1gey})
    3'b00x: t=13'b0110001010101;
    3'b01x: t=13'b1110001010101;
    3'b100: t=13'b0101110110010;
    3'b110: t=13'b1101110110010;
    3'b1x1: t=13'b1101110110010;
   endcase
   {ns,ldr1,clrr2,incr2,ldr3,muxctrl,aluctrl,ready}=t;
 end
```

`casez`: Like `casex`, except it only uses `1'bz`.

***Central ALU**: An *architecture(1)* that uses a single *ALU* for all computation. The associated *pure behavioral ASM* is usually restricted to one register transfer (*RTN* or *non-blocking assignment*) per *state*, and so algorithms designed for a central ALU architecture are usually slower than those designed for *methodical* architectures.

Central Processing Unit: see *CPU*.

CPU (Central Processing Unit): The main element of a *general-purpose computer*, besides *memory*.

Combinational: Logic which has no memory. In Verilog, *ideal* combinational logic (including a *bus* or *tri-state* device) is modeled with @ followed by a *sensitivity list* or by a *continuous assignment*.

Combinatorial: *see* Combinational

Command signal: 1. An internal signal output from a *controller* that tells the *architecture(1)* what to do. Found only at the *mixed(1)* and *pure structural* stages. 2. An external signal output from a controller to another *actor*.

Computer architecture: 1. *see Programmer's model* and *instruction set architecture.* 2. A generic term for a field of study that encompasses the computer design topics in this book along with more abstract modeling concerns not discussed here, such as networked general-purpose computers, disk drives and associated software operating system issues.

Concatenation: The joining together of bits, indicated by { } in Verilog.

Conditional: *Non-blocking assignments (RTN)* and/or *command signals* that occur in a particular *state* only under certain conditions. See *Mealy* and *oval.*

Continuous assignment: A shorthand for instantiating a hidden module that defines behavioral *combinational logic.* Allows assignment to a Verilog `wire`. Eliminates the need to declare *ports* and *sensitivity lists.*

Controller: The hardware of a machine that keeps track of what step of the algorithm is currently active. Described as an *ASM* at the *mixed(1)* stage, but as a *present state* and *next state* logic at the *pure structural* stage.

CPLD (Complex Programmable Logic Device): A fixed set of AND/OR gates optionally attached to flip flops with a programmable interconnection network allowing the downloading of arbitrary *netlist*s.

Data bus: A *bus* used to transmit words to and from a *memory.*

Datapath, *see architecture(1)*

`defparam`: An alternative way of instantiating a different constant for a `parameter`.

Dependent: Two or more computations where the evaluation of some parts depends on the result of other parts. It is hard to design *pipeline* and *superscalar* computers when computations are dependent.

Diamond: The *ASM* symbol for a decision, usually equivalent to an `if` or `while` in *implicit style* Verilog.

Digital: Pertaining to discrete information, e.g., bits. See also *special-purpose computer.*

`$dumpfile`: System task, whose argument is a quoted file name, for VCD. For a complete dump, also need `$dumpvars` and `$dumpon`. Other tasks exist for more limited dumps.

Enabled register: A *synchronous* register that has the ability to hold as well as load data.

***External status**: Information from outside the machine used to make decisions.

`enter_new_state`: A user defined task only for *simulation* at the *pure behavioral* and *mixed(1)* stages. Establishes default *command signals*. Helps with *Mealy* machines and *bottom testing loops*.

Explicit style: A finite state machine described in terms of *next state* transitions. Does not have multiple `@(posedge sysclk)`s inside the `always`. Roughly equivalent to the *pure structural* stage, except instead of separate modules for the *present state* register and the *next state* logic, both are often coded in the same module, and the *architecture* is modeled in a behavioral style, often called *RTL*. Explicit style requires the designer to think in terms of *goto*s. Contrast *implicit style*.

Falling delay: The propagation delay it takes for an output to change to 0.

`$fclose`: System task, whose argument is a file handle, that closes the associated file. See also `$fopen`.

`$fdisplay`: A variation of `$display` that outputs to a file. See also `$fopen`, `$fclose`.

Field Programmable Gate Array: *see FPGA*

Finite state machine, *see ASM*.

Flip Flop: A *sequential(1)* logic device that stores one bit of information. Used to build *register*s and *controllers*.

`$fopen`: System function, whose argument is a quoted file name, that returns an integer file handle used by `$fdisplay`, `$fstrobe` or `$fwrite`:

```
integer handle;
initial
 begin
   handle = $fopen("example.txt");
   $fdisplay(handle,"Example of file output");
   $fclose(handle);
 end
```

`fork`: An alternative to `begin` that allows parallel execution of each statement. For example, the following stores into `b` at `$time` 2 but stores into `d` at `$time` 3:

```
initial
  fork
    #1 a=10;
    #2 b=20;
  join
initial
  begin
    #1 c=10;
    #2 d=20;
  end
```

Four-valued logic: A *simulation* feature of Verilog that models each bit as being one of four possible values: 0, 1, *high impedance* (`1'bz`) or *unknown value* (`1'bx`).

FPGA (Field Programmable Gate Array): A fixed set of lookup (truth) tables optionally attached to flip flops with a programmable interconnection network allowing the downloading of arbitrary *netlists*.

`$fstrobe`: A variation of `$strobe` that outputs to a file. See also `$fopen`, `$fclose`.

Full case: A synthesis directive that causes a `case` statement to act as though all possible binary patterns are listed. May cause synthesis to disagree with simulation.

`$fwrite`: A variation of `$write` that outputs to a file. See also `$fopen`, `$fclose`.

General-purpose computer: A machine that fetches machine language instructions from memory and executes them. The machine language describes the algorithm desired by the user, as opposed to a *special-purpose computer*. Also known as a stored program computer.

Glitch: *see Hazard.*

Goto: A high-level language statement not found in Verilog. Similar to state transitions in *explicit style* Verilog. Equivalent to assembly language jump or branch instructions. Gotos are useful for implementing *bottom testing loops*. The closest statement in Verilog is `disable`, which has drawbacks when used for this purpose. Avoidance of gotos is part of *structured programming*, and is possible with *implicit style* Verilog.

Handshaking: The synchronization required when two *actors* of different speed transfer data.

Hazard: The momentary spurious incorrect result produced by *combinational* logic of non-zero *propagation delay*.

Hexagon: Equivalent to *diamond* in *ASM* notation.

Hierarchical design: Instantiation of one *module* inside another.

Hierarchical names: A path for *test code* to access the internal variables of a *module* without a *port*. The instance names of the module(s) in the path are separated with periods.

High impedance: A value (`1'bz`) that models a `wire` that is disconnected and that is also produced by a *tri-state* gate. Used in a special way by `casex` and `casez`.

Ideal: A model of a device that ignores most physical details, such as *propagation delay* and voltage.

Implicit style: A finite state machine described in terms of multiple `@(posedge sysclk)`s inside an `always`. Does not give *next state* transitions. Roughly equivalent to a *pure behavioral* stage *ASM*. Contrast *explicit style*.

Independent: Two or more computations that do not depend on each other. It is easier to design *pipeline* special-purpose computers when computations are independent.

`inout` declaration: A Verilog feature that allows the same *port* to be used for information transfer both into and out of the *module*. Corresponds to *tri-state* gates.

`input`: A Verilog feature that only allows a *port* to be used for information transfer into a *module*.

Instance: A copy of a *module* used in a particular place of a *structural* design.

Instantiation: The act of making an *instance*.

Instruction set: The set of machine language operations implemented by a particular *general-purpose computer*.

Instruction-Set Architecture (ISA): See *programmer's model* and *instruction set*.

***Internal status**: Information generated by the *architecture(1)* and sent to the *controller* at the *mixed(1)* stage so that the controller can make decisions based on the data in the architecture.

Latch: An *asynchronous* data storage device. Synthesis tools produce unwanted latches when a `case` statement is used that is not a *full* `case`.

Little endian notation: The least significant bit (byte, word, etc.) is labeled as 0.

Macro: A string of source code that the simulator or synthesis tool substitutes prior to parsing.

Macrocells: The basic unit of a *CPLD*, consisting of a fixed set of AND/OR gates and an optional flip flop.

Mealy: A finite state machine that, unlike a *Moore* machine, produces command signals that are a function of both the *present state* and the *status* inputs. Such a command is indicated by an *oval* in *ASM* notation.

***Methodical**: An *architecture(1)* where each register has an associated ALU or other combinational logic so that all register transfers may proceed in parallel. Typically allows for faster algorithms to be implemented than the central ALU approach.

Memory: Equivalent to a collection of *registers*. Has the ability to read (remember old data) and write (forget old data and remember new data instead). Often referred to as RAM. Although ROM is often used instead of RAM in parts of a general-purpose computer, because ROM cannot forget, it is not memory.

Mixed: 1. The stage of the design where the *controller* is specified as an *ASM* using *command* and *status* signals (rather than RTN and mathematical conditions), but the *architecture(1)* is specified as a structure. 2. Any such mixture of *behavioral* and *structural* constructs in a Verilog *module*. 3. A kind of digital logic *netlist* where 1 is sometimes represented as a high voltage (*active high*) and sometimes represented as a low voltage (*active low*).

`module`: The basic construct of Verilog which is instantiated to create *hierarchical* and *structural* designs.

Moore machine: A finite state machine that, unlike a *Mealy* machine, produces command signals that are a function of only the *present state*. All commands in a Moore *ASM* are given in *rectangles*.

Multi-cycle: A machine that requires several fast clock cycles to produce one result. Compare with *single cycle* and *pipeline*.

Multi-port memory: Allow simultaneous access to multiple words within one clock cycle.

Netlist: A structural design described at the level of connections between one-bit `wires` and gates.

Next state: Combinational logic that computes what the next step is in the algorithm based on the *present state* and *status* inputs to the *controller*.

Node collapsing: An optimization technique used by *place and route* tools.

Non-blocking assignment: A Verilog statement (<=) that evaluates an expression now but that schedules the storage of the result to occur later. Several non-blocking assignments can execute in parallel without delay. There are several forms, but the one used most in this book (`<= @(posedge sysclk)`) is equivalent to the *RTN* ← used in the *pure behavioral* stage for *ASMs*.

`notif0`: A variation of `bufif0` that complements its output.

`notif1`: A variation of `bufif1` that complements its output.

One hot: An approach for the *controller* that uses one flip flop for each *state*.

`output`: A Verilog feature that only allows a *port* to be used for information transfer out of a *module*.

Oval: The *ASM* symbol for a *Mealy* command.

Parallel: 1. Two or more *independent* computations that occur at the same physical time. 2. Two or more computations (*dependent* or *independent*) that occur at the same simulation `$time`. In Verilog, `$time` is a separate issue from *sequence*. 3. When one assumes physical time and *sequence* are the same, the opposite of *sequential*.

Parallel `case`: A synthesis directive that allows parallel evaluation of the conditions given in a `case` statement.

`parameter`: A constant within an *instantiation* of a *module* that can be different in each *instance*.

Pin: The physical connection of an integrated circuit to a printed circuit board.

Pipeline: A machine that requires, on average, slightly more than one fast clock cycle to produce one result, provided that each result is *independent* of other results. Compare with *single cycle* and *multi-cycle*.

Place and route: A post synthesis tool that maps the synthesized design into the limited resources of a particular technology, such as a CPLD or FPGA.

PLI (Programming Language Interface): A way to interface Verilog simulations to C software, and thus extend the capabilities of Verilog.

Port: The aspect of a *module* that allows structural instantiation.

`posedge`: The rising edge of a signal, such as `sysclk`

Present state: The register that indicates what is the step in the algorithm which is currently active.

Programmable logic: Integrated circuits manufactured with a fixed set of devices that can be reconfigured by downloading a *netlist*. See *CPLD* and *FPGA*.

Programming: 1. The act of downloading a synthesized *netlist* into *programmable logic* or a truth table into a *ROM*. 2. The act of designing software for a *general-purpose computer*.

Programming Language Interface: See *PLI*.

Programmer's model: The *registers* of a *general-purpose computer* visible to the machine language programmer.

Propagation delay: The time required for combinational logic to stabilize on the correct result after its inputs change.

***Pure behavioral**: The stage where the design is thought of only as an algorithm using *RTN*. Equivalent to *implicit style* Verilog.

***Pure structural**: The stage where the *controller* and the *architecture(1)* are both structural.

RAM: see *memory*.

`$readmemb`: System task, whose arguments are the quoted name of a text file and an array. Reads words represented as a pattern of '0','1','x' and/or 'z' from the text file into the array.

`$readmemh`: System task, similar to `$readmemb`, except for hexadecimal.

Rectangle: 1. The *ASM* symbol for a *Moore* command. 2. The block diagram symbol for most devices.

Reduction: The unary application of a bitwise operator which acts as though the operator was inserted between each bit of the word. For example, if `a` is three bits, `&a` is `a[2]&[1]&a[0]`.

`reg`: The declaration used when a value is generated by *behavioral* Verilog code.

Register: A *sequential(1)* device that can load, and for some register types otherwise manipulate a value. The value in a *synchronous* register changes at the next rising edge of the clock. Contrast with *combinational.*

`repeat`: A Verilog loop that repeats a known number of times. Very different than the bottom testing loop.

Reset: The only *asynchronous* signal used in this book, which clears the *present state.*

Resource sharing: A *synthesis* optimization where the same hardware unit is used for multiple computations.

Rising delay: The propagation delay it takes for an output to change to 1.

ROM (Read Only Memory): A tabular replacement for combinational logic. Not an actual *memory* because it does not have the ability to forget.

RTL: 1. "Register Transfer Logic." In the pre-Verilog literature, the term RTL meant the logic equations generated by the controller to implement register transfers (section 4.4.1). Today, RTL most commonly means *explicit style* behavioral Verilog.

Some vendors (notably Synopsys) also use RTL to describe *implicit style* design and *RTN*. This book avoids the use of the term RTL, in favor of the more precise terms: implicit, explicit and RTN. 2. "Rotate Two Left", a PDP-8 instruction.

RTN (Register Transfer Notation): An ← inside a *rectangle* or *oval* of an *ASM* that evaluates an expression during the current clock cycle, but that schedules the change of the left-hand register to occur at the next rising edge of the clock. Similar to the Verilog *non-blocking assignment* (`<= @(posedge sysclk)`).

SDF (Standard Delay File): A way to backannotate delay information into a netlist after place and route.

Sensitivity list: The list of `input` variables of *combinational* logic. The variables in the sensitivity list occur inside @ separated by `or`. Failure to list all variables can cause unwanted *latches.*

Sequence: The order in which Verilog statements execute in simulation. Statements in a particular `always` or `initial` block execute sequentially, regardless of *time control.*

Sequential: 1. A device that has memory, such as a *controller* or a *register*, as opposed to *combinational* logic. 2. Two or more *dependent* computations that occur in a particular *sequence*, even if they occur at the same `$time` in a Verilog simulation. 3. When one assumes physical time and *sequence* are the same, the opposite of *parallel.*

Simulation: The interpretation of Verilog source code to produce textual output and timing diagrams.

Single-cycle: A machine that requires one slow clock cycle to produce one result. Compare with *multi-cycle* and *pipeline.*

Special-purpose computer: A machine that is customized to implement only one algorithm, as opposed to a *general-purpose computer.* Special-purpose computers are often referred to simply as digital logic.

Standard Delay File: See SDF.

`strength`: Additional information about a `wire` that models its electrical properties.

State: A step that is active in an algorithm during a particular clock cycle.

Status: See *internal status* and *external status.*

Structural: An interconnection of wires and gates (or *combinational* and *register* devices) that forms a machine. Represented by a block diagram, circuit diagram, *instances* of *modules* or a *netlist.*

Structured programming: Describing an algorithm with high-level software control statements, such as `if`, `case` and `while`, but avoiding *goto*.

Subbus: A *concatenation* of a subset of bits from a *bus*.

Superscalar: A *general-purpose computer* that is able to execute more than one instruction per clock cycle.

Synchronous: A device that makes changes only at an edge of a clock signal, typically the rising edge.

Synthesis: The automatic translation of Verilog source code into a *netlist*.

System clock: The single clock used in a completely *synchronous* design.

***Test code**: Non-*synthesizable* Verilog code used only in simulation to verify the operation of other, possibly synthesizable, Verilog code that models hardware. Test code (sometimes called a testbench) gives an abstract model of the environment in which the hardware will operate.

Testbench: *see test code*

`time`: A declaration for a variable that stores the result of `$time`. Often, `integer` is used instead.

`$time`: The current simulation time step.

Time control: Verilog features that cause `$time` to advance in *simulation*: `@`, `#` and `wait`

`triand`: Similar to `wand`.

`trior`: Similar to `wor`.

`trireg`: A variation of `wire` that holds the last value when all outputs are `1'bz`.

Tri-state: A device that has the ability to disconnect itself electronically from a bus. Verilog models this using the high impedance value (`'bz`).

Turn off delay: The propagation delay it takes for an output to change to `1'bz`, associated with a *tri-state* device.

Unary code: A bit pattern where exactly one of the bits is a one. See *one hot*.

Unconditional: *Non-blocking assignments (RTN)* and/or *command signals* that are synonymous with a machine being in a particular *state*, regardless of any other conditions that may hold. See *Moore* and *rectangle*.

Unknown value: A value (`1'bx`) that models an uncertain condition for one bit, such as two fighting output `wires`, typically indicative of an error in a *structural* design. Also used in chapter 6 to model abstract *hazards*.

Value Change Dump (VCD): A standardized text file format created by `$dumpfile` and related tasks that record values of simulation variables and that is used by post-simulation analysis tools.

Verilog: The hardware description language used in this book. (See chapter 3.)

VHDL: The other hardware description language, which is more complicated for the beginner than Verilog.

`wand`: A variation of `wire` that produces 0 instead of `1'bx` when two outputs fight, which implements & without using a gate (wired AND).

`wor`: A variation of `wire` that produces 1 instead of `1'bx` when two outputs fight, which implements | without using a gate (wired OR).

`wire`: The declaration used when a value is generated by *structural* Verilog code.

Wirewrap: A technique of connecting chips together involving using a tool that wraps wire around posts connected to the *pins* of the chips. A convenient way to fabricate prototypes in an educational or hobby setting.

Worst case: A model of a device that only considers the longest *propagation delay* possible.

`~&`: The bitwise NAND operator, `a~&b === ~(a&b)`, which is not found in C.

`~|`: The bitwise NOR operator, `a~|b === ~(a|b)`, which is not found in C.

`~^`: The bitwise coincidence operator, `a~^b === ~(a^b) === a^(~b) === a^~b`, which is not found in C.

J. LIMITATIONS ON MEALY WITH IMPLICIT STYLE

Chapters 5, 9 and 10 and sections 7.4 and 11.6 discuss ASM charts and Verilog simulation for Mealy state machines, in which an operation (inside an oval of an ASM chart) is conditional. Mealy state machines are more problematic than the Moore state machines used in the rest of this book. There are limitations on using the Mealy approach with implicit style Verilog.

There are three consequences of using an oval in an ASM chart. The first consequence, which can be described with implicit style (pure behavioral) Verilog, is to allow computations dependent on a decision to be initiated in parallel to the decision. For example, the decoding and execution of a TAD instruction in chapter 9 illustrate a decision (`ir2[11:9] == 1`) and a computation (`ac+mb2`) that occur in parallel:

```
if (ir2[11:9] == 1)
  ac <= @(posedge sysclk) ac + mb2;
```

As in most of the examples in this book, the statement that carries out the computation is a non-blocking assignment, so the effect will not be observable until the next rising edge of the clock. When viewed by itself, the architecture is a Moore machine (it has registers that only change at the rising edge). Since the output (`ac`) of the complete machine (the controller together with the architecture) only changes at the clock edge, the complete machine is Moore. Only the controller is Mealy. For this reason, implicit style Verilog with non-blocking assignment can model such situations.

The second consequence of using an oval in an ASM chart arises only at the mixed stage, such as figure 5-2. Depending on how complicated the architecture is, there may be hazards created between the controller and architecture during simulation that an implicit style Verilog description of the controller will not process properly. The 1994 paper mentioned below describes a `'bx` handshaking technique with an `exit_current_state` task that overcomes this problem for Verilog simulation. This technique is an extension to the `enter_new_state` method given in this book.

The third consequence of using an oval in an ASM chart arises only when a decision involves an input to a machine, and RTN is not used to produce the corresponding output. Figure 5-7 is an illustration of such a situation. For such ASMs, the machine cannot be modeled just with implicit style Verilog (`@(posedge sysclk)` inside `always`) because the output of the machine is supposed to follow the input. In other words, if the input makes multiple changes during one clock cycle, the output should make corresponding changes during that clock cycle. The implicit style cannot model this, since the behavioral block will execute only once. Since figure 5-7 is simple combinational logic (single-state ASM), the designer uses the appropriate sensitivity list instead of `@(posedge sysclk)`. In general, Mealy machines often have multiple states, but there is no implicit style notation to describe this reexecution of the behavioral code that must take place in each Mealy state. It is necessary to use explicit style Verilog instead. The 1998 paper gives more information about this.

It is possible to use a hybrid implicit/explicit style to cope with a machine that has Mealy external outputs, such as the ASM in section 5.2.4 (figure 5-6). This hybrid approach is synthesizable. The following shows in bold the distinctions between the simulation only technique of section 5.3 and the hybrid implicit/explicit approach:

```
reg s;
always  //implicit block
  begin
      s <= @(posedge sysclk) 0;
    @(posedge sysclk) #1;
      r1 <= @(posedge sysclk) x;
      //ready = 1;
      if (pb)
        begin
          r2 <= @(posedge sysclk) 0;
          while (r1 >= y)
            begin
                s <= @(posedge sysclk) 1;
              @(posedge sysclk) #1;
                r1 <= @(posedge sysclk) r1 - y;
                if (r1 >= y)
                  r2 <= @(posedge sysclk) r2 + 1;
                //else
                //  ready = 1;
            end
        end
```

Continued

```
always @(s or r1 or y)  //explicit block
 begin
   if (s==0) ready = 1;
   else if (r1 >= y) ready = 0;
   else ready = 1;
 end
```

The hybrid approach eliminates the `enter_new_state` task with an internal state variable, `s`. A separate `always` block implements the combinational logic that computes the Mealy output based on the internal state variable and the condition(s) specified in the Mealy decision `(r1 >= y)`. Every usage of a Mealy command, including unconditional ones such as in the in the top state of this example (`s==0`), must be generated by an `always` block dedicated to that command signal.

J.1 Further Reading

ARNOLD, MARK G., NEAL J. SAMPLE and JAMES SHULER, "Guidelines for Safe Simulation and Synthesis of Implicit Style Verilog," *7th International Verilog HDL Conference,* Santa Clara, CA, March 16-19, 1998, pp. 67-69.

Index

C

E

Q

T

U

V

X

Z

For Jillian

(the cupboards were mahogany, by the way)

Cover art & logo: Jillian Ostrander
Internal art: Michelle Ostrander

With deepest gratitude to Dr. Ziad El-Khally for the Arabic translations featured in chapter eight. Thank you as well to Anwar Sabiha, for her translations.

Disclaimer

...Nach fearr leath builín ná bheith gan arán?

First printed in November 2017

1
Tosaigh

It was the rain that had brought her.

Kathleen stood on the ramshackle porch of her new home, watching the creeping mists slowly suck the colour from the hills. In the distance, sickly conifers cast their shade against an imperious sky, their trunks bright with mossy patches. A light wind stirred her dark brown hair, driving a chill into the mid-July evening.

There was so much about this place that would force a traveler away, and yet she never faltered. *Ah, sure, 'tis a fine, soft day anyway, a stóirín. A bit of mist never killed anyone, did it?* It was this one familiarity that drew her. Compelled her. And if this was as close as she could ever get...

Banks of black clouds spilled over the horizon, crushing the downy mists, and Kathleen turned her back on the outside world.

So be it.

and the RAIN it raineth Every Day

2
Amazed

The crowd was insipid at best.

Ever since the laws had ousted such frivolities as entertainment, a man courting the stage would sooner face the stocks than garner a little applause. Still, enough stultified souls had collected to make the endeavor profitable, so the show was certainly on.

When at last all were seated, the heavy double doors were pulled shut, blocking out the last traces of light. The moment lasted inexorably. The stale summer air, tinted by flaking street-lamps, grew thick, hanging and black and pierced by a thousand restless glances. Word passed of a Northerner scam. Then they heard footsteps scuffing across the stage.

The pace was slow, though whether this was from caution or insouciance, the listeners could not tell. Try as they might to peer through the gloaming, they could see nothing; only by sound could they tell that the performer had halted center-stage.

A match was struck, the rough whisk echoing off the walls. A tiny flame bleached the clinging gloom, suspended in the pool of its own light. Delicately it shone, its gossamer wings snatching at the darkness, shifting and wavering, sparkling indigo and gold.

Then the flame spoke.

"Ladies and gentlemen... I am pleased that you've gathered here tonight. However, I fear I must present a word of caution... Any who feel an irrepressible disturbance at the sight of burns, blood, or other grotesque wounds, I strongly urge you, either determine how to cope, or kindly exit the building. After all, I can guarantee nothing for my own safety... As well – to put it bluntly – to any inescapably gripped by the age-old fear of fire, this is not the place for you. You have been warned. Take it as you will. Now, without further delay – and in case of sudden emergency – I hope you all know where the doors and exits are, 'cause it's show-time!"

The flame went out, its last breath of smoke swallowed in the abyss. They realized then that this man was not like the others. But before they could wonder whether he was insane, a match was struck and flung into the air. A pulse of flame fanned out over the floorboards, a ripple on an amber pond that collected in writhing dewdrops upon a ring of rags without ever scorching the pine.

And there he stood. Thin wisps of smoke curled up around him, flickering shadows like pallid scars across his bare chest. The skeletal shade of a traveler, his arms wrapped in rags, his jeans weather-worn, a faded, red bandana tied to cover the lower half of his face. A revenant raven, an enigmatic phantom, a wraith.

In his hand he held a few make-shift, double-sided torches; deftly he snared one of the flaming rags, the fire licking up over his fingertips as he calmly lit the torch-heads, his motions fluid, almost graceful.

Finished, he tossed the rag aside, extinguishing it. Then, true to the ways of a street performer, he began to juggle. The torches cut wide spirals into the Stygian aethers, spinning and flashing in interlocking loops and knots as artfully he caught and wove the swirling chains.

Eyes locked on the whirling torches, he backed up, grinding out the last light of a smoldering rag with his bare heel. He did the same for the rest, stepping, gliding, turning an intricate dance until he had extinguished all but the torches' light. Credit to performer's calculation, he now stood closer to the edge of the stage.

In a heartbeat, the tapestry threads snapped, and all three torches were weighed in his right hand. Reaching down, he swept up a glass bottle, pulling down the bandana and taking a quick swig. Though the glass was somewhat thick, and the mahogany tint hindered visibility, the fire blazing from the torches cut through the obscurity, illuminating a smooth liquid within.

In seconds the bottle was set by his feet once more, its translucent surface resuming its unyielding demeanor. But the audience members were no longer looking at the bottle. Instead, their wide-eyed gazes were trained immovably upon the fire-eater, who was just raising the torches to his mouth.

Arching his back, he held the torches above him, bending his arm slightly so the flames were about a foot or so from his face. Then in one, great spray, he spat the fuel through the fire.

Flames shot to the ceiling in a roaring column of fury and heat, devouring the deepest shadows, threatening with tumultuous cracks to envelope the cowering world. But the fire-eater did not shy away. He lowered his torch, sustaining the blazing torrent with the power and ease of a dragon.

As the fire twisted to surrender to the darkness, a pale hand shot through the river of flame, setting his palm alight. White-gold fire pooled in his hand, dancing across his nimble finger-tips as he held it aloft like a torch-bearer beckoning ships to shore.

"Oh yeah." Once more his soft voice slid through the cracks in the silence. "*Don't* try this at home, by the way; I always forget t'mention that part. I haven't really been sued yet, but you can never be too careful...

"Right?"

He slashed his hand down to his side, and the fire spilled across the floor, smashing in a shower of sparks that burned warm purple and red. They glittered on the surface of the stage, sparkling hotly against the aged wood.

Then.

The man stepped back, spreading his stance. He stretched his arms out before him, straight, his hands limp; against the lightless backdrop, he was a mere form, a silhouette.

His fists clenched.

Fire surged up from the sparks in a wave, snapping and crackling with heat. The fire-eater shifted his weight, directing the flames in a stream before him, wrapping around him, stretching up to the ceiling at the back of the stage. The column of fire sent out shoots – delicate, coiling, reeling and melding in woven, Celtic curls. Leaves of shimmering lilac spread from lacing, russet limbs, breaking off to flutter away on glistening, egg-shell wings.

The fire-eater looked back to his audience, smirking at their shock. Laughing, he strolled out of the grip of the fire-tree's warped trunk; his laughter was like the night.

The fire continued to grow, spinning vines and spreading flowers, blossoming in whorls of red-orange light. Its many tongues severed the darkness with airy barbs, slicing open the monochrome and allowing colour to bleed in.

Few noticed the man, cased in shadows, his back almost totally to the crowd, as he pulled down the bandana to take a swig from the bottle, swallowing this time. The light barely touched the sharp lines of his countenance, concealing his identity still, loyal to the man it so willingly served.

Having taken his fill, he pulled the bandana back over his face. Then, whipping violently, he turned back, cheating out as he hurled the bottle so it burst at the base of the tree.

Apple blossoms and butterflies shattered, the plaited ropes snapped, and all was washed back down to earth like water, tidal-waving across the stage, surging towards its master.

He stood, ready to take it; what was visible of his expression was savage, bitter, and bold. And as the fire rushed to take his outstretched hand, he turned.

With agility and grace, he danced through the conflagration. Fast did he move, breaking and bending with beauty and power, an ancient, wild, controlling dance, and the fire warped to his whim, curling around him, brushing his delicate skin, caressing his arms, his legs, his face, running through his dark hair. Mesmerizing, tantalizing, electrifying, pure.

So fluid his motions, his sudden halt seemed more a dream. The colour drained from the whirlwind of fire; one smoky swirl, strained milk, stretched in a trail from the front of the stage, lifting in the air to wrap like a rearing snake, rising up to the ceiling above. Its light was faint as a fire-fly's, glowing a cool, pale blue.

The man stood center-stage, his arms outspread, defiant arrogance flashing in his eyes. "Speechless?" Something like laughter, yet caustic, more scarred, scoured through his mocking tone. He slid back his right foot, and gave a deep, sweeping final bow. *"Slán abhaile, a chairde."*

A few gasps in the crowd were audible at a distance.

Stepping out, he twisted, one hand banishing the faded, moonlit smoke, the other raising new flames from the dust – from the very wood of the front of the stage, pulled from the grave in a roaring surge. So dense, so cochineal, so searingly bright were the flames that they obscured all other forms, swallowing the world in their light.

Then, in a torque of as much unbridled power, the fire wrenched back in on itself, consuming, dispelling, expiring in a shower of flickering ashes and sparks.

It was over.

For a long moment, the villagers sat in breathless silence. For the first time in many long years, they had been amazed.

Shaking their heads in disbelief, they stood, filing out of the dim auditorium. Most were so eager to spread the word that they scurried out without a second thought; but there were a brave and wondering few who dared to glance back behind them before slipping through the door.

There was no sign of the mysterious fire-eater – not the slightest hint that he had ever existed, save for a few, thin strands of pale smoke, drifting off into the night.

The man had vanished as easily as a starving flame.

3
The Wanderer

She had never seen more green in her life.

For over an hour, Meygan had followed the cliff's edge, keeping a good ten feet back in case a sudden gust of wind should knock her over. And that entire time, she had seen no sign of anything – nothing but green hills, green treetops, green mountains – and black clouds. The trails of ropy down overhead had knotted into a dense sheet of grey, like uncarded wool, which was rapidly bearing down on her. The wind had picked up, carrying the scent sweet in a candle and on your own doorstep, dismal when miles from home: On top of everything else, it was going to rain.

Great.

Meygan bowed her head and began trudging up the steep hill before her.

Once again, she tried to go over how she had ended up here, but by all accounts it really didn't make sense. One minute they had been standing in her kitchen, Addison and Ian feeding scraps of paper to a candle as she searched for coffee mugs, a thunderstorm raging outside. The next thing she knew, they were here – stuck up on this mountain, with nothing familiar to be seen for miles, no clear way down – and no explanations. It wasn't a dream; of that much she was certain. And whatever it was, she refused to let it beat her! Sure as she lived, sure as she breathed, she was not – NOT – spending the rest of her days wandering these hills, trapped on this mountain with no one but three useless –

Three?

With a start, Meygan realized that she was no longer alone.

There. There, at about twenty yards off, the tips of his black sneakers perched just on the edge of the cliff, his long, black coat whipping in the wind, was a man.

Meygan narrowed her eyes. His hair was dark, a very deep brown, thick and long enough for the chaotic strands to hang into his eyes. He was very tall, though he slouched against the wind; his lithe body was thin as all starvation. He was looking over the edge, staring off into the distance, with a concentration no one else could comprehend.

Meygan was torn. For however unimposing a creature he may seem, he was a stranger, a man, and she was standing on the edge of a cliff.

But what other hope did she have?

Meygan dragged her foot through the grass, drawing a few inches closer. Maybe he could hear her from here? Against the wind. Against the distance. *Yeeeaaahhh...* She ran a hand back through her long, red curls. Maybe she should have brought one of the boys, after all? Then again, what use would they be? Maybe –

Jolted, the man turned so quickly that Meygan stepped back, raising her fists. For a second, they merely stared at each other across the distance, tensed. Then the man lifted a hand, restraining his hair against the wind so he could see her. "Hey!" he called.

Meygan cocked her head. "Hey..." A moment longer she surveyed him, wondering greatly. Then, decided, she moved forward; she came within eight yards, moving away from the cliff's edge – just in case. "Can you help me out?"

Now she could see him better. He looked to be in his early-to-mid thirties, yet he seemed somehow older – *centuries* older. Maybe it was the way he carried himself – broken and wearied, yet supported by an unmistakable, unconquerable pride. Maybe it was the guarded look in his dark eyes – the look of suspicion, near hostility, and nothing else – all else obscured, hidden, and utterly unreadable.

"How di'jyou get up here?" he asked, in a tone made slightly more amiable than his stare by confusion.

Meygan half-frowned. Now *there* was a question. "I'm not really sure," she said, pulling her hair back from her face with one hand and wishing she had a hair-tie. "That's kind of what I need help with."

"You need help knowing how you got up here?"

She shook her head. "No. No – I need to know how to get down."

"Why not just go the same way you came?"

"No. Focus. No –" With her palms pressed together, she directed her fingers over the edge of the cliff. "I just need to know how to get *down*. Like, it doesn't matter how I got up here. How did *you* get up here?"

He followed her gesture and said nothing.

"Ugh," Meygan complained, running her hands back through her orange curls. "Dude! Focus!"

"I climbed."

His voice, she had noticed, was not deep, but low, at times held so soft that it sounded hoarse. It was hard to hear him over the wind.

"Where?" she asked. "Where did you climb up?"

Slowly, he shrugged one thin shoulder.

Oh, he was frustrating. Meygan cast her gaze to the dank, rolling sky, interrogating God.

The man answered her. "You can get down anywhere if you're a good enough climber," he said. "But it depends on where you wanna get to. If you're after goin' south, you can't get there going down here."

"But you *can* get down here?"

"If you're a good enough climber." He turned away again, apparently disinterested in the conversation.

Meygan pursed her lips. *For a second there, I thought we were GETTING somewhere...* "If I go down here –" she pointed over the edge "— is there a town nearby I can get to?"

"Near enough."

"Which way do I go – and don't tell me it depends on what town I'm going to!!" she added harshly, pointing up at him.

He laughed. His laughter was like his voice: quiet, raspy, and briefly used; like his glare, like his smile: supercilious and taunting.

"Well enough," he relented. "You'd want the closest?"

Meygan sighed. "I'll find it." She waved him off, turning away, facing towards the open horizon below.

He again shrugged, following her gaze. He didn't seem particularly bothered by silences, by her presence – indeed, he was fairly ignoring her.

"Got anything else for me?" she asked, casting him a glance.

He shook his head.

Again, she sighed. "I didn't think so."

For a moment, the two of them stared out over the forest, studying the wide world. The woods were dark below, overcast by the shadows of the clouds, their many thousand leaves chattering in an unheard breeze like swarms of sweet-throated locusts.

"You oughtn't go into town."

"What?" She turned to him.

"You oughtn't go into town," he repeated, only slightly louder. "Especially not this close to the border, if you're alone."

Meygan put her hands on her hips. "Are you crazy?"

He looked at her. "Are you serious?" he challenged, in his cracked voice. Then he waved her off. "Go south, then; you mustn't be from around here."

Meygan lost her temper. "If I was from around here I would know how to get down!!"

"Not necessarily."

"*I* would!" she snapped. "And *no,* I'm not from around here. But the two boys with me are!"

His eyes flicked over her shoulder, then back to her, pityingly.

"*RRRRR!*" Meygan exclaimed, in frustration. "They're not here *now!!*"

"I was gonna say..."

"You're impossible, you know that?!" Meygan shouted.

He was undaunted. "*You* wanted my help."

"That wasn't help!!"

"Fine!" He threw out an arm impulsively, indicating the world below. "Go into town if you'd have it! What do I care?"

"*HOW!!?!*" Meygan demanded, dramatically.

"Go away!" Half-groan, half-snap; he turned away from her, folding his arms tightly. "I'm tired of talking to you."

Meygan scowled. "You are *so* rude."

"*Away,*" he drawled, rolling his eyes up to God.

Meygan opened her arms. "FINE!!!" She turned, stalking away briskly. She got about twenty yards; then, suddenly realizing a scathing last remark, she turned –

He was gone.

4
Begin.

"I'm sorry, Meygan."

The sun was beginning to set, casting long bands of stardust gold beneath the shielding clouds. Meygan sat on the cliff's edge, dangling her feet into the seemingly perilous abyss. She had taken up station where the wayfarer had vanished, gazing upon all he had seen. Miles of forest extended below, shot through with the glinting track of a river to the east, running north. Shadows cast a deepened mist of their own among the tall trees, blurring the varied green.

Addison sat down beside her, sliding to the ledge, but she did not turn to him; not even a glance. "I'm sorry I hit you," she responded.

"The tackling..."

"I'm not sorry I tackled you."

"Ah." He shifted, mimicking her position, and he looked out on the world below – the trees in their whispering, tumbling away in the distance. "I know you're mad. I don't really blame you, but... We have to stick together. At least – physically. It's safer for us if we don't split up."

Meygan was still mad at him. It was his fault they were stuck here. If only he had listened to her, maybe they wouldn't be in this mess now! But she recognized that what he said was right. It was a sad state, but he and Ian were all that she had in the world right now. It was safest for them to stay together.

Maybe he had a little common sense after all?

"So where do we go from here?" Ian asked; he had wisely chosen to remain a respectful distance behind them.

"You were right; Riverside Valley might not be far off," Addison responded. "At any rate, there has to be *some* sort of a town nearby. If we find it, we can get our bearings, and hopefully make it back in good time. Of course – that's assuming we can get down. I've been all over this mountain; there's nothing up here."

"Hm..." Ian frowned, folding his arms, his eyes narrowed upon the horizon. "It's just as well," he said. "I'm not sure I'd wanna meet anyone who's crazy enough to live around *here*."

"The problem is," Addison reestablished, "how do we get down?"

"We can try and weave grass into a rope," Ian offered. "Twelve years ought'a make it halfway."

"I found a way down," Meygan said.

Addison started, turning to face her. "What?"

Meygan swung her foot softly, dipping it through the calmed breeze. All things being a matter of perspective... She hooked her sneaker onto the toe-hold.

"It depends on how well you can climb."

5
Crazy Enough

Zander the Archer stared down from his perch high in the branches of an old tree. The soft light of the dawn was glancing through the drying leaves, warming his dew-damp hair; somewhere far-off, a raven sounded its call.

Zander was *not* happy. He was tired, he was hungry, and he was sick to death of sitting in this tree. Across his chest was the strap of his quiver, and a dark-stained yew bow was clenched in his fist. He had seventeen arrows in the love-battered leather quiver, each feathered in various colours and sharpened to a deadly point. And if those land-stealers stood there blocking the path any longer, he would not be above using them.

Zander hated strangers. They were nothing but trouble. It was always up to him to either assist or dispatch them, and sure it was a waste of his time. Think of all he could do if he didn't have to stand guard on this land! He could... could...

Shorn, if he had the time he could think of something to do!

Scowling, Zander shifted against the tree. They had been in the house last night. If Rahyoke found out, he was dead. *SO* dead. And shorn, he wouldn't blame him. *Hadn't* he said, confident as all arrogance, *"There's nothing to worry about; not a soul sets foot in this forest that I don't know about. Trust me!"*

Trust me.

Och, he was doomed.

But not if they left. Zander leaned forward, narrowing his eyes to peer through the curtain of leaves. The three trespassers stood below, arguing amongst themselves with intensity. The words that floated up to him were enough to give him the gist. They were looking for a direction – trying to come up with a plan. By all accounts, they might be leaving.

Eventually. He had never heard such a small issue drawn out so much in his life. He laid his head against the rough bark of the tree. It seemed a simple enough issue to him; shall we continue on our journey? Yes, we shall. What direction should we go? West.

West?!

Zander leaned off of the tree. The boy with the golden hair was going back the way they'd come. *And urging the others to follow!*

One night. One night was enough, and a failure. But *two?*

Not gonna happen.

Reaching back, Zander pulled an arrow from the quiver, knocking it on the dark yew bow. He was not going to fail a second time. He'd make sure of it.

He pulled back the gut string, the tail of the arrow clenched between two fingers, and fired. He barely cared where the arrow landed.

But Zander the Archer never missed.

Addison jumped as an arrow sank into the ground at his feet, the stiff, pale feathers whisking an inch past his sneakers. He whipped around, searching for the missile's source.

"Who's there?!"

The trees around were empty of life, stirred only by the soft morning breeze. Then, out of nowhere –

"Imighí libh, a fhoghlaithe mara! Anois!!"

Addison's ice-blue eyes narrowed, focusing on the lower boughs of a nearby tree. He could see no shape through the dense fabric of leaves, but surely the voice had come from there.

"Addison, let's go –" Meygan began.

Ian hefted a rock, glaring up at the tree. "Show yourself!" he barked, sounding surprisingly fierce.

Meygan rolled her eyes. "Yeah. Cause *that's* gonna work."

To everyone's surprise, it did.

The dry thud of boots hitting the dusty road, and there before them he stood. He was a handsome lad of about fifteen or sixteen; his eyes were a deep emerald green, narrowed and sharp with his anger, his dark brown hair ruffled and tinted with copper in the rising sun. He wore a long, light-green tunic-shirt with long sleeves, and over this a dark-brown linen vest that reached to his narrow hips. In his hands was a recurve bow, the string drawn tightly across his chest as he leveled an arrow straight at Addison's neck.

"You see me," he said evenly, his voice a low growl. "Now get gone."

"Hey!" Meygan stepped out from behind Ian. "Get that out of his face!"

"You're all trespassin' on my land," the archer replied. "I could *kill* you if I want."

Undeterred, Meygan pointed at him, more threat in her glare than a weapon held. "Listen, Robin Hood, we didn't *mean* to be here, we got lost! If you want us to go, then show us a way out! Don't just start shooting at people – what's the matter with you!?"

"Meygan –" Addison warned, but she cut him off.

"Don't 'Meygan' me, you ingrate! He should know better! I don't even know why I'm trying to save you. Jerk-face."

Very subtly, amusement crept into the archer's expression. Though the lethal arrow never twitched, he looked at Meygan in full.

"You're lost?"

"Yeah," she answered, "we are."

For travelers, they really didn't have much baggage – only the girl was carrying a backpack – a bright red one at that. "...It's a pretty long way t'be off-track."

Meygan folded her arms. "Well, it turns out some of us are really bad at navigating. I promise, we're not land-stealers, or thieves, or aliens, or whatever you keep that bow to fight against. I don't *want* to be on your land. It's very pretty, but I'd rather be on my own land. I'm over it. Now, could you please help us? Even if you just gave us directions. I would really, *really* appreciate it. I just want to go home. Can you help us out?"

He considered it, eyeing her expression. She seemed honest. In spite of what he'd been conditioned to expect of the world, this girl indeed seemed to be genuine. "...Where is it you were trying to get to?"

"Riverside Valley," she answered. She saw his brow furrow slightly, and she added, "If you can get us to Cazenovia, though, I know the way."

The boy was silent a moment. He watched Addison squirming beneath the threat of certain death as he deliberated, torn between solutions. At last, he said, "I'm guarding this land for a man. I could ask... but you'd have to come with me. He knows the land better than I do."

"Wha –" Addison began, but Meygan understood. "How far?"

"Couple'a miles, maybe; he should be headin' back."

"And you're gonna point that at us the whole way?"

"No... If you're unarmed."

Ian still had his arm cocked, poised to hurl the stone. He glanced at Meygan and dropped it.

The archer hesitated. Then, slowly, he lowered the bow. "Do we have a deal, then?"

Meygan looked at Addison, expectantly. Then she shoved him in the shoulder, and he obediently affirmed: "Yeah. We have a deal."

"This is taking forever."

They were in a stretch of forest; the floor was dirt, the shrubbery thick, and the sunlight clawed through the branches overhead. Meygan perched atop a boulder which was crowned unceremoniously with a large bush, her backpack clutched to her chest. Across the narrow way, Ian sat with his back against a tree, picking at the sparse grass. Addison paced between them, waiting without patience.

"Maybe he forgot about us," Addison admitted, with a sigh.

"Maybe he was eaten by jaguars?"

After a good, long, silent glare, Addison moved his critical gaze from Ian to the road before him. Abandoned, exhausted, crestfallen, annoyed – and now the possibility of jaguars! He continued to pace, noiselessly engraving a broad circle into the dust.

"I bet he left us," Meygan grumbled. She could so easily see him, laughing away over the hills. Leaving them to their fate, like the wanderer on the mountain. "How far did he say he was going?"

Addison kicked at a thin root that jutted from the dirt path. "He didn't. He said it was *'close'*." He turned to her. "D'you have any *food* in that bag, or what?"

She didn't like his tone. "I hadn't planned on an extended stay."

"Oh. Well *excuse me* for making a *suggestion!*"

Meygan's fists clenched. "It's *YOUR* fault we're in this mess, Addison!!"

"It's too late to do anything about that now!" Addison replied. "We need to focus on getting back. So far, *this* seems like the best option."

"We're sitting in the forest!!"

"If you have a *better* idea, let's hear it! Even better!" he continued, raising his voice. *"YOU* lead us back!"

Meygan's expression became dark. Through gritted teeth, she menaced, "I am *trying*, you weak, *selfish* little boy!"

"Oh, burn."

They turned to see the archer standing behind them, grinning broadly. He pointed at Meygan. "I like you. You're spunky." He folded his arms. "So you all remember me, and this –" a nod behind him "—is the man whose land you raided."

They looked to the man behind him. And, as one, two voices answered.

"No."

For here was the man from the mountain.

Facing an onslaught, Zander defended himself, his palms raised. "Okay! So, what it is, is –"

"You *lied* to me!" the man snarled, irate.

"I'll leave you to it," Meygan noted, bowing out. He sounded more intimidating.

"This is too far, Zander," the man continued, his low voice roiling with anger. "Do you have *any* idea how *dangerous* –"

"Trust me, I –"

"Trust you?! TRUST you!!" He dropped his arms to his sides, rolling his eyes up to God. "Zander, of all the *asinine* –"

"Listen!" Zander began trying to reason with him in the same strange language he'd spoken before, speaking slowly as he measured out the words. *"Tá... sé – tá siad – ar strae. Tá siad ar strae – le do thoil?"*

At first, the man looked startled, casting an anxious glance back to the trespassers. For a moment, he studied them – surveying, scouring. Then he shook his head. "'Le do thoil'. *In éadan mo thola a dhéanfainn é!"*

"Uh – *fan* –!"

"Ní thaitníonn an smaoineamh liom," the man continued. *"Níl áthas orm cabhrú leo. Rinne tú botún, a Shandar. A Dhia dhílis – agus nuair atá botún déanta agat féin, ní foláir leat mise a tharrac isteach ann leis, nach bhfuil?! Cén diabhal atá ort?! Tá dul amú ortsa!!"*

Poor Zander stammered, trying to find a word and a place to put it. *"Sea! Sea – ach – O, le do thoil! Níl... níl..."*

The man folded his arms. "What."

Zander struggled. *"Níl... tuig... muid..."*

"'Níl aithne acu orainn,'" the man corrected. "And what of it? I'd rather it stay that way. Oh, but you've done more than enough already to alter that, so no use in arguing it. You've made *sure* you won't be forgotten; well done."

Biting. Desperate now, Zander protested, *"Ach – ach tá – ar strae, sin an méid –"*

"Oh, *'sin an méid' anois. I ndáiríre.* Well *níl áthas orm cabhrú leo fós – nach ndúirt mé é? Ní haon mhaith a bheith liom –"*

"Look!" Zander insisted, speaking the language they all understood. "You're right; it's done. But can't we fix it? Take them back to the house, see where they're headed, and get them out of our lives."

"Back?" the man whispered, looking a little struck.

Zander bit his lips, nodding.

"Aw, shite. *Zan*-der." He was at that point of frustration that blends into disbelief.

"I'm sorry! I know – but it's just one day; that's all I ask. One day. We can afford that!"

He rolled his eyes, scowling. *"In aimn Dé –"*

"Dech do shínaib ceo."

The older man started, looking at the archer in silence. For a long moment, he said nothing, weighing it all out in his mind. The deal was done. *"Diabhal ort,"* he grumbled. He rubbed the back of his neck, studying the grass. Then, turning to Meygan, he spoke. "Alright... t'my thinking, it's no use discussing things here. And, I dunno about anyone else, but I'm starving, so... Fine. Come back with us to the house. I know this forest better than anyone; we'll figure out how to get you back. That is my offer."

The trespassers considered it, casting glances between themselves.

"First," said Meygan, "can I just do one thing?"

He looked at her warily. "Yeah?"

Turning on Addison, she exclaimed, "*NOW* d'you believe I saw him?!"

The man rolled his eyes. "It's a deal, then?"

"Yeah," Meygan answered. "Thank you, that's great."

The man sighed. "Fine, then." He hiked the sack he was carrying higher on his shoulder. "Let's go."

The trespassers collected themselves for the journey. Zander was admittedly relieved – until, in a voice too low for the others to hear, the man murmured, *"You are in SUCH deep shite."*

Zander bowed his head. "I know," he admitted. "I know..."

6
Propositions

The house was old and small and the utter accolade of neglect. The white-wash of the wattle-and-daub walls was worn and dirty, flaking and slicked with ash and mold. The thatched roof was tattered and black, overgrown with a sickly carpet of moss. The lawn of the little clearing was over-grown and wild, gnarled and blooming with weeds and wild-flowers. And the dark-painted Dutch door stood open.

"Di'jyou guys get any food?" These were the first words the man had spoken since they had started home. They were gathered in the kitchen – a room littered with dirt and junk, the counters cluttered with rubbish, and the sink filled with crusty pots and pans. At the room's center was an elliptical table which would have seated six, but there were only four chairs, undisguisedly and unabashedly mismatched.

"No," Meygan responded, in a *Well-duh* sort of manner.

"Why not?" The man swept a scattered sheaf of greasy ledger-papers off of the table and into a satchel, gesturing for them to sit before depositing it on the counter. He flashed her a condescending smirk.

"D'*you* have any food?" Ian asked, eagerly, kneeling on his seat. "I'm, like, *starving*." He splayed forward dramatically onto the table.

Meygan scowled at him. *"Really,* Ian?!"

Rolling his eyes, the man took the draw-string sack from his shoulder and tossed it to Meygan, grandly. "Bagels."

Meygan looked into the bag. "Thank you," she murmured, begrudgingly. She pulled a bagel from the sack and pushed the rest to the center of the table; the boys seized the food like a pack of emaciated jackals on a gazelle.

"Individually wrapped," the man added, pointedly. He cleared off an area of counter space by shoving everything aside and began riffling through the drawer below.

"These are good!" said Ian, astonished, as he stared down at his partly-wrapped, half-eaten bagel in awe.

But Meygan was still all business. "So," she said, folding her hands on the table in an executive fashion. "I didn't get your name."

"Nor did I catch yours." The drawer slammed shut, and he tossed a boat-knife to Zander, who caught it one-handed; in return, Zander tossed him a bagel.

Meygan scrutinized him as he leaned back against the counter, trying to open the plastic wrapper. "My name is Meygan. I'm sure my *associates*," she added, turning her glare on Addison, "can introduce *themselves?*"

"Mm!" Ian held up a finger, begging a moment. He swallowed. "Ian Burgendson."

"Addison... Mitchell."

The man nodded once they had introduced themselves. All three of them seemed to be in their early teens – probably late tweens. Not a typical traveling party by any means – and to be so far from any track or town? He wondered again what was really going on here. "I'd guess by your accents you're from the towns. Are you all from the same place?"

"Yeeaaah..." trailed Meygan, giving him a strange look.

"Which one?"

"Uh –"

"Riverside Valley," Addison answered. "It's off of route five when you're coming east out of Syracuse – before Chittenango, south of Mycenae; you turn south and keep on the main road 'til you come to this dirt road that most people think it's just a drive-way, and that goes on into the hills, but short of Caz, kinda angled back south-east –"

"Wait wait wait." The man shook his head. "That's a lot of information. What's the town?"

"Riverside Valley."

The man's eyes narrowed ever-so-slightly, before lifting to the ceiling to trace maps on its dull, plaster surface. "...Was it *always* called that?" A tinge of confused embarrassment slithered through his tone.

"What d'you mean?"

At once, Zander rounded on the man. "They're lying, aren't they?" he demanded.

The man bit his lip, slowly fumbling with the wrapper.

"They're not old enough," he decided at last, "they mightn't know." He looked back to Addison. "Which river'sit on?"

"What d'you mean?"

"I mean *what river* – it's called 'Riverside Valley', isn't it? What valley? What river? *Route five* doesn't exactly *mean* much to me."

Addison made a face. "It's not near a river."

"*...What?*"

Meygan frowned at Addison. "That's stupid."

"Yeah!" the man echoed. Then, drawing back, he resolved to try again. "Wait – so – is the town north of here, at least? Is it further into the mountains?" Crud, did they ever look confused. "How did you *get* here?!"

The trespassers glanced at each other; Meygan's look was scathing, and when she turned back to the man, his eyes steeled against hers, as guarded and indifferent as ever.

"What's your name?" she inquired, her tone low as a threat and twice as suspicious.

He cocked his head. Supposin' at a stretch it was relevant, but it did nothing to answer the question. He muttered something that sounded like "Ruh-hay-oak."

"*What?*"

He gave the slightest of smiles. "Rahyoke. Need it spelled?"

Meygan made a face. "I think I'll manage."

"Grand. How di'jyou get here?" Bowing his head, he studied the packaged bagel in his hands. "I'm only gonna help you if you tell me."

Meygan's look softened, and she, too, bowed her head, her gaze flashing back to him in shamed flickerings. "You're not gonna believe us," she quietly replied.

The man called "Rahyoke" smirked – mysterious, mocking, bitter.

"Try me."

"But –"

"*Try* me."

For a moment longer, she held his gaze. Then her eyes slid to Addison; his fault, his story.

Addison looked away. He wasn't stalling for comfort, and he wasn't stalling for lies. The story was incredible – literally, incredible; he was hardly certain he believed it himself. But he could feel Meygan's glare sharpening upon him; so, drawing a deep breath, he reluctantly explained. "We were at Meygan's house, and me an' Ian were tryin' to keep a fire goin'. All of a sudden, there was this big flash, and... I dunno – we just ended up on that mountain. Where Meygan said she – *saw* you."

The look Meygan fixed on him did not let up in the slightest. "Convenient telling, that."

Addison returned the glare, though guilt marred his own ferocity.

"So – wait. So –" Zander frowned at him. "That's it? That's all there is? A big flash of fire, and – poof – you're here?" He could not have sounded more incredulous if he tried.

"I can't make any sense of it either," Meygan quickly replied. "That's why I'm figuring, if you could just tell us *where* we are, then Mr. Fire-Blowing G.P.S. here –" An out-flung hand indicted Addison "—can get us back."

Zander was silent, but he still looked at them as if they were all morons. A fight brewed between the two of them, but before Zander could start it, Rahyoke cut him off. "Zander, stop it. The issue isn't how they got here."

Zander turned to him, aghast. *"What?!"* he half-shrilled. "It has *everything* to do with it!"

Rahyoke shook his hair out of his eyes, raising the packaged bagel to try to tear the plastic with his teeth. He said nothing as Zander continued.

"If *that's* their coming – That's *faery-stories*, Rahyoke!! Alternate worlds and Tír na nÓg – it's childish, it's pub-talk, it's the half-assed fabrication of the insane mind!"

"Hey!" Meygan shouted, sharply.

"How are we gonna find a place that *doesn't exist*, Rahyoke? It *matters* a bit, doesn't it?"

"I never said it was a different world, anyway!" Meygan challenged, her pretty teeth barred.

Zander rounded on her. *"Here's* an idea, princess: *Go* back t'the mountains. Light a fire; see what happens!"

"Do you *have* to be such a jerk right now? Just sayin'."

"Legends, Rahyoke!" the archer insisted. "Hearth-fire stories and old legends. They're playing you! And believe you me, *no* good will come of this!"

Rahyoke pulled the package from between his teeth, studying the seal with disinterest and boredom. "Why didn't you just shoot them, then?" he asked, in the same way. "You've never hesitated before."

"You're blaming me?!"

"Yeah, kinda."

Zander moved to protest, then drew back. "Okay, well, *that's* true," he allowed, watching as the man wiped the package on his jeans before trying to pull it open again.

"*You're* calling *us* crazy?!" Meygan defied, seizing the opportunity of his distraction. "Look at you! Running around in the forest like some little Merry-man, pointing arrows in people's faces!"

"Uh!" Zander began, but again he was cut off.

"She has a point," Rahyoke murmured.

Zander sneered at him, appalled. "You believe them!" he accused.

The man looked up, shrugging a shoulder. "Some things. Maybe not all... I believe they're lost," he clarified. "That much you believed yourself. And I believe they came here from somewhere; call that place Riverside Valley."

For a moment, Zander continued to stare at him. Then he shook his head, muttering crossly, "You damned *philosophers*."

"Language."

"Sorry."

Rahyoke turned back to Meygan. "What d'you want me to do? What is it you expect from me?"

"What do you mean?"

Good Lord, if he never heard that question again... "I've never heard of where you're from. At best – the best I can do for you – I can ask around town. See if anyone's heard of your – *'Riverside Valley'*. It would mean you would have to stay here at least another night. Is that acceptable?"

"Yes," Meygan answered. It was better than sleeping outside.

The man turned to Zander. "Zander, I need you to do something for me..."

Zander looked wary. "What..."

"I promised I'd run a message. But now I hafta go into town; we're out of food."

Meygan's eyes rounded. "I'm sorry – !" she began, but he waved her off, continuing to Zander.

"I need you t'run the message; can you do that for me?"

Now Zander's eyes widened. "Really?" he asked, unable to quash *all* the excitement from his tone.

That flicker of a superior smile. "I'll exonerate you of the crappy job you did guarding my house." The bagel package split. In triumph, he uttered, "A-haa! So!" With the utmost of pride, he crumpled the wrapper in his hand, still holding it in his palm even as he ate.

Meanwhile, Zander was again on the defensive, arms encompassingly outspread. "Hey! I *watched* your *house!* Look at it!"

The man shifted the food in his mouth to mutter around it, "You din' even cu' the lawn."

"Oh! Akh! *Well!*" Zander snorted sarcastically, rolling his eyes. "You were gone *two days!*"

"Hm! I as'd you *Thursday.*"

"Yeah – well – uh –"

"Yea'." He swallowed his food. "Do we have a deal, or what?"

"Oh no – yeah, totally. I'm going." Zander settled back, folding his arms. "And I'll be back within the night."

"You should, 'cause it's not far of a journey. But you move like a sea-slug."

"Auh!"

"Does that sound fine to you lot?" He turned back to the trespassers without giving further pause to Zander's offence. "That means you'll be here by yourself, though," he warned. "Zander and I both should be back by morning at the latest; can I trust you?"

"Yeah," Meygan answered, studying her bagel.

"...Fine, then." It was a big risk – but then, what could they do but harm the house? It was safer to send Zander out of it. He leaned off of the counter, turning and snagging a papery food-wrapper and pen off of the counter. "Zander'll be back before I am if he doesn't putz around."

"Who's a putz?" Zander challenged lightly, shrugging. He emptied the sack of bagels onto the table.

"Zander, these are your instructions," Rahyoke explained, scratching out a detailed message in thick, small, hurried letters.

"Am I gonna be able t'*read* that?" Zander challenged, his mouth full of bagel.

"Putz." Rahyoke turned around. "Here ya go."

Zander handed him the bag in turn. He studied the note, bringing it close to his face and squinting, twisting it this way and that. Rahyoke swatted him over the head with the empty sack.

"What!"

Rahyoke slung the bag over his shoulder, turning back to the trespassers. "Normal laws of decency and xenia apply – I trust you know how to behave and conduct yourselves. You've free reign of the downstairs – the kitchen, the living room – an' the loo's that way." He pointed vaguely to the door of the living room. "But don't go upstairs; there's nothing up there anyway."

"It's my room," Zander reported.

"Yeah." Rahyoke glanced out the window over the sink. "Ah – I'm out. See ya'll later." He swept over to the door, dropping the bagel wrapper on the counter as he passed.

"Shouldn't one of us come with you?" Meygan offered, standing.

"Yeah," Addison echoed. "Maybe we could help."

"You can help by staying *here*." Pausing in the doorway, he cast them a flash of a full-mocking half-smirk. "Besides," he added, mysteriously, "I still need someone to watch my house."

And, with that said, he was gone.

Meygan dropped back in her chair, defeated, her green eyes wide.

Addison turned to Zander. "Is house-guarder at least a *paid* position, by any chance?"

Zander the Archer merely laughed.

7
Mahallahi

He wasn't laughing long.

Zander pulled to a halt, dropping to one knee close to the promontory's dangerous edge. He was heaving for breath, his limbs shaking with exhaustion.

"Back within the – *night*," he uttered, with what little air he had. "I'll *kill* Rakyoke! *Kill*'im!"

The mountain range was narrowest near Rahyoke's property, but that didn't make the long, uphill trek through the forest, mountain passes, and up over that towering mass any less hellish. Nor did it make the prospect of what lay below any more appealing.

Raising his head, Zander glanced out over the edge.

The Ramadi Desert was the smallest desert in the world, cupped between two ranges of mountains that coiled around it like twining snakes, sucked barren by the sun and denied the pleasure of the rains that saturated the land farther north. Once, a great city had risen in its bowl, a holy city, filling the little desert with its millions. Now...

It was the dust of the city that covered the earth with sand.

Zander tied the head-scarf he wore more securely over the lower half of his face. Sand had an unpleasant way of coating your throat and lungs in this weather, and he wasn't after asphyxiating tonight.

He raised his deep-green eyes to the darkening sky. Steel-grey clouds were moving in from the south, shot through with lightning, as the sun was setting in the west, casting an orange-cream paste over the spinning sands, folding deep purple shadows over the bowl. He couldn't descend in darkness. He needed to move *now*.

He drew a gasp, trying not to choke; his sides stung dearly from exertion. He wrapped his arms around his chest tightly, and his fingers fell against the bulk of the quiver at his back. He closed his eyes, sighing. The army of Fionn Mac Cumhaill – the Fianna – could, each one of them, sprint the length of the forest without breaking a twig, without being captured, and come out of it without a hair out of place.

Time to run.

Zander slid to the edge of the promontory, lowering himself gingerly over the side 'til his feet hit the expected ledge.

"Rahyoke is *SO* dead," he grumbled, vanishing over the edge and into the darkness below.

8
City of God

In a corner, sheltered by the remains of a crumbling sandstone building, perched a creature of the night. Scraping casually at a branch with the sharpened blade of a rusty knife, he waited, savoring the sweet respite from the flow of the sandpaper wind. All around him, the shattered ruins of unkept buildings sat, golden-brown in the sunlight, but a pasty grey in the dark. Fine sand covered the earth in shifting mounds, breaking through the wall that surrounded the city and stripping the buildings to their roots. Softly, he shook his unshorn hair from his eyes, scattering a handful of their dust. He wondered why he'd been drawn to such a dismal place in the cover of the sousing night. He wondered why he had ever been drawn to this city at all.

Like a distant memory, a fevered dream, the grey mists rose again. How clearly he could still see them, hanging over the bog-lands, tumbling out over the moors, beneath the deep, teal sky. The gentle gold of the sunset, slipping over the moss-thick limbs of the hoary trees, the clear waters of the Afon Ddu shimmering in the dusk. The mountains, steep and verdant, dotted with cairns and the heather-bramble twisted with lengths of wool. How soft, how fine, the sweet summer rains, washing over the white sands of the beach at Llyn Alun, stirring the breezes among the flowers and the leaves...

Suddenly the silence curdled in the air, and he sat up with a start.

Quiet had resumed. Straining uneasily against the snake-like sizzle of the Southern winds, he listened. As ever, the scathing breeze licked headily though the broken walls, scoring the stillness, the silence.

He had lost his mind. That was it, and that was all. Any sense of that tantalizing pattern of undulating noise, without source or articulation, was nothing more than the snap of a mind worn unsound by the ever-shifting winds after their years spent in blissful torture.

He buried his face against his legs, wrapping his arms around his head protectively. The night was cold. It was the shade of these mountains, tainted deeper by the clouds which fell in from the north, which hurried from the south, carrying with them the song and whisper of better days. The stars were buried in the bank of clouds, lost in a concealing embrace, and no breath of frost or scorching sun could ever shift the night in her endless sleep, to show even a hint of that Heaven beyond to those lost souls residing below.

His fingers twined into his black hair. Escape was a sweet torment more alluring than the promise of the sun. But where was he meant to flee? Where, when all the world was desolation, and he would never be able to go back – and there was nothing to go back to.

A rustling stirred the night. He caught up his sharpened spear and rose to his feet, wielding his weapons with threatening ease, with a skill he'd been honing from birth.

There. There, and he could have cried out to God for His mercy. A flickering of motion, the whipped edge of a shadow, flashing down a crumbled side-street.

And, flipping the spear in his weathered hand, he gave chase.

Zander skulked in the shadow of a half-crumbling building, his eyes fixed on the small dwelling he sought. It was a desolate structure, and all that remained of the place was a fraction of the walls – stacks of cracked sandstone, toppled by the winds. A colourful blanket was pitched tent-style to act as a roof, its fabric woven into bright patterns that were obscured by the winds that tossed it. The door-way had been rebuilt, and a tattered, faded-purple sheet shredded at the hem had been nailed up as a door. A few pots stood sentinel at the threshold, half-buried in banks of sand.

Stealthily, he edged forward, clinging as closely as he could to the sandstone rubble. It was a ten-yard dash to the door; with a quick glance around, he darted.

He didn't make it. Startled, Zander jerked back to view the soul with whom he had very nearly collided.

A man, garbed in layers of bleached rags. His black hair was mussed and matted, standing at drunken angles. His eyes were large, a colourless-pale blue, wide and vaguely unfocused. In his hand was a knife, a rust-smudged Sgian Dubh, held threateningly close, poised to strike.

Zander had three weapons on him: the bow in its quiver strapped to his back; the boat knife Rahyoke had given him, which was hooked in his belt; and the great, ancient sword he wore belted against his left hip. But before he could reach for any of these, one word halted him.

"Kiff."

He didn't know the language, but he got the intention. Against all better instincts and judgments, he raised both palms in pacifistic surrender.

"I'm not your enemy," he attempted, though he rather wondered. "I'm a messenger –"

"Laissa ladayka ayya amalin huna." The weapon jerked closer, sending flakes of rust glittering to the earth like snow from a blood-filled sky. Zander gulped, and the guardian grinned menacingly. *"Irhal wa illa kataltuka."*

The two men stared at one another for a moment, taking each other in. Zander watched the flash of the quivering knife in the half-light as patterns curved over the blade. The speckled rust stirred maliciously on the wind. Slowly, imperceptibly, his hand began to inch toward the sword that hung at his side.

"I'm to deliver a message to Bahadur al-Esmail," he said; sanity, sangfroid. "It's super important, and it'll only be really quick. But if he's not here, I'll go –" He gasped as the damaged blade pressed against the soft skin of his throat.

At a low growl, the man offered one last time: *"Innahu laissa huna. Irhal!"*

Leave.

The orange flakes of rust melted into a pooling stream that darkened and intensified, curling over the blade. They ran in a stream that spilled over the edges, running along the thin tracks of the knife to fall glittering to the earth, soaking into the ashen sand.

Now!

Zander's grimace slurred into a swift swipe of his neatly-drawn blade. The knife retracted slickly from the side of his neck as the man leapt backwards to avoid the thick, seventeen-inch blade. The archer hurled another strike, and the sand-dweller ducked, tumbling backwards into the sand. When he rose, a pointed stick was clenched in his left hand.

Zander had to admit, it was a well-calculated move. Still, he taunted, "Ooh, a *stick*." He rotated his wrist, flipping his blade in a gleaming arc. *Come on*.

The sand-dweller's translucent eyes narrowed. Then he attacked; a swirling swoop of flittering rags brought him close enough to strike. The stick was blackthorn, a deep-north material not found this far into the mountains – and shorn, it held the archer's blade.

Zander swept back to avoid the underhand stab of the knife, the blow just missing his stomach. He twisted his blade to strike at his attacker's legs, but the man jumped, stabbing down with the pike as Zander's blow whisked clean under him. Zander leaned into the miss, falling forward so the man's pike came down in the sand and his feet came crashing down on him, destroying his balance.

The two men scrambled to their feet, throwing blows before even finding a footing.

"Da iawn," murmured the sand-dweller, slipping briefly into his native tongue. Then his lips curled in a dangerous smile. He had wanted a challenge – a reason for being here. *"Sabiyun jayid. Innaka barihun."*

He struck; lunging forward, he battered Zander's blade aside, driving the Sgian Dubh to the lad's chest.

"Holy –" Zander staggered back. All at once, his leg was hooked, and a solid, sucking mass slammed into his back, breaking him. His sword, in one, sparkling flash like a shooting star, arced above him, landing with a *shink* blade-first in the sand, a good twenty yards behind him.

Before he could look down, a second mass landed on his chest, collapsing his lungs. A blade pressed horizontally to his throat, ready to seal his fate.

"Declan!"

The sand-dweller tensed visibly, listening. A question lilted over the wind. *"A kullu shay'in ala ma yuram? Ma allazi yahsal?"*

"Laissa hunaka jundiyun huna," he said in reply, *"mujarradu kha'inin minal shamal."*

Dazed by the double impact and a mind-stirring lack of oxygen, Zander forced himself to focus. He saw an out. "Messenger to Bahadur al-Esmail, from Shivek of Nissia! I mean you no harm!" He choked as the flat of the Sgian Dubh was forced against his throat. The man above him was not content to wait for an answer from inside. And yet –

"Da'hu yadkhul."

"Maza!?"

By the sand-dweller's expression, Zander knew he was saved.

"HAH!!" he blurted defiantly, jerking painfully against the forgotten knife.

The sand-dweller glared down at him. Rethinking his course of action, Zander offered a compromise. "Keep my sword 'til I'm leaving," he said, as reasonably as if they were discussing things over coffee – though *auch*, his voice was hoarse.

"*Sawafa akhuzuhu minka aydan*," the man grated, indicating the quiver on Zander's back, and in it his bow and arrows.

"Yes, that too. That's all I have," he cautioned, to the man's distrustful look. That was all he needed – to be frisked and strip-searched in the middle of the desert.

Slowly, reluctantly, the sand-dweller got to his feet, keeping the sharpened end of his blackthorn pike pointed at Zander.

"Fast," he warned, finally in a language that Zander understood.

Sidling out of the path of the pike as he stood, Zander dusted the sand from his jeans, looking up at the guard. "Yeah, no problem." He straightened, dropped his quiver in the sand, and backed towards the door of the cavernous half-tent. "Thanks, lad; guard'em well."

The sand-dweller scowled, and Zander ducked under the blanket and into the tent beyond.

It was built like a Mongolian nomad's shelter – a yurt; lattice-work rounded out the walls, fortified with ply wood and covered in tied-down blankets, the floor swathed in layers of thick Persian rugs, intricate and rich, fit for a king. The room was sparsely furnished; a stack of crates and rolled blankets dominated one corner; a lone crate crowned with a low-burning lantern was the alter of the other.

"You are built like a stick."

Thus did the old man greet him, speaking out of the darkness just beyond the light.

Zander turned to the place, folding his arms. "I'm skinny, but I'm strong."

An aged hand dipped into the circle of the light, turning up the oil on the lantern, allowing the warmth of its red glow to saturate the room.

Bahadur al-Esmail was old – long about eighty, to Zander's estimation. His skin was dark as sun-baked milk chocolate, his eyes as black as the night. He wore a long robe of a pale colour that seemed almost to shift with the lighting, the deep V-neck and long sleeves trimmed with embroidered gold. On his head was a turban, blue as the summer sky.

Zander bowed his head. "God bless all here. And thank you for saving me."

"Ahlan wa sahlan!" Bahadur replied. "I imagine that with more training, you would save yourself." His accent was heavy and distinct, ornamented with the rolling "R"s and tongue-flicked vowels of Turranian.

Zander was stung by the remark, but evidence held it as true. "Loads more training, maybe," he replied, humbly. "If I might – are you Bahadur?"

"Almajdu lillah! That I am."

What did that mean? Zander hesitated a moment; but whoever he was, Shivek's message pertained to him, so...

Zander started slowly wringing his hands, wrangling the fingers of the left. "I was sent here to deliver a message, that... Word has been received and confirmed," he reported, "that this desert is going to be trawled, and – purged, God help us. A division from Nissia has been alerted of your position. It's been estimated you've less than a week. Does that make sense?"

"Yes."

"You must leave here by morning if you're to have a chance. Head south-west. Take your body-guard with you, and, if there's anyone else in this village..." He trailed, watching the old man's face. Bahadur's expression had darkened with thought as he considered these tidings deep within his heart.

At last, the old man broke the silence. "Once, I knew a boy," he murmured, "and he was very much similar to you." He fixed Zander with a solemn look. "You know what my answer will be."

Zander felt his heart lurch; he bowed his head. "Would... it help if I noted the world needs you?"

Slowly, Bahadur turned; a leather back-pack laid in the shadows behind him. He drew this into his hands, holding it out to Zander. "Our time on this earth is like the light of a lantern; and mine is nearly spent. The traveler brings with him fresh oil to keep his lantern steady – not the residue of the cask – not the spent crust of old matches. Accordingly, my son, it is the prerogative of the old to pass on what is theirs – their knowledge, their destiny – to the next generations. Here; give this to the young man that sent you, with my thanks. Believe me; the world got on fine before my time, and it will keep going." As the archer hesitantly accepted the gift, Bahadur offered him a wink. "Keep it safe!"

Zander allowed a small smirk. "Heh..." Just like a wise-man. "I wish you would change your mind," he responded. "Getting on or not, the world won't be the same without you."

A smile crossed Bahadur's lips, and his eyes gleamed in the lantern light. "What is your name?" he asked.

"Zander."

"Zander? *La takhaf*. Everything will turn out fine."

Zander could not help the disbelief which showed clearly on his face. "*Binnajaah*," he uttered, one of the only two Turranian words he knew. *Good luck*.

Bahadur's smile broadened. "*Inshallah*," he corrected. *By the will of God*. "Safe travels, my friend. *Rafakatka al salama. Bi aman-i-llah*."

9
Forbidden

The night shimmered cold blue, sparkling with a handful of scattered stars smeared into the cloudbank which stretched, lumbering and ominous, across the sky. Meygan sat on the slim window-ledge in the living room, watching the flickering tongues of the lightning rolling through the distance. Although the hour was late, neither the archer nor his master had returned. The intruders were in charge.

She reached into her backpack, again taking stock of what little she had, as if this could make her feel any more prepared. Some lip balm, her cellphone, one and a half chocolate-granola bars she was determined to ration, and an old journal – battered, blank, soot-stained and water-damaged. Once more, as if, she took out her cellphone, but she still couldn't find a signal. Rahyoke didn't seem to own a phone, either – in fact, the only electrical appliance on the first floor was a very small ice box, unplugged and empty save for an ice tray and a coffee mug. The longer she knew him, the less surprising any of this seemed.

"Hey, Meygan!"

Meygan rolled her eyes. Replacing her belongings and slinging her backpack onto her shoulders, she glanced to her left, and was startled to see light. A lit candle stood on the coffee table, propped in a battered tin holder, and Ian standing beside it, holding a shoe-box. Meygan stood from her perch, keeping her arms crossed. "Where did you get that?" she snapped.

Ian ignored her question. "Will you help us with something?"

Meygan frowned. "*Where* did you get that – did you go upstairs?!"

"Yeah." Ian brushed her off as lightly as anything, setting the box on the table beside the light.

Meygan scowled. "You *know* we aren't supposed t'go up there!" she scolded.

Ignoring her still, Ian made his way over to the stairs. Meygan followed.

"We had *one rule,* Ian. *Don't* go upstairs. An' wha'd'ya do?! You go upstairs!!" She stopped at the bottom step, taking the railing in her hand, watching as Ian ascended, unheeding. "And you *know* he's gonna blame *me!*" she called after him. "Are you ignoring me!? JERK!!!"

"Light a fire, will you?" Ian drawled in reply.

Meygan was scandalized. "Auh! No I will not!"

"You said you were cold," reasoned a voice from upstairs.

"Is that Addison!?" she exclaimed, in sheer frustration. "Are you guys *trying* t'ruin my life!? Rahyoke's gonna be *furious!!*"

Addison appeared at the top of the stairs. "We were trying to find candles," he said, as if this were perfectly reasonable. He, too, carried a box, a little bit larger than Ian's.

"I'm not involved with this!" Meygan covered her ears, walking away from him, retreating to the couch. "I am *SO* not involved in this!"

Addison rolled his eyes, following her. As she dropped down onto the couch – straight-backed and regal – he set the box on the table before her.

"It's just fire-wood in this," he explained, in a low voice. "I... just figured... if you wanted, you could make a fire... so it's not so cold... There's matches in Ian's box, if you want to."

"It is not *'Ian's box',*" she returned, sharply. "And I'm not setting Rahyoke's things on fire!! Although I know *you're* into that!"

Addison dropped his gaze. He said nothing, and returned upstairs.

Meygan folded her arms, well infuriated herself. Her eyes fell to the boxes on the table, framed and shadowed as they were in the flickering light of the candle. The larger box was, as Addison had said, filled with kindling. But the second box was far more interesting. From what she could see – she would not touch it, for the box, being a thing of the upstairs, had been explicitly forbidden to her, no matter *what* the boys said – the box contained a long twist of electrical wiring, a roll of duct tape, several cheap pens that may have been chewed on, a shaft-less fly-swatter, a mix of rusted-out screws and nails, and, yes, matches, among other things the light didn't touch.

Meygan drew forward, slipping down off the couch to sit on the thin carpet, bending her gaze to the side of the box. To her amazement, the surface was covered in a squiggly, sloppy handwriting, sprawling haphazardly and dizzily over the sides. Much of it was in a chaotic, sharp version of the writing system she knew, though she couldn't understand a word of it. The rest – scant though it was – was written in what looked like Ogham – an ancient text of a series of hash-marks set vertical and diagonal along a long horizontal line. The

Ogham-looking lines were cramped, small as chicken-scratch, and looked as carelessly scrawled as a grocery list. But there it was.

"Are you going through the guy's stuff?"

"Yeah, Meygan – jeeze."

Meygan turned at their voices, wrinkling her nose. Boys are impossible. "I'm not letting you get me in trouble."

Ian shrugged, setting yet another box on the table. "Suit yourself."

Meygan stood as he lifted the kindling box, turning towards the brick fireplace. "Ian – come on, don't. Auh –" He didn't seem to care for the hospitality approach; she tried another route. "Don't you guys remember how we *got* here?"

"Maybe we'll get home," Addison murmured, folding his arms.

Meygan tried one last angle. "Can't you just wait until Zander comes home before you start setting fire to things? I'm sure he's on his way."

Ian shook his head. "Meygan. It's *cold,* and it's *dark,* there's a fire-place *right here* – it's not like I'm lighting it on the coffee table – and I found the matches in the *kitchen.* We're *allowed* in the kitchen. I know what I'm doing; just trust me."

"Trust. You." Meygan shook her head. For a moment, she considered intervening – there was a good chance, in her opinion, she could take them both down in a fight. But the best course of action was to get out of the situation – away from these two fire-hungry fools.

She turned away, heading towards the kitchen. "I have no connection to you," she declared. "I will not be responsible for your stupidity." These were her parting words.

Addison watched her go. "Jeeze... Are you sure you know what you're doing, Ian?"

"Would I try it if I didn't?" Ian struck a match against the side of the match box. Once, twice, thrice... "C'mon, Addison!" he laughed. "What's the worst that could happen?"

Addison rolled his eyes, crouching beside his friend. "Yeah, 'cause *that's* not the premise of every horror film ever made." He watched Ian a moment as he continued striking the match. *Seven, eight, nine* – he grabbed the matchbox. "You're like the worst pyromaniac ever!" He struck the match – once, twice, lit. "Why do I let you talk me into things," he grumbled, lighting the kindling. The sticks were brittle and dry, and caught quickly; soon there was a cheery little blaze.

"Isn't that better?" Ian asked.

"It is." Addison moved around the coffee table to take a seat on the couch.

What was the worst that could happen?

10
The Inevitable

In the darkness of the midnight hours, a solitary flare burned bright, a candle of destruction, a beacon to the world. The trees were doused in ochre light; smoke billowed across the ground, drifting like a winter fog.

Suddenly, the smooth rippling of the flames was shattered as a streak of black shredded across the gold. Thin hands grabbed her, jerking her back from the blaze with such force she was nearly knocked to the ground, saved by his tight grasp.

"WHAT ARE YOU DOING!? ARE YOU *INSANE*?!!"

He was not gentle; Meygan pushed her curls out of her face, looking up at him as he held her by the upper arm. "Rahyoke –"

He stared at the fire, stunned, dismayed. *"A Dhia – cad a – tha-tharla – cén –"* His eyes widened. *Dread.* "Where is Zander?"

Violently, he turned on her, seizing her free arm. *"Where is Zander?!"*

"He isn't back –!"

"A Thiarna..." He looked again to the crackling frame of the house, only slightly comforted. Everything was gone, caught up in the smoke signal. *"...Dhia Uilechumhachtach."*

"Rahyoke – you're hurting me –"

With difficulty, he unclenched his shuddering fingers from her skinny arms, stumbling back a pace. *"In ainm Dé.* Wh-*why* did you –"

"Rahyoke, it was an accident – I'm so sorry!"

"A... A – 'accident'!" He scowled, his senses returning in a savage rush. "Feck me! An accident! What could you *possibly* have been doin'?! D'ya have *any* idea – Meygan, what're you looking – ?" He followed her gaze.

Addison backed up a pace, nearly colliding with Ian. "Shit..."

His cowardice made Rahyoke even angrier. "'Shit' is right." And for however terrifying his shouting had been, a thousand times worse was the tone – low, black, and disturbingly devoid of emotion – in which he demanded, "Did you think that I wouldn't *notice?"*

"Nn – w-*wait,* I – no!!"

There was a loud snap as a beam broke in the fire; the kids flinched away, but Rahyoke was too incensed to care. *"What the Hell did you do!?!"* he shouted. "*Answer* me, damn you!! What did you do?!"

"I don't –"

"So help me *God,* if you say you don't know..." Rahyoke drew a pace forward, seething, his fingers clenching like claws. But Addison was saved a gory death as a voice sounded from the woods.

"Shorn'a *mercy!"*

Rahyoke turned. There stood Zander, a headscarf tied around his neck and a leather pack upon his shoulders.

Some part of him collapsed in pure relief. But his rage was fervent, unchanneled, unrestrained and without reason; murderously, he hissed, "You were *meant* to be *watchin' them!"*

Zander gritted his teeth. Spreading his arms, he retaliated, "I just got back!"

"Well you should've been back sooner!"

"How the *Hell* could I'uve gotten back *sooner?!* You're lucky I made it back at *all!* Y'know, Mahallahi isn't exactly an *hour away!"*

"It's feckin' an hour an'a half if y'weren't such an ass!"

"My *God!* It's like you have *no* concept of time!"

"Feck off!"

"If you're blaming anyone, blame yourself!" Zander suddenly shouted, not fully certain what he meant. "It's your fault they were here alone!"

"I – *I?!* I blame YOU!! ALL of you!!" He backed away, drawing closer to the fire. "I blame them for this, and for them, I blame you!!" Fanned by the wind, the flames leapt up like demons behind him, casting wide swaths of shadow over the lawn.

The trespassers were too spread out to gather in a glance, and if he looked at Addison, he was afraid he might kill him; so Rahyoke looked to Meygan instead.

"I don't know what the Hell you're playing at," he vowed, his voice low. "No one's ever heard of your village. I asked, and no one had. They looked at me like I was crazy. Now, I don't know if you're *lying,* or if you think this sort of shit is funny, 'cause it's not. It's sick; it's fecking *sick!"* A soft wind stirred the flames behind him. He began to feel their heat, against his back and through his hair. He closed his eyes, forcing his throat to contract and swallow.

"Look." His voice shook; he opened his eyes. "I don't know what you're doing in this territory, and I don't know why the Hell you've decided it's *my* problem. I'm sure amongst the generation raised without decency or sense in this depraved, arseways purgatory of a Godless wasteland, *this* qualifies for entertainment. Grand! You've no idea what you've done. Not a clue. I'm feckin' *scuppered* thanks t'you!!!"

There was nothing to be said, nothing that could be done. Rahyoke twisted his gaze up to the sky behind him; all the stars, the moon, were blotted out, the sky filled with smoke, obliterated by ash. The whole of the soot-stained Heavens...

He paled.

"We have to leave."

Zander frowned. "Of course we have to leave."

Rahyoke turned back to him, anxiety searing through his veins. "No, we have to leave *now*." He didn't wait for an answer, sweeping away from the blaze and towards the enveloping shelter of the woods. As he passed Zander – poor stunned, half-understanding child – he touched the boy's arm. "C'mon; let's go."

"What about us?"

He froze, hunching his shoulders as he tensed. *They wouldn't* – he whipped to face them. *"YOU?!* You're the *CAUSE* of all this!! If it weren't for *you*, I wouldn't have to – to..." The last word vibrated with uncertainty.

What are you going to do now, Rahyoke? What can you possibly do?

Rahyoke settled back on his heels, looking again at the smoke-blackened sky. Heavy clouds still obscured the light, threatening more rain. The feel of the world screamed it was long about midnight, the tide of the night turning towards dawn. What was he going to do?

"Damn it all, fine." It was hard to say to whom he was speaking, but his voice was cracked and faint again. "Feck me; you can come. Sure, and why not? Maybe next you'll kill me, huh?"

"No!" insisted Ian, shaking his head.

"I'll believe it when I see it." With a wave to Zander, he turned again to the woods. "Anyone who's coming, come. If you're not, I really couldn't care less. *Tar liom, a Shander*."

Zander cast a glance back to the trespassers, not accusing, but hurt; not angry, but full of contempt. Hiking the bag he carried higher on his shoulder, he followed after Rahyoke.

"C'mon," Addison murmured to his small band.

In silence they filed forward. The last to go was Meygan, giving a long, final look to the home they had destroyed. *What could happen*.

And the moon flicked out over a seething ocean of flame.

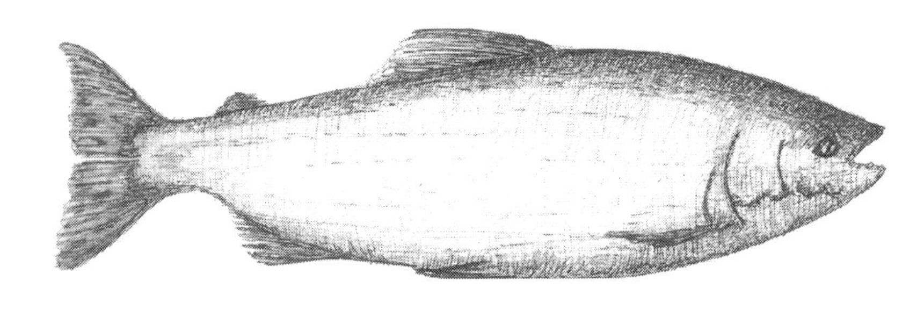

May the Road

11
Traveling

A thick, tumbling mass of soot-blackened clouds rolled across the sky. Wind whipped through the leaves of the trees, stirring massive branches, twisting them with deepened groans. A vicious evening, to be sure, cold and bleak as the ominous thundering that echoed in the distance.

Rahyoke ran a hand through his wild hair, dark eyes fixed on the reeling sky above. He was perched on a rise – an uneven hill that rose slightly on one side, then dropped, cliff-like, sloping smoothly into the forest below. The great trees strung their black limbs across the path, entwining their fingers to form a dense canopy above and below. Thin mists of dirt whipped up, tossed by the unyielding wind. The day had been dry, the scant sun baking off the rain of the night before, though the clouds loomed heavily overhead.

He folded his arms, hunching up his shoulders against the fierce wind blowing at his back. They had been walking since last night, moving ever northward, out of the shadows of the mountains. They should have been farther. But...

They dragged their feet through the tussled dust, as if they had nowhere to be in life.

Catching the glare, Zander bowed his head, trotting the last thirty yards separating him from his martinet.

"Keep up, will you?" Rahyoke demanded, as soon as he was within range.

"We need *rest*, man!" the archer protested. "We have to stop for the night."

Rahyoke scowled, though it took a bit of an effort. "We've been slowed enough as it is!" he hissed.

"We'll only slow more once they start to drop."

Rahyoke ground the toe of his sneaker into the dust, recalcitrant. "...Can't we just leave 'em?"

The archer shook his head.

Rahyoke gave an almost imperceptible sigh. Then, resolved, he straightened up, commanding sharply, "Then speed 'em up." It was a final retaliation; he turned and started down the hill.

Zander groaned. Turning back to the trespassers, he cupped his hands around his mouth and shouted, "Hey slackers! *Move* it! Let's *go!*"

Addison raised a one-finger salute, not bothering to pick up the pace. "That was rude," Meygan scolded; still, her next words to Zander were, "We're *coming*, you jerk!"

She jogged forward at her own good pace 'til she met him on the rise. "Are we *stopping* soon?" she complained.

Zander gave a low laugh, sauntering away. "Perhaps, if you keep up."

Meygan was irked. "Well, we'd 'keep up' if you *slowed* your strut a pace! What is that?!" She imitated his swagger. "Are we *fancy* fast wanderers? Really!"

Again, he laughed. "If my gait so displeases you, it may be best if you broke from the group." He walked the rest of the slope with a cocky sway that – he knew without looking – caused the girl to scowl.

"Jerk!"

That was how.

Zander ducked under the roughly-hurled stone, sprinting after Rahyoke, who was watching from the base of the knoll.

"*Iochd,*" Rahyoke groused.

"You wanted'em sped – *aiyee!*" Zander scurried out of range of a second missile. "Ach and beggorah! She throws like a boy!"

"Serves you."

Fueled by frustration, Meygan found the energy to sprint down the rest of the rise, chasing after the miscreant archer.

"What did I miss?" inquired Ian, from the top.

"We'll be stopping soon for the night if we make good ground," Rahyoke responded.

"Cool!" Ian tried to scurry down the hill, but his foot slid on the loose dirt, catching a root, and he whirled and pin-wheeled down the long slope gracelessly.

"Addison!" Rahyoke called impatiently. "Get'chyer arse down here! *Honestly...*" Without waiting, he set out again, ignoring the archer and the girl flitting through the trees.

They were far from the major foothills when they finally stopped to rest, and twilight had stolen in cool and eerie blue over the forest. The trespassers were reluctant to pull away from the invisible trail they had followed seemingly without end, but the cluster of trees that would serve as their shelter looked promising, and they were not about to argue against a break.

"Can we have a fire?" Ian asked.

"No," Rahyoke replied. Moving past them as they claimed their spots within the circle, he stood at the very edge, leaning his shoulder against a tree so his back was to them.

"Why not?"

Rahyoke folded his arms, biting his tongue. "It's going to rain," he murmured at last, a sharp underlay to his tone to suggest, *Haven't you had enough of fire?*

Meygan stood near the entrance of their circle, kicking idly at a root with slow, swooping motions. "This is where we're staying?" she asked, a lack of comfort in her tone.

"Yeah, why?" Zander returned.

"Um..." Reaching up, Meygan began untangling a length of her long, red hair, dividing the curls from each other.

Rahyoke glanced back over his shoulder. "What?" he asked, flatly. He saw her flinch. *A Dhia...* Turning to her properly, he asked, a little less harshly, "Meygan, what?"

Meygan hesitated. "What about... bugs and stuff?"

"What."

She met his gaze, wrapping her arms protectively over her head. "They'll get in my hair!"

"There're no bugs; it's fine."

She didn't at all believe him.

Rahyoke sighed. Leaning off of the tree, he slid off his long, black coat, rolling it up. "Here," he said, tossing it to her, and she caught it. "That's the best I can do."

Meygan's eyes widened. "You don't have to –"

"*Just...*" His tone spoke volumes.

"Thank you." Meygan donned the coat. She grinned; the hem of it brushed the ground. "How do I look?" she asked, posing.

"Short," Addison snickered.

"Auh!"

"Just go to sleep, the lot of you!" Rahyoke turned away again. "Honestly, for the amount of *complainin'...*"

"Okay okay okay." Meygan sat at the base of the tree she'd been eyeing so suspiciously before, dropping her backpack beside her. "Are you gonna sit down?"

Rahyoke looked away, shaking his head.

"You should."

Rahyoke said nothing; Meygan did not press him further.

The young travelers settled in, curling up on the solid earth floor. They were all exhausted, and it was not long before they were asleep.

It was then that the archer confronted him.

"You really should sit down." The soft voice from behind him caused Rahyoke to jump slightly. Smoothly, the archer shifted over to sit beside him. Maintaining that marvelously parental tone, he added, "It'd do you good."

"I'm fine."

Zander leaned back against a tree, looking up at him. "You've been going longer than any'uv us. And you haven't eaten –"

"I think I'll *survive*, archer."

There was silence. Rahyoke turned his attention up towards the high canopy of the trees. *The sky should never be green*, he thought, narrowing his eyes through the darkness. A white haze of a fog seeped slowly over the land – the only bit of the promised rain to make it through the leaves. He closed his eyes. He had lived here far too long.

Zander followed his gaze, looking out on the pale mist with wonder. "How long're you gonna run?"

The question hit him like a knife to the stomach, and he flinched in remorse. "Mm." He remained silent, listening to the distant patter of the rain.

"We can't flee forever," Zander continued, looking up at him again. "Sooner or later, we're gonna hafta face them."

"W –" Rahyoke caught himself. *He doesn't know. How could he?* He frowned, turning away from Zander. "I'd rather it be later," he muttered, crossly.

Zander rolled his eyes. Getting to his feet, he leaned again against his tree. "We should go south."

"We *can't* go south." Before Zander could speak a protest – "The terrain past the mountains is flatland – and even if we made it, they've increased the garrisons around and inroads into the mountains – *you* know that. It wasn't safe to stay there anymore. It was a mistake, to've waited that long."

Zander listened to Rahyoke, taking in his words with a heavy heart and wide, forest-green eyes. Then he bowed his head.

"It's my fault, isn't it."

Startled, Rahyoke turned to him. "Wha'd'you mean?"

"I shouldn't've left them alone," Zander said, lifting his gaze. "I should've brought –"

"It was my idea; don't worry about it. It's done now anyway..." Rahyoke paused, looking back into the forest, watching as the rain began falling harder, gliding over the leaves to spill out over the ground. "You got through?" he asked suddenly, glancing back at Zander.

"Huh? Yeah, barely."

"You can tell me about it later. We move again at dawn; make sure you get some rest."

Zander nodded, stretching. "We're going east, then, right?"

"Northeast. The terrain's nasty dead east of here; it's weird enough with all the ridges this close to the mountain."

"I still don't like it."

Rahyoke said nothing.

Zander folded his arms, shaking his head. "Tell ya what, man. You'll soon run out of land to cover."

"Ha." Rahyoke raised his eyes to the dense canopy, a strange, humorless half-smile on his lips. "Oh... I've *long* since run out of land."

For a moment longer, he watched the rain. Then, stretching, he leaned off of the red-brown tree. "Well! No sense standing in the rain, huh?"

Zander watched as Rahyoke moved back through the circle, returning the way they had come.

"Hey! And where're *you* going?!" he demanded, annoyed.

"Go t'*sleep*, Zander," Rahyoke replied, not even bothering to turn around.

Zander was appalled. "But – yer just –"

"Only fools take comfort in darkness, Zander," Rahyoke warned. "Remember that."

Zander sighed. "Fine," he relented, sulkily. "But tell me where you're going!"

Rahyoke suppressed a laugh. Standing at the circle's edge, he responded – airily, tauntingly, arrogantly – "Nowhere." He slid his hand against the tree at the edge, whipping around it and vanishing from sight.

Zander rolled his eyes, knowing by now not to bother waiting up. He moved slowly back to where his gear laid, collapsing in a heap beside it.

Gradually, the fog began to lift. The spatter of the rain became less pronounced, fading off into a weak dribble, as the Heavens spat down upon them in disgust. As the haze began to break, a feeble, goldenrod light seeped through the leaves, coating the sodden ground in faery dust, casting deep shadows on their tiny circle. Only then, as the first streaks of dawn spread into the sky, did the archer close his eyes and allow himself to drift into a shallow, broken sleep.

12
Thirteen Scrolls

The sun had risen high off the horizon, and its glaring light fell peacefully through the trees, giving the earth an almost sickly green tint. Bits of dew sparkled on the damp ground, washing away all traces of the travelers' path. It was a hot, slightly sticky morning, but the droplets still dripping like tears from the highest reaches of the trees felt cool to the touch.

It was into this world of sodden, brightened beauty that they awoke – some more abruptly than others.

"Oof!" Zander jolted upright, gathering to his chest the object which had assaulted him. Meygan's impish giggling answered him.

"You're up, are you?" a low, husky voice drawled.

Zander squinted against the light. Rahyoke leaned back heavily against one of the trees at the entrance to the clearing, his dark form standing out starkly against the red bark. His hair and clothing were still a little damp from the rain of the night before, the whole of his right leg stained with soot. At his side hung a greyish-tan canvas bag, slung so the thick strap crossed over his chest.

"Wha'sis?" Zander grunted, looking down as he rubbed at his eye with the back of a grimy hand.

"Breakfast."

"Oo!" Meygan sat up, allowing the cocoon she'd made of the black coat to fall around her shoulders. Her red curls stuck out at crazy angles, and she shook her head in an attempt to flatten the wreck. "Did you bring us bagels?"

Ian jumped up, snapping out of a deep sleep. "Bagels? I heard bagels – who said bagels – do we have bagels?!"

Since Zander had been tossed the food sack, he opened it, peering inside. "No – no bagels."

"Donuts?"

"No –"

"Artichokes?"

"Artichokes?" Zander arched a brow. "No."

"Is –"

"Maybe if you'd shut up he'd be able to tell you!" Addison snarled, woken by the discussion.

The thinnest of sarcastic grins flickered across Rahyoke's stern mouth. "Afternoon, Addison," he greeted the lad, pointedly.

Addison mumbled and grumbled, brushing the mulch out of his hair.

"Fine, Zander," Ian relented, dramatically. "I give up; what is it?"

Zander dug into the bag, pulling out a palm-sized object wrapped in wax paper. The smell of cinnamon and vanilla swirled into the early morning air.

"That smells so good!" Meygan reached forward, drawing the bag closer. She glanced at Rahyoke. "I can have some, right?"

"No; you have to watch everyone else eat."

Meygan made a face, pulling a wax-paper package out of the sack and sitting back. "I see your *sarcasm* hasn't been spoiled by a little *rest*."

"Nor has yours with all the walking!" Rahyoke noted, stretching. "Come on, lads; you can eat on the way."

"Wait!" Meygan looked up, her eyes wide. "Can't we eat first? It'll only take a couple of minutes!"

"Yeah!" Ian agreed. "Addison's not even up yet!"

"Give'm a couple minutes, Rahyoke," said Zander, unwrapping a wax-paper package. "We'll go a little bit longer tonight to make up for it."

Rahyoke resisted. "We'll go *all* night," he grumbled, slouching crossly against the tree.

"Cool with me," Meygan shrugged.

Rahyoke said nothing.

Zander settled back; in his hands was a tanned pastry, spiraled and stacked and generously smothered in sticky, white frosting. *"Oh,* mercy!" he lauded. "Have I mentioned you're my favorite person? You're a good and generous man."

Rahyoke rolled his eyes.

"Did you go into town again?" Meygan asked, all business as usual.

"Yes."

"Did you ask anyone about us?" Addison demanded, snatching a wax-wrapped package from Ian.

"Hey!"

"Yes –"

"*And?!*" Ian blurted, without waiting for a pause.

"And no one I spoke to knows where you're from."

"*No one?*" Addison repeated, incredulous. "Who did you talk to?"

"I didn't ask their *names!*"

"Well, how did you ask?"

Rahyoke gritted his teeth. Leaning off of the tree, he stood straight, his eyes cast to God as he explained. "'D'ya know how to get to Riverside Valley?' And they answerin', 'No.' 'It's a town; have ya heard of it, maybe?' 'No.' 'Do you think maybe the name's changed?' 'I think you've spent too long in Smythwicks' Pub.' 'D'ya know, maybe, anyone else I could ask?' And I asked 'em, too – 'What would *'route five'* mean to you?' 'Syracuse?' 'Chittenango?' They mean nothin' – and I'm wonderin', now – *someone* here has surely to be mad, and maybe it's me? Or maybe – *maybe* –" he pointed at Addison, accusingly "—someone's not being fully *honest* here. Is that a possibility?"

Addison jolted, sitting up straight. "Why would we lie to you?"

Rahyoke slashed his hand down to his side, taking one step forward. "I'm trying t'help you. God only knows *why*. I'll say – Zander was right. It's the sort of thing – this... People appearing and disappearing randomly, showin' up in other places, other *worlds*, even – it's hearth-fire stories. The kind of things y'tell your kids – how the *sidhe* take a man into their hills on Samhain, and he's gone a year or a thousand, and to himself it seems only a day. I mean – God help us, if you're telling the *truth – iochd*. Here it's not a man sober claimin' himself t'be 'poofed' from one place to another, by fire or any other means – in *modern* times, in *real* life – what d'you expect me t'do?!" He spread his arms wide, taking in the Heavens and all of the forest below – *his* world. "I'm a mere mortal, lads, sure as I'm breathin'. God save us, but if you *are* tellin' the truth –" Dropping his arms, he shook his head. "What d'you want me t'do?"

The terrible weight of honesty formed every ire-splashed facet of those words.

Another world... Meygan bowed her head. This couldn't be happening – he was right. This sort of thing was pure fiction – movie plots and sci-fis, not... "What if we went back to the mountain?" she suggested, with crushed hope.

"And do what, darling?" he asked. "If it was some force that brought you, wouldn't you've found it? How would that've worked?"

Her gaze dropped deeper. "Yeah..."

Rahyoke stepped back again, angling so he could view them all at once. "Look – I'm not yelling at you – I'm not trying to be a jerk – but – honest to God, lads. What can I *possibly* do for you? What if I can't get you back?" He was decent enough not to say that he couldn't.

Meygan's eyes welled with tears. "So that's it, then," she said, her voice shaking. "We're stuck here, and that's it."

Rahyoke cursed, running both hands back through his hair. "Meygan, I'm sorry. Please don't cry."

"Aw, honey." Laying his breakfast aside, Zander slid over to her, embracing her. Meygan welcomed the hug, burying her face against his shoulder.

Addison felt a spark and frowned. "We'll get back, Meygan. Don't worry."

"Yeah, Meygan!" added Ian, holding out a wrapped cinnamon-roll. "D'you want my breakfast?"

Rahyoke tossed the male trespassers a brief glare. *Do you really think you're helping?* "It's a possibility, Meygan," he said quietly. "I won't lie t'you about it. An' we need to figure what you lot'll do, in case. But I swear to you, I will do *everything* in my power to make sure you get home. And – I'll do everything beyond that to keep you safe. All four'uv you... Good enough?"

"Yes." Meygan raised her eyes to meet Rahyoke's; her eyes were a strange, pale green flecked with golden brown. "Thanks, Zander."

He allowed her to pull back, taking her by the shoulders, and smiled like a real sweet-heart. "Okay?"

She nodded. "And thank you, Rahyoke."

He shook his head, dropping back a pace to lean against the nearest tree, folding his arms again.

"Can we stay with you?" she asked.

Rahyoke jolted. "What?"

Soft-hearted Zander instantly volunteered. "Course you can!" he exclaimed, spreading his arms welcomingly.

"Until we find someplace better," Rahyoke cut in.

Zander twisted a glance back at him. "What place is better than ours?"

"We haven't even a *place*," Rahyoke reminded him. "But if you're after it –" he returned to the trespassers "—I know of plenty of good people who'll take you in – especially if you're decent for a bit of work."

"We'll see," said Addison, a bit coldly. "Right now, we're just working on getting home."

Oh, you're working, alright. Rahyoke ignored the flash of frustration. "Break's over." He hiked his bag higher on his shoulder, leaning off of the tree. "You can all very well eat on the way." Without waiting for them, he turned and moved off.

Zander glared at Addison. "Damn it, Addison! You cost us our breakfast!" Lunging, he grabbed his bag and his cinnamon-roll and scrambled to his feet. The others gathered their scant belongings and followed.

The day was sunny, and hot. Unobscured by clouds, the sun, glinting like a great, honeyed orb in the dense waters above, shone down in waves, caressing the earth with its milky, golden glow, raising the rain from the ground.

"It's so nice out!" Meygan observed, stretching out her arms. She was wearing Rahyoke's dark coat; the hem nearly trailed against the ground, and her red backpack, new and bright, stood out against it in stark contrast.

"Don't get your hopes up," Rahyoke warned. "Every so often, the clouds get blown away, but they'll be back by about five, five-thirty."

Meygan wrinkled her nose. "Well *that's* über-specific."

"You couldn't just say evening?" Addison echoed.

Zander shook his head. "He *means* five. Rahyoke's eerie-good with that sort of thing."

"Too bad he's not eerie-good with maps," Addison muttered under his breath.

Zander looked at him harshly. "With direction," he replied, "he's supernatural."

Wanting to avoid a fight, Ian spread his arms encompassingly. "There's not a cloud in the sky! Let's enjoy it while we can!"

Meygan ripped a bite-sized chunk off of her cinnamon-roll. The frosting was dry, but not solid – just firm enough to not go running down your arm when you tried to eat it. "Zander, is this our only food for the day? Should I be saving this?"

Zander cocked his head. "D'you need more food?"

Ian moaned. "How do you *live?*"

"Well enough," Zander defended, sounding bruised. "We eat better than some people, I can tell you that."

"Aren't you an archer?" Addison groused. "Can't you get us some venison?"

Zander spread his arms. "From *what!?* The forest's *bare*, Addison! There isn't a deer in the whole frickin' stretch of it, and hasn't been for years!"

"Then what good is that bow?"

"Next time I'll aim higher."

Point taken. Addison shut up, burying his snarky comments in pastry.

For a good ten minutes, they walked in silence – thick silence, heavy silence – as heavy as the humid air trapped beneath the canopy. They finished their breakfast, and Zander deposited the wax paper in the food-sack. Meygan saved half of her cinnamon-roll, wrapped in the paper, just in case.

"Aukh!" Ian complained. "I'm thirsty!"

Rahyoke rolled his eyes. "Just *couldn't* contain yourself..."

"You know when you can get water?" Zander offered, with mock excitement.

Ian fed right into it. "When?! Where!?"

"Five-thirty."

Ian groaned.

"I'm tired of rain," Meygan protested. "I at least want enough time for my hair to dry out! It's all, like, humid and – *bleh!* Is the weather *ever* nice here?"

"I'm sure God will take your demands into consideration," Zander retorted, rolling his eyes.

"Well, while He's at it, it's too hot," Ian added. "If it *does* rain, d'you think it'll reach us all the way down here?"

"Doesn't it usually?"

Ian shrugged.

"And anyway," Zander continued, "it's the sun making it so humid. You can't have it both ways. It's a vicious cycle."

"It's vicious, alright," Ian agreed. "I'm *dying* here!"

"You're *always* dying," Meygan returned.

"I am!" He spread his arms wide. "I'm starving and parched and just about *exhausted* with walking! And then with the heat, and the humidity, and it's gonna rain, and –"

Suddenly, Rahyoke whirled on them. "Oh my God!" he exclaimed, raising his voice. "*Shut* up and *stop complaining!!* If I hear *one more word* outta you, your spleen goes on the black market!! Understood?!"

Snickering, the children fell to silence, and he turned away again, shaking his head.

It was going to be a long, *long* trip.

It was five hours they'd been walking, when, all of a sudden, a bolt of lightning slit across the sky. A smattering of yelps, and the kids scattered, spiraled out of order by the fierce-driven water bullets like pool balls on the first strike. They tried to take cover under the splaying branches of the trees, but, of course, what good would it do them?

"HAAH!!" Rahyoke shouted, defiantly. Laughing, he spread his arms, lifting his gaze to the rain-shattered sky. He was drenched; his long-sleeved, grey t-shirt clung to his body, his dark hair for once lay nearly flat, streaming water over his face, though he didn't seem to care. Arching his back slightly, he pointed back at those doubters, still grinning. "Five-thirty!! Doubt me now!!"

"Lucky guess," Ian responded, waving him off.

"How can you even *tell?*" Meygan challenged. She was wearing his long coat like a veil or a cloak in a vain attempt to defend her poor hair. "And, BTW – *what* good is this coat? It's, like, *useless!*"

Rahyoke had his arms stretched straight up over his head, twined at the wrists, his face still turned to the pouring rain. "Aesthetic appeal." He uncrossed his wrists, lowering his arms to his sides with the poise of a tai-chi master. "C'mon, lads – a little rain never killed anyone."

Sighing and grumbling, they continued on their journey. But the rain only drove harder, until finally Meygan was inclined to beg, "Seriously, Rahyoke; isn't there a town nearby where we can at least take cover? Like – even for a little while!"

Rahyoke scoffed. "*You* lot wanted t'travel with me, an' *this*, my delicate flowers, is the *road!* Yer in *my* country now!"

Meygan groaned. He was right. "You're crazy."

The look Zander had cast on him suggested the same. When Rahyoke turned to him, he said, just as suspiciously, "You know, your accent *slurs* when you're tired."

Rahyoke arched a brow. Then – "I think you mean 'blurs'."

Zander remained adamant. "Rahyoke, I think we should make camp tonight."

It was his tone as much as the words that surprised him. Rahyoke offered a sort of half-smile. "Goin' soft on me?"

"When was the last time you slept?"

Rahyoke's expression was patronizing. "Aw, you're *worried* about me? Henwife! Just walk, okay?"

Stubborn as anything, Zander began counting on his fingers. "Last night you went into town, the night before we were walking, the night before *that* you went into town –"

"*Zander!*" Rahyoke hissed, trying to hush him.

"And all during the *day* never sleeping!" Zander challenged, now fierce. "And have you eaten at all?"

Rahyoke returned, "*You* are *my* concern; don't try to turn that around! If you hafta know, we have no food. I *need* t'go into town tonight."

"What town is even around here?" Zander scoffed. Then, realizing, he enforced, "What's the *name* of the town, Rahyoke?"

"Ih –" Rahyoke moved to answer, but the name slipped his grasp. Frowning, he fought to retrieve it, but only succeeded in pushing it farther away. "Crud," he murmured. "It's... it's... Fine, smart-ass; I don't know! But we still need food, and what the Hell use are you for it?"

Zander grasped his arm, halting him. "You can't do any work like this, and you know it," he whispered. "No one'll even hire you. Can't you go out in the morning?"

"Uh –" *There* was a point. Rahyoke considered it, reluctantly. He didn't much like the idea of stopping; besides the issue of food, it put them back almost ten hours. But... He had something of a solution working, though he was having a hard time forming it. *It'll work,* he kept thinking. *Whatever it is, it'll work... Whatever it is.*

He wondered if Zander was right. How many days *had* it been since he had last slept? He tried to count, but – When was Wednesday? It had been Wednesday at some point, hadn't it? They all blended into one big, long day – a big, long day during which he hadn't eaten, hadn't slept – only worked. And wandered. And adopted *this* latest hellish little undertaking.

If it wasn't a weekday, there'd be no market in whatever-the-Hell-town-they-were-nearest, anyway, and what then? Och – maybe the lad was right. No one would hire him in this state. And if they did, they would cheat him, and he'd be worse off wanting food, because he wouldn't be able to afford it, and the last of his energy spent. Either way, he wasn't eating – but at least this way, he'd sleep.

"But – but, what about – um..." He couldn't articulate a response. He shook his head, trying to clear it. "You win. Fine, we will, we will. I promise."

"We're gonna make camp?" Zander pressed.

Rahyoke shook his head. "Not here. A little later, a little farther. Slackers!" he called, turning. "If y'make good time, we're stopping tonight. Ian, that's poisonous."

Ian dropped the stick with which he'd been poking a particularly fascinating mushroom. "Right."

"Yay!" Meygan clasped her hands together, ebullient.

Rahyoke rolled his eyes. "Yay." He hiked his bag higher on his shoulder, swaying slightly. "S'is a ways off, but I'm thinkin' we can make it. Zander, don't look at me like that, I said we'd stop. Let's go."

They walked for another four hours – three and a half longer than they should have, but who cares? The rain had stopped around eight or eight-thirty, and *they* had stopped as the twilight was coming on, leading into night.

Rahyoke dropped his bag, leaning against a tree. "Here's good," he said, simply.

"*Finally,*" Addison sighed, splaying out in the sparse grass.

"I don't see what makes this clearing any better than the rest of the forest we went through," Zander muttered, annoyed for a whole different reason – annoyed at Rahyoke's defiance.

"*I'm* not complaining," Ian replied. "*I'm* just happy we stopped!"

"You seem like you've a fair bit of energy," Zander snapped. "Why don't you run as a scout? Go seek out some *dragons* to slay or something?"

Ian's eyes rounded. "*Are* there dragons to slay?"

Zander rolled his eyes, laying back and folding his arms behind his head.

"Ian, stay here," Rahyoke instructed. "Don't listen to Zander; he's just sulking."

"Like Meygan."

"*Shut* up, Ian!" Meygan threw a handful of little twigs at him.

"Lads..." Rahyoke pleaded, nearing desperate. "Go to sleep, will you? Or I'll make you walk again."

Nobody wanted that. Rebuked, they settled.

"Can we make a fire?" Ian asked.

"No." Rahyoke twisted so his back was against the tree, sliding down its rough surface. Now that he'd stopped, the whole world seemed to have caught up with him. He suddenly felt dizzy, like his body was about to give out. "Just... just get some sleep, okay? We're leavin' at first light."

"Okay." Obediently, they curled up in their respective quarters. They were not yet tired enough to fall asleep, but they were certainly exhausted enough to rest and keep quiet.

Rahyoke closed his eyes. The ground was muddy and his clothing was still damp, the bark of the tree was rough and slimy. The wind was colder after the rain, and a muggy heat rose from the earth like steam.

He would kill for a sandwich right now. Oh, Lord of Mercy – corned beef, sauerkraut – actual cheese that was fresh, and wasn't moldy. While we're dreaming, soda bread. Soda bread, black tea... whiskey. By first light, he would get a job, and by evening find the grandest little sandwich shop in town...

Rahyoke opened his eyes, just barely. He had heard footsteps. Looking up, he said, "I forgot to ask if – how did the thing go – deliverin' that?"

Zander the Archer settled down across from him, giving him a sarcastic look.

Rahyoke glowered. *"How did deliverin' the message go?"* he clarified, with attitude.

"Does this suffice?" Reaching up, Zander pulled at the scarf around his neck, revealing a long, blood-encrusted gash.

Rahyoke bolted upright, fully awake. "Why didn't you tell me?!" He reached out, inspecting the wound, railing against the darkness. "That's straight – how di'jyou get that? Who did this?" he demanded, his mind already a hundred miles at work.

"Calm down; it's fine," Zander responded, taking Rahyoke's hand. Though murderously reluctant, Rahyoke slowly settled back, his eyes never turning from the gash. "He had a guard."

Rahyoke frowned. "Like –"

"No."

"Oh. Oh... Decent someone thought to protect him; d'you know who it was?"

Zander's gaze trailed over the ground. "No, but I feel awful for him."

"Uh – oh... So Bahadur didn't agree?"

Zander shook his head.

Rahyoke sighed, sitting deeper against the tree. "I'm sorry."

"I didn't mind running the message, Rahyoke," Zander replied, raising his eyes briefly. For a moment, there was silence. Then, tentatively, he asked, "They'll be okay, won't they?"

Rahyoke said nothing; for the briefest flash of a moment, his eyes flicked Zander's way. Enough said.

Zander sighed, bowing his head deeper as he nodded. "That's a terrible situation to put a man in."

"You can't hold it against anyone; we're all pretty much in the same place."

"So we're screwed?"

"That's a word for it."

The wind whispered through the trees, speaking secrets and obscenities, telling prophecy and past; Rahyoke's gaze flashed off to follow. "*Should'a* just shot'em," he said.

Zander heaved a sigh. "I know."

Rahyoke returned to him. "I meant to ask you – where did you get that bag? You still have your bow?"

Zander nodded as he took the bag onto his lap. "He told me to give it to you," he explained, taking out his quiver and handing the bag to Rahyoke.

"Really?" Rahyoke shifted, straightening up. "Or Shivek?"

"I think he meant you."

His tone was very meaningful. Intrigued, Rahyoke looked into the bag. "What is it? Did you look?"

"Not really; I was in a hurry to get back."

Suddenly, Rahyoke's eyes widened. "Oh, my..." He dug into the bag, pulling out a set of scrolls, wrapped like sushi in a roll of leather. He laid the bundle on top of the bag, and, delicately, untied the thin strap which bound it, allowing it to unfurl. Thirteen scrolls tumbled out across its surface, all of them very old, and one clearly burnt at one end.

Rahyoke lifted the burnt scroll, carefully unrolling it. His eyes studied its surface with deep focus, widening even more.

"*Do you know what this is?!*" Energy thrilled through his tone.

"Not really," Zander drawled, examining his nails.

Rahyoke shifted so he was kneeling, laying out the scroll on top of his bag so that Zander could see.

"How can you make anything out? It's so dark!" Zander complained.

"Here." Rahyoke pulled a matchbox from his pocket, tossing it to Zander. "But, I swear to God, if you burn these, I *will* kill you."

It was not a threat he received often, and Zander knew enough to take it seriously. He had a few sickly "practice-arrows" in his quiver that had stayed dry thanks to the double-bagged protection; binding these together with a strip of cloth, he started on a make-shift torch.

Meanwhile, in spite of the darkness, Rahyoke was pouring over the scroll with an eagerness no amount of exhaustion could crush. "My God," he whispered. "D'you have any idea how valuable these are?!"

"Can we sell it?" A flash of light spilled across the page, and Rahyoke instinctively pulled the scroll away. "Hey!" Zander pouted.

"Sorry." Hesitantly, Rahyoke relented, sliding the page into the light. "*No*, we won't sell it! It's not valuable like that, anyway... Shorn; I ought'a invest in a flashlight, hadn't I?"

"So what is it?" Zander leaned forward slightly, holding the torch aloft and out of range, so he could cast its light on the scroll without danger of setting it on fire. "Oo! A map!"

The map was highly ornate, bordered thickly in gold, emblazoned with images. The ocean was patterned with whorls in a lighter blue, the mountain ranges penned in delicately and accurately, the rivers and their deltas intricately traced. And the frontier-lines – something Zander had never before seen on a map – traced in a fine, thin hand, and each country labeled, their titles inscribed in three different languages, three different hands.

"I can't read this."

Rahyoke grinned. In the lowest of whispers, he answered: "This is a map of the nations which pre-dates the war!" He pointed lightly, unwilling to even touch the finely-decorated surface. "This is Turranian," he said, indicating the first line of script. "This is Ceothann, so monks probably got it; only thing..."

"What's the last?" Zander asked.

Rahyoke shook his head. "Millah or something; I dunno."

"Can you read it?"

"That I cannot." Rahyoke pushed his hair back, out of his eyes, still studying the map in adoring fascination. "Shivek could, but I cannot."

"Are the others maps, too?" Zander asked, pointing lazily.

"I don't know!" Suddenly excited again, Rahyoke unrolled another scroll on top of the map. It was covered in a fine hand, scrunched and scrawled so the sheet was nearly black with ink. Squinting against the darkness, he began reading.

Zander sighed. "I feel stupid," he complained. He could barely read one language; how could Rahyoke understand so many, and in so many different hands?

Rahyoke looked guilty. "You're not; I'm sorry... It's – it's part of the – the Turranian Bible, like."

"That?" Zander leaned in. "How d'you know that? That it's the Turranian Bible?"

Rahyoke shrugged, pulling out the next scroll. "Bahadur *meant* for us to have these," he confirmed.

"Are all of them Turranian?" Zander asked, in hushed tones. "Are any..."

Rahyoke started digging through the texts as if inspired, pouring over their script. Each seemed to be in a different handwriting – some were neat, some illuminated, others scrawled, as the second scroll had been, in rush-jotted pen. "The Fall of Troy... Exodus with commentary... the travel log of Ibn Zala... a bit of the book of kings... another Turranian Bible excerpt... astronomical chart for moon phases... what'sit – two and – like a *half* adventures of Sindbad... a treatise on pagan idols... the travel log of Abu Hassan al-Tariq – on Alyssae, it looks like... a medical tract – uses of poppies, and... Oh!" He dropped the last page as if bitten, slipping a hand over his mouth. The final paper was smaller than the rest, its inch-and-a-quarter-wide boarder cut in fiery patterns of scarlet ink.

Zander leaned forward. "What is it?" he asked. For the first time, he recognized the letters – but the language was still lost.

Rahyoke shook his head. His eyes caught the fire-light, shining against the darkness, reflecting a deep, ingratiated sadness.

Zander squinted, studying the page; quietly, he began to recite: "*Posthāc hic ager adiungitur imperio Augustinii et subiunget legum imperii ad unum. Offensio obsequi* – "

"I remember this."

Zander's eyes flashed up at the low voice, hearing the crack and strain. "What is it?" he repeated.

"*An fógra damnaithe,*" Rahyoke answered, so quietly that Zander could barely hear him. "We called it 'the condemnation notice' – what they posted when they annexed... I guess... they were universal..." He didn't say any more, lost in a maze of memory.

Zander considered the notice, and a fury within him began to grow. "We should burn it," he determined, reaching out.

"You will *not!*" Rahyoke snarled, and he pulled the parchment back.

"Generals *frame* these for their *walls,* Rahyoke!" Zander fiercely persisted. "It's a hateful article!"

"If we burned every scrap of history we wished to change, it would make us no better than *they* are!" Rahyoke snapped. "Believe it or not, this *article, hateful* though it most certainly is, is as valuable as the rest of these!"

"Almost," Zander grumbled.

"Almost," Rahyoke conceded. "But you will not burn it."

"Fine, I swear." Twisting away, Zander stomped out the torch. Even if they had to keep it, that did not mean he had to look at it.

Rahyoke understood. He pulled the map back to the top of the pile, studying it through the dark.

For a long moment they were both silent. Then Rahyoke gasped.

"I know what to do!"

Zander was taken aback. "You do? Wait – about what?"

"It won't solve *everything,* but –" He bowed his head, leaning over the map, memorizing every scrap and every line.

"But you didn't tell me *what,* yet," Zander complained, rolling his eyes.

Rahyoke looked up. Then he smiled.

"We're going to cross the palisade."

Zander was so startled he actually pulled back. Then, leaning forward again, he demanded, *"Have you lost your mind?!"* He took Rahyoke by the arm. "Rahyoke, it's *suicide!!"*

The man was undeterred. "It's safer over there than it is here. And it's a clear path to the continent –"

"Have you forgotten how we *left?* And it lies on the other side of the greatest stretch of open land on the continent –"

"Farther North the field reaches its narrowest point; the crossing there would be maybe a week on foot, hopefully less – but I dunno about *them* lot," he added, with a clearly indicative nod.

"A *week?!"*

"Zander –"

"And we *have* them lot; what do we do with *them?"*

"Unfortunately, 'til I think of something better –"

"That's a *huge* risk, Rahyoke!"

"Oh! And wasn't it yourself after keepin' them? What *isn't* a risk, Zander?!" Rahyoke returned. "Our options right now are thin. *This* is the very best I can do."

Zander bowed his head, his eyes falling to the outdated map. No matter what way he pondered it, he was not satisfied; but he saw that Rahyoke was right. "It's a bad state we're in," he sighed, shaking his head. "How *far* North is it?"

Rahyoke began rolling the scrolls, slowly. "Zander, trust me. It'll be fine."

Zander lurched forward. *"That far!?!"*

"Zander –"

"Why couldn't we've just gone *south,* then?! End it that much quicker! Did you take me too seriously when I said we couldn't run farther? I didn't mean we should *stop!"*

Rahyoke narrowed his eyes, cocking his head as he scrutinized Zander. "D'you have a *girl* south or somethin' –?"

"Your plan *sucks,* Rahyoke!"

He pointed a rolled-up scroll in the archer's face. "We need to protect these," he challenged, sternly. "I need to protect you. And, until I find better quarter for'em, I also need t'protect those guys!"

"Your plan still sucks."

"Well, until *you* come up with a better one, mine is the plan we'll follow."

End of discussion.

Zander crossed his arms, furious. But what could he say?

Rahyoke replaced the scrolls, in their coverings, in the bag. He cast the boy a glance, and smirked. He offered the bag to Zander.

Zander was surprised. "He meant those for –"

"I can trust you, can't I?" He permitted a small smile as he asked it – one meek but full of faith and pride.

Zander accepted the bag, still stunned – and very honoured. He straightened, jutting out his chin at a haughty angle. "*Course* you can trust me!"

"Grand." Rahyoke again settled back against the tree, sliding low against its jagged surface. "Get some sleep, lad. We'll talk about it more in the mornin', 'kay? I'll work it out..."

Zander secured the drawstring on the bag, into which he'd returned his bow and quiver. "You always do," he replied, rolling his eyes. Standing, he moved to seek his own quarters. "'Night."

"Yeah..."

And within seconds, he was asleep.

13
Following Zander

"So here's the plan."

It was barely dawn; the thick shades of slate had only begun to lighten – and Rahyoke was already up.

"How could you have a plan, ya flippin' insomniac?" Zander groaned, waving him off and rolling so his back was to him. "Go back to *sleep*."

"I said dawn, didn't I?" Rahyoke asked, settling cross-legged beside him.

"No."

"Well I meant to. Anyway –" With a stick, Rahyoke began tracing patterns into the dirt. "Pay attention; I'm saying this once.

"If you cut dead Northeast from here, you'll reach the river; at that angle, you avoid the lingering foothills, and whatever caverns and what've been carved by the water. I can rendezvous with you down-river; they used to have a mill on a bend, and I think the bare foundation's still laid out there – looks like a faery-ring."

"Then what?" Zander grunted, turning his head to look back at Rahyoke. He propped himself up, pressing his forearms to the dirt, so he was lying on his stomach.

"Depends on how I fare in town," Rahyoke responded. "I might be able to travel with you tomorrow; I mightn't."

"How'll you meet with us? And how do I know what mill ruins look like?"

"You'll know, and I will." He clapped a hand on the boy's shoulder. "Have *faith*, my good lad." He stood and moved off.

Zander bowed his head with a sigh. "*When*'re we going?"

"Now."

He lowered his head onto his forearms. "Uuuhv *course* we are..." He waited a moment, then asked, ever dutiful, "Shall I wake them?"

"Got it." Two fingers to his lips, Rahyoke gave a sharp, shrill whistle.

"Jesus!" Zander jolted upright. He turned a glare on Rahyoke – as did the three trespassers.

"I was already *awake*," complained Meygan, covering her ears.

"Can you teach me to do that?" Addison asked. "I've always wanted to learn how."

"No!" Zander scolded, crossly. "That's *all* we need, is more fools that can do *that!*"

"*Some*-one's *crab*-by," Meygan sing-songed.

"Alright kids, listen up!" Surprisingly, they did. "Now: because Zander insisted on stopping for the night, I haven't any food to give you. Therefore: Zander will be leading you today, while I scrounge up enough money t'buy bread. Listen to him, and *behave*." The emphasis was deep on that word. "We'll meet up around nightfall; we'll break, but we're traveling by night."

Zander turned to him. "We're *what?*"

"Yeah," Rahyoke concurred. "So much for that; I'm going to the market –"

"You're going shopping?!" Meygan interjected, suddenly and – it seemed – inexplicably excited.

He rolled his eyes. "I'm going shopping –"

"Can I come?"

"You may *not* come," he continued, in the same tone. Meygan pouted, and he ignored her. "Zander, get the Hell goin', alright?" He returned to the tree for his canvas bag.

"Right!" Zander jumped to his feet. "Time to find a direction!"

Rahyoke paused, watching as the lad scanned and cast about, finally settling on a direction and pointing with all the cocksure and definitive confidence in the world. Rahyoke rolled his eyes, then cleared his throat pointedly.

"Huh?" Zander leaned back, glancing over his shoulder.

Rahyoke pointed Northeast – about a hundred degrees left of Zander's own direction.

Zander turned. "Right!" Though he still spoke with confidence, his cheeks burned a bright scarlet.

"I trust you'll make it," Rahyoke said flatly, walking off west.

"'Bye, Rahyoke!" the kids discordantly chorused, audibly sorry to see him go. As soon as he was out of sight, they rounded on Zander. "When are *we* leaving?" Meygan asked.

Zander slung his own pack onto his shoulders; his face was still marked with his recent shame. "Right now."

"But it's still *dark* out!!" Ian exclaimed, scandalized.

"That's not my problem," Zander retorted, walking away.

Meygan bowed her head, heaving a sigh. "I *thought* so."

Gathering themselves up, they followed.

The darkness was still heavy on the land; no birdsong rang out; the only sound was the wind through the leaves. Meygan pulled her cell phone out of her backpack, turning it on. She still couldn't get a signal. Frowning, she began waving it about, stretching out her arm and jumping around like a ballerina, seeing if maybe, *maybe* –

"What's that?"

She stopped. "What's what?"

Zander pointed to her phone, arching a brow indicatively.

"Oh. It's a Kami Fire. We got a pretty sweet deal on it, actually, because we went like the day after Black Friday. It's a couple of years old now, but it still works –" He was staring with such a look of blank confusion on his face. "Auoh," Meygan lamented. "Don't tell me you've never heard of a cell phone!"

Zander hesitated; he did not relish the feeling of seeming stupid, but... But what did he have to prove to a bunch of kids?! He shook his head. "What does it do?"

Addison bawled. "You're kidding, right? Everyone knows what a cell phone is!"

"Not necessarily," Ian pointed out.

"It lets you talk to people who are far away," Meygan explained, unlocking the screen. "So, like, if Rahyoke had a phone, and we could get service here, we could talk to him. Everyone has a number, every phone, and you enter the number of their phone into yours and it calls them. Addison, don't you dare roll your eyes at me!" She tried once more to send a text, and once more saw it fail; she flicked through to the games. "Hey; 'Fury Birdies' still works." She handed the phone to Zander. "Knock yourself out."

Zander took the phone delicately, cradling it in his hands as he closely studied the screen.

Addison disapproved. "Meygan, shouldn't you be saving your battery or something?"

"There's really no point. Besides; the Fire has a month-long battery, and – oddly enough – I charged it before we left."

"There's an intriguing sentence," Zander murmured. "'The fire has a month-long battery'..."

"Sometimes in the country you don't get service," Addison insisted, but Meygan cut him off.

"Worry about your *own* phone, Addison, alright?! Killjoy." She then turned her attention to the archer. "So Zander," she began, and, God help us, she was executive again. "Are we, like, going anywhere, or what?"

"Of *course* we're going somewhere. As long as we're moving, and in the same direction, we can't be goin' nowhere, can we?"

"That's not what I *mean*."

"What do you mean?" He cursed as his bird slingshotted backwards. "This is fun! I suck so bad at this."

"I mean *where* are we going?"

"Today?"

Meygan crossed her arms. "Never mind. You're worse than Rahyoke."

Zander grinned, reaching up to brush the low-hanging branches of a tree; he was the only one of them tall enough to do it. "Today we're going to the River Abona," he explained. "It's the biggest river on this continent. It flows South-to-North. There's a stream on the Northern Island way far from here that flows North, too, but no one ever counts that, so they say that this is the only one that does."

"We're in Egypt?" Ian asked, cocking his head, his eyes wide.

Zander raised his eyebrows at him. Honestly curious, he asked, "Are you crazy?"

Addison cut in. "Where are we?!" he demanded, fiercely. "What country is this!?"

Zander handed the phone back to Meygan. "Thank you. The southern half of the Northern Augustan Territory."

Ian gasped. "Oh my gosh! We're in Rome!!"

Zander laughed. "I'm not *that* archaic, am I?"

Addison was not amused. "What's the name of the *country?*"

"Augustinium," Zander repeated, "but I forget the name of the province; Rahyoke'll know."

Addison stared at him, agape. "Are you... That's... " There were no words; shaking his head, he snorted, "Inconceivable!" Finished.

Meygan and Ian giggled. *"Inconceivable!"* they exclaimed.

"This whole place is crazy!" Addison blurted, flicking his hands in the air above his head to dismiss them all. He stalked forward.

"Are *you* leading?" Zander smirked, as Addison brushed past him.

Meygan decided to change topics. "So," she began, returning to Zander, "how did you meet Rahyoke? How come you hang out with him?"

"Oh," Zander shrugged. "He's pretty cool, once you get used to him. He's just very private. I've known him to go weeks at a time without saying two words together to me – not in a mean way; he's just not a talkative guy. He can say a lot in a few words, though."

"Whether or not it's clear," Meygan muttered.

"Ha! That's true!"

"How did you meet?" Addison asked, the animosity surprisingly voided from his tone.

"Why do you ask?"

"Curiosity," Meygan answered. "Lots and lots of curiosity."

"Hm..." Zander raised his eyes to the canopy, wondering what Rahyoke always saw up there. A low light was beginning to creep into the trees, pale and yellow and thick with grey. *Sunrise.* "That's a long, old story," he responded, waving them off. "And Rahyoke's the better story-teller, not me."

Meygan gave him a look. "You're just gonna leave us hangin', there?"

Zander raised a palm; half a surrender. "I lost my parents when I was very young. Rahyoke... saved me. I probably owe him my life or something."

"And he tells stories?"

"Don't sound so skeptical! He's been away out of it for a while, but – yes. Didn't I tell you the man has a way with words?"

"Do you think maybe he'll tell us some stories?"

Zander shrugged optimistically. "Maybe he will yet; you never know. But enough of that. How did *you* all meet?"

Ian naturally jumped in first. "I go to school with Addison. We were gonna hang out, and he told me how Meygan'd just moved in, and I wanted to meet her... So I *did*." He wouldn't turn to Meygan, feeling her gaze, but bowed his head, clasping his hands.

Zander turned to her. "When did you 'move in'?" he asked.

"What was it? Sunday? Yeah, I guess it would be."

"Just this past Sunday? Like – what? – five days ago?" Zander's smirk was awful.

Meygan sighed. "Yes."

"Oh!" He laughed, almost cackling. "That *sucks!*"

"Your pity is lovely."

"Aaw!" He looped an arm around her shoulders, pulling her close. "I'm only teasin'; don't be sore."

"Sheesh, Zander!" She pushed him away. "Pay attention to where you're going! We have to get to the stupid river."

"Pay attention to where *Addison's* going," Ian corrected.

"Yeah."

"Don't worry!" Zander folded his hands behind his head, uplifting his face to the paling grey as if it were sun-warmth. "I'm the best navigator around!"

"Mm-hmm," Meygan responded, dubiously.

"*Trust* me," Zander emphasized, the most believable con-man. *"I'll* ge'chya there in *no* time..."

14
Challenge!

The Abona was widest at its source in the mountains. Eternally the spring high on the bowled peak of Brigdalga spilled over, its endless waters tumbling and cascading wildly down to the sea. At a calm spot close to the mountains, the old mill had been built over five hundred years prior; all that remained, as Rahyoke had said, was a circle of stones, broken and about eight yards separated from the water.

He sat on the stones of the ring, watching the trail upriver. The last rays of the setting sun reflected over the wide waters, glinting off the glassy green in sparkling bands of gold.

Where IS he?

The canvas bag at his feet was weighted with enough food to make two meals, and gripped between his hands was the last fourth of a greasy, day-old, but – Heaven help us – glorious tuna melt. He felt a little sick, but what can you expect, scarfing fast food after going a few days without eating?

Beneath the rough rolling of the wind and waters was a sharp crack, and he cocked his head, listening. *Stealth was never really his strong point.*

"So you made it at last!" Standing, he turned, wiping one hand on his jeans.

Not too far downstream, Zander had emerged from the forest, looking *dead* exhausted, surrounded by the trespassers like swarming bees. Meygan was the first to look up, and, seeing Rahyoke, jumped up and down, calling out as she waved, "Hi, Rahyoke!!"

"Hey," Rahyoke half-waved in return.

The small band descended on the rocks, scattering about like scavengers or hounds. "Did you bring us food?"

"*Did* I!" He lifted his bag by the strap, holding it out to Ian as he turned his attention to Zander. "How'jya fare?"

"Well enough," Zander sighed. "Is that tuna?"

"Yev yer own," Rahyoke replied impulsively, pulling back his fourth. "Here." He held out the bag to Zander, for Ian had been content to pick at the contents while Rahyoke held it for him rather than take it himself.

"I'm alright," Zander responded, settling down on the rocks. "I wanna rest."

Addison finally took the bag, as Rahyoke, still to Zander, asked, "How is it you're coming from due North?"

"Zander got us lost," Meygan piped, taking a seat on the stones.

"Meygan!"

"I had figured!" Rahyoke replied, mockingly impressed. "What happened? Couldn't find the river?"

"As a matter of fact, no!" Zander snapped. Amending, but still curtly, he added, "Not everyone knows the forest as freakishly well as you, Rahyoke."

Rahyoke shrugged, settling cross-legged in the grass and returning to his sandwich, his concern eased.

Taking something squishy that smelled promisingly of bacon, Addison moved to the water's edge, inspecting the broad river. The current rolled by at a moderate pace, snaking over and around a few jutting, black stones. The same great rocks built up the bank, holding back the lush, green grass from the water.

"How deep is this?" he asked.

"'Bout ten feet," Rahyoke replied. He finished off his sandwich, hoping the fact that he was still eating would detour further questions, but to no avail.

"Is the current dangerous?"

"Not if yer *standin'* there," he answered, sarcastically.

"So all that's a river?" Meygan asked.

"Mm-hm."

"Are there fish in it?"

"Yep."

"How far does it go?"

"To the sea – miles an' miles. *Mercy*, you ask a lot of questions!" No wonder Zander looked tired. He ran his sleeve over his mouth. "Okay – we break 'til just after sunset, then we cross it." To Zander, he added, "That'll solve heading too far North."

"Whoa, wait!" Meygan interrupted, signaling for a time-out. Everyone turned to her. "We're *crossing?"*

Rahyoke frowned. "Yeah."

"Like, how? Swim it?"

OH, this gets better and better! "So y'can't swim, is that it?"

Meygan's eyes rounded. "No, I can*not*."

"Mercy. Alright – anyone else?" He cast his gaze between the two boys, and, slowly, shamefacedly, Addison Mitchell raised his hand.

"Really?" mocked Zander.

"Stop it," Rahyoke scolded. "Don't shame him. So, *neither* of you can swim." Nope; he pointed to Ian. "You can swim?"

Ian grinned. "Gold medal champ of Onondaga County," he bragged.

Rahyoke dropped his gaze with the flash of a frown. "Maybe you can fish for us, then," he muttered. He got to his feet, pacing away from them a few meters, running both hands through his hair as he lifted his gaze skyward. "So that kills *that* plan."

"What do we do now?" Meygan asked.

He breathed a long sigh, keeping his back to her and his eyes on God. *You're up there, right?* "Same. We break 'til about sunset, we *walk*. There's a point farther North shallow enough to wade across."

"It's *that far* North, isn't it?" Zander lamented.

"Why would we have to cross, anyway?" Addison asked. "Everything we want is over here."

"No, it's not," Rahyoke corrected, turning back to them. "And I've been meaning to cross over awhile anyway." He smirked. "You lot just, *fanned the flame*, ey?"

"Couldn't resist?" Ian jibed.

"Saw no point." *Lord, I'm tired.* "There's apples on the other side, and fish in it; in case you haven't noticed, I can't make enough to support you all adequately; that would've been a decent supplement."

"Wait – apples?" Zander looked up, and looked sad.

"Yeah."

"Auh! *Damn* it, Addison!" Zander scorned. "You cost me food *again!*"

"We couldn't cross here anyway!" Addison snapped. "The river's too wide, and too fast!"

Zander stood. "You're just sayin' that because you're intimidated, you child!"

"I don't see *you* crossing it!"

Zander's eyes flashed to Rahyoke; though his expression remained indifferent, Zander saw him give the slightest of nods. *Bring it.* He returned to

Addison with the grin of the Devil on his mouth. "Addison Mitchell, for this slight, you shall owe me a full forty-eight hours of silence from the moment my feet again touch this shore after I *prove you wrong*."

Addison folded his arms. "Oh yeah? Or *what?*"

"Or I'll throw you in the river." Zander dropped his bag, laying a half-eaten Danish on top.

"You can't throw me in the river," Addison sneered, though he believed Zander on that one himself.

Zander pulled his shirt off over his head. His grin was taunting. "Aw, look at you," he cooed. "I could hawk you in there with one arm; but even if I couldn't, we all gotta sleep *sometime,* and you'd be hard-pressed lifting me!"

Desperate, Addison turned to Rahyoke. "Are you hearing this?! You can't possibly allow this!"

Rahyoke glanced to him lazily and shrugged. "Not *my* problem."

Addison was appalled. "You're the adult here! Control him!"

Rahyoke, disinterested, looked back to the river. He may have rolled his eyes.

Addison frowned. "I don't have to take this!"

"If you leave," Zander leered, "then I win."

"Let's see it, then." Scowling and skulking, Addison crossed his arms.

Zander's expression spoke jeers enough. He dropped his shirt on top of his bag. He was a fine specimen, lean and well-muscled, lightly-tanned; his jeans were slung low enough to just show his narrow hips. He stretched, flexing the strong sinews of his arms.

Addison rolled his eyes, looking away.

"Rahyoke – gimmie that food bag, will ya?"

Without a word, Rahyoke strode over to where he'd ditched his belongings, emptying the food sack into his canvas bag and tossing it to Zander. There was a subtle warning in his dark eyes.

"Yeah, yeah; I'll be careful." He looped the drawstring over his arm, around his shoulder, slipping off his boots. "You'll wait for me, right?" he checked, moving to the water's edge.

"Yeah, I'll wait for you," Rahyoke answered, with the subtlest flash of a smirk.

Zander leapt with a clean dive into the water.

Rahyoke waited until Zander surfaced, a fair length out, swimming forward without breaking stride; then he returned to the faery-mill ring, sitting down again.

"Is that safe?" Meygan asked.

"What?" he replied, without interest, as he picked through the contents of his bag.

"What if the water's – like – polluted?"

"By *what?*"

Addison turned. "Seriously!?"

Rahyoke's eyes flicked up to meet his, with a glare so dark that Addison almost stepped back. "Yes," he said, quietly. He had had about enough of Addison's flack.

But the sting of Zander's taunts was too fresh for Addison to back down. Stepping forward, he pulled his cell phone out of his back pocket, holding it up. "What's this?" he demanded, halting about a yard from the man.

Rahyoke cast it half a glance. "I don't know."

"Okay, now –" He put away the phone, fuming "—here *you* are, calling *us* liars –"

Rahyoke's eyes narrowed.

"—I mean, come on. It's one thing for Zander not to know; he's stuck relying on you. But *you* go into town. I don't know if you're stupid, or – why would you lie to us? I thought you wanted us gone –"

"*Believe* me I'm wantin' ya gone!" Snakes' tongues lash with less venom, less speed. "And I wanna sleep under my own roof, work f'r m'own food – not hafta defend my every bleedin' word to some pretentious, indolent little curmudgeon – but it's not gonna happen, *is* it? An' c'mere – who're *you* t'be givin' a hard time when it's your *own* claim yourself from another world? Do you have no faith in your argument, or are y'just that *argumentative?* One t'doubt so much what conflicts with your own crystalline image of the world, another to doubt what you see, *more* what y'know and've said as stark truth – and does it stand so in contention, another world to be different than your own – why wouldn't I not know all tha'chya have? So don't be shamin' me for knowin' or not knowin' as I do when I've only your own word t'go off'a, and the logic'uv yer own argument! *Think* before you lay blame, an' chill the Hell out before y'cause yourself damage pickin' fights in all the wrong places. *No,* I flippin' haven't seen such a thing. I can't *believe* I stood up for you!!"

Shaking his head, he returned to what he'd been doing.

There was no sound from Addison; cowed and rebuked and reviled, he remained silent, turning his gaze toward the river, waiting with agony to be proved wrong again.

Meygan smirked. *Ha-ha, Addison... And we DO push the poor guy to talk.* "Why is Zander so bummed out about going North?" she asked. "...Rahyoke."

"Hm?" He sat up, shaking the hair out of his eyes. "Oh; he's from the South; Zander hasn't really had much experience in the Northern region. Food's better in the South, too. But why don'chyou ask him?" With a nonchalant shrug, he ended the conversation. Though his head was bowed, all of his attention was drawn out towards the river.

There was a pause that lasted nearly a minute.

"We should build a boat," Meygan asserted.

"A boat?" Rahyoke looked at her, incredulously. "And who's the shipbuilder, here?"

"I built a rowboat once with my dad," Ian volunteered, raising a hand. "And me and my friend Joe built a raft last summer like Tom Sawyer or whatever."

"I like Huck Finn better," Meygan muttered.

"You do?"

"Well. You're just *full* of talents, *aren't* you?" Rahyoke quietly remarked, the sting subtle but there. "If you can build one before Zander's back, we'll take it."

"Auh – too late." Ian waved. "Hey, Zander."

There was a tremendous upwards splash, like a fish leaping out of the water, and the food sack, filled, rolled over the grass.

Zander came up after it, scooping it up in triumph. He carried himself with a pride to banish exhaustion, his back straight, shoulders back, rightful arrogance on his face. Water streamed into his dark green eyes as he opened the drawstring, reaching inside.

"Ha!" He held an apple for Addison to see, annoyingly close to the boy's face. "That's forty-eight hours of silence," he elaborated, pulling the apple back. Taking a huge victory bite, he strutted back to where the others sat.

"You're a bold, brazen sort of creature," Rahyoke murmured. "Satisfied?"

"For now." Zander handed the bag to him, dropping down by his own belongings.

"Are those ripe yet?" Ian asked.

"No." Zander took another great bite, saying around it, "U're sweeter, 'ough."

"Because he won. They're sweeter because he won," Meygan translated, and Zander toasted her with the apple.

She turned to Rahyoke. "D'you think you can get, like, some water bottles the next time you're in town? Do they have those here?"

Rahyoke bit his lip. "I can try."

"Wow. You really *don't* make a lot of money, do you?"

"Not really." Rahyoke stood, slinging his bag over his shoulder so the strap was across his chest. "Sunset. Let's go." He handed the apple bag to Ian.

"I'm tryin'a put my shoes on!" Zander protested, hopping around as he struggled to keep balance and pull his soft boots on over his wet jeans.

"Well hurry up!"

Meygan slipped her backpack on, uncrossing her legs and standing. "When're we gonna stop again?" she asked.

Rahyoke turned, walking backwards down the shore. "Depends. How well can you keep up?"

Ian grinned eagerly, hefting the food bag; he took out an unripe apple, eating it as he walked. "Mm!"

Zander slipped on his vest, shoving his shirt into his bag before slinging it onto his shoulders.

"Really?" Meygan denounced.

Zander began striking poses as he walked, flexing his muscles. "Really? 'Cause I think I look kinda sexy!"

Addison, trailing in last, rolled his eyes. "Jeeze."

"Heard that!"

"No!!"

SPLASH!!!

15
Flaws in the Socratic Method

"So, how far away is where we can cross?" Meygan asked.

Night had settled over the land. The trees were twisted columns of black against the navy sky, the river a shining torrent of silver, reflecting the bands of moonlight which slipped around the moon's dark rim. It was a new moon, and yet light still churned through the river's frothy waves.

Trailing at the back of their train, Zander responded with a sigh. "Far. That's why it would've been so much easier if you knew how to swim."

"You could carry me," Meygan suggested.

"I'll think about it."

Addison chucked an apple core into the river; it landed with a solid *plunk*. "Aren't there any bridges or anything across it?"

"Yeah!" Meygan agreed, and when Zander shrugged, she turned the inquiry over to their leader. "Rahyoke?"

The man, long and mercifully ignored, just managed to resist wincing at his own name. He had no desire to be involved in the conversation, so it was with great reluctance that he contributed, "There is one. But we'll reach the shallow crossing way before we'd get to it."

"There's only one bridge? Why's that?"

His aversion won out; he shrugged and said nothing.

Meygan pursued. "Do you use ferries, then?"

Again, Rahyoke shrugged, trying to buck the discussion, and Meygan took umbrage. "Rahyoke," she said, sternly, "if we're going to keep traveling with you, we're probably going to need to know more about this place."

"Not necessarily," he countered. "You could get on just fine never really knowing anything at all. 'S'how most people go through life."

"Well, I'd rather not. Could you tell me some things? I'd really like to know. Make the most of the experience."

"Why ask me?"

"Who else am I supposed to ask?!"

God, she was feisty. He made a show of fixing a wide, enervated glare on the sky, lingering until he relented, "Alright; how's this. I'll give you each three questions. And otherwise, you leave me alone. Ask Zander some things for once, maybe. Fair?"

"You're kind of short-ended, aren't you?" Addison pointed out.

"Because I don't care."

"Do you at least want one question?"

He considered as he drew the long inhalation for his submitting sigh. "Fine. Three to nine. Do we have a deal, then?"

"Deal!"

"*Grand.*"

"Can we ask you anything?" Ian inquired.

"*Fine.* That's one."

"Shoot!"

Zander laughed. "Keep *that* up and we've nothing to worry!"

"Why would you worry?" Addison asked.

"To me or Zander?"

"Never mind never mind never mind!!!"

Unseen, Rahyoke grinned. Perhaps this was a good idea after all!

The ground dropped swiftly and steeply, casting the river down. A ring of moonlight flickered through the clouds, glancing over the rushing waters.

"We're making good time," Rahyoke asserted. He jumped from the ridge, landing cleanly. "Maybe a couple more hours, then you lot can rest."

"You're going into town," Meygan guessed. "Don't count that as my question –"

"Yes." He waited for them with something near patience. "I'll go into town when we stop; at dawn, you lot'll move again, maybe break briefly at midday, stop 'til I meet you at sunset again. We might stop for the night then, depending on how we do." He caught Zander's arm as the boy, imitating his leap, stumbled a pace forward. "Sound good to you, lad?"

"Beautiful," Zander replied, coolly.

"Like that jump!" Addison teased.

"Let's see you do better then!"

"I cannot," Addison replied, haughtily, sliding and picking down the ledge. "But at least I know it."

"Oh, the shame and a half of it on you, Addison Mitchell!" Zander cursed, shaking a fist and squinting an eye at him. "I tell you rightly, if a man never ventures across his limits, he will never be made a man at all."

"That's very profound!" Ian praised.

"Yeah," chided Addison, dubiously. It was all he said.

"You got it, Meg?" Rahyoke asked, watching as the girl eyed the steep slope.

"Yeah... Zander will catch me." At first she picked at the ledge, then decided faster was better, hurtling downward in three bounds. Indeed, she collided with Zander.

"Oof!"

"Sorry!" She turned to Rahyoke. "I like being called Meg; I never get to be called Meg."

It was clear that the man didn't know what to say in response. "Alright." He waited as Ian gracelessly half-flung himself down the little cliff. Then, without a word, he led them onward.

For a merciful five minutes, they were as silent as the blackened woods. Pale and greying clouds were stretched like gauze across the moon, and the rapids were churning beneath them. Not a soul alive, not hart nor hound nor moth nor owl, flittered though the trees; all the world was a twilight of darkness, waiting in suspense.

Though the river was a little bit narrower beside them, the current was too rough for even Rahyoke to attempt. Here the apples would be thickest on the farther shore, ripest: deep ruby spheres framed with emerald shavings, weighting the ropy branches to the ground. Sweet, crisp fruits – a Heavenly bounty, tantalizingly close, but locked across that deadly void. Is it ever any different?

"I have a question," Addison asserted.

"Taking stock?"

Addison ignored him. "So, I get that we're following the river because we can eat the apples and fish and stuff, but that doesn't explain why we're *following* it."

Rahyoke waited a second; then he frowned. "That's not a question. You can't trick me."

"What I want to know is how come we're following it. There's food in town; we could go to a town just as easily."

"Seriously – that's still not a question."

"I *mean,*" Addison grated, "*if* we're following the river, we must be doing it for some reason. If we're going somewhere, where's our destination? Where are we going that we need to cross the river? Where had you been meaning to go that you couldn't get to by now? And seriously – *why* can't we just go into town?"

Rahyoke looked to Zander. "That was how many?"

"None were good enough to count."

"Rrrg!!" Addison emitted, enraged. And at last, he was able to articulate: *"Where are we going!?!"*

"Close enough!" Rahyoke decreed, waving off Zander's intended retort.

"Great," Addison dredged. "Now will you answer me?"

Rahyoke surveyed the canopy. "There is no answer that I can give that will not irritate you royally. But I'm okay with that." He half-glanced back over his shoulder. "I'll tell you tomorrow evening when we make camp."

"What?!"

Rahyoke shrugged. "The answer's very complex; I need a hearth-fire."

"Oo, yes!" Meygan exclaimed. "We can have a camp-fire like proper nomads! Can you get us s'mores stuff?"

"S'mores stuff? We'll see."

"Why can't you just tell me now?" Addison protested, thoroughly miffed.

"So I can concentrate!" Rahyoke declared, his tone just bordering grandiose. Before Addison could regroup, Rahyoke jumped down another rise with a sudden, unquenchable, god-like and puerile energy, exhorting them, "Come on, lads! There's lots of forest yet to be covered! Let's see if we can make it before light!"

16
Tales of the Sidhe

ul soir Cora Fhidhinse, 'dul soir Cora Fhidhinse,
'Dul soir Cora Fhidhinse 'sea chaith sé a stiúir,
'Dul soir Cora Fhidhinse, 'dul soir Cora Fhidhinse,
'S ní phósfadh sé anois nó go bhfágha sé bád nua.

'S nach fada uaim siar é, is nach fada uaim siar é?
'S nach fada uaim siar é, mo Cheallachín Fionn?
'S nach fada uaim siar é, is ní fheicfidh mé i mbliana é,
'S go seola Dia aniar é, mo Cheallachín Fionn."

The sun was setting over the water. It had rained for most of the day, as was usual this time of the year, and the mud was thick upon the ground, drying in the passing rays. Red and tangerine gold was the sunlight, glancing upon the quieting stream.

Zander dipped the string of his make-shift fish-pole into the water, twitching it and praying for salmon. It had been summer when Fionn tasted the Salmon of Knowledge. At least, that was how he'd always imagined it. Maybe the leaves were beginning to turn; maybe the sunlight showed through their crisp scarlet and amber foil, scattering over the Bóinne, as Finnegas, silver-haired old bard, perched with his fish-pole, waiting patiently, just like this...

"HEY!!!"

"GAAH!!!" Zander jumped, nearly loosing hold of his pole – which would have been a terrible thing indeed, for he was using the shaft of his bow.

Turning, he shouted, "*DAMN* you!! *MUST* you scare me!?!"

Rahyoke laughed. "So you didn't get them lost!"

"I did not," Zander snorted, proudly, sitting up straight.

Rahyoke dropped down beside him, cross-legged. "I wish dearly you wouldn't do that with your bow," he said, peeling the skin from an apple with the boat-knife Zander had returned to him. "Have you any idea how old it is?"

"Aah –" Zander squinted dramatically up at the gold-washed sky. "One hundred aaaannd... sixty-*two* years, would it be?"

Rahyoke grimaced. "That sickens me. Seriously –"

"Alright, alright," Zander sighed, reeling it in. He had unstrung the bow, and was using Ian's shoelace as fishing-line. "Happy?" he asked, holding out the shoelace, detached from the bow, for Rahyoke to see.

"Uh-huh. Curiosity; what did you use for a hook?"

"Paper-clip. IAN!!! I'M DONE!!!"

"*Mercy!*" Rahyoke hissed, flinching away from Zander at the piercing shout.

Ian leapt up. "Did it work?" he asked, scurrying over.

"No," Zander scoffed. "I told you it wouldn't, the *stink* of the thing..."

"Ha-ha." Ian took back his shoelace; he was wearing socks, and his sneakers were nowhere to be seen. "You're *welcome*."

"Don't let him take all yer stuff," Rahyoke advised, glancing up and behind him to Ian. "Where're your shoes? Did he take those too?"

"I fell in the river."

That was all of an explanation he gave. Rahyoke waited a moment; then, not caring enough to ask, answered, "'Kay," turning towards the river again.

"Did you get s'mores stuff?" called the voice of Meygan, from the tree-line.

"I got bread and marshmallows," Rahyoke responded. "It was the best I could do."

"Next time, chocolate syrup," Ian instructed. "We can use the leftovers to make it like those French deserts with bread and chocolate chips."

Rahyoke flicked the curled length of the apple peel into the water. "Where's Addison?" he asked.

"Oh!" Ian pointed. "Ge –"

There was a tremendous crashing and shredding and rending and shattering of leaves and branches.

"Jesus Mary and Joseph," Rahyoke muttered. "Well wou'jya get'im before he destroys the whole bleedin' forest? Honestly!"

"On it!" Ian threw a salute and scampered off.

Rahyoke shook his head. He bit into the apple. "*Uah.*" Inspecting the fruit with a look of disgust, he murmured, "This is... *Not* ripe."

"I told you so."

He turned to Zander abruptly. "D'we have firewood?"

"Yeah." Zander thumbed a vague indication, focused on cleaning his bow with the hem of his shirt. "You're welcome."

"Grand. You're doin' alright today, ya know that?" Shrugging, he bit into the apple again. "Mm – c'mon!" He got to his feet.

"*I'm* fishing," Zander declined, haughtily.

"With what?"

"I don't know!! Thanks to *you*, I have to *find* something!"

Rahyoke rolled his eyes. Leaving him to it, he moved away, ruffling Zander's hair in passing.

"Kaah!" The archer ducked, hunching his shoulders. "You *know* I hate it when you do that!" He turned, watching as Rahyoke knelt and began sorting through their meager collection of firewood. "Renegade!"

"Are we gonna have a fire?" Meygan enthused.

Rahyoke picked over the disheartening, wilted store of their fire-wood. "I'm gonna *try*."

It was nearing twilight by the time Addison and Ian made it back, carrying enough branches to build a five-year sapling.

"We can't use any of that," Rahyoke muttered, his attention on the tiny flame he'd managed to coax from the travesty of their wood store. He was building the fire close to the river, on one of the broad, flat stones of the bank.

"Why not?!" Addison dropped the branch he had been carrying – it was nearly his height, and bright with evergreen needles.

"Because fresh-cut wood is too green to burn. And it's been rained on."

"But we didn't cut it!" Ian protested, holding out two brush-like branches like wings.

"They retain moisture. Zander'll cut'em and we'll use them later," he added, mainly to placate them.

Zander folded his arms. "Oh, *will* I, now?"

"There's some food in my bag," Rahyoke continued, ignoring him. "Grab something to eat. Marshmallows are not dinner."

Childish groaning.

"They're not," he added, summarily, returning his full attention to the fire.

"How was it in town?" Meygan asked, settling cross-legged on the other side of the fire.

"Auh, same old," Rahyoke shrugged. He breathed a curse as part of the twig-tent over the fire collapsed.

"Language."

"A hundred thousand apologies." Rahyoke took one last look over their utterly pathetic failure of a fire and settled back, shaking his head. "I'm done; that's it. That's as good as it can possibly get."

"Can we try –" Ian held up one of his "wings".

"No!"

"Aaw!"

"What are the towns like?" Meygan asked. She was holding half of some sort of glazed, honey bread-knot which she had split with Zander.

"Wha'd'ya mean? They're towns; what's there to be like?"

Meygan knew he was trying to put her off. And, usually, his round-about way of addressing questions would be enough to put anyone off of an answer. But she had a weapon of her own now.

"I wanna use my first question," she informed him.

"*Augh,*" Rahyoke groaned. "Alright, fine. Let's hear it, then."

"'Kay." Meygan leaned forward, leveling him with a very intense, very serious look. "When I first met you, you said it wasn't safe for me to go into town. But you *always* go into town. But you never let *us* go. So – what's up with these towns?"

He waited, but that was it. A mocking smirk lifted the corner of his mouth. "They teach nothing of rhetoric to children in your world," he commented.

Meygan sighed. "*Seriously*, Rahyoke!" Sitting up straight, she took a length of her curls between her fingers, holding it for him to see. "In town we could get showers. Look at this! Look at how *sad* it is!" She dropped her hair. "We could get warm beds and dry off for a while! Maybe we could even help you!"

"Help me?"

"You know – we could get jobs and stuff! Help out financially!"

Rahyoke laughed – that low, demeaning laughter.

"Why not!?" she exclaimed, exasperated.

"You wanna know what's up with the towns? Fine. I'll tell you. The biggest issue – and there are *many* – is if I brought you myself, I don't have any of your papers. I could be arrested for kidnapping, for traveling without proper documentation – several counts, because you're minors. If I left you at an inn while I worked, it could be for improper supervision of minors, abandonment, some other half-assed backwards –"

"Language."

"—law, and it'd call into question Zander, who's young enough yet that they could argue against his age. An inn's gonna wanna see your paperwork, especially if you show up with a dirty old vagrant like me!"

"You're not old," Meygan replied, with an angelic-sweet smile.

Rahyoke rolled his eyes. "Ah, but I *am*."

"What if we went into town ourselves, though?" Meygan asked. "What's wrong with that?"

"Besides paperwork?" Rahyoke sighed, sitting forward, his one knee propped up and his arm draped over it. It was a tough issue to explain... He ran a hand over his face, holding it so it covered his mouth. Now on day six without shaving, he was sporting a scruffy sort of beard – a ruggedly handsome look, all told.

He answered: "It's very dangerous in the towns... *Very* dangerous." He dropped his hand, holding his empty palm level. "You're young, you're undocumented –" pointing at Meygan "—*you* especially, as a girl... Without papers, without a family to protect you, anyone could take advantage of you. It would be only too easy for the system to scoop you up, and then... Well, the workhouse and the seller's block tend to be the popular options. The towns are the kinda places where you wanna get in, do your business without attracting undue attention, and get out. Simple as."

"So it's basically like any city?" Meygan offered, summarily. "They aren't cute, homey, little country towns, but, like, dangerous cities – south-side style?"

"Yes," Rahyoke confirmed. "Yes – and there're good quarters, too – well, there use'ta be. But it's too great of a risk to take. I wouldn't put you through it, decent young souls ya are."

Meygan considered what he was saying. Did he seem trustworthy enough to go off of his word alone? Sure the towns might be dangerous – *all* towns were dangerous. At times, even the cute little country ones. But they had nothing firsthand – they had seen no towns – no works of man save the downtrodden cottage and the ruins of a mill washed out years ago. And maybe that's all there was in this world – his words, and nothing else.

Then again – what on earth reason had he to lie to them? Burning down a person's house and destroying his livelihood doesn't exactly encourage him to keep you. And Rahyoke was honest; enigmatic and cynical though he was, she could tell this one thing about him from the very beginning: at least at the very core, Rahyoke was a good, decent person. And if he said the towns were too dangerous, it was so.

"Well, what if we got paperwork, then?" she asked. "Could we get paperwork?"

Zander laughed sardonically. Noticing Meygan's responding glare, Rahyoke intervened. "No," he said, shaking his head. "That's a vast, drastic, major process, and I'd sooner die of old age. Not to mention, you'd need to apply with a valid and creditable licensee –" He frowned at the lingo "—or something."

"You aren't credible?" Addison's head jerked up, on alert.

"Look at me. Do I *look* credible?"

Zander elaborated: "They'd arrest him for vagrancy without ever checking his papers."

Meygan frowned. "There are a *lot* of ways to get arrested here."

Ian snorted. "There are a lot of ways to get arrested *everywhere*."

"If all that's true," Addison challenged, "then why haven't you been arrested yet?"

Rahyoke could've very easily given the appropriate answer – after all, it probably would've been the mature, responsible thing to do. But he was tired of answering their questions. So, with an arrogant, smirking grin like the conniving grimace of the devil on him, he challenged, "How d'you know I *haven't?*"

Meygan and Zander thought it was funny; Addison scowled, and Ian – poor, lovely Ian – eyes wide, asked, "*Have* you?"

Rahyoke did not answer; merely *looked,* with a look provocatively cunning. *Have I?*

Zander decided to move on. "Marshmallows," he said, grabbing the bag from the food-sack. When Rahyoke glanced at him, he held the bag out of reach, adding, "You can't, because you haven't had dinner yet." He then proceeded to hand them out to everyone else.

Rahyoke flashed an eye-roll and said nothing, reaching forward and snagging the bag of food.

"Are you gonna answer Addison's question now?" asked Meygan, watching as Zander skewered a marshmallow on the end of a broken arrow.

Rahyoke glanced up. "Huh?"

"I asked you yesterday," Addison clarified (sort of).

"Mmmh... Oh yeah!" Rahyoke remembered. He temporarily abandoned his dinner. Reaching forward, he took one of the sticks from their pile – a half-sturdy, arm-length branch. He began stripping the leaves and twigs from it, feeding them into the fire.

He waited until Addison demanded, "*Well!?*"

"Well!" Rahyoke grinned – supercilious and taunting. "It took me a night and a day, but at last I finally found the sense in Addison's question." To be decent, he decided to make like he didn't notice the stifled giggles that answered his mocking report, using the decorticated stick as a fire-poker. "I think I mentioned the answer would be a long one," he continued, "so bear with me... Ready?"

"Yes."

"Okay." Softly, he stirred the crumbling ashes. The fire cast a candle-like orange glow over the cold river stones, scattering shadows over the ground. It was a miserable fire, really; the hearth would not do justice to the tale.

Sin an saol.

"The whole world lies across that river, lads. Once beyond it, there's the rest of the forest. There's the fields and meadows, the moorlands to the far North-East, the end of the mountains to the south. Beyond the mountains there is desert; there is serengeti, rice fields – och, things you could only imagine. In maybe a year, travelin' as we are, we could traverse it – probably more – and come to the distant shore. And then there's the ocean beyond it, and all the world lies open to you from there. You just hafta know how to take it.

"We're not wandering aimlessly, kids... Ha – a lot of surprised looks, there. You think I'm crazy – okay, alright. Wait 'til ya hear the destination – I've been workin' on this.

"There's no town around it to give it a title. The name of it has long since been lost. Somehow, it's *so* appropriate, considering all things – considering why. No name is suitable to replace that lost, ancestral title; it's not like a dog or a town or a mountain – some fleeting thing. It's far greater, and deserves far more. You forget the name of God, you cannot call Him by any menial, temporal replacement title. Once lost, some things can never be justifiably replaced. And that loss, that vacant gap, stands as a testament for all time.

"But why. Why does anyone care about the place, let alone want to go to it? Because of what it is." With the sturdy branch, he began to trace patterns in the soil beside the stones – great, inter-locking swirls, wild and overlapping, and yet deeply and intricately patterned. "Beyond the forest," he went on, "beyond a great, dividing field like the ocean, there lies a stretch of woods. Ah – great and vast it is, far bigger than these Western woods. And you say – Oh great! *More* forest. But *shorn* – wait 'til I tell you.

"Old as the earth itself and older; at its heart, the trees are so vast that ten men cannot encircle them. Dark-wooded, ancient oaks and sinewy, red maples, a few random black conifers here and there. Big enough for to colonize a clan in one hollowed trunk, and bigger. And besides the trees – whose entwining leaves carpet the flooring of God – there is the world below. The light, golden like honey, goes flowing through the dark leaves, scattering over the ground in patterns of shining lace. And all manner of living things, great and terrible, grow up there. Thick is the grass and moss and ivy, swirling up the roots. Orchids and nightshade, snapdragons and roses and those weird little star-shaped white flowers that no one ever remembers the name of. Springs and streamlets – yes, it's a word – flow across the black soil, glancing between the bushes heavy with fruit – the pears and pomegranates, peaches and apples and black berries and mint. Salamanders hide in the running waters, minnows and frogs, and every manner of bird swarming above and below in pure harmony. In the summer, the red deer flitting through the trees, the grey wolves in the depths of winter, the great, blue heron – if you've never seen one, it's this truly *massive* bird – and *damned* if my words could ever even halfway do it justice.

"Such is the heart of the forest to which we're going. And believe me – not to put too fine a point on it, but it is safer than the towns. Wolves will generally leave you alone. People are far more ferocious creatures – and the laws of the towns are lawless. I'd explain further about the towns, but..." With the tip of the stick, he flicked a log deeper into the fire, sending up a shower of sparks. "I don't want to," he concluded. Then, looking up: "Satisfactory for an answer?"

"It was very pretty," Meygan praised.

He bowed his head again, shy. "Thanks."

Addison wasn't convinced. "Pretty, but unrealistic."

Rahyoke looked up again, his eyes narrowed. "Do you hear a lot of faery-stories, Addison?"

"What's that got to do with anything!?"

"I bet you don't." Looking down, Rahyoke scratched a thick, precise line across his drawing, breaking the mist-like swirls. "Look around. If a stretch of woods can be as utterly barren as this, is it so illogical that another could be equally as full of life? C'mon."

"You're having us chase after some – some – *fantasy!*" Addison blurted, frustrated.

The darkest smile ever wrung from the human mouth briefly flickered across Rahyoke's face, then was gone in an acerbic mire; his eyes flashed up to meet the child's, reflecting the fire's light. *"Return,"* he corrected, gratingly. "I intend to *return*."

Everyone was taken aback. Ian sat bolt upright, suddenly very excited. "You've been there!?"

Rahyoke nodded. "And I remember."

Addison shook his head. "Sometimes the mind exaggerates things," he persisted, "especially in memories."

Rahyoke was irked. "I'm not –"

"It's called wishful thinking. And while that all *sounds* great – don't get me wrong – you're building up these expectations – it isn't rational. We can't live like that."

"What?! How –"

"Look – be realistic. How do you even know it still looks like that?"

"Because I have *faith!!*" Rahyoke's voice had risen suddenly and sharply; his fists were clenched. The boy with his mindless picking had offended something deep within him. *Calm down... "A Thiarna Dé –* " half of his anger fled into the prayer "—either come or don't come, Addison; the decision's yer own. But that's where *I'm* goin'. *A Dhia dhílis...*" The rest of his anger in that. He shook his head, relaxing his tensed muscles. "Try that sass on a checkpoint sentry; they're armed."

"Just ignore him, Rahyoke," said Meygan, taking sides. "It's what *we* do."

His eyes shifted to her, peering through lengths of his dark hair. Shorn, she was angry, too. He lifted his gaze to the sky, tilting her a glance. "I take it y'don't mind chasin' down Tír na nÓg?"

She had to allow a smile; she couldn't help it. "I like my faery-stories just fine. I would rather go to your dream-place than any mean, cranky old town."

He gave a short laugh, and blushed for it. "Alright."

"What do we do when we get there?"

He shrugged. "I'm not planning your life out for you. Do what you want."

"Can we build a house there?" Ian asked.

"To burn down?" Rahyoke tossed the stick into the blaze as it caught fire.

"Ouch."

"You won't abandon us, will you?" Meygan asked.

"I? No." Rahyoke shook his head. "I give you my solemn vow."

"Thanks," said Ian, lightly, fully accepting the oath.

"I still don't understand," Addison murmured, not defiantly now, but cowed, confused, and quiet.

"You will, Addison," Rahyoke responded, leaning back on the rock behind him, his hands folded behind his head. "Someday, if you open yourself to believing and trust, you will."

That was the moral of the story, it seemed. They all pondered it a moment, in their own way.

"That was deep," Meygan mused.

"Cripes," Rahyoke muttered, his eyes tracing over the stars as they were blotted out by clouds. "You lot get some rest, okay? I'll watch over the fire."

"Fine." Standing, they retreated to their own quarters.

"When do we move again?" asked Meygan, glancing back, though by now she anticipated the answer.

Rahyoke smirked.

"Dawn."

17
Stargazers

Silver moonlight glittered through the branches of the trees, scattering in shifting patterns over the churning river foam.

He had been kneeling there for hours. His black coat, returned to him, wrapped around his shoulders, the trailing lengths of its fabric spreading a slim blemish upon the crystal grass. He was searching, scouring the wilderness for something he had lost, something he had missed – something cryptic and crucial and beautiful beyond belief. His eyes, lifted unyieldingly skyward, interrogated the moon, berating the clouds; they had loomed there for so long.

Meygan watched him from beneath a tall maple, wondering why he didn't just call it a night and turn in. She knew that he must be exhausted – they all were; yet, there he knelt. At times withdrawn, as he dreaded and cursed the dirt to which his feet were bound; at others, inclined so avidly towards the Heavens that it seemed that he would fall into the river just to be swept up in the moon's reflected glance. Twice he almost had.

There was no chance that he felt her gaze; he would never have been so open if he thought he could be perceived. Now in the privacy on which she secretly intruded, in his small shiftings, in his aspect, she could see he was upset. Not angry-upset, as he had been so reasonably when they'd burnt down his house. Sad-upset, frustrated-upset – as if he were struggling with some deep inner turmoil of sorrow, futility, angst. Maybe he was praying; she wondered what he would pray for. Whatever it was, it was consuming. Why else should he fail to notice that the fire had long ago died?

Moving as silently as possible, Meygan slid her hand onto her backpack, which she had been using as a pillow. No other connection was left to her of that lost Otherworld – a blank book, a dying cell phone, and half of a chocolate-granola bar.

Maybe we're all just crazy?

"What're you doing?"

Rahyoke jumped, turning; Meygan would not have been surprised if he had spread wings and taken flight, so startled he seemed. His eyes pierced through the half-light, adjusting slowly from his reverie, dark even against the night.

"Oh, Mercy," he breathed, laying a hand against his heart. "Why aren't you asleep? What's up?"

"I *was* asleep," she replied. She came up behind him, halting at the distance of a yard. She watched him uncertainly.

"*Was*," he quietly scoffed. He averted his gaze, studying the ragged grass with the usual air of dispassionate interest which nature seemed to invoke in him. "Were you afraid of the spiders, then?"

"They know not to mess with me."

"Huh..." Personally, he was not particularly familiar with the habits of spiders. Folding his legs, he shifted over to make room for her – more of a symbolic gesture on the banks of a river. "You're welcome t'sit, if you'd like."

The girl regarded him silently, and it was a look he was more than used to. He turned back to the river without waiting to see what she would decide.

Meygan followed his gaze. The moon was brighter in this world than in hers; although it had waned to a thin crescent, its light blanketed their clearing, dancing over the water like molten tin. The river was a little narrower here, but deeper and just as rampant; the trees were darker on the other side, the moonlight glancing strangely off of the apples in their boughs. Other than that, it was just as blank as their own side, just as covered in night.

"What *are* you doing?" she asked.

Rahyoke did not stir. A cloud flashed across the moon, throwing them into momentary darkness, and he slowly lifted his eyes.

"Navigating," he said. A smattering of stars sparkled through the oblique film of clouds, protesting as vainly as he. "It's a bit more accurate than the sun... this far North."

Meygan looked to the sky as well. She knew it had to be more than that; no one had to search that hard to find the North Star. But he had remarked it in such a helpful tone that she didn't contradict him.

"How?"

Rahyoke titled his head back, far enough this time that he could look at her. A wry, haughty smirk was on his lips. "'How'." Shaking the hair out of his eyes habitually, he returned to the moon. "Maybe I'll teach you sometime."

Perhaps it was only because he was tired, but there seemed to be something more than sardonic mystery to his tone – that same willingness of a sage and storyteller to preserve his wisdom against the mire of ignorance and time that had driven his earlier explanation to grandeur. And yet – as always – something else was holding him back; he remained reserved.

She sat down beside him, cross-legged on the bank. "We have time, right?"

He gave a short laugh, his sarcasm returned. "Strange, you children," he remarked, in that soft, scarred voice. "Always y'beg for a respite, and when you get it, never's the time that'chya use it f'r sleep."

Meygan pursed her lips. She found the archaic way he spoke intriguing – sometimes annoying, sometimes beautiful. Whether or not this was another world, Rahyoke was something else entirely. "I told you; I already slept."

"Mm-hm."

"Want me to take over so *you* can get some sleep?"

Rahyoke twisted a glance behind them, to the sad remnants of the fire. "There's not much point in a watch now." In truth, he had forgotten entirely about the fire.

For a moment, he pondered what, since not the watch, could have inspired her to come over to him. Then he faced her, for once looking at her directly, meeting those bright-green eyes.

"Look. If you want to talk, then y'may as well. I know tha'chyer... disappointed. I know that it's not much I can do for you – I can't replace what you've lost. But... I dunno, is it you wanted to talk?" He faltered for words, looking at her, confused.

Meygan shrugged; her head bowed, her face became hidden beneath a veil of fiery curls. "No... I just... it's just I wish I could've said goodbye to my mom, that's all, you know? The last thing I said to her was like, 'See you later – oh, and don't forget evaporated cream.' I don't even really know what that is, evaporated cream. And then I didn't see her."

Rahyoke's gaze dropped, flickering away. "I'm sorry."

"*You* didn't do it."

"It's... some sort of social convention that purports to make it all better. It doesn't, but... I suppose – you're meant to feel less alone."

Meygan slid him a narrow glance, peering intensely through the curtain of her hair; his voice had cracked. Slowly, she pulled a few, slender blades of grass from between the rocks, dripping them into the water. "Well, at least we have each other."

Rahyoke said nothing. His dark eyes were fixed on the river's depths, every muscle tensed, wound taut to endure that renewed inner battle; Meygan recognized that it was time to change topics.

"You know what that campfire needs?" she asked; she was glad to see him relax (as much as he ever did).

Rahyoke nodded. "S'mores."

"Well, that too." She tossed the rest of the grass into the water. "But I was thinking stories."

"Stories!" he echoed, with mock surprise. "What kind of stories?"

Meygan shrugged. "I dunno. Liiiiike – how the river got its name. Or faery tales. Or – I dunno – your *life story*," she teased.

"My life story *is* a faery-tale."

"But no scary stories!"

"It's that, too..."

He was brooding. Wringing his slim hands, his fingers a ghostly white against the black half-gloves he perpetually wore. Though he was speaking to her, Meygan knew that his heart and mind were still enwrapt in those anxious prayers.

"Does it have anything to do with the river?" she offered, a hesitant joke to distract him.

He made a face. "Does *now*."

"Have you ever crossed it?"

"Yes – did I not tell you? I've crossed over it many times."

"For your job?"

"Usually."

Meygan frowned. "If people have to cross it that much, why don't they just build a bridge? This place makes no sense... What do you do, anyway? For a job."

"*Well.*" He leaned back, shifting so he could stretch out his legs a little; his foot was going numb. "Will ya waste a question on that, now?"

"Ah, no." She waved him off. "I'll get one of the boys to ask."

Rahyoke was so startled he actually laughed – then quickly covered his mouth to keep from waking the boys. "Shorn," he whispered, a little embarrassed. "That's a thing I'd like t'see!"

"Tomorrow," Meygan promised. "Wait and see."

"Okay."

They sat a moment in silence. It was a nice night; Rahyoke could not help but think on how many miles he could've walked if he didn't have to wait. He could've been at the fields by now – by Glory, he could've crossed them. He considered again Meygan's proposal to obtain paperwork, to dwell in one of the towns. Mercy – at times he wondered if it wasn't the right decision. Who were they safer with? A town full of suspicious, cantankerous grimalkins and worse? Or him? Now *there* was a strange quandary to ponder! In the end, he always settled the matter as he had with Zander: their lives, their decision. He wouldn't hinder them if they wanted to go. But he wouldn't send them off unwarned.

"Will you tell me?"

"Huh?"

Meygan smiled. "A *story*," she clarified. "You have, like, a *way with words*. That sounds cheesy, but you do. You told the Tír na nÓg thing very nicely."

He rolled his eyes. "Thanks."

"I mean it!" she insisted. She shifted so she was kneeling, facing him, though cheating out towards the river; she would not be caught unawares and fall in. "Will you please tell us some stories? It would certainly pass the time."

"For you or me?" he grumbled. Then: "*Fine*, but not tonight."

Meygan brightened, clasping her hands. "Really!? What story will you tell?"

"How should *I* know?" He frowned, but it was a playful sort of disgruntled expression. It was difficult to be unaffected by her happiness; she was adorable. "You ought'a rest," he advised. "I wanna cover some good ground tomorrow; we've been slackin' a bit."

"Okay." Meygan got to her feet. "Are you staying with us tomorrow?" she asked; for some reason she sounded incredibly suspicious.

He nodded slowly. "Yep... Somebody's gotta teach that poor archer to fish."

18
Haul Away Joe

Dawn was a soapy whiteness down-river and to the right, choked by the clouds that thickened in the west.

Rahyoke stood knee-deep in the water – as deep as the sandy little ledge would afford – the archer's re-strung bow in his hands. His black sneakers were lying on a rock with his dark coat, and his jeans were rolled up to his knees. The archer's battered quiver was on his shoulders, the strap cross-ways over his chest. The bowstring was stretched taut, angling a deadly shaft towards the water. No fiber moved, no muscle twitched; his dark eyes were locked on the rolling waters.

Zander sat onshore, eating a cinnamon-honey bread-knot and making snarky comments whenever he could.

"I'm tellin' ya, you're never gonna catch anything that way," he said, for about the tenth time since Rahyoke had set to this "hopeless endeavor".

"I know what I'm doing," Rahyoke murmured, never breaking his concentration. Zander tore a bite off of the bread-knot; it was beginning to verge on stale. "You keep saying that," he said, his mouth half-full, "but I've yet to see the proof."

Rahyoke shushed him.

"Why don't you just get a fishing pole? Or even better, a stick I'm *allowed* to use; and I can keep trying with Ian's shoelace until we *starve*."

"The fish hear you; that's why they're not coming."

"They're underwater."

"You always mention Finnegas; d'you think Finnegas went on in this way when trying to catch the Salmon?"

"He was a bard; he probably put it in rhyme."

Rahyoke didn't deign to respond to that.

Zander swallowed; pointedly, he began to sing:

"Bád beag is bád mór, bád beag is bád mór
Bád beag is bád mór ag mo Cheallachín Fionn,

Bád beag is bád mór, bád beag is bád mór,
Agus loingis faoi sheol ag mo Cheallachín Fionn.

Óra, chuaigh sé a' rith geallta, chuaigh sé a' rith geallta,
Chuaigh sé a' rith geallta, mo Cheallachín Fionn,
Óra, chuaigh sé a' rith geallta i sliogáinín bairneach,
Is bhuach sé trí gheall, mo –"

"If you remembered your chores *half* so well as ya remember those old songs," Rahyoke denounced, under his breath.

Zander shrugged, biting into his breakfast. "*Someone* has to," he defended.

Rahyoke continued to mutter rebuke, as much to himself now as to the obstinate boy. The boy – who had woken him up lamentably early, after only about an hour's sleep, demanding, airily, "*Didn't you get me any food?*" Whoever said to teach a man to fish obviously didn't have teenagers; far better to snicker and roll over, falling back to sleep, and let him figure it out on his own. He had been three hours standing in this bloody current for his trouble – and for what?! Rahyoke tightened his grip on the wood of the bow 'til his fingers were almost shaking, gritting his teeth as the archer continued to taunt him. So many years of trawling the river bottoms had stripped its depths bare, but *damned* if he wouldn't stand here 'til *Doomsday* waiting, if he had to; if there was so much as a single fish in this bloody, wide stream to be had, he would have it – if only to *shut that brat up!!* Besides! The nearest town was five miles away, and it was even further to the next; they needed to ration their food. And the apples across the river were – nasty.

Realistically, they did need to get moving either way, no matter his stubborn pride. But if he could just wait a little longer –

Something slid against the back of his leg, brushing slickly against his calf, and he dropped his eyes, never shifting, never tensing.

Buíochas le Dia!

"What's he doing?" Ian Burgendson settled down next to Zander, watching Rahyoke with interest.

"Tryin'a fish," Zander muttered, around a mouthful of bread.

"Oh... Is it working?"

Zander snickered. "*No.*"

A sharp sploosh sounded from the stream – followed by the rending, tumultuous cacophony of shattering water. Ian winced as the spindrift spattered the rocks.

"HA!!" On his knees dangerously close to the underwater ledge, Rahyoke lifted a speared trout for the doubting Zander to see, unbalanced as he kept the ancient bow aloft. He was doused, his grey shirt clinging to him, his dark hair wild as always.

"Smooth," replied Zander, flatly. Rahyoke moved to retort but briefly lost his balance and was dunked.

"Y'okay?" the archer – still as emotionlessly – amended, as the man again jerked up out of the water, flinging himself to his feet.

"Damn it!" Rahyoke blurted, though his pride kept his tone light. "I got yer feckin' bow wet!"

"Language!" Zander sighed. "*Honestly*, child!"

"That's okay," Ian mumbled, waving him off. He was sitting cross-legged, his cand his elbows on his legs. "I hear worse on the bus."

"'"What?" Zander began to scoff, but was interrupted.

"Catch!" In an arc of water, Rahyoke tossed the fish onto the bank, shooting well over Zander's head; he didn't trust him to catch it. "Hack one of those sticks, will you? And don't use the conifer! The damn sap gets into everything."

Zander stood, moving to retrieve the dusty, bloody fish. "Auh! That's a fine, *small* specimen of a creature ya have there, Rahyoke!" he japed.

Rahyoke waded to the shore; his response to Zander was exceedingly crude.

"Now," Zander continued, regardless, "the half-drowning myself – I assume that's a crucial step; would you care to elaborate on your technique? Maybe show me again –?"

"Just get a bleedin' knife, ya smart-ass!" Rahyoke sneered, using his coat to wipe down the bow.

"Jeeze," muttered Ian; "I didn't realize you'd go *all-out*."

"You would too, if ya'd put up with *him* all mornin'!" Rahyoke snapped in reply. "*I'm* tired!"

"Are we gonna gut this now?" Zander asked, indicating the fish. His remarks on its size had been a little less than fair; the trout was the length of his forearm, big enough for breakfast, and would make a tasty meal at that.

"I will; just give me a minute." Rahyoke clambered up on shore, trying his best to keep the bow as dry as possible. "I'll need to see a knife, though."

Zander unsheathed his, holding it tauntingly.

"Ha," uttered Rahyoke, unamused. He came over to where the archer was. *"You,"* he quipped, stooping and snatching the blade, "are not *funny*."

"Well, aren't *you* talkative today!"

"Lord save us," Rahyoke responded, and didn't speak again. He managed to rekindle the fire as Zander began stripping the bark from two small but solid branches with his sword; his glare of disapproval when the boy sent Ian for firewood was dismissed with a shrug, and the archer was only spared by the arrival of Meygan.

"Why hello, Meygan!" Zander greeted her, grandly and pointedly, as she settled down beside him.

"Wha'chya doin'?" she asked.

"Rahyoke caught a fish, and we're gonna cook it for breakfast."

Meygan glanced towards Rahyoke, as he pulled the arrow from the trout. "Why didn't you use a line?"

Zander answered. "I told him his way wouldn't work; he's used to those Northern waters of twenty years ago. Up in the Isles you can dip your hand into any river and scoop out in a fist-full enough fish to feed a farmyard. But he's all –" He lowered his pitch, assuming a domineeringly authoritative demeanor as he quoted, "'If you fish from the lee side, you're *bound* to catch something'."

"What's the 'lee side'?"

"Downwind," Zander explained. "And he's crazy; everybody knows that you catch more fish from the windward side; all the bugs get knocked over there."

"Uh-huh..."

There was a tremendous crashing from the forest that caused everyone to jump, and a minute later Ian returned dragging a branch longer than he was tall. "I thought we couldn't burn these?" he said, handing the bough to Zander.

"We're *not*," Zander half-snarled, snatching the branch. "Next time could ya look on the *ground* first!? Go sit down!"

Ian raised his palms. "Jeeeeze," he murmured, suggesting Zander was quite touchy that morning. He moved off a ways and sat down in the grass, watching the proceedings.

Still using the sword, Zander stripped the bough of its smaller, leafy branches, laying them aside for future kindling. He skinned about a foot of the bark from one end, and split it down that length.

Meanwhile, Rahyoke scaled and gutted the fish with quick, deft skims of the gleaming knife, and laid aside the head, fins, and bones. The work was over in a little under a minute.

Ian gave a long, falling whistle. "Wow!" Addison – roused by Ian's pulling down the forest – had just arrived at their circle in time to watch the act, and asked, nonchalantly, "Where did you learn to do that?"

"That's what they pay him the big bucks for," replied Zander, picking the dew-damp bark from his blade.

"That's his *job?*"

"Like a sushi bar?!" Ian chirped.

Zander shrugged a shoulder. "Why ask me?"

Both boys turned to Rahyoke, expectantly. He pretended not to notice, taking the two smaller, debarked sticks from where Zander had laid them and boring them into the fish. It was an interesting tactic – one stick woven into the top half, the other at the bottom, so there were four, pencil-sized holes into the edges of the fillet.

"Rahyoke!"

"Addison." Taking the larger branch, Rahyoke slid the fish between the split halves, so the smaller sticks made a sort of cross. He then tied the top of the branch closed with a long, thin, green branch, securing the fish.

Addison frowned, watching as Rahyoke braced the other end of the bough in the crag between two rocks, so the fillet was suspended over the fire.

"Is that gonna *cook?*" he sneered, side-tracked.

"*Solabhartha,*" Rahyoke muttered, and he wiped the blade clean on his jeans.

"Eew," Meygan complained.

"It'll cook," Zander assured them, feeding the bark into the small fire; the fledgling flame was, if possible, even more pathetic than that of the night before. "If we ever get a real fire going, that is."

"If *you're* so clever," Rahyoke snapped, "then *you* give it a go!"

"*Yeah* yeah yeah..."

Meygan motioned to the refuse of the trout, which was still laying on the rock. "Are you gonna leave that there?" she asked Rahyoke. He glanced at it and shrugged, moving neither to dispose of it nor to do any more for cleaning his hands than wiping them, too, on his jeans.

"Ugh," she reproached, with disgust. She had been living for six days now with this pack of boys, with no female company, and had officially given up on trying to civilize them. Men are absurd.

May as well have fun with them. "Addison? Weren'chyou gonna ask Rahyoke something before? Or did you just wanna know if that would cook?"

"Huh? Oh yeah! What exactly do you do for a job?"

Rahyoke caught the flash of a cocky glance from Meygan, though he did not turn to her. *Well played, lass.* He flipped the knife, handing it hilt-first to Zander. "It'll cost you," he parried, "and you won't be satisfied with the answer."

Meygan pouted. *No fair.*

But Addison scoffed. "Not satisfied?" he echoed. "Why *not?*"

"You lot really don't get the whole 'three questions' thing, do you?"

"Fine; just answer the job one, then."

"Very well." Rahyoke's glance snapped to the lad – grinning, a little savage. "That's two."

Zander snickered, anticipating what on earth sort of answer Rahyoke would give. He stowed away his weapons – the bow and boat-knife in his bag, loosely, and the sword in the scabbard at his side – preparing himself to listen.

Rahyoke, as ever, took his own sweet time, checking the fish, making sure it was cooking.

"I could cheat you out of a solid answer," he noted, "by saying I do whatever's necessary. But I'll honour the bargain beyond your poor phrasing."

Addison made a face. "Again with my wording."

"You fare better in this world if you learn how to speak."

"No matter how infrequently," hinted Meygan, examining her nails.

Rahyoke frowned.

"So *anyway,*" plied Addison, rolling his eyes. "Do you teach *grammar* or something? That's your idea of an answer?"

It's taught if it's learned, and it doesn't seem to be received. Rahyoke had to bite his quick tongue, for how dearly he wanted to say it. "I do..." he began, slowly. "...whatever needs to be done. I can do readily almost anything that's asked of me. Usually, it's – I'll come to a place, and they're like – 'We're short-handed on the farm today', or, 'Can you run this message to the carpenter's?' or 'Go fetch me some tomatoes!' or something like that. In some places, I've established agreements, or at least something of a trustworthy reputation, with some employers, and I can usually find work with them, or they'll put in a good word. For example – in Osset I can almost always find a job as a dockhand, unloading ships and such – also a man called Fergus has me employed to convince people to buy things they don't need in large quantities. Like I said," he added, with a casual half-shrug, "I can do most anything. But I am typically hired for these."

"So," interjected Addison, pithily, "you just – wander around from town to town, looking for work?"

Rahyoke shrugged. "Basically."

"Wouldn't it be easier to just stay in one place? Get a steady job?" There was something condescending in it, even if the child didn't fully understand; in time, it would be intentional.

"Hey!" Meygan exclaimed, brightly. "You could work for Fergus!"

"And live in *Osset?* No." He shook his head. "Being established for so long, the towns are fairly self-sufficient, and I'd be worse in competition for wages. I fill a need when there is one, and, when there isn't anymore, I find a need someplace else. A small farm doesn't need thirty field-hands year-round; only at the time of the harvest. And maybe planting. Besides," he added, his gaze roving off to the distant trees, "I was never really one to stay in one place..."

"*No really?*" Meygan said.

"So –" Addison's own tone was flat "– you do *everything?*"

"I *told* you you wouldn't be satisfied!" Rahyoke said, with mocking cheer. Sitting forward, he again tilted a glance to the fish. "Och – breakfast!" He began dismantling the structure.

"It's done?" Addison asked, dubiously.

"Why don't you try it and find out?"

Addison's frown deepened with disapproval. "You know, you keep giving us just this random food – what if someone's allergic or something?!"

"*Are* you?"

"No –"

"Anyone *else?*"

"Nope!"

"No," Meygan answered, still examining her nails; "but I'm very picky."

Discarding the last, Rahyoke concluded brusquely: "Then it's fine!" He held out a hand, into which Zander clapped the blade of his sword rather than the more sensible knife. Wielding it almost like a pen – startlingly careless, for the length of it – Rahyoke portioned out the fish, handing Addison a chunk. "And stop being a difficult ingrate."

"Yeah, *Addison*," Ian added, haughtily; "at least be an *agreeable* ingrate!"

Rahyoke leveled the blade at him, flat. "And I don't need help from you, lad."

Ian accepted his share of the fish. "Beautifully cooked," he said, with a grin of apology.

"Eat up; we gotta get going," Rahyoke advised. Having handed out their portions, he laid the smallest slice aside for himself, endeavoring to kill the fire. This accomplished, he buried the bones of the fish in the ashes, and – finally – rinsed his hands, before settling into his own meal.

Meygan picked at her portion; it was still hot. "What kind of fish is this?"

Rahyoke shrugged, purely to annoy Addison.

"Trout," Zander answered. He turned to Rahyoke. "Right? Trout?"

Rahyoke enjoyed Addison's hateful scowl a moment longer before turning to Meygan. "Do you like it?"

"You know? I actually *do*." She allowed for a pause in the conversation. Then, rekindling the previous topic, she asked, "Interpreting 'job' as 'occupation', as an addendum, what *else* do you do? For example, do you play guitar, do you write stories – that sort of thing. I saw you can draw," she added, glancing up.

Rahyoke looked at her, very much intrigued. *Learnt*. "That was very well asked," he approved.

"*Show-off*," Addison hissed, and she answered him with a superior glare.

"Hey," Rahyoke warned them. To the boys, he explained, "She's asking because she's after me telling you all stories – hearth-fire stories and what. I don't know – are you guys into that kind of stuff? Because –"

"Yeah!" Ian exclaimed, with surprising enthusiasm. "Zander says you tell the *best* stories!"

Rahyoke threw Zander a retaliatory stare.

Ian continued. "What kind of stories d'you know? D'you make'em up?"

"I know all kinds of stories," Rahyoke shrugged.

"Faery-stories?" Addison denigrated, and Rahyoke turned at the jibe with a smirk.

"*Especially* faery-stories," he answered, something strangely feral to the mystery in his tone. His dark eyes flashed with light, and he shifted so he was kneeling before them, fixing them with that overpowering gaze. "Know what you mock, *a stórín*. Scorn that old mythology – do. Sure, and abandon all valour, all honour, every item of our past – and shame, child, to think: it's heartless ignobles like yourself, uncared-for and uncaring, who will rule the world. What cost?"

Whether or not he understood Rahyoke, Addison knew he was being ridiculed, and was ruffled. "Fine then!" he challenged, sitting up straighter. "Prove it to me! Tell us one of your worthy *faery-stories* tonight!"

Rahyoke shrugged. "Alright, then. If no one's objecting..."

"Yay!" exclaimed Meygan, clasping her hands together.

"Yay," Rahyoke affirmed, flatly. Reaching back behind him, he snagged his coat and his sneakers. "Ya'll done eating? Let's go. I won't be telling you anything until tonight."

19
Hearth-fire

It did not take Rahyoke long to start a fire. In spite of the rain which had fallen intermittently throughout the afternoon, they were able to find enough dry kindling to get a decent campfire going for once. The cheerful, golden blaze cast long blades into the darkness, playing over the rocks and slithering between the trees. The clouds above had retired with the sun, and all of Heaven was open, glittering with milky stars.

"Will you tell us now?" asked Meygan, squirming excitedly, her wide eyes watching Rahyoke.

The man drew a last, long, looping line through the ashes and dust, entwining it over and through his drawing. Since he'd made the fire, he had been tracing designs in the dirt, and had a veritable Celtic mandala before him. The line curled in a deep, winding spiral, blurring out the pattern.

"Yeah, alright."

He snapped the stick, tossing it into the fire, and sat up straight, surveying his audience. They all looked back at him with interest – whether it was open and exhilarated or covered with a film of superior cool.

He smirked. "There may be a chance that you've heard this before," he began, his low voice soft, and intriguing. "In fact, though you're from very far away, it's very likely that you have. Some things can transcend all boundaries. Time, music, legends...

"Fionn's love for Sadb was one of those things..."

20
The Flower of Almhuin

One night, as the battle season was winding down, and the leaves in their whisperings were threatening to turn; when the full moon cast her frosty light over the Bog of Almhuin, a great restlessness came upon Fionn Mac Cumhaill.

Knowing it was not a thing to be fought, Fionn arose, taking with him none save his hounds, Bran and Sceolan. Now: much has been said concerning these hounds of Fionn's, and we could be here an hour or more discussing Bran alone. Know that they meant more to Fionn than anything in the world, and were more than ordinary hounds by far. Some say they were of the *sidhe*, the Faery-people – others that they were Fionn's own kin. But whatever they were, there were no dogs to match them – not then, not before, not ever again.

Together, the three went out from Almhuin of the White Walls and into the rough darkness of the night, the two dogs walking on either side of him – Bran, who was as purely white as the snow when it first settles, and Sceolan, pearly white but leopard-specked with gold, and her ears as red as berries. The clouds were heavy on the moor, and a strange chill spiked the late summer heat. But the air of the outside world was a relief to Fionn; out here in the wilds, he could breathe.

Through the night they wandered, never heeding their path, though a path indeed they took, wrapped in the silence of the wood and the stillness of the heart and the sheer beauty of simply being alive. At that time when the first, lighting traces of the sun begin to brighten the opaque sky, and all the horizon is made silky and watery, an ever-paling blue; then it was, that – in perfect harmony – Bran and Sceolan stopped.

Fionn halted as well, a pace ahead of them, looking to his hounds in bewilderment. Strange, that they should stop at all, let alone in such unison.

Then he looked ahead of him.

Between the shapes of the trees so familiar to him, he saw a foreign stretch of land. Bare trees, stretching their white-armed limbs into the purple sky, but for the sanguine clumps of berries supported between their fingers. A golden-russet carpet of leaves scattered the ground at their feet in perfectly-ordered disarray, not a leaf of it stirred, for the wind and its whispers were dead.

Without giving tongue, Bran and Sceolan drew forward. And it was then that Fionn first saw her.

A white hind – that's the word that they use for a deer, a female one, like – a white hind stood just within the line of those spooky trees, her coat flashing and glancing in the light of the newly-opened sky. The two hounds approached her somberly, and, when they reached her, they pressed against her sides, licking her hooves and slender legs. The hind never gave glance to them; indeed, her eyes – large, dark eyes they were, deeper than the Pool of Delight, where grow the Hazels of Knowledge – her eyes were only for Fionn.

Fionn drew a pace forward; in the instant, the three in their phantom whiteness began to move, prancing and cantering, flashing and glancing, though their feet never disturbed those leaves of gold.

Fionn's did. To his dismay, that single step was a cacophony, echoing off through the trees like a shout in a sea-cave. Far at odds with his hunter's training, and even farther still from the silent beauty before him.

Ochón, was he dismayed! With a newfound strangeness, every fiber of Fionn's being, every sinew and artery and marrow and vein of him, was filled, shredded, choked, and dragged with longing. Terrible this longing – terrible, and yet more glorious than anything this world has to offer. It's something you'd have to experience for yourself; not to understand it, no – you can't. But only to feel its power. Ah, sure you will if you haven't, in time.

Anyway! In desperation, Fionn drew forward, pulled by the heartstrings to be united in that ethereal, mystic dance.

The silver beings drew off farther.

And Fionn followed.

Slow, their progression, traipsing over the dragging mire of time as if its rapid current meant nothing to them. At times the silver dancers were distant flashes in the trees; at other times, they circled back to him, as if to coax him, always moving just out of reach. And Fionn himself, as if in a dream, wandering ever farther, ever deeper into the ever-shifting wood.

In a world apart, the sun was rising in tangle-headed glory, stretching long tendrils of gold into the bed of Night. An outflung curl struck the white, stone walls, and a scattering flash dispelled the fog of dream.

Here he was. Fionn stood alone on the Bog of Almhuin. Before him stood the white-walled fortress, bathed in the light of the rising sun. His two hounds scampered around the doors in the distance, shining like gold in the dawn-light.

Fionn turned a final glance behind him. All the world was a blaze of white light, and framed within it was the hind, standing proudly, looking down on him with those deep, hazel eyes. Then she was gone, and the forest was as it had been, the rough bark of the maples and oak trees and elders a warm, cheery red in the rapidly-growing light.

Fionn shook his head to clear it. He'd have to have a word with Garbhcronan, the head serving-man, about the quality of the latest shipment of mead.

And so resolved, he plucked up his spirits, and made his way home to the Palace of Almhuin.

Now: Three things that are of the highest value to a fennid (a member of the Fianna – you'll see); and these three are: war, women, and hunting.

That might be to *any* red-blooded man, really. But if any man loved these things, then never a one among them did as keenly as Fionn Mac Cumhaill.

Fionn was commander of the Fianna of the Isle – the army of the High King, essentially. And there were many men in this organization, and deeply he loved all of them; but most of all, he loved these: Goll Mac Morna and Caílte; Madan, Cael, and Blamec; Conán Maol and Fergus; Lugaid, Donn, and Mullach Rua – which is Red Ridge. Others would come in time whom he would cherish as highly as these, but these are the men we are concerned with now. All brave and worthy souls, and great and wondrous and terrible their deeds. And it was with these men – in whole or in numbers – that Fionn loved most to go hunting.

So it was they thought among themselves to get in a final run before winter really set in. And off they set, the whole band of them, with their spears and their swords and their ready pack of hounds.

Forward-driving though Fionn ever was, his bard's heart often led him deep into nostalgia; he trailed along at the back of their train, deeply wrapped in thoughts marred by the clinging cobwebs of an unforgettable dream. Ahead of him, his men walked, talking and laughing and singing and jesting, teasing him when he fell too far behind. And he smiled at their joviality, walking along in silence, his spear on his shoulder and Bran at his side, as Sceolan raced to chase the dancing butterflies away from her master.

He had given them strange instructions that morning, and a large part of himself even wondered why: take the wild boar, take the stag, waste even a cast on a passing squirrel if you must. But do *not* hunt the red deer.

His men, hungry as they were, grumbled much about this – I'm sure you can imagine. The boar and the stag were as hard to take down as the squirrel and badger were un-filling; and wasn't the Bog fairly *teeming* with red deer?

"What about the little speckley ones?" Cael had asked him.

"Or the ones who are old and gamy and slow?" added Conán.

But Fionn had remained firm.

And on they walked, through the sunlit forest, deeper and deeper into its winds.

Fionn loved the forest. He loved the smell of it – the warm, crisp leaves, the heady earth. He loved the wind in the trees and the dew on the grass, the twists and turns of the roots and branches that mimicked the alleys of his own brain. Out in the civilized world – at Almhuin, or on Teamhair – where order and borders reigned – all was confused and confusing for Fionn; but here, where the wildness of his mind and the tumult of the forest were united – here everything made sense.

Suddenly, Sceolan halted, pricking up her ears, lifting her paw habitually as she listened. The poor, crack-winged butterfly she'd just pounced on struggled drunkenly into the air, staggering off towards the sky.

Bran had stopped as well, and Fionn along with her, though he could hear nothing beyond the pleasant chatter and clamour of his men.

Then Bran and Sceolan gave tongue!

Caílte, as ever, was leading the line, and he looked up as Bran hurtled past him, stopping short, and Madan collided with him.

"*Och*, Caílte!"

A red deer flickered away through the trees ahead of them, just as Sceolan flashed by, nearly taking out Donn.

"Ah, *come* on!" Blamec predictably complained. "*Fionn* gets to hunt deer!"

As if summoned, Fionn hurried by. Long were his strides, like the long strides of the grey-hound, his silvery hair flowing long behind him. He shouted one word to them as he flew past, and in that one word the hearts of the earth were bursting with excitement.

"Come!"

They didn't have to be told twice.

Scarcely touching the ground, the fian flew through the forest, racing after Fionn, and himself after Bran and Sceolan. They outstripped the three winds ahead of them and caught up with them again at their tails, so quickly they traveled! Sure, the whole of the world could have rolled by under the speed of their feet, traversing it crossways and lengthwise. And Fionn – he outdid them all.

Flashing through the trees ahead of him, he saw the white hides of Bran and speckled Sceolan; and – was it a dream? – the glancing sun playing over the back of the red fawn. More brilliant a creature had he never seen living, nor imagined in all his dreams.

Save one.

Fionn had never run so hard in his life. He stretched his strides 'til his muscles burned, 'til his lungs seized, 'til his heart swelled and would surely burst within him. Let it burst! He didn't care. 'Til every bound of his boundless strength had withered, died, and failed, he would not cease running, not until that glorious creature was his.

Bran and Sceolan, who would fly to the ends of the earth for their master, were beginning to fall behind. Silently had they been flyting, their feet never bending a grass-blade, their tongues never sounding a call. Now it was that they began panting, and Fionn's heart was greatly dismayed. A little farther, and they'd make it. They just had to hold out a little bit longer...

Then!

Fionn stumbled, tripping over a great root big as a sea-bed – for nothing lesser could hinder Fionn. And – och! Agony! His heart burst and his lungs fell and his sinews snapped and he fell to his knees on the ground. But if there was a greater pain within him, it was that of a heavy truth, for he knew that such a costly failure was bound to lose him the doe – and never a more costly failure had been made!

Fionn bowed his head, clenching his fists and striking the ground in frustration. If red burned the hottest fires of the world, then Fionn was the epitome of crimson. If black was the darkest pit within Hell, then ebony his demeanor. Great was the shame that fell upon him! Great the fury and anguish and –

Someone was licking his head. Fionn chanced a glance upwards and met with Sceolan's tongue. He sighed – a long, heaving, heart-heavy sigh, wrapping his arms around her. He appreciated her sweet attempt at comforting him. Even though –

By chance, his eyes lifted beyond her; and there, in a clearing, bathed in rays of sunlight, lay the faun.

Caílte was running full-tilt and had to pull up short to avoid colliding with Fionn head-on.

"Hey!" he shouted, breathlessly, taking Fionn by the shoulders and wondering how he'd out-run him. "What're ya –"

"Ssh."

Caílte was confused. Catching a second wind, he studied his friend's face, then he followed the transfixed gaze. "Och!" He twirled his spear, aiming its head at the deer to the left of him. "*There's* a strange creature, now! A wonder I didn't see it!" He glanced back at Fionn. "Shall I take it?"

Fionn jolted from his dream, seizing Caílte by the shoulders. "No!" he moaned, pleadingly.

"Alright! Okay!" Caílte raised his hands, directing the deadly spear skywards. Fionn only relaxed slightly, watching his friend with a suspicion Caílte had never before seen from those dark, grey eyes. "What do we do with her?" he asked, turning as Bran snarled Madan's hounds away from the hind's flanks.

"We follow her," Fionn replied.

"Why?" When he received no response, Caílte shook his friend. "*Why?*"

"Because," was all Fionn would answer. "Because." And so that is precisely what they did.

They picked up the other members of their fian along the way, each more winded than the next – except for Conán, who had opted out of the chase early, seeing as he could never catch up, and why expend energy on the impossible? – and they all, in a perplexed and murmuring huddle, followed their leader through and out of the woods, and onto the Bog of Almhuin.

Here the deer stopped; Fionn knowingly declared, "Now it is her own choice, whether she will follow me or won't she." And, in manner recovered but in mind, so it seemed, still totally insane, he strutted across the bog and over the threshold of his fortress, flanked on either side by his two hounds.

Caílte and the others were so struck by the strangeness of it all that they merely stood where they were and watched, as the fawn followed their leader into his home.

"Ai!" lamented Lugaid, clapping a hand to his forehead. "Our Fionn has finally lost it!"

"Well!" returned Conán – and wasn't the devil smiling? "Might as well watch."

Donn swatted his arm, but they all made their way into the Palace.

At that time, there were precious few women at Almhuin, but those few sure made a show of fawning over Fionn's tamed deer – pun not intended. And it was well agreed that they should keep her – for if nothing else, they could tell she was a special creature indeed.

Fionn was not yet as far gone as his men had feared; and he passed the evening and night with them as they always did following a hunt or a battle – in feasting. Long and merry their feasting was – as it always tended to be, whether they had a dozen good boars to feed upon and gallons of golden mead, or the slightest hare and a skin of wine to share between them. Mind you, *this* is the sort of company you ought to keep.

And when they had done with feasting – *och*, it was probably two in the morning, then – they all retired to their quarters, and Fionn ascended to his.

And that night, he dreamed.

He was following the silver-white hind again, her bright coat dappled fiery red in the light of a blood-red sun. This time, they were not running, but dancing, flying – as he had watched his hounds do so many lifetimes before. They leapt and twined in Celtic knot-work, weaving between the trees, spinning and twirling and whirling, 'til he stood at the edge of the Hill of Almhuin. Here is where his Palace had been built – but now, no sign of its white walls. The hill was bare, as it had been in the thousand years before he chose to make his home there – wild and ancient as the tangles of his heart. Fionn cast his eyes around him, staring in wonder at the stillness – the vacancy – of it all. But what was there?

A fairer creature there never was; her skin was like the fresh-fallen snow, as soft and pale as the moon when it shows as a sliver of guiding light through the dense clouds; her hair was a deep, red-tinted brown, like the coat of a red deer, and it fell nearly to her hips in cascading waves, shimmering with perfection even in the strange, mixing light. Her lips were red as the crimson haws of the bare-limbed forest, her neck like the swan, and twice as graceful, and her eyes – deep as the Pool of Delight – and staring unfailingly at Fionn.

All the world and all it contained and all his victories and achievements meant nothing. In the silence of the forest, in the ancient, gripping stillness and the eternity of the world around them, all his years had scattered away like leaves, and his life began anew. In that moment, to draw a breath had meaning, each flitting pulse of the heart; and none of it meant anything without her.

He longed to call out to her, but he had no voice. He longed to touch her, but feared with a fear greater than all the terrors of battle that his hands would prove what his eyes could not see as truth. He could only long for her, and plead with the very depths of his anguished soul:

Your name! Please – by all the gods we worship – tell me your name!!

She smiled – the slight, smirking upturn at the corner of those beautiful lips. At once, a fledgling breeze gave tongue to the leaves; the long grass swayed, the branches cracked, and she whirled about in a splay of skirts and mantle and fiery-dark curls. Fionn lunged forward with every sense of energy he had within his body, stretching out a hand, in absolute desperation she should not leave him. And the wind was whirling against his vision, tearing at his clothes, scratching his eyes and skin until all was a fog and a mire and an utter desolation, and –

WHAM!!

Fionn opened his eyes, and was at first very confused, for he found himself bound and upside-down. Then, as realization dawned on him, he groaned.

A dream. And himself, having flung his body out of the bed, cocooned in his sheets, his feet still up on the bed and the rest of him pathetically on the floor.

"I'm sorry to have startled you."

A voice like the faeries – like the tinkling of bells in its merriment – a voice softer than the winds, deeper than the twisting woods, dearer to his heart – fairly scared him half to death. Unable to distangle himself, Fionn titled his head back as far as he could – and nearly died of pain and excitement and fright and total confusion.

It was she. The woman was standing almost directly above him; a band of gold was against her forehead, restraining a torrent of dark-fire curls – the band, which was a mark of royalty, like a crown. She wore a long, flowing dress of deep green, ornamented in gold – garb of the faeries. But – sure as the breath that died within him and the heart that held for its return – it was she.

"I probably should have waited," she breathed, smiling beautifully. As she bowed closer to him, her head cocked and her curls flowing over one slender shoulder, she inspected him closer. And poor Fionn didn't have the air to speak. Like a dream, she knelt down beside him, just above his head. And she explained:

"I am called Sadb. For three years I have been in the shape of a fawn, wandering the Isle and enduring much hardship. For I refused the love of Fear Doirche, Dark Druid of the Tuatha De Danaan –" the faery-folk. "You have saved me twice this day, Fionn Mac Cumhaill," she continued. "The first, for sparing me from the hunt, and the second – if you permit me to dwell within Almhuin longer. For I am safe only so long as I dwell in a dun (a household) of the Fianna – or in the shadow of Fionn Mac Cumhaill."

Now, that was a lot of information to take in – and sure, Fionn's thoughts were just about full. Seeing his confusion, the woman drew closer; she laid one hand on either side of his head, and she leant so she might look him in the eye, although upside-down. A length of her curls tumbled down from the mass, falling against his cheek, entrancing him as desperately as the look from her fawn-dark eyes. "I ask your protection, Fionn Mac Cumhaill," she said, her tone one of utmost seriousness that belied her earlier energy. "If you can't abide by me, send me into the household of one of your other men. But I'll be nowhere safer than when I am with you."

Shorn, there was a nice thought; Fionn mulled it over like wine.

"So what do you think?" the woman asked.

And Fionn kissed her.

Men: it's the women who love the romance. And you gotta be able to use the lingo – to express the feeling that girls so readily understand. So I'll be explicit.

Imagine everything beautiful in the world. Imagine – a sunset. Rays of gold falling softly through bands of pink and purple clouds, magenta and clear, sapphire waves pouring into the gathering darkness, the swells crashing below in sparkling mist. And long the shadows of the sea-cliffs fall, over a pearl-white beach of delicate sand, the verdant hills receding into the indigo cover of night. Stuff like that.

And know, that *nothing*, for Fionn, was not dulled to a milky, gossamer blandness in comparison with Sadb. Every facet of his former lifestyle was abandoned – no more did Fionn hunt, feast, or even fight. He spent all of his time lavishing in the company of Sadb, the Flower of Almhuin.

And the Fianna were bored witless.

"Does he *ever* come out of his room?" Blamec groaned, to Mullach Rua, who was sitting guard outside it.

"No," Mullach Rua answered, tilting his head back against the wall. "And only Donn ever sees him. He brings him food."

Blamec folded his arms, looking sternly disapproving. "Some Rigfennid Fianna," he grumbled.

"He'll hear you."

"Well that he does! The irresponsible – and does he fancy his men will be training themselves?"

Mullach Rua snickered. "You. You complain when we work, now you complain when we don't. You're worse than Conán Maol."

Whom, we have hinted, was famed for his laziness.

But Fionn Mac Cumhaill was not lazy. As with any of his passions, he was devoting his utmost energy to the pursuit – and, as love for Sadb excelled all other passions a hundred-fold or more, so, likewise, increased his devotion.

But soon, Fionn's energy would be needed elsewhere.

It is now, lads, that we get to the battle.

Summer was battle season – and there were really only two seasons for a warrior.

Thanks to Fionn and his loyal army, the territorial kings of the Isle were finally at relative – at least livable – peace. Livable, but – perhaps even because of this – the summer would not be without warfare. The men of Lochlann – foreigners! – took it into their heads and upon themselves to conquer the peaceful Isle. They landed upon the southern coast, and were laying waste to it with fire and steel. Dire, it was – the lamentations rang up to the skies! And the Fianna-Fionn would answer the call.

Ah, but no graver face was there to be seen than Fionn's, for to go to war meant he had to be parted from Sabd. He chose out the best men from his company to be left at Almhuin, placing deep and serious oaths upon them: Their sole purpose in *life* during his absence was purely the guarding of Sadb.

Left in charge of them were Donn and Mullach Rua – the latter of whom, on being told of his mission, cast a dying moan of rare complaint which flooded the very ramparts with his agony.

"It's a great honour," Donn half-heartedly tried, to which Mullach Rua tastefully grumbled, "Honour my posterior."

In addition to this loyal and able, albeit vaguely irked, contingent, Fionn's druids laid spells, and prayed prayers, supplicating the gods and threatening every force known to man. Even still, for all his preparations, Fionn found it unbearable to leave Almhuin.

Sometimes dreams have a way of staying with us. And those are the dreams that *change* us – even subtly – in ways too important to understand. As he cast one final glance to the fortress, as the wind swirled through the heather on the bog and the trees rattled out their secrets, he saw Sadb turn on the threshold, vanishing within its strong, enchanted walls, her skirts flashing in the wind, and her dark-red curls a torrent swirling around her shoulders...

It was seven days, the war. Seven days, and Fionn and his Fianna broke through the ranks of the Lochlannachs, destroying them, setting fire to their ships and driving the men into the ocean, where they foundered and flapped and fought their way back towards their homeland, choosing the broad, pounding ocean over the fury of Fionn's men.

Caílte never lost sight of his leader during the battle – not even in the thick mist that fell in the morning hours – and he would swear that Fionn fought like a man possessed. A whirl of battle-lust and love-lust and exhilaration and agony and pride – *so* much pride. His energy exploded in the great wash of the enemy's blood, and he fell on them like some immortal and ravenous beast – and yet, he fought with caution, as if life was something to be more than exulted, but treasured. The great battle was all a game to him, a fantasy; his world, his reality, was Sadb.

And Lord, he nearly wiped out the Lochlanns single-handedly for the interruption of it!

The moment the fight was over, Fionn was anxious to get home.

"The king'll be after seeing you," Caílte warned, scoldingly.

Fionn shook him off. "He'll see me at Samhain."

"There'll be feasting, and you missing out on it."

"Don't care."

"It's an insult."

"Then I'll come back!"

And what more could poor Caílte do than roll his eyes skyward, answering only, "Fionn, has anyone told you that the head on you is like a forest fire?"

Stubborn and piercingly hot. Fionn flashed him one of those cunning smiles. "But pretty," he challenged.

"Ochón," Caílte groaned, but he let him go all the same.

Fionn promised again to return if there was feasting, though Caílte sorely doubted him. And, taking no one with him save Bran and Sceolan, Fionn headed home to the Palace of Almhuin.

The mist which had fallen during the battle still hung heavily over the bog, but Fionn didn't mind it; he knew the lay of the land better than he knew his own soul. He called to his dogs to keep up with him, speeding over the spongy turf – for this was a day he could overtake the gods.

In no time, the Palace of Almhuin stood strong before them; Fionn smiled to see it, though a thorn of worry entered into his heart to see the walls and the plain and the windows were bare.

Ah, she hasn't seen me yet.

Fionn lifted his pace, hurrying across the plain.

Ah, well, she mightn't expect me yet.

Fionn's feet, flying, hardly touched the springy ground.

Of course, she's modest, and fearful of crossing the threshold...

But all excuses stopped, when he stood, dead-still, before the wall, and the doors did not swing readily open, and indeed, he saw no one to swing them.

Fionn Mac Cumhaill stood at his own doorway, unable to enter his home. And his dogs stood to either side of him, waiting in the deserted yard. A cold chill ran a breath through the bestilled silence, causing the hairs to stand on the back of his neck.

"Hello?" Fionn called, feeling a dark dread begin melting into the sinews of his heart. He knew deep within him something was wrong. Dreadfully, horridly wrong...

Sceolan began whimpering, scratching at the door with her powerful nails. And Bran dropped down on her haunches, throwing her head back to give one long, mournful, haunting howl.

Fionn's dread became darker than Death. He felt the fluids of his body mix and strain and die as his heart dropped out and his intestines churned up and all of him was a shaking, torquing, torrent of agony.

"No," he stammered, stepping back from the dooring. It was at that that they opened the door.

Mullach Rua peered out, spotted Fionn in front of him, and pushed the one door open. And there he stood, the blood slowly draining from his face.

His figure gave Fionn strange hope.

"Where is she?" Fionn demanded, seizing him full by the front of the shirt. "Where is Sadb!?"

Mullach Rua started. "She was – she – we thought she was with you!!"

Fionn jolted back, taking Mullach Rua with him. "What!? How *could* she be!?"

"You came back yesterday!"

Fionn let him go. "I didn't!"

"You did! Garbhcronan saw you! We all saw you! *Sadb* saw you!"

"I didn't!" Fionn shouted, growing frantic even in his rage. *"I CAME BACK NOW!!"* And he seized upon Mullach Rua harder than before.

Mullach Rua did not fight him; but he gave a long groan, so horridly similar to that Bran had given moments before.

Fionn released him, staggering back. All he could manage, was, "Where?"

Mullach Rua shook his head. "We saw her go with –"

"She *DIDN'T* go with me!!" Fionn pushed past his friend, racing into Almhuin.

"DONN!! GARBHCRONAN!!" Name after name he shouted, rushing through the halls, pleading for an answer – any answer other than the only they could give. At last, he shouted – louder than any war cry, any lament, or any word was ever spoken, Fionn screamed out one word, and the pain of it still marks the land.

Sabh.

He had lost the Flower of Almhuin.

Seven years.

Many things happened in those years, but Fionn barely felt any of them. His world was empty. His life was merely a haze – a shadow and a mist, vacant and placid and fake. Every victory he won, every drop of blood he shed, was for her – compensation.

He never stopped looking for her. Every waking moment, and every moment of his scant and pleasureless, guilt-sodden sleep, was spent in searching for her. It was no longer Wanderlust that drove him from his bed; it was darker. No soft-calling, sweet luring of a world more real than home. It was driven, and it was driving. It was harsh and painful and controlling, violent as the evil force that had taken her away. Harder did it drive Fionn than ever he had driven his men, his hounds, himself. It was consuming. And he followed.

Compensation. Recompense. Balance in all things.

He quartered that sacred land; the footstep of Fionn Mac Cumhaill has trodden thrice over every blade of grass, every speck of dust, every particle the bards and the brehons and the Tuatha De Dannan call the Isle.

Take the wild boar, take the stag – take even the squirrel or the badger. But never, never, never – on pain of death –

Never take the red deer.

21
Oisín

"Shoot. Fire's run down."

Meygan actually jumped as Rahyoke slid off his perch, stretching to recover the fire.

"What?!" she exclaimed, staring at him with wide eyes. "You *can't* stop *there!!*"

"I'm not, I'm not," Rahyoke muttered, as he worked to stoke the failing fire. "I'm fixing this... It *is* getting late, though; are you –"

"No."

Rahyoke pressed his lips into a thin line, keeping his eyes bowed. A faint blush was in his cheeks; for all of his artistry of performance, his skill, he was shy.

"In the old days," he murmured, more to alleviate his own spirit, "they used to bait story-tellers with ample pay. You could stay seven nights, in the days of the Ard Rí at Teamhair; if you were good enough, you stayed on for life. Even in my parents' day, they at least tempted and paid the poor soul with a good meal, a warm hearth, and a little liquor."

"It was all leading up to that, wasn't it?" Zander drawled, though he could not conceal a small smile.

Having gotten the fire back to a pleasant blaze, Rahyoke sat back, leaning against his former perch. "I don't believe a little whiskey's too much to ask, do you?"

"For a full story, maybe not."

Rahyoke rolled his eyes. "Single-minded, you lot. Okay, fine."

"Did Fionn ever see Sadb again?" Addison asked – hesitantly, though he was too curious to balk over hypocrisy.

"No," Rahyoke replied, turning a glance to him before casting it into the fire. The red and orange flames leapt heartily, drawing afresh the long and distant past.

"... Not quite, anyway..."

So seven years grind past – of agony and hardship, of frantic searching and of throwing himself with desperate abandon into every fresh and deadly encounter. And through it all, Fionn's poor friends, in spite of their urgings, could do nothing but watch as Fionn drove himself deeper into the ground.

It was late in the summer, and Fionn and the closest of his men were hunting on the sides of Beinn Gulbain. There was Conán and Goll and Caílte, and, by this time, they were joined by Diarmuid, and the man known as Lugaid's son. They were hunting more out of necessity than pleasure; they were doing a circuit far from home, and were desperate in need of some sustenance.

Fionn, as he was accustomed to do, took with him no hound save Bran and Sceolan, forsaking the rest of his pack; no other hounds knew the ways of the sidhe. But, though he could order them not to take the red deer, Fionn could and would do little against his men taking along their own dogs; and, though each man of them brought no more than one or two hounds, there was a veritable pack around them.

Long were they walking, and no sight or sound that anything lived in the world save themselves. Fionn trailed along at the back of the group, listening to their conversation – their mumblings and grumblings, their teasings and tauntings. Diarmuid and Lugaid's son were young, and therefore loud, and far more jovial in spirit. They talked and sang, joked and bragged with a heartiness that Fionn could hardly remember having, though he had once been known as the best for it; in another time, in another life. Fionn looked away, spurning his gaze from his men, who – though they understood him – could never *understand*. Lord, he wouldn't want them to.

Suddenly, Fionn's footsteps dragged to a halt. There – utterly randomly – towards the base of the great hill, was a tree. Stripped utterly bare, standing a pearly white-silver even against the murky glare of the sun, save for the glistering, claret haw-berries in its branches.

As Fionn watched in wonder, Bran and Sceolan stole forward, moving calmly, regally, down the swift slope towards the tree. He watched as they purposefully, nigh nonchalantly, sniffed at the bark, circling once, twice, thrice around it. Then, of a sudden, Bran and Sceolan gave tongue, belling their cry to the winds.

Like a flash, the pair took off – and all of the hounds of the others following them.

And Fionn was hot on their heels.

Never had they remembered him to run so fast; hurtling after him themselves, his men wondered that the speed and the peril of the descent wouldn't break his neck, and they called after him, begging him to slow up.

But Fionn did not hear their voices, for they were in another world.

Down the slopes of Beinn Gulbain he chased the pack, and the men of the Fianna following after him. When at last he reached the bottom, Fionn came upon this scene:

The pack of the Fianna were fighting, quarreling as viciously as starved beasts over the marrow of a bone. Bran and Sceolan stood at the center of the mess, biting and tearing at any creature that should come near them. And there, standing behind them, was a boy.

Thin was the lad, with high looks and proud posture, drawn into himself though his gaze on the mess of gnashing teeth and wailing hounds was undaunted. His hair was long and reached almost to his feet, and was pale, a shining silver.

Fionn and the others caught up to the mess, and the little boy turned to him coolly, looking at him with big, dark eyes, dark as the Pool of Delight.

Cursing, Fionn battered through the mess of dogs, breaking up the turbulent fight. The boy's suspicion turned to curiosity, and he stood, watching Fionn with interest.

He struggled but little when Fionn scooped him up in his arms, protesting at first mutely, then with broken, poor-grammared complaints. But Fionn and his men soon calmed the child, giving him a cloak to wear – for he was utterly naked – and fed him sweet, honeyed cakes. And, if that and their kindness weren't winning enough, the outpouring of love and fondling affection from Bran and Sceolan, who whined and licked and pawed and rolled over in the dust for the lad – this the boy found pleasing indeed.

Fionn sat near the boy, keeping about three yards back – never able to bring himself closer, never able to take himself farther away. He sat with his legs crossed, his chin on his hand and his elbow on his knee, watching the boy and delighting in his every move, his every word, his every fleeting smile.

"Caílte, tell me a thing honestly," Fionn murmured, as the thin man settled beside him. He was watching as Diarmuid and Lugaid's son stationed themselves on either side of the lad, separated by only the hounds, conversing with him in a cheerful tone and words he barely understood.

"If I tell you anything, it would be with honesty," Caílte replied. Above all others, he was Fionn's confidant; he trusted no one more.

"Am I crazy?" The words were a simple murmur – not pleading or doubting; a fact of curiosity, and nothing more.

Caílte gave a slight smile. Turning his gaze, he studied the boy, ignoring all that he wanted to believe, stripping the world down to the simplicity of fact. And shorn, did his smile broaden.

"He has your hair," he murmured. He laughed as, from the corner of his eye, he saw Fionn jolt, sitting up a little bit straighter. "By Lugh and Manannan – *nobody* else has that crazy, pure hair!"

Fionn sat up perfectly straight now, turning – for the first time – from the boy, to face Caílte in full (though he still saw the lad from the corner of his eye). "Yeah?" he asked, as if astonished, genuinely intrigued.

Caílte laughed again. "Surely Grian has blessed this day!" And, lowering his voice to an astonished whisper, he intimated: "You have a son!"

Fionn was torn between belief and doubt, but he loved the child with a surety. After much convincing – and sweet-talk especially by the youths of their group, who sure could have put a love-sick bard to shame with their bejeweled tongues singing the praises to the lad of Almhuin and the Fianna-Fionn – they brought the lad back with them. And *och*, such a fuss was never made greater than that which the women of Almhuin made over him!

In time – thanks in large part to the women's insistence, I'm obliged to add – the little lad learned their way of speech. And, one night, as they sat around the fire, this is the story he told them:

For all my life, I had lived in a valley. Small it was, hemmed in with perilous cliffs that blotted out all the horizon; but the valley was filled with wonderful things. Fruit trees grew up there, and little streams ran through. And there was no season, but it was always summer, or at least the fullness of spring. And no one at all there was in that place, no other living thing, save a silvery-golden hind.

But sometimes a Shadow would come to us – a dark, towering shadow, frightening to behold. And when the Shadow came, we were very afraid.

And always when he left, we were happy again.

For many years, we lived in this way. Then, one day, the Shadow came, and I could see nothing, and feel nothing, and hear nothing, but the cries of the hind as she was driven farther and farther away.

When the darkness went away again, and I could move, I was in a very different place. And it was big, and open, and green, and the hounds all shouting around me.

And that was how he had come to the Fianna.

And Fionn knew that it was Fear Doirche, the Dark Druid, who had taken the hind, and that this lad was the child of Sadb.

Fionn called the child many loving things, each term more beautiful than the last; but the name that he gave him to have as his own was Oisín, which means "Little Deer". And for all his loss, Fionn was well rewarded with such a son as Oisín, for he grew up to be a great warrior, and a great poet, and he lived for more than three hundred years, though that mostly in Tír na nÓg. And, when he returned, he gave these stories to Saint Patrick, so it is thanks to him that we have them at all.

But that is another story entirely.

22
When in Rome...

Scattering the ashes of their humble hearth-fire, they set off the following dawn beneath a lavender sky. The farther North they traveled, the thicker the forest became along the riverbank. The grass on the uneven strip of ground between the tree-line and the rocky bank grew greener, longer, and bushes were seen about, gathered in dense, heavy clumps. Mixed in with the large, deciduous trees were occasional willows, their dragging threads dipping fingers into the water.

They were walking in silence, as, having been compelled so frequently to answer Ian's nagging – "Is this poisonous? How 'bout this? This?" – Rahyoke had finally erupted in a bout of cursing that ended with, *"Why don't you try it and find out!?!"* And at that, he'd finally gotten some peace.

"Sssoooo..." began Meygan, breaking the lull, "how long d'you think it'll be before we can cross the river?"

"Hm?" Zander looked back at her. She was walking along the big rocks at the water's edge, agile as a faery. Her inquiring glance, when raised from her footing, flashed between himself and Rahyoke, who wasn't paying any attention.

"Yo!" He elbowed Rahyoke, startling him out of his reverie. "When can we cross?"

"What?" Rahyoke looked to the river. "Couple'a days, I figure. Two or three, maybe." He glanced back to the trespassers. "All depends on you guys."

"If we travel through tonight," Meygan proposed, "do you think we could cross faster?"

"What are you doing!?" hissed Addison, his eyes wide.

Rahyoke gave a sharp laugh. "Presumably; why? Are you after it?"

Meygan shrugged. She leapt across a particularly wide gap, then across another to catch her balance. "I'd just like to get to a destination, that's all," she said.

Rahyoke considered it. He hadn't figured on traveling through the night. They were almost out of food. They had maybe, to his recollection, two apples and one croissant; not to mention he'd caught the only fish in the entire flippin' river. He would have to go into town tonight, and... Auch, but they were very far north...

"We'll see."

Meygan breathed a sigh.

"Will you tell us more stories?" Ian asked – the first words he'd ventured in over an hour.

"Maybe," Rahyoke responded; he sounded weary, and indeed he was.

Ian thought he needed encouragement. "I really liked the one you told last night," he praised. "Now we need one with more action – you know? *Really* get into the battling!"

Zander glanced at Rahyoke in time to catch him rolling his eyes. He decided to supply an answer himself.

"What?" he asked, turning and walking backwards. "Like the *Aeneid? 'To which Pyrrhus: "Therefore you will report these things and you will go as a messenger to my father, the offspring of Pelius. To him my unhappy deeds and degenerate Neoptolemus remember to relate. Now die!" This saying, to the alter itself he dragged him, trembling and slipping in much blood of his son, and he entwined his left hand in his hair, and with his right hand he raised his flashing sword and buried it in his side up to the hilt!'"* He mimed the action on himself dramatically, staggering back, and finished: "'This was the end of the fates of Priam!'"

Meygan made a face. *"What?!"*

"Your translation could use some work," Rahyoke noted lazily. Zander frowned at him.

"Translation?"

"He's makin' it up," Addison rebuked, waving them all off.

"What!" Spreading his arms, Zander recited:

"'Cui Pyrrhus: "Referes ergo haec et nuntius ibis
Pelidae genitor'illi mea tristia facta
degeneremque Neoptolemum narrare memento –"

"Memento?" Addison scoffed.

Zander continued, more pointedly: *"'"Nunc morene!" Hoc dicens altari'ad ipsa trementem traxit!'"*

"Seriously?"

"It was good enough for the emperor of your beloved Rome," Zander challenged. "Book two of the *Aeneid*."

"There's no way you could've memorized all that."

"Hey!" Zander snapped, pointing. "Don't make me *Haec finis tui fatorum* you, you *degenerem!*"

"Whaaaaat?"

"I –"

"Wait." An out-flung arm caught Zander in the back.

"Oof!"

"Quiet!" Rahyoke hissed.

Everyone stood still. They had reached a place where the bushes were very dense, twining with the vines and curtains of the willow trees. All was thick greenery about them, aside from the river to their right.

"What is it?" Meygan whispered.

There was a pause. Then, softly, Rahyoke murmured, "I heard someone..."

For a moment, they all strained, but no one else could hear anything – nothing but the sparkling stream.

Ian rolled his eyes. "I don't –"

Zander seized him, one hand across his mouth, the other arm pinning his arms to his sides. Ian rolled his eyes again, making no effort to break free.

Rahyoke glared at him a moment. Then, creeping forward, he ordered, *"Stay."* He vanished through the thicket.

Meygan got down on her knees, sliding so she could see through the bushes. She gasped.

Sitting on the bank, about ten yards off, was a man. A scarlet cloak was rolled on the rocks beside him, a leather scabbard laid over top. Leaning against the same stone was a round, leather-and-bronze shield and a long, ashen spear. The man was fishing, his rod well-oiled and gleaming in the light. He was bare-chested, a powerful, muscular man, his shins wrapped in a thick cloth to protect his skin from his hot, metal greaves.

The boys, including the released Ian, knelt down beside her. Ian had only one word to offer:

"Romans!"

Rahyoke made his way slowly down the line of trees, making as if he hadn't noticed the man. His eyes were following the canopy, tracing the dripping bands of light.

"Hey!"

Rahyoke turned as the man called out to him.

"Oh," Zander breathed, "*a Dhia Uilechumhachtach...*"

Rahyoke bowed his head in recognition of the man, clasping a fist over his heart. The man called to him again, this time beckoning him over.

Rahyoke approached the man with that casual, languid stride, talking to him the while, though his words were too low to hear. He grinned, more welcomingly, more pleasantly than his usual, mocking half-smile, gesturing at the fishing-pole, and to the reed basket beside the man on the bank.

The man gave a loud, sudden shout of a laugh, his words booming out as he exclaimed, "Ha! 'Creel'! *Maxime. Qui's tu? Esne peregrinus?*"

Rahyoke shook his head. He gestured again at the basket, and, in a lower voice, the man answered him, thumbing a direction behind him, farther down-river. Rahyoke nodded slowly.

They continued to talk for another five minutes. At the end of it, the man pointed to the creel, offering a canteen, and Rahyoke shook his head again, rocking back on his heels. The man offered a hand, they shook, and again Rahyoke clapped a fist to his heart, half bowing. This time, the man returned the gesture.

Rahyoke headed off downstream, disappearing into the forest once more.

Zander, chewing his thumbnail, mumbled around it, "He said wait... he said wait..."

It was another five minutes. Then Zander turned, nearly starting up from his place before he remembered. The sheer, adoring relief on his face was enough to assure the others.

Rahyoke, gesturing for them to remain silent, signaled for them to follow.

They cut into the forest again, trying to keep as quiet as possible, if for no other reason than Rahyoke had told them to. After what seemed like a day and a half, Rahyoke, taking a glance behind him, suddenly burst into laughter.

"Mercy! Were you worried?"

Zander did not take kindly to his teasing. "*Yes*," he scowled, darkly. "Don't you ever, *EVER*, ever do that –"

Rahyoke waved him off. "So are you lot after spendin' a night in a house?"

They were all very surprised at the offer.

"What do you mean?" Meygan asked.

He shrugged. "Just what I said –"

"You're going off *his* word?" Zander sneered, pointing back behind them.

"You go off mine."

"I'm beginning to wonder after the *sanity* of that!" Zander admitted, folding his arms and maintaining his critical stare.

"Canteen full of wine? That langer-bastard will never remember I was there!"

"Language."

"What?" He had to think hard about it. "Oh – my bad. Anyway." He turned back to the trespassers, offering a slight, but winning, smile. "Zander worries like a hen-wife; don't bide by him. How 'bout it? I can get a good place for the four of you a night. On my honour, it'll be good. Wha'd'ya say?"

They considered it.

"We'll have Zander?" Meygan asked.

"Yep."

"And his weapons?" Addison added, pointedly.

"Yes, though you oughtn't need them."

"I'm cool with it, then," Meygan decided, giving an affirmative nod. "I trust you."

Rahyoke frowned dubiously. "Thanks."

"What about you?" Ian asked, but Meygan beat Rahyoke to the answer.

"He's going into town." She turned to him. "Remember to ask about us, 'kay?"

"Always."

Zander sighed. "I still think you're flippin' nuts..."

"That's allowable."

"...but I agree."

Rahyoke grinned – cunningly, mysteriously, saucily, derisively. "Grand." Without waiting, he began walking again, turning parallel to the far-off river they needed to eventually cross. He called one last command behind him as he went.

"And next time I tell ya *whisht*, the Devil upon you ya better!"

23
The Dogs at Your Front Door

It was a modest little cottage – one floor, with a sloped roof which looked strange now that it was shingles and not thatch.

"Are you sure this is a good idea?" Zander murmured, eyeing the place with the utmost distrust.

"I'm always sure." Rahyoke turned back to the children, but to the trespassers especially. "Alright; here're the rules, kids: Take your cues from me. Pay attention. Be alert, but not enough to attract suspicion. No accents but your own."

"But –" Ian began.

"*No.*"

"...Fine."

Zander arched a brow at the wanderer. "Your own?"

"It's not gonna be hard," Rahyoke continued, ignoring him. "You're just being yourselves, but with a few minor alterations. We're probably gonna hafta say you're from someplace around here."

"Like Osset?" Meygan tried.

"Yuck. No."

"Why?" Addison asked, his arms folded. "Why do we have to be from around here?"

Rahyoke studied the sky. "Addison, if a man came t'yer house and presented you with three kids, sayin' – 'Hey! These guys are from another world!' *Tell* me you wouldn't shut the door on him."

Addison bowed his gaze. "Okay, point."

"Are we all clear? Ian?"

"Yep."

"Grand." Rahyoke gave a soft sigh, turning back to the house. "Don't blow this you guys," he murmured. "*Please.*"

"I took a theatre class in fifth grade," Meygan assured him, examining her nails. "Don't even worry about it."

Rahyoke rolled his eyes again. "Let's go."

Stepping from the cover of the forest, they made their way across the short lawn, coming up to the door.

Rahyoke leaned against the doorframe, his forearm against it horizontally at about head-height. He seemed to be thinking, hesitating for some reason, though nothing could be read on his face.

Impatient, Ian raised a hand to knock, and, without looking, Rahyoke snatched him by the wrist.

"I'm perfectly capable, thank you," Rahyoke gritted, his voice low and calm, though the rebuke was plain enough.

Ian held up his palms, backing away.

Rahyoke straightened slightly, pressing only his palm to the doorframe, his arm held straight.

He knocked.

"Alright, you win." Ian rolled his eyes. "I don't need the *look*."

"Maybe you should knock louder?" Meygan suggested, quietly.

Rahyoke held up three fingers. Two. One.

He jolted off the doorframe, clumsily clasping his hands, as the door sprang open.

Standing in the doorway was a man, about a foot shorter than Rahyoke, and very old. He wore a tattered, navy sweater, ripped around the neckline, and faded, brown dress-pants. His hair was white, and thick; his eyes a summer-sky pale blue; and *deep* was the frown upon him.

"What's this, then?" he clipped, looking Rahyoke up and down with obvious distaste.

Rahyoke ran a hand back through his dark hair, which only served to mess it more. He seemed uncharacteristically nervous – uncharacteristically fidgety.

Huh... Meygan narrowed her eyes, watching him. As he had been with the fisherman, Rahyoke was a completely different person. And, if he had an accent, it was only because he was the only one of them who could pull it off.

"Evenin', sar. Cou'jya be givin' me a moment'a yer grand, precious time, sure I'd be terribly obliged."

His speech was lilting – nervously, yet beautifully. He gave the faintest twist of a smile, his head slightly cocked, as if begging, *Please? I'm a nice lad – honest!*

The man in the doorway squinted at him. "For *what?*"

"Obliged indeed, sar!" He clasped his hands again. "An' fer what it is, sar, it's this: that I'd like t'be makin' ya, an offer."

The man arched a snowy brow. He craned his neck to look behind Rahyoke, to the four children scattered on the lawn. "I'm not in bondage." He moved to close the door.

Rahyoke pulled back – his foot in the doorway – looking appalled and astounded. "*Sell*'em!?" he cried out. "Oakh! Mercy – but I've *standards*, sar, so I do – good, moral standards! An' shar, isn't it these me own flesh an' blood, an' them two only *just* after losin' their mother –" a vague flick of an indication " – *ochón*, poor, sweet, sainted Breffni, me own an' only sister, passin' – t'poor doteens – shar how could I –?"

"What *is* it," the man broke in, tersely, "that you *want?*"

"A favor," Rahyoke eluded, suddenly calm, suddenly angling, his voice low and the prize tempting indeed. "An' it's the doin' of it, sar, ye'll be *marr* than well-compensa'ed..." Abruptly, he added, "An' don' be after accusin' me next of it's blood-money! I won't have it!"

"A favor?" The door slowly drifted to its widest extent, though the old man's lithe frame still blocked it.

"*Very* well-compensa'ed, indeed, sar."

The man in the doorway considered it. It was not a common thing, to bring in vagabonds. Especially ones toting a pack of children. Especially ones so *clearly* –

"Who's there?"

A voice sounded from within, and a woman came up behind him, ducking and weaving and standing on her toes to get a look over his barring arm. She was as old as he was, her hair as white, her eyes as clear a blue.

Rahyoke slid back a step, giving something of a very shallow bow. "Evenin', ma'am."

"Oh! And he has four children with him! Let'im in!"

The old man's eyes narrowed further, never moving from Rahyoke. At last, still full of suspicion, he decided.

"Anwen, fix the kettle," he said, slowly. The little woman clapped her hands together and moved off into the kitchen.

Rahyoke's eyes shifted, watching her go with curiosity.

"There's no sense in discussing business out here," the man grumbled, stepping aside. He gestured for them to enter.

Rahyoke stepped aside as well – one foot decidedly if casually still set on the threshold – and, obediently, his pack filed in. As Meygan passed him, he briefly put a hand to her arm, just above her elbow, protectively – a warning that was not *for* her, but on her; he followed her inside with a deterrent glance to the old man.

The man closed the door behind him, then turned to study further what he had allowed into his home.

The vagrant had his gaze slightly uplifted, surveying the room to the finest detail in a sweeping, roving glance. "*Dia anseo isteach,*" he breathed, habitually. His eyes noticed everything; so where some would have seen a dark interior and an open-concept botched by shelving, he noticed the rich tapestries, the wares gleaming upon the metal-and-glass shelves, the electric, brass sconces on burgundy-painted walls and the enormous, brick fireplace gaping to the left, the dark-wooded floorboards imported from the North.

"Come on in here," the man ordered gruffly, moving past him and almost through him. "We can discuss matters in the kitchen."

"You kids make yourselves at home!" the woman called, unseen. "Just have a seat anywhere; I'll be right with you!"

There was a discordant chorus of "Thank you"s, and the kids settled, glancing over and perching on the tasseled couch and the two arm chairs of the living-room. Throwing them a glance, Rahyoke followed the old man into the kitchen.

Although hedged by the shelving, the kitchen was stunning – granite countertops, oak cabinetry, chairs which matched one another, and a wine-rack large enough to hold maybe fifty bottles – and, God Almighty, it was full.

Feeling the old man's gaze, Rahyoke turned, smiling, and offered, "Ah, yer a right well-off man, t'be shur, an' a generous soul along with it, God prosper ya." He had spent enough time in a marketplace to know what he was dealing with.

The old man's expression remained disapproving. "We don't speak of religion in this house." He gestured for Rahyoke to sit, at the opposite end of the table.

"Ah? Ah – oh, sorry." Rahyoke shook his head, self-scoldingly, a hand to the back of his assigned chair. "Mercy – old habits. I'll try t'watch m'words, so I will." He took his place at the old man's table.

The merchant doubted the vagrant's ability to control his tongue, but he wouldn't say so. "Is the kettle on?" he asked his wife instead.

"It is," she answered. She turned from the stove, which was only about a foot or so from the magnificent wine-rack, looking again to their strange guest. "Do you take tea? Coffee?"

"What's aisiest –"

"You'll have tea," she decided, turning and standing on her toes to snare mugs from the cupboard.

Rahyoke smirked faintly, bowing his head. "Obliged."

"We also have wine," she added, dropping tea-bags into the mugs. Even their mugs were high-quality. "A nice burgundy from Lavinum's Martius Valley –"

"*Anwen!*" the man scolded – a harsh, grating breath.

The woman – evidently Anwen – cast him a glance over her shoulder. "*What?*" she defended. "The Rosea Poma is for sale, I know that. But Martius Valley?"

"Anwen!"

She rolled her eyes. "Fine, whatever."

Rahyoke flashed an unseen glance to the man. "Ah!" he half-lamented, diffusing the situation. "I amn't really a drinkin' man, anyway; so sweet an' refined a thing'ed be lost on me!"

"Alright," sighed Anwen, turning off the kettle (electric stove). "*Tea*."

As she poured the steaming water, her husband returned to business.

"So all those kids in there are yours?" he asked.

"In a way an' not. T'blond lads're uv me late sister, Breffini – God rest'er, if yu'll allow. T'eir father – a plague on'im! – hadn't t'mind t'be bothered with'em. S'that – an' me own wife dead these last twelve years – I'm after travelin' wit' ta'four uv'em. *Dech do shinaib ceo*."

"You haven't a house?" the man demanded, roughly.

"I run messages f'r a livin', most – did even before I met Roisín. I live deep north'a here, an' I amn't keen on leavin' t'four'uvem – ya know? Ta' cities c'n be..." He looked uncomfortable and darkly protective as he searched for an adjective.

Anwen jumped in. "Oh, you're right!" She set down two mugs on the table, steaming hot. "And especially how *young* they are! How old is your daughter?"

"Thar'teen," he guessed. "Thanks fer th'tea."

Anwen returned to the counter. "Do your kids take anything?"

"Ehhmmm..."

"I have hot chocolate."

"Yes."

"Okay, then." Anwen was on her toes again, fetching more mugs.

The old man again reigned the conversation back to the point. "What is it you need?"

"Ah." The vagabond turned back to him. "I'd been sent south t'run a message – grand, fine – but passin' t'rough Syene I got one f'r a man over t'river. Now – fine anuff f'r *me*, shar, but four kids an' no ferry? M'daughter's a weak swimmer, always has been. An' bad faith, t'be bringin'em across here anyway, what wit' ta'current, an' – *ochón*. I'm after a place t'hold'em," he added quickly, catching the old man's glare. "An' a man I met upriver, he said t'me, 'There's a grand lil' cottage downstream, where shar they can help a man out!' An'! – t'be *shar*, I'll pay ya *grandly* f'r th'sarvice," he swore, faithfully. "Tha' I can promise, truly."

"Well of course we could!" exclaimed Anwen, balancing a tray against her hip with the trained ease of many years waitressing.

"Anwen!"

This time the stranger didn't break the tension. He took a long drink of his tea, staying well out of it.

Anwen rolled her eyes. "Fine, then. *You* settle the *business* end of it." She moved around the table, choosing the stranger's side rather than her husband's, even though it was a longer route. "But I still say the *decent* thing to do is to help the poor man out!"

"Anwen."

"Never have I heard my name so much," she muttered, exiting the kitchen.

This tea is too hot to be drinking, Rahyoke thought absently, draining half the mug before setting it down again. But *mercy*, it was good tea!! "M'full pay fer th'river-job," he said, abruptly. He turned back to the old man. "An' shar but it's not cheap; I c'n guarantee ya *gold*, sar. In exchange f'r lookin' after me kids." He took another drink of his tea, watching the old man over the rim of his cup.

"How long?" the man asked at last.

Rahyoke lowered the mug long enough to answer, casually, "I'd be back by t'marra – afternoon at warst. An' if I'm not –" He shrugged one shoulder. "Tarn'em out. They'll find me, shar anuff." He took a long drink, allowing the old man about five seconds to contemplate the matter. "Sa' wha'd'ya say?" he asked, leaning forward on his forearms onto the table, the mug cupped between his hands. "Deal?"

The old man watched him wryly, still uncertain about this strange creature from the North. "We'll have to arrange the particulars," he said at last. "But it seems fair enough."

"Grand." Rahyoke grinned, and there was something subtly cunning in that smile. "Y'seem like a keen sart'a businessman," he gave, studying the old man shrewdly. "Let's begin."

The old man was a businessman indeed; it took twenty minutes to sort out all the particulars.

Zander and the trespassers sat in the living-room, trying not to bother anyone, and trying, if they could, to overhear. Anwen had gone back into the kitchen, leaving Zander free to relay – at the deadliest-soft whisper – anything he managed to overhear, and instructions.

At last, the two men re-entered the room, Anwen pausing to lean in the doorway. Zander stood instinctually, and Rahyoke cast him a glance.

"You'll remember our contract?" the old man asked, recapturing his attention.

"'Tis me own flesh an' blood I'm givin' ya, sar," he vowed. Pausing, he glanced back at his kids. "Speakin'a which –" He pointed first to Ian, then Addison, saying, "That's Iulius, and Adrastus. An' they're Meg, an' Setánta, m'own two."

"Mostly I go by An Damhán Alla," Zander – Setánta – added.

Rahyoke rolled his eyes. "He's all talk, tha' one; he's a good lad, I swear." He swept to the door, following behind the old man. "Again – 'tis grand merciful'uv ya t'be doin' this. I appree'see'ate it highly – an'jyer hospitality."

The old man opened the door, and the wanderer stepped out half-way, casting back to his children and their cousins, "Lads, be good as I'm gone. I'll be back t'marra –" He turned to the man. "Assured – t'marra."

"We're always good," Meygan replied.

The wanderer slipped out, backing onto the lawn. The old man stood again in the doorway.

"Before you go..."

The vagrant paused, looking back with innocent curiosity.

"I never got your *own* name. I'll need to know with whom I have entered into business."

"Ah!" He grinned. "Near slipped me mind, it did!" Then – pausing to consider – his expression became exceedingly cocky. "I'll tell yeh mine if you tell me yars."

The old merchant glowered at him, with the greatest, gruffest old-man-pout ever bourn; but, after a length, he answered, begrudgingly, "Connor."

A good name. "Mongan." The wanderer offered his hand.

The two men shook hands in a cold, business-like fashion, as often a haggled merchant will. Perhaps that was the reason the wanderer gave the old man such an annoyingly pleasant smile as he thanked him again before turning away.

He could've almost laughed at how forcefully the door snapped shut behind him.

24
Honesty and Decency

Your man Mongan came back even earlier than he'd figured, coming up to the door of the cottage at around ten-thirty the next morning, when a fine, misty rain was seeping through the air of the forest. Unshaven, full-drenched, his sneakers caked with mud – but he came, as he'd promised, bearing gold. As his pack of children collected their belongings and straightened the blankets and pillows on the armchairs and couch, he pulled from his coat a small, drawstring sack – large enough to hold a good-sized, ripe apple – and dropped it heavily into Connor's palm. Connor looked into the bag, and very nearly smiled.

Mongan corralled his kids outside with an idle warning that it was raining, thanking the old couple again for their kindness and generous hospitality. Anwen was a little sorry to be without guests so soon, but there was something about the wanderer that seemed, *off*; for that she was grateful to be rid of them.

Zander waited until they were back at the river – and the whole of the forest clean and clear around them – before breathing a sigh of relief. "Wow."

"Yeah, 'wow'," Rahyoke agreed. He dropped down on one of the great stones of the river's bank, not caring that it was wet. "I'll hand it to ya, lads, you didn't blow it; well done."

Meygan bowed, accepting credit.

"I thought we were gonna travel by night," Addison said, brushing the mist out of his hair. "What was the point of stopping?"

Rahyoke was distracted. "Shorn; I think I bruised my shin on a smelter... Sure, they say it's good to indulge in society every now and again; builds character."

"Character or *characters?*"

Then Ian stood before him, his arms folded, giving him a parentally scolding look. "*You* said no accents," he accused.

"Did I? Zander, does that sound like something I'd say?"

"I don't know which version of you we're examining."

Rahyoke gave that a silent appraisal; then, suddenly annoyed, he pointed at the archer. "Oh! And you!"

"Me?"

"*An Damhán Alla,*" he sneered. "I could *KILL* you!"

Zander grinned, spreading his palms. "*What?*"

"What does that mean?" Meygan asked.

Rahyoke shook his head. "It's a... spider..." he waved her off, too peeved to address it further.

"I thought it was fitting," Zander defended, jutting out his chin.

"Yeah – in the *future* let's just stick with Setánta, *alright?*"

Zander laughed. "Alright."

"So, did you really have a sister?" Meygan asked, and it was to Rahyoke she directed her inquiry.

Rahyoke shrugged, as if the question had not caught him slightly off-guard. "Beats me." He rubbed his shin, checking his jeans for blood. "I had to garnish a little. You're – well, your name really *is* Meg, isn't it. But I build convincing characters, do I not?"

"You lie well."

"Ohch, *'lie'* is so... I embellish."

"So, what's your name?"

He grinned – a snide, superior sort of grin. "What?"

Meygan folded her arms, cocking a hip. "No embellishment. You told the old guy your name was 'Mongan'. I want the truth."

"I use a lot of names," Rahyoke answered, shrugging. "Just as often I'll go by Diomed or Cadfael, Lochlan, Seamus, or Branigan ...Or Roland – but I'd already given my wife's name as Roisín."

"Aaw!" Meygan cooed. "Roland and Roisín!"

"That's why;" he rolled his eyes.

"*Rahyoke* and Roisín sounds more sing-y."

"That's *especially* why."

"But is your name really Rahyoke? Is it *just* Rahyoke – do you not have a last name?" Like the old merchant, she wanted to know exactly with whom she had entered into business; names were everything in this sphere. "Come on – I'll use my second question, if you want: What's your real name?"

Rahyoke sighed, lifting his gaze away to the trees beyond them, as if considering. "My name," he said, slowly, "is of no importance... But it is as I told you."

"Your full name?"

"My full name."

"How come you don't have a last name?" Ian asked.

"How." Rahyoke again rolled his eyes.

"Come on," Ian pressed; "count that from my questions, too. Why don't you have a last name?"

"Aaauughh..." He didn't like this sudden attention. *They say Mongan was actually Fionn, having discarded all he was and begun his life anew... Then he met Brótiarna, and didn't she fairly nag the bleedin' story right out of him –* "I... I don't need one," he explained. "Ah – a name, when it comes down to it, is nothin' but a word by which you're recognized, and anyone I'd want 'ur need to recognize me does. Anything else I'd need it for was erased long ago."

"Uh-huh," Ian noted, slowly; even he was unsatisfied. "Does Zander have a last name?"

Rahyoke shrugged a shoulder. "I don't know, does he?"

"Hmmm... Can I still be Iulius?"

"Y'like that? ...Shorn; it's better than *An Damhán Alla*."

"I wanna be called that from now on!" Zander exclaimed, raising a fist and a pointing finger up to the sky.

"Suit yourself." There was a sigh in that tone, very clearly.

A moment of silence passed. The rain was still a mist about them, flecked with bigger drops, and all the world that morning was grey – grey and damp and hot.

"Did you bring any breakfast?" Addison asked.

"Shorn," Rahyoke half-smirked; there was a flash of nervousness to that smile, a little dread. "Didn't they feed you?"

Ian rolled his eyes. "Not even continental!"

Zander snorted, crossing his arms. "Why would they?"

And Rahyoke's expression darkened. *"What?"* He stood, towering over them in his rage. Zander caught hold of his arm to stay him, and he jerked out of the archer's grasp. "Wha'd'ya think I *paid* him for?!" Turning his back on them, he paced off downstream, running his hands through his hair, clenching the dark, drenched locks in his fists. "Room an' board! *A Dhia Uilechumhachtach...*" He broke into a long stream of muttering in that strange language, and it was clear from the tone that he was cursing.

"Was it your full pay?" Zander asked.

Rahyoke turned back. "I'm not an idiot!" he snapped. "I gave them decent pay for room an' board!"

Zander snickered. "Did you expect them to be honest? My innocent heart, how do you suppose a man in a run-down cottage gets tapestries like that insulating his walls?"

"That's not what I'm *saying!*" Rahyoke urged, exasperated. "I mean – how d'ya not even offer – not even scraps –"

"Even *you* fed us," Meygan helped.

"That's right!"

"Don't *encourage* him," Zander frowned at her.

She continued. "Don't feel bad, Rahyoke. You're a better person than he is; that's what really matters. It all comes back around."

Rahyoke pursed his lips, kicking at a tuft of grass. "They *say* that..." They say a lot of nice-sounding things. Maybe in the second or third life... He sighed, casting off his anger. "Ah, well. There's your lesson for the day: I should've accepted the Rosea Poma."

"What's that?"

"*Very* expensive wine." He started digging through his bag. "I should have some food in here," he promised. "I'm really sorry about that, guys."

"It's okay," said Meygan. She walked forward, clapping a hand to his arm as he had to hers, in passing. "I liked being *dry*."

"There was that," he admitted. He turned after her, following with his gaze as she continued downstream. "I'll hafta go into town tonight."

"Again?" Ian rose from the rock he'd been sitting on, and, once again, they walked.

"Well, *someone* has'ta provide for you."

"We could sell The Spider," Addison offered.

"I amn't in bondage!" Rahyoke again protested – though this time he sounded amused.

"I wouldn't give him any ideas," Zander reinforced, his warning coupled with a deliciously venomous smile.

"Heaven help us." With a long sigh, Rahyoke folded his hands behind his head. *Sin an saol.* "I figure it's another day or two 'til the shallower part of the river," he informed them.

"*Good!*" Zander declared. "And hopefully we won't run into anyone on the way! I don't want any more trouble, and I mean it!!"

And Rahyoke actually laughed.

25
Waiting for the Salmon

"How do you *DO* this!?"

Zander had been standing in the water for hours, his bow in his hands, his sword at his side, his boots and shirt lying with his bag on dry land.

"Keep waiting," advised Rahyoke; "it'll come."

Weapons gripped in his left hand, Zander flopped his arms to his sides. "No it *won't*," he groaned, in complaint. "You caught the only fish in the river. *You* do it!"

"No way." Rahyoke shook his head. "I've my own job; you do yours."

Zander moaned again, as if the whole world were verging on end. "What sort of an example are you? If you got me some decent *line* –"

"Don't blame the line," Rahyoke responded, "any self-respecting fisherman knows that."

"I'm a *perfectly good fisherman!!!*"

"Prove it."

"I will!" Zander drew a shaft against the current. Defiantly, he sang:

"Oh I wish that I had me a sturdy fish-pole
'En I'd fish with a skill in this watering hole
An' shur as there'd be a great fish in the sea,
Surely no fisher is greater than me!"

"Ah!" noted Rahyoke, calling across the way. "Sure, would that y'had yer mother's voice! Even the crows are crying!"

Zander's eyes slid to Rahyoke. His response was to sing louder.

"The pole that I have couldn't carry a tune
But it shines with a beam like the light of the moon
So scoff as ye will while ye sit by the fire
But ye won't get any fish, 'cos it's mine!"

Unfazed, Rahyoke turned back to the scattered pile of twigs that was their campfire. Striking a match, he answered, "Yer scarin' all the fish away, I'm tellin' ya."

Zander scanned the water. There were no fish.

"My spirit is free and the water is wide
Roaming this world 'til the day that I die
So cast all yer nets and yer spears to the bay
And so long, my dear biddy, I'm away!"

"Yeah, far and away with ya!" Rahyoke replied, taunting.

Zander turned, placing his hands on his hips nobly, and defiantly. "How am I to catch anything if you're distracting me? Quit distractin' me!"

"Distractin' you," the wanderer scoffed, shaking his head. "Because you've caught *so* many fish before this!"

Zander clasped his hands. "Yeeeess..." he resigned, reluctantly. He bowed his head, examining the water with annoyed shame. Rahyoke had made it look so easy! If only he could...

"Scaoil!!"

A shaft struck the water sharply, causing Rahyoke to look up. "You got it!?"

Zander dove forward like a seal, capturing his prize before it could float any farther downstream.

Rahyoke cocked his head. The boy rose like the dead from the water, shaking his hair like a dog, the mangled fish clenched in one hand, the one-hundred-and-sixty-two-year-old bow sopping in the other.

"Smooth!"

Zander turned a disgruntled look to the shore, pushing his hair out of his eyes with the back of his fish-gutty hand.

Rahyoke half held out a palm. "Toss't up here, if ya would," he bid, with disinterest. "I'll clean it for you." He returned to the fire he was building.

Zander wiped his face on his wet sleeve, climbing to his feet. "I'd like to learn to do that myself," he said. With an underhand swing, he flipped the fish onto the shore, a yard and a half short of Rahyoke's reach. The man gave a look, and Zander shrugged.

"I could teach you." Not willing to get up, Rahyoke stretched forward, flicking the fish nearer with his finger-tips.

"Maybe in a bit," Zander replied, drawing another arrow from his quiver. Between himself and Rahyoke, his arrow supply was running quite low – not from the shots they'd taken, but from the diving sort of retrieval. It had occurred to neither of them just to leave the quiver on the bank. "Right now, I'm fishing."

Rahyoke said nothing, laying the fish aside until he could get a proper fire going.

Once more scanning the waves, Zander held the next arrow ready, this time taking a fighting stance, prepared to pounce. Singing had seemed to help.

"Chuaigh mé 'un aonaigh 's dhíol mé mo bhó
Ar chúig phunta airgid 's ar ghiní bhuí óir
Má ólaim an t-airgead is má bhronnaim an t-ór
Ó caide sin don té sin nach mbaineann sin dó?"

Rahyoke shook his head. Someone had to remember the old songs, aye. But that didn't mean they had to take the risk of singing them.

"So what are we doing tonight?"

Rahyoke looked up. Meygan was a few yards down the bank, picking blueberries from a bush nestled between two great rocks. The leaves of the bush were sparse and sickly, but the full, indigo berries were juicy and large.

"What do you *want* us to do?" he asked in return.

"You could tell us more stories," Ian offered. He was lying on the grass, his legs against a tree, so he was halfway upside-down.

"Oh, *could* I? I'd been plannin' on maybe getting some sleep, seeing as I had to *work* all last night..."

"You could do both!"

Rahyoke shook his head. He found himself half considering working tonight, just to find some adult company.

"What do *you* think, Addison?" Ian asked, announcing the other boy's return.

"About what?" your man asked, approaching and dropping an armload of kindling beside Rahyoke.

"What should we do?"

"About *what?*" Addison repeated, taking a seat.

"Maybe you could ask him more questions," Zander suggested.

Rahyoke's eyes narrowed. "I hate you."

"Yeah!" Ian brightly exclaimed. He flipped off of the tree, rolling onto his stomach, so he could see them better. "Go on, Addison! Ask him something!"

"Not me! I only have one question left. I wanna save it for something *good.*"

Rahyoke rolled his eyes openly. How could an attempt at gaining peace backfire so horribly?

"Ask something good *now!*" Ian whined.

"No! Why don't *you* ask something? You have *two* questions left, don't you?"

"I can't think of anything! Give me an idea."

"No – he'd know."

Rahyoke lost his patience. "I wonder how!" He glared at them both. "Couldn't you take this somewhere else?!"

Addison narrowed his eyes, pursing his lips in an exaggerated expression of close speculation. "I think I'll ask my question *now,* actually," he decided.

Rahyoke scowled, snagging the fish and his knife. Fire just wasn't doing it for him at the moment. "Fine," he grated. "As long as I can keep working."

"Give me a minute," Addison ordered, now exaggeratedly contemplative, a hand placed as if stroking his beard. "I'm working out the *wording...*"

His life was again spared by Meygan. She placed next to Rahyoke a pouch she had made of a tinfoil wrapper. "Are these blueberries?"

He glanced over her find. "Yes."

"They're for dinner."

"*Thank* you."

They were silent as he cleaned the fish, scaling and gutting it as he had done the day before. It was a trout, a little smaller than the one he had caught, but meatier. Having laid aside the head, fins, and bones, he stuffed the fillet with blueberries.

Zander waded over to the shore, folding his arms on top of one of the largest rocks. "Aw, sick!" he praised. "Do we have chives?"

"I *wish* we had chives," Rahyoke said, longingly. "You know what's good? Deep-fried rush-leeks. With cheese. And maybe bacon..." For a moment he was distracted, then he shook his head, returning. "That's all North."

"Bummer." Zander sulked for a moment, dropping his chin on his folded arms. He watched as Rahyoke unfolded Meygan's tinfoil basket, wrapping it around the fish. Using the knife as a spade, he trenched out a line in the warm ashes, very close to the fire, laying the tinfoil package within and raking the warm embers overtop.

"Will it cook like that?" Addison asked.

"Is that your *question?*"

"No." Addison crouched down near the wanderer – just out of convenient reach. "Are you ready to hear it? Are you ready for the *sheer awesomeness*, that *is*, my question? Are you all prepared!"

"In ainm Dé!" Rahyoke muttered, shaking his head.

"It is! What –" pause for dramatic effect, "—is the meaning of life?"

The other kids burst into laughter. Rahyoke looked up, absolutely appalled. *"What!?"* he complained.

Addison sat looking at him, straight-backed, arms folded, sassy and smart-aleky as anything.

Rahyoke scowled. "I'm not answering that! It's stupid, and can't be done!"

"I thought you said you could do *anything*," Addison returned.

"The answer is entirely subjective!" Rahyoke tried.

"Explain to me how, then; that's part of the answer."

For a moment, Rahyoke was simply riled. His demeanor was of the usual, dark intensity, but his eyes were flashing with agitation. At last, he muttered, "Damn you," and shifted, taking a higher position on the rock behind him.

He bowed his head, his elbows on his knees, his palms against his forehead, fingers entwined into his hair. *Stupid brat...*

"Leave him a minute," Zander advised, clambering out of the water and onto the bank. He had seen Addison about to interrupt. "It's not a *small* thing you ask, by any means, and he's not after steering you wrong."

"Can he answer it?" Addison asked, surprised and curious.

Zander arched his brows, concealing, his expression reading only, *Why ask me?* Sometimes he was a great deal like Rahyoke.

At last, the wanderer looked up. He laced his fingers together, holding his hands so they covered his mouth. But his eyes were raised to meet Addison, his gaze harsh, piercing, somber. "Damn you, Addison," he said quietly. And then, in that same tone, he explained:

"Wandering. Wandering is what we are here for, wandering is the meaning of life.

"Hold up now, let me finish; it is not *just* because I'm a wayfarer I say this, 'kay? Its specificity is subjective; everyone has their own path and purpose, interweaving, connecting – never seeing the greater Design, but always traveling farther. But the purpose of *life itself*, in *broad* terms –

"We are meant to bide our days in meaningless perambulation, wondering – for wondering is merely the wandering of thought – wondering at what we see, and thereby gaining some semblance of – recognition towards what has been created – the complexity and sorrow, beauty and convolution – and wondering at the meaning of it all. For what is the point in weaving such an intricate pattern if there is no one to marvel at its beauty?

"That's right; the point of life is to wonder at the point of life. And every God-damn time ya think y'have it figured out, you're foiled at every turn. The game shifts, the road breaks, and you founder for a direction, unanswered soundings reverberating desperate, snapping lines. We were made to wander. And that's *it*.

"That's yer higher calling for ya – and ye'll drive yerself *mad* to be thinking about it. Most mortal beings know, and settle: The purpose of life is to live. Done."

There was the fullness of his answer. And, having given it, he stretched out a hand, slipping it between the ashes to check the fish.

Addison frowned. "That's it?" Now when he folded his arms, it was crossly. "That doesn't really answer me, Rahyoke, I'm not gonna lie."

Rahyoke wiped the ashes on his jeans, his head still bowed. *"Sin an saol, a mhac,"* he replied. Looking up, he demanded, "D'you want a damn story?

"Once there was a man who wished to know everything. For years and years, he struggled to find the answers, but all he found was frustration, for years pass painfully slow. One day, yer man was offered a wish. Fed up with those years of profitless turmoil, he wished for infinite knowledge." Mid-stream, he glanced to Zander, noting, "That fish'll be ready in about ten minutes, 'kay?"

"'Kay..."

Returning to Addison: "To his dismay, above all else, he came to realize one crushing certainty: the joy is not in blindly knowing. It's in following the twisted paths of unknown wonder that is finding out, and emerging at the end of it victorious, in whatever degree. It's in wandering."

He stood; his body desperately wanted rest, but, suddenly, he wanted nothing more than to be any place but here.

"Y'want another one? There is no joy in *knowing*. In learning, yes – not knowing. The Design ain't all that pretty. And all it gets you is frustrated with the grand tangle of threads and all those feckless fools entangling them." He lifted his bag, slinging it onto his shoulder. "Some – like you – think the goal is to find and see the great Design. Youthful ignorance. The Design is maddening. The true Truth is live to exist, live to live, live for life – and *move on*." And, so saying this, he did, moving around their gathering and walking off.

Addison looked to him, frustrated and pleading. "But what does all that *mean?*"

"It means stop asking so many bloody questions!"

"Where're you going?" Meygan asked.

"To get some *chives!*" he shouted back scathingly, vanishing into the forest.

Zander shook his head.

"Could *you* make any sense of that?" Addison queried.

Zander did not look up. Reaching out, he quickly pulled the tinfoil from the hot ashes, taking up the knife and rinsing it in the river. Softly, and by way of any sort of an answer, he sang:

"I went to the market and traded my cow
For five pounds in money and a gold guinea coin
If I drank all the money and my gold I did share
Since it's no one's concern, then so no one should care

If I go to the green woods wild berries to store
Plucking apples from bowers or herding the cows
If I relax in the shade for an hour or more
Since it's no one's concern, then so no one should care

Ó caide sin don té sin nach mbaineann sin dó?"

26
A Thief, a Liar, and a Coward

Rahyoke did not come back that night, nor all the next day. He had left the food with Zander, and it was a good thing, because after four hours trying, Zander could not catch a single fish. The trespassers questioned him, worrying over what they should do. Lying on a long rock of the bank, watching the clouds, Zander merely laughed at them.

"He'll come back," he assured them, sardonically but certainly. "He always does."

Now, as the sun was setting, Zander cast off his boots once more, laying them with his bag away from the rapids. "I'm gonna get apples," he announced, shaking out the drawstring sack into his canvas bag.

"Why don't you use the other bag?" Addison asked. "It's bigger."

"There's stuff in there too big for the smaller bag, and I don't want it to get wet," Zander replied, securing the drawstring around him. "Like my bow. I'd never hear the end of it if I ruined that."

"Alright," Addison gave, with a shrug. "Anything you'd like me to do meanwhile?"

"Attend to your chores!" So many years of hearing it, and now he finally got to use the phrase himself! Jovially, he called, "I'll be back in a bit!" and jumped into the river, swimming across its rolling current.

Addison watched him for a moment, dour with envy and self-degradation. Then, turning, he made for the woods.

"Where ya goin'?" Ian asked.

"I'm gonna get fire-wood," Addison answered. "You never know when we'll need some, and you never know when it's going to rain."

"I thought Meygan was doing that?"

"She takes forever," Addison disparaged – a fine cover-up. "You wanna come?"

"Naw, you go ahead. I'll stay here in case anyone's lookin' for ya."

Addison frowned. "Alright." And he left.

Ian sat on the green, leaning back against a willow tree. The long vines of its branches swayed softly in the breeze, dragging through the water at its very edge. The day had been dry, a vibrant, cheerful blue running shots through the gull-wing sky. For the first time since they had started on this journey, he was alone.

To be honest, he had had his share of resting. He couldn't bear the anticipation. He just wanted to get to their destination, and have it over with – move on to bigger and better things. The whole world was reported to be beyond this forest, and he was ready to find his place in it. Maybe he could become a merchant like that old guy – destined to rip off poor travelers as they struggled to make their own way.

He did alright for himself, that old guy did. Kind of so much so that Ian was almost inclined to wonder if the towns were really as bad as Rahyoke made them out to be. Perhaps it was bad for a scraggly, sketchy, wandering –

Ian trusted their wandering leader, of course – but on a level. Even a fool could see that Rahyoke was concealing and enigmatic, dishonest when he had to be – or *could* be. It was clear there was a great deal he was not telling them. But was he holding back information they needed to know?

A light wind stirred over the river, rattling the leaves, and a sudden motion caught his eye. The breeze had knocked Zander's bag over, jostling it onto the grass. Through the loose drawstring Ian could just see its contents: a flash of scarlet cloth, and a papery sort of roll.

At once, a flood of emotions poured over him. Anxiety and temptation, guile, adventure, warning, and sheer audacity in its purest form. He squirmed beneath their solid weight, as they fought their conflicted war over the surface of his heart.

Curiosity. That was the word for it.

An emotion not to be refused.

Ian cast a glance around him. All the world was peace. Stealthily, he crept forward, relinquishing the shielding umbrella of the tree. Like a spider, he swept silently over the green, 'til his hand slipped over the canvas-bag. He cast a glance across the river; what he could see of it was blank. He opened the bag, dragging out its contents.

The deep burgundy scarf which Zander had been wearing had become part of a packet, bound in leather, which Ian untied, allowing it to fall open. Cautiously, he began sorting through its contents, unrolling a large scroll.

"And the son of Mirtás was Zohák, also called Bíwurasp.

One day, so they relate, a very elegant and virtuous-seeming man appeared before Zohák. And he spoke to him, weaving graceful and sweet-sounding discourse. So enthralled was the youth by this manner of speaking, it pained him when the man had made an end, and he begged him to discourse again.

The man replied that he was master of even more magnificent speech; but to hear this, the youth must first bind himself to whatever he was bid.

This oath Zohák willingly swore – innocent heart! For he did not know that the man who bound him was Iblís in disguise.

The oath being sealed, quoth Iblís: 'Thy father is old and feeble; worthless. But thou art young, valiant, and wise. Why to him, beneath his rule, pledge thy strength, when it is thee deserving well the throne? Let not thy father bar thy way; kill him.'

Hearing this, Zohák was deeply distressed, and protested. But:

'If not fulfilled your oath will be
Thine own head shall I take from thee.'

Thus beneath the crushing terror and agony of his heart, Zohák submitted. Through the wicked design of Iblís, kind-hearted Mirtás was slewn. In this way Zohák, called Bíwurasp, usurped and ascended the throne.

Iblís approached the new king, and spoke to him again.

'Thy kingdom is but slight, O lord,
And as yet incomplete
But heed me, and thy power great
Shalt envelope land and sea
And throughout every climbing age
No king shall match your power's display.'

Every form of flattery and praise poured forth from the lips of Iblís, and, raining down upon Zohák's heart, bound him more dearly in the king's favor.

Such was the extent of the demon's control, that he exerted sole and supreme authority in Zohák's kitchen, and only he prepared the young king's food. In those days, fruit and bread were the only food; but Iblís prepared grand and enticing, elaborate dishes, the like of which had never been seen, giving Zohák meat and fishes, birds and eggs.

Enamored and entranced by the delicious and savory delicacies, one evening Zohák exclaimed, 'For this, whatever I can give that thou desirest is thine!' And Iblís was glad.

'Slight,' quoth he, 'is my desire
A mere whim, though I fear to ask:
To place a kiss on your bare shoulder
Is sweeter than any gift or task.'

And Zohák, unsuspicious still, willingly complied.

No sooner had the flattering lips of Iblís pressed the unclad flesh of Zohák's shoulder than – lo! – he vanished. And from the touch there sprang forth the cold bodies of twin black serpents, writhing and hissing, bound – may God have mercy! – to the young king's skin.

For thus was his destiny.

Zohák fell into wasting despair. Then a final time Iblís appeared to him. It was he who told Zohák that the serpents might yet be killed, if fed daily on the brains of humans. Hearing with hope, Zohák believed him, and rejoiced.

This was done.

Throughout the land spread great terror. Nobles gave up their kings, kings gave up their thrones, turning all over to Zohák, in dread of his terrible power. And what he could not gain by fear, he conquered, winning a vast and insurmountable empire in this way.

And this continued for years.

It came to pass that Zohák called together his nobles and wise-men, and to these he said: 'So vast has the empire become, that it is proper that a register be kept of all the people.' This was done.

At that time there was living a blacksmith called Kavah, valiant and strong. And when the lot fell to two of his sons to be sacrificed to the brain-devouring serpents, enraged, he appeared before the king.

'Wherefore,' quoth he, 'dost thou cast upon me such cruel misery? If thy own shape holds the form of twin slavering dragons, surely the blame is not mine! Thou art unjust indeed, to satisfy thy sin with the unstained blood of mine innocent children!'

Zohák was much astounded at this speech, and, hardly knowing what he did, he released the two boys unto their father.

Kavah greatly rejoiced, and embraced his sons. But as he saw his own name being inscribed upon the register, wrathful, he turned unto the chiefs.

'Art thou men, or what, that stand here, leagued unto a Demon!' And he tore the register of blood, and departed with his sons.

When Kavah had gone, the nobles turned to Zohák, and demanded, 'How canst thou tolerate so far?'

'I know not what hath o'ercame me,' he replied. 'Nor what of this will come; for none can pierce the veil of destiny.'

As for Kavah, he rallied the people, and set off to join with Feridún; for long had it been prophesized this youth would vanquish Zohák.

Until this time, Feridún had lived in the wild climbs of the mountains, with a dervish, who taught him every art and wisdom. For at the time of his birth, a dream –"

"Find what you were looking for?"

Ian jolted. Before he could even raise his gaze, he had been seized full by his arm, just below the shoulder, and hauled to his feet by a hand as far from gentle as a cudgel. *Black sneakers...*

Brutally, his assailant shoved him, back and away from the pack and its spilled contents, forcing him to look up.

Ian tried to explain. "Okay – the thing you need to understand, is –"

"Don't insult me," Rahyoke spat. "I'm not interested in what you think I want to hear." The look in his eyes was fearsomely dark, deadly livid as black fire. He advanced a pace forward, and Ian retreated one back. "You never learn, do you? You just keep messin' with stuff that doesn't belong to you. Well, I'm sick of it; I've lost enough by you already!" His voice was not raised – not even slightly. And it was that depth, that softness, from which it dragged and snarled that made it terrifying indeed. "I'll not tolerate farther."

Suddenly, Ian didn't take comfort in being alone.

"C'mon!" Lashing out, Rahyoke seized his arm again, this time at the wrist, whipping him closer and driving him farther downstream. "Get walking!" he snapped.

"What's going on?" asked Addison, returned; he sounded curious, not concerned. Meygan was standing by his side, watching the proceeding with very big eyes.

"*This*," sneered Rahyoke, pointing at Ian, "is *your* charge. From now on, he does not leave your *sight*. Understand?"

"Yeah, no problem..."

"Keep him in line! Where's Zander?"

"Here!" he called, as Addison pointed upriver.

Rahyoke caught Ian again as he moved to slink away, grabbing him by the back of the neck. "We're traveling by night, and we're not stopping 'til we're across that river and well away from it, got it?" He cast a vague indication to the spilled scrolls. "Gather that, will ya?"

"On it," Zander replied. His usually cheerful disposition was marred by a deep frown.

"Walk," Rahyoke grated at Ian, shoving him forward. Ian did not protest.

Zander slung his re-gathered bag onto his shoulders, slipping on his boots. Only then did Addison move forward, Meygan following behind him.

"Where've you been?" she ventured to Rahyoke as she passed him, her voice timidly cheerful.

Rahyoke took a moment, trying to drain the frustration from his voice; it was not she he was mad at. "We needed supplies before we crossed."

"Okay... I guess it's a good thing I couldn't find any satisfactory fire-wood, huh?"

Rahyoke shrugged one shoulder, a little stiffly. "Kept you out of trouble."

"For now."

"True." He cast a glance behind him. "Zander, you coming? Let's go."

True to his word, he forced them to walk all night, or suffer being left behind.

It was nearing midnight when they reached the place where the river was shallow enough for them to wade across. The current was still swift, but was not so clogged with rapids here. It was deepest at the middle, where it was up to Meygan's shoulders. Being the shortest by solid inches, she was frightened, but she wasn't about to let on; after seeing Zander cross with ease, she followed.

They made their way beyond the willows, beyond the blackberry bushes and the last outliers of the orchard. They made their way through bramble hedges and gorse hangings, over ground increasingly more barren, more dusty, more tangled with roots and darker than anything they had encountered before. The trees here were densely packed together, towering conifers and thick, black-barked oak trees. It was eerie, it was haunting, and it was very, terribly dark, as the clouds filled in the thick canopy overhead.

Only when the glittering rays of dawn at last dripped through the dense ceiling of the woods, sparkling across the dusky needles of their path, did Rahyoke finally allow them to stop.

Tír na nÓg

27
Siúil a Rún

"Hey."

In the midst of the great, twelve-mile stretch between Conmaicne and Kilandubh, there is a broad set of fields, bordered strenuously on either side of the dirt-track road by a make-shift, wooden fence. To the south, the fence displays acres of well-ordered grains, still green in their growing; to the north, it hems in a vibrant array of purple flowers, their long stalks bending and swaying in the breeze. And it is here, where the fields of waving, purple stalks and budding grains and blonde grass are at their widest, tumbling and rolling 'til they reach the embrace of the ancient, chattering woods – here is where they met.

She turned, leaning off of the old fence. "Hey."

A man stood in the field behind her. He was thin as the rails of the rugged fence, and tall. *Very* tall. He was wearing a dark-green t-shirt and baggy jeans, and a black coat was tied around his narrow waist, the hem trailing down around his knees. His hair was long – longer than she had ever seen any man wear it. It reached to his hips, and was dark, a brown that was almost black, tinted red by the sun. He had tied it back in a pony-tail with the exception of a slim braid that hung against his right temple.

Without waiting, the man launched into speech. "God be with ya, love. I saw ya here an' thought that I might be askin' ya a thing, sure, seein' as it's yerself might know, and I not havin' asked ya yet thusfar." He spoke with a sort of brogue, which lent a lilting quality to his low, slightly-cracking voice. "And the thing is this, an' it's wou'jya be knowin' da' way t'Kilandubh?"

The woman arched a brow. "Sure isn't it that way?" She pointed down the road.

"Is it? Is it, now..." He folded his arms, considering. "But c'mere – di'jya come from there yerself, now?"

"No –"

"Ah, well – it's the port, isn't it?"

"No."

"No! Ahk – ah, so it's not. Oh, but da' port's Callatis, now, isn't it. Right. Jayz, but t'changin' a names t'rows me off somethin' awful."

"Ah! I know, right!?" she exclaimed, a furrow of scorn upon her. "They're fine as they are."

"They are indeed!"

"You can't get there from here, anyway."

"How?"

"The port," she explained. "With the patrols and the fences, you can't get there from here."

This was disheartening as anything. "Aaaaggghh," he groused. Crossing his arms and scowling, for a moment he pouted like a child. "Sure I didn't wanna go there anyway."

She wrinkled her nose, giving him a skeptical look. "Then why did you ask?"

"If anyone *had* known, then maybe I would've gone." Even he knew – especially with the botched *"casual"* lilt to his tone – that it sounded absurd. He gave a quick laugh. "Alright; fair's fair. I'll ask ya another thing, then – 'cause, an' I'll tell ya what –" leaning a bit forward, he lowered his voice, conspiring, "—I'm about quittin' dis unholy joke uv'a job soon's I can. Can ya believe it, no lunch pay! Sure on a ship they give ya yer rations, right anuff! Heh –" he smirked. "I'm meant t'be warkin' now, s'fact – y'won't be snitchin' on me, now, will ya?" He narrowed his eyes with mock suspicion.

"Me? I wouldn't *dream,*" she vowed, casting her eyes to God.

"Auh, grand." And he offered her a big smile.

He was an attractive man, all told. He had a narrow face, slim jaw-line, high cheekbones; his eyes were big and beautiful beneath dark brows, sort of a rounded-almond shape, defined and green as pitch. His skin was rough and tan from working in the sun, his smile shining and white and adorable. He had the look of cleverness, the look of kindness, the look of darkness, sweetness, adventure.

She arched a brow. "What did you need to ask me?"

"Ha? Oh!" He laughed. The woman had very pretty eyes – entrancing eyes, a paler green than any he'd ever seen. "Yes. So where're you from?" When he leaned forward onto the fence, he caught her perfume – citrus and energy, cloves, sandalwood, the freshness of the sea... "Havin' never seen ya before this, I suppose I must never've been there – an' sure but it can't hurt t'give da' place a go, hey?"

She pushed back a length of her dark hair, brushing it behind her ear, out of her face. She was very small – a full foot shorter than himself at least – small and very pretty. "Senona."

When he realized he'd heard, he thought that he must have heard wrong. "Senona?" he echoed. "Senona – by t'sea, Senona? Darlin'! 'tis ten *miles* from here!"

Her eyes widened with surprise. "Is it!"

"Yes!"

"Wow..." Then she smirked. "I'm *awesome*."

He gave a crackled little laugh, folding his arms atop the fence. "Sure, it must be a good set'a legs on ye, at least!"

She gave a quick raise of her eyebrows, giving him such a smile that he realized what he'd said, and blushed.

"Oh. Oh! I didn't – Jaysus, but I meant it f'r walkin' – like ya seem like a strong girl – n-not like a *big* girl – not – jus' – I'd no sight'a yer legs – not that I wouldn't – not in a *salacious* way! But – but I just – I meant – I..." By now he wasn't sure *what* he'd meant, he was so befuddled. And his tongue became so entangled that all he could do was look up at her with those big, pleading eyes, distraught and utterly disappointed in himself.

"Is it affluent or eloquent?" she teased.

He dropped his head onto his arms and groaned, "It's *neither!*"

For a moment, she studied him; a good-looking picture of defeat. Casting her eyes up to God, she assured him, "You'll do well with the patrols."

He couldn't help but laugh – a fine, snickering-snort that brewed into something hearty, albeit bitter. "Mercy." He straightened enough that only his hands were on the railing now. "Y'know, I remember days before deh' patrols," he said, sounding suddenly very old. "Before passports an' fences – before curfews an' dogs. Before every city was walled – when a man could go down t'his boat at night an' fish t'waves before a shinin' moon, an' not be afraid'a bein' shot at or arrested – which is worse?" He shook his head. "There's small chance'a true peace when we fear t'speak on politics – some politics they is, too. *Mar-can-sine,*" he emphasized, picking over the syllables pointedly. "Marcansine use'ta be called Tarvenna; m'best friend was born an' raised there. Change t'names 'uv t'territories, sure; Caesar did as much. An' ain't we all itchin' t'be stabbed in t'back twenty-t'ree times – at work, no less. But where dey get off changin' t'names'a every damn fishin' village, farmyard an' port here t'Lavinium an' back, that's what's got me." Raising a squint against the setting sun, he finished, "T'was simpler in bygone days, *Dia ár sábháil."*

She smirked. "You're very wise."

He shrugged, simply. "I've been around." He scuffed a battered work-boot against the soil, watching the dust. "Okh. 'Tis a grand first impression I make." He raised her a one-eyed squint. "Tell me a thing honest: Am I botherin' ya? Say t'word an' I'll go."

She shook her head, frowning. "No, you're not bothering me."

"No?"

"No, not at all."

"Not at all, then?" He winked. "Ye'll wish ye'd answered that different in a bit, I'm thinkin'. Well, c'mere: is Senona nice, anyway? D'yeh enjoy it, like?"

"No." She shook her head. "In fact – are you still jumpin' ship? Maybe I'll join you. I hear Alisia in the south's pretty nice."

He smirked. "Never been."

"The river runs by it; it's nice."

"Have yeh been there yerself?"

"No."

"You an'jyer hear-say information!" He shook his head. "But f'r all that – I'd commit mutiny wit' yeh in a heart-beat, right anuff."

"Oh really?"

"Ye'll need a man wit' experience – an' I've jumped a few fair ships in my day, t'be sure. T'seas were a wild place, then, darlin'. Wild and broad. I could tell ya such stories!"

Her pale eyes lit up at that. "Really?"

Her interest was strange to him; he was long used to being ignored. "Oh, *what* wou'jyeh want with an old salt's tales?"

She frowned. "You can't just bait me and leave off like that," she scolded. "Now I'm all interested."

"Yeah?"

"You don't wanna disappoint me."

"Yeah?" He considered it. This was new. And maybe he liked it. A trouble-maker's grin spread over his face, and he shot her a cunning look. In the voice of an old, salty adventurer, he goaded, "Over fifty voyages I've crossed ta' wide, vainglorious seas – an' only two an' a half wrecks!"

She arched a brow. "And a *half?*"

He grinned, mischievous and bold. "I have an excuse," he began, but was cut off by a shout from the field behind him.

"O'Rielly!"

He straightened up slowly, examining his soil-blackened nails as he continued. "I stole a jolly boat an' jumped ship deep off t'coast'a Lavinium. Got dragged inna current an' beached 'er on a reef. For t'life'a me, I can't figgur how. I was ten."

"O'RIELLY!!!"

"Gabh mo leithscéal. One moment, sorry." He folded his arms, turning back to the field. And he shouted: "Kin ya *not see* I'm havin' a *conversation?!*"

"She's too pretty for you, anyway," called another man.

The first went on. "You've been late three days, and you've got your ass in hot with the bosses –"

"'In hot?'" he murmured, not understanding.

"—and I've warned you before to quit slackin' off –"

"Slackin' off, but I'm done fer the day!" he protested, his arms flung wide. "Look t'me quarter; 'tis all quadroned off an' all an' everything's watered an' weeded an' upstandin' as anythin'! I'm *done* – 'til tomorrow." Dropping his arms, he muttered under his breath, "An' I haven't been late." To press every issue gets you fired.

"Then work Duggin's quadrant."

"Auch, I won't. I won't, 'cause y'won't pay me for it – I'm not gettin' paid half, at all, for t'work'a six men – no sir! But leave me go; I'll be back over in a minute, won't I?"

The man waved him off, uttering slander.

He spread his arms again, demanding, "Won't I?! *God...*" He turned back to the woman, shaking his head. *"A Thiarna Dhia."*

She smiled. "At least you know it's all jobs, right?"

"God save us." He rolled his eyes – and in rolling his eyes, he noticed the dimming track of the sun as it carved its downward arch, and he frowned. "Okh; but c'mere. 'Tis a long road back to Senona, so it is."

The woman cast a glance back behind her, towards the way she'd come, calculating.

"Don't take offence," he continued, "but I'd worry after yeh, yerself goin' on from here after dark. Is it back yer headed?"

She returned to him. "I *was* going back..."

"It'll be dark." Deep was his frown, serious and disapproving. He offered her a hand. "Look; name's O'Rielly – most calls me Dubhán – 'tis sort uv'an adopted name. Anyway, if ye'll trust me, I'll walk y'back safe. Or – if I don't look good t'ya – I know I amn't much t'look on – I'll fin'jya a soul yer willin' t'go wit'. But I'd rather y'not be left t'go it yer own."

She cocked her head slightly, her hand pressed against his, and she studied his face in the tangy light. His hands were thin but rough with callous, clad in dirt-caked, black half-gloves in a vain attempt against cuts and blistering; workman's hands. But he held her own so gently, with the same depth of concern in his tone, in his eyes.

She lifted her chin, casting him a hooded, cocky glare. "That's very sweet of you," she noted.

"I know yer an able girl," he defended, sheepishly. "Y'walked ten miles out here! But – 'tis for me own sake, really. Will ya do me that? After all – y'couldn't give me t'way t'Kilandubh, an' ye've got me in trouble at work f'r it all; 'tis t'very *least* yeh could do for me."

"I held'jyer hand fer ten minutes; that should be worth something."

"Mm?" His eyes dropped, and he realized he had never released her hand. He started – and *almost* let go, but thought better of it.

Sweetly, he sing-songed, "Ya *did, din'*chya? Isn't *that* grand?" Sort of swinging her hand as he said it.

She let go, but not quickly, giving him a slightly parental look. "*Well,* Mister O'Rielly. Should I let you get back to work now?"

"Naw," he grumbled, making a face. He didn't want to leave her, and didn't want her to leave. "Anyways, yeh never answered –" Suddenly he gasped, inspired. "Say! I've an idea! D'ye like soda bread, at all?"

"Soda bread? Yeah; it's like the only thing I can make."

"Make it?" His heart leapt inside him. "Well," he answered, coolly. "I dunno how well *you* make it, but Hourihan's, if ye've never been there, has t'best *I've* ever had, sartin."

"Oh *yeah?*"

"I'd be happy t'put'chya t'the test," he established, cocky. "Wha'd'ya say? Dinner at Hourihan's? On me, a'course."

"When?"

He shrugged. "Ya free now?"

"Yeah."

"Grand." His grin became devilish. "*Then* I'll walk y'back."

She rolled her eyes to taunt him, but, truthfully, she would be glad to have him along; a man is always better protection than a knife.

"Gimmie five minutes, love," he said. Turning back to the field, he shouted, "Connie... *Connie!*"

The man turned at the sharper bark. "I told you not to call me that."

Completely altering his persona, Dubhán cowed, in deference. "I'll do that thing y'were after me – Duggin's shift – I can pick it up t'marra, plus me own."

"*Simultaneous*."

"I'm a good worker; ya know I am. I'll even do it fer t'same pay-rate as always. Do I have permission t'go?"

The man in the field considered it. He cast a glance back to another man, deeper out, who was inspecting one of the quartered-off spaces. When that man looked up, he gave the thumbs-up, and the man who'd prefer not to be called "Connie" looked back to Dubhán.

"Fine."

Dubhán bowed, one foot slid a little farther back, his right fist to his heart. "May t'Lord rain blessin's on yer mercy, Conrad Egnatius." He turned back, taking hold of the fence at a flash. "That's how it's done, *a chara*," he hissed, with a wink, and vaulted the fence cleanly. Turning back to them, he called out, bowing grandly, "I'll see yez all tomorrow, then."

"Yeah, yeah."

"Shall we, m'love?" he offered her, his cheeks flushed with a tinge of crimson, his smile winsome and hopeful. He had so much energy. She surveyed him again, tallying up what he appeared to be made of. So far, the count leaned in his favour.

"We shall."

And together they set off down the long dirt road, as the red sun washed the lavender sky and the man explained the damaging effects of coral.

28
Rest

Meygan sat cross-legged atop a large, mossy stone, cutting chunks from an apple with Zander's knife. The faintest glint of sun still flickered in the sky, highlighting the indigo as it gathered. The trees stood darkly overhead, rattling their boughs with a scent threatening the coming rain, and the great, barren sea of their roots was broken only by occasional stones, and the rare twist of a dead and dying bramble.

"Hey, Rahyoke."

Although at dawn they had made camp in this place, Rahyoke had gone off immediately to travel the five miles to the nearest town and work. Now with the final traces of daylight, he retired, his anger and pride at last failing him, forcing him to rest. "Hey." He cast her half a wave and no glance, as he dropped down at the base of a tree.

"Hey!" Zander exclaimed, starting up. "Are we going?"

"Mm." Rahyoke shook his head. He looked exhausted – even more so than usual. He held out his bag to Zander, limply. "Here."

Zander took it, digging out the greasiest little wax-paper package he found within. He tossed the bag to the boy trespassers. "Jeeze," he groused, unwrapping his meal. "The shit you bring home as dinner..."

The wolf snapped. "Y'know what I could *really* go for right now? Some *meat*."

Zander glared, sullen. "Why didn't you *buy* some, then?"

"I bought m'self an archer."

Zander spread one arm. "The forests are empty!"

"Well *maybe*, if y'spent more time *huntin'*, an' less time just *sittin' around!*"

"Ouff," Addison snickered, from his safe distance. "*That* was censored."

Zander grumbled in that foreign language until Rahyoke demanded, snidely, "*What*, Zander?"

"Nothing." Zander drifted away, taking a seat closer to the boys, and farther from the man. Adults are so mopey.

Rahyoke sighed, pulling up his knees and burying his face in his hands. *Ochón*, he had *really* overdone it this time...

Meygan watched him, eating her apple slices in peace.

"Rahyoke? Are you mad at us?"

Her voice carried the sweet simplicity of the child she was, underlined heavily with the kindness and care of the woman she would become. Rahyoke shifted, running one hand into his hair, though he kept his head bowed and did not turn to her. "I'm not mad at you," he answered, quietly.

"Is everything okay?"

"Everything is *fine*."

"D'you want some of my apple?"

Rahyoke smirked – bitterly, but it was the first sign of a smile he'd offered in days. He still did not turn to her, but his tone was slightly more amiable as he asked, "*Should* I be mad at you?"

"If you should, I certainly wouldn't tell you," Meygan answered, straight as fact.

An interesting thing happened then. For a moment, Meygan could have sworn she saw a flash of genuine emotion pass over the man's typically impassive face – the same sort of emotion which had wracked him by the riverside. Pain, longing, confusion, angst, memory – and then it was gone. His eyes slid her way, one eyebrow arched. Condescendingly curious.

She returned the look. "What?"

He looked away. "For a moment," he said, "you reminded me very much of someone."

Meygan really wanted to ask who. She still had one question left, and prying into Rahyoke's mysterious social life seemed the perfect way to use it. But... she could tell that he didn't want to talk about it, and she didn't want to make him mad – or worse, any sulkier than he already was. She picked a bit of peel off of the shiny blade of the boat knife. "Do we like this someone?" she asked instead.

His head still bowed, Rahyoke nodded, shallowly. "Usually." He spoke so low that any emotion the word carried was rendered unintelligible. Still, there was something about that smile – that sorrow, that pride, that reluctance...

But Meygan kept her musings to herself. "Well," she replied, shrugging, "as long as we like her..."

Zander had noticed the exchange. And – if anyone could be said to understand at all – he understood it, and repented a little. "We're moving at dawn, right?"

Rahyoke nodded. "Mm-hmm." *If I live so long*. "Meygan."

"Yeah?"

Conspiratorially, he asked, "Is your loquacious friend still scared half t'death of me?"

"Who, Ian?" She shrugged. "Yeah, I guess so."

He grinned, vicious in spite of his exhaustion. "Good." He leaned deeper against the tree, folding his arms.

Addison pulled what looked like a cheddar croissant from the canvas bag. "Can we have a fire?"

"No." Rahyoke caught the bag as it was tossed back to him – one-handed, and barely; he had to bring up his other hand to keep what little remained within it from spilling across the soil.

Zander gave a long groan, rolling his eyes up to God. "You're stricter than a Mother Superior, you know that?"

Rahyoke shot him a look.

"What! No one's here! I'm sure *Meygan* won't think less of me!" Defiantly, he sang:

"Deir daoine go bhfuil mé gan rath is gan dóigh
Gan earraí, gan éadal, gan bólacht nó stór
Ach má tá mise sásta mo chónaí i gcró –"

"This is why I never take you into town with me."

Zander laughed.

"Yer fate's yer own," Rahyoke informed him; "that's all *I've* got to say on it!"

Zander translated the chorus, singing:

"Since it's no one's concern, then so no one should care!"

"*Och*, you're flat," the wanderer grumbled.

"I guess you'd one more thing to say."

"Tomorrow, *you're* leading them."

Zander started. "What?"

Rahyoke again flashed him a look, unwrapping some kind of greasy hotdog-hoagie thing. His expression said it all.

"Uh, *no*."

"Tomorrow," Rahyoke repeated, having done with it.

"Then tomorrow can we travel at night?" Meygan asked.

Rahyoke frowned. *"Why?"*

"I'm tired of walking," she explained. "I just wanna get to Tír na nÓg."

There wasn't much to consider. With all the trouble they usually gave him, the fact they were volunteering... "Sure," Rahyoke said, with a weary shrug. "We'll see how we do, but probably." He cast a glance to the others. "Alright with you lads?"

Not to be outdone by a girl, they answered, accordingly, "Yeah, totally; no problem."

"Grand." Rahyoke leaned his head back against the tree. Lord; he was too tired to eat, and too hungry to sleep! *This* is what pride got you! "It's long about ten." His gaze remained on the canopy as his hands absently re-wrapped his uneaten food. "You lads ought'a be gettin' some rest..."

There was that ever-recurrent parental look from Zander. "You too."

"Cinnte."

The sun dragged its dark coverings over its head, plunging them into night as the first drops of rain began to fall.

"How far out d'you think we are?" Meygan asked, carving a last hunk from the core of the apple. "How much longer do we have to walk?"

Rahyoke dropped his bag on the other side of the tree, sliding down deeply. "Uuunh – 'tween thirty-six an' forty-two hours out from the Crossing," he muttered, then clarified: "The big field it'll take a week t'cross."

"Ssssoooo..." Meygan trailed.

"About eleven days," Zander translated.

Eleven days from Tír na nÓg.

"...And *then* what?" she asked.

Already half asleep, Rahyoke rolled so his back was to them. He would give no other answer, save: "Go to sleep, Meygan."

She would have to ask him in the morning.

Damn those soft, hunter's footfalls – just quiet enough to rouse him from his sleep. The swelling dew was still trickling from the bending leaves, pattering against the spongy ground, and, somewhere far off, a lonely bird made its call. But it was those footsteps that awakened him, coming to a halt at his side.

"Zander, don't you *dare* wake me up."

The archer folded his arms. His hair and clothes were damp from the rain, and his attitude was morose. *"Really?"*

Rahyoke groaned. He had *definitely* overdone it; everything was sore, everything was tired – and everything within him knew it was time to move. Stiffly, he dragged himself upright, one palm pressed to his forehead, the other to the earth. "We've made a good pace," he murmured; "we have time for breakfast."

"I already ate," quipped Zander, stretching.

"Then it'd be good of you to get some practice in," Rahyoke replied, raising a glare to him. "When was the last time you shot, at all?"

Zander pondered it. "I suppose you're right..."

"*Yeah* I am."

"Hear that, Zander!?" rose Meygan's angry shout. "We aren't even *leaving* yet! *Why* do we have to be awake!?!"

"Ah, stop complainin' and have some breakfast," Zander grumbled, pulling his bow from his pack. "You're never satisfied."

"Travel-nazi!"

"Hey, now. Why are you so feisty this morning? Relax... Rahyoke, don't look at me like that."

Coming late to the lady's defense, Addison struck the rear guard. "Who died and made *you* leader?"

Zander considered it a moment. Bending the wood of his powerful bow, he hooked the gut string over the notch on one end, stringing it easily; it was a heavy weapon, but he did it without any thought. Flipping the bow, he turned back to the trespassers, spreading wide his arms.

"I am the morning and the evening star," he recited, in a tone that somehow managed to be both casual and grandiose.

Rahyoke looked up. "Good Lord – could yeh imagine? 'Why didn't the sun rise today?' 'What!? Oh, *darn!* I *forgot*.'"

Zander scowled at the second tone – a lighter, airier, more clueless voice. "I don't know *who* you are trying to be," he replied, with haughty indignation. "And if you're suggesting *me*, then I assure you, I would *not* forget!" Drawing an arrow from his quiver, he inspected the cracked feathers. "Might forget to bring the rains an' stuff," he added, at a mutter, "but they happen of their own accord."

"And the sun and moon don't?" Rahyoke shook his head. "Merciful Lord – I certainly wouldn't like to see the king *you'd* make."

Zander made a face. Then, resetting, he responded as only a royal could. "You dare to spurn my munificence? You will be the first to be made an example of, vagrant!"

"I usually am." Rahyoke retrieved his dinner from the night before. Under his breath, though audibly, he added, "No need f'r sun in an enshrouded, glorified marsh."

There was scattered snickering from the troops, swiftly dispelled by the royal's omnipotent glare.

"Do you even know what 'munificence' means?" Meygan asked.

"Of course I do!!"

"What's it mean, then?"

"I don't have to explain myself to you!" Stepping back into character, he exclaimed, "I'll have you beheaded!"

Meygan jumped to her feet, wielding an absurdly short stick like a baseball bat. "Not if I behead you first! Go back to your swamp, you coward!"

Zander twirled his bow, taking it between his hands.

"*No.*" In a languid tone of utmost boredom, Rahyoke scolded them from the base of his tree, taking ahold of Zander's ankle. "There will be no beheading on this journey, if y'please. Zander – I shouldn't have to tell you that. C'mon."

Zander sighed, laying his bow against his shoulder, and Meygan hid her weapon behind her back with clasped hands. "We were only playing!" he defended.

"*I* wasn't," muttered Meygan.

"How come you always take her side?"

Rahyoke shrugged, looking up at Zander innocently as he bit into the cold hoagie.

Zander shook his head, taking his quiver from his pack and tossing the bag at Rahyoke. "Got any targets for me, O wise master?" he drawled, securing the strap of the quiver across his chest. "Wanna set an apple on your head?"

Rahyoke smiled – a smart-alecky little smirk – and chose not to respond.

Zander knocked the arrow he had drawn, aiming it at a tree about fifty paces back the way they'd come. There was a soft breeze today, and it ravaged the cracked feathers, fowling the shot.

Meygan smiled, craftily. "How many times has he almost hit you?" she asked Rahyoke.

"Accidentally or on purpose?" He answered her surprised look with a smile. "Oftener than you'd think, darling... More often than you'd think." He turned to Zander. "Could I borrow yer knife?"

"For what?"

"Juggling. Wha'd'ya think! I gotta shave!"

Zander loosed an arrow; it sunk into the root of the tree, about three feet below where he'd been aiming. "Shi –"

"Language!"

"Zounds!" The archer glanced back to Rahyoke with a beaming smile, catching the man's eye-roll. He lowered his bow, taking the knife from his belt. "Meygan got apple goop on it," he warned, as Rahyoke accepted the blade.

"I did *not!*"

The argument carried the hint that it was being renewed, so Rahyoke severed it. "Thanks."

"I don't see why you should want to shave it."

This argument was stale, too. "No? Well, I do."

Zander stretched, grand in his nonchalance. "Nope. That's why I'm gonna let mine grow out. That way I will *properly* resemble a bushman."

Rahyoke narrowed his eyes. "What on *earth* are you talking about?"

"I'm growing a beard!" Zander announced. "And *you* should *too.*"

Rahyoke laughed, doubling as he nearly choked on his breakfast. He cleared his throat, and demanded, "And for what reason could you possibly want to resemble a bushman?"

"To strike fear into my enemies."

Rahyoke stifled another snort of laughter. "Well! If you'd rather be a proper *warrior*, then you'll have to grow out yer hair as well. Super old-school. In addition to your already *astonishing* beard."

"I sense mockery in your tone."

"I wouldn't dare... But yes, your scraggly, unkept appearance certainly puts me in dread." The knife still in his fist, he returned to eating his breakfast.

Zander decided that Rahyoke was just too out-of-touch and ridiculous to get it, so didn't push the issue further. He drew another arrow from his quiver, knocking and drawing it back across his chest.

"Hey!" Addison asserted, attracting their attention. "So, *I* have a question."

"No y'on't," Rahyoke replied, the knife-hand raised over his mouth. "Y've use' yurs."

"*Jeeze,*" Meygan scolded; "*swallow.*"

Rahyoke pointed at Addison mutely.

"It's not *that* kind of question!" Addison rejoined.

Rahyoke swallowed. "It's not *my* fault, anyway, you pissed away all yer questions on stupid things!"

"It's more of an observation."

Rahyoke returned to his sandwich, ignoring him.

"So," Addison continued regardless, "*Zander* gets a weapon..."

"Oh!" Rahyoke jolted, turning to him. "*Oh* no! *You* do not get a weapon!"

"If it's for protection, then –"

"No!" Rahyoke snapped. "It's cultural – Zander gets a weapon because *he* is mature enough to have one!"

"*I'm* mature!"

"He *whined,*" Rahyoke narrated, his tone so biting and so fierce that Addison knew better than to continue. "Do *not* think I'm above mentioning what you lot did t'my house. I'm not. And if I can't trust you t'follow simple instructions or reside in the same building as matches or God forbid leave off of other people's belongings, then you can bet I sure as *Hell* will not trust you with a blade."

"Ouch."

"Point. Now did you *get* breakfast?"

"No."

"Then get it."

End of discussion.

There was a snap as the third arrow skittered underneath the roots, bouncing away in the mud. Zander heaved a dejected sigh, eyeing his bow in defeat. "I just don't seem to have any luck today."

"You think too much," replied Rahyoke, biting into his hoagie.

Zander looked at him. Then he offered him the bow as a sign of truce. "Wanna take a stab at it? I think it might help me."

Rahyoke was startled. "Me?"

"I haven't shot long-range in a while. I need a refresher on technique."

The trespassers were watching the exchange with interest, especially the suddenly weapons-crazed Addison. Rahyoke debated a diversion.

"Oh! I almost forgot!" He reached into his bag. "On account of Ludi Apollinares was washed out this year, I was able to get these on discount – here ya go." He held up a bouquet of arrows, made of rough wood and fletched with chicken feathers.

Zander's eyes lit up. "Oh my God!"

Rahyoke gasped as Zander hugged him, his eyes wide, as if the gesture were utterly foreign to him.

"I know you hate this, but it's the only way I can sufficiently express my sincerest gratitude!"

"Eauh – alright! Expressed! Done!"

Zander released him at his own leisure. "There. You're a ridiculous little bushman, you know that?" He accepted the bunch of arrows, storing them in his quiver. Then, to Rahyoke's dismay: "So you gonna show me now or what?"

Rahyoke shook his head. "It's yourself who's been shooting your whole damn life. I –"

"Taught me everything I know. Take it... Go on!" Zander flicked him in the shoulder with the bow.

Rahyoke heaved a sigh, getting to his feet. "Yes; let's give the sleep-deprived man a weapon." He slipped the knife into his back pocket, shoved the rest of his hoagie in his mouth and took the bow from Zander, assuming the archer's place as the lad yielded it to him. Reaching over the boy's head, he plucked an arrow from the quiver, knocking it against the bow.

"Careful," Zander cautioned, crossing his arms. "That's a hundred and sixty-two years old, y'know."

"So'm I." Rahyoke set his stance, eyeing the stabbed pine that seemed to be the boy's target. He swallowed the rest of his breakfast, noting off-handedly, "This bow has well over a six-hundred-foot range, are you aware of that?" He took one step back, bringing into his sight a stringy sappling he had noticed earlier, which was about eighty yards off – more than twice the distance

Zander had been firing. Setting again, he leveled the bow, pulling the arrow back so the string crossed his chest.

Thwit!

"Damn," Zander muttered. *Bulls-eye.*

"Luck." Shrugging, Rahyoke handed the bow back to Zander. "Pull it back farther, settle your target. And you think too much."

"You keep *saying* that!" Zander protested, as Rahyoke dropped back down at the base of the tree. "What does that *mean?!*"

"It's all a matter of perspective. Concentration. Discipline."

"Okay. *Now* you're just being contradictory."

Rahyoke's answer was almost irrelevant. "Go ahead now and shoot the damn sapling. Seeing as now you have the proper amo."

Zander frowned. "The wind's up. It's caught the branches – it'll catch the arrow too –"

"Thinking too much!"

"Oh – go – *shave*," Zander struggled, and fired an arrow like a bolt from God straight on past his target. "Damn it!"

"If you think you will miss," advised Rahyoke, "then you will. You think too much."

Oh, that's *what he meant!* Zander scowled. "Why can't you *ever* be straight-forward?"

"But where's the fun in that?"

Addison grew impatient. "Why are we wasting our time here!?" he blurted, throwing himself to his feet like a disgruntled warrior.

Rahyoke responded calmly; taking the knife from his back pocket, he said, "Want him t'practice on you? We may have a few apples left."

Addison pursed his lips. "Point taken."

Looking up at the archer, Rahyoke recklessly nudged him with the knife. "Mercy, you're so serene; just give it another go."

Zander glanced at him, then pulled back the bow-string.

Zwit – thwack!!

"Yes! I hit it!" Zander exclaimed, too excited to be dignified. "See?"

"Now go get it."

Zander's pride deflated. He turned back to the trees, scouring the distance. "They're lost. Good thing I have more –!"

"They're not, either; you're just too lazy t'fetch'em. Go on; off with ya," Rahyoke urged, nodding in the direction of the lost arrows.

"Yea'... Don't cut yourself with that knife," Zander returned, backing up.

Rahyoke twirled the blade skillfully.

"Show-off."

Rahyoke glanced back to him only after he'd walked away – and any blind fool could see the slight twist of a proud smile he gave.

As the archer moved off, Addison stepped in, taking up station on the rock Meygan had occupied the night before. He had never seen anyone shave without a mirror, and with anything other than a razor. With an introspective glance into the reflective knife, Rahyoke eradicated his beard as deftly as he had skinned and eviscerated the trout. Addison briefly wondered if Rahyoke would have qualms wielding such skills on his fellow man. Then again – it was Rahyoke. "I have another question."

"No, you *don't;*" it was so taunting, it was nearly a sing-song.

"Why is Zander armed?" Meygan asked. "I meant to ask before; Zander has so many weapons. That guy we saw fishing had a lot of weapons. Even *you* seem pretty good with a bow." She did not miss the flash of his sardonic smirk. "Are we gonna be safe where we're going?"

"Yes."

"*Rahyoke,*" she warned.

But he remained firm. "If y'weren't gonna be safe, I wouldn't put you through it. Trust me." His little grin was winning, and very convincing indeed. "There's a difference between valour and showmanship."

Addison made a face. "You always liked her better," he grouched.

"Yes," Rahyoke drawled, without irony, as he inspected the shining blade. "I do."

"Ha!" Meygan hissed.

"You ready to go?" Rahyoke asked, as Zander returned, his quiver full, his bow looped over his shoulder, and one broken arrow cradled pitifully between his hands.

"I feel like I should bury this," mourned the archer. "I'm not still leading them, am I?"

Rahyoke got to his feet, using the tree for support. "It'll be easy," he said brightly. "And don't do that with your bow."

Zander groaned. "How will I *find* you? I've never been on this side!"

"Sure ya have – and you'll be *fine,*" Rahyoke soothed, slinging his bag onto his shoulder. "Just go straight."

"Straight?"

Taking Zander's shoulders, Rahyoke turned him east. "Straight," he repeated. He clapped the boy on the shoulder, moving off. "If I don't fin'jya, stop at sunset; sure I can hunt ya out."

Zander pursed his lips. "That's very reassuring," he noted, sarcastically.

"You kids be good for Zander, y'hear me?" Rahyoke ordered, pointing back at the trespassers. "I mean it."

"Sure thing," Addison shrugged.

"Convincing."

"'Bye, Rahyoke!" Meygan exclaimed, cheerily, waving.

"See ya." Rahyoke hooked the knife onto his belt; it was well hidden by his coat. He started to move away, then he paused. "IAN!"

The lad jolted, looking up at Rahyoke with wide eyes. "Huh?!"

Rahyoke fixed him with an intense, scrutinizing glare, his dark eyes narrowed harshly. Then, without a word, he sauntered off. But not a soul missed the evil cackle snapped back through the trees.

Zander rolled his eyes. "C'mon, trespassers," he commanded, depositing his weaponry in his bag and taking it up. "Let's move out."

And they did.

29
The Competitions of Rank

"Why are we stopping?" Meygan demanded for the fourth time, as Zander continued to ignore her.

The archer was kneeling on the ground, circling stones and propping twigs to make a little fire. "I wanna make sure that Rahyoke finds us," he at last responded, striking the blade of his great sword against a little black rock.

"He knows we were going east."

"I'm giving him time, then, to catch up," he replied tensely, raising her a glare.

"Oh-*kay*," Meygan answered, with the same effect as *Touchy!* Lifting her palms in a sarcastic surrender, she walked away, taking a seat at a distance. There's no worse company than a sulky archer (or a dejected philosopher).

"Get some fire-wood, will you?"

"*Whatever!*"

But she went, and that was good enough for him.

Though they had traveled several miles deeper into the infinite wood, they were still surrounded by those close-packed conifers and oaks, their shadows thickening as the light seeped from their spiky boughs. Change is omnipresent but slow, longed-for until it comes. Zander struck the flint rock; two more tries, and a significant enough spark leapt into the dry kindling for the beginnings of a small fire. That was life, too. Ten thousand hopeless sparks sputtered and died; few were the number who took light. Far fewer still, the impassioned fire-starters igniting the blaze.

Zander sighed. *Of these I am not.* He sat back from the kindling, legs crossed at the ankle but knees pulled up, held by the loose wrapping of his clasped arms. There was no use blowing that tiny flame into life; more likely he'd blow it out. Resigned, he watched its reflection glittering in the smooth steel of his blade.

Since he had sent Meygan off, he was left with the two boys – trespasser and traitor. Of these two, it was Addison who most affronted him.

Something about him – in his manner, his attitude, his character – abraded Zander, and this wound and his boredom demanded a hunt. He watched his quarry. Watched what he watched. And he smiled.

"You like her," he called out.

Addison jumped, turning to him. "What?!"

"Meygan. You like her; I can tell."

Addison scowled. "You're crazy."

"Then why did you respond? How d'you know I wasn't talking to Ian?"

Ian glanced up. "Huh?"

"Nothing!" Addison snarled. Shrugging, Ian went back to cleaning the bottom of his shoe.

"You're always gawking at her; why don't you just man up and ask her out already?" Zander tilted his blade, letting the fire pool. "Scared?"

"I am not –" Addison clenched his fist. "I'm not taking dating advice from a guy who's... *that*." Unable to establish a summary, he settled on the basest expression of contempt.

"Only trying to help."

"Yeah. Maybe in the Middle Ages, *'archer'*. But you don't know a *thing* about the modern world. You wouldn't last ten minutes in my world, let alone be able to pick up a girl. Actually, come to think of it, how much has the whole 'hunter' thing really done for you *here?*"

Zander pursed his lips. *So the wee fish bites...* Slowly, methodically, he began feeding more twigs into the fire, encouraging the blaze to grow.

"And how well would you have lasted in this world without me?"

"I'm tempted to see," Addison menaced.

"Just so we're clear, are you threatening t'kill me? 'Cause *that* I would like to see. Or are you threatening t'storm off into the woods and try t'go it on your own? See if you can survive the misty bogs, the holly berries, raccoons –"

"Oh! Raccoons! You mean meat that an actual hunter would catch?"

"Who eats raccoons? That's simply barbaric."

"Sounds like an excuse for sucking at hunting."

"Can't kill what's not there."

"Can't kill if you suck at it."

"I've caught more stuff than you have, haven't I? At least I contribute something to the group. What do you do, regale us with your complaints? Riveting."

"Yeah, you contribute. You 'bravely' shot at a tree. Ooh, so impressive."

"See this?" Zander returned, raising a palm. "It's called callus. It's what you get from *working*. I may not seem like much now, but I've got prospects. I have skills. And just because I haven't taken any lives doesn't mean I couldn't."

Addison pulled a face. "You expect me to believe you're a man 'cause you've got a few blisters?"

Zander smirked. A tight-lipped, condescending up-turn at the corner of his mouth – nature or nurture, that look was Rahyoke's.

"No."

From beside the fire, he lifted a stone, holding it for Addison to see. Then, taking it in the palm of his fist, slowly – slowly as all glaciers – he lowered it into the flames, depositing it in the heart of the camp-fire without so much as a wince.

Addison stared at the stone, his eyes wide, as Zander withdrew his hand, brushing his palms together to banish the smoke. "Maybe it doesn't mean much," the archer stated, his voice held low. "Maybe it doesn't prove anything. But I can do it... Can you?"

Addison was startled a moment longer. Then he gave a derisive snort. "*Chih.* Ha – you want me to stick my hand in the fire? How dumb do you think I am?"

Zander examined his nails. "Intelligence has nothing to do with it."

Addison frowned.

After a moment, Zander shrugged. "Auh. I guess I can't blame ya. If I had dainty woman's hands like that –"

Addison came forward.

He sat opposite the fire, and something like approval flashed through the archer's gaze. The hunter gestured towards the fire, openly, as if displaying the world before him. *Well?*

Addison swallowed, eyeing the flames. Then he raised his eyes to Zander. He was tired of putting up with this backwoods bushwhacker's flack.

He focused on the stone. *Mind over matter, right?* Slowly, he stretched out a hand. He passed his tongue over his lips, watching the crackling blaze. Waiting. His fingers brushed the biting tongues, and he bit his own, tasting blood.

"What the Hell!?!" A hand came down on Addison's wrist, gripping his arm and pulling back so suddenly that Addison was forced to stand or fall over.

"Uhm –" Zander's voice faltered.

"I tell ya t'look after'em and this is what you do?" Rahyoke snarled; the glare he had fixed on Zander was livid.

Zander bowed his head.

"What the Hell are you playing at? I expect better from you!" He felt Addison struggling to free himself – an annoyed wriggling – and turned on him, staying his motions. "And you! Are you daft? Do you do *every* God-damn thing you're told? Jesus – y'just can't resist setting shit on fire, can you?"

In spite of appearances, Rahyoke was strong, and Addison could not begin to break his grasp. He clenched his fist. "We were –"

"*Don't* even." He released Addison's arm. "Sit down." And – for better or worse – Addison obeyed. "*Go dtuga Dia ciall duit.*" Turning back to Zander, he commanded, "Get the Hell out of my sight."

It was a warning. Zander stood and moved off – not quickly, for he was proud, but he went all the same.

"Go and get Meygan, damn you!" Rahyoke called after him. Then he returned to Addison. "You're not a remarkably *intelligent* child, *are* you?" he demanded coldly, his arms folded. "Wanna tell me why the Hell y'let him talk you into that?"

Addison gritted his teeth. "I'm just as strong as he is!"

"Well, all that proved is that you're as dumb as he is." He assumed Zander's spot on the opposite side of the fire, crouching. "And gullible. Is that what you wanted? Probably not."

"You're as bad as he is," Addison muttered.

"I'm worse." As there was no water around (grand convenient!) he began scraping up dust to pour over the flames, making a pile on the ground. "Here's the thing, Addison. It's all about attitude. Not this sulky, self-righteous attitude," he added, shooing an indication Addison's way. "But one of *actual* self-confidence. Yes: Zander's an annoying son uv'a bitch. But if you're confident in your own strength, then you won't feel the need to prove it to him with stunts like this. Speaking of..." He paused in his scraping, raising his gaze just slightly to study the fire. Then he dipped his hand into the flames, clenching the stone in his fist and lifting it for Addison to see. "*This* is not the makings of a man; it is a display of the skills of one."

For a moment, Addison stared at the stone; then he averted his gaze. "Two."

"Don't read too much into it." He tossed the stone aside and briefly returned to scraping, but his fingers were cut and sore from work. He wiped the sides of his hands on his jeans, lifted a stick, and began breaking up the fire into crumbling embers, over and into which he poured the soil, churning and stirring and killing the fire with chained frustration. "The point is, you can't let everyone make your decisions. Use your head."

"I don't need your guys' advice."

Without looking up, Rahyoke pointed at him. "Perfect." He carried on with his work.

Addison watched for a time, wondering why it was taking Zander so long to find Meygan. Maybe he really was a terrible hunter.

"This sucks."

Rahyoke glanced up, his dark hair streaming into his eyes. He seemed surprised that Addison had spoken. Then he shrugged, scraping more dust from the ground with his stick. "Yeah, well, *sin an saol, a mhac*. That's life. Deal with it."

"That's it?"

"I thought you didn't want my advise? Make up your mind."

"That doesn't fix anything!"

Rahyoke laughed, raspily. "Do I look like I know how to make life suck less?" He sat back, studying Addison, the slightest, smirking twist to his stern mouth. "C'mere, look: Are you happy?"

Addison's brow furrowed. "What?" he demanded, as if it were the dumbest question he had ever heard.

"Don't get all defensive; I'm just wondering. Anyways, you owe me a question – and a proper man would give me a straight-forward and honest answer, wouldn't he? So answer me that. At the end of the day, at the very base of everything in your existence, all of life's seeping suckage aside, are you satisfied with your life? Are you happy?"

It was a ridiculous question. Over-simple and completely irrelevant to everything at hand, and at first, Addison simply maintained his glare. Then he answered. "I guess, yeah."

Rahyoke watched him a moment longer. Then he shrugged. "Alright." He got to his feet, brushing the dust from his hands or his jeans – the gesture accomplished little of either.

As long as the kid was happy.

Addison glowered, utterly at a loss. "Why is everyone here so weird?" he demanded, crossly.

"At least we embrace it."

"What's that supposed to mean?!"

"You need to find out what you are, and who you are – not who you're told to be – if you're ever going to have any satisfaction. And embrace it."

"Well, maybe I'm the kind of guy who just does what other people say. Ever think of that?"

Rahyoke bowed – arms outspread encompassingly, his right foot slid a little back. "Then *embrace* it."

Addison frowned.

Without straightening, Rahyoke cast Ian a menacing, side-ways look. "Wha'd'ya think, sparky?"

Ian jumped, snapping the scraper-twig against the bottom of his sneaker. "Iyuh – I – agree."

"With what." Rahyoke straightened up, folding his arms, never moving his glare from Ian.

"Ih – I – I kinda hate it when my life is controlled by other people." His voice was airy and faint; Rahyoke's unpredictable temper was still fresh in his mind.

"And so the lesson for *you* is to not put yourself in that position, isn't it?"

Ian nodded eagerly, biting his lip.

"Brilliant. We all have lessons. Zander, are you back yet?"

"You know very well I'm behind you... Fine, I've a lesson, too."

Rahyoke lifted a sarcastic grin up to God, refusing to turn. "You've learnt nothin' an' ya damn well know it." He looked at the archer over his shoulder. "Oh, di'jya, then? Alright: tell me wha'chya learnt."

"Don't get caught."

"Good boy."

30
Relics

"Are we there yet?"

"I swear t'God – I swear t'flippin' *God* –"

Zander grabbed Rahyoke's arm, staying him. They had been walking for nearly four hours; the sun had just set, lengths of rusted orange still lingering in the rapidly-darkening sky; and, although the kids had been able to stop and rest for the day, Rahyoke had *not*.

Tightening his grip on the seething man, Zander reprimanded, "Does it *look* like we're there?"

Predictably, the outburst was not enough to quell their laughter. "Fair warning," Rahyoke growled. "What you consider hilarious? To a man without sleep for a full *forty hours*, it's nagging, it's obnoxious –"

"Hey!" Meygan began, her curiosity turned elsewhere.

"The first time, allowable. The first *fifteen*, nettling. The next *thirty-five!?!*" He clenched his hands into ready and willing fists. "Incendiary! Do you *want* me to sell you?! I will!! And don't think I won't!"

"Rahyoke!" Meygan again tried, but was ignored.

"You're kinda belligerent when you're tired," Addison remarked, his tone contemptibly patronizing. "Not to mention talkative."

"Irrational!" Rahyoke corrected. "Madly, desperately, *unalterably* irrational!"

"I think you're bluffing about selling us."

"No one would *take* us," Ian contributed.

"See if I don't!" Against Zander's restraints, Rahyoke advanced a pace forward. "And believe me, it's *after* I've already cut out your tongues. I couldn't stand the route to market." He imitated: "'*Are we there yet? Are we there yet? Are we there yet –?*'"

"*Rahyoke!!*"

"*WHAT!?*" he shouted, turning.

Meygan pointed into the woods. "What's that?"

"Wha'd'ya *mean,* what's –" But he saw it. Distracted from his anger, Rahyoke narrowed his eyes, focusing through the dense needles of the trees. "Huh..."

He shrugged off Zander and followed the object of Meygan's pointing. There was a clearing some twenty or thirty yards off through the trees, and in it a large, wooden structure, a vast and dilapidated shed. Gaps crowded the walls and canopied the roof; the old boards were molded over, stained by rain and time, the stripped wood painted over with moss and decay.

"It's like a barn," Addison observed.

"It's not *like* a barn, it *is* a barn," Rahyoke corrected. "Stay here."

"Why –"

"Jist!" With a sharp, undercut stroke of his hand, he signaled for them to stay, moving forward. Silently he stepped over the lawn, taking the knife from his belt. When he heard Meygan whisper, "What, should *we* be armed?" he knew that Zander had drawn his bow, for back-up. He cast them a glance. "I'm sure it's fine; I'd just rather not be mauled by an owl."

Zander resolutely refused to lower his bow; Rahyoke just hoped that the boy wouldn't jump and accidently let slip a shot. He was too tired to dodge.

Carefully, he sidled up to the door. He could detect no sound from within, but that meant little at this time of night. He ran his thumb over the flat steel of the knife near the hilt, taking the latch of the door with his free hand and feeling the rust flake against his skin.

"Are owls vicious?"

"Maybe they're territorial."

He raised his eyes to study the chipped paint. *They* would *be the death of me*. He stuck the knife in his belt again.

Shifting, he cranked down the latch; it took both hands to slide back the thick bolt, which gave but rigidly. In a way a good sign, as it could mean that no one had made use of this place for a very long time. The door was on a track, and would slide just like the bolt, with little greater ease. He passed his tongue over his lips, wiping his hands on his jeans. *I've got this.* Bracing against the door, he readied to drag it open.

It jammed.

"Sssshhhit." The track was heavy and sticking with rust, the thick wood melded into the earth itself. He jigged the door, putting all of his force into shifting it, until at last he managed to gain an inch.

"Dhia!" he breathed, dropping back. *Well, THAT'S not working...* Again wiping the rust from his now-shivering hands, he stepped up to the gap, peering inside.

"Watch out for owls!"

"Mercy," Zander denounced, shaking his head. "You really *do* have a death-wish, don't you. Let's just leave it and get out of here; this is a waste of time." Ironically, he finally lowered his bow.

Rahyoke cast him a glance but said nothing. He slid his fingers into the gap, wrenching the door another foot – enough that he could get his skinny body between the door and the doorpost.

"Dech do shínaib ceo, a Shandar," he declared, and quickly vanished within.

Meygan turned to the archer as, rolling his eyes up to God, he returned his weapons to his quiver. "So, what; do we follow him?"

Zander sighed. "Yeah, we follow him." So that is what they did.

The interior was dim. "Are we safe?" Meygan called out, teasingly.

Rahyoke turned from where he'd been inspecting the deep shadows within the barn. "For the moment." He sat down on a crate, dropping his bag beside him in the dust.

The crate was one of a multitude of vessels stacked at the back of the barn and scattered aimlessly throughout – boxes and crates and trunks of various sizes, all sheathed in a layer of dust.

"Jeeze, what is all this?" Addison wondered aloud, nudging a crate with his foot.

"Is this someone's barn?" Meygan asked, looking to Rahyoke.

"Not anymore, I'd figure."

"Is this their stuff?"

"Maybe it was."

"Can we stay here? *Just* for a little while," she qualified, rebuffing his aggravated look.

"Nnnnnn..." Rakyoke considered it.

"Can we see what's in the crates?" asked Zander, drawing his heavy sword. "Since they don't *belong* to anyone..."

"Huh? Yeah..."

"Yay!" The kids set to riffling through the crates with a vengeance.

"What d'you think's in here?" Ian queried.

Zander rolled his eyes, wedging his blade against the lock on a trunk. "Gold. Jewels. Rare, precious items of infinite value, long since lost to the world. *Mmh.*" The lock broke with a chink. "And snakes."

"*Gyeesh!*" Ian dropped the crate he'd lifted, and the brittle wood cracked. The lid tumbled off, sending the contents spilling onto the dirty floor.

Zander looked up, startled. "What, the – ...what?" Curious now, he reached out, delving his hand into a mixed pile of satin. "Yo Rahyoke!"

"Uh?"

"C'mere!"

Rahyoke groaned. "What?" He rose, retreating to the back of the barn where they were.

Zander lifted a trailing, flowing piece of fabric, holding it out to Rahyoke. The man took it, frowning.

"Ooh!" Meygan exclaimed. "That's pretty!"

It was a shawl. "That's... random," Rahyoke said. "What's in the rest?"

"Clothes." Zander flipped open the lid of the trunk. "Is it all clothes?"

"Hm..." Deep in consideration, Rahyoke blindly passed the shawl to Meygan, who had come up beside him.

"Oh! *Thank* you!" She shook out the shawl, wrapping it like a head-scarf.

"Where did it all come from?" Addison asked.

"Likely marauded years ago," Rahyoke murmured. "It's long ditched now."

"I *liiike thiiis,*" Meygan sing-songed, twirling.

Rahyoke folded his arms. "There're seven towns within walking distance of here," he said to Zander.

"What about it?"

"We could stay here for a while."

"We could sell it."

Rahyoke turned. "Huh?"

Meygan stopped twirling. Her legs were crossed at the ankle, her arms were spread, the ends of the shawl gripped in her fingers. "What if we sold it? This stuff," she suggested. "Take what we need and sell the rest in town? That way you won't have to work so hard, and we'd be able to save up some money, ya know? I mean – it doesn't belong to anyone, right?"

"Hmm..." Rahyoke rubbed his chin. Slowly, he began pacing, his eyes on the darkened ceiling. "Not a bad idea... Shorn – a *grand* idea..."

"Can I keep this?" Meygan asked, twirling again.

"Sure." Rahyoke cast a glance to Zander. "Kenny has a 'barrow I can use."

"Who's Kenny?" Meygan again.

"Some shmuck I know," Rahyoke shrugged, waving her off. "It'll be a good way to stock up on supplies before the crossing, at least."

"People will *buy* this?" Addison pulled a face, dropping a skirt back into its box.

"Please." Rahyoke cast him an equally condescending glare. "Fergus Campbell was *fortunate* to have hired me; I could sell you your own shoes." He shook his dark hair out of his eyes. "By the way – while we're on it, what'll you guys eat? I may actually be able t'get enough now for a choice."

"Chocolate!"

"No; I mean *real* food."

"Oh."

"Bacon," Ian asserted, seriously. "Definitely bacon."

"Like a BLT!" Addison exclaimed.

They continued on down that line – shrimp scampi, lobster-tail, garlic pizza, steak – until Rahyoke, sighing exasperation, called out at last, "Never mind!"

"*Personally,*" Zander responded, slamming the lid on the trunk, "*I* think we should invest in a horse."

"A tad *gamy* –"

"Not t'*eat,*" Zander glowered; then he turned to the boys. "Could one'ur both'uv you lads grab the other end? I don't think we'll need anything out of this trunk, so..."

Addison shifted past Ian, bending to take the other end of the trunk. He scowled at the inquiring look Zander gave him.

"I'm saying for transportation," Zander continued. "On three, lad. Ready?"

Rahyoke shook his head. "Is it the tangled horse of Gilla Deacair you're after? *One horse* won't do us."

"Isn't that a drink?" Meygan asked. "Like a cocktail?"

"Huh? Oh! *Daiquiri!*" With a stifling snort, Rahyoke burst into laughter.

"The horse of the Gilla Deacair was a creature of the *sidhe,*" Zander dutifully explained, "that carried off sixteen men of the Fianna on 'is back." He and Addison dropped the trunk by the door. "Phew!"

"We should've waited to fill it," Addison realized, in retrospect.

"Boys are stupid," Meygan muttered.

Rahyoke managed to get ahold of himself. "Shorn! *I'm* sleep-deprived!" he admitted. Uncrossing his arms, he stretched as he moved over to a tower of boxes. "So!" He clapped his hands together. "Let's see what we got, hey?"

"I call this scarf thingy!" Meygan exclaimed, jumping away to begin a pile.

"Aw," Zander lamented. "But it goes so well with my eyes."

Meygan gave him a look.

"Y'want this, Zander?" Addison offered, holding up the skirt again.

Zander lifted his chin. "No sense'uv fashion on ya at all, Addison Mitchell," he remonstrated. "Why ruin an arse this nice with a shabby auld skirt? Ha?"

Meygan smirked. "You're ridiculous."

"Only a little." Zander offered her a wink. "And – *language,* by the way! How could you miss that?"

Addison dropped the skirt off to the side. "Seriously, though, Rahyoke," he resumed. "I really doubt you can make any real money off of this stuff."

"Now *there*'s a bet I long to take," Rahyoke uttered, wistfully. He knelt in the midst of the boxes at the back of the barn, inspecting the outside surfaces, as no one had thought to do.

"I bet you could make money *that* way," Meygan remarked. "Ah! Ironic..."

"What, betting? I was keen at it when I was younger. Used to *rake* in the cash – what with bets of that sort – what I could sell, what I could do – how much and what I could drink..."

"Why don't you just do that?" Addison suggested, perhaps not quite understanding the implications.

"Are ya makin' an offer?" He slit open the cracked tape on a ragged cardboard box. The box was heavily water-damaged, a little moldy at its base; likely everything in it was ruined. He reached in, pulling out a grey zip-up hoodie.

Meygan laughed, cheerfully. "We could buy a whole *team* of horses!"

"And form a happy little maurauding band!" Zander grinned like a fox. "Wha'd'ya think, Rahyoke? Uh – Rahyoke?" Curious, he clambered over the trunk, coming to the man's side. "Wha'd'ya got?"

Rahyoke jumped at his touch. He was crouching before the cardboard box, transfixed by its contents, the grey hoodie gathered in his lap. In his hands was a book, bound in blue linen, its fabric faded, the pages crinkled and yellowing, rippled faintly by water-damage.

"Can I see it?" Ian asked.

"Never you mind it!" Rahyoke snapped. Then, amending: "It's some sort of ledger... I can't *sell* it."

Ian frowned. "Oh." Shrugging – after all, it was neither gold nor snakes – he dug back into the crate he was ransacking.

Rahyoke hesitated, studying the damaged cover. They left him alone; even Zander, giving him an understanding if irritating pat on the head, returned to his rummaging.

Carefully, he opened the book and began leafing through the pages. They were covered in a small, scrunched text, slipping between languages, scrawled to utilize as much space as possible. In some places there were drawings – quaint sketches or inked designs, some labeled; in others, another hand took over, one more flowing and vine-like. Some of the entries were long; others, mere snatches: *"Call me savage, a chairde"*; *"A fine health to the company!"*; *"Harvest: damnation on all squirrels"*; *"Today, it rained"*; *"Are you still with me?"*; one, scrawled madly in a top corner: *"Ná hairr ormsa!!!"* – "Don't ask *me*".

He allowed the pages to slip through his fingers, listening vaguely to the fluttering sound they made. At the back, between the end pages, was a pressed, dry rose; a few notes were scratched in pen, and a shakily-inked *"Slán go fóill"*, in yet another hand.

He flipped open to the front, studying the first page. The message there was short – the date done out exuberantly, the first letter of the piece decorated with extreme detail, embellished lightly with coloured pencil, as were the vines, the flowers, the butterflies, the flames, dancing Celtic designs across the free space.

God bless all who view these pages! And
may He kindly look with favor upon this decent

scribe. May my humble words be pleasing in Your sight.

And may our tale endure forever.

Rahyoke ran his hand over the page, smoothing out the crumpled paper. "Can I keep this?" he asked.

Meygan glanced up. "Sure; you found it."

He rolled the notebook in the sweatshirt, slipping the bundle into his bag. Then, shaking off his trance, he eyed the door. A soft rain had started, gradually becoming more persistent, spilling over the stone threshold. He rose.

"Jayz, Meygan," Zander grumbled. "*You're* helpful."

Meygan was digging through her backpack, looking for food. "I agree with Rahyoke that you'd be more useful if you hunted."

"Auh!" Scandalized.

"'Auh' yourself," Meygan replied. "We have time, and I didn't get to eat. So sue me!"

"How could such a tiny creature need so much food?"

She pointed at him with a cheese croissant, leveling a glare. "You're gonna make me anorexic," she warned.

Zander threw out his arms. "What did I say!? God!" he complained to the Creator. "Girls are so unreasonable."

"You're *sexist,*" Meygan accused. "*That's* what *you* are. You're sexist, and jealous because *I* have *food,* and *you* were *too stupid* to *save* any."

"Why do girls always speak in italics?"

Skreeekg!!

They turned.

"What the Hell are you *doing?!*" Zander shouted, annoyed. "The door was fine enough where it is, without you messing with it!"

Rahyoke, straddling the threshold, pressed his back against the frame. "I can't get a damn 'barrow through there," he grated. Bracing, he put his feet against the door. "*Come* on." The door ground against the cement threshold, shifting a centimeter with a metallic squeal and the hum of wood on stone. "Who in the all-fecking Hell designs a *sliding* barn-door?" Rahyoke grumbled. His foot slipped, and he scrambled to catch himself, narrowly avoiding a most painful injury. "Auch – flippin' rain!"

He had been at this a solid five minutes, and was soaked through. And sure in all that time, for all that struggle, the door hadn't shifted in either direction save for that lonely centimeter and a foot.

"This is better than TV," Meygan remarked, tearing a hunk from the cheese croissant.

"Shit." Rahyoke ran a hand through his damp hair. He had thought that the rain would make moving the door easier, acting as a lubricant – it was easier to sail a ship through water...

"You're gonna hurt yourself," Zander drawled in warning, not sounding particularly concerned.

"Why don't you worry about gettin' some stuff I can sell, huh?!" Rahyoke returned. He shifted higher against the doorframe, glaring at the rusted-out steel track.

Zander rolled his eyes. "Sure."

"Are we selling all of it?" Ian asked.

"It would be unwise t'waste such a God-given opportunity," Zander pondered. He bent, using his sword as a crowbar on a crate's tricky lid. "Yet, greed is often damnation... There're only seven towns about, and not a one of us has a permit..." Almost at random, he glanced to Addison. "Do you?"

"Do I what?"

Returning to Ian: "We'll sort it at least, whether we sell it or not."

"Why?"

Zander shrugged. "For the convenience of those to come after us. Out of respect t'the poor, marauded souls. Boredom. We'll be here long enough; take yer pick."

"How long'ur we gonna *be* here?" Addison groaned.

"That would be a question for Rahyoke. But –"

There was a sudden squeak of sneakers on wet metal followed by a colossal, rattling bang.

"Aoh!"

Meygan flinched, averting her eyes. "Ooooohh..." Even the *sound* of it was painful. "Y'okay, Rahyoke?"

From outside came a whining little moan that developed into a curse.

"I *tol'*jyah," Zander admonished.

Rahyoke groaned. Unraveling from the fetal position, he reached out, grasping the edge of the door. It was an arduous mission, getting up. *"That* hurt," he rasped, doubled over as he clung to the doorframe.

Meygan opened her eyes, venturing a glance. "Leave the door alone now, will ya?"

Rahyoke pushed his hair out of his eyes; his hand, like his entire right side, was smeared with mud. "I think I loosened it up a bit, actually; I can get it." He started eyeing the track.

Meygan scowled. "Boys are *so* stupid."

Rahyoke bit his tongue. Moving inside, he dropped down beside the door, leaning back against the wall. "Is anyone else *reeeeally* exhausted?" he asked, his gaze roving the ceiling. "Hey! A loft!"

"Can I sleep in there?" Meygan asked.

"It looks *mad* rickety..."

"I want my own room."

"You can sleep in the grain room," he responded, pointing blindly to his right.

"What if there's rats in there?"

"Then we'll have a lovely, fine breakfast after all."

Meygan made a face. "Ew."

Rahyoke leaned his head back against the wall, watching her with narrowed eyes, waiting for the inevitable next question. Meygan knew that this was on his mind, so she refused to give him the satisfaction. She stood, moving over to the "granary" – an isolated stall to the right of the door, center of the long wall, boarded up completely. "It's locked," she informed him.

"So *unlock* it."

"I *can't*."

"Augh!" Rahyoke got to his feet, coming over to where she was. The door was locked with a simple bolt, just above Meygan's head, and between the height and the rust she couldn't force it open. Rahyoke couldn't either; he was a full minute trying it before Meygan, surrendering, sighed dejectedly, "I'll just sleep in one of the poopy old stalls..."

Rahyoke scowled, backing up a pace. "D'you need the lock?"

"I shouldn't –"

BANG! Rahyoke gave the door a solid kick, shattering – *one* of the hinges. Muttering, he worked off the other. "*There*, damn it!"

"Language."

"*Sor*-ry," he responded, not sounding terribly apologetic.

"Can you check it out before you go?" Meygan asked, wringing her hands.

"'Fore I go where?"

"To sleep. Away. Wherever it is..."

He groaned. "Fine." He slipped into the room.

There were a few storage bins, a cylindrical, barrel-shaped feed container, and some metallic bit of farm equipment that looked both murderous and alien and a little like a manual push-mower.

"It's too dark; never mind."

"What?! *I* can see!"

"What! I'm scared of the dark!"

Rahyoke set to a round of unintelligible grumbling. "I'm too tired t'do anythin' about it *now*," he finished, retreating out of the granary and back to the outlier crate beside which he'd left his bag.

"Wha'chya doin'?" Zander asked.

Rahyoke lifted his bag, slinging it so the strap crossed over his chest. "There're some storage thingies in there we can maybe use," he said, irrelevantly. "I'll check'em out in the morning." With his shin, he slid the crate, positioning it as he eyed the edge of the loft.

Zander smirked, throwing a glance to Ian and Addison. "This'll be good, lads; trust me!"

"You're *fortunate* a *lady's* present," Rahyoke returned. Then, without awaiting further rebuke, he jumped from the crate, grabbing the ledge. "Auh – oh, *face!* I did it! Ha!!"

"Why didn't you use more boxes?" Meygan called, as he hauled himself the rest of the way up. "If you just *stacked* them, you could've done it a lot easier."

Zander snickered.

"That might'a made sense," Rahyoke admitted, reluctantly. "Hey – I like it up here; this is my room... What's tha –?" *Crack.* "Whoa! Shit!"

Meygan rolled her eyes.

"S'okay! We're good! ...Big hole in the floor, but – we're good!"

"Boys. Are. *Stupid.*"

31
Love-seat Valour

Early the next morning, Rahyoke had headed out, taking a large knapsack he'd found and filling it with "merchandise". For their own part, the kids continued sorting until late in the afternoon, when the goldenrod sunlight was beginning to slant as it filtered through the many chinks in the walls and ceiling. Zander had been left in charge in Rahyoke's absence. And he was *lording* it.

"Addison? I figured out where we went wrong."

Addison looked up – and scowled as Zander settled beside him, sliding an arm around his shoulders. *"We?"* he half-sneered, shrugging the archer off. Bred to believe hand-me-downs were the mark of a hobo, Addison was sticking to his own duds; having momentarily removed the Syracuse University sweatshirt, he was sporting a green t-shirt with his lacrosse number (81!) on the back. Zander, however, had absolutely no problem wearing "found" clothes, and had put together the ugliest outfit possible purely to irritate Meygan. He pushed back the oversized winter hat he was wearing and explained. "I started y'off too big. Sure but y'couldn't hunt for yourself, nor even fish; how am I t'start you on snapdragon?"

"Snapdragon?"

Zander drew his sword and began cleaning it with a bit of cloth no one had wanted. "It was wrong'uv me to tease you when clearly, who is there to teach you otherwise? I figure any right man needs t'be able to defend himself. In words, aye, but with muscle."

"Aren't snapdragons flowers?"

"I could teach you."

"You?"

"Look: I'm offerin' t'help you. Are you still so Hell-bent on ignorance you would refuse it? I can teach you to fight."

Addison wanted to stay angry – he hated Zander, he really did – but he hated being called a wuss even more. Relenting, he answered begrudgingly, "Fine."

"You're welcome!" Zander got to his feet quickly, scurrying off. There was a hint of malevolence to his cheer.

Without rising, Addison turned. "What are you *doing?*" he droned.

Zander fetched a longish metal staff from atop one of the boxes; it was the shaft of the alien-killing rake from Meygan's room, which he had dismantled the night before. "Here."

Addison stood. "Don't I get a sword?"

"Blades are a little advanced for you."

"Self-righteous, condescending –"

"You wan'it so bad?!" Zander demanded. He flipped his sword with the fanciest still-in-hand move *ever*, holding the hilt out to Addison. "There y'go; take it. Strike me if you can."

Hefting the weighty blade, Addison looked up at him, startled. "Huh?"

Zander spread his arms, dropping the alien-rake shaft. "Have at me," he challenged. "You've been wantin' to, haven't you? Just *dyin'* t'take my arrogant head off? Go to! I'm all yours!"

"Don't you need a weapon –"

"Naw!" Zander assured him, with a cunning, devilish smile. "Let's go, boyo." He wagged his fingers, urging Addison on. "*Have at me*."

Addison tightened his grip on the hilt. Jeeze, it was heavy. "Okay." He slung the sword back over his shoulder, caught his balance, and hurled forward a weighted stroke.

Zander stepped wide right. "Oh, poo. Seems y'missed."

Now he was condescending.

Addison gritted his teeth. Stepping forward beneath his blade, he rolled an upwards slash at Zander's side, but Zander stepped right again, even advanced, placing his foot in the middle of Addison's wide stance. With one hand, he gave a good shove to Addison's shoulder, toppling him.

"Oof!"

"Get *up*," Zander commanded, striding and circling around the boy, making so bold as to step over his sprawling legs. He lifted the alien-staff from the dust. "You need to learn to work *with* the weight of the blade, not in spite of it."

Addison rose, taking a goal-tender's stance, ready to pounce. He lashed out again.

"Aaw. Too bad, too slow; bad blow, Joe." Zander flipped the rake, and, with it, his tactic. "Come on! Yer ma could handle a blade better than that, I'm certain!"

Addison gave a low growl, furious.

"...mine."

"Shut *up!*" Another strike sank into the floor, followed quickly by a second, a third. "How the Hell is this *training?!*"

Zander rolled his eyes, strutting lazily around as he waited for the next blow. "We learn by doing," he drawled. "C'mon, tiger-lily; while we're young."

Willingly, Addison attacked.

As they were fighting, the wanderer returned, slipping silently through the gap in the door and coming to stand beside Ian and Meygan, who were watching the battle with remarkable indifference.

Rahyoke watched too, looking intrigued. He asked, "Why are they after killing each other?"

"Why ask me?" Meygan responded.

"You seemed to make more sense. Should I ask Ian?"

Without awaiting his inquiry, Ian explained: "Well, since he was making fun of Addison the other day for weakness, Zander offered today – like a little while back – a few minutes ago, really, though it seems like longer – to teach Addison how to fight – Zander did, I mean – so that way it wouldn't be anyone's cause other than his – Addison's – own if he were still such a poor, unmanly fighter – which, *I* don't think he is, so much – he's great at *Battle Call* – but in *real* life – I don't know, I've never seen – and –"

Sighing, Meygan straightened up and clipped: "Zander's teaching Addison to fight."

Ian skulked.

Rahyoke clapped a hand on the boy's shoulder. "Welcome back, my exceedingly loquacious friend."

"And anyway," Meygan added, "'*in real life*', it's sofa valour. Anyone can be brave when they don't have to be."

"Huh... Well said, well spoken."

"But was it *exceedingly,*" Ian pointed out.

"*Eloquent* is saying something well," Rahyoke defined. "*Loquacious* is excessively talkative."

Ian pondered it. "Oh. Well, I'm that, too."

"More like 'love-seat-valour'!" Zander called, laughing as he simply dodged Addison's strike.

"*A Dhia.* Whatever it is, it's making a wreckage of your sword." Rahyoke sat down on the floor amidst the boxes with Meygan and Ian. "I got these from Kenny," he said, producing a large brown-paper sack, which he offered to Meygan.

"What is it?" she asked, opening it and peering inside. Then she gasped. "Thank you!!" Without warning, she leaned over, wrapping her arms around Rahyoke.

He tensed. "What're you doing?"

"Hugging you." Undaunted by his tone (she knew well enough he wasn't what you would call the *sensitive* type), she returned her devoted attention to the contents of the bag, pulling forth a Boston cream donut.

"Kenny's daughter's got a bakery; I'd trust her before him for anything I'd ingest," Rahyoke said, as Ian scrutinized the slightly-greasy bag.

"Sure!" Zander scoffed, grinning. "And tell me it took *all day* just to walk to and from Kenny's – even with the haggling."

"*Haggling?* Do I ever?"

"So it was just drinking!"

"Hey; when the house offers for free, it would be rude not to."

"It's a dreadful habit, Mongan," said Zander, with mock sobriety. "Shame on ya, takin' advantage of an old sot's private cellars."

Rahyoke clasped his hands, bowing his head humbly. "Ih'tis, so, playze yer Honour." He looked up, returning to himself. "Anywayz, be respectful; Kenny's not that much older than me –"

Zander arched a brow.

"Ih – *shut up!*" Scandalized and depressed, Rahyoke sulked for a moment. Then, resuming: "Ya gotta meet certain standards'uv hospitality and social grace and accord – it's a *geis*. Y'can't expect a guy t'give anythin' up for free – at least not without a proper bit of persuasion."

"Di'jya at least get the cart, ya toss-pot?"

Rahyoke frowned, looking disgruntled and proud. "I *got* it, and don't be teasin' me." He accepted the paper bag from Ian, retrieving his own dinner.

"If you didn't have to haggle," Addison chimed in, "then where's all the stuff you were gonna sell?"

"Ah – where d'ya expect?!" Rahyoke dug into the near-empty knapsack until he produced a handful of coins, slapping them down on the crate. "Ha!"

"That's it?"

"That's *gold*."

The coins were not uniformly minted; of different sizes and hues, they resembled more closely the coins of the ancients than the coinage Meygan was used to. With a glance to Rahyoke for approval, she lifted one of the coins, studying it like an archaeologist. On one side was a figure enthroned, holding what looked like a palm branch; on the other was stamped "IIKKA".

"S'from the far North," Rahyoke explained quietly.

Meygan wanted to ask him more, but they were interrupted by the fighters. Noticing Addison distracted by the gold, Zander tackled him without warning, sending the sword plopping a good yard or two away into the dust.

"What the Hell!?"

Zander had Addison's arms pinned; he pressed the flat of his knife between the trespasser's eyes, along the bridge of his nose. "Oh damn! You're dead!"

"What the Hell?!" Addison repeated, more pointedly now.

"If you think you can make in this world it without the help of a poor, bumbling rube like myself," Zander replied, just as pointedly, "then laddie, are you *ever* wrong."

"Lads," Rahyoke called over, as Zander lifted his sword, "cou'jya come eat a last meal before ye murder each other?"

There was sighing and grumbling. "Yeah, alright." The pair trudged over.

"I'm gonna go back into town tonight," Rahyoke said. "I just need a chance to eat, then I'll load up an' head off." Snagging the fistful of coins with his free hand, he got to his feet, pacing off. "I think we can have a good run of it here. There're seven towns in the area, like I said, and I think I can hit up all of 'em. They need t'be schmoozed, like Kenny."

"Why do I feel like you're a bad friend?"

Rahyoke smirked, choosing to bite into his donut rather than respond. *Yay, chocolate!* He retreated to the corner where he'd left his canvas bag, depositing the money within and settling down beside it. "Unless something drastic happens, we could be here a good while. Maybe even a week."

"I'm down with that."

"Sounds good," Zander echoed, distracted between evolving training plans and pastries.

Rahyoke took the ledger from his bag. Sure, what did they care, so long as they were provided for? Disquietude was an adult malady – and a right useless one at that. Flipping open the book, he began to read.

32
Huath

An seachtú lá déag de mhí Aibreáin

It's about the middle of the month, the middle of the night, and sleeping is boring, so I figured I'd try this. My wife has been after me a while to write my stuff down. Why? says I. Trust me, says she. I'm not after writing, really, especially about my own life. It's not I've anything against it; sure most of what we have would've been otherwise lost, but...

Okay.

But what am I supposed to write? I don't trust you with my name – I'll trust you with Sean's; he wants to be writ into history: Senan Fahy, founder of Balladarach. *Dia ár sábháil.* They call me Dubhán, the Dark One. You can call me that, too.

If you bothered making it past the first page.

Maybe you would. Maybe you think I'll divulge all my secrets, and you can turn me in for some grand reward. Isn't that lovely. You should know that I am nothing in God's eyes; I am a vessel, *a chairde.* A prideful vessel – and I ain't goin' down easy!!

What am I doing – wait. Let's try this again; I think there's a chance that we could be good friends. And for all I know, you could be my progeny reading this, if God sees fit to grant me any – och, and who else would? It serves I should be decent to you. Especially since you've been so kind as to waste yer time on me. Or are ye waiting for that reward?

Mercy. There is great anger upon me this day; it is this that inspired me to write. I fear that everything I stand for will be lost with my generation. It used to be a joke with us that we'd soon have to sign waivers to take a – to defecate, rather; at our rate of progression, that mightn't be funny a few years down the road. After all, the public waters they're thinking on closing against us, aren't they?

Here's what they did of late: The whole shipping industry has been shut down – well, for me. The "Seizure of the Ships", it's called. No one of

Ceothann, Gallic, or Odinic descent – "Northerners" – is permitted to own, captain, or crew a vessel – which runs from the big shipping rigs right down to the humble raft. Can ya stand it? If I build me a raft, I'm in contemptuous violation of the law! They'll still take on Southerners – damn, but who better can navigate those tousled waters than a Turranian?

Nobody was reimbursed when the screws took their boats, and no one was paid for the voyage wrapped up. Oh, a crew who'd just done a three-year whaling voyage was livid and nearly turned the docks over at that. I'd really like to know what was in their heads when they decided to write this law. Really. I would like to come at them with the full text-book of their laws (and hit them with it, dead) and ask them for each – "What in the all-fecking hell did you think you'd accomplish by this?"

It wouldn't be anything new. I had warned her before we were married, we couldn't get a marriage license. By law, because of what my parents were, because of what *I am*, I am legally banned from marrying. For the first 21 years of my life – who cares? Not to mention, no woman, good or bad, ever took an interest – not that I amn't sexy. I'm a damn good-lookin' SOB, *a chara,* and don't you forget it. C'mere: let's just say that the government does not sanction my copulating, (heh!) and every lass from here to the Eastern Sea and back again knows it. For I am but a lowly creature – and fire-mastery is not as great a turn-on as it was for the cavemen, let me tell you! Sometimes I wonder if my name and countenance weren't written directly into the law. It wouldn't surprise me. After all –

Whoa – where am I going? Am I going to dictate to you backwards every law and sanction ever laid-out, and that in the most incriminating way possible? Oh, maybe. Maybe, if it would help you to see where I'm coming from... alright.

New licenses have been issued – we were always meant to carry licenses which stated our name, birthplace, residence, age, nationality, citizenship and marital status. Under the new restrictions against "vagrancy" and "idle perambulation" – wandering – this was not against the law until five years ago, mind you – I see now they were gearing up for this shipping law; anyway, the license requires a residency now.

Balladarach is – strictly in the legal sense, of course – not actually a *town*. And to try to register us would jeopardize the lives of everyone here. Besides; my Orchid refuses to register. "If I can't put your name on it," she says, "then I'm not going through all that trouble. It's a load of crap anyway;

I'm not doing it!" To hell with it, then; as for myself, I'll continue under the same false name I've toted since the original licenses were issued, and see if I can't claim I live over Dermott's Pub in Marcansine. It's the only way I can get a job, and the only way I can make it there without being arrested on the road. Besides; they're harder on the men than the women, anyway – and she has me to protect her.

So! What other reams of nonsense mislabeled "justice" can I dictate to your stinging gaze? How's the Act Against Cults – that which banned our faith – which boils down everything from paganism to Daoism, and strikes it out. I bow to the subsection entitled "The Papist Laws" – ugh, I hate writing that; the skin upon me crawls of it. The – I'm not writing it again. But that's the part against Catholics, specifically; every other sect has their own, equally-insulting addendum. Care for a taste of what it entails?

The dissolution of all places of worship – from the convent to the rectory – all churches, monasteries, cathedrals, *et cetera et cetera* – and all possessions therein and land thereon to be turned over to the control and keeping of the government, not to be restored. All persons affecting religious vows are hereby disbanded. Anyone servicing a Mass (my own capitalization, the barbarians) or "harboring objects of religious intent" is subject to punishment up to and including death. Anyone offering or giving religious instruction is subject to imprisonment for up to five years; repeated offence is punishable by death. Vows of celibacy – IRONICALLY!! – remain effective. Isn't that the most bastarding thing?

All religious artifacts and icons, including but not limited to: statuary, crucifixes, (etc; it was a long list, lads, *long*) are to be turned over to the officials for destruction – And I mean everything, lads. They smashed out the windows of the churches, and cast the idols out after them. They burned the Books, melted the chalice, the tabernacle, the candle-holders; smashed the pews, the kneelers, the pulpit. They defiled the alter and all its artifacts, and – only after everything within had been utterly destroyed, one piece after another after another before the very eyes of those who'd worshiped there – then did they take the building, and pull that down, too.

And so it had been with the others – the Synagogues, the Mosques, the Temples – the oak-trees where the Northerners still plying the old religion used to worship – everything – blown up, cast down, struck out, immolated and destroyed.

That's the Act Against Cults. There are more – so many more – against which I rail in so tireless bitterness; but the hour is late now, and these are not the thoughts with which I'd sweep you off to bed. There will be enough space for ranting and enough time to accomplish it in later days; I'll tell ye a story instead.

The dirtiest little pub I know of is called Priest Rock. Bear with me, it's worth it. The place is so desolate that even the soldiers don't care to go there; they wouldn't be caught *dead* in such a lowly, scrounging place. The barkeep, Peader – *och,* he's a big lad, a little shorter than meself, but jacked; dark-hair, deep-set eyes, tattooed as all hell up his arms – Peader, *a chuisle,* is a priest. Yeah, he is; they hold Sunday Sessions on the holiest days in the backroom; use the pool-table as an alter – Christmas and Easter, and the feast day of Saint Dismas (March 25), if ye please.

It didn't matter what the government thought, so long as we were married before God; and on a glorious day in mid-winter, Peader married us, me and Orchid, before God in the back room of the bar.

I wonder if I can still get a job on the docks?

An naoú lá is fiche de mhí Aibreáin

Scant are the breaks when I may write, and scanter by the day; forgive me my long absence.

It's been a busy time of the year for me, not merely for moving into the house, but for the planting. There's apparently a lot of things to be planted in April – corn and cabbage and tomatoes and beans – and they need extra hands, now and at the harvesting. What's more is I'm just after getting myself hired at the illustrious Xena's Fish Barn. Sure but I don't mind telling ye because hundreds do work there and I don't know rightly how long I'll be at it myself. What ye do at Xena's is you're stationed on an assembly line – I'm a keen hire on account of I already know well every station – and you're gutting and scaling and preparing and packaging fish. Someone once said that it's a measure of a hard-working man – honest work being most favourable to God – that he had such dirt under his nails. But I say to you, whoever said *Cleanliness is Godliness* is a man who's been spattered in hake guts.

I often wonder if men invent sayings just to justify what they're at.

And how many men of them take authorship themselves, or try to add heightened authority with the vague and preponderous, "As a wise man once said" – the "wise" an even more ingratiating appendage. *In aimn Dé.*

And you might say to me: "Why is it you don't get yourself a more favorable position in life, lad?" – or howsoever you speak. A job more respectable like, more of a benefit to your fellow man. What a nice world you must live in; fancy that if I had the world at my feet, I would squander my time with hard labour? I'd be on a boat. The family man surrenders the transcendental life at the prow of sedentary endeavors; your freedom for your passage, sir. I work the jobs I can, for the food that I need, and try not to dwell on the feeling of being caged. So when you wonder why I work so, hear me when I say it is because I like to eat, in preference to starvation.

Oh, I got fish guts on the page. It won't come off; darn. This is why we can't have anything nice.

33
The Good Life

eygan!" The snap of a shout rang out through the glowing dawn-light, startling her. Exiting her quarters, she stood, framed in the dark of her doorway. "Yeah?"

"*You* are my favorite." Rahyoke strode into the center of the room. Over his shoulders was the knapsack he'd found; he carried a latte in one hand, and a cardboard drink carrier in the other.

"Yes!" Meygan pumped her fist.

"Others up?" he asked, setting the drink carrier on a trunk, tossing the knapsack beside it.

"Nope."

Rahyoke grinned. "Block yer ears."

"Done."

He put two fingers of his free hand to his mouth and gave a piercing, shrill whistle.

"Shit!" Zander jolted to his feet, drawing his blade. Though less prepared, Addison was just as startled; Ian, for his part, merely groaned, "Go ta' *Hell*, Rahyoke."

"Get up! I've got breakfast... Fair play to ya, Zander; that's a quick reflex."

"Did you go into town?" Zander asked, re-sheathing his blade.

Rahyoke settled cross-legged beside the trunk, digging through the knapsack. "Did I *go* to town? I *owned* that flippin' town. I rocked that market like it was nobody's business. You're welcome."

Meygan smirked. "Well, aren't *you* modest."

He bowed, arms spread. "Of course." From his bag he lifted six sporks in cellophane wrappers, four water bottles, and a mason jar filled with rich, purple jam.

"What's this?" Zander asked, removing the lid from the styrofoam container he had taken from the drink carrier.

"Uhm – she called it 'Fretters' or somethin'. It's like shepherd's pie, but it's corned beef and potatoes and onions..." His eyes dredged the ceiling for the ingredients. "Bread crumbs... Some kind'a dressing."

"It looks like Thousand Islands," Addison affirmed, squinting into his own breakfast.

"Really?" Meygan challenged. "You're an expert in *salad dressing?"*

Addison shrugged.

"Did you remember to ask about us?" Meygan asked. To her surprise, Rahyoke shook his head.

"There are seven towns in this area, and in four uv'em, I wouldn't dare ask."

"Why not?!"

"That, my dear," he replied, snagging the jam and a spork, "is a very long story. And my time here is short."

Zander frowned. "Where are you going? You just got here."

Rahyoke chose not to answer him, though he made like he hadn't heard. With his free hand – he was bent on not relinquishing the coffee cup – he started digging around in his backpack. "I got biscuits."

Zander rolled his eyes. *"Grand,* Rahyoke."

Meygan glanced between the two men. She did not like being blown off. "Well, you seemed to do alright in town for not being able to ask about us."

Zander bolstered the attack. "Where'd ya go? Or were you just unwilling to ask Kenny twice?"

Rahyoke choked on his breakfast, covering his mouth. *"Iochd."* He took a drink from his coffee, clearing his throat. With his eyes very obviously scorning God, he recited, "There are seven towns in close proximity to this area – blah-dah-something-sacred-well as the reason. I have good friends in Dryden and Ellery; Marcellus there's a wicker mill where I can usually find gainful employment and a idler called Uillien who knows a thing or two." His eyes fell to Zander, his head slightly inclined. "I can't go back to Kenny 'til I'm done with the cart."

"So where did you go?"

"Lasair – I mean, Lucisduorum," he amended, shaking his head. "Forgot the bloody name switch..."

"Why didn't you ask about us?"

"Gecsh – cou'jya calm down?" he returned, raising his palms. Then he pointed at her. "Do you have any idea what they do with foreigners in

Palmyra? God, it'd give yer *grand-children* nightmares. Do you know what they do with psychos in Misenum? I would never make it back – physically, or mentally. Do you know Paradisius is still run by Martial Law? Lasair is run by a black market – literally, by it. It's the skeeviest city of shirkers ye'd ever see – or smell, God help us all. But they make frickin' good jam – want some?" He held out the jar.

Meygan raised a brow, wondering what on earth she could possibly say to any of that. In the end, she accepted the jar, thanking him. After all – you can't pass up the chance of tasting black market jelly.

"Are we safe here?" Addison asked. He suddenly found himself slightly in dread of ending up in a Misenian institution.

"Oh yeah; absolutely," Rahyoke reassured them confidently, and shoved half a jam-smothered biscuit in his mouth.

"Classy," Meygan flat-lined.

Rahyoke swallowed. "What?" He stood, not waiting for her answer. "Di'jyou guys get the next round of stuff?"

"Wait – ho!" Zander returned, halting him. "Where're you going?"

"Paradisus," Rahyoke replied, as if this should have been obvious. "It's a five-hour walk."

"You need to sleep."

"Aaaahhh." Rahyoke waved him off.

"Rahyoke –"

"In ainm Dé!" Rahyoke complained, his voice cracking. "Fine! I'll loiter a *cúpla* hours. Ya happy?! Ya over-anxious, ungrateful hen-wife – *then* I'm going to Paradisus!!"

Meygan snickered. "Like a nap," she cooed.

Rahyoke jolted, now looking horrified as well as vexed. "D'ya see what ya did?!" he demanded of Zander.

"I thought you said Paradisius was dangerous," Addison pointed out.

Rahyoke folded his arms, cocking a hip. "Not so much so as Lasair – *Lucisduorum*. Whatever; I was stopped twice," he shrugged.

Zander gagged. "What –?!"

Rahyoke waved him off again, undaunted and unashamed. In a stilted accent, he dictated, grandly, "Alexandros Daedalus Appeninus, newly returned from the East. Where thanks to the good help of the imperial guard, our renowned culture remains unsullied by the rabble of the North."

"Daedalus?" Zander raised an eyebrow.

"Yep."

"If that makes me Icarus, then you can forget it."

"It's my name so long as we're here; deal with it."

"Do we call you that?" Meygan asked, raising her hand.

Rahyoke shrugged. "No matter t'me." He lifted his canvas bag onto his shoulder, leaving the other pack to them. The coffee was still in his hand. "Get that stuff ready; I'll be back in an hour."

"You *won't*," Zander muttered, sulking. Rahyoke ignored him; he was busy trying to figure out how to get up to the loft with his breakfast.

"Are we fighting again today?" Addison asked, scraping the last of the mashed potatoes from the bottom of his styrofoam cup.

Zander shrugged. "If y'want – if y'can *handle* it." He started to pick his teeth with the prongs of the plastic spork.

Meygan made a face. "Eew – *Zan*-der, that's disgusting!"

Zander looked up. "The corned-beef is stringy," he defended, innocently. He returned to his task.

"I'm going outside," Rahyoke announced, seeing no way of getting up to the loft without having to put either the cup or the biscuit down. "You two don't kill each other while I'm gone, alright?"

"Okay," said Ian.

Meygan, grinning, added, "Why? You wanna watch?"

Muttering again disjointedly, Rahyoke stalked outside and into the damp morning light.

34
Duir

An ceathrú lá de mhí Bhealtaine

milk
bread
two dozen eggs
shortening
tomato paste
vanilla
onion x 6
tea – "good tea" ???
cereal – "that doesn't suck"
cookies "if we can afford"
NO ketchup
gin ~~x2~~ x3

An cúigiú lá de mhí Bhealtaine

My wife says I need to stop using the book for grocery lists. It was twice, says I. But I'll stop, supposin'. It's just it was easier to remember from my own head when it was only me I was buying for! How in the hell do I know what's "good" tea and what tea "sucks"? Why the hell isn't it the same if I buy "lint-catchers" and not "dryer sheets"?

What the hell *is* a dryer sheet!?!

Aaugh!!

An séú lá déag de mhí Bhealtaine

Maeldún stood with a great stone in his hand, ready to cast, his foot braced on a burnt-black slab to steady him on the uneven ground. But just at the moment he would make his game-winning hurl, Briccre the Cleric said, "Had the youth of today any sense upon them, it would be avenging the bones upon which you stand, and not idle play you would be at."

"On whose bones do I stand?"

"The bones of thy father, Ailill."

Hearing this, a black rage and the lust for revenge came over Maeldún like the wash of the tide. He dropped the stone from his hand and made obsequiance to the priest, and he asked him on what dead fool he might take his father's blood-price.

The killer was a man whose name was unknown, and is unknown still today, but his lands were Carnmadoc, and were only able to be reached by a journey over the sea.

"Get me a boat."

There was a Druid in Maeldún's land known for his great wisdom, and it is he who warned Maeldún: on such a day shall ye depart from here, and bring with you no more than seventeen. Such a voice had he when he was saying it that Maeldún knew not to challenge the prophecy; but if he did, and no hero ever sinned or faulted, nary a narrative would we have in existence.

A currach was built for him, and a daring crew rostered out. Farewells said and blessings pronounced, Maeldún sailed out; anxiously had he been waiting the appointed day, and now it was here, not even God Himself could stay him.

Three of his foster-brothers came to the shore, their equipment all packed, and were dismayed to see Maeldún already away from the mooring. They gathered the strength within them, and shouted for Maeldún to come back; they would not rest 'til they had helped him in this quest.

Maeldún remembered the Druid's warning; no more than seventeen must be taken on this journey. But his foster-brothers would not be turned, and cast themselves into the sea.

Lest they should drown of it, Maeldún was forced to bring about, and add the three violations to the crew. Now, I do wonder for what reason he didn't just drop three of the ones he'd already had in the shallows, and thereby maintain the allotted number – but perhaps we can figure that such was the rush upon him, he didn't want to go through the arguing and the pulling of straws and all that; a ship is a monarchy, and her king was more than eager to get underway.

Under a full sail they banked around the coast and out over the open water. On they drove like a legion of demons, coursing night and day. But didn't I tell you that life is full of transgressions, and the poundings of God's wily humour? Well, I told ye now, anyway.

A storm blew up upon the third day of the voyage the like to make a man curse the day he was born. And as Maeldún worked, drenched through the flesh, at the ropes and the tiller, the sails and the steerage – anything not to have to anchor – his navigator called out, "Carnmadoc!"

There off the larboard, dipping in chaotic foam, were the high rocks and verdant turf of blood-besotted Carnmadoc.

Maeldún wrenched the tiller, his fury like that of the lightning, that of the thunder, that of the sea tenfold. Eighteen times he made to land, and eighteen times the soil rebuffed him, showered in capsizing spittle which threatened to sheer the fragile hull.

"Every man of us will be killed!" his mate implored him. "And what good to your father then, Maeldún? What good to your father then?"

Maeldún peeled his eyes from the hated turf and looked upon his men. Never before this day had he seen the fear of God nor man upon them, but how greatly it gripped them now. These were men with whom he'd fought, with whom he'd been raised, and whom he loved with all his heart. The demon of Carnmadoc was not worth their lives; they must live to sail and love and fight another day.

Maeldún's heart fell within him, and with it the fire of his rage. He gave in to the pleas of his comrades and the biting drag of the tide; the rope in his hand snapped by the Will of God, filling out the sails, and like the hounds of the underworld they were swept on over the traitorous face of the sea. The wind and the waves carried them off past the island; and the last sight that Maeldún had of it caused the need he should be restrained.

For who was it who stood on the grass of the headland, watching with calm, dark, wondering eyes, but the very man who had murdered Ailill, Maeldún's father?

35
Créde

A shadow fell across the page, causing Rahyoke to look up. "Don't rat me out to Zander!"

Ian Burgendson put his right hand over his heart, holding his left aloft. "Don't tell on me, either," he said, settling down across from him. "I'm supposed to be helping."

Rahyoke's eye-roll was brief. "Wha'd'ya need?" he asked. "Or are y'just shirking duties?"

"Are you reading that?" Ian asked, craning to look at the ledger as Rahyoke pulled it back. "Anything interesting?"

"What do you *want*, Ian?"

"I had a question to ask you. Uh –" He hesitated as he saw a flash of hardness in Rahyoke's eyes. It had been five days since the incident with the scrolls... "Is that okay?"

"That depends on what it is."

The warning was clear in his tone, and, not knowing specifically to what it pertained, Ian erred on the side of caution. Timorously, he began: "So, this isn't the question yet, this is a lead-in. There's this story. In it, there's a man who's forced to choose between two doors. Behind one, there's a hungry, vicious, man-eating tiger; behind the other is a beautiful woman he will have to marry. The only problem with that – the marrying – is that the woman he loves, a princess, is sitting in the crowd. Now, she tells him which door he ought to pick. Does he take her advice?"

Rahyoke was taken slightly aback. "Does – is that what you're asking me?"

Ian explained. "See, it has one of the most obnoxious cliff-hanger endings *ever* – it just *stops* –"

"Then why bother with it?"

"Because I need to know! I need to know whether he follows her advice, and whether or not he dies because of it!"

Really? Rahyoke gave a long and silent sigh, leaning his head back against the tree. The clouds were skittering by quickly overhead, hustling in another bout of rain. *Because I know everything...* He was halfway inclined to ask why this was so earth-shakingly important, but he didn't have the energy to cull through another of Ian's garrulous rants. "Easy enough t'tell. S'pose the lady's on the right and the – the tiger's – on the left." He halfway made an effort to raise the respective palms. "He follows her advice. He picks the one on the right."

Ian frowned, literally drawing back. "He does?" He shook his head. "But isn't the princess jealous? Or wouldn't she think, like – if she can't have him, then no one can?"

Rahyoke did not reply, but his eyes alone dropped to fix Ian an inquiring glance.

Ian sighed. "How do you know?"

Rahyoke continued to watch him a moment; then his eyes returned to the chattering leaves. "How many questions is that?"

"*Please!*" Ian begged, clasping his hands. "It's *killing* me!"

"Yes. I can see." He allowed the boy to suffer in silence a little longer. Then, passionlessly, he explained. "Art Mac Conn had gone out to Ildathach, the Many-Coloured Land, in search of Delbcháem ní Morgáin – a woman unmatched in beauty. Among the other things he came across, he met a lovely young woman by the name of Créde. This was towards the start of his journey, and it was she who advised him of all that he'd face. As the warnings progressed and the dangers piled, she urged Art the more to leave it all and remain with her; Art refused. The final danger of which she spoke was the coming to him of two of her own sisters, and they each bearing a cup. One cup was harmless; the other was filled to the brim with deadly poison. She says to him, 'Choose the one on the right'. And, in spite of her renewed pleas for his remaining, off on his journey he goes..." Rahyoke paused. He seemed to be considering something off among the leaves. "Just... out of curiosity, Ian: which would you?"

Ian frowned. "What – like, what do *I* think?"

"Not what would *he* choose, but yourself. What would you do?"

Ian was struck. He wasn't so inclined towards philosophy himself – certainly not now, when the answer seemed only a sentence away. But it was clear that this question was a rite of passage; so, impatient though he was, he settled in and gave it an honest pondering.

"Alright," he said at last, quietly. There was something in his tone – a seriousness it so usually lacked. "She would tell me the truth. I *know* she would... I would listen to her."

When he raised his eyes to Rahyoke, he caught the man's almost suspicious glare, but in a flash the look had changed; Rahyoke gave a light shrug. "In time Art, son of Conn Cétcathach, arrived at the place where he was met by twin maidens, each carrying an identical vessel. As one, they offered their chalices to him. Art took the one on the right with a word of thanks, and he drank deeply of it, draining it to its silver base. And y'know what?

"He was perfectly fine. The end."

The boy's expression was a slow-dawning but agonizing bafflement. "Uh – he was *fine!?*"

"You answered yourself, so. Better to look on the man, even he with another woman, than to look on him, and him dead. Besides; they were all polygamists in them days, anyway – *especially* the son of a king, who could afford it." Again, his eyes roved the canopy thoughtfully. "I bet *your* people got it from *my* people and left off the ending, thinking themselves clever. Thus the ambiguity."

"Is that always how it works?"

"What, oral tradition? Oh – no." Rahyoke shook his head, a bit lollingly. "No; otherwise ya wouldn't be – cliff-hanged. If there's ever the possibility of another answer – of your princess lying and betraying the man she loves – of your man the hero winding up mauled by a man-eater or struck down by Druid poison – if none of this could happen, well – you'd be back inside, helping clean out that barn and earning your bread."

Ian pondered all of this. And the best he could get out of it was: "Little shot at philosophers, huh?"

"All I'm saying is multi-task." Rahyoke leaned off of the tree. "Think we've been out here long enough for the archer not to draw and quarter us?"

"I guess so."

"Alright, then." Rahyoke gathered his belongings, slipping the book into his bag. "How do you know your own Créde wouldn't lie to you?"

Ian jumped, unexpectedly startled. "What?"

Rahyoke looked at him curiously. "Sorry... Just tracing the line of thought. Your confidence in Créde's credibility –"

"Oh. Oh – well, she loved him, you know? Like your other story – Finn would do anything for Sadb, right? If you love someone, you protect them – whatever the cost..."

"Right; alright, then." Rahyoke got to his feet, snaring his bag. "Come on; break's over."

They returned inside, and – conveniently – Zander called out on spying them, "Ah! I was just coming to get you." He held Addison at bay with the shaft of the alien-rake as Addison geared up to swat at him with his own sword. This was "training"; Mercy Above.

"Me?" Rahyoke asked, dropping his bag by the outlier crate. "What for?"

Zander knocked aside Addison's strike, grabbed his arm, and tripped him with the rake so he fell backwards into the dust, the sword skittering off like a spider.

"Is that what you wanted to show me?" Rahyoke asked, with a devilish smirk.

Addison raised Rahyoke a glare. "Is this how you trained him?!" he shouted.

Zander pinned him with the rake. "D'ya think I'm goin' too easy on'im, Rahyoke?" he asked, grinning.

Rahyoke sighed, looking up at the ceiling. "A man could not join the Fianna until he could defend himself against nine warriors wielding spears –"

"Easy enough," Addison shrugged, from where he lay.

Rahyoke arched a brow. "And himself buried up to his waist in a hole, with only a hawthorn branch t'defend him, and that less than a foot in length."

"...Oh."

"A man could not enter the Fianna," Rahyoke continued, stepping forward, "unless at a full sprint, he could jump a branch level with his chest and stoop under one level with his knee without breaking stride."

"And," Zander added, "to be able to extract a thorn from his foot at the same time."

"He had to be versed," Rahyoke went on, "in the *twelve books* of bardic literature."

"And he had to be a poet himself," Zander finished.

Addison scoffed, propping up on his elbows. "What does *poetry* have to do with anything?"

"Everything," both men replied, sharply.

Addison raised his eyebrows. "When were *you* members of the Fianna?"

Rahyoke gave a short snicker. "*Look* at me," he bid, spreading his arms. "Do I *look* like a warrior? Do I seem the man to've trained that lad?"

The replying expression was a dubious frown before even the boy had the time to consider it. Still, he did, focusing especially on the unhealthy slightness of the tall man's frame.

Rahyoke gave half a bow. "Thank you." Turning swiftly, he made his way back to the goods piled by the door and began to re-load Kenny's cart. "There's no purpose to buying an untrained archer."

Addison sat up, folding his hands between his still-splayed legs. "Then, who *did* train Zander?"

Rahyoke shrugged.

"I'm a self-made man," Zander boasted.

Leaving the barn, Rahyoke called back, with as little effort as possible: "In your training, know this: The ancient warriors used to build up skill, and savor it. It was a mark of honour. Consequently, this drive to honour made even the savviest head-hunter reluctant for the unnecessary shedding of blood. They would send out champions where they could, and let them – two men, as stand-ins for entire armies, entire *countries* – decide the war. They never relied upon numbers alone. And they sure as Hell never relied on treachery." Returned, he stalked over to the lone crate, grabbing his bag, lifting it onto his shoulder. His last words on the matter were nearly too quiet to hear: "But those days are gone."

Zander watched as Rahyoke loaded the rest of the merchandise into Kenny's cart, waiting for Addison to get up. Looking back to his squireling, Zander informed him, "In some cases, skill is born of the drive; they call this combination 'talent'."

"Auch," Rahyoke declined. "I've known some pretty lazy-ass bastards who things just come natural to them. If they practice at all – sure it might be to challenge themselves. Mostly it's just to show off."

"Yeah?" Addison responded, at last on his feet. "Like who?"

Rahyoke cast a glance between the empty trunk and the wheelbarrow outside. Then, stepping out after it, he gave a final, taunting grin, and he answered.

"Me."

36
Nion

An dóú lá de mhí na Márta

I miss the smell of peat and heather.

I have very few memories – a thing that will agonize me the rest of my days – but those that I have of my childhood I remember very well.

And I remember, always in March, my father would take me out to the moors. He was always traveling, but he wouldn't but very rarely take me. I was too young.

But in March he would take me, if only once for the whole year. And the frost and the snow had all just melted, and the rivers were swollen and clear as ice, running rapidly over their banks, the crystal lifeblood of the land. And everything was brown and green – a thousand shades, and all so vivid and powerful and eerie and bright. And the buds and the little flowers just springing from the earth, under a sky still overcast, grey and sulking, from winter.

It was the beginning of new life – everything fresh and young and wonderfully alive! He wanted to show me what that was like. How everything could be dead and cold in darkness and despair, barren famine and chilling, frigid void. And then, from it, because of it, in spite of it, the world would rise again, in energy and life.

I understand that now.

I wish I could show my own son. To let him see the earth shake off its darkness and despair, bursting forth more livid and vivacious than ever before! But it's too stupid cold here. Why the hell is it so much colder south? Ceothannanmór gets the south-born, north-bound sea and wind currents; they almost completely bypass the Mainland.

I hope that the world can wait another month – or two – or however long it takes. Ceothannanmór never waited. My *father* never waited.

Oh, how well now I understand...

An seachtú lá de mhí na Márta

What do I understand?

About thirty-nine years ago, I think it was, the destruction of the world and everything wholesome and Godly within it was put in motion. Eighteen years ago, it reached me.

We lived at the very edge of the world, on the North-Western point of Ceothannanmór where the sea-cliffs dropped steeply into the tossing waves and the road down to meet them was tough but well worthwhile. My village, as I remember, was small, and deep enough through the moors and woodlands to be fairly isolated in winter. We never saw it coming. *They* never saw it coming.

Most territories the empire conquered they merely decimated. They fought to subdue, and permanently. When the fighters were killed and the remainder sufficiently scared, they clamped down an iron hand and held it there. That was all it took.

In Gallia they tried another method; the first wipe-through was before I was born; it destroyed the clan system and territorial lands. They kept their names, a sheltered pride, but moved into the towns and kept quiet.

Somehow or other, the empire determined there was cause to wipe through again, murdering anyone of Gallic bloodline. Then again. Three times they wiped through Gallia, 'til not even blood remained, and the clinging sinews left – refugees and implants of other nations – these were calm enough that the scouring ended.

They'd gotten the idea – for the second "cleansing" – from what they'd done to Ceothannanmór. We were descended from the Gall, and – as my one friend put it – crazier. There was no attempt to reason with the island. Like a scourging fire of hell, they devoured the land, exterminating, extinguishing *everything*. Those who could fled to the Mainland, and there – God love them – they remain. My village never got the warning, never was given an option, never was granted a chance. It had been clear since the empire first set foot on the land – maybe even before – that the Ceothanns would not back down, would not compromise, would never stop fighting.

Ceothannanmór lies barren now. By law it cannot be colonized; I don't think even a garrison has been established there, for how tightly they hold the Mainland shore. A symbol of our desolation, a testament and a warning meant to stand against us for all time. But I swear, with Christ and Mary, with Brigid and Finnian and Joseph and Patrick as my witnesses, my *allies*, God willing, so long as I live and *breathe* – it won't!

This frost is not eternal, blotting out the green of the earth and the energy of its life forever. I *want* the memory of peat-bogs and heather. I want it for my son.

If my only defiance is to keep our memories, our stories, our cultures, alive in this humble village *alone*, so be it. But damned if I'm gonna let it die.

37
Night

Heavy rain poured down without, thick as the darkness through which it cut. Lightning flayed the sky above, dousing the forest in unnatural bursts of flashing light. With a tumultuous crack like the world rending open, everything flooded in white, for a heartbeat illuminating the lank shadow in the doorway. It was enough for one hazarded glance up to the treacherous heights of the loft.

And then, the world went black.

38
Red Morning Dew

"Rahyoke?"

In spite of the weak morning light, it was very dark in the loft. Thick dew sparkled eerily on the floorboards, clinging to scattered strands of straw. Zander slid forward, relinquishing the loft's steep ledge. A slow apprehension was burrowing through his body, but, with urgency, he fought it down, kneeling at the side of the motionless figure.

He had found the crates half-toppled from where they'd been stacked beneath the loft (he had tripped on one and face-planted in the dust – ehm, like a warrior, though) and those that had fallen cracked or shattered like chestnuts, the wood stained by a bleak, water-colour rust.

Reaching out, he took hold of the man's shoulder; normally, this would have been enough to wake him. "Rahyoke? ...*Donnachadh – le do thoil!*" Now he started to panic; his stomach turned as he forced his fingers to scramble after a pulse, taking the limp wrist, unsettlingly cold.

"Oh, thank God," he breathed, relieved at the faint beating. So he was alive, at least. Zander's eyes flashed to the great hole in the ceiling, wishing for more light.

He decided to try waking the man again, and this time put his hand to his side, shaking him. "Hey –"

That did it.

With a thin noise of agony, the creature writhed, struggling to breathe.

"Whoa – hey –" Zander backed off, hesitating. The man twisted, hiding his face against his forearms; he was shaking, that single heaved breathe held, released in a muted whine. Zander again took his shoulder. "Calm down – you're alright. Calm down."

It took a moment; then, with a choke, he drew a shallow breath, settling gradually. His muscles were still tightly tensed.

"You're alright," Zander soothed. *Wishful thinking?* "Look at me... Hey – *féach*. Is that right? *Féach – breathnaigh* – what happened?"

"...uh?" Rahyoke shifted, just slightly, lifting to Zander one half-narrowed eye, almost totally covered by his hair, which was lank and still damp from the rain. *"Cad... Cad atánn... tú a rá?"*

He could just make out the words – but damned if he knew what he was saying. Zander noticed a dark, slender mark running very close to that raised left eye, and, curious as he was concerned, he pushed back the clinging hair.

"Auch –" The eye closed tightly in a flinch. Lengths of blood matted and clotted into his hair, clogging the recently-gushing wound just above his temple.

"What the Hell *happened?*" Zander hissed, his horror knocking his pitch an octave higher.

"Ní – ní thuigim, ach – stad, le do thoil!"

"Damn it – sorry!" Zander let go again. He watched as the eye opened cautiously, studying the floorboards. He wondered if Rahyoke had a concussion... and what to do if he did. He was totally at a loss. And the only man around who would know what to do, who knew anything about medicine, who could *translate* this poor, damaged wretch, was lying in dust, straw and blood, slowly slipping into unconsciousness.

"Hey – where are you hurt?" he asked. The dark eye looked at him, narrowed further; he couldn't understand. Zander swore. Casting his eyes up to God, he trawled through his limited vocabulary, piecing together a sad little sentence: *"An bhfuil...* No. *Cad... Cá... gortaithe?"* He shook Rahyoke's shoulder. *"Gortaithe?"* It was the word for "hurt", and the best that he could do. *What hurts?*

"Ní – níl a fhois – 'am... Ah – Gach rud... Tá gach rud – gortaithe – go dona..."

Zander sat back on his heels, biting his lip. No, he had no idea what to do. He would only cause the man more pain, trying to figure out the problem.

"Tá... mé... Look," he said, frustrated, "I'm gonna be right back. *Ceart go leor?"*

"...togha. Ní th-thuigim thú – ar aon chaoi."

There was a *yes* somewhere in there, though he might have been insensate to what he was saying. Not caring for the stacked boxes, Zander slid over the ledge, hanging from it, dropping agilely below. He was back within a matter of minutes, having fetched the last unopened water-bottle. He cleaned and bound the wound as best he could, using a bandana he'd found in

Rahyoke's canvas bag as a bandage. Through a half-conscious haze, Rahyoke cursed at him – in that lovely, Isle-born style: *"Go n-ithe an diabhal thú!!"* – may the devil eat you. But he didn't resist.

Finished, Zander sat back, leaning against the opposite wall. "There, you defiant bastard," he murmured.

Rahyoke held his gaze a moment longer. Then, dropping back into the same position as before – one arm covering his face – he was still.

Zander breathed a long, near-silent sigh. *And now we wait.*

The sunlight still sifted through the planks and clouds, scattering waves of dust over the floor. He turned his glance to the side, following the traipsing rays, 'til his gaze again fell to Rahyoke's bag. He snagged the strap, tugging the bag closer. For a brief moment, he saw the man's eyes flash open, surveying him.

"Do you care?" Zander asked.

"...Tusa togha."

"Okay." Zander pulled the sweatshirt halfway out of the bag, lifting the ledger from between its folds. Carefully, he began to flip through the pages, looking for something he could at least halfway translate enough to understand. His eyes alit on a drawing of a dog twisted in a Celtic knot, done in cheap pen and taking up the top right fourth of a page. The detailing was intricate, the knots embedded with interlaced vines, the body coloured with knotted patterns in red pencil. He squinted at the writing that traced, jumbling, down the page. This he could make out well enough.

And he needed to get out of this place...

39
Coll

An chéad lá is fiche de mhí Iúil

"Open the door then, Diarmuid," said Fionn. "But it's not a man of us that will help you."

And with that, he, Caílte, Oisín, and Conán, turned their backs to him and were asleep.

Diarmuid opened the door.

Illuminated by a shock of lightning that cast the world in a grave of white, amid the flooding torrent that soaked puddles into his shoes, there in the doorway was the most horrific beast of an old hag he'd ever seen. The eyes of her face bulged out of her head, set in a valley of crags and threatened by the mouth of a jaw that protruded so it could swallow the whole of her squished-pumpkin face. Her thin hair hung down in matted straggles, clinging to her icy features. Humped was her back so she was nearly doubled, her knobby legs bowed to encircle a cask, and her dark clothing rags, patches, and tatters hanging upon her. Hairy and mole-ish and grotesque, and streaming with cascades of rain.

Was that a smile – toothless but for one, and that gnarled and yellow and cracked – she turned to him, looking with such hunger in those bulbulous, pale eyes?

Diarmuid gulped. He fought down a lurching sickness when, with a voice like shrieking hinges, dying gulls, rending sheet metal, she begged, "Diarmuid Ó Duibhne, let me in."

Diarmuid cleared his throat. "As a member of the Fianna, I am bound that, anything that is asked of me, I must give, so long as it's in my power. Come in."

And in the hag came, shuffling. A great gust of wind, ice, and water tumulted in behind her, pulling the door back like a receding wave, closing it with a crack.

"Diarmuid Ó Duibhne, I have wandered the depth of the oceans for these last seven years, and no one has offered me respite." The cold came from her, and more desperate a bloodless, pure cold was not found in the white flakes scattering in wind and raining hail outside.

It was all he could do to keep from stepping back; no man of the Fianna ran before less than nine men – but was it disgrace for to flee before this creature?

"What is it yer after?" Diarmuid asked.

"Diarmuid Ó Duibhne, grant me a place by your hearth."

"That is an easy thing," he said, trying to force a smile. He led her to the hearth-fire, still burning a bright yellow-gold on the stone.

And on she came – and sure but the fire grew dimmer the nearer she approached, stifled by the cold.

"Auh!" lamented Diarmuid, for hospitality is very much esteemed. "A house is but all I can give you; see, the wood is already spoilt by the damp of this place!" Though it was not entirely what he was thinking.

"True, fire cannot warm me." There was something in the way she said it; Diarmuid cocked his head, intrigued.

"And what can?" he asked. He shook his head, clearing it. "All that I have is yours."

"Diarmuid Ó Duibhne, give me warmth."

Diarmuid started, and he took a step back.

"Come and sit beside me," she said, taking a spot nearest the fire.

Frigid sickness dragged his soul down into his gut. In a voice that was faint and hollow and hardly his own, he said, "I am bound, as a member of the Fianna, I have sworn..." And, with all the power in his body – greater than the exertion of a full day's battle – he forced his foot forward. Then, with all the strength of the Fianna – more exhausting than the full length of a war – he forced the other forward. Then, with all the strength of the Isle – harsher a debt than poison – he forced the final step, and in those three steps crossed the room to her, and sank into the seat beside her.

The old hag nuzzled close to him, her skin sticky wet with rainwater, thick with the scent of rotting tide – of salt and death. He might have thrown up, but for the cold was so numbing his body, so numbing his mind. Barely he knew, as closer she pressed to him, snuggling up to, against, beneath, his mantle, drawing closer to his fair skin and chilling the marrow through and through. Dazed, through a senseless fog, he watched as the skin of his fingers paled white – a numb, yellow white, slowly slipping into blue. Dimly the life-light flickered at the back of his fading mind, and *realized* it was fading – *he* was fading! Dragged down and down to the pit of that chill, down and down into a black, encroaching nothingness that swallowed him whole!

His hand slipped to grasp the bench, but no use; his limbs were dead, and so was he.

And it was darkness and numbing and cold.

Somewhere, a fire was burning. It glowed with the radiance and warmth of a thousand, beaming suns. It fused his limbs and stirred his blood, and – slowly – slowly, Diarmuid came awake.

His soul rose, shaking and staggering, from the black chasm of dead despair, and he realized he was alive. *More* than alive.

Diarmuid opened his eyes.

Or am I?

Brighter than ever before, cheerier and more golden an amber than ever red flame flared, the fire danced and leapt and reared upon the hearth. A creature was bending over it, and – by all the gods – she was the most beautiful creature he had ever seen. Her eyes were deep a stormy grey that seemed to flash in the light, flickering with dark green, with olive, with pale blue and with sapphire. Her skin was of the whitest purity, her lips like the reddest, sweet strawberry. Her hair was thick, and dark, tumbling over her slender shoulders in undulating waves, crisp as the night air settled and stilled in the clouds that lingered outside. Her dress streamed like the running river-waters, flowing down her body in gentle folds of pale puce, and silver – glittering like a shaft of moonlight dancing over the sea-foam – was the cloak that wrapped around her.

Soft was the smile she turned on him, a gentle curve of her ruby lips.

Straightening, she came towards him, like a dream. And it was warmth that radiated from her, sparkling and gleaming and sweet as a summer dawn. She walked with the flowing of calming waters, with the grace, the silence, of a soul, and the beauty of the *sidhe*.

She knelt down beside him, and *close* was her face beside his. She smelled of summer breezes, wafting through the apple blossoms and long grasses, racing across the tumbling waves and scattering into heaven beyond; delicate and beautiful, luring and intoxicating.

"Diarmuid Ó Duibhne," she said, and sure his name never sounded sweeter than if the thousand bards across the centuries, and all the birds in their choirs, had sung out the words. "I have wandered the depth of the oceans for these last seven years, and no one has offered me respite."

Through the translucent veil of her musical voice her words shot, stinging Diarmuid's strayed consciousness. He jumped.

And she leaned forward. "Tell me now, before daylight again takes hold of this world: what would you have done for you? Nor hesitate in asking; the greatest thing which you could ask is but a trifling matter to me. For even as I was the hag who came to you, I am King Under-Wave's Daughter."

40
Pellucidity

hu're you doing?"

Zander jumped, looking up. Narrowed eyes watched him through the gloom.

"Are you alright –?!"

"Just..." Rahyoke lifted a hand as Zander scrambled to his knees, gesturing for him to settle. He did reluctantly, sinking back down, though he sat at attention, the book clenched tightly in his hand. "...uh'you have?"

Zander glanced at the book he held, wondering himself. "It's yours," he admitted. "Are you alright?"

"Nnh..." As if to evaluate, Rahyoke moved to sit up, but fell back on the instant, writhing. *"In ainm Dé!* What the *Hell* –?!"

"You're injured!"

"No shit! Aukh." He groaned, as much out of frustration as pain. Pain is too small a word; the world was edged in black, voided like the blank slate of his mind by that white-hot, stabbing discomfort. A pain that was so vicious, so voracious, as to take away his movement, his breath, his rationality... *"A Dhia, déan trócaire."*

Although these words had often been shouted at him out of sheer annoyance, Zander had a feeling the prayer was honest now. "To answer you... Are you cognitive?"

"Yeah."

"To answer you, I'm not sure what happened. You've been out pretty much all day –" He faltered briefly when Rahyoke turned to him, listening. "I – you're wounded a couple places, far as I can tell, but I'm not sure where. I tried bandaging the wound on your head, but... *you* know I'm not a trained healer. And I couldn't ask *you* about it, because I couldn't understand a damn word you were saying! What's *'Ní thuigim'*?"

"I dunno."

"How do you *not* –"

"No... no. It *means* –"

"Oh." Yeah, that made *way* more sense. After considering it, hopefully committing it to memory, he continued. "That's about all I know myself, either way. Do you remember anything that happened?"

Rahyoke shook his head.

"You understand what I'm saying to you, right?"

"Hmm." Rahyoke twisted his gaze to the ceiling. He said nothing.

Zander pursed his lips. Was that a "no", or was he just being difficult as always? Either way, it wasn't really what you'd call a *step up* from this morning. With a sigh, he sank down farther, studying the book. "There's no way you're gonna be able to work."

"I'm feelin' a lot better now, actually. Just kinda tired..." Rahyoke studied Zander's expression. "It's not catchin'."

Zander looked at him irritably. "I know head-wounds aren't contagious."

"See? Ya know more about medicine than ya give yerself credit for." Rahyoke limply checked the bandaging on his head. "Huh. Not bad. It would'a been fine, but I'm grateful for your efforts... Where did I go?"

"Paradisius."

"*That's* right... *Ar an drochuair bhí dul amú orm.* That was a poor choice."

Zander gave the ledger he held one last exploratory look, then tossed it back onto Rahyoke's bag with a gentle flick of his arm. "So which town should *I* go to?"

Rahyoke stiffened, freezing like a startled cat. "What?! No! No no no – you can't –"

"Rahyoke." He spoke the name through clenched teeth, commanding the older man's attention. "Can you for *once* stop over-protecting the Hell out of everything and be rational? Yeah, we have money now, but ya can't eat gold. And we need to stock up on supplies. Otherwise this entire endeavor is futile. I'm *going* into town. It's up to you, which one. Unless you want it to be Paradisius –"

"Oh God!" It was half choke, half agonized moan. Bracing himself, he rolled onto his right side, shifting so he could prop himself up, looking at Zander all the while in earnest desperation. "*Please,* Zander – *fan nóiméad, más é do thoil é!*"

"What."

"Rationally – okay? – rationally –" His forearms pressed to the boards, he managed to lever himself into a weak cobra pose, to better see the one whom he implored. "You shouldn't have to go into town. What if –?"

"Rahyoke, there's too much immediacy in our need."

"Zander, I can't let you." He emphasized each word. "Look. You're right. Okay? You're absolutely right. But I... *can't*."

Zander was silent a moment. In that instant, above all else, he was keenly aware of two things. First, that Rahyoke was terrified. Second, that the wanderer was utterly powerless. There was nothing he could do to prevent any of his fears from being realized. For once, that responsibility was Zander's.

"I appreciate your concern," he began, quietly, "but I'm not a little kid anymore. You trained me well. I know how to handle the merchants, the guard. And, if I have to, I know how to fight."

"You take too many risks."

"What isn't a risk – you told me that."

"Yes – and it turned out *spectacularly* for me." Rahyoke tried to sit up, but couldn't quite make it; halfway to the full cobra position, his palms pressed to the boards, he stiffened with a grunt, bowing his head.

Zander took his arm, both supporting and commanding him. "Rahyoke, I *know* I can do this," he said. "Let me prove it to you!"

Rahyoke met the boy's eyes, searching. Searching, for the ante in this game of risk was insurmountably steep. Then, his eyes never wavering, he took Zander by the wrist. "*Geall dom,*" he said. "Promise me nothing will happen. Promise."

"I swear it by all the gods our people swear by."

"That isn't funny, Zander."

"Sorry. I promise, okay?" he vowed. "Everything will be alright."

Rahyoke's eyes narrowed, sifting through his words. Then, in a low voice, he said, "If you're outside the barn, facin' the back wall, the corner at yer right. Put that to yer back, and walk straight, without deviation, and you will come to the former well of the goddess Sirona, or Saint Serena. You'd do well t'give her an auld prayer for your safety, if you can manage it. These days it's a pump-mill, which carries the water to the *thermae* in Paradisius. When you come upon the building, that pipe'll be headin' off t'yer right. You are going left, y'hear me? Dead left from that building, an' on 'til ya reach Dryden. Take back Kenny's cart. He'll be at the West end'a town, which is –"

"The opposite end from that I'm entering."

"Thank you," Rahyoke breathed, with slight relief. Maybe, just maybe, Zander could do this after all. "He'll be in Carrie's. Don't go in; have'im come out to you. Tell'im you've brought Aveline's share. Bring the nicest stuff we have left – maybe Meygan can help you pick something out. Once Aveline's paid ya and taken the goods – do make sure she pays ya – and Kenny gets his cart, head out. Don't let anybody know your name. If anyone asks, say you're a bought messenger. Don't accept credit, an' don't spend my money on booze."

"Not even a little?"

"Not even if it's free."

Zander almost sulked, but decided against it. "Okay. Where do I buy stuff?"

"Leaving Carrie's – your back to the door – head straight. There's a building across the street, cut to its left, and carry along parallel, through the woods, 'til ya get to the town of Ellery – you'll be headed due North. There's a nice farmer's market in Ellery."

"Perfect."

"At least, there use'ta be. Oh, God – I dunno. It's been a long time since I been up here. Maybe –"

"Is there any super specific secret shit I gotta do in Ellery?"

Rahyoke was silent a moment. *Maybe...* Then: "No. Make responsible use of your merchant-handling skills. And when you leave, make sure you take the east exit, and follow the road, which'll bring you back to the spring. Can ya make it back from there?"

"Yes." Regardless, Rahyoke made him repeat back the directions. "Don't worry," Zander added, at the close. "It'll be fine!"

"I do worry. I shall always worry," Rahyoke intoned. "And I know I shouldn't let you go. But..." Very slowly, he unwound his fingers from Zander's wrist, letting go. "I'm trusting you."

In spite of the intonation, Zander heard the slight, buried waver in his voice. His fear.

He gave the most confident, reassuring grin he could muster, gently clapping a hand to Rahyoke's shoulder. "I won't let you down." He slid to the loft's edge; although Rahyoke kept his head bowed, he watched the boy's every move. "I'll be back sometime tomorrow, okay?"

"Just..."

Zander paused, one leg already dipping over the precarious edge. Rahyoke raised his eyes. He looked tired, he looked worried, and, for a rare instant, he looked almost old.

"Just be careful."

The night passed excruciatingly. Zander had asked Meygan to check on Rahyoke, if she didn't mind, and accordingly, as morning was dawning, she came to the man's side. There she knelt, wondering whether she should wake him up, and if so, how; for although Zander had had the decency to mention the injury, he had lacked the sense to mention *where*.

In the end, she reached out, placing her hand on the man's shoulder. "Rahyoke?" She felt a vague tightening of his muscles, and, gingerly, she shook him. "Come on, honey; are you awake? Come on..."

He gave a weak moan, mumbling, "*Caitlín? Cé an rud – cad é atá cearr?*"

Meygan was taken aback. "What? I can't understand you... Rahyoke –" She gave his shoulder a firmer shove, jolting him.

"Huh – wha'?" He was awake.

"Sorry! Sorry..." Meygan sat back on her heels, releasing him. "Did I hurt you?"

"Ss'ee back yet?"

"Wha –"

"Is he *back* yet?" he demanded, clearly. "Is he back? He should be back –"

"No, he's not."

Rahyoke's gaze dropped. He was clearly troubled; half to himself, he muttered, "He's languorous." It always took the boy twice as long to do any errand upon which he was sent.

"He asked me to watch you," Meygan said, glossing over the fact she didn't know "*languorous*" any more than she knew "*cad é atá cearr*"; after all, she was only twelve. "I'm sure he's on his way back."

Rahyoke looked at her. The lavender scarf was tied on her head like a great bandana, covering her hair. His eyes were captivated by it. "...Meygan?" His voice was cracked and raw and sanded to near nothing.

"Yeah."

He didn't respond right away, wrapped in whatever dense thoughts he was miring through. He shook his head, clearing them. "How long s'it been?"

Meygan frowned. "For what?"

"What day is it?"

"Oh." Her frown became more pronounced with concentration, and her eyes studied the rafters. "There's a hole in your ceiling."

Mutely, his eyes flicked off to follow her gaze.

"You've been here all day, and all night."

His eyes returned to her, widened. "That long?!"

She nodded, still studying the hole that was not quite above that in the floor.

"*A Thiarna Dia,*" he grumbled.

Meygan looked back to him. "You don't take care of yourself, that's the problem."

"Ha. That's the problem, is't?" Sarcasm laced and lacerated through his tone. His eyes drifted off to the overcast sky beyond the smashed ceiling, though Meygan continued to watch him. All of his energy appeared to be going into looking strong for her sake, mimicking as opposed to striving for a real healing. And healing was what he needed; Rahyoke was typically disheveled and tired, but she had never seen him as beaten-down as this. "What happened?" she demanded at last. "How did you get hurt?"

"Hmm..." His mock consideration, much to her dissatisfaction, ended with a half-shrug. "S'nah important."

She folded her arms. "Maybe I'll use my third question, then."

"Wun' answer you."

"What! You get a pass?"

"You can have some, too."

"Are you ever gonna ask me anything, by the way?" she inquired, leaning forward. So far, Rahyoke had gotten less than nothing out of their deal. "That's not my question, but – I just feel like I owe you."

"You don't."

"*Are* you?"

He looked at her. With her vibrant hair obscured, it was easier to see her face, her fine, Renaissance features lightly tanned from the occasional sunlight, her brows and lashes dark, like those of a brunette. He looked away again. "No."

"What!? Why *not?*"

He did not answer this question either. "D'yeh see? S'gettin' light out." His voice was very faint, nearly enough so to disguise his complete departure from the topic. At the same level, he added, "Maybe he'll be back soon..."

Meygan drew a deep breath. She nodded. "Probably." The sky was a lightening, lavender slate, the sharp, black lines of the branches frothed with clattering leaves. "What's the other language that you and Zander sometimes speak?"

He looked at her but said nothing.

"Come on; you were just speaking it!"

"I've a head injury." Gritting his teeth, he huddled more deeply on his right side. "You oughtn't take anythin' I say seriously. Go down and get somethin' t'eat."

"It's too early," she said. "And anyway, Zander told me to –"

"F'rget Zander. I'm taller an' scarier than Zander, anyway."

She arched a brow. "*Scarier?* I don't think so."

"Ah? Well, go on all the same." His hand shook as he raised it to push the hair out of his eyes; the rolled bandana was wrapped a little too loosely and was starting to slip. "Don't worry about me. Zander's just –"

"A worrisome hen-wife," Meygan parroted, flatly. She snared her backpack, pulling out a biscuit. From his own bag, she retrieved the water-bottle. "Here," she said, pushing both towards him, placing them close enough to his arm that he would *have* to do something about the food. "Eat that."

His eyes fell to the offering, taking a moment to adjust. "Bread an' water."

"I'm gonna go down," said Meygan, slinging her bag onto her shoulders and edging to the loft's steep ledge. "If you need anything, you tell me."

His smirk was lost for the lighting. "Sure thing."

"'*Need*' is not the same thing as '*want*', Rahyoke!" she warned, lowering herself down onto her fortified tower of crates. "And it's *my* standard of needing, not *yours*, you absurd creature!"

The slip of a low, unheard laugh, which he quickly cut off. "Ooh..." After a moment, he recovered, summoning just enough strength to answer her: "Hen-wife!"

Meygan heard him, though she was already below. "The *thanks* I get," she muttered, shaking her head. Pilfering a biscuit and the black-market jam, she retreated to her quarters.

For now, her job here was done.

41
Muin

An chéad lá de mhí Lúnasa

There was a Moorlander who sailed with us – Aneirin was the name on him. We picked him up off the coast of Sefetaten, adrift on the choppy waters, in a pathetic vessel of his own craft; this was when I sailed under Hennighan, so my first voyage proper. The chances are steep Aneirin was an escaped slave, though he never told us either way. Since the war, Moorlanders are found in droves throughout the Mainland coasts of Gallia, Sefetaten, and even Turranbar, though in the latter two countries it's more in the grips of the slavers they are; I know first-hand. For that, it's actually or it was a concern of small Gallic commercial vessels, their crew to be plundered in Sefetaten, and their men to all be enslaved – 'cause where does that leave you as a captain, sitting with two tons of cotton and a leaking boat, and not a man to shove 'er off or clean up the pieces.

I remember Aneirin well because it was he was the first soul I'd met since leaving my homeland who could speak the same language as I, and he more or less took me under his wing. I was six at the time, and hadn't had a mutually comprehensible conversation with anyone in two years.

In addition to other talents such as science and "tab-payers", Aneririn was a remarkable piper. When he'd be off-duty and the weather nice, he'd go up on deck and play. One day I was up, swabbing the boards with bucket and rag, and himself playing a wonderful tune. Sweet was the sound of it – ideal for the clear morning, blending so crisply and finely with the wash of the waves and the creak of the wind through the ropes. I'm a hopeless romantic; what can I say?

Abruptly, Aneirin looked up – looked right at me with those piercing eyes, and he asks me, he says, "Dych chi eisiau dawnsio, gwas?" – essentially, *Wanna dance, lad?* I had long ago given over my working, not realizing, staring at the man in awe; at his words, though, I balked, answering him with a shy shake of the head. But he'd none of my excuses. "Allwedd tlodi, seguryd," he pointed out; *idleness is poverty.*

I gave him a look; even at six, I knew slavery certainly wasn't wealth – and a slave I was, or did I not mention, Hennighan having bought me in Osset. I shook my head; back to my rag with me, and, shrugging, Aneirin went back to his playing.

At length he began a song I recognized. It was a song my mother used to sing – a song my mother had danced to.

I was older now; maybe I could, too?

It's one of those hand-clapped-to-the-forehead instances when I reckon what that first dance must've resembled – nothing of my mother's grace, I'm sure – but I didn't have to look at it; I *felt* it. And it was a feeling I realized I loved.

It's – your whole body, your whole soul, is poured out and emptied for the grasp of the beholder, and in its place is the music – that driving, ancestral rhythm. It is your heartbeat, it is your lifeblood. It bends and twists your limbs with incomprehensible passion, and – twisting, writhing, soaring, flailing – Lord, but you obey. You see not the crowds, but you feel them – not as entities, but they are energy, too. An energy that you engage with – your response is your own, whether defiance or compliance to their mood, their whim, their will.

Then I felt this energy, and the energy of them was pure, gold in the darkest of hearts – in the eyes of those sinners whom I knew had pondered before how easily such a small body could be lifted up over that rail. Even the Devil can see beauty, *a stórín* – not that the dance was in excess of it; but there was some beauty in it, too, just the same.

Oakh: and how much of money, *a thaisce,* do you fancy is made from such talent? *Allwedd tlodi, seguryd;* Aneirin made out well each time we docked.

The first time we gave it a go was the Lughnasadh festival in Mabu. This was the haven in Sefetaten for Moorlanders – where they came together in great enough numbers to be safe; where they were far enough from the Augustan capitol to get away with so distinctly cultural a festival as Lughnasadh.

Lughnasadh festivals are always orange. At least, they are dock-side. Bright is the night with lanterns and trash-fires – the colours of harvest and the light of the setting sun. Shadows and gold-leaf, pumpkin and black-licorice spice, everything warm and inviting even through the mystery of night. When you get older, you watch these streams of colour whirling – churning, spinning in a reeling dance, stirred in the amber dregs of fresh mead, weaving though the fields as the great stalks are mowed down by the ever-flowing wind.

At its inception – och, *way* way back, before even the first saints walked our fair land – Lughnasadh was a great holy day, the day of the god known as Lugos (thus the name). Then it was that all the kings and their followers would gather, and – between business meetings – there would be contests of speed and strength, and afterwards everyone would party. Och, it must have been a great time they had of it, too! By my time just the partying and the merchantry remained – the politics and worship having gone away out of it – and even that was waning. Mabu was a bit of an oddball in them days; see, the only other examples I know of Lughnasadh celebrations in the past quarter-century were the somewhat-politicized festival in Lutetia and the harvest market in Callatis, both Northern Gallish ports. The Lutetia festival was shut down after a rather nasty riot in aught-six, and the format of the Callatis festival was forcibly altered I forget what year, so the cultural, not the commercial, aspect was thoroughly wiped out of it.

You can glean; the ports are a little more lawless – I think because of the vast cultural diversity and ever-shifting population. Although I haven't spent much time on land, what I know of the inland cities is they tend to be a tad more reserved about their own boorishness. Maybe because they don't have sailors to blame it on.

As for this sailor, the festivals began as and remained times of immense profit for me. I could sell my soul at those little gatherings; street performing was bread. It was also meat and cheese and whiskey. I trained myself to supply the need. Do you need a dancer, *a ghrá*? A knife-thrower? A fire-swallower? The more dangerous, *a chroí*, the more eagerly I learned it.

My debut was at Aneirin's hands. We used to pull into port for the harvest festivals and to take the produce at this time – the latter the legitimate reason, the former solely what *I* got out of it. Onto the cobbles by the docks with me; the Moorlander rounding up some other three blokes to join in. Their instruments fascinated me, like relics of a distant dream. I'd been promised a lesson at the pipes if I danced – another trade I would benefit from in time.

And so you gather, if by nothing other than my sudden urge to make the time to write, I'm feeling nostalgic again. And I told Sean, I said: "We should have a harvest festival."

He frowned at me, arching a brow – like the idea was so bleedin' crazy. "For *what?*" says he. "We haven't anything to harvest!"

I pondered this; for there was a ruddy truth. So maybe there was a point in it far beyond my sentimental urge to party. "Maybe we should," I says, right slowly.

Wouldn't it be nice if we were self-sufficient?

An deichiú lá de mhí Lúnasa

Winters are colder in these parts; Lúnasa is just a stark reminder of that. In its heat there is a staleness, bitter as dried herbs passing over a parched tongue. That gold and flaming orange and scarlet slowly ebbs, seeping into the leaves, as all the world around it crackles dusty brown and grey. The hint is there beneath this vibrance; the world can't support such energy for long.

Why am I so depressing? Shorn – to ask, you've never spent a winter here, *a ghrá*. I've complained to you before, in fleeting, of the cold, and, damn it, I'll rightly do it again. It's a cold more than the pure white-blue of the flame – it is grey. It is marrow. Cracked and caked and splitting, fracturing, gripping. And, like glistering frost, it coats everything. It is *alive* – sapping, sheeting – *in everything.*

And winter, *a thaisce,* is long.

I sit on the stoop, as I do now, reveling in the hot, burnt-mead fuchsia of the climax of the summer, trying to soak it all in. Tryin' to fortify myself with its elixir, that saturating peach ember, in the hope that if I absorb enough heat, enough of that tangerine beauty, I won't flippin' die by October.

The child is abed. My wife is making tea – fair life to her. She will be joining me in a minute. I've no further use in writing for now.

An tríú lá déag de mhí Lúnasa

I used to walk the decks at night. *Especially* during a storm. It annoyed some of'em, but I was more than expendable, truly, for however skilled at my job. And no one paid me much heed.

I remember this night:

I leaned back on the railing in front of the helmsman. I was in my teens, whatever that was. "They say on nights like this, the Ladi Wen even prowls the seas."

I had spent the last decade or so learning Augustan, and could finally manage enough for narration.

"Is that more'a yer gibberish lore?" himself asked, without interest. "Sure, you sound like me auld father."

"Sometimes ye sound like yer father yerself," I murmured, though I'd never met the man. "The mist is thick anuff tonight." I leant forward off the rail, taking hold of the helm from the other side. "See them rocks yon?" I asked, pointing off into the mists and spray. "How fast she'd break apart yer ship. An' you left, a milky corpse bobbin' on the waves 'til she comes, laughin', to collect ye. An' yer soul she'll hold forevermore. An' neither heaven nar hell'll receive ye."

"*Knock* it off," he says, pushin' me, one-handed, back from the wheel. I laughed, flitting over to the port-side rail.

"Have I told yez of the Merrows, Jamie?" I asked.

"You have, just as I've told you to stop callin' me Jamie."

But I haven't told *you, a chuisle.* Merrows are very akin to selkies, save it's they're part fish, not seal; the men, ugly as all sin, but the girls – the *girls!* – save for their scaled tails from the waist down an' the thin webs between their fingers – *them* I would crash into rocks for. And how they're alike, at all – 'cause I suppose you'd be wonderin' – is the Merrows each have something of red on 'em, a scarlet hot as the blood; usually it's said it's a feathered cap, but a sailor friend'a mine swears to the death that the Merrow took his brother had a red band tied around her shoulder, and I've heard from another man of one who'd a red sash around her waist. And if you take that red thing, and you have it or hide it, the Merrow is yours, 'til they get their red back again. And it's only the self-same stolen object which will do the trick – and it unmolested. Now, it's a helluva bit of work to molest a faery cloth, but another lad I know, called Turlach, had a wife who always wore a sad, shredded cloth, tied like a head-band, in her dirty-gold curls. And he swore up and down she was a Merrow, and he sworn only never to tell how it was he destroyed the cloth.

"Auh, it's a worse night for Merrows, anyway," I said. "*More* the sort for the Ladi Wen. The Merrow thrives on her bein' *seen, not touched,* so. And damned if a flick'uver you could see with all this fog."

"Ah. *Now* yer makin' things up. And this, from the excess of rum in yer little head. Away and sleep it off."

"No really!" I protested. "Scanter are they in water, I said, and especially so far out. But ever a gust of wind did sweep o'er the plains of Ceothannanmór, I can feel them here tonight."

He rolled his eyes. "And *what* do you feel, but the tingling of drunkenness?"

"I amn't drunk tonight," I warned him. I slowly glided down the deck, my hands trailing over the rail, my eyes lost among the waves that tossed beyond that fog. "It's a night like this they come for a man," I whispered, just loud enough for him to hear. "The whole of her pale, near translucent like the shifting moonlight on the snow. Her hair, caught and matted like damp kelp, tanglin' and untanglin' on the wind, floatin' suspended on the breeze even as it clings to the pale skin. Her dress is tatters, hiding her feet, and there she drifts. But her eyes. Her eyes are those of your lover, sparklin' and shinin' with light. And ye can look past the skin as pale a grey as chalk, and see the softness of the skin. Imagine warmth. As beautiful as it is perfect.

"It is she will lure you off the dock and drown you. She will smother you in a bog. She whose shriek sets hearts to bursting, and keens them after they've gone.

"The White Woman. The Ladi Wen. The Banshee." Thus did I end it. And – beautifully timid – an albatross wept overhead. Even I started; it was bad enough for the omen, but the blood-chilling story on top of it!

"Auch – *now* I can't sleep!" I sputtered, and I leapt back up to the helmsman, hiding behind him, clinging to his shoulder. "Singing lures them," I said. "Did you know that? Both the *ban-sidhe* and Merrows."

"Serves ye right," he said, though I saw him make the Cross, "for all yer imitations."

I might explain – I knew the superstitions about sea birds – all uv'em – and had a tendency to imitate their calls, just to freak out my fellow sailors.

"I'd not be singin' this night if ye paid me," I went on, slipping out in front of him – to where I'd been before, the other side of the helm. "Though they like a creature of their own lot." I raised a hand then, makin' a little fire dance along my fingers. I was still learning, God help me (as I am ever learning now). "But I'm afeared they'd like me *too* much, so."

"*Jesus* Mary and Joseph!" The one hand again blessed him as the other swatted my arm, breaking my concentration and expelling the flame. "Enough of your witchcraft! It's a *wooden vessel* we're on – and that in the midst of the sea! Ya charlatan..."

"It's *pyrokinesis*, Jamie," I sulked, sauntering off into the mists. "And sure but the Merrows fear it.

"There's no help for a man 'gainst the Ladi Wen."

42
Blood in the Lethe

The pages slipped slowly through his fingers. He was late. Good God of Mercy Everlasting – he was *far* later than he should ever, *ever* be. And the sun's cool rays were casting down, the shadows spinning round the dust. The evening was drawing on, rolling starkly to a close, and still, no returning footstep had graced that threshold.

He couldn't sleep. He needed to – his body begged and pleaded and deluded him towards it, but – shorn, no chance in Hell that he ever could. Not while the lad was so far away – so much in danger – so *late*.

He laid his head on his arm, too exhausted to retain his position longer. *I should never have let him go,* he thought, that one, overwhelming dread echoing through the fog. *I let him go, and now I've killed him. I...*

"Rahyoke? Yo!"

He jolted, opening his eyes. "How'dih go in town?" he blurted, before he was even fully awake. He moved to sit up, but Zander laid a hand against his shoulder, stopping him with shaming ease.

"It's fine, just relax." Zander was crouching beside him, anxiously glancing him over. "What the Hell happened to your side?"

Rahyoke gave an aggravated sigh. "Uh. *What.*" Half killed with stiffness, he had at last managed to roll over onto his other side (no one knew, but it had taken him an hour in doing it). The way he'd twisted, his shirt had caught, getting pulled up so that almost the whole of his right side was visible. He winced, swearing, as Zander pressed a hand to the bare skin of his lower ribs.

"You're all bruised!" Beneath the twisted ropes of an old scar was the thickest, blackest contusion the eye had ever beheld.

"Please stop – *Zander.*"

He jerked his fingers from the bruise. "Sorry! What happened – have I murdered you?"

God, but he even *sounded* innocent. Rahyoke drew a shuddering breath. "I don't *know,*" he reiterated, sounding vaguely annoyed. "I'm fine."

"So, you're gonna die slowly of internal bleeding."

"Mercy, Zander, I'm alright." Rahyoke shifted, gritting his teeth and hoping that Zander wouldn't notice. "It's – it's just bruised, is all. *Don't* you look at me with that Balor-esque face on ye –"

Eye-roll.

"—and I'll have none'a that, neither! Either!" Damn it. Rahyoke laid so his back was pressed against the floor, trying to pull his shirt down over the wound. Balor had been leader of the Fomorians; one eye of his had been struck with an enchantment, so that it was always closed, the lid so heavy it needed to be lifted with a metal chain. The reference was to the other eye, that which Zander cast on him in scrutiny, which bulged out as if to make up for the deficiency.

Rahyoke sighed, briefly closing his eyes. "God, I can't lay here anymore."

Zander frowned. "What are you talking about?"

Rahyoke's eyes flashed open, focusing on Zander with such glaring intensity that the archer actually pulled back. There was a clarity in that dark gaze that had been missing the last two days. Through clenched teeth, he snarled, "I – can't – *stay* – here!"

"Yeah, wha – don't!" Zander grabbed Rahyoke's shoulders as he sat up. "Hey!"

Rahyoke flinched, holding his side. But his anger – his sheer, ferocious fury – was enough to support him. "Wa' – wha'd'you – *expect* me ta' – t'do?" he struggled, drawing breath shallowly against the pain. "Just –"

"*Yes,*" Zander grated, severing the tirade before it began. "I *expect* you to sit here and rest! For *ONCE* take care of yourself –!"

"My responsibility," Rahyoke rasped, "is to you."

That tone – the lowered voice, rife with a warning – was usually enough to silence him. But Zander rallied.

"Then rest," he returned, finding a growl of his own.

Rahyoke's eyes rounded slightly. He was impressed more by the archer's boldness than his words. Slowly, reluctantly, and with an air like it was his own intention, he laid back down, never breaking from the lad's gaze. "You're not going into town again."

"But –"

"*No*. An' fine but I won't go, either – not yet. But I will'unt be sendin' you into such danger again – under no circumstances! On pain'sa *death*, you're not t'go back into town, do you hear me?"

"Alright! Fine!" Zander held up his palms.

"A week," Rahyoke grumbled. His eyes flickered away, tracing over the dense shadows as they gathered in the cobwebs above. A great deal of the vagueness had returned, and his voice was soft, soliloquizing. "I'll not have it said I lounged about, a shirker and a coward, my children languishin' an' dyin' around me."

Zander bit his lip, not sure if he should've heard him.

"Wouldn't it be perfect if there were termites through this whole bloody structure?"

"You broke your ribs," Zander decided.

"Did I use 'will'unt' a moment ago? Lord, I gotta stop readin' this..."

Zander rolled his eyes to the place that seemed to so fascinate the invalid. "Settled. You've shattered your ribs, thereby destroying and rupturing with shards yer spleen an' your pancreas, and are dying a slow and horrible death. Is that it?"

"If you could tell me their location, those organs, I'd be *most* impressed."

"Maybe I ought'a look at that head-wound?"

"I'd feel it if it were a fracture." Rahyoke's eyes returned to Zander, studying the lad's face, though the boy refused to gratify him by returning the gesture. "Zander, trust me; I'm *far* longer at medicine than you." With each new sentence, his harsh tone softened. "I promise, I'll be alright."

Faintly, Zander murmured, "I can't tell when you lie."

Guilt knotted in his throat, and this time he *couldn't* answer the boy. But Rahyoke continued to watch Zander, wishing he would return his gaze. The deep shadows of the recessed ceiling were not interesting; he knew. He'd been staring at them for the past few days.

He closed his eyes, pressing the heel of his hand to his forehead. *What the Hell happened to me?* "Di'jyou give the cart back to Kenny?"

Zander ran his sleeve over his cheek, the back of his wrist pressed below his eye. "I think so."

"Wha'd'ya mean, you *think* so?"

"He *said* he was Kenny."

Mercy. "Y'*got* food, right?"

"Yeah!" Zander scowled, leaning forward confrontationally. He saw Rahyoke's faint smile, sardonic and haughty as always, for however weak, and he withdrew his challenge. "Mm... Maybe you could at least come down."

"I can't."

"Why not?"

Rahyoke sighed, and admitted, "I screwed up my ankle."

"How?"

"I don't remember!" Dropping his hand, he met Zander's concerned gaze – with a look conversational, a tone delirious, stir-crazy, punch-drunk, insane. "I didn't notice 'til this mornin'!"

Zander sat back a little, eyeing him incredulously. "Are you okay?"

"Aauh." With this groan, Rahyoke covered his face with his hands. *This is madness – complete and utter madness – this is – A Dhia...* Trapped. The oblique net of suspension between living and death. One foot in the river – half called to life, its motions remembered with tantalizing palpability, half sucked in the drain of sweet oblivion. How sweet? *Tread or drown.* He lowered his hands so they covered only the lower half of his face, meeting Zander with wide and plaintive eyes. "I bet I could stand," he whispered.

"No!" Zander rebuked, shaking his head emphatically. "What did I just tell you?! Rahyoke, no."

"I bet I could."

His tone held the innocence of a child – the sheer and curious ignorance. Zander frowned. *No, you can't.* But Rahyoke wouldn't heed words. He was going to keep trying, no matter what anyone said. He would either stand – walk – work – or die trying. Zander's eyes fell to the bruise.

"Nnnnhhh!!" Rahyoke flinched, twisting so his back was to the antagonistic child, curling into a ball.

"Now stand," Zander ordered gravely, keeping his hand pressed to the wanderer's side.

With the last strain of a dying breath, Rahyoke hissed, *"Bastard!"*

"That gives a little; are you sure it's not broken?"

"Not *yet,*" Rahyoke grated, through clenched teeth. He lashed out with his right hand, grabbing Zander's wrist and wrenching it from his poor side. "What're you at, huh?!" he demanded, strength and fury flooding into his tone.

"You can't travel like this – *you can't,*" he emphasized fiercely, as Rahyoke rolled his eyes.

Under his breath, Rahyoke muttered a long and foreign oath.

Zander sighed. "Give it a day, Rahyoke," he begged. "For my sake."

Rahyoke narrowed his eyes, falling silent; his closest form of surrender.

"Unbelievable," Zander reviled, shaking his head.

"One day," Rahyoke sneered, quietly. "Look where *one day* got us." The implications were clear.

"I'm gonna bring you some food," Zander muttered, coldly, moving away. "Have you not been eating?"

"That's not true; I had a biscuit."

"Oh. *Well* then." Zander rolled his eyes. "I'm getting you food; I don't *care*." He slid, easing himself over the ledge.

"Zander?"

He paused. "Yeah?"

Shakily, Rahyoke propped himself up on one arm, his forearm pressed flat to the boards, for as much balance as it'd bring him. The effort had overpowered him, draining the blood from his face and the oxygen from his lungs, but he would not lay down. "T-th-*thank* you," he breathed, sincerely.

Zander averted his eyes. "I said I was with you, didn't I?"

Very faintly, Rahyoke smiled.

"Get some rest," Zander bid, shifting away again. "You look like Hell." Then he was gone, dropping into the world below.

Rahyoke listened a moment, reassured when he heard the boy swearing at one of the trespassers for not being asleep. Then he settled back, shaking near-violently from the effort of keeping steady; only when he was sure no one would hear him did he breathe, shuddering and shallow, strained with exertion.

He turned his head, studying the book in the faded light. His fingertips slid along the floorboards, slipping onto the soft, linen cover, before finally, gradually, going limp.

43
Cúiteamh

It was white.

Cleansing sea, obliterating light. Death before Hell.

Focus, focus; through annihilation, the shifting shades of steely grass. Hemmed by the mist, hemmed and swallowed. Stark were the violets – the purple petals of orchid-like violas, against the grass.

The white flowed; swirled in, snapping tendrils. There she was. Kneeling in the grass, back to him. Her slender hand sifting through the long blades, methodic, graceful, entrancing.

Please...

She wore black, her long skirt hiding her skin. Her fingers brushed the purple head of a viola, curled slowly around the stalk, breaking its sinews gently. Her hand was so white – so pale – reflecting the – swallowed by the light.

Please – just...

Her hair, brown and soft, was restrained loosely by a green band; the long strands flowed over her slim shoulders, glittering in the light. It shielded her from view.

Her name echoed soundlessly across the vacuity, dropped from an empty heart. Unvoiced, unspoken, filling the muted void.

Her hand was still, holding the viola suspended above the earth.

...turn...

Her head bowed, slowly.

That clinging gossamer white. A shade against the brightness, obscured against the fog – clarity was memory – clenching, stealing, taking, this devilish mist. The unfelt breeze slipped through her pretty hair, the long strands streaming, straining her profile.

I'm sorry... I am so, so sorry.

She turned –

Black shredded across the sky, boiling like blood across water, light, obliterating – darker than night, darker than hellfire, hashing, searing, consuming –

Gone.

Lightning ripped through the sky – deafening its thunderous crack, roiling the black. His hand struck the wall. Shadows cut like tar, oppressive, cascading thick, lethargic, molasses over vision, over air, the slapping palm that clawed the dripping wall. Drowning silence, waves of pitch – that suffocating weight, heaving, pressing, gushing, devouring, impenetrable, as death – He twisted, a cry stifled in his throat, lost beneath the roaring wind, that overpowering, crushing silence –

The sky.

He raised his eyes, drawing a shuddering breath. A far-off pulse of lightning scattered the blue darkness, illuminating the grey beyond that jagged hole. His eyes began to adjust, and he could feel the boards beneath his fingers, drenched with rainwater and stunned with cold.

"Oh, God..."

Shaking, he bowed his head against his arm.

He was still here.

As always, he was still here.

44
Ink

Meygan picked her way around the mud puddles, hugging close to the barn. A light rain was still mystifying the morning, and she wore the purple shawl draped over her head, saving her hair from the damp. It was lucky that she had found an old bucket to use as a rain-catcher; of the three mason jars she had laid out, two had fallen over – and it was the one of these that had escaped which she was hunting down.

"*Mey*-gan!" Zander called from within. "Are ya gonna be all *day* at it?"

"No," she scowled, plucking the jar from the mud and scurrying back to the door. "It would be nice if you'd help me!"

"I'm busy." What he was busy with was this: They had set in the middle of their battleground a trunk of forest-green leather to act as a table, and filled one of the old, metal feed cans from the granary with wood to act as a sort of trash fire for cooking. The archer was engaged in digging through his bag, rooting out his spoils from the other night.

"Yeah," Meygan returned. "*Super* busy. Would one of you be a dear and grab the bucket that's by the door?" She carried the mason jars over to where Zander sat, placing them on top of the trunk.

"Darn," the archer lamented. "I was gonna tease you for your scanty haul; now I have to praise you for your foresight?" He shook his head. "If you can't get the upper hand, keep your mouth shut."

"Did Rahyoke teach you that?"

Zander said nothing; merely *looked*.

Meygan uplifted her palms to the breakfast gods. "So, what did you bring us, O exulted and mighty *hunter*."

"I see the insinuation, and I look past it." Lifting out his groceries, he laid them on the table, one by one, listing them as he did. "I got tea and I got coffee. I got peanut butter, pretzels, and a box of granola-bars I may or may not have stolen. I got biscuits, scones, honey-glazed almonds; this random loaf of bread which tastes like cinnamon and honey and manna from heaven; and a green pepper."

"Mine." Ian took the green pepper; no one contested him. "Is it hot?"

Zander stared. "It's a bell pepper," he said, as if that should speak for itself.

"Is that all?" Meygan teased; no one had gone for the bucket, so she retrieved it herself, setting it down before the busy archer.

"No," Zander answered, serious as anything. "I also got jam." And he put that, too, on the table. "Thank you for the water. I'm sorry I didn't help."

"You're *welcome!* Now was *that* so hard?"

"I could've done without the patronizing tone."

"Are we gonna fight today?" Addison asked. He was feeding the fire with bits of shattered crate, stirring the slow blaze with the shaft of the alien rake.

"Of course we're gonna fight. You still suck." Zander got to his feet. "Gimmie that staff for a second, Addison; I'm gonna try t'make breakfast."

"Well," observed Meygan, "it looks like you made out well, anyway. Almost as good as Rahyoke."

"I was trained," he bowed, "by the best." Addison slapped the rake into his upheld palm.

"Do you go into town often?" Ian asked.

Zander shook his head; using the rake as a rotisserie, he looped the handle of the metal bucket over the shaft and thereby suspended their water supply within the bon-fire. "I've had my fill of civilization, personally. My place is here – with the sky and the sun and the wind in the trees." He unscrewed the top from a container of gritty coffee grounds, peering within.

"Like Fionn."

Zander beamed. "Exactly!" He dumped half of the grounds into the bucket, watching as the water came to a boil. It took a while for the grounds to be roasted to his satisfaction, and the whole time he stood over them, his arms folded, his head cocked, as if listening for a timer or a choir of coffee angels to descend. At long last, the nectar was ripe. "Meg, do you mind seeing to the toast? I trust you more with fire." He pulled the boiling coffee-water from the feed can, pouring it into the rinsed mason jars. One he set aside.

"Well, when you put it *that* way... Is that for Rahyoke?" Meygan asked. Waving off her own proffered jar, she handed Zander the lid. "Here."

"Auh, brilliant!" he said, by way of thanks, screwing on the lid as he stood.

"I know."

Zander cast a glance back at her; then, shrugging, he ascended to the alter-world via their make-shift staircase.

"Hey."

Rahyoke jumped. He was sitting against the interior wall, his left knee pulled up, his elbow against it and his hand against his temple, shielding his face. Now he turned.

He looked like a wreck. Pale as death, his eyes underlined darkly as if sleep had never touched his form; his hair was twisted and matted, clinging with a dampness beyond the dew. His black coat lay, rolled into a ball, beside him, and the grey hoodie he wore seemed to be the only dry thing in the loft.

Recognizing the boy, he offered the faintest hint of a smile. "Do I look that bad?"

Zander's eyes rounded. "Bad as what – I mean, no." He slid away from the ledge. "Did you get any sleep?"

I DO look that bad... "What more is there for me to do now but sleep?" His eyes shifted. "Wha'd'ya have there?"

"This?" Zander slid the mason jar within the man's languid reach. "It's for you. It's coffee."

"For me?" He accepted the offering, studying its contents with grave curiosity. "So that's what I smelled burning?"

"It's good; trust me."

"Di'jya try it yourself yet?" he asked, slowly twisting off the lid from the jar. "Am I gonna die?"

"Ingrate..."

Rahyoke smirked. He pushed back his hair slickly; a few, thick strands fell back into his eyes, but, for the most part, it actually stayed back. "So. How're things down in the world of the living? Goin' stir-crazy yet?"

Zander thought of Fionn – and Caílte and Osgar and Diarmuid – racing through the trees, fighting off bandits. "A little."

"You made out well in town, I guess." Raising the coffee, he took a sip. "Eauh."

"Hey; a house offers the best it has."

"Oh, sweetie, this is *so* not your best..." He cleared his throat; five copper pieces' worth of espresso plasma. "I want you to be on guard, alright?"

Zander arched a brow. "Why?"

Rahyoke frowned. He was silent for a moment, a deeply thoughtful look on his face. Inadvertently, he took another sip of the coffee – choked, swore, and answered: "Auh, it's nothin' more than usual, isn't it? I get anxious when we stay in a place for too long."

"In my experience, you run into more people when you're moving."

"Mm." He had a semi-valid point. Still, he could not shake the feeling... that something was wrong. Shorn; that stupid dream... "So we should live here, then."

Zander spread his palms. "All I'm saying, is for a little while. If you weren't so *paranoid,* then I'd be able to provide for us..."

Rahyoke considered that, though his face didn't show it.

"It's out of the way, like the house – and it's *far* away. Like the house – it's a pretty ideal location... Oh, please don't say 'too ideal' – I see it in your eyes!"

Rahyoke cast him a bitter smirk, and Zander grasped his foot, pleadingly.

"What did I say last night? Just give it a bit, alright? We'll be safe here for a little while, and it will give us a chance to get ahead for once. We'll make the crossing before winter, anyway."

He would hope so; it was August. "Well... If I'm going to stay here then I may as well be useful. Wou'jya mind bringin' up – eh – *an mála páipéir* – the sand-dweller things – scrolls! I think you poisoned me."

"The..." Zander's eyes drifted off and away. "Yeah... Yes."

"Grand. Will ya do me another favour? Can ya grab for me, like –" His eyes lifted to the hole in his ceiling. "If I had a tarp, or oilskin, or even canvas – something of that nature..."

"Why would you need it?"

Wordlessly, Rahyoke lifted his coat, wringing the worn fabric. A pint of water dribbled steadily onto the floor.

"...'Nuff said," Zander allowed. "Anything else, princess?"

Rahyoke looked at him, unimpressed. "I could do without the back-talk."

Zander laughed; having missed the earlier conversation, Rahyoke didn't see much of a reason for it, but – auh, the archer was crazy, anyway.

Rahyoke rolled his eyes. "Will ya go?!" he demanded, his voice cracking, ragged for the early hour and want of anything decent to drink. "Yeh've better things t'be doin' than annoyin' me!"

Zander grinned, sassily. "*Nothing's* better than that." In one motion, he grasped the ledge with both hands and flipped over it in a somersault, dropping below.

Rahyoke gasped, sitting up sharply – then he heard the exultation below: "*Did you guys see that?! How awesome am I?!*" He dropped back against the wall, his fingers clenching his bruise, grinding his teeth. That boy lived for nothing more than to kill him.

After a few minutes, the pain ebbed, and Rahyoke relaxed, leaning more deeply against the wall. Forgetting, he lifted the mason jar and took another drink of the coffee. *Eew! Uh!* Scant was the use for it, but a shame in a thousand to waste such a costly luxury as food...

Oh! He cast a glance about for the book, snagging it when he found it and flipping it open to a blank page; the whole last fifth of it was blank, as if the authors had merely stopped.

Taking the boat-knife, he dipped the blade into his coffee. *The lad's lack of culinary skills is better than gum arabic.*

Painstakingly, in highly-ornamented, flowing letters – the language and style of the Empire at its outset – he wrote the first thing into his head:

Domine Deus hîc omnes misere et salva.

God save all here.

45
Beith

An chéad lá is fiche de mhí na Samhna

Cathbad the Druid was asked, "What is this day good for?"

And he answered: "If a young man should take up arms today, his name shall last forever, but his life shall be very short."

Cuchulainn was playing nearby, and he overheard the words of the Druid. He immediately went to King Conchubhar.

"What is it you are wanting?" said the king. He was very busy, but the lad couldn't be easily put off.

"I want to take arms today!"

"And who put such a thing into your head?" exclaimed the king, pulling a face. His nephew had wild ideas, and this was the most absurd yet.

"Cathbad the Druid."

Conchubhar frowned to hear this, but he dared not try the Druid. So he provided Cuchulainn with arms. And wasn't it so that none were strong enough to withstand even his youthful might, but in the end he was given Conchubhar's own.

Just then, Cathbad came in. "Is it the boy is taking arms today?!" he asked, greatly alarmed and uneased.

"It is."

And Cathbad was terribly sore at heart. "*Ochón!*" he lamented. "I'd be sorry to see the mother's son that would take up arms today!"

At this, Conchubhar was surprised himself. "Did you not tell him to?"

"I did not."

He rounded on his nephew. "You lying imp!" he scolded. "Why did you tell me that Cathbad had told you so? Oh, a great shame upon you, boy!"

But Cuchulainn was resolute. "I never lied to you," he answered. "I overheard Cathbad say that whoever took arms today, his name would last forever, even though his life should be very short. And I would far rather my life could last a day and a night, if only my name would last forever!"

An chéad lá is fiche de mhí na Samhna – istoíche!!

My son will be one in a few days.

He had been little and premature when he was first born. Orchid had been worried, but I didn't understand enough to. I didn't mind that he was tiny; I loved him because he was mine, and he was hers, and he was here, and nothing else in the world mattered.

Then he kept up crying for three weeks straight and every night afterwards, and I minded a bit.

He's still small for his age – at least that's what everyone keeps telling us – but he's strong, and healthy, and very willful. And for that I am very grateful.

Orchid says he looks like me. But he *acts* like *her*, the brat.

An old friend stopped by to see me tonight. I hadn't seen him since we were kids. Through the crazy, underground grapevine of his life, he had heard mention of my name, and was able to hunt me out. No one else could; he's a clever son-uva-bitch, and shorn am I glad to see him!

It was my turn to put the kid down – it's an expedition, let me tell you; that's why we tend to switch off when we can. So my wife was the one who got the door.

As is usual in these parts, the two of them really didn't share a common language. And she's territorial, and he's quite timid at times. Finally, getting his name, she yelled out if I knew him. Yes!!

We convinced him to partake of our hospitality; he's so strict about not wanting to impose – an attribute reserved only for those whom he really favors. I've gone on the lash with him – I know.

We talked long and hard, myself often acting as translator. Though he repeatedly lied about how everything was, "Auch! Grand, fine, Donny, sure!" Eventually he admitted – "Uh... I... may no' bespeak... *such* true..." (Side-bar – my wife finds his grammar adorable.) Reluctant as he was to admit it, he naturally won't go into specifics, but what I gather is he's had the worst of the last few years, and he has nowhere left to go. Not that he's near about takin' advantage of our help. Shorn! The poor *sionnach* thinks that he's imposing because we gave him a decent meal and let him sleep on the couch (and a scragglier piece of furniture I dare you to find)!

We want to keep him on here. The stupid roof has a great, leaking hole in it, and the windows are well decayed, and shorn it would be grand to get a little more use out of the land we have – and he's a well able body. But more, for my peace of mind. I'd feel so much better if he were here when I'm gone, to help out but more to protect Orchid and my son. And anyway, a bit of honest work would not go amiss by him.

I'll give it another shot in the morning. There's no one around that I'd trust more. He'd do anything in the world for me, and if he knew how we needed him here, I can probably get him to stay.

Och! I'm glad I wrote that out. I'll try that! Ha-haa!

Och... I really need to get some sleep...

An tríochadú lá de mhí na Samhna

C'mere: What good is a tale without a proper bit of romance? Don't be making a face on it; you know it's a far better death that a warrior has when a beautiful maiden is there to be crying over him. That's all we really want of the world, isn't it: a hot girl to cry over us when we're gone? *Ochón*; maybe it's a cultural thing.

I remember the first time I made fire dance for her.

A clear night in Samhain, unseasonably warm – *buíochas le Dia*. All the stars were out that night, and the sky fairly glittered with them, casting a sheet of crystal over the dark moor.

It was dangerous to be so out in the open – or would have been, anywhere else. That field goes on for miles, and no one cares to traverse its barren ground. The whole world was ours.

"I don't usually give private shows," I told her, tossing my bag aside. She sat down on the sparse grass.

"Ooh; I feel so special."

"You should." As I had gotten older, I came to realize that what had been commonplace in my household was to the rest of the world not so; laws against it as ridiculous as if they were against mythological things, like – well, like fire-serpents. It's hit-or-miss with the general public, if they believe whether or not it's real (bear in mind of course when I say that the general public still debates the possibility the globe is lunar-centric). Some think the law is merely tied in under that against street-performance; either way, most girls are not enthused. I mean, come here: merchant, mill-owner, busker. Not enthused. The law-breaking is a deterrent to some, the low status to others. And let's face it: if it wasn't going to impress a girl, I was far less likely to break a law.

My Orchid isn't most girls.

"You're not gonna set my violin on fire, are you?" she asked, taking the sleek instrument from its case.

"No." Uncorking a wine-bottle I'd brought with me – DON'T try this at home, kids; upon my life, I swear it won't work for *you* – I slid a rag down into it; it was about a quarter full. She started tuning the fiddle. I found a stable patch of ground, and braced the bottle in a circle of stones – just far enough back from her to ensure that if – God forbid! – I *should* make a mistake, I wouldn't hurt her.

"You're right; this really isn't a first-date kind of thing."

We had by this time been courting a few months.

I held up the lighted match to her, watching her through the flame. "Trust me; I'm a professional."

My idea was that I should manipulate the fire to follow what she played; I told her to play whatever she was able, not realizing that she actually was a gifted musician – not like the "bodhran-player" I met once in Heliopolis, Lavinia – God help us all; that woman had arrhythmia in every capillary of her body. But I digress.

"What do you want me to play?"

"Whatever yer up for, *a ghrá*."

And I tell you, it was like I had questioned the very core of her being, her integrity, her intelligence, her ability; something, but whatever it was, she looked at me as vipers do: bold, ready to compete. And she played – I don't remember what, but it was *wicked* fast, glorious – a show-off after my own heart.

"Very well," I said, and started counting out the beats. Just as the little flame had burnt down the stick, climbing to my fingertips, I got it. I cast down the match, and the rag in its bottle took light, but I controlled the flame.

Because I can.

The music she played had a pulse, and, seizing it, I started to dance. Ha – there might be a time in my life yet when I'm embarrassed of the fact I can, but, when you throw fire into the mix, anything's cool. As I danced, the flames followed me, twisting up in winding chains, rearing like snakes and spreading into the sky.

"Whoa!"

"Keep playin'!"

She plays with a spirit, a fire, of her own – an energy, rich and dark and vivid, and the flames bend to it of their own will. Never have I found it easier, more natural, dancing for any other musician. Soul mates, y'know? There was no routine in it; no set pattern. I bent to her music, and the fire twisted around me, arcing and lacing into the sky. As she played, she watched me – watched as I watched her – and I saw in her eyes what I had never seen in the eyes of another.

"Mercy!" I cried at last, dropping flailingly into the grass.

She stopped playing. "The drama of a vaudeville player; are you alright?"

I sat up. "C'mere, look: Are ya sure ya don't need me t'tune it for ya?" I teased her, but she had worn me right out.

She gave me a look. "You don't know anything about fiddles."

"I'll learn. Em – yer fallin' out of time; maybe I could play awhile?"

And in the light, I saw a smirk on her – condescending, beautiful. "Whatever you're up for? Just dance." And she launched into the fastest *Geese on the Bog* I've ever yet heard.

And I loved her.

[Orchid's handwriting] You're so sweet.

[Dubhán's writing] That's why you love me.

[Orchid's] Who says I love you?

[Dubhán's] Och, you love me. 'Cause I'm so pretty, right?

[Orchid] Sure you are, Donny. Sure.

[Dubhán] NOT ENTHUSED.

46
The Fenian Oven

"Rahyoke says we can go hunting!" Zander announced, sauntering back to the fire.

"'We'?" Addison repeated, incredulously. "Who, *'we'?"*

"*'Us'* we," he clarified. "You and I."

"Not me?" Ian asked.

"You stay here and guard the court."

Ian grinned, nodding. "I'm good with that."

"C'mon, Addison," Zander bid, shouldering his quiver. "We'll be back in a while; don't wait up!"

"Whatever," Meygan called in reply.

Addison followed the archer out into the yard. There your man stood, posing triumphantly, his sword in its scabbard against his left thigh, his quiver on his back, the shining, dark shaft of the bow protruding amidst the smattering of unmatched feathers. He wore Rahyoke's black coat, the hem trailing down around his shins, and he leaned upon the alien rake like a shepherd's staff, his free hand on his hip. "Y'ready?"

"*Now* what?" Addison asked.

"Pick yer weapon, squireen, *a stór,*" Zander commanded. Leaning off of the staff, he twirled it skillfully.

"Show-off," Addison muttered. Then, "Are you gonna teach me to use that bow?"

Zander burst into laughter.

"What, *never?!'*

"No one save me touches this beauty, squire."

Addison sighed. "Then I'll take the sword."

"If you can." Zander replaced the point of the staff into the gnarled turf. "D'you want the scabbard, then, as well?"

"The what?"

"Ochón go deo! This!" He took the sheath of the scabbard in hand, indicatively. "*Mercy* above, it's the covering."

"I don't need it," Addison shrugged, his tone a little lighter than it had to be. "I'm strong enough to carry it."

Zander shook his head, holding the staff in the crook of his right arm as he drew his sword. "Rookie," he uttered, warningly.

Addison accepted the sword, hefting its weight with some clumsiness. "Alright, then!" he said, a bit – a *bit* – grandly. "Where to?"

"Hm..." Zander rubbed his chin, peering up towards the sun where it rolled through the mire of the sky. "Ah!" He snatched the staff, turning and heading back behind the barn. "Follow me, squireling! We're tracking the red deer!"

Addison rolled his eyes, trooping after the agile hunter. "*Are* there deer?"

Zander halted, though he didn't turn or tense. "Whether there are or not, we can still hunt them. It's merely a matter of success." Then, mysteriously, "There will *always* be red deer." And with that, he scampered off.

Addison sighed. *This is going to be a very, very long day.*

Zander had more skill than he made it seem. He stepped as silently through the overgrown bracken as a fish through water, a bird through the air, a doe through a weaving field. Every sense was atune – for every breath of wind, every shift, every dead note, was as sure a sign of the world and its turnings as the breaking twigs, the snap and crack of the raindrops falling – everything lived and changed and warned. And Zander heard all of it, and he knew – but he wished dearly with his all of his soul he knew where it all would lead.

A clattering, fern-crunching snap resounded behind him, and at last Zander stopped, sighing up at the clouds. "*How* are you walking so *loud?!*"

"It's not that –"

Zander whipped. "*Bears* have more stealth than you!!"

"I have stealth," Addison pouted.

"We'll catch nothing," Zander determined, gravely, moving on. "We'll starve to death."

"Oh yeah," Addison muttered, glumly. "Because *I'm* the one who scared everything off. We've seen *so* many deer already."

"Stealth does you more good than just for filling your stomach," Zander replied. "Think how you'd fare, captured, or stealing down a street past a garrison."

"*Captured?*" Addison raised a brow. "You're teaching me how to sneak in at night."

"*Addison,*" Zander sighed. "You know *so* little of this world. And so little of your own. Someday these skills may come in handy to you. But even if they don't, it's useful to know *something* – something *real.* Make your life worthwhile."

"Sssssooooo..."

Zander rolled his eyes. Stopping abruptly, he whirled, and was little more than a finger's breadth from Addison's face, filling that gap with the tip of the staff. "I'm gonna *make* you a warrior, Addison Mitchell!" he vowed, and his voice echoed through the trees.

Addison lost his balance, staggering back. "*Why?!*" he asked, honestly confounded.

"Because!" Zander shouted, stabbing the end of the alien-rake into the soil, solidly.

"You complained about my walking," Addison said; "your yelling is louder."

"Intimidated?" he teased; then, throwing his arms out wide, he thundered, "RAAH!!!"

There was a sudden clattering, rustling, a rough cawing, and the hunter whirled around as a crow jumbled out of a bush and leapt into the sky. *Twang – thwit – thud.* The shot was so fast Addison hadn't even seen Zander draw his bow.

"C'mon," the archer urged, setting off on the track of his query. "I saw him land somewhere over here."

Addison followed, picking his way through the ferns; they grew up here through the tightly-matted black of the pine trees' roots. "How did you *DO* that!?" he asked, amazed.

Zander shook his head. "It might be a game to you, Addison," he called; though he began enigmatically, he finished childishly: "Oh! There it is!" Three steps brought him to crouch at the side of a mass of black feathers. He pulled an arrow from his quiver, laying it against the creature's neck. "*Dia leat,*" he breathed, in quiet blessing.

"We're not eating that, are we?" Addison asked.

"Modern cultural convention breeds trepidation about eating the corvie," Zander replied, almost languidly. "Everyone's all – 'Oooh! It's a scavenger! We shan't eat it!'"

"'Shan't'?"

"But have ye ever seen a *pig*, Addison?" He spread the crow's wings, angling its head to the right. "Pigs lie in their own shit and would eat it, too – and even lobsters will succomb to cannibalism."

"Mm. Poetic."

Zander slit the crow's body, beneath the breastbone, from wing to wing.

"Seems kinda small."

"A warrior's meal!" Zander lauded savagely, grinning. "And I'm hungry."

"Were your parents wolves?" Addison asked, viewing him warily.

Zander laughed. "It has been said so!" And he simply left it at that. He grasped the bird by its feet in his left hand, slipping two fingers into the cut and spreading them. The noise of it was not appetizing in the least.

"Wou'jya grab us some fire-wood?" Zander asked.

"Gladly."

"Don't cut'em fresh –"

"I *know,* I *won't!*"

Zander gave a sharp laugh. "And don't be gettin' lost, now, on me!"

Addison shook his head. He followed a straight line, to make good on his word; and as he went, he swished his blade through the straggling ferns, battering the roots that crept over the dusty forest floor. Now, in video games – as in stories – this would be the perfect opportunity for something to happen. Of a sudden, you would be burst upon by drawn fighting – a hundred mobsters, a thousand wolves – auh! – *zombies*. All clamoring for your blood, and you alone, facing their masses – fight or fall! – armed with –

Addison stopped walking. He lifted the sword he carried, and for the first time, he actually *looked* at the absurd thing. The blade was as long as his arm, a semi-dark steel. By the hilt was a flourishing Celtic design of darkly-engraved swirls, which climbed down the shaft at a slant, dipping lower on the left than the right. The hilt was a simple thing, dark and murky like worn brass, the light handle wrapped in soft, dark-brown leather. No jewels, no gold, no costly silver. And yet – It was so much more. It was *so* much more than a simple strip of metal and leather. This was not a game controller. This was a *sword.* Vaguely, fleetingly, Addison began to understand. There was value in this object, there was power; if anything – if *any* of this was real, it was this.

If...

Addison shook his head. Firewood! Right! He gathered as much as he could before he grew impatient and scurried back to the archer.

"The plodding of a *Clydesdale!*" the archer called, in greeting. "I'd know it was you at ten miles!"

"Do you want this or not?" Addison asked, holding up a handful of branches.

"D'you wanna *eat?*"

"Touché," Addison gave, settling down across from him.

Zander broke the branches, building a tent for the fire on a patch of dirt. He filled the tent with poofs of tumbleweed bramble, spearing the meat on two arrows and passing them to Addison. "Hold these a sec'," he bid, pulling from his pocket the small, green box that Rahyoke used for matches.

"What, these are just for us?" Addison asked.

"It's not one of those *loaves and fishes* bits," Zander retorted, dipping the flaring match into the hearth. "I am *far* from being God."

There was an abrasiveness to his tone that ushered in a silence.

Taking back one of the spits, Zander said, "You mayn't think it – I know you look down on me, though I am your elder, and it's foolish to think that no wisdom can be gained from a man without trying him. The stupidest of all creatures at least teaches this much: not to emulate him. The true fool seeks no wisdom, nor thinks there is any to be found – himself having all of it, of course." He dipped the meat into the fire. "I can offer you much in the way of wisdom, Addison. Rahyoke, too."

"I know," replied Addison, airily, shrugging one shoulder.

"You *think* you know," Zander returned. "But for however many questions you cast to him, sure you think yourself far worthier."

"That's not true!" Addison protested; but for all his shock and injury, something lingered underneath.

"Hhm." Condescension flashed through the archer's expression, even as he held it still bowed to the fire. It was the same look of that dark wanderer, and Addison didn't like it from him, either.

"I know I can learn from you!" Addison exclaimed.

"You *think* you know."

"I *do* know!!"

"You are *learning*, squireen; but you do not yet know enough to learn."

Addison frowned, his frown deepening into a scowl, and his scowl bringing him to ask, "Are you sure that was coffee you drank this morning?"

Zander laughed, his usual haughty, hearty air returning to mix and to merge with his profoundity. "It's something hard to explain," he eluded. "But you will, God help you, in *time*. You just need to be open to that guidance."

"What guidance?"

"Aaaaahhh..." Zander tilted back his head, lifting his gaze to the chattering leaves, and the warm light that fell down through them. "Have you ever heard of the Salmon of Knowledge?"

Addison sighed. "*No.*"

"See? There you go, with that reluctance again! How will you ever know a thing if –"

"Just tell me!!" Addison broke in, glowering.

Zander grinned. "Thank you! Now was that so hard?"

Addison rolled his eyes, sticking his skewer into the fire.

"I'll tell you the *short* version, please your Majesty."

And this is the story he told:

Before Fionn ever met Sadb, his deepest love was the forest. And day and night he would go wandering, and nothing was sweeter to him than the motions of the world.

Once in his wanderings he came upon Finnegas, himself poet to the king. Fionn asked him what it was the man was doing, and him – and Fionn also – sitting on the banks of the rolling Boine. And what he was doing, he was trying all his days to capture the Salmon of Knowledge.

"What will you do after you catch him?" Fionn asked.

"I don't know," Finnegas answered him. "But after I catch the Salmon, I will."

Whether Fionn rolled his eyes or not, who's to say?

"And what is your own name?" Finnegas asked.

"It is Demne."

"It isn't," Finnegas returned – although the name he gave was the name he had been given at birth. "In the days to come you will be called Fionn, and Fionn is what I shall call you."

Your name, *a stór*, is not important, so long as it lives forever.

Fionn stayed on with Finnegas, learning poetry, and – long story cut short for an impatient, squiggly audience – Finnegas caught the Fish.

"Fionn, will you cook this?" Finnegas asked.

"Yes! And I have plover eggs too, and bitter herbs –"

"Those for yourself, and the Salmon for mine," Finnegas answered. "And don't be after tasting the Salmon!" And many a stern warning he gave to the like.

"Don't *worry*, Finnegas!" Fionn assured him. "The Salmon is rightly won, and rightly yours! You can count on me to protect him!"

Finnegas went away, and it was a great show of strength in him never to cast back a glance behind him.

As for Fionn, he spitted the fish, roasting it over the fire. Waiting was boring, and with Finnegas gone, he had no one to talk to, so his thoughts wandered. As he was distracted, a blister rose on the fish. It seemed a shame, such an imperfection on such a beautiful creature, and he pressed it down with his thumb. Fionn winced, sticking his burnt thumb in his mouth to ease the pain.

When Finnegas returned, Fionn presented him with the fish. "Long have I waited for this," said Finnegas. But – no sooner had he gotten a taste, then he lamented.

"What's wrong?" Fionn asked, alarmed.

"It is as any other fish," said Finnegas, and deep was the sorrow upon him. "Are you sure that you didn't taste it?"

"I didn't!" Fionn insisted. "Save for – a blister rose upon it, and I did not like the look of that blister, so I pressed it down, and burnt my finger. I put my finger into my mouth – and sure if that fish tastes half so good as my finger, you'll love it!"

But Finnegas shook his head. "Not I," he answered. "You."

"Me!?"

"The prophesy said that the Salmon would be eaten by a man called Fionn. I am called Fionn –" For his name was composed of *"fionn"*, which is *white*, and *"éces"*, which is *poetry* " — but so are you. It is yourself who is meant to have the Salmon," Finnegas said. "And so you shall."

So Fionn ate the Salmon of Knowledge which Finnegas the Poet had caught. And always after, if ever he needed to know a thing at all, Fionn had only to put his thumb into his mouth, biting it, and he'd know.

"Sidebar: that's how Wisdom Teeth were named. Our culture'll live on forever!" With a soft laugh, Zander took his meat from the fire, examining it; not surprisingly, it was a bit charred. "My question to *you*," he said to Addison, "is: *how* did Fionn get the knowledge?"

Addison frowned. "Huh?"

"It's not a hard question, *a stór*."

"By eating the fish."

"Ah!" Zander raised a finger. "For whatever wisdom gained of Faery got by consuming the fish, Fionn was wise before ever the *sight* of the fish – before the knowledge of it – before the word 'fish' ever crossed his mind!"

"Okay, you've lost me."

"The moral, dear Addison, is this: Fionn knew to ask questions – good, *quality* questions; not stupid stuff like *'Huh?'*" He shook his head, gingerly pulling the crow meat off of the arrow. "He asked, he engaged, he remembered. That's a worthwhile thing, *a stór*."

"What does that mean," Addison asked, "*'a stór'*?"

"Fair question. 'O treasure'." Wiping the arrow tip on his shin, Zander dropped the weapon back into the quiver at his back. "I tell the story the way I learned it. Not that I can't tell it my own way; I do. But I love the language of that version. It's not as natural for me to say, like, 'So wou'jya be after goin' it, at all?' ...'D'you wanna go?' – but I get lost in the language of those stories, and, I don't know." He shrugged. "It sticks." He bit into the crow meat, trying to give reason to why his eyes were bowed all through that speech.

Addison was intrigued. "How do you know that story?"

Zander swallowed. "Auh!" He smirked. "I loved warrior stories when I was a kid, but I used to *idolize* Fionn Mac Cumhaill. I wanted to emulate him, if I could. Save Sadb over Grania, any day of my life. No question."

Addison didn't know, but neither did he ask.

"You're burning your meat."

"Huh? Oh!" Addison pulled the crow out of the fire. "Is it any good?"

"Middling. Could've done with some seasoning." Zander shrugged. "You take what you can get."

They finished eating, and Zander put out the fire, burying the crow in the ashes and dirt, as Rahyoke had done with the fish.

"Are we going back?" Addison asked, wiping his hands off on his jeans as he stood waiting for Zander.

"It's hardly past mid-day," Zander responded. "We'll keep going a while longer." Before he could add, *Unless you're tired,* Addison answered, "Good!"

After taking careful stock of their direction – and obscuring what they left behind – Zander and Addison set off, trudging deeper into the woods. It was a quiet journey – save for Addison's plodding attempts at "stealth".

"Oof!"

There was a great clattering thud filled and frilled with the choppy rending of vegetation – an explosion of ferns and burdock.

"*Wow!*" Zander snarled, turning. "I'd be *better off* with a bear!"

Addison looked up. "I'm fine, by the way."

"You'd better hope that sword isn't damaged."

"No, I'm not impaled, thanks for asking."

"Face-plant," said Zander, grinning that savage grin. "*Literally.*" He crouched down beside Addison, to get a better look at the offending (or amusing, as you please) pit that had tripped him up. "Well, get your sneaks out of it, anyway… 'Sneaks' is not appropriate in application to you."

"*Yeah*, yeah yeah yeah…" Addison shifted, pulling his feet out of the hole, and then under him, so he was kneeling. "Are we *interested* in the hole?"

Zander looked up; oh, his grin was mocking indeed. "You don't know what this is," he said, "*do* you."

Addison griped. "Well, *you* obviously do! Lay it on me, Yoda."

Zander arched a speculative brow. Then he lifted a stone from the pit – and it about the size of his hand. The pit was about a foot and a half to two feet in diameter, its bottom coated in similar broad, flattish stones, blackened with soot, covered in ash, caked and coated with the needles and scattering leaves of last fall. "In the days of Fionn," Zander enumerated, the draw of a storyteller, the energy of a bard, in his tone, "they – the Fianna – to cook their meat, they'd dig a hole, lining it with stones like these. They'd light a fire and let it burn down, clear away the embers, and lay their meat sometimes two or three cuts deep on the stones, laying red-hot rocks o'ertop."

"And your point iiisss…"

"This is a Fenian Oven," Zander answered. Studying the stone in his hand, he murmured, "And recent enough, at that."

"Are you going to try and convince me now that Fionn Mac Cumhaill was here last night?"

"I'm loving 'try AND convince' over the usual 'try TO convince'. Very telling."

"What the Hell is it with you guys and wording?!" *To* wording? No... "Anyway – why do we care?"

At this, Zander became silent and pensive. It wasn't that he was without his assumptions; the dire frown upon him was enough to dispel such a notion. He turned the stone over in his hand and started. "What's that look like to you?" he demanded, shoving the stone towards Addison's face.

"Uh –" Branded on the bottom of the stone, incised by ash, was the letter "A". "'Ay'?" Addison frowned, watching as Zander began marauding the pit, laying the marked stones to one side, the unmarked to the other. "Is this another Fionn-ish thing?"

Zander didn't answer. Scraping the dirt bottom of the shallow pit, he turned to the marked stones, of which there were six. A, C, H, C, E, T. Unintelligible. "Someone who knows the old culture was here," he said; for although the letters were an anomaly, there was no mistaking the construction of that pit.

"What; you don't all do this?"

"No, we don't." The archer raised his gaze to the sky above. The clouds lingering in the morning had thickened throughout the afternoon, and were looming on the skyline, rapidly ushering in an early night. He set back the last stone. "Tell no one of this."

"I couldn't if I tried."

"Make sure of it." He slowly got to his feet, as if he'd aged a thousand years in the instant. "You know? Usually it's red sky at *morning* you need to worry of. But I don't like the look of those clouds."

"They are dark," Addison agreed, and dark the clouds were indeed.

"C'mon; we'll head back." He helped Addison to his feet – a favor accepted with great reluctance. "We wouldn't want to be caught up in something as ominous as that."

"As clouds?" Addison asked, following.

"There's many a story warning the misdeeds of mist. Sadb being one of them."

"Oh yeah."

"I'll tell you another to pass the time, if you'd have it."

"Anything is knowledge, right?"

"Indeed."

This story was given to Diarmuid Ó Duibhne, and so passed on down his line, though for sure he wasn't in it.

So it was that, at one time, themselves – Caílte and Oisín and bald-headed Conán, under the leadership of Fionn – were out on the hills and the valleys and the forest-places, and great was the dark mist settled over them.

It came to their minds to find a place to rest, for deep was the darkness, and night coming on rapidly. And they cast their eyes about them widely, and so it was that Fionn bespied a light. "What is that, now?" he wondered. "I never knew of a house in this valley."

"Well, then," Caílte advised, "we'd best go and check it out!"

Down three of them went, and they moving towards it; but bald Conán offered a protest. "There's no wisdom in that!" he argued. "For who's to say it won't be the death of us, surely?"

"Yes!" called Fionn in return. "Far better to be devoured by wolves!"

Conán whimpered, Conán held out – but Conán was never a man to leave his comrades unprotected, and defense they would be needing, and at least *one* reasonable mind to guide them! So he followed.

So they got up to the house, and there they stood, each waiting for the other to knock, and each not wanting to seem the coward by uttering, "*You* do it."

Finally, sighing greatly, Fionn glowered, "*I'll* do it, then," and swept forward, raising his hand to knock.

But before his flesh could touch the dooring, a great and horrible sound met their ears: three terrific screeches to curdle the blood and straighten your hair sheered out from that little shack.

"*I'm* out!" exclaims Conán, but no one else runs; and, thinking of his duty in caring for those poor fools, his comrades, bald Conán cycles back.

The moment he did, a sight to startle him back again, though he did not run: The door fell open to a dark interior, and a massive, grey man took hold of their horses. "My welcome to you, men of the Fianna! And especially to yourself, the much-awaited son of Cumhaill." And he brought their horses within.

"If ye won't heed and turn back," hissed Conán, "at least wait for the sound of crunching." But his words again went ignored.

In with them, then – and they fairly tripping over the dooring, for the black soot of an elderwood fire, and the dilapidated decay of the floor.

The only seat about the place was a board, cast over with thin, tatty furs as a sleeping-place, and here Fionn sat himself down. Only Conán did not – for well he remembered his treatment in the House of the Quicken Trees, and knew also: you always cover your exits, *a stór*.

Now: As their eyes adjusted, they saw they were not alone.

Across the fire – what they had taken for just a bundle of rags – was a hag in a grey, tattered shawl, and she having three bald and crook-necked heads on her. But worse was the man there, and him without even *one* head on him, but a great eye in the middle of his chest, pale a sickly yellow-green, a perfect circle, and with a pupil like a cat's. And worse still! One corner was peopled with a mound of headless bodies. And the other, on the far side, one of bodiless heads. And when the grey man bid them – "Rise up! And make music for the King of the Fianna!" – och, and those nine bodies and those nine heads rose up – *och*, but you can be sure that, if the music those nine heads made was worse even than the screaming before, then that of the nine bodies was worse a screeching chaos than ever the nine heads could make. And worse still, *ochón!* For the hag and her three heads joined in, and *then!* – auh, horror that ever the sound graced this earth, for if all of that "music" was awful to hear, never was worse than that of the one-eyed man. And the music went near to breaking their heads, and shattering the bones within 'em.

And, "How do you like this music?" the grey man asked.

Fionn swallowed, and uttered shakily, "When a house offers the best it has, no better could be asked for."

Then up with the grey man, and him taking hold of the ax for the firewood, and he hacked and flayed and destroyed the horses, and he pierced bits on twigs of the rowan tree, and set them over the fire. And blacker than ever the fire burned then, and the like of it to choke you a mile off. But when the meat was brought to the Fianna – all stringy and gamy – sure but it was so raw that the blood still dripped and flowed.

"Uh-uh, no way!" protests Fionn, pulling back. "Never have I eaten raw meat, and never will be driven to it."

"If you are come into our house and refuse our food, then it's a fight you are surely asking."

And up it was.

The heads and the bodies and the headless bodies and bodies with three heads all rose up and attacked them, led by that great, grey, ax-wielding man. And Fionn and Caílte, Oisín and Conán, were all driven back into a corner. And the fire was quenched – squelched and smoldered – in the ruckus, and so it was they fought in darkness, and in heavy, gagging smoke, the like they could neither see nor breathe, and they fighting the supernatural enemy.

Oh, but fierce was the raging battle, *a stór!* And the blood and the soot and the sparks from their weapons, clashing and slashing 'midst the war-cries and hellish, howling screams!

And och, they fought on, 'til the deeper black darkness came over them, drifting in the form of a mist. And no one could say but who fell to it first and who last – if truly they fell at all – but when they opened their eyes, and could see –

They were on the side of the hill. No house. No phantoms. No floating, shrieking heads. But the clear morning sun scattered away the harsh mists, glittering over the beds of heady heather. And the warriors picked themselves up, mounted their horses, and went home.

"The moral of the story is, if you get lost in a druid mist, don't go towards the light!"

"That's a ridiculous ending," Addison complained.

They were within sight of the barn now, its dark form outlined against the darkness of the trees and the greater darkness swarming up above.

"It is," Zander responded, shrugging. "But only for those fully prepared for the adventure that lies beyond."

47
Firebrand

N*í básóimid riamh!*

This was how Dubhán had ended it – these final three words, scrawled rapidly on the back of an entry about a dream concerning either moths or corvies – *dubh* and *oíche* were used so frequently to describe the antagonists, it was hard to say.

Three words. *Ní básóimid riamh!* Never will we die.

Dubhán would burn it.

Rahyoke bit the tip of the blade. The condemnation notice lay, unfolded, on the journal before him. He had folded, unfolded, read and refolded that bloody paper so many times, it no longer rolled, it no longer moved. *It no longer had to.*

Dubhán would burn it, and so would Zander. But they didn't understand.

The act of defiance was no longer in destroying it. Not when its existence meant nothing, its meaning already carried out. That was an old revolution, and futile.

But to *keep* it. No more a sarcastic, reviling gesture than to protect that which had destroyed you, and hold it for all the world to see.

Rahyoke bowed his head, twining his fingers into his hair. What *world?* What *world* would raise their eyes and see it?! Would –

Aah. He clasped his hands, wringing them, restraining them.

Posthāc hic ager adiungitur Imperio Augustinii et subiunget legum imperii ad unum. Offensio obsequi legum praecidētur cum grave poenis. Ave, Pax Augustinii. Ave, Augustinium.

Raising his eyes, he studied the paper, studying with familiarity the words so well-traced that they were committed to memory, and no longer needed to be seen.

Hereafter... This territory is annexed to the Empire of Augustinium. And, as such, it is subject to the laws and regulations of the Empire without exception. Failure to comply will be cut off with severe punishment. Hail, the Peace of Augustinium. Hail, Augustinium!

Poor Dubhán. Poor sweet, innocent little Dubhán.

Reaching out, Rahyoke folded the notice. *No more bitterly ironic bookmark was ever made.*

A battle-cry and a death rattle, never spoken, only dreamed. *Ní básóimid riamh!* Never will we die.

He shook his head, slipping the hated notice between the pages of the book.

Dubhán, a ghrá, you know nothing of the world…

48
Na Fiach

Darkness had settled on the land, and darkness there would stay. The fog rolling in from the ocean below, filling up the forest, choking with webs the black-limbed trees. Trees gnarled, spattered with lichen, barren of bark or leaves, charred. The snapped ruffle of silken feathers, a thousand snarling voices scrape against the sky with mocking laughter. Murderous, burst of agonized cackling, the raven scattered into the sky, clattering and clamoring 'til its voice was choked by the fog. Shrieking – shrieking the silence –

"Ná bíodh eagla ort, cariad."

A woman stood in the mist. Her hair fell in cascades of dark curls, dark as the night, soft as the mists, and shining. Her eyes a deep brown, like the faun's, her fair skin covered with freckles, so pale as to seem translucent in this light. A mantle red as blood around her, pinned at the shoulder; her skirts layered, slitted red, leaf-shaped green, a longish skirt of purple. Barefoot as ever, the thin scar 'round her left ankle traced gracefully with woad. She smiled – warm, comforting, beautiful, proud.

"Beidh gach rud ceart go leor."

The mists, ever-swirling, played through her night-dark hair, stirring the curls; clung to her clothes, pulling at the fringe of her mantle like sable, cleansing waters.

"Tá Dia leatsa, cariad. Go deo."

Her eyes sparkled, light beyond life. Knowing. Insistent. Glinting through the dense twilight of hazes. But the mist carded like wool, its spikes and thorns, twisting and twining like cords, spreading iron against the white, cutting the ethereal. The sky was red. Red, fibrous black, like cut sinews, like –

"Éist liom."

Her eyes, comforting, insistent.

"Rith an méid atá i do chnámha. Tuig tú? Ná bíodh eagla ort."

She was gone. A fanning of dark curls and skirts, vanished in the mist. And the red. Red flame fanned out across the valley, blurring out the grey. The world was shriveled and black, the world caustic gold, and claret and sanguine and red, tearing, ripping, goring, swallowing. One word was left echoing through the darkness, one word alone:

"Rith!"

Run.

49
Because We're Wanderers

Zander the Archer shouldered his bag. It was wrong, what he was doing. Wrong, and well he knew it. He wouldn't blame Rahyoke – he'd merited the punishment, truly – but that wouldn't stop him from going.

There was a night market in Paradisius. The items he'd selected for sale were fine – good enough to earn a decent price, but – hopefully – not enough to draw attention. There was nothing in the forest; his little excursion with Addison had proven that. And no matter what he said, one blackbird – one *half* of a blackbird – was *not* a filling meal. And they needed to make a profit.

Zander shook his head. He stepped silently through the shadows, cursing the breaking moonlight that fell through the slats, the shingles, the trees, the clouds, to scatter the ground with silver. Though full, the moon was not yet high, and with any luck – with a *lot* of luck – he could even be back before anyone noticed. With everyone confined to their separate quarters, no one would see him go – they'd know he had, of course, but maybe, if he did well enough, it wouldn't matter. Maybe –

"Zander."

"Oh – shit!" Zander staggered back, his drawn sword glinting in the half-light.

In the deep shadows to the left of the door – leaning even against the doorpost – his arms folded, his glare stern.

"What'ur yu – aah." Zander bowed his head, every muscle in his body tightening as if about to receive a blow. Rahyoke's words were hasher.

"I'd love to assume that what you're about to do isn't what I'm supposing."

His voice was low, his words dropping scathingly into the depths of the nighttime silence. For a moment, Zander's muscles screwed even tighter – as tight as they could go. Then he relaxed, sighing in dejected surrender. "How did you know?"

Rahyoke at last lowered that injurious gaze, shaking his head. "You *are* your father's son."

"We're all our father's sons," Zander responded, and he might have shrugged but for the situation. He ventured to raise his face of Balor. "How did you get down?"

Rahyoke's smirk was bitter. "Painfully. My ankle's stable now."

"What about your side?"

He uncrossed his arms, lifting his shirt just enough to show how he'd bound his lower ribs with strips of the teal-blue blanket he'd been given for use as an awning.

"It'll hold?"

"It has to."

Now Zander dreaded the tone. Looking up, he met Rahyoke's eyes. "What happened?"

Rahyoke's eyes averted. "I remembered." The enormity of pleasure he derived from that fact was probably sick. Leaning off of the wall, he swept out into the moonlight. "C'mon, we –"

"What?" Zander grabbed his shoulder, staying him. Though he spoke at a whisper, his voice was sharp. "*What* did you remember?"

Rahyoke paused. "I... I wasn't thinking. It was self-defense. I... couldn't..."

"What."

Rahyoke glanced back at him over his shoulder, before reluctantly turning to face the boy in full. "I'm sorry. Look – across the Pale we'll be safe, alright? Ih – this isn't something they want to get out; no one'll know."

"For how long?"

"One step at a time, Zander." He sounded exhausted, almost defeated. After a heartbeat, he began speaking again, gently and encouragingly. "Do you remember when we left Carteia? How they were slow in coming – we were a long time in finding the house, but we've been alright these last nine or so years, more or less – and they left us alone in the end. If they didn't expend the effort then, I hardly believe they'll go to much trouble over us now. And if they do, well I – *we'll* be ready for them, won't we?"

The small smile he gave was very reassuring. Zander bowed his head, studying the dingy floor in the moonlight. "Yes."

"And you know I'll protect you, don't you? Well, and doesn't this prove I'm able for it?"

"I know." *Jonah thought he could run, too.* Zander raised his eyes at a flash. "And you know we could always change your stupid plan? Don't we stand a better chance on our own turf? If you're able –"

Even Fionn knew insurmountable odds when he saw them. Rahyoke shook his head, supplying a lecture of reasoning in answering simply: *"Dech do shínaib ceo."*

Zander's sigh was short. "C'mon let's what?" he echoed, cycling back. He saw a look of confusion on the man's face. "Are we gonna tell them?"

"Wh... *No,* we're not gonna tell'em." Rahyoke shook his head emphatically. "Nobody needs to know. What we need now is to leave; go wake the trespassers. I'll handle it."

Zander made sure Rahyoke caught his eye-roll, then scurried to the granary and out of reach, calling for Meygan. The girl met up with Rahyoke as he retreated towards their meeting-spot at the back of the barn, amidst all that dormant, deposited, and wasted potential.

"What's going on?" Meygan asked, scurrying up beside him. "Is everything alright? Are *you* okay?"

Why did she emphasize it like that? *You.* "Get yer stuff together," he said, nodding towards her shed.

"Packed," she quipped, holding up her bag. "Are you okay?"

Stubborn... He shrugged one shoulder. "My ankle's a bit sore."

"How?"

"Easy enough to tell: Never look back when running." Before she could ask what that was supposed to mean, he began their family meeting. "Sorry for the hour, lads, but it's become necessary that we move."

"Are we in trouble?" Ian asked.

"Should you be?"

"Well – it's just that – any *other* time we've gone like this –"

"Och, no." Rahyoke shook his head. "No, you're not in trouble. But get your stuff together. Take what you can carry, nothing more. Dress against the wind."

"What does *that* mean?"

Rahyoke paused. He had never actually questioned the phrase; mentally, he ran back over every instance he had heard it, deciding in the end: "Ehm, essentially dress warm – layer, I guess."

Addison made a face. "It's *July.*"

"No it's not; it's August."

"I thought you were injured," Addison returned, fixing a Balor-esque glare of his own.

"I heal quickly," and the tone was tense enough to shut him up.

Now we at last have a wardrobe change – for which they were all a little grateful. In an act of rebellion against his former life, Addison "went native". His new shirt was a soft-linen pinnie, a whitish colour like sheep's wool. He wore khaki dress pants, rolled at the cuff, and a large, navy coat which – unlike the detail-less, robe-like, draping items that Rahyoke, Zander, and other men of their status tended to wear – was highly ornate; it looked like something a ship captain in the 1700s would have worn. He was also in charge of a hobo-esque knapsack he'd made with an old, plaid blanket, tied to the end of the alien-rake and filled with merchandise that they hoped to sell at the border. Ian took charge of the bag of food ("Worst idea *ever*, but okay"). He had been denied the use of a bright orange jacket ("Like Hell you *will!*") and had settled for a green windbreaker, a t-shirt bearing a logo they had never before seen, and a pair of Wellingtons that were too big for him ("*A Mhaighdine Muire*, that's disgusting! If you get worms, you are *out* of this group!!"). Zander had pulled a sleeveless, red hoodie on over a long-sleeved, blue v-neck. He had found a wrist-guard, and, adoring it, had strapped it onto his left arm; it was fingerless, and covered his forearm entirely, up to the elbow. Meygan, for her own part, chose to emulate the style of a gypsy fortune-teller; the lavender head-scarf completed the look.

She lifted a tweed flat-cap, studying it. "Here," she said, tossing it to Rahyoke. "That's for you."

He caught it one-handed, glancing it over before putting it on, twisting it backwards. In spite of his injuries, he was ready the fastest, having sorted out their luggage, determined and packed what of the chaos they could carry to sell, and returned Zander's bag devoid of the scrolls. He took little from the piles for himself, pulling a black t-shirt on over his long-sleeved grey one, and the light-grey hoodie under his long, black coat.

"You look like a rebel gangster," Meygan approved, the tone of the artist about her. "See – 'cause you can still see the bandana, and your hair is all – *foosh* – and – *yeah...*"

He rolled his eyes. "Great." Taking up his bag, he got stiffly to his feet, using one of the columns that supported the roof to support himself as well. "C'mon," he said, firmly, sweeping to the door. "We're about sixteen hours out from the crossing, and I think we can make it in a shot."

"Because we're wanderers," said Meygan, without any sort of enthusiasm. "And that is what we do."

Rahyoke smirked, faintly. "That's right."

He shepherded his little troop out of the barn, casting an arching glance back to all they were leaving behind. He knew a month's worth of meals was stacked up in that corner. And it was his own fault and folly they couldn't use it now.

He shook his head, leaning a moment longer against the doorpost.

This wasn't over yet.

50
Luis

An t-ochtú lá de mhí na Nollag

Our saints have become like ancient gods; nobody uses, thinks on, or hears of them, save at the cost and provocation of scandal.

The Church of St. Dismas is dead, the bar burnt to the ground and any survivor caught waiting penalty in jail.

The rest have retired to Balladarach.

"If this were an army," Senan told me, "first I would have ya flogged for this...

"Then I would make you a general."

If we do not defend to death those to whom allegiance must be owed – if we allow those wrongly accused to perish for that loyalty – in the interest of protecting masses – who may or mightn't suffer, though threatened they must be – and you *knew* that you could do it – could save them, alone, with your own bare hands – although you *know* – ungodly omen! – there *will* be consequences, though not yet manifest – *dire* – most assuredly –

Do you let them die?

Knowing the difficulty of saving those lives, do you allow them to die? Not I. I am proud of my actions, *a thaisce*. I would do it again.

I'll confess it:

I shall.

"Head-fecked," my friend called it. "You flippin', dyin', degradin' head-fecked."

"I'm a peddler of all needs, lad," I responded, haughtily. "And the need here was a warrior."

"You're an everyman!" my wife chimed in, with a teasing smile. And, supposin', so I am.

"Dubhán, we can't be so out in the *open* with this," Senan told me.

And I, my head bowed: "I know."

"If we give them a cause t'come down on us, they will, and *hard* will they punish us, surely."

And I, still-shamed: "I know."

"What y'did for Dismas was grand, aye, but we can't take 'em in so great of *numbers*, Dubhán! Two hundred people! *Dar m'anam!*"

"I..." Intrigued. Cautiously, I glanced up. "We can take'em in smaller, though."

Sean Fahy smiled.

"Can ya follow a few orders, for once?"

I can.

'Cause wouldn't you know? There are souls that we've rescued who have sisters and wives, brothers and parents scattered throughout Mainland towns. And wouldn't you know but they'd love us to save them, too?

Now, don't make that face, *a rún*; I'm not stupid. But I, inconspicuous externally, as I've been bid to be, can go into a town, and make inquiries. And I can *find* yer brother or sister or aunt, and, grand judge of character I am, read them, and determine whether or not I can bring'em home.

Nor am I the only one. Senan and I between us have hand-picked, like, fifteen lads – thirteen guys and two girls, so.

It's a start, *a mhuirnín*, and I'm excited.

It's a start.

An ceathrú lá déag de mhí na Nollag

My weird little house-guest has made it his life's mission to make certain my son's first word is my wife's name. Not "Mommy" or "*mamaí*" or "*mam*" or "mum" – but the name with which her parents baptized her.

And not a soul among us has decided yet where the scale tips on that one – whether it's annoying, or slightly hilarious.

An séú lá déag de mhí na Nollag

Once I was stranded – as far east as you can go without falling into the sea. I woke up in a cell – not, I'll admit, for the first time, and neither for the last. This cell was different, though – I saw when I rolled over. The door was a screen – nearly the whole wall. Everything in me ached, but I got up, going to the screen and, near gently, pushing it aside.

Here was a porch. Like a dock, it was, and it running, dark-wooded, all around an enclosure. And in that enclosed courtyard, the ground was all pebbles, and trees – not many, but thick – thick as a Rhanddeheuol (Moorlands) bog – and little streams feeding a dark pond, and a red, arched bridge crossing into the darkness.

I'd never seen anything like it. And I was drawn to cross that bridge.

Snow blanketed the ground, sheathing the more solid limbs of the trees, mixing with the dark moss, crowning the heavier boulders which stood, scattered like sentinels, silently watching my every move.

The walnut boards of the dock cracked and whined at my footsteps – though at fifteen (nearing sixteen, now) I was a wisp of a lanky thing. I looked down. There were a lot of boards between me and the red bridge. But jumping down onto the stones would be as much of a racket, and sure but with the shaking of my limbs, I didn't think I could make my jump. Nor could I balance my weight. The best I could do was lengthen my stride. *Skrit skrit skrit skrit* – I was scowling, furious, by the time I reached the bridge. I've always prided myself on my stealth; so why could I move, silent as a leopard, *drunk* even, across the decks of an old wreak like *The Lady Haggard* (*Delila*) – and not across a porch?

Someone must've done somethin' to me, I decided. And, *Oh! but they will pay!*

Dancing across that pondersome question – *Where am I?* – I turned my attention to the bog. Crossing the footbridge in three, angry-long strides, I stood on a sandy island, itself well-shaped and solid, rising like a cliff above the river of stones. I couldn't see the sky for the clouds – a phenomenon I called mist, for I'd never yet set foot on mountains so terribly, madly high.

I snaked my way over the island, moving between the trees. Even now, I remember the racket of clattering stones, as I staggered, light-headed with injury, altitude, anger – for sure I was sober of wine – my bones scraping and shooting, a thousand little abrasions crying out like fire. *What the hell happened?!*

Deeper I moved into the trees, wondering ever more at what I saw. Ever through their willow-branches I caught glimpses of the porch – more screens like my own, more black-walnut decking. *What* is *this place?*

"You come."

Jeeze! but I spun around so fast I fell into a stream – the stream blue and green and shaded like the depths, though it was only half a foot at its deepest, for the peaceful dark of the world around.

"You destroy."

On a sandy island, beneath a great, thick tree, cross-legged, was a man. Engulfed in light robes of honey-yellow and burnt-orange red, his head bald. A bony, skinny, leathery old creature, like a turtle, but with more nobility, more poise.

"I didn't *come*," I said, scowling, as the stream washed over and around my body. "I arrived."

The man opened his eyes. His eyes were narrow, and black, but they shone with life like the fire – a life held perfectly in balance, in control.

"Not yet," he replied.

Everything he said was calmly. And that infuriated me more.

"Then where is it I'm *'come'* to, at all?" I demanded. "Och – God help us, but if this is that usurping nation's idea of rehabilitation – mental incarceration an' madness – ye be damned but I'll get out of it!"

"You ask much."

"You *narrate*," I returned, sharply. "Do ye *impart?*"

"Rise," he instructed only.

I railed against him more, cursing and mocking and challenging, but he didn't respond. So I decided he wasn't a threat to me, and therefore deserved my respect – at least, if only enough that I got up from the stream, went over, and sat down beside him, cross-legged.

Then he explained.

Of all the monasteries that had once dotted the east, this was the only one that remained – the only one left undestroyed by the Empire, for they had never found it. High in the rugged mountains it was – And damned if I wasn't snowed in for the winter.

He then told me another thing. He said he could tell that I had power – more than I knew. But my mis-directed pride was a hindrance to control.

"What in hell's that mean?"

Pride is not bad, he explained. I was just exploding it in all the wrong places. He also told me that I was clever enough to understand, so to think before I burst out in protest.

I sighed and groaned and muttered Ceothann, but then I thought.

"I'm proud of what I am," I said.

"Good."

"I don't need any help."

"Even the mightiest need help at times."

"Thank you, Aesop." My elbows on my knees, I buried my face in my hands. It was a long moment – a long, *silent* moment, broken only by the faint rippling of the stream – before I said, "Fine, damn it. I guess what yer sayin' is... you're on my side. And so... I should... stop... tearin' ya up, like... Y'ain't gonna kill me, are ya?"

"I am here to guide you."

"Huh..." I considered it. "I like you, guy," I decided. "Sure; and as long as it isn't trickin' me yer at, sure you can 'guide' me t'yer great heart's content 'til summer."

We had two seasons in Ceothannanmór.

You can gather some of what I was. I was a real flippin' brat in them days, ***a chroí*** – a kilted terror, wild-haired and thrice as fierce as the warriors I imitated. Then I was one of the rogue crew of *Delila* – a voyage cut all too short, may God give the lost bones their resting.

What I'd since gathered of the true tragedy of the wreck of Old Murray's schooner was this: that she sheered on the rocks off the far Eastern coast, taking with her five good men and the captain – God keep them. Any fool of the rest of us with a mite of sense in his head fought for the shore, and the rocks and the hidden caves and coves made it hard for us. Myself...

Now, the monks and I *barely* shared a language, and it was this made it hard to communicate indeed. The best I can figure is I was hurled around by the waves, might've tried to swim it, and – by God's means, not my own – had the strange luck to have landed myself on the rocks, and the tide abated for a while. These were coastal mountains, very high, very bound up, and steeper on the side facing the sea. It was the goats and the monks that followed them who could scale it, and – I guess – it was one of the more agile ones that found me.

During the three or four months I remained with the monks, I was healing. Physically, of course, but also – spiritually. This latter concept I was far less receptive to. It was all hoakum. The idea that settling my spirit could do anything for my health. In a way I suspect they were trying to house-train me – to grant me some semblance of peace and self-control. It... wasn't as far-fetched as I thought.

After my talk with Bai (yon auld monk), I was at least resigned to give it a half-assed shot. I respected him. And I knew – I was clever enough to know it – that I *could've* reigned it in if I wanted to. But there was the problem: I didn't want to.

They gave me over to a man called Zheng, whom I tended to call Xiao; myself they called Hai, which meant, I was told, "sea", and was desperately easier for them to pronounce than my original name, especially as *I* pronounced it: heavy with Ceothann accent.

Zheng had a long staff, with which he would whack me – when I mouthed off, when I tried to leave, when I was being a little jerk – *thwack!* in the head. And I was made to meditate *constantly*. I was durable as I was defiant.

On a day, about the second week of my arrival, I was sitting – not quite *meditating*, because my veins were seething, my mind filled with devilish delinquent tripe – *Fine, I'll sit here – I'll show you! I won't learn anything! Ha!* The hour was getting late, and I was cross because I wanted to be outside, and Xiao would not let me. And that, *a ghrá*, was to me one of the *worst* things you could do. Now: I had thought myself pretty tolerant up 'til now. But Xiao had rapidly become a tyrant in my eyes, and, if nothing else, I would not be ruled.

I noticed that Xiao had lit a lantern – a glass orb in a strip-metal frame. By this time I had been performing with controlled fire for some years; I decided a little impromptu demonstration was in order.

Concentrating my gaze, I slipped the grasp of my senses into the little heart, filling it, cradling it, urging it to grow. The flower flared, catching and expanding and sparkling with white light. Slowly, I found the heat, as I usually did to extinguish. But instead of pulling it all at once, I began stripping it away, 'til the flame burned as red as blood. This fascinated me; I had never before done anything like it.

I began to test the boundaries, swelling the warm flame against the glass. The fire curled and coiled within itself, pressing, increasing –

A sharp crack, a glittering shatter. I started, cast into darkness.

Anyone else in the world might have exclaimed, *"What happened?"* But not Xiao. Master Zheng, our intrepid warrior, whipped me with the staff, demanding: "What you do?"

"*Aagh!* What *you* do?!" I snarled, shrilly. "Slap-crazy old man!!" I could feel the sparks in the ember ashes, but I was too mad to raise them.

He glanced behind him at the table, on which stood a number of scrolls, all covered in boxy stacks of hash-marks I couldn't comprehend. He re-lit the candle in the lantern – the motion of drawing, striking, dispatching the match, was all so fluid – like one long paint stroke – it was mesmerizing.

"Concentrate."

Good Lord. I rolled my eyes. By now I had heard that word so many times in a day I'd begun to answer to it – and that's *not* an exaggeration.

"I –"

"*Concentrate!*"

"I can't do it *now!*"

"This gift wasted on you!" he said; on other days I had wondered if the Eastern language lacked the verb "to be". But not now. Now I stared up at him, his words piercing me to the core; my anger bled, and nothing else remained.

"...What?"

Zheng shook his head. "Long I heard rumors in my travel of bygone years. The world did not see your like and kind for centuries. A true blessing. And you waste."

I waste... And my stomach turned as I thought on my parents, and the fact that *they* had been able to do this, too. Not centuries, not really.

Just not anymore.

It's an elite few, *a chumann,* who know what I can do, and who know about my life and parentage. Consider yerself privileged.

"Learn to control the self," Zheng ordered, "and you come to control this power, in time... You may go outside."

"Oakh –" It was a wordless sound I choked on, turning and splitting before he could hit me again – or worse, *ask.* I didn't care how cold it was, I didn't care that the snow blanketed the rocks and the tree-limbs were frozen. The stream was a single, drifting capillary of a hot spring, and I was sheltered there, so there I stayed the night. And, for the first time in over a decade, I cried.

I had got it; I understood.

The very first thing when I woke the next morning – I remember it clearly – a cup of tea was set in front of my face. Not too close, but dead in range, just in the perfect spot, placed with the fluidity with which God sets adrift a lotus flower. It was a plain cup, white clay, without even a handle. And, to a Ceothann, it was rather small – the size of a double shot.

I looked beyond the friendly steam. Bai.

Bai was one of the few monks I kept away from – out of respect; I avoided Ryo because he scared me, but that's another story. That dude was tough for a monk, really. Anyway – Bai I respected. Bai was saintly, and you sure as shite don't mess with saints.

Why did it have to be Bai? Aukh, he was way too wise and uplifting, made way too much sense. I sat up of some alter spirit's accord, lifting the tea as he advised, "Before the snow get it!"

My spirit was too downtrodden for tea. I stared into the cup, and the wafting aroma I associated with incense. "Why is it it's... it has fog that's in it?" I asked, picking through words.

"Ah? Ah!" Bai answered delectably. "The milk strengthen you."

"...I thought it was only focus could strengthen me," I whispered.

"Ah!" and out of the corner of my eye, I saw him wink.

You're too good to me, Bai. I settled deeper into my despair. "I can't take this; I can't take all your tea. It's yourself who deserves it."

Bai held up a hand. "Tea between friends."

Friends... auch, that made me feel both better *and* worse. To the left of Bai was a Fenian Oven dug into the sand, a little one, and a red, earthen tea-kettle suspended over it.

"Is..." I struggled to get that word out – stammering, stumbling. Finally, I managed: "Is it cold tea you have, sir?" I was closer to the stream of the hot spring; he was in his favorite spot, by the big rocks and the ancient-looking tree. I set my cup in the sand, secure; then, hesitating, I reached out my hands to Bai. He trusted me, giving over his tea.

Don't mess this up, I thought, like a prayer. *Don't screw this up. Concentrate.* If I could pull heat, I could isolate it. I could create it. The vibration of molecules, Aneirin had called it. Was I not shaking enough to bring a little warmth?

Relax. That was what my mother had said when teaching me. *Relax and concentrate. Don't be scared. God is with you. Concentrate.*

Don't screw this up. I had closed my eyes; now I opened them to slits, focusing on the tea I held between my hands. My hands were warmish – the spring was warm, too. I concentrated on that warmth – I pulled it, I twisted it, I twined its energy with the tea. Suddenly, I felt its warmth against my face; my eyes widened, and I saw the steam.

"Here." I handed the cup to Bai, quickly. He accepted the tea. "Such a good boy."

Aukh, his kindness was killing me. No, I wasn't good; I was a waste. I buried my face in my hands.

"Tea help," Bai advised. "Come come come."

I glanced up through my fingers. With a wry look, Bai sipped his tea. Tea. I pushed my hands back through my hair. I crossed my legs and took my tea; settled. For him, I would do as I was told.

His next words caused me to look up, surprised.

"I find Zheng a grumpy old man at times. Too serious." Bai made a serious face, and it made me laugh, even against my will. "What he need: More. Tea. Less *hwoot-hwt* with stick." Mimicking the thwack noise I had come to know so personally.

I smiled. The tea had a crazy aroma that filled the soul – nutmeg and ginger and cinnamon, star anise and black tea – a luxury north of Saranam. I ventured a drink. Mm; heavenly.

"The longest journey is the journey inward, Hai. This tree: it did not grow in a day."

"Thanks for the tea, Bai."

He looked at me wisely – so wisely, I knew there was more beyond those words to which disparate languages were no barrier. "Tea help."

And wou'jya know it: by the time I left, after that stay of nearly three and a half months, I could walk across those flippin' boards, silent as a monk.

Naudecametos Cantli (19ú Nollag)

Knowing as I do that this is not a language spoken, let alone written, anymore, I am using the other half of the paper for translation. See? I can be clever, darling. It is into Ddeheuol, aye, but – it is the best that I can do. And Donneen is a darling, truly; he will not rip it out. And that's all I ask. For I write so that there might be *one* record, at least, of my people's heritage, our language, and our history, if only a bit. To help shed a little light, maybe, on a near-forgotten culture.

Okay. So here we go.

My father's sister used to tell me our clan was named for Murchaidh. Now, as you can figure, it takes a mighty grand deed to get a clan named after yourself, particularly when that name is not occupational. Some gain that right by being the first of their clan to attain great-scale rulership, some by being especially proficient conquerors, others retroactively by being the common ancestor for multiple, smaller tribes. And, although Murchaidh was well able for all of that, the reason he is best known is for being the Master Architect who engineered the Castle of Ariegn.

Undoubtedly, if you've gleaned a word of Usurper propaganda, you have heard of the Castle of Ariegn. But they won't give you the whole story. They will not tell you how it stood for over two thousand years, designed by a crafty king, stacked by the hands of countless, unified tribesmen, and sealed by a white mortar which Time itself has forgotten how to concoct. They will not tell you how *we* built it; only how they took it from us.

But I will tell you, my darlings, and not a word of a lie in the length of it.

Murchaidh was king in those days. The latest in a long line which had ruled the north-westernmost territory of Gallia, as was designated to them by ancient tribal boundaries. But things change, and the lines were coming into dispute. A wooden *rath* on a flatland would no longer be sufficient to protect his kinsmen and subjects. But what can he do? It is all flat-land and trees 'til you go deeply south, and to do so would leave your entire coast fatally exposed. Coastland is a necessity to ruling – access to the coast. Food, flight, protection: coast. And amply – a veritable blessing to my people – the coastline provided that which the sapling could not: stone. You see, my darling, the answer was very simple. Mining skills, construction skills – these they put to the whetstone of human ability, and from the very land they ruled, they built up the great fortress of their salvation.

I might interject here, regarding the stone and the aforementioned mortar: Any tale you have heard of the source for the white rock of Ariegn is false, I can guarantee you. They say it was, that Murchaidh (those telling likely don't know his name), dissatisfied with the black rock of the coast, carted white stone from the cliffs of Ceothannanmór, or Sefetaten – or even Lavinium itself, depending on who you ask. And for each story, he is lugged through another set of wild feats, each crazier than the last. Dragging stones along the bottom of the ocean with a fleet of sail-boats, accidentally raising treasure and rousing gods; battling rooks as big as buildings to roll the mysterious stones out of their mythical nests and breaking them like coconuts against the desert sand; driving teams of elephants along through mountain passes, beset by phantom archers at every turn. Not so, I'm afraid. And I am sorry, but the truth is sometimes duller. This is what happened: Ariegn was built of the bald, black stones, cut in blocks and sealed with the aforementioned mortar; you could tell if you were ever inside it (I have not been, but it is my own damn family line, is it not, darling?) A great, monstrous castle, sleek and insurmountable, towering up from the high-point of the coast; straight its walls, straight and steep. Within, rooms upon rooms – I have heard tell the lie of it from various tongues, in bits and pieces, but I shall not compose it all here; too much to tell – great, open spaces at the heart of it, and rooms and halls all branching off within, so that everyone you should ever want to could come inside, and take harbour. So intricate, so *beautiful* – indeed, God Himself has to have helped in the designing of it.

The walls are white for the mortar. The recipe, I repeat, was lost, although I depravedly delight in trying to convince people the base was sea salt and gull poop. I maintain I am clever enough to figure it out, given the chance, but... I feel it would hurt to go back; to see what once was, and to know what we have become. But maybe I will yet, someday, all the same. God knows.

Goodness. So settle it: Murchaidh built up a great castle of the black stones, the first of its kind and unparalleled since, and covered it over with a hard and durable, impenetrable mortar, which helped it withstand even Time. In this castle, he kept his whole clan. Yes, *all* of them, so big was the place within, so intricate its design. And in this way, everyone was defended. *That* is why we draw our name from him. He represents the best of us; the pinnacle of our talent, the depths of our dedication to our clansmen.

And *always*, as long as men had held land in Gallia, the people of Murchaidh had held *that* land. That castle became our symbol. Never defeated. *Sego are maruos.*

A thousand-odd years that castle stood, and it was passed from Dumnorix, our king, to Saint Aídan, who had led God out of the south-east to our people in the North; at that time, the world was not so dangerous, so threatening, as it had been in Murchaidh's. Dumnorix was a strong warrior, and brave; he realized – the world around being settled, themselves being fierce – no longer had he need of such fortifications. So, the children of Murchaidh agreeing, Dumnorix gave over the castle to be a monastery, and the people moved out into their land. There we remained throughout the years, to defend what our ancestors had given us, proud in the name, proud of our culture, and strong.

That is, until forty-one years ago. When the war took everything away.

Lost to me now, my father's sister, Aveta, told me all these things, so that I would know them and remember, reciting this history over and over, so it is written on my bones and in my veins.

I am descended from Murchaidh, Master Architect and Chieftain of the Galls, father of the Morini tribe and engineer of Ariegn.

And don't you forget it.

An fichiú lá de mhí na Nollag

Keel-haul'im 'til he's sober
Ear-lye in the mornin'!

51
The Palisade

Sixteen hours *mo thóin*.

But there it stood. The great monstrosity. The sea of solid, sickly green, trampled and hard-packed more so by wind than by lunatic travelers, stretching off to devour the horizon, with never a tree in sight beyond this last dutiful line. No flower, no grain, no simpering stalk, dared raise its head above the turf – not to be seen from here anyway, though the smell of heath and pine sap lay heavily on the air. Grey the sky above, looming ominously over the caulk-slate field.

And *that* was what they were crossing.

The last of Rahyoke's strength gave out, and he collapsed, casting himself and his baggage at the base of a gnarled tree. If he felt his chest, his side, his lungs at all, sure but they were on fire.

He did not miss Zander's glare. He felt it, carving hatred over his bent form. How many times had the archer urged him – *We need to rest, Rahyoke. We need to stop*. And him answering, until all he could manage was the vaguest shake of his head.

He choked, shortly and quietly, trying to draw a breath. It came, wheezing and shallow, drawn out long. *Pathetic.*

"Y'okay?"

He jumped as her hand slid onto his shoulder blade, pressing in a friendly, encouraging way.

"Aah." He shrugged her off, and she came around, sitting against a tree three yards to his left. "Hh – I'm –" He straightened as much as he could manage, keeping his head bowed. "I –"

"I take it we're resting now?" the archer demanded, shortly. His pack was laying at the foot of a tree which marked the border, his quiver slung onto his back.

Rahyoke rolled so his spine was against the tree, sitting with his knees drawn up. He drew a long, shuddering breath. "Yes." Mercy, the word was airy. "I ha'a – ukh... I hafta – get supplies – b'fore we – so –"

"Tomorrow," Zander grated, through clenched teeth.

"*Sea,*" Rahyoke acquiesced. He swallowed; taking another breath, "Ha-have you guys ever – done anything like this – before?"

"No," answered Addison, shaking his head.

"'Kay." The stabbing pain ebbed enough that he could breathe again. "Okay... We're crossing that field starting tomorrow. Here're some things you'll need to know.

"First... A vast and terrible wind ploughs constantly across that field, stripping it barren, wearing the grass to that sad, weedy state you see it now. It parches the bogs and drives the rains into your skin; dispels the hillocks and chokes the very breath from your lungs. And it blows out of the north and east, so, lucky us.

"Second: the mist. This time of year especially, the sea-mists draw in devilish thick here – a solid and impenetrable white-grey, like a spirit's skirts, raised from the ground and dropped from the sky t'quash your reason between it. A beautiful sentiment, I'm sure.

"Lastly: the cold. I believe it was our good man Addison who mocked me so for it being summer, but the winds pull down from the Loptrian – Northern waters cold as the –"

"Hey," warned Zander.

"They're cold," Rahyoke amended, without missing a beat. "And colder by night. And, should we stray *north,* this time of year it'll get down around forty at night.

"Oh. And it also rains." A bitter contempt lathered his tone on the last – an ironic and humorless little snap.

"*Why* are we doing this, again?" Addison asked.

"Our destination's *just* across that field," Rahyoke insisted, a verging plea in his tone. "It'll take only about a week t'do it."

Meygan whirled on him. "Wait! A week?!" Her eyes were wide. At a whisper, as if the thought were too scandalously disgusting to relate aloud: "*How do we use the bathroom?*"

"How d'you *nor* – wait, don't answer that..." After a moment, he shook his head. "I don't see the problem. Look!" he added, as she moved to protest, and he drew the boat-knife with startling rapidity. "Anyone who dares t'so violate yer privacy, I'll *kill*'um. Satisfied?"

Meygan sighed, shaking her head at his absurdity.

"I'll leave you behind," he threatened, indicating her with the blade. "Don't test me; I will."

"Whatever," she grumbled, folding her arms and sulking. She didn't like the idea, but there wasn't a way around it.

Rahyoke rolled his eyes, stowing the blade again. There's no pleasing women – of that you could be damned sure. "Any other *life-shattering* qualms? No?" He leaned his head back against the tree. "Then alright." And there was the end of that.

"Does anyone care about dinner?" Zander asked, lackadaisically, and was answered with a resounding grumbling of, "No."

"Just get some rest, lads," Rahyoke bid them. "I promise ya – it looks like Hell from over this side, but, once across it, you'll be glad of the experience."

"If we don't die on the way over!" Ian trilled, and never were words more cheerfully spoken.

"Aye, Mercy... Trust me, lads. Everything will work out *fine*."

"Hell wi'*that*."

Zander looked up as Rahyoke dropped down beside him. The man had drawn unusually close, coming within a yard, and by that Zander knew that he wanted to talk. "You couldn't sell?" he guessed, and Rahyoke shook his head.

"The selling isn't the problem, but the buying." Indeed, almost all of the merchandise he had brought with him was gone, their great store reduced down to what the found camping knapsack could contain. "I can't get what I want."

"Then get something else."

"I need tack. Hard-tack – somethin' that won't expire halfway across the field – and yer man sellin' it won't sell any to me. Not unless I show him a travel visa, and I've no travel visa to show. He was quite rude to me, actually. *Iochd*, I tell you – a downright regular palisade town, they've become. They've erected walls and a gallows. It stands empty, the noose, but that's almost worse; you know it's waiting for you. D'we have any more pretzels?" He snared the food bag, pulling it nearer and rifling through it.

The day was nearing sunset, long about seven or eight in the evening. "So what are we going to do?"

"I've a plan," Rahyoke shrugged, eating a handful of pretzels; he hunched over the bag with the protectiveness of a starved wolf.

Zander's frown flat-lined with sarcasm. "Wow. Reassuring."

"Lemme ea' firs'," Rahyoke replied, crossly, around a mouthful. "Wur still leavin'a nigh'!"

Zander rolled his eyes. Taking the food bag, he began sorting through it.

Swallowing, Rahyoke asserted with annoying levity, "I was threatened by it, t'day – the gallows. 'Ah, this one scents of vagabond,' they said. 'Have y'seen our Godly stocks?' Apparently they *hang* nomads – can y'believe it? Of all the *scandalous* depths..." He shook his head. It was too unspeakable for the man of few words. "A pretty little nation – a *lovely* little town. Tell ya what, you and yer damn bush-man theory –"

"Stocks?"

"Mm." Rahyoke swallowed another mouthful. "D'ya 'member when they use'ta put vagrants in a stockade?"

"Oh! *Oh...*"

"Oh," Rahyoke acknowledged, nodding slightly. He looked down at the bag of pretzels. "I can eat and walk," he realized; *why didn't I think of it before?* He stood. "'Kay. Be packed, be ready, and have the kids the same. An' I mean it – be ready t'go the moment I come back. And if there's no sight'a me by eleven, morn, tomorrow, leave me behind."

"What are you planning?" Zander asked, well attuned to the foreboding in Rahyoke's schemes.

"Not important."

"Shall I stand watch for you, then?"

"If you can last the vigil." Keeping the knapsack, Rahyoke handed Zander the canvas bag he carried. "Guard this with your life," he ordered, and Zander slung the strap over his shoulder obediently. Rahyoke turned, lifting his gaze to the field. One line of trees separated them from that open sea, its rimed waves beaten and frozen by time. One line. "I hate this crossing," he murmured, his tone dour. "I hate this crossing, I always *have* hated this crossing, I always *will* hate this crossing." He reiterated to Zander: "Be ready."

"I will," the boy affirmed. The wanderer backed up a pace, hesitating, a flash of indefinable emotion beneath that disdain. Then, turning, he swept off, limping slightly, his black coat trailing silently against the bracken.

Zander ran a hand into his darkish hair, clenching his fist. *Should I be as worried as you?*

"Lads – let's get a fire going," he urged, getting to his feet. "We've got several hours yet 'til we leave, and, believe me, we'll need the light."

The limp fractured his stealth, as the small campfire broke theirs. But the fire was a sign of life, and readiness, and who could ask for more?

"I might'a gone a tad overboard."

Meygan looked up as Rahyoke dropped a bag beside her. "Oh my *gosh!*"

One by one, the skinny creature unslung four cloth grocery bags, two burlap sacks, the large hiking knapsack, and the black coat flung over his shoulder like a hobo backpack, piling them before the hearth.

The children started up, clustering around.

"Holy *Hell*, Rahyoke!" Zander half groaned. "What did you *do?*"

"Hard-tack!" Rahyoke proudly declared, by way of an explanation. His grin was utterly sinister. "I won't stand my dignity abused."

Meygan looked up at him with a beaming smile. "You're a very vengeful person, Rahyoke," she approved.

"That I am." He crouched down before the pile. Taking Zander's pack, he started unloading one of the grocery bags into it. "Could I press you lads into service?"

"Are you feeding an *army?*" Addison degraded, sifting quarter-heartedly through one of the burlap sacks.

"*Never* satisfied," Rahyoke grumbled, shaking his head. They emptied the grocery bags into Meygan and Zander's backpacks. The burlap sacks replaced most of the merchandise Addison had carried in the tartan, and Rahyoke moved the contents of his coat into his own bag. Ian would take the knapsack.

"But I –" Rahyoke started to protest, but Zander cut him off. "No... Rahyoke, it's fine," he snapped, when the man moved to argue again.

Rahyoke frowned, but said nothing more.

"*I'm* strong," Ian reassured them, confidently. Then, peering into the bag: "Oo! Pears!"

"Why didn't you do this sooner?" Addison complained. He was the only one left packing, and was taking his own sweet time about it.

Rahyoke's dark eyes flashed. "Because we are neither thieves nor beggars. I have worked hard for what I have. And if it –"

"Sorry! Sorry – I didn't mean to *offend* you." It didn't sound earth-shatteringly apologetic, but again we take what we can get. Addison continued to pick at his work.

"Really, wine?" Zander challenged, offhandedly, and, with the grin of a trouble-maker, Rahyoke lifted a bottle, so the olive-green glass caught the light.

"Yeah; so, I saw the hard-tack, and then I saw pears, and I saw this — Ah! *Buíochas le Dia*, the aul' bastard saved me some!" He swirled the two inches of liquid, uncorked it and tipped it back. He swallowed and elaborated. "Shorn, that's good stuff. If I'd sold the house an' all in it – including a strong lad like yerself – five times over, I couldn't afford it. *Balaena*, out of Martius Valley. I thought the bottle could be useful for storage." *Balaena* had an exclusive monopoly on the production of ambered wine – claret flavored with ambergris. The ambergris added a depth of flavour to the claret that was simultaneously earthy, musky, animal, marine and floral, warm, complex, and sweet. He took another drink. "Oh – and Addison, quit that look; I didn't drink the whole bloody thing myself. *A Dhia dhílis*." He lifted the bottle, ordering, "Get ready, will you?" before finishing it off.

Addison muttered under his breath, but nothing loud enough to bother hearing.

"Why do you swirl it?" Meygan asked, slinging her backpack onto her thin shoulders.

"It – what'sit? *Releases the flavours*, I think they say," he replied, slipping the empty bottle into his canvas bag.

"Ah... Fancy."

"Yeah," Rahyoke snorted, sarcastically. He pulled the strap of his bag over his head, trying not to wonder how he'd lift it. "I'm a flippin' connoisseur." He stood – shakily, very subtly using the tree behind him as leverage. "That'll do for now. Zander – y'wanna get the fire?" The lad set to the task. "Addison, I'm losing patience with you. If that's too heavy, I'll take it, but we're leaving *now*."

"*Right* now?" the boy nearly whined.

"*Right* now we're going."

"Come on, Addison," Zander complained, under his breath. "Yer makin' me look bad."

Addison scowled. He made no move to pick up the pace, and this time his criticism was audible. "Crazy, tyrannical old bastard."

Rahyoke frowned slightly. "Crazy." They had already spent too long on the border. They were lucky they'd gotten this far. They were cutting it deadly close as it was, and they didn't have time for this nonsense. But anger wouldn't work with this one.

At least, not on the attack side.

Rahyoke drew the unstrung bow from the quiver on Zander's back. "Fie, then!" Sweeping the bow like a broom, he swatted Addison in the side – not hard, but enough to kill a fly, and certainly enough to be annoying. "Ya wouldn't know crazy if it bit ya on the ass, son." *Hwack!* "*Çá, çá* – an' on, ye lily-blooded nyaff. Crazy is an accoutrement – an *hors d'oeuvre* – ornamentation adorned to equate quirkiness with originality." His tone remained the epitome of calm sanity, never breaking that perpetually dispassionate cadence. "*True* insanity – oh, true insanity – is the ambrosial brunch of only the most fatally brilliant. Come on, then!!" He swatted him a third time. "Wou'jye break rank, lad, ere yeh raych te' heathers? Far scar an' sixty sirens lark b'yand t'firth, a-cryin' far ye – 'Oh-roe, 'tis th'Devil's ain sport' – bu' yer honey-slip't water-blood tay them's divine ay' t'champaign-taykin' heart'sa them tha' no' cou' stand blood rich'da iron, dirt, an' whiskey – get *up*. Or hasn't my insensate rambling given delay enough?"

Meygan cracked up, adorably amused by Rahyoke's performance. Addison grimaced. "Do you just save up words so you can blurt them out like that?" he sneered. "Or do you stay quiet because you're *really that nuts?!*"

Rahyoke sighed. "Will ye no'?" He whipped Addison again with the hundred-sixty-two-year-old switch. "'Tis t'hayth an' hayther; longar y'sit, te'morn they drive ye mad. *Oop*, ye windae-lickin' caitiff! Ye'd be whip't wi' yer ain blade in a muster, y'unmannly, mewling bairn! Ye delicate mornin' glory! Ye –"

"*Knock* it off!" FINALLY, Addison jumped up, staggering back so he was out of the switch's immediate reach. "Are you *drunk?* What's your problem?!"

"It's in the bloodline." The words were slightly clipped. "Are you set t'go, your Highness?"

"Yeah," Addison responded, with all the attitude in the world.

Rahyoke returned the bow to Zander, though his gaze and his words were for Addison. "Next time Meg'll be bosun. *Do* try and see if you can prove stronger than she, *warrior*."

The insult was set forth in such a way that there would be no refuting it, and certainly no taking it back.

"A little sexist," Meygan noted, nodding. "But – oh, snap."

Addison returned to grumbling.

Rahyoke shook his head. *They'll be the death of me. I know it.* "We'll maybe stop at sun-down," he established. "'Til then, we don't. Okay?"

"Okay."

Lord, he hoped it so.

52
Fearn

An tríú lá de mhí Eanáir

It's Orchid who's gotten both tea and coffee all over this book.

I'll admit in torture it was I did the grass, the water, the sap, the blueberry jam, the mud, the scorch-mark on the top corner, MAYBE the blood –

We can neither of us agree who (*not* me) did the whiskey (she should've known I wouldn't've caught it!)

Ah, sordid monks convened, rejoice.

An deichiú lá de mhí Eanáir

When the ocean is calm – I mean dead, boring, bane-of-my-life calm, rather like this night – you can go down from the ship into the sea. It's a little like standing on a cliff's edge with your eyes closed – to me, anyway; I've a heightened sense of adventure, I've been informed. But c'mere: You have *no idea* what you're dealing with. You don't know what's beneath you – how far away is the earth. A springing wind could leave you to your doom. Not that they'd set sail and abandon you – usually. I don't know. Men are more often lost in the fishery than the merchant trade – because for all the many powers of man, you cannot bring the whale about. And I don't care who you are – my own life is not worth 300 barrels of oil, even in these electrified times. Why would I cut loose for you?

Oh, don't get sore; you know my love for you. And for all that, if you come down into the water with me, I won't set sail without you. And sure, preserved in words as I am, I can't.

Behold, the Dichotomy of Dubhán!

An ceathrú lá déag de mhí Eanáir

I got pipes! :-)

Did I – I did; I mentioned to yez before I was a piper. But never did I own a set of Highland pipes, but only my humble tin whistle. A-haa! They're beautiful, *a thaisce!* Just beautiful!

It was long about four-thirty in the morning and I came home with them. I slipped through the door, silent as anything, dropping my keys on the counter as I closed the door behind me with my foot. It was still dark in the kitchen, for the sun doesn't rise so early this time of the year. I cast a glance about me, taking the stillness of the room in with some annoyance. Resolved to be a total jerk and a nuisance, I decided to announce my presence with "Mist-Covered Mountains".

There was a tumultuous, thudding crash from the den, and the snarled squawk: *"Holy Feck!!?!"*

Laughing, I came into the room, dropping into an armchair. "Mornin'. You still a-bed?"

My friend was all out of sorts, entrapped in a cocoon of sheets, cursing and swearing in all manner of tongues. "Ye *scuttered,* ye damn fecked half-cut eejit of a bollixed whore-monger!?!" he demanded, scrambling free and to his feet. "I's a mind to flay ye!"

"Wanna see what I got?"

"I see," he snapped, begrudgingly. He's not an early riser, God love him; but neither am I, so we shan't hold it against him. "How came to ye?" He sat down on the coffee table, which I've since given up on tellin' him not to do.

"Would ye believe I stormed t'ree fortresses with naught but a pike to me name, vanquishin' armies –"

"No."

I sulked. "I didn't finish."

He yawned and stretched. "I in me cups deep – time, drink alike – you tells me, huh?"

I made a face at him. "What – yer good anuff for poetry, now, are ye?"

He dropped out of his stretch, sighing. "Go on, okay."

"Ye've ruined me flow; I need another."

He groaned, leaning back, his palms pressed to the surface of the table. "High-Ceothann," he grumbled, remarking on the make and model of the pipes.

"Near," I agree; "it's styled after it, aye, but them Highlanders don't part from their pipes for cheap –"

"Swell might," he muttered. "Kilts."

I laughed. "Ye sordid, shameful creature! But let me think of a story..." Deep in my thinking, I settled back in the chair, slouching as deeply into it; that chair's a great thing, lads; it dwarfs even me – I love it!

Finally, I started. "I was in Shifty's, an' a crowd of Turranians comes in – and they the southern, ruffian brand –"

"Holy *feck*."

"Don't tease me – pay attention! So – so, Shifty's is rife with 'em, right? An' I slanderin' desperate outnumbered. *Don't* care. I slides me to the bar, an I says to yer man the bartender: 'Gimmie twice whate'er that table swallows; sure it'll be one glass to me!'"

"Was – *if* was true, you dead." Rolled his eyes: "'Ruffians'. Goodness."

I frowned. "Me? No one'd kill me; I'm too pretty!"

"Sure..."

"Anyway – so I says it loud so's they can hear, an' a man of them – *big* man, he was, too – stands. An' he says to me: 'Tall order, Northerner. But could ye put down what *I alone* can in a draught, an' you drinkin' the whole night?'"

"Like that, he speak."

I moped. "I can't do a Turranian accent well; shut up."

He smirked. "*Go* on." Real taunting-like.

I rolled my eyes. "'Feck all,' I says: 'I can drink down what would fill that pipe bag a yorn, an' you under the table after a chanter'. Oh, och, were he mad then, lad; an' he says to me – 'Such was the claim of the Highlander what owned these pipes before. Ha! An' ye know what happened to *him?*'"

"What?"

"Yer man the sand-dweller gives a squeeze to the dronin' pipes; indicative anuff, aye?"

"*Ochón*."

"Sure anuff. An' I says to him, I says: 'I could drive the whole a yer table under an' into submission; but you alone, yer challenge I'll take; an' I gets yer luvly Highlander pipes for the bargain – the tartan, an no substitute!'"

"Sordid creature," my friend replies, picking over the words.

"Thank ye. Anyway – so we agreed." I shifted again, adjusting the pipes as I held them; I hadn't relinquished 'em yet since I got them.

"So!?" Sitting up, he spread his arms, his patience at an end. "Ye bat 'em? *'Drink 'em under'*, ye speak?"

I grinned. I love when he gets wording from my half-assed little stories. "No," I answer, confident as proud.

He arches a brow, sitting straight-backed. "How no?" Then he frowns. "Ye lose? Ye make story, ye lose! Auh!" He rolled his eyes. *"Tragic hero complex!"*

I made a face. "Ye *really* spend too much time with me wife."

He folded his arms. "An' how no' ye pissed? Aukh – you gettin' to it *sooner*, so!" he reproached, for he knew I would drag this thing out. "I hungry! I grand dry," he added, pouting to evoke my pity.

I sighed, rolling my eyes. "No mind for drama, nor grandiose, nor flamboyance –"

"I speak ye, *ending*," he challenged. Stiffly, he rises up – young little brat though he is – and goes over to the fire-place, where his guitar was leaning, and takes it up. In my own native tongue, he explains, fluently, "They measure out a quantity to fill the bag – fifteen full pitchers on the table – and your man Donny, he eyes it, and he smiles. 'Mercy, but it's a small thing!' And they measure out the same for your man the Turranian, who gives the lively count. And ye both set yourselves to drinking, and all of Shifty's chanting on. And the while, you –" He fell into Augustan for: "talkin' all matter o' smack" before resuming: "Belligerence fills your veins hotter than the drink and hotter for it, and he slights ye so you cannot stand it. And ye raise up; a noble figure ye cast, swayin' like a ship amains, the like to go down any second, but tall ye stand, right proud over him still sitting, so ye can look him level in the eye –"

"Augustan," I scold, and, like a child, he sighs and groans and grumbles and drops down on my coffee table, sitting cross-legged at its center.

"Och, yeh," he mutters. "Dis-remembered."

"Forgot," I corrected.

"For... *Why?"* He looked up at me, frowning – God help us – at the etymology of the word. "But..." He motioned, pulling towards himself, demonstrably. *"Got."*

"I don't know why the hell that's the damned word for it! You remember or you forget; that's all there is to it."

"Oukh... Well, *my* Augustinian no'... Uh." He frowned, increasingly more frustrated with his limited vocabulary. Damn, but I remember going through the same thing; it sucks when you can't express yourself, *a chairde.*

"Ah!" Brightening, he reported, as proudly as pointedly – and believe me, there was a purpose to that pride, the snot – "*My* Augustinian *shit*. Ye story." And thus passed the narrative back onto myself.

I sighed. But God forbid I should try and draw a breath, for he instantly lost patience and, in a sudden burst of energy, jumped up, standing on the table. "Auh! Feck ye!" he exclaimed, and continued in Ceothann: "The whole of the place goes up in arms and fire! Fightin' all around!"

I rolled my eyes – "Aw, hell, fine" – and joined him. "Me and the whole feckin' bar, it was, lad – the whole feckin' *bar!* And all around fightin' and warrin' and – MADNESS!!"

"*Arragh*, to've been there!"

"*I* was awesome."

"So ye were!" He began jumping around on the table, fighting off imaginary foes; these were the few times when I remembered he was young. "Punchin' and slashin' – until –" At last in Augustan: "Ye all kilt an murthered all over."

"But me and yer man!" I laud.

"Oo!" he brightens; I guess he hadn't thought so far in. "Ye *floor* him!! Yeh?"

"Get yer sneaks off me table; I'll floor *you*."

He grinned, folding his arms. "You no' can," he rejoined, childishly.

"Och, ye dubious sleeveen –"

"Anyway!" He jumps off the table, onto the couch. I've told yez before, that couch is a rugged hames; who cares if his muddy sneaks are all over it?

He deigned to Augustan as he finished, carefully as triumphantly: "Ye win pipes, glory a nation, name ever, so." He made a face. "Yeh?"

"I won the pipes, aye."

Satisfied, he crossed his legs, dropping down onto the couch. "How came to ye?" he repeated, conversationally, wanting the real answer now.

"Honestly?" I blushed. "Trade ring. Swapped a dry chicken I faked could lay twice as normal."

"Yaay, Donny! I's *pride*, so. Wanna play?" He lifted the guitar with a smile sweetly innocent. I can sell an eggless chicken like it poops gold; he can convince the chicken itself it can.

But before I could answer –

"*What* are you doing."

I jump, and my pipes gives a squawk.

My friend greets her sweetly, relinquishing the guitar as if he had never held it. "Morn, *caru, a chuisla, a stór!* Play?"

"No," she grumbles, agitatedly, moving into the kitchen. "You guys are stupid."

My friend grins, retaking the guitar. Strumming a simple tune – it is, after all, five in the morning – he sings:

"Oh, the humor is off me now,
Oh the humor is off me now,
Sayin' I'm sorry I ever got married
for the humor is off me now!"

"I can *hear* you," my wife scolds from the kitchen, unimpressed.

"It's a sign and a wonder ye didn't bother to come in sooner, darlin'," I call in reply, leaning back in the chair and folding my hands behind my head.

"You're so lucky to have me."

"I know it." And I am; scarce are the souls that can stomach bagpipes and battle at four-thirty – five in the morning. Here it's a regular occurrence.

I stay listening to her move about in the kitchen a moment, then I take up my pipes again, sort of droning on them softly, testing them.

"Where did you get those, anyway?" she asked.

"A-drinkin'!" my friend tattled and lied, grinning at me impishly. "An' him half-seas over inna donnybrook; truth's first, no' last."

"*No's* truth – I mean –" I stammer. "*Damn* it –" And I scorn his full name. "I *traded* for'em!"

"You spend too much time with Donny," says my wife to the *sionnach.*

"Iain MacFerguson make," he reported. "They a least damn good of quality. Your husband no' so damn stupid, *caru.*"

I frowned. "I don't even know how to respond to that."

"Did you drink all our money?" she asks, near disinterested; we've been married long enough.

"On the counter."

In a minute, I hear the rattling – of my keys, of the bag as she sorts it out.

"You satisfied?"

"You forgot the milk."

"Shite... I told ye it wouldn't last the whole way; I said I'd get it today."

"I wanted it *now*."

"Jesus Mary and Joseph..." I can feel him watching me, countin' backwards to the point he knows I'll get up; I'd do what I could to thwart him, but damned if I could, the *sionnach*.

Sighing, I got to my feet. "Fine, damn it." I set the pipes down on the seat and moved off towards the kitchen.

"I come?" begged my friend, clambering over the back of the couch to trail after me.

"Ask the lady."

"*Caru*, I come?" He clasped his hands in prayer. It's the eyes, however, that do it; puppies would give him scraps.

"Sure." Her tone was light. I was cross, and I was tired, but as I gritted my teeth and picked out the change to buy milk, her hands slid around my arm, and she hugged me. She reached into the wallet, pulling out a handful, pressing it into my hand. "Have fun," she said, and she kissed me.

Now I'm confused as hell; we'll *never* be long enough married to dispel that.

"A strange manner of wife I have," I murmured, appreciatively.

"Damn feckin' lucky," my friend retorts, takin' up his bag. He's out the door before I can look at him. "'Bye, *caru!*"

"Yeah," she replies; she's less of a morning person than anyone in that house, God help us all.

I empty the rest of the money on the table before her, pour in the fine bit of change she'd allowed me, and out the door I goes (I did kiss her, aye, ye saucy bastard).

T'was long about eleven-morn I'd gotten back, and eleven-night I'm writing this now. God love yez, *a thaisce*, and goodnight.

These stories I tell may seem as trifles to you, *a chairde*. But, love 'em or disregard 'em, I would rather lay them out to be scorned a thousand times over than let them be forgotten.

53
Dredging the Acheron

"You know? Whenever they tell you about stuff like this in stories, they never tell you how completely *BORING* it is!"

They were walking. The clouds had gotten heavier overhead throughout the day, and were beginning to descend to crush them. The dark wall of the pine trees filled the horizon behind them, and nothing lay ahead. It was long about three in the afternoon.

"Meg, I see you, and I raise you: *and exhausting,*" Zander agreed. He was leading the group again, praying to God he was going straight – and that Rahyoke would have the strength to correct him if he wasn't.

"We should play a game," Ian decided.

"*God,* no."

"I'll think of one, gimmie a minute... Oo! Okay! I've got one!" Eagerness was rife in his tone. "We go back and forth, and name something – like animals – that start with each letter, and if you can't think of one, you're out."

"Fine," said Zander. "But I call third."

"Why third?" Addison asked.

"I assume as creator, Ian will go first to provide an example. Second and last are stuck, alternatingly, with 'X'. I'm also assuming of course that Rahyoke'll be *too cool* to play with us."

"What?" Ian cast a Balor-esque squint to the sky. "So, twenty-four letters by four, eh? Wait – no – " He started counting off on his fingers, aloud. Uttered, "No, *that* can't be right," and tried again. Then, like a dam breaking, he gushed: "You know? It seems like somebody would've done something about that by now. It's been driving people nuts for ages – and I mean – new species are discovered every day! How could you *still* not have any animal that starts with 'X'? Well – the *microscopic* things, sure – but they're too small – I mean besides the whole 'wee beastie' thing we learned in, like, sixth grade. But those are all scientific names anyway, really. Hey! How do you spell 'Zander'? 'Cause maybe –"

"Xiphias."

Ian turned, startled at the exhausted voice behind him.

In spite of his pride (which was sure near banished to corners, weeping) Rahyoke had been reduced to the half-doubled-over shuffling stagger of an aged invalid trying to deny his state, lingering like a haunting memory at ever a greater distance from the scattered group. And he hated it.

Straightening as much as he could manage, he repeated, breathlessly, "Xiphias. It's the – genus of – of the swordfish, but – th-that's what we always – called it."

"See, though? That's still the *scientific* name; why can't they –"

"*Iochd.* Xanthareel! Xeme. Xiphosura. An' he spells it with a 'Z'. Now *please* shut up awhile, will you?"

If Ian caught the last, sure but he didn't obey. "Xiphias," he repeated, testing the word. "Can you eat it?"

Rahyoke bowed his head; he lacked the strength to sigh. "Aye, yeah. Y'can." In a feat of omnipotent malice, God in His abounding creativity had managed to divine not merely one but two additions whereby this Crossing could be made even more horrible: first, the inclusion of the unsilencable Ian Burgendson; second, these blasted injuries. He bit back a curse as he stumbled again, missing his footing. His right ankle was beginning to throb, reminding him threateningly of the barely-healed injury it had suffered. It was slowing him down, and, in turn, the lot of them.

Rahyoke cast a glance behind him like a sounding-weight – and started. The trees. The trees should be no more than a blemish upon the great rim of the horizon by now. But no – he could still see the frills of their tapered peaks, discern the soft drone of their chatter on the breeze. They were lagging. A distance of about eight hours and it had taken them fifteen. They had to have dropped far below four miles an hour. At this rate, the distance it'd take them – dear God, could it really –

"Oh!"

Distracted by what would normally be simple math, he mis-stepped, tripped and collapsed before he could even flail. If it hadn't been so agonizingly painful, he might have been more embarrassed.

"Are you alright?"

"Nnh." He managed, shaking, to get to his knees. It took him a full minute. "I, uh –" He coughed, bowing his head as he tried to draw a breath. *Is it broken?*

Zander drifted back to him. "Are you dying?"

Am I? He held a hand pressed to his rib cage, trying to steady his lungs. "I..."

"Don't you dare say you're fine, 'cause yer not."

"... tripped," he finished, not even sure himself where he'd originally been going with the sentence.

Zander folded his arms. "We should stop."

"Not yet," Rahyoke responded, his voice hoarse. Not when they were so far behind – not when he had *made* them so far behind. "I'll rest when I'm dead. Keep walking."

Zander scowled. *"Fine."* Reaching down, he grabbed Rahyoke's bag, slinging it over his own shoulder and walking off. "So come on."

Rahyoke bowed his head – ashamed, and still struggling for breath. He refused to admit that Zander had a point, and, more than that, he refused to prove it. But he had one, all the same.

Suddenly, a hand was offered. Looking up, he was honestly surprised to see – "Addison?"

"C'mon," the boy replied. "Let's go, morning glory."

Something like appreciation flashed across Rahyoke's countenance. He tried to get up on his own, but struggled so that he was forced to accept the boy's help. "Grand'uv ya," he murmured, by way of thanks. Lifting his glance beyond, he called, "Aw, y'waited for me after all? How *sweet*."

Zander responded with the rudest foreign-language phrase he knew. Rahyoke's grin broadened, flashing malicious. He would've laughed, had he been able.

"C'mon, Addison," he urged, clapping a hand on his shoulder as he moved past him. *"You're* my favorite, now."

"Aw," Meygan pouted.

Zander turned, walking backwards, his dark gaze leveled on the man. "I can slow down for you, princess."

"Hey! I don't need yer pity!" He tried to straighten up and winced. "A true wanderer stops for *nuthin'!* You think a little *bruise* is gonna incapacitate me?! Ha!!"

"An bhfuil tú ar meisce?"

"No, I'm not drunk!"

Meygan laughed – more at his tone than his denial, though his denial was marred by his tone. "You're so cute," she appraised him.

"I am *not!*" Rahyoke frowned – first at her, then at the ground. *Shorn, the slope; the territory itself is against me. "A Dhia dhílis!"*

She flashed him a cunning smile. "You know what *you* need, Rahyoke?"

"Laudnum."

"A woman."

"Oh, yes. Un-propertied, unemployed, middle-aged, four children – and kind'uva limp. *There's* an easy sell."

Meygan frowned. *And pretty sadistic.* "You've got nice hair," she offered.

A very short burst of sarcastic laughter, cut by a wheezing intake of breath. "You're a sweetheart, Meg. A hopelessly romantic sweetheart."

"I try."

The clouds were rolling overhead, driven by the hard wind. Rahyoke raised his eyes to them, watching God slowly twist the knife. "Auh!" he said, suddenly, offering a bitter smile. "Y'see that?"

"Who?!" asked Ian, looking around, frantic as a bird.

"The sky, darlings. It's going to rain."

"It's *always* going to rain!" Meygan dramatically complained.

A dampness had saturated the heavy air, battered by those tufts of wind which rolled the clouds along, gathering them in black. *And a thick fog to follow, tomorrow.*

"S'ere's my thinking," he continued; and he'd have to speak it fast, for the more energy he was forced to expend in words, the clearer he was short of breath. "Poor Zander – seems t'be nearly – *exhausted* wit' the walking... He needs his rest –" When the archer glanced back, he flashed a most sinisterly cheerful smile.

"Anything for Zander!" Meygan chirped, twirling prettily.

"Grand," Rahyoke managed. "We'll break tonight." *Thank God.*

54
Saille

2 February

Sílim nach tuirse atá ort ach leisge.
You are not tired, you are lazy, *a chroí.*

Pimpetos Dumannii (5ú Feabhra)

Yes, Donny. Because I can't reach the fridge-top, it's a safe hide for your book.

Goodness.

An tríú lá déag de mhí Feabhra

For a portion of my childhood, I was timid.

Then I grew wild.

I remembered the long walks on the moor – long to a four-year-old – and that curiosity born in me, instilled in me – and whole-heartedly encouraged by my parents.

When our wee vessel would be cooped up on the blue waves awhile, and I finished with my work (usually), I would scour every inch of the place. I would take things I liked. Mix them into other people's belongings for the pleasure of a ruckus. But mostly, I would climb.

It was forever, "Ye scurvy snake (and far worse names) get down from those ropes! Begone from the crow's nest! Ye'll break yer neck hangin' from that mast!"

And I loved it.

Worse when I'd go ashore. Never but five times did I jump ship on purpose; but often was the time when I simply missed the boat – and I never but rarely noticing – and five times I was sold off. The best thing at that last selling: "I would sell him for rat shit; I can no longer stand the absurdities of this wild child."

Fun fact: he to whom I was sold: second time I jumped ship. The bastard.

Anyway: we'd make berth. If we were in town – market days aside; that's a whole nother volume – I would entertain myself delving down every alley, scaling every building I could find. I would go into – and frequently steal from – any little shop, pub, and what-have-ye. I would mix into and learn from and love and challenge any sort of company, and many adventures – sordid and glorious – I've had.

Other times I sought the wilds. And for however grand the craic in the settled places, sure but I loved the wilds so much more. There were hills and rivers and fields and forests. Great, sea-soaped cliffs to scale, pebbled, sandy beaches to comb, waves to leap in and battle and laugh when they couldn't drown you.

It was rare on these excursions I ever got an hour from the coast. So my idea of grass beyond Ceothannanmór (for everything, always, is beyond Ceothannanmór, different from Ceothannanmór, and blander) and grass was long and yellow and pale, blondish like the tangled locks of those deck-clinging barnacle women (who were just as frequently dark). Lakes were lagoons, all full of sea water. Rivers could be grasped by the hand if a ship couldn't be run down them to port. And forests were always distant.

Standing on the rail in the crow's nest, I asked the lookout (who had bid me fall) what all of that green, chattering darkness so different on the outside could possibly look like within.

"*Trees,*" he said. "And get off the rail!"

"What are trees?" I asked. For I knew them as *draighean* and *duir*, *caorthann* and *feoras* and *iúr*; we must have had these around my home, for me to know them – for who remained to have taught me? And the language these sailors spoke was still strange.

"Are you a *philosopher?*" he scowled, and pulled me backwards off of the rail.

Sometimes in the lagoons there'd be great, wispy, willowy things. We trawled through those watery forests so you'd never imagine a tree could survive on dry ground.

I was thirteen before I ever saw a pine tree. We were unloading in – shit, *one* of the Odinic lands; it might've served me well, in retrospect, to've stopped blurting "the land of the Lochlanns!" and bursting into laughter, long enough to actually learn where the hell I was.

We were unloading in the Land of the Lochlanns. Suddenly, looking beyond the village, I distinguished the green mass around it. "Oye!" Laughing snidely, I thwacked the man next to me. "What the feck are *those?*"

He grumbled, for I'd made him lose grip on the crate he'd been about to lift. "*What?!*" He followed my pointing, and scowled. He called me a number of slanderous endearments I forget, lifting the box and telling me to carry my stupidity elsewhere; his hands were full.

My curiosity was piqued, and for a moment I abandoned my arrogance. "No, really," I insisted. "I've never seen one so tall, if it's a shrub."

In my innocent stab at not seeming a total moron, I earned the repute and consequent rebuke of one. "It's a *fir tree,* you dumbass!"

He shouldered the crate and made for the dock.

I followed after him, forgetting my duties. "What's a fur-tree, at all?!" I exclaimed, fascinated. "Why's it called so? Can I touch it? Is it soft?"

"*Fir* tree, not *fur* tree! Dumbass..."

He set down the crate, and I tripped into him, distracted by the trees. He turned, glaring brimstone, and I steadied myself by his stout shoulders.

"*Stagger* much?" I laughed. "Ye ought'a watch yer intake, lad!" My own broken way of calling himself a drunk.

He shrugged me off, pointing towards the woods. "Then off, if you'd see them! You're no help unloadin', that's certain!" He told me I could take the boatswain with me, if I was so bent, and my chances with the gunner's daughter.

"I'll unload first," I affirmed meekly, blushing at the threat of the lash. And I did.

And we disembarked ere I could see those fuzzy trees!

But I *nagged* so after a return to the Land of the Lochlanns; around my eighteenth year on this earth, I dragged my friend Cinead to the Odinic lands, back to that same port – for no other trees would do – and seized my chance. Fir trees are not hirsute, nor are they velutinous, by the way.

"Are you satisfied?" Cinead teased. "How was it?"

"Flocculent," I replied – because I'd just learned the word, and found it aboundingly hilarious – not merely because it was so pedantic, but because – of course – the resemblance it bore to –

Anyway! If you've never been so fortunate as to've seen such a creature o' God yerself as a conifer, let me describe it to you:

It's often conical, first off, if ye see it at a distance. All its branches are afrill with green, eternally – no autumn wind shrivels and achromatizes its leaves – and, if they grow dense enough, they rather *do* look soft.

But they're not. Even their bark is scratchy – scaley-like. And those frills they call needles, with reason. Each little beauty, a *cúpla* inches long, straight as sin, pokey.

And these creatures – these furry trees – they ooze a gummy sap, that – for my curiosity in seeing how far up I might climb – I got myself covered with. And dirt and needles and one great pine cone stuck to me. Pine cones are, I guess, their seeds. They look like little biddy baby versions of themselves, but made all of wood. And stuck as I was with smudge and sap, I naturally embraced Cinead, so he could experience the wonder of conifers, too.

"Yuck! You *turd!*"

I was eighteen when Captain Fitzherbert took the notion to circumnavigate the globe. I had been on one similar voyage before, when I was 14-15, under Captain Murray. Murray was notorious and dark and grand, a gritty, serious, scowling sort of adventurer. The best summary of the man I can fathom was that night we'd blown off course rounding the Cape, and ended up south – *deep* south – like as south as you can go without strikin' land or startin' north again. And Murray took the wheel from the coxswain. And it was darkness around, save in flashes of lightning, but the face – the *scowl!* – on him was darker than all sin. And the icicles froze in his great, black beard, and his blue-grey eyes, they glared fury.

Such was Murray.

And the journey around had ended, as I've told yez before, in wreckage and ruin and the crying of sea-mews, the mourning of merrows, and myself never seeing this far north-western coast again until I was eighteen. We cycle back.

What Fitzherbert proposed was a similar voyage – following the coast, tradin' at every sensible port; once as far east as one can go, re-supplying and striking out across the waves for Gallia. The thought of hitting that broad, open sea made me dizzy with excitement, for in all my years of sailing, never had I tempted it; only the desperate whalers and the massive cargo rigs ever did – it made more financial sense for them to than anyone else – and the latter I had done off with early on.

I volunteered my services to Cap'n Fitzy without a qualm, and he set me to run down a crew – me, Gaius MacIntosh, Skully, Davis, and some other guys – the scant few of us he'd gathered himself. I was given the title of boatswain's mate; we were to sign on a group and run them on a test-voyage – for there's a thing you'd never want, to be stuck running around the world with a mess of fools you can't trust.

I press-ganged Cinead, who's a year younger than I, at Lutetia, a Northern, East-Gallic port, into the fine little voyage to the Odinic land; he befriended me when I saved him from the lash. Only one man we'd gotten for the Odinic trip backed out of the circumnavigation, and, in my nineteenth year, we went abroad.

We followed the coast from North Gallia down to Southern Turranbar, then past the Straits, and along the coast of the great continent we call Alyssae, so we were down below the southern equator. Near the Cape is an island, long populated by Lavinians, although 'tis very far from their homeland. And you'll wonder when I say it why I'm after relating this story, but bear with me – there is a purpose to all things.

Docked at the town of Lōtós, most of us got shore leave, myself and Cinead included, because we were after fixin' a leak in the hull that, though small, would tear up somethin' awful and wreck us if left unattended, when rounding the Cape. A lot of the men hung around the ship – they smarter than we – because of *who* we were and *where* we were, but – I could speak Lavinian, and Cinead could scowl like the Devil, when he so chose; I declared us Augustan soldiers, and we went off into the sea-grass.

Three hours gone found us patrons of a humble home – to this day, I can't say what in all hell it was, whether diner, pub, or brothel. The lusty lasses in tight corsets and hanging, clinging dresses flowed freely as the drinks; I had never in my life gotten such attention – you'll recall I've told you, though I knew how to talk a girl up, my language, my occupation, my name were like a dirty brand. And here, wasn't every one of them but dying eager to sit on your lap, to curl up against you, to whisper in your ear. Merrows won't eat you, darling; but they will rob you blind somehow.

Now: here is the reason of my telling you (for be damned sure, if I'm sleeping on the couch for entering it in my own journal, it had *better* be with reason):

To cover my ass and Cinead's in case if we got too drunk, I said to the girls, in Lavinian: "My dears, have you ever heard of the savage Ceothanns?"

Whether they did or not, they coaxed me, chiming melodiously, "No! No!"

"Sure, chaste, pure beauties like yourselves never did. And better for it!" And with that – with a flourished air, an arrogant pride – I began to extol the horrific qualities of – sin among sins! – my own people. "They all live in the one house – if it can be deemed a house, hovel that it is, with mud for walls and floors and straw wisps for their ceilings. They go about – scandalous to relate! – unclad, and all sleeping in that same room, the men and the women and the infants, and the pigs and the cows having been brought in. Hard to determine one beast from the other; their men so dainty and frightened, their women so brutish, their beasts so comparably clean. And they feast on raw meats and beer 'til they bloat like dogs, and –" I went on. I spoke every misconception and slander of my people ever heard. I couldn't stop. I didn't want to. I loved to imbibe, to partake, even sarcastically, in the privilege of a conquering empire. I loved the scandalized horror in their eyes. Loved to see them faced with what *their own people* had reduced us to.

Cinead had paled as I spoke, the tips of his ears burning red. Shame, fury, shock – scandal. In dread of whatever the hell I was doing.

And my first reward: Wrapping her slender arms around me, draped like a mink around my neck, the lusty creature on my lap pouted, "*No* one can be *that* horrible."

Auh! Ha-haa! It was like the full intake of wine surged into my brain, and I leered, full confident.

"Oh, but my dear, they *can*. I have seen them myself, and, first-hand, witnessed how they live. My legion was long stationed in their province." The shudder had the desired effect.

"Oh!!" they chorused, innocent and alarmed.

"Were they *truly* horrible?" one nymph asked, clasping her down-soft hands.

"Oh, the *worst*." I sipped my wine, holding them all in suspension. Then, as if it were just occurring to me: "You know? I'll have done with these words, and lay them aside; they do nothing but alarm you. But – I learned their savage language – by compulsion, for they cannot understand our noble tongue, being so regressed." Smirking: "I bet you I can speak their language the rest of the night. Will you take my wager?"

"Yes." The terms were laid; and, on both sides, they were very much the same.

"And Aquillus shall speak it, too," I said, "though I doubt that he can do it."

Cinead downed the fourth glass he'd had just since I'd picked up the topic, sweating under that sea of doe-soft eyes.

One of the naiads stroked his cheek, earning a gaze of dreamy longing from eyes just visible over the cup's rim.

"*Can* you, Aquillus?" she cooed.

Cinead swallowed the last of his wine audibly. He lowered the cup onto the table.

"For you."

I shan't tell you the rest of what happened that night. Truth to God, I don't really remember. But *somebody* won that bet, to be sure – but, by the end result, I really don't think it was us.

[In the margin: 13 February. If you're wondering which lines exiled him to the couch, "lusty lasses in tight corsets" was a bad start to it, and "draped like a mink around my neck" pretty much hit it home. Dubhán, *a ghrá,* what's Lowlander for "divorce-papers"?

(Dubhán's hand): Be fair, my love.]

Over fifteen years I had traveled the sea; I had seen death, I had seen betrayal, I had seen the horrors of war and of nature and man. Maybe it hardened me. Maybe it made me bitter. I had seen good men die, and bad men flourish in prosperity; countries destroyed, and ignorant fools reaping the benefits unwittingly. And I had absorbed every word, every foul deed, every injury, holding them close in the heated, blackened banks of my heart.

Ceothanns are dirty, pathetic, groveling creatures. Choose your destiny, and God will know the truth.

Let's play.

Abair do phaidreacha agus codhladh sámh, a rún.

A cúigiú lá is fiche de mhí Feabhra

A fine mess I've dragged my house into.

Dech do shínaib ceo – the best of bad weather, mist.

55
The Merry Fields of Asphodel

"Oooooooooooo – !"

"*Knock* it off!"

"Ow! I can't see you; no fair!"

A dense fog gilded the air, sparkling and heavy and white. It was morning somewhere though that blank ice-grey, and like Hell he was sitting here longer.

"Rahyoke? Maybe we should wait," Meygan suggested.

"I can find my way."

"But I can't find *you*."

Rahyoke frowned. The obscurity of *due east* was easier to find in the thickest fog than the girl with the fire-red hair. Instinct *versus* material.

"Sneak attack!!"

"Ow! What the *Hell*, Zander!" the voice of Addison snarled.

"When Fionn and the Fianna made their attack on the Lochlanns, the mist was laid so densely upon the land that ne'er a man of the enemy escaped – though Fionn, being prepared for it, lost not a man in the fighting. Beware of the druid mists!"

"*Arraaahhh!!*"

Rahyoke listened to the manic, Pomeranian tussle going on to his left. *I remember when fist-fights were to be a source of pride.* "Lads, knock it off, or I'll discipline you the same."

"It's more the shame than anything – ouch!" Ian jumped as Rahyoke again swatted him with his cap. "See? Shame."

"*Away* outta my reach, y'eejit." Rahyoke folded his arms, habitually slipping a hand to his right side. "Oof..."

The mist was all around them, thick as a cloud, drenching the crushed grass, dampening their clothes and hair.

"I still can't find you," Meygan complained.

"Is't gonna be like this all day?" Zander asked. "Wah! Oof."

"How should – *probably*. Quit fighting one another!"

"That was a half-decent flip, Addy. But it would never work outside the haar."

"Zander."

"What?" He turned as Rahyoke finally addressed him.

Rahyoke repressed a sigh. "Noble as it is you're teachin' him your art, I think it'd be best if you waited until we're across to continue. Y'know what I'm saying? It's mostly the time and energy."

"Of course," agreed Zander, giving a shallow bow.

"I'll get out of practice," Addison sulked.

"Are you kidding?" Zander's grin was lost in the swirling mist. "If you're meant for it, it's all in the blood. Natural as anything. It'll come back to you."

"Great," flat-lined Meygan, "but I *still* can't find anyone."

"I've got an idea," Rahyoke informed her. "Zander – come here to me. I think you still have my matches; God help us if you don't."

The archer came up beside him, very nearly tripping over him. He handed over the matches, settling down beside his leader.

"Meg, *a mhuirnín*, could ya follow by sound, maybe?"

"...Maybe."

He flashed God a glare. As his hands set to his task, quietly – and with utmost reluctance – he sang:

"Oh grey and bleak, by shore and creek
The rugged rocks abound
But sweet and green, the grass between
That grows the Isle's ground
And friendship fond, all wealth beyond
And love that lives –"

He glanced up as he saw her. "Mornin', love."

"Thank you." Meygan looked down at what he was doing. He had taken the empty wine-bottle from his bag, slipping one end of a long shred of cloth down into it. "This is very dangerous, kids," he informed them, as Addison and Ian drew closer. "Do *not* try this on your own. Okay?"

"Okay."

"You sing really well," Meygan informed him.

He blushed so deeply she could see it through the fog. "Can you empty at least two of your mason jars into those water-bottles?"

"Yep." Meygan dropped down across from him, digging into her bag.

"What are you doing?" Addison asked, but Rahyoke did not answer.

He shoved the rest of the cloth down the neck of the bottle, setting it aside and twisting packaging paper into potpourri-like packets, binding them with twine – all from the food he had – *acquired*. The grass was too damp with dew to serve, so he tore more fabric from the shredded blanket in his bag, adding this to the two mason jars. "This'll do fine, Meg; save the third."

He shifted so he was kneeling, drawing his long legs up under him, putting his back to the light wind to shield his work. He struck the match, quickly and carefully lighting the contents of the mason jars before dropping the burning match into the wine-bottle, the neck held at a deep slant so the fire could slide in.

"Zander, could you puncture the lids of those jars?"

"Yes."

"What..." trailed Addison, with a frown.

Rahyoke lifted a cunning little smile. "Bottled light," he answered, quietly.

"And these will help how?" Meygan inquired.

"Yourself, Zander, and Addison each take one. This way we can see each other."

"Why can't *I* hold one?" Ian complained.

"You drop everything."

Ian gasped, scandalized. "I do not!"

Rahyoke enumerated: "The crate, the trunk, the coffee, the pretzels, the mason jar, the coffee again, the food bag – *three times* this morning alone – the pears, the – "

"Alright! Point!"

Rahyoke returned to his bag, tearing off another length of the moth-ridden blanket. "This'll help for if it gets too hot," he explained. "We'll wrap this around."

Zander twisted one of the lids, newly-perforated, onto the jar. The light shone through like fire-flies. "You know? Sometimes you're almost brilliant."

"Thank you." He wrapped the glass bases, tying the fabric to secure it. "Once we get farther out, we can use the underbrush and heathers. There's our goal for today."

"How'll we see it?" Meygan asked.

"Trust me," he said, holding out one of the jars. A tiny flame tripped and fumbled over the waxy packaging, chewing daintily on the twine. But its light flickered out from the clear glass, dispelling the gathering mists.

Meygan sighed. *"Fine."* She slipped on her backpack and stood, accepting the jar from him.

"Do we have a choice?" Addison muttered, out of the side of his mouth. Rahyoke handed him the other jar with an admonishing look.

"You're free to leave at any time, *a bhuachaill.* And good luck." His grin was malicious. As he gathered himself, getting stiffly to his feet, he gave them their morning pep-talk. "I'm afraid the trip won't be any more interesting than yesterday. Nor will it be, 'til we're across. To be perfectly blunt, this is probably the most boring place in the whole of the world. The desert has its sandstorms; likewise the ocean. But this?" He raised an accusing frown up towards the sky. "Death by inches. From boredom and chill. I literally cannot think of anything I would *less* rather be doing."

"Then why are we *going* this way?" Addison demanded.

"Because," Rahyoke responded, definitively. *And didn't I already tell you?* "For a thousand and one reasons, dear *buachaillín* Addison. And simply, summarily: *because*." He shouldered his bag, ignoring Zander's inquiring glare. "C'mon. Let's get goin' before it rains again."

Addison pulled a face. "Really? You call that an answer?"

Rahyoke threw him a cold glance. "It's in my bloodline, *a mhuirnín.* Blame it on the bloodline."

56
Tinne

An chéad lá de mhí Mheithimh

"I'm useful, Donny. I am. I want to repay you."

I looked at him. "*Everything* is useful," I replied. "There is nothing given without consequence and purpose."

"Let me show you."

An Seachtú lá de mhí Mheithimh

For long, Sean Fahy had told me: No retaliation. Your time will come.

It will come.

I restrain myself for you, *a chairde*. Often it is remarked the volume with which my race can soliloquize – sure but we'd talk you to death, and never a word of it put forth would have any semblance of sense to the listener, so round-about – Circumlocution is just another matter of the blood.

If they really listened to us, if they even listen at all, they would see that our words are not without value. On and on and on the story goes, until it is severed, and the picture traced on the binding ax is so phenomenal, so astonishing that it makes all the trailings of the telling so much the more glorious, the more beautiful.

Everything is cyclic, and we slam you back to the beginning with as much purpose as every one of those succeeding words. Yes, *everything is chosen with purpose*.

And for as much as we lavish hyperbole (I love hyperbole!), our instances of litotes – *what we do not tell you* – och, *a chairde*.

If they *listened* to us, at all.

An ochtú lá de mhí Mheithimh

A matter of the blood.

Do you find me stereotypical, *a chuisle? A mhuirnín?* Do you roll your eyes, shake your head – maybe even smile? When you read tales of my exploits. Of sitting cross-legged in a barren field, as the mists settle in the valley below, playing bagpipes to the rising sun. Of baking make-shift peat fires around which a company of three or twenty sit, raveling the epics of Fionn Mac Cumhaill through the dark hours of the night, shooting wine-glasses of whiskey, or donning and trading instruments to reel looping jigs into the twilight? Of staggerin' home from the tracks, tripping up the stair, calling a little too loudly for my wife, and accosting her most brusquely with every gentle caress from which the bloodline brings forth families of ten or twelve – to my one?

Yes. And do you think little of me indeed, *a chuisle?* Think me incapable, intemperate, incoherent – unsociable, ignorant, crude? Do ye think me an unsuitable father? An even worse spouse? A sad, creeping, effeminate, over-compensating –

The bloodline breeds savages. Ignorant and fawning, 'til kicked – like dogs.

We are dogs, *a chuisle.*

And I'm the biggest fecking mutt you've ever seen.

"Get outta the street, you savage cur!" the soldier snarled.

"Don't waste your breath," replied the other, grabbing full folds of my flesh wrapped in my dark coat, as he hauled me towards the curb. "It's in their bloodline."

Nuff said.

I was flung into the gutter.

"Lie there, you. And better know your place."

I waited until they were gone. It wouldn't do to make a scene – not now, anyway. I slithered out of the gutter, looking after them down the street and wishing to set fire to those attributes they so loved to flaunt. I wiped my sleeve across my face and realized what I'd landed in. The litter and the leaves; the horse shit, running slick as mud, through the diarrhea of the rain. The filth of all quarters clung to me, soaking me, and for a moment I just stayed there. I couldn't get dirtier lest I ate it – and I ignored the hint of a heated taste.

"Putrid, pathetic gutter-drunk."

"Well, what do you expect?"

I gritted my teeth, the low, feral noise – something sigh and moan and groan and whine – lost beneath the noise of the street. I pulled up my legs, enwrapping myself tighter, to avoid being run over and adding my blood to the stew. The most I did was shift so I was sitting on the low curb. I buried my face against my forearms – against my slightly-raised thigh – I kept my eyes down. I've put up with twenty-three years of this abuse. And small, personal rebellions mean little when you're forcibly covered in shit.

I was torn in remembering my family. I have to protect them. To protect them means bearing this scorn. Means backing down, means shaming them.

How do I look to my son? And I have raised him – small, personal rebellion, but truly all I know – in that stereotypical, Ceothann fashion. Raised him to be proud of it. And will cast him in some years – at an age when he'll so terribly resemble me – into a world that hates his very name. And they will force him off the road, call him a mewling savage – a pure spawn of the bloodline – and toss him into a pile of shit. And it's *my* doing.

It's mine.

I raised him to be good. To be proud. And this is what awaits him.

And I *raised* him to be good and proud – what sort of an example am I, taking it? Taking it to spare him...

Do you see my conundrum, *a ghrá?* It's cyclic.

Reflecting on these things, my fists clenched. But I couldn't use them.

My fury burned off those shameful tears, and, slowly, I stood – the crowd creating a wide berth around me. I felt like Moses – unstained by the river of blood, unsaturated by the sea he crossed – except *I* was the dirty one, here.

I set down the street. I've no license to utilize the "baths" – the showers at the gym, the public pool – what else? I was gonna hafta walk home – all fifty-three miles – caked and covered in shit. And the hot sun baking it to a stench.

Worse. *Worse.* I was gonna hafta *explain* this.

And the first thing I'd do was give Senan a big hug.

Patience *mo thóin.*

A dóú lá déag de mhí Mheithimh

"I thought you said you were good at this."

Sometimes my insensitivity knows no bounds. I'm okay with that.

He pressed his fists unto the edge of the table, straight-armed, leaning forward and breathing hard. He stammered: "I *get* them, didn't I?"

I hadn't heard his voice shake so much since we were kids, and that ten minutes he was terrified to approach the docks off Sycamore Row. Nearly repenting, I poured him a whiskey, a tall, generous glass. "Take this. It helps."

He shook his head. And unless you're just coming in at this point, you will know this as something astonishing. He had his forearms pressed against the table surface, fists knuckle-white, and hunching deeply he managed to scratch out, "I can't!"

I set the glass within reach anyway; there was only about my fill left in the bottle, so I rejected a glass. "Just calm down."

I'm not sure if the noise he gave was a laugh or choke; somewhere in between. *"Rucco-ti-ambi!"* he snarled, looking up on me like a wolf. *"Iexs-tu mi-sendi?!"*

This is his native language, now, that he's speaking; but it's not mine.

"What is it yer at, with yer gibbering?" I asked him, and he looked up so suddenly I started.

He stared at me – a narrow, kohl-smeared dark glare. He's young, and still full of that defiant energy, that determined energy, that unswayable, self-righteous – I don't know what. And he looked at me with that damning glare – that haughty, contempting look – damn; the look of *real* aristocracy, which not only subverted that usual teenage sulkiness, but forced it easy to his bidding. *Regal.*

At last, gracing me with Augustan:

"Quare Donny. Quare, *pity* Donny." And he shook his head; I am beneath him; I wouldn't understand.

I shook my head in turn. "Look: I don't know why you're upset. You did well to do as you did, and ye pulled it off perfect. Be satisfied, and give it a rest for tonight."

He growled, exasperated, straightening up, his fists still clenching the table. *"An-pissiusi,"* he accused, and made an abrupt shift into Ddeheuol to explain: "I have brought shame upon my entire family. I discarded my father's name, taking instead some pseudonym spawn of that hated nation as my own. I disrespected everything of my culture. Everything!"

"Roeddet ti wedi newid eich enw," I cut him off. *You changed your name.*

There is only – well, two things, *b'fhéidir,* about which he will not lie. One of those things is his name. Unfortunately, it is a very distinctly Northern name, and the name of the ruling family of North Gallia does not serve when infiltrating an Augustan guard.

"Your ancestors understand!"

All his anxiety compacted to anger. *"Craxantos!"* he snarled – which means either "toad" or "pustular", neither of which is very nice. *"Esi –"*

"Look – you got it, okay!?" I returned, riled.

He continued to swear, glaring at me side-ways with those kohl-ringed eyes.

"Move your glass," I said; setting the bottle on the counter, I lifted the rolled parchment from where it leant against the counter. Only as I unfurled the great sheet above the table did he finally accept the glass.

"Londa-adarca," he muttered, taking a drink. "Savage scum-reed" – that one I know, he reserves it for me in the depths of his fury. Ignoring him, I laid out the parchment on the table. And – *a chuisle mo chroi* – it was beautiful.

The object of our determination, our tolerance, his deceit; it covered nearly the whole of our table, a war-room document, smuggled out in the intricacies of that wee bastard's warped mind. He has an eye for architecture like I have for stories; he can look at a building, and break it down onto paper – all its angles and blocks – given only two angles, or less. He had been trained to expand this architectural mindset to city-planning, thence to landscapes. All landscapes can be broken down into floors and blocks – someone once called it topography, another cartography, and by marrying these two we get architectural landscaping. All it took was getting our boy access to the original document, and a big-ass sheet of paper.

He kicked a chair nearer and dropped into it, leaning onto the table and draining half of his glass.

"You done good, boy." I sat on the edge of the table opposite. There were their forts and their trade-routes as they stood on a full quarter of this grand old earth.

My wife stood off to the side, a quiet observer, studying the map in its intricacies. "Donn, can you mark the towns in? The names aren't on there."

Your man took a pen from his pocket, tapped the port, and asserted, "Tarvenna." He dropped the pen and finished his drink. "*Allos*."

"There isn't any more; you drank it."

"Ahkh, *caecos*," he cursed. "Go get. You owe me." Taking up the pen, he continued work on the forest, adding greater detail and trees. "I make fox-roads?" he asked, tracing in the air where he knew a secret path to be.

"Not if they don't know them."

"Aye." He paused in his drawing, looking at me. "Go get," he repeated, as if I were a stupid child. "I thirst."

"Simmer down, scrappy," Orchid bid him.

"Get Senan, will you?" I asked her. She imitated me, in a highish voice: "*Get Senan*." But she went.

"I don't sound like that!" I called after her.

"You do," my friend muttered, bending over his map. "*Carosa! Medu!*"

"You don't get hydromel for that!" she shouted, already out the door.

He frowned, shrugged, and returned to drawing trees. I watched him a moment in silence, studying the architecture of his nature. Finally, I asked, "Do you have any idea what you've done for us, lad?"

He didn't raise a glance; he was studying the outline of a forest – as far east as you can go, to that land immediate south of Ceothannanmór. Slowly, delicately, purposefully, he traced out a long line – a border many years forgotten, many years erased. In the same way, he answered.

"Yep."

57
Sanctum Sanctorum

"You said when we got to the heath. Here we are."

"No, I –" Oh.

They had reached the furze and heather as a sickly tangerine sun cast its low rays over a slick, dark-grey horizon, and – at Zander's urging, of course – had determined to stop for the night. All around them stood the skeletal shrubs, twining spindly arms of dry shoots around the even lower-growing, un-bloomed heather. A post-Apocalyptical wasteland of a sand-soiled bog.

Rahyoke snapped a twig off of the nearest bush. "I underestimated –" he began, but Zander swiftly cut him off.

"We'll leave earlier in the morning, to make up for it," he said, dropping down in the midst of the scattered, sad bushes.

"Easy for *you* to say," Rahyoke grumbled, under his breath. He dropped his bag against the nearest bush – clenching his teeth and drawing a long, hissing gasp. Suddenly lifting the terrible weight had given a painful jolt to every stiff joint in his body. "Oh, Faith," he grunted, at a near-pleading whine.

"Gonna cry?" Zander smirked.

"Good mercy –" His eyes were still Heaven-bound. *I might if it'll help.* "How *polite* ya are; who raised you so?"

"Tá mé na Mac Tíre."

"Cute. I maybe threw my back out." *Ow ow ow ow ow* – Rahyoke carefully dropped down beside his bag; the crack of the brittle furze-branches could easily have been the muscles of his spine.

"Can we have a fire?" Ian – shockingly – asked.

Rahyoke gave a short sigh. For a long moment, he was silent, his thin fingers slowly breaking the furze twig into smaller and smaller bits.

The fog had lingered throughout the day, clinging to their ankles and kicked up like whirling dust-clouds, and threatened to fall again tonight. Anything of the furze and heather would be too damp with dew in the morning. And, much as it would give him a half-twisted pleasure to deny the nagging child, the darkness was weird here on the near-heath – weird and cold.

"Y'know what? Fine." He tossed the bitty twigs unenthusiastically, his gesture speaking volumes of truth. "We'll need a watch on it, though; I'm not about to set the whole place ablaze."

Zander raised his hand.

"Not you, Phaëton. *I* will." Rahyoke began collecting branches from the dead furze-bush, preparing a place and a means for the fire.

"*Phaëton,*" Addison snarked, incredulously.

"But –" Zander began, on the verge of another parental rant – but it was a thing for which Rahyoke had no further patience.

"End of discussion," the wanderer retorted.

And so the matter was final.

The darkness curled in around the flinching fire, crackling like the burning edges of the heather. Elsewhere, the fuchsia flowers would have been in full bloom, but not here. No life here.

Rahyoke snapped a stick off of the nearest bush, bringing it into what of necessity must be called the light. The furze and the heather and the spongy, dark ground were facets of the bog – that great expanse burgeoning with the lushest green to ever shatter the warm mire of brown and grey, which had once composed his existence and still comprised his dreams. Yellow-green, amber, and red mosses and bushes towered with yellow fires, like the world aflame with life. But this – this! – this where he sat was dead and decayed. The stark, skeletal remnants of a lost love. Open and hard, bare to the pattering rains.

He drew his coat more closely about him, gathering it away from the wind. These were not light thoughts, but he could not help brooding. The dank mists crept into the tatters of the soul, sickening sharp, and dragged down into the briny depths of the gloom.

Rahyoke broke off another branch; it gave at the socket with a muted pop. The branches were all afrill with smaller limbs, and these with twigs, not a one of them budding, through they should have been, this late in the year. No better than the heather.

He slowly began to break off the twigs, tossing them into the fire, biding his time through the night.

The weather was too damp to give life to the fire; the pretty light dwindled, shivering and pale. A proper bog wrought peat bricks – fine wedges of turf, little rectangles of mud topped with green – which, dried-out, you could burn for fuel. And och, but the flame for *that* was the fullest, orangest, most cheerfully welcoming creature you'd ever seen. Burning merrily upon the hearth, maybe a pot of strew or a kettle set by, even in a stranger's house it would seem to say, "Is it yourself come home, now? Settle in, then; and there'll be a pint and a plate ready for you anon." At least, that was how it'd talk in the north. Auh; but it was a clever race indeed that could sell you dirt to burn, and sure that those trade routes had reached wide. Where were they now, he wondered, the sons of the men who had so unwittingly given life – a life worth the living – never knowing – Never knowing this cold until it swallowed them in death.

Once he had seen a peat bog burning. The forest alive with glowing embers – the life-blood of the ground, the trees, running in crackling waves, burning through in pocks to the surface. Smoldering from the inside, with a powerful, deep-orange light – the charring, black edges of the cinders chewing back the surface, spraying the air with ash. Often he wondered if it was still burning. Doubtless it was not.

All was heather and darkness – scent and sight. The slate of the unseen void pulling down to the stone blankness of the earth that decrepit veil of the all-consuming sky.

Stop. When you flinch at the unGodly chill of the raindrops, like daggers slipped in blood, you know that your thoughts have molestingly strayed. It was exhaustion that was doing it – yes. The field and its cruel expanses. *A Dhia dhílis*. Nothing – literally, *nothing* – was making him mad. And for what? They were making decent time, weren't they? He had factored it all out – indeed, he was so used to traveling with this illogical herd in tow by now, he could carve the itinerary in stone. We rest, we rise, we complain, we set out. Rinse and repeat. And they'd set out again early, before the mists rolled in.

The mists... He drew into a ball, pulling his left knee up to his chest, his right leg bent so his sore ankle laid against his left heel, his arms crossed at the wrists around his left ankle. In this way, the lanky creature was made small.

He heard the scattered raindrops slithering through the furze, a hissed chattering as the gibbering mists were called to life again. *And let it rise*. Long had the Heavens withheld their precious treasure, the earth her concealing deceit; now the hold was broken, the precious liquid spattering down with scorn, flickering across the vast blanket of the night. He heard it, more than felt it, just beneath the wind, platting over the dried grasses in bursts. As pathetic as the flame. His eyes were closed, his forehead against his upraised knee. Let that sickly ember-patch consume the world if it chose to. Damned, but he wouldn't hinder it.

What other answer for it? Sure, what could he do – raise the kids and tell them, *"Alright – we're leaving now."* *"Why?"* they would ask him, surely. *"Why can't it wait 'til morning?"* And what would he tell them then? I'm too tired? *That* would fly well. There was no excuse, dear children – none that he could offer. And even *Please – just trust me* was wearing awful thin.

And for what? *I gcuntas Dé*. No; there was nothing to do for it but wait.

How. He raised his eyes tentatively; the touch of the raindrops crumbled the ashes, disseminating the pallid flakes to the wind. He could see nothing beyond the light. The complanate sea crushed by the weight of a thousand hunts and haunting rains – all rigged and interwoven into that ancient, curling steam, blacker than night, its tendrillous, eternal, ethereal wings of everlasting, other-worldly, soul-devouring –

God! Rahyoke jolted, casting himself forward to kneel, doubled over, at the foot of the flames, pain jarring his side. God condemn this darkened world!

He put out the fire, curling into a sore and shivering roll near its draining warmth. He wanted to sleep; he would sleep.

My chosen bride is by my side,
Her brown hair silver-grey
Our daughter Rose as like her grows,
As April dawns today
Our eldest boy, his mother's joy,
His father's pride and stay;
With gifts like these, I'd live at ease,
Were I near Gaillimh Bay...

58
Muin

An tríú lá is fiche de mhí Lúnasa

So Diarmuid took the Cup of the King, filled with the healing draught from the well of Magh an Ionganaidh, and he looked on Maureen, his wife. It was no hesitation, for he intended to do as the Red Listener had bade him; there was no choice in his heart for it. But he knew well enough, what his heart said do was that very thing to destroy it.

He looked on Maureen, and, for the last time, he loved her.

He added the first drop of blood, red as the rowan berry.

He forgot all love of her.

He added the second drop of blood, red as the harvest moon.

He forgot his quest on Magh an Ionganaidh.

He added the third and final drop of blood, red as the heart of a summer rose.

And he forgot all happiness in the Land Under Wave.

Immediately his heart became a void, desperately bleak save the murky mists of longing to drift memories through its clattered spaces. Memories of the bogs lush with heather, the black oaks slicked with moss, the speckled red deer and the crying curlews that flitted in between, charging and flirting to the wide, shattered shore, the cliffs rising dizzying over white, dusty sands and broken stones, worn by the feet of the thousand warriors who made their homes above.

No more could he find pleasure in the golden halls, the flowing casks, the splendid finery of King Under Wave's palace; no more were the crystal fields, the silver orchards bright with jewels, beautiful in his sight; no more the flowing, night-dark tresses, the sea-mist eyes, the fair, pale skin of his wife and lover – Everything was faded, everything dull, compared to that far-off world he used to know; he was lost among his dreaming, the figments of another life, and reality was a notion to him unheard of the more.

Maureen knew in her heart it was a thing never to be rightened. No more could she protect Diarmuid from the fate of that mortal universe; no more could he return to her, until he had lived out the harshness of life.

She came to him at his perch on the sill of that high-up window, where he watched the windless, waving planes; the brushing willows, their trailing vines stirred languidly by a force beyond and above them. No moon, but its silver light graced everything at these depths, causing all to sparkle and shine with a soft light. The tinge of salt mixing on the full, strawberry-fuchsia, citrus air. And his eye raised, ceaselessly, to the soft-shifting, silver-blue-green-violet sky above.

Maureen came to him, and she joined him on his watch, her back to the world below. It was politeness that urged him to try and force a small smile, and Maureen smiled in turn, though it was a different sorrow which crushed and withered the flower of her heart.

"You are gone from me, Diarmuid Ó Duibhne," she said, soothingly. "And I'll no more detain you here." She stood, and he, relinquishing his post, stood also.

She took from her shoulders the great, silver mantle, shimmering in the dissolved light of the moon.

"You will forget me," she said, as his eyes, wide with dazed wonder, followed the gleaming cloak. "You will not hear my voice in the sounds of birdsong, you will not see me in the light glancing off the running streams. But for me, your aspect will live in the churning waves of the storming, your voice in the gleaming dawn."

Close to him, she was, and, looking up into his eyes, she slid her hand against his cheek, touching him, holding him, for the final time. "Goodbye, my heart."

And, stepping back, she threw her cloak around him.

All was darkness, glittering light. A whirling, rushing, roaring torrent pressed down around him, down upon him, lifting him, crushing him, hurling him with the sheer volume of its sound, tearing, grinding, and pulling; rolling – though the world around him was still.

Diarmuid opened his eyes. The water lapped against the sand-speckled shore beneath a pale, white sky. Grasses, lush and long, stirred in a breeze fragrant with lilac, as a pure, crystal sun pierced its low, morning rays through the mist.

The light! Diarmuid's heart was filled with gladness. It seemed to him the first dawn he had ever seen, and all before darkness, so sweet were its clear, white rays, flecked with the dancing cherry blossoms.

Diarmuid took the silvery cloak from around his shoulders, studying it in the light. He frowned. "What have I to do with a sea-queen's cloak?" And he tossed it onto the waves.

And the waves accepted the offering, rising up, embracing it, and bringing it down, down, to the depths where the dawn was not yet shining.

And as for Diarmuid, he never noticed, for he'd already turned, leaping up the steep sides of the grassy knoll, giving as he did the three great, joyous cries of the Fianna.

59
Perdition

It had been days.

Twilight still kicked like a greying banshee, dusting the foaming mists. Dawn was a whitish smudge separating the grey of the crushed grass from the abyss of the tattered sky, and a froth of raindrops still picked and pattered their path across the ground, the surreptitious, scornful loogies of the gods.

"Farther south, it's warmer," Zander expostulated. He was again at the head of the group, dragging his feet and sauntering slightly to prove how much they were holding him back. Rahyoke didn't know whether to be pissed off or proud.

"Cutting through the mountains," the archer continued, casting his surveying, superior glance behind him, "we'd be shielded from the winds, and the rains; we'd have streams for water and bushes for berries. And all manner of fishes and fowl to eat."

"Manna from Heaven!" Ian exulted, like an old man dying in his hunger.

Rahyoke narrowed his eyes. "Farther *south*," he grated, "more likely it's there's a girl, around seventeen, with black, flowing hair, and they call her Isadora – and wouldn't you just give your *life* to hear herself but once call *'Zander'*."

Zander pursed his lips. "Been pondering that, have you?" But sure, she was a lovely thought, Isadora.

Rahyoke rolled his eyes. This is why teenagers aren't warriors.

A thick wind was driving away the last of the morning mist, hurtling out of the east. It threatened the brume, it threatened their lanterns, it threatened their striving lungs. Meygan – with Rahyoke's begrudging assistance – had tied the purple shawl like a desert scarf, so it simultaneously restrained her hair, and wrapped around to cover the lower half of her face; the first so she could see, the second so that she could breathe against the harsh wind. She turned as she heard him coughing again.

"D'you want my scarf?" she offered, like a real sweetheart.

Rahyoke glowered. "Are you kidding?! It took fifteen minutes t'tie that, for all your complaints!"

"Ugh," she remonstrated, facing back into the wind. "You're so *stubborn*."

"I –" Rahyoke's voice collapsed into a heaving fit of coughing, muffled by the sleeve of his black coat, pulled over his hand.

"Y'okay?" Zander called back.

"Wha' did – *kuh!* – what did I tell you about wanderers?"

Meygan tipped her gaze to the sky. "Blah-dah something pity something –"

"Mercy above."

A growl of thunder rolled through the distance, the sharp snap of its jaw severing the static skyline. Rahyoke raised his eyes, watching the smudging of the grey as a black mass of clouds tumbled over the ink-stroke of the world's cold rim, shattered by flickering tongues of blue light.

Marvelous.

Addison turned to Zander. "So, what's up with this field?"

Feeling a sense of déjà vu, Zander found God through bales of cloud. "Wha'd'ya mean?"

"I mean: it's this massive, open territory," Addison explained. "Why hasn't anyone developed it yet?"

"You mean why aren't there little McMansions?" Meygan taunted.

"Look: It's a lot of unused space. I'm just saying, it's *fishy*," Addison persisted. "I mean – how many days have we been walking? And the only people we've met, are –" ticking them off on his fingers "– you, *that* guy, some old dude and his wife –"

"That guy by the river," Meygan added.

"– yeah, and him. I mean: How is there *no one?* How d'you have, like, no technology? Where *is* everything?!" He unfurled his arms wide, taking in the void. "There's nothing here! It's like the frickin' *Apocalypse!*"

"What?" Zander asked, pulling a face.

It was Rahyoke who answered. "'I looked upon the earth – and lo, it was – a void.... to the Heavens – they had no light; on the – mountains, and they were quaking... and all the hills were shaken. I looked, and – lo, there was no one – the birds of the air – had fled. The faithful land was no more – than a desert... and all of its – cities were laid – in ruins before the Lord.'"

"Kind of short of breath there?" Meygan offered.

"Do you intend to quote *all* of Revelations?"

"Jerimiah," Rahyoke corrected. "Ss'Jerimiah."

Addison scowled. "D'you hafta be such a smart-ass all the time? *Just* wonderin'."

"I'm a compulsive mnemonist; I can't help it."

"See? Smart-ass."

Rahyoke glanced up, flashing a look of frustration before his anger developed into a thin smirk. *"Tá tú glan as do mheabhair, a bhuachaillín,"* he said, briefly uncrossing his arms to slash his palm through the air above his head. *Whoosh.*

"Are you –" Addison turned to Zander. "Is he swearing at me?"

Zander shrugged. *"Ni thigim. Ach, tá tú go* hiontach, *a ghrá."*

Rahyoke audibly repressed a snicker.

Meygan rolled her eyes. "Oh-*kay*. Let's keep it *nice*, you guys."

Rebuked, they were silent a moment. Then Rahyoke murmured: "'And the sun was like sack-cloth. The full moon – like blood... And the people of earth withdrew... hiding amongst stones, and they – cried out to them, 'Fall on us; hide us from the face of God... for the day – of His wrath – has come.'"

A sudden crack of lightning shredded across the sky, ripping open the black, and the Red Sea caved in. So thick, so fast, so hard drove the violent downpour, clogging the air with grey waters.

Rahyoke gasped, standing shock-still, his shoulders raised against the drenching deluge, elbows out like an irked rooster. He heard Meygan give a little screech, ducking, her arms over her head, and well he agreed with her. The boys cursed and complained, pulling up hoods and lifting jackets over their heads for cover.

"*Now* what?" Meygan shouted, above the rain. Looking to Rahyoke for an answer. *A Dhia dhílis*. Rahyoke lifted a vicious scowl to the Heavens. *Smart-ass!!* He spread his arms, calling down the full force of Paradise, fury in every aspect of his lithe form, the fire of a different realm flashing in his eyes. "Really?!" he screamed. "REALLY!!! *COME* on!"

Meygan looked at him flatly. "Yeah. *That's* gonna help."

"Aarrrggh!" Rahyoke complained. "Damn it, *I'm* not stopping!" he shouted. Hiking his bag higher, he trudged on, through the mire and the rain and the wide gazes of his caravaners, grumbling, "Sure, the whole summer

we've been droughted, and why *wouldn't* it wait to break 'til we're halfway nearly across this flippin' bloody field? Oh, *wise* humour, that, *very!!*"

"Drought?!"

"Oh, come on; it's not that bad –" Zander gagged as a particularly wretched gust of wind staggered him back. "Mercy!"

The wind was driving now in sheets; a tearing rush surged thunderously over the grass, and Rahyoke caught Meygan's arm, shielding her until the blast subsided.

"Okay!" the girl surrendered. "This is no longer fun!"

"Yeh –" The words jammed in his throat, and he started coughing. And this time, he couldn't stop.

"Are you alright?"

Rahyoke backed away from Meygan, doubling over, one fist against his thigh, his other arm covering his face. He was coughing too hard to answer. The sound was racking, shuddering, scraped with agony, slick with lung. A sudden wrench and he clutched his side, dropping to his knees.

Oh, damn – Malachi had warned him ignoring the lung sickness would come back to haunt him! As Zander crouched down beside him, Rahyoke wrapped both arms around his lower ribs, doubling all the way, trying to force himself to breathe evenly. His aching lungs pulled shallowly with a wheezing, hollow strain. *Almighty Lord, I was only kidding! I –*

"Calm down, man –"

Calm down... Rahyoke pressed the palm of his right hand into the grass. With his head bowed, he could block just enough of the wind; he closed his eyes, compelling his lungs to open. *Concentrate, concentrate.* A little gasping, like being tossed in and out of waves. *Come on...*

"Are you okay?" Zander had wrapped his arms around him like a precautious squid.

Rahyoke opened his eyes. He nodded. *"Táim, cinnte* – yeah." Sitting up slightly, he looked back over Zander's shoulder towards Addison, surreptitiously using the boy's strong shoulder to block the wind. "Addison – or maybe Meygan – one of you has some'a the merchandise I didn't sell – there's a reddish wrap thing – d'you –?"

Addison dropped his bag, digging through until he found and pulled out a deep-red cotton shawl.

"I'd love you eternally if you'd give it t'me."

Balling it up, Addison tossed it, and the wind caught and unraveled the bundle, and Ian had to make a grab for it.

Rahyoke's expression spoke volumes. *"Thanks."* Taking the cloth, he tied it so it covered the lower half of his face; not as intricately as Meygan's scarf, but it was all he could do for the wind. "Aukh." It was already damp with rain. "Alrighty, now. Vote of: *this sucks*."

"Seconded," Addison moved.

"Aye," Ian agreed.

"It's settled." Rahyoke stood carefully, trying to ignore Zander's hesitant support. "Let's get on, then. The sooner we're done with this – well, we'll be done."

"Good logic," Meygan approved. "Are you okay, though?"

"I'm always okay."

Her brows knit. "Well that's just not true."

"You know, you probably have asthma," Addison, son of a doctor, observed. Rahyoke looked at him mutely, and Addison shrugged. "Just the way it gets wheezy like that."

"...I'll try an' keep that in mind." To the group as a whole: "Come on. If it stops rainin', we'll stop, but I'm not settlin' in this rain."

He would make this crossing if it killed him.

The rain had slowed to a soft clatter on the crushed grass, and the wind at last fell still. It was night. Night of the ninth day, and it was dark, and it was muggy, but the moonlight was still clinging to life overhead.

They sat in silence; really, what did they have to talk about? Polite conversation had long since been exhausted, and deep revelations and philosophical musings were not on the top of anyone's list of desires. Beyond the customary "So it'll be dawn, then?" and "Can we have a fire?" they had left Rahyoke completely alone. And he was fine with that.

Zander stirred the ashes of a watery fire with his sword. This one was his own creation, and it shamed him rather a bit, how sad it was. He had already taken jibes from Addison and Meygan, and given a rousting exchange. Now, to break the weighty silence, he sarcastically, sweetly, softly, sang:

"Down in the willow garden
Where me and my love did meet
T'was there we sat a-courtin'
My love fell off to sleep
I had a bottle of burglar's wine
My true love did not know
And I did murder that dear little girl
Down on the banks below."

"Zander?" returned Meygan. "You are a *terrible* child."

He laughed. "Y'know, Meygan; it occurs to me, we never got to have our duel."

She sighed. "Not right now; I don't feel like it."

He pouted playfully. "How're you going to progress? Hone your warrior skills, y'know?"

"I was *born* a warrior; I *wanted* to be a bridal consultant," she replied. "You know – like on TV. 'Are you saying "yes" to this dress?' 'Yes.' It's so happy and great." She left it at that.

Zander was silent a moment. Then he shrugged. "You could do both –"

"Damn it!" The sudden hiss was from Rahyoke, and he got stiffly to his feet. "Damn it, damn it, *damn it!!"*

"What?" Zander inquired, lazily.

Rahyoke turned; and if he couldn't hold himself fully upright, sure there was command enough in his stance. "This is no life for you!"

He said it as if it explained everything, which of course it did not. "What do you mean?"

"Ih – juh – Look. You – you wan'a future, don't you?"

A disconcerting silence was broken unsatisfactorily by Addison's disinterested, "Yeah"; by Ian's shrugging, "Not really"; by Meygan's scrutinizing, *"What do you mean?"*

God, give me strength. "Maybe not now, y'don't see it," he said, with the terse patience of the world. "But there are people who can give you what I can't – id – *prospects,* and – *life,* more than survival. Sustenance living – slow starvation – running – *constantly* – ah – d'you *want* that? That's life with me. This!" He spread his arms wide. *"This* is life with me! You deserve more than this!" Then, pointing at Zander: *"You* deserve more than this."

Tension. Tension snapped tighter than a drawn bow.

"I'm not leaving!" Zander insisted, fervently. "I've told you a *thousand times!*"

"Oh, *come* on, Zander!" Rahyoke's voice cracked as it was raised. "Things've been shot to Hell for me – you have a chance now! West of the Pale it was too dangerous for you to be on your own. But it's better where we're going – people won't be as keen on your paperwork. You could have a normal life! You would be safe, and cared-for – Live in a house, get a job, meet a girl – start a *family* someday – things you can't do on the run!"

"I'm not on the run; I live in the wild."

God, the immaturity of that simplistic, rationalizing tone! Far in contrast was Rahyoke's outcry: "You are not a Fenian! Zander, those were just *stories* I –"

"This is the life *I* want to live!" Zander returned. "I don't want to be confined to normality! I want the freedom of the skies and the birds and the trees – I want my stories – the life you *can't* live in a town. And besides!" Having reached a climax of volume, his voice abruptly dropped, becoming poised. "Things'll pan out for me. I'm stunningly good-looking; any woman would leave the settled life to join me in the wilds."

Rahyoke cocked a brow.

"It's true," the boy boasted, candidly. Any words with which Rahyoke might have rebuked him caught painfully in his chest, sundered in the face of that confidence, leaving him mute. It was a moment before he spoke; when he did, he was subdued. "So yer in with me, then, are you?"

"I'm *always* with you."

"You'd be better off if you weren't." He turned and walked off, settling a fair distance away. And there, at his distance, through night and rain, he remained.

Meygan bit her lip. Her heart was sad for so many reasons, but it broke to see him there. "Stay here," she urged her companions; then, standing, she crossed the space between the fire and their leader.

It had been almost a month since they had come here; almost a month since she had stood on that high cliff, studying this man. She barely knew any more about him now than she had then, save that he deserved the trust she gave him. Even if she couldn't explain why.

She settled down beside him – a fair six inches away, though it was close enough to make him tense up.

He was sitting loosely cross-legged, his elbows on his knees, his head in his hands. It was a ten-minute battle with his long-dormant sense of hospitality before Rahyoke pulled down the red scarf that covered half his face and demanded, "*What.*"

Meygan, irritatingly undaunted as always, folded her hands in her lap, cocking her head to assess him as she noted, "We wrecked things for you, didn't we."

It was more of a statement than a question, and over a decade of bitterness and misanthrope was only checked with difficulty as he reminded himself again, *she has no idea.*

He didn't answer.

Meygan hesitated. Anyone else she would have embraced; but everything within him, down to his aura, was so strained, so forbidding, so closed, so brutally black – Meygan thought better. "Is your life ruined?" she asked, bluntly, driving straight to the point.

A flash of emotion registered in the fiery darkness of his form. Faintly, he drew a shaking breath, looking up and meeting her gaze with a look of sarcastic pleasantry. "Is yours?"

Meygan unwound the part of her scarf that covered her face, studying the slateish sky. She, too, said nothing.

Rahyoke smirked, turning away again. He didn't really want to talk; frankly, he wished she'd go back to the fire. It was better for her to be over there – with the light, if nothing else – than wasting her time on him.

"Meygan..." He spoke her name slowly, ponderingly, as if lingering over a thought. "How..."

She glanced at him. "Yeah?"

"How... did you get down from that mountain?"

Meygan was surprised. Of all the issues and all the complexities in the world, that was the one he had chosen to address. Then, her eyes glittering, she leaned against him and teased, "Is that your question?"

He continued to study the grass. "...No."

Quiet was his voice, small, rebuked. Meygan pulled away, slowly, studying him out of the corner of her eye. His dark hair hung, covering his eyes, stirred vaguely by the breeze. She could tell, though, that his teeth were clenched, his mouth a sullen line. In pain from *something,* though how much was physical, how much deeper, was anyone's guess.

She tilted her head, the corner of her mouth lifting in a small smile.

"I climbed."

And she turned away again, lifting her gaze to the far horizon, searching for something of land.

Rahyoke mulled over her answer. And at last, though he still did not turn to her, he could not help but return that smile. *"Dech do shínaib ceo."*

Meygan turned. "What?"

"Sometimes it takes distance, to know why a thing happens," he said. He folded his arms, low, against his stomach. "Some people are better at it than others. I don't know, and – I suspect – *you* don't know, why this has happened. But it's for a reason. And, I can tell you... if *anything's* right in this forsaken world, it *all* is. It's destiny, y'know?" His voice was so soft on the last line, she wasn't certain she had heard him. He shook his head, straightening up slightly to point off to his left. "That's North." He folded his arms again. "Sure as I know that, I can tell you, *it's right*. D'you trust me?"

Meygan had been studying the way he'd wrapped his arms – so much tighter this time than before. "Huh?" She looked up, fixing him with those wide, light-green eyes.

Rahyoke met her gaze, leveling her an almost equalizing stare. "D'you *trust* me."

Meygan pursed her lips. The rough wind caught at her orange curls, which had been escaping from the shawl-wrap all day, and she pushed back the red tresses with annoyance. "Yes," she answered. "And are you okay, there?" She lifted a limp indication to his side.

"Uh? Ah." He slouched deeper. "You're always sore when you're old. I fell on it weird, earlier; the lung-hacking; this sulking position..." Meygan was giving him a look. "Ukh. Off with your aporetic glances," he urged, turning away with an air of dismissal. "So that's as far as you trust."

"No, like; I trust you to help me, I don't trust you to help yourself. Rahyoke, are you –"

"I've made it this far, darling, and I'd thank you to go back to the fire. It's pretty late, now, anyway; you'll wanna rest."

"But –" She broke off. He wasn't going to talk any more, and he *certainly* wasn't going to tell her what was the matter.

"Fine." She stood. "Get some rest, though, okay?"

"Mmh." He heard her walk off, but found little comfort in it. Sometimes – *sometimes* – she was so alike, it nearly killed him.

"Oh – hey, Meygan?"

She cast him an arched brow. "Yeah?"

He lifted an arm, pointing high above the horizon to where a smattering of stars glimmered through the cloud bank. "That's Aquila. Eagle of Zeus – carrier-pigeon of the gods, essentially. I don't know what good he is for navigating, but he's an August constellation; he'll be following us the remainder of our journey."

The girl raised her eyes to the sky, concentrating on the dim constellation. She smiled. "Cool."

He turned his gaze back to the fire, finding the archer poking at its dim light. "Zander." When the boy turned, he continued. "How long d'we have, ya think?"

"...What day is it?"

"It's Tuesday." A bitterly saccharine smirk. "Happy Tuesday."

Zander lifted his narrowed gaze up to the milky moon. "Day *ten*." A hint of acridity coiled beneath that usually-pleasant tone. The black night was streaked and corded through with omnipresent veins of white, draped in a haze of iron grey, rising like the Holy Ghost from plains of solid water. The air smelt of dust and rain, the crispness of fall in muted swatches beneath the lush of the near harvest. Phantoms hung in scattered darkness, blocking the horizon, hanging like a curtain over the blue. "I think we can make it by morning," he shrugged, trying in all his power to sound casual. "Depending on how often you decide to stop."

Rahyoke grinned, fiendishly, a bold light flashing in his eyes. "Yer an audacious one, *a chumann*. Let's give it a shot."

60
Coll

An naoú lá déag de mhí Iúil

Dubhán Ó Raghallaigh, who has walked drunkenly into the sea at Bheith i Gcrunachás and is being brought home to Bailey to be buried.

That's what my epitaph will say. Because that is how I'm goin', *a chuisla*. That way, and no else. So don't worry, me to be killed in the scraping genocide – the eugenic raid on our blood to purify the line; don't worry me to be killed in battle, brought down by an enemy's hand. Long they've been trying to get me; in the end, they never will. An eccentric, tired old man, I'll walk into the sea, let it take me if it will, let alone my life (if I'm any more than that), and let it do or it won't do, reject me or murder me, carry me back over to whichever shore, and maybe I living still. God decide. God decide *then* if all that I did was worth it.

Meanwhile, I'll keep on, the same.

They say the Ancients lived on the sea – so close were their homes to the promontories and tides. At times, a great storm would whip up. It would threaten their homes, their families.

They wouldn't leave. They would draw their weapons, they would wade out into the sea, to stop it in its course. Never fearing the tearing thunder; the lightning that split open the sky; the waves that fought back to drown them. Naw. They would laugh in the face of unescapable danger. They would laugh at the roaring Heavens, the beating wind, the shredding waves, and the tumultuous, swallowing ground. They would beat back the waters with their blades as the sea rose up around them, the sand stirred to sucking around their ankles as the water rolled over their thighs, their stomachs, their heads. And they'd come up *laughing*. *Defiant* to the last breath.

A lot of people – a *lot* of people – and anyone who'd relate it – did so to say that these were savage fools. Look at this impious dumbass, who doesn't know when to back off, when to back *down*, when to go in out of the rain.

But I – oh, how I understand them.
I am standing against the waves.

An tríú lá is fiche de mhí Iúil

"Where's Orchid?" I asked him, full breathless as I was.

He thumbed an indication. "Damn vineyard. Waste time," he muttered, around a mouthful of cereal; he's forever complaining over that vineyard, *a stór*.

For the last – nuh? – three years, now, I've been working a factory packaging job in a town that's a three-to-four hour walk from here. And about eight months ago we got in an order all the way from Lavinium – large the order, short the time-frame. I was workin' the line when I hears yer man the executive says to the other: "Look! We just can't do it! There's not enough *time*." I knew well what a financial loss would be suffered; and while I hate those ezeks dearly, the financial well-being of the company is in my own best interest, employee as I am.

Without turning:

"Actually, if you cut through what they now call Silva Poma goin' south-east, you can dart dead south at Briva Isara an' cross the Senones mountains at Isca Vinovium. There's a path through the valley that angles further east-east-south from there an' goes to Durocortorum – I mean Argentoratum." That was all I gave them, and that got them as far as Lavinium.

There's a bit of a silence to go with my words.

"Turn around."

"Yeah." I turn.

We argue it out. Silva Poma is deceptively named (because "Apple Forest" is so damn cute an' cheerful) an' turns innavigable in the patch I'd lead us dead through. Furthermore, it's not the coming down from Isca Vinovium that's the issue; it's getting *up* there. And isn't the path but so steep that not even a fool would try?

"But you know a way?" Cross-armed and dubious.

I nodded, shrugging one shoulder. "Yeah."

And recruited I was.

It's five months down and back for us – a big gang of Augustans, peppered with Lavinians, the Gauchammi camel-driver Rajesh, the financial genius Tariq, and the lowly path-finder, me. *Dia ár gcumhdach*. Now I'm back.

Miss me?

I'm glad they didn't ask me how I'd known; I'd never been there myself, you see. But I have seen it, done out all pretty-like on a map. Which returns us to the subject at hand.

To resume! at your-man-with-the-cereal's bidding, I went back outside and around to the garden. Our garden is a fenced-off enclosure to the side of our house which I hired the food-stealing son-uva-bitch to build. We built up lattices to enclose it in like a great box, with rungs as a ceiling, and all the rungs and the lattice flecked and laced with vines. There's a great gap between roof and floor where the little gate is, and I leaned onto the wood structure – biddy, white-washed, board structure it is, the gate; it matches nothing – and I leaning onto it with my hands, I called within, "Woman of the house!"

And there she was. Shafts of sunlight fell down through the ceiling, slipping over her hair, her clothes, her bare feet. She was standing on her toes, poking at the ceiling with a rake to mend a fallen vine. She turned at the sound of my voice. "You're back!"

"I am, so," I said, as she came over to me.

"Why didn't you tell me!" Without giving me chance to answer, she again went up on her toes, wrapping her arms around my neck and kissing me.

"Shorn," said I, and moved for more, but she stepped back.

"Can you help me?"

I frowned a bit.

She clasped her hands, straight-armed, twisting slightly and lookin' up at me with that cute face I cannot resist. "'Cause *you're* so *tall* and you *love me...*" She trailed.

"Yeah, yeah." I entered.

The dear girl scurried over to where she'd abandoned the rake, pointing up at the ceiling. "See how it falls down?"

I lifted the vine, reaching up to re-tie it. "Ye know ye won't get grapes this way?"

"It's pretty; I told you before. It's for the aesthetic appeal."

I return with a shortish half-laugh. And I'm twinin', I'm twinin'. "I missed you," I try, meekly.

She smiles. She hugs me, wrapping her arms around me as I work. "I missed you too."

"Your man inside says we gotta go over the map thingy again." A-ha, see?! So back inside with us. Within, yer man has himself seated at the table, and Orchid demands of him: "Hey, mooch; where's my baby?"

His eyes flash up, and he answers, mouth filled, "*Faoin mbord.*"

"I trust you to watch him and he's under the table?!"

"*Níl sé* ar *an mbord*," he pointed out. At least he's not *on* the table.

Sighing and grumbling – "*'Íosa, 'Mhuire's a Iósaif!*" – she stooped and lifted the bairn from under a chair. "You're irresponsible," she scolded.

"Aye, lad," he redirected, "Jeeze." Standing, he went to the sink, depositing the empty cereal-bowl.

Dissatisfied with his petulance, Orchid rounded on me. "Do you know what your son's been doing?"

Uh-oh. She's disowned him.

There *is* something to a feral bloodline, *a thaisce;* he gets his stubbornness from herself – and, I'm afraid, a good strain of the wildness from me.

"He climbs up on everything," she extolled. "He's all – jumpin' off of trees and tables and chairs!"

"What do ye want *me* to do about it?" I asked, with the dull boredom of a real *prize* of a husband and father.

Her eyes flashed with that darkness – a look I could read by now.

"Ope! 'Kay!" Coming forward, I lifted the squirmy *buachaillín* from her arms, putting my free hand to her shoulder and kissing her soft hair. "Hi, sweetie."

"*Mmmmmm-hhmmm.*"

Every woman wants the bad boy. Every mother does *not*. Lying in bed one night, I demanded, "Look: when ye married me, what did ye expect yer children to be? They get half of their blood from their father." She hit me with a pillow and told me to, "Shut up you drunken bastard, it's three o'clock in the morning."

I hope the red cells of sensibility passed from herself down to the poor child. I couldn't bear it if he was all me. She says the same of herself: "He's shy. I hope he'll be more like you."

"A careening loud-mouth? Oh, goodie."

I don't know what he is, *a chuisla*. He's his own person. What more could you want than that?

Oh. A well-behaved child... yes...

Orchid went into the other room to get the map. Your man turned to me, arms folded, a smirk on him. "Yer trip good?" he asked, tauntingly. "I no' miss you."

I rolled my eyes. "Why do I keep ya?" He moved around to the other side of the table – across from the sink, the long side – and sat down, a good few feet back from the table. I knelt one knee onto the chair he had formerly occupied, at the head and closest to the door. "It wasn't some pleasure trip, ye know."

"You," he replied, indicatively. "*All* 'trip' wi' you 'pleasure', vagabond."

"*A Dhia dhílis*."

My wife came back. "Ready to do this?" she demanded, pulling the great map out of the meter-long tube we kept it in. "I'm making tea."

I couldn't help but smile. "*A chuisle mo chroí*," I told her, "begin."

The world can turn a blind eye, easy, to the disappearance of six hundred souls; this I've learned. But even the blindest of eyes can find them all again if they are all gathered in one place.

We've gone over the map, again and again and again. Where we're at – that is, the location of Balladarach – is ideal – so ideal, it's miraculous, God help us all.

We've had the map for a little over a year, and in all that time it has done us wonders. We know where the major forts and army routes are, and – the saving of our lives! – we can avoid them. And, when we have needs of shadier deals – how do you figure there're six hundred of us? – we can go about them unmolested.

But I see the track we're on.

"Y'canna no' – aukh –" a few snaps "*—sustain exponential growth*," my friend had explained to Sean. "Yen split, yen start t'refusin'. Ye fro' me, yen no' start refusin' now. S – someone is – *mad*, someone is – is – give you up, y'know?" And he's right.

If we overpopulate, we're found.

If we overpopulate, we lose control, thereby jeopardizing all our lives.

If we overpopulate, lose control – we become an army, and this would be the death of us all.

Why?

We are outnumbered, we are unarmed, we are not *trained* – and so many of our numbers are women and children. And all of us, *all of us*, young.

This is down the road, but – *exponential growth* – more and more rapidly approaching... There are six hundred of us now. A good, fine number. I can control six hundred people. Sean can control six hundred people. I don't know what either of us could do with an army; I am not a trained soldier.

Do we stop helping them?

Then they feel abandoned. *Then* they get angry. They turn against us and turn us all in. Or worse. Then they're abandoned – then they languish and perish, sacrificed before our eyes, because of our own –

But what else can we do?

"What if we split it?" I asked, as we were again studying that fine-drawn map. "The town, I mean."

"'En do what?" yer man asked, with a frown.

"What, the halves? One we keep here, yeah? But the other's for a new town."

His frown deepened. "*Where?*" He spread his hands. "I's – *shocked* – us no' be find *here*. Lucky. Damn *grand* lucky. Where *you* ken, Donny boy?" he asked, at a taunting sneer.

I pursed my lips. "Wherever the pipes are callin'."

But he had a point. We are an oasis, *a ghrá*. Even if we managed to find another sanctuary in this great, vehement desert – how in all of hell do we get to it?

"What if we went west?" I offered, folding my arms. "Right to the heart of it? No one would suspect a thing."

"Huh?"

I pointed. "Frith Gallia, the forest south of Ariegn."

"Aukh! Hah!!" A sharp, cackling laugh. "You *damn* stupid, boyo! *Damn!*"

I spread my arms. "You can get two hundred –"

"No' this day," he warned, shaking his head. "No' a s-*stagnant* town. Them – *three times*." He made a sweeping gesture over the map, brushing north-west from Lavinium. "A four be them *happy* as hell, truth."

Did I – I *did* tell you, *a ghrá*.

The look on his face was stone, but I could see how deeply it pained him. As far as he was concerned, Gallia was dead. As Ceothannanmór is to me.

"He has a point," my wife observed, breaking that stringy tension. "They do seem to go through Gallia an inordinate number of times."

"What you, so?" he asked, looking up.

"Go down more towards Lavinium," she replied. "That incorporates Dubhán's idea about going up under them, but it's in a territory less suspicious."

"Aukh!" Gesturing to Lavinium with both hands: *"Tri-border, caru!!"*

"If we can pass *you* off for a *Lavinian captain* –"

Pointing to various obstacles forcibly: "Imperial roads, forts, trade-roads, walls –"

"We get them in small enough numbers –"

"Train' ground. *Military train' ground.* Military guys. Young – want t'prove somethin'. *T'ree* camps!!"

"Mountains!" my wife insisted, forcefully, leaning onto the table, her arms straight.

"Ha?"

"Mountain. Ranges," she emphasized, tracing with her finger the two mountain ranges – the first at the very northern border of Lavinium, and the second about two inches – or two hundred miles – north.

He watched her, flatly, his mouth a small and emotionless line.

"So," said I, "we build a town in the mountains, yeah?"

"Yep."

He shook his head. "HOW, *a chuisla mo ghrá?"*

"So you at least give me it'll work?"

"Ish, aye, *a thaisce.* Ish. *Place* work, aye. More so'n Gallia. Yet!" Again, now wordlessly, he pointed out the obstacles indicated before.

"What if we went north?" I observed. My arms were folded, and I looking thoughtfully on the map over which they contended.

"Ha?"

"Goodness, boy!" my wife teased him. "You need to learn your directions! *An tuaisceart."*

Glancing up at her, he traced a line with the flat of his hand, moving upwards from Lavinium, towards her. "North?"

"Yes!"

To me – a snide snork of a snicker. "You *damn* half-seas, Donny." In Ddeheuol: "*Our* side of the Ran –" the traditional name for the inlet-sea that separates the Land of the Lochlanns from dear auld Gallia "— might be tame, but *theirs* is so heavily guarded –"

My wife asked what he was saying, and I translated as he spoke, a little behind.

"—they have steel fortifications, a towering wall, and hundreds of men, and dogs, and weapons – barbed wire – *just* to protect the *coast*. Any boat we try to get into the water would be destroyed. That's assuming we can cross form here to there on land *with* boats; it's tight up there, as you can plainly see. Whole swaths of structured, coastal towns. Beyond that, boats are illegal for us to possess – assuming even that a one among us could build one –"

"He's pretty talkative," she noted, giving him a smile.

"He says for you to pay attention." I turned to him, cutting him off. "And anyway: what's *your* plan, then? You seem content to be doomin' us, sure anuff!"

"Are you still translating?"

"No; that's me."

"*I*' no' have plan f'r ye," he haughtily, a tad angrily, replied. "Me own job, t'get map." He slapped both palms to the table. "Done!"

I rolled my eyes. "Maybe we should ask Sean –" I began, but before I could finished, the two of them chorused: "No."

The *sionnach* expanded the thought: "Senan no' so after expansion t'first. 'Less you make *damn* good plan, he stop expansion."

"Yeah," she agreed.

"Look. Senan – he – he make laws. You – you is doing – all *outside* town, he inside. Is..." He trailed, making an exaggeratedly thoughtful face at the ceiling until he settled on: "*Division of labour*. Inside, outside. Business, popular. Two kings. Divide." The foreign and the domestic rule. "Way I see it, this be all your call right now."

"Yeah," she agreed again.

Having done with me, he turned back to the map, quietly singing as he thought. I for my own part bore an expression of soul-fatigue. "So I'm king, then," I allowed.

"Point's I's *no'* one," he added, anticipating my question.

I prevailed. "It's yourself's my vizier, though."

He groaned, leaning onto the table, his forearm pressed to the surface. I had been illuminating the edges of the map with a pen, and this he took up, drumming it on the table. *Click click click click* – "Them has whole world, really, yeah?"

Click click click.

"Yeah." I watched the tapping of his pen – no real rhythm other than anger. His arms now folded, he tapped the pen against the table's edge, leaning onto that great swath of desert land so many miles south of here.

Lavinium isn't far; I said, I can make it there and back in five months by chariot, and that stopping near every night, and stopping over a few hours when we get there to unload. But it's not the distance that concerns me. *Everything*'s far when you're running.

I pray you never know what I mean.

It occurred to me then I'd never seen a map like this until I'd had one myself. One with borders and trees. I'd seen nautical maps all over – sea-charts and coastal charts for every province, with only the ports and the main towns we cared about marked in.

There was more in Lavinium than Isca Vinovium; there was more in Submisso Gallia than Osset... There were *forts*. There were *roads*. There were *walls*. Everything ripped down, re-drawn, rebuilt by Augustinium.

Looking at the map, I followed the Western coast, up to the northern-most tip of what had once been Gallia. There was a little island north of it – no label drawn, no city marked, no grand fort guarding the coast. Blank; for so they'd made it.

I have a son now, and so I can never forget him. But at times, he does split my heart and mind. I want to fight for him, but I want to protect him. And both must happen; both are opposed. I've told ya before: it's for saving his life that I keep quiet – his and his mother's and all those around. But life is not worth living without that fight, and all it's for.

They own the world, *a mhuirnín*.

They own the world.

Click click click click – clatter. He cast the pen so it rolled across the table. "*Ochón*; I's fight'em; it better. *Is* better, so. Fight 'em *then* West. And North. And... un-North –"

"You *always* wanna fight," I grumbled. *And damn it, I do too.* But there stands the problem, *a ghrá*: that six hundred souls can love peace easy where six hundred and one souls *cannot*. And all I've said thus far beyond.

"I agree with him," my wife vouched, raising her hand.

"You, too!" I snapped. For they are a rebellious house.

"*Rraaaaagh*," he complained. He stood, pointing at me. "You after to ken what I do, here be this: all ways is death! If you still do a thing, you go all out. All places them find you, 'cause all places them circle you. An' squish!" He clapped his palms together in a grinding, mutilating way. "You be runnin' and you be runnin', and really you no' better off anywhere than –"

Suddenly, he cut off, and he clapped both hands over his mouth, giving a muted squeak.

"What?" I asked.

He turned away, moving over to the sink, keeping his back to me. Nervously, he started singing – "Wayfaring Stranger" – softly and beautifully, his eyes tracing the counter by the sink. There was more in the tone than meant by his speaking the words, but none of it was an answer to me.

"C'*mon*, man!" I complained, stepping forward. "If ye know a thing, then say it! What was it ye were thinkin' on before?"

I could tell he was biting his lip, stalling, weighing it out. Then he turned – and him giving one word, harsh and definitive, before sweeping out.

"Ceothannanmór."

An séú lá is fiche de mhí Iúil

What is Ceothannanmór?

I was working – awful close to the overseeing man – who felt the need to inform me, "Don't think you'll have special treatment of me, *Ceothann*. Just because you can navigate doesn't make you an asset; you're a pack-mule."

"*Aye*, sir; *thank* you, sir." I rolled my eyes behind his back, as he walked off.

A lad near my age, a little younger, who worked next to me, elbowed me for attention, and asked, lowly, "What's 'Ceothann'?"

I laughed – then cut off as he cast me a look. It was then that it struck me.

I suppose, with a breaking heart, it'll be of necessity to explain to yez – as it's a thing, God help us, you more than like mightn't know – and so often it comes up in these pages. *What* is Ceothannanmór.

Pay attention, dear hearts.

To the north of what was once called Gallia, now Submisso Gallia or more common the Northern Augustan Territory, and God only knows what by the time you get this – lies the Isle of Ceothannanmór; on a clear day, if ever there is one, you can just see the line of it from this bloody shore.

Ceothannanmór is divided into three parts. A lot of it's moors and bog; some fleckin's of forest, but the hills are mainly bare – the mountains, too; many the streams, the little lakes – *lochs* – and a couple of teeming rivers. As to division, picture it as an inverted triquetra; and though there's hills and mountains and moors and marshlands in all of it, we tend to think of the southern "third" as the Moorlands, the north-eastern coast as the Highlands, and that smallest of sections quadroned in the north-west as the Lowlands, *Dia ár sábháil*.

The third of the Moorlands is the biggest of the three. The land is softly rolling and barren, nearly unarable in much of it, but for the heather and thistle and whatever weeds. Some of the lower swaths, the vales, are thick with bog, and the bog is where the trees are – bare, eerie trees, ghost-like and haggard, washed out and blending into greater depths so you can't tell what's crowned with fog, mossy blue-green leaves – and what's water. The coast is not horrendously steep; for the most part, there's good ground and good fishing, so the largest-stretching port cities you'd ever imagine. They protect the coast and the trade – trade is *huge*. Farther north, in the thick of the rolling, lichen-coloured hills, they make their money in sheep. Seriously – wool, meat, knitting, theft. It's damn near barter economy; they could live on their own means, if not for their love of tea. *Real* tea, from the continent; not that piss-water made from flowers (I'm obliged to add here – I'm sure chamomile's very good; I just can't be bothered to try it. Dandelions are for sheep).

Anyway: they – yer people, not the sheep – are a Gallic people brought up from the souls who first migrated over – the dark son, the bard, who wedded in with a golden-eyed *bean sidhe*. They call their land "Rhanddeheuol", and this encompasses the Moorlands; they call themselves Deheuwyr, which is "southerners"; and they speak a language called Ddeheuol. I can too.

Again: all of the Isle has its rivers and faery forts, but the eastern coast is where it becomes most mountainous. Cutting a diagonal from *maybe* about the middle of the top of the island, the range slices down until the last fifth, where it meets the coast. All that terrain is high, so the cliffs leave fairly no access to the sea save by that which is got by cleverness. This is the territory of the Highlanders, who adhere most strictly to the bonds and divides of clans – I guess ye have to, split so by mountains – and as such, they've no real unified title for themselves; I grew up saying "Àrdians" 'til I found out that itself is a blend of languages that makes no sense as a title; so let "Highlander" serve, I suppose.

So for their language, which I've always heard as "Highland-Ceothann"; again, not our own word, but I don't know any better.

What I do know is this: By the turn of last century, it was they who were the warriors of the Isle. Kilted terrors, little in the way of armor, much in the way of weaponry. Big, wild-haired lads, mostly. Alright; maybe a stereotype, but it's one I've yet to see wrong. And a man with a big heart is a big enough man.

And on. The last third – and the smallest – is yer northwestern corner, and that has everything. Rolling hills and mountains, forests, bogs and lochs, high sea-cliffs and hidden coves; the most isolated section of Ceothannanmór – that's the Lowlands.

The Lowlander territory is itself divided into five parts: Dubhachta, Cuaintír, Iarus, Mailachta and Dúnadach, but for some reason we've taken (or been given) our name from Cuaintír, and are called Cuanceothanns. Dúnadach is the farthest north out of all Ceothannanmór; Bailecolgáin (my town) is right at the edge, near to the cliffs. In the old days, powerful chieftains ruled in each quarter – as more powerful chieftains ruled in the older five divides. I should mention here, before Ceothannanmór conglomerated into three sections, then one, it had been more greatly divided. Into clan holdings, yeah, but there were five great sections overall – the Lowlander and Highlander remaining largely the same, the Moorland territory further divided; over each was a king, and a king among kings at the island's center. What good beyond internal affairs he could do there I don't know; he must have been very powerful indeed.

Anyway; the Lowlands was split into five clan holdings; I don't know who had what or what I had, if anything. But there you are.

We spoke a language most similar to that of the Highlanders, and no one knows where the hell the Moorlanders came from, but we liked them anyway, mostly. So call our language "Lowland-Ceothann", because that's what it's called today.

My memories are of the Lowlands; in vague shades wracked with darkness and fire drift my knowledge of the rest – visions of fear, of grey as ash and red of blood and pain, sifting through the moonlight. White the frost as gossamer, black the stealing crow, and where it rests, I shudder to recall. I'd rather any fate than left to be picked over, gasping, by crows.

What memories are not gouged, shredded, shielded, shattered, shrouded, and destroyed by that fear? Och; precious few I'll venture after. The moors, yes – the moors in March. The sea curling over the black cliffs by our home. A smattering of stories, precious few names. *"An bhfeiceann tú, a Dhonn?" Do you see, Donn? Do you see?*

Much have I seen in my lifetime; Ceothannanmór is my world, *a ghrá – m'anam* – my soul. The white sheep grazing on the hillsides; the red deer flickering through the wood; the pale gulls nesting on the cliffs, sea-splashed, the choppy foam rising against the black stone; I hardly know what's memory, what's dream; it's all the same to me. I've felt the peat, I've felt the salt-spray, I've watched the island slowly slipping away under full canvas – and silently slipping by, a serpent through the dank mist, crested with purple hills.

Is airiúi. The hour grows late, and I spurn lighting a candle. There are not enough hours, enough pages, enough *wicks* for me to extol to you the half of my – sure, all I have for that island. It is my soul, *a chara*. It is my soul.

Standing that day in the factory, I looked at the man beside me, and wondered: how do I explain? How could I ever explain?

"A little less than two thousand years ago, it was, a man looked across the water and said, 'I could take that island with a legion; but I don't think it's worth the bother'; so no one ever did. Nearly two thousand years later, men of his nation crossed over and did – with three times as many forces. So cocky, the conquering race – themselves believing that so long as they hold the coasts around, the lands around, the sea, the ships – as they do – nary the foot they need set on shore. And the fact there are generations of men growing up and living and *dying*, now – *now* – without knowing our languages, without knowing our mythology, without knowing *our name* proves it: ours is a race that has been wiped off the planet, verily. A strip of land – a barren strip of land – is what remains. Do ye follow?"

"No."

"No one would." Deep inside, all of me felt sick; I sighed. "There's an island that used to be up north... called Ceothannanmór; a 'Ceothann' is a creature of that island."

"Ooooohhh..."

As if he understood the world now, God love him, God bless his little heart.

My soul is a barren rock in the sea.

An t-ochtú lá is fiche de mhí Iúil

Rannsuigh mo rún
Múch mo teine
Bristear mo saol
Is tusa m'fhíorghrá

Tá mé caillte
Is tusa mo sholas
Tá mé caite
Is tusa mo neart

Doirtear m'anam
Glan mo chroíse
Dóigh mo dhúchas
Seas an fód

61
Druid Mist and Ivy

Dawn.

Vivid orange and watered fuchsia-peach, the light against the whipping mists, which would not – in spite of all nature – be dispelled by the roaring breeze. Their whiteness scattered everything, glittering and opaque, and the sun was cracked and crisp as if droughted itself beneath its hazing, milky veil. All the world was washed-out and livid, dyed in shades of water-coloured snapdragons, pimpernels, and slate-grass, and their forms were like raven-feathers against that ethereal heat.

"Flippin', feckin', bloody day *ten*. I didn't swear." Rahyoke hiked his bag a little too forcefully and winced. "*Shit.*"

"*Now* he swore," Ian tattled, having been near enough to hear it.

"A chariot can cross this field in twenty-four hours," Rahyoke replied. "A man alone walking, a week. A *week. Stupid bloody fog!!*"

"You know that doesn't do any good," Meygan ventured, repressing a smile.

"It does me!" The wind was rigid with the scents of the field – drained of heather, it was a meadow, drained of salt-spray, it was sea-mist; a cold and watery smell, hinted with dead grass and peat moss and sap and dirt-clods and –

"Sap," Rahyoke said, suddenly, halting.

They turned to him.

"What? Huh?"

"Sap!" Rahyoke repeated. "It definitely smells like – bark-moss and lavender – and dirt and water and rain, but I can –" He cast a glance around. "You *can't?*" he exclaimed.

Meygan frowned, pondering. What does sap smell like? It's sticky and golden and white, like honey and beeswax... "Are you saying we're almost there?" she asked.

"I'm saying it smells like sap," Rahyoke responded; but when he started walking again, his pace was just the little bit faster, quickened with a dash of restrained excitement. "What do I know, what it's from?"

A collective eye-roll, though by now they expected no better.

"You know," Meygan observed, following, "it's not very romantic."

Rahyoke frowned. *Contrary; poets have field-days with the lure of wandering, the hardships of the trail – the gouging wind, the lashing rain – the very colours of the dawn – the roughness of such beauty, murderous – scathing, to dare and temp behold* – "What's not?" he asked, his soft voice drowning out his thoughts.

Meygan gave him a look; apparently, it should have been obvious. Spreading her arms, she said, with bardic gravity and flamboyance: "And Oisín asks Niamh: 'Oh! What's Tír na nÓg like?' And she goes: 'Ah! It smells like *sap*.'"

Rahyoke made a face. "Who talks like that?"

"*You* do. And that's not the point!"

"Me?! I just –" He severed. Something had caught his attention – something deep beyond the obscuring blanket of the mists. "Hey!"

The fog was shades of white and grey and darker grey and whiter white – and black. Black only in front of them by the illuminating light of the dawn, the slick shadows shifting in the sun-kissed and suspended dew. And all that silence, all that pressing void, was softly shattering in the distance, cascading and breaking a thousand times over in great sheets of waterfalls, glaciers shaving, severing off. Chattering, battering, tumbling down, a thousand billion stones rattling, clattering – a muted roar beneath that silence, distinct as memory, filling the shades of the distance. Zander heard that chattering, and, recognizing it, he smiled.

Trees.

Ian gasped. They were here! Throwing his arms in the air, he sprinted, shouting for all the world to hear: "WWHHHOOOOOOOOOO!!!"

A Shíorghrá, a Dhia dhílis – they were here.

The children scattered like so many butterflies, running between black-barked oaks and paper-pale birches, strong cypress and maples with jagged leaves. Shrubs of holly and broom and purple rhododendron bordered the blankness of the field, spilling out onto the parched grass. The trees, the ground, the massive roots were wrapped in moss and ivy, studded with morning glories just opening to the milk-pale sun. The dirt, rich and dark, was hidden from sight, for the over-abundance of ferns, silver-bush, red valerian and campion. Though the richness of the August honey-light was hidden in the froth of silver mist, no darkness could obscure the beauty, the serenity, the *truth.*

This was Tír na nÓg.

Rahyoke cast Addison a grin, as haughty as it was pleased, and the boy rolled his eyes.

"Fine, you *win*. You were *right*."

Rahyoke's smile broadened. Turning away, he called out to his caravan, "Hey, lads! We're there!"

Over-dramatic little sunflowers, they made their exclamations – "*Fine-ally!*" – and dropped into the green.

Rahyoke settled down at the base of a maple tree, the trunk of which was three times the width of a man. Through the screen of the rhododendrons and flickering, pale butterflies, he could still see the grey of the field. He hadn't realized until now – beyond that silver stream were miles and miles of open ground – harsh and flat terrain, and they'd crossed it. Beyond that sparkling curtain of steel was the invincible bond of the Palisade.

And they'd *crossed* it.

This was the beginning of something; whatever it was, he knew it was right. Not good, not awful – but right. *Sin an saol. Sin an cath.* And it was safe – at least – to rest.

"No fire."

62
Gort

Tritos Equi (3ú Meán Fómhair)

Let the wind blow high, let the wind blow low
Through the streets in my kilt I'll go
All the lassies cry "Hello!"
Donny where's yer trousers?

[Dubhán's hand] You're not funny.

An naoú lá de mhí Meán Fómhair

The *sionnach* and I were sat on the stoop, I with my back to the doorpost, the both of us furnished with fine, summer ale, which I had bought with some of my harvest-earnings. It's only the middle of the season, but I'm doing well so far, thank God.

For his own part: having finished the addition on the back of the house, he's started trying to tame our make-shift garden, out of sheer spite. He apparently promised a ransom to one of the neighbor girls if she can bring him, alive, the leader of the squirrels – a task with which her daddy, auld Senan, made me acquainted with a measure of remonstrance. I don't know; I think it's nice they for once have a game which doesn't end with a bathtub full of squabbling cats or broccoli stuck to the ceiling.

It's almost two years that he's lived with us, now, but it feels like he's always belonged to this household. And no matter what Sean says, about the stories his daughter's picked up off of him, or the carousal and carry-on he's often dragged into our house, Senan knows he's proved his worth to this village a thousand times over, and we were right for letting this noble stray in.

"Hey." So thinking and saying, I poked the elegant cur with my foot, thereby gaining his attention. "You know what?"

"What?"

"I'm happy you're living with us," I said. "It's like I've a proper family. No matter where ya go or how long you've been away, you'll always be welcome with us. You know that, right?"

He was silent a moment then, not looking at me. Then, quietly: "You just no' wanna cook for the house."

"Hey: to be fair, I'd no knowledge you could cook when I brought ya in."

He said nothing to that, but I noticed him smirk. He took a drink, watching as the sun sank in bands and veils of yellow gold. Then, with sincerity, and in Lavinian, he wound this bit of poetry:

"In the north... and, I am told, far west... the ocean rises up in shifting arrays of mulberry and slate and verdant turquoise, crowned with silver foam. It cascades in rivulets over the black rocks, scattering thick mists across the rolling hills. And the sun, setting, sends a thousand jets of fire glittering through that fog, catching every droplet like snowflakes, spiraling in sparkling rays like the power of God, settling over the mossy, black limbs of the great oaks which still withstand the pressage of time – and yes, I mean that word." He took another drink, as our own sun slipped the bright, goldenrod disk of its brassy perfection down beneath the tree-line, casting shafts and shadows long. "This is a near daily vision," he went on. "And if you stand on the tops of the buildings, clinging like a last autumn leaf to the girders and brick, you can see it. Maybe it's the danger of the height that helps it, you know? Makes it all the more beautiful. I don't know." A shadow flashed across the light, and he turned to me, admitting, "Of all these things, I'd rather be right here, with your family. At least," he added, turning away, "that is, for now."

For a moment, I watched him. Knowing him so many years, *a chairde,* I know he's not as stupid, as useless, as lazy as the world seems to think. I hope, *a rún,* you know that as well.

The last corner of the sun's bold orb dipped beneath the horizon, dragging the traipsing, ethereal rays of its light garment.

I smiled. "I love it when you wax poetic."

"Feck you."

An dóú lá déag de mhí Meán Fómhair

The opiate of the masses is force. They took our religion, they throw us out of bars – all they have is force. But it works.

Let me address the bars. Och – they can't take them away, *can* they? Sure but the soldiers themselves would riot. And we're complacent drunk, aren't we? Och – *sure* I am! Surely! *A Dhia dhílis.*

So they control us by force, toss us a drink now and again – all the respectable bars forcibly devoid of our presence, the shitholes into which we crawl terminated for conspiracy and supposed treason – but they still sell at markets, and sure I can get drunk just as easy on cheap wine. It's in the blood, *a ghrá*. We are – *sigh* – a lowly sort. An irrational race. A savage, reckless, intemperate people.

A Dhia dhílis.

And it's easy. Most people are content to comply – the vast majority of people are followers, *a thaisce*. Remember that. Stupid, sinful, complacent followers.

Most mutineers are depicted as fools, because they must be fo'c'sle hands, and not Captains, for a reason. Intelligence does not depend on your station.

We could build an empire. I know the mechanism, inside and out. I don't know what good it would do.

But I could.

An ceathrú lá is fiche de Meán Fómhair

Here's another thing I remember.

There was no warning when they came to us, *a ghrá*. In the south, they had seen the ships rising out of the ocean, and had had time to run for their scant weaponry, casting their children into boats. But so quick, so thorough, their decimation, that we never heard a word, the smoke of their fires caught and blended into the mists, blown south.

I remember that night – cold, as the frost lay thick upon the ground, and black with the ash that blotted the skies above. We were awakened into a world of burning thatch which cast crows, cackling, their wings scattered against the grey smoke. And the blood. Blood matted the dark grass, sputtered in screams and pain and flooding from wood and steel.

And they took him. As he fought with a pike and chains of their own fire. And they tore the pike from him. Gave it back into his ribs. And cast him down. And brutally they finished. And – there for you, fire-eater, into the flames you befriended.

And his wife they took as well. Barefoot, she stood in the dust, and from her distance, pulled the flames, flooding them back on the men who had made them. They both could do what I can, my parents, and they were murdered for it. Through cruelty. *By law.*

She suffered the same fate as her husband. Cast into the flames. Her scream was defiance, not pain, though they killed her. Anger and defiance.

And I saw everything.

I saw Bailecolgáin cut down one blade-stroke at a time. Vicious. Merciless. Helpless and kind-natured souls, who had never heard the name "Augustinium", dragged from their beds and cast, mutilated, into their graves. I was held – held so I could watch my parents, a blade angled so, when plunged, it would sever my shoulder from my neck, and I'd die a little slower. I still have the scar, for it wasn't at all kindly I was held.

I only live because I obeyed my mother's final order – slipping near through that blade to accomplish it and leaving Ceothannanmór behind.

"Run."

But sure as the breath and blood they died for thrives still within these bones:

I'm not running anymore.

63
When Oisín Returned

Meygan's hair was sad. A full month without Moroccan Argon oil and anti-frizz conditioner. A full month of wind and rain and dirt. *Dirt!*

She was sitting, cross-legged, at the base of some huge tree. Beside her, a crick lined with smooth stones carved its way through the green, a trench filled with glass-clear, quick-moving water. She was watching Rahyoke in his attempts to make himself remotely presentable; he had not shaved in nearly two weeks, and so had a full beard with which to contend. He had slept the entire day – a fact he had realized upon waking up as the sunlight streamed red through the hanging mist – and was trying to work by the light of an absent sun and a few dozen, soft-glowing fire-flies.

"Do you have shampoo in this world?"

"Huh?" A last slice killed the beard, and he turned her a grin. "What are you insinuating?"

Side-tracked, Meygan giggled. "I forgot what you looked like without a beard!"

"Oh." He wiped the knife he had been shaving with on his jeans. "*Beautiful*, right? Absolutely gorgeous?"

"Totally." The sarcasm and sincerity were so well-balanced, her words needed no supplement.

"Yeah?" He gave a slight frown, trying to find his reflection in the blade of the knife; it was too musty and grimy to use as a mirror, so he spat on it, wiping it clean on the inside of his coat. "Hopefully my clients will share your opinion," he mumbled.

Meygan made a face. "Just don't do *that* in front of them."

"You still look like a wild-man!" the ever-pretentious of Addison Mitchell announced, from stage-left.

"A wild-man!" Ian beamed, in agreement.

Having made it across the field, the "warriors" had been amusing themselves with training for the past couple of hours, fighting staves with the battered sword and the alien rake. Ian, inducted into the game as Addison's page, was keeping score. Addison wore the tartan they had used as a knapsack, knotted at his left shoulder for lack of a brooch – *"The proper way,"* as Zander had quaintly put it. The way of a real warrior.

When Zander quipped, "Well, I'd hire you to scare crows," Rahyoke cried, "I'm *workin'* on it, a'right? *A Dhia dhílis...*" He took off the burgundy scarf, the paddy cap, and – carefully – untied the red bandana, flinching as it pulled a bit. Taking the scarf, he dipped it in the clear water, soaking and wringing it out. Bent over the stream, he toweled his hair as one might to dry it, scrubbing at the dirt, the grime, the clinging blood.

"Does that actually work?" Meygan asked. "Maybe I'll try it."

Rahyoke pulled the scarf off, shaking his head like a dog in the rain. "Does it?" he returned, sitting back on his heels. He didn't care much how effective it was, truthfully; it was the effort that counted, right?

Meygan laughed. "Your hair's all sticking up."

"I'll take that for a 'yes'." He laid the scarf to dry on the bank, along with the bandana he'd washed. Then, smoothing back his hair (not all of it stayed), he put the cap back on, sitting back from the waters. "I think that's as good as it gets."

"Go bhfóire Dia orainn," Zander invoked, rolling his eyes piously.

"God save yerself kindly indeed," Rahyoke quipped, liltingly. Then it was down to business.

It had been a long time since he'd been on this side. A lot can change in a decade – and where did that leave them? Could it be that the land was ravaged – mutated to pretty little palisade towns? Ugh; he hated the sound of that – the thought that they had gone through so much, only to be worse off than they were before... He shook his head. What was, what wasn't, what may or might never – "Damn it," he muttered, irritated.

Mossy roots twisted their way through the glossy carpeting of ivy, flecked with the sheaths of morning glories and the white stars of those flowers no one can name; in front of him, wound into this patchwork tapestry, was a bounty of shamrocks of varying sizes – all with three leaves. He sifted through them lightly with his hand, watching their tousled motions. *A map.* He scooped a handful of stones from the river, shifting so he was kneeling again. The look he fixed on the shining clovers was one of deep concentration, intense focus, near trance as he sifted through their stalks, sweeping them aside to place stones amid their clusters in a great, pattern-less grid.

"The fury of the restless," Zander mocked, observing.

"Aaauh, sure. And *act* like ya didn't spend every occasion t'traverse my land and its surroundings like a hawk. I'd be glad'uv ya t'spend two days at a time in the house, but wou'jya? Nope." Studying the diagram, he hesitated, a rock upheld in his hand.

"You can't have it both ways, can you?" Zander returned, fighting Addison with almost mechanical ease; it was slow. "I mean, yer land was tiny as Hell, but y'can't watch the back yard the same as the front for the house; it's the limit of mortal vision."

"How tiny is Hell?" Ian wondered, a philosophic murmur.

"It's the same as everything," Zander went on. "Even the man with the keenest of vision – introspect or external – cannot see the future for the present, cannot see the west as he sees the east –"

"And can't mow the lawn 'cause he's busy climbing trees," Rahyoke finished. "I *have* farther to run, Zander. It's only a matter of when the Lord sees fit to break the road."

Zander frowned. He was so put off by this answer that he nearly missed his block. "Gaahh!" he complained, stepping back to avoid a solid scrape to the arm. "I might have to re-think how I'm training you."

"I'm doing fine!!" He turned to Rahyoke, waiting for Zander to re-set. "So what's the plan?"

"Shorn. Is it always we need to plan fifteen steps ahead? Is it so impossible in your own culture to simply enjoy where you're at?"

Meygan and Addison chorused: "Yes."

Rahyoke pursed his lips. "I'm tired of being the plan guy. Addison? Where do your budding warrior instincts tell you we should go?"

Addison was startled. "Uuuuuhhh..."

"So there." Rahyoke cast a tight frown on his diagram, fingering that glossy stone. At last, he snapped, "Hellfire!" And blindly chucked the rock into the crick with a splash.

"Two points!" Meygan exclaimed, dropping her hair to give a touch-down signal.

"We'll move camp when I get back. You lads stay here – don't kill each other while I'm gone. And no fire – y'bleedin' pyromaniac." He added the last in a mutter lost beneath Ian's groan of protest. "I wanna get more – what? – inland. *Aaauuchh*." He had stood; now he drew a hiss through his teeth, shifting his weight uncomfortably.

"Y'okay?"

"Sure I'm fine. Auh." He straightened, folding his arms. "Ya gotta test the waters before you shove off; I'll see what this Paradise has to offer, and then I'll get back to you." He took up his canvas bag, limping off in a general north-east direction. "Don't wait up."

There *was* more ground to cover. A lot. And through the shifting waves of time, he didn't remember *any* of it.

Which was infuriating. Memory was the one thing he had, for better or worse – had and still actually liked – perhaps, even, took pride in. But everything had changed. That was the way of the world, though, wasn't it? Compensation. Recompense. Balance in all things. Fionn had had Sabh for the shortest of whiles. He'd received Oisín in return. And Time carves new wounds, scratching over the old. Ancient, gnarled trees had fallen. Wells and springs had gotten clogged. Cairns and dolmens had been toppled – at the violence of what hands they'd never tell. And the town, when he found it. Sure, hadn't they cleared away the trees for farmland twenty miles around? And itself a *perfect* little township. A perfect little Mainland town. They knew no better; it had been too long.

"Anyone up for tree-climbing?" inquired the unsinkable Ian Burgendson, to whom Meygan responded, "*Eew*, no."

Rahyoke couldn't help a slight smirk. They had moved camp when he'd gotten back – late, *late* in the evening yesterday. He had found them something of a clearing; a spring welled up a seven-minute walk north-north-west, and the open grass of the clearing was well-shaded by a loose canopy overhead. Zander was building a Fenian oven from stones he had found near the spring; Meygan and Ian were just getting back from refilling their water supply, and Addison was watching Zander work, eating a bagel and pretending like he was too cool to care.

Rahyoke lay on his back in the longish grass, his hands folded behind his head. It was maybe about nine-twenty, nine-twenty-five in the morning, but he didn't feel like going to work just yet. It was rare that he got the chance to bask anymore, and, by God, he was going to make the most of the opportunity now.

"So!" the archer broke in, with the promptness of command. "What's our itinerary?"

Rahyoke's smirk broadened. They boy had been granted begrudging permission to show his skill with that fireplace, and this small indulgence had made him the High King of the world. Pride, *a chairde. A Dhia dhílis.*

"Yourselfs? You lads can stay here. Behave. Don't kill each other and don't burn the forest down. *An seanscéal céanna go deo.*"

"That sounds *enthralling.*"

"You can climb a tree an' be look-out for me," Rahyoke proposed, off-handedly. "There're towns south and east and north-east; there's the ruddy field off west."

"I'm an adult; I don't need to play crow's-nest."

"So there for your training opportunity."

Zander gave an over-dramatic, moaning sigh. "*Fine.*" For a moment, he sat, despondently looking over his half-finished fire-pit. "You know, if I'm gonna be top-side, should I even bother with the fire? 'Cause –"

"Finish the fire," Rahyoke directed. "One'a your proud squirelings can watch it. Teach the quintessential pyromaniac how t'keep a half-decent vigil."

Zander frowned, dubiously.

"Auh!" Addison sat up ram-rod straight, throwing his arms out wide. "You doubt me! Me! Your own – trainee – student –"

"Shmuhgegee," Rahyoke helped. Addison jerked him a glance.

"Sure, I'll trust ya for it," Zander shrugged. "I'll be in the tree anyway, so I'll know if you screw up."

Addison dropped his arms. "*Thanks.*"

"Any time."

It was a nice morning. Sunny, warmish – the light scattering down to grace them through the leaves. A great bush of white roses was growing over by an oak; vines of ivy sparsely spiraled up sycamores and cypress. Some sort of bird was singing its sorrows of the coming winter to a love in another tree.

"So, what are *you* going to do today?" Meygan asked Rahyoke.

"Me? Well, it's *Thursday,* so – I suppose I'll get up, walk to town, find a job, work 'til I'm fired, an' trudge on home. You know – change it up a bit." To go from the mock perkiness of the revelation to the flatness of the final line prompted Zander to note, in a similar way:

"I sense some sarcasm, there."

"You get fired?" Addison asked.

Rahyoke waited until they knew he wasn't responding; then, as he watched the light shifting through the trees, he quietly sang:

"Siúil, siúil, siúil a rún
Siúil go socair agus siúil go ciúin
Siúil go doras agus éalaigh liom
Is go dté tú mo mhuirnín slán."

"Hey, Rahyoke!" Ian now. "How are the towns?"

"Yes!" Meygan echoed. "Worth sailing, or do they suck?"

"All told, the Isles didn't suck when Oisín came back to them; of course not! Saint Patrick was there!"

"But Oisín was sad."

"Don't read too much into it, sunflower." Gritting his teeth, he sat up, gingerly; then he turned to them. Sometimes it struck him how young they all were. "I went to a town called Vinovia last night. Newer town; still developing. Loads of young people moving in, so ideal for you lads. There's a slightly older town nearby I can't remember the name of I'm gonna check out today, but it looks promising. Really, with luck and grace, it'll just take a bit of searching so I can find someone willing to take you."

"Why can't we just live on our own?" Addison demanded, as if this were the most idiotic of oversights.

Rahyoke hesitated; Zander chucked a pebble at Addison, who didn't flinch until impact.

"Ow!"

"That's why," Rahyoke answered, grateful for the example. He moved on before Addison could muster an argument. "Look: Wou'jya do me a favor an' stay in camp? *Oakh.*" He was stiff and stung, and it was hard for him to get up. "I don't even want you going for water. I haven't had a chance yet t'scope things out, really, and it would – I'd be grateful."

Addison made a face. "*Where* would we *go?*" he sneered.

Rahyoke said nothing. Getting his bag, he went back to singing, still in the same soft way.

Zander shook his head; he knew Rahyoke's concerns. "Ah, you're a worrisome hen-wife."

"Are you not done yet?" Rahyoke returned. "When Donn Mac Duibhne cooked for the court of Cormac son of Art and the whole of the Fianna besides, when did it ever take him more than three minutes to prepare a place for the fire?"

Zander frowned. "That's the most random example ever; when did that even happen?"

Rahyoke shrugged. "When y'can't dazzel'em with style, baffel'em with bull-shit. Are you done with my matches? I need those."

With the air of a performer, Zander took a match from the green-coloured box, struck it magnificently, and dropped the flame into the pit. Rahyoke watched as it took light.

"You know, that won't happen every time you do that," he warned, coming over to take the box.

"Oh, sure. And why not?"

He snatched the box from Zander's upheld, open palm. "You take *far* too much for granted."

"But you love me anyway, right?"

"Love you." Rahyoke rolled his eyes. "Get to your post."

"Aye." Zander stood, saluted, approached a great foe of a chestnut, and gave a phenomenal Salmon Leap to its upper branches.

Rahyoke winced. "Please. At least pretend you're trying to be careful. For my sake; it does my soul good."

"Look; if you're too old and fragile to climb up here, leave me to it my own way."

"...I'm too old and fragile to catch you. Enjoy your own way, then." He started off; then, pausing: "Hey Zander. Heads up."

Zander caught what he flipped into the tree – a folded leather wallet that looked like it had belonged to Noah. "Wha –"

"See if you change yer mind." That was all Rahyoke offered of an explanation; then he was gone.

Zander looked at the wallet in his hand. Very faintly, the Imperial Seal was stamped onto the coloured leather. Opening it, he realized it was the covering for a license and travel visa; both documents bore the red boarder demarking official Imperial papers; both documents bore the name "Fionnius Aleczandros Arāneolus, Vinovia". It was not an alias he had ever heard Rahyoke use, although he used many – oh of course. It was meant to be his.

There was another sheet of paper tucked in with the official documents – yellowed, water-stained, and folded sloppily into fourths. He opened the note and read:

-- Zander

scroll #7

And when there comes to them information about [public] security or fear, they spread it around. But if they had referred it back to the Messenger or to those of authority among them, then the ones who [can] draw correct conclusions from it would have known about it. And if not for the favor of Allah upon you and His mercy, you would have followed Satan, except for a few.

So fight, [O Muhammad], in the cause of Allah; you are not held responsible except for yourself. And encourage the believers [to join you] that perhaps Allah will restrain the [military] might of those who disbelieve. And Allah is greater in might and stronger in [exemplary] punishment.

Whoever intercedes for a good cause will have a reward therefrom; and whoever intercedes for an evil cause will have a burden therefrom. And ever is Allah , over all things, a Keeper.

4:85

Zander grinned. Rahyoke had translated the scroll for him. To the side, he had copied out the original, scimitar-esque script (as well as he could), with the transliteration. Another tale to add to his mental store – another keepsake of the life the Empire had destroyed.

Zander folded the wallet and all it contained, sticking it in his back pocket and leaning back against the strong trunk of the tree. Two pieces of paper could be the foundation for a new life in the right hands; he had his mother's determination, his father's manipulative skills... The cleverness of ravens, the voracious thirst for life of the wolves. Here was home; here, with the scent of the rain and wild scallions and honeysuckle, with the sunshine glittering through the leaves. Not in a town, with its smog and its fleas and its taxes. He thought briefly on his upbringing – his parentage, wild and true.

Changing his mind was not in his nature.

64
Ngetal

An deichiú lá de mhí Dheireadh Fómhair

He didn't sneak off in the night. He stayed for supper, as we asked him. And, when my son cried at his leaving, this is the conversation they had:

"Bairn, is it me dead?"

"... No."

"Is it me loving you?"

"Yes!"

"So no crying of you!" He had taken the child up, as he was accustomed, half-balancing him on his hip; and he spoke to him as easy as anything, in the best of the language he could offer. "It ain't the last of seeing you – goodness! I ain't dead – do no' you keen me! Hear me?"

The lad nodded, and my friend smiled. "When going with himself from a feast, was it Cuchulainn crying? No. No' him, no' Fionn, no' Achilles, no' Camulos – no hero. Cause they'd no' for a reason. You?"

The lad nods, and my friend rolls his eyes.

"*No!* You'za hero – grand, *grand* hero!" He shrugged. "And I come back. No keening me!"

Whether or not the lad understood him , he took great pleasure in this speech, and loved him greatly. *Go n-éirí an bóthar leis*. And damned if the house isn't an inch bigger and three times as quiet without him.

An tríu lá déag de mhí Dheireadh Fómhair

I need to lose this brogue.

Greatly it pains me. I feel like I'm abandoning my past, and that's a thing I never wanted to do. It's betrayal.

But I *need* to lose this brogue.

They stop me twenty times at least going into a city. ID? Papers? Registration? And I picked a hard way of finding out they were "expired". Och!

They heard me talking to a merchant, and when I'm leaving it's: "Hold there, Ceothann."

And like a moron I stop.

I'm surrounded.

"Registration?"

"Yep." I start digging though everything, affecting that nervous air they expect from the less bellicose of my race – wondering what the hell I'd done with the damn thing – 'til I find it and show it. I never hand it over lest I'm made to – it's a good moral. Otherwise people steal your stuff.

He did to me anyway, and, when he saw it, he gave a laugh. Believe me, *a chairde,* it was not a laugh you're wanting to hear.

"This is expired."

"I just renewed –"

"What's in the bag?" he asks me, random, totally throwing me off-guard.

"'Tisn't against the law to carry things; you saw yourself what I just now got."

As for himself, he gets menacing. "What's in the bag, you little savage?" Which, I suppose, he's seen fit to deem synonymous with "*Ceothann*".

I haven't a good hold on my tongue when I'm angry, and I haven't a good hold on my anger. So I *tell* him "what's in the bag".

Och. And I can tell yourself in retrospect, *that* wasn't the best of ideas. Live and – oh, God knows that I'll never learn. Not in the heat of the moment, anyway, and that's why I need to talk like them. To take that stupid, nasally, nail-scraping caterwauling accent into my soul like a second nature.

To be fair, 'tis my wife's own accent, but in her it's attractive – a soft, low-toned music, so different from the gritting, grating babble of the usurpers. If I'm to take on any version of their speaking, it'll be only hers... But a little more manly, of course!

I'll see if she isn't willing to be teaching me, especially after this... I will see if she *will* teach me??? Oh, Lord, I've a long way to go yet.

It does feel like an act of treachery, but... "Your ancestors would understand." A virginal sacrifice to stop the volcano, to slake the dragon. Collateral damage. Friendly fire. Change your name. As I have done, time and time again. Change your background. Your accent. Pharaoh can even take your very lives. But they cannot take your soul. What are you willing to martyr for the greater good. What irons are you ready to don to win back your freedom. Will you sell your father to buy back everything he loved?

Oh, Christ. What am I thinking?

An naoú lá déag de mhí Dheireadh Fómhair

In the beginning, there was Chaos. And from the time that Chaos was demulsified and sedimented into the great, elemental parfait of

fire

air

earth

and water

– From the beginning, Prometheus had a fatherly love for man. It was he who is said to have fashioned Man out of the elements, to have modeled him after the gods. It was he who tricked Zeus into giving man the best portion of the meat-kill for his dinner, while the gods got only the bones. And it was he who took pity on man for his defenselessness and pathetic nature in the face of all the world's dangers, and sought to bestow upon him the gift of fire. This would be the greatest kindness ever done for us, and it was also our benefactor's downfall.

Prometheus went to the upper strata, the highest sphere, where dwelt the gods and their coveted element, fire. There he asked Zeus to spare but a spark, an ember, a mite's coal, for to man this small gift would be the greatest treasure. But Zeus turned him down cold.

There were three reasons for Zeus' niggardliness: 1) He had had about enough of Prometheus' generosity, which was leaving the gods penurious. 2) He reckoned that man had had enough of Prometheus' generosity, and t'was best to deny him ere he got to be thinking himself on par with his masters. And 3) Zeus saw fire only as a weapon, designed purely to mete out divine retribution. But Prometheus saw things differently. Aye, fire could be a weapon, and could be used for making greater and more terrible weapons besides. But with fire, man could keep himself warm on the long winter's nights. With fire, man could cook food that was nutritious and soul-loving and cake! With fire, man could weld idols, work and stain glass memorials – he could create, an art Prometheus loved best of all. Fire was dangerous and beautiful. Fire was life and death. Fire was the chief element of Chaos, and fire would give Man a fighting chance.

Prometheus knew the risks, and he knew that the prize was worthy. He would take care of his children, no matter the cost.

Prometheus stole fire from the gods.

No one can quite agree on how he managed it, but no one can deny the damage was done. Man was happily grilling hamburgers, and Zeus' patience had reached its end. Nor would he be content merely to cast Prometheus into Tartarus with the other Titans, turning his immortality against him. No; he would make him an example to Man.

Prometheus was accordingly bound in heavy chains upon a mountain, where each day a great bird would feast on his liver and heart. Each night – and just as painful – the organs would regenerate, and the excruciating cycle continued.

But Prometheus could bear the unimaginable pain of his torture. For he'd see the light of their fires, he would smell the wisps of their smoke, and he knew what he'd done was right. And he would smile.

Man had attained mastery over fire, and this could never be undone.

An seachtú lá is fiche de mhí Dheireadh Fómhair

The sun falls lightly through the slats on the window, scattering and seeping across the kitchen. Och, the hour's but early yet, and somehow, its rising seems late. Late, as it scours and scrapes through the thick bulkhead of clouds still gathered just beyond, lurking over the forest, around the dark horizon. The trees are still wet and dripping with the rain that's hounded us these last three months, nigh incessantly; the sound of it drumming through the night, soaking the thatch and wearing down the trees. And the great silence of its strangled pulse – reduced to a pettering of slow-sliding drips that plash so loudly, so vibrantly, on the waterlogged ground – that silence is overwhelming.

Everything slips and drains and drags – the black trees into the pitch-dark ground, the spreading indigo into the pale white below, the heavy shadow of everything else around – blurring like delirium tremens.

And yellow – a slick, muted amber – is the white light slipping through the slats. It casts grey shadows around the kitchen, reflecting quarely through the bottles on the sill, dribbling faintly the blue, the brown, the green, creeping, across the floor, moving towards the table like sneaking dogs. The daylight is unreal after so long a storming, and foreboding clings to the shredded strips of light, promising another wave of washing rain, harsher than that before as punishment for the lull. But now...

Now the greying light spills over the clear glass on the table, scattering through the clinging foam. The lingering, gold-brown leaves mutter drunkenly in the thick breeze. The pooled rain drips steadily through the broken pane, puddling on the counter.

Should I be more worried than I am?

An tríochadú lá de mhí Dheireadh Fómhair

65
Tides

The fire had burned down by the time Rahyoke came back, and Addison and Ian were sitting near it, poking at it with sticks.

"The town's called Savaria, if anyone cares." He stopped, looking down at the sad, smoldering ashes of the fire-pit. "Wow," he said, flatly.

Addison looked up. "It's windy."

"Uh-huh. Where's Zander?" Rahyoke strolled to the base of the tree the archer had climbed. "Zander! Clear?"

"Yes! Hello, by the way," the ragged chestnut replied.

"Hiya. Are you lookin' only one direction, or is it every way, y'dosser?"

"Every way!" Zander defended, but the crack in his offended tone was more than youth. "...Now."

Rahyoke rolled his eyes. "Hey Addison! Where's yer darlin'?"

Addison actually blushed. "I don't know; I haven't seen her."

"Haven't... haven't *seen* her?" he murmured, sounding mildly bewildered. He shook his head, turning to face the boy in full. "Seriously, Addison; don't kid around. Where is she?"

"She –" Whatever was in Rahyoke's expression broke the confidence in Addison's own. "She – she went to look for Ian."

Dread snapped his patience. "Ian's right here!" Rahyoke shouted, gesturing to the boy.

"Ian had gone to the well; Meygan wanted to get him before you got back!"

"Shit. *Shit!* How long ago? Never mind – Zander!" Long enough for the child to look nervous; long *enough...* "Keep on it – I'm getting Meygan! *Feck...*" He cast the male trespassers an eviscerating glare, and, with all the dark-toned, low-pitched ferocity of Hell, he warned them, "Stay, *here.*" And, dropping his bag, Rahyoke took off at a sprint.

66
End of the Run

Long shadows were beginning to stretch across the grass. Twilight was coming on, and soon the sun would set. Darkness would once more spread across the land, a harsh and cloud-filled darkness, and by then it would be too late...

"Meygan!"

The sudden grip on her shoulders was so tight, the force with which he collided with her so sharp, and the sound of that voice so familiar, she could not cry out. He took her by the wrist, left hand to left arm, and stepped back, shifting to her side as he half-spun her around.

Rahyoke.

"Ouch! You scared me!" Her eyes were wide.

"Meygan!" he gasped. Relief. And pain; it was hard to breathe – nearly as hard as when he had first injured his ribs. He took in a wheezing breath, deep as he could, and demanded, "What are you doing?!"

"I was coming back – are you alright?"

"Hy – uhh –" He bowed his head, struggling to get his breath back. He was nearly doubled over, using her frail arm for support. "I – *ran* here's the – matter, s'what!" he retorted; the snap was lost in the raspiness of his voice.

"D'you wanna sit down?"

He shook his head, swallowed with some difficulty, and said, "I t – I told you not to – not to leave camp! You can't just – go wandering off like – that – What if something happened to you?! You don't know what's – wolves, bandits – I dunno! *Damn* it; I don't know *what* it is with you kids and disobedience –"

"But I needed to find –"

"Save it; come on." Like a blade-stroke, his tone sought to sever any more conversation, but it lived on.

"Wait!" If he could breathe, he would have shot her a look from Hell; as it was, he could do no more than obey. "I need to show you something."

Rahyoke dropped an inexact curse far too vulgar to have been spoken in front of her – then covered his mouth quickly. "Id – uh – Look: priority-wise –"

"It'll be quick." Her eyes dropped; he was holding his side again. He released her arm, backing to lean against a nearby tree. "Seriously, Rahyoke – what's wrong?"

The gasp he drew was strained. "My ribs – are bruised. I'll be fine. Now –" He straightened to his full impressive height. Meygan could see she was losing ground, and so she appeased: "You're totally right; you're absolutely right, I never should have gone off. I'm sorry. Can I please just show you this one thing? I think it's really important –"

He groaned, leaning against the tree again, casting his eyes very obviously to God. "*A Dhia liom* – fine! It better be close –"

"It's at the spring; it isn't far." Having evaluated him, she offered him a hand, teasing, "Want a hand? Don't be all moody; the sooner you come see, the sooner I'm back at camp."

He rolled his eyes, leaning off of the tree under his own support. He followed her to the well.

The well no doubt had once been sacred to the Galls, and that was what had preserved this grove of trees. At the top of a knoll, in the bottom of a wide indentation at the summit, a natural spring bubbled up from deep in the heart of the earth, creating a dark pool. Large, disparately-coloured boulders were strewn like toppled monuments or benches within the hollow; a few trees straggled up from the holy ground, their branches occasionally tied with rags which were decades – perhaps even centuries – old, hung as offerings, their fabrics washed and dyed by nature.

Young and healthy, Meygan was quicker than he was, and she darted down into the bowl of the well, returning to hand him a largeish, flatish stone.

"Doesn't it look like something was written...?"

Rahyoke pursed his lips, completely prepared to be annoyed at this egregious waste of time – but there was indeed a rubbed soot marking on the stone: two vertical lines joined horizontally. He turned the stone, studying it. "Is it an 'haych' or an 'ih'...?" The stone was warm, as if it had sat in the baking sun all day.

"There are a bunch."

"Are there?" he asked, absently. He was inspecting the rock with the dire depth of scientific observation with which he studied everything that he found odd, turning the rock this way and that. The soot had been rubbed off of the rock; what remained was the scorch.

He allowed Meygan to take his wrist and lead him down to the spring. A few feet back from the water's edge she uncovered more stones which had been hastily scattered among ashes and dried leaves. They seemed to be broadly gathered in a roughly-marred circle, which Meygan was gently repairing.

"I think they make words," she explained, lifting the stone Rahyoke held and replacing it whence she'd taken it. "It looks Latin-y. Do you think it's Latin?"

Rahyoke cocked his head, studying the stones from her angle. *I think it looks like Lavinian.* "Missing a few?"

"Yeah; I think someone tried to cover it up. They might be in the spring...?" She looked at him. "Isn't it weird? What's it mean – do you know what it means?"

H – C – S – V – space – D – R – A – C – O – N. Third declension – not nominative – plural subject? "'Hīc sunt dracones': 'Here there be dragons,'" he breathed. "That's very weird..." Suddenly he felt all of his hair standing on end. He at last raised his eyes, meeting hers. "Did you see anyone around here?" She shook her head; he did not know whether or not the denial was reassuring. "Mmh; I don't want to take chances. I don't think we want to be running into the cryptic loon who makes fire-offerings to dragons. Maybe we'll spend the night in Savaria?" He offered her his hand, helping her around the fire-pit and away from the spring. "You did well to show me."

"Does this mean that other people are around here?"

His eyes narrowed slightly. "Perhaps Ian knows." Without another word on the matter, he led her to the crest of the hill.

Meygan looked up at him, scurrying to follow his long if slow strides. "There aren't really dragons in this world, are there?"

His smirk was for God and the clouds and the secrets only they shared. "If there are, it's not myself who's heard of it... Although that *was* how Crazy Steve always claimed he'd lost his leg. He had this great peg in place of it that looked like it had been gnawed on." He looked at her. "What?"

She returned a judgmental stare. "You hang out with a dude called 'Crazy Steve'?"

He shrugged. "I've had an odd life."

Rather than ask about that – although interested, she had had enough weirdness for the hour – she asked instead, "Do you really think it's an offering site?"

Rahyoke was serious again, pondering. "It was at one time, whatever the intensions of that fresh... thing – I lack a noun. Hearth?" His eyes slid her way. "It has a creepy feel about the place in general either way; do you feel so?"

"Maybe we should have stayed on the other side."

He glanced at her again, his expression clever. "Wanna cross back over?"

"No!"

He started to laugh, but stopped abruptly as a shout sounded from the direction of their camp. He lashed out, blocking Meygan with his arm.

He was momentarily relieved as he recognized the voice as Zander's, and greeted the archer with some surprise; and then: "What's the matter?"

The archer had come bolting out of the cover of the brush with the speed and agility of a panther. He carried nothing; his bow and quiver had been left back at the camp. His eyes were wide with anxiety, his face pale in spite of the run. "I'm sorry – I know, I – it's just – I was keepin' watch as ya said –" The words poured out with an intense urgency – a harsh and deadly urgency.

Oh God...

" – ya know – up in the tree – I left my bow – I said I would be –"

"Zander!" Sharp; not loud, but *sharp*. Rahyoke took a step forward. "I won't be mad, just spit it out!"

"They – through the woods – I don't know how to stop them – they –" Terror pulsed through the archer's very bones, and he uttered the fatal words: *"They're coming!!"*

Rahyoke paled. "My God," he breathed. "Oh my God oh *God!*"

Meygan looked back and forth between them. A thousand questions flooded her mind – questions that would never be asked.

"How close?" Rahyoke demanded, his voice hoarse despite his firm tone.

Everything within Zander wanted to seem brave, but he was panicking. "We need to leave now! I'm sorry –"

"How many?!" Lord – his voice waxed shrill! "How long – how much time do we have?!"

"I'm sorry, I –"

Rahyoke spat a single word – she didn't have to know the language to realize its meaning.

"What's going on?"

Rahyoke jolted; he had legitimately forgotten she was there. He looked at her, and, in a voice that was strained, he said, "We're at war."

This was so far from anything she would have expected him to say. It made sense in a way – in so many ways. The sentries, the curfews, the paperwork, the palisades and the Romans. It made sense. And yet...

"What?"

The innocence. That's the most painful part, is destroying that innocence. "Meygan, our world is at war. And I was born on the wrong side. If they catch me, they will kill me. Do you understand? We have to go now."

Her eyes rounded. Not waiting for her answer, he stepped forward, seizing her arm. "You need to come *now* –"

"Christ –!"

A single word shattered the confines of everything he had known, everything he had tried so hard to protect. He turned at the sound of Zander's voice, and froze.

The archer clutched at his heart, red with blood, his shaking fingers flitting over the shaft of the arrow imbedded there. He had fallen to his knees as if in prayer, though his left hand reached for the sword at his side. For a moment, his eyes were fixed on the blood that stained his fingers. Then, he pulled the arrow from the wound. And collapsed.

The shadow of a man stood upon the long grass. Stripped of his soul, all indifference was gone, replaced by anguish, dominated by shock – paralyzed by a nightmare turned real. After everything – he couldn't even catch him; hold him while his life bled out across the soil. One faltering step forward –

She screamed. The shrill cry cut through his thoughts like a swordsman's blade, flooded with terror, pleading, despair. Dazed, he turned, and was startled to see they were no longer entirely alone in the clearing.

A man had Meygan by the arm – a man with a gleaming sword and a quiver of black-feathered arrows not yet stained with blood. She struggled desperately against his grasp, like a faery caught in barbed wire, but in vain. He was dragging her towards the woods, taking her away.

"RAHYOKE!!!"

In a pulse, Rahyoke drew the boat knife from where it had been concealed. He wrapped his right arm around Meygan, his hand clasping the wrist that the other man held. In the same motion, he pulled back, tearing her from his grasp and plunging his blade into the soldier's neck. The demon fell without a sound, and Rahyoke's knife dropped to the grass beside him.

They stood in silence.

Meygan buried her face in Rahyoke's arms. The light of the sun would dispel nothing, irradiating the truths beyond hope's shroud. This was life, its true nature. Horrible, beyond credence, and cruel. She wished she had never come here.

"I'm sorry."

He was rigid, stiff as death himself, and shaking. Shaking as if every glass fiber of his body had been cracked, and would shatter beneath the weight of his grief, as soon as his mind could comprehend it. As if in another world, he gently took her by the shoulders, pushing her a few steps back. His eyes were on the Heavens, questioning his infernal God.

"Rith an méid atá i do chnámha... Ná bíodh eagla ort."

Meygan looked up at him, studying his face. His dark eyes were wide with a question, his mouth twisted into a thin line of pain. He was fighting back what could not be fought. And it was killing him.

"Rahyoke..."

He closed his eyes.

He heard the noise before she did; heard the distant crackling of twigs as distinctly as the whisper of fire. *Rith...*

In a flash he turned back to the girl, taking her firmly just below the shoulders. In his grief-torn heart he wanted counsolance, but she couldn't help him. Her beautiful eyes bore frangible emotions, of virtue and naiveté, and it pained him to know there was no time to comfort her.

"Trust me once more," he enjoined her. "What happens next I would prefer you didn't see; it's the sort of thing unfit for anyone's eyes."

They were drawing nearer, the snapping roar that betrayed their footsteps swelling with each fleeting heartbeat. He urged: "Do not remain hidden, but never surrender. There is nothing for you to fear... Now go!"

With that he released her, propelling her with gentle exigency towards what he desperately hoped was the safety of the woods. In a single glance, Meygan knew there was no way she could argue; she obeyed.

He waited until she disappeared within the line of trees; when he turned back, the first soldier had already broken cover.

Swiftly as he could, Rahyoke grabbed the murderer's sword from where it lay in the grass, getting it up just in time to deflect a crushing blow.

The two blades had barely connected when his attacker raised for another shot.

Locked in that unstable position – back arched, knees bent, legs apart, upraised sword parallel to the earth – Rahyoke saw his chance. The inexperienced fool thought to batter the sleek steel into submission and was aiming to repeat the downward slash.

Right.

The steel flashed in an arc, and Rahyoke twisted, straightening up, bringing his right foot closer and meeting his opponent's blade with a stroke from the side as the two blades scraped past each other. With a jerk, he stepped into him, slamming the force of his elbow and lowered left shoulder into his enemy's chest, knocking the wind out of him. One step backwards, and a neat slash – blade held in his left hand, parallel to the ground – slit his attacker's throat. The would-be assassin hit the ground, his clean sword hitting the dirt beside his body.

Two.

Breathing a little harder than he was proud of, Rahyoke cast his glance around the clearing. There were more – he could see their shades shifting amongst the trees, held at bay by a show he feared would soon prove no more than just that. His eyes briefly fell to the still form of the archer behind him.

Taking a wide stance, he brandished his sword like a baseball bat, over his right shoulder, the hilt gripped in both hands. His gaze narrowed on the camp of his enemies, he placed himself between them and the boy. *Come on...*

...I dare you.

One by one, seven warriors slipped from the cover of the trees. All, like the first two, wore scaled armor the dull grey of an overcast twilight, and bore the insignia hated by all rebels. Each carried a sword in one hand and a small, circular shield for hand-to-hand combat in the other. Their helmets resembled the un-crested helmets of the fabled ancient Greeks. They did not charge, but strode forward, outspread to deter a retreat, inflamed with battle fury.

Seven... This should be interesting.

They leveled their swords.

His mouth twisted in a defiant sneer. He uttered but a single word:

"Cowards."

This drew forth the first combatant, and like a shot he was upon him. Unlike his predecessors, this man was ready, and far better trained.

Rahyoke was forced into defense, focused on catching the savage blows and avoiding any thrusts of the shield, which was reinforced with an iron boss and a razor's edge. Each crash of the soldier's blade against his, each tweak of his muscles as he turned to avoid a blow, sent shudders through his brittle ribs, making him dizzy. His eyes darted, searching for a break in his adversary's defenses – watching as the other men stood by, and watched him in turn. Content to save their energy, if the outlaw could be conquered by one man. Contentment. He had forgotten the feel of it himself.

"Auh –" he stumbled back as the shield forced his own arm into his ribs, and caught one of the bystanders looking behind him, to what he so frantically guarded. His eyes narrowed.

Well, you'll have to kill me, too.

At once, he settled into a firm stance. He swung back his blade, hilt gripped in both hands, and hurled all his strength into a twisting strike.

"Auh!" His harrier was startled as the crushing blow fell against both sword and shield, pinning his blade to the boss. Laying his palm against the flat of his blade, Rahyoke shoved forward, pushing the man back. His bruised ribs panged terribly; not by force, then, but by skill.

He stepped forward with his right foot, passing his sword between hands as he did, so that his right hand met the shield that had been raised to repulse it, and the left hand plunged a spurting blow from above.

Three.

Swords bristled like hackles as he jerked the blade free. He looked at them beneath the shelter up his upraised arm, his eyes flashing with cold fire.

"Adjuvāminī."

They attacked, but suddenly the vagrant was doing far more than defending himself – he was fighting back. Nothing to lose, he was a rabid wildfire, tearing, clawing, biting with a strip of steel and an uncontrolled rage. He fought like a demon, anticipating their blows, and they, like a starving wolf pack, were closing slowly in.

They had him surrounded. Without turning, without thinking, Rahyoke flipped the sword in his hand, stabbing back behind him productively. But as he tried to yank the blade free, one of his assailants saw his chance. It was a wide, clumsily-made blow, but it found its target nonetheless.

"Auuhh!!"

The shimmering blade reddened slickly as it slashed across his stomach.

Rahyoke doubled over, one arm around his waist, his sword gripped loosely in his hand. He felt his own blood running along his fingers. He was totally at their mercy, and yet no blade lifted for the final blow.

He looked up, stunned with pain, his dark hair trailing into his narrowed eyes. They were... waiting.

He just managed to get his sword up, his weak, half-hearted slash quickly battered into the ground. He tried to straighten up, and failed. "*God –*" He staggered to the right, slinging the sword up into both hands. His breathing was heaving now, the taste of iron in his mouth. He gritted his teeth, willing his body to fight. With a downward hurl, he lurched his force into a heavy stroke, beating his opponents' blades to the ground and digging his into the soil.

"Unnh..."

He lifted his blade, shaking.

C'mon...

He swung back, rolling slightly to the side, nearly tipping as he avoided a swipe to his ribs.

C'mon!!

Teeth bared, he flung his force forward, his blade connecting with his attacker's, and locking. The two men held in a contest of strength and will, separated by that slitting edge. Rahyoke pushed forward, forcing his opponent's sword back, bending the sharp equilibrium towards his enemy's throat.

Then.

A pommel raised. It was driven into the side of his head with one, sickening crack, and it was over.

The wanderer would roam no further.

And as the pale light of the sunset dyed the sky a bloody shade of red, from the safe shelter of the forest, she watched as her world was destroyed...

67
War

The opiate of the masses is force. They take, and they assume that we will not fight back. Generally, they have been right. But not necessarily so.

Do not remain hidden, but never surrender. Fight.

There is no point in running anymore.

Meygan shouldered the battered quiver, slipping the strap over her head so that it crossed diagonally over her chest. Ten shots were all the trusty quiver had to offer; ten discordant, rustic shafts. In one hand she took up the ancient, powerful weapon; in the other she held the string.

"*What* are you doing?"

She raised her gaze; Addison. "I'm following them," she replied. Bowing her head again, she hooked the string to one end of the bow, placing that end on the ground so she could string it. "They took them, so I think that they might still be alive, because why would you take dead guys? They took them in chariots. I've never seen a chariot in real life, but I bet they leave tracks, so it'll be easy to follow them."

"And what if they're dead?"

Meygan looked up, startled at the insensitivity of his tone. She searched Addison's expression, but compassion was not among the emotions she found there. "They deserve a proper burial," she said.

Addison shook his head. "What are you gonna do? You're gonna get yourself killed."

Meygan stared at him, appalled and confused. "What... so you're just going to abandon them? How can you do that – after all they did for us?!"

"Meygan, those two guys were probably criminals. From the sounds of things, Rahyoke was on the wrong side of the law; at the very least, we know he lies, he steals – he *kills*, apparently –"

"He saved our lives!" Briefly she dropped her gaze to the bow; she was not strong enough to string it. "Help me string this," she both commanded and begged, twisting a glare to Addison. "You're the one who's supposed to be a warrior; help me!"

"Meygan, listen to reason –"

She shook her head. Settling on the ground, she knelt one knee onto the wood of the bow, holding it steady between her legs so she could bend it. "You're so selfish!" She could feel the tears stinging her eyes. "I can't believe how selfish you are. You would rather starve to death in the woods than try and save the two guys who have helped us all this time! Why? Because the guys trying to kill them say they're criminals? This is literally life or death, Addison. Maybe you didn't like Rahyoke and Zander, but they were there for us – they saved us – that should count for something!" At last the gut string hooked into place. "Auh!" She let go, her hands shaking from the effort.

Addison moved again to dissuade her, but she severed.

"Think of it this way, Addison," she bid him, getting to her feet. "What are we here for? Are we here only for ourselves? Or are we here to *do* something with our lives? I want my life to count." She lifted the bow.

"Meygan, this is a *horrible* idea," Addison pleaded, a slight waver to his voice.

"Then don't come with me," she responded, a sudden chill lacerating her tone. "You were never there for me before; why should I expect you to be now?"

Her words were so cold, they stung him to the heart.

Meygan slung her backpack onto her right shoulder. It would be harder to shoot with it on, but she wasn't about to leave it. Addison would give it up in a heartbeat.

"Don't wait for me," she finished, cutting off whatever statement he was about to make. "And don't follow me," she added, walking away.

"Meygan – ah –" Addison stood in the clearing. Maybe *she* was willing to die for a couple of ragged, reclusive thieves. But *he* wasn't.

He simply watched her go.

They had departed just as the last light of the late summer sun sank into the parched grass of that open field. And, true to his namesake, Sinon Terfel had been left behind. On one level, sure, the guys had found the irony amusing. But in reality, they had not yet gotten what they wanted.

Sinon combed the grass and roses like a bloodhound, searching every inch of ground. He was lightly armed, a standard-issue *gladius* belted at his waist, though he doubted he'd have to use it. In fact, he doubted he would even find what he was searching for; he did not put much weight on the Cunīculus's rushed assurances. At least they wouldn't be going back empty-handed – and the prize they had captured was grand.

The sun was beginning to rise; they'd be back soon, and without him. Damn it; and they'd probably take all the credit, too! Figures; you're finally assigned a *real* mission, and you're left for reconnaissance and all the guys go back and say you were sleeping it off at a brothel or something – which is only any fun if it's true –

Thwit!

Sinon jolted as the dirt leapt up beside his foot, a sting slashed his calf, and a solid *thunk* sickened the air. He drew his sword as he turned, narrowing his eyes on the trees. "Show yourself!" he commanded, in the modern language of the Empire. "By order of Augustinium, cease fire and come into plain sight!"

An arrow jutted from a sap-bleeding root beside him; the edge of the feather had grazed his leg.

"Tell me what you did with them!" shouted the forest, with the voice of a little girl; she spoke with the clear, unaccented Augustan of a town-born citizen.

Point for Cunīculus. Sinon straightened, settling on his heels. "So it's a misunderstanding, then." He sheathed his sword. "You're Meygan, am I right? Come out, and I'll explain everything. I mean you no harm."

Concealed in the protective hold of the forest, Meygan narrowed her eyes. *How do you know my name?* If anything else before had been weird, this sure beat it all. "Yeah right."

He would not have been surprised if she'd drawn another arrow. "Still don't trust me, huh? You're a smart girl, Meygan. Alright, then." Let her hide; he would explain things as they stood. "We don't want any trouble. We're just here to get something of ours back."

"That doesn't interest me."

Sinon repressed a smirk. She sounded young, but there was fire in this one. *Very well.* "Meygan, let's negotiate. You tell me what you want, and then I'll tell you what I need. We're civilized people, are we not?"

Meygan chewed her tongue, biting back the tide of recalcitrant responses that leapt readily in her young and feisty mind. Silently, she drew another arrow from the bag.

"I want the two men your men took."

Huh... Sinon pursed his lips, considering. "What... could a nice, young girl like you want with two traitorous fugitives?"

"Nice try."

"Sweetheart, if only you knew what you were defending... I'll show you where they are, Meygan. Come back with me," he prompted. "You won't have to worry about your safety if you're with me. I can get you through the patrols and customs, and I can protect you from the marauders who roam abroad; and it'll be faster than going on foot."

"The only way I'm coming with you," she said, before he could continue, "is if you give me that sword... And the scabbard thingy."

Sinon smirked, bowing his gaze. Chuckling mutedly, he shook his head, unbelting the sword. "This?"

"Yeah."

He wrapped the belt around the scabbard, tossing it forward. She watched, noticing the side it was strapped to, the arm that he threw it with, and how well he could throw.

"Okay," she said.

"You'll come with me?"

"I promise," she vowed. "But one more thing..."

Sinon looked up. "What?"

She fired.

Master

# of Flames

68
Fire-Torture

As far north and west as you can go without falling into the ocean is the castle of Ariegn. It stands, clinging, at the top of the highest cliffs, as the tumultuous sea rolls and roars below. Above ground, Ariegn was both a palace and a military base. But beneath Ariegn was a vast maze of holding cells, a notorious prison and torture chamber, akin to the corridors of Hell but darker – much darker, by far. The deeper one traveled, the more barbaric the accommodations; at its very depths, the cells were hewn from the cold, black stone itself. Only once in the prison's history had every cell been filled; now that great labyrinth was home only to one.

And it was the task of Judah Basilicus to retrieve him.

Normally, he would consider such a chore to be beneath him; he was, after all, second-in-command to the most powerful Empire in the world. But this prisoner was different. He wouldn't trust the job to anyone but himself.

Judah stood at the foot of the stairs, sheltered by the doorway. His torch cast a flickering orange light about the room, exaggerating every shadow. The silence was profound here, in spite of the echoing stone walls, in spite of the room's close proximity to the world above. It was the kind of silence that carried the absoluteness of eternity, powerful, revered, and frightening. He stepped towards the cell, focusing his attention on finding the right key in the dim light.

"If you value your God-forsaken life, before you put that key in the lock, you will *tell* me what you've done with him."

He froze, his eyes locked resolutely on the place from which that Hellish soft voice had sprung. Deep in the farthest shadows of an isolated corner, where no light touched. Of course. "So at least I won't have to drag you."

A pair of eyes flashed open, narrowed, catching the amber light. No words were spoken; no words were needed.

A trademark sneer curled the soldier's lip; leaning forward, he took hold of the bars. "You wanna know what we did with your boy?" he asked, malign pleasure scoring though his tone. "Fine. He's dead."

Nothing in the world was more sharp or mordant than those insensitive, casually-spoken words. It felt like a thousand knives had been driven straight through him – through his stomach and throat, mind and heart. The eyes closed.

A black shadow slid out of the reach of the light, and Judah grinned with triumph, hoisting his torch aloft to send its glow into the gloaming. "A good job they did of it, too," he continued. "One straight shot to the chest, clean through the heart. Now *that's* marksmanship."

"What..." Barely more than a breath. There was no venom left in that haughty voice now; all that remained was faint, hollow, and faded; perplexed and uncertain. *Pain.* "What... did you do... with... with – the body..."

Judah gave a derisive snort, stepping back from the bars to swat the question away with a dismissive flick of his hand. "What else do you do with a useless object? We gave him a proper fisherman's burial. Good enough for your people, isn't it?"

"You... Ya couldn't'uv..."

Guilt twisted those winding daggers, seizing and shredding the heart too tattered to take it. All from a truth too horrible to be believed, a death too fresh to be comprehended, a pain too old to be ignored.

The moan of a dying soul in agony grated through the silence.

Judah wedged his torch into a sconce; he didn't need it. "A touching show of emotion, fire-eater," he commented, mildly. "I didn't know you had it in you. *Bravo.*"

Fire-eater... Shorn, he hadn't been called *that* name in a while...

Clearing his throat, Rahyoke the Fire-Eater leaned his head back against the wall, studying the guard in the torchlight, unseen. A bitter smile flitted across his thin lips, supercilious, indifferent. "So you remember my occupation but not my name? Interesting..."

"All you're good for, fire-maggot."

"And here you went to all that trouble to capture me, Bassus... Or should I say, *re*-capture me?"

Judah – "Bassus" – ignored him, sliding the key into the lock. "Get it in while you can, torch-bait."

"Your men still travel in chariots. I never really took you for a traditionalist, Bassus, but I see I stand corrected."

That got him; for a moment, Bassus was surprised. Then he shook his head. "I told them they should've hit you harder."

"Probably," Rahyoke agreed, though he didn't see how they could've without killing him.

The key clicked in the lock.

He was on his feet before he could think better of it, and immediately he stumbled back into the wall, dropping heavily against it. Pain like torrid fangs ripped through his bones, springing from myriad fountains and gushing along his veins.

"Damn it," snarled Bassus, "would you be careful?!"

Rahyoke choked, wrapping his arm around his stomach as he fought against the blackness slithering across his eyes. It was then that he remembered the swart contusion spilling over his side, the fractured ribs searing with a thousand white needles, the gaping, red sword-wound athwart his abdomen. "I'm bleeding," he noted, studying his hand; a fine trickle of blood ran across it, seeped from the open wound. His raised gaze was feral. "Not shy on wasting your precious commodity, *are* you?"

Bassus pursed his lips. "A little bit more enthusiastic than necessary, I suppose..." He shrugged one shoulder. "No matter. He will be dealt with accordingly."

The words were as cold as the hostile night; the fire-eater felt a shiver of foreboding cross his spine.

"Don't be fooled, fire-worm," Bassus went on. "That was a skilled strike, much as the one that killed your boy. I sent only my best men to capture you. Of course, now thanks to you, three of them are dead."

"I'm sorry." It was the only response in his wide repertoire that could have struck so deeply, though he meant it with some sincerity. Some. Supported by the sand-stone wall, he made a noise akin to contumelious laughter. "Would've made less'uva mess if you'd just run me through."

Slipping the keys into his pocket, Bassus drew his knife from his belt, examining its shining beauty in the light of the ensconced torch. "Believe me," he muttered, "I'm tempted." Taking a step back, he swung the door wide. "And," he added, brightly, "should you try anything, I will! Now, come on."

Everything within him urged Rahyoke to resist. But then he heaved a sigh. "What reason have I to struggle?" he realized, his voice faint. "After all, thanks to you and your fine company, the last of my children is dead."

He carefully eased himself off of the wall, moving forward at a deliberately slow pace. The moment he was within range, Bassus seized him by the arm, dragging him out of the cell. The door slammed closed, and the knife was pressed to his back.

"Why did you take my shoes?" the fire-eater asked, bewilderment in his tone. "My knife, yes. Sword, naturally. But my *shoes?*" He shook his head. "That's just mean." He knew that they had been looted; he just didn't think it was fair.

Bassus swore vulgarly. Rahyoke stepped forward before the knife-slinger could vent his anger.

He was lead at knife-point down vacant halls, through winding corridors as familiar as a forgotten dream. He had never known Bassus to be so silent; usually his malicious sense of amusement would concoct all kinds of detestable commentary detailing the graphic end to the stay of his unfortunate captive. But this night, as he drove his enemy at the point of his blade, Bassus uttered not a single word.

At last they came to a small, pine door, standing open like a gaping bird's mouth. Realizing where they were headed, Rahyoke stopped short, and was rewarded with a painful jab to the back.

"What do you *want* with me?" he demanded in annoyance, refusing to go any farther. He tried to turn but was stopped by the sharp edge of Bassus's blade at his neck. Just as well; better a knife to the back than the heart. "I'm of no use to you! Why don't you just kill me, too?"

"Not yet, fire-eater," responded the soldier, in a sickeningly soft voice. "We need you alive."

"*Need* me," Rahyoke snarled, making a face. "And what, *pray-tell,* makes you imagine I'll cooperate? You've nothin' on me – not now."

Bleak contentment carved the soldier's lips. "I'm not so sure." And with that, he pushed Rahyoke into the room.

Of all the rooms in Ariegn's castle, above or below ground, none was more famed, more beautiful, or more lamented than the Great Hall. Easily as big as a small church's sanctuary, the room had once held feasts boasting four- to five-hundred guests. The ceiling was arched like a grand cathedral's, disappearing high into the shadows, its beams of stone leaping and twining like

Celtic designs. At one end, towards the south, two oaken doors joined together in the shape of a great arch, inscribed with elaborate patterns. Far at the other end, the floor was raised in a stage-like stone dais, upon which the long-vacant thrones of the king and queen still stood. About a decade ago, if one had stood on the dais, the wall to the right would have been open; the solid layers of stone would have started perhaps twenty feet above the ground, supported by arched columns every ten feet. Beyond was the massive courtyard, paved with stone; guests had been able to stream in and out, dancing in the courtyard, feasting in the hall, with freedom and ease. The wall to the left was solid, a formidable structure, once hung with the most beautiful tapestries in the world. Now...

"*Mercy,*" Rahyoke breathed.

The space between those open columns had been covered by a great wall of stone, laid in solid brickwork, so that only a gap of a finger's breadth remained at the very top. The three long tables that had snaked in a horse-shoe around the room had been removed, replaced by smaller ones akin to oversized picnic-tables. The walls were bare, but for a few empty sconces at the doors. All was dark, naked stone, clad only in the shadows of the heavy night.

"Familiar territory, ey fire-blower?" Bassus gave Rahyoke a shove. "You recognize your lovely palace?"

"The wall's new."

"Like it?" Bassus glanced emotionlessly at the new fortification. "Built that beauty after you left us. No surprise."

"I thought... you heathens preferred your victims treated in the old monks' quarters..."

Bassus shrugged, *"inadvertently"* digging his knife's tip between the fire-eater's shoulder-blades. "Oh, we still use that old chapel for executions, not to worry. But you're *far* too remarkable a find to be left to any old killing, aren't you?"

"Hhm. Super."

They stood on the dais, ahead of the two ancient thrones. The only light in the room rose like pale ghosts from the torches of two guards. Decked out in ceremonial armor, they stood five feet in front of the dais, near a stone brazier filled with driftwood. Standing between them, bathed in the red torchlight, was a woman. She was medium height, and a few years older than Rahyoke; her hair was black, relatively straight, and reached just past her shoulders; her eyes were an intense greyish green, like the sea before a storm, and their mien was just as harsh. Rahyoke narrowed his eyes in turn.

"Don't even *try* pretending that you don't know one another." Though he spoke to Rahyoke, Bassus' message was meant for them both. "It would save us a lot of time and trouble if you just comply, and I have plenty of other things I'd rather be doing right now!"

I'll bet you do, Rahyoke thought, incensed. "She had no part in this! Why bring her in now? She doesn't know what I am – let her go!"

Bassus allowed a quick and pompous laugh. "Easy, matchstick-eater! Though a wise choice, I must say. You should realize why she's here! After all, aren't you feeling a little more inclined to cooperate, for her sake?"

Rahyoke exhaled slowly, his gaze sinking to the dusty floor. *"Sea, is mise a dúirt é."* Though there was something in the timing of all this that seemed off... "So it seems you have an innate talent for throwing my words back at me. Skillful oration. Explain to me, then, with all your words: are you proud? You can capture and kill both women and children. Am I meant to cower in fear?" He gasped slightly as he was seized by the shoulder, the edge of the knife finding his throat. Oh – *yes;* that was all Bassus wanted: to watch the wretched fire-worm bleed! "Threaten me all you want, Judah," he warned, his voice malevolently dropped. "You dare harm her, and my cooperation will be the least of your concerns."

A silence fell, sharp and cold as the blade pressed to his neck. It didn't always take a knife to win an argument. Raising his head as much as he dared, he returned his attention to the woman, calling over, "It's been a long time, Michelle. How's life?"

The woman leveled him a stern glare. "Child, it must be some mess you've gotten yourself into."

Ill-humoured and spiteful were his smirk and the slight laugh that bolstered it. "Woman, you've *no* idea." Before she could ask anything, he continued, "I've missed you. They haven't hurt you, have they? You can tell me. 'Cause I swear t'God –"

"They haven't yet, no. Mind telling me what's going on?"

"A little. Would ya happen t'know the time, love? I got conked on the head and seem to've lost track..."

"Evening."

"Really! And on what day, pray-tell?"

It was Bassus who answered him, impatiently cutting off their conversation. "You've been here since yesterday, fire-eater," he snapped.

"Have I really? Well, now..." Though he did not turn, he knew his mockingly indifferent tone caused Bassus to suffer great irritation; his fingers had twitched on the dagger's hilt.

Before another word could be spoken, a heavy pounding sounded on the other side of the oaken doors. Out of the corner of his eye, Rahyoke watched as one of the guards moved to answer the call, marching away down the long center aisle. If he could only see –

A cruel voice whispered in his ear. "If you make any move, fire-eater, my knife is through your back and into your lungs. And *this* time, I will be justly praised for it."

Rahyoke clenched his teeth. These words could only mean one thing... And if it was true, he knew that little could stop him from chancing the speed of that blade.

Bassus had to have known this; Rahyoke was startled as he was pushed from the dais so abruptly that he nearly fell. He only just managed to catch himself, stumbling into the side of the stone brazier and grabbing the rim for support. He was just turning to snarl at Bassus when the slamming of the door re-captured his attention.

A procession had entered, consisting of four guards, and a woman older and more evil than the Devil himself.

Once they had called her beautiful. Scarcely more than five feet tall and bird-bone thin, she was small, but carried herself with regal dignity. Her features were delicately shaped, and her eyes were an icy blue, as pale as the light of the moon and as frigid as the frost on the winter plains. In her youth her hair had been the colour of sunlight, so blond it was almost white, and had only lightened with age; tonight the straight, silver-white strands were pulled back into a sheer bun, to match the severity of her light, black gown. Like a phantom of limbo, she had no name; those who served her called her Regina, Imperātrix – empress and master. Those many who despised her had other names for their ruler – many, many names. Rahyoke himself called her Murrain, for to him she was no woman, but a plague – a stinking shadow of black death creeping across the land.

Murrain, Ruler of Augustinium, Queen of Ariegn.

Bile like snakes' venom welled upon Rahyoke's tongue, and for a heartbeat it seemed as though he would attack her then and there. But he forced the venom down, into a voice as brittle as the voidish darkness around them, into words.

"By the blood of my line, I swear that you will regret this."

The pale eyes narrowed with disgust. "So. Duncan O'Rielly makes his triumphant return to Ariegn, and that is all he has to offer?"

A curious expression passed over Rahyoke's face then, and his eyes widened in near surprise. Twelve long years had passed since he had last answered to that name, yet he would reclaim it in a heartbeat. "The usual formalities, then?" Deliberately, he gave a sweeping bow, right foot slid back, right arm half out-flung; the challenge was in the gesture, in his tone, in his eyes as he straightened and cast that fearsome gaze around the room. For – true as the blood that coursed through his veins – *this* was his real name. "*Dia anseo isteach*! I thank ya for the hospitality ye've shown me in bringin' me to Ariegn. After all, it's *so* much better than last time," he added, rubbing his left wrist with a meaningful glare.

The old woman's mouth tightened, and she half turned away, flicking a dismissive hand at Bassus as if to say, "*You* handle him!"

Bassus was only too happy to oblige. Roughly, he shoved his prisoner in the back, instantly commanding the attention of the room; then he moved away, repositioning himself so as to be able to look at the fire-eater head-on, a judge before the criminal. "Do you know why we brought you here, fire-raiser?"

Duncan rolled his eyes. "I couldn't imagine..."

"You're here to finish what you started twelve years ago."

"Oh. And here I was thinking the only reason I wasn't completely run through was for lack of skill –"

"Our precious commodity."

That got his attention.

"Hm. Guess I threw the right words. You stole that book, fire-eater, which is cause enough to believe you figured out what it is we wanted. And either you give us what we ask of you, or you will die trying."

Duncan's eyes rounded, and he paled. He did not care if he died; he just did not want to die in that way. "I can't do what you're after," he breathed.

"Give it a good try. For your friend's sake," Bassus returned. "All good things come in time."

For a moment, it seemed like he would choke on the absurdity of the bargain. Then he relented. "*Dhia.*" He stretched out his left arm and began rolling up his sleeve. "You'll make her watch, will you? She's not like me, you know – she has a family, and they'll miss her. You need a reason to detain her, and – shouldn't you keep this a secret? 'Cause if –"

"*Save* it, flare-fingers," Bassus replied. "By aiding known insurgents, she has made her choice. She knows that if she wants to protect her family, she will stand here quietly throughout the demonstration you will generously provide."

"Unless," hissed their ruler, "you are willing to make a trade?"

Duncan said nothing. He rolled up the sleeve of his black coat with the slow, methodic motions of a moving meditation. Then, pushing up the long sleeve of his t-shirt, he revealed a strip of beige cloth that wrapped all the way from his wrist to his elbow. He dropped his hand to the basin again, watching his own fingers as they slid through the soot on the edge.

"The Fire-Eater's Chronicle is no longer in existence."

All eyes were on him; he, uncaring, dragged his finger through the carbon matter dust. "I should know," he added, pointedly, "seeing as I'm the one who destroyed it."

At first, a stunned silence was the only response his words received. Then the calm shattered in outrage.

"He's lying!" accused one of the men; Duncan recognized him, but did not know him by name. "That's all his breed knows as a defense."

"As long as we have *him* –"

"There's still what he stole. With that, we wouldn't need him *or* the chronicle. He's more a nuisance than a use –"

"Destroyed that, too," Duncan uttered boredly, stirring them anew.

"Lies!"

"The little pyrophile was keen enough to risk stealing it; no *way* he'd do all that just to destroy it!"

"It means as much to him, too, since –"

"That's right!"

"I don't buy a word of it," Bassus spat in agreement. He gestured to Duncan with his knife, thumbing the blade wistfully. "This runagate knows *exactly* where they're hidden."

Duncan was unfazed. "Search all you want, it makes no difference to me. I'm just letting you know, you won't find it."

All the fury they had shown before was nothing compared to that which filled the following silence. What made them angrier, Duncan wondered. The sheer audacity he was showing towards them? Or the apparent truth of his words?

When at last someone found the tongue to answer him, it was Murrain who spoke, frost cutting into every word. "Well, *fire-eater*. Then it seems we will be needing your talents after all... Unfortunately, thanks to you we have no more books to burn."

His thin fingers curled on the basin's edge, digging into the soot. "I was beginning to wonder if you'd ever join in." Holding out his other arm, he began rolling up that sleeve in a like manner. "Aiding known insurgents, huh? Regardless, Michelle, I'd thank you t'keep out of this."

"Too late," the woman muttered, but said no more. Though there was much that she did not know about Duncan, she knew the Empire all too well.

Murrain made a sign to Bassus, who grabbed the fire-eater's outstretched wrist. Duncan shot him a dirty look, but did not resist; he simply pushed his sleeve up the rest of the way. Wrapped around his wrist and halfway up his forearm was a strip of faded black cloth, tied tightly against his skin. Bassus forced him to hold his arm straight out, bending back his hand so that his wrist was exposed. Holding the flat of his blade against the fire-eater's veins, he slipped it under the cloth. When sawing the binding failed, he twisted the blade, breaking the band with a snap and catching it in his outstretched hand.

Duncan's fingers curled into a loose fist. He bowed his head, studying what the knife-wielder had done as Bassus cut the knot securing the other cloth and unwound it from Duncan's arm.

"What are you looking at?" Bassus demanded, seizing Duncan by the arm and yanking him closer to the torchlight.

The dim light played across the man's bare arm, illuminating a gruesome sight. The skin that had been covered by the band was as pale as a dead man's and as warped as if it had been shredded and pieced back together. The surface appeared both withered and raw; it was the mark of countless burns, healed and re-healed until it could not heal properly anymore. Fresh gashes crossed the damaged flesh, spilling blood into the dried channels of older wounds.

Duncan looked at the knife-thrower with an expression of chilling indifference. "Admiring your handiwork?" He rubbed his other arm; as bad as his scarred and bleeding right arm looked, it was not nearly as damaged as his left.

"The fewer times I have to dirty my blade on your blood the better." He cleaned his knife on the tattered rags; they were useless for anything else now. "Go on."

Duncan gritted his teeth. He dug in his pockets and was somewhat surprised to find his matchbox was still in his possession. He drew a match. *"Amaideach,"* he muttered. With a defiant air, he struck the match against the side of the basin and cast it into the bowl with the same dispassionate grandiloquence for which he had so often scolded Zander. The match's feeble glow found the surface of the drywood, igniting it so that fire leapt up over the basin's edge, casing a ring of light around the room.

He looked down on the old crone as she stepped forward, a black shadow marring the scarlet gold. *So it's yourself come to play, is it?* Without hesitation, he stretched out his right hand, plunging it into the flames.

Fire licked up between his fingers, cackling at the rattled gasps from around the room, but he did not waver.

For a moment, Murrain looked somewhat surprised. Then, reaching between the flames, she quickly plucked his hand towards her, giving his scarred forearm to the blaze. A drop of blood fell from his wrist, startling the fire and causing it to flare. A cruel smile twisted the old woman's lips, and this time when she looked up at him there was no sign of the disgust that had been in her expression before.

"Both wrists, if you please."

Duncan's muscles tightened, and for a moment he hesitated, stubbornness freezing in his hot blood. Slowly he raised his hand, his eyes flickering over his disfigured left arm as he contemplated tempting Scylla or Charybdis. Then, bracing himself, he reached through the fire.

Murrain took hold of his wrists, holding his veins to the flames she so carefully kept her own hands clear of.

"Do it."

The fire-eater bit his lip. *No,* everything within him screamed. *No, don't!* They wanted the power, and to give that over to them – even for the briefest of instances, was – abhorrent – betrayal – profane. Nothing good could possibly come from this endeavor; the only guarantee in this was pain. Pain... *Nothing good came from resisting, either, did it, Duncan*? And still, in blood-soaked memories, that defiant voice whispered...

The fire between them gave an impatient snap.

"Do it *NOW*, fire-wretch!"

Duncan bowed his head. Closing his eyes, he drew a deep breath. "*Maith dom é le do thoil*," he breathed.

Undertoned, a barely-audible voice; he began speaking so rapidly that no one could understand what he was saying. A torrent of words flooded together, meshing into a single churning, lilting, grating breath which mimicked the singing of a flame. His words became like fuel, feeding it, coaxing it, demolishing his last shields. The fire bit him, sinking its needle-like teeth into his flesh. Hot tongues seeped into his open wounds, drawing out his blood. The fire swelled with unsatable lust; the more it drank, the more it craved. The blood coursed with a thousand untold emotions, and the fire fed off them all.

And there was nothing that the fire-eater could do about it.

He felt the fire surge, billowing in great gusts like a sail in a violent storm, its many flickering tongues snapping at the air above his head. The dust of embers brushed against his cheek, singeing him viciously. Barely contained, it threatened to overflow the slim stone edges of the bowl. *This is what you want me for?* he wondered, as instinct fueled his commands. *To set the whole world ablaze?*

The fire gave a hiss, and Duncan felt the terrible heat of its crooked fingers running through his hair. The flow of words on which the fire fed was stifled as quickly as it had begun. Head still bowed, his eyes flicked open.

The fire shone like a tawny prism, sparkling like glass, rimmed with cochineal, and as unnervingly still as if God held time suspended.

Maybe... this time... it could work.

Suddenly, Duncan heaved a hard gasp, his body shuddering and shaking as if each pulse of his heart upset his system. Each breath he drew dragged through his clenched teeth, ragged and rapid and wheezing. One sharp convulsion wracked him, and Bassus was forced to support him to keep him from collapsing. Fever-heat burned through his pale and clammy skin as all of his blood was drained – dying the fire red.

When the fire-eater started choking, asphyxiating on agony, Bassus addressed his master. "Domina –"

"You're holding back," the plague sibilated.

"Ih –" It was so hard to speak. "Ih – *in* – t – *ter* – *cē* – *dit... ea*," he gasped, piecing Lavinian like an ignorant Ceothann. But his words were more interesting than his grammar.

Intercēdit ea. She intervenes.

The two Augustans who held him there exchanged a glance. Bassus moved to ask what he meant, but the empress commanded, "Show her to me."

He closed his eyes. He didn't want to. After so many years of yearning, craving – yet he had never wanted it to be like this. She never would have wanted it to be like this.

The sanguineous cracks of the fire receded, healing, 'til the flames were of the most cheerful orange-tinted transparency. Gradually, O'Rielly's body calmed as well, the chaotic trembling settling to a steady shiver as the fire withdrew from his veins. He kept his eyes closed, his teeth gritting into a pained and angry scowl as he heard that satisfied snake's drawl: "You won't be able to escape from it that easily, matchstick-eater."

The fire-dancer felt as if all the breath and life had been sucked from his chest, stopping his heart cold. He had waited so long. And yet... Slowly, slowly, his eyes opened – and no matter how hard he tried, he could not hold on to his indifference any longer.

"Mei Dominus..."

There she was. In the depths of the flames, the ghostly figure of a woman had appeared. Her long hair fell about her shoulders in loose, dark-coloured waves; even now he recalled the feel of them running smoothly through his fingers, flower-scented and soft. Her head was bowed deeply, her face turned from everyone, turned from him, shielded by her hair as it always was in those haunting dreams. Her right hand was pressed to the ground where she sat, legs half-folded beneath her as if she'd been kneeling; her hand was a fist, pale white.

She was even more beautiful than he remembered.

"And here you doubted yourself."

Duncan's fingers clenched tightly. Beyond the flames, beyond his light, in a world far more caustic and tenebrous than the one he longed to hold on to, that hated creature spoke. He closed his eyes tightly, opening them to slits.

"You've missed her, haven't you." The devil's tongue; the claws tightened sharply on his wrists.

Of course. Gently – so as not to incur maltreatment – he straightened, standing under his own support. Of course he had. Every heartbeat of every day and every breath he'd drawn had ached a painful reminder; *of course* he missed her.

"So that's what distracts you in your work? Your loneliness?"

Duncan ground his teeth, wringing blood. He could almost feel the soft brush of her fingers as the spirit relaxed her hand – almost, but never again. She was nothing more than a vision – the memory of a dream – projected by longing. In reality – stark reality – she –

"Or is it merely the fact that it's your fault she is dead?"

His eyes dropped, widened, to stare through the fluttering flames. Something about hearing the words aloud made them so much more potent.

"So that's why you distain to indulge your talents, even though you have at last discovered what it is you are capable of, your true potential. All these years, hiding, living a worthless existence. Groveling and scrounging, too afraid to give your true name. Your guilt binds you, as love once did. Guilt brought by the knowledge that you failed her. She saw you, watching, as she was led away, as her children were torn from her, and she called your name, and you did nothing. You let her die."

It wasn't true. As stoically as he had before, he tried to hold her gaze, but a sliver had rived the surface. They could say nothing – they knew, *nothing*. Not his soul, not her thoughts – not how he'd been living these past several years...

"You failed her. And yet you comport yourself a paragon. The blameless victim. As if your hands are so clean. Tell me: what *really* holds you back, Duncan? Is it your fear? The fear that you *are* the monster that the old laws made you out to be? The fear of your inadequacy? The fear that stems from the knowledge that all your morals are empty boasts, and it is not virtue that constrains you. It is cowardice. Your selfish cowardice brought about her death."

Slowly, like vipers, the words tore into his soul. He had forgotten that just enough of his heart remained intact to feel the horrible sting of pain. Involuntarily, he leaned against the side of the basin, lifting his gaze to the fire-spirit once more. Twelve years' worth of unanswered questions – of rived hopes and shattered dreams, of unavailing searching and endless, *endless* longing – twelve years' worth of prayers bled out by that shredded heart – twelve long and wretched years poured out everything once suppressed into his imploring gaze. If anyone could answer him and bring his silent suffering to an end, it was she – the mute ghost illuminated by flame. *Please...*

"If only, Duncan. If only, if not for that fear. You could have saved them, and you know it. You could have saved your son. And yet –"

"Leave him alone!"

Michelle's voice cut through the haze, severing the dark tendrils that wrapped around his mind. Duncan's eyes widened, and, for a moment, the image in the fire flickered, panicking him. But she remained, thank God; and that flagitious voice spoke again.

"I know it seems cruel." Tenderly the devil spoke, consoling the aching heart. "But it is important that you realize, this fight is futile. She is dead. And dead she will remain. Let go of your trepidation. Let go of your guilt, your love. These things confine you. You need to move on.

"We can offer you *life*, Duncan. More than sustenance – dignity, occupation. Salvage your respect; stop resisting. You *know* that it's right...

"Or do you still think it's worth it to risk your life cavorting after a departed dream?"

Duncan closed his eyes, bowing his head. She had been dead for twelve years now. Twelve years was a long time – a *very* long time. And every word that had been spoken to him – *nearly* every word – stung with a ringing truth. No matter all his extravagant powers, he could not resurrect the dead. His wife, his son, his daughter – they were all dead. And they were dead, because... because –

The blaze of his fury rekindled in his bloodied veins, and he raised his gaze, tilting back his head to look on that conjured vision – that image – that lie.

But sixteen years was longer.

She turned.

Duncan's heart stopped. Ivory, her soft skin, regal and thin her features, black and long her lashes, shielding those eyes – bright and pale a green, flecked with copper, rimmed with iron – eyes that pierced to the depths of his soul, hard as steel, brave as fire, sorrow traced in every gleaming ray. He remembered every detail, every line of her form, sharp as reality, and those eyes most of all. Those eyes, which ensnared him, luring him into her world, and no word, no cry, no fatal wound could ever tear him away from her. It never had.

Bowing her gaze – the heart lurched inside of him, pulling him closer – she reached out a hand to him – comforting him, welcoming him, beckoning him away. Her left hand; slender and lined it was, from years of hard working; glinting with silver bands. She was a siren – an erinyes – a white woman – a banshee; he longed for her more than anything else known to man, and yet he feared her just as desperately – her, and her irresistible summons to Hell.

As he watched, she twisted her hand palm-up, curling her fingers into a loose fist. Slowly, she pulled her hand back towards her, drawing him in. And she smiled, as daring as it was beautiful, her white teeth glinting against the shade.

The fire-eater understood. And, once more, she was gone.

"The choice is yours, Duncan," the vile demon averred, as the fire-eater looked away.

A dream... With iron hardness, he recovered his inveterate impassivity, the familiar vague frown returning to his lips. Again, he was as unreadable as a burned book; as supercilious as a king. There was a *reason* for all that suffering. A reason he lived in fear – a reason his soul was rent and torn – a reason why he must keep his expression dispassionate even as his heart ached and bled and his flesh was seared from its bones. A reason his sweet wraith had looked so sad. Duncan felt his fingers clench. It was nothing more than a dream, was it!?

The fire-eater raised his eyes to meet Murrain's, his gaze dark with a chilling animosity. A horribly vicious grin contorted his lips – and he turned his hands palms-up.

Yes, he understood.

White fingers clasped the old woman's wrists, seizing them tightly. And, with a motion so slow it was nearly imperceptible, he pulled back.

Realization drained the self-satisfaction from Murrain's face. Though she fought against any show of weakness, the fire's touch upon her skin sparked a light of panic in her eyes. She felt its hot-blooded heart wrap itself around her fingers, its many tongues dancing and coiling as if welcoming back an old friend. A heat like no other began to creep into the flames.

As soon as she let go he released her; the action was so instantaneous it was nigh impossible to tell who let go first.

Though Murrain quickly tore her fists from the fire's grasp, Duncan did not move. The blaze dwindled until there was only a small fire, big enough to fill the palm of his hand, dancing on the charred surface of the stone. Its smoke curled up around his arms, brushing against his newly-rawed skin. The ghost of a smile flickered at the corner of his mouth as he saw her cradling her scalded hands.

When the old woman looked back at him, she had regained her expression of proud disgust. "Well," she remarked, sharply, "it seems you can be of some use to us after all." With a quick nod she directed her guards to the door, the others following her train as obedient as dogs; no doubt they were eager to be out of the presence of the iniquitous fire-dancer.

"Leave the girl; it'll give them a chance to catch up. I'm sure our little fire-spitter has a lot of explaining to do." Reaching the door, she paused, casting a final glare back over her shoulder to take in all the damage she had done. Her eyes locked on Duncan, and, in a voice more pitiless than any she had used yet, she said, "Think carefully about where your loyalties lie, Duncan O'Rielly. You may have been a tremendous help to us this time, but be forewarned. Try again what you did last time, and you will find that our tolerance has reached its end. Remember: there are worse things that we can do to a man than take his family away."

And then she was gone, vanishing into the night as the echo of her words permeated the stones.

Duncan watched as the door slammed shut, the heavy board which barred it falling into place with a taunting crack. He listened for footsteps in the corridor outside, though he knew he would hear none. These doors were well made; they let nothing in – or out.

They were gone.

For a heartbeat longer, Duncan stood firm, his narrowed eyes locked on the door, bidding them every curse on their retreat. Languid, his attention raised to the ceiling, and he wondered if Heaven ever still watched over the once-blesséd castle of Ariegn.

He closed his eyes, faltering.

The fire in the brazier went out, vanishing in ash and ember. Michelle only just managed to catch him as he collapsed, hooking her arms under his from behind. Although he was slight, the force with which he fell made it hard for her to hold on; quickly she knelt, supporting him as best she could.

"Rahyoke?"

Moonlight flickered over his still form, highlighting the black of his clothes, the claret blood, the yellowing cerise of his burns, the deathly-pale whiteness of his skin. His pulse when she felt it was lurching, fevered hot.

"O'Rielly!"

His eyes opened, found her, and focused. Faintly, hoarsely, he demanded, "Are you alright?"

Michelle stared. "Yeah," she answered, shortly. "Are you okay?"

Duncan grimaced, his eyes falling to his fresh wounds. Gingerly, he lowered his arms, setting his elbows close enough to his shoulders that he would be able to push himself up without putting too much weight on his forearms. Then, carefully but quickly, he sat up before Michelle could stop him.

"Easy!" Michelle took hold of his elbow, halting him, as he got to his feet. "Sit down, will you? You need to rest!"

He gave the light note of a humorless laugh, pausing to glance back at her. "I? Shorn, Shelly, since when do I rest?" He drifted out of her grasp, but she caught him again, holding him there. He looked at her, but she said nothing, merely looked at him in turn. *Knowing,* as she always did; knowing and wondering. Knowing and wondering and being too damn kind and decent to have deserved any of this – it was his fault she had been torn from her family – as so many others before...

"I don't want to explain!" he blurted. He jerked his arm free, backing out of the light. "It isn't your burden to bear, Michelle. It isn't fair on you – none of this is fair on you. You should have just told Alan to man up and kill me like a stray dog the day I came begging at your door – a flea-ridden, stray dog with rabies and mange. You would've been better off... *Everyone* would have been better off." He looked away, forcing his hand back through his hair; the dark strands were clinging with blood and sweat. When had his hands last been clean?

Keeping that raised palm pressed to his forehead, he turned his eyes towards her. "But I owe you because of it," he said, in a voice so quiet that it lacked tone. "At the very least, I owe you an explanation, 'cause it's my fault... I'd understand of course if you didn't want to listen." He dropped his hand. "If you'd rather I just keep quiet and go away, I can do. You've every right. And it isn't your burden. It's not, and..."

"Rahyoke?"

There was that name again. "Uh?"

Michelle stood. She was so little – as the lass in the flames had been. But when she opened her hands, spreading her arms to either side of her, he felt like a forlorn child.

"I'm listening."

O'Rielly bit his lip. Long he was silent, torn sickeningly through and through. "Gimmie yer hands."

Michelle scrutinized him warily. There were a lot of questions that stood in the space between them now – strangely, even more than had the day they'd met.

She took his hands.

Duncan held her fingers delicately, feeling every one of those piercing questions. He still wore his black gloves; the cracked leather pressed against her palms. "...That little trick with the fire," he said, his voice hushed, and urging, as if begging to be believed. "I can do *so* much more than that...

"Watch. I will not move, I will not speak, I will not *breathe*..."

All the world was still, and time itself dared not to stir. He held her gaze steadily, his dark eyes filled with such intensity that they cast away the night, though his mind was deep in foreign lands. There were no words this time – no fast-paced outpouring of long-abandoned tongues. There was only energy, harsh, rigid, and powerful.

There was light.

Michelle turned. A fire burned brightly in the heart of the brazier. It fed on nothing, and yet its light seemed to fill the room, its radiant glow as welcoming as the hearth on a cold night. Its flames of golden amber were scarred with garnet, flashing and glittering like faery dust. They seemed almost to whisper, feeding on the stories of the air where wood and ash had failed them.

The fire-eater studied the flames impassively, his expression as unimpressed as it was sorrowful. He let go of her hands, dropping his left back to his side. For a moment he seemed entranced, his eyes fixed upon the golden waves. One sooty hand reached up to touch something that hung around his neck. And then the levies broke.

"If you'd have it," he whispered, "I will tell you everything. Anything you wish to know, I promise." He averted his gaze, and, with rare candor, he finished profoundly, "The truth."

Michelle looked on him – this tortured, broken man, damaged and bleeding and on the verge of tears, who could so carelessly bring light out of pure darkness.

"Okay."

69
The Fire-Dancer's Confession

High above the tossing waves, the black clouds twisted in their troubled sleep. A silver thread of moonlight drifted over the wall, dividing the abyss like a blade-stroke, flickering and flaring like beating moths' wings; like the ice-blue heart of a flame.

Michelle sat with her back against the brazier, savoring the warmth of its blaze. Across from her, his head bowed in wresting contemplation, sat Duncan. The fingers of his left hand were held loosely to his neck, hooked slightly, though she could not see what it was he held.

As if sensing this, he glanced up. His fingers clenched. "I've never explained this," he admitted, with an apologetic half-smile. "Half of it I only figured out after –" He cut off. Cleared his throat. "I'd better start at the beginning...

"Among those jobs listed under the Unsavory Occupations Act I have used to shape my résumé, there's one that, when they'd listed it, I'm sure they considered a mere nuisance. Not realizing until later what it was that occupation encompassed.

"I am a fire-eater," he urged, meeting her gaze. "And that means *so* much more than you think."

He unwound his fingers. In the soft light, Michelle noticed the silvery chain around his neck, from which hung a band of white gold. He hid the band again beneath his t-shirt, lowering his hand stiffly.

"There are... different *levels*, to what I can do. The most basic I could teach people. Torch juggling. Quenching. Blow outs. Y'know," he half-shrugged, "tricks. For that, we fell under the same rule that kept penny-whistlers off the streets... Kinda ironic, really.

"Other things... other things I don't think can be taught. Sustaining dragon's breath without fuel. Bending fire into shapes..." He nodded towards the brazier. "Spontaneous combustion... That's what you might call the second level. All it requires is a little communication with the flames. A little will.

Something like concentration. And there is *much* to be done within this level. Though most of it I haven't attempted in a long time...

"Which... I suppose... takes me to the third and – God save us – final level. That which you've so recently borne witness to... It's the only level that causes pain. It saps the strength from your very blood. *Feeding* on it. Devouring it... Certain things – certain things – like *that* – require sacrifice. Blood. The blood of a fire-eater." He spoke these last words with dignity and contempt, spitting them like poison from his tongue.

"Now... As to why they want me.

"I think the old bitch mentioned the books, right? – though Bassus certainly *eluded* to them enough." He glanced at her. "Remember the massive book-burnings? You might have been too young. The books were destroyed for the same reasons as the people: predominantly, for the knowledge they contained, the words they expressed, and what they could *do*." He bowed his head again, studying his fingertips in the flickering light. "There were books – *certain* books – that could perform on the aforementioned third level when burned. They were destroyed too, but – shorn! Just imagine what possibilities that could have opened up! The amount of power ordinary individuals could control. But by the time they realized it – through the grace of God, it was too late. A shelf's worth of books, maybe, remained. And all the fire-eaters were dead. That was an Act, too, you see – what was it called? Something... for the detainment and... of kinetics..." He shook his head. "I was violating laws purely by being alive. Regardless of what I'd chosen to do with that life, which –" He smiled, and there was something very dark, very bitter, and very, very proud in that smile. "That's another story.

"When they became aware of the power they were missing out on, they sought to gather it in any way possible. I was the last pyrokinetic. I *am*, the last pyrokinetic."

"They wanted you to replace the burned books and murdered fire-eaters."

He nodded. "Aye."

"It's not something someone else could learn?"

He shook his head, chewing at his thumb as if to cut a hang-nail. "It's inherited," he muttered around it. Then: "My parents were able for it, too. It's... something in the blood."

"What is it exactly that they need you for?" she asked. "What does the third level do?"

He hesitated. Blanched. "I'm not... Nobody told... I cah... I can't do it. I tried – believe me, I tried, but I can't do it! They say that I can, but I can't, and I always fail, and fail dramatically! You saw yourself, didn't you! I can't –"

"*What*, Duncan?" she intervened.

He stared at her. Lies were comfortable, lies were easy. So why was he putting himself through this torment of truth now? Indeed, the purposelessness of everything gleamed in finding him in the same place, the same position, that twelve years of tireless fighting had sought to help him evade... So secrets had proven just as pointless, hadn't they. "...But I know from a good source that the books can." He took a deep breath, and declared, "Transportation. That's my term for it. Transportation between space. The fire creates a rift in the space, *through* it, and – poof – yer object's sent, gone, and arrived. Simple as." After a beat, he laughed nervously, adding, "I know; sounds mad."

"All of this sounds mad," Michelle responded. "But I've just seen what you can do with my own eyes."

"Believe me, I won't press you to spread the word."

"May I ask another question?" When he nodded, she asked, "What's the 'Fire-Eater's Chronicle'?"

"Ah." He allowed a relieved smirk. "It's – really, all enigma aside, it's just my journal. For some reason they thought it'd contain some monumental store of vast worldly power and knowledge – that I'd write down all my secrets – of rebellion, but more importantly, of how I do stuff with fire." He shrugged. "*I* don't know how I do it; and if I did, what use in writing it down?"

"You said it's destroyed now," Michelle observed, "didn't you?"

"I did." A dragging lilt. And if she studied the look on his face for a thousand years, sure she could never read it.

So instead she prompted, "You've been here before."

"Ah." He nodded. "Ya wanna know how I got *their* book, then? It's long, so – *I* find it long."

But Michelle nodded. "Get it all out."

They say it helps. Duncan stifled a sigh, bowing his head. He shifted, sitting so his knees were pulled up almost to his chest, his arms folded as loosely as possible around them. And he said: "Sixteen years ago, I married the most beautiful woman in the world. We moved to a refugee town, outside the knowledge, and so outside the laws, of the Empire. We had two kids... an occasional third... And for four years, my life was perfect.

"Then one day... It was – over a decade ago. Very early in October. I was traveling, and – that was the sheer brilliance of it, really. I had always been in the habit of wandering off; no one would think to look for me. Not 'til it was too late, anyway... They took me. The Empire captured me. And Ariegn so terribly close... The journey took less than a night. And then...

"The first night was easy. They beat me until I could not stand. Over and over they made their demands, and... when I would not deliver what they wanted..." A shiver crossed his spine, and at once he flinched. In the heavy darkness, even the fire's nervous chatter was hushed.

He narrowed his eyes. "How can I put this simply?"

Placidly, he got to his knees, a weary warrior set to make his final prayers. His dark coat slid from his shoulders, slithering onto the cold floor to gather like a black spirit on the stone. Then, gritting his teeth, he pulled his torn shirt off over his head, trying not to wince as the soft fabric scraped over his abraded skin.

He turned, putting his back to the warm light for her to see.

"Oh my God..."

The firelight played across his skin like sunshine, highlighting a massive web of overlaid scars more brutal and grotesque than those on his arms. The marks etched across his entire back like ivy laced with thorns, some deep and thick, others pale white threads, still others whole patches clearly the result of burns. Nor were they confined only to his back; they spread in blooming tendrils up over his shoulders, down his arms, and no doubt onto what remained covered, what the light did not touch.

"*This* is what they did to me," he said. "I am the longest-staying occupant of that hole to rival Tartarus, the deepest chambers of this prison. I am the only soul to've survived the thousand and one tortures of Ariegn's depths. I am the only prisoner to've escaped from Ariegn."

He hiked his coat onto his shoulders as he turned back to Michelle, his eyes scanning the flagstones as if some answer could be found in the red-tinged dust. "I endured four months' torment," he continued. "That was nothing; I see that now. Because there is always something greater, always something more that can be taken away. And *that* is the problem with resilience... They will *always* find a way to break you...

"Always.

"By the end of January they'd run out of things to throw at me – they can do more than one torture at once; they're very resourceful like that. And, at the end of it..." His eyes raised to the ceiling, his expression one of scorn, disbelief, agony and accusation – his tone slit with ironic laughter as humorless as Hell as he spat, "I thought I'd be released. God! That I'd know *so much* of the world, and still be so utterly *stupid!* And sure – it was that ignorance that had spared me. That ignorance that would ruin me in the end... Someone had betrayed us," he said, "they *had* to've. Someone had given over Balladarach, and, sure enough, late in January... they brought in my family. My oldest was almost four; youngest wasn't even a year yet – eight months," he amended, quickly. "She was eight months...

"And they killed them."

It was almost sick, how little emotion ran into those last words. How light, how faint, how utterly indifferent. A shadow passed across the moon, staunching her silver light, and, uncontested, the fire cast its crimson mantle over the darkness, illuminating everything within reach.

"I was interred in the cell-structure they call the Crypt – a damp, cold, horrible place in the bed of the earth, more wretched than Hell; the Devil himself would cry, being caged there. Being *here*... But the single blessing in that inundation of atrocity was that it gave me time to think – something I hadn't had in over a month. It gave me time... to realize...

"I won't dwell on the details. But I'd known about the books – from a very reliable source," he was careful to repeat, "and I realized at length what they'd wanted me for. There was only one book left then, and they were keeping that here as well – somewhere in the old cathedral. After a month in the Crypt, when it was nearly March, I made my escape. And I took their thrice-accursed book with me. I took it, and I wrecked the bloody thing. I burnt it to nothing – not even a fleck of ash remained. Then I ran.

"I wandered for about a year, honestly somewhat crazed. I don't remember a lot of it, I just remember – being – *afraid*. And..." Now he looked at her. And, in a voice barely more than a whisper, he finished: "Then I found you."

Michelle bowed her head. *There* was a story she knew well. He had offered her husband his labor in the fields in exchange for a loaf of bread – or socks, if they could spare them. There had still been snow on the ground, and there he stood, tensed like a young deer, skinny as all starvation, with nothing but that thin, long coat to protect him against the cold. They had hired him as a field-hand, and he had proved a good worker, strong and industrious in spite of his timid ways. And he had worked for them for nearly a year until, one day, he went into the hills and simply vanished, as inexplicably as he'd come.

"I miss Alan," he said. "He'll be pissed when he learns you're here."

"He could've used you in the fields."

"There's still the harvest. Has that all been taken in?"

"Stop evading." She looked up in time to see him turn away. "You still have ten years to account for. The spring planting was beginning; why did you leave?"

"I had gone into town... I had forgotten that they would be at festival. Even after two years, the crowds still made me *so* nervous, but... it's in the blood, the wandering." The only explanation he gave was a shrug. "The only way out was in; I tried to skirt through the crowd, keeping my eyes down. A detachment of Ariegn's guard had been stationed there – temporarily, they said, but *shorn*, there were a lot of them. I was tryin' t'bypass a knot of them when I noticed... They had a little boy with them – aged about six or seven, standing in their midst. Just the *look* on him – that weird mix of defiance and fear you see on abused slaves. That he was, and standin' there, wearing rags, covered in filth and mud, and holding his arm, 'cause it was broken. But it was the look on him – the *face* on him – that made me realize...

"He was mine."

Even now, the words felt heavy. So heavy, they caused his voice to scrape, as his throat was closed against them. He stood, but Michelle caught his hand, causing him to flinch.

"Wait. What do you mean?" she asked him. "I thought you said –"

He looked down at her. "I couldn't bring him back to you – no, they hadn't killed him. They'd hoped he could – *replace* me. But he can't do what I can..." He shook his head, clearing his throat. "I owed you – I owed you that much, the not bringing him back – that's why I never returned t'you. It would've been too dangerous..." He trailed. "I was trying to *protect* you," he said. His eyes flashed away; and the sheer, riling venom that entered his sneer was astounding. "*Shorn* an' didn't I know it would've been better for him if I'd've just given him up? Hadn't I learned after the *first* round of this Godforsaken game that he'd be better off with*out* me and I – I..."

His voice faltered. He clenched his teeth, casting his glare on the sky – on God, Whom he knew *must* be up there – with a look so furious, so anguished, so pained. And, after twelve long years and a lifetime of endurance before it, he could feel the hot tears begin running down his cheeks.

"A Thiarna Dé. Ormsa an locht. Agus – agus tá sé – a Dhia!!" He collapsed, doubling over, his knees bent beneath him, his face buried in his hands. And he wept.

Michelle knelt beside him. There was nothing she could do. Nothing could repair that damage. She slid her hand onto his back, a small comfort for what could not be salved. "It's alright, Duncan," she said. "It's alright."

He shuddered under her touch, choking sharply, curling up as much as his many wounds would allow. He shook his head, drawing a breath. *"Mharaigh siad é!"* he sobbed. *"Mharaigh mé é..."*

It's my fault. I killed him. He's dead because I failed him. He is dead.

Tears were never any use; they brought nothing back, forgave no sins. With vast difficulty, Duncan settled; he ran his sleeve over his face. He whispered, "How many times can I fail her?"

Unbidden, and to no one, he explained.

"Three times. *Three times*, I tried. And what did it get me? I knew I had limits, but I didn't understand them. I didn't care. When they took her away from me, I tried *everything* to get her back. Everything... I'd always hoped that she'd made it – back... But when I nearly died of it, that was when I knew. She was dead. And following her would only bring death to me as well... Because I *can* do it, can't I? I mean – the amount of energy – of sacrifice – it takes is horrifying, but... So she's dead – she's dead, she must be, or –"

He stood. The tiny, golden flame still perched upon the bloodstained ashes, leaping and prancing like a hound trying to recapture its master's torn affection, and Duncan drew forward until he stood at the side of the brazier. He reached out a hand to the flames, his fingers brushing the curling smoke, and, for a heart-beat, the red-orange glow seemed to ebb.

Taking hold of the sooty rim, he bowed low over the fire, and admitted the words that broke him.

"Or maybe I didn't try hard enough."

The tears came again, now silently, bitterly, filled with the desperation he had never been able to dispel. Drawing a breath, in a voice surprisingly steady, he confessed, "I am weak, and a coward. Every time I felt my strength blending into that cold, dark abyss, fear – what remained of my feckless *sanity* – would *pull me back*. I would wake up. Alone. Still *here!"* On that final word he struck the rim of the basin with a closed fist, savoring that deserved sting, wishing it were more. "I failed her! *Ba chóir go mbeinn marbh.* I should be dead.

Agus..." And what? They were gone. They were all gone, and only he was left. And nothing he could ever say or do would change that.

Duncan pulled back slowly, out of the sting of the smoke, away from the razor edge, lifting his gaze once more. Vacant the world and vacant the skies, and deaf to his violent pleading. "*Agus tá mé críochnaithe,*" he breathed, his voice thin and raw.

I'm done.

Michelle turned to look up at him, her eyes wide with alarm. "What?"

"*Auhssh...*" He hadn't meant for her to hear, not really. He raised his fist, allowing the mixing light of the fire and moon to play once more across his damaged skin; charred blood was staining his sleeve. His eyes rounded.

"I have nothing left to lose."

He walked away, still studying his sleeve. His wayward pacing brought him near the tattered wreckage of his discarded clothing. Bending as little as possible, he swept up the cloth, dropping down a few feet from the brazier. He had layered two shirts to keep warm in the crossing; now, taking up one ruined shirt, he tore it into a long strip, feeling the distorted pleasure of being able to release his anger in destruction.

"What are you doing?" asked Michelle, watching as he again shrugged off his black coat.

"First aid," he grumbled, struggling slightly with the coat; both arms were stretched back behind him, and he was glaring almost viciously as the fabric rasped his burns.

"D'you want help?"

"I don't *need* help!" he all but snapped, finally getting free of the tangling cloth. Carefully, he began to unwind the tight binding from around his ribs, trying to ignore his discomfort. A thousand curses – a hundred thousand billion curses on the imprudence of those fools! He laid the bloody cloth off to the side; then, sitting up as straight as he could manage, he began bandaging his own wounds – something he was becoming quite proficient at, he noted darkly. It only took him a moment to tie the make-shift bandage around his stomach and lower ribs, knotting it securely in place. The tourniquet applied a light pressure, alleviating some of the pain; hopefully it would help the trickling wound to clot as well.

Taking the other shirt in hand, he tore this too, shredding it into two thin strips. He held out his left arm; already the reddened skin had begun to blister and peel. The burn was much worse this time; he tensed, muscles tightening in agonized anticipation, and dutifully he laid the strip of cloth over his outstretched wrist.

"Aauhh!"

He sucked in breath through his teeth; a sharp pain like prickling nettles seized his arm each time the fabric slipped. Again and again he tried to wrap his burned arm, but it was impossible to do so without the fabric sliding even a little. And each failed attempt only caused his frustration to grow.

He felt a hand alight on his shoulder. Startled, he looked up.

Michelle gently took the tattered cloth from his shaking fingers. "You suck at first aid," she said quietly, calming the feral beast. "Hold out your arm."

For a moment he hesitated, his eyes flickering over his burns. Then, reluctantly, he extended his arm. As delicately as if she were handling a bird with broken wings, Michelle bandaged his wounds. Duncan was conscious of every fiber of cloth that touched his fragile skin like a scarab's pinch, but he offered no complaint.

Michelle knotted the first bandage. "Other arm," she said, and he shifted obediently. "So, what do we do now?" she asked.

"What do you mean?"

"They got what they wanted from you, didn't they? I don't think that they're coming back tonight. Do you think we can escape?"

He shook his head. The cloth pressed against the place where Bassus's knife had gouged him; cauterized, the wound was as ugly as a convict's brands. If only everything could be so easily hidden away.

"We wait."

Michelle made a face at him that, at a pang, reminded him of Meygan. "For what?"

"Hm? I dunno." He shrugged the shoulder of the arm she wasn't attending to. "An opportunity, maybe? Armageddon perhaps?"

Michelle rolled her eyes. He had a truly weird sense of humor. "We're waiting for them to come get us, you mean." She knotted the bandage.

"Essentially."

"Who?"

Duncan frowned. "Who what?"

"Well, do you mean like *rescue* us, or what."

The slightest of grim smiles tugged at the corner of his lips. "I'm afraid *we're* going to be the only ones doing any rescuing, darling." He shifted, rolling his shoulders back in an exaggerated stretch in an attempt to cover his flinch. "I'll take first watch."

Michelle rolled her eyes. Although there were a lot of things she did not know about him, the one thing she knew beyond doubt was not to waste her time trying to drive any sense into his stubborn brain. "Fine." She got to her feet. "You'd better put that out."

"Hm? Oh, yeah..." Like puppies or fireflies, the dancing flames chased each other around the great, sullen bowl, still vying gleefully for attention; standing beside the basin, he passed his hand slowly over the surface, settling them, wiping the fire away. A voice in the darkness gave solemn praise:

"Cool."

Duncan laughed. "Now it's too dark to see you!"

"Good! Maybe you'll rest!!"

Duncan sighed rather dramatically. A faint flash of moonlight darted into the room, helping his eyes to adjust.

"This is good," she heard him murmur, though she was not sure who he was trying to reassure. "Believe me, it's good. If their goal was to keep me tame..." He trailed. Gingerly, he folded his arms, cradling the broken skin. Though the darkness hid it, on his face was an expression of bitter determination, far blacker, far more ominous than all of the shadows gathered in the room. Shaking his head, he murmured, "They should've just left me alone."

70
Daring

To his credit, he had tried to keep watch, but physical and mental exhaustion had dominated his willpower in less than twenty minutes. He lay, curled like a hound beside the basin as if it could offer any warmth, totally asleep. Michelle sat beside him, her head bowed as she fingered a set of wooden Rosary beads. The murky, grey light of twilight was gradually beginning to peel back the ebony night. They would be coming soon, and when they did...

A snap of fire flashed suddenly in the brazier.

"You hear that, I presume?"

Michelle glanced to Duncan, who was sitting, almost crouching, a few meters away. His body was drawn into a sprinter's starting position, one knee folded beneath him and the other raised. Though he had regained his composure, even in the darkness his face was pale.

"They're coming," Michelle translated. "So what's your master plan?"

He remained silent, mulling a thousand things over until he mused, "Which way do those doors open, I wonder?"

"What?"

From his pocket he pulled a ragged, faded red strip of cloth – the sad remnants of a bandana. Handing it to her, he ordered, "Send this back t'me when yer safe."

Michelle withdrew. "Rahyoke – no –"

"Please." He turned his head to look at her fully. "The two of us together would never make it out alive. I'll hold off the guards; you keep to the shadows, and the first chance ya get, run for it." His eyes searched hers, and a flicker of alarm crept into his tone as he pleaded, "*Trust* me!"

"Are you *nuts?!* You'll be killed!"

"No I won't."

"Do you have *any* sense of self-preservation?"

He looked offended. "What would I want with that? *Tá mé dubh dóite de féinchaomhnú.*" Anywhere else, the pride in his tone as he said it would have been humorous. Here it was merely truthful.

Michelle scowled; taking the bandana, she tied it as a headband, rebel-style. "So who's gonna rescue *you?*"

"God willing?" Duncan actually smiled, a heartbeat of hope lightening his expression. "You!"

Michelle shook her head. "Just promise me you'll be careful, Duncan. Make sure there's something left for me to rescue."

Duncan pursed his lips, returning his attention to the door. *Be careful, Duncan. Stay safe.* His eyes narrowed, and he thought of the absolute uselessness of words.

"I can promise nothing anymore..."

There was a noise outside – a solid crash – and his eyes flashed towards Michelle.

She understood. With half a wave, she swept into the shadows. "God go with you, Duncan O'Rielly!"

"*An bhail chéanna ort.*"

There was another crash, louder than the first, and this time against the door; its metallic echoes rang chillingly against the walls.

Here we go.

Eyes locked on the door, Duncan rose slowly, retreating 'til he felt the basin at his back. Reaching behind him, he took hold of the rim, bracing himself. The fire went out.

There was a scraping beyond the door as the heavy board that barred it was pulled away. One door swung open, just far enough for a single man to enter. The fire-eater smiled. They had sent *one man.*

Fools...

The guard's torch-light barely made it beyond the doorway. For a moment he hesitated, eyes unaccustomed to the gloom, ears wary of the silence. The deep, still silence...

"Took ya long enough."

The guard drew his sword with a *shing!*

There was a laugh, cold, humorless and scornful. "You draw your sword against an unarmed man?"

The guard could just see him now in the grey of the twilight. Even as the vagrant spoke his challenge, he slumped down a little further against the brazier, sinking his lower back against the rough stone. Still, the guard tightened his grip on his sword, and he felt more than saw that the fire-eater noticed this with a smirk.

"Come get me."

And of course, he had to.

Though his injuries slowed him down, it was a simple matter for Duncan to drop and disarm the guard. Unfortunately, his few seconds' slowness meant that the guard had just enough time to cry out –

There was a soft, sickening crack, accompanied by Duncan's low curse.

As he listened to the chaos in the hallway incited by the guard's scream, he again leaned back against the brazier with a short, overwhelmed exhalation. He'd forgotten – he'd forgotten about his ribs, about his wrists, about the gaping wound congealing across his stomach – about exhaustion, hunger, age – though, he was glad to admit, his ankle did seem better. He almost laughed out loud.

At last, the doors swung wide.

It would have made more sense to have sent all three guards in the first place. No doubt they had drawn lots – or wrestled, which would certainly explain the crashing. But there the two men stood, their figures black against the light from the hall. One man carried a torch; the other, a long, shimmering blade.

Duncan put a hand briefly to the binding around his ribs. He could handle two.

"Heus, juvenes," he greeted them languidly, his hands casually taking the rim of the basin behind him.

The guard with the blade – one Silvanus Praetolis, also called Silvius – took in the scene at a glance, his eyes flashing briefly to the fire-eater before falling upon what he had done. *"Inferi!"* he spat. "Did you kill him?"

"Hm. Did I?" Duncan inquired; there was more honesty to his tone than sarcasm. "He does *look* dead, doesn't he?" The other guard moved towards the fallen man cautiously, but he halted when Duncan lifted a sword. Duncan laughed. "Aaw; *cad é sin? An bhfuil eagla ort?"*

The guard erupted in anger. *"Taceve, squalus Ceothann perfide, excarnificabō te infere –"*

Silvanus held up his hand, stopping him; on his face was an amused grin. "Do you intend to fight us?"

"I intend to *resist*."

"And you really think you can oppose us with a blade?"

A dangerous light flared in his eyes. Straightening up, he lifted his arms, holding them outstretched like a bird about to take flight. The silver blade glinted in his left hand like a flickering snake's tongue. Laughingly, mockingly, scornfully, he retorted, "*No*."

Not yet.

The caustic clatter of metal on stone sliced through the air, its reverberations echoing off the thick walls; the stolen sword was lying at his feet. And, slowly but deliberately, he twisted his hands so his open palms faced the ceiling, cupping the faded light of the moon.

The two guards froze, causing his devilish smile to broaden. They barely noticed the slightest of nods he gave, his head bowed, his dark eyes level with theirs, daring them; he alone saw a fleeting shadow across the light of the door. *Perfect.*

"So you remember me," he said, his low voice a strange complement to the rising dawn. "...Good."

71
Consequences

"I don't care *what* he says he can do!"

The voice roared through the corridors, echoing against the walls.

"You are a disgrace to your comrades, your rank, and your country. He is nothing more than a mangled, *unarmed* vagrant. He is not dangerous, he's clever – for God's sake, the wretch figured out how to escape from the *Crypt*, and you morons leave him unguarded in a *first-floor holding cell?!* My God! Do you have any clue as to what we are dealing with here? The consequences alone for if you lose him – oh. You think you're scared of *him*? You're gonna be wishin' he'd boiled you alive.

"Now, if you two slobbering pansies don't get down there and grab that rotting gadabout, I'll have you both executed as traitors and cowards. Is that understood?!"

In another army, the argument may have continued; but Judah Basilicus was among the most respected of history's generals, and so Silvanus Praetolis, the best among soldiers, clapped his right fist to his heart, gave his general a short bow, and settled. The displeased glare with which he fixed his leader was for once allowed to pass – though whether this was because he recognized Silvanus's point or because handling the fire-breather was enough of a punishment, who's to know?

Silvanus turned on his heel, making a brisk exit; the younger, lower-classed, foul-mouthed soldier, Flavius, trailed behind him, this time armed with an assortment of glittering weapons. Lot of good it would do him, the show-off.

The holding cell was on the first floor, just a few short corridors from the Southern Gate. Once it had been used as a storage closet; now it was barred and locked with black iron, the temporary cell for those who were to be quickly dispatched or were too weak to resist. The interior was dark and narrow, its stone walls barely three feet apart. It went straight back, about fifteen feet deep, with a seven- or eight-foot ceiling. Reaching the cell, Silvanus impatiently unlocked it, flinging the door wide. And he did not welcome the sight that greeted him.

"Back so soon, are we?" the epitome of disrespect sounded quietly through the gloom.

Silvanus scowled. *"Paedor."*

An escape artist indeed. He had been restrained in the fashion customary to the holding cell: chained to the far wall, arms and legs spread. The manacles were tight, as if they had been formed to fit a young boy – and yet he'd already managed to worm his left hand free. Apparently, the everyman had been a contortionist in his past life.

Caught in the middle of trying to wring free his right hand, Duncan merely grinned, cocking his head at the insult. There was nothing he could say, and even less that he wanted to.

For all their boasts and taunts and threats, every soldier of Augustinium knew the legends of this fiery serpent, and the truth of these legends was something not many of them were willing to test. It was certainly for this reason that Flavius, usually so bold and eager for action, did not offer protest when Silvanus ordered him to remain outside of the cell as a back-up guard... just in case.

"One on one. That worked s'well last time."

Armed with his keys and his short sword, Silvanus stepped into the gloom. "You're more trouble than you're worth, if you ask me. The sooner we're rid of your filth, the better."

Like every other Ceothann, the maggot was hot-tempered and proud, and his cynical-sounding drawl was typical form: *"Mar sin féin, aireoidh tú uait mé 's gan mé ann."*

Though Silvanus, being a soldier of the higher class, understood Turranian, Lavinian, and the three dialects of the Odinic lands, Ceothann was simply not taught. "Cute. It's a shame they never teach you classless Northerners how to speak."

"And yet," growled the prisoner, speaking Lavinian in an open refusal to humor the Augustan guard, "*I* am the one who understands *you* without qualm."

His next sound was a strangled gasp as Silvanus pushed him against the wall with the flat of his blade to his throat. *Speaking of hot tempers.*

"Remember, *flame-eater,*" hissed Silvanus, far too close for comfort; he took hold of the chains that still bound the fire-eater's right arm. "I could crush your hand to release this bond. Or cut off a few of those dainty little fingers. No one would bat an eye. You are not worth as much as your blood, and as long as you're living there's a *surplus* of that." He leaned in, just enough to make his prisoner squirm; the vagrant was very aware of the blade at his neck preventing him from turning away. "So I think it's time you start treating me with the respect demanded of my class, *lowlife*. Isn't that so?"

"I'm sure," sneered the wretch, through gritted teeth. He tried to swallow against the blade and felt the cold steel tighten its grip on his neck. "Say! You're a little bit braver now that I'm chained, aren't you?"

Silvanus looked down on him with disgust. He gave the blade a harsh jerk, a quick application of hard pressure, before stepping back.

The vagrant doubled, choking; in the twenty seconds it took him to recover, he was aware of one fact above all others, and that was Silvanus's laughter. He drew in a shaky breath, his eyes scouring the guard with an expression ruthless, feral.

Silvanus had half turned away, so he was taken by surprise when the fire-eater lashed out with a wild left hook.

The chains snapped taut, like a whip, and he realized he had lunged forward. He also very suddenly remembered his ribs, and with a moan he collapsed against the wall, his free arm wrapping around his waist as he rolled to the side.

Silvanus – unharmed, uninjured, *untouched* – folded his arms, looking at his charge with contempt. "Hot-headed *Ceothanns*," he remarked, slowly, *tuttingly*.

Duncan O'Rielly hacked, spat, and looked up. "Yes," he replied, hoarsely. "I can see how I'm a threat."

Silvanus grabbed him by the hair.

"Dar Dhia!" complained the fire-eater, swatting out as if the guard were no more than an especially obnoxious gnat. Silvanus seized his wrist, slapping it in handcuffs. *"Damnú air!"*

He was made to raise his arm enough that his right wrist could be shackled as well before he was fully released from the wall. The metal was cold as death and pinched his wrists as the shackles locked with an echoing soft click. He hissed through his teeth, grateful for the cloth between his raw burns and that unforgiving cold.

Silvanus never sheathed his blade, and it was his use of this as a prodding shepherd's staff, and his constant, harsh tugs on the fire-eater's chains, that finally dragged the rebellious prisoner from his cell. And it was by these means – whipping the chain violently forward, jabbing or slapping him with the blade – that Silvanus Praetolis brought his captive to their destination.

Bassus awaited them in the atrium – a grand, broad entrance hall between the Great Hall and the Southern Gate. The ceiling here was incredibly high, sloping upwards from the Great Hall wall until it disappeared somewhere in darkness at the heights of the Southern Gate's massive wall. The Southern Gate itself was a commanding structure; twin doors of dark cherry wood, a foot in thickness and elaborately carved; beyond this was an iron portcullis, and a drawbridge made of solid white oak, fortified with black iron, that sealed the castle closed. Above the gate – too high to purposefully shoot an arrow through – a long row of arched windows extended down the wall. They were narrow and empty, allowing great streams of muggy dawn light to trickle through.

Though the hour was early, Bassus was decked out in his usual military attire: chain-mail and tunic, a broad sword in a golden scabbard at his side. He cast the fire-eater the briefest of all glances, turning a critical eye on his two guards. "I've seen *rocks* move faster than you," he remarked, scornfully. "What the Hell took you so long? Did you chisel a path through the wall?"

Silvanus cast his prisoner down at Bassus's feet, forcing the vagrant to his knees, keeping a tight hold on the chain that bound his wrists. "The prisoner was resisting transport." On impulse, Duncan gave a derisive snort.

Bassus gave him a once-over. "Not much," he commented, and though Duncan's pride was jabbed he could not help but agree with him.

Silvanus's near eye-roll was not missed, nor would it be tolerated with the enemy so near at hand. Bassus turned on him like a cobra. "Are you suggesting that two soldiers of the Imperial Household are *not* enough to handle one half-dead, waste-of-bone beggar? Are you insinuating that I miscalculated? That I did not send enough men to do the job?"

"Not at all," Silvanus responded coolly. "I'm merely indicating that the act took a little longer than anticipated."

Augustan soldiers had a unique aptitude for political rhetoric, in consequence of which they could talk to death the slightest of all issues, and still they'd carry on. Growing preemptively impatient, the fire-eater heaved a sigh. "Is this what you fetched me for?" he griped, fixing Bassus with a hard stare.

Bassus returned his glare with a scowl, knowing all too well what he meant. "It'll *wait,*" he snapped. He turned back to Silvanus with renewed vigor. "In the future, you would do well to heed my orders with the utmost promptness. Is that understood?"

Silvanus and Flavius bowed their heads, respectfully and obediently. "Yes sir."

"As for you..." Looking down on the fire-eater, he regarded him as if viewing un-breaded calamari for the first time. The obstinate creature met his gaze with defiance. Bassus narrowed his eyes. "You seem to be of the belief that it is *you* in control of what transpires here," he observed, coldly.

The fire-eater's expression became a sneer. "It's mutual."

And suddenly the knife-slinger smiled, a chilling, murderous little smile. "Of course. You take pride in what you consider to be your own private rebellion. Yeah? Setting your little flower free right under our noses must seem like a *great* victory in the war you wage against us. But let me tell you: you have accomplished nothing. We will find her, and we will kill her, too. Happy? *That* is all you have done for her. Her blood will be on your hands.

"You are going to have to learn, O'Rielly, one way or another: there are consequences to your actions. And I suppose it's going to be up to us to teach you. So lesson one: *You* are not the boss here. *I* am. And as long as you remain within Ariegn's walls, I am the one you answer to, I am the one who controls you. Keep hold of your pathetic rebellion if you must. But I will get what I'm after, in the end. I will break you, Duncan O'Rielly. You can be sure of that." He turned back to his soldiers, his scowl retained. "Flog'm," he spat, with a decisive flick of his hand. "We'll see how resolute his will-power stands."

Flavius began to finger a shimmering dagger, a pleased curve to his grave lips; the two guards bowed, only too eager to begin.

At a dark murmur only Duncan could hear, Bassus concluded, "There is no escape from justice."

The fire-eater bowed his head, studying the hazy morning light as it wakened in the stones. Under his breath, he answered, "That's two."

The sharp slap echoed long after he was gone.

72
Survivors

He had inherited it from his father, no doubt. Some called it Resilience, others called it Durability; still others, at daring whispers, called it Immortality.

Sully called it Dumb Luck.

It had been Sully who noticed the strange novelty coughed up by the tide. He had been out on what he called an "unofficial border patrol" – his excuse for taking a sea-side stroll dangerously close to Ariegn. He walked just within the line of trees with a confident, easy stride, hidden by the shadows at the forest's edge. The western sun was casting its rays over the castle's high walls, setting the sea alight, reflecting boldly across the white-washed surface of the walls. The cliffs here were shallow, the drop between the flat, grassy lawn and the raging sea filled with large, craggy stones. The rocks were as black as an angry storm-cloud, pounded brittle by slate-grey waves. A fine sea-mist was suspended over the low waters like a ghostly haze, collecting like dew-drops upon the path of the tide. And there, tar-black and foreboding against the surface of the stones, was the object.

Sully paused beneath the shelter of the trees. He was not one to go picking through Lady Ocean's refuse, yet for some reason this particular wad of ragged flotsam had ensnared his attention. He came forward, to the very edge of the shade's covering garment, clinging to the rough bark of an elm like a straggling rope of ivy. He narrowed his eyes against the hazy sunlight, wondering about his next move.

There was a solidity to that object, wasn't there? A solidity and a weird call from God to investigate. God and His sense of humor! Shaking his head, Sully relinquished all safety, his black boots shifting silently over the dry twigs of the bramble. He drew his blade smoothly and silently; it was a short sword with a slight curve to it, bought off a South Turranian almost twenty-five years ago. Sadly, it was starting to look its age.

Sully smirked grimly. *Not unlike its current owner...*

"Hello?" he offered, dubiously, scoffing at his own suspicion. Wouldn't they just love to see him talking to a bundle of old rags? "I will warn you, flotsam, I am armed."

Advancing with the slow and careful footwork of a beginner's waltz, he cast an anxious glance back at the castle. Ariegn, in all its wretched glory, was looming all too close for his liking. It was one thing to be audacious; stupidity got you killed.

Reflecting on the absurdity of it all, he decided to try an old ploy that had gotten him through the barricade at the channel of Alyssae back in '92. Scavengers rarely approached Ariegn, but every so often one madman would tempt fate, so Sully took the risk and feigned it. Slinging off his long coat, he stooped low, his legs spread wide, knees bent. With a crab-like, side-ways, lurching hobble, he made his way over to the rocks at the water's edge. *And I am but a poor, washed-up sailor, rum-soaked in my youth and now haggard with old age! A harmless old coot!* He almost laughed aloud. God help us, but boredom sure has an odd effect on the mind of a man, does it not?

Mercy. Sully picked his way across the lawn, keeping his head bowed, hunched between arched shoulders, and his cap pulled low over his eyes. When he reached the rocks, he clattered down among them, picking a spot a few boulders to the west of his chosen prize so as to avoid the suspicion of a goal.

Ah, tide pools. Sully discreetly sheathed his sword, wondering if he'd be missed if he skipped out on the next meeting. Probably not. There was nothing to discuss other than whether or not to light a candle because of the new moon. Reconnaissance was much more akin to their greater purpose, and why be mundane when you could take a crack at being useful and score some souvenirs in the process? If nothing else, Aila had a fascination with Ariegn sea-trash – believed it could be used to determine the greater soul of an army. Personally –

Midway through clambering over a great rock, the mad, trash-picking beggar straightened stiffly. For a little too long, he stared with his head cocked at something he'd spotted, his eyes wide.

"Mother Mary, deliver us."

There was no mistaking the prone figure for anything other than what it was now. Perhaps it was the sad remains of a degenerate servant, an insolent stable-boy or useless cook. Perhaps it was some long-washed-up fisherman or a seafarer's bones, or a beggar much like the one he'd portrayed.

Eew. Suppose it was dead?

Sully pursed his lips as his hair stood on end. Dead things he could not do. Suppose its ghost became attached to him, following him everywhere like an evil shadow?

Then again – could he really just leave it? Its bones blistered, whitening on some discarded shore. The decent thing to do was bury it – the decent, humane thing – which would, of course, consequently placate the ghost. And then he would burn his coat.

Like a swooping vulture, he stooped, throwing his coat over his treasure and hauling it from the rocks like a fisherman dragging a net. In under a minute, he had lugged his find across the short, lush lawn and into the dark cover of the woods. The moment he was out of range, Sully lifted the bundle into his arms, as gently as he could quickly, and walked off.

He carried his burden back to his camp, which was only about two hundred feet from the forest's edge – what some would consider daringly close to Ariegn's walls. As it was unsafe to light a fire here – even Sully had to admit that was just asking for it – his "camp" was more "where he'd dropped his bag." And now, where he dropped the bundle.

He laid his find at the base of a pondersome beech, sitting across from it. He supposed he ought to see what on earth he'd picked up. He rolled his eyes; the chances that thing was alive were as slim as his actually being hanged. But, he reasoned, leaning forward with a sigh, might as well check...

Sully pulled back the shroud – and dropped it with a gasp. His gaze shot to Heaven, and he blessed himself with the sign of the cross half a dozen times.

What had he gotten himself into?

It was an hour before the young man stirred. An hour in which Sully kept a silent vigil, his wide eyes fixed on the motionless figure, the *Salve Regina* fresh on his lips. The sun was beginning to set, a faint tinge of gold mixing with the thick grey of the twilight clouds. Already beneath the heavy canopy the shadows were looming long, and soon it would be dark.

The only sound was the wash of the tide against the rocks of the seashore; faint though it was, Sully could feel it. Each slap of its heart-beat was even with his own. Each dragging breath that tugged at the shoreline reminded him that God had put him on that turf for a reason.

His eyes raised to the darkening Heavens. The sky was the colour of steel, a cold blue-grey. Impassive. He had seen nights such as this at sea, when the sun had slipped silently behind the clouds without so much as a warning or a goodnight. God wasn't giving any prophesies tonight.

Sully frowned. *And Fate was wide open.*

A soft murmuring like the whispering of the wind or the gibbering of spirits caused Sully to jump. His eyes focused on the source – his flotsam – but he did not relax. The young man had uttered something wholly unintelligible, his expression becoming strained, his brow furrowed with discomfort and anger. He was waking up.

Sully listened as the young man muttered another mindless strain, wondering what language that was trying to be. What kind of stress had this young body been through? And what new dangers did his survival bring?

Not for the first time, Sully considered throwing him back.

The lad's eyes flashed suddenly open, and instantly he sat up, putting a hand to his waist, then his shoulder, then blurting a hoarse and strangled curse. Sully knew then that he was in for a rough time.

The boy demanded something in some other language, and Sully did not bother to translate.

"Do you speak Augustan?" he asked, nonchalantly, completely unfazed by the young man's fury. "Might be easier to communicate..."

For a full minute, the youth merely stared at him. Then, grudgingly, he answered, "Yes."

"Good!" That was all he said.

It was the youth's turn now, though he was loath to speak. Pressing his back against the tree, he drew his knees up slightly, eyeing Sully with all the suspicion in the world. At last:

"If you're gonna kill me, just get it over with."

Sully studied him with an expression as unreadable and coarse as the steely sky above. "Y'look a little too scrawny to eat."

The young man stared, his dark green eyes flashing. "I'm a lot stronger than I look."

"Even worse. Tough meat is deplorable."

Slowly, slowly, the anger began to drain from his expression in spite of himself. "Who are you?" he asked, in his split-bark voice, and the older man gave a laugh.

"Me?" He offered a hand as calloused as the face of the earth. "Name's Joe Sullivan. Most calls me Sully, but you can call me what you like."

Hesitantly, the boy accepted the offered hand.

"Jeeze, you *are* strong, aren'chyew?"

The faintest flicker of a smile.

Sully leaned back against a tree opposite the boy, watching as he tried to make himself more comfortable; the youth moved with the ginger caution of an ancient old man. He was a wiry, sinewy creature, and would probably be tall for his age, if he could stand. His face was set in an expression of harsh determination, even for the gentleness of his young features. The darkness in his deep green eyes spoke of tribulation, and a will that would never break.

"You're a lot like your father."

Every fiber of those sinewy limbs stiffened as if doused in ice, the unconcealing dark eyes grown wide.

"*...what?*"

Sully once more studied the boy. "Your father was a sailor, wasn't he? I sailed with him back in aught-nine when I was quartermaster for the *SV Evangeline*. Every morning he threatened to jump ship." Sully saw a glimmer of recognition in the boy's eyes and set the matter aside. "What's your name?"

The young man's eyes narrowed, and he studied the stranger in turn. Sully was a tall man, about 6'1", and of strong yet slim build. Everything about him – his tattered, salt-stained clothing, his sun-bleached, straw-blond hair, his calloused, tanned, leathery skin – spoke of a long life of sea-faring, a life full of mystery and hardships you'd have to see to even imagine. A thousand stories clung to him in a thick veil, hiding all their secrets behind pale blue eyes. There was nothing to trust, and nothing to mistrust. With unbridled honesty, he was what he was, take it or leave it.

And the youth decided he rather liked that.

"Zander," said the boy, extending a hand with such haughty moxie that the redundancy of the gesture was completely banished in his poise.

Sully gave a nod of acknowledgement, shaking the boy's hand. "*Zander*," he repeated, contemplatively. A wry smile found its way onto his usually-unmoved lips. "Now *there's* a name."

"Joe," the youth returned, flatly. "There isn't."

"It's Joseph, actually."

"Even better."

Sully glanced up, and gave a quick bark of a laugh. "Hah! Jeeze, you're a bit of a smart-ass, aren'chya? We may well be good friends yet!"

The young man – *Zander* – smiled. A short silence drifted in, though neither man was particularly bothered, accustomed as they were to the solitude. Putting a hand to his left shoulder, Zander grimaced slightly. "Well, if you bound my arm, I'm guessin' you aren't gonna kill me."

"You'll wanna get that looked at when we get back to camp."

Zander furrowed his brow. "Camp? What camp – isn't this camp?"

Sully shook his head. "Base camp." Drawing up one knee, Sully laid his arm across it, letting his hand hang limp. "Deep in these woods there is a secret camp. The camp of the resistance fighters, the Company. You may know – but actually, why would you? – so, the Company evolved out of the small town of Balladarach, at the lead of a single man. A small, unified tribe of varying peoples, all joined under one cause. They are the remainder, the survivors, of our history. They are the sad fragments of the lost nations, the stories, the talents, the beliefs – the very soul – of their lands. These are the ones who will carry the fight, who will make a stand, once and for all, against the Empire, to win freedom or die trying, as it were. That's the romance of it, anyway," he tagged, with a shrug.

"The romance?"

Sully smiled grimly. "I'll let you see for yourself when you get there." Reaching up, he adjusted his cap, bringing it lower over his gleaming eyes. "Now: before we proceed any farther – how ya doing?"

Completely thrown off-track, it took Zander a moment to understand the question. "Huh? Fine."

"Fine enough to walk?"

The young man frowned with concentration, bracing himself to rise. He got about half a foot before dropping back down. "No," he hissed, through tightly gritted teeth. "Not yet."

"It's alright, it's alright," Sully reassured him, motioning for him to settle. "We aren't in any kind of hurry. We'll wait a little, huh?"

Zander was reluctant, until he realized that he wasn't breathing. He cleared his throat, unclenching his jaw and surrendering hoarsely, "Alright." He eased back into his place. After a moment's evaluation, he pressed a palm tentatively against his wound; it stung. "Sully?" Shorn, one could not help but feel scrutinized under that piercing, blue stare. "T...tell me more about camp. I've lived in this forest for, like, my whole life, and I've never heard anything about a resistance force."

"We move around a lot," Sully shrugged. Then, leveling that pounding glare again: "You're not a spy, are you?"

This brought a keen scowl to the boy's noble lip. "The Empire tried to kill me – twice. And they killed my father –" His voice snapped, and he found himself unable to speak any more.

"Your father is dead?" Sully asked, and the boy could do no more than nod. "That's a shame..." The sailor did not know what else to say. He had never had children, and, though young Zander was certainly on his way to manhood, Sully in his fifty-four years certainly considered him something of a child. "So... the Camp of the Traitors, current home of the Company, is divided into four quarters." He was gratified by a slight lift in the boy's glance. He continued. "The Northern quarter is home to the First Ceothann branch, which is mainly Highlanders and whoever else falls in. To the East is the Second Ceothann branch, which tend to be Lowlanders, Moorlanders, and whoever didn't already fall in the first branch. In both of these camps, although predominantly in the Second Branch, there are Balladarach survivors. About fifty in total.

"To the South are the Calchians – highly regimented, they are. They're Lavinium's proud rebels, scant though they be. The soldiers – and their sons and their grandsons and daughters – who defected, choosing death rather than violate their vows by expanding the Empire through truly needless violence and bloodshed.

"And finally, the Western quarter is devoted to the Turranian faction, under the leadership of their nineteen-year-old deposed king, Shrikant al-Rajiv. Not that too many people heed the boundaries," Sully added, "especially the Ceothanns. There are enough in and around camp who are floaters; the exchange is worst between Ceothann camps, as it were. There's no real danger in it – we're all the same army, for lack of a better word. But I will say that if you're a Ceothann alone drifting into the Turranian camp, you'd best have your wits about you and a reason. Other than that..." He shrugged. "The divisions are more guidelines for organization than anything else."

Organization... By the sound of things, Zander doubted there was anything "organized" about it. "I'd never heard that anyone even survived Balladarach."

"I'm surprised you would have heard about it at all. That was way in the east, over the Palisade and even past Marcansine."

"I was born there," Zander responded, determined not to let a desire for anonymity keep the strands of his future at bay. "I only survived because the Empire wanted my father."

The conviction in his tone would play tug-of-war with your heart. By his looks, there was no mistaking the son of the dark-haired sailor, and that sailor whom Sully had known certainly had had an aptitude for insubordination and trouble-making which could have put him at odds with the Empire – but not to the weird extent Zander described. Troublemakers and their families were not ransomed. Then again, that tonal conviction – no Augustan spy, no matter how well trained, could feign such intensity. So either the young man had conked his head hard enough on the way down to reinvent a history, or... Or there was more to true history than there appeared.

"Do you have a surname?" Sully demanded.

The child looked taken aback and confused. "I don't know that it would make much difference. My dad has a tendency to compulsively lie about his name; I could list off all the ones I know, but it might not be the one he gave you decades ago. I have his real last name, and he hardly ever uses that –" Catching Sullivan's glare: "O'Rielly."

Sully said nothing for a long time, his expression impassive, staring at Zander. It was a hard look to endure, and after a while, sighing, Zander bowed his head, clutching his injured shoulder. "It's probably dangerous for me to come to the camp with you. If the Empire finds out that you have me, they'll come down hard on all of you."

"You're coming to camp."

There would be no questioning that matter.

"You sure?" Though he tried to sound casual, the boy perked up visibly.

"Yeah," Sully snorted, dismissively. "Ya think you're gonna make it, Flotsam?"

Zander shifted tentatively, ever more reluctant to move. Out of deep concentration rose his definitive, "Yeah. Of course."

Sully rolled his eyes. *A true warrior, this*. He stood with enviable ease, slinging his sea-bag onto his shoulder and offering Zander a hand. Rather than take it, the boy handed back the sailor's long, black coat. "Thanks for letting me borrow that," he said, putting both hands to the bark behind him, determined to stand on his own. "It's cool; I'm kinda jealous."

Again, Sully was compelled to roll his eyes at the youth's unfaltering drive towards independence. "Sure." He slipped on his coat as he waited for the young man to stand. Zander noted Sully's seemingly boundless patience with some curiosity, wondering how one went about achieving that.

When at last the young man stood, Sully demanded, "Ya ready?"

"Yeah."

"Let's go."

And, without word, without question, Zander followed.

73
The Messenger

Evening was long, and the ceaseless walk adhered well to its arduous length, insufferable yet necessary as the semi-reverential silence and the inescapable, disconnected thoughts it churned. Birds, sea-ravens, cackled overhead, speaking keen and cadence in the same grating, wailing breath. All mourners weep for their own loss, and some wounds slowly, some wounds never, clot. Was it the gnawing of infection or the renewal of complete solitude that wormed so poignantly, so inconveniently, so relentlessly into his heart? If heart there still remained...

"Okay! This may seem like a stupid question, but why is the camp *so* far away?"

Sully turned, the hem of his coat flowing about his shins, and continued walking, backwards. "Ya holdin' up okay, kid?"

"Yeah." He tried his best to pick up the pace a bit, ignoring his shortness of breath. "*All's* I'm sayin', is it seems like a wasted effort, puttin' your army so far back from the enemy's walls. By the time they get there, they're too tired to do anything. How much farther?"

Sully took two more steps back, then turned smoothly, slowing his pace to fall in beside Zander. "Oh, I wouldn't call them much of an *army*," he replied, his hands deep in his pockets, his eyes on the sky above. "That's the romance again. No; their main goal is to stay alive. It's all boasting and bluffs – empty words, and nothing more. About fifteen minutes, give or take."

"Fifteen *more?!*"

"I'm afraid so."

The young man groaned dramatically. Then, returning to the previous topic, he asked, "But why, though? Are they being hunted?"

"Eeehh." Sully rocked his hand like an uncertain teeter-totter. "So-so. Many of us have warrants out – some more than others – but the bounty's not yet high enough to encourage a real active pursuit beyond personal vendettas. It's not even high enough for us to turn traitor."

Rather than ask about that (did he really want to know?) Zander hung onto one of his earlier words. "Wait – *'us'?*" he echoed, grinning deviously. "What do *you* have a warrant for?"

Sully sighed. "'Defiance of Seizure', I think it's called? Have you heard of the dismantling of the fleets?"

"Oh, *God*, yeah," Zander groaned, lifting his eyes to the Named.

There was a glint in Sully's eye as he said, "Yeah, I should've figured."

"So, does that mean that you have a boat, then?" Zander inquired, and this time Sully's sigh held more sorrow.

"No," he answered, wistfully, his eyes scanning the same expanse of hidden sky. Then he winked at Zander. "But *they* don't have her, either!"

Zander grinned.

They walked on, each man resuming his thoughts. Sully was another one whose face could not be read with any amount of precision other than the coldness of his pale eyes and the fraction of a smile he gave. Overall, he was reserved, scornful, and rough, his glance as scourging as a trawler's net.

Zander wondered about his own expression, and realized he was no longer so different. Lifting his gaze to the sky, he saw the steely, harsh grey of the twilight had given way to an even more impenetrable deep violet. Not a single star, no chink of moonlight, slipped between the clouds – a perfect night for stealth; a terrible night for loneliness. Zander pressed his lips into a tight line. By God, he was beginning to understand Rahyoke.

"So!" he began again, wanting to fill the silence with something other than thought. "I guess no one else must have a death warrant out, then."

Sully thought about it, pondering over a few sticky names.

"Ariegn doesn't often declare war on an individual," he at last decided.

"Lucky me," Zander groused, shoving his hands deep into his pockets as he continued walking along. In a voice barely more than a dejected mutter, he remarked, "Not that it really matters much anymore."

"No!" Sully protested, in an intriguing tone. "Don't ever say that! It matters, Zander – more than you know."

"Because I'm O'Rielly's son."

"Because you're *Zander*," Sully corrected. "If God didn't need you, you would've been left to drown."

"Uh-huh..." Zander was too tired for philosophy and fate. "So what now?" he asked. "We go to camp; then what?"

Sully made a face. "How should I know?"

Zander rolled his eyes, though he couldn't help a soft smile. Following dutifully after the sailor as he picked up his pace, he decided to let it rest. After all, Sully had a point.

A long silence drifted in, and this time Zander did not stop it.

He had never questioned God before in his life. Not when his father had been taken by Ariegn's soldiers, not when his village was destroyed and his mother and sister murdered. Not ever, through good or bad, for he trusted God to guide him to what was right. But now, covered in blood and salt water, his eyes raised to an indigo sky that was starless, moonless, cloudless, and blank, he wondered.

The trees loomed, vast and ghastly, to every side of him. Their slim, straight silhouettes held up the sooty sky like columns, separating the black from the hazy blue of the world beneath. A fine, faint mist like ghosts' pale hands began to stretch across the ground, gripping at the thorns and brambles, roots and dry dust of the forest's ancient floor. A soft chill was settling with it – the eerie, foreboding chill of a deceptive August's night. All the world was gathered in, splinter fingers and pitch-dark cloaks swirling 'round as shadows seeped in like the tide. It was the sort of night when Death-spirits gathered, banshees sounded their calls, and the swift hands of unnamed spirits led your soul away. The wind's cry, no longer a story, was an omen. And as the mist choked out the wind's dying breath, he could still hear them, feel them, clinging to him like cobwebs, with whispered words upon his flesh and cold hands on his heart.

God, what now?

"Zander!"

The boy stopped dead in his tracks. He did not recognize the voice that had spoken his name – or did he, but could not remember? A vague recollection stirred within him, and slowly, indecisively, he turned.

And what a sight met his eyes!

Perhaps twenty yards off, tied uncomfortably to a large oak, was a man. His arms had been bound tightly so that they were folded behind his head, and the rope from which he hung was just short enough that he was forced to balance on his toes where a thick root sloped up to join the tree. He was a small man, tough but nimble. His shoulder-length hair was pale ginger, straight, and had the vague wildness of having been ignored; it lent a sort of fierceness to his otherwise disheveled appearance. Under a brown vest of simple cloth he wore a light linen tunic of dark blue which was too big and half-tucked-into his jeans. And he had some really nice, if scuffed, brown dress shoes.

The man's eyes locked with Zander's, narrowing as he studied the boy's face. "Zander O'Rielly?" he tried again, this time with more uncertainty. Offering a sheepish smile, he noted, "F'yain't, chya look jus' like'm."

Zander hesitated. Then, taking a step forward, he demanded, "How do you know my name?"

The man allowed a grateful sigh. "Gad – thought Ah wuz goin' nuts," he said; then he answered: "M'naym's Murphy. Daíthi Murphy. I wuz'a good friend'uv yer da's. Ah me'chya once – y'were abou' – uunh – two den. Ya won' r'member me, y'were so small. Bu' – aah! I'd know yeh anywhere! *Jus'* li' yer fa'der, righ' – well – mos'ly. *Clayn'd-up* version'a yer da'. Deyr's anuff'uv'im in ya, anyhow, righ'?" He smiled weakly, the slightest upturn at the corner of his mouth.

Intrigued, Zander drew nearer still. "Enough," he agreed, struggling to comprehend the man's bizarre accent and slurring words. "But I would appreciate it if you kept it to yourself – you know... *Instead* of calling it out to the whole woods."

"Un'erstood, yeah." The man gave a nod of acceptance. Then, casting an indicatively curious glance to the ropes that bound him to the tree, he began, lightly, "Say. Sin'syer in t'*neighborhood* an'all, f'yain't too *busy* ur nufin'... Ya see dees ropes –?"

"Alright." The harsh voice of Sully cut him off as the sailor moved to stand just behind Zander. "That's enough, Barfly."

The man's expression was one of annoyed disgust. "Ah, *tha'* again? *A chiall* – hain'*chyou* high an' migh'ee, *Sullivan*." He hiccupped slightly. "Ugh. Uh – *dan'* gimmie tha' *look*, Ah gatta bad stomach. Judgin' auld –"

"What did you do *this* time?" Sully drawled.

"*Dis* time?!" Daíthi protested, his raised voice cracking to a higher pitch. "D'ah do 'is *off'en?!*"

Sully regarded him stonily. "What d'you think, Zander?"

"Uh –" Zander hesitated. "A con man always gives a little too much information."

"So does a drunk one." Sully maintained his disapproving glare on Daíthi. "Maybe we should just leave him."

Daíthi noticed Zander's slow nod. "No – c'mon! T'aint *fair*, nee'ver!" He sounded the tiniest bit desperate. "Ah cain't feel m'legs!"

"Oh, qui'chyer cryin' Barfly."

"I hain't *cryin'!!* Ah'm *parched,* man!"

"So is that, like, a punishment, then?" Zander asked Sully. "Like a pillory?"

"Mahr li'a keel-haul," Daíthi explained grudgingly, "an'ihs uncom'ferbul az *Hell.* Please?"

Shaking his head, Sully turned away, but he paused as the child took a step forward. "I'd leave him," he warned.

"We *can't* just leave him!" snorted Zander, though he wasn't quite sure why. He held out his palm to Sully, who, with an eye-roll, slapped his blade into the youth's hand.

"*I* can." Still, he stuck around.

With murmured thanks, Zander moved to within two feet of the tree from which the man hung, studying the fine curve of the blade.

"*Thaynk* you," breathed Daíthi, with relief.

Zander cast him a brief glance. He was now close enough that he could see Daíthi's face through the gloom. He had somewhat fox-like features, dark brows in spite of his pale hair, his cheeks scarred with sun damage like freckles. His eyes were dark with the black lashes of a Lavinian, almost as if outlined with kohl, and were the strangest colour Zander had ever seen. A deep amber, maybe with green mixed in – emerald, or sapphire, flecked with chocolate. The colour seemed to shift with the lighting, the day, his mood. Personally, Daíthi called it hazel.

Overall, he was not a bad looking guy – just savagely unkept.

Zander raised the blade, ready to cut the rope, but then he paused. An interesting look of rumination passed briefly over his face, and, much to Daíthi's consternation, he flipped the blade, laying it over his shoulder, at rest.

"*Why*'re ya lookin' a'me li' tha'?"

"I wanna know why you're tied to a tree," the boy pronounced. "And –" stepping close to Daíthi, fingering the blade, "—I wanna know how you know my father."

Daíthi stared at him in shock. "Ih – Tha – tha's *torture!!*"

Sully came up beside Zander, clapping a hand on his uninjured shoulder. "I like the way you think, kid."

Daíthi cursed. "Ah cou'na tell ya better'un f'Ah wuz stan'in on t'ground! Jus' be dacent an' cuh'me down, will ya?" His foot slipped. "*Perfide!*"

Sully glared at him. "Then tell us why you're up there. Go on."

Daíthi's face was an expression of pure outrage; in the end, though, he saw that there was no way around it. "*Anderos* – look! Ah may'a had a bi' t'drink an' harangued d'guard."

"What were you doing at the gate?"

"Wuz comin' back'm Carteia – a fin may'ee you'd know, f'yever came back now'nn again yerself –"

"Alright, alright. Just answer the boy and he'll cut you down; what brought you back?"

Daíthi looked at them both resentfully. "Girl as't me ta' deliver a message. Z'issa feckin' inquisition? Wanna as' who Ah pray teh, Sullivan?"

Sully looked at him flatly. "I do *not* wanna know who you pray to."

Daíthi snickered. "Yea'," he sighed, leaning his head back as far as he could manage, his gaze wandering the canopy. "Da' Company weren' interested much, nee'ver. Shaym..." His eyes flicked back to Zander. "Dey're all nuts, Zander – Sullivan tell you? Tay're ah' scared, bored, an' bloodthirsty, an' all ou' f'r *number one* – whi' sure'az Hell ain't *me*. S'funny," he added, with a dark-humored grin. "'Ss d'very same thing f'r which dey strung *me* up for, 'sentially."

Zander strove to keep up with him; this was turning out to be a very overwhelming night, and the man's thick accent *alone* wasn't helping. "Wait – so – you're here to deliver a message?"

Daíthi nodded deeply. "Na, cou'jya *playze* un-*tie* me?!"

Zander and Sully looked at one another, and Zander shrugged.

"Yeah, alright."

He twirled the sword with a flash of silver, inspiring sudden realization and regret in Daíthi, who gasped just before the blade sank into the wood of the tree, slicing the rope. Suddenly released, Daíthi collapsed.

"Shit! Ah mayn – *thaynks*."

He got to his knees with some difficulty, pulling his bound arms over his head so at least they weren't tied *behind* him. Zander grabbed his elbow to hold him steady, then, with a samurai's precision, slit the tight binding.

Daíthi found he could breathe again once the boy returned the blade to its owner. He rubbed his wrist as the feeling flowed back into his limbs, looking up at Zander from the base of the tree. "Ya migh' be ge'in' in trouble now," he cautioned.

Zander's only answer was a shrug; what matter?

Daíthi smirked. "Sa wha' abou' *you?* Where d'Hell *you'd* come from, an' af'er so many years?"

Zander's attention was on Sully. "If they did this to *him* –" he began, but Sully shook his head.

Daíthi beat him to it. "Naw. Y'ent *narely* notorious anuff t'be strung up, hon." Ignoring Sully's glare, he struggled to his feet, leaning heavily against the tree. It took him a moment, but he managed it. "A'ywayz," he continued, shaking his hair out of his eyes, "you'llav'nn in wiv da' sen'ahmen'alists, won'chya?"

"W-what—?" Zander began, but Sully intervened, demanding impatiently, "Have you been to camp?"

"Eh? Me?"

"No, Cleopatra."

"Yeah, jus' t'th'edge, why?"

Sully's expression darkened. "I wanna make sure before I go draggin' *two* sodden wretches back with me."

Both men were offended, though neither could deny it.

"Hey – Ah *been* ta' *camp!*" protested Daíthi, raising his voice perhaps a little too much. "How'a ya think she as't *me* an' nah one'a them *uver* bastards'ss messenger, huh!?"

Zander coughed and muttered, "Only one willing."

"Huh?" Daíthi glanced at him as distractedly as if he'd forgotten he was there. "Oh. Hell. Din' ayv'n think'a tha'."

Rolling his eyes, Sully walked away. "Fine then. Come if you're comin', or arn'chya?"

Zander and Daíthi glanced at one another. Then, with a muttered, "'Kay," Daíthi used the tree and a rough grasp on Zander's injured shoulder to gracelessly ease himself onto his own two feet. This done, he patted the youth gruffly on the back. "Thaynks, hon."

Zander rolled his eyes, which made Daíthi smile. He had a strange, lopsided smile, the smirk of a real trouble-maker.

"No problem."

Scrutinizing him now with a confused disgust, Daíthi swayed slightly, and in a drunk, accusing tone, he demanded, "Ahr yew *taller* 'en me?!"

Zander looked him up and down (mostly down), and leered. "Yeah. I am." He had a full four inches on him, in fact.

Daíthi shook his head, kvetching in some other language.

"So you're *not* coming?"

With a mutual eye-roll, the two "sodden wretches" followed.

Sully strode on ahead, sometimes turning and walking backwards as he urged them to hurry it up in no uncertain terms. Zander looked at him imploringly, his arms folded. He missed his own weapons so badly at this moment, it almost hurt. He didn't trust any of them. Not even the man who'd known his name.

Daíthi almost laughed when the lad finally lost patience, demanding – still as politely as possible, *a chiall* – "Are we *there* yet?"

"Yu'll feel't."

"What?" Zander frowned, turning back to Daíthi.

"Yu'll know," Daíthi explained, with bright innocence. Lightly, he put a hand on the boy's shoulder. "Liss'n. Liss'n *closely*," he urged, keeping his voice low. The boy looked fed up and tired, but genuinely intrigued. He listened. "You'll hear't first, if y'know wha'chyer lookin' for. T'very wind speaks ihs naym, wrappin' t'whole'a da' place in legend. D'ya hear't? Ih's their history 'at gives'um away.

"Nex', if ya strain, yen see t'ligh'. S'fain' as a haun'in bog-fire, lurin' as a banshee's song. Faint's t'ligh', an'jya see't, jus' ahead'uv ya, like t'horizon a' sea, somewheres off'n d'trees. An' all da'time, t'ligh' whisperin' t'ya, tellin' ya stories y'ev always longed t'hear.

"An' then, b'far ya know't, yer upon it. There, ou'a t'dead a'nigh', risin' from da' fog'uv darkness li' fabled Brigadoon. An' then – well, deyr's a sight you'll hafta see yerself, hain'it?"

The camp of the rebel warriors.

74
Fire's Son

"Uuuuhh – Joseph?" Despite his uneven gait, Daíthi had folded his arms tightly across his chest, his gaze darting constantly, shifting from tree to tree. "A-are we –? K'we go t'rew anu'ver gate, may'ee? Please?"

Sully scowled. "D'you think you'll get any better a reception at another gate?"

"No."

"Then come on."

"*Ooohh...*" Daíthi worried, tremulously, but said no more. Zander watched him closely, his eyes narrowed. He was just about to ask what his deal was when Sully announced, at a whisper, "We're here."

His two followers stopped dead in their tracks.

A thin fox-path, barely visible, wended its way through the trees. And beyond a bank of sea-mist, beyond the clustering branches and shrubs, beige tent-canvas like sail-cloth breached the darkness, outlined in the hallucination of fire-light. The half-hidden phantasmagoria of freedom; the edge of the Highlander quarter.

The Northern Gate was guarded by a single sentry, but it was guarded well. The man was built like a linebacker – big as a house, every inch of him strong muscle, tough as steel. His eyes and thick, curly hair were as black as a raven's, and he was garbed as Zander had always pictured the heroes of old: a short-sleeved, chain-mail tunic over one of navy cloth; a heavy cloak pinned at the shoulder, a plaid kilt; and armed with knives and swords, a sharp, menacing broad-sword ready in his hand. As he surveyed the strange band of night-visitors with a well-trained eye, the warrior's gaze fell on Daíthi, and he smiled.

"They se'jyeh whar an escape artist, Barfly," he noted, in a hefty brogue. "Ah niv'ver dreamed t'was only on account'uv yeh had *resarvs* t'be helpin' yeh!"

Sully gave a derisive snort.

"Okh, wait – Sullivan," the sentry corrected himself. "Mah mistake."

"*Hi*, Tam." Daíthi's voice shook *hard*. "S-so – y'ent off-duty yet, huh?"

"Did Ah nae just tell yeh yer no' welcome aroun' this camp?"

"C-come on, Tam!" Daíthi implored. "Ah know'ih weren't e'zackly da' bes' re-inner'duction, but uh – uh – I – Ah'm 'ere on business, a'ywayz – "

"Tae deliver a message," Tam confirmed. "Ya kin leave it wi' me an' go."

"Nnnnnnn – tha' – tha'ss on'ee *part*'a mah job, though."

"Uh-huh. An' the other half?"

"I – Ah'm accompanyin' dees guys. Th-tha' one –" he said, thumbing an indication to Zander " – Ah'm meant ta' guide 'im aroun' – new recruit – real classified an' shit, Ah wun' e'spect you'a been told – bu' – Ah mayn, jus' look't 'im – Look't how bloody trus'worthy 'ee looks, don'ee! You c'n lemme in on 'is accoun', can'chya? C'mon."

Tam regarded the three of them, his expression stony and unimpressed almost unto sorrow. "*Tell* ye whah'," he said at last. "Ah'll let *yew* pass," to Sully, "Ah'll let *yew* slide," to Zander, "an' Ah'll let *yew* hang," he said, and the indicated went, if possible, even paler.

Daíthi gulped. "Ach – n-*n'armal* Ah'd be up far'a challenge, Tam, bu' – Ah bein' drunk, an' unarmed, an' – look, c'we jus' call't a draw f'r t'nigh' an' –" He staggered backwards, his progress halted as Sully impatiently grasped his shoulder.

"I'll look after him, Tam," the sailor growled, and Tam gave a good-humored smile. "A'right, Sullivan," he relented. "Bu' be shoor yeh do!"

Sully gave a nod.

With a word of thanks, Zander slipped past and onto the well-tread path, Daíthi trailing after him. Tam shot Sully a pointed glare, but the sailor merely shrugged. At his own good pace, he followed.

The moment they were out of reach, Daíthi breathed a long sigh of relief. Zander turned to him, one eyebrow arched. "Should I have listened to Sully about you?"

"*Ah* niver do," Daíthi boasted, smirking impishly.

The sailor gave him a hefty push in the back as he passed, grumbling, "*Walk*, ya reprobate!"

Daíthi only just caught his balance; he voiced his objections as he draggled at the back of the group, but Zander couldn't understand a word.

In spite of the late hour, the camp was active – quiet, but busy. In the midst of the tents and campfires, Sully paused, bringing his small troop to a halt. "I don't know when the next meeting is," he realized.

"Auh – s'rye *now*, innit? May'ee?" Daíthi's interjection was so slurred it was almost unintelligible.

"Uhm," Zander asked, "could you try that again?"

"Uuh –" His eyes flashed up to the canopy, and he struggled to enunciate, slowing down his fast dialect and stilling his lolled tongue. "T'mee'in – huh – *meeeeT-iiinG*. Ih' still's may't – be –" He gave up with an impatient sigh and implored Sully, at nearly a whine, "S'one *t'nigh*! T'was goin' on 'en Ah firs' came!"

Zander just stared at him blankly.

Growing extremely frustrated, Daíthi moved to explain again, but Sully held up his hand. "Alright, alright, I got it," he blew him off. Daíthi looked a little confused, but said nothing. Sully stuck his hands in his pockets, casting Daíthi a contemptuous glare. "You already have a hard enough time with the language; you shouldn't be allowed to drink on top of it!"

The spat insult earned an easily-dismissed scowl from Daíthi. This time the man's response *was* another language, one that Zander did not understand.

Sully was unfazed. "Like I said." He turned to Zander. "I'll see if it's still going on or if it's already shattered in chaos. You gonna be alright here?"

Zander nodded.

Sully looked at them both with a stern eye. "Wait here," he commanded, and strode off.

Daíthi made a face. "'*Wai' here*', 'ee sayz," he quietly sneered. Still, folding his arms, he deigned to stay put. It wasn't like he had anything *better* to do.

"Hey Barfly!"

Daíthi stiffened sharply. Zander stopped pacing and stood in front of him, raising a curious stare. Never once looking at the boy, Daíthi's eyes slid to the place whence the voice had come. Then, reluctantly, he turned.

Near at hand was a tent and a campfire, around which sat clustered six or seven men. Most had their gazes bowed and were paying adamant attention to the fire. One, confident pose, malicious smile, flask in hand, was looking directly at Daíthi. Twm O'Rourke.

Daíthi hesitated, then moved slowly towards the camp, motioning for Zander to follow.

"Yeah?"

"Where's your wife?" Twm demanded, callously. "I hear from my good friends in Marcansine that she's *very* friendly. With men. I was hoping to enjoy the pleasure of her company!"

There was some snickering from the group. Daíthi narrowed his eyes. "Her company 'ur her thighs?" The second snap was even coarser; she was with Sergestus Carmellus.

"What?" asked O'Rourke, taken aback; apparently this was not the answer he'd been expecting.

"Wha' – Din' ay'one tell ya, Twm?" Daíthi calmly took the flask from O'Rourke's hand. "Ah'm a free man! Far love'a Sergestus Carmellus, low-rankin' gen'ral in t'Augustan army bu'a gen'ral none'aless, she'd ahr marriage d'clared null an' void. On t'groun's 'at Daíthi Murphy'za low-born *'Ceothann'* vagrant, an' worse, orphan *minstrel,* t'marriage's claimed illegal an' made s'if t'ad ne'er eh'zisted. Sa wha'z Ah doin' t'las', oh, *eighteen years,* was't?" He took the cap off the flask slowly; mercifully, it was quite full. "Nah. *Thaynks* ta *Sergestus Carmellus,* Minor Imperatores, *perfide! Seems* Ah'm single! Hain' yeh' figgered how t'feel 'bout tha'." With a toast, he kicked back the flask, drawing half of its contents in a long, savoring draught.

"Um –" O'Rourke began, half-reaching a hand, but he changed his mind, waving it off, letting it go. After all, it was his fault anyway.

Daíthi noticed the motion and briefly lowered the flask, listing slightly. His gaze was far off and unfocused as he mulled a broken stream of dark thoughts. At last, he muttered, "Z'a stupid naym, hay'wayz..."

Before he could empty the flask: "What is?"

Daíthi swallowed what he'd taken, casting the speaker a vague glare. "Candice (*hicc*-uh) *Carmellus,*" he sneered, swaying. "H'esspecially considerin' – on'a (hicc) 'coun'a she goes by *'Candy'* – (hucc). *Perfidy...*" He raised the flask again, this time to drown.

"Candy Carmellus," muttered a voice, testing her name. "Yer right – that is stupid."

"Is it much better'n *Candy Murphy?*" A too-loud whisper.

Daíthi finally lowered the flask, hesitating now between offering it back or keeping it for himself. He deserved *something* in this world, didn't he!? Looking to the speaker, he began, "Han wun'chyeh f'r *once* min'jyer *own* – hey." He glanced to his side as a new hand gently lifted the looted treasure from his own. O'Rielly's son.

With mild discomfort, Zander returned the flask to its owner; it was hollowly light and smelled as strongly of liquor as the man he'd taken it from. He took Daíthi by the arm. "C'mon. Let's ditch."

Daíthi didn't protest. He lobbed Twm an aloof "*Diolch*" and walked off.

They met up with Sully as he was coming back. He greeted them with a curse. "Can't you two stay in one place?!"

Daíthi pointed at Zander.

"Is the meeting still going on?" Zander asked.

"We're in luck," Sully reassured him. "They're just getting back from respite, so we'll be just in time." He started to move, then cast Daíthi a glare. "Keep close if you're coming; I'm not about to go looking for you if you get lost."

Daíthi rolled his eyes and followed, his hands shoved deep in his pockets. Though he said nothing, he made sure he clipped Sully's heel, stepping a little too far, shrugging when he turned and smiling *Oh well*. There was no mispronouncing *that*.

It was not a far walk to the designated meeting place: a clearing surrounded by trees at the center of the Company's camp, where all four factions could come together. Around the edge of the clearing, like a gap-toothed wall, was piled natural debris – fallen trees and immense boulders, over-grown roots, stubbly bushes and moss-covered stumps – which served as stadium seating. Generally, whoever wished to attend the meetings was welcome, unless it was a closed meeting or you were some utterly degenerate soul like Daíthi. Tonight, however, attendance was sparse – which was fine enough for Zander. About ten warriors in various garb were kicking around the clearing, waiting for the others to arrive. Most wore regular street clothes, over which they'd belted their weaponry. Scant though it was, the sight of their weapons – claymores and keris, crossbows and javelins – was marvelous to behold, and with a pang of longing and envy Zander thought of his own polished bow, which would have stood a modest beauty among these treasures. Knowing being armed certainly would have come in handy, he felt almost sick with disgust to think of some Augustan wretch handling his long-cherished weapons, melting down his battered sword, unstringing his trusty bow.

Still...

"*Iochd*, this is awesome!"

Sully cast Zander a subtle smile. "You haven't seen anything yet." A great, mossy stone stood solemnly to the boy's right among some smaller cousins, and Sully leaned forward against it, his hands clasped, his forearms pressed to the stone.

"So, who *are* all these guys?" Zander asked, coolly; he contained himself with grace and poise, although inside he could've fairly died with excitement. "Do you know any of them?"

Sully squinted at the group as if searching the horizon for an island in the fog. "That I do.

"See him?" Sully nodded towards a man at the other end of the clearing. He was built like Tam, a broad, muscular form, though he was shorter. His hair was short and pale, his eyes a deep coffee-brown. He wore a breast-plate, greaves, and a Tartan cloak. "That's Eoghan Fisher, leader of the Northern camp. He's as Highlander as you get, and a hard man to sway. See the man next to him?"

Sitting beside Eoghan was a tall, slim man with pale brown-blond hair. His features were thin, his expression one of haughty contempt. Most striking were his eyes: a washed-out teal in colour, and as piercing as a dagger to the back. He held himself with an air of superior impatience, surveying the crowd with a look of boredom, his mouth a flat-line frown. "That's Caomhan MacAdara. He's Eoghan's right-hand man. He's a Runner, a messenger, and the fastest we've got. Professionally, he's a weapons specialist, which means that any weapon you can think of, he's trained in it, and you can best be sure it's worth some skill. He's not a man you want against you."

"A fact I'm sure he never tires of reminding anyone," Zander murmured, studying Caomhan's superior glare. Pointing discreetly, he asked, "And who are they?"

At a distance were three men dressed in what Zander took to be Augustinian garb. They had remained seated, and were chatting with lowered voices in a smooth, silk-soft language.

Sully followed his gaze. "They? Well, the one wearin' the full uniform, with the gold-hilted blade – that's Astyanax Caeculus Tyndareūs. Leader of the Lavinian clan. Beside him is Acates Glaucus Sarpedon, his second-in-command. The third – what's his name? – Antoninus. One of his generals. They're good men, all of them.

"Now, see there?" Surreptitiously, Sully pointed to their right, where a crowd was trickling in. They all bore what Zander was beginning to see as a *distinctly* Ceothann look. "They're all from the Second Ceothann branch. They get bored and flit in and out of meetings."

"Who's that, leading them?"

"Amergin O'Shea," Sully remarked, sounding intrigued. "He represents the Second Ceothann branch when its usual leaders are away." Upon further examination, Sully smiled. "He looks bored as *Hell*, doesn't he?"

"And who's that, talking to him?"

"Well, that would be the *reason* he looks so bored, isn't it?" He straightened slightly, folding his arms. "That there is Moesen O'Rourke."

"T'little squint!" spat Daíthi, right on cue. Zander glanced at him. He was leaning back against a thick tree, his arms crossed and shoulders hunched, eyes narrowed irately.

"What camp does he belong to?" asked Zander, slowly and reluctantly removing his gaze from Daíthi, who was no longer paying any remote attention.

"Ours, if ya see it that way," Sully answered. "He's often making himself heard at these meetings. For some reason he's regarded with some esteem – as Eoghan Fisher's third."

"Ah."

There were others besides filing in – just enough to form a ring around the edge of the clearing, with some large gaps in between. A fine cluster formed where the three men were standing, shielding them slightly, adding to the protection of the shadows.

Zander watched absently as everyone settled in, amazed by their numbers. "What are they all here to talk about?" he asked, and was answered again from behind.

"Nuthin'!" the voice spat, and Zander turned back to Daíthi. He was still leaning against the tree to Zander's left, heavier now, as if the effort of standing upright was becoming too much for him; clearly Twm O'Rourke's liquor hadn't done him any bit of good.

Catching Zander's look, Daíthi realized he'd spoken aloud, and so explained. "Tay 'ont talk abou' nufin, an' all 'ey *do* iz talk! T'aint any more use'n any *government*, r'ally. 'Tis a *fine* bunch'a *cowards*, them lot!" With that, he returned to staring at nothing in particular.

Zander looked at Sully, who shrugged as if to say, *Well, he's right on that one, I'll give it to him.*

"Barfly!!"

Startled, Daíthi slipped against the tree, uttering what was undoubtedly a curse in his native language. For a moment, it looked as though he would blend into the tree or bolt, for as harshly as he pressed his back against the bark. Then, to the sure amazement of everyone, he stepped forward.

"Aye?"

His accuser stood in the center of the clearing, anger in his stance. Unrelenting, he continued. "I thought we'd run you out of here?"

Daíthi swallowed and tried his best to stand upright without swaying. "Me? Sure Ah on'ee jus' got 'ere – you sure 'at warn't someone else? *You've* a taste f'r barley, hain'chya? P'raps yer confused."

Under his breath, Sully murmured, *"And his head upon a pike they speared..."*

If he heard him, Daíthi ignored him.

"You better watch yer mouth, vagrant!" the speaker snarled, "or it'll be more than a pillory we're stringin' you up for!"

"Ih' zair'a *reason* j'ya called me ou', now, 'ur d'ya jus' enjoy t'sound'a me voice?"

Zander wondered whether Daíthi had found his courage or if the liquor had just kicked in. He hoped Daíthi would stay alive long enough to find out – a sentiment strengthened as the accuser stepped forward and Daíthi stepped back – a motion on the latter's part embarrassingly more like a lurch.

"What makes you think you can just come waltzing back here – and to council, no less! You –"

Not waiting for the insult, Daíthi cut in. "Ah still haf'a message t'deliver, an' I'ent goin' back 'til Ah do! I'ent on'a go back on me word."

"I'm sure."

"Jus' lemme say m'piece an' I'en go back'a Carteia an' –" His impatient but reasonable compromise was cut off like a descending ax.

"Go back to Carteia *now!*"

Daíthi stepped forward. "How'z whut Ah gotta say anyfin 'genss whu'chya'll talk abou' – I'ent takin' away fru' nufin!"

"Any word from *you*'s not worth the listening to, nar the waste of our time!"

"*Wha*'re yew *afraid* 'a!?!"

"Now!!"

Irate, Daíthi shrilled something in his own language, cut off as Sully restrained him from behind, hooking his arms around Murphy's elbows.

Confused, Daíthi struggled a protest, but Sully pulled him back, whispering, "*Leave it!* It's alright! Just leave it for now – *come on!*"

It took a moment, but Daíthi relaxed, relenting with extreme reluctance. He cast a glance at his accuser, and the whole of the gathered tensed as he jerked his arms out of Sully's grasp. But he did no more than throw them a glare full of hatred, turn on his heel, and stalk off.

Brushing past Zander, he hissed, "Yer turn," giving him a slight shove in the shoulder towards the clearing.

Zander watched as Daíthi settled a fair distance off, dropping cross-legged at the base of a tree (whether by accident or on purpose), waiting and watching how the meeting progressed. If for no other reason than sheer stubbornness, he was going nowhere.

The man at center stage continued, bolstered by his victory. "See? This is exactly what I was talking about! The state of the army is being decayed by the fragility'uv our borders. If rot like that is able to come waltzing in – and dead drunk, no less – what's to stop an army from crashing through?"

"Lack of desire?" prompted Amergin, with a broad grin.

The childish giggling of the Second Branch warriors was lost in the more serious debate that followed.

Sully returned to Zander's side, shaking his head. He settled back against the stone, leaning as he had been before.

Very soon, Zander saw the truth in Daíthi's earlier words: they really *did* talk about nothing. And, quickly growing bored, his attention began to wander.

He looked over at the Lowlanders, watching as they whispered among themselves. For so many years, he had listened to his father tell stories of Ceothannanmór. Of the Company. Of Balladarach. These were *his* people, in blood and in bondage. And, for whatever reason God had seen fit to bring him here, he was grateful.

Across the way, Caomhan looked up as a woman knelt beside him, gesturing with close-drawn hands as she whispered to him ardently.

Sully jolted as he was whacked in the arm.

"Who's that?!"

He gave the boy a sidelong glare. "*Jeeee*-sus," he muttered, shaking his head. Then he followed the boy's stare – and recoiled. "She?"

Every serpentine emotion thrived within the word.

Seeing the youth unrelenting, Sully gave a sigh. "Caomhan's cousin. They call her Roscanna, whatever her original name may've been. She's a mysterious creature, that – like a moonless night, so it's said. A fearsome temper, and rancor to boot. Like the goddess Hera incarnate."

Zander nodded absently, watching the woman from a distance. She was not remarkably tall, but she *seemed* taller. Slim and graceful, she carried herself with rigid pride, her intransigent wrath visible in every motion, every curve of her form. Her hair was black as night and bologna-curled, reaching to her hips. Her eyes were bright as ember, an almost electric blue. Though her face was delicate, her hard expression was intimidating enough to cow any lingering stares – though that never stopped them from looking. She was a forbidden beauty, a tantalizing, acrid, and powerful flower.

"She hates you, by the way."

Zander turned as if slapped. "What?!" he gasped, instantly hurt.

Without looking at him, Sully explained, "In short, she blames O'Rielly for the destruction of Balladarach, and her husband an' son with it. By extension..." He shrugged. "She hates you."

Zander was shock-silent a moment, too many words crowding the tip of his tongue. At last, he swallowed them, breathing, "This night just keeps getting better and better..."

And it wasn't over yet.

"Oh, *hello*," voiced Glaucus Sarpedon, leaning out to get a better view. "What else has Sullivan netted for us today?"

Zander turned to Sully, suddenly anxious. The sailor laid a hand on his shoulder. "He's just a boy," he said, protectively.

"Bring'm forward!"

Zander stepped into the clearing, free of Sully's grasp. "I can bring myself just as easily," he announced, no longer particularly inclined to go out of his way to be polite. He had nothing to lose.

"Ooh," taunted Caomhan, his voice low, like a snake's. "Kid's got a little moxie, doesn't he?"

"What's your name, son?"

As if inspired, the general company murmured the question in eager reverberation.

Zander hesitated. How many times in his life had he lied? *Setánta Fahy.* The familiar pseudonym stirred at the back of his mind, ready and able. But this time, something held him back.

He was done. He had lost everything – *everything!* – and for what?

He closed his eyes, feeling the soft touch of the cool night against his cheek, listening to the silence of the world.

"Well?!"

He was tired of running, damn it! Tired of existing on faith alone. Tired of answering to a world that despised him, tired of being beaten down. Tired of praying for a better day and watching his prayers go unanswered. For too long he had accepted the way of the world, and now, his waiting was *over*.

God – how much of Destiny is made?

He opened his eyes. And, with pride, with anger, with undeniable strength, he made this answer:

"My name is Zander Mac Donnachaidh Ó Raghallaigh, son of the Master of Flames!"

75
Bad Company

Zander O'Rielly walked through the lines of tents, trailing after Daíthi and wondering what on earth he'd gotten himself into.

The dead silence following his introduction had erupted into disorder louder, angrier, and more chaotic than anything he'd ever seen. Some declared he was a liar and a con man, others said he was a spy, a madman, a drunk. Some averred to have recognized him in an instant – as a fraud, as a loon, as the man he claimed to be. Many were emphatically for making him their king, their sworn leader, then and there. Many more wanted to kill him.

Those calling for his death had been particularly persuasive, though they could not agree on how, or why. Treason, said some, audacity, said others; he knows too much! Some wanted to shoot him, to drown him; others were for the noose or stoning. The more they talked, the more they reasoned, the more they justified their own claims. In the end, they were arguing for the killing of each other, vying heatedly for the position of executioner.

That was when Daíthi had stepped in – literally. He had strode silently to the center of the mad vortex and put an arm protectively around Zander. And then, glaring murder at them all, he admonished, "Wou'jya kill ta' on'ee man 'at can save yer coward arses?!" And before they could kill them both, he led Zander away.

The boy's life was now entirely in his hands.

The moment they'd been out of ready earshot, Daíthi had taken to grumbling about the stupidity of the situation, and Zander had left him to it. He could not understand most of what the man said anyway, as it crossed over five languages and two or three dialects, he was muttering, he was staggering drunk, and even when he *did* speak a language Zander knew, his accent was so bizarre that it alone rendered most of his speech incomprehensible.

So Zander followed like a dutiful puppy-dog, his hands jammed deep in his pockets, considering his own furious rant long after Daíthi had fallen silent.

"Where are we *going?*"

Daíthi glanced back behind him, slowing up a little bit as he realized Zander was falling behind. "Lookin' f'r a campfire," he explained, shortly.

"Don'chya think we should be back at the meeting?" Zander demanded, surprising himself. "Ya know – defending our honor and whatever?"

"Ah'yent got a'y honor *lef'* t'defend," Daíthi responded. His eyes were scanning the campfires they passed, as if looking for a face not hostile. "An' anywayz, dey're mahr lye'lee'a behea'jya *lookin'* achya. An' Ah fancy me life, ih spi'a hss'warth."

Zander bit back a request for a translation – *was that "in spite of"?* – holding his silence.

"S'look – I 'ent much f'r runnin'a nigh', so I'd 'ppreciate it *ih'mensely* f'yed jus' stick aroun', a'kay?"

"I'm not a captive," Zander replied, curtly. "I came here of my own will."

Daíthi evaluated this. "Hah... J'yah know, Ah li' t'think t'same thing, bu' Ah'd run sure's Hell soon's Ah'm done here. Hain'a place ya wanna stay long lest yeh've some kin'a power. An' – *shar* – I ent gat none." He cast Zander a more serious glance. "Truss' me, hon. Don' run."

"Ih –" Zander paused, considering the man's words rather than glossing over them this time. "I won't run..."

Daíthi heard and recognized in the trail Duncan's particular brand of dishonesty – that quaintly open half-truth toying with mistrust. "Yeah? Good..." He glanced at the boy from the corner of his eye. "An'?"

"Nothing."

"Uh-huh... Dey wanna kill you, aye, boyo? Fink'uv ih dis way, f'ss mahr t'yer Gad-damned juvenile *sa'hissfaction*: Ah ent guardin' dem fra' you... Ah'm guardin' *you* fra' da bleedin' *caymp*."

"Maybe... Maybe I should be armed?" Zander could have flinched at the amount of childish hope that fell into his tired voice as he spoke these words. And, accordingly, Daíthi answered, "Auh, shit; dan' warry abou' tha'. Auh, an' shar, dey coun' kill deyr way outta a' empty room."

Zander found this notion more disconcerting than the barrage of death threats he had just endured. "But – they're our *soldiers* –"

Daíthi laughed, a bitter and raucous sort of giggle, spreading his arms. "Yur army, mah prodigal son!" Speaking as if blandishing God, he added, "An' damn'tt, ain' *Ah* th'one owed, talk abou' slaugh'ered calfs an' *loyalty*, a'twain – uh!" He tripped, falling against Zander, spiting his smaller stature by nearly taking him down. Zander was forced to support him as best he could, the strain on his shoulder nearly killing him; it had to have been the twelfth time that night Daíthi had done something stupid to aggravate Zander's wound, and the youth was beginning to wonder if it wasn't on purpose. Not merely a messenger, but the worst assassin ever.

What with the muttered swearing and dizziness, Daíthi was somewhat slow in asking: "Hey. Y'arrite?"

"...Yeah." Zander's voice cracked. "Yeah; I'm fine."

"Uuh..." It took some effort to get himself back on his own feet, all the while keeping a hand firmly on Zander's shoulder for balance. With an apologetic half-smile he observed, "S'funny, huh, t'timin'a thin's? Y'hain' never ready..."

Zander stared.

"Ahr..." Daíthi's gaze returned to the distance, squinting against something in his head. "May'ee... *yar*. An'jya cann'a realize *why* 'til after."

Zander softened. "Is it you or the whiskey that's profound?"

"Ha! O'Rourke's brew packs *wondars*, don' it?" Trying to focus, he bowed his head; his body was limp, limbs heavy, head whirling. "Alwayz hi'sya la'er. R'membar tha'. When ya leas' need'ur e'spect it."

"'Kay. D'you need a hand there?"

"*Naa*." Daíthi half waved him off and had to catch his balance again. "A'righ. C'mon, le's go."

Zander followed, sticking close in case his guide conked out. Though he wouldn't say it, he was grateful for the staggering slow pace. He was exhausted, and his bones were beginning to ache, his wounds beginning to really sting. *It always hits you later. When you least need or expect it.*

"Hey, Daíthi –?" The man turned a little too swiftly, and Zander grabbed his wrist to steady him – and did not let go. Fixing the man with a stern glare, he demanded, "Are ya cognitive, or do I hafta wait until you're sober?"

Daíthi managed to work out a comprehension. "Yeh'n *try*."

"Okay." And, before he could change his mind, Zander began, "What did you mean when ya said *I* was the only man that could – *'save them'*? Save them from what?"

"Sa' who?"

"What will I save the Company from? What did you mean?"

Ah; *there* was a sobering question! "Uh – Ah say'a lah'a things –"

"C'*mon*, Daíthi!"

"*Lis'n*, kid – I – Ih c'n wai' 'til mahrnin' –"

"Don't blow me off!"

"LOOK!!" Daíthi lost his temper. "*Yew* know'uz well'uz *I* do an' az well'uz Joe does an' az well'uz *ev'rybody else!* A'twain t'two'uv us 'ere'z anuff change t'disrup' t'whole balance'uv things! Bu' hain' *nufin* gonna be done t'nigh'. Nah c'mon!!" He pulled his arm, jerking Zander forward. "Da' pro'llum wi'ddis camp is t'pro'llum 'uv you. Tay 'ont rah'lize whu' I tol'jya ahrlier. Ya hain' *never* ready. Gad'iz own plan. An' all ye c'n do iz hang on, hope f'r t'best, an' try li' Hell t'do righ'. Righ'?"

"...I *guess*."

"Hain' no guessin' abou' it, nee'ver. Ya *can't*. Ya never know!"

Zander considered this shockingly wise revelation; releasing Murphy, he resumed following in brooding silence. Maybe Daíthi had a point. *When you least expect it...*

"Murph, ya fool! Come here 'til I see ya!"

Daíthi looked up, and his initial concern broke in a smile. They had reached a tent at the back outskirts of the Highlander camp, before which eight men sat in a packed circle. The man who had summoned them stood waiting. He wore a tweed cap over his short, vividly red-orange hair, a waistcoat, white shirt, black dress-pants, a slim, green tie – and a great big, welcoming smile.

"Red!" Daíthi greeted him, clapping the man's outstretched hand in his. "Man, ihsa relief an'a half t'see you!"

"Murph! Come'an si'chyerself beside ahr water-logged fire!"

The men were seated on logs arranged in something of a circle, but for one man on the side opposite who sat on a stump. The ground between them, like much of the ground at the outskirts of camp, was all mud and roots and wood-chips and dead leaves. Most of the men had drinks in their hands – either bottles or pub glasses – and one man (Zander would come to know him as Kelly) sat absorbed in trying to light a pipe. They made a shuffling effort to make room, but they only succeeded in creating a gap of about a foot's space.

"Who's this, now," inquired a man of their circle – a dark, wiry man called Rhys, "that ya brought back with ya?"

Daíthi swung a leg over the log, straddling it. "Ah'm t'wach'm."

"Huh..." Stroking his sparse beard, Red tipped Daíthi a sly glance. "Who's his mammy? He dasn't look a wee mite like ya."

Daíthi scowled. "If yer gonna make room a'tall, den be well'n shar yer makin' room far his'self, 'cause *tha' righ' there*, gen'almen, is O'Rielly's on'ee son." He slung a hand on Zander's shoulder, moving him forward ever-so-slightly. "Hon, this rabble'uv chaos'iz t'big'ess mess ya ever turn'jyer eyes agains'."

"Ra'lly, Daíthi?" drawled another man. "An' that's all t'introduction ya give us, an' after all these years?"

Daíthi rolled his eyes. "Fine." And, in no particular order, recited: "Here's Rhys Daukyn, Shaw MacLochlan, Mike Flannery, Faolán Kellog – we call'm 'Kelly' – an' tha's Matty Calhoun da' seminarian, Gerry – Gerard – Masters – sometimes we call'm 'Mast' – an' Sheeran, who hain' nare tellin' ya his full name on accoun'a he don' wan' ya t'know ihs *Lemoine O'Sheeran*."

Sheeran, the man on the stump, pouted fiercely. "Ya lil' tattle-tale!"

Ignored in the disorderly introduction, Red puffed himself up proudly, offering Zander a hand. "No sense in 'is head, that one. I'd be Padraig O'Donnell, but there's those that call me 'Paddy' an' others still who call me 'Red'. At yer leisure." He then introduced his comrades in the correct order, purely to show Daíthi up. "Ya got it?"

Daíthi received Red's glare and corrections with a shrug. "Hey, Ah'm *slammed*. Seven ou'a eigh' hain't bad." Shakily, he swung his other leg over the log, dropping down beside Gerry, another big, Highlander man.

Red rolled his eyes. "And wou'jya have a farst name, *'O'Rielly's son'?*"

"My name's Zander," he said; he had never had to introduce himself so many times in his life – certainly not in one night.

"*Zander*," repeated Rhys Daukyn, appraising him. "Foreigner?"

"Balladarach."

"Ooch!" They all jolted back, arms outstretched, overwhelmed, it seemed, by the immensity of the answer, and –

"T'*insensitivity!*" exclaimed Red, scandalized at Rhys' imprudence. "T'*shayr* insensitivity!" And, in agreement, Shaw punched Rhys in the shoulder.

"I don't really mind –" Zander began.

"Ach, how *pih'lichya* ahr!" Red whacked Kelly's shoulder indicatively with the back of his hand. "Move over."

Kelly jolted, his concentration momentarily jarred. *"Jayz,* Paddy!" He shifted to his left, and Red took a seat beside him.

Taking Zander by the arm, Red said, "Si'chyerself down, Zandereen Óg." When he had, Red continued, "Don't be mindin' Rhys, now. He hasn't a mite'a good culture, but he means well anuff."

Rhys frowned. "A man c'n take offence t'that."

"But he means well anuff!" Red enunciated pointedly, and Rhys said no more. "Now tell us, Zander – because I wouldn't believe a word of it from Murph – not a word – but you seem to have fallen outta the woodwork, so ya have."

"O'Rielly's only son," said Shaw, ponderingly, anglingly. "That *is* a powerful interestin' patronymic..."

Zander bit his lip, but once again Daíthi came to his rescue. "Gents, layve'im rest th'nigh'. 'Eez been f'rew anuff Hell'nn interrogation wivout you ninnies addin' needlessly t'the mess. Mah throat's dry – Fa'der?" He clicked his fingers at Matty, reaching across Gerry to demand his drink.

"You would do well teh arn it every now an' again," Flannery grumbled.

Daíthi was affronted. "Eah? Wha' – an' how, like?"

"Och! Ah forgot!" And just like that, Gerry rose and entered the tent. The noise of scrounging about – clattering and thuds and rufflings and one loud smash followed by an unenthusiastic "Whups" – rang out for the next five minutes, and no one felt particularly inclined to do anything until he came back. When at last he did, he was carrying a sleek, maple guitar, which he handed over to Daíthi.

Daíthi ran a hand over the maple wood of the guitar; for the first time, Zander saw the beginnings of a genuine smile on his face. He strummed the strings – and flinched. "An' no'a *soul* among ya could tune 'er?"

Red laid a hand on his heart, holding the other palm-up as he vowed, "'Pon me life, t'same wards I spoke meself. Isn't it so? An' wouldn't I'uv tuned it far ya – had I but known how to meself?" He added the last line with a mischievous grin, and the others all groaned, *"Aaauuggh!!"* as the man next to him moved to push him over.

Daíthi permitted a sort of irrepressible giggle; then it was back to business. "Ah deman' sustenance!! Fa'der Calhoun – ee'ver *be* da vessel'a God an' gi' drink teh he who tharsts ahr pu' down yur beer an' homilize me on *intemperance."*

"Ah –" Matty was initially lost for words, unable to choose between sermon and barley. "Rough night?" he guessed.

"Try *week*." Having roughly gotten the instrument in tune, he strummed up a nicely-chorded ditty and sang:

"So come all ye jolly young fellows
I'd have ye take warnin' by me
Whenever yer out on the liquor, me lads
Beware of the pretty colleen
She'll fill ya with whiskey an' porter
Until yer not able to stand
And the very next thing that you'll know, me lads,
Yer landed in Van Dieman's land..."

Even with the sheer amount of attitude he poured into the lyrics, his voice was beautiful; like a barroom-hero angel.

Zander looked up. "That was nice."

"Ya luh' surprised. Ain' nobody ever think Ah can do shit." He turned to Gerry, showing the back of his left hand, fingers splayed; a slim, white mark was at the base of his ring finger, but no ring. "Spaykin'a shit I ain't doin'..."

Gerry pondered the transition, but then he realized what Murphy meant, and he grumbled a sigh. "So she got ya a'last. Ah'm sarry, lad."

"*I* ent," Daíthi answered.

"Naw, but doesn't it suck, so," observed Flannery.

"Aye," Shaw chimed in. "Though I never did like'er."

"Ya never did *meet'*er!"

"But I don't like her, jus' t'same."

"An' how wou'jya be knowin' such an opinion on someone, Shaw MacLochlan, at all, an' yerself never havin' once met her?"

"Fer –"

"Guys, *guys!*" Daíthi cut them off, irritated; he had reviewed the situation enough without their help. "Sumfin else, please? Ha' some decorum."

"*Decorum?*" Shaw repeated, mockingly impressed.

"Aw, Jaysus," Sheeran complained. "So we can't talk about castin' off the gold-digger'uv Babylon, an' we can't talk about how you conveniently acquired th' last blood 'uv the O'Rielly –"

"Gimmie a beer'un I'll pass ou' an' den you can talk abou' wha'ehr da feck you wan'!"

"Tolerable," Lemoine O'Sheeran enunciated in reply, savoring his own six-gold-piece word. "Poor, tolerable Daíthi."

From a bag near his feet, Shaw pulled a somewhat-loved and grimy bottle. Lifting a recently-emptied pint-glass from beside his sneakers, he poured himself a generous serving, raising it high. "I'd like t'be makin' a toast, gents, if no one's objectin'." They weren't, as he well knew, and so he continued without more than a second's pause. "'Tis both a farewell, an' a beginnin'; an end to a' old life, an'a hail to'a new an' better one. An' it goes as such –"

In no shortness of obscenity, Shaw gave them a veritable lecture, bidding farewell to Murphy's now-ex-wife – whose name he could not recall, having never known it, though he had no lack of grossly expressive and highly colourful titles for her – and ended with an appraisal of life as a single man, which was, if possible, even more raunchily descriptive. His audience listened with a mix of appalled shock and amused horror, too stunned to break him off.

"Sláinte!" the homily ended; Shaw shot Daíthi a smile and a wink, tipping back his glass with poise.

Daíthi bore an expression both pleading and mildly disgusted. *"Gad,* Shaw! K'we –" thumbing an indication to Zander "— 'ee ain't narely mahr'ena child, like."

Rhys pulled Sheeran's bag closer, taking out another glass. "Well, *yar* wasted," he pointed out.

Daíthi was indignant. "Oi han't *plann'* on *comin'* hayr, ya *spineless* traitor ya!"

"Alright then!" Red reached across Zander, grabbing Daíthi's arm as he leaned forward. "Settle yarselves, why don'chya? We need no bloodshed t'night."

Daíthi was reluctant, but he sat back, and Red released his arm. "Hanyways, Ah'm soberin' up mos'... Oh! Sully said ee'd be by a'nigh', on accoun'a d'boy an' d'mee'in, an' – He *will!"* he added, in sudden protest to their dubious, pitying glances. "Shite! Gad-damn –" he nicked a beer can from beside Matty's feet, cracking it open *"—Siaxsiou tiopritom; se-eia tarī-caecos eti exuertina. Luge!"* Of all the languages into which he lapsed, this seemed to come most naturally. A smooth, silken, serpentine language, like Lavinian, but punctuated with heavily rolled "R"s like Turranian and guttural "ch"s, like Ceothann. It was a weird sort of language – beautiful, but weird.

Red rolled his eyes. "Ya know no one else can understand that tongue, Murphy. Pick another!"

"If yer gonna keep proper watch," Matty decided, "one drink'll suffice t'warm ya!" He moved the rest of his beer supply out of Daíthi's reach.

"Skivers, the lotta yeh," Daíthi grumbled, before dipping into his much-awaited drink.

"Skivers!" echoed Shaw. "Ya know, I think we teach a good lesson; maybe ta' rest'a camp ought'a learn from us? Stop buildin' houses an' start buildin' an army!!"

"What do you mean?" Zander asked, speaking up at last.

"Hoy!" Shaw exclaimed. "New guy! Ye have a tongue in ya after all!"

It was Red who actually answered him. "Ya see, Zander, ta' idea towards ta' start was to be *mo-bile*. We need trainin', right anuff – most of us, we're *fighters* but we ain't *warriors*, right? A mobile camp keeps us out of danger – 'til we're ready to be established."

"As it is," continued Flannery, "ta' camp as a whole is kinda stuck in a rut."

"Invisible Fortress Syndrome," scoffed Gerry.

"I'm not sure if they think we're immortal ar' just didn't *think*," Matty agreed.

"Hain' it whu' Ah tol'jya?" Daíthi slid a glance to the boy, his eyes narrowed slightly. "Tay think ehree'thin's par'minent an' unchangin'. Warse, *unchangeable*."

"I tell ye, gents," said Shaw, leaning straight-backed, his arms folded. "*Ahr'ganization*, tha'ss the key. And Zander, m'boy –" a few stiffened, worrying what knowledge he would impart "— yer lookin' on ta' most *organized* part'a this whole damn camp."

"We hain't nee'ver *ahrganized* nahr *mo-bile*," Daíthi scoffed, "yer bleedin' kiddin' yerselfs. 'Specially Gerry."

"Wha –!" the Highlander began to protest, but Daíthi cut him off.

"Yer feckin' tent. 'En Ah lived in a house Ah din' have nare as much possessions in'tt as you."

"Which house?"

Daíthi ignored the inquiry. "Hain' no use fer a novelty shop a'y longer, Ah'd say."

Gerry looked at him flatly, unmoved by the younger man's ideals. "Would yeh."

"S'yer shit," Daíthi shrugged, turning away again and leaving it at that. He tipped a toast to the gathered. "A health t'the Company," he muttered, taking a drink.

"I fergot how contrary ya are," said Red, rolling his eyes at Daíthi. "Go on an' do us a song, why don'chya? A full one, now."

Daíthi groaned. "Kohr, *Jaysus*, Ah'm too drunk now," he said, a roll slipping into the "R". He finished off his drink, then slouched forward, setting the drained can on the ground and wiping his mouth on his sleeve in one stroke. "Wha' d'Hell d'yeh do wivout me when Ah'm gone? Da whole feckin' camp goes ta' *shit*." A jerk of one peg knocked the instrument into tune, and he strummed the guitar softly, thinking. "*Ah'm* da feckin' *ahrganization*, ninnies... A'kay, Ah got one." Without further eloquence, he looped his thoughtful pickings into a song. One man who instantly recognized the tune hailed it with a hushed, "*Yesss!*" and Daíthi smiled, and started to sing:

"I am a young fellow that's aisy and bold
In Castletown Conners I'm very well known
In Newcastle West I've spent many a note
With Kitty and Judy and Mary

Me father rebuked me for bein' such a rake
For spendin' me time in such frolicsome ways
But I ne'er can forget the good nature of Jane
Agus fágaimíd siúd mar atá sé."

True to his word, he gave the full rendition of the conquests of the Rake. The song ended on a slow chord; purely to show off, he strummed a quick, gorgeous little Celtic melody after it. His reward:

"Show off."

Daíthi's grin was cocky. "A reg'lar pub-singer y'are, Fa'der." Matty, the accused, merely shrugged. "Dass werff anuver drink, ah?"

"*Sober,*" Padraig emphasized. "*I* won't be intercedin' wit' the Captain on yur behalf, so I won't!"

"Khh'man, man, dan' mayk me go cold!" Daíthi moaned, in insistent complaint. He turned to Zander, surprisingly anticipating his question. "Ah gatta as' Sully wha' t'do."

"...About what?"

"Oakh, 'ee'z stayin' back'a'da mee'in, aye?"

That one eluded Zander entirely; he stared at the man, wracking his brains frantically until Flannery translated: "Something about Sully and a meeting!"

"Oh –"

"S'eez gatta tell me wha' t'do!" Daíthi quipped. "Ih'dey still wanna kill ya we gatta move ya."

Matty started, alarmed. "Kill –?"

Daíthi waved him off with a flick of his hand. "Trifles. Ah gat ih *aaalll* un'er control!" He got to his feet – too quickly, for he stumbled and had to grab his balance again. *"Deuos, immi mesco,"* he uttered.

"It's nice you can be so flippant," Zander clipped.

"You'll be safe with us, Zander," Matty reassured him, with genuine concern in his soft, brown eyes.

"See? Wha'd Ah tell ya!" Daíthi quipped.

"Ah, go on ya," Flannery complained.

"Play us a tune; it'll keep your blood warm."

"A chiall. Hain'chyou demandin'." Natural as anything, he again set to tuning the guitar. "Zander, honey, wha'd Ah tell you? Hain' nufin gonna be done a'nigh'. Okay?"

Zander narrowed his eyes slightly. Beneath the levity, debauchery and sauce, there was in that line something solemn, something paternal, something of a guardian. "Alright, Barfly."

Daíthi raised his eyes to God. "T'thanks Ah get," he muttered. Then, to his avid audience, he declared, "Fine. Ah'll do ya one more, then someone else's gatta tayk over, a'kay?"

"Let me get warmed up," Shaw replied, flashing a smirk as he took a drink from his glass.

"Deuos, uoretos." Drawing a quick chord across the strings, Daíthi turned, and something caught his eye. With a raptness he had abandoned, he focused through the gloaming. *Delirium Tremens... No sweeter death could you have brought me...*

A slim figure was cutting through the trees at the edge of the Northern camp, heading for the woods. Her skin was so pale, her grace so flowing, that at first he took her for a Ladi Wen, a spirit. But fear was not the feeling she evoked. In spite of the darkness, he knew she was beautiful. And, as he beheld her, he knew a feeling he had never truly experienced before...

"Ah'righ'..." Absently, his fingers arched over the guitar strings, caressing a few, fine notes. A song began to take shape, the plinking notes became a soft, smooth strumming, and, in his rough, clear, beautiful voice, Daíthi sang.

"Oh, all the money that e'er I had, I spent it in good company
And all the harm I've ever done, alas it was to none but me
And all I've done for want of wit, to mem'ry now I can't recall
So fill to me the parting glass. Goodnight and joy be with you all.

Oh, all the comrades e'er I had, they're sorry for my going away
And all the sweethearts that e'er I had, they wish me one more day t'stay
But since it falls into my lot, that I should rise and you should not
I gently rise and softly call, 'Goodnight and joy be with you all.'

If I had money enough to spend, and leisure time to sit awhile
There is a fair maid in this town, that sorely has my heart beguiled
Her rosy cheeks and ruby lips, I own she has my heart enthralled
So fill to me the parting glass. Goodnight and joy be with you all..."

76
Scars

The sun never shone in Ariegn – at least, not for a prisoner.

For the thousandth time in the last four hours, Duncan tried to shift. It had been humiliating, being so paralyzed with pain and exhaustion that he could not even utter retaliation as he was abused. Helpless! Had he been able, the accursed word would've made him shiver.

With a massive effort, he at last succeeded in rolling onto his back, his re-broken scars pressing against the cold stone. He gasped, but, in a moment, the numbing chill was soothing enough. How long, he wondered, did they expect him to last if they kept him like a piece of raw meat – bloody and open to the flies?

He was somewhere on the first floor, a hall off of the great courtyard. The room was a long hallway that, if he stood and faced the bars, dead-ended to his right. To his far, far left, a short flight of three steps, set a foot's gap from the bars of the first cell, led to an ancient little door clearly unaccustomed to its more modern purpose; it was a feeble, but noble, figure. The room was long, but relatively narrow, thanks in part to the addition of the cells. They were made of iron bars, each set about two inches apart. And everything else – the wall at your back, the floor at your feet, the ceiling hidden far above your head – like nearly everything else in Ariegn, was made of stone.

And this, for the moment, was his world.

Gritting his teeth, he pulled up his right knee, elevating it as much as he could manage. When he could get his foot flat on the floor, he surrendered. *Good enough.* He closed his eyes.

A lock churned somewhere deep through the darkness with the slow squeal of metal on metal. He listened to the soft, brief protest of the ancient hinges and braced himself, though he did not yet bother to open his eyes.

Footsteps on the short flight of stairs; he recognized the gait and waited until it was right outside his door.

"No one else will come and visit me? Shame..."

The footsteps came to a halt, and the voice of Judah Basilicus made an answer. "You always have to get the first word in, don't you, matchstick-eater?"

"... And the last."

"Uh-huh." Unimpressed. "You know, for once it might *behoove* you to be nice to me. I am, after all, the one holding your life in my hands."

Duncan's eyes opened slowly, but he made no move to reply; merely folded his left arm back behind his head, blocking Bassus, however slightly.

"What, no back-talk?"

The fire-eater shot Bassus a side-ways glare. He said nothing.

Bassus's smile broadened. "The great *Fire-Master* won't talk to me!" he exclaimed, grandly. "Now there's a loss!"

The fire-eater turned away.

"Perhaps we've been going too easy on you." His fingers curled around the bars. "Son, if you can remember what we did to you last time – Lord only knows how addled your brains must be – by God, that's *nothing* compared to what we've been developing these last ten years. Haven't yet been able to find a body willing to endure the test, but... Flagellation and immolation don't seem to have had much impact on you. And you're strong. Expendable. Tell me: can you envision the process by which, using dull tweezers, one might be able to castrate an octopus from the *inside* –"

"What do you *want*, Judah?" Duncan demanded, in a tone as gruffly intolerant as it was bored. Still, he shifted, drawing up his other knee.

Bassus's grin was patronizing. "A little conversation never hurt anyone, O'Rielly," he observed, leaning back. He didn't expect a response, but he allowed for a slight pause anyway before he revealed the answer. "I'm here to deliver food."

Duncan frowned. *Hm. Yum.* "A little honest work never killed a soul, either," he grumbled.

Now Bassus frowned. "Feel better, now that you've had your say?"

Duncan shrugged one shoulder uncomfortably. "Forty-seven lashes. I feel nothing."

"Forty-seven, huh?" Basilicus cut. "You *are* a strong body."

For a moment, Duncan was content to remain silent. Then, weakly, he mused, "I've had practice." Whipping for slaves, whipping for seamen, whipping for landsmen likewise confined. He had first tasted the lash as a child, and Ariegn had before now given him his surfeit. "They went easy on me this time, though the formula was the same. You know the drill: Strip. Spread eagle. Extend your arms. *Pray.*" He swallowed and, in a voice even more faint, he breathed, "*One.*"

You could almost hear the whip crack at the end of the word. Bassus shook his head. "You have a real talent for creeping people out, you know that?!" he spat. "A *talent.*" Bending down, he pushed his delivery through the bars, sliding it across the cold and grimy floor. Duncan watched the motion silently. There were worse things to be talented at.

"Here," Bassus glowered, grudgingly, as he straightened up. "*Food* for you, you little scuzz-ball."

Duncan turned his head, viewing the offering at the soldier's feet with disinterest. "What is it?" he drawled, rather shortly.

"Your rations."

Duncan made a face, then returned his gaze to the occult ceiling above.

Bassus scowled. "Don't you *dare* go on hunger strike again!" he warned, his voice raised menacingly.

"I don't trust you!" Duncan snapped, his voice raised in return.

Bassus swore violently. Trying as hard as possible to sound reasonable, he explained, "Look. There's no point in druggin' you, and there's *certainly* no point in poisoning you."

He had to say it, for, addled or not, Duncan remembered well the tortures of Ariegn. His stomach turned, and it was all he could do to keep his voice even as he demanded, "*How* does that make me trust you?"

Bassus uttered another curse. For a moment he stood in contemplation, his head bowed. "Y'know what?!" he blurted at last, dropping down to seize upon the vagrant's rations. His voice, his motions, his expression – everything about him was the picture of a man who'd reached his patience's end. He held up the bread and water for the ingrate to see, then, breaking the bread in half, he muttered, "You are *such* a child," and ate a small piece from the middle, chasing it quickly with some of the dingy water. Holding up the food again, he swallowed and demanded, "*There!* Happy?!" And set it back on the floor.

Duncan sat up slowly, his eyes fixed on what would be his first meal in nearly four days. He felt almost faint with hunger, but he would rather be whipped a hundred more times than let Bassus know it. Sliding on his dark coat, he cast Bassus a brief glance; apparently he was now important enough to have a taste-tester. Tensing, he dragged himself over to the bars, gingerly but smoothly – certainly more smoothly than it felt.

The bread was wickedly stale, but at least it wasn't moldy; he could tell by the taste it wasn't poisoned – insipid, but not poisoned. He took a drink from the cup Bassus had passed him and pulled a face. "Ugh. When I said give a man a drink, I meant a *man*, not a flower," he complained, quietly.

Bassus made a face in return. *Lousy Ceothann drunkards.* "And you're worried about us *poisoning* you," he snorted.

"*God*," he rasped, taking another drink of the bland, clear, bitter liquid. "Just about!"

"Don't make me take that off of you, you ungrateful slug!"

"What good would *that* do?" he responded lazily. He glanced around him briefly, then stuck his free hand through the bars. "Well, *help* a man up, will you?"

Bassus looked at him with disgust and took a step back.

Duncan rolled his eyes. "*Fine*," he hissed, under his breath. Setting down the cup, he took hold of the bars with both hands and began the laborious task of dragging himself to his feet. It took him a full minute to accomplish it, all the while muttering a steady stream of Ceothann in a bitter, angry tone.

At last, he stood! Well, he had to cling to the bars pretty tightly for the moment, just until he got his bearings, but – he stood! He looked at Bassus, a broad grin on his face. "Ha!" he triumphed, with pride. He bent shakily, just enough to snag the cup, keeping one arm firmly wrapped around the bars. Then, rising, he toasted the knife-slinging guard. "*Thank you, Basilicus*," he said, in the most obnoxious sing-song he could muster.

The soldier's expression darkened. "It's more than you deserve," he muttered in reply.

Duncan offered his usual slight, enigmatic smile. "And how much longer," he inquired coolly, "can I expect to be incarcerated at her majesty's pleasure?"

"Just so," Bassus rejoined, and that was all.

"Huh... Do you think it'll change anything?"

Bassus smiled, that awful, condescending smile. "In time," he murmured, *knowingly*. "I have faith in you." Reaching a hand through the bars, he clapped his palm on Duncan's fragile shoulder in what was meant to be a quick but annoying gesture – and flinched. He drew back, looking at his hand in the dim light. It was covered in blood.

Duncan hid his grimace with a sneer. "Wash your hands'uv *that*," he taunted, his voice rough.

Bassus was silent for a long time, staring at the blood on his hand. His eyes were wide with a kind of fascinated shock, as if he could not believe that so much blood could soak through anything in so short a time.

"Jeeze," he said at last, his voice quiet because of the unfamiliarity of such honesty. "You know... maybe... if you're gonna be gettin' blood all over the place –"

"No," rasped the fire-eater, harshly, glaring at the knife-thrower with hatred in his eyes. "If we're going to be wasting this blood, then let us waste it. I *will not* save it for *you*."

"We want you *alive*, fire-rat!" Bassus nearly shouted, moving a step closer to the bars. "Whether you believe it or not! What use to us is your blood clotted in frozen veins?!"

"That's no concern of mine!" snarled the fire-eater, through clenched teeth.

And what could he say to that?

Bassus looked a little disturbed; it took effort not to glance at his hand. Drawing the cloth that he used to clean his sword, he wiped away the blood. "Fine," he growled, his glare harsh. "If you want to bleed to death, then that's your business."

Duncan gave a slow and solemn nod.

Bassus clenched his teeth. He would have his coward's legion scourged as well – how he wished in that moment they could revive the custom of killing ten as an example to the army! To let a spineless little garter like *this* scare them...

The faintest of soft smiles twisted the fire-eater's treacherous lips. "What makes you so sure I'm gonna stick around, anyway?" he asked, setting his half-empty cup on the stone floor.

Bassus laughed aloud. "For the love of God, *face* it, fire-eater! There is nothing left to fight for! Your forces are scattered, your people are destroyed. Ceothannanmór is a ruin, and ruin it will remain."

"Nothing is ever destroyed if its memory lives on. Do you know the story of Troy? They say the ashes of decimated Troy scattered to the winds, and from that dust sprang many nations. From that dust sprang Rome."

"You put too much faith in children's stories. But all the same: Rome is said to have fallen, too."

"The remarkable thing about enduring children's stories is the amount of truth they hold. Not every off-shoot of Troy took root; fledging nations have short lives. You are but the spume and sputum of Lavinium, and Lavinium's name is upon all aspects of what will be preserved of this kingdom when it falls. Lavinium is what will endure – not Augustinium."

"How fertile have Ceothannanmór's ashes proven, O'Rielly?"

"As long as I have not died of hypovolemia, they remain."

"You're alone." Bassus pressed his left hand against the bars, feeling the cold, solid iron. "Any other 'spark' like you has been dampened by his own cowardice – or wisdom, if you like. One cinder does not a phoenix make; one rebel rebellion. The ashes of Ceothannanmór know better than to rise. I'll say it again: Ruin and ruin remain."

"How many people had said that to Aeneas?"

"I don't know. Care to go ask him?" His flashing grin matched his briefly unsheathed blade. "And if you're choosing to play that card, O'Rielly, let me just remind you that our Roman hero, Aeneas, possessed above all one quality that you Ceothanns fail to encompass: *bravery*."

"You've said so," muttered Duncan.

"And I'll say it again. I see it in you, fire-eater, though you wouldn't care to admit it. I know your Ceothann legends and superstitions as well as any so-called *seanchai*; don't think that I don't know you. You, Duncan O'Rielly, Master of Flames," he spoke the title grandly in his soft tone, ridiculing him. "You could lower this whole place to the ground and us along with it if you wanted to, but you won't. And it's for the same reason you haven't escaped yet. The reason you will not use your powers – not for anything beyond illusions or juggling. It's because you are a *coward*, Duncan O'Rielly. It's because you are scared."

Duncan pursed his lips. But, beyond all of Bassus's words, what made him angriest of all was that there was nothing he could say in reply.

Because it was true.

Gently, he leaned forward, taking hold of the bars almost a foot above Bassus's own hands. The knife-thrower stepped backwards, casting him a look of disgust.

"*My people*, you say," remarked the fire-eater, at a murmur. "Did you forget that half of Lavinium rebelled against the Empire as well? Or do you deny that little bit of treachery, too?"

"*My* people are right here," Bassus responded, on edge. "Augustinium. The future lies with the Empire, vagabond. It's time you learned to accept that."

"*Never*." Duncan straightened proudly to his full height as he spoke the indignant challenge.

Bassus sneered. "Then you've spoken your own damnation."

"And you yours."

"Look around you, fire-eater! Again, I don't see *anyone* here to help you!" He stepped forward, lowering his voice to mock, "Tell me, then. Where are your followers – your precious sparks – now?"

Duncan frowned. "I... I never had any followers."

"You don't believe that," Bassus accused. "Half the world lives on legends of what you can do."

Duncan was honestly confused. He let go of the bars slowly, grateful that the shadows would hide his face.

"You know, Bassus," he said, quietly, "you raise an interesting point. I'm tired of being afraid."

Bassus' own expression was clear: he looked amused. "And let's say that you are. What do you intend to do about it? What could you *possibly* do?" He leaned into the bars, his face the most exquisite portrayal of malicious joy the eye had ever seen. "*One* blow was all it took last time, as I recall," he hissed, avarice in his tone. "I suppose the last thing you heard was her scream?"

In spite of it all, Duncan paled.

Yes. Screaming his name.

And the sound of her cry still echoes in my mind, tears apart my heart, haunts me every night and day. It still runs veins through my name, the memory of that...

He winced, just barely managing to repress the urge to cover his ears.

And Bassus smiled, a taunting, vicious, satisfied smile. Tapping the bars of his enemy's cell lightly with one finger, he walked away. Leaving the fire-eater with nothing but an unnatural darkness, the blood streaming down his back, and the ghost of a scream still ringing in his ears.

77
Breakfast with Outlaws

They were deciding what to do with him.

Zander sat at the base of a large oak tree, staring up at the sky. The morning mists had cleared, leaving a fine dew upon the whispering leaves and a heavy silence over the woods. He had never felt so anxious, so helpless, so *bored*.

Daíthi leaned against the tree beside him, so deeply that he was almost lying flat. His arms were folded loosely across his chest, and a tweed cap was pulled low over his eyes, so that all Zander could see of his face was his tight frown. Looking at his straight-legged, arms-folded, laid-out position, Zander had commented that this was a rather dry wake. Daíthi had told him to shut up.

It was still early in the morning. The sun had barely risen a foot above the horizon, but its pure light streamed brightly through the trees, spreading soft, white rays through every leaf. Daíthi had "procured" a few apples, which he had given to Zander, muttering that he wasn't hungry.

Zander sighed, leaning his head back against the tree. Between the cool, August air, the crisp, sweet apple, and the unusual break of sunlight, it was almost easy to forget that he was probably bound to hang.

"Fer th'*life*'uv me... Ah cain't rightly reckon how ya b'caym *mah* responsibilih'ee..."

The groan could've come from the depths of Hell, it was so rough, so low, so dragging, so hostile.

Zander turned. "Because I'm so much fun t'be around, isn't that it? Ya love me an'd miss me were we apart..."

"Yeah," Daíthi grunted, sarcastically. "I'm *shar* tha's t'*ezact* thing..." He shifted vaguely, and closed his eyes against the sun, instantly and overwhelmingly regretting the move. "*Gad.* An'ihs prob'ly Monday too, *hain'it*. Sunday whiskey, fer da' sin, alwayz taysts bett'ar, bu' ihsa steep price fer Hell in t'mahrnin'. Righ'?"

Zander smirked arrogantly. "I wouldn't know." He bit into one of the apples; it was just shy of ripe, but he was so hungry he didn't care.

Daíthi responded with disgust. "Aukh! Taller *and* temperate!" he enunciated. *"A chiall – nuair a thig cioth tig balc."*

Ceothann... Zander was intrigued. "Where are you from, exactly?" he asked.

"Me? Nowhere."

"No whe – *come* on! That has'ta come from *somewhere,"* Zander said, pointing a finger at him.

Daíthi lifted the brim of his cap. "Ah tayk offence t'tha'..."

"I mean your *accent!"* Zander huffed.

"Mine?"

"No; Saint Peter's."

"Well, ihs *mine, hain'*it?"

"Murphy!"

Daíthi laughed. "Sharn, yar aisy, ain'chya! Ol' Donny wuz quick teh rile, too. Tell ya wha': Assen' hain'a fin t'go on, fars' off. Language nee'ver. Ih changes."

"Is uh – 'ass'n' – accent?"

Daíthi smirked. "Comes aroun', f'ya pay atten'in."

God – I'm beginning to understand him. Zander addressed him flatly, "Where're you from."

"Mo' Deuos! Gid *Gad,* ye Ceothanns're all alike – full naym han place'a birth, *a chiall*. Oh, an' damned hif it ain't rubbed off on me, too!" He shifted again, sinking deeper into his position. "Gallia. City'uv Tarvenna." It was the *way* he spoke those words. As if cradling something holy, something fragile, something beautiful. And at the same time with a familiarity, gracing the word with something more natural than his own name. There was power there. Ancient and engulfing power.

Zander stared at him with wide eyes. Did he dare to read into that speech?

"I've never heard of it."

Sharply, Daíthi returned, "Dey changed t'naym." When the boy didn't speak again, he realized he may have been a bit harsh. "Heard'a Gallia?"

"Yes."

"D'language'ss a lot li' Lavinian. Li' Ah said – y'cain' often tell be d'language. Ah do lapse inna ih when vexed. Also t'annoy Padraig. Feckin' hilarious."

"Do you know Lavinian?" Zander asked, and Daíthi nodded. "And you know Ceothann."

"Fashionin' dialects, Ah'm fluid in five, plus ha'-dacent in four, languages... *Fluen'*," he corrected himself, peeved. Then, glibly, "Ah also dabble in dis here Augustan. Badly." He tipped Zander a jaded smile. "P'raps ya noticed?"

"Uh –" Zander felt a blush "—so, nine languages, huh?"

Daíthi shrugged, pulling his cap lower over his eyes. "Yea', mos' people're s'prized t'larn Ah know stuff. No shaym."

"No – sorry, I –"

Daíthi lifted a hand, the slight motion cutting him off.

Zander pressed his lips together. *Whups.* "I didn't mean to pry," he insisted, meekly. "Really. I'm –"

"Ah *like* you," Daíthi replied, enunciating the point. "Yen nufin'a warry'a me."

The boy said nothing. Daíthi could not think of anything in particular to offer either – teenagers weren't exactly his demographic – so he allowed a long silence to drift in.

The boy must've followed a long train of thought to get to his next point: "You never answered my question."

"Ha? Yea' Ah did –"

"No, from last night."

Daíthi was silent a moment. "J'you as' sumfin?"

"I asked you how you knew my dad."

"Nn." Something about that sounded familiar; sure it was a natural enough thing to wonder. He sighed, lifting his eyes to the brim of his hat; he felt cross-eyed. "Ent hard t'tell," he murmured, with a shrug. "Shar we go *way* back – Ah met'im on t'docks in d'place dey call *'Marcansine'*." He spat in the dirt.

"Thank you for not spitting on me," said Zander, rolling his eyes.

"Anytime! Here – imagine if dey naym'd – Ah dunno, sumfin y'cayr abou'. Jus' *tayk* d'name righ' from under yeh. *We* cayr jus' az much abou' nayms as Ceothanns, Ah hafta tell ya."

"'We'?"

"Galls." Bowing his head, he put his hands to the sod on either side of him, hoisting himself up so his shoulder-blades were braced against the tree; he was still mostly lying flat. *In stages...* "Ach. S'thayrs *mah* life story, eh? Nah how's abou'chyerself?" He squinted at Zander, appraising him. "Yar abou' fifteen –"

"*Sixteen*."

Daíthi held up his palms, giving it to him. *The little lies with which we entertain ourselves*. "Yar'a Balladarach survivor. Yer ahlone, unarmed, polite, naïve – an' tall," he added, with an *"innocent"* smile as Zander turned on him.

"And we're just gonna let them decide our fate, is that it? We're just gonna *sit* here, *eatin'* breakfast, while *they* decide whether or not to be killing us."

"Yep." Daíthi pulled his cap over his eyes, folded his arms, and settled in against the tree's rough bark. "Pre'ee much."

Zander frowned; perhaps if he were older he could have pulled off a shudder-inducing dark look; as it was, he only succeeded in appearing sulky.

Daíthi snickered. "Don' warry yer pretty head, hon. When dey wan'us, tay'll come f'r us. Gad'll send 'Is messenger. Be assured, hon: f'tay whar gonna kill us, we'd ah'ready be dead."

"Marvelous."

"Aukh, lad," he began, heaving himself up the rest of the way with his right hand while pushing his hat back with the left; he would get no rest with the child bent on talking to him, anyway. "Din' I tell ya las' nigh'? *Trus' Gad*, hon. He'll take care'uv ya, He will." Dropping his hat, he added, "*Thayrs* yer way, hain'it?"

Zander gave a tight-lipped sigh, studying the apple in his hand. He had a point.

"Wan'it?" He held out the last of his breakfast to Daíthi. "S'the last one. Might be your last meal. Ever. Just sayin'."

"*Nuuuuhh...*" Daíthi moaned, turning away.

Zander smiled.

"Ihs *nah* amusin'!!"

"It is a little."

"Yar *mayn*," Daíthi sneered, tugging his cap even lower as – for once – the sun broke through the clouds. "Yar jus' *mayn!*"

"Is that *'mean'* you're saying?"

"*Uuunnh*. Shut up."

Zander clapped a hand on Daíthi's shoulder, causing him to flinch. "Always hits ya later, Murphy, remember that."

Daíthi scowled. "*See* hi'fy tell ya a thing again, *Zander O'Rielly*. Tha's t'*thaynks* Ah get f'r bein' *dacent* t'ya!"

"Hospitality, Daíthi," Zander enunciated, eyes on the sky and half-bitten apple to his heart. "*Sheer* hospitality."

Daíthi wrinkled his nose. He turned to Zander, intent on giving him a piece of his mind (really, he could *take* it!), when he was distracted.

"Ach – yonder."

Zander turned.

Daíthi cast a dubious, hazel glare to Heaven. "Rahlly? Ya cou'n do bett'ar?"

"Hi, Sully!" Zander called, waving the remains of his breakfast. For, sure enough, there was Sully, striding towards them through the trees and dawn light from the direction of camp.

"Greetings, Flotsam." Sully came up and stood before them, a few inches from Daíthi's feet, which he pointedly did not move but to cross them at the ankle.

"Joe *Sullivan*." The arrogance was fluid in his tone as Daíthi raised his eyes to the sailor. "*You* look *well-rested*."

Sully's expression remained impassive. "You look like Hell, Murphy; what happened?"

Daíthi's glare was murder. "*Ten ahrs* f'r yeh t'*relieve* me, Joe?! Ih wuz bleedin' *dawn!* Damn fine watch-guard, you!! An' – F.Y.I., Joe? Ya wanna wake a man up, ya don' kick'm t'th'kidneys, a'righ'? 'Pon me life, I ough'a throttle yeh."

"If you'd've woken up the first – oh, five or six times, I wouldn't've had to, ya great drouth." Sully nudged Daíthi's foot with the toe of his boot. "Hey – move your feet and I can sit."

"Move'n inch; yah'll block t'sun."

"A fine welcome to you, too." Sully sat down opposite Zander, bracing his arm on Daíthi's ankle until he got himself situated, exaggerating the move. Daíthi's expression was vitriolic, but he remained silent.

Sully clapped a hand on his ankle. "Thanks, Barfly." He drew a flask from his coat pocket, holding it out to Daíthi. "Here."

Daíthi stared. "Wha' t'Hell iz'at?" he asked, without emotion.

"Flannery said t'give it to you."

Daíthi wrinkled his nose. "I'ont trus' nufin'a Flannery," he snorted, reaching out to take the flask. He looked down at it in serious examination. "Looks full," he remarked.

"You're welcome."

"Huh?" With slow suspicion, Daíthi uncapped the flask.

Sully rolled his eyes for Zander. "Never mind."

Zander smiled. "Were you five minutes earlier, you could've caught breakfast." He held up what was left of his apple. "Unless you want this?"

Sully held up a hand. "I'll leave ya to your last meal."

A sudden choking re-captured their attention, and they both turned back to Daíthi. Doubled over, he twisted away from the others, holding the flask so as not to spill it as he gagged. "*Gad!*" he managed, still choking. "Wha' t'Hell *is* that!?!" He sounded more offended than anything else.

Sully could not help a smile. "Buttermilk and coffee. Seems your boys thought you'd be in need of some double-strength Cure."

"Gad, Ah'm gonna puke."

Sully leaned back, putting his hands to the earth behind him for support. "What's the matter, Barfly? Can't stomach it?"

Daíthi hacked, trying to catch his breath, then gave up and fell back against the tree, the flask still held aloft like a torch. He looked at it warily. "Tha'ss why I *never* thank ya fer a thing 'til Ah see't meself." Then his eyes flashed back to Sully, narrowed. "Dis is t'kind'a thin' ya warn a man first, Joseph. Flannery's Skimagig's known to've killed people – probably!"

Sully rolled his ice-blue eyes.

Daíthi did the same in return and raised the flask to his lips again. Zander noticed him tense slightly but sharply before taking the liquid, and shudder faintly before swallowing. Zander couldn't blame him. "That must be *spectacular*," he commented, in a low, sarcastic tone.

Daíthi lowered the flask, clearing his throat. "Sour'uz t'Devil's tongue," he replied, his voice even more hoarse than before.

Zander turned to Sully. "So, why aren'chyou at the meeting? Is it over? What did they decide?"

"Yeah," Daíthi echoed, tipping back more of the evil brew.

Sully leaned his head back, squinting up at the clouds; still overcast from the night before. Zander thought nothing of his ponderous silence – but Daíthi clearly did, because suddenly he fixed Sully with a wide gaze. "Joseph, if they ain't done – bu' weren'chyou s'posta –" He closed his eyes with an aggravated groan. "Aouh! *Don'* tell me ihsa closed trial!"

"Ya knew as much would happen, so why waste yer breath complainin' about it?" Sully returned, but his dismissal sparked Daíthi's rage.

"Ih's been a'most seven years'a this!!" Daíthi shouted, imploringly. "Dey keep playin' games – ih's gonna be da' death uv'em – of *all*'uv us! We ain't got time fer dis nonsense – an' we all know dey'll vote ta' hang me, anywayz!" His tone shifted subtly from urgency to impatience – a subtle shade.

"He's right," Zander said, quietly, and both men turned to look at him. He had his head bowed, his eyes scanning the apple core in his hand. There was no more to be had from it; he tossed it away like the others, this time aiming back towards camp. "Murphy's right."

"'Preciate da' vote'a confidence –" Daíthi began, snidely, wondering why everyone always went straight to the noose with him – until Zander seized his arm.

"They're not gonna talk about anything else until they've decided what to do with us, are they?"

Daíthi eyed him warily. "Yeeaaahh..."

Zander looked at him sidelong. "Murph – God helps those that help themselves, doesn't He?"

"Y-yeah –"

"Then what are we waiting for!" Letting go of Daíthi's arm, Zander jumped to his feet, forgetting his weariness, forgetting his pain, knowing nothing but destiny – and the pure moral outrage needed to fuel it.

Daíthi looked up at him lazily, pushing back his cap. A few, pale ginger strands of hair straggled into his eyes as he squinted against the light. "*Arglwydd da!* I do b'lave yer beginnin'a understand t'concep'a *Fate!*"

Sully turned to him. "It does seem a decent idea, representin' yourself at your own trial, don'chya think?"

Daíthi nodded slowly. "T'would seem downrigh' indacent *nah* to..."

A mischievous smile slowly took hold of Sully's rough features, and he leaned forward. "Five silver they throw ya out again – ten it's bodily. Wha'd'ya say? Think you can stack against those odds, Barfly?"

Daíthi considered it, taking a long drink from the flask.

"Yeah, ah'righ'."

78
Sunburstry

"Destiny starts now, hon."

"Finally! ...Does it ruin it that I just thought of Shaw's speech? *'A beginning and an end'...*"

"Aukh. *Yesss.*"

They had tried to make themselves as presentable as possible – *"pretty us'selves up, huh?"* as Daíthi had put it. Pressed for time as they were, it didn't show much; Daíthi had finally tucked in his shirt and Zander had hurriedly finger-combed his dark hair, hoping to tame the unruly beast. When Sully had remarked on their "useless vanity", Daíthi had retorted, "Dacent *some'*a us's got *dignity*." And now here they were, standing before the full mass of the Company, highly aware that "sodden wretch" did not count for "having bathed."

And quite the audience they had, hadn't they? Word had spread quickly about the night before, and, lured by curiosity, it seemed that the entire countryside had turned out for the show. The three vagrants lurked at the edges of a crowd in constant motion. The frayed edges mixed and shifted like solar flares off the sun, surging over the toppled trees, large rocks, broad stumps, and high branches like swarming bees and climbing ivy. They bickered for position, seating themselves like a packed stadium. *A coloseum,* thought Zander, frowning, *and we the lucky gladiators.*

"Lord," muttered Daíthi. "Closed mee'in *mo thóin*. Ih's like dey're es'pectin' free food ahr summat."

Zander jumped. "*Is* there free food?!"

Daíthi scowled. "Lord – j'ya jis' ate! Yer makin' me look li'a bad parent, ahr –" He turned to Sully abruptly. "Wha'ya say? Give'm a fair route an' run li' Hell 'fore dey unleash de'tigers?"

"Tigers?" inquired Zander, but Sully waved him off.

"It is a turn-out indeed..."

"A fine, non-commi'al answer, Joe." Daíthi moved forward, slipping into the shadows of the trees nearest where a bunch of Lowlanders had gathered. Zander followed.

"D'you really think they'll listen to us, Murph?"

"Naw," Daíthi answered, dishearteningly. "Bu' wha' da Hell, righ'?" He drank from the flask, and this time didn't flinch.

"*That* won't help anything!" Quick as an adder, Sully snatched the flask, dropping it into the pocket of his long coat.

Daíthi's cheeks turned red. "Quick, ain'chya?" He paused, then – "Okh; dan' warry sa' much! It ain't li' dey es'pect any better! C'mon – Ah ent done; give't back."

Sully rolled his eyes. "*Pretend* t'be professional, Barfly."

"*You* be professional."

Sully was mature enough to not say anything in response; Zander murmured, "It's gross anyway; why do you want it?"

"Hhhmmmmm..." Daíthi's gaze toured the canopy. He was about to turn with his rebuttal when something way more important entered onto the scene – something fair and shining and fresh from the world of his dreams. His gaze had found that stunning creature – the woman from the night before – and he found himself entranced. He saw the soft cascade of her night-black curls, the delicate beauty of her smooth, freckled skin, the graceful fall of her long skirt, the vibrant fire of her ice-blue eyes... Shorn, a better bard may have made a poem of it, but he found his mind choked of words, his heart choked of blood, his body choked of any strength, all at the very sight of her.

"Her name is Roscanna," said Zander.

Daíthi turned, faintly and fleetingly. "Huh?"

"The woman you've been staring at. Her name is Roscanna."

Though Daíthi tried to feign indifference, he was betrayed by a scarlet blush. Zander could not help but notice that, as Daíthi bowed his head to examine his fingernails, dirt-smudged and jagged-short, he was mouthing the name for himself.

The last of the crowd arrived – and arrive they did! Clustering on the thick slope of the great, fallen tree were the assembled ranks of the Turranians – a group of about twenty men, their skin as dusky as un-ground cinnamon, their eyes as black as a moonless night, their posture as imposing and as pompous as a king's; the Turranians were a proud and elegant people, and the poorest man among them wore embroidery on his tunic; even in rags they were glorious. At their center – there was no mistaking the dark-haired youth – was their leader, Shrikant al-Rajiv, a haughty squint to his inky eyes, a pouting frown to his thin-drawn mouth, a sun-kissed brown to his youthful skin, and a slightly-worn resilience to his golden tunic. The crowd from the Turranian camp was small, but the young king himself was a formidable enough creature. If his superior glare was any indication, this youth would not be easily swayed.

"Farget playin' t'th' *marcy*'a their *hearts*," Daíthi murmured, eyeing the crowd. "Par'aps we 'en beat it int'a tay'r thick skulls."

Sully made a face. "*Lord* you're sulky when you're hung-over."

Daíthi shrugged. "Nufin'a good donnybrook never did solve." He smacked Zander's arm. "Stay here, shut up, watch Joseph." And, just like that, he glided silently into the eye of the storm. There was something so commanding in his poise that they turned to him; something so strange in the calm way he drew his switch-blade, his head bowed as he used the blade to clean under his nails, that they fell silent, listening. And, with the arrogant boredom of a prince, he inquired, tonelessly, "D'jya decide whe'ver ahr nah we'd be killin' 'im?"

No one responded, so he continued, in the same indifferent tone. "Like, c'we ate 'im, d'ya think? Migh' bay'a bi' *gamy*, bu' he'll cook fine anuff."

Zander frowned. *Why does everyone want to eat me?*

"*If*;" Eoghan spoke first, addressing him in measured words. "If we *are* tae kill 'im, it'd be on th' groonds 'uv him trespassin', and him comin' from the direction o' the Castle himsel'."

"Trespassin' –" Daíthi murmured, but was cut off; the one word he got out was spoken with some surprise.

"Banishment would be a more suitable punishment," noted one of the Calchians, "but he knows where the camp is."

"So the only option *apart* from killing him –"

"And of course we haven't the means –"

"Would be a sort of house-arrest."

Zander was stunned. Never before had he seen anything blown so far out of proportion (though he had seen nothing yet). Stepping forward, he protested, "I'm on your side!!"

Daíthi turned to him with a glare he knew well; it was the *Shut UP, Zander* glare, the one tinged with the undertones of *Don't worry, son; I'll take care of you*. It was one he had seen so often shot from Rahyoke's eyes that instantly his tongue was bound to silence.

"Well," said Daíthi, shortly, turning back to the crowd. "A poor lot yeh'd be – t'lott'a yeh!" A current of anger flashed through his tone, a spark of electricity. "Kill'im, den! Prove yerself t'be all dey say we are – t'rebels're dumb, barbaric, *drunks*, who slaughter tay'r children, who cain't fight, who steal an' hide an' cower – prove'um righ'!" He raised his arms in a helpless gesture, snapping the switch-blade shut and dropping his arms back to his sides.

"You would have us kill him?" asked Caeculus, cocking his head.

"To assume he is the man he claims to be," noted one of the Turranians, "he *is* a danger to us." At this, Daíthi and Zander exchanged glances across the way, both thinking the same thing. *Yes; because when you lie about your name, "Zander" is the natural go-to.*

"He's a stranger," responded a Highlander, in a thick, rolling brogue. "Who 'ed ken what he is an' isn't?"

"God-*damn*; ih was sarcasm," Daíthi groused, but the Highlander had sparked a thought.

"We could ransom 'im," offered one of the Lowlanders. Amergin turned to him with a withering look, scolding, "*Dude!*"

"The man there has a point," agreed a Turranian from the top of the fallen tree. "We might hold him as prisoner, and sell him to the Ariegn!"

"Which would give up our exact location..."

"A decision on the matter needs to be reached quickly," added Caomhan, at his usual casual drawl. "The longer we keep him, the more he's in the know."

"Very well then," announced a Lowlander, jumping up, "we'll hold it to a vote!"

"Ah'll vouch far'im!"

Everyone turned to Daíthi in surprise; his raised voice had overpowered their rabble, wavering though it was. He cleared his throat, looking at no one in particular. "I... I'll vouch far'im," he repeated, praying to God that that was the right word.

Eoghan was watching him with an accustomed look. "Yeh'll vouch fer yer ain guy."

"...yes."

The gathered looked to one another, for once debating with their eyes instead of their mouths.

It was then that Shrikant spoke.

"What I want to know, it is this: Who is *this* man?" he demanded, and the one to whom he pointed was Daíthi. His coal-dark eyes turned from Amergin to Eoghan, as if demanding that one of the leaders take responsibility for the impetuous rogue.

The name "Barfly" was dropped, and Daíthi knew he'd better take over. "M'naym's Murphy. Ah'm one'a Eoghan's men. Been doin' field-work. Uv' come back wid a message tha' hain' no one sayms t'cayr abou' hearin'."

"Field-work," Eoghan repeated, in his same parental tone.

Daíthi looked to him. "Uh-huh."

"In a bar."

"Ah gat hungry. Look, why'um Ah – Look!" he emended, suddenly. "Look – a'da *very* base'a thin's – a'da very base!" Without looking, he flung an indication, pointing steadily at Zander. "*That* is a *boy*. An' *if* 'ee wuz gonna feckin' run – run t'the enemy, as yer all sa' dang scay'd he'll do – *don'* ya fink 'ee'd've done ih las' nigh' while Ah wuz passed ou' drunk?!" His voice cracked as he raised it in annoyance – a sharp contrast to the reasoning, calm logic of his closing statement: "S'it too much t'ask t'jus' leave 'im live?"

Eoghan leveled Daíthi a very serious stare. "Murphy. If yeh've an honest fleck'a blood in yer body, let it be used in givin' claim tae this lad."

Daíthi acquiesced with a deep nod. "By mah life an' very naym, he is da spawn'uv Donn O'Rielly."

They all turned to one another, debating in lowered tones. "Perhaps, then," observed Caomhan, "if we acknowledge the boy as O'Rielly's, we might consider who we'll entrust him to."

"What d'ya mean?" asked Amergin.

"We can't kill the only son of our founder, can we? And to let him leave would be a threat to all of us. So it's the only viable option that, to ensure the protection of the lad, we place him under some care. And it seems only right that we entrust the son of our leader and founder to someone of some, *reputable*, character."

A look of slow understanding dawned on Daíthi's face, and his expression darkened into a scowl. "*'Scuse* me?"

Caomhan merely shrugged. "Sullivan can't be expected to tether himself to the camp," he reasoned, as if this should obviously explain all.

Daíthi's pale cheeks burned red with anger. "Caomhan MacAdara, be *plain* when yer callin' me out!" he growled, his voice not loud, but fierce.

Caomhan readily accepted his challenge. "Come on, now. Don't pretend like *you're* the ideal guardian, *Barfly Murphy.*" He turned to the rest of the crowd. "Should we really entrust the son of O'Rielly, or any mother's son at that, to the rake king of a band of motley thieves, rogues, and drunks?"

Amergin nudged the Lowlander beside him. "I take offence t'that," he murmured, with a smile.

But Daíthi wasn't amused. "Say it t'*me*, Caomhan!"

"Very well, then, I will..."

Once more the rogue was set to defend his slighted honour, yet even as he argued, Zander could not help but realize the truth of Caomhan's words. Sweet as they were, Daíthi and his *contubernium* were the rugged sort of rabble your parents warned you about. Tattered, homeless, haphazardly-employed, they smoked, they drank, they stole, cheated, and cursed. Daíthi alone was strange, quick to anger, irreverent, enigmatic, and never without the three-and-a-half-inch switch-blade that he tended to use as a bottle-opener. A white-hot mess and proud of it – not, to be true, your ideal guardian.

And yet...

"Look! He c'n damn well protect his'self, f'r Chris'sake!" Daíthi snapped, grown impatient. "He's adult anuff – he don't need a guardian! He ain't in any danger *here*, is he?! An' he ain't a danger ta' us!!"

"And what makes us believe *you?*" Moesen O'Rourke, entering into the fray, stood, and stood with a power and command that proved why he was seen as Eoghan's third. "What makes us believe some phantom-spawn wanderer that this is legitimate and isn't a con?"

Daíthi chewed his tongue, bowing his head for a moment to handle his rage. And, in a tone as amiable and level as it was – Lord help us – professional, Daíthi Murphy replied, "O'Rielly gave *us* ta'benefit'a th'doubt. Why cain't we do t'saym f'r his son?"

Oh, did he ever have a point there. Like scolded children they hung their heads, considering his words. But Moesen O'Rourke remained resolute.

"O'Rielly," he snorted, with an unsavory level of contempt. "O'Rielly is dead."

Murphy gritted his teeth. And, with fire, he gave up his best gambit.

"Is he?"

They stared at him, all ten thousand eyes, wondering. He shifted, sliding his hands into his pockets, his own eyes sweeping over them at a glance. "'Cause if 'ee is, den so are we.

"A ten-min'it chariot run fra' Ariegn, we sit, waitin', in a 'temp' camp we lef' standin' f'r over a year. Damn't all, dey ain't blind. T'a *moment* dey tayk it inna deyr heads t'sarch fer us, we're dead. An' you warry abou' *Zander!?*" His voice became sharp, rebuking, and filled with disgust.

"What would *you* be havin' us do?" a Highlander – Nollaig – broke in. "Shift t'entire camp, as we used to?!"

"Nah," hissed Daíthi, shaking his head. "No mahr. We ain't runnin' no mahr."

"Well, what yer *say*-in' –"

"F'anyfin, we've run outta time f'r tha'." He paused, only long enough to draw a breath. "We have *no* time. T'time far preparation, f'r hesitation, f'r dreamin' an' idle boastin', far *runnin'*, is over. Dey'll be on us soon, shar anuff. An' tay'll kill us wivout givin' us a chance. T'way *Ah* see't, ihs damn well time we start livin' up t'all'a them boasts we made o'er t'years. Prove yer warth teh' Gad A'mighty, an' prove we ain't jis' cowards hidin' in t'woods. Prove, once an' f'r all, dat we ain't gonna settle. Tha' we're a powerful force t'be reckoned wiv. Tha' we're *marh* t'en dissenters – mahr den rebels. We *are* dis land, han we're takin' it back! We must *fight!!*"

His voice cracked under the strain of raising it. Daíthi watched the crowds with charged defiance in his hazel eyes; so adamant was his audacity that scarce was the soul that would realize that slightly, slightly, he tensed up. He knew these people well enough to know what was coming next.

And he wasn't disappointed.

Murphy winced as if he'd been swung at, for the sheer force of their rebuttal – their banshee screams and screeching cries of fury, the howls of anger and thirsty roars for his blood.

"Deuos, trougetos," he pleaded, in his own tongue. But his confidence remained. "Yea', *tha'ss* righ', avert t'a discussion! Ihs yer own blood yer wastin'!!"

The shrilling voices separated, becoming intelligible strands. And one line rang above them all.

"WHAT GIVES YOU THE RIGHT TO DECLARE WAR?!"

What gives you the right to declare war? Daíthi pursed his lips. "I *ain't*. Jayz – if y'knew me at a glance, y'd know sich'd be far from me own wantin', wun'chya?! Good *Gad* –"

Still on his feet, Moesen glared down at him. "Yeah?" he demanded. "Then what, then? What."

"I ain't declarin' war," he repeated carefully, defining every word. There was that look again – the plaintive eyes of a treacherous dog as he hesitated, steadily wringing his hands. Then – at long last –

"O'Rielly is... He's alive."

Treis... Allos... Oinos... Gaah! He couldn't make a word out of it but that it was noise – angry noise – noise and anger directed at him, directed at the world, directed at the Lord on high Himself for Fate, for toil, for rage, dejection, disbelief. And ach – *Ach!* – Lord have mercy – it hurt. Any second now, surely, his brain was bound to implode.

He was seized.

"Uh –!"

"WHY DIDN'T YOU TELL ME!?!"

The boy was stronger than he looked, *a chiall*. He had seized Daíthi so harshly by the shoulders that he lost his footing, and his knees went out from under him, making him a tad dependent on the lad's violent support. For a moment, he was surprised. Then Murphy responded with a scowl.

"Yew never asked," he sneered, his voice low.

Zander tightened his grip with a jerk. "I did! Yes I did! What the Hell's wrong with you you didn't just *think to tell me!?!"*

"Li' five minutes ago Ah said 'ee warn't dead," Daíthi hissed, "why din'chya –"

"Because I can't understand a word you say!!"

Daíthi smirked. "Oh. Den may'ee Ah *did* tell ya –"

"Listen, Barfly, you damn traitorous little liar – !"

Daíthi got his footing under him and stood with his own force. "Trust me," he quipped, severing the boy's costly rant before it could get underway. There was not an ounce of pleading to that tone.

"Why the Hell –"

"'Cause Ah'm t'on'ee one ya gat," Murphy snarled.

Zander bit his lip. But slowly, with more reluctance and misgiving than his young body had ever given into, he relaxed his grasp on Murphy's shoulders, letting the man go.

Daíthi stumbled back a pace, just in case the boy regretted his generous show of trust. But before either of them could utter another word, the hostile deluge was shattered.

"F'r feck's *sake."*

Daíthi turned.

Amergin had stood, and by that act alone had silenced them. And, in a tone which wove steely command from an abundance of over-worn faith, he demanded, "*How* d'ya propose we go about it?"

One of his men began a protest – "You actually *believe* –" – but Amergin severed him with a hush.

Daíthi bowed his head. *So NOW we're wanting to hear the story, are we?* "Tha's fer yerselfs t'be decidin' righ'ly," he answered, giving Zander a generous shove to the shoulder that urged him back to his seat. "Bu', whar it so ya had a leader, t'en t'matter 'ed bay d'tarmined mee a bit fast'ar, if y'would."

"*What,*" snorted a Lowlander, leaning forward. "Li'a *king?*"

"This is a democratic organization," furthered Antoninus, with that infallible Lavinian political flare.

"Yea'! That's right!"

"We make our *own* decisions!"

And what right had he, they reasoned, or any other soul, for that matter, even a soul like the fabled O'Rielly, to impose any judgment upon them? Daíthi listened with a tight frown, his eyes flashing from one speaker to the next. He didn't know what a *democracy* was, but it sure as Hell wasn't anything he wanted a part of.

They continued to democracy the place up; pretty soon they'd be calling for blood again. *And since we've already been THERE,* he reasoned, discretely rolling his eyes, *looks like it'll have to be ME to be breakin' it up.*

"F'wur all goin' at it," he offered, raising his voice just enough to be heard, "migh' I be havin' a say?"

There was some disgruntled murmuring, but they allowed it.

Daíthi dipped his head in a shallow nod of polite thanks. Then – "If ya ain't gatta leader an' ya all follow on'ee yer own decisions, an' ihs aich decidin' 'is own way an' followin' it, lemme ask you. Wha' t'*Hell* kind uv'an *army* izis?"

"What would *you* know about any'uvit?" defended a man, jumping to his feet.

"A damn well better'un *you!*" Daíthi returned. "Look a'chya! Yer'uz *dis'ahrganized* 'uz –"

"And you think throwin' a leader int'a this'll make us any more battle-ready?"

Another man – Rankin Tomlinson – stood. "There's nothing wrong with the leaders we have!"

Kiss-up. In a tone of tense boredom, Daíthi responded, "Tayr *ineffective*." He hit every syllable perfectly – unfortunately. Oh; their outcry was like a field of daggers to his aching brain, and he was again forced to steady himself, but this time he'd had enough of their nonsense.

"SHIT!!" The single word shredded through their argument like a switch-blade through water. "Gad-damn it! We cain't ayvin converge in counsel wivout tearin' each'uver apart! An' damn fine we'd be in war – SCATTERED!! Y'ent led?! Ye've no direction!! F'r feck – ye hide in t'forest b'hind nayms an' legends – *use*'um, f'r Chris'sake! Live *up* t'yerselfs! Hain' *none*'a you so-called 'leaders' 'is done *anyfin* an' you *know* it! When war comes – an' ih'll come whe'ver *you* like it ahr nah – ye'll *all* be *killed* fer yer ain flippin' stubbornness! Bu' *enjoy* yer feckin' *'democracy'* – while ih lasts!"

"We aren't ready," Rankin hissed, advancing to join Daíthi in the circle. "No matter *how* many leaders you put over us, it won't change that!"

"*One* leader, dipshit!" Daíthi clarified, holding the indicated number of fingers in the other man's face. "*One* leader. Righ' now, we're divided – we are *clearly* divided," he enunciated, pointedly. "I ent sayin' git rid'a ahr kings. Far from it; keep yer damn kings! Bu' add someone who can unite dem – unite *us*. Someone who ain't just loyal t'their ain faction, like – someone wivout..."

"Unbiased," Amergin provided.

"Yes! Someone unbiased! Someone yeh can direct yer rulin's t'rew, an' he can make sure ihs divided out equal. An' *tha'* way –"

But before he could get out another word, Moesen rebuffed, "And let me guess. *You*'d have us nominate the boy."

Daíthi's eyes narrowed. "Aza ma'her'a fact –"

"That was really your purpose for bringin' him up to us at all, *wasn't* it!?" With all the confidence in the world, Moesen now sauntered to the center of the circle. "You wanna put your *own* man forward – a malleable little boy, with nothin' but a name, by means of whom you'd take control of this organization – and send us to our deaths!"

Oh. LORD. "Now tha'ss a very interestin' proposition, Moesen. Thaynk you fer bringin' it inn'a me mind."

Moesen scowled. "He'll abandon us like his father, the coward."

"Damn yer tongue!" Daíthi snarled, his hand balling into a fist. "Dan' be makin' comments whar ya ain't gat business, O'Rourke! Y'never did meet 'im yerself; ye've gat no right!"

"I've as much right as *you*, a random beggar saunterin' into council – *who* d'you think you *are?!*"

"T'Gad-damn son'uva-bitch tryin'a *save* yer sarry ass, is who! Ah gave me warnin'; tha's me finished. Ye'd be best t'tayk it, but if ya won't – Hell, screw ya. Screw ya all! It ain't *me* who'll suffer!" He backed up one solid pace. "I ent gatta tayk none'a dis abuse, nee'ver – an' me on'ee messenger!"

"A *fine* messenger," Moesen retorted, "pushing yer own gain!"

"A fine *warrior* – deaf t'any sense an' too *COWARDLY* t'liss'ena *TRUTH!!!*"

Without warning, Moesen grabbed him – by the throat, forcing him to look up with a vicious jerk. "You *slanderous* little bastard!"

Daíthi noticed out of the corner of his eye that a fair half-dozen men had risen, in solidarity with the mob. More eager to spill the blood of their own than to risk spilling that of their oppressor. His glare became smug. "Ya ain't *followers*..." One man actually sat back down, in embarrassment.

"Shut up!" His assailant pulled up, yanking Daíthi onto his toes. Being very small in a rapidly-retrogressing society clearly had its draw-backs. "You're nothing but a chauvinistic deadbeat; you don't know the first thing about war."

Daíthi managed with the last of his breath to choke out two words as far from an apology as any words can get.

As Moesen again tightened his grip, Eoghan intervened, apathetically, "O'Rourke, you're proving his point."

Moesen hesitated, then released Daíthi abruptly. The latter had no footing, and had to stagger back a few paces to find it, messily catching his balance. "Damn," he rasped, rubbing a hand against his throat to play up the injury.

"It's smart-mouthed war dogs like you who slaughter armies, you nepotistic little snake," Moesen reproached him, "pushing untrained men to the front-lines using scare tactics. It's disgraceful."

"It ain't a scare-tactic," Daíthi said, huskily. He cleared his throat. "Ihs a warnin'. Dey're comin' fer us, an' soon. An' we'll fair better facin' 'em as a unified force den if we a'ready done picked each u'ver off!!" He took a step towards Moesen as he said it, and if Moesen hadn't had half a foot on him, he would've been right up in his face.

Moesen waved him off, returning to the outer ring. Rankin assumed his place.

"Fair better?" he scoffed. "Moesen's right. Whether they're coming or not, it does nothing to change what we are. We are not prepared to take on Augustinium."

"We'll *be* ready," Daíthi returned. "An' if you can't hack ahr stomach it, princess, may'ee y'ought'a give up da' charade, crawl back ta' Augustinium an' start settlin' up yer back taxes!"

"Hey!" Rankin grabbed Murphy – a two-handed hold this time, one hand gripping his shoulder, the other his collar, preparing to throw him out. "You've had *more* than enough of your say."

"Wha', gonna pick up where Moesen left off? You ain't got da' balls, cupcake."

That did it. *Fist cocked back and...*

The name "Daíthi" means swift, and he was when he had to be. Rankin threw a point-blank punch; Murphy dodged, grabbed the outstretched arm, yanked its off-balance owner closer, and twisted his arm behind his back, shoving the fool into a steep bow, pinning him. "Ah gat all day," growled Daíthi, through clenched teeth, and the man quit struggling for the moment. "Now! Are ya gonna bay dacen'a me an' shut t'Hell up? Ahr do I gotta break yer feckin' arm off, ya lil' shit?"

"He has a point."

A woman's voice broke his imposed calm – a beautiful, melodious, *dissenting* woman's voice.

Daíthi raised his eyes to God. "Auh. *Lord* wha' now?" He turned, and even though she opposed him, he found his tongue gagged by her beauty.

Roscanna, standing between Nollaig and Caomhan, folded her slim arms defiantly. "Thae'ss ah' fine *pub-talk*, bu' we're in the *real* world here. An' t'would be our deaths, goin' tae war now, ne' matter *who* lead us." She had a good Highlander brogue, did Roscanna.

Daíthi rolled his eyes; in one, smooth motion, he spun Rankin away from him with a twist and a shove. The man got his balance with a jolt, cast them all a glare of bruised dignity and sauntered his wounded pride back to his seat at the far front of the crowd.

"Hard ta' talk 'real world' wiv an Augustan sword a'chyer throat, love."

"We're nae at tha' point yet, rogue." Like a goddess she paced forward, gliding over the ground, stopping beside him.

His eyes flashed over her in a lingering glance. "You gonna figh' me too?"

She dismissed him with a frigid glare, then turned to face the gathered. "We can'na take on the Empire as we are. We haven't teh trainin' nar weaponry, an' throwin' a leader in front'uv us won' change tha' – nae matter the fame'uv his da', an' *especially* nae some milksop bairn who can'na tell a dirk from a claymore! Even *if* the Empire's wrath is imminent, we're *nae* out of time yet. We can train. We can plan. We can get ourselves *ready*, an' pu' ourselves under the command'uv an *able* lead'ar. *Then* we go after the Empire."

Daíthi shook his head.

And she, turning back to him, finished, "We *don'* jump in'tae it at th'command'uv some half-slammed bard and 'is infant lackey!"

Daíthi frowned. "Lovely." He offered a hand pointedly. "M'naym's Daíthi, by d'way. Ya know; jus' in case da' creative flow'a literary pejoratives starts ta' go stale... An' *you* are?"

Roscanna's eyes narrowed. Daíthi turned his hand palm-up with a *Well? I'm waiting* glare. Then, with a shrug, he stuck his hand in his pocket.

"So, what is it yer suggestin' meantime?" Amergin asked, trying to reign them in. "Seein' as it's safer t'assume he's right in the *'they'll find us soon'* bit."

"Backtrack," Roscanna advised, with a graceful shrug.

"*Back*-track," Daíthi snorted, contemptuously.

She glared at him. "*Move tae a safe location*. 'Til we're ready teh fight."

"Tripe."

"Bloody immature." Roscanna shook her head.

Daíthi gritted his teeth. "Arrite, den, sweet'art. Lemme ask you dis: Where we gonna go? Can't go North; tha'ss Ariegn. Can't go west, 'cause deyrs Vasio. T'da east, we 'ave Northambria, Thysdrus, Ulcisia, Sirmium – an' a feckin' river. An' da' troop tha' jus' culled Mahallahi's right up ahr asses – i'fact, how did Mahallahi fare? Tha' worked ou' *real* nicely as a southern outpost, *din'* ih? We've a'ready had *two* repeats'a Balladarach; you ready fer a third, darlin'? 'Cause *tha'ss* how ihs gonna go!"

"Cou'jya *be* any more boorish in yer diction?"

"Oh; Ah'm *sarry* Ah cain't *e'sspress* meself in t'*tarms* 'a yer *nobility! Immi sui trougo caxto, suados vanos!*"

Roscanna's glare was livid, and for a heartbeat she merely held his gaze. Then she answered.

"Dusios ex-anderos, incors tó gulbion eti gabi in mantalon!!"

To Daíthi's absolute shock, Roscanna began berating him in his native tongue. For a moment he stared at her dumbly, taking the verbal beating; he would have been less shocked if she had actually hit him. He had known no one living who knew Gallic outside of his own teaching, and he was – fascinated. Then, realizing what she was saying, his complexion darkened, and he retaliated.

Everyone watched the exchange in mesmerized silence, not a soul among them understanding a word. Finally, Amergin decided to cut them off.

"HEY!!!" The pair of them turned to him in unison, bloodlust sharpening their glares. "Kill each other on yer own time! Ta' Turranians think us barbarians as it is!"

Shrikant nodded, shrugging. True.

"Can ya conduct yerselfs proper, or have ya spoken yer sides'uv it?"

The pair glanced at one another, evaluating. And Roscanna threw in the towel. "Ye've heard sense, now Ah give ye back tae him!" And that was that; she returned to her place at Caomhan's side.

Daíthi pursed his lips, momentarily furious, but then he shrugged, allowing his anger to melt away – if for no other reason than he knew it would irritate Roscanna.

"Ah'll summarize quick, den. I ent sayin' we gatta go t'war righ' dis *secon'* ur nufin – all's I'm sayin's we gotta decide *soon*. Think on't. When ya realize t'er ain't no u'ver way, we'll vote on't. How long'll y'need t'be makin' a decision, d'ya think?"

Amergin glanced around at the gathered, then favored the little messenger with a shrug. "I meself could be makin' a decision by tomorrow if I'm t'be speakin' for m'people, as I hafta be consultin' with 'em. Otherwise I'd say I'm ready t'be goin' at t'word, save I'm needin' two nights t'be mobilizin' m'camp far th' women an' childer, like."

"Livestock," added Shrikant.

Amergin was piqued. "I'm like t'be takin' *offense* ta' –"

"Are not your people herders?"

"Oh." Amergin blushed. "So they are. My mistake and yer pardon."

"I don't like the rush for a decision myself," noted Caomhan, returning to the subject at hand. "I'd give it two nights of thought." They all conceded.

Daíthi let slip a vulgar and indiscreet curse, earning their attention. With a shake of his head and palms raised, he noted, "Ah *concede.* But t'end'a two nigh's, ye'll be givin' me an answer, aye?"

"...Aye."

"*Awesome.*" With an eye-roll serving as the meeting's closing formalities, Daíthi walked out.

The boy took him by the arm as he moved past, briefly halting him. Daíthi repressed a sigh. When he turned to the lad – and Joe Sullivan, who was still hanging around, tied to the camp or not – he met them with a forced smile.

"Well!" he said shortly, clapping his hands together. "*That* warn't warth t'gettin' up dis marnin'!"

"What d'we do now?" asked Zander, scrambling slightly to keep up with Daíthi as he continued walking.

Daíthi shrugged. "Damned hif *I* know." He turned to Sully. "Sa – ye'll be shovin' off nah, so t'spayk? Seein' as y'ent gat need'a be watchin' over us, so."

"As it's two days 'til a decision's made by the lot'uv ya?" Sully walked with a casual stride, his hands deep in the pockets of his great coat. "Aye, I'm supposin' I will... But I'll be back in time for the meeting. I'm curious myself to know whether your urging had any effect on'em."

"Yeah, you an' me both."

"I'm assuming you'll be keepin' the boy with you, then?"

"Supposin' I *have* to, huh?" Daíthi returned, an edginess to his tone.

Sully ignored it. "Alright, then." He turned to Zander. "I'll be seein' ya, Flotsam."

"'Bye, Sully."

He tipped his hat to them both and delved as smoothly into the crowd as a dolphin in the waves.

Daíthi breathed a slow sigh, watching him go from beneath the brim of his cap.

"He's gat me damn Cure."

Turning back to the boy, he said, "Ah'm gonna be headin' off meself, if y'don' mind. Don' leave camp in th'mayn-time, a'righ'? Jis' be stickin' aroun'jyer own people, so's t'be sure tayr's sumfin left'a yaz when I come back, 'kay? I'll be back by a'nigh'." He backed up a few paces, moving towards the uninhabited woods.

Zander's eyes narrowed. "Let me come with you."

"Hah!" Daíthi's sharp laugh banished the notion.

"There're still some things I need answered, Murphy!"

Daíthi's eyes flashed off to the distant Heavens as he drew in a breath through clenched teeth. Ah, that was true, wasn't it. But sure he'd had enough fighting for the one morning, thank you very much! For a moment he merely stared at the piercingly overcast sky. Then at last he settled on the most suitable answer.

"'Kay bye!" he blurted, and, turning, he was gone.

79
Drownin' Foudroyant

The sun was beginning to set by the time Zander started heading back. He had been left on his own at the heart of the Company's camp – a group, *shockingly*, meant to be his allies. In truth, they had done little to maintain let alone cement the bond of kinship he had felt for them before. And he could not help but sense the truth in Rahyoke's familiar dictum, "Other people just get in the way."

Rahyoke... If Murphy could be believed, Rahyoke was still alive, and that meant he was most likely being held as a prisoner at Ariegn. If he was dead, well... all the more reason to give the old empress a call. This, however, raised more problems than it solved. Ariegn was impenetrable. Alone, he did not have the means to conquer it. And although he doubted it was something that friendship and teamwork could overcome, for now he could do nothing but sit tight and monitor how the situation progressed.

He suddenly recalled Murphy's response when he'd claimed that he wasn't a prisoner: *Ya know, I like to think the same thing...* But they were victims of their own designs.

Dredging a curse he hadn't realized he'd known, Zander turned his steps back towards "home" – the camp of the rake king and his band of motley thieves, rogues, and drunks – wondering where those steps would take him next.

As he approached the camp, Padraig O'Donnell looked up with a smile. "Hey! 'Tis himself, ol' Zandereen Óg!"

"Murphy's nae back yet," added Gerry, who stood nearby holding a floor-lamp, "but yer welcome tae sit awhile, if it pleases yeh."

Zander remained silent, staring at them in wonder. A roaring fire blazed at the center of their circle. Gerry was in the process, with Matty Calhoun's help, of clearing out his tent, as Shaw, Paddy and Rhys loitered about, picking on them, scraping through the belongings, and being generally of no use whatsoever. A mass of items that seemed almost too many to have filled such a small tent cluttered the area of their camp like a deranged yard-sale, and Gerry and Matty ever bringing out more to add to the pile.

"Aye!" Red pushed a large knapsack from the log beside him carelessly; it fell to the ground with a clattering crash. "Si'chyerself here, m'lad! Take a load off!"

Though admittedly hesitant, Zander moved forward until he came level with Gerry's tent. "What are you doing?" he asked, watching as Gerry tried to steady a lopsided pile with the floor-lamp like a staff.

"Sellin'," Gerry answered with a shrug; as if that single word should explain everything, he withdrew into the tent.

Red rolled his eyes. "He's bent on t'notion t'make t'camp *mo-bile*. Y'got any int'rest inna traffic light?"

"A what?"

"We heard that they won't be killin' ya!" Shaw exclaimed. "Congratulations!"

"Thank... you," Zander responded, not really certain what to say. "They're also thinkin' about goin' t'war."

"HAH!!" Gerry re-emerged from the tent just long enough to lob a triumph.

"Plannin' an' doin' are two different things!" Red returned, glaring after Gerry. With a more pitying look to Zander, he explained, "We've heard it all before, kid; ye'll see soon anuff. Sit down."

Now Zander obeyed, taking a seat on the log beside Red O'Donnell. "Murphy seems to think something'll happen," he muttered.

"Murphy?"

Zander's eyes flicked up from the blaze of the campfire. "He proposed making me leader in a war on Ariegn."

Red and Shaw glanced to each other with wide-eyed looks. Then they burst into laughter. "Da' day *Daíthi Murphy* takes an honest interest in politics – "

"Find an honest bone in 'is body!" Rhys challenged.

"In that case, Zander," Shaw reassured him, "don't think a *thing* on it. The day they listen t'*him*'s farthest off."

"As much as I hate teh agree," said Gerry, "it may very well be true."

Zander pondered this deep within him. It was in no way funny to him.

"Guys, you've upset him!" Matty complained, compassionately. "C'mere, Zander: let it sit for a night. No sense worryin' in what's a ways off."

"Aye!" Red agreed, with a smart-alecky grin. "An' if they decide that it's t'war they'd be wantin' t'go, we'll worry about it at t'decidin'. As it is – rejoice!" He lifted high a clean and empty glass. "Yer alive, an' 'tis *Murphy* who'll hafta deal wit' their choice!"

Zander smirked. "I'll give ya that."

"Ah ne'er knew ya had sa lil' fait' in me."

Everybody turned. There stood Murphy, wearing a night-black hooded-sweatshirt and the same tweed cap from before, a full satchel slung over his shoulder, a sea-bag on his back, and an uncorked wine bottle in his hand.

"Murphy!" Shaw greeted him, warmly. "We were just talking about you!"

"Where yeh been?" Red inquired. "And what di'jyeh bring back wit' ya?"

"Hm? Ah wuz o'er in *Parvum Lavinium*, talkin'a Aila... Rav'er, tryin'a ou'-bid 'er on dis." With his free hand, he indicated the sea-bag on his back. "Hey, Gerry – why ya gat half da' country in goods emptied ou'side yer door?"

Gerry and Matty were busy trying to maneuver a large chest out of the tent. "Yer crap's bee the tree, there," Gerry replied.

Daíthi followed the nod. "Thaynks," he said. Beside a tree near the campfire circle were two battered bags that looked like they'd traveled the world and his beloved guitar. He added the sea-bag to the pile, supplementing, "She's flippin' furious o'er it, too. Means Ah gatta git a new axel fra' someone else."

"Maybe you can get one from Morann?" Shaw proposed. "He's back in camp; says he needs ta' see ya."

Daíthi's brows knit. "Morann Doyle?"

"Do you know another Morann?"

Daíthi took a drink, mulling it over. "Wha'see wan'?"

"Says you gotta do a delivery." When Murphy groaned in reply, Shaw added, "Hey! Yer a messenger! All part'a the gig!"

Daíthi rolled his eyes, choosing to take another drink rather than respond. "Mmh. *Foudroyant,* tha'."

"Is that wine?" Shaw asked.

"Did ya bring us food, Daíthi?" Matty added, stepping through the clutter. For whatever reason, he was wearing an oversized sombrero that fairly dwarfed him.

Daíthi grinned, answering them both summarily: "Yup!" In one fell swoop, he tipped back the bottle, unslung his bag, and tossed it into Rhys's arms with a clatter and a thud.

"Oof!" Rhys cupped the bag like a football as it hit him in the chest. Peering within, he reported, "Bread an' beer."

"T'substance'a life! Dayr's some u'ver stuff, too." He took a seat beside Zander, setting the wine-bottle between his feet.

"Yer gonna get drunk."

"I ain't, nee'ver. Ah'm tryin'a bring su' class t'dis place." He reached into the pocket of his hoodie, producing a pack of cigarettes. "You lot ain't got na' mind f'r flamboyantry, fair anuff; stay a low-bred, indacent, classless bunch'a ninnies."

"Flamboyantry?"

"Aye." Murphy slit the cigarette package open with his switch-blade. "Give tha' sack t'Paddy, will ya? Is his nigh' t'be cookin' dinner."

"Oh *is* it!?" demanded Paddy, haughtily, even as he reached for the sack. "Can I be *sarvin'* ya, Princess?"

Daíthi's eyes flickered up, his expression one of accustomed boredom. He lit his cigarette from the fire.

Paddy continued. "Since yer an eavesdropping shirker, no doubt yer aware of our surprise in findin' that *you* of all people decided to make Zander a war-lord, and us 'is supportive cannon-fodder. And I'm wonderin', now, if you could answer me: *why*, an' *what*, possessed ya, in faith, t'be makin' such a proposal as *war?"*

"Di'jya *wonder*, Paddy?" Daíthi remarked, his exaggerated expression of consideration but a thinly-veiled eye-roll. He stowed his knife, taking up the bottle again. It was a wide-open question, and he could have gone anywhere with it. "Well? Boredom..."

Red rolled his eyes. *"Boredom* me arse."

Daíthi raised the palm that held the cigarette. "Jus' deliverin' a message, tha'ss all."

"An' t'rowin' Zander int'a t'bargain," added Red.

Daíthi smirked, motioning a toast. "Ez'akly righ'." He tipped back the bottle.

"Fer who?"

"Hm?" He didn't lower the bottle, but he looked Paddy's way.

Paddy leaned forward, emphasizing, "Daíthi Murphy deliverin' a message uv his own good will an' kindness? Not a lick'a good payment?" He snorted. "Not likely. Fer who di'jya get that message?"

Daíthi swallowed slowly.

"It must be some sorry, piss-poor son-uva-bitch to have ended up with you!" Shaw laughed.

"Aye," agreed Rhys, through a mouthful of bread filched from the bag. "Usually r'sarved f'r *fast* souls, tha' job."

"Uh!" Daíthi glared at him, scandalized. "Y'know wha', layve off! Ah give! Attend a damn mee'in ahr two yerselfs, if yer wantin' a'know, a'tall!"

"Don't take offence," Shaw urged him. "We're just all very curious to know what the Hell happened to bring us to this point."

Daíthi sighed. Then he shrugged. "Li' ya said, Shaw: Ah'm a messenger. All part'a da gig. An' anywayz, da amoun'a bitchin' yez all do pinin' after a figh' yerselfs, ya shun' be so quick ta' condemn me. Ya know wha' else ya said, Shaw? 'Stop buildin' houses an' start buildin' an army!' Tha'ss wha'chya said. So we're uprootin' – an' da first step'a formin' a fightin' force is givin' 'em a leader aroun' which ta' rally!!" *Wait*. "Whom," he corrected himself.

"Invisible Fortress Syndrome!" Gerry called from the tent, relishing his favorite original term. Daíthi toasted him indicatively.

Shaw rubbed his chin, thoughtfully. "Ya know? Ya do make the meetin's sound powerful interestin'; maybe I'll go. When's the next one?"

"Two nigh's fra' now."

"So it's tomorrow night? Or d'ya mean the night after?"

Daíthi's eyes rounded. "Uuh..." When the first trickle of giggling started from his companions, Daíthi let out a vulgar curse.

Shaw clapped him on the shoulder. "Wha' t'*Hell* kind'uva messenger are *you?!*"

Daíthi's cheeks burned red with frustration and embarrassment. "Ya know wha'?! *You* give'tt a try, smart-ass!! Ihs a lot teh keep track of –!"

"It's a very detailed con."

He was surprised by the soft, sardonic remark from Zander. The boy was very clearly not in a good mood; Daíthi suspected it had something to do with a combination of irritation from being talked *about* instead of *to*, and anger over not being fully informed about his father.

"Are ya cross? Aw, don' be sore, Zander!" He smiled. "Ih'll all wark ou'! *Luge eti ueidu*. T'mee'in ain't 'til two days fra' now; le'ss jus' layve't t'nigh, 'kay?"

Zander looked away. "Convenient."

"Tru'ss me," Daíthi insisted.

Zander whipped like a viper. "I don't even know you!"

"Ah know *you*." Grabbing the satchel from Red, Daíthi pulled out a loaf of bread, pressing it into Zander's hands. Soda bread. "Here! *Dinner*, dammit! Stubborn son'uva..." It seemed as though there should have been more to the sentence, but he added nothing else.

Zander chewed his lip, vacillating. How did a loaf of bread – even *soda bread* – make it all better? How was that an answer to his questions? How did that erase the fact that Daíthi was – as it really seemed – using him to his own advantage?

Then again, he was hungry. "Thanks," Zander muttered, breaking off a piece of the bread.

Daíthi tossed the satchel back to Paddy. "Anywayz," he continued, still talking to Zander, "y'done a good job t'day. Be sa'isfied! We'll deal wiv it t'marra!"

Zander bit into the soda bread, saying nothing. *Refuel...*

Daíthi permitted a silence between them, finishing off his cigarette as he thought and flicking the spent remains into the fire. Then he addressed Zander. "Look. Ah'm sarry Ah left ya. I ought'a re-innerduce ya ta' Aila; she's from Balladarach, too. She wuz very friendly wiv yer family. Ah'chully, deyr's a couple people fro' Balladarach who'll wanna meet ya. Heh – don' wan' ya thinkin' Ah'm jus' tryin'a get back on Aila's good side!" Zander rewarded him with a side-long stare, and Daíthi grinned. "Da' mee'ins are hostile, bu'chya ain' wiv'out friends here. Ah'll tayk ya ta' mee' wiv some. Promise."

Zander dropped his gaze, considering, but there didn't seem to be anything less than genuine in that vow. But what pushed him to allowing a miniscule smile was Daíthi's response when Rhys mocked, "Well, ain't you sentimental, *Barfly*." Uncurling one finger from the neck of the wine-bottle, Daíthi flipped Rhys off with both hands, saying, "'Ere's me sen'iments towards you!"

"Behave," Paddy warned, and Daíthi turned to him.

"Christ. Wha' da' *Hell* 'ur you makin'?" he asked disparagingly, studying whatever Paddy had going in a suspended stew-pot.

"Holy shit," Shaw said, by way of an answer.

Red pursed his lips. "Don' be makin' me express sentiments, now!" His critics snickered and relented for the moment.

Daíthi wasn't so lucky.

"Although," Rhys murmured, "I think the young O'Rielly has a fair point."

Tension drew a taut silence. Daíthi lowered the wine bottle, meeting Rhys' eyes. A blush that was more than passion had bloomed on his sun-damage-freckled cheeks, but his confrontational gaze found a steady target.

"Do ya."

Rhys prevailed. "Come on. When was the last time you spoke in one of these meetings? I think you've done it *once* –?"

"More'nn you."

"You're an attention-whore, that's all it is. What; stage shows gettin' a little sparse, rock-star? Ya gotta drum up some action here?"

Daíthi jumped, but Shaw moved faster, holding him back. "Easy, Murphy –"

"*You,*" Daíthi snarled, pointing at Rhys with as much of his free hand as he could raise. "Ah've just abou' *had* ih wiv you!"

"I'm just makin' conversation," Rhys began, subversively, but Gerry cut him off.

"Rhys! Murphy! That's enough from both'a yeh! There'll be no bloodshed here tonight, mark me!"

There was a frictive moment, then with reluctance both men relaxed, casting one another dark glances before settling.

Daíthi sank onto the ground, leaning his back against the log on which he'd been sitting, lounging. He took a long swig from his bottle, breaking off suddenly when he realized he had another question for Zander. "Mmh! Zi'yer shoulder better?"

Zander jolted. "What?"

"Yer –" Daíthi frowned. He laid his hand against his heart, his fingers just brushing the clavicle. "S'at t'righ' word?" His eyes had fallen to Zander's wound. "Sul' – Sullivan sai'jyur s'possa get 'ih looked at; sai'jyer *damn* lucky ih din' kill ya, where ih went t'rew. Lot'ta kill-points deyr, y'know? Gotta make sure ih' don' get infected, now. Di'jya?"

"...yeah," Zander lied. He didn't feel the need to trouble a healer with it – or himself with a healer.

"Good!" approved Daíthi, though he audibly muttered into his drink, "Ih'd be shittin' hard t'r'place ya."

At this point, Zander merely rolled his eyes. It was the point most eventually reached with Daíthi.

But their exchange had won the attention of the others, and, sitting forward to better examine Zander's shoulder from across the way, Shaw observed, "Ooh, there's blood! How di'jya do *that?*"

Zander pressed his lips into a thin line. Too many memories surrounded that puncture – far too many to explain. "Uhm..."

Daíthi came to his rescue again. "A *lotta* people wan'im dead. Some'uv been nearer su'sessful den u'vers. Wha'd Ah tell you? Trifles." He lazily flitted a hand over his mouth, suppressing something like a hiccup. "Shar wun'chya fi'ggur, him bein' O'Rielly's son an' all?"

"You know," Matty said, thoughtfully – and the pensive nature of his expression nearly made Murphy groan. "The timin' of things is all a bit peculiar, isn't it? Here we've been, waitin' around for years, and now out of the blue we find Zander, and we get an order to march from some secretive sender, and – well, it's all very mysterious! We aren't even an army yet – everybody knows it! So why now?"

Murphy noticed Zander's expression – that devious little smirk tugging at the boy's down-turned lips. Smug. He suddenly had an inkling of what Zander had meant by calling his humble if keen message delivery a con – an inkling that rapidly grew into an oil-spill as Zander pronounced, "Funny, the timing of things, isn't it? Always seems to hit you when you least need or expect it."

Murphy rolled him an unimpressed glare. "Gad'ss mysterious, Matty," he quickly undercut, but Zander rallied.

"Did he tell you my father's alive?"

Daíthi cursed. "Wai' –"

Too late. "O'Rielly?" Matty gasped, his eyes wide. "O'Rielly's alive? Is he coming to help us?"

Daíthi shook his head slurrily. "*Naw* 'ee ain't," he grumbled. "'Ee cain't."

"See – how the Hell d'you *know* all this!?" Zander demanded. "How d'you know he's alive? Did you *talk* to'im?! How!?"

"Ah tol'jyah a'fore. Da' girl'nn Carteia tol' me."

"'Kay, how did *she* know?" Zander pressed. "How d'you know she's not lying? And who the Hell is she, anyway?!"

"*I'*uh – wha'a'you *want?*" he complained, taking another drink.

"I'm sick'uv haf'in to piece everything together!" Zander exclaimed suddenly, his voice cracking. "I want someone to answer me! For *once!* I want someone to give me a straight answer!! I will *not* blindly fight somebody else's war!!"

Daíthi choked on a lungful of wine. "Somebody *else's* – y'fink dis'uss *my war!?!*" He turned on Zander, anger nearly making him sober. "When dey as'tt, you cou'da said *'Zander'*, plain an' simple. Bu' *you?* You took yer fa'der's name – da' full, Ceothann naym, *din'*chyou? An' *warse,* yuss'slap on *'Master'a Flames'*. J'yuh know, 'ee ne'er *once* used tha' title hissel'? On'ee *legends,* Zander Óg – an' damn well d'jya know it, too."

"What are you saying?"

"Wha'd'you *want,* Zander?" Murphy challenged. "Wha'd'*you* gain fr'uh dis war, huh?! Why j'you stay? You – you took'tt fer gran'it tha' *Duncan O'Rielly* were sumfin uva legend – j'yuh knew *ezact* whu'chyou meant, *han* da' consequences. You knew tha' wiv Donny tayken, d'Company b'caym d'ness target, positioned *so* well wivin grasp. An' you wuz jus' *sick*'a runnin', *warn'*chyou? Ya used d'fame a' his naym ta' join us."

Zander's fists clenched. "You –"

Daíthi iced him. "Dey won' move wid'outta leader."

"Why me!?" Zander protested, in agony – horrified at the suggestion he would use his dead – not dead – father's name for his own gain. *"Why me!?"*

Daíthi grabbed Zander's arm in an iron grasp. "Yer dad's mahr den a catalyst, Zander!" he insisted. "Dey'll *follow* you!! An' *ONLY* you!" His grip tightened. "Y'fink dis ent yer war? 'Un yew *NEVER* should'a taken on d'naym."

His words were cutting, biting, true.

"Zander, *wha'* d'you want, huh? Wha' d'you wan'?"

Zander lost his temper – finally. *"I WANT MY FATHER BACK!!!"*

To his surprise, Daíthi smiled. Releasing him, he noted, "We' talk in d'mahrnin'."

Zander was shaking – every muscle in his body, shaking. Paddy announced that dinner was served, and with good, hearty eagerness, they collected their rations for the night. As Zander reached forward, he did not miss as Daíthi, openly relieved, tipped back the wine-bottle again. He couldn't help but envy him. "Foudroyant, that," he whispered.

Daíthi snickered, lightly. "Tha'ss wha'chya get wi' undiluted, ain'it?" he quipped.

Zander elucidated. "You were right, you know. They'll come for us no matter what. Now that O'Rielly's out of the way... I – I just wanted to help, but... I've made it worse, haven't I? This is all my fault –"

"*Jay*-zus, Zander, yer a feckin' buzz-kill, j'ya know tha'?" Daíthi hiccupped; he took a shot of wine to settle his stomach. "Who *cayrs* whose fault wuz wha'? Ihss over! Wha' matters now'ss wha' we're *gonna* do. An' t'be hones', *most'*a tha'ss outta ahr control, too. Stop worryin'." He raised the bottle again, adding, "Hun' anywayz, yer – fifteen? Hain' nobody gonna blame you in da' ballads, hon."

"But..."

"Hanywayz too, ya gat me! (Hucc!) Ugh... Wha'ever *tha'ss* warth."

"Promise?"

It was the way he asked it – unnaturally meek and uncertain, his voice as small as a child's – that struck Murphy, causing him to turn. "Huh?"

Zander bit his lower lip, hesitating. "You... Promise not to bail on me!" Even though he tried to sound confident, Murphy still found the desperation in his tone. "I may've gotten myself into this mess, but you made it worse – so don't ditch me, okay?! I can't do this by myself; I need your help. Promise you won't betray me!"

Daíthi's hazel eyes narrowed slightly as he searched the boy's face. The fear of being abandoned... Like the promise to O'Rielly, to the girl in Carteia, no one would hold him to this but God.

He rolled his eyes. "*Fffyyne*, whatever. Ah promise Ah won' jilt 'chya; yeh have mah word... Yeh suck da' fun ou'a eh'ree-fin, j'ya know tha'?"

Still, out of the corner of his eye, he noticed the boy finally relax enough to manage a small smile.

"Hey." Daíthi smacked Zander's shin, lazily grabbing his attention. "Yuh still wanna help, righ'?"

"Of course."

"Den tha'ss *e'zactly* wha'chyou'll do." He raised Zander a knowing grin, and finished: "Tomorrow."

80
Vertigo

"Up, y'alcoholic son'uva-bitch!"

"Nnuh..." With a long, dragging, whining moan, he pulled the blanket more firmly over his head. The light pierced through the dense fabric, stinging bolts of fire into his enfeebled mind. "...Kah go back t'sleep?"

There was a shuffling – the painfully loud crunching of mulch gravel as his killer got to his feet. "Nope."

"Uuh..."

The footsteps moved off, but did not go far. As he began to come back around, he placed the progress of his killer somewhere about the fire, near the smell of food which was somehow both nauseating and alluring. The very air tasted of moth-balls and vinegar; his fingers closed on his cap, and he tugged it onto his head, low over his eyes. *One, two... two and a half... two and... three.* Pulling the blanket over his head like a lady's shawl, he dragged himself up from the depths of black comfort and into the lancing sun, sitting up slowly. "Aukh." Repressing a wave of sickness, he swallowed and admitted, "Ah'd say *never again*, bu'... We'd all know ihsa lie."

Zander was kneeling beside the fire, stirring a pan which he'd suspended over the low blaze. His grin was cloying. "Morning, sunshine."

"Yeah yeah yeah," muttered Daíthi, snaking his hand to his shoulder-blade, which ached almost as much as his head. "You kicked me, din' you. *Son*'uva –"

"It wasn't all that hard... Anyway, we got stuff to do today, and I didn't know if you wanted breakfast before we go."

Breakfast. Daíthi's stomach turned at the mere thought. "Mmh... Yer feckin' *bossy*, you are."

"Well, he *is* meant to be a war-leader, isn't he?" Camp was sparsely populated that morning, with only Matty, Flannery and Faolán Kellog joining himself and Zander. Snagging a can of beer from a fishbowl, Kelly concluded, "Isn't that part of the gig?"

"Yeah," Matty laughed, "if he can get *you* to move, commandin' an army will be no problem!"

"*Eh'*ree-body's a smart-ass," Daíthi grumbled. "'Ey, Zander. Wanna grab me tha' bag, there?" He stretched, his right arm bent behind his head to grasp his elbow, left arm raised, straight but for the finger he bent towards Shaw's satchel. Zander hesitated but obeyed, tossing Daíthi the bag.

"Oh, 'tis a grave vice indeed," Matty intoned, as Daíthi pulled out the whiskey bottle.

Daíthi tipped him a winning smile. "Ah'm internally damned, Fa'der; will ya save me?"

"Baking soda or prunes would do for that," Flannery grinned.

Daíthi looked at him flatly, a little loathingly. "Shut up."

"Moral fiber," Matty snickered.

"Shut *up!!*" Taking his switch-blade out of his pocket, he flicked it open, working it under the cap to lever the cork. He intended to finish the bottle. "Flannery, wha'ss da' logistics on Gerry's miscellany? He's sellin' it, ain'ee?"

"Well?" Flannery was studying an astronomical globe which had been stabbed through with a wide array of darts. "Would ya have a suggestion?"

Matty chimed in. "We've been arguin' out transportation an' location. Carteia's closer, but the markets in Emona are better."

Daíthi nodded slowly. "Tha'ss true; Carteia's all farm goods an' hookers." Bracing, he pulled the cork from the bottle, dropping it into the satchel, snapping his switch-blade closed. "Bu' unless y'can get a cart, it'd hafta be Carteia."

Flannery set the globe on top of a tilted pile, steadying it with an artist's delicate poise. "Where would we get a cart?" he scoffed.

"Ah dunno; do a jig an' see if one falls outta da' sky." Daíthi took a drink, muttering, "Why'mm Ah da' on'ee one who can *think* in dis' camp?"

Zander slid him a sidelong glance, still stirring whatever he had going in the pan – a thing like a broken omelette, all peppers and onions and three eggs. "Oh, yeah; you aren't orchestrating *anything*."

Daíthi grinned. "Ah dunno wha' tha' mayns." He took a drink, judged Zander's expression, and hazarded a guess. "Oh!" He laughed. "Yer generous, Zander; Ah ent no mastermind genius social engineer. *A chiall!* Nah; Ah'm on'ee tryin'a help."

His smile was so genuine, so sweet. But Zander was the son of Orchid O'Rielly, and looked at him flatly. "I'm sure."

His skepticism was delightful. "A'rrite; layve'tt aside, Ah'll tell ya straight. Yer fa'der's captured at Ariegn, righ'? So, since'eez there, he'll pro'lly try an' make d'mos'uv d'essperience. Knowin' him – well, *you* know him, righ'? Shar ya do. We're gonna hafta save 'is courageous ass b'far he's gat all the good killin' over an' done wiv. Bu' ha' d'we do it? By doin' wha' he mahr or less as'tt of us!" A shot allowed sufficient time for that to sink in. "T'on'ee way t'help him is t'seize Ariegn. T'on'ee way t'seize Ariegn is wiv an army. T'on'ee way t'get an army's be givin' dem someone dey'll follow. Dey followed yer fa'der. Dey'll follow you. Tah-dah!"

Zander arched a brow, watching in silence as he pondered Daíthi's proud answer.

"Yer all sworn t'silence, be t'way – un'erstan'?!" Murphy demanded, paranoid, swishing an indication around them all with the bottle.

Matty raised a palm. "Priest."

"Faolán don't talk anyway."

"Ih'll all wark ou', Zander!" Daíthi chirped, apparently satisfied with that answer. "Ah wanna help yer da'. F'Ah din' fink dis were d'bes' way t'go abou' ih, Ah would'a bailed on 'is stupid-ass idea."

"What does that say about your loyalty?"

"A'have mah leader's bes' interest in mind, even when he don't. Wha's *tha'* say?"

Point Daíthi.

Zander looked down at his breakfast – and cursed, pulling the pan from the fire. "Aw; you made me burn my food!"

Burnt onions... Daíthi resisted the urge to vomit, taking another shot of whiskey. "*Yar* food," he repeated, at a mumble. "Yar food I bought, ya lil' ingrate."

"I'll share," Zander offered.

Daíthi made a weak little noise of pain, shying away, and Zander emptied the pan onto a plate he'd found, settling cross-legged beside the fire. Matty took his place, suspending a campfire coffee pot. "Ya want some, Zander?"

"Sure." He looked to Daíthi. "Do we have time?"

Matty pursued the target. "Would you like some coffee to spike yer hooch with, Murphy darlin'?"

"Eh? *Gan bhláthach, b'fhéidir –*"

Flannery was affronted. "*Without* buttermilk?! That was quality buttermilk, I'll have you know!"

Daíthi rolled his eyes. "Sorry, Matty; we ain't got time ta' wait for it. Zander, eat it ahr leave ih; le's go."

Zander picked through his omelette, pushing the onions to the side. "So, how do you know they're holding him at Ariegn?"

"Hm? Oh, yer da'? Well, Ah gat good intel, bu' – Ah mayn, where else cou' dey hold'im, righ'?" Murphy shrugged easily as he said it, but Zander paused, looking at him through the fringe of bangs shorter, neater, and of a somewhat lighter brown than Duncan's, but the effect of the glance was the same. Noticing, recognizing, Daíthi smirked, a teasing laugh to his tone as he responded, "Ain'it?"

Zander nodded, returning his attention to his meal. "So," he began again, changing tack, "I mean, since you make it sound like it's the only option, how are we gonna get them to follow me?" He stuck a sporkful of omelette in his mouth, shifting it like a chipmunk to add, "Maybe some of 'em will, but you heard Roscanna. I don't think they'll all follow me just 'cause of my dad."

"Ya don' need ta' convince all'a dem, Zander. Ayvin a mob, if ya look for it, has a hierarchy. Nah; invokin' yer fa'der's mahr fer da Balladarach survivors den anyfin else."

"Aren't the Balladarach survivors the minority?" Zander asked, around another mouthful.

"Yes!" A quick swig. "Bu' an important minority. Sometimes d'minority rahlly has d'control. An' shar – s'like kings, ain'it?"

"The Balladarach minority's not in control," Matty denied, frowning. He was digging around in his bag. "Where's my soup?"

"Dey ain't in upfront control, bu' deyr say counts. If Balladarach were agains' ih, ih wun't go." When Matty shook his head, Daíthi insisted: "If history weren't important, nah one'd give a feck abou' Zander. He's yer litmus test. 'Ee resonates wiv Balladarach, who cling t'the original cause, amplified by deyr own losses. Balladarach remains, while not king, d'life-blood uv *our* empire."

"So I hafta win over Balladarach?"

Daíthi pointed to Zander. "No!"

"Then what?"

"We hafta meet wiv Morann." Daíthi stood – swayed, looking somewhat confounded – then smirked. "*Lord;* Ah think Ah'm still drunk."

"I'm still going, right?" He recognized Daíthi's glare – it was the *Wrap it up or I'll leave you* glare – and he obediently dug into his food, scarfing down the rest. "Mmh! Hey!" He swallowed, setting his spork down on his now-empty plate. "Finished!" He got to his feet as well. "Boy; *you* bounce back fast."

Flannery snorted. "Son, you ain't seen *nuthin'!*"

"Wait up!"

"Yo! Morann, ya son'uva bitch!" It might've been the happiest those words were ever spoken.

In spite of the early hour, nearly everyone in the Lowlander camp seemed to be out, if not about. Great droves of people huddled around campfires, wrapped in cozy blankets and the mixed, savory scents of food. Herders drove flocks of sheep and even a few cattle dead through the heart of the camp, accompanied by packs of dusty dogs and ruffled chickens. The entire camp was at once a market, a public house, a tenement square, and a gorgeous, never-ending hooley.

They found Morann standing atop a simple, flat-bed wooden cart, surrounded by tarp-covered crates and barrels. He was busy working a tap into one of the larger barrels, and he raised his head at Murphy's informal greeting, giving the tap a final, definitive whack.

"Is that Murphy!"

"It is." Daíthi offered his hand. "Ihs been a while; how da Hell are ya?"

"Ah, it's too early to tell yet," Morann said, accepting the handshake. "Too early to tell entirely. Wouldn't ya say?"

"Aye."

Having dispensed with the formalities, he returned to his barrel, taking a clean pint-glass from an open crate.

Murphy lifted onto his toes, folding his arms atop the cart rail. "Dayr wuz rumors you went ta' Osca," he noted.

Morann sighed. "'Tis always straight to business with you." Lowering the glass, he pulled a pint of dark, rich stout from the barrel.

"Hey; f'Ah'd me choice, I'd be sleepin'. You summoned me fer a reason, Ah reckon?"

Morann looked up, considering ponderously; though he didn't watch it, he knew by feel exactly when to cut the tap. He raised the glass to his line of vision, inspecting the pint in the morning light; it was the most picturesque pint of foam-crowned black stout that the eye had ever beheld. "I did, I did."

He said no more, and Murphy could not help but groan, burying his face against his folded arms. "Damn it, Morann. Yer as thick as manure and half'uz useful, ya know tha'?"

"Now, do you want to see what I've brought back with me, or no?"

"Ah swayr ta' Gad, Mo –" He gasped, seeing through the mystifying murk of Morann's words. He raised his gaze. "Wai'."

Morann smirked behind his upraised pint-glass.

"You got ih?! Lemme see!"

Morann shrugged. "As you may." He set his drink upon a chair and began futzing around with a shock-blue tarp. The tarp was tied down – to itself, to other objects, to the cart, and now to Morann's shoe – with a wide and tangled mess of bungee cords, and it was all Morann could do to get them unstuck. Daíthi hoisted himself up, clambering over the rail slowly, a bit clumsily, but successfully, coming to stand at Morann's side.

Layers of wadded-up, squiggly cords were peeled away, snapping and hissing like a nest of angry snakes. *I wouldn't untangle that racket if you paid me,* Daíthi thought, as Morann dropped the last of the half-braided cords onto the deck. The tarp was a war and a half as well, but at last Morann threw it aside triumphantly (it crumpled over the edge of the cart, still with one corner caught under a barrel) to reveal his worthwhile treasure.

"What is it?" Zander asked, taking up Daíthi's place at the rail. He had never seen such an object in his life. It was like a big, oddly-shaped box made of grating and metallic plastic, almost the size and shape of a barrel. It was dusty and cob-webby and entirely alien, but Daíthi seemed to know what it was in an instant.

"*A chiall.* Morann Doyle, you *are* as good as yer word," he murmured, crouching down beside the novelty to better inspect it. "Ah'm damn nayr impressed!"

Stout in hand, Morann settled into a chair beside the strange barrel. "Is that a *compliment* I hear, now?"

"S'it function, ahr have yi'nah tested ih?"

Morann shook his head. "Damn busted up."

"Mmh." It was hard to tell if Daíthi was listening, so intent was the gaze he had narrowed on the strange plastic barrel. "Luh's li' ihs been ou'a commission f'r 'bout twen'ee years, this."

Zander shifted so he was sitting on the edge of the cart, leaning over the rail. "Murph, what is it?"

Daíthi glanced up. "Morann, Zander."

"Hey, Zander," Morann greeted the boy. "D'you go by that?"

Zander shrugged; before he could think of a word to say, Morann continued, "D'you drink?"

"He does not," Daíthi intervened. Turning to Zander, he answered him, "Ih'sa *generator*." It didn't matter that he'd enunciated the word perfectly; Zander had no clue what it meant. "A damn good'un, a'that."

Zander frowned. "What's that?"

"Ah jus' – oh." Daíthi hung a hand over the top of it, drumming his fingers tunelessly as he thought. "Ih mayks –" What? "Mayks – mayks f'ins *wark*, so." That was a terrible explanation. Murphy huffed a sigh, wondering why Morann had to be so damn useless in conversation. "Ih creates – a force. *Leucos*. Energy – an' ih mayks thin's *function*, like. Li' t'grea', flower-lookin' post Gerry has in'is tent – this'll mayk ih give light." The floor-lamp. The magical power of the odd barrel would make the floor-lamp operate. It made electricity.

Zander brightened with a spark of recognition. "Oh!"

"So tell me, Murphy," Morann said, languidly. "D'ya think you could fix it?"

Daíthi's head jerked up. "Yew insult me!"

Morann raised his hands. "Had to ask," he said, by way of a defense. He finished his drink and stood.

Daíthi pursed his lips, returning his gaze to the generator. "Ah'd give ih... Gimmie t'ree days, tops."

"*That's* ambitious."

"Supposin'," answered Daíthi, lightly. How on earth he was meant to get it back was the real difficulty, but – hey! What better use for pesky children? Clapping a hand to the top of the generator with a solid, plastic *snap*, he got to his feet, turning to face Morann. "Ah'll pay ya when Ah come back fer ih. Ah cain't carry ih – ahr –" his eyes scanned the generator "—Ah ent *gonna*."

"You are a special kind of skiver."

Murphy shrugged. He noticed a box of almond flour sacks among the crates, and, motioning to it, he half-asked, "Ah'll deliver tha'?"

Morann followed his gaze. "Mmh? If it goes missin', I'll tell her you have it," he said, by way of assent.

Daíthi smirked, tilting his head slightly in a way that was both compelling and conniving. "Perish da' thought, hon."

They returned to camp, where Daíthi picked up one of his bags and successfully finagled Flannery into retrieving the generator; then they made their way back to the Lowlander camp, the box of flour sacks in hand.

"What are we doing?" Zander asked, tasked with carrying the box.

"Ah had'a pick tha' up from Morann," Daíthi explained. "Ah'm a messenger."

Zander rolled his eyes. He was about to pursue the issue when – most unexpectedly – a strange beast erupted from the forest and tackled Daíthi to the ground.

The great creature was covered in black and white fur, its eyes as blue as ice, its fangs as white as the moon's merciless dagger, claws sharp and blackened with mud. It darted back and forth over the fallen man, settling on his chest when he rolled onto his back with a curse, pinning him to the soil.

"*Amadán!*" Complaining loudly, Daíthi shoved the beast off of him with one arm and sat up, wiping his other arm over his face. This delighted the creature, who began running around in tight circles, body held close to the ground.

"It –" Zander's voice was as strained as if sieved through the hand of Death "—ih – it's a wolf!"

"Dis?" Entirely insensitive to the boy's obvious fear – and, indeed, because of it – Daíthi burst into laughter. "*Mo' Deuos* – she'za dog!"

Zander bit his lips. "Yeah. Dog like Cerberus." He watched in nervous fascination as the beast ran up to Daíthi and resumed licking him in the face with overflowing enthusiasm, sniffing his hair and pawing at his arm as he ran his fingers through its fur. Daíthi looked up at Zander with a smile and explained, "Her name's Cuchulainn. I, uh – we don' do much 'animal husbandry' in da' city, so I din' raylize a' first she warn't a boy. Gave 'er a nice, manly name. Go on an' pet 'er! She's nice – aren't you, *a Chulainn, a stór?*"

"Ih – is she your dog, then?"

"Well..." Daíthi raised his eyes to look beyond the happy wolf, and his smile softened. A woman's voice greeted him, "She missed you."

Daíthi rose, hooking the dog's collar with his thumb to ensure she didn't run off. "Di' you nah?"

A young woman stood nearby, her brown-black curls tied into a loose bun that spilled a shower of escaped locks down her back. Her skin was splashed with a flurry of freckles, her eyes as green as emeralds, and her own smile sharp and devious. "No. What's in the bag?"

With his free hand, Daíthi again hefted his fallen bag. "Will ya do me a solid?" She gave him a look. "Auh, please? Dayr ain't skivvies in 'ere – an' Ah promise dey ain't rank." He chewed his lower lip. "A dozen *pizzelle*."

"Two dozen."

"*T'ank* you!" He handed over his bag.

"Oof." She made a show of almost dropping it because of the weight, although it was clearly only half full. "Jeesh," she strained, playfully, "you're wardrobe's bigger than mine."

"Take wha' ya like," Daíthi shrugged, now taking the box of flour sacks from Zander and offering this to her in turn. "An' dis is yours. From Morann, wiv love."

"Ooh!" She dropped the bag and cradled the box. "*Thank* you!" Her gaze then found the one who had carried it, and her brow furrowed with thought.

"Fahy, d'ya remember Zander? O'Rielly's lil' bairn? 'Ee lived in Balladarach wiv us..."

When the name hit home, her face lit with excitement. "Oh my gosh! The little Zander we played burgle-cats with?"

"Da' one you used fer cat-bait, aye."

"Oh!" She set down her crate atop the laundry bag and pulled Zander into an embrace.

Zander blushed. "Oh – um – hi."

She took him by the shoulders, stepping back; even Daíthi was a little taller than she was, so she had to look up to meet Zander's eyes. "You probably don't remember me," she said. "I'm Deirdre Fahy."

"Senan Fahy's daugh'er," Daíthi said.

Deirdre released him, studying Zander almost like a proud mother. "I'm so happy to see you, sweetie!" Then, turning a glare on Daíthi, she admonished, "So it's true you're sending him off to war!"

"Don' look'a me li' tha'," Daíthi half-sneered. He released his hold on the dog, folding his arms as he continued. "It ain't li' Ah'm t'rowin' a puppy inna den'uv snakes. He's able for it. Truss' me."

Deirdre continued to look at him scoldingly. "I know what you're doing," she said, in a low voice. "He isn't his father."

"An' you ain't yours; Ah know. If ya wuz, we would'a been able ta' put dis whole thing ta' bed a Hell'uva lot sooner, wun' we? Bu' yeh've aich got yer roles. Believe me, Deirdre; 'eez young, but 'eez clever. An' ya don' bring down an Empire' uv dis size wiv dees numbers on brute force alone. We need someone smart. From wha' Ah hear, Ariegn had'im in deyr grasp, an' were made ta' give'im up li' ih was deyr own idea."

"Twice," Zander uttered, tensing as the wolf began sniffing his shoes.

"Twice!" Daíthi looked at him in surprise. "Holy shit!" Then, to Deirdre, "Talent!"

Deirdre sighed, bowing her head, and Daíthi took her by the shoulders, encouraging her to look up. "Fahy," he said, with surprising tenderness, "we're gonna do ih dis time. Okay? I promise. We can do dis."

"We have to, don't we?" She lifted a hand, crossing it to hold his wrist as she searched his eyes. "You better take care of him," she warned.

"Took care'a you pretty damn good, din' Ah?" And at that she finally smiled.

"I'm serious!" she laughed, pushing him back so she could heft her crate again. "You protect him, ya hear?"

"*Chih!* Why'a ya think Ah been takin' da' licks'a oration in 'is stead? It ain' li' Ah'm hard-up fer altercations. Bu' t'way Ah figgur, better me den him, righ'?"

"I heard." Deirdre folded her arms around her box. "Only two fist-fights? Not much of a debut."

Daíthi winced at her barb, laying a hand over his heart. Then he thumbed an indication back to Zander. "He's a killer, nah a brawler, an' we're short-staffed as it is. Ah'm simply da' Aaron ta' his Moses. Da' Caomhan to his Eoghan, if you will."

"You certainly think highly of yourself," Deirdre muttered, making Zander laugh. "Speaking of Caomhan, watch out for him, alright? You seem to be swimming against the current of the Highlanders, and he'll be the one with the net."

"Rah'lly? A fishin' metaphor? Yer such a cheese-ball, Fahy."

She gave a humble dip of her head. Then she resumed. "You know I can't protect you from your own camp. And Caomhan's a cleaner – the guy they bring in to make guys like you disappear."

Daíthi rolled his eyes, then took her by the shoulders again, this time more insistent than consoling. "Lissen ta' me. Are ya lissenin' ta' me? Ah know. An' Ah won' let anythin' happen ta' Zander. Ih'll all be fine. Ah've gat da' next great war-lord on me side, don' I?"

Zander stretched, folding his hands behind his head. "So long as you keep feeding me."

Daíthi arched a brow at him, but Deirdre laughed. She patted Murphy's arm as she shrugged off his loose grasp. "We'll discuss it over *pizzelle*. I'll be here." Stepping back, she picked up his bag. "Are you taking Coolie?"

"Nah yet; Ah feel better knowin' she'll guard ya 'til Caílte ge'ss back – whatever good he'll do."

She grinned. "I *like* Caílte!"

"*A chiall...*" He stepped away, motioning for Zander to follow him. "Stay outta trouble, Fahy!"

"Two dozen *pizzelle*, Murphy!!" Deirdre whistled for the dog. As she walked away, she called back, "Good luck!"

That night, it rained.

Most of the camp had retired, surrendering to the weather and the hour, but Daíthi, Gerry, Red and Flannery were still awake, keeping watch on a sputtering campfire.

"Everyone's asleep, are they?" Paddy asked, casting a glance about. He and Flannery were playing cards, each holding a hand of five, with one card and the deck on a travel trunk between them.

Daíthi, seated on the ground with his back against a log, breathed out the smoke from his cigarette, watching it fade into the night. "Yer so damn antsy," he drawled. "Jus' say wha'chya gotta say."

"Then I will." Red laid down a card. "Now, I've kept mum in front'a yer boy, 'cause yeh've clearly got an angle, and who am I teh spoil a good game outright? But I thought I should ask ya, just t'be clear, at what point does it end?"

Daíthi's eyes narrowed with feigned curiosity. "Hm?"

"Ya can't possibly expect ta' convince these people ta' go ta' war, Daíthi," Flannery reasoned, playing a card before shuffling through his hand. "*We* know tha'chyer a bright anuff fellow, but... Well, what else are ya?"

"Barfly Murphy," Padraig agreed, summarily.

"Aye, not a soul'd take that seriously."

"Yer nameless ta' most, an' nonsense t'da' rest."

"Aye."

"Take it from me, Murphy," Paddy urged, drawing a card from the pile. "I'm older an' wiser den you. Guys like us, we have aspirations. But ya gotta know your place. *Ná tarraing an anachain ort mura bhfuil tú i riocht é a chur díot.*"

Don't pull calamity on yourself unless you're in shape to get rid of it. Daíthi took another long drag on his cigarette, unfazed. "Are ya willin' ta' wager on tha'?"

Red and Flannery both paused, looking at one another with concern. Then Red laid down another card, watching the board as he asked, "An' what then, Murphy? Let's say tha'chya do get us ta' war. 'Tisn't us who da' odds currently favour – not a bit. Why should we t'row our lives away?"

"Ya keep tellin' Zander it's all fer da' sake'uv his father," Flannery pointed out, laying down a card in turn. "But what's so great about dis O'Rielly chap, anyway? I mean, I get that he's a founder, and a powerful fine rebel and all, but –"

"But a lot'uv people would be givin' up deyr lives in dis fight, Murphy," Red explained. "What makes dis O'Rielly so –"

"'Eez da' on'ee thin' keepin' dem from us, Paddy. *Tha'ss* why 'eez so damn important."

A moment of silence fell as they pondered what this information did for the odds. "What –"

"Jus' trust me, okay? Gad willin', you'll see."

Red's smirk was a sneer. "God willin', he says." He threw down his cards. "Well, that's about all the hope we have of winnin' a battle anyway, Murpheen." He piled the change on the table into a stack, depositing it into his pocket as Flannery collected the cards. "'Tis a dangerous game yer playin', and though I pay little enough heed ta' Rhys's antagonisms, it does make me wonder what yer tryin' ta' prove. Deyr's easier ways ta' gain respect than slaughterin' yer comrades –"

"No one's makin' ya stay, Red."

"Aye. But no one's makin' me leave, yet, either." He stood. "Tell ya what. Twenty gold says ya can't convince them ta' go. Two hundred against us winnin'."

Murphy bowed his gaze, and smirked. "Ah don' need yer gold, sweet'art. F'Ah win, Ah wan' a favour. On mah askin', no questions, no complaints."

Paddy grinned. "Deal." Waving for Flannery to follow, he walked away, calling back, "When I win, I get da' gold!"

Daíthi lifted a hand in farewell, waiting until they had gone before sliding down against the log with a groan. "Ihs true wha' dey say abou' loyalty an' thieves," he muttered. After a moment, he turned his gaze to Gerry, and offered a forced grin. "Wha'd'you say, Gerry?" he asked, quietly. "'Ur all mah past failures evidence'uv some indefatigable pattern – an' will dis be dee' apogee? Da' apotheosis an' da' zenith 'uv a tragic if sparkly career?" He raised the cigarette to his lips again, pretending he didn't care about the answer.

Gerry had been spending his time whittling; a little girl from a neighboring campsite had commissioned a rabbit to play with. The tiny figurine was barely two inches in height, yet it was so detailed it seemed almost alive.

Gerry held up the figurine, examining it in the low light. "Let me ask yeh somethin', Murphy," he said.

Daíthi looked at him expectantly. "Yeah?"

"Do yeh have any other option?"

Murphy bit his lips, dropping his gaze, as Gerry returned to his task, carving a delicate "V" so his little creature could breathe.

"...No."

"Then ya have yer answer, don't yeh?"

Murphy was silent a long moment, studying the woodchips and listening to the sound of the rain. It was beginning to fall harder, sinking into the black earth.

He lifted his gaze again to Gerry, a subtle if avid flick. "Yes."

81
The Price of a Fire-Eater

The rain that had fallen late in the night clung in glistening pools to the courtyard's paving-stones, shining softly in the milky morning light. The world was quiet, sweet with a cheery summer dampness and troubled only by the distant cries of seagulls as they scavenged along the shore.

This was the place Bassus had chosen for the meeting. Not for the impending gloom of the down-grey sky, nor even for the salty, metallic air that blew in off of the sea. He had chosen this place because it was Ariegn's center. And the man he was due to meet was a bit skittish.

The courtyard, once in the time of kings the most vibrant of places, now was as dark, as bare, as black, and as empty as the vessel of a merciless heart. A covered walkway lead all the way around at this, the ground, level, fenced with arched pillars like those that O'Rielly had so missed in the great hall. Above – higher up those towering, slate-grey walls – was the catwalk, the place from which patrols could see to the ground far, far below.

In all, it was a great, empty space, paved with cobblestones and crowned with a sky more often than not of the same colour, and had remained so, quite frankly, because no one could figure out what to do with it. After all, once you remove all the trappings and finery, what more is a room than a room?

Still, it was daunting, this vast, empty space, and *that* was why Bassus had chosen here.

The wretch he awaited was late, though it wasn't particularly surprising; this peasant rabble was useless for doing anything in a timely fashion. At last a door at one end of the courtyard opened, and Bassus turned, as a figure stumbled out onto the covered walkway.

"Sorry. This whole place is stairs and archways."

Bassus narrowed his eyes. "Cunīculus, I do not exaggerate when I say it is *fortunate* for you your employment has reached its termination."

His words were like a barrier and halted the Cunīculus in the door.

Bassus's expression developed into a scowl. "I will not conduct this business yelling across a courtyard, you coward! And you do *not* want *me* to come to *you!*"

"Okay, I'm sorry," answered the Cunīculus, raising his palms and bowing his head as he started forward, slipping into the pale morning light. "Haven't really gotten used to all this yet."

All this. He said it as if there were anything to get used to. It was a conscious choice – one decision on the path of life. Join up with the Empire, or go down with the sad, rustic *curragh* of a dying, faery-tale state of ignorant sinners and modest, wide-eyed virgin saints. Less than a choice. It was *survival* for these leeches. One little slip of treachery. And then they were done.

There was really nothing to get used to at all.

"I will try not to waste your time," vowed the Cunīculus. For whatever reason, stupidity or courage, he halted closely before the general, easily within a knife's reach. Shame, in a way, that he was so useless. He was clearly bold, and conveniently lacked a conscience when the matter wasn't too close to home. But he was driven by self-interest – a fact that made him delightfully easy to control for the moment, but only so long as you held the right cards. He had an expiration date. It was time to show his hand.

"I suppose you want your thirty pieces of silver?" From the satchel at his side, Bassus pulled a roll of parchment, bound with the royal seal.

The Cunīculus's eyes widened, aglow. In his response, however, all excitement was veiled as he asked, "If I may be permitted to read it?"

Bassus's lip curled in a taunting smile. "You learned *something* from us, at least."

The Cunīculus watched as he held forth the scroll, scarcely able to believe his eyes. There it was. Three long years of work, justified at last! And he could finally go home. Back to his country. Back to his own people, his calm, quiet life. Back to –

"I wonder, though." His gaze roving the sky, Bassus repealed the magnificent treasure, laying it against his strong shoulder. "How wise is it to simply turn you lose?"

The Cunīculus paled.

"After all," Bassus continued, stinging him with his glance, "once a traitor, *always* a traitor, eh?"

"But – but I'm a traitor *for you,*" the Cunīculus persuaded, with surprising poise. "Why would I want to work for anyone else? Loyalty aside, even if I was *scum,* I know no one could make a better offer!"

"Loyalty?" Bassus held the scroll out over the Cunīculus's head, watching the man's eyes follow it like a puppy follows a steak. "Your loyalty is to this; don't insult my intelligence."

The Cunīculus's eyes returned to him. "Why would I betray you? There is nothing besides the Empire; any fool can see that."

Bassus allowed a brief pause in the conversation purely because he could. Then, pointing the scroll at the youth's thin chest, he accused, "You told Scipio Evander of the fire-eater, his brat archer of a son, and the girl with the book. When would it have been convenient for you to mention the boy?"

"I was bringing him with me – "

"Uh-huh."

"There were – uh – *complications,*" the Cunīculus admitted, taking a step backwards.

"*What* complications."

Ominously calm. "See – we – everything was going fine – all according to plan – but then, last second – like, honest t'goodness, we'd only about a day left of travel – we were attacked." He released the last of his tensed breath in a short sigh, as if relieved at having gotten that much out.

Bassus looked disturbingly unimpressed. "Attacked."

"Ih – by one of those roving bands of forest rats," the Cunīculus explained, obligingly. "I'd been provided with neither guards nor weapons, and –"

"And now the gutter-sucking gypsies have the one element to give away what's happening here." Bassus advanced a step forward, menacing. "Is that about right?"

"There's really no danger –" the Cunīculus began, a little half-heartedly, his eyes on the blade at Bassus's side.

"No. The rebellion is simmering so far underground that the dregs could take centuries to boil it. That does *not* mean we are entitled to leaving out loose ends. This is a loose end, Cunīculus, however inconsequential. Rather like yourself."

"Ih – *wait,* now!" the Cunīculus protested, holding up his palms. "What could they possibly find out? As far as they'd manage, the guards arrested three vagabond criminals – thieves and rejects! Killing outlaws is nothing new! The fire-eater flaunted himself under a false name – to be honest, it was Scipio who identified him, not I! And besides," he added, convincingly, "even if they *did* know, and – even more unlikely – did take it as a spark for rebellion, what are they gonna do?" A flash of genuine, cruel, and mocking humor crept into the Cunīculus's tone, echoed in his eyes and smile. "What could they *possibly* do?"

Bassus shook his head.

The Cunīculus insisted. "Yes, they were able to overcome *me*. But their weaponry is crude, their skills even cruder. They fight with rustic clubs, make-shift arrows for cracked, ancient bows. Rakes and fish-nets. *Knives* as opposed to *swords*. And there were *twelve* of them," he added, meaningfully. "What are twelve men against an empire? I tell you – even if they'd twelve *thousand* men, they could not take an army of Augustinium at half that in the field, let alone against the impregnable fortress of Ariegn. Believe me," he added, in confident conclusion, "if anything those ignorant rakes'll probably kill'm before a word's said. There is *nothing* for you to worry about."

Silence met the Cunīculus 's speech, but the young man did not waver, even beneath the force of Bassus's crushing glare.

At last, the general answered.

"If I do not contest you, do not think it is because, by any means, I believe you," he said, holding out the scroll, "but because you are a colossal waste of my time."

The Cunīculus's eyes lit up at the sight of the proffered scroll. Sure, he could be a fine waste of time, if that was what it took. He was utterly, totally cool with that.

"Thank you, sir; surely, sir, yes –" He reached out for that coveted treasure, but just before he could take it in his grasp –

There was a sudden crash as a wooden door was thrown open against the stone wall.

Bassus turned. Two guards entered the courtyard, traveling along the far wall, dragging a man between them. Their captive was doubled over, resisting every step fiercely with the last of his strength. The guard at his right had his sword drawn and an expression to suggest it was taking all within him not to use it; that on his left had a tight grip on the base of his neck, clutching his hair and forcing him to stay down. He staggered beneath their driving punishment, and more than once tried to break free and run for it, but they held him, sure and fast.

Bassus smirked. "Right on time..." Stepping forward, he lifted a hand, halting the struggling guards. "Hold, Aesculus. Eryx."

The two men paused, their captive still writhing between them.

"Subdue him."

A hard blow to the stomach did the job.

"Cunīculus. If you would..."

With a haughty grimace, the Cunīculus stepped dutifully forward until he stood beside and a step behind Bassus.

"Surely you would observe the fruition of your several months' hard work."

And as he stood there, so he did. The wild-haired prisoner raised his head, squinting against the dawn light after his long stint underground. The remains of his clothes hung from his broken body, stained with broad, greasy patches of dried blood. He was frail, his defiant sneer marked by the distinct weakness of hunger, exhaustion, and pain, betrayed in the dull, creeping confusion in his dark eyes and the unusual paleness in his skin. Yet still, here was O'Rielly, and that alone was something to be feared.

Bassus made an indicative motion with his hand. "Raise 'im."

"Wha – aauh!" An angry hiss dragged breath through his teeth as he was hauled upright by the hair, his hands twisted behind his painfully arched back. A look of animalistic ferocity crossed his face.

"Should I be *honoured?*"

Another punch rewarded his boldness, though they did not allow him to crumple.

"We're shiftin' his cell," explained Aesculus, who looked rather pained that he hadn't gotten a turn to punch the vagrant yet.

"Keep him within ready range," Bassus commanded; "we'll be needing him soon enough."

Duncan O'Rielly opened his eyes, catching enough of a breath to snarl, "F'r *what?*"

Aesculus looked eagerly to Bassus with all the excitement in the world; Bassus had barely given a sign when a sharp blow caught Duncan to the kidneys.

The vagrant groaned, a slight, involuntary noise. His slitted eyes glared down at the soldier, his chin raised proudly, resiliently, and wearily.

Then he noticed the Cunīculus.

Bassus watched the vagrant's eyes, noting well the subtle change that came over his face. "Ah. Yes... I'm sure you remember our unfortuned friend?"

He did. Duncan gazed upon the forlorn figure of the traitorous scum that had turned him in, a form that had become so unsavorily familiar to him across the space of these past few weeks. The lad was dressed in more imperial garb now, the scarlet cloak of the Empire wrapped Ceothann-style, clasped at the shoulder. His childish air had gone, and he stood looking at all the world with arrogant contempt, as if all of it lay beneath him. But a look did little to change what he was.

"Hello, Ian."

His voice was void of emotion save the silvery hint of unsuppressed strain, solid and slicing and cold.

A silence fell, disrupted only by a soft morning breeze stirring through the courtyard – until it was disrupted by Aesculus.

"Wait – he knows him?" He cast his gaze between Eryx and Bassus, as if hearing the story second-hand and not being present in it himself.

"Oh, aye," answered Duncan, lightly, his voice now soft and eerily equable. "He sucks."

Bassus rolled his eyes. "Y'know, most people get wise after the *first* betrayal. What are *you* waiting for?"

O'Rielly would not deign to respond to that. That was the way with these Ceothanns; they'd either talk you to death and madness, or fall into a silence far worse.

For his own part, Bassus allowed the moment to linger. It was fitting, this scene, for the bleak, iron dawn. When he at last spoke, he kept his eyes on the obstinate fire-eater, though his raised words were clearly meant for a far broader audience's ears.

"You've done well, Cunīculus. Remember this sight. Those who would seek to destroy the peace this Empire has fought so long to achieve – those who would drag down the nation into anarchy and savage rule – this is all they are. Foolish parasites, hungry for power, bolstered by greed and bloated notions of pride and noble sacrifice, breeding and festering beneath the streets 'til they scrape up from the gutter, casting their feckless allegiance behind an idolic name, slurring and clamouring for the destruction of their life-source. *This* is all they are – and for that, they will not prevail. The harmony, the justice, of this Empire is unshakable. And even the most *fearsome* of rebels can be tamed.

"Isn't that right, Duncan?"

The fire-eater looked at him; and though his face remained impassive, something had entered his eyes – something stunned. "I..." For once, he found himself utterly lost for words. *Peace? ...Harmony?* He shook his head. "Is that what you tell yourself so that you can sleep at night?"

The Cunīculus – Ian – turned to Bassus. "Are ya gonna kill 'im?" he asked, speaking with remarkable apathy of the murder of the man who had fed and cared for him for nearly two months.

"Where the Hell do you come from?" O'Rielly marveled, studying him with furrowed brow. Then – "You're not twelve," he accused, a humorless laughter blackening his tone.

"No," agreed Ian.

"No," echoed the man. "S'how old *are* you? 'Cause damned if ya don't look ten."

"I'm eighteen."

Duncan's smile broadened. "Ah. And did it intimidate you, the fact that you're smaller than a woman and as unbearded as an infant when my own young son already looked like a man?"

Ian glowered, a dark look so unfamiliar on that young face. "And look how far it got him."

Duncan pursed his lips, though the word of the traitorous maggot did not particularly affect him. "I liked you better as an obnoxious, over-talkative and under-educated twelve-year-old." He turned to Bassus. "Didn't you?"

"You done?" Bassus clipped, annoyed.

O'Rielly shrugged. "Just savin' ya answerin' his question. Wouldn't do for yer subjects to know ya plan t'keep the *'last rebel'* here, bleedin' him dry f'r yer insane and insidious purposes." Grandly, he proclaimed: "The peace of the golden age! Built, fueled, and sustained from the veins of a fire-eater!"

Ian glanced to Bassus, curious. "He's useful?"

"No," Duncan growled, fire in his eyes. *"Dar m'anam* – I *can't do* what yer *after! Nothing you do* can change that! No matter how long you keep me here! No matter what you do to me! No matter – what – big – long – speeches you make!" He was so exasperated, he was nearly stammering. "Observe *that, 'Cunīculus'*! *This* is the fate of those whom the Empire uses!" A sharp jerk as the guards tightened their grip silenced him but for a last, muttered, "Fine."

Bassus leered. "Save your energy, fire-eater. You'll be needing it."

"I suppose you've been waiting a while t'use all those catchy lines? *Ouff!*" Duncan doubled over as another shot sank into his gut, nearly bringing him to his knees but for the supporting arm of his attacker. *That* little beauty of a blow had caught him across his wound; he was honestly surprised to find his guts weren't spilling across the courtyard paving stones. He spat onto the man's shoes – thin froth flecked with blood – and the next blow caught him in the head.

"What can he do?" Ian asked.

"That is no concern of yours," Bassus nearly barked; Ian wears on everyone's patience.

"Th' *feck*'ss not," O'Rielly grumbled, his words slightly slurred.

Bassus's eyes narrowed. *Damn stereotypical Ceothanns.* "Pull up on 'im," he ordered, and Duncan was yanked forcibly upright again. Though he tried feebly to resist, it was easier just to lean back into the man that had him by the hair and give over.

Bassus advanced until he stood less than a foot from O'Rielly, glorying in every minute.

"Comfortable?"

Duncan's lip curled in a scowl.

"I know that you will never see my way," Bassus informed the fire-eater. "You are too stubborn for it. And so this must be done. For the good of the people, this must be done."

Duncan could have said a lot of things; indeed, he could have sacrificed the last of his saliva teaching boundaries to those who drew the lines. But he knew it would do no good.

"'This,'" he said, calmly, defining the argument; "'this' is murder."

"Consider yourself the slaughtered lamb. Be reassured that in that sacrifice you will have been globally useful. Whatever you must, O rebel. Take him away."

This time the fire-eater did not resist the dismissal, and was even subdued as he allowed the guards to lead him out of the courtyard, back into the darkness of Ariegn's core.

Again the courtyard was still, as the last rays of dawn slipped beneath the cover of the clouds. A storm was coming, to be sure.

Best finish this matter now.

Bassus considered the scroll in his hand through the gathering darkness. Funny, what souls would do for a little scrap of paper...

The Cunīculus cleared his throat, a harsh note beneath the rising wind. He would not dare speak up, yet neither would he simply drift away in surrender.

Bassus turned, his mockingly thoughtful gaze tossed up to the sky as he tapped that coveted scroll against his chin. "You know?" he began, and Ian felt his soul drop right out of him. "Perhaps it's best we hold onto you a little longer?"

He leveled his gaze at the informer, a warning in his eyes and under his tone. "You've done well. You ought to take a break. Relax. Stay here for a while, and enjoy the spoils and comfort this mission has brought you. What's coming next for sure you will not want to miss."

And, scroll in hand, he strode back towards the door.

"Ah – but – wait!" The Cunīculus stumbled, stretching out a hand. "The – the scroll was all I really came for, I – I really don't want to *impose* –"

A simple smirk was all it took to shut him up – a small, serpentine smirk. "It's a freaking *castle,*" Bassus returned. And with that he was gone, leaving the young fool of a traitor standing alone in the middle of an open courtyard, as the last of the morning birds gave up their sweet songs.

82
Murphy's Law

The moment the rain had stopped, Zander had applied himself to collecting firewood with a vengeance.

Since his old clothes were torn and irreparably blood-stained, Daíthi had given him a new t-shirt (an oversized bar-promotional reading "Schmidtty's") and a navy zip-up hoodie. He had also re-bound the boy's wound, recognizing that Zander had lied about seeing a healer. In spite of his professed squeamishness, Daíthi had actually done a good job, even using proper sterilization and dressings. It was the fresh bandaging that had confined Zander to the campsite while it'd rained, for Daíthi had warned against getting the binding on his shoulder wet, "Ahr it'll get ruined, an' Ah'll *beat* 'chya." Whether he would or not was for many a reason a matter which Zander did not want to test.

He was nearing home now for the fourth time, as the pathetic sun sank low in the sky, and under his arm a pathetic bundle of twigs that would starve even the most pathetic of fires, and who should he chance to run into but the man himself?

"Auh! Speak of the devil!"

The lithe figure stumbled at his call where he'd been weaving his way through the trees and tent legs, looking up as if struck. He straightened when he saw Zander approaching, dropping his hand from where he'd been holding his side loosely and offering a weary smile. "Hiya, Zander," he greeted him, pausing a moment to wait. "Wha'd'ya got?"

Zander came up level with him. "Firewood – ach! *Jeeze!* What happened t'*you?!*"

"Hm? Ah!" Daíthi gave a short, triumphant laugh. It was clear he had been fighting; his lip was bleeding, his left eye was blackened by a harsh bruise that spread over his cheekbone, and he could barely stand upright. He wiped the blood from his mouth with his sleeve and asserted, "Wiv honest pride Ah c'n say: Ya should'a seen dee' u'ver guy!"

"What... What happened," Zander repeated.

Daíthi waved him off. "Tiff. Nufin'a warry 'bou'. Ahr consarn now is da mee'in'; you ready?"

"I hafta bring back this firewood."

Daíthi looked over his stock. "Aw, hon, it ain't warth da' trip."

Zander pouted. Daíthi rolled his eyes at him.

"C'mon, Zander; le's go." He headed out, waving sweepingly for the boy to follow. "Wha'sis, round two?"

"Round three."

"A chiall."

A heavy rain started when they were halfway to the meeting place, and showed no signs of letting up. Zander made sure to emphasize that the fact that his shoulder dressing was getting wet was not entirely his fault.

"Jaysus – ihs comin' down!" Daíthi squinted up at the sky through the downpour. His long hair was soaking wet, clinging to his damp face from under his backwards-turned tweed cap. He turned to look at Zander, then, reaching up, pulled up the boy's hood. *"Mo' Deuos;* you'd die wivout me, wun'chyou?"

Zander tilted his head back, trying to view the shorter man from beneath his low-pulled hood. "Of what?!"

"...Drownin'."

They stood at the edge of the great tree-ring where their meetings were held. The crowd was scant – drenched, scrawny figures flitting between the trees, holding tarps and shawls over their heads; a handful even had umbrellas.

Zander nudged Daíthi. "It's like they've never seen *rain* before."

Daíthi smirked. "Damn southerners." That was all he said.

Folding his arms, he leaned his shoulder against a tree, surveying his audience. The "democracy" was small tonight because of the rain. And that would serve them well. It would be far easier to win over a more intimate number. Not to mention a number dying to get out of the rain! He laid his head against the mossy bark, pushing a strand of hair out of his eyes. Through the steel sheets of driving rain, he watched the people as they settled. Clear, crinkled ponchos, mud-caked shoes and spattered jeans, thick cloaks raised against this chilling "northern cold"...

Zander nudged him in the ribs a bit harder than necessary, noting in a sing-song, *"I* know who you're *lookin'* for!"

Daíthi jolted and snarled, *"Dún do chlab!!"*

Zander just snickered.

Daíthi sulked, pressing the sleeve of his hoodie to his mouth to try to staunch the slow trickle of blood still dripping from earlier.

At last the crowd was assembled. Highlander drifters settled in around where Daíthi and Zander stood; one, called Cesan, greeted them simply with, "Hey! I know ye guys!" before turning away again.

Last of the Highlanders to arrive (certainly from Murphy's perspective) was Roscanna. Over her dark curls was a blue-and-black checkered scarf – a great swath of fabric that draped over her head and fair shoulders, trailing artfully over her hips to frame her graceful body. Poor creature was soaked through, yet she stood as determined as ever, and it was clear that come Hell or high water she would see this thing through to its end.

Zander turned to Daíthi, whose unwavering attention was fixed adoringly on the woman. He leaned into him annoyingly, taunting, "You *liiike heeer* –"

"Go ta' Hell, Zander," Daíthi replied, though he didn't – and wouldn't – deny it.

Unfortunately, Roscanna heard him, and turned. "Yer back."

"So're you."

Although he still held the cloth over his mouth, she noticed his bruised eye. "Wha' happened t'you?" she asked, unfeelingly.

"Ah gat hit in da' face," Daíthi shrugged. "You gonna pu'llikly shame me again?"

"As if ya had shame."

"Fair. Ya gonna *try*'nn pu'llikly shame me?"

Roscanna was silent a moment. Then she responded, "Yer lucky mah cousin is so fond'uv you an' tha' scoundrel O'Rielly. Yer lucky Ah love 'er enough teh stand down for 'er. Bu' yer *unlucky*," she added, her gaze sharpening, "'cause Ah've no cause t'believe yer oration skills're half so good as yeh think. An' while yeh may have *her* support, Caomhan will no doubt have ya beat!"

"...Lit'rally?"

She looked at him flatly. "Yer an idiot," she declared, and turned away again.

Daíthi pursed his lips. Then he twisted his cap forward, pulling the brim low over his eyes, cutting off Zander with a gruff, "Don' say any'fin."

The meeting began with what Zander supposed was a shorter version of *the usual proprieties*. Amergin O'Shea moved to the center of the circle, wearing a drenched, knee-length, sleeveless black tunic over a long-sleeved canvas-coloured linen shirt and dark jeans, a broadsword sheathed and strapped at his waist. Raising his voice above the rain more than over his soggy audience, he announced, "Right! Let's get this damn thing started! We're here t'discuss ta' possibility'a goin' t'war, an' t'be decidin' whether 'ur not we'll have a go at it. Any first takers? – an' keep it *short*," he emphasized, clasping his hands, *"please!"*

Having said as much, he returned to his place. Zander nudged Daíthi, muttering, "Means you."

"Mayns yer *face*."

Zander rolled his eyes. "Clear: savin' your *fine rhetoric* for later."

"Mmh." Daíthi's attention had wandered again. *More* clear was the fact that nobody had discussed anything beforehand; this was going to be a long-winded ordeal, to be sure. No one wanted a war; it was exactly as Gerry and Red – and, to be fair, Sully – had warned him, and it was going to take all of his fairly considerable skill to work this. Right now, it was purely a matter of timing...

They began to talk logistics; how many fighters were in each camp? How many weapons? Who actually knew how to use them? Daíthi gave a low but audible groan, leaning his head back against the tree and twisting his gaze to the sky. *How* could he have thought they would be ready? How could they *ever* be ready?

Man, democracies *sucked*.

His attention fell to the trailing hem of an azure-and-ebony shawl, and he must have lost track of time, because suddenly he was met with a beautifully livid glare, crystal-lightning-blue eyes narrowing beneath a shower of errant curls. *"Stop it!"*

Only at the sound of her voice did he realize he'd been blatantly staring at her. "Huh?"

Roscanna was very annoyed. "Quit starin' at me!" she hissed. "Why're ya always starin' at me?!"

Daíthi blushed. For a moment he was utterly tongue-tied, wondering how best to explain. Finding he lacked better words, he stammered, "'Cause... 'cause yer – pretty."

His breathless admission was received with an eye-roll.

"Sure." She turned away again.

Daíthi's blush faded slightly; he was more comfortable with confrontation than kind words anyway. He shifted against the tree so he was more directly facing her. "Shar *yar*." She wasn't buying it. Taking hold of the tree-trunk, he insisted, "*Insinde se: bnanom bricton*."

She turned. "What?"

"D'reason Ah stare." He drew back slightly having said this, re-settling against the tree. "Ah'll stop if I can," he promised, and that was that.

After a moment, Roscanna turned away as well. But she continued to watch him from the corner of her eye, wondering somewhat at the novelty of this creature.

His eyes trained on the clearing, Daíthi murmured to Zander, "Twen'ee secon's. Time me, f'ya want."

"What were you sayin' to Roscanna?"

Daíthi gave a tight frown. "In those days, Math son 'uv Mathonwy cou' on'ee live while 'is feet were in t'lap uv'a maiden – 'less t'turmoil 'a war preven'ed him."

Zander wrinkled his nose. "What on *earth* does that mean?!"

"...when yer older."

Zander opened his mouth to demand a better answer, but, honest to God, he was stumped.

Daíthi cast him a glance. "Tha's twen'ee." With a cocky smirk, he tipped the boy his sodden cap and stepped into the ring.

"*TWO FECKIN' DAYS!!*"

Again, your man was not a loud soul, but he easily overpowered them. Their arguing fell silent, and Murphy stood at their center, a furious fire in his glare.

"Ah gave you, *two feckin' days*," he growled, "an' you sai'jye wou' have a decision far me."

Silence reigned in the gloom, broken only by the steady torrents of rain.

Daíthi twisted his cap backwards, pushing his hair out of his eyes. "Wha'sit, den?" he asked. "D'ya think we c'n do dis *peacefully?* Ah tell ya, I gat *fifty-four years* dat says *uver'wise*."

"It's more *complicated* than that!" Amergin got to his feet from where he'd been sitting on a stone.

"Aye?" Daíthi stepped forward, unintimidated. "How lang'sit take f'r Ariegn t'swoop a campaign t'crush us?"

"Ariegn has *resources,*" Amergin returned, moving forward as well. "They have – weapons an' armor an' – *people,* Murphy! People an' numbers – *trained soldiers* – tha' we frankly *don't have!*"

"We'll wark wid wha' we have!"

"We have *nuthin'!*"

"Han we'll wark wid it!!" he declared, spreading his arms. "Look! We *gatta* get dis thin' movin' a'fore dey fin' word'uv it! We've *gat* to!" Addressing now the seventy-or-so of his crowd, his voice carried, almost echoed, through the rain. "When dey *realize* we ahr here, dey *will* kill us! When dey realize we're plannin' tae move, an' we ent? We're *dead.* Plain an' simple. If *we don' attack!* We ahr *dead!*" His gaze returned to Amergin, and he demanded, "*Is* tha' complicated!"

Amergin scowled, standing stiffly to his full impressive height. He said nothing.

"Dayr ahr *two* options lef' t'ya now," Murphy continued, his tone calm though his eyes were still ablaze. "Fight *now,* ahr surrender."

For as long as he lived, Zander never met a better finagler than Daíthi Murphy. Standing on the sidelines in the pouring rain, he studied the man's manipulative orations like an apprentice, knowing that all too soon his own turn would come to ply his hand to the trade. And *mercy* he felt stupid.

"Yes, we're ou'-numbered by Augustinium," Daíthi said. "Bu' dis battle ain' *on* Augustinium, *is* it? We need ta' take on da' capital. Cut off da' head. We need ta' take ou' Ariegn. An' we ou'-number *dem,* four-ta'-one."

Amergin shook his head. *"Is fearr rith maith ná drochsheasamh*: a good run is better than a bad stand."

"T'run ends here."

Amergin gritted his teeth. Although he kept his voice low, there were more than a few who heard him hiss, *"I'm on yer side, dumb-ass!"*

Daíthi lifted his chin, glaring up at him defiantly. "Den *give* me a feckin' *answer,* Amergin."

Amergin pursed his lips. For a moment, he did nothing but seethe; so hot was his temper that the rain nearly rose off him as steam. Then, turning sharply, he called out to the woods:

"North-West Ceothannanmór d'clares war on ta' Empire of Augustinium!"

Venom, the grin that Daíthi gave. Taking liberty, he clapped Amergin on the shoulder. He charged them: "Tha'ss one."

"How do you believe this will work?" Glaucus Sarpedon stood. "Show us a plan, give us an outline – we must not rush blindly into war."

Agreement reverberated.

Daíthi turned to him with a scowl. "Tha' warn't part'a dee' agreement."

"It is the counsel we have determined nonetheless," noted Balaji, Shrikant's Vizier.

Daíthi turned again, starting to feel caged. "Tha' ent a *decision!*" he insisted.

"An' though *Amergin,*" voiced Eoghan, "may've made a decision f'r *his* people –"

"—in Rael's stead –" Shrikant condescended.

"—we do not believe it wise to make any hasty declarations before a plan is fully thought out," finished Caeculus, the only one among them to sound even slightly sympathetic.

All eloquence was gone, choked in a fury and mire of rage and shock. Daíthi looked between them all, gagging out only: "I – ukh –" But there was nothing taunt-worthy in his stammering; not for the thunderous poison flashing in his eyes.

Now Caomhan stepped forward, and Daíthi turned again. The Highlander wore a long, black trench-coat, black T-shirt, black boots, and bore an expression as cold and heartless as steel in his pale-teal eyes. Concealed beneath that trench-coat were undoubtedly all manner of loathsome weapons, from the heavy long-sword to the fabled lock-pick he utilized in extracting confessions...

The weapons specialist stopped a few yards away from Daíthi; and, in that patronizing low drawl, mocked, "Do you have *any* idea what you're in, *bard?"*

Daíthi looked at him frigidly but said nothing.

"You *are* in for a rush-job," Caomhan went on. "You talk of glory and suffering, war or die. Run or live. It's all black and white to you."

"Don' insult me," Daíthi cut in. "I ent *half* 'uz stupid az you think. I un'erstand dayrs complications –"

"They are not trained," Caomhan emphasized, "and vigilantes die in war – people like you, who think they can get by on sheer determination – wit and faith alone – *die."*

"Ah'd ra'ver die fightin'," Murphy murmured, to which Caomhan responded with a disdainful snap of laughter, shaking his head at the man's innocence. He noticed the contusion worsening beneath Daíthi's left eye, and commented, "Huh. That's quite the mark ya have there, Barfly."

Daíthi's eyes narrowed. "*Nat'rully* tha' has bearin' on dis."

"You never think about the consequences," Caomhan replied. Folding his arms, he began to circle around Daíthi; the man refused to turn. "Sure you may *arm* them, but that will not change what they are. We are not like the Turranians and Calchians, trained warriors. You are pitting the sons of farmers and waitresses against a well-oiled war machine built on dynasties of highly-skilled soldiers – a mechanism which dominated the world in less than sixty years. This is not some silver-tongued bard's ritzed-up rendition of a fully winnable war. This is reality. Your childish imaginings and liquid-courage drive *will* be the death of them. And yourself. Two *thousand* lives, all sent to the slaughter. Their blood upon your hands. Cold. Dead. And *glorious*.

"But sure. Have Eoghan sound the call to war."

Daíthi's expression was an interesting one. Intractable and shaken, embarrassed and hateful. "Yer a *democracy*," he grated. "*Vote* on it."

"Be reasonable!" Caomhan nearly yelled.

"We'll se'ull all in one go!" Murphy shouted in reply. He flung a snap of an indication out to Zander. "If t'son'uv O'Rielly can come up wid a plan, Ah'll tayk d'blood, he tayks d'blame, an' y'give'im overall leadership. Shri, Eoghan, Caeculus, an' whom-e'er-d'-Hell as legionnaires – four kings, one advisor. Whoe'er commits t'war, dey follow. Whoever commits ta' his plan, follow." He looked to Eoghan with narrowed eyes. "D'choice is yurz."

They agreed. But by God, did they have their reservations.

Daíthi listened to them—all seventy-whatever souls, reasoning out their acceptance – and a shiver crossed deep within him. This was it. This was the end of all their arguing. Of all their empty words and useless planning. Of all their shifting camp and playing soldier. This was the beginning of the end; this was history being made. And Lord, it was as intimidating as Hell.

Inaudibly, he cleared his throat. Unanimously, Fate was sealed. And, when at last all answers had been given, he said:

"Ah ask in 'is stead one nigh' f'r consideration; we re-convene Friday nigh'?"

Some low but amiable murmuring; Shrikant shrugged: "It is a reasonable request."

"Zander O'Rielly," spoke Amergin, for once addressing the lad instead of his guardian, "do you accept these terms?"

Zander nodded; Daíthi could've hugged him for his nonchalance. "Aye."

Satisfied, Daíthi started to walk away.

"I move for a closed meeting."

"Seconded."

"Third."

Daíthi turned, and in one of the most *un*-diplomatic, uncalculated moves he ever made, shouted impulsively, "*SONS OF BITCHES!!!*"

"Ooch," murmured Cesan.

"Daíthi Murphy to be henceforth banned from meetings until otherwise moved."

"Second."

"Third."

"Fourth."

"Fifth."

This time on purpose, Daíthi restrainedly enunciated, "Sons. Of. *Bitches*." He folded his arms.

Eoghan looked at him. "Daíthi Murphy, Highlander Camp, yeh ans'ar teh *me*."

Daíthi's gaze dropped to the mud at his feet. "Ah concede t'me leader. Move t'provide a second f'r Zander?"

"Denied."

"Ih wun't be me."

"Denied."

"Formal protest."

"Denied."

"Na one as't you, Glaucus."

Eoghan rolled his eyes. "*Denied*."

Amergin stood, watching Daíthi; he himself had given no judgment, but by vote, he was overruled. He faced his fellow leaders, who so outranked him. "I won't be supplyin' bouncers for teh maintain a closed meetin', an' Faith my people're free t'come an' go as dey desire."

With childish "Auh!"s of delight, his "people" high-fived extravagantly.

"Denied; as a stand-in leader, if you want your vote to count, you must follow the rules of the camp."

Amergin looked to his nation. "Sorry, lads." He rejoined them, resuming his seat upon the damp stone, amidst plaintive groans. *"Tá a fhios 'am, tá a fhios 'am,* I know..."

Daíthi raised his eyes. "Move fer popular representation; limited body'a peers, albeit chosen be leaders, as demanded un'er subsec'in 'G', Martial Council. Da'ss righ': Ah helped write t'bloody constitution, an' ya'll can kiss my ass."

Rolling his eyes, Caeculus formally summarized, "We re-convene Friday night before a limited council; no more than ten individuals per camp plus the leader and his second shall suffice."

"So, *twelve* then," murmured a Lowlander, and, reaching back, Amergin struck him subtly in the shin. They agreed.

"Mee'in *adjourned,"* Daíthi snarled, claiming a right that was in no way his. He turned and walked away, and the meeting collapsed gratefully beneath the force of the driving rain.

He grasped Zander tightly by the arm as he passed him, forcing him to a conspiring level. *"Kill me if yeh'd have't,"* he hissed; the boy answered only, "Not yet."

Daíthi laughed. Letting go of Zander, he called congenially, "Cesan!"

The addressed turned; noting the rogue's seemingly-genuine smile, he set aside his misgivings, turning his steps back. "Yes?"

"Do me a favour?" Murphy raised two fingers, and between them a stack of gold pieces.

Cesan raised his hand in a similar way. "Peace."

Murphy's fist closed over the gold, and he came up to Cesan, angling. "Send me Morann Doyle, an' tell'im t'bring Joe Sullivan. Tell 'em ta' come tomorrow mornin', an' tell 'im dey'll bot' be well-compensa'ed – will ya do that?"

"That's it?" Cesan was wary.

"Tha'ssit!" Murphy took Cesan's wrist in one hand and clapped gold into his palm with a hand-shake of the other. "I'll be in m'usual place – wit' Gerry an' Red an' them."

"I know it."

"Good lad. T'marra mornin'!"

"Alright, then," Cesan shrugged. Easiest twelve gold-pieces he'd made in his life. "See ya around."

"Brilliant." Daíthi watched him go. Zander looked back and forth, between Cesan and the smirk on the manipulative little bastard's face.

"What the Hell?" he asked, flatly.

Daíthi looked back at him over his shoulder. "We're armin' t'sons 'uv farmers."

83
Trial by Fire

For once, it was a clear morning.

Judah Basilicus stood on the rampart, looking south. There were the trees, green with the fullness of life, their dense canopies broken in swaths by villages and farmyards. To the right, the sea blanketed the horizon, shifting and sparkling blue-steel and silver. Far away to the south rose the Nóbergusia mountains, and beyond them, the world. To the east, the Abona River, its wide mouth pouring into the waves, dipping beneath the Sky Bridge. A soft breeze stirred the hazy morning sunlight, and seagulls voiced their annoyance on the rocks below.

"You summoned?"

Judah smirked. Turning, he beheld Athanasius, accompanying a thin little girl with fiery red hair, who clutched a slightly-more-red backpack tightly to her chest. It was she who had spoken, fixing the commander with a dark glare.

"I did indeed. Having had other matters to attend to, I'm afraid I have been delayed in making your acquaintance. I am Imperator Judah Basilicus, first in command of the Augustan army and second of the Empire."

Meygan nodded slowly. *If you're first in command of the army, you're first in command of the Empire as well, no matter what your business cards say.* But she didn't say that.

He went on. "As for yourself, I hear you're perfectly resolute on anonymity... Very bold."

Meygan was biting her lips, considering. She thought of Rahyoke. "My name is easy enough to tell," she responded; "I'm Meygan." She was content to leave it at that. Beyoncé. Adele. Meygan.

A smile crossed the soldier's hardened face. "It's a pleasure, Meygan. Welcome to the castle of Ariegn."

It was the sixth day since she had arrived here. Sinon had kept his word, effortlessly bringing her through customs and patrols, even though she had shot him in the crook of the arm, rendering his sword-arm useless. She had been given her own personal guard and a very nice private room complete with an ocean view, a fireplace, and an en-suite, where she had bided her time, awaiting an audience with the general.

She watched him warily now, tightening her grip on her backpack. There was nothing deceitful about this man – nothing scheming, nothing dishonest. Although she still wondered which of the soldiers would toss her off of the wall.

"I appreciate how you guys have treated me so far," she said, quietly. Was it good or bad that her voice sounded so small? "You've been very nice."

"You don't have to sound so frightened, Meygan," Imperator Basilicus said. "We're not going to hurt you."

"You killed my friends."

The general's gaze dropped before the child's.

"Yes... Legatus Iovita has informed me regarding what happened. I am truly sorry for your loss; no matter what he was, I know he made quite an impression on you."

"He was a good man. I don't care that he was a thief – that wasn't justice!" Her voice rose. If they were going to throw her from the castle walls, then they were going to have to face their wrongs. "And Zander never did anything to anyone! Why should I *not* be afraid of you? You killed a little boy because he – what?! Because he knew too much?! How –"

"Meygan," the general interrupted, calmly.

"Zander never hurt *anyone*," she emphasized; she could feel the tears welling in her eyes. "I want him back. Why did you take him?"

Athanasius felt direfully uncomfortable; he looked helplessly at his general. But Judah was watching Meygan.

"Zander's death was unfortunate. Although young, Zander was a fugitive as well – evading the law – but that merits no sentence as severe as capital punishment. The marksman missed his shot, and he has been dealt with for his transgression."

Meygan's look hardened. "So what are you going to say when you kill me?"

After a moment, the general allowed a faint smirk. "Hm. Legatus Iovita was right, you are a stubborn one. Athanasius, head back down to the mess; it's about time to shift sentries." Meygan watched as her guard departed, leaving her and the general alone. As Athanasius left, Imperator Basilicus asked her, "How long did you live with – what was the name you called him?"

"Rahyoke. You killed him and you didn't even know his name –"

"Hold up, Meygan. He may have called himself 'Rahyoke' when he was with you, but he called himself by many other names, too."

"You still killed him."

"Meygan." By mercy; this was exactly why women were not allowed into the army. Judah frowned, looking out over the landscape again as he considered.

"Meygan, do you see that?"

"What?" she asked suspiciously, following his gaze.

"*All* of that. The trees – the forest they call Frith Gallia, Silva Cervae. Where they break are the cities – there's Carteia, and you can just see Emona there. Off to the west, the ships sailing between Osset and Northambria, their sails white against the sea. The mountains in the distance – Brigdalva, and the Abona flowing down from it. All ordered, all peaceful.

"Some – a very small, select, few – say that this tranquility rests on blood – that it was built on the backs of a gore-spattered past, the ancestors slain. That we've erased the flame of life. They cannot see that we've rekindled it anew; that we've staunched the flow of blood, rebound the wounds of these countries, allowed life to flourish in peace."

"Rahyoke said this country was at war."

"The only ones who consider it a war are the agitators who would make it so. The war ended thirty years ago. There's always a fight." He cut her off, but not sharply. "Name a single country without combat, conflict, or strife. No; the insurgents – like Rahyoke – seek to 'restore the old ways'. Regression – backwardness. Risking thousands of lives, for their own devolution. They think that they can do a better job."

Meygan frowned. "I don't see Rahyoke as an agitator bent on world domination."

"Well, that's because he's very good at what he does."

She gave him an incredulous look. *"World domination?"*

"How much do you truly know about him, Meygan?"

By now, she really didn't know.

Imperator Basilicus sighed, turning a gaze up to the sunlit sky above. "I'll level with you, Meygan. Your Rahyoke, he wasn't horrible. Just horribly misguided. He'd fight to the death for what he believed in, no matter how many times he's proven wrong. A quality both admirable and abominable in a man. Unfortunately, he was one of those crazy souls convinced that the way to 'save the world' was by resurrecting a war that had ended decades ago – before you were even born. He has had a strong impact on your worldview, but I urge you to at least try to see past this image. He *was* a good man, Meygan. But he was an extremely dangerous man as well. Can you see that?"

Meygan shifted, gaining a better grip on the bag she held clutched to her chest. She said nothing.

Imperator Basilicus turned to her. "The important thing to determine now is what to do with you. Do you resent us too much to remain here?" he asked. "Where are you originally from?"

Meygan had still been looking out over the parapet, in the direction of Osset. She wondered if Fergus would miss Rahyoke.

"I have no home in this world."

The general was silent, looking at her. But before another word could be spoken, a new soldier joined them on the rampart, giving the accustomed salutary bow.

"Hastatus Prior Rosarius Marcus Leo." Name rank and number.

"Proceed."

"In request of your immediate presence by Tribune Scevola."

On Judah's (rather casual) command, Meygan accompanied the two soldiers back inside, down through the twisted climbs of labyrinthine staircases. They arrived at a room on the east-facing side of the castle, a floor below the calefactory. The room was called the hypocaust; inside were rows of square-stacked columns, and a great fireplace burned on the far side. Once this room had been used to heat that above; now it was one of the many rooms where the soldiers conducted their business.

Flanked by lower-ranking soldiers, a man stood where a long, wooden table stretched beside the fireplace. His hair was blondish-brown and stuck up from his head in spikes, and his side-burns were long, though he sported no beard.

"Tribune Scevola," Judah addressed him. "What is so urgent that it requires my immediate presence?"

Scevola looked up as the arrived gathered around his table. Without a word, he lifted a bag from the floor; and if there had been any doubts as to its ownership, they were dispelled as he flipped its contents onto the table.

Thirteen battered scrolls clattered rustlingly onto the wood.

"These belonged to the fire-eater," he explained. "They were found where he had been camping. The Cunīculus brought them in."

"Is that all he had?" Judah asked.

"The Cunīculus claims there was a second bag, but that was taken in the ambush."

"Figures," the general scowled. "Lay'em out."

One by one, they unrolled the scrolls, spreading them out on the large table, pinning their corners with candles, goblets, knives, and stones.

Meygan narrowed her eyes, straining to make out the words beyond the design. She jumped as Scevola drew his sword.

"He crossed the Palisade with these articles in deliberate violation of intranational transport laws – well, you know. They're somewhat cryptic; the contents in anyone else's hands would be weird, but when it comes to the fire-eater, everything's suspect." Using his blade as a pointer, he indicated two colourful scrolls. "A map of the old borders, and the Annexation Decree." He pointed again. "The tale of a peasant uprising led by a God-sent against a snake-ridden imperialist." Again. "Various religious tracts – on Moses, on Christ."

"Savior material."

"Manuscript material at the very least; what use this charting the phases of the moon does is beyond me."

"Perhaps it impacts his powers?"

"Oh, *that* would be interesting!"

"So this foreboding..."

"With his wit and golden tongue, these objects certainly would have proven galvanizing to the Eastern Turranians. Once clear of the mountains, he could get to the rest of the territory itself."

"What good is that to him?" Marcus asked.

The Imperator looked to him. "He could pass almost flawlessly for a Turranian. It could begin a revolt." Turning to Scaevola, he wondered, "But is it plausible?"

"*Nothing* about this man is plausible. And yet, there he is."

"But –" Marcus ventured "—if it isn't too bold, I do wonder why he headed east?"

"Easily answered," Judah replied, congenially. "He probably knew about the border raids. The Cunīculus said he first encountered the fire-serpent in the mountains. You know, Scaevola, it's a crazy idea, but he just might have tried it had he gotten the chance. And anyway – possession of these documents is incriminating enough." For a moment, he was in his own thoughts, looking over the remarkable collection of Turranian manuscript material. "Fascinating," he murmured.

Or horrifying, Meygan thought. Those were Rahyoke's scrolls. Rahyoke – the *fire-eater's?!* – scrolls.

"Rahyoke – wasn't a rebel leader," she corrected. There had to be a mistake. Rahyoke was a hoarder of paper and culture – of course he would hang on to these; they were beautiful. He – he had transported them across the Pale; guarded them with a vengeance. As he had the towns, the book, the world.

"He's not," she insisted.

"Not anymore," said one of the soldiers, as another reverberated, "Not yet."

Rahyoke... *"I have farther to run, Zander. It's only a matter of when the Lord sees fit to break the road."* For a heartbeat, Meygan pulled back, and allowed herself to see him as they – as the rest of the world – did.

"Look at me. Do I look credible?"

"Why haven't you been arrested yet?" "How d'you know I haven't?"

She remembered the violence with which he had grabbed Ian for touching the scrolls –the speed and ferocity, near skill, with which he had fought with and dispatched three of those soldiers... He had *killed,* three soldiers. Servants and citizens of the Empire. An empire which seemed to run on a peace so savagely absent from him, even when he was still.

She remembered Zander's panic when they'd beheld that lone soldier. How he'd pleaded with Rahyoke not to attempt what he had again. Anxiety and fear. And dread.

How Rahyoke had laughed him off easily, adopted his character, and passed the soldier off as a "langer-bastard". Disrespectful and proud. Hot-tempered, serious, a-social, mysterious – and dark. Bitter and near-humourless. Quiet.

Vengeful.

She had watched him steal out of anger, lie for his own personal gain.

Compensation. Recompense.

"I won't stand my dignity abused."

Rahyoke the Fire-Eater. Traitor. The more she considered it, the more sense it made.

Meygan bowed her head. "You never said what you wanted me for." Lifting her gaze but slightly, she met the general's eyes. "Sinon told me that you want something I have. But I don't have anything. What do you want from me?"

Marcus and Scaevola glanced about the room, looking at everyone in it in turn. But Imperator Basilicus kept his eyes on the table, lingering over the scrolls – the relics of a lost world. He answered.

"A decade ago, the man you call Rahyoke stole something from this castle – something very valuable. An informant of ours knows that this object fell to you, and he claims you carry it with you always... In that little red knapsack of yours."

Meygan squeezed her bag a little tighter, feeling its protection against her heart. The only thing of value she had in her backpack was her cell phone, and she knew that he couldn't mean that. It was useless to them here – and it certainly hadn't come from their castle.

"But I don't have anything," she repeated, her voice barely more than a whisper now.

And then the Imperator said the last thing she would have expected them to want from her:

"You have a book."

A book she had indeed. But Meygan's book was blank. And old. And broken. Even if she could accept a little willing suspension of disbelief and refrain from adding *And I found it on my kitchen table,* she could simply not see what anyone, let alone an *empire,* would want with such a decrepit, unworded book.

But wait.

In the barn, Rahyoke had found a book – that blue, even-more-damaged ledger. *That* book had words in it. Were those words worth killing for?

He stole something very valuable... She hadn't seen him steal either book – but she had seen the recognition in his eyes when that blue book was first in his hands. Maybe he *had* taken it, years ago. Forced to run – by what? – he had come back for it, where it had been stashed in the barn. But where was it now, she wondered; wasn't that his stuff, spread out all over the table? And if a map and a note about Moses could be so incriminating, surely that book warranted a place upon their table – the crucial evidence clinching his conviction.

So they didn't have it. And neither did she. But if she was to have any hope of surviving this adventure, they really didn't need to know that.

"If you truly have no place in this world," Imperator Basilicus said, "then I believe we can work out a deal. This is where the book belongs, and this is where it will stay. But we can make room for you here as well. You like the room you were given? It's yours. Three meals a day, the freedom of the courtyard and North Wing – free tea service – even an allowance, maybe, depending on the budget. We can work out the details. In exchange, you would act as the guardian of that book. You would make sure it stays where it belongs, and that only the elite few who have clearance could have access to it. This is a unique chance to serve your Empire at the highest level. Do you accept? Or will you simply turn the book over to us, and return to the countryside, and whatever foraging you may find there?"

Meygan chewed her lip. There were many pitfalls in this bargain, not the least of which being she knew they could kill her either way. If she stayed, did she sacrifice her freedom, her future? Her last clinging chance of ever getting back? If she had really lived anywhere else at all... If she left, would they hunt her, hound her, as they had Rahyoke? What choice did she truly have? What hope...

One thing was certain in this terrible mire: she could not give them the book. Whatever it took – whatever the reason – she must not give them the book.

"How can I trust you?" she asked.

Imperator Basilicus folded his arms. "No matter what the rebels say about us, even they know that we always keep our word."

But how could she manage it? How could she keep both her book and her freedom? She needed to buy more time.

"Alright," she answered, stretching out her slender hand to shake and seal the deal. "I agree. Thank you for giving me this opportunity... to serve the Empire."

84
Lessons in the Art

Rather than bring Zander the slightly-wending fifteen minute walk down to the nearest offshoot of the Abona River, as a good and sensible guardian would do, Daíthi had shrugged, claimed he was busy, and sworn that, "as Paddy always says", when a great enough need arose, the "damn stream" would "call to him", and he'd find it. Which, as it turned out, was true, but that didn't make it any less absurd.

He really was an odd creature, Daíthi. But he had a plan, which was more than Zander could say for himself at the moment. He had spent all night and all morning wracking his brain, but not even the soothing waters of the stream could help his straining mind. He was stuck. He felt so utterly useless, so terribly stupid – but it was nice to be clean. In a way, he felt almost content; he was clean, he was clothed, and for the first time in a while he was nearly well-fed. *Almost* content; almost.

Ceothann law was built wholly on the idea of recompense; every offence demanded repentance, retribution, revenge – blood demanded blood. It was an idea so inherent to the culture that it had been central to every lesson from his youth. And the way Zander saw it, he was owed. Augustinium had taken away his mother and sister; they would not take his father, too.

But Glaucus had a point. You cannot rush blindly into war.

Zander breathed a long sigh, dragging the toe of his boot through the soft earth. *Lord, what do you want me to do?*

Thick, morning sunlight slanted through the trees, the damp wood and mud steaming beneath its rays. Before the campfire sat their expected guests, Morann Doyle and Sully; Murphy, the master of all manipulators, stood leaning his back against the tree, the brim of his cap pulled low to half-heartedly hide his bruised eye.

Sully folded his arms, clearly disgruntled about whatever knowledge Daíthi had just imparted. "Seems fishy."

Daíthi giggled.

Sully rolled his eyes. "Kian's Lowlander camp; what has that got to do with you?"

"Ah git t'job done. Don' Ah?"

Sully folded his arms and dredged, "Alright..."

"Look: go inna Brigdavios, ask fer Christine Olliveri. See *nae one else,* un'erstan'? Buy rakes."

Sully drew back, a look of surprised skepticism on his face. "What?"

"Rakes. Farm-tools. Long-shafted, sturdy – think turf-cutters, yeah? Abou' seven 'undred."

"Yer insane," Morann protested.

"Ah'm jus' tryin'a help; Chrissie owes me one. Ahr d'ya ra'ver go ta an'undred different places scourin' 'em ou' one-by-one?" Clearly they didn't. "Good men. Christine Olliveri – gat ih?"

"Yes we do," affirmed Sully, getting to his feet. Surprisingly, he offered Daíthi a hand. "No worries, Barfly. Take care'uv yerself."

Morann shook his head, murmuring disbelief. He, too, shook Daíthi's hand, and he and Sully moved off.

"Ahoy, Flotsam!" Sully smirked, ruffling the boy's hair as he passed. "Good luck today."

"Thank... you?" Zander followed Sully and Morann's leaving with a confused frown, watching as they departed. He jumped when a hand clapped his shoulder.

"Ooch – sarry!" Murphy laid his other hand to Zander's forearm. "Fargat, tha' was t'wound – *a chiall.*"

"What was Sully –"

"C'mere wiv me, will ya?"

"I'm hungry," Zander complained. Was a little breakfast too much to ask? Had he thought he was well-off too soon?

"Ah'll gi'chya sumfin. C'mon!"

Rolling his eyes, Zander followed.

Daíthi led him through the Highlander camp, lacing between tents. Sauntering as if he hadn't a care in the world, answering those few who called out to them in kind.

"K'you spayk Turranian?"

Zander arched a brow at him. "Like, *five* words."

"A'rrite. Well, kayp'em t'yerself, den." That was all he said.

"Why... why?" Zander turned to him, frowning his confusion.

Daíthi shrugged. "Hey – y'like scones?"

"Yes... why're you so –" Zander's confusion only increased as Daíthi handed him a scone. "Where –?"

"Sa' how much d'you know abou' Turranians?"

Zander inspected the scone for signs of witchcraft. "I dunno. Ish."

Now it was Daíthi's turn to roll his eyes – as if Zander was the crazy one here. "It ain't poisoned; Ah procured ih. Nah lissen up.

"Wha'er ya know abou' Turranians, above all ya gotta know two thin's: Firs', dey *hate* Ceothanns. Hate 'em. Ya ha' several – *t'ousan* years'a border disputes wiv Gallia ta' thank fer tha', so... yer welcome. Secon': day're very – Godish. Very livin' Judgment – Fear'a Gad – Sa' here's gonna be some rules.

"Above all else: Check d'language 'uv religion bu' kayp da' values. Don' mention Gad, bu' eh'ree statement you make, yeh mayk before His'self. Ergo, you do *not* say: abou' gamblin', abou' theft, abou' smokin', abou' drinkin' – no unarmed combat – street-fightin' an' sich – ihs beneaff dem. No beggin' lest a mendicant or cripple. No MacLochlan-zian *conquests* – basically, anyfin Shaw wou' do or say, be ih absent fra' yer consciousness. Gat ih? Shaw Rules apply!!"

"Shaw Rules –?"

"Ahr dee' *Anti-Shaw!*" He was very fervently insistent on this point. "You talk abou' d'war ahr da' weather, an' dass' *it*. In fact!" With sudden energy, he swished his arms wide as if separating the Red Sea. "Don' talk! Le' *dem* talk!" He tipped Zander a glance. "Separates da' 'tards fro' the wise-men, Zander: Look b'far ya speak. Tha'ssa life lesson righ' there."

"You are so not politically correct."

Daíthi continued regardless. "Words; now language. Dey'll spayk Turranian; jus' wai' ih out. Don' rebuke 'em; dey won' stop, an' ih'll on'ee piss 'em off. Again, don' give'tt a go yurself; dey mock an' revile beginners, 'specially if dey ain't ayvin half-good. Wait ih out – wait an' lissun. Be respec'ful. Sometimes in life ya get farther wid some by bein' a flippin' gombeen-man den a terrorist."

Zander rolled his eyes at Daíthi's version of the phrase. "Honey an' vinegar; but why the Hell are you telling me all this?"

Now the look Daíthi cast him was upsettingly pitying. "An' don' swear."

"S-swear, like –?" After so long a time in Daíthi's company, it was far easier said. Zander groaned. "What counts?"

"Heh – damned hif *Ah* know."

"Aw, shite-waffles!"

And that was when he realized they'd arrived.

The Turranian camp, in short, made that of the Highlanders look welcoming. Rows of tents clustered together, turning their backs on the other camps like a woolen palisade. The inner gate was guarded by a tall, fit, typically-Turranian man armed with two shamshirs (curved swords) in the bright red sash at his waist.

Daíthi snickered at Zander's deplorable timing. The archer again looked at him flatly. "You're *not* orchestrating this?"

"Ah mayn, no one *forced* you ta' walk wid me..."

Zander shook his head, turning back to the guard. *Alright...* Ten to one, what guesses Daíthi's purpose in not only bringing him here, but priming him in etiquette on the way?

"I hafta speak with Shrikant," he announced, coming to stand before the gate guard. "Z'he in?"

The sentry's look moved from disapproval to disappointment. "Your accent shows you as Ceothann," he said.

"Really?" Zander cocked his head, feigning contemplation. "I'd always thought'a myself as Turranian at heart..."

"As does your bad sense of humor."

Zander grinned. "Only a sense. Say – is yer chief in then? Or what. I was meant to meet with him. Top-secret war-councilly stuff."

The sentry's eyes narrowed. He straightened proudly, folded his arms barringly, and responded curtly, "Were he 'in', it would make no difference; we do not admit Ceothanns beyond this point."

"Oh, sweet'art, how mistaken you are!"

Zander turned back to Daíthi, who held a half-eaten potato in his hand. Rather than ask about his remark, he demanded, "Where did you get that?"

Daíthi looked surprised. "Auh! Ihsa bleedin' *miracle!*" He took another bite, parked the solid chunk out of the way like a tobacco plug, and explained, "Dey a'mit leaders, an' messengers on business, an' you're both, Zander." Returning to the guard, he drew forward, almost strolled, the potato upheld at a jaunty angle, and continued, "Ihs amazin' how quickly you sand-dwellers've forgotten. It ain't *Ceothanns* you got a beef wiv, love. Ihs Galls. D'you know wha' a 'Gall' is, sweet'art?" His grin became delectably sinister. "Ihs me."

The guard was not impressed. "And who are you?"

"Da' *true* heir of dis land. Guardian uv'a culture which t'rived a'fore Turranbar wuz anyfin mahr den a collective'a Bedouins all ponderin' shared well-rights – an' da' guy who can mayk yer life *exceedingly* difficult t'day." He stole one, sauntering pace forward. "Ya know... mayk me own way inna yer fine camp. Raise da' cry'a Hell an' lead astray yer young men, seduce yer women, an' devour yer livestock wid da' blood-lust uv'a t'ousand vampires! An' in so doin' give da' last seeds'a mah longin' culture some fertile ground, an' corrode yer lovely people li' ivy corrupts brick. *Or,*" he shrugged, "you cou' let Zander in, an' Ah'll divert mah plague ont'a Lavinium."

"*You* are not gaining admittance –" the sentry began, maintaining his gruff but reasonable tone; he broke off as he noticed Daíthi had edged closer to the border than he liked, grasping the Gall's shoulder and shoving him back towards Zander.

Daíthi was unfazed, and carried on the direction of his own accord, stepping behind Zander and placing his hands on the boy's shoulders. "So you'll let Zander in, den?"

"Unaccompanied," the sentry growled, tiring of these antics.

"Grand." Daíthi gave Zander a gentle nudge forward, stepping lightly to his right. Clasping his right wrist behind him, he began backing towards the Calchian camp. "Hey; say 'hi' ta' Shrikant for me, will ya? Remind 'im wha' king colonized Brigdalga. Ah'll give you a hint: it ain't called *Jabal-al-Nār, is* it?"

"*Leave,*" the sentry ordered, sounding more bored than bothered.

Cackling contumeliously, Daíthi sashayed away, leaving the wreckage to Zander.

Zander closed his eyes a moment, an elongated sort of flinch. When he dared to glance at the sentry, he was jolted by his harsh glare.

"For the sake of this union," the Turranian warned, "it is hoped that you will prove nothing like the sordid rabble of your people."

Zander's eyes narrowed. "And which 'people' would that be?" Straightening to his full height – he could nearly look the sentry in the eye – he pointed towards the heart of the Turranian camp and stated: "Now. As my current position as messenger and *leader* within this Company permits me – lead on."

And – marvelously – it worked.

85
Fishing

Fifteen minutes. It could not have been more than fifteen minutes, and Zander stood again outside the camp of the Turranians, feeling utterly cheated.

See, it had all gone down like this:

He had been admitted to Shrikant's tent by the gate guard, who had not only led him there but had stuck annoyingly close in the doing it purely to prove a point. The Turranians had cooler tents than everyone else, built of blankets, dyed and patterned with rich, bright desert colours. Shrikant's was no bigger than those of his people, nor more elaborate. Three large blankets made up his circular tent; one blue-and-burnt-orange stripped, one red and coconut-flesh white, and the third like a tapestry taken from a Turranian palace, embroidered with gold and silver.

Zander had to duck low to enter the tent; he found the inside as dark as night but for the warm, red glow of a kerosene lantern. He shuddered.

The lantern was standing on a low table – like a coffee table, but half the length; this was the only furniture in the room. A great blanket, like those that made the tent, covered the earthen floor. In all, the tent occupied about as much space as a king-sized mattress.

Zander's eyes quickly adjusted to the dim lighting. He found himself near face-to-face with Shrikant al-Rajiv.

The deposed Turranian king had that same hard, angry look of disapproval that he had almost perpetually on his face. His raven-black hair was uncovered, and boyishly ruffled – the *only* sign he was no more than three years older than Zander.

"God bless all here," Zander murmured habitually, but intentionally. "I thank you for agreeing to meet with me."

Shrikant said nothing. He moved to a place on the opposite side of the table, placing it and its light between them. He gestured with an open palm to the cloth floor, and, with a polite half-bow, Zander took a seat. Shrikant echoed him.

"Tahir has said that you wish to speak to me," Shrikant said, the words clear and clipped beyond his slightly-rolling "R"s; the Turranians had always done better with this language than the Ceothanns. "He also said your guardian made feeble threats against the sanctity of my people."

"Yeah, I apologize for Murphy. He's really... *astoundingly* rude. He grows on ya, though. But I didn't come here to talk about him."

"And what is it," Shrikant demanded, icily, "that you came here to discuss?"

Zander was silent a moment, studying the great web of the situation he was in. "Murphy wasn't very clear when he told the Company my proposed role. I'm not looking to take over the army; I have no right. *You've* been training for this your whole life. I'm a hunter and a survivor, but that doesn't make me a warrior or a leader. So what I am – what I really should be – is the acting general for the four of you leaders. The fall-guy, maybe, if it comes to it. But definitely the guy who takes your ideas, your advice, your concerns, and your warnings, and turns them into a plan you can all agree with."

"Is this just because you cannot come up with a plan of your own?"

"No. I have plans. I have plenty of plans. But what good are they to me if everyone resents me too much to follow them?" He folded his arms on the table. "I want to work *with* you, not, like, in spite of you, or against you, or whatever. This isn't an army if we're all fractured – Ceaothanns all doing their own thing, and Calchians all doing their thing, and Turranians another thing. *That's* where I come in. I don't want to be in your way. I want to help."

"And I want what is best for my people," Shrikant returned. "Plans are not weighed by quantity, and victories are not built upon good will. When the alternative to victory is only the destruction of my people, I will not take the decision so lightly as to risk their lives on good intentions."

"Then don't," Zander replied. "Believe me, I understand your concerns. And, believe it or not, I care about these people *just* as much as you do. I do!" he snarled, at Shrikant's doubtful gaze. "I care about their safety, their futures, their protection – their freedom."

"A Ceothann –"

"I'm not a Ceothann," Zander snapped. "I was born in Balladarach. A refugee town carved into the former territory of Gallia, carved *into* the Empire of Augustinium, built by immigrants and the children of immigrants from all nations. I am as much a Lavinian as a Gall, a Ceothann as –"

"A Turranian?"

Zander's fists clenched against the tabletop, and suddenly something came into his mind: the parchment Rahyoke had translated for him. And, to even his own surprise, he found himself reciting:

> *"Man yashfaAA shafaAAatan hasanatan yakun lahu naseebun minha waman yashfaAA shafaAAatan sayyi-atan yakun lahu kiflun minha wakana Allahu AAala kulli shay-in muqeetan."*
>
> *Whoever intercedes for a good cause will have a reward therefrom; and whoever intercedes for an evil cause will have a burden therefrom. And ever is Allah, over all things, a Keeper.*

A moment of silence reigned. Then, very quietly, Shrikant said, "One does not simply bandy about the Word of Allah."

Zander raised his eyes, meeting the king's gaze. "My father taught me a deep respect for your culture. I don't expect you to feel the same about me. But I do expect you to know that when I say I care about these people, and will give my all to help them, it's the truth." He got to his feet – now looking down on the king. "And I dunno about you, but I feel that the best way to defend them is to take up arms and stay the blade that's poised above them. Because those sons-of-bitches are ready to take us, Shrikant. And I *refuse* to give up on these people without a fight.

"I'll be in the Highlander camp with Murphy if you decide that you wanna help, too."

And that was how he left it, turning and striding out without waiting for a response – although he did listen, and none was offered.

And now? Well...

"I am not letting you back in," warned the Turranian sentry behind him.

Without turning, Zander raised his hands in surrender and walked off. He wasn't entirely certain whether or not he had accomplished anything. He wanted to call it a calculated risk, but, truth be told, he wasn't a manipulator. He was too brutally honest – not a bit of his father's devious finesse; he hadn't even picked up any rhetorical flares from Daíthi. He was just being honest; he had said what he felt was right, and... Maybe he'd ruined everything? He should probably find Daíthi...

"O'Rielly!"

Zander's last footstep dragged to a halt. "I'm not used to being called that."

"You will be." It was the voice of Caomhan that spoke to him.

Zander turned. "Hey; long time no see, huh?" His eyes flicked over the massive sword Caomhan was carrying, its tough blade resting against his shoulder. It was a claymore, a beautiful, two-handed, cross-hilted broadsword.

Caomhan noticed the appraising look the lad gave the weapon, a look not wholly of awe, but of expertise. He smiled. "Heading back to camp? So am I. I'll accompany you."

Pleasantly though he said it, it was not a question, not an offer, and Zander was sure to make note of that. But he smiled, having at least learned enough from his mentors to reply just as cheerfully, "Grand! I'd be glad of the company!" Even though he wasn't.

"You seem reluctant," Caomhan observed, his teeth glinting white like a serpent's venom-glazed fangs. "That's good. Trust is man's mortal flaw. Trust in the goodness of others. Trust in the strength of a fortress. Trust in numbers. How's your planning coming?"

Zander glanced at him, then shrugged. "Oh, grand."

Caomhan nodded. "Ya know, I've a lot of expertise in the area..." he began, and trailed.

Zander turned to him. "I thought you were against the war?"

"It's not so much I'm against the war," he corrected. "I just recognize a pending failure."

"In *what?*"

"Come on, kid. Even a guy as green as you can see we're unprepared. And moreover..." He lowered the hand still holding the sword's hilt, so more of the blade was cradled against his body, the tip now angled straight up in an effort to avoid severing any strayed tent ropes or laundry lines. "You simply cannot rush a war. It isn't a bar-fight, it's a blood-feud. Let's say you win. Let's even say you win with *no* casualties, 'cause I'm feeling particularly generous today. Does the war end with the fall of Ariegn? Farthest from it. If Ariegn falls, if we beat them, you are setting in motion a changing of the guard which will alter the face of our world. You are plunging the current order into chaos, and you are setting these people to push it back into a place of peace.

Ariegn falls. Then you have to take Thysdrus, Vasio, Northambria, Brigdavios, Danaum – and you're six battles in, and still west of the Pale. Let's, for a moment, say you win. Are you prepared for the long-haul, O'Rielly? Are they?"

"Let me ask you a question in turn," Zander responded, sticking his hands in the pockets of his hoodie. "Where did *you* learn how to fight?"

"I imagine from a place very similar to you. I found the tools suitable to my trade. I studied them, and, in due time, I mastered them."

"'These people' aren't like the other 'peasants' – the sons and daughters of more common waitresses and farmers," Zander said. "'Cause they're *here*. They've already shown they have more guts, more drive, than everybody else on the planet, just by being here. Is it so hard to believe they can learn weapons too, same as you and me?"

"Can you teach the art of war in a month, O'Rielly? A week? A day?"

"Then you've answered your own question."

He said no more than that, and that simplicity forced Caomhan to take pause. "Hm?"

Zander cast Caomhan a sidelong glance. "You don't win a war in a day, but you don't fight it in a day, either. It will take days, and weeks, and months, to battle Carteia, Baeterrae, Isurium, Lutetia – and through every victory – before, during, and after it – they will be trained."

"So for the long-term. But you need to get there first, don't you? Maybe the Turranians can hack it. Maybe the Calchians can. But the rest of us? Try fitting into formation a rabble as hot as this, and they'll break ranks before you even get to the gate. Not necessarily through cowardice. But, having taught ourselves, we work better on our own. And that won't cut it in the field. There is a stark line between disciplined mercenaries and –" He paused. They were about halfway through the Highlander camp now, deeply mixed in the crowded, disorganized tents. It was then that Zander heard it – the cacophony of voices, loud and eager and barbarous, like spectators at a sporting event. He turned to Caomhan.

"What is that?"

Caomhan took his arm, halting him. "As I said to your so-called guardian, kid, no amount of faith can make warriors out of untrained vagabonds and peasants. Do you really think that they're gonna stand any kind of a chance?"

"I have every bit of confidence in them," Zander replied.

Caomhan's smile became a smirk of superior pity. Confidence had nothing to do with it. "Suit yourself," he said. Speaking his last words on the matter, he took hold of the lad, propelling him around the last corner. "But don't say I didn't warn you."

He was gone. Leaving Zander at the back of a great crowd all clamoring joyously for blood.

Casting a glance behind him, Zander moved forward, snaking his way through the jostling crowd. And naturally, who should he find at the heart of the commotion but Rhys Daukyn and Daíthi Murphy?

The two were locked in full combat, a fight without pause, without restraint, without respite. They fought with furious agility, grappling and breaking like angry dogs, cursing each other out in a language foreign to most of the crowd. Rhys seemed to favour a fighting style somewhere between wrestling, kick-boxing, judo, and cheap-shots; Murphy was straight boxing, holding the slightly-flattened stance of a left-hander used to going against righties. He landed a nice uppercut as he dodged Rhys' roundhouse kick, and a familiar voice called out, "Any takers on the even odds, now?"

Zander looked to his left and spotted Red, his cap pushed back and a few strands of vibrantly-red hair falling into his eyes as he counted bills, hand-written pledges, and IOUs, recording them in a little leather notebook. Zander sidled up beside him. "What's all this, then?"

Red turned, smiled in greeting, and returned to counting his notes. "Och, it's a long time brewing, this," he answered, eagerness veining his low voice. "If you ask me, 'tis a wonder they haven't killed themselves at it yet."

As if on cue, with a strangled curse, Murphy was flipped, slamming onto his back in the soil.

"Another five on Daukyn, if ya please, against old Murphy." A strayed Lowlander handed Red a note.

"Gladly." Red took the note and the rascal slunk away. With a wink, he murmured, "Poor bastard won't know what hit 'im."

Zander smiled.

Murphy was back on his feet, and stood eyeing Rhys, his fists raised. He jolted when he heard Red call out, "I'm givin' five t'one on Rhys!" "What?!! You *git* –!!" Rhys launched first, tackling him, prompting Red to declare, "Nine t'one! I'm givin' nine t'one on Rhys Daukyn!"

Zander watched for a few minutes. It wasn't procrastinating – no! Indeed, he was busy weighing the very viable tactic of simply placing Rhys at one end of Ariegn and Murphy at the other and telling them to find and kill one another through the fort. The Empire wouldn't stand a chance.

"Zander turned to Paddy. "How long'll they be at it, do ya suppose?"

"Days... weeks..." He shrugged.

"I don't doubt it..." Zander pressed his lips into a thin line. Much as he hated to break off this epic fight, he needed a word with Daíthi.

"Hey – is this to the death, or what?"

Daíthi raised his gaze. Breaking form, he had Rhys in a headlock and was aiming back a punch when his eyes found Zander. "Hey!" he said, with a grin. "Zander!"

"*Not* buyin' it," grunted Rhys, and rewarded Daíthi with a jab to the knee that crippled him.

"Auh! *Shiiiitt!*" complained Daíthi, wrapping both hands around his injured leg as he lay on his back in the dust.

Rhys looked up. "Oh! It is Zander – hi, Zander!" To Daíthi: "My bad."

Daíthi muttered pure fire, getting stiffly to his feet. "Hows'yer –" Rhys turned to him. "Hold on – hold *on*." The second time was through gritted teeth. "Zander, how'jyer mee'in go? You talk t'Shrikant?"

Zander nodded. "Talked t'Caomhan, too."

"Oh yea'? Bet tha' was int'restin'. Ih went well, though?"

"I... I don't know."

To his surprise, Daíthi smirked, a flash of impish malice in his eyes. "Ya cursed at 'em, din'chya?"

Zander's blush deepened. "Well – I mean – contextually..."

Daíthi laughed. "*Flattery* ain' exactly yer strong point, huh? Auh, well; we'll talk long an' hard'a da' details la'er if ya fancy. Bu' fer now – dees walls got ears – and *mouths*," he added, casting a pointed glare which drew mocking complaints from the crowd. "Go ge'chyer scran, Zander; yeh've earned yer food!"

"Really? That's it?"

Rhys interrupted impatiently. "Could we get *on* with it now, your *Highness?*"

Daíthi rolled his eyes. "Ah'll mee'chya la'er, Zander; deyr's sumfin Ah gotta take care of first. Cain't say how long Ah'll be; mules're notoriously hard ta' train."

"Why, you –!"

"Have fun," Zander enabled, with a wave, as he dropped back into the crowd. Dearly as he wanted to stay and watch the fight, he had his own business to attend to. So, with a sigh, he tapped Paddy's shoulder with a fist, urged him, "Tell me how it ends," and drifted away from the thronging crowd and the whirlwind it surrounded.

He had bigger fish to fry.

86
The Dirty Glass and Her Piercing Gaze

Roscanna stole though the undergrowth and the dimming twilight, making her way back to her camp. She lived apart from the rest of the Company, in a modest vardo parked at the outskirts of the Highlander territory. And every night, in the small clearing that stood in the space between her camp and theirs, she would meet up with her daughter, and they would walk home together.

Tonight she was late, because Caomhan and his apathetic punctuality had seen fit to keep her waiting while he discussed the validity of serrated shield edges with his pack of lackeys. *They* did not have to worry about their children wandering off – or the dangers that the forest hides.

She picked up the pace; she knew this stretch of ground better than anyone, and could navigate it as quickly and as silently as a hunter. Dipping beneath the last sheet of over-hanging vines, she stepped into the clearing, and stopped.

There was Molly. A young woman of twenty years, sitting upon the great stone, her ragged black curls falling about her shoulders and trailing long bangs into her eyes. On her face was a frown of deep concentration, her brown eyes narrowed intently as she picked at the strings of a maple-wood guitar. Her hands moved across the strings as if it were a harp, the neck supported in the crook of her skinny arm. And there, crouching at her feet, was none other than Barfly Murphy.

Molly looked up, and immediately stopped playing. Cradling the instrument with angelic gentleness, she stood, placed the guitar upon the stone, and flounced over to her mother.

Daíthi turned, and his eyes rounded. He seemed surprised to see Roscanna, but he said nothing, merely watched as Molly wrapped her arms around her mother's upper arm, whispering in her ear. Her mother, meanwhile, never moved her icy glare from Daíthi.

Roscanna laid a hand against Molly's black curls. "Go on home now, Molly," she said, in even tones. And, with a beaming smile, Molly obeyed.

Murphy watched her go with a subtle flick of his eyes, using the support of the rock to rise slowly to his feet. He waited until she had vanished into the woods before offering, amicably, "Sweet kid." His eyes met Roscanna's. "Ya look angry."

She was. "I'd appreciate it if yeh stayed away fro' mah daughter in the future, rake."

Daíthi was startled. "Ah din' mayn harm. Ah'm sarry."

"I don' know *what* possessed yeh, rake," she sneered, "bu' if yew *ever* come up tae her again –"

"Whoa, hey! Ah din' come up t'her, arrite? Ah wuz here first; Ah din' rahlize ye *owned* dis particular stretch'uv real-estate –"

"What were you doin' out here?"

"Tryin'a catch a small break from baby-sittin', ironically!" Before Roscanna could berate him further, he raised his palms, commanding the floor for a speech of surrender. "Look: Ah'm terrible sarry to've troubled yeh. Goodnigh'." Grabbing his guitar, he moved to duck into the woods, but – shock among shocks – she called him back.

"Wait."

He halted with vague reluctance. Sighed. Inquired, "*Yess?*"

"'*Aoidh Na Dèan Cadal Idir*'?"

He turned to her in full. "Yes. Da' melody's simple ta' teach."

"Hm... Mah husband used ta' sing tha' to her when she was small."

"...Husband?"

His voice was slight and shaky. He cleared his throat. Willed himself not to sound bitter. "'Carse... A'carse!" How could he have been so stupid? Of *course* she was married – they were discussing her daughter! And rare flowers like this are not left to whither on the vine.

Roscanna studied the same patch of grass that he seemed so utterly fixated on. "It was sweet of yeh to teach her," she continued, softly. "I... niver had the heart to."

"Uh –? Oh, so now ihs sweet, is it?" He shook his head, tightening his grip on the neck of the guitar. "Look, darlin', I ain't got time fer yer mind-games, arrite? Do you hate me? Den say ya hate me an' have done, 'stead'a laydin' me on in'a these traps an' –"

"No!" Roscanna's protest was honest; she took a step forward as if to physically restrain him, her slim hands closed into raised fists. But there she stopped. Unable to continue. Wondering why she wanted to. Daíthi watched her in silence, offering no answers, only a shared question. *Then what, darling? Now what?*

Roscanna pushed a few long strands of hair behind her ear, reading his expression easily and anxiously. "Yeh dohn' understand," she breathed. She searched him, uncertainly, biting her rose-red lip, unwilling to trust what she found. Then she started backing towards the woods. "I... I appreciate th' kindness yeh showed mah daughter," she said, almost stiffly. "People dohn' usually – *do* things like tha'... *Tìoraidh*." She turned.

"Roscanna."

She paused, looking back at him.

"Yih – leave if yeh hafe to, bu' don' layve outta awkwardness. You 'uv all people ha' d'layst reason t'be shamed. Ah know you know; Ah'm jus' sayin'."

She stared at him. He waited, but, as he could not understand her harsh expression and she wasn't offering any answers herself, he shrugged. Guitar in hand, he sat down upon the stone, perching lightly on the edge. He indulged in a brief imitation of Molly's harp-posture before settling into the correct position, running his fingers gently over the strings.

When Roscanna sat down beside him, he did not react, did not turn. He simply sat there, playing through a rich, dulcet air.

Roscanna bowed her head, closing her eyes as she listened. "Wha' key is tha'?" she asked.

"Hm? E minor."

She nodded. "Ah; that's it, then. I've only heard it in A or B... Yer very good."

"Heh... Tha'ss very kind'uv ya." He slid his left hand over the frets, fingering A minor. "Ya play harp, don'chya?"

"How do yeh figure?"

"D'way Molly held d'guitar; wha' you saw wuz'a compromise... Ha' long yeh played?"

"It's been a very long time since I played."

"D'you wanna –"

"Okh; yer better than me; you keep goin'."

Shrugging, he slipped into "*An Fhideag Airgid*".

The ease and beauty of the combinations in the set, let alone the skill of the pieces themselves, spoke of a level beyond the professional. She watched his fingers, traipsing so delicately over the strings as he plucked out the ancient Highlander tune. "They call you 'bard'," she said, quietly. "Do yeh play fer a livin'? Well – suppose it would be hard –"

"Ah play when Ah can, aye. Mos' da' lustier bars, on accoun'a d'laws an' shit. Dey're ra'ver gran' company, though, f'yed see't." He grinned. "Dey know all t'wards – ayvin t'dis. Li'..." His fingers were mid-chorus, so he joined at the next verse.

"Fàilte, fàilte, muirn is clià dhut
Ho ro hu a hu il o
Hi ri liu hill eo, hi ri liu hill eo
Fìdhleireachd is rogha ciuil dhut
Ho ro hu a hu il o
Hi ri liu hill eo, hi ri liu hill eo
Co a chanadh nach seinninn fhìn i?
Ho ro hu a hu il o
Hi ri liu hill eo, hi ri liu hill eo.

"Ah swayr t'Gad," he insisted, "da' toughest bastard ye'd meet knows all d'wards ta'layst one *sean nós* tune."

"Show-off."

"Oakh, I ent, nee'ver," he replied, proudly.

"Oh?" She shifted, turning towards him. Her dark curls fell forward as she tilted her head, the tip of a long strand brushing over his fingers and causing a shiver to race over his spine. "Yer the picture 'uh modesty, is tha' it?"

"Ah'ma mere strummer, lass. Divine inspiration an'a wee bi'a stout now an' again –" He met her eyes and his mind faltered. Her eyes were ensnaring, piercing, enthralling, and so *very* bright a blue. And ach they dared him, scarring him, wounding him, yet he was unable to pull away, for strength and for weakness, forever drawn in. He cleared his throat. His own eyes flitted briefly away and at last found focus on the trailing tendrils of her hair, soft and black even against the darkness of the night. Without realizing it, Daíthi smiled.

"Aye, tha's a bard if I ever heard one," Roscanna affirmed, having given up on ever receiving the rest of his sentence.

"You shou' sing."

The remark was so soft that it was hard to know whether he'd intended to say it. But the words were there all the same.

Roscanna drew back. "Ah dohn't sing."

"Shaym."

Claws out. "I haven't since mah husband died."

Daíthi jolted. "*Jay*-sus, yer quick t'offend! Tayk a compliment!"

Roscanna pressed her lips into a tight line. He had a point. She averted her gaze.

Daíthi did likewise. His fingers found a new melody, and he stroked the vibrant strings. At a murmur, he added, "Ah'm sarry, by d'way."

Roscanna responded with a long sigh, raising her eyes to the blackened stars above. She wanted to tell him that it was alright – that she appreciated the compliment – that she knew he'd meant no harm – but the words were not there. To be given, or received. And she knew it was better to just leave it be.

Once more that strange sort of silence flowed in, broken only by the notes of the minstrel's guitar. Though overcast, it was a gorgeous night; pleasantly warm, thankfully dry, and long, as if God had held still the moon.

"So, why now?"

Daíthi perked up. "Hm?"

Gently, she nudged him with her arm, achieving his attention without throwing him off. "Why *now?*" she insisted. "We've been encamped here over a year, the Company's been around over a decade an' a half, an' who the Hell are you? Where di'jyeh emerge from suddenly, an' why pick *now* t'start a war?"

Daíthi glanced at her and shrugged. "Messenger –"

"Dohn' try tha' wi' me," she warned.

Daíthi realized now how thin was the ice of her warm disposition. One small fracture and she was lost. He stopped playing, folding his arms atop the guitar, considering.

"Deyr warn't neever spark nar fuel," he said, slowly. "Ya whar righ', dey ent trained figh'ers... *we* ain't," he admitted, quietly. "Wivout spark, wivou' fuel, yeh canna get farmers' sons t'war. D'you b'lave in fate, Roscanna? F'r a long time, ahr fate wuz sit ahr die. We've one chance ah' dis now."

"What: sit an' die or *fight* an' die?"

"...ahr win."

Roscanna made a dismissive noise, shaking her head. "You're crazy."

"Ah'm *sober*," Murphy corrected, making his eyes wide. "See wha' though'ss happen in a man's head when 'ee ain't drunk? 'Tis wild dangerous –"

"Weren't yeh drunk on Sun'dee's meetin'?"

"...*Touché*." He picked at some high non-notes. "Nah answer me one: you ent a combatant, ahr ya? Well, den why'a you stick aroun'?"

"Healer." She drew very close to him suddenly, meeting his eyes with about three inches' distance. "The *best*."

Oh, I love you. Daíthi attempted to tease her, but he couldn't manage it; his tongue was a dead weight. Oh well. She moved anyway, returning to her own space.

"Zander has a wound on 'is shoulder 'ee won' ge' checked proper. Ah wuz gonna tayk anuver a look at ih, but may'ee you should?"

"Squeamish?"

Murphy spread his palms. "Wha' luv'a *pus* y'all Ceothanns have, Ah'll never understan'!"

"Yeh care a great deal fer tha' O'Rielly boy, don'chyeh?"

"Hm? Oh." Murphy shrugged. "'Eez li'a nephew t'me, s'all."

"So you *are* a nepotist."

"Mus' we *label?*"

"Idiot."

"Wanna mayk ou'?"

She whacked his shoulder playfully with the back of her hand. *"Idiot."*

He laughed. "Ah'm glad Ah met you! Ah ne'er know wha'chyer gonna say!"

"Mercy," she grumbled.

"C'mon, cheer up!" Leaning sideways, just barely against her, he tried,

"For I'm a rover, seldom sober
I'm a rover, o' high degree
It's when I'm drinkin' I'm always thinkin'
How to gain my love's company!"

"D'you *only* know songs in bits an' pieces?" she teased.

"Dass all songs *is*, is pieces. Yer welcome'a taych me beh'er." He grinned, mischievously, leaning a little farther.

She put both hands to his shoulder, pushing him to his own side. "Get *off*, ya shameful yeh!"

"Heh! Yer da first – *ooookh!*" He lifted his eyes to God. *"Ochón,* Ah nayr said tha'!" Strumming up another tune, he sang:

"Oh dinna lippin laddie,
I canna promise laddie
Oh dinna lippin laddie
In case I canna win O!"

She rolled her eyes. "Yer so *strange*."

"Hey; if ih gets me d'attention uv'n beau'ihful lass, den Ah fink I'm doin' okay."

Roscanna gave him a dubious look. *"Bnanom bricton,* righ'?" And his cheeks went a vibrant crimson.

"OOOkh," he trailed, a hard Highlander note. "Women give reason anuff, t'be shar..." His voice wavered slightly, either warningly or nervously, but she knew it was in jest. He played a fine rhythm on his guitar, then breathed a soft and reluctant sigh. "Damn," he lamented, at last. "Ihs been fun, bu' I ought'a go. Me men're *helpless* wivout me." He stood – then doubled with a sharp gasp, catching at his right knee. "Mmh; crud."

Roscanna straightened. "Are ya alrigh'? Si' down –"

He shook his head, waving her off, although he was gritting his teeth. In a few seconds the pain subsided, and he answered, "Happens a lot; me leg's banjaxed, an' Rhys punchin' it din' help. Don' worry; ih'll heal."

Roscanna looked at him disapprovingly. "You *warriors*. We *have* an infirmary, ya know!"

Daíthi grinned, and this time there was a little spice in the expression. "Zat'un invitation?"

Roscanna pursed her lips.

He pulled the strap of his guitar over his head to free up his hands, shifting the instrument so that it lay against his back, upside-down, and out of his way. "If ih starts painin' me too much, Ah'll swing by fer a kyphi drip, 'kay?"

Roscanna rolled her eyes. "Away off." She got to her feet. "I need teh get back tae Molly."

He nodded. "Ah'll sing one f'r you, hey?"

Roscanna could not help but smile – a small smile, granted, but a beautiful sight all the same. *"'Til a' the songs are sung,* righ'?" she said, quoting lyrics.

Daíthi blushed. "G'nigh'."

"Goodnight."

With a pace somewhere between limp and stroll, he headed back into the forest. Before his shadow vanished out of sight, she heard him sing:

"If e'er I led you from the way
Forgie a minstrel anince for a'
A tear fah's wi' his parting lay
Goodnight and joy be wi' you a'..."

Roscanna smiled, shaking her head. Lord, he was certainly a strange one. She turned away, following the path her daughter had taken, drifting farther from the wild rake and on towards her makeshift home. And, at the softest of whispers, she allowed:

"Co a sheinneas an fhideag airigid
Ho ro hu a hu il o
Hi ri liu hill eo, hi ri liu hill eo..."

87
God and Murphy

"Alright, it's flippin' hopeless."

Zander flopped down purposelessly in the mulch, his eyes fixed on the sky. He had been seated, kneeling on the ground before a trunk spared the selling, using bits of twigs and leaves and an empty can of tomato-paste to map out something – *something* – that even had the beginnings of a plausible battle plan. And he was losing the fight.

He put his fists to his forehead, closing his eyes. By sea was, of course, out of the question. Not only did the Company lack a navy, but any attack to the west, north, or north-east would have to contend with the cliffs, the rocks, the crushing waves – it was a well-fortified spot on the coast, to say the least. Nor could they meet in the field. Caomhan had said that the men of the Company would break ranks in a formal battle, and he was right. They couldn't face, let alone withstand, an Augustan army in the field. Although the Company outnumbered the men in Ariegn four to one, Ariegn housed the apex of Augustinium's strongest soldiers. And as Caomhan had also said, trust in numbers was a fatal flaw.

So not by land and not by sea. The only option that remained was to take the castle itself – to take them, by surprise, on their own grounds.

How?

Daíthi had told Sully to buy rakes, not shovels, so presumably tunneling was not on the table. Zander reckoned tunneling would take far too long, anyway. The walls of the castle could not be breached – they were too thick. The door was too secure; it could not be smashed, it could not be burnt – and even if it was, there was a second door and a portcullis to contend with beyond. Air drops were impossible. The walls were too high to scale a full assault – not enough rope in the province, and ropes were so easily cut. And beyond all that – beyond the gates and walls – remained the soldiers. Many, powerful soldiers, who were better trained, and better armed.

For as right as Daíthi had been in saying that Ariegn would soon destroy them, Caomhan had been equally correct in noting that they were totally unprepared for war.

"Hard a' wark?"

Zander propped himself up on his elbows. There stood Murphy, his arm braced against a tree for casual support.

Zander grinned. "The triumphant return!"

"Heh. Nah quite." He unslung the guitar from his back, setting it amidst his belongings, which were all roughly obscured by a tarp. "Rhys caught me a devilish blow. Broke mah shin an' knocked me ou'. Ah'll get 'im nex' time, though. Damn, ihs cleared ou' here; Gerry sell 'is stuff, den?"

"Uh –" Zander really would have preferred to talk about the fight. He glanced behind him to the empty camp. Eerily similar to how it had appeared the day he had first arrived here, although the half-toppled tent and lack of companions made the whole place feel abandoned. "Yeah. They managed to get a cart; they're going with Sully and Morann to Brigdavios. Did you –?"

"Good; dey'll be back fer da' war, den." Carefully if awkwardly, Murphy eased himself onto the ground beside his belongings. "Le'ss hear wha'chyou ha' planned," he urged, as he pulled the tarp away from the generator.

Now Zander definitely preferred to talk about the fight. "Uhm? Well..."

Daíthi cast him a look over his shoulder that was definitely parental. "Ah wuz kiddin' a'fore; di' you really do nufin?"

"I did!" Zander protested. "I've done a lot of things!!"

Daíthi started digging around in one of his old bags. "Alrigh', den." He pulled out a tool that was either part of a crowbar or part of a flathead screwdriver; with this he set to removing a back-panel on the generator. "Infiltration plan?"

Zander frowned. "How do you know everything?" he muttered.

"Wou' you grab me a beer?"

Zander stood, moving to fetch one from the tent. "That's the part I'm stuck on," he confessed. "The infiltration is... necessary, but impossible."

"Den skip infiltration an' come back to ih."

"I can't! Without infiltration, nothing else matters!"

Daíthi allowed a posh sort of groan of disgust. Only when Zander handed him his beer did he answer. "Den Ah'll gi' you a hand."

He set the can on top of the generator, using the same machine to haul himself back onto his feet. When Zander offered to help, Daíthi told him to grab one of his bags – one of the two Zander figured was filled with clothes. "...Okay, then."

Daíthi moved to Zander's war-table – slowly. Zander noticed that he was all but dragging his right leg, and his left didn't seem quite up to making up the difference. He was wondering whether to ask Daíthi if he was alright, but Daíthi spoke first.

Looking over the boy's plans, he smirked, folding his arms. "Heh. Visual learner, huh?"

"It's a map." Zander joined him at the table, kneeling on his own side. "These are some guys," he explained, indicating the leaves, "and these are some guys," the twigs. "And the can's Ariegn."

Daíthi lifted one fist to his mouth, half-stifling a snicker.

Zander raised him a surly glare. "I didn't say it was a *good* map!!"

"Ah notice ih'sa map'a da' *outside* 'uv Ariegn. Were you tryin'a map a field battle?"

Zander shook his head. "I saw them train when I was a kid. Hoplite phalanx, marching *testudo*. Even their horses are armed; the chariots have spiked wheels."

"Ah know."

"We can't beat that. I was starting with where we are, so I can know where we can move. These are us –" the sticks and leaves. "The leaves are warriors. The sticks are – farmers." He uttered the last bit dejectedly, bowing his head to look over his map. So he missed as Daíthi's taunting smirk vanished into genuine curiosity.

"Huh." Daíthi uncrossed his arms and lowered himself to the ground across from Zander, on the other side of their platter of war. "Hain't a bad start, hon. Truly." Zander looked up, watching as Daíthi hooked his bag, dragging it closer. "Aw, damn; mah beer –"

Zander retrieved it from atop the generator. As he did, Daíthi pulled from the bag a very ornate little box – of the sort Zander would have expected to contain an heiress' jewels. "What's –" he began, but he was answered as Daíthi opened the box. "Paper!"

Not just any paper. Not the recycled, waxy wrappers or homemade sheets of lint and plant fibers that even the upper middle classes tended to use, but clean, white, military-grade paper.

"Oh, you definitely stole that," Zander said, holding out Daíthi's beer for the second time. "From Ariegn?"

Daíthi shook his head. "Nah from Ariegn." He accepted the beer, cracking it open in one hand, with a chef's finesse. "Ah had dis a long time. Stuff lasts longer when ya ain't got cause ta' use ih. D'you mind...?"

"Nope." Zander pushed his "map" aside, gesturing for Daíthi to utilize the space. Taking a drink, Daíthi used his free hand to set a clean sheet of paper on the table. Before shoving the box back into his bag, he removed from it a pen made of lapis lazuli.

Zander couldn't resist; reaching out, he half-sneakily ran his finger along the paper. "It's so smooth," he breathed.

"*Chih.*" Daíthi grinned. "Ah'll give ya some, s'long as ya don' go wavin' it aroun' in front'a dee' Augustan guard." He shook the pen to get the ink flowing. "A'rrite. Tell me where we stand."

"Um, well... as far as fighters go, we have... Turranians there're about three hundred. Calchians I think six hundred. Ceothanns about... Fourteen hundred. But they don't all have weapons. I think they said we can arm one thousand men. I don't think they took into account how many men have, like, plural weapons, though. I think we can stretch it. But we've got, like, zero shields, so... Yeah."

That about wrapped it. Daíthi smirked again. "Aw; ya *did* pay atten'in in da' mee'in's." Taking the pen up in his left hand, be began writing numbers in the top-right corner of the sheet. "Though ya rounded."

"Well, it –" Zander frowned. "You know exact numbers, don't you."

"Ah'm good a' math." Daíthi turned the paper towards Zander. His handwriting was spindly and compact, like Duncan's, but he spaced everything much farther apart – which Zander was grateful for. "Dis is total camp size," Daíthi said, pointing to the first column. "Dis is warriors, dis is noncombatants, an' dis is da' difference. Yer farmers. Turranians an' Calchians 're well-defined, so ihsa column tha' on'ee applies ta' the Ceothann branches. Bu' dis..." He tapped the remainder for the First Branch Ceothanns, the Highlanders, with the pen. "Dis includes me, dis includes you, an' Paddy an' Rhys an' Shaw..."

"Why!"

"'Cause *self-trained* is *very* differen' fro' a hereditary line a' bred-fra'-birth warriors." He cast Zander that parental look again. "When you wuz two, yer parents started teachin' you ta' write yer name. When Caeculus

Tyndareûs's son, Aeneas, wuz two, he learnt da' difference 'tween a Hoplite phalanx an' a Macedonian one. Consider us untrained figh'ers, if ya will."

Zander frowned. It didn't set well with him, but deep down he knew that he didn't hold his bow quite correctly, and that the man who had taught him swordsmanship, for however talented, was the busking son of a fisherman. Untrained fighters it was.

Looking over the middle column, he asked, "Why do we have noncombatants, out of curiosity? Especially when some people still live in the towns. Shouldn't the war camp be for fighters?"

"Well, some's healers, li' Deirdre. Some... Ihs li' a replacement Balladarach. Some people don' wanna live under da' Empire, some people *can't* live under da' Empire. Dis is deyr home. Li'... Take Tam MacCay, fer instance. His wife an' kids're here. Now, MacCay wuz a big-deal clan in da' old country, fine, bu' da' real rub wiv Tam is he stole 'is wife on da' eve uv' a forced marriage wiv an' Augustan *praetor*. Deyr kids were born in outlawry – two uv'em in dis very camp. Ta' dis day, deyr's a five t'ousand silver boun'ee on Tam. If *any* a' his family's seen in town, dey'll be killed." He took a drink. "Kilmeny ain't a warrior; she'll stay back wiv da' kids – deyr's four non-combatants." He started writing another column onto the page, totaling the trained and untrained fighters. He did the math in his head so quickly that there was no pause, his hand moving methodically down the page.

Three hundred and thirty-six Turranians. Six hundred and forty Calchians. Six hundred and sixty-one Lowlanders. Seven hundred and sixty-three Highlanders.

"Twen'ee-four 'undred, total."

"Exactly?!"

"Neat, huh?" Daíthi drew a line to separate the work, caging off their numbers. "So now ya got intel. Wha'chya gonna do wiv ih?"

Zander rubbed his chin, studying the paper, the twigs, the tomato paste can.

"We'll send the Highlanders first," he said. "The front line will be made all of warriors. Each untrained fighter should be accompanied by a warrior – they'll fight side-by-side, so no one runs, no one's pitted against anything he can't handle. We'll attack in waves. Each wave drives further into the castle, leaving others behind to make sure what we've taken is held. After the Highlanders, the Turranians should come. They have a lot of archers. One hundred and fifty, right?"

"Mm-hm."

"And they can take and hold the parapet. The Lowlanders should follow, and the Calchians should closely follow them. I want the last lines to be very well-trained. They won't bolt while they're waiting, and they'll make sure no one else bolts, too. And just when Ariegn thinks we've run out of good fighters – we'll give 'em Aeneas to deal with." He raised his gaze. "That's a cool name, Aeneas."

Daíthi arched a brow.

"Anyway! That's my idea. And the camp leaders – Shrikant and Eoghan and them – they can assign their guys to what rooms to hold, and tell them when to advance and stuff. They know their guys better than I do. Plus too, then they don't feel as much like I'm bossing everybody around."

Daíthi nodded slowly, thinking it over. "Ah like it."

"But there are still a few things to be sorted," Zander pointed out. "Like how we get in. How are we gonna get in?"

Daíthi paused in drinking to counter, "Deyr's a gate."

"We can't just walk through the gate!" Zander responded, rolling his eyes.

"Why nah?"

"It's locked and stuff. We don't have time to wait out a siege – everybody would end up dead. We need to do it in one fell swoop. But if we try to break in – we can't hit them from the trees, and we can't get up close because they'll chuck stuff down on us until we're dead."

"On'ee if ya go in loud."

"How do you not go in loud!? It isn't like you can pick the lock –"

"From dee' ou'side."

Zander stared at Daíthi, demanding an explanation by sheer force of expression. Daíthi smirked and took a drink.

"What are you saying?" Zander enjoined. "We can't stay here an' we can't run, we can't engage them in the field and we can't enact siege warfare – I'd assumed when you signed me up for this that it wouldn't be *impossible*."

Daíthi grinned. "Tha' wuz yer firs' mistake."

"Murphy!!"

"Alrigh', a'rrigh'."

Zander watched as Daíthi began to draw – not sketching, but solid, free-handed bold lines. He drew the outside portion of Ariegn that could be accessed by land – the southern and eastern walls, replete with windows and gates. "Guards're heaviest on da' sout'. Most'a da' eastern wall'z obscured by sea, an' sheer anuff ya won' have nothin' ta' grapple to – windows're too high. Where da wall slopes a'da' base is called a *talus*. Ih strengthens an' supports, an' gives 'em a nice surface ta' drop stuff onto. A boat'd be smashed on da' rocks, Ah'm told. All tha' deters entrance, so fewer guards. Further, da' walls ain't solid stone, an' Augustinium ain't been doin' upkeep. Da' mortar's cracked here –" he indicated a place on the eastern wall, drawing in the cracked mortar near the sea. "An' you can see where da' joints a'twain da' stones are. Da' blocks're mathematically uniform," he explained, with a hint of excitement that Zander could not understand. He then explicated, using a lot of very specific numbers, how the angles on the building's corners, comparisons of the foundation stones and merlons and crenellations, and an appraisal on the weight of the stones necessary to reach such a height, could all be used to determine the necessary size of the blocks. "An' da' break in da' plaster's a good check fer tha' – see? As well as ta' confirm da' fact tha' da' wythe, a'layst, is a basic running bond."

Zander had no idea what any of that meant, nor what he was meant to do with this information.

"Super."

Daíthi looked up, cocking his head. He had thought it was all pretty straight-forward. "See? Ih le'ss you... *extrapolate*." He drew in first the mortar lines that could be seen through the cracked plaster. Then, sticking to scale as best he could, he drew in where he figured the remainder of the joints would run. "Wiv da' righ' tools, you can ge'chyerself up deyr. Ihs easier ta' pierce da' joints den da' bricks. A stout crossbow'll jam a metal bolt in far anuff tha' – reckon ya use t'righ' weapon an' bolt, it'd support upward'sa two-forty pounds. If ya get da' right angle, you'd on'ee need two sturdy bolts – big, steel. One grapplin' hook. An' a load'a cable." He studied Zander's expression. "Wan' me ta' explain angles an' physics?"

"I'm good. Won't they notice if you start shooting at their fortress?"

"Time da' waves. Da' tide comes in a'nigh', an' da' waves'll be loud."

Zander was silent a moment, pondering. "Is... what you're proposing... that one guy goes in, and opens the door for the rest of us?"

"He'd hafta. One on five 'undred ain't great odds. Dey'll've gone full capacity wiv yer dad there."

"That's all very nice in theory," Zander said. "But mathematical perfection and the reality of a shot are two very different things. Do you know anyone who can make that shot? I mean, that's gotta be like a hundred yards... Can *you* do it?"

Daíthi jolted. "Me!? Wha –"

"You seemed pretty confident in your answer."

Daíthi's expression was an intriguing blend of thoughtful alarm. "Id – uh – I – I..." His wide eyes fixed on his beer can. He said nothing for long enough that Zander had time to wonder whether the beer can made a better model for Ariegn than the tomato paste can.

"Daíthi."

"Ah wun' put ih all on me; ain' no one gonna like tha'."

Actually, Zander had a feeling that that might be the perfect angle to work. "But *could* you?"

Daíthi realized his hands were shaking; he finished off his drink, pausing to calm his nerves. "Mmh. Ah cou' be understudy hif ya really need. Bu' find someone a lil' better, won'chya?"

"Like who?"

Daíthi shrugged. Zander gave him a look, and he reasoned, "Hey, I ain't plannin' da' whole shebang for ya, am I? Spaykin'a which..." He leaned onto the table, inspecting his drawing in the scant light of Zander's campfire. "Wou' you grab me a beer?"

Zander smirked. "I'll fly if you buy."

Daíthi raised him a sullen glare. "You suck." Bracing on the trunk, he moved to stand, but Zander flagged him down.

"I'm only kidding; here..." Zander again retreated to the tent, this time bringing the whole case. "Are you gonna get smashed again? And if you do, will you be okay for the meeting?"

"Helps me concentrate," Daíthi replied. "An' I ain't goin' t'the mee'in, remember?"

Now it was Zander's turn for dread. "Oh yeah."

Daíthi noticed his expression. Cracking open the beer, he said, "Aaauhh, you'll be fine, hon."

"Easy for *you* to say; you're *made* for this stuff!"

Daíthi laughed. "Cor, yer dramatic. Ih could work t'yer a'vantage." Zander dropped dejectedly into his place, and Daíthi continued. "Here, look: ever forget da' words to a tune 'fore a crowd'a ten-t'ousand?"

"No."

"If ya done a good enough job, dey'll sing for ya ta' help ya to yer place." He took a drink. "Start wiv a good hook, carry wiv a good premise. Yer golden. If ya respect dem, dey'll respect you, too. Dass anuver life lesson for ya." He motioned a toast, watching Zander as he drank. Noticing the boy's unease persist, he sneered, "If yer gonna lead a *war*, Zander, yer gonna hafta be able'a *talk* ta' dem!"

"You were my Aaron," Zander said, lightly tossing Murphy's words back at him.

Daíthi was not amused. Pursing his lips, he tapped the expensive pen against the table's edge, measuring out his ire. Then, flipping the pen, he quipped, "Work ih out." He returned to planning.

Zander decided his stage fright was a matter best left for the day of. He folded his arms on the edge of the table, resting his chin on top.

"So your guy goes in stealthy, and manages to get to the gate," he said. "There's a drawbridge, a door, and one'a those iron thingies."

"Portcullis."

"The door's really heavy. One man can open it, but not without drawing attention."

"If 'ee can break da' drawbridge an' portcullis, yer front line'll take care'a da' doors."

"How's he manage that?"

Daíthi pointed at him with the pen. "You better come up wiv *some*'a dis plan, mister!" He took a quick drink and explained. "Dayr're two gate-houses, one ta' ee'ver side'a da' main entry, a floor up fro' da' Great Hall. Makes thin's mahr spacious an' aesthetically pleasin'; also makes breakin' da' defenses harder. Da' gate an' portcullis each is on a counter-weight system. Each respective winch is in a gate-house... Make sense?" He squinted uncertainly; after all, he had thought the maths explanation abundantly clear.

"So, the winch for the drawbridge is in one gatehouse..."

"An' tha' fer da' portcullis is in da' u'ver, yeah. Nah, theoretically, ya kill da' bridge-winch, da' coun'er-balance system fails an' da' bridge opens. Yay. Bu' ya kill da' portcullis winch, portcullis stays down, yer screwed. In da' old days, you cou' lift da' portcullis pretty okay, bu' nah dis one; ihs too big, an' ihs steel. So ya gotta kill da' bridge-winch, an' raise an' stall da' portcullis."

"That sounds tricky."

Daíthi snickered. "Fih wuz easy, eh'reeone'd do it!" He took a drink. When he pulled out a cigarette, Zander took it and lit it from the fire so he wouldn't have to get up. "Aaw, thanks."

"How do you break a winch?" Zander wondered. But Daíthi seemed to be paying more attention to enjoying his cigarette than war plans. It was Zander's turn to calculate.

"Alright..." He bowed his head, studying the glowing campfire. "So the drawbridge and stuff's probably raised with chains, not ropes, right? I don't like the idea of relying on the crossbow to break the winches. The shot's too precise, and too easily pulled out... You'd hafta *really* wreck the chain..." He watched the flames eating away at the main log in the campfire, ash whittling away its center. He smirked. "Like metal in a forge." Raising his gaze to Murphy, he explained, "Fire can be used for both fusion and fission. You can use fire to weaken the chain and break it, or you can weld something into the chain so that it'd stick. Solves both the portcullis and the draw-bridge, doesn't it? But..." His gaze flashed away for a moment, as he made a quiet comment on a subject Rahyoke had once forbade him from ever bringing up. "I can't do what my father can."

A curious pause followed this admission, but when Daíthi returned his glance, his expression spoke only of cocky approval. "Ah can help ya dayr." He motioned a toast with the cigarette, leaving Zander in suspense.

"What," Zander asked, flatly, "are you a fire-master, too?"

"Cou'jyah imagine!" Daíthi laughed. "Naw, hon. I ou'-bid Aila fer an oxy-acetylene torch. Meant ta' use ih on *tha'* piece'a shit," he added, pointing towards the generator. "Anywayz. You can both weld an' cut wiv it, so ihs pretty damn handy. That'd do ya, sure." He took a long drag on the cigarette, using its sweet smoke to cleanse his lungs, breathing out. "So tha'ss infiltration sorted, ain'it?"

Zander pursed his lips. He gave a hard nod. But they were still short on weapons, and shields. Then again... Anything could be turned into a weapon, couldn't it? Broken bottles and chair legs, fishing hooks and... His eyes rounded.

"Rakes."

He looked at Daíthi, who was busy trying to figure out if he could use his expensive pen to crack into a new beer can to shotgun it, the cigarette held between his lips. Zander wasn't sure whether he should be impressed with or afraid of his little friend – or if both, then to what degree.

He watched the smoke from the cigarette go rising off to join the clouds and stars.

"Breaking winches isn't the only thing fire's good for," he said, quietly.

"Nnh?" Daíthi looked up, and Zander met his gaze.

"In the battle against the men of Lochlann, Fionn Mac Cumhaill and his men took the victory because they were shielded by sea mists. If we play it right... I mean, I know we don't know the terrain in the castle, but... if we play it right..." He nodded over his own thoughts. "Maybe... this could work."

"Articulate," Daíthi muttered around his cigarette.

Zander's first impulse was to insistently defend his honour, but he decided to try his hand at being crafty. "Murphy. With all your phenomenal powers –" Daíthi rolled his eyes "—do you think you can get me the means for a smoke screen?"

Daíthi was silent a moment, pondering a variety of things.

"We don't have enough shields to go running in there," Zander explained. "I think that some kind of a smoke screen might make up the difference. Could give us cover 'til we can get shields off of enough dead guys... Or – I don't know; is it stupid?"

"It ain't stupid, hon," Daíthi replied. Shakily, he hauled himself to his feet, again waving off Zander's help. "Gatta suck up ta' Aila, bu'... Ah can ge'chya wha'chya need," he said, the cigarette clenched in his teeth. "Da' rest is up ta' you." Tossing Zander the pen, he limped back over to the generator, dropping down beside it much more heavily than he would have liked. He snagged his other bag, digging for something he could use as a brace. "Shit." He took the cigarette out of his mouth. "Grab me beer? Na' – tha' one's empty. Sa' tha' gives ya men, weapons, cover an' strategy, don' it? Yer abou' set, den?"

"Yeah. Catch!"

It was a bad throw, and Daíthi couldn't nab it with his right hand, so the can flipped into the dirt. "*Jeeeee*-sus. Nice *spiral* on ih." Daíthi wiped the dirt off the can on his pant leg. "War-strategy. So aisy, a kid can do it!"

Zander rolled his eyes. He took one last look over the drawings – the leaf matter, the fancy pen, the discarded cans – his future – and dropped back into the mulch again. "Wake me up in time for the meeting, 'kay?"

"Sure thin', hon."

There was a crack, a hissing spray, and an outpouring of curses to which Zander, thoroughly exhausted, fell fast asleep.

88
Waiting

Duncan was being punished now because he had refused to speak in anything other than Ceothann for the last three days – that was how he'd been made to understand it, anyway. Every question, every threat, had been answered with innocently widened eyes, a soft shake of the head, and *"Ní thuigim; arís, le do thoil?"* Three days without rations was his premium. *"An bhfuil tú ag pleanáil do mo choinneáil as seo go dtí sin?"* he'd asked. Are you planning on keeping me that long? *"Dar leat: An mairfidh mé beo?"* Think I'll survive? It was worth it, though, to his reckoning; after all, if we're going to be stuck here, we may as well have a little fun.

Closing his eyes, he leaned his head back against the wall. They had buried him deeper this time; his cell was one of many in the vast network of winding tunnels and chambers cut beneath the castle basement, carving their way into the belly of the earth – the vast labyrinth known fondly as the "catacombs". Aptly named. But for all the macabre bleakness of this place, he knew there was a worse place he could be. Beneath this winding, subterranean hive, at the very base of it all – at the very center of the earth – was the Crypt. Darker than Tartarus in its blackest corners – a darkness so unimaginably thick that it seemed to clog your lungs; more silent than the grave, and cold. Cold, and damp, and stale, and rotting. And the silence – suffocating – with every sound muffled, every sound magnified. A man imprisoned would strive to dig *down* for an escape, preferring the heat and the torment of Hell to that maddening, crushing abyss.

Duncan opened his eyes, and at first felt sick when he could not see the difference in the greys. He bowed his head, forcing himself to breathe. This was the catacombs, not the Crypt. Solitary confinement would not drive him mad; sure it was a relief after traveling with four teenagers for so long. If he just kept reminding himself of that, he would not mind the silence. Lord, he may even embrace it!

He raised his eyes to the shadowed ceiling. Mercy, it felt close. He wondered if it was damp down here, and how damp was it? Enough to wring a drink for a starving soul? Enough to drown it? Bracing himself, he lifted a hand – his left. Even though he'd been burned more badly on that arm, the dull pain of his right side made movement even more unwelcome. He held his hand up in front of him, his arm just shy of fully extended, and could just see the outline of its shape. He flexed his fingers, wondering what he had the strength left to do – and if his weakness wasn't more emotional than physical. He again questioned why he was still fighting. Your veins are dry when you're dead; wouldn't that piss them off more? But no – because there's nothing more aggravating than knowing that what you're after is in your grasp – right in the palm of your hand – and you can't have it. Like a vision beyond the veil of a dream, freedom beyond the gap in the bars, water beyond the darkness...

"Oof – *ssshhhh...*" He doubled, wrapping his arms around himself. He couldn't reach the ceiling. He didn't know why he'd tried.

They had not yet tried to burn him again, he reflected, as he settled against the wall. He wondered what they were waiting for. That's what they wanted, wasn't it? That was the real reason they beat him and starved him. They wanted to tap that most precious vein, to mine it for all it was worth. That greed would drive them again to his side soon enough, he knew. But for now...

Duncan cast a glance around him. He was alone. He was tired, he was beaten, he was worn out beyond belief – but he was alone.

Wincing, he dragged himself into the darkest corner of his cell, away from the bars that kept him from the outside world. With a soft sigh, he laid his back against the wall, suppressing a shiver at its stabbing cold. He pulled his knees up, as near to his chest as he could manage. And God willing, God willing, he could sleep.

And start it all again.

He sighed, closing his eyes. "Aah... *Ach, tá mé réidh... an uair seo.*"

He was ready.

89
The Wolf Nest

"Why do you only come around when you need something?"

Daíthi dropped into one of the wooden lawn chairs that stood outside of Deirdre's tent. "How d'you know I need anyfin?" He handed her a box filled with *pizzelle* which Zander had watched him hand-make earlier that morning. No end to the talents of Barfly Murphy.

"Because you always need something," Deirdre replied, taking the box and plopping Daíthi's laundry bag unceremoniously into his lap.

"Aw, yer a dear," he said. "Actually, wou'jya mind givin' Zander's shoulder a check?"

Deirdre had started to roll her eyes, but at Zander's name cut her gaze to Murphy. "What di'jyou do to him!?"

"Ah din' do nufin! Zander –"

"I have an old shoulder wound," Zander assuaged, pulling the collar of his shirt so she could see the bandaging. With a nod to Daíthi, he added, "He wrapped this for me the other day, Sully did before."

"I'll take a look at it." Putting a hand to Zander's arm. she guided him into the other chair, twisting a look back to her tent to call, "Sophie, do we have any more tape?"

Zander had pulled his t-shirt and hoodie half off; he paused to glance over his shoulder, and was surprised to see Roscanna emerge from the tent, a roll of medical tape hooked on her finger. She looked at Daíthi with a malevolent smile. "Yer everywhere, aren'chyeh?"

Daíthi instinctively moved to stand, but Roscanna waved him down. "Stay off yer leg," she ordered, and he obeyed.

"What di'jyah do you to yer leg?!" Deirdre exclaimed, in more annoyance than concern, but he ignored her.

"Ah – Ah din' e'spect ta' see you here," he admitted, a little breathlessly.

"Ya don' know the camp very well," Roscanna accused. She tossed Deirdre the roll of medical tape. "Deirdre's mah cousin; Ah'm often here."

Daíthi looked at Deirdre, nettled that he hadn't been made party to this information, but now she ignored him. "I got the sheets together there," she said, pointing to a pile of sheets atop a platform of crates.

"Great." Roscanna retrieved a laundry basket from the tent, set it on the ground beside the crates, and began folding blankets.

Daíthi was silent, watching as she folded a sheet into a tidy rectangle, dropped it into her basket, and then lifted another.

He shrugged, settling back. "F'Ah'd've known, Ah would'a brought more *pizzelle*."

Deirdre frowned. "I would share," she complained, but Zander whispered, "I think he's trying to flirt." She turned to him, intrigued. "Is he?"

Daíthi cast them both a withering glare.

Roscanna followed. "Wha'ss *he* doin' here?" she demanded, of Zander.

"Be nice teh him," Daíthi warned. Roscanna was struck by his tone; not even Caomhan dared to speak to her like that. She snapped a sheet and folded it, saying nothing. A little less sharply, Daíthi explained, "Deirdre's tapin' da' wound Ah tol'jyah he has. What're ya at here yerself?"

"Laundry." Roscanna dropped the folded sheet and grabbed another. "New concept?"

"Tha'ss dirty?"

"Yes."

"Den why'ur you foldin' ih?"

"It's easier to carry."

"Y'have a *basket*."

"... Shut up."

He smirked, flashing her a mischievous glance. "Sartinly buys ya *time*."

Roscanna glanced up. She dragged a chunk of the pile into her basket. "Wou'jyeh be opposed teh helpin' me wi' this down t'the river?"

Daíthi shied a little. "Mmh; wha' for?"

Roscanna hefted the basket. She was bored with answering dumb questions. "T'harvest leeches."

Murphy frowned, his expression blank. "What'ur they?"

She steadied the basket against her hip. "Great slimy crayturs who look like tar bubbles that'll stick pincers in yeh an' draw ou' all yer blood. We use 'em teh' bleed out the 'bad blood' from the sick. The pure survive, bu' sinners usually get killed off."

Daíthi's eyes were wide with shock – the strangest green-amber she had ever seen – and he'd paled as if struck with leeches himself. "Wai', ra'hlly?"

Roscanna snickered. "No! Tha'ss *barbaric!*" Lord, he was an innocent creature. "Ah'm doin' the wash! Are ya comin'?"

"Mmh... Nah a big *fan*'a d'river, rahlly," he trailed.

"Alright, then," she shrugged. Turning away, she said over her shoulder, "See ya, rake. If Ah find any leeches, I'll bring some back teh ya for a taste, 'kay?"

"Tay –? Uhm – id – be careful down be da' river, 'kay?" he blurted, sitting up. "Da' currents are... jus' be careful, okay?"

She paused, looking back at him in full, one pretty brow cocked. "Okay, rake," she said at last. Then she walked away.

Daíthi sat back slowly, watching the place where she had vanished beyond the tents.

"Wow. That was some *quality* flirting."

"Shut up, Fahy!" he protested. His wolf came over to him, and at his invitation jumped into his lap, making a nest of his laundry bag. "A'layst *some*'uv us'ss makin' an effort."

"*'Some of us'* is a coquette," she replied. Then, switching topics before he could retaliate, she asked, "The meeting's tonight?" Daíthi nodded. "You ready?"

Daíthi looked to Zander. The boy's attention drifted back and forth between Deirdre and the bird singing in the canopy of a nearby tree – never alighting on the conversation at hand, let alone the future they discussed. Daíthi frowned. "Ya know, Ah'd've hoped."

Deirdre traced his gaze, but before she could ask, a bright war-horn cut through the clear morning, startling the bird away.

Zander flinched, instantly alert. "What's that –?!"

Deirdre clasped her hands. "They're back!"

Daíthi rolled his eyes. "Good; Ah git mah hellhound back." He ran his hand into the dog's thick fur as she licked his forearm and hand.

"Who's back?" Zander asked.

"Let me finish," Deirdre bid him, "and then you can go see."

She finished bandaging his shoulder, he put his shirt and hoodie back on, and, with the promise to send Caílte back to her, Daíthi and Zander headed out, accompanied now by the wolf.

"So who's back?" Zander asked again, not having the patience to wait and see and not convinced that he'd recognize even if he did.

"Margan an' Rael. Dey went on a supply run a'fore we got here; now dey've come back wiv food an' stuff from da' towns."

"Is that what the war-horn was for? Doesn't that seem a little dangerous?"

Daíthi shrugged. "Does *anyone* in dis camp strike ya as particularly sensible?"

Fair point.

They continued on through the Lowlander camp, the wolf circling around them and giving Zander unnerving and loving nudges all the way. At the heart of the camp was a grand commotion, centered around a convoy of carts piled with goods and supplies. The atmosphere was festive and jovial as people flocked to meet their friends, to unload the carts, to merely be a part of the action.

"Damn," Daíthi murmured, in admiration. "Dey done good."

"Murphy!"

He and Zander turned and beheld a skinny, sun-burnt man with straw-blond hair, who, without further ado, took Murphy by the arm and led him over to one of the carts, explaining en route, "There's somethin' you'll wanna know, I'm thinkin'."

"Damn it, Farney," Daíthi groused.

Farney looked to Zander. "Dis guy wid you?"

"*Yes*. His name's Zander, an' 'eez been declared *imperator* in yer absence. Wha'uh yeh wan' tha'ss so urgent ya cain't wait t'rew proper salutations?"

Farney gave Zander a last scrutinizing look. Then he launched: "So get dis, right? It all seemed a normal trip, us doin' our usual split – us wid Margan, we hit up a *cúpla* towns. Bu' two sort'a *oddities* occurred, hear me? Two anomalies, so ta' speak."

"Yeah?"

Lowering his voice, Farney conspired, "So, even though we were in splits, we each had, *encounters*, ya know what I mean?"

Daíthi looked at him, disturbed. "Encoun'ers? Li'..."

"Yes an' no. *We* came out on top, so we did, but..." Farney cast a furtive glance around. "Word is Rael's whole party was captured."

Daíthi stared. "*Wha'?*"

"We missed 'em at ta' rendezvous point, an' word about town uv'a big arrest fit deyr description alright. We reckon dey must'a been taken teh Ariegn. It'll be rough goin' a bit, but – I mean, we're gettin' *really* good at dis raid business, so we can pro'lly just get 'em back when dey move 'em teh another prison, right? Shu'nt be long –"

"If dey're still alive."

Murphy's low voice was so bleak that it shocked Farney out of his confidence, and he looked to his comrade with wide eyes. "Wh... what?"

"We ent had a loss dis big in o'er a decade. If dey're willin' teh –" He noticed Farney's expression, and softened his own with a reassuring smile. "Sarry; Ah worded tha' wrong. Ah meant if dey took 'em alive in da' firs' place, deyr's a good chance dey're jus' lookin' fer compensation – we'll be able'a get 'em back alright. 'Specially if Rael's wiv'em; he's crafty anuff ta' get 'em out on 'is own, ain'ee? Ihs jus' a minor set-back, is all. Anyway, ya mentioned ya had some luck yerself, din'chya? Wha' did comin' ou' on top gain ya?"

"That's right!" Farney climbed onto the cart and began sifting through his haul. "So, me, Diarmuid, Sencha an' some guys jumped a supply cart on its way east – got maybe fifty bows, with arrows, and t'ree-dozen swords, give'ur take."

"*Moni Deuos,*" Daíthi sent a quick prayer to God; at least *something* had gone their way. He leaned onto the cart. "Ihs quality, den, is it?"

Farney smirked with pride. "*Look* a'this." He straightened up, holding in his hands a sleek, mocha-coloured hunting bow.

Zander's heart gave a leap within him. Shorn, it was a beautiful piece of work. "Can I – can I please see that – please?" His voice cracked, but he didn't care.

Daíthi followed his indication, then cast him a glare. "Ya see it fine."

"*Come on*, Murph!"

"Auh, what harm in it!" Farney exclaimed. Daíthi glowered at him as he handed Zander the bow.

"Jus' be *cayrful,*" Daíthi sighed. "Please?"

Zander nodded, clambering aboard the cart; Daíthi knelt on the edge, arms folded atop the rail. Farney handed Zander an arrow bag with a languid

"Here", and Zander placed it on top of a crate, drawing a straight-shafted weapon from the bag. The arrow appeared to have been fletched with a blue jay's feathers, which would give the arrow a different weight than the chicken-and-crow feathers he was used to. *Cool.* He knocked the arrow, readying his stance as he found a target tree out of the range of the camp. But as he lifted the bow and drew the arrow in the same motion, a sudden pain stabbed through his shoulder, and he was forced to lower, clamping the arrow in his left hand as he pressed his right to his wound.

There was more than mere disappointment in the curse the child uttered. Daíthi studied the boy for a moment – the crumpled, confident stance, the practiced grip that held shaft and bow locked.

"Y'ever shoot a crossbow?"

Zander looked at him, not even bothering to turn – the raven-like glance of his father. "Hm?"

Daíthi lazily motioned to Farney over the railing. "Farney, give'im one'a them – naw, the black one – *do* wha' Ah say, damn you – yes, tha' one. Tryin'a give 'im tha' piece'a shit –"

Zander laid the beautiful recurve bow on the crate atop the arrow bag, the fingers of his left hand still flitting over the arrow uncertainly as his right hand accepted the foreign, heavy crossbow from Farney.

"Yer arrow shou' fit ih fine," Daíthi said. "Nah, dis is gonna work a lil' diff'rent from tha' bow, 'cause ya cock ih first, then load..." With calm instructions, he briefly led Zander through the process of spanning and loading the bow, watching carefully as the boy followed along. At the front, there was what resembled a miniature recurve bow, like Zander's, and this end he placed against the floor of the cart, his foot on top to brace it. Then, grasping the string with both hands, he pulled up on it, sliding it along the barrel until it locked. "'Kay; tayk da arrow wit' the odd-feather down, slide ih in tha' track – yep." Zander docked the arrow in the flight groove as he was told, then lifted the bow, to aim. "Righ'-handed? Thank Gad; brace'tt agains' yer righ' shoulder – bandage holdin'? Good – aim –" everything after that was a gel of mixed curses and prayers. Zander took aim; he was used to aiming down his left arm, but the difference was slight. There was a metal trigger on the underside of the contraption, and, a tree in his sights, Zander gave it a twitch.

"Whoa." That was a powerful weapon indeed!

Having cringed, Daíthi ventured to cast a following glance. Thank God, no one was dead – and, praise Jesus, the lad seemed to have hit his target;

the arrow was wedged irretrievably in a tree a good forty paces off. *"Moni Deuos,"* Daíthi breathed.

"Foof!" Farney gave a sharp, falling whistle. "Yer a keen sight wid'a bow deyr, lad! An' you!" He nudged Daíthi with his foot. "I didn't know ya knew anythin' about weaponry, Murph!"

"People are inexplicably always surprised he knows anything," Zander uttered. He cast Farney a glance. "And yet, he does."

Daíthi repressed a laugh. "Do 'at again," he ordered, his calm, commanding tone maintained.

This time without instruction, Zander spanned, loaded and fired the crossbow, matching his prior shot with the difference of about an inch – he didn't want to risk breaking the first arrow. Crossbows were easier weapons – they required less skill than a recurve – but that aim was something harder to teach.

"It's a lot easier on my shoulder," Zander admitted, marveling somewhat. He was still, by the placement of his body, physically barring Farney from retrieving the recurve bow, although he did not seem inclined to reclaim it.

Daíthi noticed. "Farney, deyr ain't many who'd ha' use fer dah' shit. Tell ya wha' – since dey ain't givin' him a sword, let 'im tayk both, will ya? Come on – d'you know ha' long ih wou' tayk ta' train someone on tha'? Ee'ver'uv dem?"

Farney looked at Daíthi, disinclined; but the look of desperate hope on the boy's face was too much for any soul to bear, let alone deny. Farney rolled his eyes. "*Fffffyyne*. They're yours. Take ta' damn arrows, too – ya may as well. Can even have them from the tree, if you c'n get 'em out. You owe me, Murph."

Zander nearly died, he was so overwhelmingly happy. He was an archer once again. An archer with two beautiful weapons – one new and stunningly powerful, one sweetly familiar, both striving amiably and admirably to fill the void of his father's stolen bow. Taking up the recurve, he ran a hand along the smooth wood. He felt as if he had just been handed a part of his soul – a light, and he wandering in darkness through the longest week of his life. Here was one glimmer of his identity – something to cling to – to remind him who he *really* was, beyond the great chaos of names.

"Breev, Zander," Daíthi urged. "Y'hafta –"

He hugged him. The little war-leading killing machine, still armed with the crossbow, *hugged* him! At first, Daíthi was utterly startled. Then he snickered. "Wha'sat far?" he asked, as the lad released his hold.

"Everything," the boy answered, definitively. "Flippin' *everythin'*, Murphy." He turned to Farney, who raised his hands in halting surrender. Coolly, Zander thanked him: "*Go raibh míle maith agat, maith fhear*."

"Hey! A Ceothann after me own heart!" And Farney embraced him.

Daíthi rolled his eyes. "Losh," he grumbled loudly, jumping down from the cart, "Ah'm gonna go drink beer and punch stuff; *coun'eract* you *lassies!*"

Zander flipped his – *his!* – recurve-bow onto his shoulder. "Ky join ya?"

Daíthi ignored him, devoting all of his attention to Cuchulainn the Wolf, who was enthusiastically showering him with affection and kisses. Farney cast a disdainful glare on him, snorting, "Punch stuff."

"C'mon, Zander, le'ss –" Daíthi looked up in time to see the boy notice something, and stiffen. His narrowed eyes were locked on something deeper in the crowd, and beyond his vague shock was a look Daíthi'd had yet to see on his face.

Rage.

"Zander?" he asked, concerned. "Ya'll righ'?"

Zander frowned, a look on him to kill. "I'll be right back." Grabbing the rail, he vaulted down – recurve bow and three arrows in hand – disappearing into the crowd.

"Shit," Daíthi griped.

Farney had followed Zander's gaze, and now, recalling something with a curse, leapt clumsily from the cart and ran after him. And, if that weren't enough, the blasted dog ran after the both of them, too!

Swearing profusely, Daíthi rose from his crouch, using the cart for support. There was no way he was running; muttering oaths in Gallic, he followed.

Cuchulainn, being a dog, caught up to Zander first; he stood, rigid as a tree, his white-knuckled fist clenched around the arrows and bow, his eyes locked on his target with a vast, overpowering, murderous loathing. The target was scrawny, his hands bound behind his back, a short rope limiting the ability of his legs. And he was staring back at Zander, a look of dumbfounded astonishment on his face.

Sensing his fury and hatred, Cuchulainn gave a low growl that developed into a harsh bark, one sharp note of warning for the enemy to back off. The captive cast her a terrified glance, which only fueled Zander's anger. Clenching his teeth, he took the bow between his hands, dropping two arrows to the dust.

"I should've killed you when I had the chance," he growled, raising the bow.

For the prisoner was Addison Mitchell.

90
Traitors

"Hold up, lad! Hold up!"

A metal shaft descended between them, breaking Zander's concentration. Only then did he notice the dark-haired man guarding his traitorous enemy. But the arrow had been drawn and aimed, and a few words and a broken gardening tool would do nothing to dissuade him.

The man with the stick tried again. "Wou'jya be leavin' off a mo' on t'*killin'* him, lad! Settle down!"

The archer turned on him a glare filled with pain, with anger, with fire. "*Cén fáth?!*" he demanded. *Why?*

There's a question. No words had he heard spoken with such desperation. But – by God's mercy – the bow was lowered before he could answer, the boy shaking as if turning away that arrow were the most difficult act of his young life.

"*Buíochas,*" the man thanked him. "Yer a good lad, right anuff."

"How did he come here?" the archer asked, stiffly.

"Zander –" Addison began, and the arrow was raised again.

The man with the stick turned to him as well. "You're so *stupid,*" he complained. "*Why* would ya do that?"

"What, you think you've a right to last words?! After all you did to me?!" Zander demanded, his voice manic. "Don't think I haven't pieced it together, *Addison* – how conveniently they found us once we'd found you! I should've known after Rahyoke was attacked in the town. You betrayed us!"

"No –"

"Shut up!!" Zander's voice cracked in ways he hadn't known it could. "It's thanks to *you* we're all in danger now, isn't it?! Rahyoke was the last thing standing between all of us and the Empire – I bet they sent you to take him down, huh?! I knew your story of how you got here was stupid – I should have known that you would be crazy enough to betray us!!!"

"Zander, I didn't – I swear! *Ian* –"

"Coward!!!" Zander pulled the bowstring taut. But right before he took his shot – took his sweet, deserved revenge – a hand came down, clamping the arrowhead and bow-shaft together and jerking his aim to the dirt.

"Zander, tha'ss anuff."

Zander met Daíthi's gaze. How dare he speak so calmly. How dare he put himself between him and this traitor!!

Never breaking eye contact with the boy, Daíthi said, evenly, "Diarmuid, we ain't supposed ta' take prisoners."

The man with the stick – Diarmuid – answered. "He isn't mine – sure, I haven't a care for 'im save Margan had me guard 'im. It's herself and God only who could explain why."

"Secure 'im," Daíthi ordered. When Diarmuid gave a second's pause, Daíthi turned him a grating glare over his shoulder. "Diarmuid Cahillane, *secure* him."

Diarmuid sighed. "Feckin' bossy," he muttered, grabbing Addison by the collar and hauling him away. He was joined by Farney, whom he immediately sent to fetch rope.

Before Zander could see where they went, Daíthi forced him into the opposite direction. "Wait – damn you!"

"*Enough*."

Daíthi directed him away from the crowd, away from the tents, into an area of the forest between the river and their camp that was bright with sunlight and white birch trees. There they stopped, and Daíthi turned to him, arms folded.

"You had no right!" Zander snapped, pointing the tip of the bow at Daíthi's face.

"Nee'ver d'you."

"Aw, *screw* you!" Zander launched into a sort of frenzied, pathless pacing, covering the same large patch of ground in a million different ways. "I could've killed him!" he shouted. "He was right there!"

"Aye, tha'chya could'a, but ih wou'nna solved nufin. Believe me; dayr's a time an'a place fer everythin' – revenge included. Righ' now, you've other consarns –"

"It would'a taken *two seconds!!*"

"Zander, ya gatta know when'a strike, an' when'a let it go. Zander!" He grabbed Zander's wrist, halting him. Making sure that the boy met his eyes, he said, "Ah know ih seems on impulse tha' shootin' da' boy in da' face wuz da' best option –"

"Don't patronize me."

"Bu' yer smarter den tha'. Yer *better* than tha'; Ah know you are."

Zander gritted his teeth. "I am so, *sick*, of having to be better."

"An' yer righ' in that. Morali'ee sucks. Bu' ihs da' *on'ee* thing tha' separates us, from Ariegn." He released Zander's wrist, folding his arms again. "Dayr's purpose in eh'reefin, Zander. Figgur ou' how t'use 'im. If he did betray yer dad, who wuz he runnin' messages to? *How* was 'ee? Wha' does he know? You figgur how ta' take a'vantage of him. You make him work fer you. You wai'. Bu' save yer anger fer Ariegn, Zander. *Tha'ss* where ye'll get yer real revenge."

Zander looked away, to the bristled ground at his left and Murphy's dog sniffing at the roots of a tree. He said nothing.

An O'Rielly surrender. "D'ya wanna brood on'ih an' get back t'me?"

Zander pursed his lips. He didn't want anyone to be right, or to make sense or – "Fine." Either way, it wasn't really Daíthi he was cross with.

"Yer good wiv'at yoke," Murphy admitted, as Zander looped his bow over his arm in the way to make all real archers cringe. "Bu' Ah din' give't t'ya so's ye'd be shootin' eh'ree bloke 'at looks a'chya sideways."

Zander rolled his eyes. Then, setting aside his anger, he nudged the little man with a wrangling grin. "In life," he said, "I was an archer."

Murphy arched a brow at him; then he shook his head. "Come on," he said, waving Zander – and the wolf – on. "We gotta find Caílte an' send 'im back ta' Deirdre."

They rejoined the crowd, and saw at a distance a tall, beanpole of a man, holding a great sheaf of parchment in his hands. As he sorted through the varied "papers", he called out names, yelling for Runners to the other camps as he needed them.

"Who-e'er gave Seamus mail-call, is a moron." Murphy paused at the edge of the crowd, restraining his dog as a girl moved to pet her. He was only half-listening as the names were called, until strangely, all of a sudden –

"Anyone know a 'Morini'?"

Daíthi cursed. Leaving Cuchulainn and the little girl to Zander, he slid easily through the crowd and was level with Seamus in seconds. "Me."

Seamus looked up, his brown eyes narrowed. "You know 'im?"

"Ihs me... Ihs my 'regularized' name. Give'tt here."

"David?"

"Don' ask." He accepted from Seamus a folded sheet of military-grade paper. When he opened it, it was almost two feet long and a foot in width, covered with a scrawling black script on one side.

Murphy frowned. "Wha's dis?" he asked, running his finger up and down the page. Nudging Seamus, he repeated, "Ah cain' read; wha' *iz* this?"

Seamus glanced over. "Oh!" His grin returned. "Congratulations, Daíthi Murphy!" He slapped your man on the back so heavily he moved. "You are now ta' *second* most wanted man in ta' world!"

"...Huh?"

Seamus pointed to the paper, slinging an arm around Murphy's shoulders. "So, ya go'chyer name, ya go'chyer residence –"

"Tha' sayms li'a lot'a wards f'r jus' m'naym an' township."

"That! Auh! That's yer *crimes*, that is!"

Daíthi's frown became darker. "That."

"Yea'."

"*All'a* that."

"Yeah!" Seamus grinned. "Ye'll be wantin' the whole long thing?"

"Is't long?"

"Aye."

"Have ya other business?"

"I'll read t'whole thing, then." And he began:

"Notice: one David Morini, Marcansine, Northern Gallia Province, for outlawish behavior, acts of treason, and other motions presenting a clear danger to national security. Nice! Ah – such acts are herewith inscribed and are as follows: violations of laws established under the Act of the Partition of Gallia, the Annexation Act of the Gallic Provinces, the Northern Security Act, the Papist Laws – hey! I didn't know you were Catholic! – and the Labor Embargo. Additional crimes – Lordie! – include: forgery; extortion; fraud; money laundering; tax evasion; perjury; traveling without proper documentation; obstruction of justice; resisting arrest; assaulting an arresting officer; criminal contempt of court; arson; poaching; public indecency; invalid permit, fifteen counts; public urination – really?"

"One time – ih was *one time*."

"Loitering; public intoxication; drunk and disorderly; gambling; vagrancy; disturbing the peace, fifty-one count; destruction of public property; destruction of personal property with malicious intent; vandalism – wow, redundant much? – evasion; conspiracy; criminal possession of a weapon; murder, first degree, two counts; assault, *massive* counts; theft, including burglary, robbery, and larceny, in the following... Hoe-lee *shit*."

Daíthi's brows knit, and he raised his glare to Seamus. "Wha'?"

"'Tis a very prolific stream of thefts, so it is. By God, dey're peggin' ya fer the theft 'uv the Scarlet Emerald 'uv Sefetaten, are they!" He twisted the page out of Murphy's grasp, and Murphy made a jump for it in spite of the splints. "What's 'Aurora's Tanzanite'?"

"Damn'tt, Seamus!!"

Seamus held the page so Daíthi could see, but not reach. "Look at that! That's all thefts that's pinned on you, so it is!" He ran his finger up and down an impressive paragraph of work; it took up half the page.

"*Give* me tha'," Daíthi snarled. He snatched the paper from Seamus, his eyes scanning down the text; he may not have been able to understand Augustan letters with much confidence, but he could tell their numbers, sure enough. "Figures, the timin'," he muttered, as he searched. "*Convenient* – t'self-righteous whore! Tha' finaglin' bitch. So *tha'ss* da' price'uv imperial annulment, is it?!"

Then at last, he found what he was searching for.

His eyes trailed over the number, making sure he was reading it right. And suddenly, he laughed. It was a quiet, humorless, chilling sound. And in a voice cracked with something akin to madness, he exclaimed:

"Ihs high anuff t'tarn traitor!!!"

91
Geasa Droma Draíochta

Zander followed Daíthi through the forest, content to trail along quietly and watch the rain-clouds filling in the space between the leaves. As Seamus had agreed to send his brother Caílte back to Deirdre, they were accompanied by Murphy's wolf. She scouted the territory, sniffing at the trees and underbrush as they headed out from the Lowlander camp, navigating deeply south to skirt the Calchian camp, *Parvum Lavinium*. As if on patrol, they kept well away from the tents and gates, turning west and then north, following the ebbs and flows of some invisible boundary Daíthi seemed to know by instinct. Whenever the wolf would stray too far, Daíthi would summon her with a snap of his fingers, never speaking a word. The longer he held his silence, the less often the dog left his side, until she stopped leaving him altogether, walking beside him, nudging his arm and looking up at him with a piteous expression she must have learnt by copying.

At last, Zander could bear the suspense no longer, and demanded, "Where are you going –"

"Ah don' *know*, okay?!" It was an outburst of anxious fury, and Daíthi halted, clasping his hands over his mouth too late. His wolf looked up at him, her ears flattened against her head. "Ih – um – I..."

Zander pursed his lips. He came up level with Daíthi, reaching out a hand to brush the wolf's encouragingly-waving tail. "What's wrong?" he asked. "Is this 'cause you've got a rap-sheet? I though everybody had one of those."

"Tha'ss 'cause you were born wiv one," Daíthi muttered, the lower half of his face still hidden behind his sleeves.

Zander rolled his eyes. "*Chih*. Well, if I was, then it was a pretty boring one." The wolf turned to see who was touching her tail, and he bent down to pet her, ruffling her silky fur. "So... all those thefts, right? Were they, like, heists? You know – like where you sneak in and steal art and diamonds and stuff?"

Daíthi lowered his hands, his expression one of reluctance. "About 'aff."

"That's cool," Zander murmured, with far more admiration than Daíthi felt the situation deserved. "Are you, like, a master thief, then?"

"No."

"Oh. So..." Zander looked up, cocking his head. "...you *didn't* steal all those things?"

"Zander, a *master* don't get caught." Daíthi moved to walk away, but before he could, the boy spoke again.

"Maybe it's good that word got out. Now people will know what you can do."

Daíthi tensed. "What?"

"Well..." Zander straightened up, rubbing the back of his neck. "That ransom poster was a pretty solid résumé, wasn't it –?"

Daíthi's eyes widened. "Damn it, Zander! Ah tol'jyou ta' get someone else fer yer infiltration plan, din'Ah?"

Zander spread his palms, reasoning, "Look, Daíthi. You said that you can take the shot, and I really don't think that anyone else can –"

"You can!" Daíthi took one step forward with eager apprehension. "Ah seen your marksmanship! You can do it!"

"Not with my shoulder like this!! And anyway, I'm not good enough at math to make a mark that I can't see." He pointed at Daíthi. "You know as well as I do that if somebody else could do it, it would already be done, wouldn't it?"

It was more complicated than that – after all, once through the gate, there were guards – but Daíthi could not bring himself to fight off the vestige of his own words with the argument Caomhan had given him. Caught in this struggle, he merely stared, mutely – and stumbled back one pace. "D-don' make me da' crux 'uv yer plan, hear me? Ye'll regret it."

"Why?"

Daíthi looked away, drawing a long breath through his teeth. He said nothing.

"Daíthi. *Why?*"

The wolf began nuzzling her master's hand, looking up at him, her tail still lolling hopefully. Both men watched her for a moment. And then, very quietly, Daíthi said, "One jink iz a quick, evasive turn. Two jinks iz disaster."

Zander narrowed his eyes. "What?"

"Ah wreck things by mere proximity. Sure, ih all goes swell fer a while – Ah'm clever anuff ta' make sure 'uv it. Bu' if Ah linger, if Ah fail ta' back down when Ah should – sure it *all* goes ta' Hell. Any language Ah ever took to's dead. Any home Ah ever lived in, demolished. A'yone who ever loved me, has been killed. Ah'm da' plan-guy, da' messenger – Ah'm an asset in design. Ah shou' be *anathema* ta' da' execution." He turned to Zander, a pleading look in his eyes. "You wanna stake t'ree thousand lives on me?!"

His tone was just on the verge of frantic. Zander stared for a moment, too startled to know how to respond. Then he shook his head. "So what, you set me up for failure? I'm about to pitch an idea that can't possibly work, because of some – banshee complex?!" he blurted, at a loss for any other designation.

Daíthi gritted his teeth, lifting his gaze to the sky. "Ah know ih sounds stupid," he breathed, "but you *hafta* believe me –!"

"You're blowing this way out of proportion," Zander dismissed. "All you hafta do is get us in! After that, we're golden –!"

"An' if we fail?"

Though asked frankly, his voice was so quiet that Zander wasn't necessarily sure he'd heard him. "Huh?"

"After we're gone," Daíthi echoed, his voice louder, still blunt, but cracked, "who carries da' fight? After we're gone, who carries on da' fight?"

"...what?" Zander didn't have an answer.

"Caomhan wuz righ', Zander," Daíthi hissed. "Ihsa small camp, ain'it. 'Specially compared to da' world. Millions – billions, may'ee – 've had deyr lives impacted by Augustinium. People who lost families, friends – wives, fa'ders, siblings, children. An' nary a *damned* one'a dem will take up da' mantle an' fight. Dey surrendered long ago, content ta' sit beneaff a foreign rule an' gripe – pay deyr taxes, an' never shake da' boat. Ihs immutable. Do you hear me? *Miletu!*" He felt his heart constrict, railing against his bones. "After us, wivout us, deyr ain't jus' no *army*, Zander. We – we ain't warriors! Mahr den *half* uv us ain't warriors! Yet we'll go, an' we'll fight, because we *have* to. We'll go, an' we'll tayk on tha' *dunom – ison isarnon cambion rate* – an' we'll die. An' Ceothannanmór, Alyssae, Turranbar, Gallia, Lavinium, Sefetaten, Soonekk – *ehreefin dies wiv us*. We are de' *end*, Zander. An' Ah don' *want* t'be dee' end."

His voice dropped abruptly, as if it'd collapsed under the weight of this realization, though his tension remained high. Very high.

"Ah'm too much uv'a liability, Zander!" Daíthi exclaimed, pleading and insistent. "When ya scale a high-rise an' nick an heiress' engagement ring, nobody dies – nobody's ever in danger'a dyin'! When da' safe yer crackin' open'sa military base, ya don' get diamonds, you get snakes – you get them sea urchins're all spiny an' deadly, bu' dey jump! Ya get – I'uh – *DAMN* dis Usurper language!!!" His claws tightened on the air, and his gaze dropped further. His dog pressed herself against his legs like a cat, but he didn't seem to notice. "Ah dunno how ih would happen – maybe Ah'd set off some silent alarm, maybe Ah jus' miscalculated deyr numbers – bu' *somethin'*ll happen if you leave it up ta' me. Somethin'. Somethin' tha' *won'* happen if ya leave it ta' somebody else. *Anybody* else! You hafta believe me!! Dey're all gonna be dead because'uv me! *Everyone* will be dead! *Everything* we ever worked towards –"

Zander grabbed his arm. *"Daíthi!"*

Daíthi looked up at him. Something had snapped – vaguely he realized it himself. *"Anderos* – nah now..." *Not now!!* Bowing his head deeply, he pressed his clasped hands to his chest, trying to quell his rapid heartbeat, trying to remember how to breathe.

"Are you gonna be sick?"

"Shut up; ihs fine!" Daíthi blurted, his voice cracking. *Smooth, dumbass.* "Ihs fine. Ihs fine... It'll *be* fine. We hafta fight, an' we hafta attack first; eh'reethin' stands as Ah told ya. Bu' *ev'rythin'* stands as Ah told ya. *Dey* hafta fight. *You* hafta fight. But I – I..."

Zander maintained his grip on Daíthi's forearm, even though Daíthi was faintly pulling, leaning away from him, as if he wanted to bolt and yet couldn't quite work around to it.

"A-ah jus' wan'eda help," Daíthi protested. "An' may'ee – t'make sure my – tha' they din' die in vain. Ta' mayk sure Gallia had stood fer somethin'. Tha' was defiance, weren' it? Ah got ambitious. Yer daddy says such convincin' things. Ah – Ah thought Ah had a chance. Teh take back what Ah wuz owed. Ta' take back wha' dey took from me. I... Such nice, convincin' things – an' Ah reckoned we had a chance, bu' if I die – ih meant nuthin' didn't it? It meant *nuthin'*. An' I drag ehree'one down wiv me. Red wuz righ'; Caomhan wuz righ'; ehree'one wuz righ' bu' me..."

Zander looked on in silent wonder. In part, pondering whether Murphy had always been this crazy and, if so, how it had taken him this long to notice. In part, trying to figure out what Murphy meant.

He was scared, yeah – spooked at the very least. Afraid of some nameless, nonsense curse, afraid of dying, afraid of letting everyone down... *Everything dies with us...*

Murphy was a Gall; his country, his parents, everything he'd ever known and loved were destroyed. He was banned by law from his religion, his occupation, his homeland. Versed in nine languages, he carried those cultures – mythology and history – and for Gallia, the burden was on him alone. Bard, warrior, king – in all this was power. This was why Eoghan was wary of him, why Caomhan hated him. He was more than a messenger, more than an orator. He had been with O'Rielly from the beginning, and he was made of the same stuff. At least, O'Rielly had thought so. They had both had nothing – and everything – to lose.

"Trust God."

Daíthi looked up sharply. "Huh?"

"*God,*" Zander repeated, as if Murphy may have forgotten who He was. "He always has His own plan, and you're always more ready than you know. If it's His will that we're wiped from the earth now, then it will happen, without pause – no matter *what* we do. As for me, I'm goin' out in style. I would rather live only a single day if it means my name will last forever.

"Besides; death is like a dog; it comes when you call to it."

That's – ridiculous. Absurd. Totally, utterly mad. So it must be true. Daíthi bowed his head. The sleeves of his hoodie were pulled to cover his twisted hands, and he raised them over his mouth. Muttered against them, "Tha' won' please da' Company."

"Then screw the Company!" Zander snapped. "Nobody's forcing them to join me! I've gotten this far without them, haven't I?! I'll save O'Rielly myself. And if you were any kind of friend to him, you'll join me." He lifted his hold on Murphy's arm at a jerk, forcing him to look up. "You promised you wouldn't ditch me, you little punk. You still wanna help? Then help me now. Help me make everything we've lost stand for something. Help me bring Augustinium down. We've wasted years – *decades* – on empty vows and – meetings! Oh my *God,* meetings! What good has it done them?! This war isn't gonna be won with talking, and this meeting – like ALL the others – is pointless! I don't care what they think of me, I don't care *who* they are – what they say or what they decide to do! *I'm* fighting! And if they had *any* sense, they'll fight alongside me! *Why* are you smiling!?"

It was in the gaze that he had raised him. The most satisfied and pleased of sly grins was hidden behind that hoodie sleeve – which he appeared to be biting like a dog chews a blanket. "Ye'll regret havin' me," he mumbled around it.

Zander lifted his eyes to the tormenting Heavens. "I regretted havin' ya five minutes after I *met* ya."

Daíthi gave a short laugh, soft and helpless. "Well, Ah warn'jya, anywayz. R-remember y'had yer chance t'get outta ih, will ya?"

"Are you gonna do the stupid infiltration plan, then?"

Daíthi nodded. Only then did Zander release his grasp on his arm, instead pointing at him warningly. "If you try to bail, I swear I'm gonna kill you."

"I ent gonna run." Finally taking his hands away from his mouth, Murphy dropped one and raised the other – his left. "Ah *swayr*." There was something halfway sarcastic about the tone. He crouched down, running his fingers through his dog's soft fur, embracing her as she licked his neck. "*Maith, a chailín*."

Zander frowned, his niggling suspicion grown. "That was on *purpose*, wasn't it?" he demanded. "You're playing me."

"Wha'!?" Daíthi started, alarmed. "Paroxysmal anxiety's really painful f'r me; wou' Ah induce tha' jus' t'prove a point? *A chiall*."

"You did, didn't you – if only because you *say* that –"

"Low opinion y'have," Daíthi droned. "C'mere; ye've a mee'in' t'get to, an' Ah'm gonna go self-anesthetize an' see if Ah cain' mayk up f'r yer gran' dearth'a pity."

Zander looked up at the sky, trying now to judge the time beyond the clouds. "There's one more thing I need to check before we go to the meeting," he observed, ponderously.

Daíthi smirked. "Hm. 'Course ya do." He ruffled his dog's soft fur, then, using the tree behind him for support, got to his feet again. "Le'ss grab some grub first, d'ough, arrite? Ah don' want you famishin' in da' middle'a da' bloody conclave."

"...Alright, fine."

Daíthi gave a conciliatory nod. Without further eloquence, he walked off, giving a sharp whistle for his dog – or Zander? – to follow.

92
Old Ghosts

Duncan woke with a jolt.

A warm light poured through the darkness, blaring from a magnifying glass lantern. It was the door that had awoken him – the door as it slammed against the thick stone walls.

A guard pushed his way through the doorway, dragging a prisoner by the neck; the poor creature wore the long garb of a sand-dweller with Ceothann flare – a knee-length tunic with long sleeves of an off-white fabric, jeans beneath, a sleeveless sort of long vest like the one Zander had worn, battered leather sandals, and a long, bulky, red-tartan scarf wrapped loosely around his neck.

Duncan closed his eyes. Not his problem.

"Hey *fire*-eater!"

Oh shit.

"Got a cell-mate for ya."

Duncan opened his eyes, watching as the guard pulled the unfortunate whelp down the stairs. "Not to the chapel?" he asked, quietly. "Whatever could you use him for?"

The guard dragged the sand-dweller up to the cell adjoining Duncan's, slamming him into the bars as he dug for his keys. "He's a rebel-leader," he answered.

"A rebel *leader*?" Duncan scoffed, smirking. "Don't tell me the ashes and parasites 'slurring and clamouring for the destruction of their life-source' have actually scraped up from the gutter and *organized!*"

"What; don't you recognize him, fire-licker?"

Duncan furrowed his brow. "Should I?"

"Chih! He's part of your army."

Duncan laughed – a loud, sharp, echoing laugh. "If I had an army," he challenged, satirically, "what the *Hell* would I be doing here?"

The guard lolled him a glance. "I didn't say they were a *good* army."

"Mmh," Duncan responded. He watched as the guard flicked through the keys with one hand, the other holding the sand-dweller easily restrained. "Is that why I've been forgotten down here? You're squashing small fires out in the real world?"

"So you're done speaking that bog-trotter language, are you?"

"Yes. Now bring me a sandwich."

"Hhm." Jamming the key in the lock of the cell next to Duncan's, the guard unlocked the door and discarded the sand-dweller inside, caging him.

Cast off-balance, the sand-dweller collapsed to the floor. After a two-second recovery, he found the energy to send his jailer a glance – a look so full of hatred and defiance that it undeniably confirmed his identity as a dissident.

"They're like hornets, Ixion," the fire-eater said, coldly. "Kick them, it'll only piss them off."

The guard turned to him, his smirk quashed.

"That's your name, isn't it?" the fire-eater drawled, adopting the fallen expression for himself. "Ixion Briareus. You were always behind the scenes – the man who pulled the switch once the others had secured me. Never could raise a hand against me yourself, though, could you, sunflower?"

"Hey!" Ixion pulled a short sword from its scabbard, thrusting it savagely through the bars. "One more word like that and I'll cut the clever tongue right out of your mouth, you bastarding vagabond!"

Duncan merely grinned – a sweet flash of that old, indifferent smile. "Well, my invertebrate nebbish," he answered, softly and smooth, "isn't it a convenient thing indeed that your master ordered you not t'harm the bastarding feckin' vagabond?"

Ixion's eyes narrowed. Angling the sword at Duncan, he fairly growled, "Mark my words, flame-spitter. When the time comes for killing you – and mark me again, it *will* come, and quickly – then *I* will be the one there. Not as a bystander, not throwing the switch, but with the blade *in my own hand!*"

"Get in line."

Duncan settled back into his corner, satisfied. Ixion refused to be baited, sheathing his sword and double-checking the sand-dweller's door was secured. "Sleep tight, boys," he bid, still unflaggingly leering. Then, without further ado, he swept up the stairs and was gone. The heavy wooden door fell into place, and the bolt upon it slid.

Lethargically, Duncan rolled his head against the wall, turning a glance to the rebel warrior. The lad had slid all the way up to the wall, and was striving to shift so he could sit up by leaning against it. Duncan sighed. *Damn it, I'm responsible again.* Pushing off of the wall as much as he could, he half-straightened his legs, turning to the sand-dweller in full.

"Rael."

The young man stiffened, tensed as an alert deer.

"Do you remember me?"

The sand-dweller narrowed his eyes. "I can't say I do."

"You named me." He shifted, clutching at his ribs. "It was years ago; you were young at the time. You made up a word for me 'cause you couldn't think uv'a fitting title – not content to say 'Field' like everyone else. You told me I deserved more out of life – and I told you to stop pulling Margan's hair for attention. Rael al-Tâjir –"

The boy looked up suddenly, and O'Rielly smiled.

"—you *know* me."

"*Rahyoke?!* What the – no – yes?" Rael slid closer to the bars that separated their cells, coming within a foot. "Oh my – what the *Hell!*" He laughed, overjoyed and helplessly surprised. "I thou'chyou were *dead!*"

"Yeah..." He sank against the wall a little. That one was always odd to respond to.

"What are you doing here!" Rael exclaimed, his voice wisely hushed, but excited. "Jeeze, it's been a decade!"

"Aye; you've grown up, *rebel leader*."

Rael became pensive. "I suppose we both have some explaining to do... you first."

"I –"

"Your friend there already said what *I* was," Rael deplored, impatiently. "He took all of the fun out of it."

"I'm interested to learn more of this fabled 'rebel force' from someone who is neither leaving nor desperate to *kill* me," Duncan returned. "D'you know much?"

"Yep."

"Out with it!"

Rael sighed, rolling his eyes. "Rael al-Tâjir, leader of the Second Ceothann Faction of the Company –" He saw the jerk Rahyoke gave as he stiffened at the name. Rael continued a little more cautiously – but his bravery and pride would not allow him to quit. "I am one of four primary leaders; we lead a force of thousands –"

"Thousands?"

"Thousands! Okay, *two* thousands, but –"

"Two thousands. The Company."

"Why do you sound so disbelieving? It's nothing new – surely you've at least seen our wanted posters. No; we were started like nearly two decades ago – like a decade and a half – out of a town called Balladarach –"

"Whu – " He had started from the wall at a jolt, but doubled, his left arm around his stomach, right fist to the floor. "Aaaugh..."

"Are you alright?"

To his surprise, Rahyoke began laughing, shaking his head. *"A Dhia Uilechumhachtach."* He looked up. "Rael, you're my successor."

"What?!"

Carefully, the man shifted, sitting up, away from the saving wall, his legs half-folded beneath him. "Don't freak out."

"Never a good start..."

There was a sharp snap, the clicking of fingers – and a heart of flame flashed to life in the farthest corner of Rael's cell. Rael jumped, clapping a hand to his mouth. "Hm! Sheesh..."

"Smooth."

Rael watched the flame fanning its tongues like butterfly's wings, skittering over the dry stone like a rolling candle. He passed his tongue over his lips. "They talk of you," he said, his voice faint. "They speak of you, but I thought surely it was a myth. The stories of old, retold so many times that... that..." He swallowed, turning to the conjuror in a darkness blacker for the light. "The Fire-Spirit."

Duncan pressed his lips into a line. He was holding his hand out towards the flames, cupped as if it was in his palm that they rested and not an inch above the cold, stone floor. "Rael, that is *what* I am," he said in reply. He closed his fingers, bringing the tips together. The fire went out quickly, as if it pained him to keep it going. "Not *who* I am. My name is O'Rielly. Donnachadh Mac Thomais Ó Raghallaigh."

Rael stared. That was a lot to take in – and he was still wondering if maybe he'd been beaten to death and this was God's way of putting some fun into Purgatory. He drew slowly up the wall 'til his back was rail-straight, never breaking his gaze from his really weird friend.

The man – the Fire-Spirit – O'Rielly – smirked. "What."

Rael waved a hand up by his temple. "Mind-blowing."

O'Rielly gave a snort – something like laughter and something like derision.

"Don't tease me! That was a lot!" He pointed with his good arm to the place where the flames had been. "How did you do that?"

"I dunno."

"You're mean. But – seriously, okay? Seriously, now – are you really O'Rielly?"

"A Dhia dhílis." Like anyone would pretend to be me – how many fire-eaters are out there – why is my name so blasted important – I wonder how much thought went into its creation? His eyes were raised to the ceiling. "C'mere – what's this about the Company, now?" His gaze slid to Rael. "And if you're in here, what good does that do?"

"Someone'll hold my place."

"As you mine. It's a small movement that habitually loses its leaders."

"That's life; don't judge a movement on numbers. Hope is alive, and that's what counts, isn't it? Besides: we're not dead."

"That's a *small* comfort, Rael."

"Stop bein' so emo, Rahyoke – O'Rielly – gah! Are y'able t'come closer? How 'bout I yell my confession t'the whole hall?"

He sighed, unable to do anything else with the emotion within him. "You're *so* impatient," he hissed, looking down to refocus. "Here, God willing; be *patient*."

Gritting his teeth and bracing himself against the wall, he slid until he was within a foot of Rael's cell. Everything in his body was as stiff as dried blood, but the rest he had had had thankfully restored him to a workable point.

"Are you alright?"

...If not fully. "Mm-hmm." His attempt to make it sound convincing failed, and, forcing a small grin, he carried on. "Anyway. How di'jyou get here?"

Now Rael sighed, becoming very much like a defeated old man. He leaned against the bars. "Myself and nineteen men had gone up to Danaum. We were inadvertently singled out. They cut down Connla, my messenger, and brought us into Sirmium Fort. My men remain there, alive or dead. And I'm here."

O'Rielly swore empathetically.

"It's worse, I'm afraid," Rael went on, raising his eyes. "I *will* be missed back at camp. My wife can handle the army, but – " Then, quick as that, he shifted gears. "Did I tell you about my wife?"

O'Rielly was startled. "No."

"*Ah*, my friend!" Rael grinned, bold and saucy. "Tell *me* not to pull girls' hair!"

"Auh – Margan!"

Rael smiled, and it was a look filled with young love. "Yes!"

"Oh, you're too young to be married," said O'Rielly, of the twenty-five-year-old Rael.

"And how old were *you?*"

"Ah – d'wha –" *Twenty-one.* O'Rielly shook his head, offering a hand through the bars. "Congratulations, anyway. *Go maire sibh bhur saol nua.*"

"That sounded cool; what was that one?"

"'May ye enjoy yer new life.'"

"I'll hafta remember that one!" Rael shifted somewhat awkwardly, giving O'Rielly his right hand; his left arm was too damaged for use.

"You're banged up pretty bad, aren'chya, kid?"

Rael waved him off. "Hey; if I stay here, I'll be dead and it won't matter. If by a miracle God sends Gabriel or the Archangel Michael to open this door, then I can get back to camp where we've fine healers and sanitation."

"That was an odd sentence." Duncan pulled back again. "But a good answer."

"I try."

O'Rielly leaned his head back against the wall. Somehow he didn't think they compared to Saints Peter and Paul.

"Do you know a lot about this place? Have you been here a long time?"

Without turning, O'Rielly tottered a hand. *Ish.*

"Maybe I'll ask, then."

"Sure, fine."

"Why would Ariegn need a book?"

Duncan jumped. "Aw, *shit!*" Wrapping his arms around his stomach, he grunted, "I hafta quit doin' that..." Looking up, he strained, "What book?"

Rael settled back, his eyes tracing the ceiling as he tried to remember what he had overheard. "I'm not sure. But they said about a – what was the word they used? – a sibyl, guarding some book. It's important enough that they gave her a constant guard, to protect her. Why would a –"

"They have another book!!" He had sat up too quickly again, but, not heeding, threw himself forward and to his feet. "Glory to God," he breathed, as he set off pacing. "Glory be to God!"

"O'Rielly –"

"Do you understand what this *means!?*"

"*No.*" Rael was watching him warily, fairly confident that O'Rielly had snapped. "Are you okay?"

"Rael!" He turned so abruptly that he nearly lost his balance, though even as he rightened himself he didn't seem to notice. "I will get you out," he vowed, pointing at him commandingly. "I swear on m'life I will! I just need t'figure – how..." There was the fix, wasn't it? He put his hands on his hips, standing now in the middle of his cell as his gaze roved around him, wondering what on earth he should do.

"Rael, di'jya bring any food with ya, by chance?"

"Would have, could have, *should* have, didn't." He watched as the man resumed his loopy, deliberating route. "C'mere – why is this book so important?" he quizzed, and was honestly surprised to receive something – albeit unsatisfactory – of an answer.

"I wan'it," the man answered, like a shot. "I wan'it, and *they* wan'it because *I* want it."

"...Ooohh-kaaayy," Rael said, dragging out the word in confusion. Again there was a pause. "Rahyoke, quit pacing around like that! You're driving me insane!"

He did. "Alright, then!" he said, brightly. "What d'you propose we do?"

"*Sleep!*" Rael laid down, curled up as he was. "That's what *I'm* doing."

Duncan sighed, feeling a little deflated – or maybe just light-headed. He resumed his place at the bars of Rael's cell, but he did not sit down. "That's it?" he asked, taking hold of the steel. "Sleep? Some rebel-leader, givin' up."

"I am *not* giving up. I'm going to *sleep*."

"But –"

"Maybe we'll get an opportunity. Maybe there's a chance when they're moving us. But for now? *Is é taoscadh na farraige é.*"

It is like bailing the sea. Duncan waited, but he said no more. Turning, he pressed his back against the wall and slid, dropping slowly and painfully back to his place. The cold and the darkness were no longer claustrophobic – they were an obstacle.

He glanced to his side and realized Rael was already asleep; God love the poor lad, he must be exhausted. *Alright, then; my watch it is.* He stretched, then settled again into a sort of roost, studying the place where he knew the door of his cell to be. *Maybe...* Methodically, he began massaging his hands, working the blood into his cold fingers, circulating heat. He could feel his heart fluttering like a caged bird, nerves and excitement battling for control. *Just once couldn't hurt, could it? Maybe just once...*

He snickered. "From time to time a fish is caught when bailing."

93
Discourse Again

They had dealt with Addison in the usual manner; it made life at least halfway easier for Diarmuid. A soft rain had started, and the twilight mists were seeping in; they gathered around his sneakers like bog-earth as he leaned against a tall oak, smoking a bummed cigarette. He was still in charge of the traitor, but he needn't stick too close to him. Every half hour or so he would switch his position, shifting to another tree, just to let him know he was still watching. No one had come to relieve him yet (surprise surprise), but when they did, he would definitely recommend the strategy; it was the only thing to do to keep from getting horrendously, murderously bored.

Diarmuid took a long drag on the cigarette; he didn't smoke often, but when he did, it was savory. When he looked back into the forest, he saw a hooded figure lacing through the trees. He recognized the gait and clothes and knew that it was Zander.

"Oye!" he called out. He waited until Zander was nearer before he continued. "Found yer way here, boyo?" He smirked, pulling on the cigarette again.

The lad looked up – deeply up, for the hood was pulled low over his eyes and Diarmuid was very tall. "Farney gave me directions."

"Oh, *Lord*. 'Tis lucky ya found me at all!"

Zander pushed back his hood a little so he wouldn't put a strain on his neck. "So you're still guardin' him, are ya?"

"Yea'." Diarmuid rolled his eyes. The pale smoke he blew into the drizzly night dissipated eerily in the darkness. "Yer unarmed."

"If ya want, I could relieve you for a bit. Even if ya just wanted t'get food an' come back."

Diarmuid narrowed his eyes against the rain. "If they decided you could lead this army, no doubt ye could strangle a man with yer bare hands, right?"

"On my honour, I won't."

Diarmuid contemplated the lad as he smoked, his arms folded against the unusual August cold. "A solid fifteen'll do ya?" he asked at last.

Zander nodded. "Fifteen'll be beautiful, thanks."

Diarmuid took from where it leaned against the tree the shaft of the broken gardening tool, pressing it into Zander's hands. "Just in case," he said. Flicking the ashes from his cigarette, he added, "I'll be back. Don't kill 'im."

Zander nodded. "See ya." As soon as Diarmuid was off, he moved.

It was not hard to find his target – look for the only man tied to a tree. Zander realized when he found him that they had been kinder to Daíthi; Addison's tree was taller, thinner, straighter, the branch to which he was tied higher up, and no rest for his feet; he was perched painfully on the tips of his sneakers, his arms bound together behind his head with chains.

Zander's anger had not dulled in the slightest, and at the sight of the traitor he felt it flare again. But he had promised too many people he would not let it get the better of him in order to gain this chance – Diarmuid, Murphy, Farney, Gerry, two gate-guards, Deirdre, a kid selling pottery... And he was, whether he liked it or not, a man of his word.

Zander pulled his hood lower over his eyes; the rain was coming down harder.

"Bastard!"

He used it like a name; Addison hated that his head automatically jerked up. Zander's voice was the first he had heard in hours, for Diarmuid seldom spoke, and never to him directly.

"Zander, before you say anything –"

"Shut it!" Zander pointed the staff at him like a blade. "You said Ian was the one who betrayed my family," he said. "Well, you better make a *damn* compelling case for how that's so."

Addison was silent a moment, measuring out his words.

"Go on."

"Ian totally ditched me," Addison blurted, begrudgingly.

"So? I would gladly ditch you, too." In truth, he probably wouldn't – stupid honest streak!! – but he *wanted* to, and that counted!

"He stole Rahyoke's bag," Addison tried again. "All he cared about was that stupid notebook and scrolls; he ditched me for those. Although I guess you'd probably do the same."

Zander wondered what value those papers could have to Ian. Rahyoke had told them they weren't worth any money, and Ian didn't particularly strike him as the scholarly type. Still... "That just makes him a looter."

"What do you want me to say?! It wasn't me who betrayed you, alright?! Who would I even go to? Everybody in this entire stupid, psychotic world is in this camp! There was *no* one I could have told about you guys. And why would I want to? You're crazy hermit people! Who cares!? Why would anyone care!?!"

Above all else, Zander realized one thing about this answer: Addison was speaking truthfully. Even now, he had no idea who or what an O'Rielly was.

Zander settled reluctantly onto his heels. There was another question he was loathe to ask.

"Where's Meygan?"

"I could ask you."

"What's that supposed to mean?"

"She went after you, to rescue you – or avenge you, but I guess that's not the case. Did you do this to her, too? Did you kill her?"

"I never saw her. How am I to know you didn't kill her yourself? Wait; dirtying your hands isn't really your style, is it." Zander briefly entertained the idea that Meygan had betrayed them. But her temper did not lend well to subtlety. He was fairly certain that if Meygan had gotten it into her head to kill him, she would have accomplished it in a moment of rage, quick and simple... and probably would have felt guilty about it later. "When did you last see her?"

"I dunno. Like right after you left."

"Did you see where Ian went?"

Addison balked. "I'm not stupid. I know that as soon as you have the information you want, you're gonna kill me."

"You *are* stupid. I could kill you either way. I could leave you tied up here and let the crows pick out your eyes and your pancreas and your liver. Some Prometheus *you'd* make. I don't *need* the information you could give me. I'm just extending a courtesy for you to make a defense for yourself. If you *are* actually smart, you'll take it."

Addison pursed his lips. Much though he hated it, well – his hands were tied. "I don't know directions, man; I'm not all weird and psychic like Rahyoke."

"Coming back across the field, you were headed West."

"Then he went North. North?" He thought it out. "Yeah, North I guess. But I don't know where he was going. I mean, the plan as far as I knew it was to go after Meygan and save *your* ass if we could. But, like I said, he ditched me, so I don't know what he was thinking."

What was North? Ariegn was North. Northambria, much as the name suggested, was Augustinium's northern naval port. All good places to rendezvous with the enemy. Did Addison know this? Probably not.

Much to his vexation, Zander didn't think Addison knew anything else that was useful. So what now? Addison was Margan's prisoner; Margan's responsibility. But – stupid honest streak – Margan did not know Addison like Zander did. She had not spent the past two months traveling with him across West Gallia; she had not spent the past two months training him to be a fighter.

Zander pursed his lips, tightening his grip on the neck of the staff, resigning himself to his own idea.

"Addison, tell me a thing honest," he said, his voice quiet, a little cracked. "Do you want to help Meygan?"

Addison knew Meygan's idea had been stupid. What were they supposed to do against trained soldiers, an army, a government? But he had to weigh his options. And, as Zander had pointed out, he was dead either way. Either these crazy guerilla warriors would slit his throat as he hung from this tree, or he could die with – what? – a sword in his hand? Addison sighed.

"Yes."

"Compelling." Zander folded his arms, the staff cocked in the crook of his arm. "I'll give you a chance. You can accompany us into battle. You are not as skilled as my men, so don't try anything; you want them with you, not against you. As for the battle itself, let God decide whether He'll save or desert you. I will accept His choice."

"That's –" Archaic? Barbaric? Insane? "Th-thank you."

"I'll let Diarmuid decide where you belong 'til then." With a feral sneer and a mocking half-bow, he concluded, "Welcome to Purgatory." And with that, he departed, the staff gripped firmly in his hand, restraint gripping his every, seething bone.

He was surprised to find Diarmuid back at his post – clutching a bag of potato chips and a container of canned coffee that looked like a heavy metal poster. He was even more surprised to see Diarmuid was accompanied by Murphy and Deirdre.

Zander spread his arms. "Did you not trust me?!" he demanded, piqued, and Daíthi replied by sniggering and pointing at Deirdre, who started, recovered, and punched him in the arm. Zander shook his head. "Where's your wolf?"

"Back a' camp. She don' like rain," he explained, almost interrogatively, squinting up at the weather. "Strung up mah blanket ta' make a tent for 'er. What're you carryin'?"

Diarmuid, accepting his proffered weapon, answered, "Used'ta be a rotary cultivator. Took it as a war prize."

"Huh! Alrigh', den." Returning to Zander, he asked, more pragmatically, "Mee'in'll be abou' ready now; how 'bout you?"

"Yeah. You're not allowed to come, right? Then can you do me a favour and relieve Diarmuid?"

Daíthi smirked at Diarmuid. "Yea', 'cause 'eez so feckin' put out."

"No chips for you!" Diarmuid declared, regally, cradling the bag as he snapped open his coffee.

Deirdre rolled her eyes. "Come on, Zander," she urged, stepping forward and hooking her arm through his.

Zander was quite surprised. "Oh! You're –?"

"Ev'ry leader needs a second, ey?" Murphy said, by way of an explanation.

"Now *I'm* your Caomhan!" Deirdre boasted, turning a bright smile up to Zander.

Zander was pretty sure he had a distinct memory of the Company denying him a second. But... well, who was in charge here, anyway! If he wanted a second, he would damn well have one! And only the most profound sort of idiot would turn away someone as pleasant as Deirdre.

"Alright, deputy," he said, unlinking his arm from hers and looping it instead around her shoulders. "Let's go start a war."

Daíthi snickered. But his tone was one of confident pride as he declared, "Yer ready."

94
Precious Sparks

The crowd that had gathered to decide the fate of nations that evening was small by Murphy's standards; he would have considered the gig "intimate". But to Zander, who would be making his public speaking debut, they numbered like the stars – or the slew of fireflies that had taken their place this overcast night. Compounding the issue was the fact that Margan and Amergin had flouted the twelve-man restriction, arriving defiantly with an entourage of at least thrice that number. "You never could behave, Amergin," Deirdre had teased him, to which he responded, "And who invited yerself here, Miss Fahy?"

Deirdre was the single addition to their numbers for which Zander was grateful. Something about the presence of one guranteed ally made the whole thing seem manageable. Because, apart from Deirdre, there was Caomhan and Moesen, Shrikant and Balaji, his second, Glaucus and Caeculus, who had brought with him his cool-named son, Aeneas, and wife, Ursula Fallon, whose eyes pierced like talons, making Zander feel more like a sparrow than a war-hawk. Wouldn't it be nice to fly away?

He was startled as Deirdre took his hand.

She smiled up at him. "I know that you're not your father, Zander," she said. "But you are his son. You can do this."

Zander was still unsettled, but he took courage from her words. He set his eyes on the modest crowd – his allies, whether they liked it or not. He gave her hand a quick squeeze before, letting go, he stepped into the ring. They had surrounded the circle with a number of torches, sticking one in the middle. It was beside this center torch that Zander stationed himself, taking hold of it at its neck.

"At your leisure."

They hushed with unnerving speed, all their eyes upon him. He suddenly had more respect for Murphy.

If you're gonna lead a war, Zander, you're gonna have to be able to talk to them!!

He cleared his throat. Cleared it again. Spoke. "Thank you... for giving me this chance to prove myself to you. Does anyone mind if I start in on it, then?" His eyes found Shrikant, who gestured, saying, "By all means, *yā isfahsalār*."

Zander cocked his head at the appellation. Then he launched, laying out his plan in as even a pace and tone as he could manage, his gaze roving with his thoughts, trailing the dust, scouring the canopy, jumping from face to face, lingering occasionally on Deirdre whenever he felt particularly pressurized.

"We will take Ariegn," he finished, "as Augustinium took her empire, country by country, township by township – room by room. I need you to help me arrange your fighters. You know them better than I, and it will take all of us together to get this thing accomplished."

It wasn't the strongest note to end on, but it was all that he had left. He stood, his breath bated, waiting for their blow.

Not surprisingly, Caomhan was the one to strike it. "That's a very pretty plan."

"Thank you."

"But whoever you send to breach the gate would have to be a man of some skill."

Zander shrugged. "Just need a man who's good with a crossbow." Boy, was that ever an understatement.

"I hope yeh have a man in mind," Eoghan remarked, bearing the patient smile of a parent to a child who has just suggested shipping himself to the moon.

Amergin's grin was more undermining. "Caomhan, *a stór*, you're a weapons specialist, aren'chya?"

"Aye," Eoghan affirmed, "and if yeh wanted someone tae creatively put tha' fear o' God intae the wall, then Caomhan 'ud be yer man. But ta' take a shot like tha'?"

"It would take another man entirely," Caomhan agreed, stretching.

"And I think," Zander resumed, "that you can all readily agree with my resolution. I believe it would be best if whether or not we go ahead relies on the man who campaigned so hard for this war." The warmth here drained from his tone, the youth, and he spoke his verdict with all the cold, heartless indifference of Caomhan or Roscanna. "Let it all rest on Murphy's head. If he dies, there's recompense. If he succeeds – we take it as a sign from God to go forward."

As expected, the immediate response was scorn.

"Barfly?!"

"The success of your plan hinges on *Barfly Murphy*."

"You're looking at it all wrong," Zander responded, patiently. "You're acting like your lives are depending on him; the only life here at risk is his own. Look: the reality of your results is this: You either become the first warriors in history to breach the walls of Ariegn, and take down an empire doing it, or, at the very least, you get rid of Daíthi. Win-win."

He would have felt worse saying it (especially when out of the corner of his eye he saw Deirdre's expression) but for the fact that he knew that, in the same position, Daíthi would have suggested it himself.

"Can we *rely* on Murphy?"

"He wants this war bad enough," Zander angled. "If he bails on us – and he *won't* – then I could always give it a shot myself. But I'm telling you, he's not bailing on us." *I'll kill him first, and he knows it.* "If he dies, we run. Ariegn is otherwise impenetrable. At worst, we'd lure them into *our* territory, meet them on our own grounds. But that isn't likely to happen."

"Is Murphy able for it?" Amergin challenged, with genuine concern in his voice. "Or are we just sendin' da' man to his death? Does –"

Seamus McCarthy tentatively raised a hand. "Uh – if I could note?"

Margan welcomed him kindly. "Go ahead, by all means!"

"Well... as far as vouchin' for 'im... yer man Murphy is a man of high skill, insofar as bein' a thief goes. I've it in writin' an' everythin' – well. Me own copy I gave t'himself, but it's there, anywayz. And I don't mean a shop-liftin', pick-pocketin'-type thief. I mean..." He turned to a man beside him. "You were around t'hear it – an' di'jya see, they're pinnin' the high-rise heist in Fabius on him – the theft of the Athene Emerald – the Brigdavios Races –"

One of the Turranians actually paled, his jet eyes wide. "That was a real thing?"

Seamus continued. "So far as Augustinium's concerned, he's a threat fer just t'exact skill we need."

"I'd like t'second," one of the Highlanders voiced. "I've sayn'im fight; he's a braw, bold fighter, too."

"If nothing else," added – shocking – Moesen O'Rourke, "he's small. Less weight on the cord and less attention drawn if he's sneaking by the guards."

Margan patted Seamus on the head. "Well done, Seamus; I'm glad we brought you!"

"Yeh have confidence in him," Eoghan said, looking directly at Zander. It was not a question – more of a call to observation. Although Caomhan and Moesen did much of the talking for the Highlander camp, and Eoghan was more of a warrior than a diplomat, he was by no means a fool. He knew Zander was not sending his greatest advocator out to die.

Zander gave a firm nod. "I have full confidence in his abilities, yes."

"But how are we on weapons?" Moesen demanded. "That was among the many problems we faced not a week ago. You can get a man to the battlefield, but what's the point if he's got nothing with which to defend himself?"

"Wait." Shrikant had been thinking matters over meticulously, and now he raised a hand, commanding the entirety of the floor. And, in as careful a manner as his considerations had been made, he said, "Most of my men have multiple weapons. Keris and scimitars and bows. I have spoken with them, and they are willing to share their weapons out amongst your unarmed men, provided their return is assured."

Zander was downright stunned by this generosity. "Wow... Does that mean you're in with me, then?"

Still in that drawn-out way, Shrikant established, "It seems to me that this is a likely plan, and one that is best suited to my people... Therefore, Turranbar is united with its current ally, Zander O'Rielly, in declaring war upon Augustinium."

This caused a stir among the ranks, which was quelled as Moesen again demanded, "But is it enough?"

"My men and I brought in another shipment," Margan said.

"Aye," Seamus agreed. "About fifty bows and t'erty swords, give or take, we stole, an' another twenty bows we bought, like good little Cat'lics, an' a goodly supply of arrows to boot."

"There's still a lack among the Branch warriors," admitted Kian MacManus, Amergin's rarely-recognized second.

"I'll take care of it," Zander said, waving them off.

"Oh, will you."

He wasn't partial to the tone. "The weapon that's easiest for any man to pick up and use is a pike, isn't it? Some of my men are off making that a thing. We'll be able to supply the rest of the camp with pikes."

"Your men?"

"Barfly's men; do you want weapons or not?"

Amergin badly stifled a laugh.

Before they could ask any more stupid questions, Margan rose, positioning herself in the torchlight. She was a very pretty woman of twenty-four, with flowing black curls and soft green eyes, arrayed in colourful layers of skirts and scarves like a gypsy-woman. "I know Amergin spoke for me the other day," she said," and I agree with him. Shrikant is right to say that this is the best plan for our people, and I believe that this boy's plan is the best one we can hope for in the time-frame we have. As you've probably all heard, Augustinium has come after my family, they've come after my forces, and they will come after you, too, soon. My men can wield pikes. And it's better to make a stand, instead of just letting them wipe us out. I'm sure Rael would agree, we are officially allied with Shrikant and Zander in declaring war on Ariegn." She sat back down. "And I would welcome any of your fighters who feel that way, too."

"So that leaves two of you," Zander said, casting a glare first on the Calchians, in their fine, Lavinian attire, then on Eoghan, his own camp leader. "We have weapons, we have fighters, we have a plan – and a very promising one at that. I see no need for further discussion before a decision is made: So come on. Will you have me and these ideas of mine, or won't you?"

Ursula pursed her pale lips. "This is not how our decisions are made."

Zander found he was losing patience – and this, in turn, made him brave. "What, 'cause you're a 'democracy'?"

Eoghan looked squarely at Ursula. "Truthfully, hen, we've a'most eigh'teen hundred wi'out yeh. It wounae matter a whit if yeh j'ined, or if yeh took off back tae Lavinium an' yer grapes." To Zander: "Tha' First Branch Ceothanns 'uv tha' Company agree teh declare war on Augustinium and Ariegn. Wha' – wou'jyeh do it tomorrow?"

"Ih – tomorrow night," Zander found himself saying.

"Aye! And tomorrow night it is!" Eoghan slapped a hand to his knee, having done with it. Caomhan shrugged, a small and eager grin lighting his eyes like Hellfire.

Balaji turned a cold glare on the Calchians, and in heavily-accented Augustan, demanded, "Caeculus, doth thy woman speak for thee?"

The little nerdy bit of Zander that recalled one of his father's linguistics lessons thrilled to hear the old, wartime Turranian dialect of Augustan. The more prevalent part of him reveled in not being the target.

Caeculus' fist had clenched, but Glaucus was whispering to him, reasoning with him – as much to his wife's annoyance as the insult had been. Why had they brought her?

At last, Caeculus gave a hard nod, and said, "You have our support in this." He said no more.

Amergin sat up from his lounging on a stone. "So that's all of us, then?" he asked, sounding rather surprised.

"That's it," Margan confirmed.

"Call your people together," Zander ordered, with quiet clout.

Amergin jolted. "Which *ones?*"

Zander rounded on him, and, fiercely, commanded, "*All* of them!"

"We – we need t'send messengers –"

"Fine; I'll take care of it," the boy said. "Gimmie five."

Leaving Deirdre behind to ensure the status quo, he headed out for the Lowlander camp, returning to the tree-prison. There he found Murphy and Diarmuid taking their ease on the damp ground, the latter still smoking, the former lazily interrogating Addison. "C'mere; ya answer me honest, Ah'll ge'chya sumfin ta' stand on so it ain't so hard on yer ankles –"

"Murphy."

Daíthi turned at his name; when he spotted Zander, he teased, "Oh, good! Da' cavalry!"

"Hey, Zander," Diarmuid greeted him. Thumbing an indication to Addison, he asked, "What's this one's name?"

"Addison... what was it? Mitchell, I think."

"Addison Mitchell," Daíthi repeated, an appraising scoff. "*A chiall*, da' nayms dees days. Where'jyou pick dis guy up?"

"Oh, he's feckin' nutters," Diarmuid assured him. "Thinks 'eez from another world. Like an alien, like."

"Does 'ee, now!" Daíthi marveled, turning again to Addison. "Cor, yeh'd hafta be born in a vat'a beer t'be tha' head-fecked, wun'chya? C'mere, Zander – ya gonna make me ask or guess how yer mee'in's gone?"

"We're only half through."

"Ge' kicked ou'?"

"No. I need your help to summon everybody else that's in the Company. I wanna announce that we're goin' to war."

"Ah!" Diarmuid exclaimed, with a pleased and wicked grin.

"Eh'reeone? Ya do rahlize ya jus' increased yer audience 'bout thirty-five times over?"

"Yes," said Zander quickly; he didn't want to dwell on it.

"Ah like yer ambition." He swatted Diarmuid in the arm. "Get Sheeran, an' get Aila."

"Me?"

"Naw! Da' guy tied to a tree!"

Diarmuid sighed dramatically even as he got to his feet. "Sure, is it myself goin' around *claimin'* to be a messenger? Just 'cause yer too lazy to run, Murphy..." He pointed at Zander. "Don' be startin' widout me, now. I'll be seriously agitated, and I after fetchin' dee' *entire* army." With that, he jogged off, heading south.

"Come on," Zander urged, as Daíthi used a tree to drag himself to his feet. "They're waiting, and I don't want them to change their minds."

"Ya know, ayvin though 'ee iz a head-fecked sart'a git, Ah feel kinda bad layvin dis guy tied to a tree. Wha'd'ya say? Ah find a good leash, an' we can bring 'im, li'a puppy?"

Zander folded his arms. "No. He'll have his freedom soon enough." He ignored Addison's outcry of *"You can understand what he's saying!?"* "Until then, I'd leave him on the tree, so he doesn't get too full of himself."

"Ya rule wid an iron fist, *a chiall*." He looked at Addison, and enunciated in an accent more like that of the Turranians or Lavinians than his own, "Do you want somefin' to stand on, or what?"

Zander frowned. The rolled "R" was telling. "So that's what an actual Gallish accent sounds like, is it?"

"Aye," Daíthi confirmed. "An' a'fore ya ask, Ah hate ih. Ah don' use ih wiv you, 'cause... well, yer keen anuff ta' ken, ain'chya?" he said with a wink, then returned to the baffled Addison, flitting an indication. "Lift your feet."

"Is this a trick?" Addison demanded, tentatively obeying the command.

"No." Daíthi pulled his sweatshirt off over his head. Beneath, he was wearing a black, sleeveless t-shirt which showed off the edges of the warrior's tattoo which wound its way around his left shoulder, over his heart and shoulder-blade, and down his left bicep to his elbow in curling, woad-blue whorls. Rolling up the sweatshirt strategically, he set it under Addison's feet, explaining, "Seems small, but it does help. Trust me; I know." Folding his arms, he gave the prisoner an appraising look. "Better?"

"Yeah, actually."

"Grand." Gratefully resuming his usual drawl, Daíthi bid the captive farewell with a hearty and mocking, "'Til yer freedom, Ah guess," and accompanied Zander back to the meeting place.

"Time!" Amergin jokingly called out, as they rejoined the circle. Someone hesitantly complained about Murphy being there, and the insulted responded by covering Deirdre's eyes with one hand and gesticulating rudely at the objector with the other.

They waited. In a matter of minutes, four distinct signals rang out, calling the Company to the meeting place. And they answered. They poured into the clearing in droves – all three-thousand souls, fighters and non-fighters – filling the meeting place as they had not done since they had first come to this place. And when all had arrived, and Diarmuid rejoined them, Daíthi nudged Zander forward. "Arrite, hero. Prepare yer troops for war."

Zander looked at him a moment – at the encouraging smile Deirdre gave – at the casual thumbs-up Diarmuid offered.

"Okay."

Zander strode into the flickering light of the center torch. Three thousand people ringed the ancient clearing. Mothers with sleepy babies, herders with old sheepdogs – a little girl who led a chicken on a leash of string. Hereditary warriors. Farmers' sons. The seedlings of the future generation, and the remnants of the last. His soldiers. His people.

Zander smiled.

"I thank God every day for your presence," he told them. "In my childhood, I wanted nothing more in life than to meet you. To a kid, you were like ancient gods – a secret force of spirits who reigned beyond man's grasp, man's control. When I got older, I realized: you are so much more than that. And you don't even know.

"Maybe you think that you're just trying to stay alive. And that the cultures you carry represent the relics of a forgotten world – even something that holds you back. I know what it's like to be denied meals and fair pay just 'cause of my ancestry. It's a hard life you have lived, and I know that. But to people like me, who tried and failed to live by their laws, and who wished for a better day – whether the freedom of the past, or the freedom of a waiting future – *you* guys gave me hope. The fact that Balladarach had at one time, for a brief and beautiful moment, been a real thing, meant *everything* to me. It meant that there could be another way. A better way. How much greater, then, to find you alive now – and what's more, to be permitted to walk among you. Thank you for that, sincerely. You are our hope. And now, it is time for you to become our salvation.

"We... by the joint proclamation of Eoghan Fisher, Shrikant al-Rajiv, Amergin O'Shea, Margan of the Single Arrow, Caeculus Tyndareūs, and myself... Zander O'Rielly, if you're like, 'Who's this guy?' I'm O'Rielly. Anyway, by our proclamation, tomorrow night we're gonna raid Ariegn, and take it or die trying with the utmost of glory. We have declared war on Augustinium, and we are gonna win."

A murmuring began and quickly died; they listened to him, as patient as young children during their favorite story.

"The plan is basically a cattle raid, only the castle is our cow. We're seizing Arieng. As you know, that will both cut off the head and give us a base. The specifics of the plan will be laid out to you by your commanders. But right now, that isn't what's important." He paused. They all looked at him curiously, an excited energy beginning to stir through the crowd. "You have all come *so far*. You have done so much – to honour those lives lived and lost before you. Your ancestors look on you with pride. But while surviving in the face of adversity is a commendable rebellion, it doesn't win the war. For too long we've been rocks, standing immovable against their waves – lurking just beneath the surface. It's time we realized that together, we build a dam! And we will stop the tide – the flood – that is the Empire. We, together, can not only survive – as we have done all these years, in spite of them! But we CAN take them down!

"They think that, in a matter of days, they will march into these woods and exterminate us – like so many bugs, as if we didn't matter. They're wrong. They have always been wrong.

"The time has come for those of us who would to stand against them – to protect everything we have fought this long to keep alive. Now, know: no one can force you to fight. I won't make you. That's literally the exact kind of savagery we are fighting to *prevent*. But the decision needs to be made now. *Right* now. And you *need* to let me know, because if you just go off, that puts all the rest of us in peril. To desert is an act of cowardice that betrays and endangers your countrymen. To decline responsibly is the warrior's option. But if you're thinking of taking that option – and it's *not* 'cause, like, you're a healer or a mammy or, like, three years old – there's a couple of things you need to know.

"First: we fair a lot better if we stand together. Maybe it sounds cheesy or cliché or something, but generally it's a lot harder for one guy to go against an empire than three thousand. And the more of you run off, the less likely we are to come out of this thing on top. And if we lose... they'll hunt down the stragglers. And they will find you. Take it from a guy who has spent his entire life on the run, they will find you, they will figure you out, and they will slaughter you, and anyone you know and cared about. If we stand together, we survive. We win.

"Second... you are all *far* more ready than you know. Believe me. You are able for this. God provides. And ya know what?"

Zander O'Reilly – son of Donnachadh, who had led them before – looked on that beautiful crowd. And with all the casual, proud, bold confidence in the world, he smiled.

"We can take them."

How much of destiny is made?

95
Snapdragon

Dumb luck and soda bread. That was all that had kept him alive.

Less than half an hour had passed, and already the castle was on high alert. The place was like a giant anthill – filled with twists and turns, and through all of them men running, trying to salvage their jobs. And well that they should, for Rael – leader of the Company and mastermind behind it all, surely! – had gotten free, and was on his way to avenge himself and his comrades on Sirmium Fort in Danaum. Grand.

Duncan O'Rielly leaned his head back against the cold stone wall, a half-hunk of black olive bread clenched avariciously in his thin fingers, held close to his lips. He could hear their footsteps scouring from the sweet privacy of his fortunate alcove; he could almost have laughed. *There, now YOU know how it feels to be running!*

He knew the castle well, but the journey had been confusing. Every time he sensed a guard, he was forced to duck into another hallway, another room, another darkened niche – and he'd follow it or linger until he was forced to turn or hide or duck again. Long and short of it was: now he didn't know where he was.

No worries, though. Duncan took a bite of the bread, savoring the first food he'd tasted in days. Something ravenous in him urged him to swallow it in a bite, but he held back. After all, he might not quickly see food again.

Suddenly, he stiffened. The footsteps in the hallway had stopped. He felt the presence of the guard – *guards* – felt their eyes staring into the blackness of the hall – *his* hall. And shorn, they were tensed as well.

Swallowing as silently as he could, he eased off of the wall. A grand blessing, now, to be barefoot. Edging sideways, he urged all of his woodsman's senses, all of his hunter's soundlessness, all of his warrior's inner strength – and every ounce of piety from here to High Heaven.

A torch threw him in a pool of gleaming light, and swords bristled like a forest.

"Shit!!"

They were on him, and, whirling, Duncan was off like a shot, sprinting down the hallway at full tilt.

He had always been a good runner. Hell – it kept him alive more than *luck*, if you could believe it. His lithe body and long legs made it easy for him to quickly outstrip his pursuers – who were mercifully weighted down by their "light" infantryman gear.

Of course, he was reminded yet again, he had difficulties of his own; after only fifteen yards, he was gasping – wheezing like an old man on his deathbed – an old man with a crushed rib cage – and after three hallways of this he was near doubled with the effort of drawing breath. The pain and lack of oxygen made it even harder to concentrate on where he was going – and it had been a *long* time since he was last here. And the dark, and the –

He couldn't run. His ribs were threatening to bite into his diaphragm – he *could not* run. He staggered into the wall, looking back as they advanced upon him, and weakly held up a hand.

"Brón orm."

The torch rebelled, striking at its bearer like a snake. With a cry the man dropped it, and fire screened the hall, barring them from their prize.

"Now *stay* there," Duncan hissed, and slipped off down the hall with as much speed as he could manage, one arm clenching his injuries to keep his innards where they belonged.

They *all* knew where he was now, and they knew the castle better than he. Duncan drew a hard breath as he reached a crossroads, only half-hesitating before hurling himself down another hall.

The shouts of the guards and his speeding flight were attracting more pursuers, and he was beginning to doubt the sanity of this decision he'd made. He drew in a choking gasp, forcing everything into lengthening his stride, increasing his speed, until surely his lungs would fail, surely his heart would burst of the effort.

He skidded around corner after corner, barely controlling his path, through another hall, through a battered monks' door, and –

Outside.

The door slammed open, crashing against stone as he hurtled through it. It moved a bit more easily than he'd expected, and he tipped forward, stumbling with and against his momentum as he was thrown forward. He tripped dangerously and suddenly pulled short, lurching backwards and staggering to a halt.

He had found his way to the Western Balcony – a structure more north-west than anything, but whatever. It was a wide place – maybe thirty-some yards across. The old stones were weathered and battered, chewed and stained by the elements, slick with the ever-present sea-mists – but *Lord* they spoke of grandeur. Once princesses and kings – high chieftains and their queens – had stood out on that balcony; once complacent, wondering monks had gazed out over that sea, marveling at that farthest, bright horizon. Now the railing at which they'd stood was gone. Two thousand years of pounding waters had washed the cliffs away, pushing them farther inland, and kept eating, washing the retaining wall out with it. The edge of the paving-stone floor was jagged, and abrupt. Duncan swayed, looking over that ledge. He didn't trust his own balance any more than he trusted those tilted stones.

Far below – far, *far* below – black and grey waves were biting at the night-dark cliffs, scratching and lashing and wailing like a creature encaged. He felt their power even at his height – felt them tearing at the foundation of a house of God gone profane.

He shuddered, casting a glance behind him, as the door whipped and banged like a reed in the wind, before turning his gaze to the horizon. The sky was dense with clouds, boiling and black as they were whipped by the salty winds.

Behold, the face of God.

Duncan almost smiled. "Lord, how small we are..."

A wind rose, and he stepped back from the cliff again, farther now, fearing he'd be whisked away. Whatever the size of the world, there was little here to help him make his escape. Raising his head slightly, he narrowed his eyes against the lashing salt, glancing to either side of him. Dead end.

"There'll be no escape for you this time, fire-eater."

Duncan turned half-way, and what he saw was Bassus. The man filled the doorway, one hand clenched on the sword at his belt, the other holding open the door. The Augustan general held every advantage: the only weapon, the only exit... And he was a knife-thrower.

Duncan gulped as he saw Bassus draw something like a paring knife – disposable but deadly in the right hands. He would have to think fast if he was going to get out of this alive; he knew he was too tired to dodge.

"Yer rebel-leader's free," he remarked quietly, his voice barely audible through the wind and the waves. "Rael al-Tâjir, Lord of the Forest Thieves and leader of the Company. He's free."

Something flashed in Bassus's eyes, like lightning. He lifted the knife, taking aim. "For now." His voice was silky with vehemence, like a feral beast savoring its cornered prey. *Like a madman,* Duncan thought. The blade's shimmering tip caught the ethereal light of the outside world, gleaming like the morning dew.

Just as Bassus readied to make his throw, fire licked up from the bare stone, spiraling around Duncan, both weapon and shield.

Bassus smirked. "Well. Look at you."

"I told you I'm tired of being afraid." His voice cracked distinctly as he spoke these words. *Of course.*

"Are you?" Bassus closed the door behind him. "I will not run like my men before you, O'Rielly. And I will not leave you with the power to run again."

"I know." Again, his soft voice wavered. "I'm not gonna go easy."

"But you will go, nonetheless." His eyes flickered to Duncan's bare feet as his heel broke a chunk of rock from the edge, causing him to gasp. Bassus laughed at his fright. "Your only other choice is suicide."

"Is it?"

Duncan tried to settle his stance, hoping he could look stronger than he was – ignoring how his bare heel hung over the cliff's edge. *I can't take on the whole castle with the fire. I shouldn't – I can't. It's too – it's too much – it's too much.* He didn't want Bassus to know what he could do...

Duncan looked the knife-thrower in the eye, considering. He could still hear the waves, crashing, thrashing below, deep in their booming. He smirked. "Hey Judah... That's four."

Like a flash, he turned, pushing off from the ledge and arching his back in a dive.

And the sea, in all its glory, rose to meet him.

96
For We May or Might Never

Zander sat on the low stone wall that ran on the south-east side of the Lowlanders' camp. The wall was made of flatish grey stones, flecked with moss and lichen, worn by wind and rain and time. From his perch upon the wall, he watched the heart of his camp. In the common area they had strung lights in addition to the torches, constructing a make-shift stage out of carts on which their musicians played. Besides the music and light, there was food. *So* much food – tables covered with baked goods and fruits, fresh vegetables and meat – *real* meat! Great kegs of every liquor imaginable were rolled out. There was coffee and tea. Cans and bottles and pints were flowing. And the feasting – Lord'a Mercy! Zander had never eaten so well in his life. A far cry from the meal of an outlaw.

"Hey."

Zander looked up as Murphy sidled up beside him, his gait a slow and casual saunter. The little rogue looked vaguely disheveled, his faithful tweed cocked at an angle halfway between smart and slovenly, and in his hand was a half-empty pint of rich, dark beer. He leaned onto the wall next to him, his drink cupped between his hands and his elbow on the wall. The wall stood at about four feet, but his low slouch made it seem higher.

"Hey yerself!" Zander greeted him.

"Why ya ou' hayr? Go enjoy yerself – ye've me blessin' t'day."

Zander smirked, following Daíthi's indication towards the sea of vibrancy and life. "Can I get a beer?"

"Ha! Nat a drop, nat a *chance!*"

Zander sulked, his air at odds with the frivolity before him. He winced as a bagpipe gouged a bad note into the night. "Are you sure that this is any form of a good idea?"

"Dis?" Flicking his hair out of his eyes with a toss of his head, Murphy turned to view what Zander saw, putting the wall to his back and both elbows on it casually. He took a drink, contemplating. Then, "Aye, yea'."

Zander frowned at him. *"This."*

"Shar, give 'em dis, Zander. Personally, Ah can naym ya easy a'layst seven reasons to *encourage* it." He could feel the boy's glaring gaze upon him as he took another drink. "Ya wanna know da biggest reason?" he asked, extending a toast towards the festivities. "Look a'that. Notice sumfin diff'rent from when ya first came? Ihs called 'solidarity'. On tha' stage t'nigh's been Mainlander guitarists, Lowlander pipers, Turranian drummers, tha' fiddler-woman Aludra, Lavinian lyre-players – same fin wit' the crowd. Dey're one army. Fer da' *first* time – Ah c'n tell you – da' very first time. Da'ss a *Hell* 'uva lot mahr ahrganized den when ya came here b'fore, ain' it? ...An' wha'ss more, Ah fink dey've finally remembered dey're alive."

Zander was quiet. But at least he had returned his gaze to the crowd, allowing himself to view the truth of Daíthi's words. *The Company...*

"Mmh, ya ha' some kind'uva a point, though." Grunt in tone, though dragged like a drawl. Daíthi elbowed Zander slightly, lazily jockeying for attention. "Zander... You ha' a lot for t'prove in a short time?" His voice was lowered, lilting – purposely delicate over a very grave threat. "An' yer fortunate all far-five leaders've pledged t'back ya. Armies follow *dem*, 'ey do not follow *you*. Sa' yer gonna hafta prove yer *ass* off dis battle."

Zander nodded slowly, pressing his lips into a thin line. "Sobering, that." Then, opposing himself, he groaned. "We're all gonna die."

Daíthi just laughed. "Oh, one mahr thin'."

"Yeah?"

All of a sudden, Daíthi reached out with one hand, pushing Zander forward off the wall with ease.

"Oye!" Pin-wheeling, Zander landed on his feet.

"Ya mayk a feckin' *appearance!*" Daíthi forsook the wall as well. "Shake yer mammy's reserve off ya an' le's go! It ain't a warrior nar hero scay'd an' cryin' o'er be da' wall on accoun'a da' crowd. Man up! Dis'us sumfin ya gotta es'perience. Ya ent lived 'til t'nigh', an' ihs much I gotta show ya. You in, hero?"

Zander hesitated a moment – in part because he was trying to figure out if Daíthi had just called him a mama's-boy. He was about to move forward when a voice from behind startled them both.

"That's very responsible."

They both turned quickly, but Daíthi's startled flinch was so dramatic that the arrived commented flatly, "Wow, much?"

Margan. She stood on the other side of the wall, in the eerie stillness of the untouched forest, holding a dim lantern. She was dressed in her usual, colourful array of skirts, jewelry and scarves, her black curls loose about her shoulders.

With an intriguing amount of affectionate disregard, Murphy scolded, "Dan' sneak up on me, you brat."

Margan continued to stare at him, her large, green eyes like the depths of the ocean, capturing her lantern's light. "Do you know what intrigues me? Half of the camp knows you like you're an old friend. The other half's never seen your face."

Daíthi half-folded his arms atop the wall, quipping before a casual drink, "One-fift', ra'hlly."

Margan looked at Zander. "I have something for you."

Zander found himself taken aback. "You do?"

Margan drew close, holding her lantern aloft. The light struck his face, and, two feet from it herself, she paused. "Huh." She smiled. "I didn't notice who you looked like before. That's very curious."

Daíthi allowed a hiatus in his drinking. "Great. Ahr best shot's nah on'ee a pacifist, bu' goin' blind."

The smile she turned on him was odd against her words. "My eyesight's never been good."

"Sa' pacifism's a good 'scuse fer a bad shot."

"I sense it. I sense the heart," Margan said, cocking her head and maintaining her grin as she laid an indicative hand. "That's all you need."

"Ah've na' idea wha' t'make'a tha'." He raised his beer again, ending the dry pleasantries.

Margan turned back to Zander, and he wondered if he would be surprised every time she did. She reminded him very much of a butterfly – flickering and colourful, yet you could never tell how much of the colouring was camouflage, and for what. And the fluttering.

Margan handed him a tin whistle.

Yep; Zander was even more bemused. The pipe was a thin metal tube almost a foot in length with a mouthpiece of green plastic, old, though it still it shone brightly, almost ethereal in the weird light of this final night.

"Um... Thank you."

Sensing the rise of a barrier of confusion otherwise incorrigible, Daíthi looked to see what had been given. "Mmh. Shiny," he commented.

Zander found it a lot easier to challenge Daíthi than Margan, and so did. "And what do I do with this?"

"Play't."

"God made me a warrior and not a bard for a reason."

"Jayzus. Y'en use it ta signal da army – durin' ta' fightin', like."

Zander looked at him dubiously. "This. This tiny little flute-thingy. I can't even play it –"

"Ihs called a tin whistle, it ain't hard t'manage –" he leaned against Zander annoyingly, his eyes glinting green-chocolate in the marsh light "—an' small thin's ha' a way'a maykin' *quite* d'biggest ruckus."

Zander arched a brow. Then: "Hey – you're a musician. Can you teach me to play it?"

Murphy pulled a face, leaning away from the boy again. "I ent swappin' saliva wit' ya, bu' yen match pitch what Ah've seen. Tell ya wha': t'marra mornin' wake me up ahrly. I'll taych ya den. Deal?" He offered Zander his hand.

Zander's eyes flashed down for a moment, then he accepted. "Deal."

Daíthi turned to Margan. "Zander's an archer, too. Yen give'im some tips, can'chya, Single-Arrow?"

Margan perked up. To Zander: "I think I have, if your daddy is who I think he is. He taught you?"

"Uh – yes!"

"You're welcome!" Returning to Daíthi: "I feel like we need to speak."

"Mmh? Mmh." Daíthi had drained the last of his beer. "Ah need anuver one'a dees. Follow me, bot'uv ya, an' we'll ge'chya all sorted out."

Margan vaulted the wall, and she and Zander followed Daíthi through the mist and trees. As the light cast away the sea-fog, they soon found themselves threading into a crowd, stepping over tent ropes and around long tables, trying to avoid being crushed. The crowd was a living beast, its breath filled with the heady scents of liquor and sweat and fire, its sinews and veins pulsating with the rhythm of its throbbing core that bled with light and music and sound. The crowd was tightest at its heart, fed to frenzy by music, flavored with the savory spices of guitars and bodhráns and pipes.

Zander hung back, and Daíthi, noticing, halted beside him. "Sca'yd?" he taunted, with a cocksure smirk.

"I'm not scared!"

Daíthi's smile broadened, half-pitying, and half-teasing. "Aw, honey! 'Ur yew *shy?"* He looped an arm around the lad's shoulders. "Go on! – ye've yer pick'a lassies t'night! Yer a feckin' *war*-leader!" At a conspiratorial murmur, he added, "Use't up t'da *hilt*."

Zander was just working on a stammering protest when Daíthi released him, taking one step to the side. A girl was walking past; Daíthi hooked her arm, and she spun, pivoting back to them.

"Hi! Who're you?" she asked, affably.

"Da' question ain't who *I* am, sweetheart," Murphy corrected, gently putting his arm around her shoulders to re-direct her. "Da' question is, who's *tha'* guy. An' tha' guy wou' be Zander."

Zander had been staring at the girl, blatantly and unabashedly. She was nearly Murphy's height, her golden-amber curls shining like the drops of a late summer sunrise, cascading over shoulders thin and white as a swan's. Her eyes were large, shining and gorgeous, a deep and endless emerald green.

He blushed.

Ah, young love. "Funny yer timin'; Zander wuz jus' tellin' me ya seem t'prettiest'a all girls – sumfin abou' how yer hair sparkles wiv d'radiance'a twisted gold – I ain'a poet, he'd tell ya better." Zander wondered if Murphy could read minds with those faery-blood eyes of his or if he'd accidentally spoken aloud. "Wha'ss yer naym, lass?"

"Merilee."

"D'ya dance, Merilee?"

Merilee shrugged one graceful shoulder. "Middlin'."

"Grand." Daíthi nudged Zander painfully in the side. "Dance wiv'er, wou'jya?"

Zander jolted. "D'yuh – I – uuhm –"

Merilee's eyes brightened and she perked up like a sunflower. "Yes!" she exclaimed, grasping Zander's hands between her own. "Do you want to dance, Zander?"

"I? I – uh –"

"T'isn't hard, most, the dancin'," Merilee added, explaining. "Really we just jump around. If you c'n keep jumpin' f'er t'ree hours straight, yer golden."

"I –" What warrior couldn't jump? Zander's look became confident; it was no longer a question of style so much as of physical endurance, and *that* he surely had. "Course I can!" he defended, aloofly.

Daíthi gave him one last push from the nest. "Grand; you two have fun." He turned back to Margan. "C'man; I'll spot ya."

Merilee seized Zander by the arm. "Yay! Let's go!" She darted back towards the crowd; Zander cast a last glance at Murphy, looking as shocked and pleading as if he'd just been handed a firebrand and told to light his hair on fire. Then they were gone, vanished into the sea of bodies and mist and light.

Margan sniggered, warmly. "That was nice of you."

"I ain't nice!" He led her back the way they'd come, procuring a dram for himself and a coffee for her en route to the wall.

Margan assumed Zander's forsaken seat, crossing her legs daintily.

"Ah fink Ah know wha' yer at, bu' go on," Daíthi leaned his back against the wall, slouching low as before, cradling his drink in his hands.

Margan looked down on him with a kind smile. "It's nice to finally get to talk to you. Seamus speaks highly of you."

"Hmm... Ha' any bu' you said ih, I'd call lie." He took a drink. "Ha' you allocated yer haul? Ah don' – dayr shouldn't be much weapons distribution so late as t'marra, I don't think. Layst not in ihs entirety."

She arched a brow. "Aside from the pikes you promised? My camp's taken care of. And they all got their floor assignments with their weapons. 'Cause I'm efficient like that."

"Hmm." That was all of an answer Daíthi gave. He was watching the flitting colours of that closely-confined chaos of jollity, like dragonflies in a mason jar, thinking.

Margan was thinking too. At length, she asked him, "Are you worried about Zander?"

Daíthi took a cigarette and a lighter from his pocket, trying to light the former without relinquishing his hold on his beer. "Are ya wonderin' why I don' stop 'im fightin'?"

"No." Margan took the cigarette from his hand and lit it from her lantern. "My daddy taught me to use a bow when I was just a little girl. I think they're proud of me and Rael – my parents." She handed Daíthi the cigarette. "You're very fond of Zander."

"Ah've known'im a long time. His parents wou' be proud'a him, too." He took a long drag from the cigarette, savoring the charred tobacco vapor, holding it in his lungs before breathing it out again. "T'be hones', Ah'm worried *for* him, nah abou'. Bu' Ah don' have th' heart t'hold'im back. Go ahead an' ask me; Ah know ya wanna."

Margan looked down at her coffee, watching various lights reflecting on its surface like a shower of false stars.

"Augistinum 'cleansed' Gallia t'ree times; Ah lived t'rew two uv'em. Ah can tell ya, dey're grearin' up ta' do it again. Yes, dey know yer parents. Dey know all yer families. It ain't yer fault – ya jus' cain't hide nothin' from a government tha' powerful. Sa yes. Whe'ver ahr nah we figh', they'll come fer us. You ha' mahr questions." He didn't ask, he confirmed. Having taken a drink, he continued, "Go on; ask me. Ihs on'ee one freebie, an' tha' wuz ih."

Margan tilted a raised stare to the source of the reflected light, working out the wording. She had a thousand questions, but they did not have the time for all of them to be asked.

"O'Rielly..." she began, trailing. She allowed the pause to linger for so long that Daíthi had time to finish his cigarette, flicking the spent remnants into the dirt.

"Uh-huh," he prompted, before taking a drink.

"Is..." Her pretty eyes narrowed, focusing her thought, condensing that sea of questions into one. "Is he... real."

Daíthi slid her a glance, continuing to drink. He knew what the question meant; and although he enjoyed avidly spreading rumours like wild-fire, working his bardic magic to the frenzy of all around – he didn't feel like it.

"'Ee'za man," he shrugged. He again set both elbows back upon the wall, tilting his head back and enjoying the gentle summer-night breeze.

Margan looked at him. She wanted more of an answer than that, but she refused to seek it in words. She watched him, determined to suss out satisfaction to her query herself.

Daíthi smirked. Almost in refusal to grant her request, and yet almost as if it were an answer, he sang:

"Kind friends and companions come join me in rhyme
Come lift up your voices in chorus with mine
Come lift up your voices all grief to refrain
For we may or might never all meet here again.

So here's a health to the company and one to my lass
Let's drink and be merry all out of one glass
Let's drink and be merry all grief to refrain
For we may or might never all meet here again..."

97
Strange Twists of Fate

T*á mé beo.*

Impossible to reach by land, too dangerous to attempt by sea, but somehow, *somehow,* by the warped Grace of God, he had made it, cast onto that ledge like a dead and dying fish. A last desperate grab through his drowning secured it, and – Mother Mary! – he was saved. For as quickly as the waves had caught him and spewed him up, they ebbed slowly from the cliff-side, and that's what spared him; the rough ledge was at the right height and the right angle that the wave curled down onto it, shattering – and not so low that it could drag him out to sea.

For better or worse.

Duncan choked, fighting against the water clogging his lungs, his mind filled with the roaring of waves. Somewhere far distant from his damaged body, he began to take stock of himself.

He was alive.

Drawing a sweet, savory, water-logged breath, Duncan cast his gaze to the sky.

I'm *alive!*

He felt the faint pull of the receding water as it washed over and under him and glanced back behind him as another, lower wave sprayed foam over his body.

"Whoa."

He smiled, then laughed, his voice as hoarse as the sea's. "Next time a bit more shallow a dive, eh Duncan? *Lordie!*" His laugh was lilting and nervous.

He was alive.

Facing forward, his eyes fell to his hands. Laying flat, he'd managed to grasp a vertical twist of iron – an external vine of the inner cave network. His fingers, wrapped in his dark gloves, were strained white, bent and shaking, caked in salt. Not unlike the rest of him.

The tide crashed against the rocks; as the mist rained down, it made a gentle hissing sound, blending with the never-ending rhythm of the sea. Bracing himself, he dragged his body forward an inch.

"Augh –" His outcry of pain died in his throat, doused in the volume of another sharp wave. He dropped his head against his arms, coughing and choking and muttering, "*Sáile, sáile, sáile...*"

Salt in an open wound. Small price for a great mercy. He didn't dare stand, for to do so would require him to release his hold on the bar, and he knew he would slip and be swept out to sea. He needed to get closer, use the bar as a brace. He gagged harshly, gasped, and – without hesitation – jerked his body forward, pulling as close to the cave's mouth as possible.

"*Uuaih!*" He writhed, and, whipping desperately, scrambled to his feet, grasping the rocks and iron like death, using them to lift his shattered weight.

In spite of the destroyed shaking of his body and the slight dry-heaving it and his half-drowning induced, he was able to seize an understanding of where he had ended up. The battering ocean had broken through the dungeon wall; he was standing at the mouth of a half-natural cave, clutching and leaning onto the dooring for support. Breathing deep, he controlled himself as best he could, fully stunned.

But alive, he reminded himself, raising his eyes to God. And perfectly positioned for nothing better than to reenter the castle, unseen, and totally unexpected.

Duncan heaved a sigh, running a soaked sleeve across his brow. "I'm goin' with ya on this, Lord," he vowed. He caught himself as he swayed, slipping slightly on the refuse of another washing wave. "Sure though I don't understand it."

The sky above – far, far above him now – was rolling and dark, and he couldn't tell whether it was night or day. The waves collided with the skyline, lifting and spraying like salty rain, blocking out whatever light or answer God might have offered.

Shaking his head, Duncan stepped into the shelter of the mountain, muttering as he did: "*A Dhia, cuidigh liom*; I'm *goin'* with ya on this!!"

She stood in the darkness of the Western Tower, her long hair dripping over her shoulders like sea-foam, every strand as straight as a shining blade. The first light of dawn was beginning to flicker along the horizon, and a cold wind growing with it.

"You are certain he is dead?"

Far below, the waves were thrashing the rocks, the grey-green waters drawing back from the black cliffs and receding into the distance.

"Like a candle in a pool; there is no way that he could have survived."

Slow, soldier's footsteps drew forward, and Murrain turned. In the half-light of the dark room, she could just make out Judah Basillicus, standing at its center.

"That's what you thought last time."

Judah's eyes narrowed. He felt the coldness in her tone. "I have increased the guards at all the gates and on the parapet. I have posted a constant double watch. If you'd like, I'll send out a patrol, but the way the sea-cliffs stand, he couldn't've reached Gallia's shore. The currents are against him. My men are soldiers; they are not sea-trawlers."

"For now, he is superfluous, but that makes him no less of a threat. Killing his son may have empowered him. And a rebellious O'Rielly empowered is a danger to the Empire."

"The fall from the cliffs alone would have killed him. That's well over four hundred feet. If a rising wave swallowed him, the pressure behind it would've smashed him into the rocks. The weight of the water crushed him. The sheer volume of it filled his lungs and drowned him. But if there is some ungodly chance that O'Rielly may have survived, my men will find him. He wouldn't be able to make it so far as the green."

"Make sure it's so." Murrain's eyes glinted in the twilight, a threat lingering unsaid.

"It will be done," the soldier affirmed. He made his way to the door.

As Murrain turned back to the window, Bassus paused, and vowed: "Duncan O'Rielly is as good as dead."

98
Sláin(t)e

It was early that morning they began preparations.

Breakfast had been provided by Zander, who somehow managed to burn pancakes black while the middle was nice and runny-raw. "*Jayzus*, yer a menace," Daíthi had grumbled in complaint – but he ate a handful of the palm-sized pancakes anyway. God love him, he wasn't picky.

The day had begun with the disruptive arrival of the rest of their *contubernium*. The cart they had brought with them to Brigdavios stood parked where Gerry's tent had been, piled high with bungee-corded sheaves of rakes surmounted by the crumpled tent-canvas. The party had largely scattered, many returning with Morann to "check on his wares", although Paddy and Sheeran had stuck around for breakfast. They were lured less by Zander's cooking and more by curiosity, for, since dawn, Murphy had been engrossed in a complex mechanics job, the majority of which he seemed keen to accomplish using the oxy-acetylene torch.

"Shun't you wear gloves wi' that or somethin'?" Sheeran asked, around a mouthful of pancake. He was eyeing the torch warily, squirming like a disturbed child. "Maybe some *goggles* –"

Daíthi swore at him – then swore at the blow-torch as it shut off again. Pulling out the lighter he had "borrowed" from Kelly, he answered, "Jus' stay clayr'a me an' qui' bein' distractin', an' ye'll be fine!"

"What're ye at there with yer tinkerin', Murphy?" Paddy asked. He was sitting a few feet away, tasked with keeping Murphy's dog away from the mechanics job. The wolf was licking his face in an attempt to earn pancakes for sweetness.

Murphy narrowed his eyes, re-opening the acetylene needle valve a tiny bit so the torch would light. "Dis type'uv generator is an army prototype fer a semi-perpetual-motion storage/conductive generator, but ihsa Model Gamma-Seven, in all'a which da' magnetic fields were installed a tick off, so da' torque ain't maximized, meanin' it stores less over time. Tha' some'uv da' generators lost energy – so ta' speak – as heat meant tha' six'ur ten uv'em caught fire, an' da' Gamma model was discontinued. But, if ya just shift Permanent Magnetic Field B tha' lil' bit, works like a feckin' charm."

"Ah! So if it's as simple as that, then why aren'chya finished with it yet?"

Daíthi lifted a pained glare to the Lord. "Why ain' Ah finished with it yet," he echoed, at a mutter. "Padraig O'Donnell –" he addressed the man by name, then launched into an explanation of "magnetic dipoles" and "sinusoidal voltage" and all kinds of stuff like that; they let him go for about five minutes before Deirdre, their guest that morning, mockingly condescended, "*Sure* Murphy. Just, *sure*."

She was sitting audaciously close to the man with the torch, perched at the end of a log, dipping *pizzelle* in a beer mug filled with hot cocoa. Murphy looked at her flatly, as if she were an expected betrayer. "We lissen'a you go on fer twen'ee minutes abou' hair-cuts wiv *'side-bangs'*..."

Deirdre smiled and offered him a slice of a *pizzella*, and he accepted the truce readily. Popping the *pizzella* slice into his mouth, Murphy returned to the task at hand, laying the torch into whatever inside the machine had earned such a punishment. "Catch any races i' Brigdavios?" he asked, around the mouthful.

Paddy grinned, pulling the squirmy wolf into an embrace. "Why, Murphy, *a chara*, yeh know deyr's no bettin' at t'races in Brigdavios! Not in five years! Vica Pota won da' day-four."

"*Perfide!* Well, Ah guess she weren't named after a vic'try goddess fer nuthin'."

"Indeed!"

Suddenly, Cuchulainn started up, wriggling out of Paddy's grasp and darting forward. Everyone turned to see Diarmuid Cahillane stooping to pet the wolf as she jumped and danced around him. "God save all here," he greeted them, shyly.

"Wha' da' Hell, Diarmuid!" Murphy returned, cheerfully, shutting off the many valves that fed the scorching flame and safely setting the torch aside. "Wha' brings ya ta' our little corner'a Hell, mate?"

"I heard ya have t'worst pancakes in d'world."

"Ugh!" Zander pointed the spatula at a widely-grinning Sheeran. "It was you, wasn't it!? Just because you're the one I sent f'r water –"

"Why, whatever cou'jya mean, Zander, my dearest, *a chuisle*?"

"I'm hungry," Diarmuid continued.

Zander pointed to a plate by the hearth. "Help yerself." Returning to Sheeran, he threatened, "And *mark* me, Lemoine O'Sheeran –" (Daíthi laughed childishly at the use of his full name) "– they made me leader, so that makes me the best'uv you lowly lot! So you'd better show me some respect!"

"Wait –" Deirdre half-raised her hand. "So yer the best of the worst men?"

"...Yes."

Diarmuid watched as Daíthi rummaged through his bag for electrical tape. He looked over the campsite, taking everything in – the rakes, the machines, the long coil of cord half-wrapped in white tape. "What're ye at?"

"Me," Daíthi corrected. He snagged the black electrical tape, unrolling a great line with a tug. "Ah'm da' on'ee one doin' any work, here."

Zander pouted. "I made breakfast."

"Fine; me an' Zander." Rising to his knees, he snaked the tape into the generator. "Stick aroun', Diarmuid, if yer wan'in ta' see our salvation unfold."

Diarmuid frowned. He wasn't entirely sure what any of that meant, but he didn't have any place better to be. "Alright." He grabbed a handful of pancakes. "By da' way, Paddy, you owe me five bronze; we're goin' ta' war."

"So we're doin' this, are we?" Paddy asked, lackadaisically.

"Aye, Red," Daíthi answered, with a tone so serious, so monumental, it was gruff. He cut the tape with his teeth like a wolf, turning Paddy a glare to match his tone. "In ev'ry meanin' uv' da word, we are doin' dis!"

Far in contrast, Red sighed, tossing Diarmuid a few coins. "Insider knowledge," he grumbled. Diarmuid quietly pocketed his winnings.

Zander turned to him. "Diarmuid, how opposed would you be to keeping an eye on Addison during the fight? Don't go out of your way for him or anything, but... He's kind'uva wuss, but he wants to help, and I don't wanna keep him with me 'cause the crossbow'll be too tempting."

"Margan mentioned y'were lettin' him fight wid us," Diarmuid acknowledged. "So long as I can kill 'im if he goes rogue, I've no objection to it. He better not get me killed."

"Damn, Diarmuid," Murphy jeered, "yer sa' bloodthirsty."

Diarmuid turned to him, and abruptly changed topics. "Oh, shoo – 'tis a fine guitar, there. Maple?"

Daíthi tipped him a glance. "Ihs mine. Try't?"

Diarmuid's eyes rounded. "Can I?"

Murphy shrugged. "Be all mayns. Have fun."

"Ooh! A hundred t'ousand blessin's on Daíthi Murphy, an' his children! May yer generations be long! An' prosperous! An' rich, if ye'd have it!" Diarmuid finished the rest of his pancakes and took up the guitar. He ran a few test chords, savoring the harmonic notes, then jumped into an old, sassy banjo tune.

"*Thayr*, damn it!" Slamming the panel closed, Murphy stood, using the generator for support. He spat in the dirt before the machine, muttering oaths, curses, and warnings against it. Then, turning aside, he reached out to start it. "Gad help us, lads," he said, and then flicked the machine into life.

There was a bit of a stall, and then a loud whirring as the thing started up. Murphy waited, listening intently, measuring the rhythm of the generator and praying it wouldn't blow up. Then –

"Yes! Alrigh'!" Dropping down, he began sorting through wires, binding some with tape, connecting a power-strip to the generator, and, to that, another thick chord.

"How is *this* gonna win a war?" Paddy drawled.

"Ah'm *gettin'* dayr!" But he made no move to illustrate just yet.

Diarmuid was full in the trad zone now. Deirdre flicked lazily at a tuft of his hair, gently commanding, "Sing for me, Mister Cahillane."

"Aye, lass." Bending over the guitar, leaning into the music, he looped into a variation of "The Maid that Sold Her Barley", and he sang:

"It's cold and raw, the north winds blow
Black in the mornin' early
When all the hills were covered with snow
Then it was winter fairly
As I was riding o'er the moor
I met a farmer's daughter
Her cherry cheeks and coal-black eyes
They caused my heart to falter."

"Ye've a good voice, Diarmuid," Murphy complimented knowledgeably. "Surprised I ent heard ya."

"I use'ta play a bit around, it's true," Diarmuid explained, falling into the rhythm of another old tune. "Unfortunately what stopped me was, some years back, a raid of ours went a little awry, and I was seen without my knowin' it. And who else would have t'bad luck that me next performance they should get a name t'me face as well? After that..." Diarmuid shrugged. "A warrant was put out for me. Now t'only way I go into town is under t'name of Demna Muldoon. An' poor Demna can't play guitar." Looking up at Daíthi, he added, "Seems it'll be so for you now, too, lad. Now you've gotten yer own lengthy little note."

Murphy's expression darkened. "Gad willin', mate, tha'll soon anuff change."

Diarmuid nodded expressionlessly, swaying into his next tune; he played them in fourths and halves, ranging from one song to the next without completing the thought. Murphy smirked; *familiar style*. Turning aside from the generator, he began working around the tarp which covered the smaller object sitting beside it.

"What *is* it?" Deirdre asked, watching him work.

"We dunno," Paddy shrugged.

"And it's *killin'* us, Deirdre!" Sheeran insisted, his eyes grown wide.

"Be *patient, a chiall!*" Daíthi braced against the generator, to lower himself in front of the tarped object and all of its cords, but stalled halfway, forced to shift his leg again before he could settle because of the splint.

Paddy witnessed his movement with some concern. "Lad –"

Daíthi shook his head. "Won' be da' firs' time Ah climbed wid a broken leg, by any means. Don' think I ent figgered eh'reefin; Ah wuz injured a'fore I volunteered. Besides, t'ain't mahr'una wee climb! 'Tis t'battle *after*'ll be t'death'a me!" His laughter took on a strange quality halfway through as he realized his words might be the truth. He cleared his throat. "Hhm... Hey! Wha's un'er dis tarp!" He uncoiled a last cord from the pile, plugging it into the power strip. "Sheeran, grab me one'a dem sticks, will ya?"

Sheeran looked perplexed, but he stood and moved without complaint.

"*Should* I move?" Diarmuid asked, looking to Paddy, and Deirdre echoed, "Should *I?*"

"Gad love ya, *a gcroíthe*," Murphy taunted, rising and accepting the rake from Sheeran.

"D'we get t'see what that is now?" Sheeran asked eagerly, settling now next to Zander – and as far from the machinery as he could get.

"Yes."

"Ooh!"

Murphy stepped over the cords, moving behind the machine. "T'see it ya wun't know it," he warned, for once breaking from his performer's drama. And with that, he pulled the tarp.

"Yer right," said Diarmuid, who sat closest to the machine and still had no idea what it was.

Murphy half-smirked. "Don' jump." Wielding the rake shaft, he slid one end into a 1 ½ -inch hole at the top of the machine. There was a loud, crunching sheering and whirring – Diarmuid jumped – and Daíthi pulled out the shaft.

"Poin'-eight secon's," he declared, with subtle pride, holding the rake vertical like a shepherd's staff.

"Jayzus!" Sheeran blurted. "Ye've a pike!"

"So *that's* how yer gonna do it!" exclaimed Zander, getting to his knees.

"Shame tha'chya e'er doubted me," said Daíthi, wrapping his hands around the shaft and leaning against it. The formidable tip glistened, smooth as ice, in the early morning light.

"Swear on me life," vowed Paddy, crossing his fingers over his heart, "never will we doubt ya again."

"Yeah righ'." Turning to their guests, Daíthi said, "Yer help is welcome, bu' nah required."

Diarmuid enthusiastically threw a salute, answering for them both. "Anythin' f'r me country, lad! Where d'we start?"

Murphy smiled. A bold, cocksure, lopsided smile. "Grand."

99
Sever

Meygan paced up and down in front of the altar, a blade held in her hand. It was heavy, made of shining, silvery steel, like a shaft of moonlight. So pure as to cry it had never been used, though here in the chapel its aura was dark as deceit.

Meygan halted before the altar, setting the point of the blade on the ground, her hands folded neatly over the top. The place looked like a monastery after a Viking raid – stripped bare of all gold ornamentation, down to the doors of the empty tabernacle. All the chalices and plates and crucifixes – all and any signs of faith – were gone. The altar was all that remained.

She leaned back on her heels, cocking her head to study the altar's stone side. All four sides had been carved with intricate and beautiful Celtic designs in the La Tène style – great, thick, looping vines dancing around one another, never crossing, yet filling the space with their intricate patterns. On the side facing the congregation were four rectangular cells, each containing the symbol of one of the Four Evangelists, the frame emblazoned and built with Celtic designs.

She raised her eyes to the tabernacle. No candle burned beside it, no Host sat within. But God was here, she knew it. Different than He had been once, assuredly, but He was still here. In every corner, every shadow, every speck of light that filtered through the damaged ceiling. And now, it was up to her to defend Him.

Bowing her head, she tipped the sword so it was balanced between her hands, measuring the weight of its sturdy blade. Not for the first time, she wondered why they had given her such a heavy weapon. Why leave her as guard? Being a guard was boring. Meygan leaned back, raising her eyes to a particularly large gap in the ceiling that was casting a beam of light onto the altar's bare stone. She knew instinctively there was a reason for her being here. She just didn't know...

The door opened.

Meygan turned.

A flash of silver, and it was closed. The shadows were heaviest in that farthest corner, but she knew, by feel, exactly where he was. Slipping along the wall, along the columns, creeping towards the light.

"They're looking for you."

He froze as if struck dead; it was not a presence that surprised him, but the fact that it was – instead of a soldier – in spite of it all –

"What?" he asked, weakly, his voice so faint that it didn't echo.

"There's no use in you hiding in the shadows. They're gonna find you soon, anyway."

He stared, unable to speak a word. He had seen hallucinations before – plenty in his lifetime. But not even they had made as little sense as this.

"Or you can just turn around and go right back out there," she said, when the pause she allowed him yielded no results. "You're not getting anything from me. It's safer for you if you just go, actually, because they don't just put anyone as official guards, okay? So just go –"

"Meygan?"

Now it was her turn to be surprised. Because when they had said that a rebel prisoner had escaped and had stationed her to guard this room, she never could have imagined that prisoner would be –

"Rahyoke?"

How. How, in the name of sanity...

"H-how di – *where* di –" His shock clouded. "Have they hurt you?"

Meygan said nothing. She was happy he was alive. After all, that was what she had come all this way to ensure. But there was something – off. Something was wrong.

Maybe it was what they had said.

"No," she said, her volume soft but her tone sharp. Her voice was eerie, like a winter-morning fog. "But they might have been telling the truth."

He felt the cold cut into him, slicing to his core. He forced himself to swallow – Lord, his throat was parched! – and ask in as calm a voice as he could muster, "The truth about what?"

She still could not see him, but she swore she could feel his gaze, scouring, searching. She flexed her fingers on the hand-grip of her sword. "You didn't know I was here," she said. "So why did you come here?"

For the book. The chapel was the end-all for Ariegn – the place where prisoners were executed, the place where the one thing they valued above all else would be kept. It was the most defensible position, so safely away from the entrance gate that nothing could threaten invasion. If Rael had been right, and there was another book, here was where it would be kept. Here...

His eyes fell to the backpack on the altar.

Softly, he slipped through the shadows that rimmed the room, until he was close enough to see her as more than a spirit framed in that ethereal morning light. She was wearing the red military tunic of an Augustan soldier, belted with her hoodie, which was tied around her waist; her black leggings and sneakers. The purple scarf was again tied like a kerchief, covering her hair almost entirely.

A weird, chilling feeling swept through him at the sight of her – the blade she held – the book she hid. A feeling of absolute dread.

His voice was strange when he managed to ask, "Meygan. What are you doing here?"

She narrowed her eyes through the gloom. He had come far enough into the light for her to see him, too. Covered in brine and blood, his hair wild, slick and spiked and mad. He was pale, his eyes underlined with a haggard darkness, and he was badly in need of a shave. Bare-chested but for a torn, stained cloth wrapped around his stomach, and bare-foot. A long sword glittered in his right hand.

Her voice was both frigid and caustic as she answered: "Protecting my book."

She watched as shock dominated his features again, wiping the slate clean.

There IS another book...

Suddenly, his look hardened, becoming determined, intense, almost feral. "Give it to me."

"What?! No!!" She stepped back, shielding the altar with her body. "It's mine!"

"Come on, lass," he breathed, on the borderline of patience. "Ya don't realize what yer playin' at, here... Give it over t'me."

Meygan clenched her jaw. She wasn't protecting this book because some crazy Empire had told her to. She would protect it because it was hers. It was her last connection to the Otherworld, and if an Empire could not tear it from her grasp, what hope had he?

She lifted her sword, and with equal determination, she said, "No."

Duncan gritted his teeth. "Meygan, I swear t'God, I don't have time for this. That book is a danger to our world. *And* yours, if you'd have it. You've no idea what it's capable of –"

"And I should give it to *you?!*"

His voice raised to a shout. "Better off with me by far than with the Empire that would destroy us –"

"Destroy *you*," she defied, overpowering him with her small voice. "And *I'm* a better keeper of this book than *any* of you! You want it? Then you're going to have to take it from me." She hefted the heavy sword like a baseball bat, her eyes locked on his. "And I have a *way* cooler sword than you."

Duncan's eyes narrowed. "Very well." A glint of silver caught his blade, and he vanished into the shadows of the columns that ringed the room. Meygan turned right, holding her ground.

"Don't even *think* of sneaking around. Coward."

A dark snicker rolled beneath the shadows. "I wouldn't dream..."

There was a flash, and Meygan cried out, swatting his blade away with a heavy stroke of her sword before he could knock the bookbag from the altar. *"Really?!"* He struck again towards the bag, but was forced to step back as she nearly gouged his ankle.

She again put herself between him and the altar, lifting her sword. He was moving slowly, and she realized that the blood staining his clothing was his own. She could take him.

He watched the dragging sling and pull with which she wielded her weapon, bringing it back over her shoulder. He smirked. "Mine's lighter," he offered; "d'you wanna switch?"

Meygan wrinkled her nose. *"No."*

"Hm... Stubborn," he muttered, just audibly. He tried to side-step her again, but she swung at him, making him stumble back.

"If you change yer mind," he shrugged.

There was no way he was going to get past her while she was armed. She had far more energy than he did; her wrath was inexhaustible. If he was to ever have any hope of getting his hands on that book, he needed to disarm her.

Without warning, he attacked, pounding back her sword effortlessly and with enough force to make her fingers sting. Every blow she tempted, he swatted away with an ease that made her scowl in fury. He was aiming for her blade; she began aiming for his body.

She had read somewhere once about this sort of sword; it was all a matter of balance. A matter of finding the right terms.

She spun out of the reach of his blade, using the momentum to swing up her sword, hurling a chopping blow. O'Rielly caught it, but he gasped.

Meygan grinned, but – *och*, that grin was savage!

Balance.

She stepped forward, driving the force of her own body through her blade, thrusting it forward, up O'Rielly's own.

He flipped his sword, dragging hers off to the side, crashing them both against the surface of the altar.

Enlightenment.

The blow had been too clumsy, and she freed her sword with ease, whipping to her right to dislodge the blade and carrying through the full 360, aiming to slash him while his weapon was down. Duncan jerked backwards, sacrificing his footing to avoid the chopping blow.

He fell to the ground beside the altar, landing on his back. "Auh... Shit!" He was forced to splay his legs to evade her next strike. "Hey!" He knocked her sword aside, staggering to his feet.

She threw forward another down-slashing arc; he jumped back, allowing the heavy blade to connect with the stone floor. At once, he swung his weapon into hers, clapping them both against the side of the altar.

He stepped forward, leaning his full weight against the stone, the length of his leg pinning the locked blades.

Meygan jerked the hilt of her sword, trying to pull it up and away from his, but it wouldn't budge. She was stuck.

Her fingers uncurled from the hilt, and he grabbed her, taking her wrist in an iron grasp.

She gasped, looking up at him, but he only had eyes for the book. Letting go of his sword, he reached for the altar, jerkily opening the backpack and emptying its contents – then stopping cold.

That book. *The* book. The same book he had destroyed all those years ago. The same exact book – he recognized the water-damage, the pattern that it warped. It was the same book. *How* could it be the same book?!

Frantic, he looked at her, meeting her eyes like a hawk. Her eyes, that soft, pale green, flecked with amber, filled with defiance beneath brunette brows...

"A Dhia..."

She felt his grip slacken to nothing, watching the blood drain from his face. Seeing her chance, she took it.

Summoning all her strength, she pulled up on the hilt of her sword, breaking his grasp on her arm and freeing the trapped weapon. She twisted, lashing out viciously with the flat of her blade, striking him hard across the face.

"Uah!"

His blade clattered to the floor and she kicked it away, hefting her own again like a baseball bat, poised to make the final blow.

He was totally at her mercy. He had fallen to his knees, both hands pressed to his right cheek. Blood speckled the stone floor, streaming from between his bone-white fingers.

"Do it."

The words were so soft, so low, so *unbelievable,* that she scarcely fathomed she heard them.

"What?"

Slowly, he looked up, meeting her eyes with a dark and hateful glare. His fingertips were slick with blood, the fine drops trailing down his wrists and sticking to his gloves. He scowled, burning with violent rage.

"Take me *NOW,* if you're going to take me at all! For I *swear,* you will *never* have this opportunity again."

Meygan hated his tone. She hated that evil look in his eye – that look and tone so abrasive, so savage, so supercilious and sneering and cold. She hated how he disregarded her, belittled her, made her out to be something weak, far weaker than she was. She *hated* him for it. She hated him!

She adjusted her grip on her sword, steadying it. Her eyes shifted; his blood was making patterns on the floor, small drops dappling the ancient, grey stone, sparkling in the faint light.

Do it. Take me now.

He spoke the words so defiantly, and yet there was something so tragic in hearing them said. How awful, to want so little out of life that you begged a little girl to kill you. How awful, to want to kill a man over a book.

She met his eyes – the same dark green as Zander's – shattering her anger to study him anew. She knew all too well what ran so slowly, like a tear, down her finger, across the back of her hand, burning like a spark.

Duncan flinched as her blade clattered to the floor.

"I'm sorry!" Meygan buried her face in her hands and started to cry. "I am so, so sorry..."

His bloodied hand slid onto the altar, bracing so he could stand. Only the tips of his fingers remained pressed against his wound – a deepish gash to the cheekbone, just below his right eye.

"Don't be." He lifted the book in his free hand, holding it up. "I stole your book; it's only fair."

She looked up at him, her hands still hovering to hide the lower half of her face. "What are you gonna do with it?"

He turned to look at the book he held, allowing his other hand to drop from the cut on his face. *Keep it,* his heart pled, *use it.* And then what? Accidentally leave it behind, and the Empire would conquer the galaxy? That was a risk he shouldn't – *couldn't,* take.

"I have to destroy it," he breathed, his voice shuddering.

"But..." Meygan twisted her hands together, wringing her delicate fingers. "But..."

"I don't want to," he said, "but if they get it... *Is airiúi,*" he moaned, as if making this choice was like cutting out his own heart. He put the book on the altar. He swept her tumbled belongings back into the bag. "Believe me, dear, I want it as much as you. And... we... I..." He was vaguely holding out her backpack to her. She wiped her eyes on her sleeve and took it.

"I'll make a deal with you," she said, her voice low. "I'll let you destroy it if you get us out of here." He looked at her, surprised, but she remained firm. "*Us*, Rahyoke."

He stared at her a moment, torn – not over her proposition, but over the possibility that maybe, if he was careful, he could use the book to get them out. Out of this castle. Out of this country. Out of this world...

He shook his head quickly, snapping back and offering her a hand. "Deal."

They shook.

He turned his attention back to the book, barely noticing as Meygan instinctually backed up, the backpack once more gripping her slim shoulders. He voided his mind, banishing all those treacherous thoughts – of the consequences of what he was doing. He could torment himself after this was done. He pressed his palms against the cover. He could feel its energy, its *promise.* An unnamable aura of unadulterated power, its moral for its user to decide. He could decimate the world if he desired – all worlds – he was clever enough to know how. It was easy. As easy as burning a page...

He *had* to destroy it.

Duncan bowed his head. His hands were cold, his body shaking.

"...*Tine.*"

It began to burn.

Slowly, the black spun its web out from the center of the cover, consuming all the pale green, curling back the pages in shavings of night-wing ashes. Flecks of burning ember flashed along the wave, like ruby rust, glittering like stars amid the roiling, engulfing false night. Softly, a flame licked up between his fingers, shining first as pure as yellow gold, then darkening as it recognized its master.

He held his hands to the cover a little longer than he should have, but he didn't care. The bite was not sharp enough – not violent enough by *far* – to so much as hint towards the pain of his suffering. So long had that fire of agony burned out, the conflagration consuming his heart and soul; so long, so long...

He stepped back, folding his arms, and watched.

Meygan looked on in silence as the fire engulfed the book, swallowing it in its hold, feeding on the ashes as greedily as it fed on the unused pages, melting it all away. Her last real connection to her former life – to everything she held dear. And now it was gone forever.

"I'm sorry, Meygan," he whispered, his voice, faint as a dying breath, pulling out the last of his desecrated heart. "I'm so sorry..."

Meygan looked at him, wondering what crazy faery knoll he had fallen out of. "How... did you do..." She didn't bother finishing the question. After all, this was Rahyoke; there was no way he was going to answer.

As the last of the flames and ashes vanished in a twist of ghostly smoke, dying this time for good, he allowed himself to wonder how in the name of sanity this could be the same book. He'd destroyed it, hadn't he? With fire. With bloodied hands, and fire. And...

His eyes rounded.

"...Rahyoke?"

"I... I think..."

"...Rahyoke."

"I think I know how to get you home."

If he'd said he could also fly, it would have shocked her less than that statement – after all, she had just seen him fire-bending. There was no end to the weirdness of Rahyoke the Vagabond.

"What?!"

"I – I th-think I can – I can try, at least." He was shaking. Staring at his hands, as if marveling at a great weapon he had stolen. But he had done it before, hadn't he?! The proof was – well, he had just destroyed it, but – "I can, but –" He needed time. His expression darkened. His voice was just above a whisper as he asked, "What did you mean, they'd find me soon?"

Meygan jolted, looking towards the doors. "Oh no."

His fingers curled into fists. "I need time!"

And God, in His infinite irony, denied him. At that moment, the doors burst open, and he had just enough time to hear the words "Disarm him" before he was stunned by a searing pain.

Meygan gave a little cry – startled, not hurt, thank God – but she had the sense to remain where she was. That was all they needed, the Empire realizing she'd switched sides...

She turned her eyes to the door. There stood Judah Basilicus, flanked by two very big, very threatening, very well-armed soldiers. One, Meygan would know as Pyrrhus; the other, Eryx, knew Duncan.

A snap of fire skittered over the floor, slamming the doorframe, but quickly died, and Duncan dropped heavily to his knees. He drew his hand from where he'd clenched his now-injured shoulder, pulling a very small, coated knife away with it.

"You... s-said you wun'... drug me..."

Recognizing the voice, Eryx started. "Good *God!* He's *immortal!*"

All the contempt in the world saturated those words. "God is," O'Rielly responded, "but I am not He."

Blocking the doorway, Bassus snapped his fingers, and both soldiers unsheathed their swords. "I knew you would come back here, wretch," he growled, satisfaction blending into the hatred in his tone.

O'Rielly shook his head. *Come on; you've had worse. Remember Lōtós? You're fine.* He grinned. "I *destroyed* your feckin' book," he boasted, at almost a sing-song. He was proud he managed to sound so much more cocky than drunk.

Still, Meygan decided that she made a better actress. "I was *bait!?*" she shrieked, turning on the general. "*Nobody* told me he was here. You told me he was dead! And a rebel psychopath! And you *used* me for *bait?!!*"

"Did he get the book?" Bassus asked, sounding almost disinterested.

"*Yeah,* he got the book! He makes fire with his mind! *What* was I supposed to do with that?!" Calming suddenly, her look hardened. "I was *promised* a *guard.*"

"No matter," Bassus responded, mildly; it was that mildness that made his tone terrifying. "We won't be needing the book now."

Meygan again clasped her hands, backing against the altar. In a voice that was very quiet, and very timid, she asked, "What do you mean?"

Duncan looked between them. When he saw Eryx step into the room, he moved to rise without thinking – to put himself between them and her, no matter what ruses it destroyed. They weren't buying it anyway. But his legs gave way beneath him.

"Okh; *Dhia...*"

"Are you familiar with the term 'collateral', Meygan?"

"Please," he heard himself beg.

And the general spoke but one command to his soldiers, and eagerly, dreadfully, it was obeyed.

"Take them."

100
Hear It Knell

Sunlight seeped into the Great Hall, flecked with shadows and dust. It crept through the thin crack at the top of the wall, flooding through the open door, where it spilled from the high windows into the foyer. It laid like a gossamer veil across the grey, its harshening rays laced with the scent of rain.

The Empress stood on the great stone dais, watching the sunbeams crawl along the walls like spiders. She was wearing a black silk gown, simple, floor-length, and plain; if there were sleeves, they were hidden beneath the trailing lengths of a dusty-sapphire shawl, which covered her hair in a fine web of closely-woven lace.

She had sent Bassus for the girl. For the past week, the young woman known as Meygan had been residing within these walls as little more than a rumor, guarding a book of as much veracity. The time had come to see whether she was worth wasting their table-scraps on, and for what length of endurance; personally, Murrain could not envisage the need for the child was great, nor even what purpose she had served thus-far. But the appointed hour of their meeting had since whittled away, and General Basilicus was never tardy without reason. And the sounds ringing down the hall were promising indeed.

The girl was the first to enter, her fingers laced like she might wring them, dressed as an aspiring fashionista's attempt at reinnovation. She cast a glance behind her, stepping quickly out of the way and into the room as Bassus entered behind her, taking her arm at the elbow. Her light-green gaze surveyed the room, finding Murrain, and on the instant, her demeanor changed. Calmly, she lowered her hands to her sides; her back straightened, her chin was raised, her eyes were narrowed, royal. Maybe she had learned something from the fire-eater after all.

The girl walked down the center aisle, guided by the soldier. Stopping before the dais where the old crone stood, she offered a bow, sweeping off to the side.

It was Bassus whom the Empress watched. He raised a hand – as a greeting, as a signal, one finger raised, three loosely bent; *wait for it...*

In came Pyrrhus and Eryx, escorting a man between them. Doubled-over, his arms twisted behind his back, each fully secured in the grasp of a soldier. Not an unfamiliar scene, though one this time which made the Empress of Augustinium smile.

"Back again from the grave, are we, Duncan O'Rielly?"

The wretch twisted, writhing against the brutal grasp with which he was borne, jerking upright to spit in her direction. The crack of the saliva hitting the floor almost echoed off the walls; it was nearly as loud as the punishing slap he received.

The drugs had worn off.

"We caught the sticky-fingered fire-monger in the midst of attacking the girl," Bassus reported. "The book has been destroyed."

Meygan sucked on her lower lip. *Looks like you're out of a job...*

"Are you certain he didn't transport it?"

"*Search,*" sneered O'Rielly, daring them. This time Pyrrhus lowered the blow, punching him just below the broken ribs, knocking the breath, the sass, and nearly the senses, right out of him.

"He said he'd come to destroy it," Meygan ventured. "He said that, 'for all they've taken' – if he couldn't have it, then no one could. He said it was – 'recompense '." She nodded, remembering the word from the story of Oisín. "'Compensation'."

Balance in all things. Duncan drew a breath, choking it in as quietly as he could. Looking up, he angled his gaze to his right, where she stood, his blood staining her fair hands, watching her, protective hawk, through the matted lengths of his dark hair. *O Dhia...*

"He destroyed it."

Meygan nodded. "It's a pile of ash on the alter."

Grand, a chuisle.

Murrain turned her gaze to Bassus. "She's out of a job now, isn't she?" she asked, a mere conversational curiosity.

Yup, exactly.

"She's served us well," Bassus responded, his voice a shade low. "She was able to detain the fire-spitter long enough for us to get to her. Gave him that nice little gash to the face."

Murrain looked to the fire-eater, and Eryx obediently yanked his hair, pulling him so that the bloody mark could be seen. Duncan hissed through his teeth, and, all being satisfied, Eryx released him.

"Mm. Perhaps we could find something for her," Murrain allowed, to Basilicus. "Archer. Stable guard. Scullery maid."

You never want to hear your rewards listed in such a down-grading way.

"And what shall we do with *him?*" Murrain continued, droll, flipping an indication to O'Rielly; that was all her enemy had become. "He *has* been a naughty child, running away so."

Duncan gritted his teeth, only clenching them harder as he listened to Bassus speak.

"I-225 can be used for amputation," the soldier said, so vaguely that Duncan knew perfectly well what he meant; the memory of Inquisition Two-Twenty-Five nearly made him throw up. "He couldn't get very far with one leg."

"All of I-two-hundred through three-eleven could potentially be amputation, when you think about it," Pyrrhus offered.

"That's a good point."

Murrain smiled down on the fire-eater as a black widow smiles at her lover. "Make certain he's docile. Whatever it takes. We have one more task for the hot-blooded mongrel before we can rid the Empire of his degradation and delusion. I would make sure he completes it."

"What do you need him for?" Meygan asked, with an almost maddening similarity to Michelle.

Bassus turned to her, and was on the verge of answering when O'Rielly snapped.

"Don't involve her in this!" he shouted. He twisted his gaze to Bassus, looking up at the soldier imploringly. "Please – you ignored me with Michelle. Michelle I can see. Michelle was rebellious. But Meygan, she – she's only a little girl, Bassus – one who has served you well. I – I never told her – she doesn't know what I can – she doesn't *have* to know – and you can just let her go. Let her go on serving the Empire in citizenry. The Empire is kind to its citizens, is it not? She's only a little girl, Bassus. *Please* leave her go!"

By some warped grace of God, his words for once seemed to be having an effect: the soldier agreed with him. He could see it on Bassus' face – the manifestation of that imperial kindness the boast of which Duncan had heard pontificated for years. Idealism and propaganda at its base, beautiful, simple yet so rare in presentation. The Empire protects its citizens; the Empire maintains its peace, rewards loyalty, shepherds the impressionable and young. But as far as the Empire was concerned, Judah Basillicus was only second in command.

"It may be too late for that."

Something in Murrain's voice cut O'Rielly to his very core, and dread, black and silencing, was the blood that welled from that severing wound. "What do you mean?" he hollowly asked.

Murrain watched what little colour remained in the fire-eater's face drain from it, savouring that leaching ebb. "I think you know what I mean."

If the ripping burn of dread had been eviscerating, it left nothing but the cavern of horror in its wake. Oh, how well he knew.

His legs gave out and he collapsed, so suddenly that Eryx nearly dropped him.

"Something to *add*, fire-wretch?"

Duncan couldn't breathe; head bowed, kneeling with only his captors' support, he fought against that dizzying agony of mordant panic. He fought into resignation, his despair clenching like knives in his veins and gut. He fought into the dark corner of passivity in which he'd made his abode.

The question hung in an untapped silence, ominous and strange, tinged with a fate, looming, more horrible than death. In the sarcastic jibes of our enemies, opportunities lie. Terrible and great. They did not see it yet.

"Yes..."

He would make them see.

Duncan looked up, meeting Murrain's cold gaze, attracting it with the force of his own. The fear had left his face – as had the tiredness, pain and defeat. If ever defiance had adopted the pose, likeness, and determination of capitulation, this was the emotion reflected in his eyes, in the sure set of his mouth. Perpetually remorseless intent. So at odds with the words that he then spoke.

"I surrender."

The world came to a screeching halt at those calmly-spoken words, and even Murrain was stunned.

"What?"

Duncan straightened his back as much as he could, trying to mimic in body the authority which gleamed in his eyes. "You would torture her," he said. "Bleed her pretty veins 'til she gives you what you want. That's the way you run things here, is it not? Well, she has nothing more to offer you. And I will not let you kill her. I will trade my body for Meygan's – take me; I'm yours. But you agree to let her go, and think no more on her afterwards. You do not harm her, you do not touch her, you have *nothing* to do with her, here out. I will make sure of it, dead or alive. My bloodline has a way of never really dying out." He raised his chin slightly, haughtily. The subtlest twist at the corner of his mouth spoke no longer of concession; it was daring. And, clearly, he repeated the condemning words. "I surrender."

Now the silence filling the Great Hall was emptied, all potential used up, possibilities displayed, answer clear. It had taken nearly thirty-eight years to break Duncan O'Rielly. And all it had taken was the tears of a little girl.

"Done."

O'Rielly bowed his head in submission.

"You know, technically, we don't need him." Bassus turned to address Murrain. "It's clear he's a danger to the Empire, and a growing threat. A decade ago he was a flighty little idealist – scarcely more than an agitating nuisance. He has since made a mockery of our rule, and would stand as a statement of our weakness if left unpunished. He needs to be wiped out."

Duncan looked up. There was nothing in the gaze which he fixed on Basilicus – a dark and vacant void.

"We only need his blood."

Duncan cocked his head. Still his dark eyes were fixed, wide and emotionless, upon his captors. Tonelessly, he asked, "Are you gonna kill me?"

Meygan gasped, her own pretty eyes, so full of life, grown wide. "Wait – no!"

"Meygan, *a mhuirnín*, it's done." The last thing he needed was her throwing some childishly heroic wrench in the works. "Trust me, Meygan. Please." His voice fell to a whisper, unable to support the weight of that final word.

Meygan nodded, fighting the tears welling in her eyes. She wanted to argue for his life – the usefulness of his talents, the merits of a prison stay – anything to save him from death. But she was forced to trust him... after all, he had promised...

"Oh, that's damn near touching," Bassus grumbled, heartlessly.

Promptly, Murrain addressed the business now at hand. "We will maintain your girl until after the execution. To *ensure* you keep your word. In the meantime: Basilicus? Would you care to escort our traitorous fire-master down to the depths of the catacombs?"

"The Crypt?"

Murrain's smile broadened, in answer and in triumph; and she truly *was* the devil.

Duncan looked up as their hands came down upon him, seizing him, hauling him to his feet. *Oh, God* – His face had gone so pale it was as if Death had already clenched a gelid fist around his heart. And down he would be taken, into the crushing darkness, the killing silence, the cold and the damp and the stale, rotting madness, never to see the light again, never to hear another voice, murdered by that gaping void, murdered –

"Wait – *wait!*"

His voice sounded like that of a frightened child, weak and protesting. He tensed, dropping into a fighting stance as he struggled against their grasp. He did not notice the swords drawn; he didn't *care.* Let them kill him now! Far better to die now than –

Twisting violently, he caught sight of Meygan, and froze.

No... No, he couldn't do it. There were no options left to him, now; he had no choice in the matter.

It would not be *him* they would punish.

Duncan untensed slowly, and immediately he was restrained, his arms pulled behind his body and blades prodding him like meat. He looked at Meygan, and he tried to force a smile. "It's gonna be okay." God, how his voice was shaking! "I *promise,* Meg!"

She bit her lip, silent tears gleaming on her cheeks. "Okay."

"C'mon!" Pyrrhus hiked his captive up, ushering him towards the hall.

"Maybe we *should* break his legs," Eryx observed. "Just in case."

"Take a good twenty men with you," Bassus advised his soldiers. "Believe me; you'll need it."

"Sure..." Duncan allowed himself to be led – half-dragged – to the door. But on the threshold, he paused, casting his gaze back into the perverted bowels of that once-mighty castle. And, quietly, he said, *"Ní básóimid riamh. Ceothannanmór go deo."*

101
Fear

"I don't know – maybe you remember? It used to be a long way to the depths of Ariegn – even longer to the Crypt. It made sense when they made it, of course – storage and security, sure. But was it ever an inconvenience for us! And what use to you, if your soldiers spend all their time in the great labyrinth of their own prison? So they built the elevator shaft – that was a little while after you left. It took about eight years to engineer – eight years for what miners can do in a fortnight, right? Now, I say that, but you gotta understand – this is no ordinary hole in the ground. You know what I'm sayin'? Like a quaint little anthill – Daedalus' masterpiece inverted – do you know Daedalus? Yeah, you would, Everyman. Sheesh. I won't bore you with the details of the engineering – it's pretty complex, and if you can understand it at all, then you already know, am I right?

"Anyway, the unfortunate thing in it, though, is that – for security purposes, again – the shaft begins on the tenth floor. So you've still gotta walk down ten flights of stairs and corridors before you get to *ride in style. We're* used to the walk, of course – dark halls and winding passages, a rickety shaft and all that – no problem. I don't know how far a wanderer like *you* can go. But, once we get there, it's a straight shot down to the depths. Imagine a fall like that! Sheesh. It's still a long ride down, but... the time difference is incredible. Cut the trip to fractions. Increased efficiency vastly. I tell y'what: *that's* a miracle of ingenuity. God bless technology, right?"

An elbow to the ribs prodded the weak breath of an answer: "Yes."

The procession was a rare one, even for Ariegn. Typically one guard would be enough – at most two – but two were merely in his forward guard. These two led the procession, another two took the rear guard, and two more escorted the prisoner, one on each side, keeping a firm hold on his arms though his wrists were bound in heavy chains. One more man traveled behind them all as a sort of safety; he would move up front with the floodlight once they were below. In all, they had taken seven men in the delivery of this particular captive. Bassus had warned them that seven wouldn't be enough, but they would not believe him. Who would?

Their captive kept his eyes on the ceiling, which, like the floor, like the walls, like the light, like the path that lay before him, was hidden in the gloom. For the most part, he had offered no resistance, but that didn't mean they trusted him. He had not spoken a single word – no pleas, no jeers, no prayers, no snarky, witty comments. Only silence. And in this business, silence was disconcerting; you never assumed stupidity or a mute. Always assume plotting, always assume the most devious secrecy. For it was the silent ones that got you.

And he had been silent for a while.

Not that he really posed any great threat. As a result of sleep deprivation, starvation, dehydration, torture – his strength was beginning to fail him. He had fallen once, and when they blamed his weakness on fear he had cursed at them vehemently. But what else could he do? He was an unarmed, half-dead vagrant against seven highly-trained, armed professionals.

And seven was enough.

For the most part, the guards had remained as silent as their prisoner. It had been entertainment that sparked their inevitable jeers. It had been boredom that sparked the more recent Tour of Hell. A tour much shorter than he'd expected; it was true that the elevator had been built in the time between his visits, and he was counting floors. Each staircase rang with echoing footsteps, another drawn chime to toll that looming hour; another shovel in the grave.

"I tell y'what, though," the guard continued; any sound seemed sudden in that gloom; "fractions or not, this trip still bites. We'll only just get down there and back and we'll hafta bring you up again."

Come back up... He found no comfort in the notion his mind would not allow him to comprehend. What mattered was the now – but the now was bleak and horrid – and cold and dark and unending – never ending...

He swallowed, trying to re-direct his thoughts, trying to focus as he stood his weight on shaking legs, forcing himself to remain stable. But what he heard next almost caused his heart to stop altogether:

"Well, this is the last floor, anyway; the elevator's just at the end of the hall."

Oh God! His mind began to race, and he staggered, jerked impatiently forward. Oh Lord, oh God, oh Mercy!

His eyes were never still. With bird-like, jerking motions, he surveyed the close walls – the faint torches – the abyss above and below. He was not aware that his captors were half-supporting him. He knew his muscles tensed, his sinews strained, and a sword blade pressed into his back. He was aware of how slight he was, how helpless. And down – ever down they went, their progress agonizingly slow and ominously quick.

He swallowed, hard. He forced his eyes ahead of him, deep into the gloom.

It was irrational to fear the Crypt – the crypt. Absurd to fear the darkness. Ridiculous to fear a cell of iron and stone, mildewed walls which snuffed all flame, murderous silence and a blackness thick as Night's own nightmarish dreams. There was nothing to hurt him there. He was a grown man. He didn't have anything to be afraid of. He... uhm...

With a deafening soft crack his foot struck stable ground. The descent was over. This was the end.

And blind terror overtook him.

102
The Leaving of Frith Gallia

The sun was setting, its amber-fire wings spreading gold across the sky, dying the early mists a soft pinkish orange. Perhaps it was the task at hand, or the imminent threat of his own mortality, or simply the ripe crispness of the breeze, but Daíthi could not help but remember other sunsets, both secreted and shared, on the other side of the Pale.

He had kitted himself out for heisting. His hands were half-wrapped in black cloth for protection (half, for he'd no concerns about finger-printing), his poor, damaged shins wrapped in white splints. Ariegn was a weird animal to heist, because its walls were white enough to out a spider, its grounds dark enough to betray a louse. To compensate, he would make a shield of the dark blanket he had used as bedding for nearly a decade, and a white hoodie. The hoodie was a little tight around his biceps because it was Deirdre's; he comforted himself with the ill fit and the fact that it wasn't sparkly. He'd cinched a satchel around his waist for use as an arrow bag, and, when he felt like it, he would strap the sea-bag containing the oxy-acetylene torch onto his back.

On the boulder before him sat his crossbow. A massive beast of a beautiful weapon, built by himself to suit his unique purposes. She could shoot far heavier bolts than the crossbow he'd given Zander, and at a far greater range. She had a heavier draw, consequently, but that should only be annoying once in this endeavor – ie, when shooting a grappling shot while hanging from the first bolt, dangling a billion feet above the hungry sea. Other than that...

This was gonna be fun.

Daíthi lifted the crossbow, hefting her familiar weight, running his fingers along the string. How long the thread that Clotho spun... and could Atropos be conned into sparing it?

"There you are."

He tossed Zander a grin over his shoulder, indicating the bow. "Ha' ya met Cassandra?"

Zander cocked his head. Like Daíthi, he was without armor, putting all of his faith in the two bows he carried and the quiver of arrows on his back. "Hi. You're gonna make us late."

Daíthi rolled his eyes. "Ah ain't gonna mayk us late." He set Cassandra aside long enough to arrange the torch on his back. "If ya could'a started ih'tall wivout me, ya wun've come lookin' far me, wou'jya? Ff'you din' need me, ya would'a jus' let me..." He trailed, his back to Zander, his eyes contemplating the deep woods, and something a little beyond their physical realm.

"Murphy?"

"Zander..." There could be no more putting off the question. When Daíthi turned back to Zander, his expression was grave, somber almost to the point of remorse. "Have you ever killed someone?"

Zander pursed his lips a second. "No."

Daíthi said nothing, so Zander spoke again. "But I've killed other things. Crows. Fish. Even a couple rabbits and squirrels, which you almost *never* find in the forest –"

"Killin' a man's a bi' different fra' killin' a fish," Daíthi said, quietly.

Zander looked at him; it seemed that the man had more to say, even more that he didn't want to.

Daíthi faced him squarely. "When you figh'," he said, "kill 'em li' fishes. Kill 'em li' rodents – rodents dey are. Dey took yer grain, dey threatened your bairns, dey spread dolor li' disease t'rew dis land. Kill 'em like rats. Dass all dey are."

Zander nodded, not sure what else to do. To him, it seemed he had enough anger in him, enough resentment towards the enemy, to obey without qualm.

"Good." He hesitated, even more uncertain as to whether or not he should say the next part – at least, right now. But he may have no other chance. "On'ee after... *After* da' battle, when da' killin's done, an' your people are saved... on'ee den remember humani'ee. It ain' easy killin' a man," he said, his eyes searching Zander's face, measuring every honest word. "Bu' on'ee victors ha' da' privilege 'uv rememberin' deyr enemies as such. Kill 'em li' rats. Remember 'em afterwards as men. Make sense?"

Zander nodded slowly. The more he considered Daíthi's words, the more disconcerted he was to realize how difficult a time he had in recognizing his enemies as anything more than cockroaches infesting his domain. "That'll be the hard part," he breathed, truly a little horrified at himself.

Daíthi's smirk was bitter and pitying. "You'd be surprised."

He allowed a moment for that to sink in. Then his smile became encouraging, and he clapped a hand below Zander's – non-injured! – shoulder. "Ye'll be fine." He walked past Zander, toting all his gear. "Come on; yer makin' us late."

Rolling his eyes, Zander followed.

"So eh'reeone's armed?"

"Yes!" Zander beamed, with pride.

"Eh'reeone's in position?"

"Yes."

"Eh'reeone knows da' plan?"

"Yes! I'm a really great war-leader!" Zander declared, spreading his arms both defensively and confrontationally.

Now Daíthi rolled his eyes.

"Matty and Faizel are gonna say blessings over the army before we go," Zander reported.

"Hm. Hope Caomhan don't burst in'ta flames."

Zander laughed. They were nearing the place where everyone had gathered together – those who would stay behind guarding their carts and wagons, the warriors farther ahead, standing – or sitting, as the case may be – in their ranks. Orderly... maybe they could pull this thing off after all.

"Murphy!!"

He looked up. "Fahy, Ah a'ready tol'jyah: ya *cain't* ha' mah guitar if Ah die!"

Deirdre joined them, making a poor road-block with her thin frame. "Hi, Zander."

"Hi."

She put her hands on her hips, returning to Murphy. "Well, if you're dead, you don't get much of a say in it, do you? Anyway, we needed to say goodbye to you."

"Ya a'ready – we?" He lifted his gaze over her shoulder, a little further down the path, and saw Roscanna.

Deirdre stepped aside, taking Zander's arm and whispering in his ear, "Watch this; it's gonna be amazing."

Daíthi lifted a hand, rubbing the back of his neck. "Hi."

"Hey..." Roscanna twisted the toe of her black flat into the dust, her hands clasped behind her back, her black hair loose about her shoulders. "Yer gane off t'yer war now?"

"Yeah..." He drifted one step closer to her. Before he could say any more, she spoke.

"So, there's a custom among mah people – maybe yeh have it, too, I don' know – bu' – Ah don' know if yeh – d'yeh wan' a token?" She blurted the last bit, impatient with herself, her own perceived silliness, and the prospect of his scorn.

But scorn he did not have for her. In fact, he blushed like a shy child, answering softly, "Yes."

"Okay." Businesslike now, she stepped forward, pulling from behind her back a navy-blue ribbon. "Yer hair's too lang," she muttered, reaching out and tying his hair back with the ribbon, in a low pony-tail. "Yer gonna fall an' die 'cause'a yer hippie rock-star hairdo."

He snickered. "Thaynk you." He could smell the rose scent she used in her hair, delicate and sweet.

"Hm." She picked a few fussy adjustments, ignoring his eye-roll. Then, satisfied, she briefly put her hands to his shoulders, then stepped back, clasping her hands before her. She met his eyes. "Return wi' yer shield or on it," she said.

Daíthi gave a nod. His eyes searched hers, finding the sorrow beneath that no-nonsense calm. The worry in the blue. He was not the first man she had sent off to war... but she was the first girl he had ever met worth dying for.

"Aw, Hell. *With* it!" He kissed her on the cheek.

Her eyes went wide with shock, and when he pulled back, taking her wrist, she pressed her free hand to the kiss. He smiled. *"Carumi si,"* he said, with a laugh. And then he was gone, vanishing around the back of a cart.

Roscanna stared after him – and Deirdre and Zander, who scuttled in his wake – her expression still one of surprise. *"Carumi si,"* she breathed, in repetition. Reciprocation.

I love you.

103
Bloodline

The last rays of the setting sun tumbled, glittering, onto the stone alter, vanishing like embers in dust. The only light left came from torches – flickering on the walls, sparkling in the hands of soldiers – but even for all their ranks, a great shadow still overcast the room.

Meygan stood among them, her hands bound before her. Having been traded for the fire-eater, she was no longer a prize, but a prisoner – a bargaining piece, to make sure that he upheld his end of the deal. She still wore that fine, Roman attire; the purple shawl was now slung around her neck, leaving her fiery curls loose and free. They had taken away her heavy sword, and at last provided her with a guard – a hulking, muscular man, armed with a spear that was twice her height and a sword broader than her bicep.

"Are you gonna kill me?" Meygan asked her soldier, almost conversationally. Not surprisingly, he didn't answer. "My name's Meygan," she continued. "What's yours?"

The guard scowled.

Meygan shrugged. "Just thought it would be useful."

They were still waiting for everyone to arrive. It was a spectacle, but a reserved one. The guards of higher rank, soldiers of greater skill, were in attendance. As a reward for their years of dedication. And in case.

She was about twelve yards from the alter, and, behind it, the old crone stood, defiling the Seat and House of God with her presence. Her long, silver hair was tied back into a bun, one graceful tress curling down from the sheer snare. Over her hair was a veil of red lace, and her dress, as usual, was black, long-sleeved, floor-length, the sleeves and part of the bodice etched with lace. And oh, the expression upon her was *satisfied* indeed. Meygan gritted her teeth as she looked on her, feeling her anger boil. She had never hated anyone more.

For the most part, the other guards were equally as well-armed, some with crossbows, others with swords unsheathed, tensed, as if waiting on a battle. Crimson crests and shining breastplates, polished chain mail and round shields – even a few of the long, full-body shields of the Galls, the paint cracked and chipping. Most wore cloaks – military standard-issue red – lighter, as it was summer – clasped at the neck – save for one hesitating man who hovered in the back. He wore no armor, carried no weapon, and his face was anxiously, not excitedly, set upon the crowd. His cloak was heavier, more battered, and tied at his shoulder, Ceothann-style.

Meygan narrowed her eyes, straining against the dark distance. Limp, flaxen hair. Wide, brown eyes. *"Ian?"* She could scarcely believe the word, but of course you knew it to be true. There he stood. Risking the annoyance of her guard, she called out to him. "IAN!!"

He turned like a shock, and his eyes found her. And steady was his gaze, firm, unmoved – barely a flicker of surprise. Older, in a way, no sign of joy, but a satisfaction, not directed at her – not at all – but evident all the same.

"Shut up," her guard snapped – not merely irked, but anxious. This was not the place to fail when your captive was no more than a little girl.

Meygan glanced at him. "But that's –"

"Some scum half-breed informant."

"—Ian..." Now Meygan looked at her guard in full. She did not want to understand. It was as horrible a thing to understand as Rahyoke being a rebellious murderer. It was the most plainly-spoken of all twists, but the turn in the plot was too vile for her to accept. "That's Ian; he's my friend," she insisted, as if this should have meant something. Once, she had thought that it had.

"Some friend," muttered Eryx, who stood nearby – saying as little as he dared, because he was only just senior enough to be included in this muster, and he wanted to be able to rub said inclusion in Aesculus' face later. Meygan and her guard both turned to him, but a snap from a higher officer brought the room to silence. Ian and his dastardly treachery were nothing compared to what was about to follow.

The tell-tale bang echoed down the hall, ringing through the open door like a midnight bell. In its wake entered Bassus, hauling a lanky, black stain of a creature. No struggle – not the slightest effort to move. He was a dusky form, a shattered mass, a listless, lifeless shadow – and nothing more.

The raven himself is hoarse that croaks the fatal entrance of Duncan beneath my battlements.

Meygan closed her eyes briefly. Thank God he had never read Shakespeare.

The soldier half-dragged, half-carried the derelict to the front of the room, dumping him before the alter at the center of a platformed box covered with an iron grate.

"On your knees, filth."

The poor wretch emitted a small moan, almost pleading; just barely, he shifted, pressing a palm to the grate.

"Up!" Bassus grabbed him by the back of the neck, hauling him upright.

"Tá a fhios 'AM, a –" his angry retort was cut into the hiss of a cat or a trampled snake as Bassus cast him to his place, and he was left, sitting with one leg folded beneath him, one palm pressed to the grating, and his body bent as if the force of his degradation could cripple him.

"So this," sneered the Devil, from above, "is the almighty *Duncan O'Rielly.*"

He bowed his head deeper, and shame like a drowning net cast away his scarring fear, his searing anger.

O'Rielly should be so much more.

He jolted slightly as he felt a shin against his back, and something slipping over his head.

"I told you I would be here." Even through his daze, he recognized the voice of Ixion Briareus. His eyes fell to the object circling his neck: a device fondly known as the scythe-noose. It was as it sounded: a tear-shaped blade, the sharpest edge pointed inwards, so the bite of it circled all the way around the neck of the unfortunate soul wearing it. The point of the tear was attached, in this case, to a long pole. It rendered him nothing more than a wild animal.

"Imperator Basilicus, the charges."

Bassus grinned. How long he had wanted to do this! In a clear, commanding voice which filled the chapel and all the world around, he began:

"We are on this day, the twenty-sixth of August in the fifty-fourth year of the Empire, gathered at Ariegn, world capital and chief seat of the Empire of Augustinium, to witness the bringing to justice of the convicted. For contemptuous disobedience and refusal to abide by every law set forth by the State; violations of the most egregious sort, including but not limited to: conspiring against the government, high treason, the brutal murder of eight soldiers of the government, including two high-ranking and one middle-ranking officers; assisting two dangerous fugitives to escape from Ariegn prison, and twice, in contempt of justice, escaping Ariegn prison himself. For these and myriad lesser but by no means less heinous crimes, the convicted, Donnachadh Ó Raghallaigh, Bailey, Ceothannanmór –"

"Bailecolgáin, if ya please," he corrected, quietly, without raising his head.

Meygan could almost have smiled. *That's* my Rahyoke...

Judah Basilicus scowled, yet he obliged. "Duncan O'Rielly, the Fire-Eater of Bailecolgáin, Dúnadach, Ceothannanmór, you have been sentenced on this day to perish for crimes against the crown. This is the decree of the highest court of Augustinium, and cannot be overturned."

The end.

So he'd broken every law set forth by the State, huh? God, he hoped so.

"Just get it over with," he breathed.

When he opened his eyes, he saw Bassus approach the alter, lifting a blue-tinted wine bottle filled with a strange-clear liquid – and a match-box covered in faded green linen.

Duncan groaned. "Oh, you *wouldn't*," he agonized, as Bassus uncorked the bottle.

The imperator grinned. "Honour your bloodline." He stepped forward.

Duncan hunched his shoulders, closing his eyes as he felt the liquid pour over him, slick and oily to the touch. The acrid smell sparked long-buried memories of little cottages with their thatch roofs all aflame. Black ravens gushing from the sacred oaks and the humble belfry of the chapel, their rough screams mingling with the chaos below. And the sight – good God, the sight – as they were ringed with fire and cut down. *"Rith, cariad! Rith!"*

And thirty years beyond them, he would perish in the same way. What good was their sacrifice? What good was his?

The kerosene ran slickly through his hair, dripping down his back, drenching his dark coat. He opened his eyes slowly, watching the thin drips fall from his bangs onto the cold grating. Slowly, his eyes widened, as he realized what that platform was for.

"Oh, God," he breathed. "God... Is that what the grate's for? To catch...?"

He heard the soft scrape as Bassus opened the matchbox and raised his head. "You'll let her go if I do this, aye? Judah, I know you're a man of yer word." In spite of his hopeful tone, his voice was almost shuddering with fear.

Bassus didn't answer him. He was busy with the match, striking it against the worn grit – *once, twice, shoot.*

Breathing a quavering sigh and averting his gaze again as a false-hearted flame flickered into being from the phosphorous, Duncan closed his eyes, rocking as gently as the scythe-noose would allow. *My Lord, and my God.* Under his breath, he began to pray:

"Pater noster, qui es in caelis,
sanctificetur nomen tuum.
Adveniat regnum tuum.
Fiat voluntas tua,
sicut in caelo et in terra.
Panem nostrum quotidianum da nobis hodie..."

Bassus had been advancing, and now stopped with little more than two feet between them, and stood, holding the match.

"...et dimitte nobis debita nostra
sicut et nos dimittimus debitoribus nostris.
Et ne nos inducas in tentationem,
sed libera nos a malo...
Bheirim suas duit, a Íosa,
Tré chroidhe rógheanmnaidhe na maighdine Muire
Agus le h-intinn do chroidhe ró-naomhtha féin
M'uile smaointe, mo bhriathra, mo ghníomhartha
Agus gach a bhuil i ndán dom d'fhulang ar feadh an lae seo."

Murrain lost patience. "*Kill* him," she ordered, darkly. But, for once, her men did not obey.

"Admhuighim do Dhia uile-chumhachtach
Do Mhuire naomhtha atá riamh 'na hÓigh
Is do na naomhaibh uile..."

Without turning, Bassus replied, "We are not barbarians." He crumpled the spent match in his fist, drawing another from the box.

"Gur pheacuigheas go ró-mhór
le smaoineadh, le briathar, is le gníomh
Trí mo choir féin
Trí mo choir féin
Trí mo mhór-choir féin."

"You are letting him make a martyr of himself," Murrain hissed. "My men will not be swayed by his heathen supplications."

"Go ndeanaidh Dia uile-chumhachtach trócaire orm
Go maithidh Sé mo pheacaidh dhom
Agus go dtugaidh Sé seilbh dom
Agus go dtreoruighidh Sé mé chum na beathadh síorraidhe."

He could have held them off forever, stalling them in this way; he knew enough Lavinian, knew enough prayers. But that was not a reason to pray. Summoning every last bit of willpower his shattered soul contained, he uttered the final word.

"Amen."

Donnachadh Ó Raghallaigh visibly shuddered, drawing a shaking breath. Gone was the arrogance, the resilience, the spirit, the pride; these he had surrendered to God. He was meek, he was resigned. He bowed his head. *God, grant me dignity, please...*

"Shit." The match broke, forcing Bassus to draw another. Meygan watched as he struck it, closing her eyes as the small flame sprang up from the wood. She could not watch him die. Nor was she willing to stand by quietly, no matter what he had told her. She stepped forward as Bassus raised the match. "Wait –" Eryx grabbed her shoulder, and when she dealt him a blow with her elbow, her guard struck her, half to her right shoulder, and half to her neck.

"Damn it, Tarquin," Bassus hissed.

O'Rielly looked up, anger coursing through him so ravenously it was dizzying, but before he could retaliate, something absolutely bizarre happened.

A low, rumbling bang shook the walls to their very foundation, the resounding crack above it echoing down the halls, filling every room and corner, banishing every shadow. The tensest of pauses followed it; then they heard the roar. A swelling tide of abstract noise, building, growing, crescendoing, racing down the halls. The soldiers started, glancing, fretting as they went without orders.

A new sound pierced through the chaos outside – clear as the day and familiar as the oldest love-song. O'Rielly cocked his head, listening. And, slowly, a strange smile spread across his face.

By God, that was a tin whistle.

Leering viciously, he cast his gaze to Murrain, and in time with the wildly out-of-place tune, he sang:

"Is go dté tú mo mhúirnín slán!!"

He laughed – not the usual, stifled laughter, but hearty, almost mad. "You're too late!" he shouted, and no words had ever been so beautiful, so sweet, so marvelous to speak!

"KILL HIM NOW!!"

Reaching up, Duncan took hold of the scythe-noose where it laid against his neck, curling his fingers around its sharp point. Twin flashes of heat ran around the blade, slicing up its shaft.

"Shit!" Ixion let go, staggering back, staring at his burned hands, and the fire-eater un-slung the restraint from his neck.

"Secundus Ventidius," Bassus ordered, taking the now-empty wine-bottle in his hand. Men snapped into action at the code, forming ranks, setting to bar the door – and Meygan was seized from behind.

Duncan dove out of the way just in time, and the glass shattered against the grate, Bassus's body arching, exposed, above him.

"Meygan, run!"

"I can't!"

Duncan rolled into Bassus's legs, and the soldier staggered upright and out of the way. The fire-eater had lost the scythe-noose, and now Bassus lifted it, wielding it like a lacrosse-stick.

Duncan was on his knees, his back to the alter, with all the world before him. At a flash, he held up his palms.

"DÓITEÁN!"

There was a loud crack, and the door was engulfed in flames. A screen of orange fire consumed the ancient wood as rapidly as gasoline-soaked paper, reducing the patient work of countless engravers into charred and blackened ash.

Everyone turned, stunned. And Meygan felt Tarquin's grip upon her slacken.

"Meg!"

She took off at a sprint. The flames fanned out to either side of her, burning a crystalline white, barring the soldiers from their escaping captive. The fire before her did not move.

She tensed, she trusted him, she leapt like a sprinting hurdler, slicing though the flames, reaching the other side untouched.

"Ha!" Duncan exclaimed. He staggered to his feet, swaying like a drunk, grinning like a madman, filled with a reckless, audacious defiance.

Now you will know what I can do.

Abruptly, a beam of white slashed across his vision, striking him full to the side of the head.

The quartz fire shattered in a thousand glittering stars, and the red tongues that clogged the doorway collapsed, crumpling in on themselves like dying lilies.

Duncan opened his eyes. Thank God it was the flat of the blade that had hit him!

Far across his inverted vision, he saw the scattering, black shadows flitting through the doorway, leaping over the flames. The last shade to tempt it was lanky, smaller, wrapped in a scarlet cloak, Ceothann-style –

Flames shot up across the door, double-ringing the deserted, burning so hot that their mere breath would pierce the skin.

Oh, you're not going ANYWHERE!!

Duncan looked up, just barely, his dark hair and a slick line of blood running into his eyes. He was lying on his left side, his back to the alter, and Bassus stood above him once more, the scythe-noose raised like an executioner's ax, the blade gleaming in that Hellish light.

Iochd!

The blade sliced through the air. He only just rolled in time, putting his back to his attacker. Reaching behind him, he grasped the shaft, trying the same trick he had pulled on Ixion, but Bassus was too swift, sacrificing the scythe and pulling out his sword.

"Auh!" Duncan was forced to turn his back again, holding up the shaft of the scythe as a defense. His left arm, which held it, was weaker, and crumpled under the blow, but the ends of the scythe caught, the blade on the ground and the end of the shaft on the alter's edge, separating him from that cleaving blow.

Now what?

Bassus raised the sword again, concentrating the force of his blow, aiming to shatter the staff and sever the fire-eater's clever fingers in the process.

Seizing his chance, Duncan sat up, diving forward, hearing the sword strike the ground just behind him. He struck, left-handed and half-blindly, towards the noise, and met the soldier's chain-mail.

"Rah!" Lashing out with the blade, Bassus knocked the scythe out of Duncan's hands.

"*Shite!*" Duncan groaned – and twisted to avoid another crushing blow.

Sprawled, he turned a glance back to Bassus, as the soldier drew his knife in his left hand. The general was double-armed – and Duncan was left foundering like a fish!

He rolled into the alter, grasping the top and dragging himself up before he was forced to drop for another descending blow, clinging like a leaf with his back to Bassus. Without thinking, he grabbed the wrist of the soldier's sword-arm, and Bassus plunged the knife to scrape O'Rielly's shoulder-blade.

Duncan swore, writhing as Bassus twisted the blade in dislodging it, readying a second stroke. But before he could make it, Duncan lunged. Springing against the alter, he hurtled himself backwards, colliding with Bassus and taking him down. The knife skittered away, clattering across the cold, stone floor. Duncan made a jump for it, retaining his hold on the soldier's wrist, but Bassus grabbed his legs, high around the thighs, rendering the attempt futile.

Bassus rolled, pinning the surprised fire-eater on his back beneath him. His free hand lashed out, seizing the vagrant's wrist. As the wretch struggled, Bassus got to his knees, trapping O'Rielly with the force of his weight.

Duncan twisted and writhed, but there was nothing he could do; he was too drained, too weak. His only advantage was his grip on Bassus's sword-arm, though even that was beginning to abate, the gleaming blade slipping ever closer to his head...

Oh wait.

A sheet of pale flame spiraled over Bassus's wrist, lacing up his arm. Bassus gritted his teeth, resisting until the fire bore into his flesh, cutting through the guards and lacerating his skin.

"Gah!" He jerked back.

Releasing his advantage, Duncan twisted, making a wild, stretching dive, the tips of his fingers slapping down on Judah's prized blade. Frantically, he reeled it in, turning just in time to catch Bassus's left-handed downward-slash.

To re-capture the fish, Bassus had knelt one knee on his shins, leaning his weight into that precarious balance. Summoning his strength, Duncan jerked his leg, wiping the ground out from under Bassus so he fell, hard, on his right side. And he dove.

Bolstered by fire, the knife shattered the sturdy chain-mail, and the sword fell to the floor with a clang.

Duncan scrambled, pulling himself forward and restraining his enemy as Bassus had restrained him before. The blow had not been enough to kill him, though blood was welling from the torn skin above his heart.

Bassus looked up at the fire-eater, feeling the weight of his own knife in his chest and his blood pooling thickly around it. He smiled.

"Do you have it in you, fire-eater?" he hissed. "Many people have tried to kill me. Some have failed in the fight, but most, through their own cowardice. But all have failed." His grin broadened, *daring* him. "What makes *you* any different?"

Duncan narrowed his eyes, glaring down at the demon who had stalked and tormented, jailed and haunted him for so many years. And he felt nothing but hatred, dire and pure.

He leaned in slightly, making sure his captive would hear.

"I could ask the same of you."

One quick, twisting pulse shattered the heartstrings, bursting the vessel in a gush of red blood. In his final moment, Bassus was almost impressed.

Duncan threw himself upright, staggering back from the body, his hands dripping with gore. He knew it was wrong that he smiled, even smugly, but – damn it, he didn't care.

Judah Basilicus, Summum Imperator of the Augustan army, was dead.

Duncan turned, narrowing his eyes on the doorway. *Now...*

Ian had sunk to a crouch, cowering, burying his face against his arms. The heat of the fire assailed him, singeing his cloak, blackening his hair, snipping at his skin like playing cobras, gloating over their prey. He was dead now for sure – he knew it. For why would a man who could summon fire on a whim spare the one who had sent him to his death?

He was seized, dragged to his feet.

The flames burned as pure a white as the eternal sun, snapping and flashing and shrieking with the voice of a thousand wars, a thousand misdeeds, a thousand lost lives and the lies that had betrayed them. And behind that fury was Duncan.

"You owe me one question's answer," the man growled, in a voice that was lower than Satan, darker than sin. He tightened his grip on the young man's neck, feeling the rapid pulse and longing to let it bleed. "Why did you do this to me?"

The lad looked at him with fear in his eyes – muteness and utter fear.

"Why did you do this to me!?" he nearly shouted, raising his voice to be clear. "What could you *possibly* gain? I wanna know." He added, "I *need* to know."

And Ian answered. "Th-they were gonna kill my family," he stammered, struggling to raise himself above the fire-eater's grasp so he could breathe. "Th-they took my fiancé – and her family – banished them over the Ran, and they said – they said they would kill them, unless – you can understand, can't yu–"

"And *what,*" Duncan spat, sneering, "you thought that when you were done, they would let you live out your life with your sweet bitch in eternal harmony and never trouble you again? You thought that you could *save* her? Fool! What are you to them? Some worthless sheep-herder from a pristine province, driven by desperation. The moment they'd finished with you, they'd crush you. Your guilty, coward's blood would flow into the earth, and no one – *no one* – would shed a tear, you lying, deceitful, ungrateful wretch! Whatever on earth made you think they would spare you?" And then, quietly as death: "What makes you think *I* will?"

Ian stared in horror, feeling that tight grip upon his throat, feeling the heat of the all-consuming flames; and there was no doubt in his mind now that this was surely the end. He tensed, ready to feel the sting of the well-earned blow. But to his surprise, it never came.

"Listen to me now, traitor, if you value your life," Duncan began. "If I ever, *ever*, see you again, without hesitation, I *will* kill you."

He thrust Ian away from him, forcing the traitor to stumble back through the curtain of dying flame. For a moment, Ian looked at him in awe, the fresh, stinging caress of sparks dancing over his skin. There was no mercy in vagrant's eyes.

Ian backed a few hesitant paces, then turned, bolting through the open door, choosing the fury of unarmed combat over the fury of the man he had wronged.

Duncan O'Rielly stood in the wreckage of the chapel of Ariegn, and waited, giving the treasonous bastard a head start. His own business was only halfway finished. And, apparently, they were at war.

A cúig, a ceathair, a trí, a dó...

A haon.

104
...and Glorious

The stoic order of the fort of Ariegn had erupted, shattering into the manic chaos of war. Rowdier than a Ceothann hoolie, fiercer than a Turranian war-horse, more tactile than the sturdiest Lavinian merchant, the great torrent of rebels ripped through Ariegn's halls, causing mayhem and destruction wherever they went.

Duncan stood outside the chapel, leaning against the stone wall, one arm wrapped habitually around his injuries. There was a terrible beauty in the scene before him... Like the plagues God couriered through Moses. Terrible. Necessary. Gorgeous.

He started as a young woman nearly fell into him, giving ground just before she plunged her sword into her enemy's throat. A white bandana was tied on her head, and the lengths of her extremely long, straight black hair fell from the bun she'd restrained it in, trailing down her back. She wore baggy overalls, a white, silk tank-top, and a set of goggles on top of her head; a rapier glistened with blood in her left hand.

"Aila?"

"Whoa!" She turned abruptly at the name. "What'er ya –" She stopped, peering at him closer. "Whoa!"

"Aila, *a mhuirnín!* I thought'chyou were perished for sure!"

"*Ochón, mo bhuachaill!* I'd hug ya, but I'm all covered in gore! I'd thought we'd hafta do way more to rescue you, t'be honest – or you were dead and Murph's a rotten liar – but I didn't expect you t'be wanderin' around, like – why d'you smell like kerosene?"

He smirked warmly. "Listen, *a stór*, we'll talk long and well afterwards, but right now I'm looking for someone. Have ya seen a little girl with red hair? I'm meant to protect her."

Aila looked back over her shoulder at the battle to make certain they weren't ambushed as they were speaking. "I forgot how odd you are," he heard her say. Then, "Yep. She's with Seamus."

"Who?"

A loud bang sounded from down the corridor, and a set of doors burst open, heavy black smoke billowing through. A number of shadows flitted out, one of them doubled over, coughing. "Aila!" the poor lad called. "Sixth floor tower – get where yer s'posta be!"

"Right!" She tossed Duncan a brief salute. "Come back t'me alive – I desperately need a good chat. Stay safe!" And she scurried off down the hall.

"Aila – AILA!! Oh, damn..." O'Rielly shook his head. Of all the people to run into, Aila was – God love her – the least helpful. He cast his gaze around the hall, wondering how he would know Seamus.

His eyes fell to Aila's victim: one of Ariegn's rare archers. Dropping to a crouch, he began looting the soldier, taking the sleek, black-wooded recurve bow and the single arrow in his quiver.

He stood a little too quickly and was forced again to lean against the wall. He was fairly positive pushing it was going to come back to bite him. But adrenaline is a marvelous drug, and he was always more the sort to go out in style.

The screening smoke being used by the invaders would make finding Meygan all the more difficult; as he'd learned in the Pale, even with vibrantly red hair, she was easy to miss. Lucky for her now, an unneeded challenge for him.

Suddenly, an arrow shot past his face, slamming into the wall directly beside his hand. He whirled, following the shot, bringing his own bow up like a sword, and what he saw across the chaos was a short man with a big crossbow, a bloodstained white hoodie, and a taunting, lopsided smile.

Murphy raised two fingers of his left hand to his temple and gave a cocky salute, fully enjoying the look of absolute shock on his friend's face.

After a moment, O'Rielly recovered. Clasping right fist to heart, he raised his bow and gave a sweeping sort of bow. Gratified, Murphy shot an assailant away from O'Rielly, pointed a commanding direction down the hall, and vanished after Aila.

Duncan turned, looking to the arrow still jammed deep in the mortar of the wall. Pierced onto it and clinging with dutiful defiance was a faded red bandana. He fingered the fabric, thoughtfully. Then, tugging the cloth from the shaft, he grabbed the arrow and was gone.

"If I get me arse handed t'me over this, it's all back on you, *a chailíní*."

They were crouching by the door of the first-floor holding cell – Meygan, her wrists finally freed of their bonds; the tall, skinny man with the straw-blond hair and chocolate eyes; and the woman with dark red hair entrapped by her scarlet cloak. The man – Seamus – was armed with a pike and a bow, the latter of which was tucked into the empty quiver on his back. He held the door open an inch, watching the chaos outside. Their goal was just down the hall, and yet – through the tide of battle – so very far away.

"Here's t'plan," Seamus whispered, without turning his gaze. "Fintan and Maurizio'll be by soon anuff; we make a dart down t'hall t'the courtyard, then up the parapet –"

"Sure 'tis a narrow staircase," the red woman cut in. "Yer better goin' up t'rew the foyer."

"Woman, 'tis madness – 'tis sheer madness!"

"How many'a yer lads are holdin' it?"

"There's da' problem – 'tis t'*gateway*, like, woman. T'gateway – sure we'd never get t'rew."

"We'll blend in –"

"*Chih!* You gingers'll blend in!" he scoffed.

The red woman flicked the back of his head.

"*Ow.*"

"I don't think you guys're with the whole concept of *lying low*," Meygan hissed.

Seamus turned. "An' it'll be no sass from *you*, missy, neither! Now pay attention. On my count, you girls run with me, got it? Ready in t'ree, two –"

One. Zander unleashed an arrow, watching as it zipped over the crowd – and cursing as it shattered against his target's tough armor. That was the way of this battle; it was like one of those stupid rip-off games they would have at the fairs – tunic is ten points, chain mail is fifty, and don't waste your shot on the plate- or – Heaven help us! – *scale*-armor!

He dug into his quiver, pulling out another shaft, spanning and knocking it for another shot. He was standing on the parapet that overlooked the courtyard – one of the few of the Company to be holding that position, as yet. He was waiting to be relieved so he could continue on to the fifth floor, where he was meant to be stationed. They were taking over the castle in waves, and taking it they most certainly were!

Down below, the battle was a mosh pit of blood and fire. Many of the Company's troops were playing through, using the courtyard as a route to their appointed stations. Most had armed themselves with trophies; forbidden though they were from taking cloaks and armor, all else was fair game, and they exercised that right to the fullest, wielding stolen swords and bows, shields and spears, and whatever else they fancied could make a useful weapon.

Zander fired down the parapet, striking an Ariegn soldier in the wrist, knocking the sword from his hand. "Yes!!"

He re-loaded the crossbow, turning again. Somehow, *brilliantly*, he had wound up in the middle of the walkway, and was being threatened grievously on both sides, with no relief in sight – and only seven arrows to his name.

He was too proud to signal for assistance; if worst came to worst, he had a stolen short-sword strapped to his side – not to mention the weapons of those he was rapidly taking down. Zander grinned, at home in the thick of battle, confident in his strength, and full certain that he was immortal. He fired another shot, turning again to his right.

Damn.

He was met with a Gallic long shield, which covered entirely the body of his advancing enemy. On his left, he was met with the same.

"Oh, shite and onions..." He'd picked a bad time to remember Murphy's note on Augustan battle prowess.

The crossbow was spanned. "Alright!" he shouted, glancing a challenge to either side. "Let's do this!"

The shields advanced quickly, covering each a line of Ariegn archers, all of whom had their arrows trained on the raging crowd below.

It's not just my own fate we're deciding, Zander thought, and, realizing it, his worry vanished into rage. Gritting his teeth, he fired towards the sky-blue shield at his left.

The ancient shield caught the bolt like it was a rain-drop, holding strong. Damn the Galls and their over-abounding cleverness! Zander scrambled a hand over his near-empty quiver, grabbing another bolt as he was steadily sandwiched between the shields.

He spanned the crossbow and loaded the bolt, but he was out of room. He tried to raise the weapon, but an enemy blade flashed; Zander jumped back, narrowly missing its deadly bite, landing on the parapet's high wall. He wobbled, swinging his arms crazily as he attempted to regain his balance.

It was easy work from there. The sanguine shield thrust forward, knocking into Zander in full, over-balancing him and barring him from any hope of a return.

"Whoa-what!?" His clawing fingers missed the rim, and he toppled over the edge, dropping into the madness and fighting below.

Endurance. Immortality. Dumb Luck.

Duncan O'Rielly wrenched his bow away from the soldier, stabbing him under the breastplate with a long, blood-stained pike. He twisted left, slashing an oncoming attacker across the face with the bow, whipping it back as he aimed a second strike. And that was when he noticed – a sword, raised, gleaming in the light of the flames, at the soldier's feet a warrior stirred, unable to rise.

"Nope!" Duncan knocked an arrow, firing it in a burst of strengthening flame. The sword clattered to the ground, its sound lost in the raging mire of the fight. Duncan stepped forward, wading towards the man he'd saved. The poor kid had sat up, his attention seemingly on his inexplicably-downed opponent. He was probably injured, definitely unarmed; thanks to the lingering courtesy of Bassus' nice, farewell stab-wound to the shoulder-blade, Duncan no longer had the strength for his bow, and figured it would be better-wielded by the youth.

"Hey, lad?" Duncan began, coming up behind him. The boy turned, and Duncan's heart stopped.

"Rahyoke?"

His hands fumbled blindly to find something to hold on to; his bow clattered to the ground, and he fell on his knees before it.

The spirit rose before him, a vision, demanding if he was alright – those eyes – *his* eyes – so dark a green – rounded and urgent. *Are you alright?* Was *he* alright!?

A Dhia.

It had been ten years since Zander had seen that look in the man's eyes – that pale, mute, disbelieving terror. At that time... He had been just old enough when he was taken to remember his father when he saw him again, but shorn, his look had changed. There had been none of the affection in his eyes, none of the joy and youth and rapture. Instead, there was fear and uncertainty, incredulity and guilt, and such an overwhelming desperation, hope, and sorrow – like a prayer he didn't dare speak or believe. When he touched him, if he ever did, he no longer gathered the boy into his arms, lifting him high, tossing him into the air, sweeping him into an embrace that not even an army could hinder. If he touched him at all, it was the brush of his fingertips, the timid, begging half-touch as he merely tried to get his attention without having to use too many words. And the words were gone. Though always a soft-spoken man, he had never exhausted his words for his son. Constantly it was stories, by day and by night – grand, beautiful, bardic stories that filled up and bolstered the boundless expanse of a little boy's imagination, always urging him to dream more, and dream better.

Ariegn had destroyed all that. Ariegn had taken that from him. They had taken his father, as they had taken his grandparents before. And, when he had found what remained of the man, two years after and ten years before, all that was left of him was fear. The fear that his child would be taken again.

And now, as the man he had been torn from yet again knelt on the floor before him, looking up immovably, Zander looked down at his face, and he saw nothing. He saw pity, he saw fear, he saw pain. He saw the steel blackness of his pitch-dark green eyes. He saw *nothing*.

"D— *Rahyoke!*" he cried, now desperate, seizing the man by the shoulders. "J—just say something!!"

Sympathy was the emotion that won out; with a soft shake of his head, O'Rielly said only, "*Dodhéanta*."

Dodhéanta. Impossible.

Zander felt himself shaking. "It isn't, though." His mind raced, as he tried to figure out how to reason with the irrational. "It isn't. I'm durable – remember? Uh – I – the time I was bit by a snake. Or when I jumped off the roof. Or when I accidentally shot myself. Or – ten years ago – last week – a *lot* of stuff happens to me! And I'm always alright! I'm – *please!*" He couldn't bear that expression anymore – that vacancy – the idea that that vacancy was all he'd been left with – not now, not after everything. He dropped, flinging his arms around the man's shoulders in a tight embrace. "Ih – *níl sé dodhéanta*," he said, like a plea. "*Tá mé anseo!*" I'm here.

Dodhéanta. He had seen him die. Seen the life-blood stain his chest. The piercing shaft embedded in his heart, reddening. And they'd dumped him in the sea. Like an unwanted fish. Like garbage. A cat only has so many lives; a man... A boy. And all those years of training, of warning, had done nothing to defend him. He had still spent his days wandering the forest. He had still thought he could conquer an enemy beyond his comprehension. He had still slung his bow over his shoulder like an idiot... again.

"Zander..."

"Yeah –" Zander moved to pull back, but without warning, the man returned the embrace.

In one smooth jerk, he hugged him – on his knees, without the strength to stand. He took him in his arms, far tighter than he maybe should have, wanting to know that he was real.

"Don't *ever* do that to me again!" To Zander's surprise, it sounded as if Duncan was fighting back tears. "Don't ever... Oh, *a Dhia Uilechumhachtach*."

Zander was alive. Nothing was sweeter than this simple miracle. If everything seemed too good to be true – if all this was just a dream – his strength, his salvation, his living son – well, he'd enjoy it while it lasted. Let the fire-spirit take him; to live in this reality was worth it. His son was still alive. His boy was still alive.

He felt Zander tense suddenly and turned, raising a hand and a fire-shield that devoured the inbound javelin.

"Whoa."

Duncan was less impressed. "Damn; I'm rusty," he mumbled, studying his hand.

Zander shifted so that the arm Duncan had kept around him was braced on his shoulders, so he could help the man up. He had his mother's practicality; as sweet as reunions were, they were still very much in the middle of a battle.

Duncan allowed himself to be helped to his feet; it was nice for once not to be hauled. "I hope I'm able for it," he murmured.

"For what?"

Still muttering to his rusty hand, he answered, "I think I figured out a way to send Meygan home."

Zander halted again, this time staring at him until O'Rielly turned.

When he did, he smirked. "I know; it's been an odd day."

"Tell me about it," Zander uttered, shaking his head. "Alright, then," he said, as Duncan shakily eased himself onto his own feet, "but I gotta get Addison."

"Addison?!"

"I know, right?" Zander laughed. "Who'da thought I'd ever make a warrior outta that mollycoddled prat?" Then he frowned, realizing: "Or he's dead." For a moment he pondered that quandary, deciding on a shrug. "Either way. Are we doing this now, then?"

Duncan momentarily muddled in the mire of his son's train of thoughts; then he shook his head. "Fine; yes. Do you want me to come –?"

"No." He had damaged the recurve bow by falling on it – perhaps that was why he wasn't supposed to loop it over his shoulder? – so he lifted the bow Duncan had dropped. "You get Meygan; it'll be faster. D'you want this?"

"Keep it." Duncan eyed him. Having only just found him, he was reluctant to send the boy out into the battle fray again. "...Meet me at the calefactory," he said, speaking slowly. "It's –"

"Got it," Zander replied with a nod. With great eagerness, he raced off in search of his traitor.

"Uhjjh – be *careful*, damn you!" O'Rielly shouted after his son. Then, shaking his head, he made for the parapet. Height aided vision... and, out of practice or not, even the sturdiest of wooden shields meant nothing to him.

There was a sick, bubbling squish as Diarmuid dispatched his last victim, running a sword into his throat. He let the corpse drop, wiping his blade on the shin of his gore-spattered jeans. "First's t'hardest, Mitch," he remarked, offhandedly, as he studied his latest work. "Auh, I'm kilt all over. C'mere – Alex, d'ya want a shield?"

"I've one here, thanks."

Diarmuid sheathed his blade. The small courtyard before the barracks was cleared. A slime of red coated everything – the walls, the weapons, the warriors both living and dead. He and Alexius had had the run of the work, once Tyrus and Ai, Kelsey and Caomhan had gone off. Mitch was the apex of apprenticeship; green as the spring leaves. But he wasn't dead, so good for him.

"I see ya fared."

The lads turned. "General," was the polite and cordial, professional greeting of Alexius; Diarmuid smirked broadly and exclaimed, "How's she cuttin', Zander?"

"You guys alright down here?"

Diarmuid spread his arms; he'd half a mind in him to be scandalized. "Are we – c'mere, *look* at us!"

"S'all clear," Alexius established.

"Clear it is!" Diarmuid emphasized. "You're right welcome, aren'chya?" He folded his arms.

"Brilliant job," Zander admitted, in honesty. "I'm proud and impressed."

"That's more like it."

"Diarmuid – Alexius – you stay here and safeguard the supplies in the barracks. Keep the area clear. Addison – you come with me."

Diarmuid saluted, two fingers of his blood-caked right hand to his forehead; likewise Alexius, though his was a Lavinian salute of a clasped fist to the heart and a subservient bow, with the added compliment, "Hang tight, General. You've got this."

"We *got* this, Zander!" Diarmuid echoed, with confident swag.

Zander put a hand to Addison's shoulder; the boy had managed to arm himself with an Augustan short-sword, fairly clean, and he was shaking. Supporting him, Zander ushered the youth away as Diarmuid began complaining for cigarettes.

"Keep focused, lads," Zander bid himself and them all. "This thing ain't over yet."

105
Thy Will

"We don't have a lot of time."

Unceremoniously, he dragged Zander into the room, Addison following, before quickly shutting the door. The room was moderately large, half the size of the Great Hall, and dark. The far wall was graced with thin, Romanesque windows coupled in three sets of three which, for the hour, let in no light. Arches like the flying buttresses of Gothic architecture laced across the ceiling, ribbing the walls and arching down into columns throughout the room.

"Why is it empty?" Zander asked, every hunter's muscle tensed. He was jolted as Meygan pulled him into a hug.

"It was locked," Duncan replied, snatching Addison's sword from his hand and slipping it between the door handles.

"Meygan!" Zander exclaimed, returning her embrace. "What are you doing here?"

"I came to rescue you!"

"Well done!" He felt her go rigid as she noticed Addison over his shoulder, but he reassured her. "Don't worry. He's cool. You know – loyalty-wise."

"And Ian's staying here with his girlfriend," Duncan quipped. He shot heat through the blade, twisting and bending it to bar the door. "Sorted. Now on to the main event."

"What are we doing?" Meygan asked him, letting go of Zander.

"God willing?" He turned to them, rubbing his poor, blistered hands. "You're going home."

Meygan's eyes widened. "Wait, right now?"

"That was the idea. I – don't have time to explain the methodology," he apologized. "But – do you trust me?"

Did they trust him.

"Alright," Addison replied, his voice soft, but resigned.

Meygan nodded, confidently. "Let's do this!"

Duncan looked away. "Okay, then." Reaching behind him, he slit the soft pad of his right forefinger on the blade, causing the blood to well.

There was a blank wall on a space to the left of the door, adjacent, that must have once held some great tapestry. This he approached, his eyes studying it to depths that no one else could comprehend. In a languid, fluid motion, he reached up, drawing a line along the wall, perfectly straight and horizontal, in blood. The red gleamed eerily in the half-light, a deep, warm ruby-wine. Slowly, painstakingly, he carved a series of hash-lines across it, vertical and diagonal. He lingered on the final stroke, bowing his head. One long and insurmountable word.

Stepping back a pace, he pressed both hands against the wall beneath it. The stone was cold – cold as death and a fireless winter night. Blood ran thinly from the deep cut, slipping over his hand.

Head bowed, he closed his eyes, trying to summon strength. He passed his tongue over his lips, vaguely aware that he was shaking.

Can you do it?

"Tine."

His voice was merely a breath, lost in a world apart, a world far more beautiful, more precious, than anything he could be offered here.

There was a reason...

His eyes flashed open. *"TINE!!!"*

Energy flowed, rippling over the wall, delving down into its surface, a dark, red, livid energy.

Duncan's hands drew together 'til they touched, shifting over the harsh grain of the stone. He rocked back a step, dragging his fingertips from the surface.

Released, flames spider-webbed the ancient stone, lacing gold-veined burgundy-orange over the black rock, twining and weaving and drawing together into a liquid pattern of light.

He leaned back, throwing his arms out at a snap. An explosion of yellow-rust flames seethed up the wall, over the rafters, down the windows and columns, following the brickwork, flowering with a thousand tongues of flagrant fire. And as he raised his out-flung arms, the fire rose with him, swelling, growing, towering over the floor.

His wrists met over his head, fingertips cupped and just barely touching. The fire roared around him, glistening, white, and savage, tearing and surging, billowing as it destroyed the clinging darkness.

Then he dropped, pulling his arms down to his chest and collapsing forward onto his knees. The fire gave one last magnificent pulse, then all was still, contained.

Meygan stared in amazement. All was dark save for the tapestry wall, and this burned as if from within. White flame flickered in vines along the plain of sanguineous heat, which itself seemed to swirl and writhe with effervescent, living Celtic whirls, gleaming like blood, like amber, gold.

This was what they had wanted him for. *This* was what he could do.

She brought her eyes to the man who had conjured the flames.

Duncan still knelt, doubled over all the way. He was breathing heavily; every intake scraped and scathing, the release a dry-heaving gag. He rocked, shaking tremulously, sweating and pale as the dead.

Dimly, he felt a hand slide onto his shoulder, and he stammered, "St-still – tr-ust – me?"

The hand rubbed his numb shoulder, bringing back the feeling, and the flat voice of Zander dredged through the mire: "Foudroyant, isn't it?"

Low, humorless, slurring: "You've *no* idea."

The blackened blurring of his sight began to clear, and his body calmed. He reached out, half-blindly taking Zander's arm. "I did it," he breathed, proud, excited, stunned. "*A Dhia Uilechumhachtach,* I *did* it!!" With a crazy half-smile, he laughed in exultant disbelief – a choked sort of cackle.

"Give't a minute," Zander warned, but helped him to his feet anyway.

"No time!" Duncan replied, airily. His veins were on fire, his marrow humming with a billion stingers. *Steady...* He stood, hesitantly releasing Zander's support, casting a glance to the door before returning his attention to Meygan and Addison. "*This,*" he said, pointing languidly to the fire, "this will g – get you – home." He bowed his head, trying to breathe.

The trespassers looked to the spiraling mass of fire. Flashing like a thousand snake-tongues over the musty rock. Stained with his blood, the dire torrents of Hell.

"Alright."

Duncan hesitated. The weight of his heart was like a stone. At a rough murmur, he added, "Zander will be your guide."

Zander turned to him as if struck. "But – but – they have no need for me in their world!" he protested. "With their cars and their mobile phones and – and – and I'm needed here! Look at what we've started! The Empire can't –"

Duncan embraced him, drawing him close and enveloping him in the sort of hug brought only by distance and despair. "Please do this," he begged, his voice muffled and faint. "*Please.* I need you to protect – them. And I... I – I can't – if you –" His voice broke, and Zander understood.

Heart shattered, Zander returned the embrace. "I promise."

His reward, the frailest of all whispers: "I'm so proud of you."

It was worth it.

Duncan stepped back, with a reluctance of agony. Trying to accomplish in a moment a lifetime of looking on the son he would never see again.

"*Bíonn súil le muir ach ní bhíonn súil le tír,*" he whispered, his voice raw and dry. "*Go n-éirí an bóthar leat... Mo sheacht mbeannacht ort.*"

Zander wracked his brain. "*An... An bhail – ort.*" *The same blessing on you.*

"Smart-ass."

Zander smirked; then he turned back to the wall, shaking out his arms, bouncing on the balls of his feet as he eyed that perilous wall of flames. Then he settled.

He reached out, dipping his fingers through the fire, through the *wall*. He grinned – broad, bold, devilish, enthralled. "*Slán go fóill,*" he bid Duncan, in the clearest Ceothann he had ever spoken. And, stepping forward, he vanished into the flames.

Duncan drew a long, sharp gasp through his teeth, clasping his right arm to his chest. His capillaries pulsed with paresthesia, stealing away his sight. "*A Thiarna – trócaire –!*" He understood now why they'd invented the books.

"Are you alright?"

"*Táim* – uh – yeah, fine." He straightened slowly, still cradling his right arm. "Still a few kinks in it... Anyway – Addison, you're up."

Addison cast a glance to Meygan. Then, stepping forward: "Ever hear that riddle? You have a fox, a chicken, and a bag of seed, and you can only carry one across the river at a time..."

Duncan rolled his eyes. "Yer truly a marvel. I'm glad to've met you. Now go away before they break the door down an' kill us."

Addison frowned. "I'm just sayin', if you look at this *logically* –"

"*Armageddon*, Addison!"

"—one could realize the potentially *dangerous* –"

"Over-thinking!" He grasped the boy's shoulders. "Safe home, my friend. Go!" And he pushed Addison so he fell through the glancing curtain of flames.

"Ooh –" The first punch, the first drink, the first tuna-melt, are all nothing compared to the last.

"Rahyoke?"

He opened his eyes. He had to look up to meet Meygan's gaze.

"I don't want to hurt you."

He straightened up, drawing a long hiss.

"...Again."

"I'm durable as all Hell, *a mhuirnín,*" he replied, hoarsely but casually. "When will you learn?"

Meygan looked up at him. Rahyoke, Mongan, Duncan. Master of Flames. She smiled.

On her toes, she wrapped him in a tight hug, so suddenly that he stepped back.

"Thank you for *everything!*" she said. "I'm gonna miss you so much! Can't you come with us?" And she raised to him that look – that *look.*

He averted his gaze. "No," he replied, at a low murmur. "Not yet."

She moved to let him go, but, to her surprise, he actually returned her embrace. "Meygan – *thank* you," he said, with desperate insistence in his quiet tone. "You have *no idea* how much you've done for me. Thank you."

He was right; she *didn't* know. She had destroyed his property, shattered his family, almost killed him at *least* twice. But...

Sometimes it takes distance, to know why a thing happens... But it's for a reason. And, I can tell you... if anything's right in this forsaken world, it all *is. It's destiny, y'know?*

Releasing him, she pulled the purple scarf from around her shoulders, holding it out to him. "Here. I want you to have this so you'll remember me," she informed him. "And to thank you. Because *you've* done for me more than you know."

He smirked. From his pocket, he pulled the damaged, red bandana. "We'll trade," he said, offering it to her in turn. Thanking him, she took it, tying it so it wrapped around her wrist and hand. His smirk broadened into a smile. "I *will* miss you," he admitted. "*Slán abhaile, a chuisle.* 'Til we meet again."

"Take *care* of yourself, okay?" she instructed, and turned to study the wall of flames. She hesitated a moment, fearing a little. Then, with her right hand, bound by the bandana, she reached out, pressing her hand through the reddened gold. It was not hot, but sweetly warm, caressing like the depths of an August night. She tossed him a grin, her light green eyes flashing. Warrior, princess, faery, rogue. "*Slán go fóill, a mhuirnín*. Good luck!" And she stepped into the flames.

He stood alone in the calefactory, staring at the fire. The heart, the soul, the mind within him were deeply rent and torn, filled with an anguish and shredded with agony, his only comfort the timid voice whimpering at the back of his mind – *I did the right thing, though. I know I did the right thing...*

However horridly unfair.

Closing his eyes, he bowed his head, trying desperately to steady himself. *Compensation, recompense, balance in all things...* A small price, his suffering; they were safe now—safe, in Riverside Valley. In her world.

Lifting his gaze to the swirling flames, he reached out a trembling hand, pressing his palm against the wall, solid and cold. The flames spiraled away from his touch, fanning and vanishing, as if they had never been, taking everything away...

Lord, but was it worth it?

...until only the thin scorch mark of her name remained.

Duncan looked upon it, studying the thin, soot-less, bloodless, branching vein – the only mark to prove that any of this – anything in his *life* – had ever existed. Had ever been more than a flickering dream, the haunting, hunting vision wrung from a shattered heart. To prove that it all had been *real*.

His fingers twisted around the wedding band at his neck.

... yes.

106
Flame

The darkness was punctuated by their raging fires, the silence by their screams and battle cries. But the tower was impenetrable.

The tower could never fall.

Somewhere beyond its stone-firm walls, the sea raged against the shoreline, the wind tore across the open fields, wearing the rocks and turning the stream-beds against the ever-flowing current of time. The summer winds were fading, their scathing day-time heat dying in the crisp air of the wintry night. And all around them, moonlight glowing, drifting through the soft banners of ever-present rain, falling upon the carpets of moss and orchids as they glistened 'neath the trees. Eternity had held this balance, woven its rapid turmoil into discipline and peace. Compensation. Recompense. The night cascaded down, sifting in its silver-shot folds to gather 'round the palisade, but soon the dawn would come. And everything would change.

Light darkened the room. White fire, pure as ghosts' wings, streamed across the windows, melting the very stone so it streaked and ran together, blending, sickly and seamlessly and grotesquely as sin. Fire consumed the dooring, fire ringed the room, harsh fire whose light was self-contained, glowing within, casting its light on no shred of darkness save that itself consumed.

She turned, her silvery hair flowing into the black abyss between her and the smokeless, ravaging fire.

There in the doorway, a shadow against the darkness, the skeletal shade of the traveler, Donnachadh Ó Raghallaigh, slouching, one hand casually raised, his palm tilted and upheld to his God.

"I knew I'd find you here."

His voice was soft, cold, indifferent. Completely and utterly, *indifferent.*

A flame flashed to life in his upraised palm; like all the fire, it did not cast its light beyond his hand. Slowly, the fire began to thicken, as every bit of his fury, every bit of his pain and desire and sorrow, solidified in its heat. And the red glow of it was angry, shifting, carnal, volcanic and pure.

There was light. Faint was the light, but it cast its glow upon his face, upon the outline of his broken body. His face was impassive, but his *eyes*.

The look in his eye was death.

She stood there, robbed of speech, unable to move, unable to breathe, locked in the power of those dark, green eyes.

The fire in his palm snapped, cracking sharply, each sharp snap sending out pulses of light, flickerings of tongues, and jets of congealed orange-red-gold that dripped from his fingers, spilling and oozing onto the floor like blood – like the blood of the heart of the flame.

His eyes narrowed, catching the light of the fire, burning again that hot russet-gold over and through the vibrant midnight green. "When you killed my parents," he said, "you should have killed me too."

He took one step forward.

And the curtain of flame flowed dutifully in behind him.

107
Slán

Dawn was a thin veil over the horizon, spreading roots into the canopy of clouds. The last of the rain was scattering over the black mud, slipping down the willow-boughs, dropping rhythmically from the clogged gutter.

Kathleen sat on the rickety steps of the front porch, a mug of tea between her hands. Even outside, with the full length of the house between them, she could hear them arguing in the kitchen, beneath the din of pots and a shrilling kettle they seemed determined to ignore. They had been separated for nearly thirteen years; they certainly knew how to make up for lost time.

Reaching up, she pushed back the bandana that restrained her dark, flowing hair. The sad little cloth was tattered and faded, permanently stained with the hearty scent of fire. She smiled.

Dech do shínaib ceo.

Appendix

Pronunciation Guide

(Well, I tried, anyway, *a chairde!*)

In honour of Morgan

A Dhia dhílis: uh YEE-uh YEE-lish (meaning "O, dear God!")

Aila: eye-lah

Amergin: AVE-ur-ghin (AVE like in "have"; ghin hard "G", rhymes with "sin")

Aneirin: ah-nAY-rin (slightly-rolled "R")

Ariegn: arr-ee-ayng

Bailecolgáin: ball-yah-coal-gone ("coal" and "gone" like the English)

Balladarach: ball-uh-DAR-uH (dar like "car"; slightly-rolled "R")

Caílte: KwEEL-tchya

Caitlín: kat-leen

Caomhan: KWAY-vin

Ceothann: K^{e}yo-han (han like "man")

Ceothannanmór: K^{e}yo-han-ann-more

Daíthi: DAH-hee

Diarmuid: DEER-mid

Deirdre: DEER-druh

Donnachadh: DONE-uh-kuh

Dubhán: Doo-awn

Eoghan: Oh-in

Faolan: fwail-awn

Padraig: PAW-drig

Rael: rye-ELL (rye like the plant)

Rahyoke: ruh-HAY-oak (hay and oak also like the plants)

Seamus: SHAY-muss

Tá a fhios 'am: tuss ahm. (*tá a fhios agam*: tuss ah-gum; meaning "I know")

Tír na nÓg: Teer nah nogue (nogue, like "brogue")

Twm: TOOM

Gallish translations

Since apparently this ain't a language you all'd know, the author commissioned my fine and humble self to give ya this (and you are WELCOME) :

Chapter 14 (Challenge!)

Abona	"River"; thus "Abona River" and its inverse are both tautological at best.
Brigdalga	*briga* is "hill" or "mountain"; *dalga,* conveniently enough, is used for both "pine" and "torch". Thus, "Pine Mountain" or "Torch (as enflamed!) Mountain". Remember that when you hit Chapter 84, for your convenience.

Chapter 50 (Luis)

Sego are maruos.	"Victory before death"

Chapter 56 (Tinne)

Rucco-ti-ambi!	"Shame surround you!"
Iexs-tu mi-sendi?!	"You're saying that to me?!" (Literally: "Say you me that?!")
An-pissiusi.	"You do not see."
Craxantos!	"toad" / "pustular"
Esi	"You are"
Londa-adarca	"Savage scum-reed!"
Allos.	"Another"
Caecos.	"Crap."
Carosa!	"Luv!"
Medu!	"Mead!"

Chapter 73 (The Messenger)

Anderos.	"Hell."

Chapter 75 (Bad Company)

Siaxsiou tiopritom	"I will seek revenge."
Se-eia tarī-caecos eti exuertina.	"This is bull-shit and unjust!"
Luge!	"By the oath!" / "I swear!"
Deuos, immi mesco.	"God, I'm drunk."
Deuos, uoretos.	"God, relief / help / support"

Chapter 77 (Breakfast with Outlaws)

Mo' Deuos!	"My God!"

Chapter 78 (Sunburstry)

Deuos, trougetos.	"Lord, mercy."
Treis... Allos... Oinos...	"Three... Two... One..."
Immi sui trougo caxto,suados vanos!	"I am your miserable slave, sweet killer!"
Dusios ex-anderos	"Demon out of Hell"
incors tó gulbion	"Shut your mouth / beak"
eti gabi in mantalon!!	"And take the road!"

Chapter 79 (Drownin' Foudroyant)

Luge eti ueidu.	"I swear and I pray."

Chapter 80 (Vertigo)

Leucos	"lightning"
Mo' Deuos	"My God"

Chapter 82 (Murphy's Law)

Insinde se: bnanom bricton!	"Behold: the magic of women!"

Chapter 83 (Trial by Fire)

Nóbergusia	Literally, "Our mountain". In the course of the long history of border disputes between the Turranians and the Galls, the Galls had pointedly if pettily renamed the entire mountain range simply "our mountain".

Chapter 89 (The Wolf Nest)

Moni Deuos	"My God"

Chapter 91 (Geasa Droma Draíochta)

miletu	"destruction / ruin"
dunom	"fort"
ison isarnon cambion rate	"that iron, twisted fort"

Chapter 102 (The Leaving of Frith Gallia)

Carumi si	"I love you"

Months in Gallic

January
Samonios

February
Dumannios

March
Riuros

April
Anagantios

May
Ogronnios

June
Cutios

July
Giamonios

August
Simiuisonna

Semptember
Equos

October
Elembiuos

November
Aedrinios

December
Cantlos

About the Author:

C.C. Ostrander graduated from Le Moyne College with a dual-major Bachelor's degree in History and English, with minors in Religion, Irish Literature, Latin, and Medieval Studies, and from University College Cork with a Master's degree in Celtic Civilisation. In 2012 she published a book of poetry entitled *Wanderlust*. This book which you are currently perusing is her first novel. Thank you for your time!

God bless!